Lecture Notes in Computational Vision and Biomechanics

Volume 28

The research related to the analysis of living structures (Biomechanics) has been a source of recent research in several distinct areas of science, for example, Mathematics, Mechanical Engineering, Physics, Informatics, Medicine and Sport. However, for its successful achievement, numerous research topics should be considered, such as image processing and analysis, geometric and numerical modelling, biomechanics, experimental analysis, mechanobiology and enhanced visualization, and their application to real cases must be developed and more investigation is needed. Additionally, enhanced hardware solutions and less invasive devices are demanded.

On the other hand, Image Analysis (Computational Vision) is used for the extraction of high level information from static images or dynamic image sequences. Examples of applications involving image analysis can be the study of motion of structures from image sequences, shape reconstruction from images, and medical diagnosis. As a multidisciplinary area, Computational Vision considers techniques and methods from other disciplines, such as Artificial Intelligence, Signal Processing, Mathematics, Physics and Informatics. Despite the many research projects in this area, more robust and efficient methods of Computational Imaging are still demanded in many application domains in Medicine, and their validation in real scenarios is matter of urgency.

These two important and predominant branches of Science are increasingly considered to be strongly connected and related. Hence, the main goal of the LNCV&B book series consists of the provision of a comprehensive forum for discussion on the current state-of-the-art in these fields by emphasizing their connection. The book series covers (but is not limited to):

- Applications of Computational Vision and Biomechanics
- Biometrics and Biomedical Pattern Analysis
- Cellular Imaging and Cellular Mechanics
- Clinical Biomechanics
- Computational Bioimaging and Visualization
- Computational Biology in Biomedical Imaging
- Development of Biomechanical Devices
- Device and Technique Development for Biomedical Imaging
- Digital Geometry Algorithms for Computational Vision and Visualization
- Experimental Biomechanics
- Gait & Posture Mechanics
- Multiscale Analysis in Biomechanics
- Neuromuscular Biomechanics
- Numerical Methods for Living Tissues
- Numerical Simulation
- Software Development on Computational Vision and Biomechanics

- Grid and High Performance Computing for Computational Vision and Biomechanics
- Image-based Geometric Modeling and Mesh Generation
- Image Processing and Analysis
- Image Processing and Visualization in Biofluids
- Image Understanding
- Material Models
- Mechanobiology
- Medical Image Analysis
- Molecular Mechanics
- Multi-Modal Image Systems
- Multiscale Biosensors in Biomedical Imaging
- Multiscale Devices and Biomems for Biomedical Imaging
- Musculoskeletal Biomechanics
- Sport Biomechanics
- Virtual Reality in Biomechanics
- Vision Systems

More information about this series at http://www.springer.com/series/8910

D. Jude Hemanth · S. Smys
Editors

Computational Vision
and Bio Inspired Computing

 Springer

Editors
D. Jude Hemanth
Karunya University
Coimbatore, Tamil Nadu
India

S. Smys
RVS Technical Campus
Coimbatore, Tamil Nadu
India

ISSN 2212-9391 ISSN 2212-9413 (electronic)
Lecture Notes in Computational Vision and Biomechanics
ISBN 978-3-319-71766-1 ISBN 978-3-319-71767-8 (eBook)
https://doi.org/10.1007/978-3-319-71767-8

Library of Congress Control Number: 2017959542

Printed on acid-free paper

This Springer imprint is published by Springer Nature
The registered company is Springer International Publishing AG
The registered company address is: Gewerbestrasse 11, 6330 Cham, Switzerland

Contents

Genetic Algorithm Based Hybrid Attribute Selection Using Customized Fitness Function

C. Arunkumar[1]([✉]), S. Ramakrishnan[2], and Siva Sai Dheeraj[1]

[1] Department of Computer Science and Engineering, Amrita Vishwa
Vidyapeetham, Coimbatore, India
c_arunkumar@cb.amrita.edu, dheerucr9@gmail.com
[2] Department of Information Technology, Dr. Mahalingam College of
Engineering and Technology, Pollachi, India
ram_f77@yahoo.com

Abstract. Attribute selection is an important step in the analysis of gene expression for cancer or illnesses in general. The huge dimensionality of gene expression data that includes many insignificant and redundant genes reduces the classification accuracy. In this study, we propose a hybrid attribute selection method to identify the small set of the most significant genes associated with the cause of cancer. The proposed method integrates the advantages of filter and a wrapper to perform attribute selection by devising a customized fitness function for the genetic algorithm. Three data sets are used that includes leukemia, CNS and colon cancer. Results of our technique are compared with the other standard techniques available in literature. The proposed hybrid approach produces comparably better accuracy than the standard implementation of the genetic algorithm.

Keywords: Attribute selection · Information gain · Genetic algorithm
Region of characteristic

1 Introduction

Cancer is a deadly disease prevalent across the globe. As per the reports of World Health Organization, 13% of all deaths was due to cancer that accounted to 7.6 million in 2008. By 2030, it is projected that it might rise to over 13.1 million [1]. Cancer that's diagnosed at an early stage, before it's had the chance to get too big or spread is more likely to be treated successfully. If the cancer has spread, treatment becomes more difficult, and generally a person's chances of surviving are much lower. Accurate predicting cancer at an early stage can save thousands of patients' lives. To achieve this, data mining algorithms can be used for gene classification. DNA microarray technology has had a huge impact on cancer research. Using the technology, large numbers of gene expressions can be simultaneously measured. Gene expression levels are analyzed in order to detect the altered genes and thereby detect the type of disease

© Springer International Publishing AG 2018
D. J. Hemanth and S. Smys (eds.), *Computational Vision and Bio Inspired Computing*,
Lecture Notes in Computational Vision and Biomechanics 28,
https://doi.org/10.1007/978-3-319-71767-8_1

[2]. Generally, only a small number of genes play a significant role in a disease. Identifying these most relevant genes can help us predict if a person is infected with the illness. The major setback for this analysis is—the number of patient data available is limited whereas the number of gene expressions for each patient is comparatively very large. Moreover, the data available is noisy. We intend to propose a new hybrid gene selection method that would reduce the large dataset to a small set of most significant genes and removal of redundant genes. In order to achieve this, the raw dataset is normalized in the range $[-1, 1]$ and subsequently the information gain filter and genetic algorithm wrapper is applied and the accuracy of the genes are analyzed.

The main contribution of this paper is to combine the filter and wrapper approaches and customize the fitness function of the genetic algorithm wrapper. Information Gain filter method is applied to the normalized dataset and the reduced subset is fed into the genetic algorithm wrapper method that uses a customized fitness function for the genetic algorithm. The selected subset of attributes using the proposed hybrid approach is fed into various classifiers like Random Forest, Random Tree, Naïve Bayes, J48 to compute the statistical accuracy of the proposed method.

2   Proposed Hybrid Approach to Attribute Selection

The main objective of this paper is to reduce the dimensionality of the dataset by removing the redundant genes and thereby increasing the classifier accuracy. The removal of the redundant genes is achieved by using a hybrid approach that combines the filter and wrapper approaches. Information Gain filter is applied and the genetic algorithm using a customized fitness function is used as a wrapper technique. The reduced gene subset is used to compute the accuracy using traditional classifiers available in literature.

2.1   Attribute Selection

Attribute selection is the process of picking relevant data, or rather the process of refining data with the idea of removing the irrelevant and redundant data. The aim is to find the smallest subset of features that can represent the dataset accurately [3]. Attribute selection is important in classification as it reduces the effects of disadvantages of dimensionality, minimizes cost of computation and other resources and helps in achieving good accuracy. Attribute selection has two major approaches namely filter and wrapper methods. Filter methods uses the concept of statistical scoring. A score is assigned to each feature, the features are ranked based on the score and a decision is taken whether to preserve or remove the feature from the reduced dataset [4]. Wrapper methods consider the feature selection as a search problem and the different combinations are prepared, evaluated and compared to other combinations. A predictive model is used to evaluate a combination of features and assign a score based on model accuracy [5]. The advantages and disadvantages of filter and wrapper approaches [6] are given in Table 1.

Table 1 Advantages and disadvantages of filter versus wrapper methods

Model	Advantages	Disadvantages	Examples
Filter	Highly scalable, faster, independent of the classifier and better computational complexity	Interaction with classifier ignored	Chi-square, information gain, correlation based feature selection
Wrapper	High interaction with classifier, simplicity, dependent, better classification accuracy	Computationally intensive	Genetic algorithm, sequential forward selection

2.2 Information Gain Based Entropy

The process of feature ranking is performed using Shannon entropy. Information Gain filter is applied on the raw dataset by adopting the algorithm given below. The Information Gain is determined using the following algorithm: Given a decision system, $DS = \langle \bigcup, C \cup D, V, f \rangle$, where C is the condition attribute set and D is the decision attribute. $B \subseteq C, \forall a \in C - B$, the gain of attribute 'a', Gain(a, B, d) can be defined as in Eq. (1) below

$$Gain(a, B, D) = I(B \cup \{a\}; D) - I(B; D)$$
$$Gain(a, B, D) = H(D|B) - H(D|B \cup \{a\})$$

(1)

If $B = \phi$, $Gain(a, B, D) = H(D) - H(D|\{a\}) = I(\{a\}; D)$. When the value of Gain(a, B, D) is higher, it implies, under the known condition of B, attribute a is more important for decision attribute D. Actually, Gain(a; B; D) can be used to evaluate the significance of attribute a in attribute set B relative to D. Consequently, a mutual information gain based algorithm for attribute selection can be described as follows:

Input: Let Z represent the raw dataset obtained after the process of normalization. Z is subjected to dimensionality reduction using Information Gain filter. It produces the set of all conditional features A and the set of decision features B (nth column in the dataset that determines the type of disease).

Step 1. Let $B = \phi$
Step 2. For every attribute, $a \in C - B$, compute the significance of condition attribute a, Gain(a, B, D);
Step 3. Select the attribute which maximize the Gain(a, B, D), record it as a;
Step 4. If Gain(a, B, D) > 0, then $B \leftarrow B \cup \{a\}$ goto Step 2, else goto Step 5;
Step 5. The set B is the selected attributes that possess entropy value >0 [7].

Output: The reduced feature subset that is ordered based on the ranking of features. All features with an entropy value of zero are eliminated.

2.3 Genetic Algorithm

Genetic Algorithms (GA) is a search heuristic method that mimics the process of natural evolution. Initially, GA initializes a population of chromosomes and iterates over it using selection combined with functions like mutation and crossover to get the next iteration of population. The idea for this process is to mimic the evolution of living things in nature [8]. Comparative terminology between GA and human genetics is given in Table 2.

Table 2 Comparative terminology between GA and human genetics

Human genetic	GA terminology
Phenotype	Decoded string
Genotype	Encoded string
Genes	Attributes/features
Chromosomes	Bit strings

2.3.1 Methodology

The core steps in any genetic algorithm are

1. Initialisation: Initial population is created by generating individual solutions randomly. The entire population is represented as an M × N 2D matrix, where M is the number of chromosomes and N is the chromosome length [9].
2. Selection: The aim of the selection step in GA is to make sure the population is being constantly improved. The selection mechanism helps the GA in discarding bad designs and keeping only the best individuals. For this, Individuals are selected using a fitness-based process using a fitness-based process.
3. Reproduction: Subsequent generations of population are generated from the previously selected parents by applying genetic operators like mutation and crossover. This is repeated until the required population size is obtained. The idea behind this step is that the new solution created would inherit the characteristics of its parents. The average fitness of the population generally increases by this.
4. Termination: This is the endpoint for the iterative generation process. Termination conditions are according to the requirement, but some common terminating conditions are finding a solution with minimum satisfying requirements, maximum number of generations reached, allocated resources used, no improvement in the result even after successive iterations or any combination of the above conditions [7, 10].

Steps of a genetic algorithm are given below.

1. Initial population is created
2. The individual's fitness is evaluated
3. The following steps are repeated until termination:

   (a) The fittest individuals are selected for breeding
   (b) Crossover and mutation of selected individuals are used to generate children
   (c) The fitness of new individuals are evaluated
   (d) The previous generation population is replaced with newly created population [8, 11].

2.3.2   Modifications in Fitness Function

The finesses of the chromosomes are calculated using a function called as fitness function. Lesser the value given by the fitness function, fitter is the individual. A very successful modification in creating a fitter population is to include a few fittest individuals from the existing generation to the next, unaltered. This is known as elitist selection and the individuals are called elite children. It can be quite effective to combine GA with other optimization methods. The hybrid approach to attribute selection is proposed in [12]. This paper uses Information Gain as Filter method, and Genetic Algorithm as Wrapper method. The two fitness functions defined are defined as fitness function 1 [13] and fitness function 2 [14]. The fitness function 1 and 2 are used on flavia and ionosphere dataset as represented in [13, 14]. This paper tunes the genetic algorithm parameters as represented in Table 4 to apply the fitness functions on binary cancer microarray gene expression dataset and evaluate the classifier accuracy on the microarray dataset.

Fitness function 1

$$f = \alpha/N_o - N_s \tag{2}$$

where α represents kNN-based classification error, N_o represents the cardinality of the original dataset and N_s represents the cardinality of the selected features

Fitness function 2

$$f = \alpha/N_f + e^{\left(-1/N_f\right)} \tag{3}$$

where α represents kNN-based classification error and N_f represents the cardinality of the selected features

2.4   Proposed Hybrid Approach

The following are the steps in the proposed method:

1. The raw datasets are pre-processed to range from [−1, 1]. This step ensures that all the gene expression values in the dataset are brought to a common scale. Moreover, it makes it simple to identify the effect of the gene on the illness. For example, 0 represents that the gene is equally expressed in tumour and normal tissue [3].
2. Information Gain based filter method is applied on the normalized dataset. This is the first step in the hybrid approach. The threshold entropy value is taken to be 0.
3. Attribute selection is done using Genetic Algorithm wrapper using a customized fitness function
4. Finally, to evaluate the usefulness of the proposed method, the accuracy is computed using various classifiers like Random Forest, Random Tree, Naïve Bayes, J48, ADABOOSTM1 and REPTree.

The parameters used in the genetic algorithm and the method of implementation of the genetic algorithm wrapper are as in [15, 16]. For the purpose of fitness functions, two fitness functions are designed and implemented. The constraints that are considered for the design of fitness functions were the loss of information while selecting attributes and the number of attributes should be less. According to these constraints, the fitness function is directly proportional to the loss and number of reduced features. A factor called re-substitution loss, provided by the KNN classifier gives the value of loss computed for the data. This is represented as α. Also, since the number of reduced features should be less, the number of unselected features is taken to be inversely proportional. This is done to reduce the fitness function. The lower the fitness function the higher the fitness of the child.

3   Results and Discussion

3.1   Dataset Description

Three binary datasets are used for the study and the details are tabulated in Table 3

Table 3   Description of the dataset

Dataset name	Initial samples (samples $\times$ features)
Leukemia	7129×72
Central nervous system (CNS)	7129×60
Colon cancer	2000×62

The above datasets consists of raw data. They are normalized in the range $[-1, 1]$ using min-max normalization.

3.2   Parameter Settings

The hybrid approach to attribute selection using the information gain filter and genetic algorithm wrapper is evolved by tuning the various parameters as indicated in Table 4.

Table 4 Parameter tuning for the proposed hybrid approach

Parameter name	Values assigned for proposed method
Entropy	Shannon
Population size	100
Genome length	100
Population type	Bit strings
Fitness function	KNN-based classification error
Number of generations	300
Crossover	Arithmetic crossover
Crossover probability	0.8
Mutation	Uniform mutation
Mutation probability	0.1
Selection scheme	Tournament of size 2
Elite count	4

3.3 Experimental Results

Suitable experiments are performed on three binary cancer microarray gene expression datasets namely leukemia, CNS and colon cancer. The number of attributes obtained by applying the information gain filter and genetic algorithm wrapper using two fitness functions are tabulated in Table 5.

Table 5 Attribute selection using proposed hybrid approach

Dataset	Number of genes in raw dataset	Number of attributes using information gain filter	Number of attributes using genetic algorithm wrapper—fitness 1	Number of attributes using genetic algorithm wrapper—fitness 2
Leukemia	7129	924	181	183
Central nervous system (CNS)	7129	73	10	2
Colon cancer	2000	135	15	2

It is evident from the above table that the proposed fitness function selects 2.56% of the total genes in the leukemia dataset, 0.03% genes from the CNS dataset and 0.1% genes from the colon cancer dataset. The redundant genes are eliminated the informative genes are preserved. The reduced gene subset is subjected to classification using traditional classifiers available in literature. The comparison of the accuracy of different classifiers is shown in Tables 6 and 7.

Table 6 Classifier accuracy for proposed hybrid approach with fitness function 1

	NB	J48	RF	RT	AdaBoostM1	REPTree
Leukemia	95.833	88.889	95.833	83.33	93.05	91.67
Central nervous system (CNS)	76.667	63.33	80	66.66	76.66	71.66
Colon cancer	77.419	80.6452	80.645	75.806	79.023	80.64

Table 7 Classifier accuracy for proposed hybrid approach with fitness function 2

	NB	J48	RF	RT	AdaBoostM1	REPTree
Leukemia	94.44	76.383	93.056	91.667	81.944	79.166
Central nervous system (CNS)	70	78.33	70	71.66	80	70
Colon cancer	74.193	66.129	69.354	80.64	66.129	64.51

The Region of Characteristic (ROC) is plotted using the False Positive Rate (FPR) on the x-axis and True Positive Rate (TPR) on the y-axis as shown in Figs. 1, 2, 3 for leukemia, CNS and colon cancer respectively.

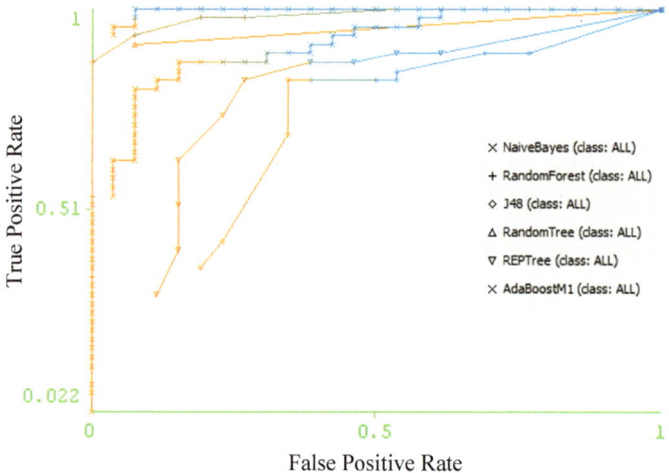

Fig. 1 ROC for Leukemia dataset with fitness function 2

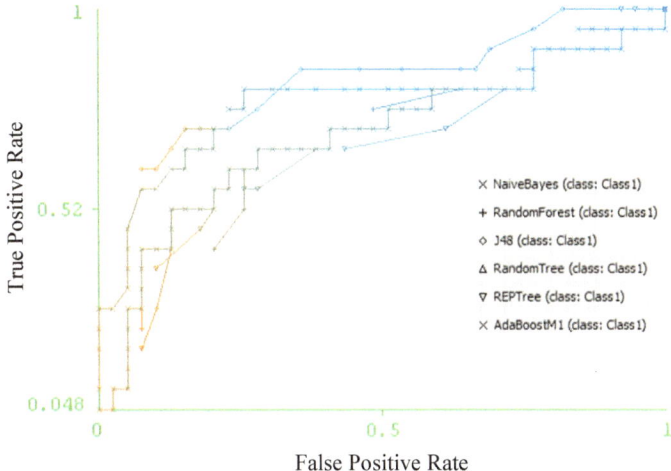

Fig. 2 ROC for CNS dataset with fitness function 2

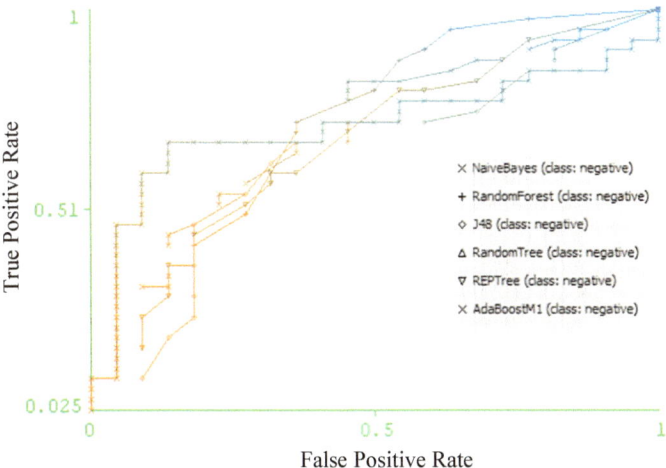

Fig. 3 ROC for Colon cancer dataset with fitness function 2

4 Conclusion

In this paper, a hybrid gene selection method to identify the small set of the most significant genes associated with the cause of cancer is proposed. The proposed method has achieved the goal of integrating a filter and wrapper method to perform attribute selection. The main contribution of this work is devising a fitness function for the genetic algorithm for three binary cancer microarray gene expression datasets namely leukemia, CNS and colon cancer. The results of our proposed technique were compared with the other standard techniques available in literature and our approach produces comparatively higher accuracy in majority of the cases. The fitness function can be further customized for obtaining better results. Different classifiers can also be used to improve the statistical parameters for our research work.

References

1. Eseyin, O.A., Satt, M.A., Rathore H.A.: A review of the pharmacological and biological activities of the aerial parts of Telfairia occidentalis. Trop. J. Pharm. Res. **13**(10), 1761–1769 (2014)
2. Latkowskia, T., Osowskia, S.: Data mining for feature selection in gene expression autism data. Expert Syst. Appl. **42**(2), 864–872 (2015)
3. Huy, P.Q., Ngom, A., Rueda, L.: PAFS—an efficient method for classifier-specific feature selection. In: Proceedings of the IEEE Symposium Series on Computational Intelligence (SSCI) (2016)
4. Pashaei, E., Ozen, M., Aydin, N.: A novel gene selection algorithm for cancer identification based on random forest and particle swarm optimization. In: Proceedings of the IEEE Conference on Computational Intelligence in Bioinformatics and Computational Biology (CIBCB) (2015)
5. Hoseini, E., Mansoori, E.G.: Selecting discriminative features in social media data: an unsupervised approach. Neurocomputing **205**(12), 463–471 (2016)
6. Saleha, A.I., Rabiea, A.H., Abo-Al-Ez, K.M.: A data mining based load forecasting strategy for smart electrical grids. Adv. Eng. Inform. **30**(3), 422–448 (2016)
7. Dai, J., Qing, X.: Attribute selection based on information gain ratio in fuzzy rough set theory with application to tumour classification. Appl. Soft Comput. **13**, 211–221 (2013)
8. Dietterich, T.G.: Approximate statistical tests for comparing supervised classification learning algorithms. Neural Comput. **10**(7), 1895–1923 (1998)
9. Sreepada, R.S., Vipsita, S., Mohapatra, P.: An efficient approach for microarray data classification using filter wrapper hybrid approach. In: Proceedings of IEEE International Advance Computing Conference (IACC) (2015)
10. Das, A.: Digital communication-principles and system modelling. ISBN 978-3-642-12743-4 (2010)
11. Ye, S., Chen, Y., Hu, T.: Evolutionary algorithmic deployment of radio beacons for indoor positioning. In: Proceedings of IEEE Congress on Evolutionary Computation (CEC) (2016)
12. Hsu, H.-H., Hsieh, C.-W., Ming-Da, L.: Hybrid feature selection by combining filters and wrappers. Expert Syst. Appl. **38**(7), 8144–8150 (2011)
13. Oluleye, B., Leisa, A., Leng, J., Dean, D.: A Genetic Algorithm-Based Feature Selection. Int. J. Electr. Commun. Comput. Eng. **5**(4), 899–905 (2014)
14. Oluleye, B., Leisa, A., Leng, J., Dean, D.: Zernike moments and genetic algorithm: tutorial and application. Br. J. Math. Comput. Sci. **4**(15), 2217–2236 (2014)
15. Arunkumar, C., Ramakrishnan, S.: Hybrid information gain based fuzzy roughset feature selection in cancer microarray data. In: Proceedings of IEEE International Conference on Innovations in Power and Advanced Computing Technologies, Vellore Institute of Technology, Vellore, India, 21–22 April 2017
16. Arunkumar, C., Sooraj, M., Ramakrishnan, S.: Finding expressed genes using genetic algorithm and extreme learning machines. In: Proceedings of IEEE International Conference on Advanced Computing and Communication Systems, Sri Eshwar College of Engineering, Coimbatore, India, 6–7 Jan 2017

Application of Evolutionary Particle Swarm Optimization Algorithm in Test Suite Prioritization

Chug Anuradha$^{(\boxtimes)}$ and Narula Neha

University School of Information Technology, GGSIPU, Delhi, India
anuradha@ipu.ac.in, nehanarula665@gmail.com

Abstract. Regression testing is a software verification activity carried out when the software is modified during maintenance phase. To ensure the correctness of the updated software it is suggested to execute the entire test suite again but this would demand large amount of resources. Hence, there is a need to prioritize and execute the test cases in such a way that changed software is tested with maximum coverage of code in minimum time. In this work, Particle Swarm Optimization (PSO) algorithm is used to prioritize test cases based on three benchmark functions Sphere, Rastrigin and Griewank. The result suggests that the test suites are prioritized in least time when Griewank is used as benchmark function to calculate the fitness. This approach approximately saves 80% of the testing efforts in terms of time and manpower since only 1/5 of the prioritized test cases from the entire test suite need to be executed.

Keywords: Particle swarm optimization · Benchmark functions
Prioritization · Regression testing

1 Introduction

Software testing is the one of the most crucial phase in Software Development Life Cycle (SDLC). For the success of any software it is very important that the software is tested thoroughly with intent of finding maximum number of defects while testing. Software testing consumes approximately 50% of the software development efforts [1]. Changing business requirements and market trends urge the need for the software to undergo changes even after the software is delivered and becomes operational. Regression testing ensures to verify the correctness of the changed software and its corresponding affected parts after it becomes operational. Testing the whole software again demands a huge set of resources which is a big constraint in terms of time and money.

Test case prioritization (TCP) ensures that the test cases are arranged using priority value so that the test cases with highest priority are executed first. The test cases can be prioritized either randomly or based on the branch or statement coverage [2] of the software. Prioritizing test cases ensures that the time required to reach a performance goal is optimized [3]. The potential benefit of prioritizing test cases is that it ensures to rank the test cases rather than reducing or discarding the test cases from the test suite [4].

© Springer International Publishing AG 2018
D. J. Hemanth and S. Smys (eds.), *Computational Vision and Bio Inspired Computing*,
Lecture Notes in Computational Vision and Biomechanics 28,
https://doi.org/10.1007/978-3-319-71767-8_2

Particle Swarm Optimization (PSO) is a stochastic population based optimization technique given by Kennedy and Eberhart in 1995 [5] that is motivated by social behavior of fish schooling and bird flocking. The population in PSO called as particles that fly through the problem domain or the search space to reach an optimum solution. PSO is widely used because of its implementation simplicity, dimensions scalability and performance in terms of empirical solutions for global search problems [6]. It is used to solve a number of complex problems like permutation-flop shop sequencing problem [5], real-time embedded systems [7], capacitor-problem [8], soft-sensor [9] and generating test data in the basis of data flow coverage [10].

In this paper, we have implemented PSO with three different Benchmark Functions (BFs) for calculating the fitness value of the test cases. BFs are used to find optimal solutions for optimization problems and to compare and analyze the effectiveness of different optimization algorithms. In this work, we have used three single-objective BFs that are Sphere [5, 11], Rastrigin [5] and Griewank [5, 11, 12]. In current work, for a set of fifteen JAVA programs we first calculate the number of independent paths by drawing control-flow graph (CFG) and decision to decision (DD) graph. For every independent path a set of fifty test cases are generated that are further prioritized using PSO. The test results are driven based on the time taken and the global best value obtained while prioritizing test cases with each BF. The results proof Griewank works best with PSO as it takes approximately 8.90% of the average time taken by Sphere function to prioritize the test cases. It is also proved that Griewank takes approximately 64.17% of the average time taken by Rastrigin function for prioritization.

The paper proposes a way to minimize the testing efforts invariably by substituting different fitness function for calculating fitness of the test cases based on their position. This work suggests of using Griewank as BF with PSO to prioritize test cases as it would help in saving a lot of time, effort and cost during critical project deliveries when resources availability is a crunch. In this work, the test cases are prioritized and not reduced or removed from the test suite.

The rest of the paper is organized as follow: in Sect. 2 literature reviews is discussed. Section 3 covers the concept of test suite prioritization using branch and statement coverage. Section 4 explains PSO algorithm and BFs in detail. Section 5 covers a case study on TCP. In Sect. 6 results are derived based on the case study. In Sect. 7 different threats to validity are discussed. In the last section paper concludes and suggests future work.

2 Related Work

An extensive research work is carried out on test suite prioritization, selection and minimization using different nature inspired meta-heuristic algorithms [13]. An extensive work has been done in the field of test case prioritization, selection and test data generation using PSO. In 2014, Mor [4] has evaluated the effectiveness of different test suite prioritization techniques using Average Percentage of Fault Detected (APFD) metrics. Walcott in year 2006 [14], has given an algorithm for time constraint aware prioritization. Sharma [15] has modified the algorithm for prioritizing the test cases based on timing and APFD metrics in 2014. Hassan et al. worked on Genetic

Algorithm (GA) and PSO and proved that PSO works best amongst the two in year 2006 [12]. Elbeltagia et al. [6] worked on GA, Memetic Algorithm (MA), Ant Colony Optimization (ACO) and PSO and proved that PSO works the best amongst the four algorithms. Nayak [16] has used PSO for automatic test data generation for data flow testing in year 2010. In 2014; Chawla [17] has given a hybrid algorithm for test data using soft-computing technique with PSO and GA.

Harman [18] has done a survey work in his paper in the area of regression testing minimization, prioritization and selection and suggested potential areas and scope for future work in them. In 2011, Kaur [19] has blended PSO with cross-over operator to avoid convergence of population to local best. In May, 2011 Arora [20] has given Hybrid Particle Swarm Optimization (HPSO) that is based on combination of PSO and GA to increase the search space. Routing problems, job-scheduling, task-scheduling problem [21] have been solved using PSO by many researchers in a distributed environment to solve tasks in efficient and cost-effective manner. El-Sherbiny [5] has given an algorithm for particle swarm optimization without using velocity for calculating position of particle. In 2004, Yang [22] has been used PSO to solve NP-hard problems like knapsack. In 2017, Kumar [23] has used PSO to solve NP-complete problem like test suite generation. In year 2007, Hendtlass [24] has worked on PSO algorithm with focus on counting total number of evaluations for calculating fitness.

As far as the literature survey is concerned El-Sherbiny [5] in his work has given an algorithm inspired by PSO for optimization problems that work without calculating velocity at which particle moves in the search space. He has efficiently reduced the number of iterations for calculating the best solutions for an optimization problem. In 2014, Mor [4] has used APFD metrics for measuring the rate of fault detection while prioritizing the test cases based on the coverage criteria provided by different prioritization techniques. He has concluded that higher values of APFD metric provide a better rate of fault detection. In 2006, Walcott [14] has given an algorithm for prioritizing test cases using GA in a time constraint environment for systems like PlanetLab and MonetDB that uses the concept of nightly builds and unit testing of the software. In 2010, Nayak [16] has simulated GA and PSO to generate automatic test data for data flow testing. His work proofs that PSO outperforms GA by 100% in def-use coverage. In 2014, Chawla [17] has proposed a hybrid algorithm based on PSO and GA to automate test data generation and the effectiveness of the algorithm is confirmed using percentage of fault coverage against unit of time and percentage of fault detected by generated test cases.

In 2011, Kaur [19] has given a Hybrid Particle Swarm Optimization Algorithm (HPSO) that combines the techniques for PSO and GA for widening the search space. In HPSO the initial population of PSO is mutated before performing rest of the steps of PSO algorithm to improve average percentage of fault detected. In 2010, Harman has performed both detailed analysis and survey in the field of regression test prioritization, selection and minimization. And his work suggested how the three terms are related to each other in terms of implementation. In 2012, Ming [25] et al. has combined mutative scale chaos and PSO to mitigate the problem of slow convergence and local optimum points of PSO algorithm. In 2007, Hendtlass [24] has used three fitness functions Sphere, Rastrigin and Schwefel's function in different number of dimensions. He has proved that best fitness evaluation is found using Schwefel's function. In the paper, we have referred the research work done after the year 2000 on test data generation, prioritization and optimization during software testing.

3 Test Case Prioritization

To make regression testing cost-effective we need to prioritize the test cases in such a manner that after executing the prioritized test cases it's ensured that the changed part of the software is tested effectively. Test case coverage is a way to describe how well the code is covered by a given test suite. Therefore, while testing the software it becomes very essential to design an effective test suite that uncovers maximum number of defects in the software. To save the testing effort there is a need of ranking the test cases based on criteria that fulfills the prerequisite of uncovering maximum defects in the updated software. Different prioritization techniques are used for prioritizing test suite but in this work we focus on two aspects of test case coverage [7] as described below.

3.1 Prioritizing Based on Statement Coverage

Software coverage (SC) is based on the concept of executing all the statements in the code at least once for a given set of requirement. SC can be achieved with a basic knowledge of code structure and by using flowchart for the software workflow. SC is represented by a metrics that defines the number of statements covered by the test case in a code block. In any program SC is calculated using the formula specified in Eq. (1). For a given procedure P as shown in Fig. 1a the number of statements covered by a set of three test cases is represented using statement coverage metrics in Fig. 1b.

$$\text{Statement Coverage} = \frac{\text{Number of executed statement}}{\text{Total number of statements}} \times 100 \tag{1}$$

```
Procedure P
1 int a=5, b=0;
2 while (a>0)
3   If(a%2==0)
4         Print "Even"
5   Else
6         Print "Odd"
7   End If
8 End While
9 Print "Completed."
```

1(a) Procedure P

Statement Coverage	TC 01	TC 02	TC 03
1	X	X	X
2	X	X	X
3		X	
4		X	
5	X		
6	X		
7	X	X	
8	X	X	
9	X	X	X

1(b) Statement Coverage for Procedure P

Branch Coverage	TC 01	TC 02	TC 03
Entry	X	X	X
2 True	X	X	
2 False			X
3 True	X		
3 False		X	

1(c) Branch Coverage for Procedure P

Fig. 1 For a program P shown in **a** that consists of two branch nodes the corresponding matrix for calculating the statement and branch coverage of the program is depicted in **b** and **c** respectively

3.2 Prioritizing Based on Branch Coverage

Branch coverage (BC) is also known as all-edges coverage or decision coverage. BC focuses on the aspect of covering all the branches or simply all the true and false conditions in the code. Prioritizing test cases on the basis of BC focuses on ranking the test cases on the basis of total number of decision or branch nodes covered by the test case. In this study we have designed the test cases in such a manner that all the branch conditions in a program are covered by the test cases. A test suite that covers all the decision nodes in the software would automatically give 100% of SC. The BC for any given software can be calculated using the formula specified in Eq. (2). For the procedure P in Fig. 1a the BC is represented by listing all the branches and the corresponding test cases that cover these branched is shown in Fig. 1c.

$$\text{Branch Coverage} = \frac{\text{Number of decisions outcome exercised}}{\text{Total number of decision outcomes}} \times 100 \qquad (2)$$

4 Particle Swarm Optimization

PSO is an evolutionary process derived on the basis of social behavior of birds migrating to a destination that is currently unknown [1]. Each bird in the swarm is called a particle and for each particle we need to optimize its position in the swarm. In PSO every particle in the swarm has a position and velocity associated with it in an n-dimensional space. Every particle in the swarm strives for two best values, the first one called as pBest that is the best fitness value retrieved by particle so far and another is called gBest that is the best fitness value achieved so far amongst all the particles.

In PSO the size of the swarm population is denoted as S, $S \in R$ (S belongs to set of real number) and the position and velocity vectors are defined as follow: $X_i = (x_{i1}, x_{i2}, x_{i3} \ldots x_{ik})$ and $V_i = (v_{i1}, v_{i2}, v_{i3} \ldots v_{ik})$. On the basis of position of the particle in the swarm the fitness is calculated using fitness function and is denoted as $F_i = (f_{i1}, f_{i2}, f_{i3} \ldots f_{ik})$. The fittest particle found at a given point of time is denoted as $F_g = (f_{g1}, f_{g2}, f_{g3} \ldots f_{gk})$.

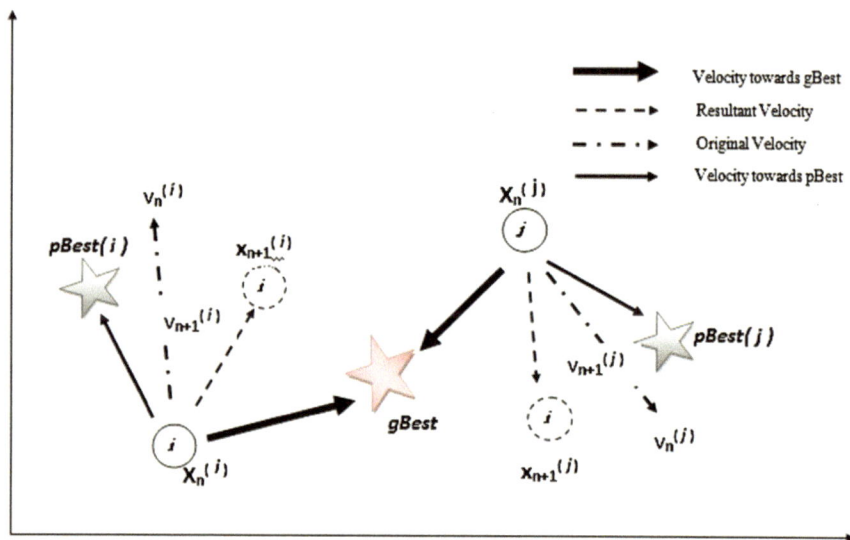

Fig. 2 The typical movement of two particles $X_n^{(i)}$ and $X_n^{(j)}$ particles in a swarm is depicted that changes its position and velocity to reach the global best value

The typical scenario in PSO is represented in Fig. 2 for two particles i and j that fly in the search space to reach a gBest. To understand the movement of a particle in swarm towards gBest lets understand how particle i move in search space. The particle *i* have an initial position and velocity denoted by X_n and V_n. It is moving towards the gBest by updating its position to X_{n+1} and the resultant velocity to V_{n+1}. The best position of the particle i attained so far is maintained in variable pBest.

A pseudo code for a generic PSO algorithm is depicted as follow. The algorithm begins by initializing the swarm population with velocity, position and fitness value. After the population is initialized the particle's velocity and position is updated for a number of iterations. The algorithm checks and updates the pBest of particle if it is greater than its current fitness value. Similarly if pBest is less than gBest the algorithm updates gBest value with pBest value of current particle.

Generic PSO Algorithm:
Input: A swarm of particles that fly in search space to reach a position that is currently unknown.
Output: The global best position in the swarm is found.
Begin
1. For i = Total number of particles
2.   For j=1 to Total dimensions
3.   Initialize the velocity, position and fitness of the particle.
4.   Initialize the particles personal best cost and personal best position same as initial position and velocity.
5.   End, End
6. Iteration = 0
7. While (iteration < Maximum Number of Iterations)
8.   For i = 1 to Total number of particles
9.   For j=1 to total dimensions
10.   Update particle's position and velocity.
11.   Calculate particle's fitness value
12.   If (Fitness (*particle*) < Fitness (*pBest*))
13.   Update pBest with current position of particle.
14.   If (Fitness (*pBest*) < Fitness (*gBest*))
15.   Update global best position i.e., gBest with pBest value.
16.   End, End, End, End
17. Until iteration= Maximum Number of Iterations

4.1   Mapping PSO to Test Suite Prioritization

In this work, we have used PSO for prioritizing the test cases to save testing efforts during regression testing. To understand how PSO fits in Test Suite Prioritization (TSP) problem; in this section mapping of PSO to TSP is explained. In the test suite the test cases are equivalent to particles in the swarm. For every test case we calculate initial velocity and position using formula's defined in Eqs. (3) and (4) respectively and the initial velocity of the test case also incorporates the average value of the test data. The aim is to reduce the cost associated with the test case that is equivalent to the fitness value in generic PSO algorithm specified in pseudo code Generic PSO Algorithm. The initial fitness value for the test case is calculated using formula specified in Eq. (5) that takes position of the test case and dimension as an input.

$$v_{ik} = (maxX - minX) * rand + minX + (mean) \qquad (3)$$

$$x_{ik} = (maxX - minX) * rand + minX + (mean) + v_{ik} \qquad (4)$$

$$f_{ik} = CostFunction(x_{ik}, dimension) \qquad (5)$$

The process to find candidate solution is repeated for a set of iterations and during each iteration particle updates its velocity and position on the basis of formula given in Eqs. (6) and (7) [7] respectively.

$$V_{ik} + 1 = w.V_{ik} + c_1.rand.(f_{ik} - x_{ik}) + c_2.rand.(f_{gk} - x_{ik}) + (mean) \qquad (6)$$

where

V_{ik}	velocity of particle i at iteration
w	weighing function
c_1, c_2	weighing factors
f_{ik}	particles current best position
x_{ik}	particles current position
f_{gk}	global best
rand	random number in range [0, 1]

$$X_{ik+1} = x_{ik} + v_{ik+1} \qquad (7)$$

The value of weighing function, w is calculated using formula given in Eq. (8). The weighing factor suggest how well the particle is moving towards the best solution.

$$w = minX - (t/MaxGeneration) * (maxX - minX) \qquad (8)$$

where
minx	Initial weight
maxX	Final weight
MaxGeneration	Total number of iterations
t	represents current iteration.

The parameters and variables that are used in implementing PSO for test suite prioritization are listed in Table 1.

Table 1 PSO parameters

Weighing factors	$c_1 = 2.0, c_2 = 2.0$	Dimension	1
Initial weight (minX)	0.0	No of particles/test cases	50
Final weight (maxX)	2.0	Number of Iterations	20

4.2   Fitness Calculation in PSO

The fitness or the cost value of a particle in PSO can be calculated using different single or multi-objective functions. The objective function is also known as benchmark function and is used for evaluating the effectiveness of optimization problems based on factors such as performance, convergence rate, precision and robustness.

In this work, the BFs are used to calculate the cost value associated with the test case based on the position and velocity of test case in the test suite. In this paper, we have used three single-objective functions as shown in Table 2 [11, 24] to calculate the fitness of the test cases using formula specified in Eq. (5). The performance of the three functions is compared with each other in terms of time to identify which work best with PSO for prioritizing the test cases.

In this work, the BFs are used to calculate the cost value associated with the test case based on the position and velocity of test case in the test suite. In this paper, we have used three single-objective functions as shown in Table 2 [11, 24] to calculate the fitness of the test cases using formula specified in Eq. (6). The performance of the three functions is compared with each other in terms of time to identify which work best with PSO for prioritizing the test cases.

Table 2 Benchmark functions

Function	Characteristics	Dim	Range
Sphere $f(x) = \sum_{i=1}^{d} x_i^2$	Continuous, convex, unimodal	1	$[-100, 100]^D$
Rastrigin $f(x) = 10d + \sum_{i=1}^{d} \left[x_i^2 - 10\cos(2\pi x_i) \right]$	Local minima, highly multimodal	1	$[-5.12, 5.12]^D$
Griewank $f(x) = \sum_{i=1}^{d} \frac{x_i^2}{4000} - \prod_{i=1}^{d} \cos\left(\frac{x_i}{\sqrt{i}}\right) + 1$	Continuous, differentiable, non-separable, scalable, multimodal, regularly distributed local minima's	1	$[-600, 600]^D$
Optimal Value for all functions is 0			

4.3 Proposed Algorithm

The proposed algorithm for prioritizing test cases using PSO is given in pseudo code as follow [25, 26]. The process of prioritizing test cases begins by identifying independent paths in program by drawing CFG and DD-graph. For the identified paths the test cases are generated based on decision nodes in the code that guarantees 100% BC and SC [7].

Proposed PSO Algorithm:
Input: A number of test suites based on number of independent paths in the program.
Output: Test cases prioritized on the basis of cost or fitness value.
Begin
1.   Identify independent paths of the program under test.
2.   For every independent path generate test data based on branch condition in the program.
3.   For paths= Number of independent paths
4.   For i = Total number of test cases
5.   For j=1 to Total dimensions
6.   Initialize the velocity and position for each test case and calculate the cost associated with test case using a benchmark function specified in table 2.
7.   Initialize the test case personal best cost and personal best position same as initial cost and position.
8.   End
9.   End
10. Iteration = 0
11.   While (iteration < Maximum Number of Iterations)
12.   For i = Total number of test cases
13.   For j=1 = Total dimensions
14.   Pick random number in range [0,1] and update position and velocity of the test case.
15.   Update cost value of the test case using updated particle's position for the selected benchmark function at step 3.
16.   If (Fitness (*particle*) < Fitness (*pBest*))
17.   Update test case best known position i.e., pBest with current position of test case.
18.   If (Fitness (*pBest*) <Fitness (*gBest*))
19.   Update global best position i.e., gBest with pBest of current test case.
20.   End
21.   End
22.   End
23. End
24. Until iteration= Maximum Number of Iterations
25. End //Repeat steps 4 -24 for every path in program.

The algorithm begins by initializing velocity, position and fitness using formula 3, 4, and 5 for each test case on the path. The initial value for pBest position and pBest cost of the test case is same as that of the test case position and cost. At step 11, the counter iterates for a pre-defined number of iteration. For every iteration the velocity and position of the test cases are updated using formula 6 and 7. The algorithm verifies if pBest fitness value of the test case is greater than current fitness value of test case; the pBest is updated with current fitness of test case. Similarly, the algorithm validates for fitness value of gBest and pBest; and updates gBest with pBest fitness of the test case. The algorithm also incorporates the dimension variable while iterating the test cases.

4.4 Flow Chart of Proposed Algorithm

The flow chart for the algorithm proposed in this work is shown in Fig. 3. The algorithm begins by initializing the test cases or swarm population with random values. The global best is assigned a maximum value and aim is to minimize the value for gBest after performing a set of iterations. The result achieved at the end of the algorithm is a set of prioritized test cases sorted on the basis of increasing order of pBest.

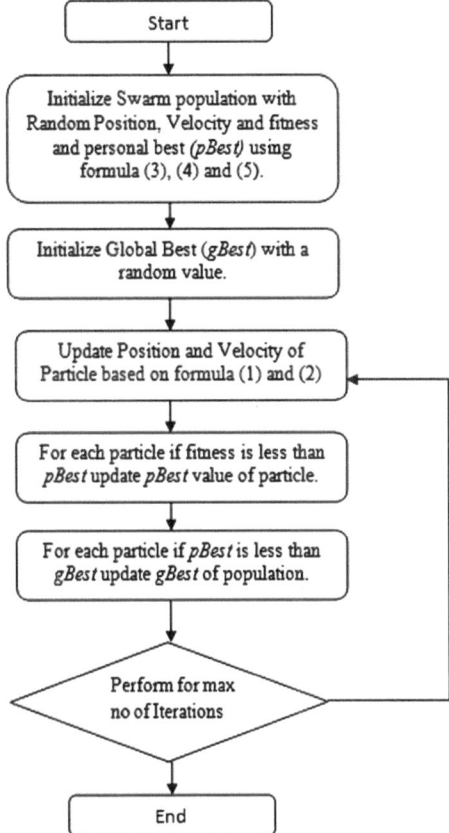

Fig. 3 Flow chart for PSO prioritization

5 Analysis and Evaluation

This section includes the steps of data collection followed by implementing PSO for a case study to derive results and showcase how the test cases are prioritized on the basis of fitness value using different BFs.

5.1   Data Collection

In this work, we have implemented PSO for a set of fifteen JAVA programs using Eclipse as an Integrated Development Environment (IDE). The test programs along with their number of input and their independent paths are listed in Table 3. The PSO algorithm is performed for twenty iterations for every test path and the dimension value used is 1.

Table 3 Test programs

Program no	Program	No of inputs	No of paths
P1	A calculator program	3	6
P2	Program to find leap year or not	1	4
P3	Calculate area and circumference of circle	3	4
P4	To check number is a Armstrong no	1	3
P5	To calculate factorial of number	1	2
P6	To swap two numbers	2	2
P7	To generate Floyd triangle	1	2
P8	Student classification problem	3	6
P9	To compare two numbers	2	3
P10	To generate fibonacci series	1	2
P11	To generate inverted triangle	3	3
P12	To check whether a number is palindrome or not	1	3
P13	Roots of a quadratic equation	3	3
P14	Triangle classification problem	3	5
P15	Find square root of number	1	2

5.2   Case Study: Calculator

In this work, calculator program is used as a case study for understanding PSO implementation and the code for which is shown in Fig. 4. The program has 32 Line of Code (LOC) and expects three inputs from the user; input 1 specifies the operation and input 2 and input 3 are integers on which operation needs to be performed. The DD-graph is drawn as shown in Fig. 5 that serves the basis of generating the test cases based on branch nodes.

```
Calculator.java
1. public class Calculator {                              16.        switch (operation)
2. public static void main(String args[]) {               17.          {
3. Scanner in = new Scanner(System.in);                    18.          case 1:
4. int num1 = 0, num2 = 0, operation = 0;                  19.              result = (double) num1 + num2; break;
5. double result = 0.0;                                    20.          case 2:
6. System.out.println("Please enter a choice value:" + "\n" + "1.Add" +   21.              result = (double) num1 - num2;break;
   "\n" + "2.Subtract" + "\n" +     "3.Multiply"+ "\n" + "4.Division");    22.          case 3:
7. operation = in.nextInt();                               23               result = (double) num1 * num2;break;
8. if (operation == 1 || operation == 2 || operation == 3 || operation == 24.          case 4:
4)                                                         25.     result = (double) num1 / num2; break;
9. {        System.out.println("Enter 1st Number in range of 1-100");     26.}
10.         num1 = in.nextInt();                           27.          System.out.println("Result: " + result); }
11.         System.out.println("Enter 2nd number in range of 1-100");     28.}
12.         num2 = in. nextInt();                          29. else
13.         if (num1 < 1 || num1 > 100 || num2 < 1 || num2 > 100)         30.          System.out.println("Invalid operation selected");
14.         System.out.println("Invalid input values entered");          31.}
15.         else {                                         32.}
```

Fig. 4 Calculator program implemented in JAVA

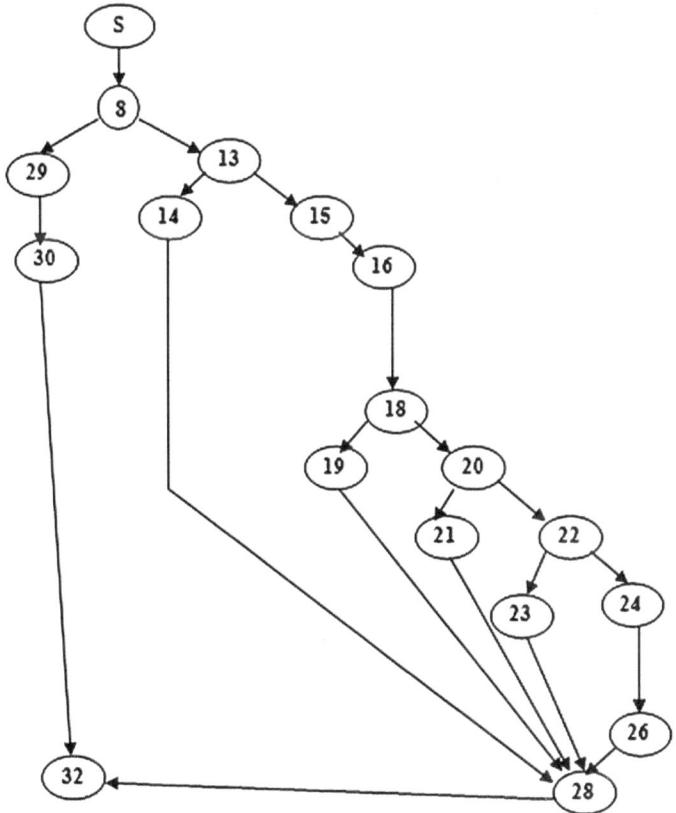

Fig. 5 DD-graph for program calculator.java

In Table 4 the independent paths for the DD-graph drawn in Fig. 5 are shown. For every identified path in Table 4 a set of 50 test cases are generated using decision nodes as boundary criteria while generating test cases.

Table 4 Independent paths for calculator.java

IP1 S 8 29 30 32	IP4 S 8 13 15 16 18 20 21 28 32
IP2 S 8 13 14 28 32	IP5 S 8 13 15 16 18 20 22 23 28 32
IP3 S 8 13 1516 18 19 28 32	IP6 S 8 13 15 16 18 20 22 24 25 26 28 32

In Table 5 two test cases for each path based on the boundary condition is shown along with the average value for the test data is given. The mean value is valued for initializing the velocity of the test case using Eq. (3). The values for input 2 and input 3 are in range of [0, 100].

Table 5 Test data for the independent paths

Path	Boundary condition	Test case	Parameters			Mean (Input to PSO)
			Input 1	Input 2	Input 3	
IP1	Input 1 <> {1, 2, 3,4}	1	5	7	8	6.66
		2	8	8	9	8.33
IP2	Input 1 = {1, 2, 3, 4}; Input 2 > 100 || input 2 < 0; input 3 > 100 || input 3 < 0	1	1	−25	25	−16.33
		2	3	24	101	42.67
IP3	Input 1 = 1; 0 <= input 2 <= 100; 0 <= input 3 <= 100	1	1	1	2	1.33
		2	1	3	4	2.67
IP4	Input 1 = 2; 0 <= input 2 < = 100; 0 <= input 3 < = 100	1	2	1	2	1.67
		2	2	3	4	3
IP5	Input 1 = 3; 0 <= input 2 <= 100; 0 <= input 3 <= 100	1	3	2	1	2
		2	3	4	3	3.33
IP6	Input 1 = 4; 0 <= input 2 <= 100; 0 <= input 3 <= 100	1	4	2	1	2.33
		2	4	4	3	3.66

5.3   Empirical Results

The PSO algorithm specified in Fig. 4 is run on all the independent paths of Table 4. The algorithm takes mean value of the test data generated for calculating the velocity, position and fitness of the test cases. At the end of the algorithm we get test eighteen prioritized test suite that includes three test suites for each path based on three BFs applied on a test suite. The results drawn are not repeatable since calculations for PSO variables uses random numbers. The time taken for prioritization may vary depending on the test environment and machine configuration.

In Table 6 the prioritized test cases for six paths in the calculator program using PSO for three BFs is shown. The results show top ten prioritized test cases amongst the fifty test cases for each path. The gBest value achieved for all the paths using the BFs is listed in the table and the gBest for each path matches the pBest value of the first prioritized test case. The total time taken for prioritizing all the test paths using Sphere, Rastrigin and Griewank is summarized in the last row of the Table.

Table 6 Test suite prioritized for calculator program using PSO

Path	Sphere		Rastrigin		Griewank	
	Prioritized test cases	gBest	Prioritized test cases	gBest	Prioritized test cases	gBest
IP1	1, 6, 5, 20, 16, 12, 0, 19, 10, 2	0.004726	1, 16, 0, 32, 30, 9, 2, 26, 5, 35	0.000465	43, 16, 1, 0, 27, 47, 20, 23, 22, 6	0.000991
IP2	45, 16, 44, 15, 3, 13, 18, 25, 1, 30	0.009869	0, 28, 43, 4, 29, 21, 20, 18, 19, 25	0.009865	48, 28, 27, 43, 3, 35, 33, 14, 1,24	0.009865
IP3	28, 49, 45, 4, 18, 43, 8, 41, 44, 38	0.000003	11, 5, 7, 4, 48, 34, 26, 49, 42, 44	0.000004	6, 38, 37, 39, 47, 7, 15, 11, 36, 0	0.001366
IP4	39, 38, 47, 3, 36, 23, 16, 34, 42, 7	0.000015	1, 5, 35, 8, 4, 27, 31, 42, 32, 3	0.001010	38, 13, 8, 25, 49, 18, 12, 6, 4, 23	0.009917
IP5	26, 49, 48, 9, 5, 47, 2, 25, 32, 23	0.001612	12, 26, 22, 40, 27, 34, 7, 14, 47, 1	0.002611	45, 40, 2, 5, 49, 31, 38, 1, 41, 21	0.001228
IP6	30, 18, 39, 21, 49, 14, 4, 2, 37, 27	0.000010	4, 45, 42, 36, 39, 43, 21, 31, 49, 33	0.000611	42, 43, 9, 46, 25, 37, 41, 35, 0, 5	0.001267
Time (ms)	52		8		6	

6 Results and Interpretation

In order to prove the effectiveness of PSO with the specified BFs we ran PSO on fifteen JAVA programs under same test environment to produce unbiased results. In Table 7 we have listed the results derived during this work in terms of the time taken by the algorithm and the global best value reached for every independent path. For implementation simplicity we have shown the value for global best value up to five decimal places for all the identified paths in the program. The last column of the table i.e., the time taken for prioritization is calculated by addition of time taken for prioritizing each independent path for the program. The least time taken by PSO amongst the three functions is highlighted as bold in the table and where a path doesn't exist the entry is shown as **X**.

The time taken by Griewank is approximately 8.90% of the average time taken by Sphere function and 64.17% of the average time taken by Rastrigin function. The results prove that Griewank works best amongst the three functions with PSO by taking into consideration the execution time and value of gBest achieved in comparison to the optimal value of the benchmark functions as specified in Table 2.

Table 7 Results (Time taken by the algorithm) and the global best value for every independent path

P	BF	Global best (gBest)						Mean time
		Path 1	Path 2	Path 3	Path 4	Path 5	Path 6	
P1	Sphere	0.00231	0.00075	0.00000	0.00254	0.00115	0.00042	59
	Rastrigin	0.00995	0.00986	0.00000	0.00077	0.00161	0.00034	8
	Griewank	0.00988	0.00986	0.00000	0.00992	0.00063	0.00687	**6**
P2	Sphere	0.00236	409.565	9.69788	X	X	X	63
	Rastrigin	0.00996	0.30067	0.00002				4
	Griewank	0.00044	0.42755	0.41175				**3**
P3	Sphere	2.35024	0.07074	0.00686	0.03865	X	X	55
	Rastrigin	0.00314	0.01003	0.00085	0.00336			12
	Griewank	0.05675	0.01127	0.00637	0.01421			**9**
P4	Sphere	0.00001	0.00061	0.00024	X	X	X	49
	Rastrigin	0.00036	0.00060	0.00021				4
	Griewank	0.00002	0.01452	0.00159				**2**
P5	Sphere	0.00001	0.02694	X	X	X	X	47
	Rastrigin	0.00036	0.00222					4
	Griewank	0.00128	0.00055					**2**
P6	Sphere	0.68587	0.00076	X	X	X	X	42
	Rastrigin	0.03897	0.00740					5
	Griewank	0.00987	0.00330					**3**
P7	Sphere	0.02826	0.00869	X	X	X	X	43
	Rastrigin	0.00099	0.00411					**6**
	Griewank	0.08740	0.00019					**4**

(continued)

Table 7 (*continued*)

P	BF	Global best (gBest)						Mean time
		Path 1	Path 2	Path 3	Path 4	Path 5	Path 6	
P8	Sphere	0.01021	0.09533	0.00620	0.00073	0.02167	0.00011	50
	Rastrigin	0.02753	0.00848	0.00794	0.01016	0.00005	0.00019	8
	Griewank	0.08981	0.16629	0.03268	0.00710	0.00053	0.00115	5
P9	Sphere	0.00018	0.01118	0.02522	X	X	X	38
	Rastrigin	0.00612	0.01024	0.00026				8
	Griewank	0.00990	0.01149	0.01509				7
P10	Sphere	0.09492	0.01024	X	X	X	X	43
	Rastrigin	0.02038	0.00714					5
	Griewank	0.00604	0.01190					2
P11	Sphere	0.64828	0.02668	0.00068	X	X	X	44
	Rastrigin	0.06147	0.00174	0.00982				4
	Griewank	0.01014	0.00000	0.00001				4
P12	Sphere	0.00068	5632.89	159.076	X	X	X	49
	Rastrigin	0.04709	7.91535	0.16338				5
	Griewank	0.00077	14.2083	0.78569				3
P13	Sphere	0.00015	0.05450	0.00595	X	X	X	50
	Rastrigin	0.00279	0.00298	0.01003				4
	Griewank	0.00001	0.01432	0.01013				2
P14	Sphere	0.00204	0.10044	0.00246	0.08367	0.02341		58
	Rastrigin	0.00036	0.01834	0.01393	0.01018	0.00992		7
	Griewank	0.00378	0.00989	0.00042	0.00090	0.01061		4
P15	Sphere	0.03100	0.49967	X	X	X	X	40
	Rastrigin	0.00818	0.00747					3
	Griewank	0.01020	0.00146					2

The bar graph for the results driven in Table 7 is shown in Fig. 6. In the graph x-axis represents the 15 programs that have been taken in this work and y-axis represent the mean time taken by PSO for prioritizing test cases for the selected programs using three functions: Sphere, Rastrigin and Griewank. The results drawn are not repeatable as time may vary based on system configuration and the programming language on which the experiment is performed.

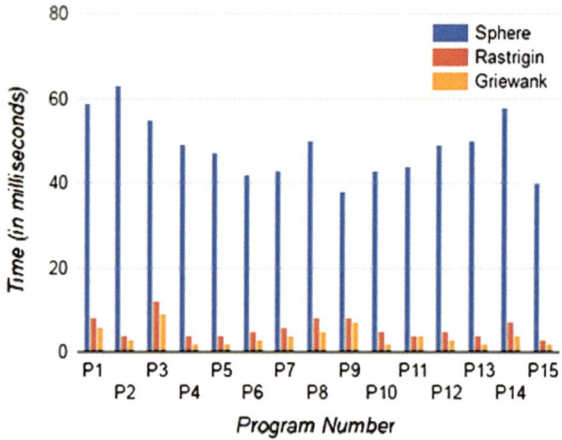

Fig. 6 Time taken by benchmark functions with PSO

7 Threats to Validity

This empirical investigation is carried out on Eclipse that may not be the correct representative for all the software's available in the market but we have tried to the best of our ability to generalize our results for different scenarios. For making our results unbiased and generalized we have followed the basic concepts of object-oriented architecture while implementing fifteen program in JAVA. The results presented in this empirical investigation would not be complete without the discussion of threats of validity: External, Internal and Constraint.

External Validity means the degree of to which the results can be generalized and includes the factor that impact our ability for generalizing the results. The main treat to external validity in this work is that the test programs are small and medium with similar fault patterns and that might not truly representing large scope of programs.

Internal Validity is defined as the degree to measure the consequences on change of independent variables on our dependent variables. In this work, we have minimized this effect by selecting the range of random number between [0–1] for calculation of PSO variables.

Constraint Validity is the degree to which the results are appropriately captured using independent and dependent variables. The way in which random numbers generated for calculating PSO variables may vary depending on the framework used for JAVA programming. In this work, we have tried to minimize this threat using eclipse for JAVA development that is widely used in many organizations for large and complex program.

8 Conclusion and Future Work

In this work, we have used three BFs to calculate fitness of test cases and then sorted test suite based on fitness value. We have verified the effectiveness of Griewank function with PSO in terms of global best value achieved and time required for prioritization of test cases. The results are derived by running fifteen java programs with three BFs and it is proved Griewank works best with PSO in terms of average time required for prioritizing test cases.

In future we plan to run the algorithm for programs with 1000 and more LOC and verify the results under the same test environment. The effectiveness of the approach suggested could we further improve using more BFs or either by combining one or more BFs while calculating fitness of particles in PSO algorithm. The algorithm can also be blended with other meta-heuristic algorithm like GA, MA and ACO. A hybrid PSO can be designed that uses a combination of the three functions for test suite prioritization.

References

1. Joseph, A.K., Radhamani, G.: A hybrid model of particle swarm optimization (PSO) and artificial bee colony (ABC) algorithm for test case optimization. Int. J. Comput. Sci. Eng. (IJCSE) **3**(5) (2011)
2. Suri, B., Singhal, S.: Implementing ant colony optimization for test case selection and prioritization. Int. J. Comput. Sci. Eng. **3**(5), 1924–1932 (2011)
3. Rothermel, G., Untch, R.H., Chu, C., Harrold, M.J.: Prioritizing test cases for regression testing. IEEE Trans. Software Eng. **27**(10), 929–948 (2001)
4. Mor, M.A.: Evaluate the effectiveness of test suite prioritization techniques using APFD metric. IOSR J. (IOSR Journal of Computer Engineering) **1**(16), 47–51
5. El-Sherbiny, M.M.: Particle swarm inspired optimization algorithm without velocity equation. Egypt. Inf. J. **12**(1), 1–8 (2011)
6. Malhotra, R., Khari, M., Molga. M., Smutnicki, C.: Test suite optimization using mutated artificial bee colony. In: Proceedings of International Conference on Advances in Communication, Network, and Computing, CNC, Elsevier, pp. 45–54 (2014)
7. Hla, K.H.S., Choi, Y., Park, J.S.: Applying particle swarm optimization to prioritizing test cases for embedded real time software retesting. In: IEEE 8th International Conference on Computer and Information Technology Workshops, 2008. CIT Workshops 2008, pp. 527–532, IEEE, July 2008
8. Oo, N.W.: A comparison study on particle swarm and evolutionary particle swarm optimization using capacitor placement problem. In: Power and Energy Conference, 2008. PEC on 2008. IEEE 2nd International, pp. 1208–1211. IEEE, December 2008
9. Wang, H., Qian, F.: An improved particle swarm optimizer with behavior-distance models and its application in soft-sensor. In: 7th World Congress on Intelligent Control and Automation, 2008. WCICA 2008, pp. 4473–4478. IEEE, June 2008
10. Singla, S., Kumar, D., Rai, H.M., Singla, P.: A hybrid PSO approach to automate test data generation for data flow coverage with dominance concepts. Int. J. Adv. Sci. Technol. **37**, 15–26 (2011)
11. Jamil, M., Yang, X.-S.: A literature survey of benchmark functions for global optimisation problems. Int. J. Math. Modell. Numer. Optimisation **4**(2) (2013)

12. Hassan, R., Cohanim, B., De Weck, O., Venter, G.: A comparison of particle swarm optimization and the genetic algorithm. In: 46th AIAA/ASME/ASCE/AHS/ASC Structures, Structural Dynamics and Materials Conference, p. 1897 (2005)
13. Yang, X.S.: Nature-inspired metaheuristic algorithms. Luniver Press (2010)
14. Walcott, K.R., Soffa, M.L., Kapfhammer, G.M., Roos, R.S.: Time aware test suite prioritization. In: Proceedings of the 2006 International Symposium on Software Testing and Analysis, pp. 1–12. ACM, July 2006
15. Sharma, I., Kaur, J., Sahni, M.: A test case prioritization approach in regression testing. Int. J. Comput. Sci. Mob. Comput. **3**, 607–614 (2014)
16. Nayak, N., Mohapatra, D.P.: Automatic test data generation for data flow testing using particle swarm optimization. Contemp. Comput, 1–12 (2010)
17. Chawla, P., Chana, I., Rana, A.: A novel strategy for automatic test data generation using soft computing technique. Frontiers Comput. Sci. **9**(3), 346–363 (2015)
18. Yoo, S., Harman, M.: Regression testing minimization, selection and prioritization: a survey. Softw. Test. Verification Reliab. **22**(2), 67–120 (2012)
19. Kaur, A., Bhatt, D.: Particle swarm optimization with cross-over operator for prioritization in regression testing. Int. J. Comput. Appl. **27**(10) (2011)
20. Kaur, A., Bhatt, D.: Hybrid particle swarm optimization for regression testing. Int. J. Comput. Sci. Eng. **3**(5), 1815–1824 (2011)
21. Kong, X., Sun, J., Xu, W.: Particle swarm algorithm for tasks scheduling in distributed heterogeneous system. In: ISDA'06. Sixth International Conference on Intelligent Systems Design and Applications, 2006, Vol. 2, pp. 690–695. IEEE October 2006
22. Zhi, X.H., Xing, X.L., Wang, Q.X., Zhang, L.H., Yang, X.W., Zhou, C.G., Liang, Y.C.: A discrete PSO method for generalized TSP problem. In: Proceedings of 2004 International Conference on Machine Learning and Cybernetics, 2004, Vol. 4, pp. 2378–2383. IEEE (July–August) 2014
23. Kumar, S., Ranjan, P.: A comprehensive analysis for software fault detection and prediction using computational intelligence techniques. Int. J. Comput. Intell. Res. **13**(1), 65–78 (2017)
24. Hendtlass, T.: Fitness estimation and the particle swarm optimisation algorithm. In: IEEE Congress on Evolutionary Computation, 2007. CEC 2007, pp. 4266–4272. IEEE, September 2007
25. Chen, M., Wang, T., Feng, J., Tang, Y.Y., Zhao, L.X.: A hybrid particle swarm optimization improved by mutative scale chaos algorithm. In: 2012 Fourth International Conference on Computational and Information Sciences (ICCIS), pp. 321–324. IEEE, August 2012
26. Molga, M., Smutnicki, C.: Test functions for optimization needs. http://www.zsd.ict.pwr. wroc.pl/files/docs/functions.pdf

An Autonomous Trust Model for Cloud Integrated Framework

C. K. Shyamala[✉] and Ashwathi Chandran

Department of Computer Science and Engineering, Amrita Vishwa
Vidyapeetham, Coimbatore, India
ck_shyamala@cb.amrita.edu, cb.en.p2csel5001@cb.
students.amrita.edu

Abstract. Cloud integrated frameworks (CIF) lay the ground work for controlling and managing data gathered from various sources. Data quality trust addresses detection of manipulation, concealment and rollbacks on data exchanged in CIF. The work presented here proposes a trust enhanced CIF that comprehensively takes into account processing, transmission, privacy, data quality and recommendation in addition to availability, bandwidth of the servicing. A weighted average computation using weights derived from Service Level Agreement given by CSU is proposed over a moving window on the performances maintained by Trusted Centre Entity (TCE). TCE monitors the system for malicious behaviours of CSPs that may change the behaviour of the system. It assures data quality using Merkle hash tree (MHT) and manages the trust and reputation scores for the CSPs in the trust enhanced CIF. It performs public auditability to detect data roll back, data concealment and incorrect data. The work proposes a scheme that provides for autonomous trust management in CIFs.

Keywords: Cloud integrated framework · Trust · Data quality trust
Merkle hash tree

1 Introduction

Integrating the dominant abilities of cloud computing and data assembling ability of data networks, cloud integrated framework (CIF) received much attention in many areas. CIF is a foundation for enabling control, monitor, and management of data gathered from data gathering networks. The explosion of Cloud integrated frameworks played an important role in academia and industry. The emergence of such a framework unlocked many capabilities for networked processing. The integrated framework typically provides for service invocation, management of data streams, processing of data, and analysis of datasets for decision making. A typical CIF needs to address the challenges of interoperability, data validation security, data consistency, privacy and heterogeneity. Interestingly, interactions and processing in CIFs demand adequate levels of trust among its networked entities. The emergence of big data technology

© Springer International Publishing AG 2018
D. J. Hemanth and S. Smys (eds.), *Computational Vision and Bio Inspired Computing*,
Lecture Notes in Computational Vision and Biomechanics 28,
https://doi.org/10.1007/978-3-319-71767-8_3

permits plenty of data sources; this advancement brings with it the need for verification of data, for a trustworthy data storage and delivery. Characteristics of data have undergone a radical change in terms of volume, velocity, variety and veracity when associated with Big Data. Trust for Big Data and Big Data for Trust are the two threads of research for CIF. Trust in big data describes the chances and challenges in creating big data trust. It outlines the issues such as trust in nodes, CSP trust, data quality, evaluation of trust in big data, etc. Data quality trust is an issue that is central to trust in big data. Metrics and measures are needed to assess data quality. CSP provides services to cloud users. Users need not implicitly trust the service provider. Guaranteeing trustworthiness or confidence of nodes is very important. Equally important is assuring quality of the data exchanged in cloud environments. The work presented in this paper focuses on providing autonomous CSP trust management while assuring data quality for big data in CIFs. The paper is organized into the following sections: Details of threads of research in cloud integrated framework, big data and trust quantification and models is given in the literature survey section. Section 3 describes the proposed work; it presents the methodology for establishing trust for CIF. Section 4 presents the Simulation and results along with analysis of enhanced trust in cloud integrated frameworks and finally Sect. 5 concludes the work. The work contributes towards (i) autonomous trust management in CIFs (ii) data quality trust metric (iii) trust modelling for handling behavioural oscillations along time and (iv) SLA based weighted averaging scheme with trust, servicing and reputation parameters.

2 Literature Survey

The CIF in [1] introduces a body cloud architecture that monitors and manages data from body sensor data streams while addressing the CIF challenges. Specifically, in CIF network providers (NP) contribute information assembled by the data gathering networks to the cloud service providers (CSP). CSPs make use of cloud, to store and process data and then deliver in an on-demand manner to the cloud service users (CSU). CSUs can access the data in a request—reply paradigm. Accordingly while, NPs play the role of data sources, CSPs play the role of service providers and CSUs are the data requesters. Scientifically and economically valuable data can be gathered, processed and utilised by incorporating new features for authentication and access rights [2].

A significant thread of research is directed and prevalent in modeling trust among networking entities. Quantification of trust in CIFs permit networking entities to assess confidence/trust levels and interact only with the most trusted ones. CIF environments generate security breaches in various stages; node capture during data generation, black hole, Sybil and worm hole attacks during data transmission-node compromise in network processing. In the cloud itself there is breach in confidentiality, availability and integrity. More importantly the trust management for a CIF may be under attack. An adversary may attempt change the behaviour periodically (in time) or misbehave towards subset of nodes (in space). Exploiting weak identification (Sybil attack, newcomer attack, white washing attack) and weak authorization is another area of concern. The cloud integrated framework in [3, 4] is modeled around NP, CSP and

CSU. The work proposes a trust and reputation management system for CIF. Trust can be described as a positive expectation from a trustor to a trustee on its outcomes from an interaction and reputation is the overall judgement of participants in a network. The trust model for CIF is strictly defined for three trust metrics namely, privacy trust, processing trust and transmission trust. Typical attacks like, collusion, white washing and good/bad mouthing attacks targeted on recommendation systems are also handled.

Work in [5] proposes a trust system which collects information that are significant for service provider's estimation. The trust model is constructed based on the belief that certain proposition is correct. A consensus metric is derived to estimate the aggregated opinion for a particular statement from various data sources. While a discounting metric suggest weights based on trustworthiness of data source. One approach to set up trust for CSPs is through Service Level Agreements (SLA) that guarantees offerings and promises that a CSP may able to meet. The clauses for compensation are noted by CSPs, so that customer gets benefits on SLA violation. SLA violations are monitored so that CSUs can request for compensation policies mentioned during agreement. Diverse auditing standards such as FISMA, SAS 70 II are adopted in CIF to guarantee CSUs on their offered application and services. Work in [6] contributes a trust model based on average performance (reputation) while [7] considers additionally short-term performance (risk) computations.

While integrating cloud to sensor networks on one hand enhances the abilities of WSN, on the other hand the integration introduces security issues in both cloud and WSNs. [8] proposes a group based trust management system for wireless sensor networks, where each group contributes a trust value. The trust estimation is carried out in three phases, at node level, at cluster head and at the base station. Trust is primarily quantified based on past interactions and peer recommendations.

Work in [9] presented a pre-standardised approach for trust and reputation models that are useful for distributed systems. A survey of trust/reputation mechanisms for multi agent systems, peer to peer systems, ad hoc networks and wireless sensor networks is presented in the work. Recursive estimation of reputation providing higher weight for recent ratings has been proposed in Sporas [10]. The three aspects of individual, social and ontological for reputation computation has been employed in Regret [11]. Fuzzy concepts have been used in AFRAS [12]. MTrust [13] handles cooperative interactions using Bayesian network concept. DWTrust [14] assigns dynamic weights which adapt themselves based on feedbacks and trust policies. AntRep [15] models a system based on swarm intelligence. EigenTrust [16], assigns trust based on performance history. RRS model [17] models two values, a reputation value and a trust value. PTM [18] defines a decentralised trust methodology in pervasive environments. A thread of research is being directed in formulating trust models for choosing uncorrupted data by effective reconfiguration and assignment of resources [19].

Measuring trust in cloud database services can be based on extrinsic evidence and empirical evidence is introduced in [20]. A relative positive extrinsic evidence based expectation of the trustor on his trustee and positive empirical evidence based expectancy at different levels of database is used for computing trust. Further SORT enables peers to figure out the trustworthiness of other member peers based on past recommendations and interactions [21].

The features of data have changed in terms of volume, velocity, variety and veracity when associated with Big Data. Research work in [22, 23, 24] highlights the issues associated with big data, trust and cloud environments. Big data for trust specifies how huge a data must be for trust estimation. It focuses on creating a reputation score for a particular trustor. It takes into account the issues in collection and preparation of data from multiple sources, computation and finally storage and communication. Homomorphic linear authenticators detect data integrity through public auditability. Random masking may be provided so that third party that verifies the data for integrity is actually unaware of the data itself [25]. Work in [26, 27] employ Merkle Hash Trees (MHT), an integrity technique designed for big data; it is an authenticated tree like data structure, a binary hash tree. If the exchanged data is incorrect, data quality is lost. A trusted CSP assures data quality. CSPs need to be assessed for trustworthiness with respect to data roll backs, data loss concealment and data modification attacks. A trust quantification aligned to the above stated would give an indication of the extent to which the CSP can be trusted to provide the data without performing these attacks on data quality. This forms the basis of the work presented in the paper. The proposed trust model for CIF centres on data quality, a very important issue that is fundamental to trust in big data. The work presents a comprehensive autonomous scheme for trust computation of CSP and management in CIFs.

3   Proposed Work

3.1   Architecture for Trust Enhanced CIF

Figure 1 illustrates the trust architecture proposed for CIFs to autonomously compute and manage CSP trust; here, CSPs are data sources and CSUs as data requesters. CSP receives data from NP, processes the raw data and stores in cloud. On demand from CSU the processed data is transmitted by CSP to the CSU. A TCE (Trusted Center Entity) is a trusted third party that manages the system in a centralised manner. It monitors the system for malicious behaviours of CSPs that may change the behaviour of the system. It assures data quality using Merkle hash tree (MHT) and manages the trust and reputation scores for the CSPs in the trust enhanced CIF. The TCE does public auditability and detects data roll back, data concealment and incorrect data. It is responsible for managing the following:

(1) SLA of CSU
(2) CSU feedback on CSP services
(3) CSP Trust scores
(4) CSP Performance
(5) Hashed root of MHT for files.

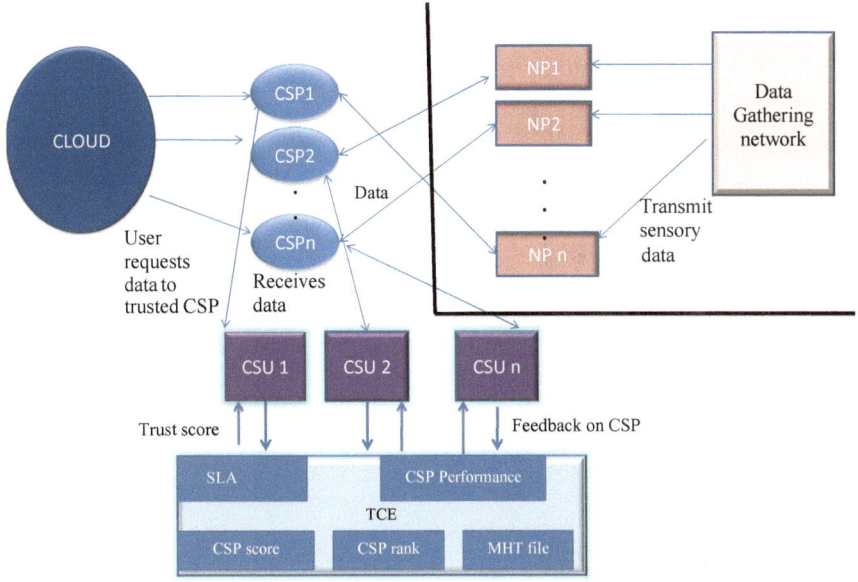

Fig. 1 Trust architecture for cloud integrated framework

3.2 Data Quality Assurance

Public auditability is performed by TCE in an enhanced CIF with trust modelling capability. Maliciousness in CIF may result in false data being serviced to CSU. It is essential that data quality need to be stringently assured. The verification token of a file is the hashed root of the file in its MHT. A comparison of the CSU generated verification token with the TCE stored token verifies the file for attacks performed by CSP. Our work employs hashed root of file MHTs to detect attacks on data quality. Data roll back and data concealment are affected on frequently updating data. While roll back attack, rolls back the data to its previous updated state, data concealment attack on the other hand hides the updated file and provides user another file or a file updated much earlier. Whereas, data modification attack compromise the integrity of the data by changing its contents. Trust on a CSP should increase as long as it gives correct and consistent data to the users. A malicious CSP may attempt to perform attacks compromising data quality. A CIF enhanced with a trust model is equipped to detect maliciousness among CSPs. Data quality compromise at malicious CSP is discussed as follows:

(a) *Roll back attack*

   A malicious CSP rolls back the data to its previous updated state and supplies this previous updated version to its users. A CSU may consume the data if no verification token is available with it regarding the latest state of updation.

(b) *Concealment of data loss*

A malicious CSP attempts to conceal data loss by supplying a much earlier version of the same data or some other data to its users.

(c) *Data modification*

Integrity of the data can be compromised by a malicious CSP by modifying data. Further processing may be performed within inconsistent/incorrect/modified data which may disrupt user expected behaviour or results. Mitigation of these attacks in the proposed model is performed by using MHT algorithm; TCE verifies incorrectness, inconsistency and correctness of data through hashed roots of a file. Verification performed on the serviced data using MHT is depicted in Table 1. Hash of the data/file chunks forms leaf nodes of a MHT. Files are divided, creates hashes for the leaf nodes then each data chunks are hashed and a hex code is generated for the file. The leaf nodes are concatenated to form the internal nodes which are again hashed. The procedure repeats till it gets a hashed root value, which is used for data quality verification.

Table 1 Merkle hash tree

1. Divide ()-Divide a file, F in to chunks by setting a PART_SIZE.
2. Hash ()-Hash each chunks, {m1, m2......mn) using any hashing algorithm.
3. The resultant hashed data parts, {h (m1), h (m2)...h (mn)}   are the leaf nodes of the MHT algorithm.
4. Set index= 0.
5. for each element in the list,
do
5.1. if index<list_size   then
5.1.1. left=getElement(index)
5.1.2. right=getElement(index+1)
5.1.3. parent=hash(left‖right)...
//(concatination)
5.1.4. add parent to the list
6. return new list

3.3   Trust Modeling for Trust Enhanced CIF

3.3.1   Trust Metrics

The model uses four trust metrics, data processing trust, data privacy trust, data transmission trust and data quality trust. The first three trust metrics have been adapted from [3]. The work introduces data quality trust metric. Data processing trust (PT) is the measure of whether the data is processed with error at the CSP.

$$PT = \frac{S_{pt} + 1}{S_{pt} + F_{pt} + 2} \tag{1}$$

where, S_{pt} is the non-error number and F_{pt} is the error number of the processed data. Data Privacy Trust (PRT) is the measure of whether the stored data is accessible by unauthorized entities.

$$PRT = \begin{cases} \mathbf{1}, if\ \mathbf{F_{prt}} = 0 \\ \mathbf{0}, if\ \mathbf{F_{prt}} > 0 \end{cases} \tag{2}$$

where, F_{prt} is the number of data accessed by others for each service from a CSP to a CSU. Data Transmission Trust (TT) is the measure of whether data transmission from CSP to CSU is successful.

$$TT = \frac{\mathbf{S_{tt}} + 1}{\mathbf{S_{tt}} + \mathbf{F_{tt}} + 2} \tag{3}$$

where, S_{tt} is the success number and F_{tt} is the failure number of transmitted data for each service from a CSP to a CSU.

Data Quality Trust (DQT) is the measure of whether data quality of the data given by CSP to CSU is assured.

$$DQT = \begin{cases} 1, if\ D_q = 0 \\ 1/D_q, if\ D_q > 0 \end{cases} \tag{4}$$

where, D_q is the number of data given by CSP for which data quality is not assured.

3.3.2 CSP Trust Score

The trust score for a CSP is computed as the weighted average of {PRT, PT, TT, DQT}. The proposed trust model uses SLA for computing CSP trust scores. SLA is a service agreement between two or more participants, which defines a service contract including serviceability, reliability, availability, performance, etc. A CSU gives its priority for each of the trust parameters in the SLA shown in Table 2. The weights for the 4 trust metrics are derived from these priorities.

Table 2 Trust parameters in service level agreement

Data privacy	Data processing	Data transmission	Data quality
W_{prt}	W_{pt}	W_{tt}	W_{dqt}

Apart from the above parameters the SLA additionally includes bandwidth and availability metrics [in Eqs. (5) and (6)] for evaluating CSP trust score

$$\text{Bandwidth(BW)} = (\text{total Bytes of file size})/\text{elapsed TimeSeconds(in bps)} \quad (5)$$

$$\text{Availability(AV)} = (\text{Uptime taken by a CSP for its service/Total time of the system}) * 100 \quad (6)$$

The trust score of CSP is comprehensively computed by taking into the reputation a CSP enjoys in the CIF. Reputation for a CSP is computed as:

$$R(S_i) = 1/n \sum_{i=1}^{n} f(i) \quad (7)$$

where f_i, is the feedback from a CSU to a CSP for a transaction (store /retrieval) in the range [0, 1], n is the number of transactions, and S_i is the service or transaction from a CSP to CSU.

The work presented in this paper introduces a comprehensive SLA format (Table 3). The SLA has 7 parameters $T_{i=1 \text{ to } 7}$: 4 trust parameters, 2 servicing parameters and 1 reputation parameter.

Table 3 Service level agreement format

Data privacy	Data processing	Data transmission	Data quality	Bandwidth	Availability	Reputation
W_{prt}	W_{pt}	W_{tt}	W_{dqt}	W_{bw}	W_{av}	W_{rep}

The proposed work computes trust score for a CSP systematically as the weighted average of the 7 SLA parameters {PRT, PT, TT, DQT, BW, AV, REP}. For each transaction/servicing of a CSP with a CSU, the performance in the form of T_is are maintained at the TCE for the CSP. CSP trust score is a weighted average of T_is. Each T_i is computed over a window of length 'l' as a weighted average for the corresponding SLA parameter (Fig. 2). While the latest entry in the window is given the highest weight, the last entry is given the least weight. The work introduces a computation of T_i over a moving window of length 'l'. The window moves over time, accommodating only the most recent performance/behaviour of the CSP. Significantly computation removes behavioural oscillations [9] of CSPs along time.

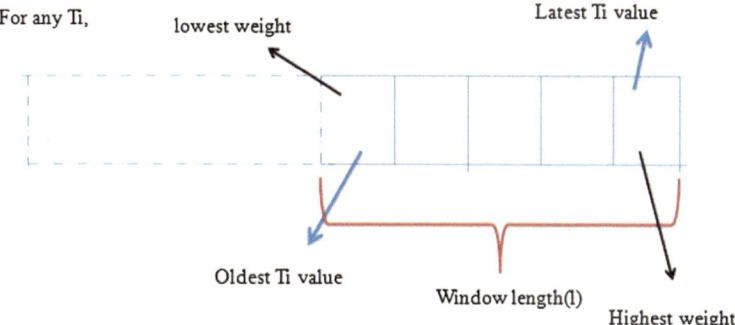

Fig. 2 Weighted window for a T_i

4 Simulation and Results of Trust Enhanced CIF

The Trust Enhanced CIF was simulated using CloudSim 3.0 API in Java Eclipse IDE platform. Actors in CloudSim are mapped into the three CIF major players; Data Center-CSP, Broker-TCE and User-CSP.

Experimentation was performed with 10 CSPs, 10 CSUs and one TCE for a total of 150 transactions. The transactions constitute 90 reads (R) and 60 store (S) servicing requests. The transactions were performed for text and image files. The CIF was simulated and the experimentation was conducted in 4 phases (Table 4) with 10 CSP and varying CSU, read and store servicing.

Table 4 Phases of experimentation

Phase	Number of CSUs	Number of CSPs	% of Reads	% of Stores
I	5	10	22.22	16.6
II	5	10	22.22	33.33
III	8	10	44.44	33.33
IV	10	10	11.11	16.66

The initial Phase I, is a warm up phase that experiments for 10 stores and 20 reads with 5 CSUs. Phase II experiments for 20 stores and 20 reads with 5 CSUs. Phase III experiments for 20 stores and 40 reads but with 3 new CSU. The final phase, Phase IV experiments for 10 reads and 10 stores with additional two new CSP. SLA for the 10 CSUs is given in Table 5.

Table 5 SLA for 10 CSU used in the 4 phases of experimentation

	Privacy	Processing	Transmission	Data quality	Bandwidth	Availability	Reputation
CSU1	1	2	4	3	6	5	7
CSU2	7	6	4	5	2	3	1
CSU3	4	3	7	5	1	6	2
CSU4	1	5	3	4	7	2	6
CSU5	3	6	7	2	1	5	4
CSU6	4	3	6	1	5	7	2
CSU7	3	5	7	6	2	1	4
CSU8	1	7	6	3	5	4	2
CSU9	6	7	4	3	2	5	1
CSU10	4	2	3	1	7	6	5

T_i's are captured and maintained in the TCE for each read/store servicing provided by CSPs for CSUs. CSUs enable the choice of CSPs based on the ranking computed by TCE on CSP trust score. Figure 3 confirms quantification of CSP trust with attack simulation for roll back, concealment and modification attacks that compromise data quality in the CIF. The attack simulation includes attack on data privacy as well. CSP trust quantification for the experimentation conducted in 4 phases with 90 reads (R) and 60 store (S) servicing requests are shown in Fig. 3.

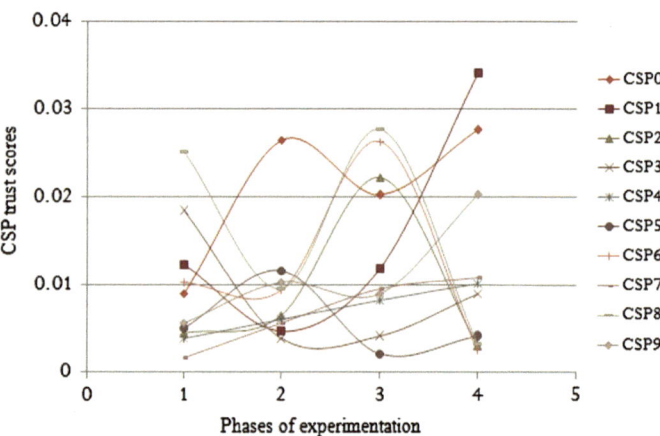

Fig. 3 CSP trust score in 4 phases of experimentation

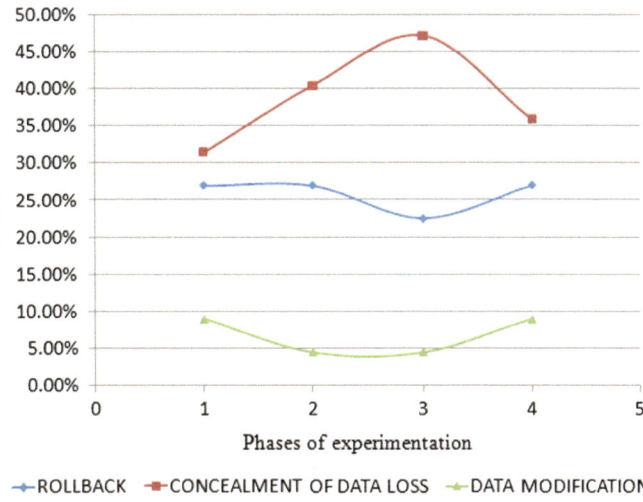

Fig. 4 Attack model for experimentation

Experimentation was carried for read and store transactions with 3 attacks (rollback, concealment of data loss and data modification) discussed in Sect. 3.2. Figure 4 shows the percentage of attacks in each of the experimentation phases and Fig. 5 depicts the attacks performed by the CSPs.

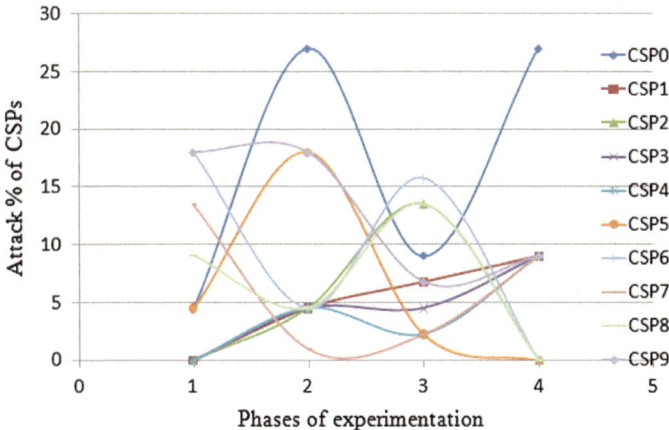

Fig. 5 Attack percentage in 4 phases of experimentation

Experiments were conducted by taking varying SLA for a CSU to observe and validate the proposed trust quantification. Figure 6 depicts the choice of CSP provided by the trust enhanced CIF for varying SLA. In Phase II, CSP0 performs almost 27% of attacks and its rank drops down from 1 to 5. In Phase III, CSP2 performs almost 13% of attacks and its rank drops down from 3 to 9. In Phase III, CSP0 performs almost 9% of attack and its rank increases from 5 to 2. These observations validate the quantification of trust(using privacy trust, processing trust, transmission trust, data quality trust, bandwidth, availability and reputation) proposed in this work.

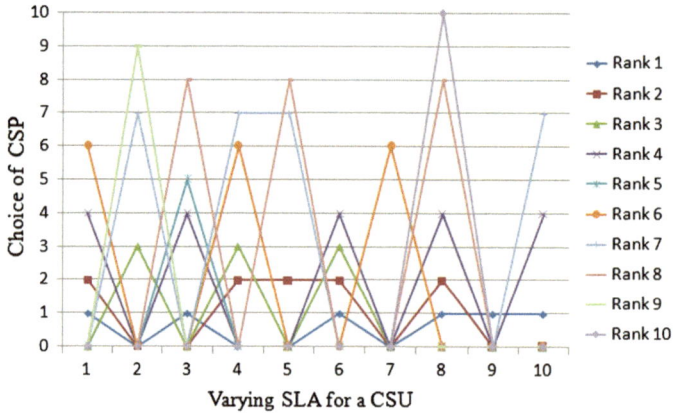

Fig. 6 Choice of CSP for varying SLA

5 Conclusion

Cloud Integrated Frameworks should be designed to comprehensively handle big data for data quality issues in addition to other factors such as breach of privacy. Servicing with data quality has a profound and direct impact on the confidence placed on a CSP by CSU. This work explores trust modelling for data quality in addition to transmission trust, processing trust and privacy trust for evaluation of trust in CSPs. The trust modelling presented in the work addresses importantly the issue of oscillatory behaviour of CSPs using a weighted averaging in moving window over the performances— T_is. Detailed discussion and analysis of CSP trust quantification is presented and evaluated. The work proposes an autonomous scheme that provides for CSP trust modeling in CIFs. The trust enhanced CIF comprehensively quantifies trust of CSPs for trust,servicing and reputation factors (4 trust parameters, 2 servicing parameters and 1 reputation parameter). Results from simulation and phases of experimentation show that the trust model is effective in scoring CSPs for trust, recommending a ranked list to CSU in accordance with its SLA.

References

1. Fortino, G., Pathan, M., Di Fatta, G.: Bodycloud: integration of cloud computing and body sensor networks. In: 2012 IEEE 4th International Conference on Cloud Computing Technology and Science (CloudCom), pp. 851–856. IEEE, Dec 2012
2. Ahmed, K., Gregory, M.: Integrating wireless sensor networks with computing. In: 2011 Seventh International Conference on Mobile Ad-hoc and Sensor Networks (MSN), pp. 364–366. IEEE, Dec 2011
3. Zhu, C., Nicanfar, H., Leung, V.C., Yang, L.T.: An authenticated trust and reputation calculation and management system for cloud and sensor networks integration. IEEE Trans. Inf. Forensics Secur. **10**(1), 118–131 (2015)
4. Yuriyama, M., Kushida, T.: Sensor-cloud infrastructure-physical sensor management with virtualized sensors on cloud computing. In: 2010 13th International Conference on Network-Based Information Systems (NBiS), pp. 1–8. IEEE, Sept 2010
5. Habib, S.M., Ries, S., Muhlhauser, M.: Towards a trust management system for cloud computing. In: 2011 IEEE 10th International Conference on Trust, Security and Privacy in Computing and Communications (TrustCom), pp. 933–939. IEEE, Nov 2011
6. Shyamala, C.K., Padmanabhan, T.R.: A trust-reputation model offering data retrievability and correctness in distributed storages. Int. J. Comput. Appl. **36**(2), 56–63 (2014)
7. Savas, O., Jin, G., Deng, J.: Trust management in cloud-integrated wireless sensor networks. In: 2013 International Conference on Collaboration Technologies and Systems (CTS), pp. 334–341. IEEE, May 2013
8. Sheikh, R.A., Jameel, H., Lee, S., Rajput, S., Song, Y.J.: Trust management problem in distributed wireless sensor networks. In: Proceedings of the 12th IEEE International Conference on Embedded and real-Time Computing Systems and Applications, Sydney, Australia, 16–18 Aug 2006
9. Mármol, F.G., Pérez, G.M.: Towards pre-standardization of trust and reputation models for distributed and heterogeneous systems. Comput. Stand. Interfaces **32**(4), 185–196 (2010)
10. Zacharia, G., Maes, P.: Trust management through reputation mechanisms. Appl. Artif. Intell. **14**(9), 881–907 (2000)
11. Sabater, J., Sierra, C.: REGRET: reputation in gregarious societies. In: Proceedings of the Fifth International Conference on Autonomous Agents, pp. 194–195. ACM, May 2001
12. Carbo, J., Molina, J.M., Davila, J.: Trust management through fuzzy reputation. Int. J. Coop. Inf. Syst. **12**(01), 135–155 (2003)
13. Songsiri, S.: MTrust: a reputation-based trust model for a mobile agent system, Autonomic and Trusted Computing. No. 4158. In: LNCS. Third International Conference, ATC 2006, pp. 374–385, Springer, Wuhan, China, Sept 2006
14. Huang, C., Hu, H., Wang, Z.: A dynamic trust model based on feedback control mechanism for P2P applications, Autonomic and Trusted Computing. No. 4158 in LNCS, pp. 312–321, Springer, Wuhan, China, Sept 2006
15. Wang, W., Zeng, G., Yuan, L.: Ant-based reputation evidence distribution in P2P networks, GCC. In: Fifth International Conference on Grid and Cooperative Computing, IEEE Computer Society, pp. 129–132, Changsha, Hunan, China, Oct 2006
16. Kamvar, S., Schlosser, M., Garcia-Molina, H.: The eigentrust algorithm for reputation management in P2P networks. In: Proceedings of the International World Wide Web Conference (WWW). Budapest, Hungary, May 2003
17. Buchegger, S., Le Boudec, J.Y.: A robust reputation system for P2P and mobile adhoc networks. In: Proceedings of the Second Workshop on the Economics of Peer-toPeer Systems, Cambridge MA, USA, June 2004

18. Almenárez, F., Marín, A., Campo, C., García, C.: PTM: a Pervasive Trust Management for Dynamic Open Environments, Privacy and Trust. First Workshop on Pervasive Security and Trust, Boston, USA, Aug 2004
19. Kim, H., Lee, H., Kim, W., Kim, Y.: A trust evaluation model for QoS guarantee in cloud systems. Int. J. Grid Distrib. Comput. **3**(1), 1–10 (2010)
20. Priyadarshani, W.E., Wikramanayake, G.N., Ekanayake, E.P.: Measuring trust and selecting cloud database services. Adv. Comput. Sci. Int. J. **2**(5), 114–120 (2013)
21. Can, A.B., Bhargava, B.: Sort: a self-organizing trust model for peer-to-peer systems. IEEE Trans. Dependable Secure Comput. **10**(1), 14–27 (2013)
22. Sänger, J., Richthammer, C., Hassan, S., Pernul, G.: Trust and big data: a roadmap for research. In: 2014 25th International Workshop on Database and Expert Systems Applications (DEXA), pp. 278–282. IEEE, Sept 2014
23. Kanagasabapathi, K., Deepak, S., Prakash, P.: A study on security issues in cloud computing. In: Proceedings of the International Conference on Soft Computing Systems, pp. 167–175. Springer, India (2016)
24. Sangeetha, K.S., Prakash, P.: Big data and cloud: a survey. In: Artificial Intelligence and Evolutionary Algorithms in Engineering Systems, pp. 773–778. Springer India (2015)
25. Wang, C., Chow, S.S., Wang, Q., Ren, K., Lou, W.: Privacy-preserving public auditing for secure cloud storage. IEEE Trans. Comput. **62**(2), 362–375 (2013)
26. Liu, C., Yang, C., Zhang, X., Chen, J.: External integrity verification for outsourced big data in cloud and IoT: a big picture. Future Gener. Comput. Syst. **49**, 58–67 (2015)
27. Venkatesh, M., Sumalatha, M.R., SelvaKumar, C.: Improving public auditability, data possession in data storage security for cloud computing. In: 2012 International Conference on Recent Trends in Information Technology (ICRTIT), pp. 463–467. IEEE, April 2012

Combined Classifier Approach for Offline Handwritten Devanagari Character Recognition Using Multiple Features

Milind Bhalerao$^{(\boxtimes)}$, Sanjiv Bonde, Abhijeet Nandedkar,
and Sushma Pilawan

Shri Guru Gobind Singhji Institute of Engineering and Technology,
Nanded, India
mvbhalerao@sggs.ac.in

Abstract. Offline handwritten character recognition is the process of recognizing given characters from the large set of characters. OCR system mainly focuses on the recognition of printed or handwritten characters of a scanned image. The proposed system extracts features that are based only on gradient of image which is helpful in exact recognition of characters. A technique to recognize handwritten Devanagari characters using combination of quadratic and SVM classifiers is presented in this paper. Features used are directional features that are strength, angle and histogram of gradient (SOG, AOG, HOG). Using a Gaussian filter, the strength and the angle features are down sampled to obtain a feature vector of 392 dimensions. These features are finally concatenated with HOG feature. Applying these to the combination of quadratic and SVM classifiers to obtain maximum accuracy of 95.81% using 3 fold cross validation.

Keywords: Devanagari handwritten characters · SVM classifier
Quadratic classifier · Feature extraction · HOG

1 Introduction

Character recognition is a system that can be used for identification of characters. It is the recognition of either handwritten or printed text by computer. There is one more possibility of classification of characters which is based upon the way in which the characters are being acquired. The two systems classified are offline recognition systems and online recognition systems. Handwritten character recognition (HWCR) having various applications has become a popular research area [1]. The bank cheque processing, postal automation, automatic data entry, retrieval of text from old documents are some of the important application areas. Researchers have proposed many recognition systems and many approaches towards the handwritten character recognition [2, 3]. An extensive research is going on in this field since more than three decades; hand written character recognition is at nascent stage in the Indian context [4].

© Springer International Publishing AG 2018
D. J. Hemanth and S. Smys (eds.), *Computational Vision and Bio Inspired Computing*,
Lecture Notes in Computational Vision and Biomechanics 28,
https://doi.org/10.1007/978-3-319-71767-8_4

1.1   About Devanagari Script

Devanagari is an Indian script, followed by many Indian languages. It consists of 13 vowels and 33 consonants [5]. With varied structural features their shape is complex. Every character is written below a horizontal line, called as 'header line' which is also termed as 'Shirolekha'. The vowels mentioned in Fig. 1 come in various shapes and are connected to the consonants in different manners which are termed as modifiers. These modifiers are placed above the Shirolekha or may be put at the bottom of the characters. In certain case a modifier may be placed at the mid-way between the characters [6]. The hand written characters also include the 'Jodakshar' or compound characters formed due to joining the two or three characters. Group of either simple or compound letters forms a word. Group of these words forms a sentence. The sentences are written from left to right. The Devanagari script is used by 120 languages including Marathi, Hindi, Pali, Nepali, Bodo, Konkani, Maithili and Sindhi. It makes it one of the most used and accepted writing scripts in the world. Figure 1 shows all the basic vowels and consonants used in Devanagari script.

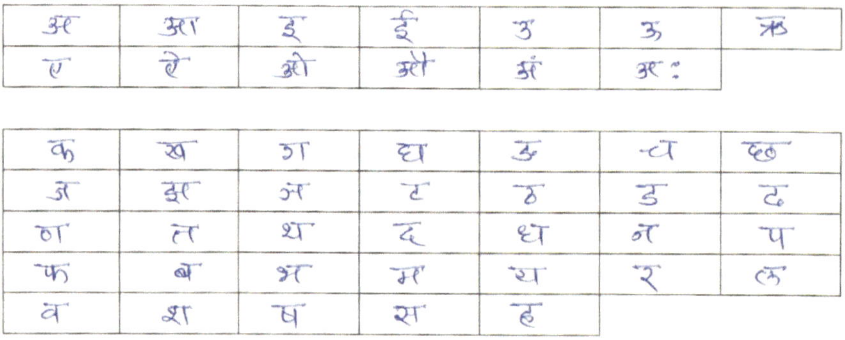

अ	आ	इ	ई	उ	ऊ	ऋ
ए	ऐ	ओ	औ	अं	अः	

क	ख	ग	घ	ङ	च	छ
ज	झ	ञ	ट	ठ	ड	ढ
ण	त	थ	द	ध	न	प
फ	ब	भ	म	य	र	ल
व	श	ष	स	ह		

Fig. 1 Basic vowels and consonants

1.2   Related Work

Different researchers have applied different techniques over printed and handwritten characters as well as numerals. Arica [7] have provided detail overview of how the character recognition systems have evolved. He also detailed most commonly used methods and processes involved in character recognition. Most of the Indian languages such as Hindi, Marathi, and Pali are being written in Devanagari script. So for the recognition of Devanagari characters, it is important to properly extract the different features of Devanagari characters. Roy [8] have used five different types of features and HMM classifier for the recognition of Bangla words. Vaidya Madhav et al. [9]

proposed the combination of horizontal and vertical projection profiles along with diagonal summation vectors to reduce the feature space. Kumar and Singh [10] have used Zernike moments features and obtained accuracy of 80%. Sharma et al. [11] have used quadratic classifier for Devanagari numerals and character recognition. They used directional chain code and contour points as features to obtain recognition accuracy of 80.36%. Pal et al. [12] have used quadratic classifier for the recognition of Devanagari characters. They have used strength and angle of gradient as feature and obtained accuracy of 94.24%. Shi [13] have used character recognition model for the detection of stroke and structure. They applied this method on numerals and obtained accuracy of maximum 95%. Pal et al. in [14] obtained the result up to 95.13%. In [15] Pal et al. used the feature set consisting of curvature and gradient information obtained from binary as well as gray-scale images. Also they used twelve classifiers and computed different results. Out of which using MIL classifier they could obtain a result of 95.19%. Holambe, Thool et al. considered 5000 basic characters and used Sobel (3×3) and Robert (2×2) operators separately. With SVM classifier they got the accuracy of 94 and 94.45% respectively [16]. Mahesh Jangid used 12,240 handwritten Devanagari samples with 314 dimensional vector and 10 fold cross validation, obtained 94.89% recognition rate [17]. Gunjan, Lahiri with 1000 handwritten samples, employing backpropagation neural net obtained a total recognition accuracy of about 93% [18]. Discrete cosine transform is used for feature selection to reduce the number of comparisons for classification [19].

Database used for this work is created at our institute written by different individuals. It consists of 29,440 data samples. It approximately consists of around 640 samples of every Devanagari character. A gray tone image with 300 dpi resolution in TIF format is used. All these images are subdivided into three groups and testing is done on each group. The rest of the paper is organized as—Sect. 2 deals with the proposed method used in this work. Section 2.1 describes the preprocessing, Sect. 2.2 gives details of features, Sect. 2.3 explains classifiers used. Section 3 discusses the results obtained and Sect. 4 provides the conclusion.

2 Proposed Approach

In order to increase accuracy, the proposed system extracts features that are based only on gradient of image along with the combination of two classifiers namely Quadratic and SVM which is helpful in exact recognition of characters. The proposed system is shown in Fig. 2.

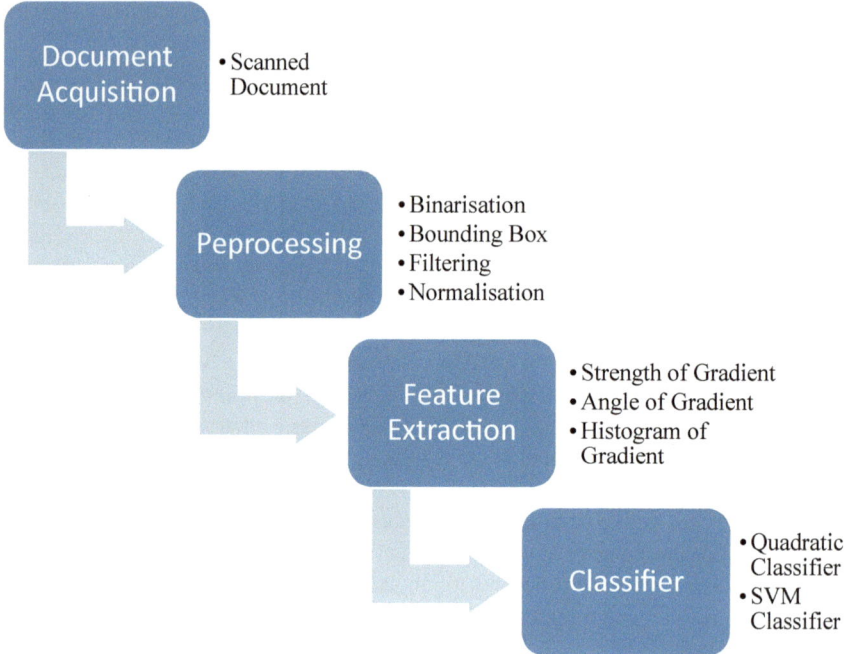

Fig. 2 Proposed approach

For the recognition of handwritten characters, a set of feature vectors need to be calculated for correct classification. After the acquisition of the image document, the image need to be preprocessed so as to reduce it to a suitable form for further process. The preprocessing involves image binarization, bounding box, filtering and normalization. Wiener filtering followed by image normalization is carried out in preprocessing. Three features mainly strength, angle and histogram of gradient are considered in the feature extraction step. Thereafter we have used the combination of quadratic and SVM classifiers and final results are obtained.

2.1 Preprocessing

To remove the noise and unwanted distortion in the script, preprocessing is an essential step [20] in handwritten character recognition. A number of tasks are being performed in the preprocessing like binarisation, thresholding, slant, detection of skew, removal of skew, detection of baseline, smoothing, resizing and so on. After normalization of the characters, segmentation and classification accuracy is increased [21]. There is a huge variability in the handwriting style of every individual. The main purpose of the preprocessing tasks is making this writing style as uniform as possible [22].

The preprocessing techniques which are carried out in this work are explained below.

- The OSTU [23] method is employed to convert the gray scale image into binary image. The advantage of using this method is that it chooses threshold automatically to reduce the intra class difference between black and white pixels.

$$\sigma_w^2(t) = w_0(t)\sigma_0^2(t) + w_1(t)\sigma_1^2(t) \tag{1}$$

w_0 and w_1 are the probabilities of the two classes separated by a threshold t and σ_0^2 and σ_1^2 are the variances of these two classes.

The class probability w_0 and w_1 is computed from the L histogram

$$w_0(t) = \sum_{i=o}^{t-1} p(i) \tag{2}$$

$$w_1(t) = \sum_{i=t}^{L-1} p(i) \tag{3}$$

- The edge and the bounding box of the binarized image is computed. There are various ways of obtaining edge of binarized image such as Robert, Sobel, Laplace and Canny. We have used Canny edge detection method in our system. This is because it can obtain background information more effectively (Fig. 3).

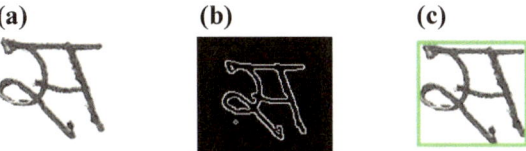

Fig. 3 Preprocessed image **a** Original image. **b** Edge of image. **c** Bounding box

2.2 Feature Extraction

Gradient image is obtained by using Robert filter. Two features mainly strength and angle of gradient is calculated. The Direction of gradient is calculated as given in Eq. (4). These directions are quantized into 32 directions and each direction is accumulated with strength of gradient.

$$\text{Strength of gradient: } g(u, v) = \sqrt{(\Delta x)^2 + (\Delta v)^2} \tag{4}$$

Gradient image is obtained by using Robert filter. Two features mainly strength and direction of gradient is calculated.

$$\text{Angle of gradient: } \theta(u,v) = \tan^{-1} \frac{\Delta y}{\Delta x} \qquad (5)$$

where

$$\Delta x = f(x+1,y+1) - f(x,y)$$
$$\Delta y = f(x+1,y) - f(x,y+1) \qquad (6)$$

$f(x,y)$ is intensity value at point (x, y).

After calculation of these two features, a separate feature called as Histogram of Gradient (HOG) is obtained. HOG is a feature descriptor which counts total number of occurrences of gradient i.e. depending upon the direction of pixel intensity it counts intensity value. A feature vector of $1 \times 36,864$ dimension is taken which is resized to obtain a feature vector of the dimension 1×392. All of the above features are combined to form a feature vector of dimension 1×392. An array of feature vector is given to classifier for final recognition.

2.3   Classifier

The combination of quadratic [11] and SVM classifier for the recognition of individual characters is used. The discriminant functions used for quadratic classifier is given in Eq. (7)

$$g(x) = \ln(P1) - 0.5 * M * ((incv) * M) \qquad (7)$$

where $P1$ is the priori probability, M is the sample mean and $incv$ is the inverse covariance matrix.

To obtain this discriminant function following steps are followed

1. Divide the given feature vector into three classes.
2. Total number of samples of each class is calculated. Let $n1$, $n2$, $n3$ be the number of samples in each class and N be the total number of samples.
3. Calculate mean vector M of samples which is given by $M = \frac{1}{N}\sum_{i=1}^{N} Ai$
4. Subtract mean from each class and calculate the covariance matrix
5. Obtain the inverse of covariance matrix denoted by $invc$.
6. Calculate the priori probability $P1$.

The features are applied separately to both the classifiers and finally the results are combined.

A discriminative classifier which is used for separating hyperplane is called SVM (support vector machine). Classification, regression and other tasks can be effectively performed using SVM. SVM is capable of handling a large number of data. The advantage of using SVM is that it can be used to find best possible margin between two classes. SVM can be used to achieve robust performance. SVM classifies the hyper plane having maximum margin from given two classes in the feature space. This is achieved by navigating the input data on a higher dimensional feature space [24].

In SVM different types of kernel function can be used such as radial basis function, linear function, sigmoid etc. Figure 4 shows the recognition result of individual character 'स'.

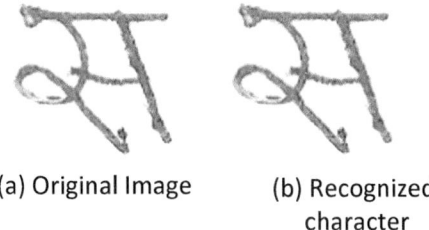

(a) Original Image   (b) Recognized
                               character

Fig. 4 Recognition result

$$z(x) = \exp\left[\frac{\|x - \mu\|^2}{\delta^2}\right] \tag{8}$$

3 Experimental Results

For the experiment we have used 29,440 samples of different individuals obtained in our institute. Approximately 640 samples of each individual character with 300 dpi resolution is considered. We have used 3 fold cross validation that is we divided the data samples into three parts. One part is used as test samples and remaining two are used for training. From the experiment we have obtained the accuracy table of top choices as shown in Table 1.

Table 1 Recognition result of top choices

Top choices	% of Accuracy
1.	86.71
2.	89.68
3	91.02
4.	92.52
5.	98.99

The overall accuracy of 95.81% was obtained with zero percent rejection. The accuracy of 88.19% was obtained by considering top two choices of the result. From the experiment it is also observed that maximum accuracy of 98.99% was obtained for the Devanagari character 'ह'. Second highest accuracy was obtained for 'थ'. Lowest accuracy was obtained for character 'घ'. Table 2 shows the top accuracies of individual characters. Table 3 gives comparison of results.

Table 2 Recognition results of individual character

Character	Accuracy	Character	Accuracy
ह	98.99	झ	98.31
फ	97.87	हा	95.05
क	97.92	आ	97.43
ड	97.62	औ	98.42

Table 3 Comparison of results

Sr. No.	Method	Features	Classifiers used	Data size	Accuracy (%)
1.	Kumar and Singh [10]	Gradient	SVM	25,000	80
2.	Sharma et al. [11]	Chain code	Quadratic	11,270	80.36
3.	Pal et al. [12]	Gradient, Gaussian Filter	Quadratic	36,172	94.24
4.	Arora et al. [25]	Shadow, CH	MLP	4900	92.80
5.	Pal et al. [14]	Gradient	SVM and MQDF	36,172	95.13
6.	Pal et al. [15]	Gradient	MLP	36,172	95.19
7.	Holambe, Thool [16]	Gradient	SVM	5000	94
8.	Jangid [17]	Zone density feature	SVM	12,240	94.89
9.	Gunjan and Lahiri [18]				93
10.	Kumar and Jindal [26]	–	–	300 documents	93.51
11.	Kamble and Hegadi [27]	Rectangle histogram oriented gradient	SVM	8000	95.64
12.	Proposed method	SOG, AOG, HOG	SVM and Quadratic	29,440	95.81

4 Conclusion

India being the country of multi script and multi languages, the recognition of each and every character is an important but a difficult task. Devanagari script is used by most of the Indian languages. We have used combination of quadratic and SVM classifiers to obtain overall recognition accuracy of 95.81%. We have used 3 fold cross validation and accuracy of each individual character is averaged to obtain the overall accuracy. This work can be used by researcher to recognize more complex script such as compound characters. The number of features can be increased or may be changed.

References

1. Plamondon, R., Srihari, S.N.: Online and off-line handwriting recognition: a comprehensive survey. IEEE Trans. Pattern Anal. Mach. Intell. **22**(1), 63–84 (2000)
2. Pal, U., Chaudhuri, B.B.: Indian script character recognition: a survey. Pattern Recognit. **37**(9), 1887–1899 (2004)
3. Liu, C.-L., Suen, C.Y.: A new benchmark on the recognition of handwritten Bangla and Farsi numeral characters. Pattern Recognit. **42**(12), 3287–3295 (2009)
4. Vaidya, M., Joshi, Y.V.: Handwritten numeral identification system using pixel level distribution features, Smart Innovations, Systems and Technologies, vol. 2, pp. 307–315. Springer publishing (2017)
5. Gupta, D., and Madhu Nair, L.: Improving Ocr by effective pre-processing and segmentation for Devanagiri script: a quantified study. J. Theor. Appl. Inf. Technol. **52**(2), 142–153 (2013)
6. Shelke, S., Apte, S.: Performance optimization and comparative analysis of neural networks for handwritten Devanagari character recognition. In: Signal and Information Processing (IConSIP), International Conference on. IEEE (2016)
7. Arica, N., Yarman-Vural, F.T.: An overview of character recognition focused on off-line handwriting. IEEE Trans. Syst. Man, and Cybern. Part C (Applications and Reviews) **31**(2), 216–233 (2001)
8. Roy, P.P., et al.: HMM-based Indic handwritten word recognition using zone segmentation. Pattern Recognit. **60**, 1057–1075 (2016)
9. Vaidya, M.V., Joshi, Y.V.: Marathi numeral recognition using statistical distribution features. In: Information Processing (ICIP), 2015 International Conference on. IEEE (2015)
10. Kumar, S., Singh, C.: A study of zernike moments and its use in Devanagari handwritten character recognition. In: International Conference on Cognition and Recognition (2005)
11. Sharma, N., et al.: Recognition of off-line handwritten Devanagari characters using quadratic classifier. In: Computer Vision, Graphics and Image Processing, pp. 805–816. Springer, Berlin (2006)
12. Pal, U., et al.: Off-line handwritten character recognition of devnagari script. In: Ninth International Conference on Document Analysis and Recognition, 2007. ICDAR 2007, Vol. 1. IEEE (2007)
13. Shi, Cun-Zhao, et al. Stroke detector and structure based models for character recognition: a comparative study. IEEE Trans. Image Process. **24**(12), 4952–4964 (2015)
14. Pal, Umapada, et al.: Accuracy improvement of Devnagari character recognition combining SVM and MQDF. In: Proceedings of 11th International Conference on Frontiers Handwriting and Recognition (2008)

15. Pal, U., Wakabayashi, T., Kimura, F.: Comparative study of Devanagari handwritten character recognition using different feature and classifiers. In: Document Analysis and Recognition, 2009. ICDAR'09. 10th International Conference on. IEEE (2009)
16. Holambe, A.N., Thool, R.C., Jagade, S.M.: Printed and handwritten character & number recognition of devanagari script using gradient features. Int. J. Comput. Appl. 2(9), 975–8887 (2010)
17. Jangid, M.: Devanagari isolated character recognition by using statistical features. Int. J. Comput. Sci. Eng. 3(2), 2400–2407 (2011)
18. Singh, G., Lehri, S.: Recognition of handwritten hindi characters using backpropagation neural network. Int. J. Comput. Sci. Inf. Technol. 3(4), 4892–4895 (2012)
19. Vaidya, M., Joshi, Y.V., Bhalerao, M.: Marathi numeral identification system in Devanagari script using discrete cosine transform. Int. J. Intell. Eng. Syst. 10(6), 78–86 (2017)
20. Rehman, A., et al.: Simple and effective techniques for core-region detection and slant correction in offline script recognition. In: Signal and Image Processing Applications (ICSIPA), 2009 IEEE International Conference on. IEEE (2009)
21. Blumenstein, M., Cheng, C.K., Liu, X.Y.: New preprocessing techniques for handwritten word recognition. In: Proceedings of the Second IASTED International Conference on Visualization, Imaging and Image Processing (VIIP 2002), ACTA Press, Calgary (2002)
22. Pastor, M., Toselli, A., Vidal, E.: Projection profile based algorithm for slant removal. Image Analysis and Recognition, pp. 183–190 (2004)
23. Otsu, N.: A threshold selection method from gray-level histograms. IEEE Trans. Syst. Man Cybern. 9(1), 62–66 (1979)
24. Sinha, R.M.K., Mahabala, H.N.: Machine recognition of Devanagari script. IEEE Trans. Syst. Man Cybern 9(8), 435–441 (1979)
25. Arora, S., et al.: Combining multiple feature extraction techniques for handwritten devanagari character recognition. In: Industrial and Information Systems, 2008. ICIIS 2008. IEEE Region 10 and the Third international Conference on. IEEE (2008)
26. Kumar, M., Jindal, M.K., Sharma, R.K.: Segmentation of isolated and touching characters in offline handwritten gurmukhi script recognition. Int. J. Inf. Technol. Comput. Sci. (IJITCS) 6 (2), 58 (2014)
27. Kamble, P.M., Hegadi, R.S.: Handwritten Marathi character recognition using R-HOG Feature. Proc. Comput. Sci. 45, 266–274 (2015)

Comprehensive Study on Usage of Multi Objectives in Recommender Systems

M. Sruthi[(✉)], Sini Raj Pulari, and Ramesh Gowtham

Department of Computer Science and Engineering, Amrita School
of Engineering, Coimbatore, Amrita Vishwa Vidyapeetham, Amrita University,
Coimbatore, India
cb.en.p2csel5023@cb.students.amrita.edu,
{p_siniraj,r_gowtham}@cb.amrita.edu

Abstract. Recommender systems have changed its purview from prediction accuracy oriented to finding more relevant and useful recommendations to user. "Usefulness" of items are different in different applications. This paper summarizes the works that have been done in this direction. Personalization, context awareness, multiple objectives of recommendations and evaluation metrics are reviewed in this paper.

Keywords: Context-awareness · Multi-objective · Personalization
Recommender systems

1 Introduction

Recommendation systems provide users with personalized content and assistance after searching, filtering and ranking of pieces of information from lively generated huge volume of information. This is helpful for addressing the problem of information overload. Initially the recommender systems were accuracy focused and by maximizing the similarity between items user liked and items to be recommended gave good accuracy scores. Then apprehending that recommending unpopular yet relevant items will improve user satisfaction, the novelty factor came into picture. Identification of user preference from user behavior, context and sessions can result in diversified, fascinating and amazing recommendations still preserving usefulness of items. Factors like trust, robustness, privacy and scalability is also playing an important role in user satisfaction.

2 Objectives of Recommendations

Recommendations can be made more valuable if they are able to address more than one objectives of a user. The objectives of recommendations which are addressed in various research works are summarized as.

© Springer International Publishing AG 2018
D. J. Hemanth and S. Smys (eds.), *Computational Vision and Bio Inspired Computing*,
Lecture Notes in Computational Vision and Biomechanics 28,
https://doi.org/10.1007/978-3-319-71767-8_5

2.1 User Preference

Comparisons of recommender systems can be done how one system outperforms other in terms of user satisfaction. User satisfaction is imperative to quantify, it can be measured by measuring components that contribute to user satisfaction. The recommender system should be designed to enhance these components.

2.2 Prediction Accuracy

Prediction engines are the base of the recommender systems and prediction accuracy is the utmost explored property of recommender systems. Recommender systems predict the inclination of user on items and chance of purchase. Users tend to prefer recommender systems giving accurate recommendations. Accuracy can be measured by offline experiments, but the system accuracy may vary according to user preferences.

2.3 Novelty

Novel recommendations are items which are relevant to user, but haven't seen before. So important part of novel recommendations is to filter out already seen or purchased items. When the information about purchase or preference is unavailable, novel recommendations are difficult to generate.

2.4 Serendipity

Serendipity measures how amazing the fruitful recommendations are. For instance, if user likes books from certain authors, recommending a book of same author may be a novel recommendation. But it may not be qualifying as a amazing one. Obviously, arbitrary recommendations might be extremely amazing, and we hence need to adjust serendipity with exactness. Recommending a document with new information is called both serendipitous as well as novel in text retrieval. To restricting the human interaction in recommendation procedure, separation estimation between items based on its content is used. In the case of collaborative filtering, a separation based score is calculated as separation from an arrangement of already seen items. Content-based filtering state, the separation score is the separation from the user profile. Serendipitous recommendations will be having a higher separation score.

2.5 Diversity

Recommendation of comparable things may not be constantly useful for the user. Then the concept of diversity is introduced. Diversity is largely characterized as the inverse of closeness. Sometimes items related to previously purchased or seen or reviewed and relevant to user is a better choice to offer. Consider for instance a proposal for a vacation, where the framework ought to prescribe vacation packages. At initial stages of recommendation different places with different types of tourists spot, lodging and amusement qualifies as diverse recommendations. Recommending one place with different options for lodging and amusement is not a diverse recommendation. At that point, we could quantify the differing qualities of a list in view of the total, normal, min, or max separate between item sets, or measure the benefit of adding every item in the suggestion list as the new item's assorted qualities from the items as of now in the rundown. The item similarity estimation utilized as a part of assessment can be unique in relation to the similarity estimation utilized by the calculation that registers the recommendation lists. Diversity may require sacrifice of some other properties like prediction accuracy, then for evaluation of recommendations a plot of these properties will be helpful.

2.6 Trust

Trust in recommender systems refers to the trust in its calculated ratings. Ratings are predicted grounded on the ratings of neighboring items or users. Extrapolating ratings for eliminating sparsity have an adverse effect in trust. Trust is evaluated by user interaction by asking the user about the effectiveness of recommendations. Thus offline trust evaluation is not possible as it involves user collaboration.

3 Literature Survey

The survey is mainly based on the objectives of recommender systems, how approaches are made to achieve these objectives, areas focused by various researchers and evaluation of recommendations in different perspectives.

3.1 Approaches Dealing with Combined Objectives

Novelty and Serendipity are introduced as a key aspect of recommendation in approach by Herlocker et al. [1]. Users can be suggested with alluring and unexpected recommendation with serendipity oriented approaches. In a study conducted by Vargas et al. [2], two new matrices ranking sensitivity, and relevance-awareness for computing novelty and diversity respectively are suggested.

The algorithms which gives higher accuracy and relevance are mainly based the characteristics of popularity and similarity of items. The approach presented in [3] solves the predicament of accuracy-diversity by presenting a hybrid algorithm by combining two existing algorithms—HeatS and ProbS. With an adjustable degree of hybridization, the proposed algorithm achieves optimization of both of the parameters. ProbS algorithm contributes to accuracy and HeatS tend to recommend diverse items.

When recommendation task is modelled as rating prediction task, ranking the items according to the predicted rating delivers good accuracy and shows a deprived performance with respect to diversity of items. The approach suggested by Adomavicius et al. [4] improves the diversity by employing new ranking techniques with a little compromise to accuracy, These ranking technique are flexible enough that the parameters can be changed and reused in any rating prediction algorithms. They suggest further improvements can be achieved by making use of optimization based approaches. Finding a single metric which captures both the characteristics of accuracy and diversity is an interesting research domain. User preference, acceptance of diversified recommendation and user satisfaction can also be considered.

Castells et al. [5] identified novelty and diversity as vital quality of recommendations. Discussions on defining these characteristics from different dimensions, designing grades and procedures for evaluation are presented. The approach suggested in [6] discussed about adding novelty and diversity along with relevance in tag recommendation domain. Novelty of item in recommendation catches how distinctive the item is from every single other items seen in a given context. In a rundown of recommended items, assorted qualities of items can be characterized implicitly or explicitly. The previous alludes to how distinctive the suggested things are from each other. An explicit meaning of diversity as a rule adventures a scientific categorization, for example, an arrangement of classifications or areas. A diverse recommendation list is the set of assorted items which involves distinctive topics. In addition, pertinence, novelty, and diversity are imperative viewpoints for the viability and utility of tag recommendations. While diversity advancement gives better coverage of various conceivable topics, novelty may advance more uncommon, supportive and particular tags to enlarge the expansion of the content depictions. In this manner the findability of items is easier in tag based search.

Mouzhi et al. [7] suggested two novel metrics for assessing recommender systems namely coverage and serendipity. The quality and "usefulness" of recommendations are interpreted by this metrics even though they contrast with accuracy measures. Coverage is defined as measure of space of items over which the framework can generate recommendations. The portion of items in the list for which the framework produce recommendation is called prediction coverage and the fraction of effectively suggested items to the user is called catalogue coverage. Serendipity is characterized as a measure of the degree to which the prescribed items are both appealing and astonishing to the users. The serendipity metric is calculated as the usefulness of the unexpected recommendations. Kotkov et al. [8] talked about difficulties of serendipity in recommender frameworks. Serendipity is challenging to research, as it incorporates a passionate measurement, which is hard to catch, and serendipitous experiences are exceptionally uncommon, since serendipity is a mind boggling idea that incorporates different ideas. Content-based unexpectedness, collaborative unexpectedness and primitive recommender-based serendipity are the three categories of serendipity evaluation metrics. The principle hindrance of content-based and collaborative unexpectedness metric is that they measure unexpectedness independently from relevance, which may bring about missteps.

The fundamental inconvenience of primitive recommender-based serendipity measurement is that they are sensitive to recommender systems. Sugiyama et al. [9] proposed a scholarly paper recommender system addressing serendipitous and relevance perspectives. Researchers whose interests differ from a target user might guarantee possibility to produce fascinating and amazing recommendations. Therefore, profiles from users that are maximally not quite the same as our target user are selected and formed a dissimilar users (DU) set. Inverse of similarity value between target user and candidate user are used for ranking candidate users. This approach favors serendipitous recommendations. The work presented by Lathia et al. [10] concentrates on the temporal dimension (inter-list diversity) and whether precisely the same items are being requested to users more than once; not the semantic relationships or popularity of items. The aim is to develop a recommender framework from rating data that offers various outcomes that change according to each user's taste. Additional information such as clicks, sessions, reviews will help in improving the system.

Yamaba et al. [11] suggested a serendipity oriented procedure for book recommendation. The items (books) are labeled with tags that and impressions that the user felt are extracted as concepts. These concepts are used for selection of items for recommendation to users. Traditionally collaborative filtering frameworks have depended vigorously on matches between the ratings profiles of users as an approach to differentially rate the predicted value of different profiles [12]. O'Donovan et al. [13] suggested trust as one factor which influence recommendation. This approach considered trust in reference to how much one may trust a particular profile with regards to making a particular rating forecast. Two trust models for profile and item level is developed. In a period of time, accuracy of predictions made by a profile is monitored and then trust is the rate of right predictions in profile or item level. Sparsity of rating matrix has an adverse effect on recommendation accuracy. For finding similar users, a trust weight obtained by propagating trust score along a trust network of users can be used in place of similarity score [14]. This algorithm works well in cold start problem areas while preserving a good coverage. Addressing the user's shifted and dynamic inclinations for novelty, a versatile procedure are needed for recommendation. Kapoor et al. [15] build up a regression model to foresee these changing novelty inclinations of users utilizing data from their former associations.

3.2 Context Awareness

Various studies conducted illustrates that the use of contextual information in recommendation process will improve the user satisfaction. Contextual information is collected explicitly from sessions, browser history, interactions in social media, reviews provided by user and interactions with user.

A method suggested by Woerndl et al. [16] uses a collaborative filtering for getting an initial list of items for mobile users for the context-aware recommendation. Information base of mobile applications is used to make sense of items which are important to the user in the present circumstances. Here context is portrayed as a vector of context features, just like for users and items. The approach by Adomavicius et al. [17] uses contextual data at different phases of the recommendation procedure, including at the stage of pre-filtering and the post-filtering.

Nasraoui et al. [18] proposed a two-step recommender system exploiting the session data and mapping it into a pre-found profiles and afterward utilizes one of a few profile-particular URL-indicator neural systems to give the final recommendations. This method achieves higher level of context awareness than from the collaborative filtering technique. The survey conducted by Pearl et al. [19] reviewed how users see and assess recommender systems. The review resulted in some usability and user interface design strategies.

3.3   Personalization

Providing custom-made services and content based on the user fondness and behavior is called personalization. Personalized recommendations should satisfy user by delivering items which are useful to him. "Useful" items comprises of relevant, novel and varied ones.

An methodology for recommending items in internet marketplace suggested by Cho et al. [20] makes use of clickstream data for learning user preferences. A decision tree induction is then employed for choosing users who are probably going to purchase recommended items. Then, the express support of the advertisers and the formal utilization of background learning, for example, the item taxonomy are additionally presented in the recommendation procedure. At long last, business-proficient items are selected from candidate items. A semantic extension to hybrid tag based recommender system is proposed by Durao et al. [21]. Data from ontologies of open linked data and WordNet dictionary are used for identifying semantic relations between tags. Usage of the semantic source for identifying relationships between tags exhibited an enhancement in precision in resultant recommendations.

The approaches formed by Khribi [22] is used to achieve personalization using mining of the web and intelligent search engines. Content profiles are constructed by performing scalable search on massive online content and techniques like association rule mining and clustering are used to assemble user profiles. Combination of content based and collaborative methods are utilized as a part of the recommendation phase. Chen He et al. [23] laid out a method for recommendation making use of visualization methods to empower users of the system to pick up knowledge about recommendation procedure and support them to control it. The paper discussed about 24 existing interactive recommender systems and came out with a conclusion that the visualization techniques influences in accuracy and improves user acceptance.

Assigning personalized hybrid parameter to each of the user will significantly improve the recommendation accuracy. Guan et al. [24] found that the customized hybrid parameter will vary from user to user and is positively correlated with average item degree gathered by the user. Experiments conducted with benchmark dataset exhibited improved accuracy, but estimation of customized hybrid parameter still exists as a challenge. In survey papers [25, 26], they audited different impediments of the present recommendation strategies and talked about conceivable augmentations that can give better recommendation capacities. These augmentations incorporate, among others, the enhanced demonstration of clients and items, joining the relevant data into the suggestion procedure, strengthening the multi-criteria assessments, and arrangement of adaptable and less noisy suggestion procedure.

3.4   Evolutionary Algorithms

Multi objective optimization considers enhancement issues including more than one objective capacity to be streamlined all the while. Multi objective enhancement issues emerge in many fields, for example, building, financial aspects, and coordination, when ideal choices should be taken within the sight of exchange offs between at least two clashing objectives. For instance, building up another objective may include minimizing weight while expanding quality or picking a collection may include augmenting the normal return while minimizing the risk [27].

Commonly, there does not exist a solitary arrangement that at the same time upgrades every objective. Rather, there exists a (perhaps vast) arrangement of Pareto optimal solutions. An answer is called Pareto optimal if none of the objective capacities can be enhanced in esteem without corrupting at least one of the other objective qualities. Without extra subjective inclination data, all Pareto optimal arrangements are considered similarly great.

The areas using evolutionary algorithms in recommendation systems are (a) approaches in which GA or EA are utilized to improve weights of various components or recommendation procedures, (b) approaches using genetic algorithm (GA) or evolutionary algorithm (EA) for clustering and (c) hybrid and other approaches [28]. In the first approach GA is utilized to weigh features comprising of qualities of items and users, ratings provided, preferences obtained from social networks and so on. Next method is grouping of users and items with similar attributes and using GA to find ideal cluster center and ideal subset. The methodologies regarding latent models are utilizing GA to determine ideal weights for hidden parameters of users and items, EA to enhance matrix factorization of utility matrix and EA based clustering relying on latent parameters for neighbor discovery. As future research in evolutionary computing (EC)-based recommendation is identified as complex recommendation states, to attempt to tackle the long-tail issue and to expand the differing qualities, novelty and serendipity of recommendations.

Smeaton et al. [29] explored that needs for future research are partitioned into five noteworthy regions—Personalization, User Modelling, User communication, Evaluation and Social impacts. Ribeiro et al. [30] suggested a personalization approach for recommendation based on hybridization achieved by dynamic adjustment of accuracy, diversity and novelty parameters. One disadvantage of the approach is the cost of evolutionary algorithm. A diversified recommendation framework based on MOEA/D is proposed in [31]. MOEA/D algorithm can create different recommendation list for every user at one time without giving up accuracy. The algorithm works in two steps—an item based collaborative filtering and multi-objective evolutionary algorithm for generating series of Pareto optimal solutions.

The method proposed by Boumaza et al. [32] uses evolutionary algorithm for searching neighbors shared by all users, alluded to as Global Neighbors (GN). The aim of the algorithm is to minimize the size of Global Neighbors and thereby reducing complexity of the computation of recommendations. Exploiting the Pareto efficiency idea will optimize objectives (accuracy, novelty, and diversity) of recommendation without essentially harming each other [33]. The Pareto-efficiency concept can be done by considering items in N-dimensional space or hybrid algorithms with 3—dimensional co-ordinates. Recommendation of unpopular but novel items will degrade the accuracy of recommender systems. The approach by Shanfeng et al. [34] formulate multi-objective framework for Long tail recommendation with minimal accuracy loss. The proposed algorithm MORS is effective in suggesting novel and relevant items.

The multi objective recommendation framework focusing on accuracy and diversity is proposed by Bingrui et al. [35]. Multi-objective optimization is used to enhance the matching function satisfying diversity and accuracy or recommendations in a candidate list formed by performing collaborative filtering. An approach by Zuo et al. [27] modelled personalization as multi-objective optimization problem which optimizes accuracy and coverage. They proposed a hybrid algorithm consisting of ProbS (Probability Spreading) and MOEA (multi-objective evolutionary algorithm) for addressing accuracy and diversity respectively.

4   Evaluation of Recommender Systems

The Evaluation of recommender system in user's perspective involves how useful recommendations are with respect to user. This involves an online evaluation procedure. But for evaluating recommendation algorithm based on the factors can be done offline also (Fig. 1).

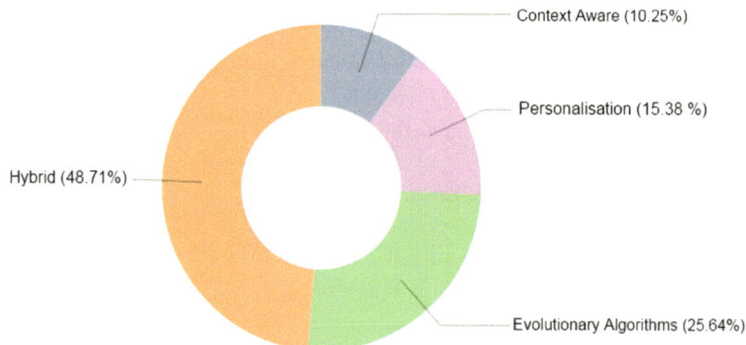

Fig. 1 Algorithms explored for addressing different objectives

Table 1 shows the evaluation metrics used in different recommender systems in different scenarios. The parameters appearing in the equations are tailored to certain approaches, still it can be modified according to the scenario.

Table 1 Evaluation metrics used in different recommender systems

Factor	Equation	Description
Accuracy	$Precision = \frac{N_{rs}}{N_s}$	Fraction of number of relevant items to number of items selected
Accuracy	$Recall = \frac{N_{rs}}{N_r}$	Recall describes probability of a relevant item get selected in recommendation list
Accuracy	$Fscore = \frac{2 \times Precision \times Recall}{Precision + Recall}$	Harmonic mean of precision and recall is the Fscore
Diversity	$Prediction coverage = \frac{|I_p|}{|I|}$	Ratio of available items to which prediction can be made is the prediction coverage
Diversity	$Catalog coverage = \frac{|U_{j=1...N} I_L^j|}{|I|}$	Let I_L^j represents the items returned by jth recommendation in a list L.I is the set of available items and N is the total number of recommendations
Accuracy and Diversity	$Fscore = \frac{2 \times Precision \times Coverage}{Precision + Coverage}$	Harmonic mean of two contradicting parameters
Accuracy	$\bar{E} = \frac{\sum_{i=1}^{N} |p_i - r_i|}{N}$	Mean absolute error measures deviation from predicted and actual user ratings
Diversity	$F_D(R) = \frac{1}{L(L-1)} \sum_{i \in R} \sum_{i \in R, i \neq j} d(i,j)$	Average dissimilarity of all pairs in a recommendation list R with length L Distance function $d(i,j) = 1 - sim(i,j)$
Novelty	$EPC@N = C \sum_{i_k \in R}^{i_N} disc(k)[1 - p(seen|i_k)]$	Amount of novelty in top N recommendation list R can be calculated as EPC@N $disc(k)$ is the rank discount and C is a normalizing constant

(continued)

Table 1 (*continued*)

Factor	Equation	Description
Novelty	$Novelty = \frac{1}{mk}\sum_{u=1}^{m}\sum_{i\in L_u} d_i$	Novelty is the ability of a recommended system to suggest items which user haven't seen before L_u is the top-k list of user u, d_i represents total number of users rated item i and m is the total number of users
Diversity	$EILD@N = \sum_{i_p\in R, i_q\in R}^{N} C_k disc(p)disc(q|p)d(i_p, i_q)$	A distance based model for diversity
Diversity	$H_{u_a u_b}(k) = 1 - \frac{Q_{u_a u_b}(k)}{k}$	Inter user diversity using hamming distance
Diversity	$D_u(k) = \frac{1}{k(k-1)}\sum_{p,q} s(i_p, i_q)$	Intra user diversity of recommendation list of two users p and q
Serendipity	$SRDP = \frac{\sum_{i=1}^{N} u(RS)}{N}$	Calculate an unexpected set by taking elements which are not in primitive model but present in recommendation list. N is the total number of unexpected elements
Serendipity	$unexpec_u(i,u) = \frac{1}{|I_u|}\sum_{j\in I_u} 1 - sim(i,j)$	Content based unexpectedness
Serendipity	$unexpec_u(i,u) = \frac{1}{|I_u|}\sum_{j\in I_u} -\log_2\frac{p(i,j)}{p(i)p(j)} / \log_2 p(i,j)$	Collaborative unexpectedness
Serendipity	$ser_{pm}(u) = \frac{1}{|R_u|}\sum_{j\in(R_u/(E_u UPM))} rel_u(i)$	Primitive recommender based serendipity
Trust	$Trust^P(p) = \frac{|CorrectSet(p)|}{|RecSet(p)|}$	Profile level trust means the fraction of apt recommendations contributed by the profile
Trust	$Trust^I(p,i) = \frac{|\{(c_k, i_k)\in CorrectSet(p):i_k=i\}|}{|\{(c_k, i_k)\in RecSet(p):i_k=i\}|}$	Item level trust is represented as fraction of apt recommendations made for an item

5 Discussion and Conclusion

The researches in improving the quality of recommendation involve improvement in various stages of recommendation process. From data collection to forming the final list of recommendation items refinements are possible. As the survey mainly focused on objectives of the recommendations, designing a unified factor which depicts the importance of factors according to the scenario is a research problem. In the area of multi-objectives, hybrid algorithms works better than content or collaborative based algorithms. Incorporating contextual information favors the user satisfaction to an extent. Personalization and related areas (especially user modelling) will pave an effectual way to better recommendations. Some recommendation scenarios needs making use of the emotional characteristics of user which makes same recommendation useful at one point and irrelevant at some other point. Instead of conducting an offline analysis and evaluation, we are highly recommending online evaluation with user involvement (Fig. 2).

Fig. 2 Factors discussed in recommendations for the year 2005–2017

References

1. Herlocker, J.L.: Evaluating collaborative filtering recommender systems. ACM Trans. Inf. Syst. (TOIS) **22**(1), 5–53 (2004)
2. Vargas, S.J., Castells, P.: Rank and relevance in novelty and diversity metrics for recommender systems. In: Proceedings of the Fifth ACM Conference on Recommender Systems. ACM (2011)
3. Zhou, T., et al.: Solving the apparent diversity-accuracy dilemma of recommender systems. Proc. Nat. Acad. Sci. **107**(10), 4511–4515 (2010)
4. Adomavicius, G., Kwon, Y.: Improving aggregate recommendation diversity using ranking-based techniques. IEEE Trans. Knowl. Data Eng. **24**(5), 896–911 (2012)
5. Castells, P., Hurley, N.J., Vargas, S.: Novelty and diversity in recommender systems. In: Recommender Systems Handbook, pp. 881–918. Springer, US (2015)

6. Belém, F.M., et al.: Beyond relevance: explicitly promoting novelty and diversity in tag recommendation. ACM Trans. Intell. Syst. Technol. (TIST) **7**(3), 26 (2016)
7. Ge, M., Delgado-Battenfeld, C., Jannach, D.: Beyond accuracy: evaluating recommender systems by coverage and serendipity. In: Proceedings of the Fourth ACM Conference on Recommender Systems. ACM (2010)
8. Kotkov, D., Veijalainen, J., Wang, S.: Challenges of serendipity in recommender systems. In: WEBIST 2016: Proceedings of the 12th International Conference on Web Information Systems and Technologies, vol. 2. SCITEPRESS (2016). ISBN 978-989-758-186-1
9. Sugiyama, K., Kan, M.Y.: Towards higher relevance and serendipity in scholarly paper recommendation. ACM SIGWEB Newsletter (2015)
10. Lathia, N., et al.: Temporal diversity in recommender systems. In: Proceedings of the 33rd International ACM SIGIR Conference on Research and Development in Information Retrieval. ACM (2010)
11. Yamaba, H., et al.: On a serendipity-oriented recommender system based on folksonomy. Artif. Life Robot. **18**(1–2), 89–94 (2013)
12. Venkataraman, D., Gangothri, V., Saranya, S.: A comprehensive review of recommender system. Int. J. Appl. Eng. Res. **10**, 13909–13919 (2015)
13. O'Donovan, J., Smyth, B.: Trust in recommender systems. In: Proceedings of the 10th International Conference on Intelligent User Interfaces. ACM (2005)
14. Massa, P., Avesani, P.: Trust-aware recommender systems. In: Proceedings of the 2007 ACM Conference on Recommender Systems. ACM (2007)
15. Kapoor, K., et al.: I like to explore sometimes: adapting to dynamic user novelty preferences. In: Proceedings of the 9th ACM Conference on Recommender Systems. ACM (2015)
16. Woerndl, W., Schueller, C., Wojtech, R.: A hybrid recommender system for context-aware recommendations of mobile applications. In: 2007 IEEE 23rd International Conference on Data Engineering Workshop. IEEE (2007)
17. Adomavicius, G., Tuzhilin, A.: Context-aware recommender systems. In: Recommender Systems Handbook, pp. 191–226. Springer, US (2015)
18. Nasraoui, O., Pavuluri, M.: A context ultra-sensitive approach to high quality Web recommendations based on Web usage mining and neural network committees. In: 3rd International Workshop on Web Dynamics in Conjunction with the 13th International World Wide Web Conference New York City, New York, USA, 18 May 2004
19. Pu, P., Chen, L., Rong, H.: Evaluating recommender systems from the user's perspective: survey of the state of the art. User Model. User-Adap. Inter. **22**(4), 317–355 (2012)
20. Cho, Y.H., Kim, J.K., Kim, S.H.: A personalized recommender system based on web usage mining and decision tree induction. Expert Syst. Appl. **23**(3), 329–342 (2002)
21. Durao, F., Dolog, P.: Extending a hybrid tag-based recommender system with personalization. In: Proceedings of the 2010 ACM Symposium on Applied Computing. ACM (2010)
22. Khribi, M.K., Jemni, M., Nasraoui, O.: Automatic recommendations for e-learning personalization based on web usage mining techniques and information retrieval. Advanced Learning Technologies, 2008. ICALT'08. Eighth IEEE International Conference on. IEEE (2008)
23. He, C., Parra, D., Verbert, K.: Interactive recommender systems: a survey of the state of the art and future research challenges and opportunities. Expert Syst. Appl. **56**, 9–27 (2016)
24. Guan, Y., et al.: Preference of online users and personalized recommendations. Physica A Stat. Mech. Appl. **392**(16), 3417–3423 (2013)
25. Adomavicius, G., Tuzhilin, A.: Toward the next generation of recommender systems: a survey of the state-of-the-art and possible extensions. IEEE Trans. Knowl. Data Eng. **17**(6), 734–749 (2005)

26. Yedugiri, K.B., Chandni, S.P., Raj, S., Souparnika, S.: Recommender systems—a deeper insight. Int. J. Appl. Eng. Res. **9**, 28521–28531 (2014)
27. Zuo, Y., et al.: Personalized recommendation based on evolutionary multi-objective optimization research frontier. IEEE Comput. Intell. Mag. **10**(1), 52–62 (2015)
28. Horváth, T., de Carvalho, A.C.: Evolutionary computing in recommender systems: a review of recent research. Nat. Comput. 1–22 (2016)
29. Smeaton, A.F., Callan, J.: Personalisation and recommender systems in digital libraries. Int. J. Digit. Libr. **5**(4), 299–308 (2005)
30. Ribeiro, M.T., et al.: Pareto-efficient hybridization for multi-objective recommender systems. In: Proceedings of the sixth ACM conference on Recommender systems. ACM (2012)
31. Wang, J., et al.: Diversified recommendation incorporating item content information based on MOEA/D. In: 2016 49th Hawaii International Conference on System Sciences (HICSS). IEEE (2016)
32. Boumaza, A., Brun, A.: From neighbors to global neighbors in collaborative filtering: an evolutionary optimization approach. In: Proceedings of the 14th Annual Conference on Genetic and Evolutionary Computation. ACM (2012)
33. Ribeiro, M.T., et al.: Multiobjective pareto-efficient approaches for recommender systems. ACM Trans. Intell. Syst. Technol. (TIST) **5**(4), 53 (2015)
34. Wang, S., et al.: Multi-objective optimization for long tail recommendation. Knowl. Based Syst. **104**, 145–155 (2016)
35. Bingrui, G., et al.: NNIA-RS: a multi-objective optimization based recommender system. Physica A Stat. Mech. Appl. **424**, 383–397 (2015)

Visual Analysis of Genetic Algorithms While Solving 0-1 Knapsack Problem

B. P. Sathyajit$^{(\boxtimes)}$ and C. Shunmuga Velayutham

Department of Computer Science and Engineering, Amrita School
of Engineering, Amrita Vishwa Vidyapeetham, Coimbatore, India
`cb.en.p2csel5016@cb.students.amrita.edu`,
`cs_velayutham@cb.amrita.edu`

Abstract. This paper presents heat map based visual analysis of Genetic Algorithm (GA) solving 0-1 Knapsack Problem (KP). The current work is a preliminary investigation to understand the search strategy of GA solving KP through visual means. A simple GA has been employed to solve 50, 100 and 500 items 0-1 KP. Heat map based visualization of best chromosomes shows clearly the explorative and exploitative search strategies of GA in conjunction with convergence characteristics. This paper demonstrates the potential of visualization to analyze and understand Evolutionary Algorithms (EA) in general.

Keywords: Genetic algorithm · Heat maps · 0-1 knapsack problem
Exploration and exploitation · Visualization · Search strategy

1 Introduction

Evolutionary Algorithms (EAs) are stochastic search algorithms modeled on the processes of natural evolution. EAs typically start with a population of random solutions for a given problem. The population is then subjected to the iterative crossover-mutation-selection cycle to eventually result in a population of high quality solutions [1]. The hidden mechanisms behind population evolution, roles of the above-said three operations and the knowledge of search space are all still elusive to a large number of researchers who are working towards understanding EAs. A thorough understanding of EAs is crucial to harness their full potential.

Towards this, the EA research community primarily employs empirical and theoretical analyses to understand the evolutionary mechanisms behind EAs [2]. While empirical and theoretical analyses hitherto have provided valuable insights about the working of EAs, they still fall short in exposing the complex evolutionary mechanisms that underlie the evolving population. This necessitates alternative analyses techniques that should be complementary to the empirical and theoretical analyses and should offer insightful perspective to the problem of understanding evolutionary dynamics. Visual analysis is a potential alternative capable of providing perspectives, the EA researchers have had never access to. Consequently, visual analysis of EAs is gaining more attention in the EA research community as an alternative method of analysis [3–8].

© Springer International Publishing AG 2018
D. J. Hemanth and S. Smys (eds.), *Computational Vision and Bio Inspired Computing*,
Lecture Notes in Computational Vision and Biomechanics 28,
https://doi.org/10.1007/978-3-319-71767-8_6

This paper is intended to do a preliminary investigation on visual analysis of Genetic Algorithms (GAs) solving the simple 0-1 Knapsack problem. Heat map visualization is employed to analyze the best chromosomes of GA runs to understand the strategies employed by the algorithm solving the knapsack problem. A thorough visual analysis of GA run for a given benchmark problem to gain valuable and deep insights about evolutionary dynamics is not only a feasible but natural extension of this work.

Accordingly, the paper has been organized as follows. Section 2 discusses related works in the context of this preliminary work. Section 3 provides the experimental design for the GA runs and Sect. 4 gives the subsequent visual analysis of those runs. Finally, Sect. 5 concludes the paper.

2 Related Works

Visualization of Genetic and Evolutionary computation is now an emerging research field with researchers attempting multidimensional visualization of EA data. The research trend in visual analysis of EAs has either been in the development of versatile visual tools or in proposing novel/existing suitable visualization techniques for EA analysis.

VIS [8], GAVEL [5], EAVIS [6], GraphDice [3], ELICIT [4] etc. are some of the visualization tools proposed for visual analysis of EAs. The VIS [8] system facilitates visualization of EAs with most detailed levels but with ad hoc representations. GAVEL [5] on the other hand, is an effective tool to visualize and trace the history of generational Genetic Algorithm (GA) run even to the level of custom selected alleles of GAs. It also helps to understand the role of crossover and mutation in the working of EAs. GraphDice [3] is a multidimensional visual exploration tool that allows interactive visual analysis of scatter plot matrix with relative ease. ELICIT [4] is a recent interactive tool that allows visualization from a single run to aggregate multiple runs of EAs. It is also capable of visualizing genetic heritage and handling multiple representations. While these versatile tools enable effective visualization of EAs, effective usage of these tools to gain insights about the EA dynamics has not been taken up well in literature.

Bedau and Bullock [9] proposed visualization of both alleles and genotypes using activity wave diagrams that helps identifying significant mutations across generations. Pohlheim [10] provided a set of visualization techniques for all levels of EAs as well as multidimensional techniques for handling larger dimensions. Romero et al. [11] proposed self-organizing maps for visualizing various aspects of GAs. Pryke et al. [12] employed heat maps to visualize the population of multi-objective algorithms especially the parameter space. Collins [13] has presented an introductory review of the existing techniques for visualizing evolutionary computation. McDermott [14] proposed a technique for search space visualization where distances between points are projected into two dimensions. Analyzing search spaces is yet another direction for the visual analysis of EAs. With the proliferation of tools and techniques, the stage is set for effective use of such tools/techniques to carry out visual analysis of EAs to gain deep insights. This paper is a preliminary effort in this direction.

In EA (specifically GA without loss of generality) literature a number of test problems have been employed to study and analyze various properties and behavior of EAs. Knapsack problem (KP) is one such classical combinatorial optimization problem. A few representative examples presenting the research trend in employing KP to investigate EAs follow. Shao et al. [15] employed a greedy GA using improved selection strategy to solve 0-1 Knapsack problem. Shen et al. [16] proposed a dual population GA for solving 0-1 KP and demonstrated the superiority of proposed GA over traditional GA. Changdar et al. [17] proposed an improved GA to solve constrained knapsack problem in fuzzy environment. As can be seen from above works, the recent research trend primarily focuses on proposing novel variants of EAs to solve increasingly complex variants of KP. There are very few works that focus on deeper issues involved in employing EAs to solve KP. By way of an example, He et al. [18] carried out theoretical investigation on the KP solution quality of EAs. Amidst the proliferating improved GAs to solve KP variants, the fundamental knowledge as to how (using what strategies) an EA solves KP (also applicable to any typical optimization problem) is still elusive. This paper is a preliminary investigation towards this direction.

3 Design of Experiments

Knapsack problems are NP-hard combinatorial optimization problems that have been intensively studied because of their applications but more for theoretical reasons. These problems require that a subset of some given items (each with a capacity and a profit) is to be chosen such that profit sum is maximized without exceeding a given limit of knapsack capacity. Depending on the knapsacks and distribution of items, there are different types of knapsack problems. The 0-1 knapsack problem, focus of this paper, is the simplest of all were each given item is either picked once or not. The problem may be formulated as follows

$$\text{Maximize} \sum_{j=1}^{n} p_j \tag{1}$$

$$\text{Subject to} \sum_{j=1}^{n} w_j x_j \leq c \tag{2}$$

$$x_j \in 0, 1, \quad j = 1, \ldots, n$$

where n is the number of items, p_j and w_j are respectively the profit and weight associated with a jth item, c is the knapsack capacity and x_j is the binary variable that decides the inclusion (if equals 1) of jth item in the knapsack or otherwise (if equals 0). We assume all coefficients p_j, w_j and c are positive integers.

In this paper, we have considered 0-1 Knapsack problems of three capacities viz 50, 100 and 500 to be solved by GA. The data for the above problems have been adopted from [19]. In fact, the data provided in [19] contains 1000 sets with different capacities for each of the three-considered number of items. For the sake of easier analysis, we have adopted, from [19], one set each of items 50, 100 and 500 with capacities 16,264, 34,092 and 194,586 respectively.

Genetic Algorithms are population based search algorithms. They typically start with a population of randomly initialized solutions appropriately represented for a given problem. After having calculated the fitness value of each population member (i.e. candidate solution), the population is subjected to selection-crossover-mutation-fitness calculation cycle (termed a generation) for a maximum number of generations (other termination criteria are possible too!).

In this paper, for all simulations (involving 50, 100 and 500 items cases) we have employed a simple GA with the following parameters. The binary representation which is a natural fit for 0-1 KP has been used to encode the solutions. While the solution (chromosome) size is decided by the problem (50, 100 or 500), the population size has been kept fixed as 200. The total profit $\sum_{j=1}^{n} w_j$ serves as the fitness function. However, if $\sum_{j=1}^{n} w_j$ of a solution exceeds the stipulated capacity c, the fitness value of the solution is made zero. Being a maximization problem, the solution with zero fitness will automatically be weeded out. The tournament selection of size 2 and fitness proportionate selection are employed for parent selection and new population selection respectively. Single point crossover with probability 0.5 and single point bit flip mutation with probability 0.1 is used for all simulations. GA is run for a maximum of 100 and 500 generations for 50, 100 items each and 500 items respectively. By virtue of GA's stochasticity, 30 such runs of GA have been carried out for each of the three problems.

The visual analysis of GA runs, solving the 0-1 KP, is intended to understand the strategy, if there is any, employed by GA. Since the candidate solutions (chromosomes) are in binary representation in the knapsack, a suitable technique to visualize a table of numbers is needed. This paper employs heat maps for the visual analysis of GA runs solving the 0-1 KP. The best solutions (chromosomes) from each generation of an entire GA run are visualized as heat map. When the best solutions of a GA run are stacked one over the other (thus forming a table) the sum of each column refers to the frequency of picking each item for that run. This, in turn, conveys the relative importance on each of the items shown by GA while solving 0-1 KP. Since the best chromosomes of a GA run (as it evolves towards the optimal solutions) are considered, any pattern in the relative importance of each item for solving 0-1 KP will give valuable clues about the working of GA for the given problem. In fact, the above said heat map visualization of best chromosomes is extended to all 30 runs of GA thus resulting in a 2D heat map.

4   Simulation Analysis

As detailed above, 30 runs of GA solving 0-1 KP has been carried out to do a preliminary visual analysis of working of GA. Figure 1a–c show the convergence graphs of 30 GA runs for the items 50, 100 and 500 respectively. The convergence graphs are obtained by averaging the best fitness values of each generation for all the 30 GA runs. As can be seen from the figures, GA runs display a common characteristic of convergence—an initial rapid convergence phase followed by an improvement phase that leads to a flattening phase. It is intended to visualize the best solutions, in each of the three phases for all 30 GA runs, as heat maps.

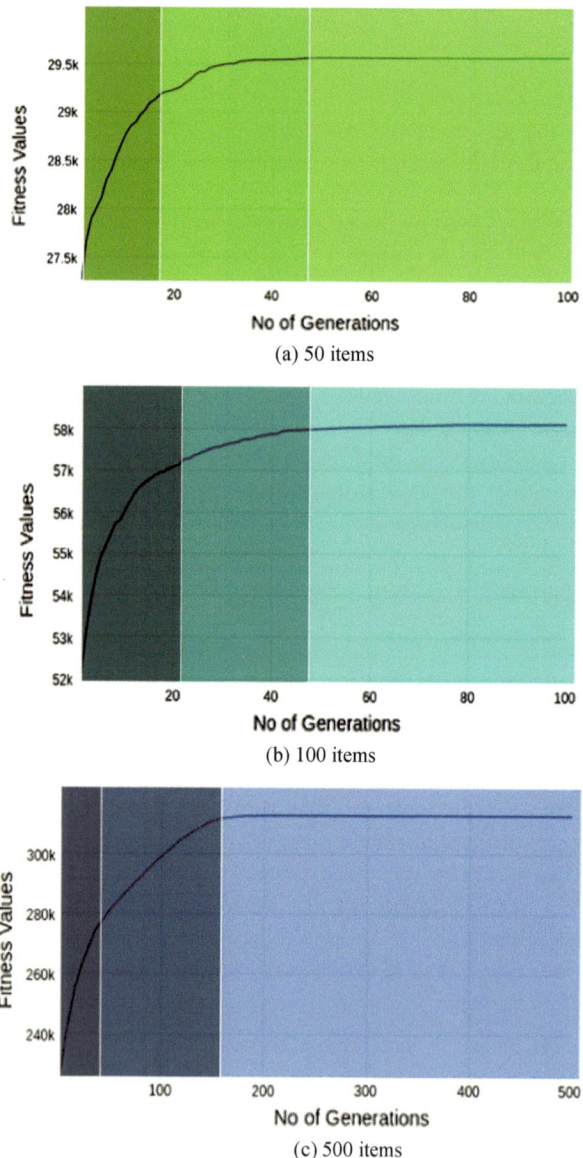

Fig. 1 Convergence graphs of 30 GA runs for the items **a** 50, **b** 100 and **c** 500 showing three phases

Figure 2a–c show the heat maps of GA runs solving 50 items 0-1 KP respectively for each of the above identified phases of convergence. Each row in the heat map is obtained by aggregating the best chromosomes in an identified phase (for a particular range of generations) of a single GA run to take an allele wise sum resulting in frequencies of picking each item by the best solutions in that phase. These frequencies are color-coded with the rarely picked item as blue and the most often picked item as red. The transition between these extremes (i.e. the different frequencies between minimum and maximum) is color-coded suitably with corresponding color gradient between blue and red as shown in legend to the right of each heat map. The 2D heat maps are obtained by stacking the color- coded frequencies of each item for all 30 runs. Thus, while rows represent the frequencies of each item picked by best solutions in a single GA run's phase, columns indicate the frequencies of each item picked by best solutions in all GA runs' phase. The items in the x-axis are arranged in an increasing order of their value-to-weight ratio. Thus, a greedy algorithm is expected to focus strictly on the right most items in the x-axis as they are characterized by larger value-to-weight ratio.

The rapid convergence phase of GA, visualized in Fig. 2a, does not seem to display a greedy characteristic. In fact, almost all the items have been picked by either one or other best solutions during the rapid convergence phase. However, the dense bluish color in the first one-third of x axis (first 7–8 items) in Fig. 2a shows that the GA, by virtue of the fitness function, has figured out the futility of picking items with very low value-to-weight ratio during the rapid convergence phase. Interestingly, though the GA runs did not choose to ignore very low value-to-weight ratio items altogether—a mark of absence of strict greedy characteristic. Similar arguments hold good for the remaining two-thirds of the heat map (Fig. 2a) where pronounced reddishness shows the relative importance of the items with large value-to-weight ratio.

As can be seen in Fig. 2a while GA employs an exploratory search (trying out every item) during the rapid convergence phase, the remaining two phases (improve-ment and flattening phases) require more exploitative search that improves the existing solutions. Figure 2b, c very well reflect the exploitative search characteristics of GA. The absence of most of the intermediate colors in the color gradient in both Fig. 2a, b marks the focused exploitative search strategy adopted by the GA. This strategy becomes little more acute as we move from (Fig. 2b, c) improvement phase to flat-tening phase. In fact, the heat map in Fig. 2c is characterized by the presence of only extremum colors that is red and blue with very few stray intermediate colors. Thus, all the best chromosomes in the flattening phase either choose to include an item or not to. Interestingly however this choice is not uniform across the 30 runs showing the dif-ferences in the quality of the best solutions obtained by each of the runs.

All the above observations and arguments in case of 50 items 0-1 KP extends to the

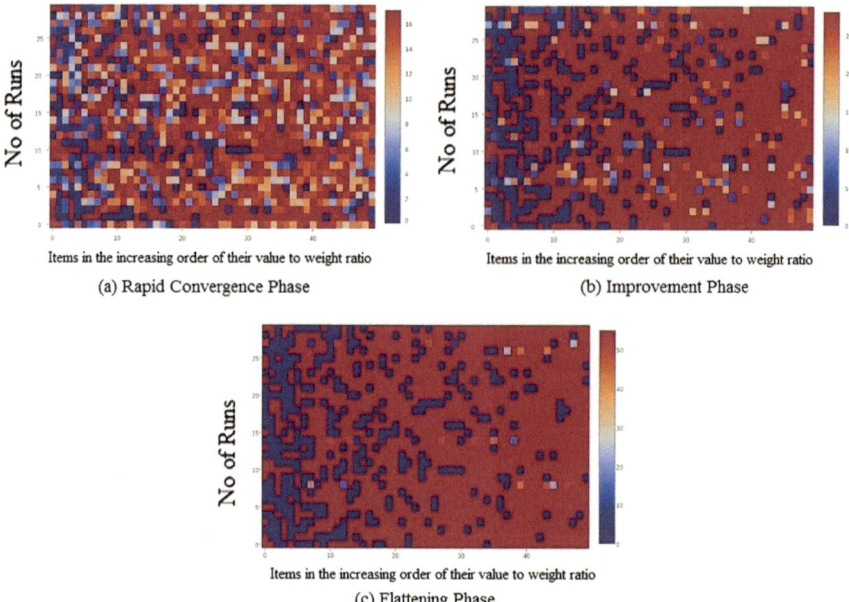

(a) Rapid Convergence Phase   (b) Improvement Phase

(c) Flattening Phase

Fig. 2 Heat maps of GA run solving 50 items 0-1 KP

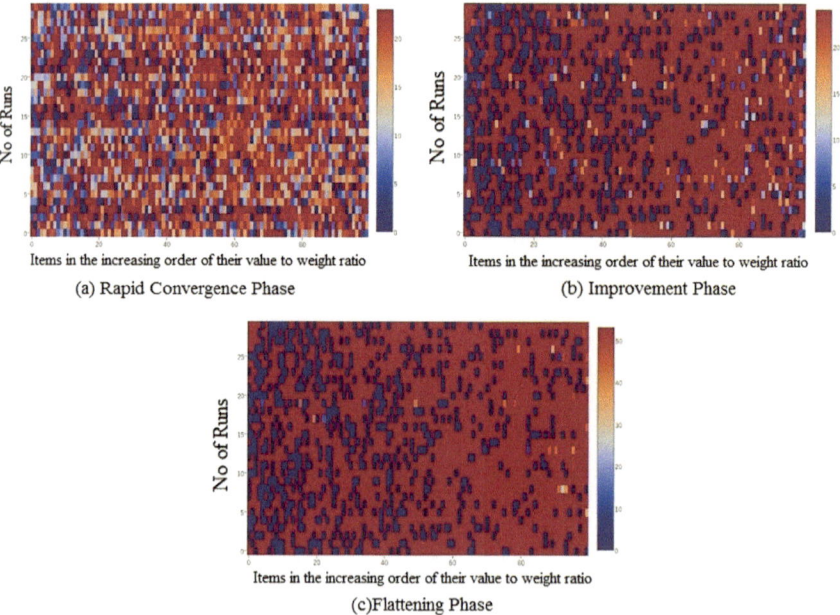

(a) Rapid Convergence Phase   (b) Improvement Phase

(c)Flattening Phase

Fig. 3 Heat maps of GA run solving 100 items 0-1 KP

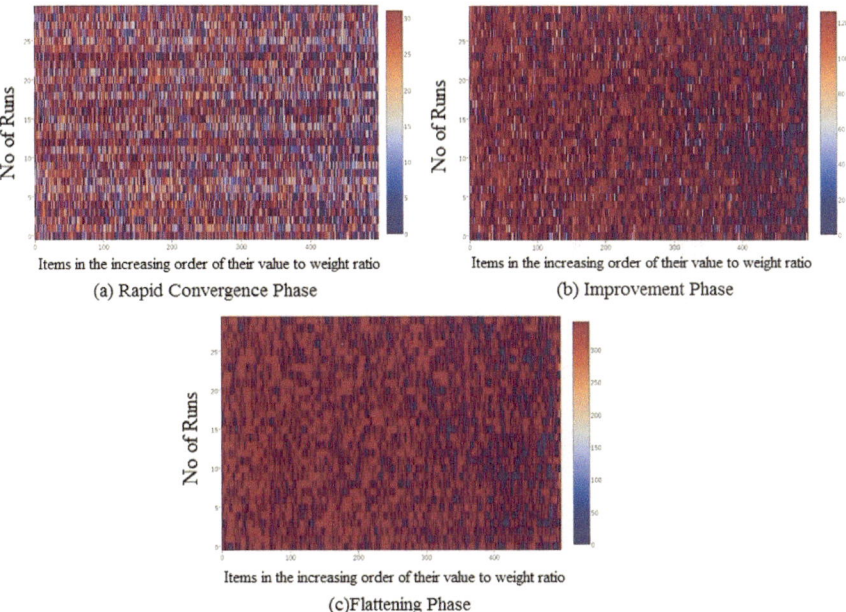

(a) Rapid Convergence Phase   (b) Improvement Phase

(c)Flattening Phase

Fig. 4 Heat maps of GA run solving 500 items 0-1 KP

case of 100 items 0-1 KP too. A visual comparison of Figs. 2 and 3 confirms the case of similarity. However, in case of 500 items 0-1 KP, the GA runs do not seem to have solved the problem effectively. Consequently, the heat maps in Fig. 4 deviate from our earlier observations. While the rapid convergence phase (Fig. 4a) shows GA's experimentation with every item as has been observed earlier, the relative importance of items with larger value-to- weight ratio is not prominent in improvement phase (Fig. 4b) and flattening phase (Fig. 4c). This reflects on the low-quality of solutions obtained by GA runs for the 500 items 0-1 KP. However, the exploratory and exploitative search strategies are still clearly visible in Fig. 4.

Figure 5a–c show the heat map visualization of worst chromosomes (in terms of their fitness values) during the three phases of convergence respectively while solving 100 items 0-1 KP by way of an example. The explorative and the exploitative search characteristics as observed in best chromosomes are observed in the case of worst chromosomes' evolution. The improvement in the quality of worst chromosomes while the algorithm converges (as can be seen in Fig. 5) strongly supports the fact that in case of GA (as well as typical EA) population is the unit of evolution and the algorithm strives to improve the quality of the population as a whole across generations.

The above observations show that GAs, by virtue of their operators (crossover-mutation-selection) and fitness function, inherently employs explorative and

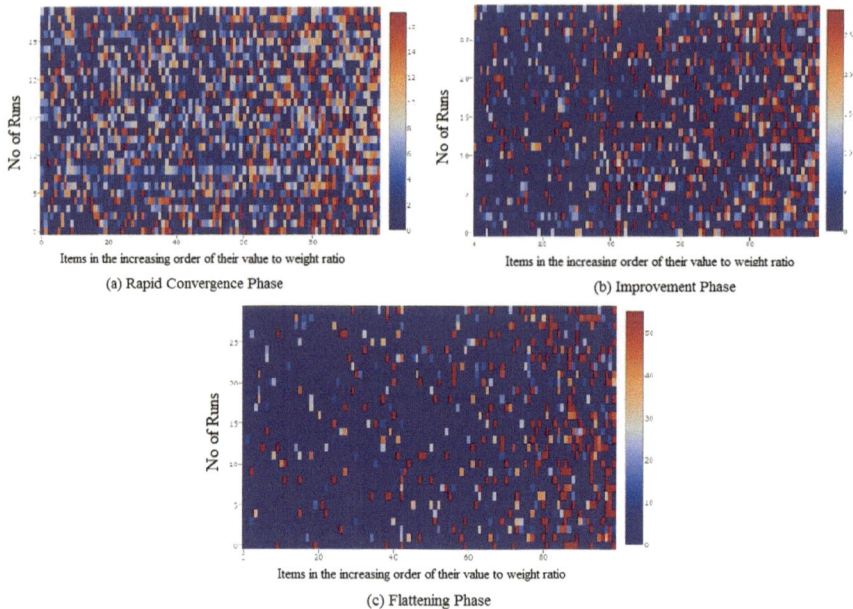

(a) Rapid Convergence Phase

(b) Improvement Phase

(c) Flattening Phase

Fig. 5 Heat maps of GA run exploring the worst candidates while solving 100 items 0-1 KP

exploitative search strategies. However, balancing the above strategies so as to reach high quality solutions depends on careful design of algorithms. Towards this, the following facts about the current work are worth admitting. Since the focus of this work is primarily on the visual analysis of GA solving 0-1 KP, algorithm design (in terms of the choice of operators and associated parameters) has not been given much importance or rather is ad hoc. However, the nature of analysis in this work is affected by the optimality of final solutions. Also, the potential of visual analysis as an alternative analysis technique to understand GA dynamics is demonstrated in this paper. Though the current work is more of preliminary in nature, it does present scope for effective extensions for robust analyses. In fact visual analyses focused towards the role of operators and their parameters would give us more insights about the working of GAs specifically (EAs in general) and hence this would certainly become part of our future work.

5   Conclusion

A heat map based visualization and analysis of GA solving 0-1 KP has been presented in this paper. A simple generational GA has been employed to solve 50, 100 and 500 items 0-1 KP. The best chromosomes obtained across generations of a GA run and across different GA runs have been visualized as a 2D heat map. Heat maps corresponding to best chromosomes from different phases of convergence, are analyzed to understand the working of GAs solving the 0-1 KP. The explorative and exploitative search strategies of GA have been clearly observed. The visualization showed GA's relatively high importance to high quality solutions (as dictated by fitness function) while not completely ignoring low-quality solutions (as against greedy characteristics) during the exploration phase. The exploitative phase is characterized by the uniform inclusion/exclusion of items by each and every best chromosome in that phase. These observations present the potential of visual analysis of EAs and also present scope for extension leading to rigorous analysis of EAs. Visual analysis of the role of operators and their parameters, to cite an example, will give more insights about EAs.

References

1. Eiben, A.E., Smith, J.E., et al.: Introduction to Evolutionary Computing, vol. 53. Springer, Berlin (2003)
2. Jeyakumar, G., Velayutham, C.S.: A comparative study on theoretical and empirical evolution of population variance of differential evolution variants. In: Asia-Pacific Conference on Simulated Evolution and Learning, pp. 75–79. Springer (2010)
3. Bezerianos, A., Chevalier, F., Dragicevic, P., Elmqvist, N., Fekete, J.D.: Graphdice: a system for exploring multivariate social networks. In: Computer Graphics Forum, vol. 29, pp. 863–872. Wiley Online Library (2010)
4. Cruz, A., Machado, P., Assunção, F., Leitão, A.: Elicit: evolutionary computation visualization. In: Proceedings of the Companion Publication of the 2015 Annual Conference on Genetic and Evolutionary Computation, pp. 949–956. ACM (2015)
5. Hart, E., Ross, P.: GAVEL-a new tool for genetic algorithm visualization. IEEE Trans. Evol. Comput. 5(4), 335–348 (2001)
6. Kerren, A., Egger, T.: Eavis: a visualization tool for evolutionary algorithms. In: 2005 IEEE Symposium on Visual Languages and Human-Centric Computing, pp. 299–301. IEEE (2005)
7. Radhika, P., Velayutham, C.S.: Visualization-a potential alternative for analyzing differential evolution search. Intell. Syst. Technol. Appl. 1, 31 (2015)
8. Wu, A.S., De Jong, K.A., Burke, D.S., Grefenstette, J.J., Ramsey, C.L.: Visual analysis of evolutionary algorithms. In: Proceedings of the 1999 Congress on Evolutionary Computation, 1999. CEC 99, vol. 2, pp. 1419–1425. IEEE (1999)
9. Bullock, S., Bedau, M.A.: Exploring the dynamics of adaptation with evolutionary activity plots. Artif. Life 12(2), 193–197 (2006)
10. Pohlheim, H.: Visualization of evolutionary algorithms-set of standard techniques and multidimensional visualization. In: Proceedings of the 1st Annual Conference on Genetic and Evolutionary Computation, vol. 1, pp. 533–540. Morgan Kaufmann Publishers Inc. (1999)
11. Romero, G., Merelo, J., Castillo, P., Castellano, J., Arenas, M.G.: Genetic algorithm visualization using self-organizing maps. In: International Conference on Parallel Problem Solving from Nature, pp. 442–451. Springer (2002)
12. Pryke, A., Mostaghim, S., Nazemi, A.: Heatmap visualization of population based multi objective algorithms. In: International Conference on Evolutionary Multi- Criterion Optimization, pp. 361–375. Springer (2007)
13. Collins, T.D.: Visualizing evolutionary computation. In: Advances in Evolutionary Computing, pp. 95–116. Springer, Berlin (2003)
14. McDermott, J.: Visualising evolutionary search spaces. ACM SIGEVOlution 7(1), 2–10 (2014)

15. Shao, Y., Xu, H., Yin, W.: Solve zero-one knapsack problem by greedy genetic algorithm. In: International Workshop on Intelligent Systems and Applications, 2009. ISA 2009, pp. 1–4. IEEE (2009)

16. Shen, W., Xu, B., Huang, J.p.: An improved genetic algorithm for 0-1 knapsack problems. In: 2011 Second International Conference on Networking and Distributed Computing (ICNDC), pp. 32–35. IEEE (2011)

17. Changdar, C., Mahapatra, G., Pal, R.K.: An improved genetic algorithm based approach to solve constrained knapsack problem in fuzzy environment. Expert Syst. Appl. **42**(4), 2276–2286 (2015)

18. He, J., Mitavskiy, B., Zhou, Y.: A theoretical assessment of solution quality in evolutionary algorithms for the knapsack problem. In: 2014 IEEE Congress on Evolutionary Computation (CEC), pp. 141–148. IEEE (2014)

19. Pisinger, D.: Where are the hard knapsack problems? Comput. Oper. Res. **32**(9), 2271–2284 (2005)

Securing Image Posts in Social Networking Sites

M. R. Neethu[(⊠)] and N. Harini

Department of Computer Science and Engineering, Amrita School
of Engineering, Amrita Vishwa Vidyapeetham, Amrita University, Coimbatore,
India
`cb.en.p2csel5011@cb.students.amrita.edu, n_harini@cb.`
`amrita.edu`

Abstract. The most unbeatable technology, Internet brings to people for communication is social networks. Starting from exchange of text messages it goes up to posting of images and videos in social networking sites which are viewed by many people. When these social networks are available to all the users for free, it will lead to various types of security issues. Image security has been a topic of research over decades. Enhancements to individual techniques and combinations proposed till date have offered different levels of security assurances. This paper aim to present a technique for secure sharing of image posts in social network. The significant feature of the scheme lies in the selection of security technique based on image content, evaluation of peers with whom the image can be shared based on text classification, transliteration and tone analysis. The proposed scheme a cost effective solution as it does not require any additional hardware. The utility of the model is demonstrated by mapping the scheme with Facebook and analyzing its performance through simulation.

Keywords: Image security · Peer recommender · Online social network
Secure sharing · Tone analysis

1 Introduction

With online social networking sites gaining popularity, in recent years, security and privacy concerns related to information sharing through these sites is becoming a key research area. In the past few years, the social networking sites such as Facebook, Twitter, and instagram has gained tremendous popularity. Social networking sites facilitates connectedness among people enabling their interactions through virtual space facilitating a platform for collaboration, experience sharing and trust formation. It is important for one to understand that the information about the users in social networking sites that is available in the users' profile can be searched based upon different criteria. Thus it can also be accessed by the strangers. Most of the people expose real identity information; thus it comes up with privacy and security issues [1]. Unfortunately many users are not aware of this. The images on user profile and timeline are available to public as per present privacy settings of most of the

© Springer International Publishing AG 2018
D. J. Hemanth and S. Smys (eds.), *Computational Vision and Bio Inspired Computing*,
Lecture Notes in Computational Vision and Biomechanics 28,
https://doi.org/10.1007/978-3-319-71767-8_7

OSNs. These images can be hacked using techniques like, Dragging the image, Right Click and save as image, Snipping tool in windows OS etc. With Security risks becoming more pressing and attacks becoming more daring there is a clear need for a scheme that ensures security on sharing of images in social networking sites. The rest of the paper is organized as follows, Sect. 2 presents motivation for the work, Sect. 3 discusses the related literature, and Sect. 4 provides details on the proposed scheme for secure image sharing. Section 5 discusses the results that validate the suitability of the scheme on real time online social network (Facebook), finally the inferences and scope for future work are presented in Sect. 6.

2 Motivation

The statistics on users of online social networking sites are showing a clear evidence of increase in number of users per day. The text, images, and videos posted by these users on to these social networking sites are also increasing day by day. The misuse of posted images in OSNs is also increasing and it creates many issues in the society. Many of these kinds of problems lead to the threat of human life also. Even though these things happens in the society and most of the people are aware of this, posting of images in the social networking sites are growing. As the technology grows, misuse of it also grows. Hacking of images is done in a wide range and problems are created in the society. A proper security scheme to protection to the images shared in the social networking sites is desirable at this era of growing technology.

3 Literature Survey

Literature witnesses several contributions from researchers on securing images. A survey on the literature was conducted to understand the techniques adopted for image security. The understandings from the literature laid the guidelines for selection of strategy in the proposed system.

3.1 Image Security Techniques

Cryptography, steganography, watermarking are widely used technique to protect the images. Variations and combinations of these schemes have found to offer improved results.

Visual Cryptography Cryptography system can be broadly classified as private key systems and public key systems [2]. A technique of transforming a message to unreadable format to protect it is called cryptography. The unreadable format of the message is the cipher text. Authentication, confidentiality and identification of user data are provided by it. Visual cryptography, to decrypt the encrypted images uses the human vision characteristics. Visual cryptography uses simple algorithm.

Watermarking Critical and challenging issue which is prevailing now a days is copyright protection of intellectual properties Digital image watermarking is one of the important solutions which include embedding owner's information or logos into digital image to be protected [3]. Major quality attributes of image watermarking techniques include robustness, imperceptibility, information hiding capacity and number of security levels achieved. Thus, robustness is a measure of immunity of watermark against attempts to image modification and manipulation like compression, filtering, rotation, scaling, resizing, cropping etc. The technology guarantees prevention of images from being misused by unauthorized users [4].

Invisible Watermarking A good quality watermarking technique should be capable to hide maximum watermark information in host image. Simultaneously, it should be difficult for attacker to detect the watermark. Invisible watermarking techniques include extraction techniques to retrieve hidden copyright information from host media. It also required that watermarked must be resistant to different image attacks to ensure to extract hidden watermark after any addition, alteration deletion [5].

Digital Image Encryption Digital image has vividly, abundant information, so it was widely used in communication. Along with the digital image using universal, its safety attracted people's attention increasingly. Firstly, the interactive image is encrypted [6] by pixels scrambling; secondly, add watermark to the scrambled image; at last, choose a camouflaged image to vision or the pixels of the interactive image. The core parameters are encrypted [7] by Elliptical Curve Cryptography.

3.2 Recommender Systems

Recommendations given by expert system modules based on input data are powerful alternatives that can aid one in decision making process. The relevance of the recommender system on the context of this work is justified as follows. The expert system module provides recommendation based on emotional analysis performed on posts that have been shared by peers, to suggest the degree of belief one can have with the peers with whom the post need to be shared.

2P Recommendation The 2P Recommendation system which comprises of 2P recommendation primary and 2P recommendation social classifies the image based on its content and then, based on the meta data. After the classification a proper security scheme is predicted based on policy normalization followed by policy mining and prediction.

Trust in Social Networks STBAC (Trust based access control for social networks) is an access control method that helps the user in sharing of data to their peer in social network by the computation of trust value. It determines to whom any data can be shared to [8]. This is done by assessing the user's foregoing interactions with its peers which helps in classifying the peers into various classes [9].

3.3 Summary of Findings

Social networks have evolved as the most important means for interactions. The rapid growth the social networks clearly illustrates that it has become an indispensible item of our daily life. The present system exposes the resources shared with a level of privacy and security as specified under terms and conditions (different for different

OSNs.) [10]. With the escalation in the number of security threats and privacy violation attempts the existing system does not appear robust enough to address [11]. This brings out a clear need for a new scheme that enhances security.

4 Proposed Work

Proposed system facilitates secure sharing of images to peers with different groups in social networks. Security offered is based on the content of the images being shared with the peers, and the evaluation of the peers as done by the recommender engine using statistical measure based on emotion analysis. The system evaluates the final credibility of the peer based on the conversations done on the social networks. The system also includes the transliteration process (tanglish to Tamil) to perform analysis on the posts that don't follow strict norms English vocabulary. Figure 1 presents the architecture diagram of the modules that participate and contribute to meet the overall objective of the scheme.

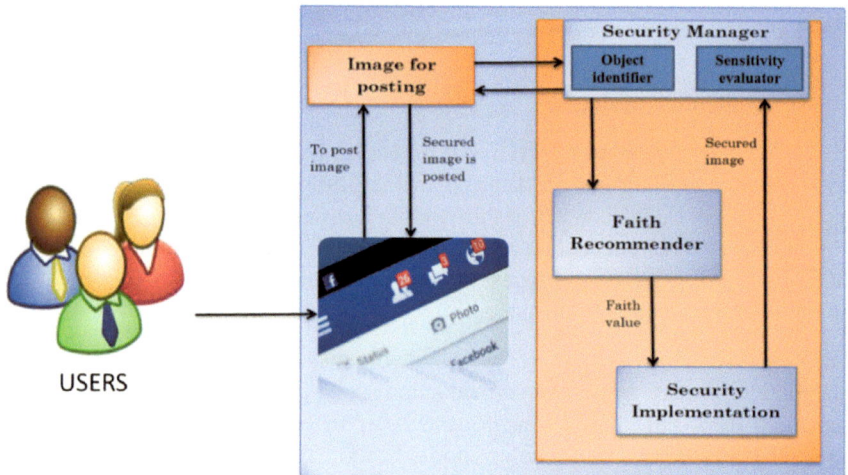

Fig. 1 Proposed architecture diagram

The user selects an image to be shared in social networks which is then evaluated to find if any sensitive information is carried by it. Any portion in the image can be considered to be sensitive. Human faces identified in images using OpenCV haarcascaders are used for evaluation of sensitivity in the image. An appropriate security technique selected by security evaluator component based on perfect secrecy is applied to the identified sensitive parts. The remaining of this section discusses the working model of the important modules in the proposed architecture.

5 Evaluation Based Selection of Security Technique

The effort a hacker has to put for cryptanalysis is the measure of the strength or weakness of a cryptosystem. To make the cryptanalysis more difficult, the system must provide a perfect secrecy. Generally probabilistic or information theory based evaluation is carried out for the assessment. The proposed system uses information theory based perfect secrecy algorithm [12].

5.1 Algorithm 1: Perfect Secrecy Algorithm

Let, m, k, c—set of discrete random variables
Let, M, K, C—respective possible set of values the random variable can take.
Given that, $M = (m_1, m_2, m_3, m_4)$, $K = (k_1, k_2, k_3)$ and $C = (c_1, c_2, c_3, c_4)$. The key and the messages are equiprobable (Table 1).

Table 1 Representation of cryptographic scheme to be evaluated

	m_1	m_2	m_3	m_4
k_1	c_1	c_2	c_3	c_4
k_2	c_5	c_4	c_2	c_1
k_3	c_4	c_1	c_2	c_3

Evaluation of crypto text probabilities is done using probability measures as follows:
Let C_1 be observed: M_1K_1, M_4K_2, M_2K_3
Probability of $C1 = P\{m_1*k_1\} + P\{m_2*k_3\} + P\{m_4*k_2\}$
Similarly one can find the probability of occurrences of all the encrypted /secured image or subject.
$P\{M_1/C_1\} = P(M_1, C_1)/P(C_1)$
...... to all possible combinations.
Any securing algorithm is perfectly secure only if
$P(M/C) = P(M)$.

5.2 Selection of Peers

It is equally important to share right images with right peers. An expert system module that uses comments collected from Facebook with applying language transliteration, classifies the text based on semantic analysis. The results obtained undergo tone analysis using IBM Bluemix Watson service Tone Analyser, to predict the tone associated with a particular post. The results obtained are fed to a recommender engine that predicts the amount of belief one could hold on the peer. These results help in identifying the right peer with whom the image can be shared. The detailed working of the sub modules related to language transliteration, text classification, tone analysis, trust calculation and recommendation are presented in Table 2.

Table 2 Peer selection process

(continued)Collecting of Facebook comments using Graph API Explorer

- Open https://developers.facebook.com/
- Go to Graph API Explorer
- Under GET tab, select version
- Give image_id/comments or page_id/image_id/comments
- All the comments will be retrieved.

When you collect the comments from Facebook, the attributes will be:

- Created time
- Person's id
- Comment/message
- Location details
- Name of the person

User can use these attributes to perform various actions*Language Transliteration (Tamil)*

To create a basic transliteration engine

Step 1: Create the array of string for possible letters in English to the corresponding Tamil letters

Step 2: Create another array of string for representing Tamil Unicode characters

Step 3: These two arrays are compared for each character entry in English*Text Classification*

Text classification is mainly involves two steps

Representation of the databases/datasets for enabling it to learn and train it as a classifier

To use this classifier to new incoming text comments

To read the dataset, Arff Reader object is used and to classify String to word vector filter combined with a naïve Bayes classifier is used*Tone Analysis*

Done using IBM Bluemix Watson service, Tone Analysis. The steps are as follows

1. Create a bluemix account and create a tone analysis service
2. Connect to Bluemix with command line tool
3. Create and retrieve service keys to access the tone analyser service
4. Update the .env file with the information retrieved in step 4
5. Install the dependencies needed for the application
6. Start the application locally

This way the analysis of a comment using bluemix tone analysis can be done. To take the analysed result for further evaluation, the following command can be used

1. Copy the credentials (user name and password) of the tone analyser service that has been created
2. To analyse the tone and get the analysed output for further evaluation, curl command can be used

   curl -u "{username}":"{password}" -H "Content-Type: application/json" -d "{\"text\": \"text to classify \"}" https://gateway.watsonplatform.net/tone-analyzer/api/v3/tone?version=2016-05-19>filename.json

3. The analysed results will be stored in filename.json that will be created in the root file directory

6   Trust Calculation and Recommendation

Table 3 Notations

Sl no	Notations	Particulars
1.	R	Reliability of the user
2.	N	Average negativeness of the comments
3.	P	Average positiveness of the comments
4.	n	Negativity of the comment
5.	p	Positivity of the comment
6.	C_n	Number of comments posted by user in the wall of the peer
7.	C_i	Number of positive comments
8.	C_{n-i}	Comments are negative
9.	u	User
10.	T	Transaction
11.	T_{in}	Transaction from peers to user
12.	T_{out}	Transaction from user to peer
13.	Cr	Credibility of the peer
14.	Tr	Trust value of the user
15.	p(u, v)	Transaction between the user 'u' and a friend node 'v'
16.	p(u, i)	Total number of transaction of u with all the friend node 'i' in the social network
17.	p(v)	Friend node 'v'
18.	T_{Th}	Trust Threshold which is input as per the user

6.1   Reliability of the Peer

Consistency of the peer to the user and by the user is very much important in the case of any online social network. For each user, negativity and positivity of each comment is found (Table 3)

Average negativeness of the comments:

$$N = \frac{\sum n}{Cn - i} \tag{1}$$

Average positiveness of the comments:

$$P = \frac{\sum p}{Ci} \tag{2}$$

If N > P, | 0, Then negativity is more towards the user by the peer, thus reliability is 0. Else ‖ 1, positivity of the user is more, thus reliability is 1.

6.2 Credibility of the Peer

$$S = T_{in} - T_{out} \tag{3}$$

If $S > 0 \mid 1$ denoted as Sp => incoming is more than outgoing
Else if $S = 0 \mid 1$ denoted as So => incoming and outgoing transactions are equal
Otherwise $\mid 0$ denoted as Sn => incoming is less than outgoing

$$Cr(p(u, i)) = \left\{ \left\{ \frac{1}{i} \left\{ \sum Sp + \sum So \right\} \right\} \right\} * 100 \tag{4}$$

If $Cr(p (u, i)) >= 75 \mid 1$ => friend node 'i' assures that the user 'u' has high credibility
Else if $Cr(p (u, i)) > 50$ and $< 75 \mid 0.5$ => user 'u' has low credibility
Otherwise $\mid 0$ => user 'u' has very low credibility.

6.3 Trust Value of the User

Let, user 'u' perform n number of transaction with another peer 'v'. Thus trust (T) of the peer 'u' with respect to 'v' will be calculated as follows:

$$Tr(p(u, v)) = \frac{Tout(p(v))}{\sum (p(u, i))} * 100 \tag{5}$$

$Tr(p(u, v) > = T_{Th} \mid$ Notify Allow access to be authorized friend
Otherwise $\mid$ Access denied, remain as unauthorized friend

6.4 Recommender Engine

The proposed system uses content based filtering mechanism for recommendation of peer group with whom the image can be shared based on the reliability, credibility and trust value of the user.

The recommendation offered is based on the proper mix of the above three mentioned attributes.

```
Recommendation of the peers:
If (R == 0||Cr ==0)
            Then access denied.
if (R == 1)
    Take all the credible users. (Sort based on Credibility percentage)
    Recommendation of the peers is done based on the top values of Tr.
```

7 Experimental Setup and Results

The system was tested with a dataset containing images requiring security to be applied to varied portions of the images. This ensured that the object recognition part was functioning correctly. The dataset for peer recommendation was collected from Facebook accounts of author's friends. The performance analysis demonstrated that the presented scheme was able to achieve the stated objectives in a wholesome manner. Tables 4, 5, 6, 7, 8, 9, 10, 11, 12, 13, 14, 15 and 16 present the results of Tone Analysis, Positive and negative values when separated, Emotional values are separated, Positive and negative average of emotional tone, Positive and negative average of each comments are found for a peer, Total positive and negative average of the user is found, Reliability of the peer, S value calculation, credibility calculation, trust score calculation and final recommendation.

Table 4 Tone analysis

Tones/0/score	Tones/0/tone_id	Tones/0/tone_name
0.018973	Anger	Anger
0	Analytical	Analytical
0.719038	Openness_big5	Openness
0.042955	Disgust	Disgust
0	Confident	Confident
0.279874	Conscientiousness_big5	Conscientiousness
0.013415	Fear	Fear
0	Tentative	Tentative
0.613473	Extraversion_big5	Extraversion
0.794777	Joy	Joy
0.622743	Agreeableness_big5	Agreeableness
0.165241	Sadness	Sadness
0.316637	Emotional_range_big5	Emotional range

Table 5 Positive and negative values when separated

Positiveness		Negativeness	
Openness	0.719038	Anger	0.018973
Conscientiousness	0.279874	Disgust	0.042955
Extraversion	0.613473	Fear	0.013415
Joy	0.794777	Sadness	0.165241
Agreeableness	0.622743	Emotional range	0.316637
Positive average	0.605981	Negative average	0.1114442

The engine segregates comments based on positivity, negativity and neutrality, as an initial step following which a detailed tone analysis is performed for correct classification. The tones of social context like openness, agreeableness etc. is not considered as they tend to skew towards positiveness always. Thus only the emotional tones are taken for the evaluation of the comment (Table 3). Positive average is taken as the value of joy in the emotional tones. If total positive average is greater than negative average, then reliability is 1, else 0. This is shown in Table 10.

Table 6 Emotional values are separated

Tone	Score
Anger	0.018973
Disgust	0.042955
Fear	0.013415
Joy	0.794777
Sadness	0.165241

Table 7 Positive and negative average of emotional tone

Positive average	0.794777
Negative average	0.060146

Table 8 Positive and negative average of each comments

Comment	Negative average	Positive average
1	0.060146	0.794777
2	0.29849425	0.001159
3	0.2916865	0.018558
4	0.29505325	0.009106
5	0.28697825	0.001063
6	0.0047445	0.96323

Table 9 Total positive and negative average of the user

Total positive average	0.297982167
Total negative average	0.206183792

Table 10 Reliability score

	Negativity	Positivity	Reliability
Peer 1	0.206183792	0.297982167	1
Peer 2	0.24647165	0.1649326	0

.

Table 11 S Value calculation

	Tin	Tout	Difference	Sp/Sn/So	S
Peer 1	1	1	0	So	0
Peer 2	1	2	−1	Sn	−1
Peer 3	5	3	2	Sp	1
Peer 4	2	9	−7	Sn	−1
Peer 5	10	5	5	Sp	1
Peer 6	3	4	−1	Sn	−1
Peer 7	6	6	0	So	0
Peer 8	7	4	3	Sp	1
Peer 9	2	3	−1	Sn	−1
Peer 10	9	2	7	Sp	1

Further computation of credibility is performed using Eq. (4). Credibility of each user is calculated based on this calculation of S value. The credibility in percentage thus found is depicted in Table 12.

Table 12 Credibility

Sum of all Sp	4
Sum of all So	0
Credibility	0.4
Credibility (%)	40

Table 13 p(u, i) Calculation

	p(u, _)
Peer 1	10
Peer 2	11
Peer 3	5
Peer 4	6
Peer 5	7
Peer 6	17
Peer 7	1
Peer 8	2
Peer 9	4
Peer 10	5
Total	68

Table 14 Trust score

$T_{out}(p(v))$	14
Trust score	0.2059

Table 15 Integration of all scores

Peers	Credibility	Reliability	Trust score
Dhanush	**40**	**1**	**20.59**
Chaithra	75	1	21.45
Nileena	54.2	1	23.67
Dharani	**36.4**	**0**	**34.44**
Meera	64.3	1	55.67
Achu	61.5	0	57.78
Sruthi	**45.78**	**0**	**43.79**

Table 16 Final recommendation

Recommendation
Achu
Meera
Nileena
Chaithra

The initial threshold of 50% was assigned to study the behaviour of the system. The acceptance and the rejection rates of the users with whom the image can be shared is presented in Table 16. The computed value show that it is safe to share image posts with achu, meera, nileena and chaithra and unsafe with dhanush, dharani and sruthi.

8   Conclusion

The current security measures available in social networking sites for image posts are not sufficient to provide ample freedom for its users. These security schemes have to be enhanced to provide a proper security to the social networking sites. These schemes provide different types and levels of security for social networking sites. The system in the present form does not include analysis on sarcasm included in comments. And also does not consider the relationship of the user with peers across other social networks. Incorporation of these could lead to the development of more generic scheme.

References

1. Echaiz, J., Ardenghi, J.R.: Security and online social networks. In: XV Congreso Argentino de Ciencias de la Computación (2009)
2. Harini, N., Surya Prabha, K..: 'CHATGAURD'—a system that ensures safe posting in social networking sites. Int. J. Eng. Technol. (2016)

3. Metkar, S.P., Lichade, M.V.: Digital image security improvement by integrating watermarking and encryption technique. In: IEEE International Conference on Signal Processing, Computing and Control (ISPCC), pp. 1–6. IEEE (2013)
4. Liu, T.Y., Tsai, W.H.: Generic lossless visible watermarking—a new approach. IEEE Trans. Image Process. 1224–1235 (2010)
5. Harini, N., Padmanabhan, T.R.: 3C-Auth: a new scheme for enhancing security. Int. J. Netw. Secur. **18**(1), 143–150 (2016)
6. Saini, J.K., Verma, H.K.: A hybrid approach for image security by combining encryption and steganography. IEEE Second International Conference Image Information Processing (ICIIP), pp. 607–611. IEEE (2013)
7. Padiya, I., Manure, V., Vidhate, A.: Visual secret sharing scheme using encrypting multiple images. Int. J. Adv. Res. Electr. Electron. Instrum. Eng. (2015)
8. Omanakuttan, S., Chatterjee, M.: Experimental analysis on access control using trust parameter for social network. In: International Conference on Security in Computer Networks and Distributed Systems, pp. 551–562 (2014)
9. Ceolin, D., et al.: Trust evaluation through user reputation and provenance analysis. In: Proceedings of the 8th International Conference on Uncertainty Reasoning for the Semantic Web-Volume 900. CEUR-WS.org (2012)
10. Gao, H., Hu, J., Huang, T., Wang, J., Chen, Y.: Security issues in online social networks. IEEE Internet Comput. 56–63 (2011)
11. Joe, M.M., Ramakrishnan, D.B.: A survey of various security issues in online social networks. Int. J. Comput. Netw. Appl. **1**(1), 11–14 (2014)
12. Harini, N.: A system to screen posts that minimize user frustration. Int. J. Appl. Eng. Res. **11** (6), 3944–3949 (2016)

A Types of Multi-granular Nanotopology and Its Applications

K. Indirani[1] and G. Vasanthakannan[2(✉)]

[1] Nirmala College for Women, Red Fields, Coimbatore, India
indirani009@ymail.com
[2] RVS College of Arts and Science, Sulur, Coimbatore, India
vasanthraj.002@gmail.com

Abstract. In this paper the difference, relationship and the means of calculating the significant measured, nano degree of dependence of decision attributes in practical application of Multi- Granular nanotopology and Multi*-Granular nanotopology, which are the types of nanotopological models are studied. Moreover an example is taken to highlight the Multi*-Granular nanotopology which are examined for the practical implementation of this hypothesis.

Keywords: Multi*-granular nanotopology · Multi*-lower approximation
Multi*-upper approximation · Nano degree of dependence · Core

1 Introduction

In 1980s Pawlak [1–3] introduced the concept of rough set model. A multi granulation rough set [4, 5] is an extension of rough set model. Lellis Thivagar [6, 7] introduced the concept of nanotopology. He applied nanotopology [8] to get a nutrient model for leading healthy life. Considering the main nutrient present in various food stuff alone [9]. Devised the idea of granular computing which is based on single classical equivalence relation. Further Lellis Thivagar [9] introduced the concept of Multi-Granular nanotopology and Multi*-Granular nanotopological [10] model where the set approximations are defined by using multi equivalence relations on the universe. He has also investigated the difference and relationship based on their approximations and also found the significant measure in practical problems. In [11] along with food, water and air also included to find the ideal combinations for a healthy life. In this paper, we investigate the difference and relationship among Multi-Granular nanotopology and Multi*-Granular nanotopology based on their approximations and also by means of calculating the significant measure named as **nano degree of dependence of decision attribute for the problem** [11].

2 Preliminaries

Definition 2.1 [8] Let U be a non-empty finite set of objects called the universe and R be an equivalence relation on U named as the indiscernibility relation, elements

© Springer International Publishing AG 2018
D. J. Hemanth and S. Smys (eds.), *Computational Vision and Bio Inspired Computing*,
Lecture Notes in Computational Vision and Biomechanics 28,
https://doi.org/10.1007/978-3-319-71767-8_8

belonging to the same equivalence class are said to be indiscernible with one another. The pair (U, R) is said to be the approximation space. Let $X \subseteq U$.

(i) The lower approximation of X with respect to R is the set of all objects, which can be for certain classified as X with respect to R and it is denoted by $L_R(X)$. That is $L_R(X) = \bigcup_{x \in U} \{R(x) : R(x) \subseteq X\}$, where $R(x)$ denotes the equivalence class determined by x.

(ii) The upper approximation of X with respect to R is the set of all objects, which can be possibly classified as X with respect to R and it is denoted by $U_R(X)$. That is $U_R(X) = \bigcup_{x \in U} \{R(x) : R(x) \cap X \neq \emptyset\}$ where $R(x)$ denotes the equivalence class determined by x.

(iii) The boundary region of X with respect to R is set of all objects, which can be classified neither as X nor as not X with respect to R and it is denoted by $B_R(X)$. That is $B_R(X) = U_R(X) - L_R(X)$.

Definition 2.2 [8] Let U be the universe, R be an equivalence relation on U and $\tau_R(X) = \{U, \emptyset, L_R(X), U_R(X), B_R(X)\}$ where $X \subseteq U$. Then $\tau_R(X)$ satisfies the following axioms:

1. U and $\emptyset \in \tau_R(X)$.
2. The union of the elements of any sub collection of $\tau_R(X)$ is in $\tau_R(X)$.
3. The intersection of the elements of any finite sub collection of $\tau_R(X)$ is in $\tau_R(X)$.

That is, $\tau_R(X)$ forms a topology on U called as the nanotopology on U with respect to X. We call $(U, \tau_R(X))$ as the nanotopological space.

Definition 2.3 [12] Let U be the universe, P, Q be any two equivalence relation on U and

$$\tau_{P+Q}(X) = \{U, \emptyset, L_{P+Q}(X), U_{P+Q}(X), B_{P+Q}(X)\} \text{ where } X \subseteq U.$$

$$L_{P+Q}(X) = \bigcup_{x \in U} \{[x] : P(x) \subseteq X \text{ or } Q(x) \subseteq X\}$$

$$U_{P+Q}(X) = \bigcup_{x \in U} \{[x] : P(x) \cap X \neq \emptyset \text{ and } Q(x) \cap X \neq \emptyset\}$$

$$B_{P+Q}(X) = U_{P+Q}(X) - L_{P+Q}(X)$$

That is, $\tau_{P+Q}(X)$ forms a topology on U called as the Multi-Granular nanotopology on U with respect to X. We call $(U, \tau_{P+Q}(X))$ as the Multi-Granular nanotopological space.

Definition 2.4 [12] Let U be a non-empty finite set of objects called the universe and P, Q be an equivalence relation on U named as the indiscernibility relation, elements belonging to the same equivalence class are said to be indiscernible with one another. The pair (U, R) is said to be the approximation space. Let $X \subseteq U$.

(i) The Multi*-lower approximation of X with respect to P and Q is the set of all objects, which can be for certain classified as X with respect to P and Q and it is denoted by $L_{P*Q}(X)$. That is $L_{P*Q}(X) = \bigcup_{x \in U} \{[x] : P(x) \subseteq X \text{ and } Q(x) \subseteq X\}$, where $P(x)$ and $Q(x)$ denotes the equivalence class determined by x.

(ii) The Multi*-upper approximation of X with respect to P and Q is the set of all objects, which can be possibly classified as X with respect to P and Q and it is denoted by $U_{P*Q}(X)$. That is $U_{P*Q}(X) = \bigcup_{x \in U} \{[x] : P(x) \cap X \neq \emptyset \text{ or } Q(x) \cap X \neq \emptyset\}$ where $P(x)$ and $Q(x)$ denotes the equivalence class determined by x.

(iii) The Multi*-boundary region of X with respect to P and Q is set of all objects, which can be classified neither as X nor as not X with respect to P and Q and it is denoted by $B_{P*Q}(X)$. That is $B_{P*Q}(X) = U_{P*Q}(X) - L_{P*Q}(X)$.

Definition 2.5 [12] Let U be the universe, P, Q be an equivalence relation on U and $\tau_{P*Q}(X) = \{U, \emptyset, L_{P*Q}(X), U_{P*Q}(X), B_{P*Q}(X)\}$ where $X \subseteq U$. Then $\tau_{P*Q}(X)$ satisfies the following axioms:

1. U and $\emptyset \in \tau_{P*Q}(X)$.
2. The union of the elements of any sub collection of $\tau_{P*Q}(X)$ is in $\tau_{P*Q}(X)$.
3. The intersection of the elements of any finite sub collection of $\tau_{P*Q}(X)$ is in $\tau_{P*Q}(X)$.

That is, $\tau_{P*Q}(X)$ forms a topology on U called as the Multi*-Granular nanotopology on U with respect to X. We call $(U, \tau_{P*Q}(X))$ as the Multi*-Granular nanotopological space.

Example 2.6 Let $U = \{a, b, c, d, e\}$ and $U/P = \{\{a, b\}, \{c\}, \{d, e\}\}$ and $U/Q = \{\{a\}, \{b\}, \{c, d\}, \{e\}\}$ be the equivalence relations on U and let $X = \{a, b, d\} \subseteq U$. Then

$L_{P*Q}(X) = \{a, b\}, U_{P*Q}(X) = \{a, b, c, d, e\}$ and $B_{P*Q}(X) = \{c, d, e\}$, hence the Multi*-Granular nanotopology $\tau_{P*Q}(X) = \{\{U\}, \phi, \{a, b\}, \{a, b, c, d, e\}, \{c, d, e\}\}$.

Definition 2.7 [12] Let $(U, \tau_R(X))$ be nanotopological space and let $X \subseteq U$. Let $U/D = \{D_1, D_2, \ldots D_K\}$ be all decision classes induced by decision attribute D and A is divided into a set C of condition attributes, then the nano degree of dependence is defined as $\gamma[P, D] = \frac{1}{|u|} [|L_P(D_1)| + |L_P(D_2)|]$.

3 Analysis and Comparison

Example 3.1 Now, consider the problem of finding the difference and relationship based on approximations and to find the **nano degree of dependence** in all the cases, of the ideal combination of the triplet <air, food, water> for a human being in order to live a healthy life.

Consider the following Table 1 giving information about the ideal combinations of triplet <air, food, water> for a human being in order to live a healthy life. **Water and Air, Food, Processed water and Junk food, Machine-driven Air are the conditional attributes of the system**, whereas the decision is the decision attribute.

Take a_1, a_2, a_3, a_4 and D will stand for Water and Air, Food, Processed water and Junk food, Machine-driven Air respectively. The domains are as follows,

V_{a_1} = {Natural water, Natural air, Pot water, Copper vessel water}, V_{a_2} = {Raw vegetables, Sprouted Grains, Vegetarian Food, Non-Vegetarian Food, Fruits}, V_{a_3} = {Mineral water, Hot water, filter water, Beverages}, V_{a_4} = {Fan Air, Air Conditioner Air}.

Table 1 Information table for the ideal combinations for a healthy life

Students	Group I (a_1)	Group II (a_2)	Group III (a_3)	Group IV (a_4)	Decision
S_1	$\{NW, NA, P, C\}$	$\{R, S, V, N, F\}$	$\{M, B\}$	$\{FA\}$	Healthy
S_2	$\{NA\}$	$\{V, F\}$	$\{H, FW, B\}$	$\{FA\}$	Unhealthy
S_3	$\{NW, NA, P, C\}$	$\{R, S, V, F\}$	$\{M, H, FW\}$	$\{FA, A\}$	Healthy
S_4	$\{NW, NA\}$	$\{R, S, V, F\}$	$\{M, H, FW, B\}$	$\{A\}$	Healthy
S_5	$\{NA, C\}$	$\{R, S, V, N, F\}$	$\{M, H, FW\}$	$\{FA, A\}$	Healthy
S_6	$\{NA\}$	$\{R, S, V, F\}$	$\{M\}$	$\{FA\}$	Unhealthy
S_7	$\{NA\}$	$\{V, F\}$	$\{M, H, FW\}$	$\{FA\}$	Unhealthy
S_8	$\{NW, NA, P, C\}$	$\{R, S, V, F\}$	$\{MF, W, B\}$	$\{FA\}$	Healthy
S_9	$\{NW, NA, P, C\}$	$\{R, S, V, N, F\}$	$\{H, FW, B\}$	$\{FA\}$	Healthy
S_{10}	$\{NW, NA\}$	$\{N, F\}$	$\{M, H, FW\}$	$\{A\}$	Unhealthy

The entries in table are the attribute values. The columns of the table represent the key factors evaluated in the ideal combinations of human beings in order to live a healthy life and the rows represent human beings called as the objects. The above table renders a information system where $U = \{H_1, H_2, \ldots H_{10}\}$ and A = {Water and Air, Food, Processed water and Junk food, Machine-driven Air} be the set of attributes from the Table 1, we can find that

$$U/a_1 = \{\{H_1, H_3, H_8, H_9\}, \{H_2, H_6, H_7\}, \{H_4, H_{10}\}, \{H_5\}\}$$
$$U/a_2 = \{\{H_1, H_5, H_9\}, \{H_3, H_4, H_6, H_8\}, \{H_2, H_7\}, \{H_{10}\}\}$$
$$U/a_3 = \{\{H_1\}, \{H_2, H_9\}, \{H_3, H_5, H_7, H_{10}\}, \{H_4\}, \{H_6\}, \{H_8\}\}$$
$$U/a_4 = \{\{H_1, H_2, H_6, H_7, H_8, H_9\}, \{H_4, H_{10}\}, \{H_3, H_5\}\}$$

From Table 1, we have, $U_D = \{D_H, D_U\}, D_H = \{H_1, H_3, H_4, H_5, H_8, H_9\}$, $D_U = \{H_2, H_6, H_7, H_{10}\}$

where $U/D = \{\{H_1, H_3, H_4, H_5, H_8, H_9\}, \{H_2, H_6, H_7, H_{10}\}\}$.

From the table, to find the nano degree of dependence in all the cases:

Case 1:

$$L_{a_1+a_3}(D_H) = \{H_1, H_3, H_4, H_5, H_8, H_9\}; L_{a_1*a_3}(D_H) = \{H_1, H_8\}$$
$$L_{a_1+a_3}(D_U) = \{H_2, H_6, H_7\}; L_{a_1*a_3}(D_U) = \{H_6\}$$
$$\gamma[a_1+a_3, D] = \frac{1}{|u|}[|L_{a_1+a_3}(D_H)| + |L_{a_1+a_3}(D_U)|] = \frac{1}{10}[9] = \frac{9}{10}$$
$$\gamma[a_1*a_3, D] = \frac{1}{|u|}[|L_{a_1*a_3}(D_H)| + |L_{a_1*a_3}(D_U)|] = \frac{1}{10}[3] = \frac{3}{10}$$

Hence, from the above it can be found that,

$$\gamma[a_1*a_3, D] \le \gamma[a_1, D] \le \gamma[a_1+a_3, D], \gamma[a_1*a_3, D] \le \gamma[a_3, D] \le \gamma[a_1+a_3, D]$$

Case 2:

$$L_{a_1+a_4}(D_H) = \{H_1, H_3, H_5, H_8, H_9\}; L_{a_1*a_4}(D_H) = \{H_3, H_5\}$$
$$L_{a_1+a_4}(D_U) = \{H_2, H_6, H_7\}; L_{a_1*a_4}(D_U) = \phi$$
$$\gamma[a_1+a_4, D] = \frac{1}{|u|}[|L_{a_1+a_4}(D_H)| + |L_{a_1+a_4}(D_U)|] = \frac{1}{10}[8] = \frac{8}{10}$$
$$\gamma[a_1*a_4, D] = \frac{1}{|u|}[|L_{a_1*a_4}(D_H)| + |L_{a_1*a_4}(D_U)|] = \frac{1}{10}[2] = \frac{2}{10}$$

Hence, from the above it can be found that,

$$\gamma[a_1*a_4, D] \le \gamma[a_1, D] \le \gamma[a_1+a_4, D], \gamma[a_1*a_4, D] \le \gamma[a_4, D] \le \gamma[a_1+a_4, D]$$

Case 3:

$$L_{a_1+a_2}(D_H) = \{H_1, H_3, H_4, H_5, H_8, H_9\}; L_{a_1*a_2}(D_H) = \{H_1, H_3, H_5, H_8, H_9\}$$
$$L_{a_1+a_4}(D_U) = \{H_2, H_6, H_7, H_{10}\}; L_{a_1*a_4}(D_U) = \{H_2, H_7\}$$
$$\gamma[a_1+a_2, D] = \frac{1}{|u|}[|L_{a_1+a_2}(D_H)| + |L_{a_1+a_2}(D_U)|] = \frac{1}{10}[10] = 1$$
$$\gamma[a_1*a_2, D] = \frac{1}{|u|}[|L_{a_1*a_2}(D_H)| + |L_{a_1*a_2}(D_U)|] = \frac{1}{10}[7] = \frac{7}{10}$$

Hence, from the above it can be found that,

$$\gamma[a_1*a_2, D] \le \gamma[a_1, D] \le \gamma[a_1+a_2, D], \gamma[a_1*a_2, D] \le \gamma[a_2, D] \le \gamma[a_1+a_2, D]$$

Case 4:

$$L_{a_2+a_3}(D_H) = \{H_1, H_3, H_4, H_5, H_8, H_9\}; L_{a_2*a_3}(D_H) = \{H_1, H_4, H_8\}$$
$$L_{a_1+a_4}(D_U) = \{H_2, H_6, H_7, H_{10}\}; L_{a_1*a_4}(D_U) = \phi$$

$$\gamma[a_2+a_3, D] = \frac{1}{|u|}[|L_{a_2+a_3}(D_H)| + |L_{a_2+a_3}(D_U)|] = \frac{1}{10}[10] = 1$$

$$\gamma[a_2*a_3, D] = \frac{1}{|u|}[|L_{a_2*a_3}(D_H)| + |L_{a_2*a_3}(D_U)|] = \frac{1}{10}[3] = \frac{3}{10}$$

Hence, from the above it can be found that,

$$\gamma[a_2*a_3, D] \le \gamma[a_2, D] \le \gamma[a_2+a_3, D], \gamma[a_2*a_3, D] \le \gamma[a_3, D] \le \gamma[a_2+a_3, D]$$

Case 5:

$$L_{a_2+a_4}(D_H) = \{H_1, H_3, H_5, H_9\}; L_{a_2*a_4}(D_H) = \{H_5\},$$
$$L_{a_2+a_4}(D_U) = \{H_2, H_7, H_{10}\}; L_{a_2*a_4}(D_U) = \phi.$$

$$\gamma[a_2+a_4, D] = \frac{1}{|u|}[|L_{a_2+a_4}(D_H)| + |L_{a_2+a_4}(D_U)|] = \frac{1}{10}[7] = \frac{7}{10}$$

$$\gamma[a_2*a_4, D] = \frac{1}{|u|}[|L_{a_2*a_4}(D_H)| + |L_{a_2*a_4}(D_U)|] = \frac{1}{10}[1] = \frac{1}{10}$$

Hence, from the above it can be found that,

$$\gamma[a_2*a_4, D] \le \gamma[a_2, D] \le \gamma[a_2+a_4, D], \gamma[a_2*a_4, D] \le \gamma[a_4, D] \le \gamma[a_2+a_4, D]$$

Case 6:

$$L_{a_3+a_4}(D_H) = \{H_1, H_3, H_4, H_5, H_8\}; L_{a_3*a_4}(D_H) = \phi,$$
$$L_{a_3+a_4}(D_U) = \{H_6\}; L_{a_3*a_4}(D_U) = \phi$$

$$\gamma[a_3+a_4, D] = \frac{1}{|u|}[|L_{a_3+a_4}(D_H)| + |L_{a_3+a_4}(D_U)|] = \frac{1}{10}[6] = \frac{6}{10}$$

$$\gamma[a_3*a_4, D] = \frac{1}{|u|}[|L_{a_3*a_4}(D_H)| + |L_{a_3*a_4}(D_U)|] = 0$$

Hence, from the above it can be found that,

$$\gamma[a_3*a_4, D] \le \gamma[a_3, D] \le \gamma[a_3+a_4, D], \gamma[a_3*a_4, D] \le \gamma[a_4, D] \le \gamma[a_3+a_4, D]$$

Observation 3.2: It is that by using the **significant measure named as nano degree of dependence** on X it's all the cases we infer that the Multi-Granular nanotopology becomes finer than the Multi*-Granular nanotopology. From the above contribution, it is found that when two attributes sets in information system possesses an inconsistent relationship.

4   Real Life Application in Multi*-granular Nanotopology

Definition 4.1 [12] Let (U, A) be an information system, where A is divided into a set C of condition attributes and a set D of decision attribute. Intuitively, some attributes are not significant in a representation and their removal has no real impact on the value of representation of elements. If it is not significant, one can remove an attribute for further consideration called as reduct and it is denoted by Red_A.

Definition 4.2 [12] Let (U, A) be an information system, where A is divided into a set C of condition attributes and a set D of decision attribute. Then a core is a minimal subset of attributes which is such that none of its elements can be removed without affecting the classification powered attributes. And it can be found by $Core(A) = A - Red_A$.

Definition 4.3 [12] Since the M*GNT model mainly considers the Multi*lower approximation and the Multi*-upper approximation of a target concept by multiple equivalence relations in the following, we introduce a measure of importance of condition attributes with respect to decision attributes in a decision system.

Definition 4.4 [12] Let (U, A) be an information system, where A is divided into a set C of condition attributes $C = \{a_1, a_2, a_3, \ldots a_n\}$, then the inner significance measure of C is defined as

$$Sign_{inner}(a_i, C, D) = S_{C-\{a_i\}}(D) - S_C(D)$$
$$Sign^{inner}(a_i, C, D) = S^{C-\{a_i\}}(D) - S^C(D) \quad \text{for all } i = 1, 2 \ldots m.$$

where

$$S_C(D) = \frac{1}{|u|} \left[|L_{a_1 * a_2 \ldots a_n}(D_A)| + |L_{a_1 * a_2 \ldots a_n}(D_R)| \right]$$
$$S^C(D) = \frac{1}{|u|} \left[|U_{a_1 * a_2 \ldots a_n}(D_A)| + |U_{a_1 * a_2 \ldots a_n}(D_R)| \right].$$

Definition 4.5 [12] Let (U, A) be an information system, where A is divided into a set C of condition attributes $C = \{a_1, a_2, a_3, \ldots a_n\}$, then the inner significance measure of C is defined as

$$Sign_{outer}(a_i, a_j, D) = S_{\{a_j\}}(D) - S_{\{a_j \cup a_i\}}(D)$$
$$Sign^{outer}(a_i, a_j, D) = S^{\{a_j\}}(D) - S^{\{a_j \cup a_i\}}(D) \quad \text{for all } i = 1, 2 \ldots m.$$

4.1   An Attribute Reduction Algorithm in the M*GNT

Step 1: Given a finite universe U, a finite set A of attributes that divided into two classes, C of condition attributes and D of decision attribute, an equivalence relation R on U corresponding to $a_1, a_2, a_3, \ldots a_n$ represent the data as an information table, columns of which are labeled by attributes, rows by objects and entries of the table are attribute values.

Step 2: Generate the M*GNT, $\tau_{a_1*a_2\ldots a_m}(H_H), \tau_{a_1*a_2\ldots a_m}(H_U), \beta_{a_1*a_2\ldots a_m}(H_H),$ $\beta_{a_1*a_2\ldots a_m}(H_U)$.

Step 3: First we find the Multi*-lower approximation reduct.

Step 4: Now, find the inner significance measure for each a_i; $i = 1, 2 \ldots .m$.

Step 5: Put a_k into Red, where $sign^{inner}(a_k, A, d) > 0$, choose $Max\{sign^{inner}(a_k, A, d), sign^{inner}(a_r, A, d)\}$. If there is a tie choose arbitrary.

Step 6: Now. Compute the outer significance measure $a_i; i = 1, 2 \ldots k - 1,$ $k + 1, \ldots .m$. Then

$Red_A = Red \cup \{a_0\},$ where $sign(c_0, Red, d) = \max\{sign^{outer}(a_k, A, d), a_k \in C - red\}$.

Step 7: Next, by using the Multi*-upper approximation find the Multi*-upper approximation reduct by repeating steps 3, 4 and 5.

Step 8: Now we can generate the Multi*-Granular nanotopolpogy and its basis with respect to Red_A, where $i = 1, 2 \ldots .nn < m$.

Step 9: Then Red_A is the Multi*-Granular reduct iff $\beta_{a_1*a_2\ldots a_m}(H_H) = \beta_{a_1*a_2\ldots a_n}(H_U)$ and $\beta_{a_1*a_2\ldots a_m}(H_H) = \beta_{a_1*a_2\ldots a_n}(H_U)$.

4.2   Solution

From Table 1, based on the decision attribute

$$U/D = \{D_H, D_U\}, D_H = \{H_1, H_3, H_4, H_5, H_8, H_9\}, D_U = \{H_2, H_6, H_7, H_{10}\}$$

where $U/D = \{\{H_1, H_3, H_4, H_5, H_8, H_9\}, \{H_2, H_6, H_7, H_{10}\}\}$.

$$L_{a_1*a_2*a_3*a_4}(H_H) = \emptyset; L_{a_1*a_2*a_3*a_4}(H_U) = \emptyset; L_{a_1*a_2*a_3*a_4}(D) = \emptyset, \emptyset;$$

$$U_{a_1*a_2*a_3*a_4}(D) = \{\{H_1, H_2, H_3, H_4, H_5, H_6, H_7, H_8, H_9, H_{10}\},$$
$$\{H_1, H_2, H_3, H_4, H_5, H_6, H_7, H_8, H_9, H_{10}\}\}.$$

And also Multi*-lower approximation reducts: To find the inner significance measure a_i; $i = 1, 2, 3, 4$

$$S_C(D) = \frac{1}{|u|}[|L_{a_1*a_2*a_3*a_4}(D_H)| + |L_{a_1*a_2*a_3*a_4}(D_U)|] = \frac{1}{10}[0 + 0] = 0$$

$$S^C(D) = \frac{1}{|u|}[|U_{a_1*a_2*a_3*a_4}(D_H)| + |U_{a_1*a_2*a_3*a_4}(D_U)|] = \frac{1}{10}[10 + 10] = 2$$

$$Sign_{inner}(a_1, C, D) = S_{C-\{a_1\}}(D) - S_C(D) = 0$$

$$Sign_{inner}(a_2, C, D) = S_{C-\{a_2\}}(D) - S_C(D) = 0$$

$$Sign_{inner}(a_3, C, D) = S_{C-\{a_3\}}(D) - S_C(D) = \frac{1}{10}$$

$$Sign_{inner}(a_4, C, D) = S_{C-\{a_4\}}(D) - S_C(D) = \frac{1}{10}$$

$$Sign^{inner}(a_1, C, D) = S^C(D) - S^{C-\{a_1\}}(D) = 2 - 2 = 0$$

$$Sign^{inner}(a_2, C, D) = S^C(D) - S^{C-\{a_2\}}(D) = 2 - 2 = 0$$

$$Sign^{inner}(a_3, C, D) = S^C(D) - S^{C-\{a_3\}}(D) = 2 - \frac{19}{10} = \frac{1}{10}$$

$$Sign^{inner}(a_4, C, D) = S^C(D) - S^{C-\{a_4\}}(D) = 2 - \frac{19}{10} = \frac{1}{10}$$

By algorithm, we select the attribute 'a_3' to find the outer significance:

$$S_{a_3}(D) = \frac{1}{|u|}[|L_{a_3}(D_H)| + |L_{a_3}(D_U)|] = \frac{1}{10}[3+1] = \frac{4}{10}$$

$$S^{a_3}(D) = \frac{1}{|u|}[|U_{a_3}(D_H)| + |U_{a_3}(D_U)|] = \frac{1}{10}[9+7] = \frac{16}{10}$$

$$Sign_{outer}(a_1, a_3, D) = S_{\{a_3\}}(D) - S_{\{a_1 \cup a_3\}}(D) = \frac{4}{10} - \frac{3}{10} = \frac{1}{10}$$

$$Sign_{outer}(a_2, a_3, D) = S_{\{a_3\}}(D) - S_{\{a_2 \cup a_3\}}(D) = \frac{4}{10} - \frac{1}{10} = \frac{3}{10}$$

$$Sign_{outer}(a_4, a_3, D) = S_{\{a_3\}}(D) - S_{\{a_4 \cup a_3\}}(D) = \frac{4}{10} - 0 = \frac{2}{5}$$

$$Sign^{outer}(a_1, a_3, D) = S^{\{a_1 \cup a_3\}}(D) - S^{\{a_3\}}(D) = \frac{17}{10} - \frac{16}{10} = \frac{1}{10}$$

$$Sign^{outer}(a_2, a_3, D) = S^{\{a_2 \cup a_3\}}(D) - S^{\{a_3\}}(D) = \frac{19}{10} - \frac{16}{10} = \frac{3}{10}$$

$$Sign^{outer}(a_4, a_3, D) = S^{\{a_2 \cup a_3\}}(D) - S^{\{a_3\}}(D) = \frac{20}{10} - \frac{16}{10} = \frac{2}{5}$$

We select the next attribute 'a_4'

$$S_{a_4}(D) = \frac{1}{|u|}[|L_{a_4}(D_H)| + |L_{a_4}(D_U)|] = \frac{1}{10}[2+0] = \frac{2}{10}$$

$$S^{a_4}(D) = \frac{1}{|u|}[|U_{a_4}(D_H)| + |U_{a_4}(D_U)|] = \frac{1}{10}[10+8] = \frac{18}{10}$$

$$Sign_{outer}(a_1, a_4, D) = S_{\{a_4\}}(D) - S_{\{a_1 \cup a_4\}}(D) = 0$$

$$Sign_{outer}(a_2, a_4, D) = S_{\{a_4\}}(D) - S_{\{a_2 \cup a_4\}}(D) = \frac{2}{10} - \frac{1}{10} = \frac{1}{10}$$

$$Sign_{outer}(a_3, a_4, D) = S_{\{a_4\}}(D) - S_{\{a_3 \cup a_4\}}(D) = \frac{2}{10} - 0 = \frac{1}{5}$$

$$Sign^{outer}(a_1, a_4, D) = S^{\{a_1 \cup a_4\}}(D) - S^{\{a_3\}}(D) = \frac{18}{10} - \frac{18}{10} = 0$$

$$Sign^{outer}(a_2, a_4, D) = S^{\{a_2 \cup a_4\}}(D) - S^{\{a_3\}}(D) = \frac{19}{10} - \frac{18}{10} = \frac{1}{10}$$

$$Sign^{outer}(a_3, a_4, D) = S^{\{a_2 \cup 4\}}(D) - S^{\{a_3\}}(D) = \frac{20}{10} - \frac{18}{10} = \frac{1}{5}$$

and also

$$\beta_{a_1 * a_2 * a_3 * a_4}(H_H) = \beta_{a_3 * a_4}(H_H) = \{U, \emptyset, \{H_1, H_2, H_3, H_4, H_5, H_6, H_7, H_8, H_9, H_{10}\}\}$$

$$\beta_{a_1 * a_2 * a_3 * a_4}(H_U) = \beta_{a_3 * a_4}(H_U) = \{U, \emptyset, \{H_1, H_2, H_3, H_4, H_5, H_6, H_7, H_8, H_9, H_{10}\}\}.$$

Hence $Red_A = \{a_3, a_4\}$ = {Processed water and Junk food, Machine-driven Air} is a redact of these granular in the Multi*-Granular Nano topological space. Thus Core $(A) = \{a_1, a_2\}$.

Observation 4.8: From CORE (A), we conclude that **(Water and Air) and (Food)** combinations are ideal combinations for the human beings to live a healthy life.

5 Conclusion

Finally we conclude that the result obtained using Multi*Granular nanotopology is same as the result got in problem [10] using Multi*-Granular reduction method. Thus we conclude that the result of the same problem considered in both the paper is consistent under the both methods.

References

1. Pei, D.: On definable concepts of rough set models. Inf. Sci. **177**, 4230–4239 (2007)
2. Pawlak, Z., Skowron, A.: Rudiments of rough sets. Inf. Sci. **177**(1), 3–27 (2007)
3. Pawlak, Z.: Rough set theory and Its applications. J. Telecommun. Inf. Technol. **3**, 7–10 (2002)
4. Qian, Y.H., Liang, J.Y., Yao, Y.Y., Dang, C.Y.: A multi-granular rough set. Inf. Sci. **180**, 949–970 (2010)
5. Yao, Y.Y.: Information granulation and rough set approximation. Int. J. Intell. Syst. **16**(1), 87–104 (2001)
6. Lellis Thivagar, M., Richard, C.: Note on Nanotopological Spaces, Communicated
7. Lellis Thivagar, M., Richard, C.: On nano forms of weekly open sets. Int. J. Math. Stat. Invention **1**(1), 31–37 (2013)

8. Lellis Thivagar, M., Richard, C.: Nutrition modelling thorugh nanotopology. Int. J. Eng. Res. Appl. **4**, 327–334 (2014)
9. Lellis Thivagar, M., Sutha Devi, V.: On multi-granular nanotopology, South East Asian Bulletin of Mathematics. Springer Verlag, Accepted
10. Lellis Thivagar, M., Priyalatha, S.P.R.: A heuristic approach to establish an algebraic structure on multi star granular nano topology. Ultra Scientist **28**(1), 33–42 (2016)
11. Vasanthakannan, G., Indirani, K.: Ideal Combinations of Triplet <Air, Food, Water> Through Nanotopology, Communicated
12. Lellis Thivagar, M., Richard, C.: Computing technique for recruitement process via nanotopology. Sohag J. Math. **3**(1), 37–45 (2016)

Early Detection of Lung Cancer Using Wavelet Feature Descriptor and Feed Forward Back Propagation Neural Networks Classifier

R. Arulmurugan[1] and H. Anandakumar[2(✉)]

[1] Department of Information Technology, Bannari Amman Institute
of Technology, Sathyamangalam, Tamil Nadu, India
arulmr@gmail.com
[2] Department of Computer Science and Engineering, Akshaya College
of Engineering and Technology, Coimbatore, Tamil Nadu, India
anandakumar.psgtech@gmail.com

Abstract. A Computed Tomography (CT) scan is the most used technique for distinguishing harmful lung cancer nodules. A Computer Aided Diagnosis (CAD) framework for the recognition of lung nodules in thoracic CT pictures is implemented. A lung nodule which can be either benign or malignant can be easily classified as the better support for treatment. The proposed work is based on using Wavelet feature descriptor and combined with an Artificial Neural network for classification. The computed statistical attributes such as Autocorrelation, Entropy, Contrast, and Energy are obtained after applying wavelet transform and used as input parameters for neural network Classifier. The NN Classifier is designed by considering training functions (Traingd, Traingda, Traingdm, and Traingdx) using feed forward neural network and feed forward back propagation network. The feed forward back propagation neural network gives better classification results than feed forward. The proposed classifier produced Accuracy of 92.6%, specificity of 100% and sensitivity of 91.2% and a mean square error of 0.978.

Keywords: Lung nodule detection · Feature extraction · Wavelet transform
Texture feature computer tomography · ANN

1 Introduction

Lung cancer is the fundamental driver for cancer death around the world, and it is hard to recognize in its early stages since side effects seem just at advanced stages. Many people die due to lung cancer rather than any other types of cancer such as blood, colon, breast and prostate cancers. The Lung Cancer remains a primary cause of mortality [1]. Moreover, mortality from the lung cancer disease is required to keep ascending, to be turned into around 17 million worldwide in 2030. Early recognition of lung disease is profitable. There is a better result that the early detection of lung tumor will diminish the death rate. The early stage can be performed in tenants screening.

© Springer International Publishing AG 2018
D. J. Hemanth and S. Smys (eds.), *Computational Vision and Bio Inspired Computing*,
Lecture Notes in Computational Vision and Biomechanics 28,
https://doi.org/10.1007/978-3-319-71767-8_9

A lung malignancy that duplicating and creating of anomalous cells because of this reality, the lung cancer growth survival rate is 15% in five years [2] A lung nodule is a round lesion with a breadth littler than 3 cm. It can be either benevolent (non-cancerous) or dangerous (cancerous), and is found in 1 of every 100 CT scans of the chest [3]. In a CT output, the lung malignancy is seen as round white shadow nodules, thus it is critical to identify and group those nodules for the screening and analyzes of cancerous purposes.

The probability of the lung nodule may be malignant around 40%, but their risk factors can vary based on the several conditions. For instance, in individuals with age under 35 years, the possibility of lung nodules can be cancerous is less than 1%, though the half of lung nodules in individuals more than 50 are dangerous (malignant) [4]. To acquire an exact interpretation from CT scans requires a major effort by the radiologist. The analysis of cancer identification turns out to be more complex if the level of the disease is still not visually detected (early stage) [5]. Computer Aided Diagnosis has been conferred a new step in the early diagnosis of lung tumor [6]. CAD systems expect to enhance the sensitivity, specificity, accuracy, and cost-effectiveness of lung cancer screening programs.

This paper focused on the design of early detection of cancer method using wavelet and ANN for identifying cancerous lung nodules and without cancerous lung nodules. The first section discusses various related work for the lung cancer detection. The second half of paper describes proposed work which nodules were processed by calculating the texture features by using gray level co-occurrence matrix (GLCM) from wavelet domain. Finally ANN classifier is used for categorization of cancerous nodules from non-cancerous nodules.

2 Related Works

Many CAD approaches have been proposed for the undertaking of the order of lung nodules characterization from CT scans [7, 8]. In 1963, first digital computers were used for detection of lung cancer nodules. Most of the techniques comprise of four stages for Lung cancer classification. They are (i) preprocessing, (ii) lung segmentation, (iii) nodule candidate detection and (iv) nodule classification. The fourth module is classification used to differentiate malignant lesions from normal lesions using their inherent characteristics [9].

Hybrid based classification scheme is introduced by [10]. For characterization of lung nodules inside the CT scans, a feature extraction based nodule was implemented. Due to this causes that some blood vessels were classified as lung nodules. At the second stage, the surface elements were computed keeping in mind the end goal to separate the veins. The approach utilized for groupings SVM with a rule-based framework. This method produced result of 84.39% sensitivity.

In [11] the framework of CADs system was constructed in two stages that can detect and diagnose automatically from CT scan lung nodule with cancerous or non-nodule. The first stage used for preprocessing and segmenting region of the nodule and in the second stage is mainly for diagnosis of malignant nodule using a fuzzy system based. The 50 testing images containing 685 slice images were used. This proposed system obtained was 90% sensitivity.

A novel approach proposed using active contours with SVM by [12]. The region of lung area segmented by using active contours and masking methodology was applied to modify non-isolated nodules into isolated. Then by using SVM, nodules were recognized in 2D form and anatomical 3D features. Here 4 datasets are used in the testing. By this novel approach, sensitivity was 89%. Table 1 shows various CADs system performance based on sensitivity.

Table 1 CADs systems sensitivity comparison

Various classifiers	Sensitivity (%)
Ruled-based SVM	84.39
SVM	89
Fuzzy system	90

3 Proposed Methodology

The proposed methodology for designing the Intelligent CADs system are: (1) Segmentation, (2) Applying Wavelet transform, (3) Feature extraction from CT images, (4) Feature and sub-band selection and (5) Classification using ANN. Figure 1 shows the detailed flow chart of proposed methodology.

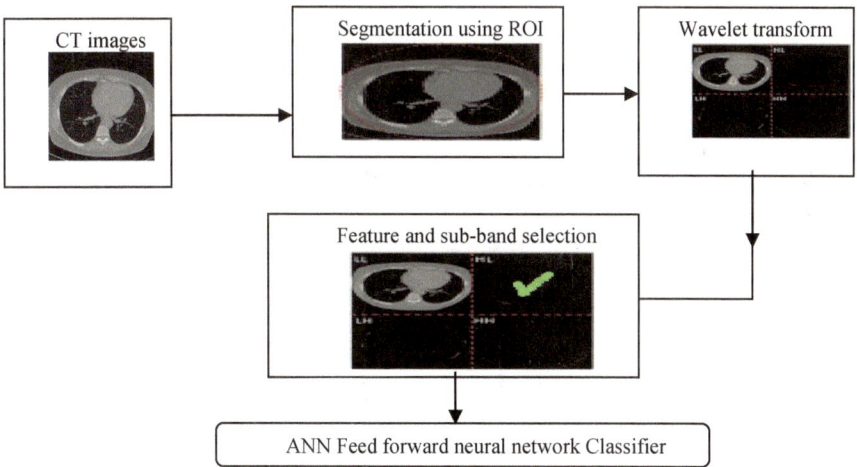

Fig. 1 Proposed CAD system for early detection of lung cancer

The data set from Lung Image Database Consortium (LIDC) [13] is used for our proposed work. Around 150 CT scan of lung images are retrieved which contain both male and female. The patients average age is 64 years (younger age is 18 years and oldest is 86 years). The CT lung images with diameter ranges from 260 to 400 mm with the layer thickness of 0.75–1.25 mm. Here 110 nodules of size less than 3 mm are used. The different stages cancer nodules such as circumscribed nodules, pleural nodules and vascular nodules are considered for the research work [14]. The lung cancer image in CT scan shown in Fig. 2.

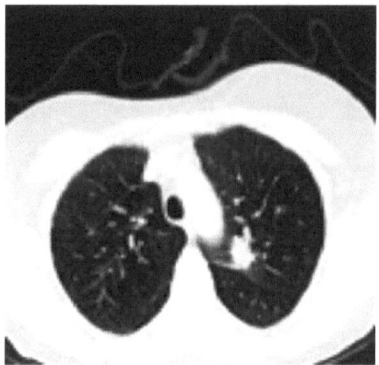

Fig. 2 Lung cancer CT scan image

The CT scan taken from LIDC contains a few slices and contrasts between them. Some scanner generates different nodules information and contains different shape. To remove difference, a Region of Interest (ROI) for every CT picture was applied. Hough transform is used for computing ROI for all CT scan images. Next stage is to perform segmentation for separating the lung region from CT picture. ROI extracted images are changed to transformed domain i.e. spatial domain images convert to the frequency domain. Discrete wavelet transform is applied on CT images which represent high-frequency values.

The Daubechies wavelets can have much impact on the achievement of texture characterization that filters positively the quality descriptors [15]. The Daubechies wavelet transforms db1, db2 and db4 were applied. However, other orthogonal wavelet families can be used. The transformation is achieved by convoluting columns and rows of CT images with low-pass and high-pass filter.

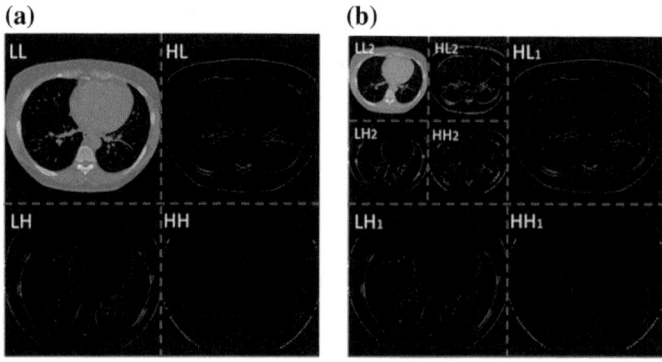

Fig. 3 First level decomposition (**a**), Second level decomposition (**b**)

From Fig. 3a, b, calculation of LL, LH, HL and HH done using Daubechies wavelet transform db4. After obtaining quality descriptors, textures information are extracted from lung nodules using gray level co-occurrence matrix (GLCM). From GLCM, 8 gray quantized values are used.

Autocorrelation, Entropy, Contrast, and Energy are parameter that is computed from transformed descriptors using Eqs. (1)–(4)

$$Autocor = \frac{\sum_i \sum_j (i,j) p(i,j) - \mu_x \mu_y}{\sigma_x \sigma_y} \tag{1}$$

$$Entr = -\sum_i \sum_j p(i,j) \log(p(i,j)) \tag{2}$$

$$Con = \sum_i \sum_{i,j=0}^{N-1} p_{i,j} (i-j)^2 \tag{3}$$

$$Eng = \sqrt{\sum_{i,j=0}^{N-1} p(i,j)^2} \tag{4}$$

p_{ij} are normalized parameter after GLCM for image rows and columns. μ_x μ_y are mean and σ_x σ_y are standard deviation for x row and y column.

Final stage is classification using artificial neural network using computed features from wavelet transform. The neural network is designed by using feed forward and feed forward back propagation neural method with layer input and output units. The feed forward neural networks use Adeline, Hebb, Perception, and Madaline. The neural network passing of information from the input layer to hidden layer and reach the output layer. There is no feedback in this neural network. In feed forward back propagation, back propagation is used to train based on error correction. So this neural network is called as error back propagation algorithm.

4   Experimental Results

The CADs system is trained with feed forward back propagation by using 150 images that contain single slices of 2 lungs. In that 110 images are cancerous and 40 are non-cancerous images. The training is done using 70% images and testing done with 30% images from total images. The sensitivity, specificity and accuracy are calculated for both networks. Training stops when any of these conditions occurs:

- When maximum number of epochs is reached
- The maximum amount of time is exceeded
- When minimum Performance is achieved
- Minimum gradient falls below
- Performance for Validation increased more than maximum fail times when in the last time it decreased.

4.1   Sensitivity

It is used to identify the probability of actual positives from Eq. (5). That is the percentage of segmented slices containing cancerous nodule is correctly classified as cancerous.

$$Sensitivity = \frac{T_p}{T_p + F_n} \tag{5}$$

True Positive (T_p): cancer nodule is classified as cancerous.
False Positive (F_p): without cancer nodule is classified as cancerous.
True Negative (T_n): without cancer nodule is classified as non-cancerous.
False Negative (F_n): cancer nodule is classified as non-cancer.

4.2   Specificity

It is ability to measure the actual negatives through Eq. (6). The percentage of segmented slices without cancerous nodule is correctly identified as non cancerous.

$$Specificity = \frac{T_n}{T_n + F_p} \tag{6}$$

4.3   Accuracy

Accuracy is a statistical measure of how well a classifier correctly identifies or excludes a condition calculated from Eq. (7). The accuracy is the proportion of true results (both true positive and true negative) in the population.

$$Accuracy = \frac{T_p + T_n}{T_n + T_p + F_p + F_n} \tag{7}$$

The result obtained from neural network for feed forward back propagation is comparatively high with feed forward network. The classification accuracy of feed-forward networks and back propagation network is shown in Fig. 4.

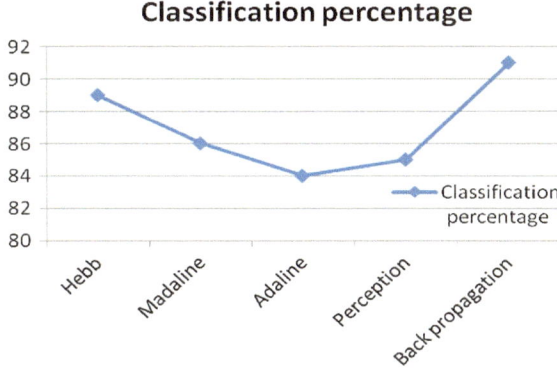

Fig. 4 Classification accuracy of feed forward and back propagation network

Four training functions of the back propagation network are trained with different momentum factor and learning rate from 0.1 to 0.9. Keeping one of these parameters constant and varying the other, the performance of the network is studied.

The best classification accuracy and least mean square error of different training functions for different momentum and learning rate are shown in Table 2. Table 2 shows that the training function Traingdx gives the maximum classification accuracy of 92.61%.

Table 2 4 Training function for classification in NN

Training functions	Momentum	Learning rate	Classification accuracy	Mean square error
Traingd	0.4	0.6	85.6	0.126
Traingda	0.3	0.5	86.79	0.128
Traingdm	0.4	0.7	87.88	0.125
Traingdx	0.3	0.7	**92.61**	0.978

5 Summary

A CADs system to classify lung nodules done using Wavelet feature descriptor which computed from the gray level co-occurrence matrix of a Daubechies wavelet transform and feed forward back propagation neural networks classifier. The region of interest is obtained from the segmented single slices containing 2 lungs. The neural network that constructed using four training functions (Traingd, Traingda, Traingdm, and Traingdx), the traingdx training function gives the maximum classification accuracy. The proposed NN feed forward back propagation classifier produced Accuracy of 92.61%, specificity of 100% and sensitivity of 91.2% and a mean square error of 0.978. The sensitivity is calculated by evaluating the percentage of segmented lung nodules containing cancerous nodule that is correctly classified as cancerous. From this new approach, radiologists can use our CADs for lung cancer detection accurately and easily in the early stage of cancer.

References

1. Takiar, R., Nadayil, D., Nandakumar, A.: Projection of number of cancer cases in India (2010–2020) by cancer groups. Asian Pac. J. Cancer Prev. **11**(4), 1045–1049 (2010)
2. Patil, S.A., Udupi, V.R., Kane, C.D., Wasif, A.I., Desai, J.V., Jadhav, A.N.: Geometrical and Texture Features Estimation of Lung Cancer and TB Image using Chest X-Ray Database (2009). ISBN 978-4244-4764-0/09
3. Diciotti, S., Lombardo, S., Falchini, M., Picozzi, G., Mascalchi, M.: Automated segmentation refinement of small lung nodules in CT scans by local shape analysis. IEEE T. Bio-Med. Eng. **58**(12), 3418–3428 (2011)
4. Farag, A., El Munim, H., Graham, J., Farag, A.: A novel approach for lung nodules segmentation in chest CT using level sets. IEEE Trans. Image Process. **22**, 5202–5213 (2013)
5. Choi, W., Choi, T.: Automated pulmonary nodule detection based on three-dimensional shape-based feature descriptor. Comput. Meth. Prog. Bio. **133**(1), 37–54 (2014)
6. Magesh, B., et al.: Computer aided diagnosis system for the identification and classification of lessions in lungs. Int. J. Comput. Trends Technol. IJCTT (2011). ISSN 2231-2803
7. Li, Q.: Recent progress in computer-aided diagnosis of lung nodules on thin-section CT. Comput. Med. Imag. Grap. **31**(4), 248–257 (2007)
8. Ambrosini, V., Nicolini, S., Caroli, P., Nanni, C., Massaro, A., Marzola, M., et al.: PET/CT imaging in different types of lung cancer: an overview. Eur. J. Radiol. **81**(5), 988–1001 (2012)
9. Van Ginneken, B., Schaefer-Prokop, C., Prokop, M.: Computer-aided diagnosis: how to move from the laboratory to the clinic. Radiol. **261**(3), 719–732 (2011)
10. Jing, Z., Bin, L., Lianfang, T.: Lung nodule classification combining rule-based and SVM. In: Li, K. (eds.) Proceedings of the IEEE Fifth International Conference on Bio-Inspired Computing: Theories and Applications, pp. 1033–1036, 23–26 Sept 2010, Changsha, China. Piscataway, NJ: IEEE Computer Society (2010)
11. Kumar, S.A., Ramesh, J., Vanathi, P.T., Gunavathi, K.: Robust and automated lung nodule diagnosis from CT images based on fuzzy systems. In: Manikandan, V. (eds.) Proceedings of the IEEE International Conference on Process Automation, Control and Computing, pp. 1–6, 20–22 July 2011, Coimbatore, India. Piscataway, NJ: IEEE Women in Engineering
12. Keshani, M., Azimifar, Z., Tajeripour, F., Boostani, R.: Lung nodule segmentation and recognition using SVM classifier and active contour modeling: a complete intelligent system. Comput. Biol. Med. **43**(4), 287–300 (2013)
13. Armato, S., McLennan, G., Bidaut, L., McNitt-Gray, M., Meyer, C.: The lung image database consortium (LIDC) and image database resource initiative (IDRI): a completed reference database of lung nodules on CT scans. Med. Phys. **38**(2), 915–931 (2011)
14. Kostis, W.J., Reeves, A.P., Yankelevitz, D.F., Henschke, C.I.: Three-dimensional segmentation and growth-rate estimation of small pulmonary nodules in helical CT images. IEEE Trans. Med. Imaging **22**(3), 1259–1274 (2003)
15. Singh, R., Khare, A.: Fusion of multimodal medical images using daubechies complex wavelet transform: a multiresolution approach. Inf. Fusion **19**(1), 49–60 (2014)

A Study on Fuzzy Weakly Ultra Separation Axioms via Fuzzy $\widehat{\mu}$B-Kernel Set

J. Subashini[1($\boxtimes$)] and K. Indirani[2]

[1] Department of Mathematics, RVS College of Arts and Science,
Sulur Coimbatore, India
subashini.j@rvsgroup.com
[2] Nirmala College for Women Red Fields, Coimbatore, India
indirani009@ymail.com

Abstract. In this paper, we introduce a new class of fuzzy closed sets called fuzzy $\widehat{\mu}\beta$-closed sets also we introduce the concept of fuzzy $\widehat{\mu}\beta$-kernel set in a fuzzy topological space. We also investigate some of the properties of weak fuzzy separation axioms like fuzzy $\widehat{\mu}\beta$-R_i space, i = 0, 1, 2, 3 and fuzzy $\widehat{\mu}\beta$-T_i-space, i = 0, 1, 2, 3, 4.

Keywords: Fuzzy $\widehat{\mu}\beta$-closed sets · Fuzzy $\widehat{\mu}\beta$-kernel set · Fuzzy $\widehat{\mu}\beta R_i$-space, i = 0, 1, 2, 3 · Fuzzy $\widehat{\mu}\beta T_i$-space, i = 0, 1, 2, 3, 4

1 Introduction

The fundamental concept of a fuzzy set was introduced by Zadeh in 1965, [1]. Subsequently, in 1968, Chang [2] introduced fuzzy topological spaces. Later on, Chang's idea was developed by Goguen [3], who replaced the closed interval I = [0,1] by a more general lattice L. In 1985, Kubiak [4], and Sostak [5], in separated works, made topology itself fuzzy besides their dependence on fuzzy sets. In ordinary topology, semi preopen sets were introduced and studied by Andrijevic [6]. Subashini and Indirani [7] introduced $\widehat{\mu}\beta$ set and continuity in topological space. In this paper, we introduced the concept weakly ultra fuzzy separation of two fuzzy sets in fuzzy topological spaces using fuzzy $\widehat{\mu}\beta$-closed set and using this concept to define the fuzzy $\widehat{\mu}\beta$ kernel set of a fuzzy set A of a fuzzy topological space. We also formulate some of the properties of weak fuzzy separation fuzzy $\widehat{\mu}\beta R_i$ space, i = 0, 1, 2, 3 and fuzzy $\widehat{\mu}\beta T_i$ space, i = 0, 1, 2, 3, 4.

2 Preliminaries

Throughout this paper, X represents a nonempty fuzzy set and fuzzy subset A of X, denoted by A $\leq$ X, then it is characterized by a membership function in the sense of Zadeh [8]. The basic fuzzy sets are the empty set, the whole set, and the class of all

© Springer International Publishing AG 2018
D. J. Hemanth and S. Smys (eds.), *Computational Vision and Bio Inspired Computing*,
Lecture Notes in Computational Vision and Biomechanics 28,
https://doi.org/10.1007/978-3-319-71767-8_10

fuzzy sets of X which will be denoted by 0_X, 1_X, and I^X, respectively. A subfamily τ of I^X is called a fuzzy topology described by Chang [2]. Moreover, the pair $(X, \tau) = (I^X, \tau)$ will be meant as a fuzzy topological space, on which no separation axioms are assumed unless explicitly stated. The fuzzy closure, the fuzzy interior, and the fuzzy complement of any set A in (X, τ) are denoted by cl (A), Int(A), and $1 - A$, respectively. A fuzzy set which is a fuzzy point [1, 9] with support $x \in X$ and value t $(0 < t \leq 1)$ is denoted by x_t, and $Pt(X)$ will denote the family of all point fuzzy sets $x_t \in I^X$. For any two fuzzy sets A and B in (X, τ), $A \leq B$ if and only if $A(x) \leq B(x)$ for each $x \in X$.

Definition 2.1 [10] Let A be a fuzzy set of a set X. The support of is the elements whose membership value is greater than 0, i.e., supp(A) = $\{x \in X: A(x) > 0\}$.

Definition 2.2 [11] Let A and B be any two fuzzy sets in X. Then we define $A \vee B: X \to [0,1]$ as follows: $(A \vee B)(x) = \max \{A(x), B(x)\}$. Also, we define $A \wedge B: X \to [0,1]$ as follows: $(A \wedge B)(x) = \min\{A(x), B(x)\}$. By $A \vee B(A \wedge B)$, we mean the union (intersection) between two fuzzy sets A and B of X.

Definition 2.3 [10] Let A be any fuzzy set in a set X. The complement of A, is denoted by $1_x - A$ or A^c and defined as follows: $A^c(x) = 1 - A(x)$, for each $x \in X$.

Remark 2.4 From definitions (2.2 and 2.3), we have, if $A, B \in I^X$, then $A \vee B$, $A \wedge B$ and $1_X - A \in I^X$.

Definition 2.5 [12] A fuzzy point x_t in a set X is a fuzzy set defined as follows:

$$x_t(y) = \begin{cases} t \; if \; y = x \\ 0 \; otherwise, \; and \; 0 < t \leq 1 \end{cases}$$

where $0 < \lambda \leq 1$. Now, supp(x_t) = $\{y: x_t(y) > 0\}$. Then supp $(x_t) = x$, so the value at x is t, and call the point x its support of fuzzy point x_t and t is the height of x_t. That is, x_t has the membership degree 0 for all $y \in X$ except one, say $x \in X$.

Definition 2.6 [2] A fuzzy topology on a set X is a family τ of fuzzy sets in X which satisfies the following conditions:

(i) 0_X, $1_X \in \tau$,
(ii) If $A, B \in \tau$, then $A \wedge B \in \tau$,
(iii) If $\{A_i: i \in J\}$ is a family in τ, then $\underset{i \in J}{\vee} A_i \in \tau$.

τ is called a fuzzy topology for X and the pair (X, τ) (or simply X) is a fuzzy topological space. Every element of τ is called τ-fuzzy open set. A fuzzy set is τ-fuzzy closed, if its complement is fuzzy open set. As ordinary topologies, the indiscrete fuzzy topology on X contains only 0_x and 1_x (i.e., $\emptyset$, X), while the discrete fuzzy topology on X contains all fuzzy sets in X.

Definition 2.7 A fuzzy subset A of a fuzzy topological space (X, τ), is called a fuzzy $\hat{\mu}\beta$ closed set (briefly $F\hat{\mu}\beta$-closed set) if $\hat{\mu}cl(A) \leq U$ whenever $A \leq U$ and U is fuzzy β open in X.

3   Fuzzy $\widehat{\mu}\beta$ Kernel and Fuzzy $\widehat{\mu}\beta R_i$ Spaces, i = 0, 1, 2, 3

Definition 3.1 The intersection of all fuzzy $\widehat{\mu}\beta$-open subset of X containing A is called the fuzzy $\widehat{\mu}\beta$ kernel of A this means $F\widehat{\mu}\beta\text{ker}(A) = \wedge \{G \in F\ \widehat{\mu}\beta O(X): A \leq G\}$.

Definition 3.2 In a fuzzy topological space (X, τ), a fuzzy set A is said to be weakly ultra fuzzy $\widehat{\mu}\beta$ separated from B if there exists a fuzzy $\widehat{\mu}\beta$-open set G such that $G \wedge B = 0_X$ or $A \wedge \widehat{\mu}\beta\text{cl}(B) = 0_X$. By definition (3.2), we have the following: For every two distinct fuzzy points x_t and y_t of X,

(i) $F\widehat{\mu}\beta\text{cl}(\{x_t\}) = \{y_t: y_t\}$ is not weakly ultra fuzzy $\widehat{\mu}\beta$ separated from $\{x_t\}$.
(ii) $F\widehat{\mu}\beta\text{ker}(\{x_t\}) = \{y_t: x_t\}$ is not weakly ultra fuzzy $\widehat{\mu}\beta$ separated from $\{y_t\}$.

Corollary 3.3 *Let* (X, τ)*, be a fuzzy topological space, then* $y_t \in F\widehat{\mu}\beta\text{ker}(\{x_t\})$ *iff* $x_t \in F\widehat{\mu}\beta\text{cl}(\{x_t\})$ *for each* $x \neq y \in X$.

Proof Suppose that $y_t \notin F\widehat{\mu}\beta\text{ker}(\{x_t\})$. Then there exists a fuzzy $\widehat{\mu}\beta$ open set U containing x_t such that $y_t \notin U$. Therefore, we have $x_t \notin \widehat{\mu}\beta\text{-ker}(\{x_t\})$. The converse part can be proved in a similar way.

Definition 3.4 A fuzzy topological space (X, τ), is called fuzzy $\widehat{\mu}\beta R_0$ space (F $\widehat{\mu}\beta R_0$ space) if for each fuzzy $\widehat{\mu}\beta$ open set U and $x_t \in U$, then $\widehat{\mu}\beta\text{cl}(\{x_t\}) \leq U$.

Definition 3.5 A fuzzy topological space (X, τ), is called fuzzy $\widehat{\mu}\beta R_1$ space (F $\widehat{\mu}\beta R_1$ space) if for each two distinct fuzzy points x_t and y_t of X with $F\widehat{\mu}\beta\text{cl}(\{x_t\}) \neq F\widehat{\mu}\beta\text{cl}(\{y_t\})$, there exist disjoint fuzzy $\widehat{\mu}\beta$ open sets U, V such that $F\widehat{\mu}\beta\text{cl}(\{x_t\}) \leq U$ and $F\widehat{\mu}\beta\text{cl}(\{y_t\}) \leq V$.

Theorem 3.6 *Let* (X, τ) *be a fuzzy topological space. Then* (X, τ)*, is* $F\widehat{\mu}\beta R_0$ *space if and only if* $F\widehat{\mu}\beta\text{cl}(\{x_t\}) = F\widehat{\mu}\beta\text{ker}(\{x_t\})$*, for each* $x \in X$.

Proof Let (X, τ), be a $F\widehat{\mu}\beta R_0$ space. If $F\widehat{\mu}\beta\text{cl}(\{x_t\}) \neq F\widehat{\mu}\beta\text{ker}(\{x_t\})$, for each $x \in X$, then there exist an fuzzy point $y \neq x$ such that $y_t \in F\widehat{\mu}\beta\text{cl}(\{x_t\})$ and $y_t \notin F\widehat{\mu}\beta\text{ker}(\{x_t\})$, then there exist an U_{x_t} fuzzy $\widehat{\mu}\beta$ open set, $y_t \notin U_{x_t}$ implies $F\widehat{\mu}\beta\text{cl}(\{x_t\}) \nleq U_{x_t}$ this contradiction. Thus $F\widehat{\mu}\beta\text{cl}(\{x_t\}) = F\widehat{\mu}\beta\text{ker}(\{x_t\})$. Conversely, let $F\widehat{\mu}\beta\text{cl}(\{x_t\}) = \beta\text{ker}(\{x_t\})$, for each fuzzy $\widehat{\mu}\beta$ open set U, $x_t \in U$, then $F\widehat{\mu}\beta\text{ker}(\{x_t\}) = F\widehat{\mu}\beta\text{cl}(\{x_t\})$ U [By definition (3.1)]. Hence by definition (3.4), (X, τ) is a $F\widehat{\mu}\beta R_0$ space.

Theorem 3.7 *A fuzzy topological space* (X, τ) *is an* $F\widehat{\mu}\beta R_0$ *space if and only if for each G fuzzy* $\widehat{\mu}\beta$ *closed set and* $x_t \in G$*, then* $F\widehat{\mu}\beta\text{ker}(\{x_t\}) \leq G$.

Proof Let for each G fuzzy $\widehat{\mu}\beta$ closed set and $x_t \in G$, then $F\widehat{\mu}\beta\text{ker}(\{x_t\}) \leq G$ and let U be a fuzzy $\widehat{\mu}\beta$ open set, $x_t \in U$ then for each $y_t \notin U$ implies $y_t \in U^c$ is a fuzzy $\widehat{\mu}\beta$ closed set implies $F\widehat{\mu}\beta\text{ker}(\{y_t\}) \leq U^c$ (By assumption). Therefore $x_t \notin F\widehat{\mu}\beta\text{ker}(\{y_t\})$ implies $y_t \notin F\widehat{\mu}\beta\text{cl}(\{x_t\})$ [By corollary (3.3)]. So $F\widehat{\mu}\beta\text{cl}(\{x_t\}) \leq U$. Thus (X, τ) is an $F\widehat{\mu}\beta R_0$ space. Conversely, let a fuzzy topological space (X, τ) be $F\widehat{\mu}\beta R_0$ space and G be fuzzy $\widehat{\mu}\beta$ closed set and $x_t \in G$. Then for each $y_t \notin G$ implies $y_t \in G^c$ is fuzzy $\widehat{\mu}\beta$ open set, then $F\widehat{\mu}\beta\text{cl}(\{y_t\}) \leq G^c$, so $F\widehat{\mu}\beta\text{ker}(\{x_t\}) = F\widehat{\mu}\beta\text{cl}(\{x_t\})$. Thus $F\widehat{\mu}\beta\text{ker}(\{x_t\}) \leq G$.

Corollary 3.8 *A fuzzy topological space* (X, τ) *is* $F\widehat{\mu}\beta R_0$ *space if and only if for each* U *fuzzy* $\widehat{\mu}\beta$ *open set and* $x_t \in$ U, *then* $F\widehat{\mu}\beta cl(F\widehat{\mu}\beta ker(\{x_t\})) \leq$ U.

Theorem 3.9 *Every* $F\widehat{\mu}\beta R_1$ *space is a* $F\widehat{\mu}\beta R_0$ *space.*

Proof Let (X, τ) be a $F\widehat{\mu}\beta R_1$ space and let U be a fuzzy $\widehat{\mu}\beta$ open set, $x_t \in$ U, then for each $y_t \notin$ U implies $y_t \in U^c$ is an fuzzy $\widehat{\mu}\beta$ closed set and $F\widehat{\mu}\beta cl(\{y_t\}) \leq U^c$ implies $F\widehat{\mu}\beta cl(\{x_t\}) \neq F\widehat{\mu}\beta cl(\{y_t\})$. Hence by definition (3.5), $F\widehat{\mu}\beta cl(\{x_t\}) \leq$ U. Thus (X, τ) is a $F\widehat{\mu}\beta R_0$ space.

Theorem 3.10 *A fuzzy topological space* (X, τ) *is* $F\widehat{\mu}\beta R_1$ *space if and only if for each* $x \neq y \in X$ *with* $F\widehat{\mu}\beta ker(\{x_t\}) \neq F\widehat{\mu}\beta ker(\{y_t\})$, *then there exist fuzzy* $\widehat{\mu}\beta$ *closed sets* G_1, G_2 *such that* $F\widehat{\mu}\beta ker(\{x_t\}) \leq G_1$, $F\widehat{\mu}\beta ker(\{x_t\}) \wedge G_2 = 0_X$ *and* $F\widehat{\mu}\beta ker(\{y_t\})$ G_2, $F\widehat{\mu}\beta ker(\{y_t\}) \wedge G_1 = 0_X$ *and* $G_1 \vee G_2 = 1_X$.

Proof Let a fuzzy topological space (X, τ) be $F\widehat{\mu}\beta R_1$ space. Then for each $x \neq y \in X$ with $F\widehat{\mu}\beta ker(\{x_t\}) \neq F\widehat{\mu}\beta ker(\{y_t\})$. Since every $F\widehat{\mu}\beta R_1$ space is a $F\widehat{\mu}\beta R_0$ space [By theorem (3.9)], and by theorem (3.6), $F\widehat{\mu}\beta cl(\{x_t\}) \neq F\widehat{\mu}\beta cl(\{y_t\})$, then there exist fuzzy $\widehat{\mu}\beta$ open sets U_1, U_2 such that $F\widehat{\mu}\beta cl(\{x_t\}) \leq U_1$ and $F\widehat{\mu}\beta cl(\{y_t\}) \leq U_2$ and $U_1 \wedge U_2 = 0_X$, then U_1^c and U_2^c are fuzzy $\widehat{\mu}\beta$ closed sets such that $U_1^c \vee U_2^c = 1_X$. Put $G_1 = U_1^c$ and $G_2 = U_2^c$. Thus $x_t \in U_1 \leq G_2$ and $y_t \in U_2 \leq G_1$ so that $F\widehat{\mu}\beta ker$ $(\{x_t\}) \leq U_1 \leq G_2$ and $F\widehat{\mu}\beta ker(\{y_t\}) \leq U_2 \leq G_1$. Conversely, let for each $x \neq y \in X$ with $F\widehat{\mu}\beta ker(\{x_t\}) \neq F\widehat{\mu}\beta ker(\{y_t\})$, there exist fuzzy $\widehat{\mu}\beta$ closed sets G_1, G_2 such that $F\widehat{\mu}\beta ker(\{x_t\}) \leq G_1$, $F\widehat{\mu}\beta ker(\{x_t\}) \wedge G_2 = 0_X$ and $F\widehat{\mu}\beta ker(\{y_t\}) \leq G_2$, $F\widehat{\mu}\beta ker(\{y_t\}) \wedge G_1 = 0_X$ and $G_1 \vee G_2 = 1_X$, then G_1^c and G_2^c are fuzzy $\widehat{\mu}\beta$ open sets such that $G_1^c \wedge G_2^c = 0_X$. Put $G_1^c = U_2$ and $G_2 = U_1$. Thus, $F\widehat{\mu}\beta ker(\{x_t\}) \leq U_1$ and $F\widehat{\mu}\beta ker(\{y_t\}) \leq U_2$ and $U_1 \wedge U_2 = 0_X$, so that $x_t \in U_1$ and $y_t \in U_2$ implies $x_t \notin \beta\text{-}cl(\{y_t\})$ and $y_t \notin F\widehat{\mu}\beta\text{-}cl(\{x_t\})$, then $F\widehat{\mu}\beta\text{-}cl(\{x_t\}) \leq U_1$ and $F\widehat{\mu}\beta\text{-}cl(\{y_t\}) \leq U_2$. Thus (X, τ) is a $F\widehat{\mu}\beta\text{-}R_1\text{-space}$.

Corollary 3.11 *A fuzzy topological space* (X, τ) *is* $F\widehat{\mu}\beta\text{-}R_1\text{-space}$ *if and only if for each* $x \neq y \in X$ *with* $F\widehat{\mu}\beta cl(\{x_t\}) \neq F\widehat{\mu}\beta cl(\{y_t\})$ *there exist disjoint fuzzy* $\widehat{\mu}\beta$ *open sets* U, V *such that* $F\widehat{\mu}\beta cl(F\widehat{\mu}\beta ker(\{x_t\})) \leq$ U *and* $F\widehat{\mu}\beta cl(F\widehat{\mu}\beta ker(\{y_t\})) \leq$ V.

Definition 3.12 Let (X, τ) be a fuzzy topological space. Then X is called:

(i) Fuzzy $\widehat{\mu}\beta$ regular space ($F\widehat{\mu}\beta r$-space), if for each fuzzy point x_t and each fuzzy $\widehat{\mu}\beta$-closed set F such that $x_t \in 1_x - $ F, there exist disjoint fuzzy $\widehat{\mu}\beta$-open sets U and V such that $x_t \in$ U and F $\leq$ V.
(ii) Fuzzy $\widehat{\mu}\beta$ normal space ($F\widehat{\mu}\beta n$-space) iff for each pair of disjoint fuzzy $\widehat{\mu}\beta$ closed sets A and B, there exist disjoint fuzzy $\widehat{\mu}\beta$-open sets U and V such that A $\leq$ U and B $\leq$ V.
(iii) Fuzzy $\widehat{\mu}\beta R_2$ space ($F\widehat{\mu}\beta R_2$ space) if it is property fuzzy $\widehat{\mu}\beta$ regular space.
(iv) Fuzzy $\widehat{\mu}\beta R_3$-space ($F\widehat{\mu}\beta R_3$ space) iff it is $F\widehat{\mu}\beta R_1$ space and $F\widehat{\mu}\beta n$ space.

Remark 3.13 Every $F\widehat{\mu}R_K$ space is a $F\widehat{\mu}\beta R_{K-1}$ space, k = 2, 3.

Theorem 3.14 *A fuzzy topological space* (X, τ) *is* $F\widehat{\mu}\beta r$-space ($F\widehat{\mu}\beta R_2$ space) *if and only if for each fuzzy* $\widehat{\mu}\beta$ *closed subset* G *of* X *and* $x_t \notin$ G *with* $F\widehat{\mu}\beta ker(G) \neq F\widehat{\mu}\beta ker$

$(\{x_t\})$ *then there exist fuzzy* $\widehat{\mu}\beta$ *closed sets* F_1, F_2 *such that* $F\widehat{\mu}\beta\mathrm{ker}(G) \leq F_1$, $F\widehat{\mu}\beta\mathrm{ker}$ $(G) \wedge F_2 = 0_X$ *and* $F\widehat{\mu}\beta\mathrm{ker}(\{x_t\}) \leq F_2$, $F\widehat{\mu}\beta\mathrm{ker}(\{x_t\}) \wedge F_1 = 0_X$ *and* $F_1 \vee F_2 = 1_X$.

Lemma 3.15 *Let* (X, τ) *be a* $F\widehat{\mu}\beta r$ *space and* F *be a fuzzy* $\widehat{\mu}\beta$ *closed set. Then* $F\widehat{\mu}\beta\mathrm{ker}$ (F) = F = $F\widehat{\mu}\beta\mathrm{cl}$(F).

Proof Let (X, τ) be a $F\widehat{\mu}\beta r$ space and F be a fuzzy $\widehat{\mu}\beta$ closed set. Then for each $x_t \not\in$ F, there exist disjoint fuzzy $\widehat{\mu}\beta$ open sets U, V such that $F \leq U$ and $x_t \in V$. Since $F\widehat{\mu}\beta\mathrm{ker}(F) \leq U$, implies $F\widehat{\mu}\beta\mathrm{ker}(F) \wedge V = 0_X$, thus $x_t \not\in F\widehat{\mu}\beta\mathrm{cl}(F\widehat{\mu}\beta\mathrm{ker}(F))$. We showing that if $x_t \not\in$ F implies $x_t \not\in F\widehat{\mu}\beta\mathrm{cl}(F\widehat{\mu}\beta\mathrm{ker}(F))$, therefore $F\widehat{\mu}\beta\mathrm{cl}(F\widehat{\mu}\beta\mathrm{ker}(F))$ F = $F\widehat{\mu}\beta\mathrm{cl}(F)$. As $F\widehat{\mu}\beta\mathrm{cl}(F) = F \leq F\widehat{\mu}\beta\mathrm{ker}(F)$ [By definition (3.1)]. Thus, $F\widehat{\mu}\beta\mathrm{ker}$ (F) = F = $F\widehat{\mu}\beta\mathrm{cl}$(F).

Theorem 3.16 *A fuzzy topological space* (X, τ) *is* $F\widehat{\mu}\beta r$ *space* ($F\widehat{\mu}\beta R_2$ *space) iff for each fuzzy* $\widehat{\mu}\beta$ *closed subset* F *of* X *and* $x_t \not\in$ F *with* $F\widehat{\mu}\beta\mathrm{cl}(F\widehat{\mu}\beta\mathrm{ker}(F)) \neq F\widehat{\mu}\beta\mathrm{cl}$ $(F\widehat{\mu}\beta\mathrm{ker}(\{x_t\}))$, *then there exist disjoint fuzzy* $\widehat{\mu}\beta$ *open sets* U, V *such that* $F\widehat{\mu}\beta\mathrm{cl}$ $(F\widehat{\mu}\beta\mathrm{ker}(F)) \leq U$ *and* $F\widehat{\mu}\beta\mathrm{cl}(F\widehat{\mu}\beta\mathrm{ker}(\{x_t\})) \leq V$.

Proof Let a fuzzy topological space (X, τ) be $F\widehat{\mu}\beta r$-space ($F\widehat{\mu}\beta R_2$ space) and let F be a fuzzy $\widehat{\mu}\beta$ closed set, $x_t \not\in$ F. Then there exist disjoint fuzzy $\widehat{\mu}\beta$ open set U, V such that $F \leq U$ and $x_t \in V$. By lemma (3.15), $F\widehat{\mu}\beta\mathrm{cl}(F\widehat{\mu}\beta\mathrm{ker}(F)) = F\widehat{\mu}\beta\mathrm{cl}(F) = F$, in the other hand (X, τ) is a $F\widehat{\mu}\beta R_0$ space [By theorem (3.9) and remark (3.13)]. Hence, by theorem (3.6), $F\widehat{\mu}\beta\mathrm{cl}(\{x_t\}) = F\widehat{\mu}\beta\mathrm{ker}(\{x_t\})$, for each $x \in X$. Thus, $F\widehat{\mu}\beta\mathrm{cl}(F\widehat{\mu}\beta\mathrm{ker}(F)) \leq U$ and $F\widehat{\mu}\beta\mathrm{cl}(F\widehat{\mu}\beta\mathrm{ker}(\{x_t\})) \leq V$. Conversely, let for each fuzzy $\widehat{\mu}\beta$ closed set F and $x_t \not\in$ F with $F\widehat{\mu}\beta\mathrm{cl}(F\widehat{\mu}\beta\mathrm{ker}(F)) \neq F\widehat{\mu}\beta\mathrm{cl}(F\widehat{\mu}\beta\mathrm{ker}(\{x_t\}))$, then there exist disjoint fuzzy $\widehat{\mu}\beta$ open sets U, V such that $F\widehat{\mu}\beta\mathrm{cl}(F\widehat{\mu}\beta\mathrm{ker}(F)) \leq U$ and $F\widehat{\mu}\beta\mathrm{cl}(F\widehat{\mu}\beta\mathrm{ker}(\{x_t\})) \leq V$. Then $F \leq U$ and $x_t \in V$. Thus, (X, τ) is $F\widehat{\mu}\beta r$ space ($F\widehat{\mu}\beta R_2$ space).

Theorem 3.17 *A fuzzy topological space* (X, τ) *is* $F\widehat{\mu}\beta n$ *space if and only if for each disjoint fuzzy* $\widehat{\mu}\beta$ *closed sets* G, H *with* $F\widehat{\mu}\beta\mathrm{ker}(G) \neq F\widehat{\mu}\beta\mathrm{ker}(H)$ *then there exist fuzzy* $\widehat{\mu}\beta$ *closed sets* F_1, F_2 *such that* $F\widehat{\mu}\beta\mathrm{ker}(G) \leq F_1$, $F\widehat{\mu}\beta\mathrm{ker}(G) \wedge F_2 = 0_X$ *and* $F\widehat{\mu}\beta\mathrm{ker}$ (H) $\leq F_2$, $F\widehat{\mu}\beta\mathrm{ker}(H) \wedge F_1 = 0_X$ *and* $F_1 \vee F_2 = 1_X$.

Proof Let a fuzzy topological space (X, τ) be $F\widehat{\mu}\beta n$ space and let for each disjoint fuzzy $\widehat{\mu}\beta$ closed sets G, H with $F\widehat{\mu}\beta\mathrm{ker}(G) \neq F\widehat{\mu}\beta\mathrm{ker}(H)$ then there exist disjoint fuzzy $\widehat{\mu}\beta$ open sets U, V such that $G \leq U$ and $H \leq V$ and $U \wedge V = 0_x$, then U^c and V^c are fuzzy $\widehat{\mu}\beta$ closed sets such that $U^c \vee V^c = 1_X$ and $F\widehat{\mu}\beta\mathrm{ker}(G) \wedge U^c = 0_X$, $F\widehat{\mu}\beta\mathrm{ker}(H) \wedge V^c = 0_X$. Put $U^c = F_2$ and $V^c = F_1$. Thus, $F\widehat{\mu}\beta\mathrm{ker}(G) \leq F_1$, $F\widehat{\mu}\beta\mathrm{ker}$ $(G) \wedge F_2 = 0_X$ and $F\widehat{\mu}\beta\mathrm{ker}(H) \leq F_2$, $F\widehat{\mu}\beta\mathrm{ker}(H) \wedge F_1 = 0_X$. Conversely, let for each disjoint fuzzy $\widehat{\mu}\beta$ closed sets G, H with $F\widehat{\mu}\beta\mathrm{ker}(G) \neq F\widehat{\mu}\beta\mathrm{ker}(H)$, there exist fuzzy $\widehat{\mu}\beta$ closed sets F_1, F_2 such that $F\widehat{\mu}\beta\mathrm{ker}(G) \leq F_1$, $F\widehat{\mu}\beta\mathrm{ker}(G) \wedge F_2 = 0_x$ and $F\widehat{\mu}\beta\mathrm{ker}$ (H) $\leq F_2$, $F\widehat{\mu}\beta\mathrm{ker}(H) \wedge F_1 = 0_X$ and $F_1 \vee F_2 = 1_X$ implies F_1^c and F_2^c are fuzzy $\widehat{\mu}\beta$ open sets such that $F_1^c \wedge F_2^c = 0_X$. Put $F_1^c = V$ and $F_2^c = U$, thus $F\widehat{\mu}\beta\mathrm{ker}(G) \leq U$ and $F\widehat{\mu}\beta\mathrm{ker}(H) \leq V$, so that $G \leq U$ and $H \leq V$. Thus (X, τ) is $F\widehat{\mu}\beta n$ space.

Theorem 3.18 *Every* $F\widehat{\mu}\beta R_3$ *space is* $F\widehat{\mu}\beta r$ *space*.

Proof Let F be a fuzzy $\widehat{\mu}\beta$ closed and $x_t \not\in$ F. Then $F\widehat{\mu}\beta\mathrm{ker}(\{x_t\}) \neq F\widehat{\mu}\beta\mathrm{ker}(F)$, then for each $y_t \in$ F there exist fuzzy $\widehat{\mu}\beta$ closed sets G_{y_t}, H_{y_t} such that $F\widehat{\mu}\beta\mathrm{ker}(\{y_t\}) \leq G_{y_t}$,

$F\widehat{\mu}\beta\mathrm{ker}(\{y_t\}) \wedge H_{y_t} = 0_X$ and $F\widehat{\mu}\beta\mathrm{ker}(\{x_t\}) \le H_{y_t}$, $F\widehat{\mu}\beta\mathrm{ker}(\{x_t\}) \wedge G_{y_t} = 0_X$. let $\beta = \wedge \{H_{y_t} : x_t \in H_{y_t}\}$, so we have $\beta \wedge F = 0_X$. Hence (X, τ) is $F\widehat{\mu}\beta n$ space, then there exist disjoint fuzzy $\widehat{\mu}\beta$ open sets U, V such that $F \le U$ and $x_t \in \beta \le V$. Thus, (X, τ) is $F\widehat{\mu}\beta r$ space.

4   Fuzzy $\widehat{\mu}\beta T_i$-Spaces, i = 0, 1, 2, 3, 4

Definition 4.1 Let (X, τ) be a fuzzy topological space. Then X is called:

(i)   Fuzzy $\widehat{\mu}\beta T_0$ space ($F\widehat{\mu}\beta T_0$ space) iff for each pair of distinct fuzzy points in X there exists a fuzzy $\widehat{\mu}\beta$ open set in X containing one and not the other.
(ii) Fuzzy $\widehat{\mu}\beta T_1$ space ($F\widehat{\mu}\beta T_1$ space) iff for each pair of distinct fuzzy points x_t and y_t of X, there exists fuzzy $\widehat{\mu}\beta$ open sets G, H containing x_t and y_t respectively such that $y_t \notin G$ and $x_t \notin H$.
(iii) Fuzzy $\widehat{\mu}\beta T_2$ space ($F\widehat{\mu}\beta T_2$ space) iff for each pair of distinct fuzzy points x_t and y_t of X, there exist disjoint fuzzy $\widehat{\mu}\beta$ open sets G, H in X such that $x_t \in G$ and $y_t \in H$.
(iv) Fuzzy $\widehat{\mu}\beta T_3$ space ($F\widehat{\mu}\beta T_3$ space) iff it is $F\widehat{\mu}\beta T_1$ space and $F\widehat{\mu}\beta r$ space.
(v)   Fuzzy $\widehat{\mu}\beta T_4$ space ($F\widehat{\mu}\beta T_4$ space) iff it is $F\widehat{\mu}\beta T_1$ space and $F\widehat{\mu}\beta n$ space.

Theorem 4.2 *A fuzzy topological space (X, τ) is $F\widehat{\mu}\beta T_0$ space if and only if either $y_t \notin F\widehat{\mu}\beta\mathrm{ker}(\{x_t\})$ or $x_t \notin F\widehat{\mu}\beta\mathrm{ker}(\{y_t\})$, for each $x \ne y \in X$.*

Proof Let (X, τ) be a $F\widehat{\mu}\beta T_0$ space then for each $x \ne y \in X$, there exists a fuzzy $\widehat{\mu}\beta$ open set G such that $x_t \in G$, $y_t \notin G$ or $x_t \notin G$, $y_t \in G$. Thus either $x_t \in G$, $y_t \notin G$ implies $y_t \notin F\widehat{\mu}\beta\mathrm{ker}(\{x_t\})$ or $x_t \notin G$, $y_t \in G$ implies $x_t \notin F\widehat{\mu}\beta\mathrm{ker}(\{y_t\})$. Conversely, let either $y_t \notin F\widehat{\mu}\beta\mathrm{ker}(\{x_t\})$ or $x_t \notin F\widehat{\mu}\beta\mathrm{ker}(\{y_t\})$, for each $x \ne y \in X$. Then there exists a fuzzy $\widehat{\mu}\beta$ open set G such that $x_t \in G$, $y_t \notin G$ or $x_t \notin G$, $y_t \in G$. Thus (X, τ) is a $F\widehat{\mu}\beta T_0$ space.

Theorem 4.3 *A fuzzy topological space (X, τ) is $F\widehat{\mu}\beta T_0$ space if and only if either $F\widehat{\mu}\beta\mathrm{ker}(\{x_t\})$ is weakly ultra fuzzy $\widehat{\mu}\beta$ separated from $\{y_t\}$ or $F\widehat{\mu}\beta\mathrm{ker}(\{y_t\})$ is weakly ultra fuzzy $\widehat{\mu}\beta$ separated from $\{x_t\}$ for each $x \ne y \in X$.*

Proof Let (X, τ) be a $F\widehat{\mu}\beta T_0$ space then for each $x \ne y \in X$, there exists a fuzzy $\widehat{\mu}\beta$ open set G such that $x_t \in G$, $y_t \notin G$ or $x_t \notin G$, $y_t \in G$. Now if $x_t \in G$, $y_t \notin G$ implies $F\widehat{\mu}\beta\mathrm{ker}(\{x_t\})$ is weakly ultra fuzzy $\widehat{\mu}\beta$ separated from $\{y_t\}$. Or if $x_t \notin G$, $y_t \in G$ implies $F\widehat{\mu}\beta\mathrm{ker}(\{y_t\})$ is weakly ultra fuzzy $\widehat{\mu}\beta$ separated from $\{x_t\}$. Conversely, let either $F\widehat{\mu}\beta\mathrm{ker}(\{x_t\})$ be weakly ultra fuzzy $\widehat{\mu}\beta$ separated from $\{y_t\}$ or $F\widehat{\mu}\beta\mathrm{ker}(\{y_t\})$ be weakly ultra fuzzy $\widehat{\mu}\beta$ separated from $\{x_t\}$. Then there exists a fuzzy $\widehat{\mu}\beta$ open set G such that $F\widehat{\mu}\beta\mathrm{ker}(\{x_t\}) \le G$ and $y_t \notin G$ or $F\widehat{\mu}\beta\mathrm{ker}(\{y_t\}) \le G$, $x_t \notin G$ implies $x_t \in G$, $y_t \notin G$ or $x_t \notin G$, $y_t \in G$. Thus, (X, τ) is a $F\widehat{\mu}\beta T_0$ space.

Theorem 4.4 *A fuzzy topological space (X, τ) is $F\widehat{\mu}\beta T_1$ space if and only if for each $x \ne y \in X$, $F\widehat{\mu}\beta\mathrm{ker}(\{x_t\})$ is weakly ultra fuzzy $\widehat{\mu}\beta$ separated from $\{y_t\}$ and $F\widehat{\mu}\beta\mathrm{ker}(\{y_t\})$ is weakly ultra fuzzy $\widehat{\mu}\beta$ separated from $\{x_t\}$.*

Proof Let (X, τ) be a $F\widehat{\mu}\beta T_1$ space then for each $x \neq y \in X$, there exist fuzzy $\widehat{\mu}\beta$ open sets U, V such that $x_t \in U$, $y_t \notin U$ and $x_t \notin V$, $y_t \in V$. Implies $F\widehat{\mu}\beta\ker(\{x_t\})$ is weakly ultra fuzzy $\widehat{\mu}\beta$ separated from $\{y_t\}$ and $F\widehat{\mu}\beta\ker(\{y_t\})$ is weakly ultra fuzzy $\widehat{\mu}\beta$ separated from $\{x_t\}$. Conversely, let $F\widehat{\mu}\beta\ker(\{x_t\})$ be weakly ultra fuzzy $\widehat{\mu}\beta$ separated from $\{y_t\}$ and $F\widehat{\mu}\beta\ker(\{y_t\})$ be weakly ultra fuzzy $\widehat{\mu}\beta$ separated from $\{x_t\}$. Then there exist fuzzy $\widehat{\mu}\beta$ open sets U, V such that $F\widehat{\mu}\beta\ker(\{x_t\}) \leq U$, $y_t \notin U$ and $F\widehat{\mu}\beta\ker(\{y_t\})$ V, $x_t \notin V$ implies $x_t \in U$, $y_t \notin U$ and $x_t \notin V$, $y_t \in V$. Thus, (X, τ) is a $F\widehat{\mu}\beta T_1$ space.

Theorem 4.5 *A fuzzy topological space* (X, τ) *is* $F\widehat{\mu}\beta T_1$ *space if and only if for each* $x \in X$, $F\widehat{\mu}\beta\ker(\{x_t\}) = \{x_t\}$.

Proof Let (X, τ) be a $F\widehat{\mu}\beta T_1$ space and let $F\widehat{\mu}\beta\ker(\{x_t\}) \neq \{x_t\}$. Then $F\widehat{\mu}\beta\ker(\{x_t\})$ contains anther fuzzy point distinct from x_t say y_t. So $y_t \in F\widehat{\mu}\beta\ker(\{x_t\})$ implies $F\widehat{\mu}\beta\ker(\{x_t\})$ is not weakly ultra fuzzy $\widehat{\mu}\beta$ separated from $\{y_t\}$. Hence by theorem (4.4), (X, τ) is not a $F\widehat{\mu}\beta T_1$ space this is contradiction. Thus $F\widehat{\mu}\beta\ker(\{x_t\}) = \{x_t\}$. Conversely, let $F\widehat{\mu}\beta\ker(\{x_t\}) = \{x_t\}$, for each $x \in X$ and let (X, τ) be not a $F\widehat{\mu}\beta T_1$ space. Then by theorem (4.4), $F\widehat{\mu}\beta\ker(\{x_t\})$ is not weakly ultra fuzzy $\widehat{\mu}\beta$ separated from $\{y_t\}$, this means that for every fuzzy $\widehat{\mu}\beta$ open set G contains $F\widehat{\mu}\beta\ker(\{x_t\})$ then $y_t \in G$ implies $y_t \in \wedge \{G \in F\widehat{\mu}\beta O(X): x_t \in G\}$ implies $y_t \in F\widehat{\mu}\beta\ker(\{x_t\})$, this is contradiction. Thus, (X, τ) is a $F\widehat{\mu}\beta T_1$ space.

Theorem 4.6 *A fuzzy topological space* (X, τ) *is* $F\widehat{\mu}\beta T_1$ *space if and only if for each* $x \neq y \in X$, $y_t \notin F\widehat{\mu}\beta\ker(\{x_t\})$ *and* $x_t \notin F\widehat{\mu}\beta\ker(\{y_t\})$.

Theorem 4.7 *A fuzzy topological space* (X, τ) *is* $F\widehat{\mu}\beta T_1$ *space if and only if for each* $x \neq y \in X$ *implies* $F\widehat{\mu}\beta\ker(\{x_t\}) \wedge F\widehat{\mu}\beta\ker(\{y_t\}) = 0_X$.

Theorem 4.8 *A fuzzy topological space* (X, τ) *is* $F\widehat{\mu}\beta T_1$ *space if and only if* (X, τ) *is* $F\widehat{\mu}\beta T_0$ *space and* $F\widehat{\mu}\beta R_0$ *space.*

Theorem 4.9 *A fuzzy topological space* (X, τ) *is* $F\widehat{\mu}\beta T_2$ *space if and only if*

(i) (X, τ) *is* $F\widehat{\mu}\beta T_0$ *space and* $F\widehat{\mu}\beta R_1$ *space.*
(ii) (X, τ) *is* $F\widehat{\mu}\beta T_1$ *space and* $F\widehat{\mu}\beta R_1$ *space.*

Corollary 4.10 *A fuzzy topological* $F\widehat{\mu}\beta T_0$ *space is* $F\widehat{\mu}\beta T_2$ *space if and only if for each* $x \neq y \in X$ *with* $F\widehat{\mu}\beta\ker(\{x_t\}) \neq F\widehat{\mu}\beta\ker(\{y_t\})$ *then there exist fuzzy* $\widehat{\mu}\beta$ *closed sets* G_1, G_2 *such that* $F\widehat{\mu}\beta\ker(\{x_t\}) \leq G_1$, $F\widehat{\mu}\beta\ker(\{x_t\}) \wedge G_2 = 0_X$ *and* $F\widehat{\mu}\beta\ker(\{y_t\})$ G_2, $F\widehat{\mu}\beta\ker(\{y_t\}) \wedge G_1 = 0_X$ *and* $G_1 \vee G_2 = 1_X$.

Corollary 4.11 *A fuzzy topological* $F\widehat{\mu}\beta T_1$ *space is* $F\widehat{\mu}\beta T_2$ *space if and only if one of the following conditions holds*:

(i)   *For each* $x \neq y \in X$ *with* $F\widehat{\mu}\beta cl(\{x_t\}) \neq F\widehat{\mu}\beta cl(\{y_t\})$, *then there exist fuzzy* $\widehat{\mu}\beta$ *open sets* U, V *such that* $F\widehat{\mu}\beta cl(F\widehat{\mu}\beta\ker(\{x_t\})) \leq U$ *and* $F\widehat{\mu}\beta cl(F\widehat{\mu}\beta\ker(\{y_t\})) \leq V$.

(ii) *For each* $x \neq y \in X$ *with* $F\widehat{\mu}\beta\ker(\{x_t\}) \neq F\widehat{\mu}\beta\ker(\{y_t\})$, *then there exist fuzzy* $\widehat{\mu}\beta$ *closed sets* G_1, G_2 *such that* $F\widehat{\mu}\beta\ker(\{x_t\}) \leq G_1$, $F\widehat{\mu}\beta\ker(\{x_t\}) \wedge G_2 = 0_X$ *and* $F\widehat{\mu}\beta\ker(\{y_t\}) \leq G_2$, $F\widehat{\mu}\beta\ker(\{y_t\}) \wedge G_1 = 0_X$ *and* $G_1 \vee G_2 = 1_X$.

Remark 4.12 Every $F\hat{\mu}\beta T_K$ space is a $F\hat{\mu}\beta T_{K-1}$ space, k = 1, 2, 3, 4.

Theorem 4.13 *A fuzzy topological $F\hat{\mu}\beta R_1$ space is $F\hat{\mu}\beta T_2$ space if and only if one of the following conditions holds*:

(i) *For each* x $\in$ X, $F\hat{\mu}\beta\mathrm{ker}(\{x_t\}) = \{x_t\}$.
(ii) *For each* x $\neq$ y $\in$ X, $F\hat{\mu}\beta\mathrm{ker}(\{x_t\}) \neq F\hat{\mu}\beta\mathrm{ker}(\{y_t\})$ *implies* $F\hat{\mu}\beta\mathrm{ker}(\{x_t\}) \wedge \beta\mathrm{ker}(\{y_t\}) = 0_X$.
(iii) *For each* x $\neq$ y $\in$ X, *either* $x_t \notin F\hat{\mu}\beta\mathrm{ker}(\{y_t\})$ *or* $y_t \notin F\hat{\mu}\beta\mathrm{ker}(\{x_t\})$.
(iv) *For each* x $\neq$ y $\in$ X, *then* $x_t \notin F\hat{\mu}\beta\mathrm{ker}(\{y_t\})$ *and* $y_t \notin F\hat{\mu}\beta\mathrm{ker}(\{x_t\})$.

Remark 4.14 Each fuzzy separation axiom is defined as the conjunction of two weaker axioms: $F\hat{\mu}\beta T_K$ space = $F\hat{\mu}\beta R_{K-1}$ space and $F\hat{\mu}\beta T_{K-1}$ space = $F\hat{\mu}\beta R_{K-1}$ space and $F\hat{\mu}\beta T_0$ space, k = 1, 2, 3, 4.

Theorem 4.15 *Let* (X, τ) *be a fuzzy topological space and* $F\hat{\mu}\beta\mathrm{ker}(\{x_t\}) = \{x_t\}$ *for each* x $\in$ X *then* (X, τ) *is* $F\hat{\mu}\beta T_3$ *space if and only if it is a* $F\hat{\mu}\beta R_2$ *space*.

Theorem 4.16 *Let* (X, τ) *be a fuzzy topological space and let* x $\neq$ y $\in$ X, *implies* $F\hat{\mu}\beta\mathrm{ker}(\{x_t\}) \wedge F\hat{\mu}\beta\mathrm{ker}(\{y_t\}) = 0_X$, *then* (X, τ) *is a* $F\hat{\mu}\beta T_3$ *space if and only if it is a* $F\hat{\mu}\beta R_2$ *space*.

Theorem 4.17 *Let* (X, τ) *be a fuzzy topological space and for each* x $\neq$ y $\in$ X *either* $x_t \notin F\hat{\mu}\beta\mathrm{ker}(\{y_t\})$ *or* $y_t \notin F\hat{\mu}\beta\mathrm{ker}(\{x_t\})$, *then* (X, τ) *is a* $F\hat{\mu}\beta T_3$ *space if and only if it is a* $F\hat{\mu}\beta R_2$ *space*.

Theorem 4.18 *Let* (X, τ) *be a fuzzy topological space and let* x $\neq$ y $\in$ X, *then* $x_t \notin F\hat{\mu}\beta\mathrm{ker}(\{y_t\})$ *and* $y_t \notin F\hat{\mu}\beta\mathrm{ker}(\{x_t\})$, (X, τ) *is a* $F\hat{\mu}\beta T_3$ *space if and only if it is a* $F\hat{\mu}\beta R_2$ *space*.

Remark 4.19 The relation between fuzzy $\hat{\mu}\beta$-separation axioms can be representing as a matrix. Therefore, the element refers to this relation. As the following matrix representation shows:

And	$F\hat{\mu}\beta T_0$	$F\hat{\mu}\beta T_1$	$F\hat{\mu}\beta T_2$	$F\hat{\mu}\beta T_3$	$F\hat{\mu}\beta T_4$	$F\hat{\mu}\beta R_0$	$F\hat{\mu}\beta R_1$	$F\hat{\mu}\beta R_2$	$F\hat{\mu}\beta R_3$
$F\hat{\mu}\beta T_0$	$F\hat{\mu}\beta T_0$	$F\hat{\mu}\beta T_1$	$F\hat{\mu}\beta T_2$	$F\hat{\mu}\beta T_3$	$F\hat{\mu}\beta T_4$	$F\hat{\mu}\beta T_1$	$F\hat{\mu}\beta T_2$	$F\hat{\mu}\beta R_3$	$F\hat{\mu}\beta T_4$
$F\hat{\mu}\beta T_1$	$F\hat{\mu}\beta T_1$	$F\hat{\mu}\beta T_1$	$F\hat{\mu}\beta T_2$	$F\hat{\mu}\beta T_3$	$F\hat{\mu}\beta T_4$	$F\hat{\mu}\beta T_1$	$F\hat{\mu}\beta T_2$	$F\hat{\mu}\beta R_3$	$F\hat{\mu}\beta T_4$
$F\hat{\mu}\beta T_2$	$F\hat{\mu}\beta T_2$	$F\hat{\mu}\beta T_2$	$F\hat{\mu}\beta T_2$	$F\hat{\mu}\beta T_3$	$F\hat{\mu}\beta T_4$	$F\hat{\mu}\beta T_2$	$F\hat{\mu}\beta T_2$	$F\hat{\mu}\beta R_3$	$F\hat{\mu}\beta T_4$
$F\hat{\mu}\beta T_3$	$F\hat{\mu}\beta T_3$	$F\hat{\mu}\beta T_3$	$F\hat{\mu}\beta T_3$	$F\hat{\mu}\beta T_3$	$F\hat{\mu}\beta T_4$	$F\hat{\mu}\beta T_3$	$F\hat{\mu}\beta T_3$	$F\hat{\mu}\beta T_3$	$F\hat{\mu}\beta T_4$
$F\hat{\mu}\beta T_4$	$F\hat{\mu}\beta T_4$	$F\hat{\mu}\beta T_4$	$F\hat{\mu}\beta T_4$	$F\hat{\mu}\beta T_4$	$F\hat{\mu}\beta T_4$	$F\hat{\mu}\beta T_4$	$F\hat{\mu}\beta T_4$	$F\hat{\mu}\beta T_4$	$F\hat{\mu}\beta T_4$
$F\hat{\mu}\beta R_0$	$F\hat{\mu}\beta T_1$	$F\hat{\mu}\beta T_1$	$F\hat{\mu}\beta T_2$	$F\hat{\mu}\beta T_3$	$F\hat{\mu}\beta T_4$	$F\hat{\mu}\beta R_0$	$F\hat{\mu}\beta R_1$	$F\hat{\mu}\beta R_2$	$F\hat{\mu}\beta R_3$
$F\hat{\mu}\beta R_1$	$F\hat{\mu}\beta T_2$	$F\hat{\mu}\beta T_2$	$F\hat{\mu}\beta T_2$	$F\hat{\mu}\beta T_3$	$F\hat{\mu}\beta T_4$	$F\hat{\mu}\beta R_1$	$F\hat{\mu}\beta R_1$	$F\hat{\mu}\beta R_2$	$F\hat{\mu}\beta R_3$
$F\hat{\mu}\beta R_2$	$F\hat{\mu}\beta T_3$	$F\hat{\mu}\beta T_3$	$F\hat{\mu}\beta T_3$	$F\hat{\mu}\beta T_3$	$F\hat{\mu}\beta T_4$	$F\hat{\mu}\beta R_2$	$F\hat{\mu}\beta R_2$	$F\hat{\mu}\beta R_2$	$F\hat{\mu}\beta R_3$
$F\hat{\mu}\beta R_3$	$F\hat{\mu}\beta T_4$	$F\hat{\mu}\beta T_4$	$F\hat{\mu}\beta T_4$	$F\hat{\mu}\beta T_4$	$F\hat{\mu}\beta T_4$	$F\hat{\mu}\beta R_3$	$F\hat{\mu}\beta R_3$	$F\hat{\mu}\beta R_3$	$F\hat{\mu}\beta R_3$

References

1. Wong, C.K.: Fuzzy points and local properties of fuzzy topology. J. Math. Anal. Appl. **46**, 316–328 (1974)
2. Chang, C.L.: Fuzzy topological spaces. J. Math. Anal. Appl. **24**, 182–190 (1968)
3. Goguen, J.A.: The fuzzy Tychonoff theorem. J. Math. Anal. Appl. **43**, 734–742 (1973)
4. Kubiak, T.: On fuzzy topologies, Ph.D. thesis. A. Mickiewicz, Poznan (1985)
5. Sostak, A.P.: On a fuzzy topological structure. Rendiconti del Circolo Matematico di Palermo. Series II, 11, 89–103 (1985)
6. Andrijevic, D.: Semi preopen sets. Mat. Vesnik. **38**, 24–32 (1986)
7. Subashini, J., Indirani, K.: On $\widehat{\mu}\beta$ set and continuity in Topological Spaces (Proceeding) (2012)
8. Zadeh, L.A.: Fuzzy sets. Info. Control **8**, 338–353 (1965)
9. Wali, R.S., Benchalli, S.S.: Some topics in general and fuzzy topological spaces, Ph.D. Thesis, Karnataka University Dharwd (2006)
10. Klir, G.J., Clair, U.S., Yuan, B.: Fuzzy set theory. Foundations and applications (1997)
11. Balasubramanian, G.: Fuzzy β open sets and fuzzy β separation axioms. Kybernetika **35**, 215–223 (1999)
12. Sarkar, M.: On fuzzy topological spaces. J. Math. Anal. Appl. **79**, 384–394 (1981)

Selection of Algorithms for Pedestrian Detection During Day and Night

Rahul Pathak[1,2] and P. Sivraj[2,3(✉)]

[1] Department of Electronics and Communication Engineering, Amrita School
of Engineering, Coimbatore, India
rahulpathak0212@gmail.com
[2] Amrita Vishwa Vidyapeetham, Coimbatore, India
p_sivraj@cb.amrita.edu
[3] Department of Electrical and Electronics Engineering, Amrita School
of Engineering, Coimbatore, India

Abstract. This paper presents an image processing based pedestrian detection system for day and night contributing to Advance Driver Assistance System (ADAS). The process, Histogram of Oriented Gradient (HOG) with Support Vector Machine (SVM) as linear classifier is compared and analyzed against Convolution Neural Network (CNN) for performance selection of best algorithm for pedestrian detection during both day and night. Performance analysis was done on standard datasets like INRIA, ETH, etc. and locally created datasets on Intel Processor with Ubuntu operating system. Implementation of HOG-SVM algorithm was performed, using DLib, python (2.7) and OpenCV (3.1.0) and accuracy of 96.25% for day and 96.55% for night was obtained. The implementation of Convolution Neural Network was performed using Anaconda3, TFLearn and python (3.6) and the scheme achieved an accuracy of 99.35% for day and 99.9% for night.

Keywords: ADAS · CNN · Histogram of oriented gradients · Pedestrian detection · Convolution neural networks

1 Introduction

The continuous development in science, technology and industrialization has led to the rapid growth of motorization, which in turn has increased the volume of vehicles on roads [1]. Skyrocketing growth in the usage of vehicles in India is illustrated in Fig. 1. Safety is a critical factor in the design of automobiles and to ensure the safety of both driver and pedestrian it is important to be complaint with the standard safety regulations. At present the trend is on upgrading ADAS which provides safety to the driver and pedestrian. Considering safety as a critical factor, ADAS is being developed for blind spot detection, lane departure warning, recognition of traffic lights, adaptive cruise control, driver weariness or intoxication and pedestrian detection [1, 2].

© Springer International Publishing AG 2018
D. J. Hemanth and S. Smys (eds.), *Computational Vision and Bio Inspired Computing*,
Lecture Notes in Computational Vision and Biomechanics 28,
https://doi.org/10.1007/978-3-319-71767-8_11

According to the statistics released by Ministry of Road Transport and Highways (Transport research wing 2015) fatalities of the pedestrian corresponds to 9.5% of the total accidents in India [3]. The main aim of pedestrian detection system is to minimize the number of fatalities in road accidents by informing the driver about the presence of the pedestrian in advance. As compared to day time, the majority of fatalities occur during the night due to inadequate illumination [4].

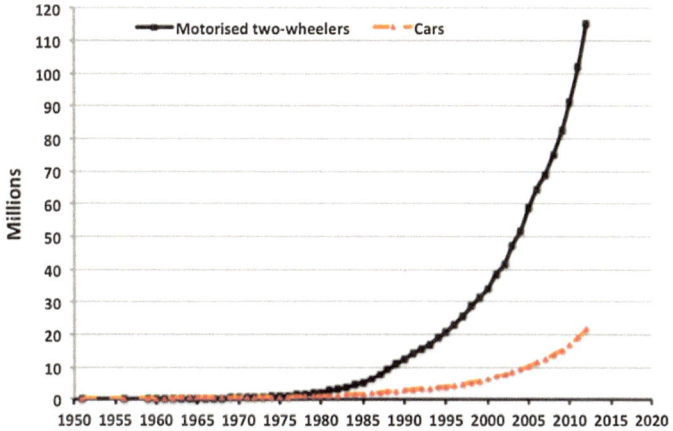

Fig. 1 Rapid increase in vehicles registered in India [3]

Pedestrian detection for ADAS identifies potentially unsafe situations involving pedestrians and inform the driver or actuate appropriate protective measures (e.g., automatic vehicle braking) on time [2]. Generally such system uses passive camera (i.e. those based on thermal vision) or active camera (i.e. equipped with illuminators and near infrared cameras) for night vision and visible camera for day vision [1].

The design of a robust pedestrian detection system must address the following challenges [5].

- The different posture of pedestrians.
- Pedestrian in different clothes carrying different objects.
- Variation in the background.
- Different appearances of the objects present on the roads like statues, parked vehicles.

This paper illustrates a novel approach for pedestrian detection during day and night conditions with the implementation of the technique using two different algorithms, analyses the performance and selects the best algorithm. The adopted methods for implementation are HOG and SVM as classifier and Convolution Neural Networks (CNN) [6]. The datasets selected for implementation consist of combination of standard and locally captured images. The implementation was done on Intel core processors and the software's used were python, OpenCV, Dlib and TFLearn.

A detailed literature survey of different schemes for pedestrian detection can be found in Sect. 2. Section 3 presents the methodology for implementation of the proposed algorithms. Finally, the conclusion is presented in Sect. 4.

2 Literature Review

2.1 Vision Oriented Approach for Pedestrian Detection

Vision based pedestrian detection is a preferred choice over LASER or RADAR as vision based can acquire more information about the environment [5]. The vision system uses a camera installed in the vehicle to capture the images of pedestrian. On the basis of electromagnetic spectrum cameras can be classified into different categories. The covering range of visible camera is 0.4–0.74 μm [5]. Near Infrared camera (NIR) has a range of about 0.75–1.4 μm. Visible cameras are used during day time whereas thermal and NIR cameras are used during the night time.

The first night vision system was launched in the year 2000 by Cadillac using FIR sensor. Honda was the first company to incorporate night vision system with pedestrian detection by using dual thermal imagers [7]. In 2005 Bosch developed the most successful night vision system and introduced it in Mercedes. Thermal cameras cannot be blinded from the headlamps of the incoming vehicles, which help drivers to see the incoming vehicles. NIR-Night Vision System (NVS) uses near infrared lamp for illumination and it consists of a camera based on CCD or CMOS technology [4].

Considering image acquisition methods into account, the night vision systems in automotive can be divided into passive or active systems. Active system emits infrared light which is near to the visible region of electromagnetic spectrum, also called as near infrared region. The advantage of active system is its high resolution which helps the driver to clearly see the objects appearing in front of the vehicle while driving. Camera used in the active system is of low cost and smaller in size. Detection range of this type of camera is shorter than the passive system ranging up to 150 m [1].

In passive system, a thermal camera is used for image acquisition. This system detects the electromagnetic radiation with wavelengths in the range of 3–30 μm (far infrared or FIR for short), but the cameras that are used for pedestrian detection uses a narrower range, i.e. 8–14 μm. Living beings with a body temperature of more than 0 K emits radiation. This system captures the infrared radiation emitted by the living objects. Detection range of passive system can reach up to 300 m for high quality camera which is higher than the active system [1, 4].

Both active and passive vision system are useful in automotive application especially in pedestrian detection where the detection range can go till 100 m [4].

A generic flowchart of pedestrian detection system is represented in Fig. 2.

The first stage is image acquisition and in further steps image pre-processing (noise removal) and object segmentation (finding the region of interest) is done. The next stage involves feature extraction which minimizes the amount of image data that describes the object. In the final stage the object is validated by classifier. The classifier categorizes the detected object pedestrian or non-pedestrian (Fig. 3) [1, 4].

2.2   Dataset Review

Over the years multiple pedestrian datasets are generated like Daimler, INRIA, ETH, TUD-Brussels, Caltech-USA [8]. The most popular datasets are Caltech Pedestrian and TUD—Brussels, which are more realistic and demanding.TUD Brussels has 1326 annotated pedestrian with a resolution of 640×480 pixels which were captured from a car [9]. Caltech—USA and KITTI are predominant datasets for pedestrian detection [8]. INRIA is the oldest and widely used dataset and it contains high quality pedestrian images taken at diverse environment [8].

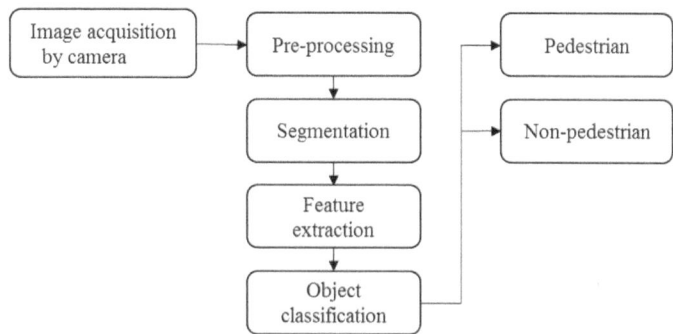

Fig. 2 Steps involved in pedestrian detection system

2.3   Features Extractors

Detection based on features: Histogram of Oriented Gradients (HOG): Histogram of Oriented Gradient (HOG) is well known feature extractor methodology which was first proposed by Dalal and Triggs [5, 9]. HOG has been an important descriptor in the field of object detection in which gradient orientation plays a significant role. It has been proposed that performance of pedestrian detection with HOG is better as compared to other features sets [10]. The scheme works gathering the gradient information in local cells and creates a histogram by using tri-linear interpolation. The overlapping blocks that contain neighbouring cells are normalized. Methods like interpolation, normalization, histogram binning makes the method powerful against the varying lightning conditions [9]. In the implementation, image is divided into small spatial regions called cells, for every cell histogram of gradient direction is computed. The HOG feature is capable of detecting the pedestrian for both day and night conditions [11]. Four types of HOG blocks techniques are Rectangular HOG (R-HOG), Circular HOG (C-HOG), Bar HOG and Center-surround HOG [5, 11, 12].

Texture based detection: Local Binary Pattern (LBP): LBP method is used for feature extraction in the training process of object detection and has shown high detection rates. Ojala proposed this method for texture classification [11]. With visible spectrum cameras and NIR cameras facial detection technique is implemented using LBP. The main advantage of LBP is that it is computationally less complex [5]. The LBP features for a particular region of interest are determined by giving one pixel at a time along with a structuring element. The LBP value for a structuring element calculated by thresholding operation of the outer pixels with the center pixel of the structuring element. Result of the 8-bit binary number, is converted to an integer. A histogram is generated after the image has been traversed and integers have been computed for all pixels [11].

Other feature extractors that are used in pedestrian detection system are Scale Invariant Feature Transform (SIFT), Haar-like features, shapelets, and Center symmetry—LBP (CS-LBP) [10, 13].

2.4   Classification

Image classification focuses on categorizing the detected images as positive samples or negative samples. In the context of this work if the pedestrian is detected then it is called as positive sample otherwise negative. Various classifiers used for pedestrian detection are described below.

Support Vector Machine (SVM): Support Vector machines (SVM) with linear kernels are popular pedestrian classification algorithm due to its performance and speed [9]. It is majorly used in the area of facial recognition and gesture recognition [11]. It functions by calculating optimal hyper plane separating the classes in higher dimensional space. Formation of a hyper plane divides two classes on the basis of "maximum margin". The two bounding planes are parallel to the classifier and they passes through adjacent input vectors. Input vectors nearest to the bounding planes are called as support vectors. This method is based on a supervised learning algorithm. SVM performance is better than other classifiers especially when compared to Adaboost [10]. Mathematically the Classifier's hyper plane equation is given by the equation

$$v^T r + k = 0. \tag{1}$$

where parameter vector is denoted by v, r represents input data vector and k denotes the bias term. Bounding planes are shown by Eq. (2).

$$v^T r + k = \pm 1. \tag{2}$$

Data which is not linearly separable can be classified by non-linear SVM. In this method the input feature are mapped to higher dimensional space using kernel methods.

Adaboost Classifier (Adaptive Boosting): In this classifier, a method is adopted to combine many weak classifiers into a strong classifier with the help of weighted sum whose weights are learned by the samples that are misclassified by the current classifier [14]. Adaboost can be combined with different detectors [10]. All the weak classifiers are combined to form a stump which is a single variable decision tree. During training process, stumps forms its classification decision from the input data and learns the weight for its count on the accuracy of the data. This methodology is suitable when the training data is in huge amount [14].

2.5 Deep Learning Based Feature Extraction and Classification

Deep Neural Networks (DNN): is now the state of art for pedestrian detection systems and they have exhibited improvement in accuracy and fast detection in real time applications [15]. In recent year's deep learning techniques have evolved as a powerful machine learning technique for pedestrian detection [6]. The main aspect which makes deep learning technique different from other technique is that, in deep learning the features are automatically extracted from the raw data. The branch of deep learning used for image specific features for pattern recognition is called Convolution Neural Networks (CNN). It consists of different kind of layers such as convolution layer, pooling layers, and fully connected layer with nonlinear activation functions. The data is given to the lower layers and it gradually transforms to the upper level layers. The lower layer extracts the fine features such as line, border and corners whereas the upper layer extracts the features of pedestrian parts [16]. In this technique features are extracted from the filters which are present in the hidden layers and their weight are optimized, so that the total classification error is minimized. The optimization technique used is Adam based. CNN gives high accuracy at low computational cost in many image recognition problems like face detection, face recognition, etc. [6, 16].

Among the above discussed feature extractor's and classifier's, we have selected HOG-SVM and Convolution Neural Network for implementation and have come up with a performance comparison of both algorithms for pedestrian detection method under different environmental conditions.

3 Methodology for Algorithm Implementation

3.1 Pedestrian Detection by HOG SVM Method

The first algorithm presented in this work is HOG-SVM. Methodology adopted for pedestrian detection in HOG-SVM is shown in Fig. 4.

Dataset creation: The implementation of this algorithm was carried out with indigenously created dataset during day, evening and night duration and also the algorithm was tested on standard dataset like ETH. The total dataset was divided into set of training images and testing images. Table 1 shows the number of images used for training and testing for HOG-SVM method.

Algorithm Implementation: Image acquisition is the primary step in image pro-

Fig. 3 Images of the pedestrian taken during day and night *Source* Images have been taken indigenously inside the college campus

Table 1 Number of images used in day and night

Test condition	Day	Night
Training images	392	391
Testing images	80	87

cessing algorithms. The acquired images consist of noise due to the imaging sensor. Noise removal from the image can be done by processing the captured image with suitable methods. The process of removing noise is called as denoising. In this work the noise removal is done using fastNIMeansdenoising method.

The implementation was done for both day and night condition. At first the captured images are resized to 255 × 255 pixels. Then the pedestrians are annotated by creating bounding boxes for training. This annotated pedestrian file is saved in xml format and given for training to SVM detector. The test images are given to the classifier which detects the pedestrian on the basis of trained data. The detected pedestrians are highlighted with bounding box in Figs. 5 and 7. The respective HOG descriptor is shown in Figs. 6 and 8.

Experimental Setup: The execution of the experiment was done on Intel i3 processor with 4 GB of RAM. The software's used are open CV 3.1.0 and python 2.7.12 and machine learning library Dlib.

Experimental Results: The experimental results performed using HOG-SVM for day and night are tabulated in Tables 2, 3 and 4.

Accuracy = (true positive + true negative/total images) [17]

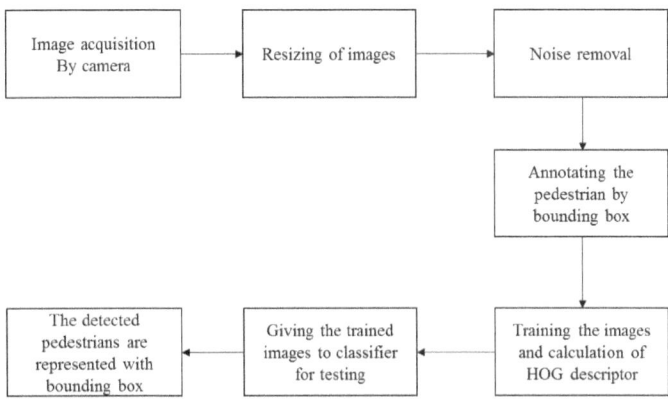

Fig. 4 Methodology adopted for pedestrian detection in HOG-SVM

Table 2 Confusion matrix for day

Actual\predict	Pedestrian	Non pedestrian
Pedestrian	57	3
Non Pedestrian	0	20

Table 3 Confusion matrix for night

Actual\predict	Pedestrian	Non pedestrian
Pedestrian	63	3
Non pedestrian	0	21

Table 4 Performance metrics

Parameters	Day	Night
Accuracy	96.25%	96.55%
Miss rate	5%	4.5%

Miss rate = (false negative/true positive + false negative) [12]
TP (True Positive) = Actual pedestrian; predicted as pedestrian
FN (False Negative) = Non-pedestrian; predicted as pedestrian
FP (False Positive) = Actual pedestrian; but predicted as non-pedestrian
TN (True Negative) = Non-pedestrian; predicted as non-pedestrian

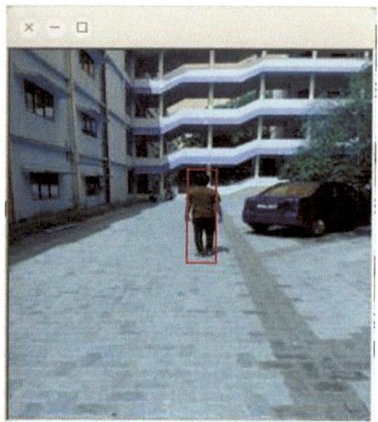

Fig. 5 Pedestrian detected in day

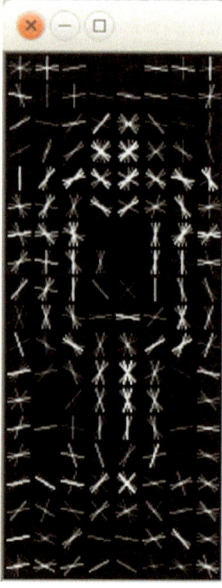

Fig. 6 HOG descriptor for day

Fig. 7 Pedestrian detected in night

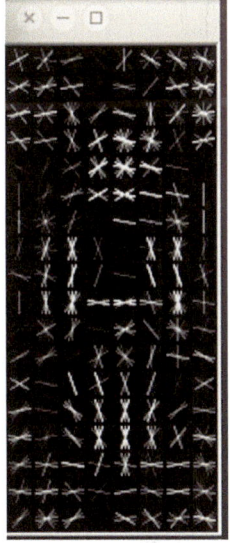

Fig. 8 HOG descriptor for night

3.2   Pedestrian Detection by Convolution Neural Networks

The second algorithm presented in this work is CNN.

CNN is a feed forward multilayer neural networking which the output and input layer consist of set of image arrays. In this method both positive and negative samples were taken from the locally created dataset and standard datasets for both day and night conditions. The captured images were resized to 32×32 pixels and then given to the network. The CNN architecture used in this work has two convolution layers, two pooling layers and one fully connected layer. The CNN was trained for RGB channels. The specifications of the parameters used in the implemented architecture are briefly explained in Table 5.

Dataset creation: In the implementation of this algorithm, the dataset was created indigenously by capturing the images in day, evening and night conditions and also images from standard dataset like INRIA and ETH was taken. Figure 9 shows the images from ETH database. Created dataset was subdivided into two classes, training and testing. Table 5 depicts the number of layers, filters. Activation function and the classifier used in the implementation. Table 6 shows the number of images used for training and testing in CNN method. Table 7 shows the results obtained from CNN.

Table 5 Description of the number of parameters used in CNN implementation

Name of layer	Parameter	Specification
First convolution layer	No. of filters	32
	Activation type	Relu
	Pooling layer (max)	2
	Drop out	0.5
Second convolution layer	No. of filters	64
	Activation type	Relu
	Pooling layer (max)	2
	Drop out	0.5
Fully connected	No. of filters	700
	Activation type	Relu

Fig. 9 Shows the samples image from ETH dataset [18]. *Source* Images have been taken from publicly available ETH dataset

Table 6 Description of the number of images used for pedestrian detection for CNN

Test condition	Day	Night
Total samples	11,959	14,297
Training samples	8370	10,000
Testing samples	3589	4297
Positive samples	5888	9297
Negative samples	6071	5000

Experimental Setup: The environment used for experimental setup consists of a PC with Intel i5 3.2 GHz, 8 GB of RAM. The software's used are Anaconda 3, TFLearn and IDE used was SPYDER. The programming language used was python version 3.6.

Table 7 Evaluation parameters and specification

Evaluation parameters	Day	Night
Accuracy	99.35%	99.9%
Miss rate	0.1%	0.04%
No. of epoch	50	60
Learning rate	.001	.001
Batch size	500	400

Table 8 Performance metrics comparing two algorithms

Parameters	HOG-SVM			CNN		
Conditions	Day (%)	Night (%)	Average	Day (%)	Night (%)	Average (%)
Accuracy	96.25	96.55	96.40	99.35	99.9	99.62
Miss rate	5	4.5	4.75	0.1	0.04	0.07

It is evident from the performance metrics tabulated in Table 8 that CNN out performances HOG-SVM in terms of accuracy and better prediction. The average accuracy variation between HOG-SVM and CNN is around 3.22 and the miss rate variation is 4.68 which are considerably high as the pedestrian detection application is quite critical in nature. There is not a big variation in accuracy but a considerable change is observed in miss rate performance of individual algorithms across day and night, but there is still scope of improvement, in CNN, during day time. Also as the number of images for training is increased the computational requirements are more for HOG-SVM. Compared to the weights adjustment process in CNN, the HOG-SVM method requires the annotation of the pedestrian with bounding box for training the system, which is a complex and time consuming process.

4   Conclusion

In this paper, two different methods for pedestrian detection are implemented and analyzed to select the best algorithm, in terms of accuracy, computational requirements, and miss rate, for pedestrian detection implementation during day and night. The algorithms were trained and tested on locally created datasets as well as standards datasets on an Intel processor and Ubuntu operating system. CNN gives better results than HOG-SVM in terms of accuracy, better prediction and training and run time computational requirements.

The CNN method can be validated using more images and the performance can be improved, especially for day. The selected algorithm can be ported to a microcontroller platform, like raspberry pi with a camera interfaced to capture images in real time, helping in real world implementation and validation.

References

1. Piniarski, K., Pawłowski, P., Adam, D.: Video processing algorithms for detection of pedestrians. Comput. Methods Sci. Technol. (CMST) **21**(3), 141–150 (2015)
2. Sandeep, A.K., Nithin, S., Ramachandran, K.I.: An image processing based pedestrian detection system for driver assistance. IJCTA 7369–7375 (2016)
3. Mohan, D., Tiwari, G., Bhalla, K.S.: Road safety in india status report, 2015 [online]. Available: http://tripp.iitd.ernet.in/road_safety_in_India_status_report.pdf
4. Luo, Y., Remillard, D., Hoetzer, D.: Pedestrian Detection in Near-Infrared Night Vision System. In: IEEE Intelligent Vehicles Symposium, San Diego, CA, USA (2010)
5. Ramzan, H., Fatima, B., Shahid, A.R., Ziauddin, S., Safi, A.A.: Intelligent pedestrian detection using optical flow. IJACSA **7**(9), 408–417
6. Ucar, A., Demir, Y., Guzelis, C.: Moving Towards in object recognition with deep learning for autonomous driving applications, 2016 International Symposium on Innovations in Intelligent Systems and Applications (INISTA), Sinaia, pp. 1–5 (2016)
7. Cho, H., Rybski P.E., Bar-Hillel, A., Zhang W.: Real-time pedestrian detection with deformable part models, 2012 IEEE Intelligent Vehicles Symposium, Alcala de Henares, 2012, pp. 1035–1042
8. Benenson, R., Omran, M., Hosang, J., Schiele, B.: Ten Years of Pedestrian Detection, What Have We Learned? Max Planck Institute for Informatics, Saarbrücken (2014)
9. Walk, S., Majer, N., Schindler, K., Schiele, B.: New Features and Insights for Pedestrian Detection. Saarbrucken (2010)
10. Vasuki, P., Veluchamy, S.: Pedestrian Detection for Driver Assistance Systems. In: Fifth International Conference on Recent Trends in Information Technology (2016)
11. Hurney, P., Waldron, P., Morgan, F., Jones, E., Glavin, M.: Night-time pedestrian classification with histograms of oriented gradients-local binary patterns vectors. Galway (2014)
12. Dalal, N., Triggs, B.: Histograms of Oriented Gradients for Human Detection. In: IEEE Computer Society Conference on Computer Vision and Pattern Recognition (CVPR'05), Montbonnot (2005)
13. Neethu, A., Athi Narayanan, S., Bijlani, K.: People count estimation using hybrid face detection method, IEEE conference of International Conference on Information Science (ICIS) (2016)

14. Gandhi, T., Trivedi, M.M.: Pedestrian Protection Systems: Issues, Survey, and Challenges. IEEE Trans. Intell. Transp. Syst. **8**(3), 413–430 (2007)
15. Jiang, X., Pang, Y., Li, X., Pan, J.: Speed updeep neural networkbased pedestrian detection by sharing featuresacrossmulti-scalemodels. Neurocomputing 163–170 (2015)
16. Szarvas, M., Yoshizawa, A., Yamamoto, M., Jun Ogata, J.: Pedestrian Detection with Convolution Neural Networks IEEE (2005)
17. Han, J., Kamber, M.: Book on Data Mining Concepts and Techniques. Second Edition
18. Ess, A., Leibe, B., Schindler, K., Gool, L.V.: A Mobile Vision System for Robust Multi— Person Tracking, IEEE Conference on Computer Vision and Pattern Recognition (2008)

Hybrid User Recommendation in Online Social Network

A. Christiyana Arulselvi[1](✉), S. SendhilKumar[2],
and G. S. Mahalakshmi[3]

[1] Anna University, Chennai, India
christina.jeyakumar@gmail.com
[2] Department of IST, Anna University, Chennai, India
[3] Department of CSE, Anna University, Chennai, India

Abstract. Social web and recommendation system has become an indispensable part of today's e-world. In such environment, most of the communications and transactions happen with unfamiliar persons. Any communication with unfamiliar persons is problematic as it lacks trust. Thus trust is an important factor to integrate recommendation system with the social web. In current scenario recommendation of trustable individuals with the similarity in behavior or trusted products is what is essential. The work reported in his paper aims towards recommending a trustable similar individual. The current work proposes user-user recommendation methods. Since the proposed approach is a hybrid approach it uses content-based technique and collaborative technique for the recommendation. Similar users are identified using content-based technique with Formal Concept Analysis (FCA) and Jaccard index and trustable users are identified using collaborative based technique. The trusted similar users are recommended using decision tree algorithm. The recommendation provided by this hybrid system is evaluated for its accuracy.

Keywords: Hybrid recommendation · Content-based technique
Collaborative technique · Trusted similar users · FCA · Jaccard index
Classification · Decision tree algorithm

1 Introduction

Most of the population today started using the social network to make friends or sharing information who are similar in terms of opinion or interest rather than persons with different interest. In such a vast environment there is a possibility that the users are genuine or fake. In this situation identifying genuine persons for accepting or making friends is the most critical issue. The online social network (OSN) is a medium that grows enormously day by day where there is a need for safer communication. Such safer communication can be achieved by identifying (a) users of similar interest and (b) trusted users and (c) recommending trusted users of higher similarity.

Recommender system provides an individualized recommendation to guide the user to identify the trusted users of similar interest by which the user can gain more trusted

© Springer International Publishing AG 2018
D. J. Hemanth and S. Smys (eds.), *Computational Vision and Bio Inspired Computing*,
Lecture Notes in Computational Vision and Biomechanics 28,
https://doi.org/10.1007/978-3-319-71767-8_12

information, knowledge, and suggestion and so on. "People are more willing to make friend with the one who has the same opinion or interest than the person who holds different attitude. At the same time, a user may more likely to receive information from those he likes" [1]. In general recommender system recommend individual or product but doesn't highlight intentional scammers. Thus there is a need for recommender system with an inbuilt component for trust.

Recommendation for users may require two components: 1. Similarity among the users 2. Trust on individual. Hence content-based filtering technique is used for identifying similar users based on their interest and collaborative filtering technique is used for identifying trusted users. Hence there is a need for hybrid recommendation system. The recommendation of trusted similar users is done through the decision pattern classification algorithm as a hybrid process.

Thus the proposed approach to the hybrid system is a parallel design technique of content-based filtering and collaborative filtering. The First step towards identifying trusted user is content-based filtering technique with an application of FCA and Jaccard index for computing similarity between users. The next step is the collaborative filtering technique where the trusted users over the period of time are identified through the stochastic process [2]. Finally trusted similar users are recommended using decision pattern classification technique.

2 Related Works

Recommendation system can be defined as a decision-making process when there is a possibility for choices. Decision making and its quality are an essential part of any circumstances, which can be achieved through recommendation system [3]. The information overloads in the web environment are sorted out by the recommender system [4] and the relevant information specific to the particular user are recommended according to the user's preference, interest or the observed behavior about the item [5].

Several approaches that have been proposed for recommendation system are collaborative filtering, content-based filtering and hybrid recommendation [6]. The purpose of collaborative filtering for the present work is to find out the similarity between users instead of finding similarity between items as in user-item based collaborative filtering [7]. Generally speaking, collaborative filtering methods consider only user rating (preference) for unknown item/user prediction. User's ratings are compared with those of other users who have the similar interest [8, 9] through which the trust network is generated. Trust network using collaborative filtering results in trust-enhanced recommendation [10, 11]. Such trust network can be deduced using explicitly or implicitly (latent) collected trust information from the user behaviors [11]. User behaviors can be represented via several types of patterns, such as frequency pattern, sequential patterns, neural network models and graph models [12], out of which the current research made use of frequency patterns. The latent features once identified can predict the rating of the user (trust value) by the stochastic algorithm [11].

Content-based filtering predicts recommendation from the information provided by the same user, not on the suggestion from the other user [13]. The recommendations are normally made based on the profile information of the user [14] which consists of the user preferences, needs, topics or the subjects the user is interested [15, 16]. The user profile is referred as a set of concepts associated with the user [17], in which the user preference or interests are represented as a concept vector for analyzing user interest and it can be represented via concept that reflects the content of the given user's interest in a hierarchical structure [18–20]. The hierarchical relationship that exists between concepts is described by super-concept and sub-concept. Every member of a sub-concept is a member of super-concept [19]. Jamil M. Saquer [21] explains the concept of lattice theory and application of FCA for clustering. Formal concepts that are generated out of voluminous datasets are so large and hence similarity or dissimilarity metrics are used for grouping such concepts. Jaccard index is one among the similarity metric which can be applied to the asymmetric binary variable to compute similarity between concepts [22].

A hybrid recommendation system is composed of two or more diverse recommendation techniques including collaborative filtering, content-based filtering, knowledge-based recommendation, demographic technique, etc. [17, 23–26]. Hybrid recommendation overcomes the limitations of individual recommendation system and improves the performance of the system [27]. Many hybrid techniques are based on traditional collaborative filtering merged with content-based filtering [28, 29]. Thus the proposed system makes use of hybrid recommendation technique to identify the trusted users of similar interest.

3   User–User Recommendation Scenario in OSN

The OSN is a virtual platform through which its users interact with one another on the basis of assumed trust and with those who have similar interest. The recommendation proposed in this paper consists of following three steps:

1. Identifying users of similar interest
2. Identifying trustable users in OSN
3. Recommending similar interested users.

3.1   Identifying Users of Similar Interest

Similar interested users in OSN can be identified through the interest specified by the users in their profile. If the two users are of similar interest, then there is a good chance that they both can become friends. Thus the friend suggestion can be generated among the similar interested users. But having similar interest alone can't be a criterion to accept the friend request because the requesting person should also be a trustable person.

3.2   Identifying Trustable Users in OSN

To compute the trustability of the user in OSN, a plausible scenario is depicted in Fig. 1 in which the reputation of the mutual friend is conferred upon a friend where the interaction information is the main consideration. Most of the Facebook users today are willing to accept a friend request from an unknown entity. The unknown entity may be a FOAF or stranger. People in the social network tend to trust FOAF rather than stranger [30] and also friends are the medium to propagate friendship in social networks. Though the OSN providers are showing so much of concerns on privacy and security issues for their user, there is no guarantee for the genuineness of the person hosting friend's request. In the case of FOAF's request, the suggestion for accepting or rejecting the request are given in terms of trustworthiness of the requester based on the reputation of the mutual friend who has enough interaction with the requester.

In OSN, people communicate via posting pictures, exchanging messages, giving responses as comments and likes and they do chatting in order to have private and personal conversation and to gain new friends. A naïve acceptance of friend request of someone who is the friend of a mutual friend won't be safe because the depth and the closeness between the mutual friend and the requester are unknown and inaccessible because the requester may have just recently become the friend of the mutual friend or may not have enough interaction. Paradoxically the requester may even be the stranger to the mutual friend. At times a faker may claim himself to be a friend of a mutual friend because of the ignorance or reluctance of the mutual friend while accepting the faker's friend request. The topology depicted in Fig. 1 will save us from the trap of any intentional scammers, who may try to join one's friendship community through the mutual friend.

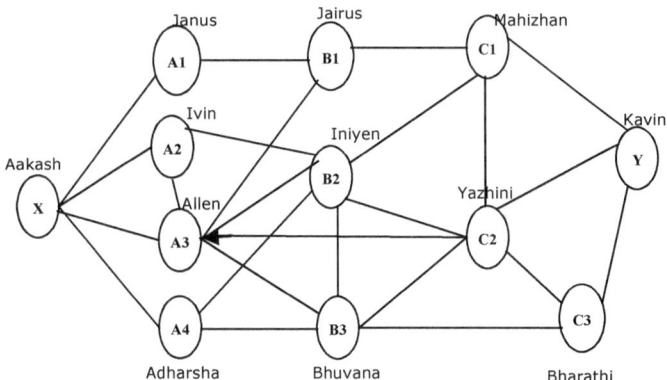

Fig. 1 Social network scenario

In the Fig. 1, B2 is a prime user with A2, A3, A4, B3, C1 and C2 are the members/friends of B2. The user C2 is generating friend request to the user A3, which is shown by the arrow line in Fig. 1. It is assumed that there is no direct contact between C2 and A3. So C2 could be a FOAF or a stranger. It is observed that C2 is having mutual friends say B2 and B3 with A3. So C2 is related to A3 as a FOAF category. Now A3 has to decide either to accept or reject C2's request. In this situation, C2 and A3 have B2 and B3 as a known_persons (mutual friend) to them. But just because of having mutual friend alone can't accept C2's request as in current social network's trend. So, A3 may look for suggestions (reputation) from mutual friends to take a decision. In order to take decision for A3, the following factors need to be considered and analyzed.

(i)   Is C2 recently became a friend of B3 or B2 who are mutual friends?
(ii) What is the closeness of C2 with B3 or B2 in terms of behavior?
(iii) Does B2 or B3 directly know C2?
(iv) Was C2 introduced as a stranger to B3 or B2?
(v)   Was C2 intentionally become a friend of B3 or B2 to reach A3?

All these queries will be answered if the interaction behavior of B2 or B3 with C2 is analyzed as in [2]. Thus the reputation of a mutual friend through the interaction on the requester (C2—A person who is generation/sending friend request) will help the requestee (A3—A person to whom the requester is sending friend request) to take a decision. An informed view of such possible interactions through social network will safeguard the requestee against the danger of unwanted friendships. This reputation value predicts the trust of the requester. This trust value is computed based on the number of likes, comments and chats the prime user had with his/her friend over a period of time. As these interaction details change over a period of time, the stochastic differential Equation which measures the randomness over the period is applied to compute the trust [2]. The trust value that is computed is used to rate the friends of the mutual friend (prime user). This process is called collaborative filtering method where the trusted users are filtered from the untrusted users.

3.3   Recommending Trusted Users of Similar Interest

The similar interested users and the trusted users identified from the above two sub-sections to find the trusted users of similar interest using decision tree classification algorithm as a hybrid recommendation process. The processes of hybrid recommendation to find the similar interested users are shown in Fig. 2.

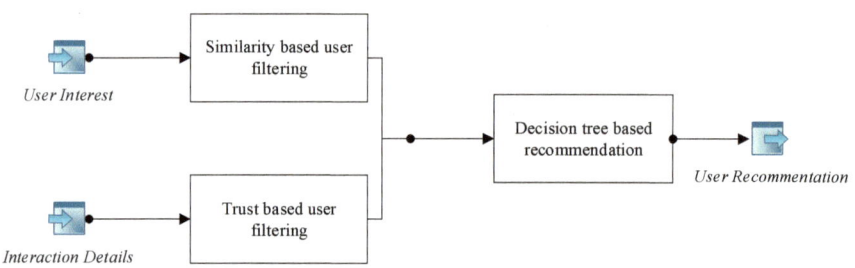

Fig. 2 Proposed hybrid recommendation scenario

4   Proposed Work

The proposed hybrid recommendation system aims to recommend the trusted similar friends to the friend of the mutual friend (prime user) based on the closeness of the mutual friend with every one of the member in the community along with the interest based similarity between friends using the hybrid recommendation technique as shown in the Fig. 3.

To recommend a user(s) to another user in a social network, the following steps are essential.

1. Finding the similar users based on the interest
2. Identifying the person either known or unknown
3. Finding the trustability of the friends w.r.t the prime user in the community
4. Identify the trusted similar users
5. Recommend the trusted similar users.

4.1 Recommendation Techniques

Hybrid recommendation system involves more than one recommendation techniques like collaborative filtering, content-based filtering, knowledge-based technique and demographic technique [31]. The main goal of the recommendation system is to enhance the performance of the recommending system. The use of complete and perfect recommendation technique is essential for any recommender system to provide a good and suitable recommendation to its users.

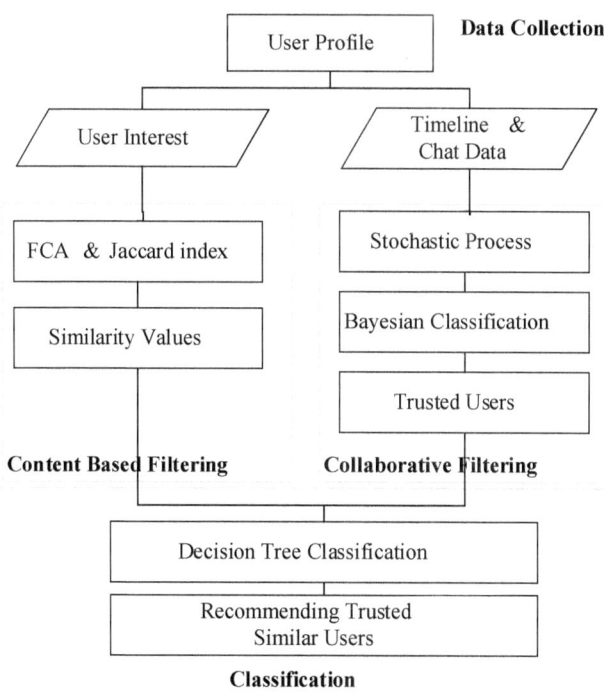

Fig. 3 Hybrid recommendation

Content-Based Recommendation. Whenever the decision has to be made based on the content analysis, content-based filtering technique is used such as recommending similar web pages publications, news, etc. In content-based filtering, the recommendation is made based on the user profiles using the features extracted from the content of the items the user has evaluated in the past [32, 33].

The proposed system aims to find the similarity between users in OSN based on the interest. FCA is one of the methods that computes the similarity among the users in OSN [34] because the user's interest is defined in terms of sets that are fully or partially similar. Hence the proposed system adopts FCA model to find the users of similar interest.

Apply FCA. The main purpose of FCA model is to use the formal context of objects and attributes to describe the formal concept of a domain. The FCA model explores the formal context between the interests of users, extract the formal concepts and construct the concept lattice.

Definition 1 A formal context of users and interest is a triplet $K = (O, A, R)$, where

$O = \{o_1, o_2, o_3, \ldots, o_n\}$ is a set of users
$A = \{a_1, a_2, a_3, \ldots, a_n\}$ is a set of interests
R is a binary relation between O and A where $R \subseteq O \times A$.

Table 1 shows a sample formal context matrix of the formal context (O, A, R).

Table 1 Context matrix of user-interest

Objects	Sports	Movie	Travel	Books
User 2 (U_2)		X		X
User 3 (U_3)			X	X
User 4 (U_4)				X
User 5 (U_5)	X			
User 6 (U_6)	X	X	X	
User 7 (U_7)		X	X	
User 8 (U_8)	X	X		

From the formal context, the formal concept and concept lattice are extracted using FCA. The formal concepts are the possible set of concepts that can be derived from the formal context and concept lattice is graph structure that defines the hierarchy between the concepts with subconcept-superconcept relations. Figure 4 shows the concept lattice for the context of Table 1.

Definition 2 A pair (U, I) is a formal concept of formal context (O, A, R) iff $U \subseteq O$, $I \subseteq A$ and $R \subseteq O \times A$. (o, a) $\in$ R is an object o has attribute a.

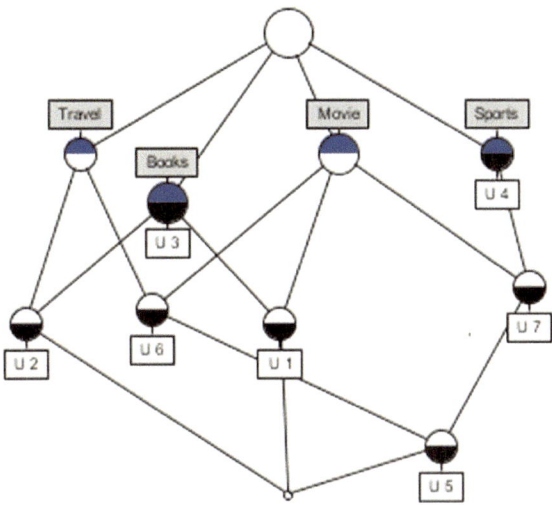

Fig. 4 Concept lattice using FCA for context in Table 2

In the formal concept, the set of users U is referred as an extent of the concept (U, I) and the set of interest I is referred as the intent of the concept (U, I). For the context (O, A, R) in Table 1, the formal concepts that are generated is shown in Table 2.

Table 2 Formal concepts of concept lattice

Concept No.	Concept set
C_0	$\langle\{U_5, U_6, U_8\}\{sports\}\rangle$
C_1	$\langle\{U_2, U_6, U_8\}\{movie\}\rangle$
C_2	$\langle\{U_3, U_6, U_7\}\{travel\}\rangle$
C_3	$\langle\{U_2, U_3, U_4\}\{book\}\rangle$
C_4	$\langle\{U_6, U_7\}\{sports, movie\}\rangle$
C_5	$\langle\{U_2\}\{movie, book\}\rangle$
C_6	$\langle\{U_3\}\{travel, book\}\rangle$
C_7	$\langle\{U_6\}\{sports, movie, travel\}\rangle$
C_8	$\langle\{U_6, U_7\}\{sports, travel\}\rangle$

The Table 2 shows that the concept sets generated are partially similar. Thus the similarity between concepts needs to be deduced in order to find the similarity between users. One of the famous similarity measures called Jaccard index is used to compute the similarity between concepts.

Similarity measure. The Similarity measure is a function that is used to quantify the similarity between two objects. Any similarity measure should satisfy the following properties.

Property 1   According to set union and intersection theory S(u, i) <= 1
Property 2   S(u, u) = 1
Property 3   S(u, i) = S(i, u).

The current work adopts the Jaccard index as a similarity measure to compute the similarity between two users based on the users' interest. As the proposed context is represented as an asymmetric binary variable, the Jaccard index is the best measure that suits for the computation of similarity between users in the concepts. The Jaccard similarity between the concepts C_1 and C_2 is defined as

$$J(C_1, C_2) = \frac{|C_1 \cap C_2|}{|C_1 \cup C_2|} \qquad (1)$$

Table 3 shows the similarity between concepts for the list of concepts in Table 2 using Eq. (1).

Example $J(C_0, C_4) = \frac{|C_0 \cap C_4|}{|C_0 \cup C_4|} = \frac{1}{2} = 0.5$

The property that is defined for the similarity measure is satisfied for the Jaccard similarity measure, which is proved from Table 3.

Property 1 S(u, i) $\leq$ 1—The similarity index is the Jaccard Similarity (C_i, C_{n-i}) in every row of column 2 is proved to be $\leq$, for every J(u, i) in column 2. So property 1 is proved.

Property 2 S(i, i) = 1—The row 1 of column 2 shows that Jaccard Similarity (C_0, C_0) = 1, Thus property 2 is proved.

Table 3 Similarity index between concepts using Jaccard index

S. No.	Concept combinations	Jaccard index	Similar users
1.	(C_0, C_0)	1	U_5, U_6, U_8
2.	(C_0, C_1)	0	–
3.	(C_0, C_2)	0	–
4.	(C_0, C_3)	0	–
5.	(C_0, C_4)	0.5	U_5, U_6, U_8
6.	(C_0, C_5)	0	–
7.	(C_0, C_6)	0	–
8.	(C_0, C_7)	0.33	U_5, U_6, U_8
9.	(C_1, C_4)	0.5	U_6, U_8
10.	(C_4, C_5)	0.33	U_2, U_6, U_8
11.	(C_4, C_7)	0.67	U_6, U_8
12.	(C_4, C_8)	0.67	U_6, U_7, U_8

Property 3 S(u, i) = S(i, u)—The Jaccard Similarity (C_i, C_{n-i}) in any row of column 2 remains the same as Jaccard Similarity (C_{n-i}, C_i). The row 10 shows that the Jaccard Similarity (C_4, C_5) = 0.33 = Jaccard Similarity (C_5, C_4). So property 3 is proved.

Thus that the similarity metrics is satisfied for the Jaccard index that has been used to compute the similarity between concepts of the proposed work.

Collaborative Filtering Technique. User-based collaborative filtering technique calculates the similarity between users by comparing their ratings on the same item/user [35] and the rating is considered as an important factor in collaborative filtering technique to form the neighborhood [31]. Christiyana et al. compute the trust rating of the prime_user (mutual friend) based on the interaction (implicit information) the prime user had with other users in the community [2]. The implicit features once identified can predict the rating of the user (trust value) by the stochastic algorithm [11] and also model based collaborative filtering approach uses probabilistic method to learn user preferences [31]. Thus the present work considers the work done by Christiyana et al., the reputation value of the prime user to compute the trust rating on other users in the community using the stochastic differential equation, a probabilistic method [2]. This trust value defines the closeness of the mutual friend with other friends. Trust improves recommendation, by combining similarity and trust between users.

Table 4 shows the sample set of trust computation w.r.t mutual friend (U1) (ego user) with every other users in the community as computed by Christiyana et al. [2] through Bayesian classification.

Table 4 User Trust Index

S. No.	Users	Trust index
1.	U_2	A
2.	U_3	R
3.	U_4	R
4.	U_5	A
5.	U_6	R
6.	U_7	R
7.	U_8	A
8.	U_9	R

The values 'A' and 'R' in the trust index of Table 4 represents the Acceptance and Rejection the requestee may do on the requester based on the reputation of the mutual friend on the requester. The reputation is computed over the period of time from interaction the mutual friend had with the requester using the stochastic Differential Equation [2].

Hybrid Recommendation. The proposed hybrid recommendation system uses parallel hybrid technique where two recommendation methods are performed parallelly and the outputs of both are combined to give a final recommendation. Content-based recommendation is done using FCA and Jaccard index to compute the similarity between the community members of the mutual friend. The similarity between individual users is deduced from the grouped users in each concept. Table 5 shows the similarity between users U_6 with every other user in the community using the Table 3.

Table 5 Similarity of User 6 with other users in the community

S. No.	Users	Similarity index
1	$S(U_6, U_2)$	0.33
2	$S(U_6, U_3)$	0
3	$S(U_6, U_4)$	0
4	$S(U_6, U_5)$	0.33
5	$S(U_6, U_7)$	0.67
6	$S(U_6, U_8)$	0.67

Using collaborative filtering technique as in Sect. 4.1 trustability of the user w.r.t prime user is computed for every member in his community as 'Accept' or 'reject'.

The outputs of content-based filtering and collaborative filtering techniques are combined as in Table 6 for further processing.

The value 'R' and 'A' in the trust index represent the acceptance and rejection as referenced by the mutual friend [2]. Table 6 shows that the user 6 shall reject user 7 and accept the user 8. Thus the user 8 is recommended as a trustable similar user. For the larger data set, the same process is implemented using classification process, where the

Table 6 Similarity of User 5 with other users with trust index

S. No.	Users	Similarity index	Trust index
1	$S(U_6, U_7)$	0.67	R
2	$S(U_6, U_8)$	0.67	A

classes are labeled manually based on the similarity and trust index.

Classification. A classification is a form of data analysis that is used to extract models describing important data classes [36]. The current work adopts decision tree classification technique to classify the trusted similar users in a social network community. Classification is of the two-step process:

(i) Learning phase
(ii) Testing phase.

The following subsections will explain the classification process in detail.

Learning Phase. Decision tree learning is the inductive learning process, which is the first step in the classification process where J48 the classification algorithm builds the classifier (a classification model) using the training set. The training set is the combination of database tuples with a predetermined set of classes. The sample training set D is shown in Table 7 where 'A' and 'R' in content-based technique is changed to 'T'-Trust and 'U'-Untrust respectively.

Each tuple in the database is a collection of user similarity index based on user interest and the trust index evaluated through the mutual friend for the particular user based on the mutual friends' interaction information. The class label 'rd—recommend' and 'nrd—not recommend' is assigned manually based on the higher value of similarity and trust index.

Table 7 Training set (D) for Decision tree classification

S. No:	Users	Similarity index	Trust index	Recommendation label
1	$S(U_6, U_2)$	0.33	T	nrd
2	$S(U_6, U_3)$	0	U	nrd
3	$S(U_6, U_4)$	0	U	nrd
4	$S(U_6, U_5)$	0.33	T	nrd
5	$S(U_6, U_7)$	0.67	U	nrd
6	$S(U_6, U_8)$	0.67	T	rd

```
Algorithm: Generate_Decision_Tree
Input: Database D, set of training tuples and its
        associated class labels;
        Attribute_list, set of attributes;
        Attribute_selection_method, a procedure that
        selects the sequence of attributes in order to
        partition the data tuples into individual classes.
Output: A decision tree
Main steps of J48 Algorithm:
 1. Select all the attributes
 2. Apply attribute_selection_method
    a. Compute information gain, an attribute selection
       measure
    b. Split_attribute    attribute with largest information
       gain          <--
    c. Attribute_list    attribute_list - split_attribute
    d. For each value of split_attribute branch the tree
       i.  If (split_attribute is discrete_valued) then
           1. Branch the tree for each value of split_attribute
       ii. Elseif (split_attribute is continuous_valued) then
           1. Find the split_point, a midpoint of two adjacent
              values of split_attribute
           2. Branch the tree as
              value(split_attribute)<=split_point and
              value(split_attribute) > split_point
 3. Repeat the attribute_selection_method till the
    split_attribute vanish or there is no more tuple in D
 4. The resulting decision tree is obtained
```

The J48, decision tree induction algorithm for the current work is explained below.

Implementation of Decision Tree Algorithm. The database D for the proposed work consists of 60 tuples with two split_attributes (similarity index and trust index) and one class_label attribute (recommendation label) similar to the samples in Table 7.

Application of attribute_selection_method. The expected information is an important measure that is needed to classify the tuple in D, which minimizes the information overheads needed to classify the tuples in the resulting partition and reflects the least impurity in that partition. The expected information can be computed using the formula in Eq. (2) [36]

$$info(D) = -\sum_{i=1}^{m} p_i \log_2(p_i) \tag{2}$$

where p_i is the tuple in D belongs to the class C_i, where i = 1,2, …,m. The attribute with the highest information minimizes the information needed to classify the tuples in the resulting partition and reflects the least impurity in that partition. The Info(D) is also known as Entropy (D).

Info (D) for the proposed work is computed as follows:

Number of tuples = 60; Tuples with (recommendation label = 'nrd') = 51; Tuples with (recommendation label = 'rd') = 9. According to the Eq. (2) expected information of D is:

$$info(D) = -\frac{51}{60}\log\left(\frac{51}{60}\right) - \frac{9}{60}\log\left(\frac{9}{60}\right) = 0.05999 + 0.12359 = 0.18358$$

The expected information for every attribute A, $Info_A(D)$ also computed in order to compute the information gain.

$$Info_A(D) = \sum_{j=1}^{v} \frac{|D_j|}{|D|} * info(D_j) \tag{3}$$

The attribute A, a split_attribute split D into v disjoined partitions $D_1, D_2, \ldots, D_v$ based on the v distinct values of A. D_j contains the tuples in D that have the unique value of A. If the split_attribute is discrete_valued, partition is done based on the distinct category of A and if the split_attribute is continuous valued, a midpoint is computed between every adjacent value a_i of split_attribute using the Eq. (4), which computes the midpoint between values a_i and a_{i+1}.

$$midpoint = \frac{a_i + a_{i+1}}{2} \tag{4}$$

Using the Eq. (4) (v − 1) possible midpoint are obtained. The split_attribute, 'similarity index' in the proposed work is a continuous_valued attribute having 7 distinct values as: 0, 0.2, 0.25, 0.33, 0.5, 0.67 and 1 that computes 6 midpoint as: 0.1, 0.225, 0.29, 0.44, 0.585 and 0.835 using the Eq. (4).

Example For the adjacent distinct values 0.33 and 0.5 the midpoint is computed using the Eq. (4) as shown below.

$$midpoint(0.33, 0.5) = \frac{0.33 + 0.5}{2} = 0.44$$

The threshold (split_point) for each midpoint is the lower value a_i of the each midpoint. Example, the threshold of the midpoint 0.44 is 0.33, which is the split_point. The expected information for the split_attribute, 'similarity index' is computed for each threshold. Then the best split_point is identified as the split_point that produces minimum expected information for the split_attribute. The best split_point that is obtained for the split_attribute, 'similarity index' is 0.33.

The expected information $Info_{similarity\ index}(D)$ is computed as below. According to the Eq. (4) the split_point that is obtained for the split_attribute, 'similarity index is 0.33. Thus the number of tuples (similarity index <= 0.33) = 47 and number of tuples (similarity index > 0.33) = 13.

The tuples ((similarity index <= 0.33) and (recommendation label = 'nrd') = 47) and tuples ((similarity index <= 0.33) and (recommendation label = 'rd') = 0). According to the Eq. (2), the expected information for the split_attribute 'similarity index' is:

$$info(similarity\ index \le 0.33) = -\frac{47}{47}\log\left(\frac{47}{47}\right) - 0 = 0,\ since\ \log(1) = 0$$

The ((similarity index > 0.33) and (recommendation label = 'nrd') = 4) and ((similarity index > 0.33) and (recommendation label = 'rd') = 7).

$$info(similarity\ index > 0.33) = -\frac{4}{13}\log\left(\frac{4}{13}\right) - \frac{7}{13}\log\left(\frac{7}{13}\right) = 0.30$$

The expected information for the split_atttribute 'similarity index' is computed using the Eq. (3)

$$Info_{similarity\ index}(D) = \frac{47}{60}*0 + \frac{13}{60}*0.30 = 0.065$$

The information gain is defined as the difference between the information requirement based on the proportion of the class (Info(D)) and information requirement after partitioning the attribute A. i.e.,

$$gain(A) = Info(D) - Info_A(D) \tag{5}$$

The information gain for the attribute 'similarity index' and 'trust index' is computed according to the Eq. (5) as:

$$gain(similarity\ index) = Info(D) - Info_{similarity\ index}(D) = 0.18358 - 0.065 = 0.11858$$

$$gain(trust\ index) = Info(D) - Info_{trust\ index}(D) = 0.18358 - 0.15436 = 0.02922$$

The split_attribute is the attribute with highest information gain. From the gain (similarity index) and gain (trust index), it is observed that the attribute 'similarity index' is holding higher gain value. So the attribute 'similarity index' is taken as a first split_attribute and is considered as a root node. The resultant decision tree that is obtained using WEKA tool is shown in Fig. 5.

The predictive accuracy of the classifier is estimated in order to use the model for new data sets. The classification accuracy obtained from the WEKA tool is given in Fig. 6.

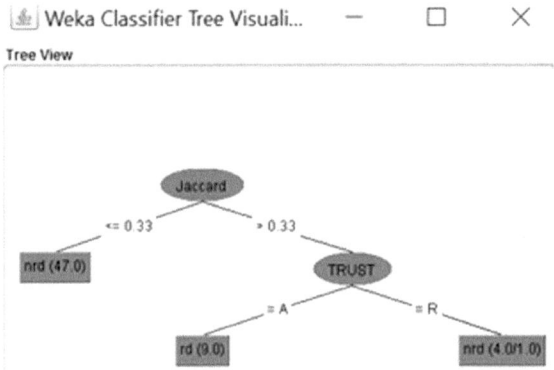

Fig. 5 Decision tree classifier

```
Correctly Classified Instances       57              95    %
Incorrectly Classified Instances      3               5    %
Kappa statistic                       0.8393
Mean absolute error                   0.0997
Root mean squared error               0.1986
Relative absolute error               34.9725 %
Root relative squared error           53.277  %
Total Number of Instances             60

=== Detailed Accuracy By Class ===

              TP Rate  FP Rate  Precision  Recall  F-Measure  MCC    ROC Area  PRC Area  Class
              1.000    0.060    0.769      1.000   0.870      0.850  0.970     0.769     rd
              0.940    0.000    1.000      0.940   0.969      0.850  0.970     0.990     nrd
Weighted Avg. 0.950    0.010    0.962      0.950   0.952      0.850  0.970     0.953
```

Fig. 6 Recommendation classifier accuracy

Testing Phase. The classifier obtained from the training phase is used for further classification. The Fig. 6 shows that the classifier is obtained 95% accuracy. Thus the classifier is considered as optimistic and hence the recommendation classifier can be used for further classification of tuples with unknown class labels. The proposed system uses 20 tuples to test the classifier and the accuracy of the testing process is given in Fig. 7, which shows 100% accuracy for the test set.

```
Correctly Classified Instances          20            100    %
Incorrectly Classified Instances         0              0    %
Kappa statistic                          1
Mean absolute error                      0.0375
Root mean squared error                  0.0968
Relative absolute error                 13.6765 %
Root relative squared error             27.0367 %
Total Number of Instances               20

=== Detailed Accuracy By Class ===

                 TP Rate  FP Rate  Precision  Recall  F-Measure  MCC    ROC Area  PRC Area  Class
                 1.000    0.000    1.000      1.000   1.000      1.000  1.000     1.000     rd
                 1.000    0.000    1.000      1.000   1.000      1.000  1.000     1.000     nrd
Weighted Avg.    1.000    0.000    1.000      1.000   1.000      1.000  1.000     1.000
```

Fig. 7 Recommendation classifier with testing set accuracy

Evaluating the Recommendation Classifier. Evaluating the accuracy of the classifier is an important part of the classification process. Thus the accuracy of the proposed recommendation classifier is evaluated using the K-fold cross validation, an accuracy evaluating measure. The accuracy of the classifier by 6-fold cross validation is obtained to be 99%. Thus the accuracy of the recommendation system is evaluated and is considered optimistic.

5   Conclusion

Recommender system has become the inevitable part of today's social network as it filters the relevant information from the large amount of data and predicts the users who are of great interest and in particular trustable to the recommender based on the profile and behavior (frequency of interaction, type of interaction, etc.) of the user in social network. The proposed system uses the conceptualization and machine learning technique in order to recommend the trusted users of similar interest. In this paper, necessary steps has been taken to identify the similar users of recommendee (person to whom the recommendation is done) through the content-based technique and trust value obtained from the interaction information of the mutual friend is taken as a reference in order to identify the trusted users through the collaborative filtering technique. Both these methods are combined to obtain the trusted similar users by the hybridization technique using the decision tree classification process and its accuracy is evaluated to be optimum. Thus the recommendation of trusted similar users will be of more help to the users in OSN.

References

1. Nie, Y., Jia, Y., Li, S., Zhu, X., Li, A., Zhou, A.: Identifying users across social networks based on dynamic core interests. Neurocomputing **210**, 107–115 (2016)
2. Christiyana Arulselvi, A., Sendhilkumar, S., Mahalakshmi, G.S.: Provenance based trust computation for recommendation in social network. In: Proceedings of the International Conference on Informatics and Analytics, ICIA-16, ACM Article No: 114
3. Pathak, B., Garfinkel, R., Gopal, R., Venkatesan, R., Yin, F.: Empirical analysis of the impact of recommender systems on sales. J. Manage Inf. Syst. **27**(2), 159–188 (2010)
4. Konstan, J.A., Riedl, J.: Recommender systems: from algorithms to user experience. User Model. User Adapt. Interact. **22**, 101–123 (2012)
5. Pan, C., Li, W.: Research paper recommendation with topic analysis. In: International Conference on Computer Design and Applications IEEE, vol. 4, pp. 264–268 (2010)
6. Jalali, M., Mustapha, N., Sulaiman, M., Mamay, A.: WEBPUM: a web-based recommendation system to predict user future movement. Expert Syst. Appl. **37**(9), 6201–6212 (2010)
7. Yao, G., Cai, L.: User-Based and Item-Based Collaborative Filtering Recommendation Algorithms Design
8. Cho, Y.H., Kim, J.K.: Application of web usage mining and product taxonomy to collaborative recommendations in e-commerce. Expert Syst. Appl. **26**(2), 233–246 (2004)
9. Herlocker, J., Konstan, J.A., Riedl, J.: An empirical analysis of design choices in neighborhood-based collaborative filtering algorithms. Inf. Retrieval **5**(4), 287–310 (2002)
10. Shambour, Q., Lu, J.: A trust-semantic fusion-based recommendation approach for e-business applications. Decis. Support Syst. **54**(1), 768–780 (2012)
11. Lee, W.-P., Ma, C.-Y., Enhancing Collaborative Recommendation Performance by Combining User Preference and Trust-Distrust Propagation in Social Network. Knowl. Based Syst. **106**, 125–134 (2016)
12. Park, D.H., Kim, H.K., Choi, I.Y., Kim, J.K.: A literature review and classification of recommender systems research. Expert Syst. Appl. **39**(11), 10059–10072 (2012)
13. Min, S.H., Han, I.: Detection of the customer time-variant pattern for improving recommender system. Exp. Syst. Appl. **37**(4), 2911–2922 (2010)
14. Bobadilla, J., Ortega, F., Hernando, A., Gutiérrez, A.: Recommender systems survey. Knowl. Based Syst. **46**, 109–132 (2013)
15. Gauch, S., Speretta, M., Chandramouli, A., Micarelli, A.: User Profiles for Personalized Information Access. In: The Adaptive Web, pp. 54–89. Springer, Berlin (2007)
16. Lops, P., de Gemmis, M., Semeraro, G.: Content-Based Recommender Systems: State of the Art and Trends. In: Recommender Systems Handbook, pp. 73–105. Springer, Berlin (2011)
17. Jannach, D., Zanker, M., Felfernig, A., Friedrich, G.: Recommender Systems: An Introduction. Cambridge University Press, Cambridge (2010)
18. Nanas, N., Uren, V., de Roeck, A.: Building and applying a concept hierarchy representation of a user profile. In: Proceedings of the 26th Annual International ACM SIGIR Conference on Research and Development in Information Retrieval, ACM, pp. 198–204 (2003)
19. Singh, S., Shepherd, M., Duffy, J., Watters, C., et al.: An adaptive user profile for filtering news based on a user interest hierarchy. Proc. Am. Soc. Inf. Sci. Technol. **43**(1), 1–21 (2006)
20. Yu, J., Liu, F., Zhao, H.: Building User Profile Based on Concept and Relation for Web Personalized Services. In: International Conference on Innovation and Information Management. Citeseer (2012)
21. Jamil, M.: Saquer, Formal Concept Analysis Based Clustering, Southwest Missouri State University, USA. (Source Title: Encyclopedia of Data Warehousing and Mining, 2nd Edn. Copyright: ©2009, p. 6. https://doi.org/10.4018/978-1-60566-010-3.ch138

22. Radeleczki, S.: Classification systems and their lattice. Discussiones Math. Gen. Algebra Appl. **22**(2), 67–181 (2002)
23. Melvilleand, P., Sindhwani, V.: Recommender Systems. In: Encyclopedia of Machine Learning, pp. 829–838. Springer, Berlin (2010)
24. Pathak, D., Matharia, S., Murthy, C.: Nova: Hybrid book recommendation engine. In: IEEE 3rd International Advance Computing Conference (IACC), pp. 977–982 (2013)
25. Adomavicius, G., Tuzhilin, A.: Toward the next generation of recommender system. A survey of the state-of-the-art and possible extensions. IEEE Trans. Knowl. Data Eng. **17** (6), 734–749 (2005)
26. Adomavicius, G, Zhang, J.: Impact of data characteristics on recommender systems performance. ACM Trans. Manage Inform. Syst. **3**(1) (2012)
27. Murat, G., Sule, G.O.: Combination of web page recommender systems. Expert Syst. Appl. **37**(4), 2911–2922 (2010)
28. Amini, B., Ibrahim, R., Othman, M.S.: Discovering the impact of knowledge in recommender systems: a comparative study. arXiv preprint arXiv:1109.0166 (2011)
29. Bhowmick, P.K., Sarkar, S., Basu, A.: Ontology based user modeling for personalized information access. IJCSA **7**(1), 1–22 (2010)
30. Anuradha, Y., Chakraverty, S., Sibal, R.: A survey of implicit trust on social networks. In: International Conference on IEEE Green Computing and Internet of Things (ICGCIoT), pp. 1511–1515 (2015)
31. Nadee, W.: Modelling user profiles for recommender systems. Ph.D. Thesis, Queensland University of Technology (2016)
32. Burke, R.: Hybrid recommender systems: survey and experiments. User Model. User Adap. Interact. **12**(4), 331–370 (2002)
33. Bobadilla, J., Ortega, F., Hernando, A., Gutiérrez, A.: Recommender systems survey. Knowl. Based Syst. **46**, 109–132 (2013)
34. Alqadah, F., Bhatnagar, R.: Similarity measures in formal concept analysis. Ann. Math. Artif. Intell. **61**, 245 (2011). https://doi.org/10.1007/s10472-011-9257-7. (Springer conference)
35. Isinkaye, F.O., Folajimi, Y.O., Ojokoh, B.A.: Recommendation systems: principles, methods and evaluation. Egypt. Inform. J. **16**(3), 261–273 (2015)
36. Han, J. Kamber, M.: Data Mining Concepts and Technique, 2nd edn. Elsevier, USA (2006)

Geometrical Method for Shadow Detection of Static Images

Manoj K. Sabnis[1(✉)], Kavita[1], and Manoj Kumar Shukla[2]

[1] CS&IT Department, JVWU, Jharna, Jaipur, Rajasthan, India
manojsab67@yahoo.co.in, drkavita@jvwu.ac.in
[2] Amity School of Engineering, Amity University, Noida, India
mksukla001@gmail.com

Abstract. The current applications consist of a number of services, of which there are these basic services to which additional services are added, called as the add on services, as per the requirement of the basic services. The application domain of tracking has object identification as one of its basic services. QoS conditions then specified, identify the objects of interest and a number of false identifications. Thus the object identification with false identification may further lead to false tracking. This problem basically arises due to other entities having the shape similar to that of the object of interest. One such entity identified is the shadows of the object of interest. In this paper, removal of these shadows forms the objective which is presented in detail utilizing geometrical method.

Keywords: Tracking · Shadow detection · Geometry · Shadow regions
Angle of orientation

1 Introduction

Having decided the broad domain i.e. shadows, the next stage was to examine their types, methods, taxonomy and algorithms. From all these, the working domain identified was the geometrical based method, which is further presented in this paper as a proposed method over the existing one. For justification, this method is represented through an algorithm which is then implemented for evaluation of its obtained results.

2 Shadow Types

The images, restricting them for static images, the shadows are classified into the cast shadows, self-shadows and attached shadows. The shadow is the dark region which is formed on the background where the light is occluded by the object [1–3]. The thick and uniformly dark region of this shadow is called as the umbra region [3, 4]. From the boundary of umbra region till the outer edges, the intensity of the shadow gradually decreases, this region is called as the penumbra region [3, 4].

© Springer International Publishing AG 2018
D. J. Hemanth and S. Smys (eds.), *Computational Vision and Bio Inspired Computing*,
Lecture Notes in Computational Vision and Biomechanics 28,
https://doi.org/10.1007/978-3-319-71767-8_13

The umbra and the penumbra regions together forms the shadows region [3, 4]. This shadow has different types as represented below as.

The cast shadow is the hard and dark shadow of the object of interest which is casted on the background. This shadow has its shape similar to the object and if the object is less illuminated then the dark colour of the shadow closely maps to the object colour [5, 6].

The self-shadow is the dark area created on the object itself due to some parts of the object being occlude by the light [7, 8].

The attached shadows can be of two types, the shadow of the object of interest attached to the shadow of the nearby object or the shadow of object of interest attached to another nearby object itself [9, 10].

Examining these three conditions, Cast shadow is identified to be the best candidate selected for its detection and elimination. The set of cast shadows have a number of parameters like, illumination types, their number and direction, reflection, direction, colour and its types, background, orientation of the shadows on the background, the point of attachment of the shadow to the background, texture and intensity of background and shadows etc. [8–10, 11].

From such types of varying parameters, only a selected few like intensity, colour geometry and texture are selected in order to form standard methods for the algorithm development [8, 9, 14, 15]. This selection is standardized by defining the taxonomy of shadow detection methods [12–15].

Fig. 1 Cast shadow

Fig. 2 Self shadow

Fig. 3 Attached shadow

3   Taxonomy

The shadow suppression approach can be statistical or deterministic [14–16]. Images in statistical class have different parameter. Of these parameters, it is possible to apply probabilistic function to classify the parameters into different classes. Two sets of these classes are proposed, first one which depends only on the colour of the image as a whole i.e. SNP [16, 17]. In the second case, parameters prominently present are detected, first region wise and then pixel wise, i.e. SP [16, 17].

In case where decisions can be taken at ON/OFF level, deterministic approach can be used wherein in which case, depending on the illumination conditions either model development is DM based mechanism or else if features like colour, Texture, histogram gradient are possible to be extracted from the forming image the DNM i.e. nonmodel based approach is developed [14, 16, 17].

4   Geometry Based Method

The geometry based method is considered for shadow detection and its elimination. First the existing approach is examined image wise, algorithm wise, result wise and then evaluation wise. On the basis of the short comings so observed, improvements are suggested which are presented in the proposed approach as the improved algorithm for better output conditions. This is demonstrated by comparing the various evaluation metrics.

4.1   Existing Approach

In image wise approach, the image selection considered should be such that the object shadow pair should be able to form a single BLOB [14, 16]. The shadow should be completely available at the background, attached to the object only at a single point and preferably at the bottom. With all these conditions Fig. 7 named as *Image 4* and Fig. 8 name as *Image 5* forms the most suitable candidate.

Fig. 4 Image 1

Fig. 5 Image 2

Fig. 6 Image 3

Fig. 7 Image 4

Fig. 8 Image 5

As for other images, Image 4.1 named as *Image 1* though have the object shadow pair joint, the shadow is not completely on the background. Figure 5 named as *Image 2* has shadow object pair disconnected similarly, some shadows in Fig. 6 named as *Image 3* do not have their objects to form the object shadow pair.

In case of single shadow images, first BLOB is formed for an object shadow pair. In case of multiple shadow images, first regions are formed and then within the regions BLOBs are identified. Thus, two separate algorithms are required for the above mentioned cases [14, 16]. However, the proposed algorithm utilizes, a single algorithm for both single and multiple objects representation. Other basic conditions related to illumination, camera conditions, image shadow connection etc. are reflected in the proposed algorithm which are already considered in the existing one [13].

The resolution of the results depends upon the display method selected. In the proposed algorithm however, advanced display mode 1 is selected to get better result representation.

The results so obtained have to be justified, for this evaluation techniques are used namely, qualitative and quantitative. The quantitative evaluation is done at the accuracy level and the qualitative one is done at the metric level. In case of the proposed algorithm the quantitative along with accuracies are represented one level down in detail in terms of successful and unsuccessful shadow pixel determination. This is further represented in a tabular and graphical form. Similarly in case of qualitative, the metrics are further drilled down to absolute and relative measures which are further quantified into the graphical representation.

4.2 Proposed Approach

The reference algorithm considered here is with respect to dynamic images. Therefore, as per the requirements of the dynamic images, the algorithm is implemented in four stages viz background subtraction, histogram projection, object moment, orientation stage and finally the Gaussian model representation [14, 16–18].

The reference algorithm being only for static images works only two of the existing stages i.e. histogram projection and object moment and orientation stage. This is further added with operations like global and local thresholding, region labeling and forming, morphological operations, indices mapping and masking along with advance display methods.

All this is implemented in the proper sequence which is further represented in the next stage i.e. algorithm representation.

5 Algorithm Representation

The algorithm is implemented in the form of two module. The first is the detection model in which shadow is detected and the second is the elimination model in which the detected shadow is eliminated.

5.1 Module I: Shadow Detection

This stage of implementation is further divided into a number of working program modules namely.

Image Selection: The image so selected should not have attached shadows to insure that BLOBs of object shadow pairs can be formed. These BLOBS should have a orientation of their regions such that a single angle of orientation can be obtained with the single point of content of object and shadow pair.

Binary Image: For the image, calculate the global threshold value and mark the image pixels above it as one and below it as zero. The marked one are then converted to logic one and others to logic zero. Inversion operation is then performed on this binary image to get its inverted image which now gives the shadows and dark object as logic one and objects and background as logic zero.

Morphological Operations: Opening operation is performed to remove small white areas. Filling operation is further done to fill the remaining holes. This is similar to fine noise removal that merges the pixels with background [18].

BLOB Formation: Region labelling is performed so as to get the largest BLOB which is the required object shadow pair in case of single shadow or multiple object shadow pairs in case of multiple shadow [19].

BLOB Marking: The image can have a single object with single shadow that forms a BLOB or it can have multiple objects and shadows forming a region which will then be divided into BLOBs

Now first histogram projection is done, which, if gives a single vertical peak, represents a single object. If the projection gives more than one vertical peak means there is more than one object, thus forming a region. Let this vertical point for a single image be **P1** [14, 16].

Now the farthest horizontal point is found along the ends in both left and right direction and marked as **P2**. This is the peak point of the shadow. Region **R** is the region of the BLOB where **R1** is the object region and **R2** is the shadow region [16]. This is as shown below in the Fig. 10 named as *Histogram*.

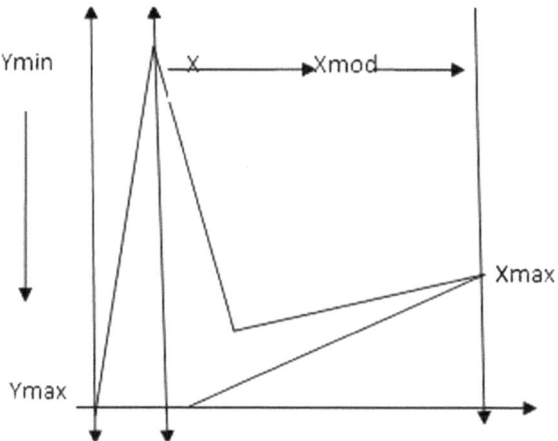

Fig. 9 Max, min

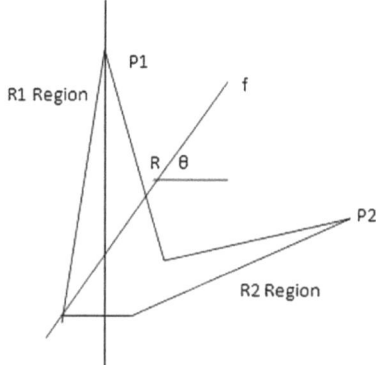

Fig. 10 Histogram

Indices: The indices of the image are obtained in the form of [I, J]. This helps to mark the maximum and the minimum points as shown in Fig. 9 named as *Max, Min*. As seen from the figure, X is the maximum vertical peak of the object, Xmax the farthest horizontal point of the shadow, Ymin, point X is mapped as initial Y point on the y axis, Ymax, measuring from Ymin in the downwards Y direction, the final point is marked as Ymax. The indices values for these points Ymax are marked as [Imin, 0], Ymin as [I, Jmax] and Xmax as [Imax, J].

Orientation angle: Referring to Fig. 11, for the region R, $(\overline{X}, \overline{Y})$ is the Center of Gravity where,

$$\overline{X}, \overline{Y} = (1/|R| \sum\nolimits_{(X,Y) \in R} x 1/|R| \sum\nolimits_{(X,Y) \in R,} Y) \tag{1}$$

where $\overline{X}, \overline{Y}$ is now considered as the Center of Region **R**.

Let **θ r** be the Orientation of Region **R** which is obtained as,

$$\theta_r = \arg \min \sum\nolimits_{(X,Y) \in R} [(x - \overline{x}) \sin \theta - (y - \overline{y}) \cos \theta]^2 \tag{2}$$

By using $\overline{X}, \overline{Y}$ center of gravity and θ r, angle of orientation, the edge contour of region **R** divides the region into two i.e. **R1** the object region and **R2** i.e. the shadow region. However, the shadow pixels cannot be detected due to irregular shadow contour around line **fR (x, y)** [14, 16].

After finding the angle of orientation, its direction also has to be determined. This also gives the direction of the shadow casted on the background. For this the process is as follows.

Let Xm and Ym be the mode values of X and Y respectively. Then, Y = Ymax, i.e. select the maximum vertical pixel value of the foreground region as the Y coordinate. This is selected for drawing the angle of orientation.

Then find, (Xm − Xmax) & (Xm − Xmin). Now find the absolute value of both as, Abs (Xm − Xmax) & Abs (Xm − Xmin). Abs (Xm − Xmax) is less then Abs (Xm − Xmin) then the object and its shadows orientation is as shown in Fig. 11 is named as *Left Orientation*. The point of orientation is selected at Xmin. In the reverse case, the object and its shadows orientation, as shown as in Fig. 12 named as Right Orientation where the point of orientation angle is represented as Xmax.

Fig. 11 Left orientation

Fig. 12 Right orientation

Let θ_r be the angle of orientation, it is calculated as

$$\theta_r = \frac{1}{2}\tan^{-1}\frac{(2\mu 12)}{(\mu 20 - \mu 02)} \times 100 \tag{3}$$

Mask Formation: Two regions **R1** and **R2** are formed where **R1** is the object region and **R2** is the shadow region. Now region labeling is done i.e. R1 = 0 and R2 = 1 i.e. R1 is added to background so remaining R2 of the shadow is only available and it is used as the mask.

Shadow Detection: The binary inverted image is then multiplied with the generated mask to give the shadow detected image.

5.2   Module II: Shadow Detection

Input: The output of stage 1 i.e. shadow detected image along with its respective input image.

Binary image: The input image is converted into grey scale and its global threshold is calculated and applied on it to mark the pixels above threshold as one and those below as zero. The marked image is then converted into binary by making pixels logic one which were marked as one and the others with logic zero.

Inverted image: Create inverted binary of the binary image so as to have the shadows and dark object as logic one.

Morphological operation: Perform the erosion operation on the inverted binary image.

Mask smoothing: Convolution operation is performed.

Averaging: Finding the averaging pixel intensity in the shadow and non-shadow areas for Red, Green and Blue. Calculate the ratio of non-shadow and shadow areas in each i.e. Red, Green and Blue. To this apply the shadow removal formula to get the required resultant.

$$\text{Result} = \text{ratio}/(1 - \text{smooth mask}^{*}\ \text{ratio} + 1) * \text{image} \qquad (4)$$

As this equation is a generalized equation, for the original image respectively RGB, three separate equations are written one for Red, green and blue.

Display method: Of the three available display modes, advanced lightmode 1 is preferred for the display of the output images.

6   Results

For results, representation of two images with single object and two images with multiple object conditions are considered. They are represented below as shown.

Fig. 13 Input 1

Fig. 14 Output 1

Fig. 15 Input 2

Fig. 16 Output 2

Figure 13 named as the *Input 1* is the input image. In this the object and its shadow are well oriented. Figure 14 named as *Output 1* is the output. At the bottom side of the image the object comes ahead of the shadow which is shown as the discontinuity.

Similarly, Fig. 15 named as *Input 2* is the input image and Fig. 16 named as *Output 2* is the output image showing only the detected shadow.

Fig. 17 Input 3

Fig. 18 Output 3

Fig. 19 Input 4

Fig. 20 Output 4

Figures 17 and 19 named as *Input 3* and *Input 4* respectively act as the input images with multiple objects. The same algorithm is applied to these inputs. The output images with only detected shadows are identified and represented through Figs. 18 and 20 named as *Output 3* and *Output 4* respectively.

7   Evaluation

The evaluation of the algorithm is performed are quantitative and qualitative basis. The quantitative is to quantify the output numerically whereas the qualitative evaluation is to justify the user's acceptance.

7.1   Quantitative

Quantitative evaluation metrics are used to evaluate the accuracy of the algorithm. Three types of accuracies are defined. These accuracies are Producer Accuracy, User Accuracy and Combined Accuracy [20, 13].

Producer Accuracy: This consist of two parameters, ηs and ηn, where these accuracies are represented as

$$\eta s = TP/(TP + FN) \tag{5}$$

$$\eta n = TP/(FP + TN) \tag{6}$$

ηs is related to detection of true shadow pixels from the shadow region.
ηn is related to number of shadow pixels detected from the object region.

User Accuracy: This is used to measure the precision of the shadow detection algorithm. It consist of two parameters Ps and Pn Where,

$$Ps = TP/(TP + FP) \tag{7}$$

Ps is ratio of number of correctly detected true shadow pixels over the of total detected true shadow pixels.

$$Pn = TN/(TN + FN) \tag{8}$$

Pn is the ratio of number of correctly detected non-shadows pixels over the of total detected non-shadow pixels.

TP: No of true shadow pixels identified correctly.
FP: Non shadow pixels identified as shadow pixels.
TN: Non shadow pixels identified correctly as object or background pixels.
FN: True shadow pixels identified as nonshadow pixels.

Combined Accuracy: This combines the accuracy of the user and producer and is called as the combined accuracy

$$T = (TP + TN)/(TP + TN + FP + FN) \tag{9}$$

TP + TN: correctly detected true shadow and non-shadow pixel.
TP + TN + FP + FN: Total pixels in the image.

The producer accuracy, user accuracy and combined accuracy are represented in a tabular form in Table 1, where the table is named as *Quantitative Evaluation.* This table is represented below and the Justification for the table contents is as follows.

The producer accuracy gives the number of rightly detected shadow pixels, therefore this value has to be on the higher side. Figures with serial number 1, 2, 3 and 4 show values on the higher side, whereas Fig. 4 has no orientation forming, multiple over lapping shadows is shown over a value. Figures 2 and 3 gives better results than Fig. 1 as these shadows are well represented on a coloured background.

The user accuracy detects the true shadow and non-shadow pixels. Ps is for true shadow and Pn is for true non shadow pixels. Figures 1, 2, 3 and 4 show these values on higher side and for Fig. 5 Ps is not applicable as no shadow is attached to the object and object pixels are detected without self-shadow as Pn. The combined accuracy gives the number of true pixels. The true pixel detection is seen on the higher side for all the images.

For the results obtained it can be concluded that only those images which have their shadows oriented with their objects in a single direction can only give higher accuracy values.

Table 1 Quantitative Evaluation

Sr. No.	Input image	Producer accuracy		User accuracy		Combined accuracy
		Ns	Nn	Ps	Pn	T
1.		0.3074	0.0141	0.7780	0.9687	0.9654
2.		0.8692	0.0293	0.9698	0.9955	0.9947
3.		0.8522	0.0293	0.9698	0.9955	0.9947
4.		0.4802	0.0238	0.7196	0.9742	0.9661
5.		0	0	–	0.7611	0.7611

7.2 Qualitative

The qualitative metric defined and used are *Robustness to Noise* (RN), *Object Independence* (OI),

Scene Independence (SI), *Computational Complexity* (CC), *Shadow Independence* (SID), *Penumbra Detection* (PD), *Illumination Independence* (ID), *Detection/ Discrimination Tradeoff* (DT), *Chromatic Shadow* (CS), *Shadow Camouflage* (SC) and *Surface Topology* (ST).

For the geometrical method, for the output available, these metric are evaluated individually and in groups as the Absolute measure and Relative measure and are further quantified in the range of one to five as shown in the Table 2 named as *Quantifying Values*.

Table 2 Quantifying values

Grades	Absolute measure	Relative measure
1	Excellent	Best in the group
2	Good	Better than the average
3	Fair	Level in the group
4	Poor	Average level in the group
5	Very poor	Lowest in the group

Table 3 Quantifying table

Metric	RN		OI		SI		CC		SID		PD		ID		DT		CS		SC		ST	
	E	P	E	P	E	P	E	P	E	P	E	P	E	P	E	P	E	P	E	P	E	P
AMV	3	2	3	2	3	2	2	3	4	3	2	2	2	2	3	2	4	5	5	5	2	2
RMV	3	2	2	2	4	4	2	2	1	1	2	2	4	4	2	2	5	5	4	3	3	3

The metrics defined above and the values specified in the table for absolute value and relative values are represented in Table 3 called as *Quantifying Table* [21, 22].

8 Result Analysis

The results obtained by qualitative and quantitative evaluation is now represented in a graphic form for detail Analysis.

8.1 Quantitative

The quantitative analysis is done through visualization of five graphs representing User accuracy. The user accuracy is further split into two graphs one for Ps and one for Pn. Similarly, in case of producer accuracy, two graphs, one for ηs and another for ηn. The fifth graph is of the combined accuracy.

The graphs are plotted for five images of which images I1 and I2 are plotted for single object while I3 and I4 are for multiple objects respectively images with I5 as an attached shadow image.

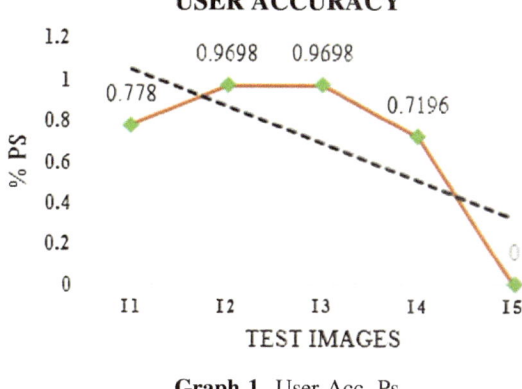

Graph 1 User Acc, Ps

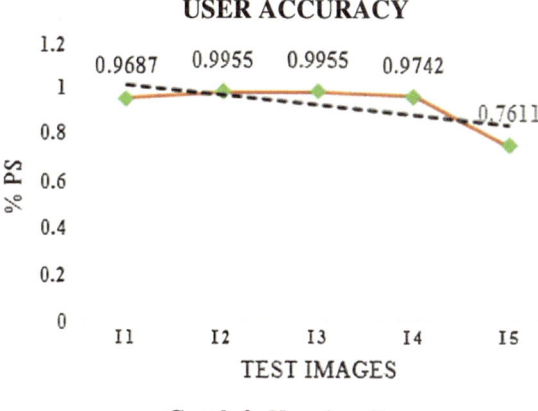

Graph 2 User Acc, Pn

For Graph 1, named as *User Acc, Ps* represents the cup image I2 gives better performance as compared to the girl image I1. This is so because the cup object's colour, background colour and shadow colour are quite distant from each other as compared to the girl object which is gray and comparable in colour with the back ground and shadow. The same relation is also seen between image I3 and I4. Image I5 has the object and its shadow not touching clearly to each other. Thus it cannot give a proper orientation angle. Therefore, the value of Ps is almost zero.

For Graph 2, named as *User Acc, Pn*, all the values are on the higher side as the objects are Coloured. Therefore detection of non shadow pixels are seen to be on the higher side.

Considering Graph 3, named as Producer Acc, Ns, true detection of shadow pixels

Graph 3 Producer Acc, Ns

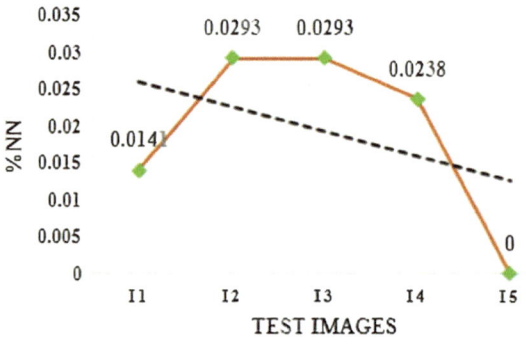

Graph 4 Producer Acc, Nn

form the shadow region is found to be on the higher side in both the single and multiple object images. Now, within these images, the one with difference in colour of object, shadow and background, the detection is found to be on the higher side. As for image I5, it cannot relate between the shadow and the object region. Therefore it has a zero value.

Considering Graph 4, named as *Producer Acc, Nn,* the object being more clear and Coloured, the detection of their pixels for all the images are on the higher side except for image I5. As for combined accuracy, of Graph 5 named as *Combined Accuracy,* the total number of true detection of pixels are found to be on the higher side for all the type of images considered.

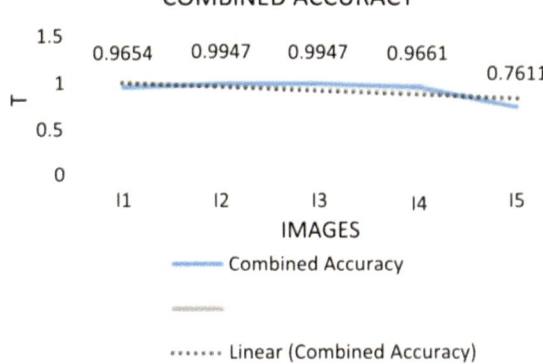

Graph 5 Combined accuracy

8.2   Qualitative

All the eleven metric, are then assigned numerical values as per their absolute measures and relative measures. Accordingly, two graphs are plotted for comparing the existing method with that of the proposed one.

The benchmark used for comparison is taken to be two, so as to find out how many metric are above two. This would enable to represent an improvement from existing to proposed algorithm.

For Graph 6 which represents the absolute value, from total eleven metrics, improvement is seen for five metrics, while other three, maintain their level same. The problem is identified only in computational load, shadow independency and chromatic shadows accordingly.

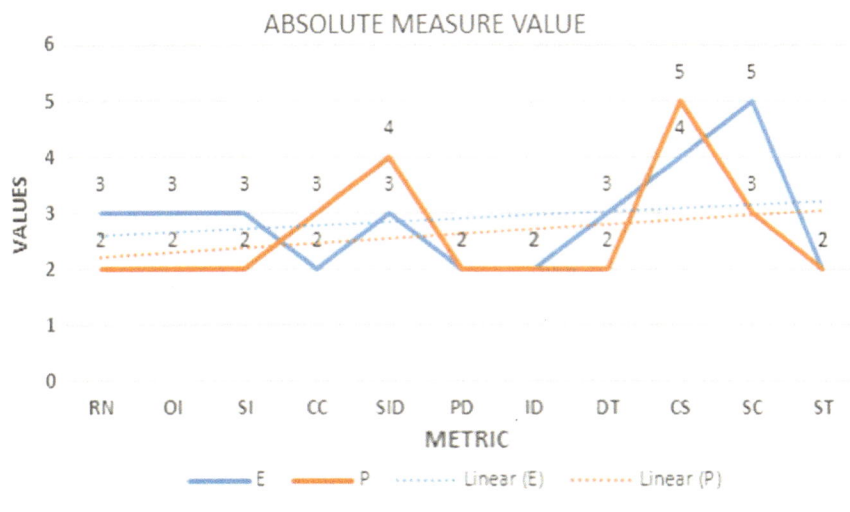

Graph 6 AMV

For Graph 7 named as RMV, out of total eleven metrics, improvement is seen in seven while three maintain their level same i.e. level two.

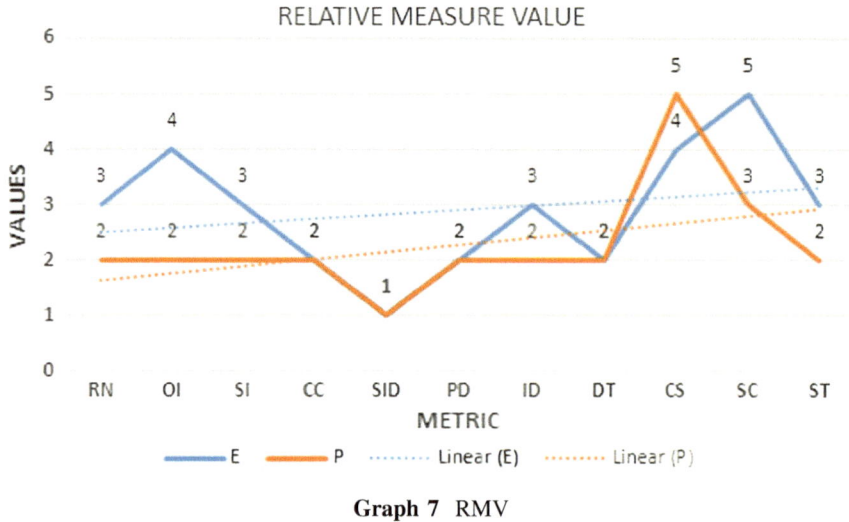

Graph 7 RMV

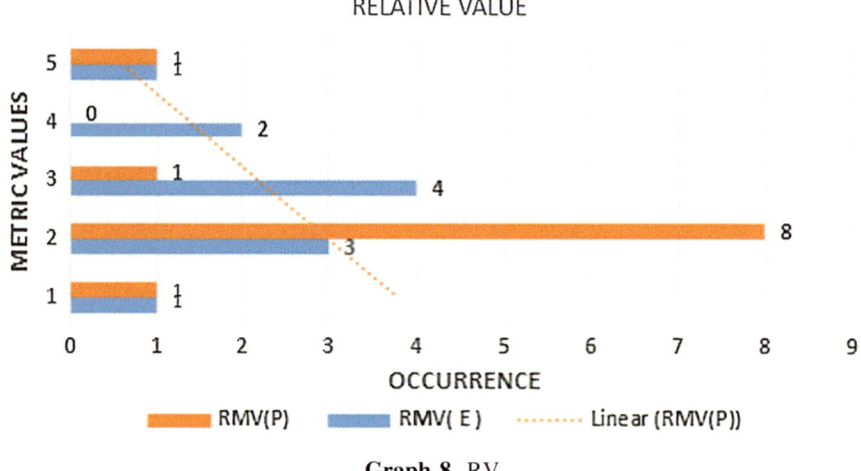

Graph 8 RV

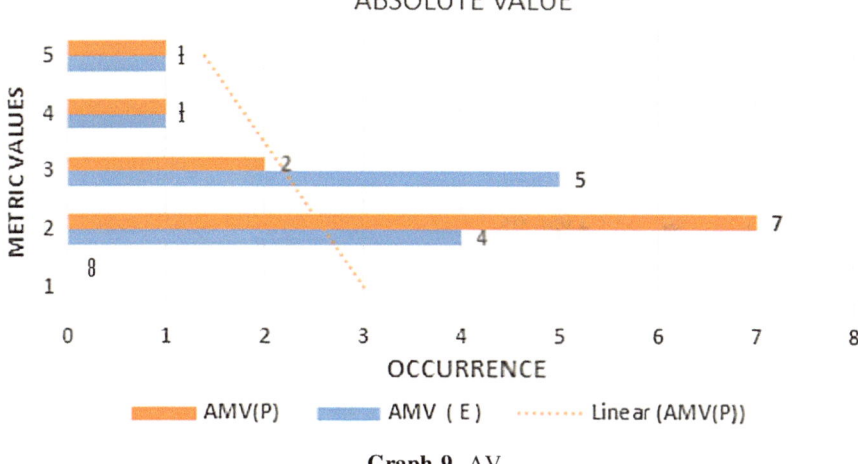

Graph 9 AV

Graphs 8 and 9 the occurrence of metric value two for absolute and relative value respectively. As for the case of relative value the occurrence of two in the existing algorithm was three which is now improved to eight in case of proposed algorithm. Similarly, in case of absolute value this increase is from four to seven.

9   Conclusion

In case of reference algorithm, two separate algorithms are present one for single object and one for multiple objects. The proposed algorithm is the single algorithm which being based on region labeling, it is possible to have the same algorithm for both, single and multiple objects. The proposed algorithm also takes care of shadows on left and right of the object due to four extreme point conditions, Xmax, Xmin, Ymax and Ymin. The proposed algorithm also gives accuracy results very near to the accuracy results obtained for comparison from reference algorithm.

10   Future Scope

Proposed algorithm has limitations i.e. it works on single object with single shadow i.e. light orientation is only in one direction. In case of multidirectional light conditions there can be instances of multiple shadows of the same object.

The algorithm works on four extreme point conditions. Suppose a edge represents a straight line to give any of the max/min edge points depending on object or shadow edge then this algorithm will find it difficult to adapt to it.

References

1. Blajovici, C., Jozsef Kiss, P., Bonus, Z., Varga, L.: Shadow Detection and Removal From a Single Image. In: 19th Summer School on Image Processing, pp. 1–6, Szeged, Hungary
2. Wren, C.R., Azarbayejani, A., Darrell, T., Pentland, A.P.: Pfinder: real time tracking of the human body. IEEE Trans. Pattern Anal. Mach. Intell. **19**(7), 780–785 (1997)
3. Haritaoglu, I., Harward, D., Davis , L.S.: W^4: Who? When? Where? What?—A Realtime System for Detecting and Tracking People. In: Proceeding of International Conference in Automatic Face and Gesture Recognition, pp. 1–6, Nara, Japan (1998)
4. Paragios, N., Deriche, R.: Geodesic active contour and level sets for the detection and tracking of moving objects. IEEE Trans. Pattern Ana. Mach. Intell. **22**, 266–280 (2000)
5. Sanin, A., Sanderson, C., Lovell, B.C.: Shadow detection: A survey and comparative evaluation of recent methods. Pattern Recogn. **45**(4), 1684–1695 (2012). ISSN 0031-3203
6. Jyothirmai, M.S.V., Srinivas, K., Srinivasa Rao, V.: Enhancing shadow area using RGB colour space. IOSR J. Comput. Eng. **2**(1), 24–28 (2012). ISSN 2278-0661
7. Tao, X., Guo, M., Zhong, B.: A Neural Network Approach to Elimination of Road Shadow for Outdoor Mobile Robot. IEEE International Conference in Intelligent Processing Systems, pp. 1302–136, Beijing (1997)
8. Onoguchi, K.: Shadow Elimination Method for Moving Object Detection. In: Proceeding of Fourteen International Conference of Pattern Recognition, vol. 1, pp. 583–587 (1998)
9. Stauder, J., Mech, R., Ostermann, J.: Detection of Moving Cast Shadows for Object Segmentation. IEEE Trans. Multimedia **1**(1), 65–76 (1999)
10. Ivanov, Y., Bobick, A., Liu, J.: Fast lighting independent background subtraction. Int. J. Comput. Vision **37**(2), 199–207 (2000)
11. Chang, C.J., Hu, W.F., Hsieh, J.W., Chen, Y.S.: Shadow Elimination ForEffective Object Detection. In: 15th IPPR Conference on Computer Vision and Image Processing, pp. 185–193 (2002)
12. Prati, A., Mikie, I., Trivedi, M.M., Cucchiara, R.: Detecting Moving Shadows: Algorithms and Evaluation. California Digital Media Innovation Program & California Department of Transportation, pp. 1–9
13. Prati, A., Mikic, I., Trivedi, M.M., Cucchiara, R.: Detecting moving shadows: algorithm and evaluation. IEEE Trans. Pattern Anal. Mach. Intell. **25**(7), 918–923 (2003)
14. Moving Cast Shadow Detection, Vision Systems: Segmentation and Pattern Recognition, pp. 47–58
15. Leone, A., Distante, C.: Shadow Detection for moving objects based on texture anaylisis. J. Pattern Recogn. **40**, 1222–1233 (2007). ISSN 0031-3203
16. Chang, C.C., Hu, W.-F., Chen, Y.-S.: Shadow elimination For Effective Object Detection. In: 15th IPPR Conference on Computer Vision and Image processing, pp. 185–193 (2002) (geometrical method)
17. Blajovici, C., Jozsef Kiss, P., Bonus, Z., Varga, L.: Shadow Detection and Removal From a Single Image. In: 19th Summer School on Image Processing, pp.-1–6, Szeged, Hungary
18. Horprasert, H., Harwood, D., Davis, L.: A Statistical Approach for Real time Robust Background Subtraction and Shadow Detection. Computer Vision Lab, pp. 1–19, University of Maryland
19. Vinitha Panieker, J., Wilsey, M.: Detection of Moving cast shadows using edge information. IEEE 817–821 (2010). ISBN 987-4244-5586-7/10

20. Prati, A., Mikie, I., Trivedi, M.M., Cucchiara, R.: Detecting Moving shadows Formulation, Algorithms and Evaluation. Technical Report-Draft Version, pp. 1–39
21. Hybrid shadow restitution techniques for shadow free scene reconstruction
22. Book of advance in signal processing and intelligent recognition system, 264, Springer, Berlin

Medical Diagnosis Through Semantic Web

P. Monika[1,2,4(✉)], M. R. Vinutha[2,4], B. N. Srihari[2,4],
S. Harikrishna[2,4], S. M. Sumangala[2,4], and G. T. Raju[3,4]

[1] CSE Department, R&D Centre, RNSIT, Bengaluru, India
monikamanjunath@gmail.com
[2] Department of CSE, DSCE, Bengaluru, India
[3] Department of CSE, RNSIT, Bengaluru, India
[4] Visvesvaraya Technological University, Belagavi, Karnataka, India

Abstract. Semantic web work towards mining of the semantics of data and
further processing from the collection of current web resources rather than
pattern matching during the information extraction process, there by leading
towards the automation of knowledge extraction procedure. Healthcare is one
among the major domains, where huge data production happens on daily basis.
There is no specific technique or model to successfully utilize the available
information during the course of diagnosis. The key to upgrade is to raise
awareness among the people. This paper aims at developing a model with the
usage of Semantic Web, Ontology concepts and Apache Jena reasoner to
improve and refine the basic clinical skills required to provide effective and
efficient primary care. The proposed work—*Healthub* is being evaluated with
respect to correctness and accuracy of diagnosis. Results obtained using Apache
Jena reasoner show promising responses approximately much nearer to expert
conclusions.

Keywords: Semantic web · Ontology · Healthcare · Automated knowledge
extraction · Reasoner · Apache Jena · Protégé · Mining

1 Introduction

Semantic Web deals with extraction of the meaning of data rather than its structural
representation. The World Wide Web (WWW) being the container of enormous
amount of data and information, it is also known for its disseminative nature. Semantic
Web is the reinforced genre of WWW which can be referred as the organised version of
the web. The purpose of semantic supported document in healthcare domain is to
mitigate medical errors, revamp physician efficiency and enhance patient safety and
satisfaction in medical practice. Semantic Web technology assists in achieving these
objectives in an ontology driven process that involves ontologies, automated semantic
annotation of documents and rule processing [1].

All of these are procured basically by executing rules on semantic annotations and
relationships among ontologies. The work presented here utilizes these technologies to

© Springer International Publishing AG 2018
D. J. Hemanth and S. Smys (eds.), *Computational Vision and Bio Inspired Computing*,
Lecture Notes in Computational Vision and Biomechanics 28,
https://doi.org/10.1007/978-3-319-71767-8_14

build a robust framework that supports a wide spectrum of applications. Healthcare domain is one of the most data intensive and data driven industry in the world. As mentioned earlier vast amount of data is fostered from health care providers, customers and researchers. The challenge is just not storage and access but also to make it usable by rendering appropriate outcome. Even though data is retrievable, it may be irrelevant with respect to each person. If this information is provided to doctors, patients or commoners, it leads to disastrous health implications. By providing dedicated web portals for such diagnostic intelligence, effectiveness of the medication can be boosted up and further undesired healthcare hurdles can be avoided.

An attempt is made here to build a web application—*Healthub* which reads symptoms from the users and responses the diagnostic conclusions based on the provided set of input symptoms with suggestions for the possible disease or the situation of the patient who is under diagnosis. The crave for relevant opinion by the healthcare experts has increased drastically since the number of diseases and its disastrous outcomes have intensified gradually. Hence it is important to make use of emerging technologies in healthcare domain to assist people with eventuating diseases and its repercussions.

The rest of the paper is organized as follows: A brief on related work is presented based on the literature survey. Methodology employed with the required tools and technologies is described in detail followed by data set generation, proposed model architecture and implementation details. Further, Experimental results are reported. The paper is concluded with conclusion and references.

2   Related Work

Health of each and every individual need to be monitored regularly to certain extent in order to avoid the occurrence of serious consequences in future. Lee et al. [2] have proposed eHealth Recommendation Service System (eHeaRSS) to recommend the health services and information to the needy, especially patients who need emergency medical services. The proposed technique suggests the likely suspected diseases, proper departments, and recommendable doctors. The system is implemented using cloud computing environment. Data has been classified into doctors, departments, symptoms and diseases ontologies; employed Case-based reasoning for semantic recommendations. Authors have conducted experiments using ontology-based data processing there by concluding that the proposed system reduces the recommendations by 33% when compared to other DB-based system. Remote data monitoring also plays a vital role in ubiquitous patient monitoring. Christopoulou et al. [3] have projected vhMentor based on a solid ontological schema system which employs mobile agent technology for encapsulation of remote sensing and medical device signals in an ontology-supported healthcare information system.

Data collection for semantic conclusions is also equally important during the automation of healthcare process. Gai et al. [4] suggested ontology based approach called Secure Ontology-based Self-diagnosis (SOS) Model for the representation of medical knowledge for alarming potential risks caused due to improper diagnosis. Writers comment that the experimental research on the proposed model has shown good performance results. Healthcare service providers face a general challenge of data integration in the present heterogeneous web of information. Health records in different formats need to be integrated prior in-order to semantically conclude the correct decisions by the recommendation systems. To support the data interoperability in diverse environment, Jung and Yoon [5] suggests the usage of worldwide information standards and terminologies such as HL7/SNOMED CT. The authors have recommended integrated system architecture using Web Ontology Language (OWL) for data collection generating intelligent biomedical knowledge base, automatic knowledge procuration method for auto correction and knowledge updates.

In order to support Community Health Workers in decision making, Pappachan et al. [6] have suggested Rafiki system for mobile and wearable computing devices which infers the treatments and diseases based on the symptoms and the patient context with the help of reasoners working on ontology files. Authors also note that knowledge can be easily shared using the semantic representation of data especially in healthcare domain. A functional semantic data exchange model to integrate the patients data throughout has been proposed by Xiao and Wei [7]. Medical group discussion with the experts also benefit to certain extent in the health care automation system [8]. The data in Semantic Web is being represented in ontological format for automation of knowledge extraction. But interactions among various ontology agents in respective domains play a major role while drawing semantic conclusions [9]. Hu et al. [10] presents architecture for a Personal Health Recommender System (PHRS) which works on person's Personal Electronic Health Record (PEHR), followed by augmentation by combining crowd-sourced data mined for treatments, diseases, symptoms and best practices together with the concept of domain ontologies. Authors conclude that as the PEHR gets populated with more data, the ontology may evolve in the Semantic Web. Krishnamurthy et al. [11] have explored the trends in Mental Health and Behavioral Studies (MHB) by presenting a novel approach where personality traits from unstructured text of patients and general social users are compared via statistical analysis using Psychiatric Disorder Determination (PDD) algorithm with the use of concepts of ontologies.

3   Methodology and Tools

In Healthcare, almost all existing ontologies and designs that incorporate Semantic Web concepts are packed specifically with particular abnormalities related to certain diseases. The proposed work represents the semantic conclusion methodology for generic symptoms and diseases which are frequently occurring. Ontologies impart the Semantic Web vocabulary which is employed to elucidate websites in order to extract meaningful machine interpreted facts by correlating the entities and their properties. Data collection was done through case studies based on often occurring diseases and couple of real time data from domain experts. The data sets in the ontological form was created using *Protégé* [12]—a well know ontology editor developed by Stanford university. The protégé tool maps the symptoms to its respective diseases, with the aid of object properties, data properties and class properties eventually leading to graphical representation of related data in Resource Description Framework (RDF) format— which is a framework for portraying information about resources in a graphical layout using Universal Resource Identifier (URI). RDF format manifests data as triplets: subject, predicate and object as shown in Fig. 1.

Prominent feature of OntoGraf is that it favours synergistic navigation relationship of ontologies. Figures 2, 3 and 4 shows OntoGraf: An ontological representation of the given input data. Symptoms like headache, insomnia, shortness of breath, fatigue, chill & sweats, pain behind eyes, blueness of lips, dark urine, nasal congestion, pale stools, bloody stools, convulsions, diarrhea, lethargy, rashes, swollen lymph nodes, hallucination, recurrent nose bleeds, bone pain or tenderness etc. were included in the ontology. Diseases like asthma, chikungunya, emphysema, dengue, hepatitis, malaria, typhoid, tuberculosis, dementia, leukemia, etc. were successfully diagnosed using the proposed model.

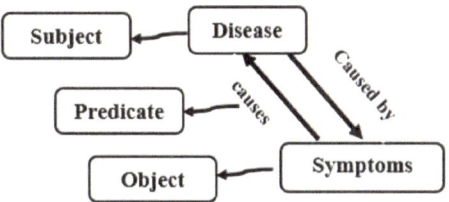

Fig. 1 RDF Triplet format

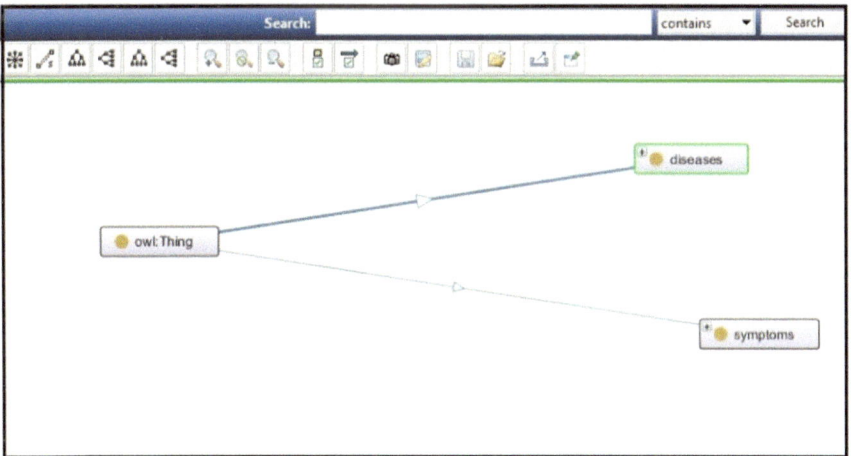

Fig. 2 OntoGraf of base classes

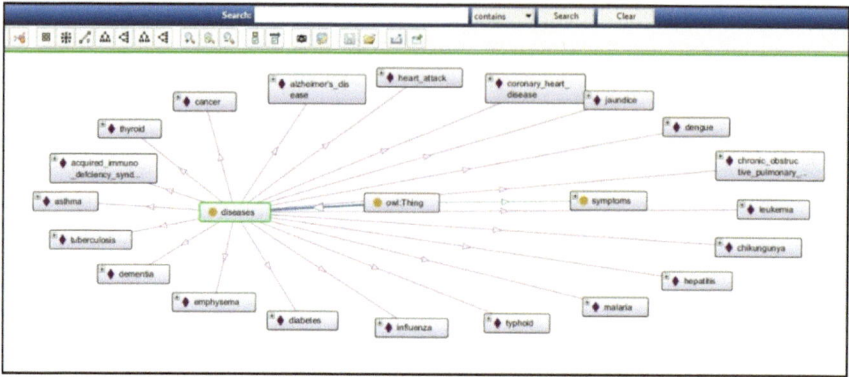

Fig. 3 OntoGraf of diseases

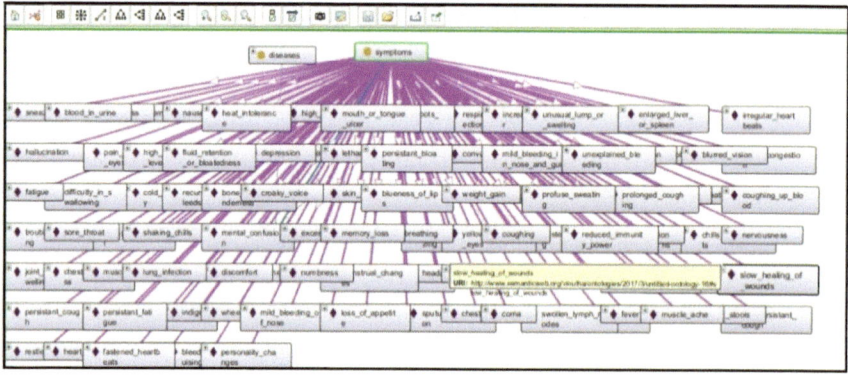

Fig. 4 OntoGraf of symptoms

According to Google survey, the healthcare domain produces billion Giga bytes of data globally, which is expanding exponentially. Therefore usage of graphical data structure to represent data help us in expressing amplified dataset. The input file which was created for the demonstration of the proposed work was designed with the following metrics: 2 classes, 2 Object properties, 137 Individual count, 489 Axioms and 348 Logical axioms.

Apache Jena plugin is an open source Semantic Web framework which is most wide used in Semantic Web applications built using Java language. A Reasoner is software that infers logical conclusions or outcome from the given set of input data. Jena framework renders plenty of its inbuilt reasoners which can be utilized to encompass few rule based inferences to accomplish reasoning on the ontologies presented in RDF format. Amidst these inbuilt reasoners, we have conducted majority of experiments using Transitive Reasoner module supported by Apache Jena.

SPARQL is a semantic query language for fetching information from the semantic knowledge bases using which we can store, retrieve and manipulate data presented in Resource Description Framework (RDF) format. In *Healthub*, SPARQL is used to extract the required data from the semantically mapped knowledge base.

The *Apache tomcat* server is used to provide the service. Scripting languages like *HTML, JSP, Java Script, CSS, J Query* have been used to build the Graphical User Interface (GUI). *Bootstrap* is used for the development of front end template. *Property file* is used to save description of the symptoms and the diseases.

4 Proposed Model Architecture and Implementation

The architecture of the proposed Model is as shown in Fig. 5. The model is depiction of Semantic Web which includes the components such as front end scripts for user interaction; server to handle the requests, requirement processing module—which works at the backend for knowledge processing is implemented using the concepts of core Java, Reasoner module for semantic conclusions using the data presented in ontological format. Figure 6 explains the interaction of the Model Factory Application Programming Interface (API) and reasoner at the back-end.

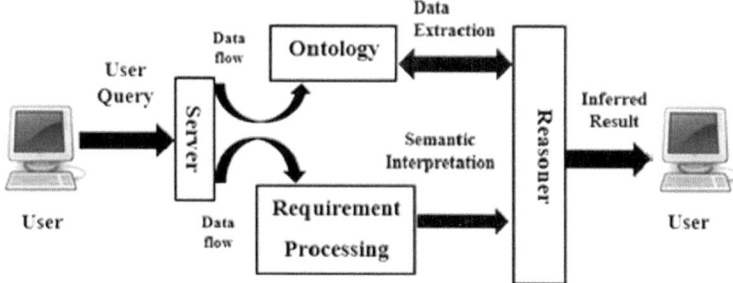

Fig. 5 Proposed architecture

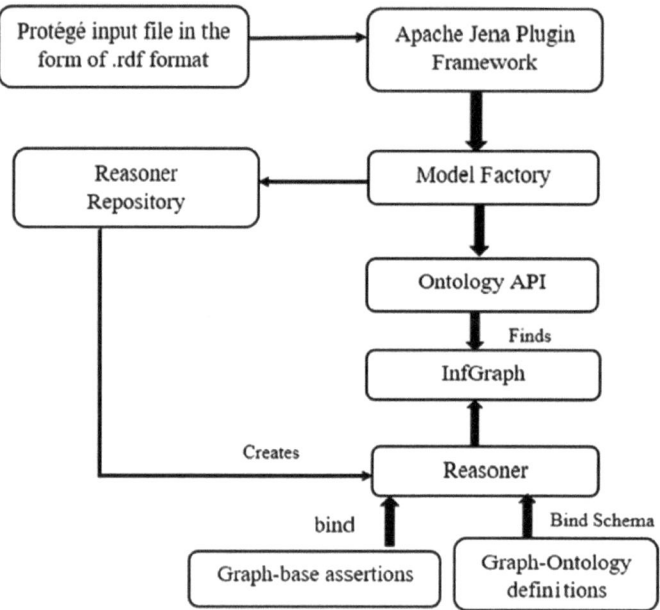

Fig. 6 Structure of inference machinery

Healthub is a web application that provides a simple GUI that connects the core semantic processing model with the user. The system reads the input symptom(s) from the user and process the input semantically at the backend using Model factory API. The API communicates with the ontology file which is in RDF format for drawing inferences using Apache Jena built-in reasoners. SPARQL query will be triggered on to the reasoner conclusions in order to extracted knowledge of user interest and the same is displayed as response to the user query with appropriate description. Based on the symptoms provided by the user, the list of the diseases caused due to the given symptoms will be displayed back in the output screen with the support of GUI modules.

5   Experimental Results

The Figs. 7, 8, 9, 10, 11 and 12 show the output screen shots of the proposed working model.

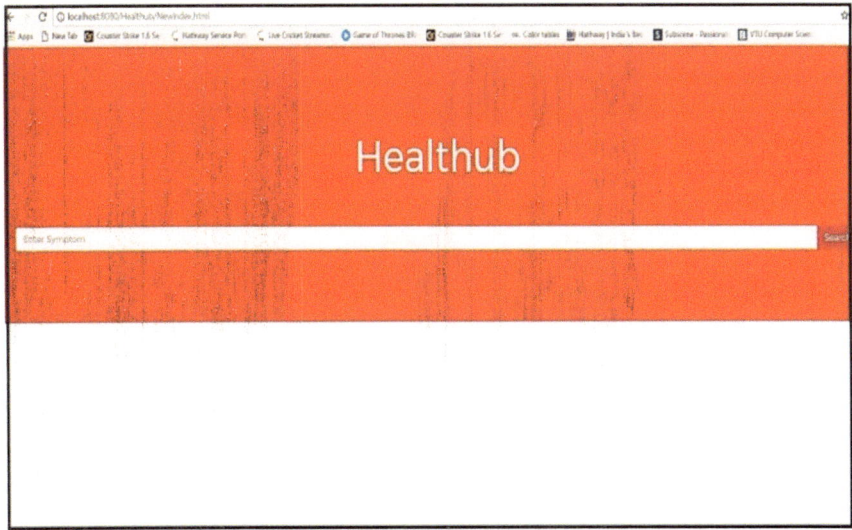

Fig. 7 Healthub home page

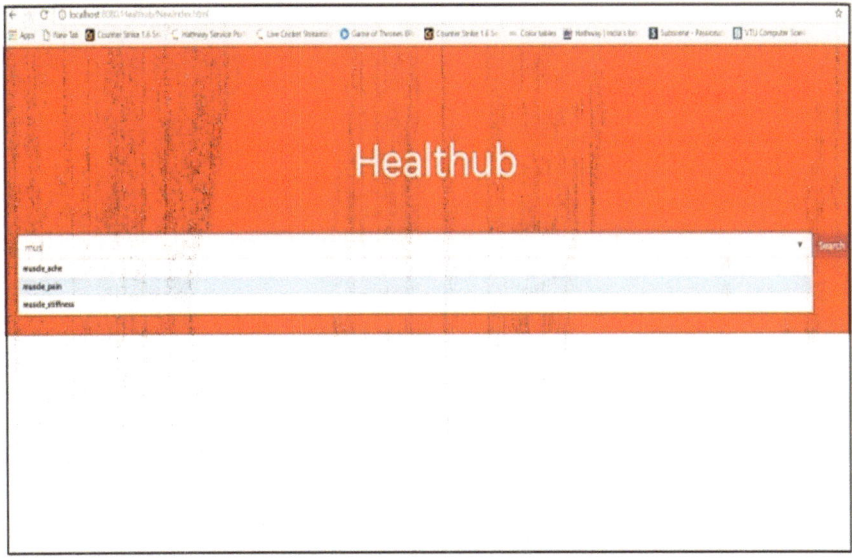

Fig. 8 Suggestion of symptoms

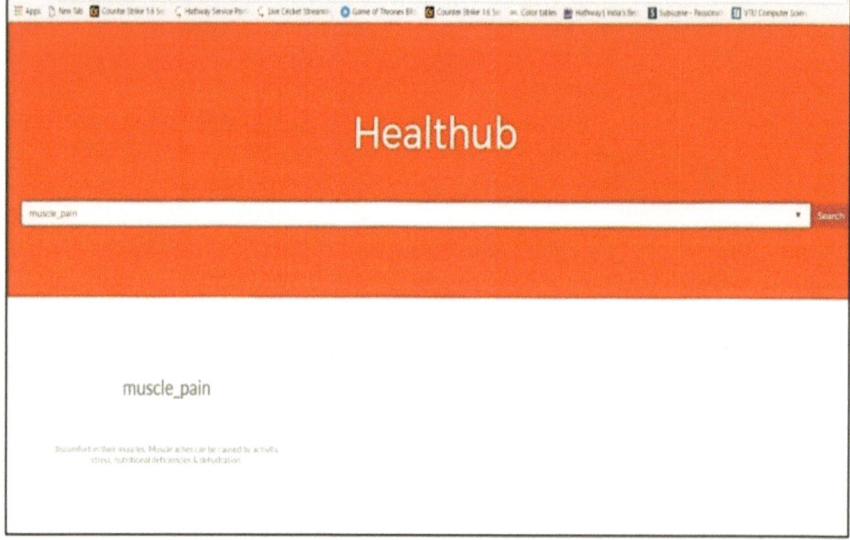

Fig. 9 Symptom description

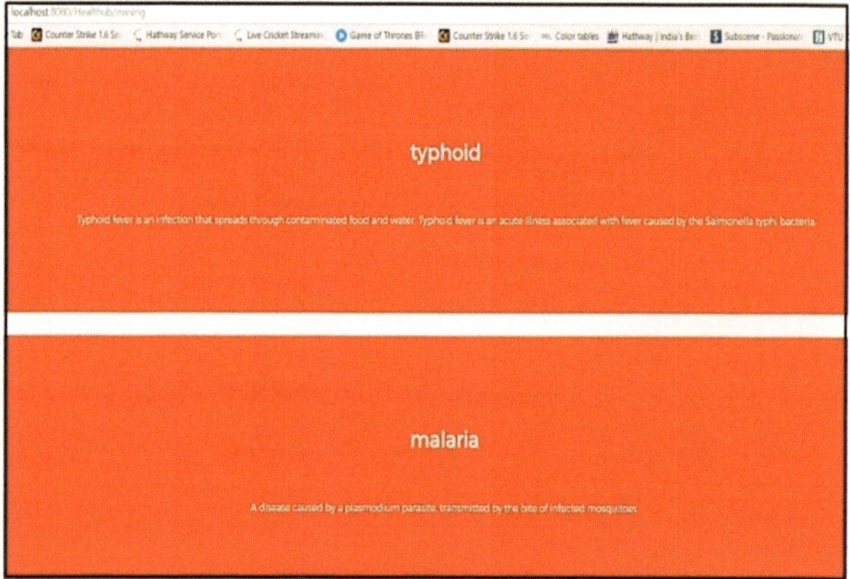

Fig. 10 Inferred result for single symptom

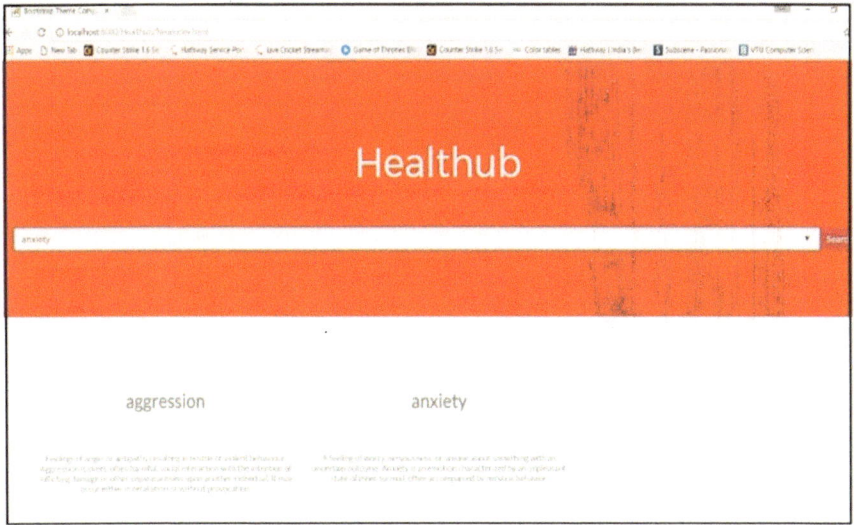

Fig. 11 Multiple symptoms description

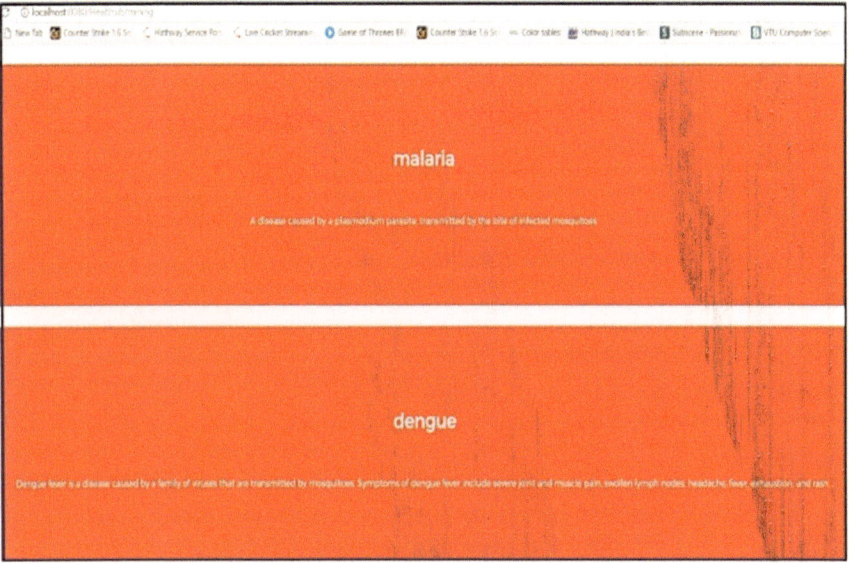

Fig. 12 Inferred results for multiple symptoms

A series of tests were conducted using the set of symptoms like as headache, insomnia, shortness of breath, fatigue, chill & sweats, pain behind eyes, blueness of lips, dark urine, nasal congestion, pale stools, bloody stools, convulsions, diarrhea, lethargy, rashes, swollen lymph nodes, hallucination, recurrent nose bleeds, bone pain or tenderness etc. which were included in the ontology. Diseases like asthma, chikungunya, emphysema, dengue, hepatitis, malaria, typhoid, tuberculosis, dementia, leukemia, etc. were successfully diagnosed using the proposed model. The domain experts were approached and their diagnosis results for the same symptoms were tabulated to evaluate the correctness of the proposed model. Table 1 represents the semantic conclusion of diseases drawn by the proposed model based on the provided input symptoms.

Table 1 Test cases and decisions

Test case	Symptoms	Possible diseases
1	S1, S2, S3	D1
2	S4, S16	D8
3	S1, S4, S5	D2
4	S8, S9, S10	D5
5	S1, S7	D3
6	S16, S18, S19	D10
7	S11, S12, S13	D6
8	S1, S6	D4
9	S15, S14	D7
10	S2, S17	D9

6 Conclusion and Future Work

As the information available in the web is heterogeneous and scattered in nature, it is difficult to obtain appropriate results and sometimes it may be faulty or misleading. Such risks are not affordable in healthcare domain. Hence, it is important to procure genuine and precise suggestions and results, assisting the patients and commoners. The main goal of this work was to develop a model called Healthub with the usage of Semantic Web, Ontology concepts and Apache Jena reasoner to improve and refine the basic clinical skills required to provide effective and efficient primary care. The proposed model, being evaluated with respect to correctness and accuracy of diagnosis, the results obtained using the inbuilt Apache Jena reasoner show promising responses approximately much nearer to expert conclusions. In future, we are working towards enhancement of the dataset to suite the real time dynamic requirements along with better semantic ontology and suggestive system. Further plan is to develop an android app for the proposed model.

References

1. Monika, P., Raju, G. T.: Hybrid Architecture for Rule Based Automated Decision Support in Healthcare. In: IEEE International Conference on Telecommunication, Power Analysis & Computing Techniques (ICTPACT-2017) (2017). ISSN 978-1-5090-3381-2
2. Lee, H.J., Kim, H.S.: e-Health Recommendation Service System using Ontology and Case-based Reasoning. In: IEEE International Conference on Smart City/SocialCom/-SustainCom together with DataCom 2015 and SC2 (2015). https://doi.org/10.1109/SmartCity.2015.217
3. Christopoulou, S.C., Anagnostopoulos, I., Kotsilieris, T.: A Health Care Monitoring System That Uses Ontology Agents. In: 11th IEEE International Workshop on Semantic and Social Media Adaptation and Personalization (SMAP) (2016). https://doi.org/10.1109/SMAP.2016.7753382
4. Gai, K., Qiu, M., Jayaraman, S., Tao, L.: Ontology-Based Knowledge Representation for Secure Self Diagnosis in Patient-Centered Telehealth with Cloud System. In: 2nd IEEE International Conference on Cyber Security and Cloud Computing (2015). https://doi.org/10.1109/CSCloud.2015.72
5. Jung, Y., Yoon, I.K.: Data Integration for Clinical Decision Support. In: Eighth IEEE International Conference on Ubiquitous and Future Networks (ICUFN) (2016). https://doi.org/10.1109/ICUFN.2016.7537008
6. Pappachan, P., Yus, R., Joshi, A., Finin, T.: Rafiki: A Semantic and Collaborative Approach to Community Health-Care in Underserved Areas. In: 10th IEEE International Conference on Collaborative Computing: Networking, Applications and Worksharing (CollaborateCom 2014). https://doi.org/10.4108/icst.collaboratecom.2014.257299
7. Xiao, L., Wei, Q.: Developing a Standard Protocol for Clinical Data Exchange and Analysis. In: 6th IEEE International Conference on Software Engineering and Service Science (ICSESS) (2015). https://doi.org/10.1109/ICSESS.2015.7339086
8. Lee, H.J., Sohn, M.: Health Service Knowledge Management to Support Medical Group Decision Making. In: 10th IEEE International Conference on Innovative Mobile and Internet Services in Ubiquitous Computing (IMIS) (2016) https://doi.org/10.1109/IMIS.2016.78
9. Monika, P., Raju, G.T.: Semantic web with ontology agents for improved search results—a survey, scopus indexed. Int. J. Appl. Eng. Res. (IJAER) 10(86), 264–270. ISSN 0973-4562
10. Hu, H., Elkus, A., Kerschberg, L.: A personal health recommender system incorporating personal health records, modular ontologies, and crowd-sourced data. IEEE/ACM International Conference on Advances in Social Networks Analysis and Mining (ASONAM) (2016). https://doi.org/10.1109/ASONAM.2016.7752367
11. Krishnamurthy, M., Mahmood, K., Marcinek, P.: A hybrid statistical and semantic model for identification of mental health and behavioural disorders using social network analysis. In: IEEE/ACM International Conference on Advances in Social Networks Analysis and Mining (ASONAM) (2016). https://doi.org/10.1109/ASONAM.2016.7752366
12. Musen, M.A.: The Protégé team: the Protégé project: a look back and a look forward. AI Matters 1(4), 4–12 (2015). https://doi.org/10.1145/2757001.2757003

An Algorithm for Text Prediction Using Neural Networks

Bindu K. R.[1,2(✉)], Aakash C.[1,2], Bernett Orlando[1,2],
and Latha Parameswaran[1,2]

[1] Department of Computer Science and Engineering, Amrita School
of Engineering, Coimbatore, India
{j_bindu, p_latha}@cb.amrita.edu,
{aakashchan24, bencube.ben}@gmail.com
[2] Amrita Vishwa Vidyapeetham, Coimbatore, India

Abstract. Neural networks have become increasingly popular for the task of language modeling. Whereas feed-forward networks only exploit a fixed context length to predict the next word of a sequence, conceptually, standard recurrent neural networks can take into account all of the predecessor words. In this paper, an algorithm using machine learning which, when given a dataset of conversations, is able to train itself using Neural Networks which can then be used to get suggestions for replies for any particular input sentence is proposed. Currently, smart suggestions have been implemented in chat applications. Google uses similar techniques to provide smart replies in Email through which the user can reply to a particular email with just a single tap.

Keywords: Recurrent neural networks · Text suggestions · LSTM
Machine learning

1 Introduction

The core component of automatic text prediction that incorporates syntactical and semantic constraints of a given natural language is the Language Model. In recent years prediction uses mainly backing off models referred in [1]. In the rescoring stage feed-forward neural network LMs, first introduced in [2], have become an important technique. In [3], the vanishing gradient problem was analyzed in detail. The gradient of error function gets scaled by a factor when it is propagated back through a unit of neural network. This factor is either greater than one or smaller than one in practically relevant cases. The authors re-design the unit of a neural network in such a way that its corresponding scaling factor is fixed to one to avoid the scaling effect.

In solving the problem of vanishing gradient in recurrent neural network deep learning is giving good performance. Neural nets can remember the context with the introduction of gated cell. Long short term memory [1] is used to give suggestion to user, based on the context [4, 5] in the text prediction model. The process begins with

D. J. Hemanth and S. Smys (eds.), *Computational Vision and Bio Inspired Computing*,
Lecture Notes in Computational Vision and Biomechanics 28,
https://doi.org/10.1007/978-3-319-71767-8_15

collection of data, building the neural network and training the network. Once the training is done, the neural network is tested with real world use cases.

2 Proposed Algorithm

Corpus chosen contains 2,20,579 dialog conversations. Several conversations on varied topics were required, and this corpus served all the needs. Conversations in the dataset are mostly between 9035 characters with gender of the character mentioned. Dataset is pre-processed to remove stop words. This took a giant leap in performance of neural network [6, 7]. The reason for choosing this data is because movie dialogues cover a wide areas. This dataset seems to be the right fit as the intention of algorithm is to generate proper suggestion.

3 Work Flow

The architecture has been outlined in Fig. 1 starting from the collection of data till we get the final test model.

Step-1   For any machine learning problem, the most important part is the data. So, the very first step is to identify the dataset that we need to use for training and efficiently extract it.

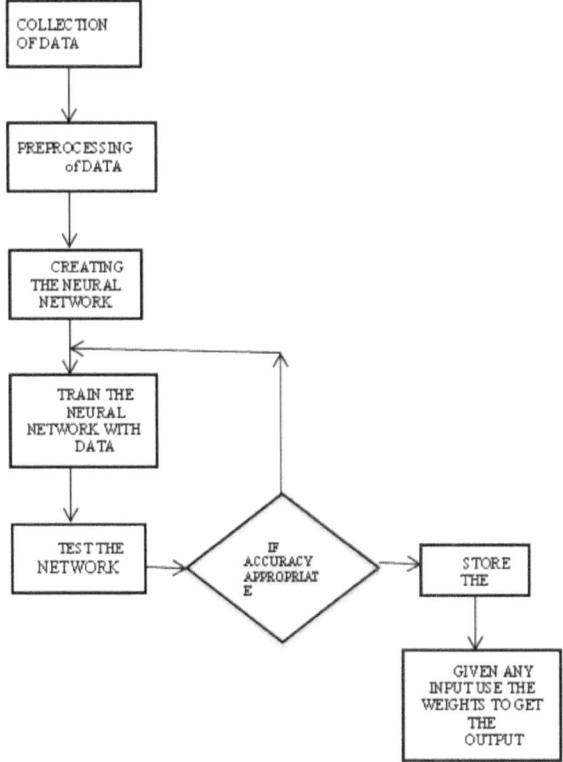

Fig. 1 Architecture diagram

Step-2 The data that we extract from the Internet will be in raw format and so processing it according to our needs will be extremely difficult in the long run. So, the next step is to preprocess the downloaded data according to requirements.

The preprocessed data is exported and we get four files which are:

    train.enc train.dec test.enc test.dec

"enc" denotes the data for the encoder and "dec" denotes the data for the decoder. We segregate the data for training and testing.

Step-3 Next step is to create the neural network that we will be using for training as well as testing. This is the core step in the entire process as this step defines how well the algorithm works in the end. Define the number of hidden layers, the number of neurons in each of these layers, etc. these are called the hyper parameters and are tunable. These parameters can be changed to get the best accuracy possible.

Step-4 This is the step where the actual training takes place. train.enc file from the data preprocessing step and give it as input to encoder and get the corresponding output for the current weights in the decoder. The output got from "train.dec", compare the two results and change the weights of the network using backpropagation algorithm.

Step-5 Once all the training data is used in the network, the testing phase will start. During this phase, the contents from "test.enc" are sent into the encoder and the output as before will be taken from the decoder. "test.dec" file is used to test how accurate the network is performing. Accuracy to be achieved needs to be mentioned.

Step-6 Once the testing is done, accuracy of the network is checked whether it is greater than or equal to the accuracy mentioned. If it is greater then, stop the training process and go to the next step. Otherwise, training is done again to improve the accuracy till we reach the required amount.

Step-7 The weights obtained during the training must be stored to be used elsewhere in an application. Otherwise, the training must be done every time the application is run which is not a feasible solution as it will take a lot of time.

Step-8 A GUI application is built to get the input sentence and the number of suggestions required and using the weights obtained during training in the above steps we generate the requested suggestions.

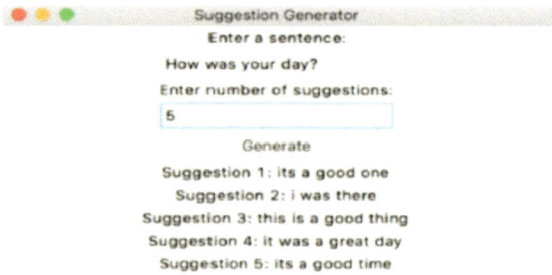

Fig. 2 Suggestion 1

4 Discussions

Existing text prediction models are based on frequency of usage and the suggestions are linked with the previous words used by the user. This can sometimes lead to irrelevant suggestions when the user uses the same words in a different context. Therefore, we have decided to study the effectiveness of using machine learning for text prediction [8].

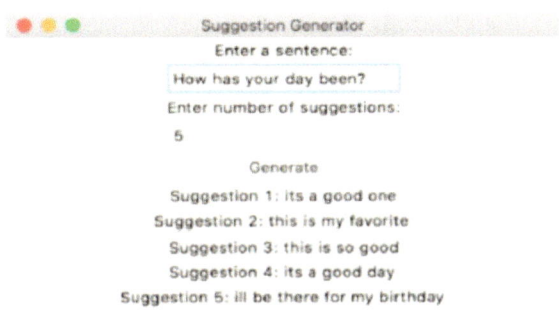

Fig. 3 Suggestion 1

Most of previous research has been done in areas related to language modelling using machine learning algorithms and "Recurrent Neural Networks" are found to be the most effective [1]. Out of the various RNNs available the most widely used model is known to be the LSTM—Long Short Term Memory [2] which is used to develop a suggestion generator which takes in a input sentence and gives several suggestions that can be used as replies to test how well the algorithm is able to give relevant suggestions.

Once the data is processed and is available in the needed format we can move on to training a LSTM based RNN for this data. We have two parts in the LSTM network which we call the "Encoder" and the "Decoder" which are both used during training as well as testing [9–12].

Once the training is done, we can use any input sentence to get a output sentence based on the given data. But since we will be needing more than just a single output for a given sentence we need to modify the encoder-decoder system [9] a little bit.

Softmax

In the existing model, the last layer is taken as the softmax and the output is provided based on the highest probability output given by this layer. Since we might require more than one suggestion, this will not for work for us and so we need to extend it. The idea that we incorporated is based on the fact the original algorithm only takes the highest probability for finding the next word in the output. So, if we use the next highest probable output, then we can extend it to one more output. Similarly, by taking the different outputs one by one, we can get any number of suggestions from the decoder.

5 Experimental Results

To test the efficiency of the algorithm, we built a GUI where the user can input a sentence and the number of suggestions that he requires and on pressing a "Generate" button, the suggestions will be generated and displayed on screen.

The two Figs. 2 and 3 show an example of how the algorithm can suggest replies for a given input sentence. Also, to understand how well the algorithm can understand the sentence given, two results are compared with each other. One input is "How was your day?" and the other is "How has your day been?". Both mean the same thing but the words used are slightly different. But the algorithm can understand that the question means the same thing and the suggestions generated are like each other.

To test how good the suggestions are we invited 75 people to test the GUI and rate the suggestions on a scale of 1 to 5 where,

    1—sentence has no meaning
    2—sentence irrelevant to the given input
    3—sentence has meaning but not to the point
    4—sentence has meaning and relevance. Good suggestion
    5—the best suggestion.

Since, the suggestions are completely based on the data used for training, the review phase is split into two halves. During the first phase, the users could give any input sentence that they feel alike and there was no restriction on it. The result from the test was (Table 1).

Table 1 Result from the first phase

Rating	No. of users
1	6
2	18
3	40
4	10
5	1

During the second phase, the users were specifically advised about the type of dataset used for training and to give the input sentence like it and test the GUI.

The result from the second phase was (Table 2).

Table 2 Result from the second phase

Rating	No. of users
1	0
2	3
3	26
4	38
5	8

Average rating during first phase-2.76.
Average rating during second phase-3.68.

From the above result, when the input data is like the ones used for training, then the result as shown by the user ratings are more relevant [10, 11].

From the results obtained above it is understood that more the data is used, more the area covered and thus better the suggestions will be. It can scale up the model and fit in a wide variety of dataset and can keep on improving it.

6 Conclusion

Specific domain based datasets can be collected and used for training to get suggestions for that specific domain. An API can be created using which this application can be open to all developers who can use the suggestions given as the output in their own chat bots and chat applications.

Computation power is the main limitation for any machine learning based application as the training phase takes huge amount of time based on hardware capacity. Also, the suggestion is completely based on the trained dataset and so any input given outside its scope can give irrelevant outputs.

References

1. Sundermeyer, M., Schlüter, R., Ney, H.: LSTM Neural Networks for Language Modeling. In: Interspeech, pp. 194–197 (2012)
2. Mikolov, T., et al.: Recurrent neural network based language model. In: Interspeech, vol. 2, pp. 1045–1048 (2010)
3. Anlauf, J.K., Biehl, M.: The adatron: an adaptive perceptron algorithm. EPL (Europhysics Letters) **10**(7), 687 (1989)
4. Bindu, K.R., Parameswaran, L., Soumya, K.V.: Performance evaluation of topic modelling algorithms with an application of Q & A dataset. Int. J. Appl. Eng. Res. **10**, 23–27 (2015)
5. Bindu, K.R., Parameswaran, L., Nambiar, S.R., Chandran, J.: Performance evaluation of algorithms for expert finding on an open email dataset. Int. J. Appl. Eng. Res. **10**, 71–75 (2015)
6. Bebis, G., Georgiopoulos, M.: Feed-forward neural networks. IEEE Potentials 13(4), 27–31 (1994)

7. Jarmo, I., Kamarainen, J.K., Lampinen, J.: Differential evolution training algorithm for feed-forward neural networks. Neural Process. Lett. **17**(1), 93–105 (2003)

8. Baddeley, A.D., Thomson, N., Buchan, M.: Word length and the structure of short-term memory. J. Verbal Learn. Verbal Behavior 14(6), pp. 575–589 (1975)

9. Funahashi, K.-I., Nakamura, Y.: Approximation of dynamical systems by continuous time re current neural networks. Neural Netw. **6**(6), 801–806 (1993)

10. Hochreiter, S., Schmidhuber, J.: Long short-term memory. Neural Comput. **9**(8), 1735–1780 (1997)

11. Jizhou, J., Zhou, M., Yang, D.: Extracting chatbot knowledge from online discussion forums. IJCAI 7, pp. 423–428 (2007)

12. Jiyou, J.: The study of the application of a web-based chatbot system on the teaching of foreign languages. In: Proceedings of SITE, vol. 4, pp. 1201–1207 (2004)

A Deeper Insight on Developments and Real-Time Applications of Smart Dust Particle Sensor Technology

Shreya Sathyan[1,2] and Sini Raj Pulari[1,2(✉)]

[1] Department of Computer Science and Engineering, Amrita School
of Engineering, Coimbatore, India
`cb.en.u4csel3457@cb.students.amrita.edu`,
`p_siniraj@cb.amrita.edu`
[2] Amrita Vishwa Vidyapeetham, Amrita University, Coimbatore, India

Abstract. Sensors are the major technological advancements which have contributed abundantly in the field of IoT, Big Data etc. Sensors are used to sense, detect and respond to various electrical and optical signals. In earlier days, sensors had very large size with high power consumption and with poor computational capabilities. Hence the smart sensors in the size ranging from tiny dust particles with higher computational, processing capabilities were introduced. This paper takes you on to a tour of various developments in the field of Smart dust technology and the various real-time applications that use the concept.

Keywords: Micro-electro-mechanical sensors · Motes · Sensors
Smart dust · Technology

1 Introduction

Smart dust is one of the hot topics in today's era. A lot of real-world applications rely on the imminent concept of smart dust. From the information in Gartner's 2016 hype cycle for emerging technologies, the innovation trigger of smart dust is one of the most recent with fewer than 10 years [1]. In the near future, smart dust will emerge as one of the most disruptive classes of technologies which will contribute to the perceptual smart machine age [2]. The Smart dust has capabilities like sensing, computing, and wireless communication. Smart dust is made up of very small, wireless sensors called 'motes'. Smart dust consists of micro-electro-mechanical (MEMS) sensors. Smart dust particles are tiny enough that they can be suspended in the atmosphere so that various

© Springer International Publishing AG 2018
D. J. Hemanth and S. Smys (eds.), *Computational Vision and Bio Inspired Computing*,
Lecture Notes in Computational Vision and Biomechanics 28,
https://doi.org/10.1007/978-3-319-71767-8_16

environmental factors can be measured. Various real-world problems related to industries, agriculture, natural disasters etc. can be solved with the integration of smart sensors [3].

2 Related Works

First man-made sensor was developed in 1883. This was a thermostat and the size was huge compared to the present time sensors. As years passed by a lot of sensors have been invented of varied sizes and specifications. More and more technologies were incorporated in these sensors.

There has always been a large fluctuation in the size of the sensors. Motion Detector Sensor has dimensions $5.75 \times 3.5 \times 7.75$ inches [4]. These are sensors which look for a Doppler shift in waves it returns to a detector, which would indicate wave has hit a moving object. If a Doppler shift is detected by this, it activates to show it has detected motion. Infrared (IR) Sensors are of $1/2$ and $2/3$ inches according to the change in focal length. These sensors are used for operating television remote. In the case of DSLR cameras, the sensor size has increased for increased exposure of light. It ranges from 22.2×14.8 mm, 23.5 to 23.7×15.6 mm and 36×24 mm [5]. Speed sensor dimensions are $33.5 \times 36.9 \times 8.1$ mm. Electro-chemical sensors are 20 mm size [6]. The vibrational sensor is of dimensions $15 \times 15 \times 15$ mm module. Vibration sensors can measure and analyze displacement, linear velocity, and acceleration [7]. Board-accelerometer sensor is of dimensions 20×20 mm. Tilt sensors are of 4 mm diameter and 12 mm long. Analog sound sensors are of dimension 22×32 mm [8]. Tilt sensors are of 4 mm diameter and 12 mm long. Digital humidity sensors are used to measure the level of the humidity in the air. Also, moisture and air temperature are also found out by these kinds of sensors. Digital humidity sensor dimensions are $2 \times 2 \times 0.75$ mm^3 [9]. Temperature micro-sensor has a tip diameter of 200 micrometers. Then there was introduction of sand-sized sensor particle called smart dust. Smart dust is of size 120 millionth of a meter [10] (Fig. 1).

3 Developments in Smart Dust

According to Gartner Hype cycle for emerging technologies released in July 2016 (Fig. 2) [1], Smart dust is one of the major emerging particle sensor technologies.

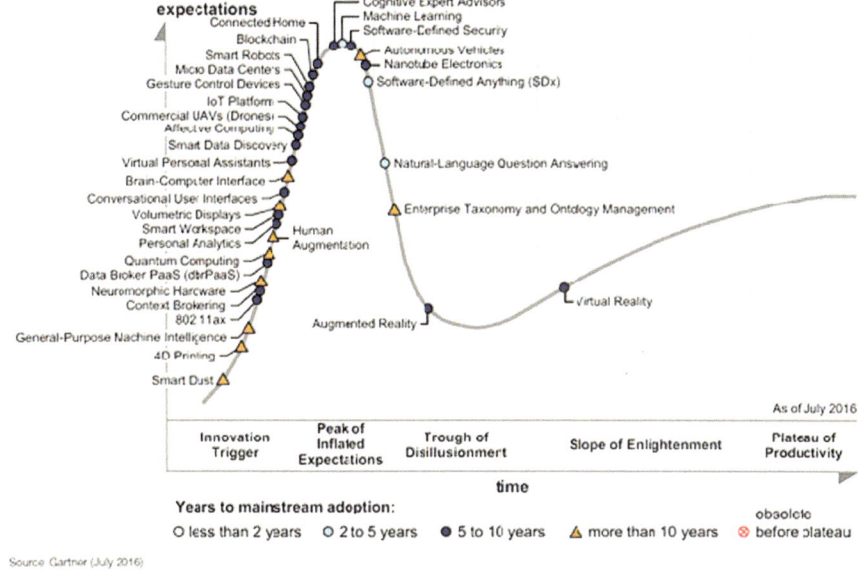

Fig. 1 Gartner hype cycle for emerging technologies (July 2016)

3.1 Smart Dust Architecture

Measurement of the entire world can be made much easier if the sensors were as tiny as grain particles. This dream is made possible with the help of smart dust. Smart dust not only includes sensors but also other micro-electro-mechanical (MEMS) devices with computational ability [3].

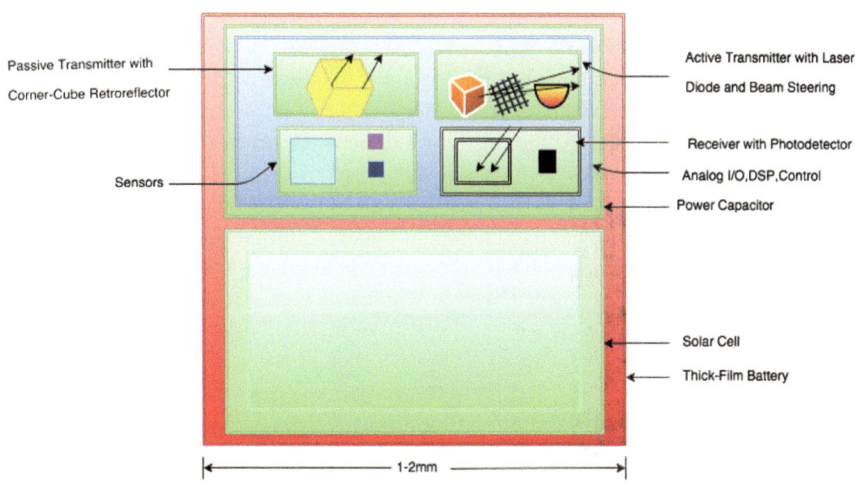

Fig. 2 Architectural diagram of smart dust mote

Smart dust motes are run by microcontrollers. These microcontrollers consist of tiny sensors for recording data. These sensors collect data and are stored in memory for further interpretations. Motes encompass nano-structured silicon sensors for the spontaneous reporting of a local environment. Active optical transmission is done using a semiconductor laser diode and beam steering. Passive optical transmission is done using Corner Cube Retro reflector (CCR). Solar cell and thick-film battery are provided as the power source [11].

3.2   Advantages and Disadvantages of Smart Dust

Table 1   Merits and de-merits of smart dust

Advantages	Disadvantages
1. Smaller size	1. Privacy is a major concern. Due to the smaller size, people are scared that they would be spied by companies
2. Reduced weight	
3. Reduced system and infrastructure cost	2. Security cannot be assured. Hackers can steal or manipulate data
4. Increased productivity	3. Once the smart dust particles are suspended in the environment, it is very difficult to retrieve due to the tiny size
5. Consume low power	
	4. Can cause unemployment, since it takes over human tasks

See Table 1.

4   Smart Dust Applications in Real-Time

4.1   Lightweight Authentication Protocol (LAP) for Smart Dust Wireless Sensor Networks

Wireless networks [12] are always prone to security problems. The main reason for this is the limitation of a number of resources send and some environmental factors [13]. Since from the introduction of smart dust in these wireless sensor networks the usual authentication algorithms have become insecure and unreliable. Hence, there is the introduction of Lightweight Authentication Protocol. Here, LAP is treated as a key management protocol [14]. Key management protocol is the protocol for the management of cryptographic keys [10]. LAP uses an unlimited number of sensor nodes. Computation cost, data stored and exchange of data is minimum in LAP. Data re-coded is processed by the sensor nodes itself. This is made available to smart dust nodes due to the global shared key approach [14].

4.2   Micro-Robotics

Smart dust plays a major role in making tiny robots. Wings and legs are attached to the smart dust so as to develop these micro-robots. All the functionalities of smart dust are included in this. Sensing, thinking, and communication are performed by this micro-robot. These micro-robots can fly and walk. It consumes power less than 10 mill watts. This power is provided by solar cells. It can also lift 130 times weight more than the actual weight of the micro-robot [15]. One of the most interesting tasks that can be done with micro-robots is finding survivors in a collapsed building due to the earthquake, plane crash etc. [16].

4.3   Ultra-Low Energy Analog to Digital Converter (ADC) for Smart Dust

ADC is developed for the Smart Dust mote. Most of the ADC needs a full microprocessor. Here a finite state machine (FSM) controls this mote which cannot be reprogrammed. This ADC is embedded on the CMOS chip along with a photo sensor, the FSM, and an optical receiver. The optical receiver is for downlink communications. Along with this, there is another sensor in the form of a capacitive MEMS accelerometer which is multiplexed by the FSM to the ADC input. A four-quadrant corner-cube retro reflector (CCR) takes care of the uplink communications. A MEMS solar cell array supply power to the system. Solar cells are connected in series for providing the variable supply of voltage which is isolated from each other [17].

4.4   Zero Power Design for Smart Dust Networks

The power consumption of the network is reduced by optimizing the clock frequency. Two types of processors are compared and provide the clock rate accordingly. In addition to this, a zero power communication is also proposed. Node to node communication can be done mainly in two ways. One is radio frequency which has many disadvantages due to the small size of smart dust. Optical communication is another way to communicate. It can accommodate small sized smart dust for communication [18].

4.5   Tracking Real-World Phenomena with Smart Dust

Novel techniques for target location estimation, node localization, time synchronization and message ordering are being proposed for the real-world phenomena tracking using smart dust. Remote controlled toy car is taken as sample target. Proof-of-concept system for tracking the location of real-world phenomena with Smart Dust is already proposed. This system can have a wide range of possible target types. System implementation uses Commercial off the (COTS) dust. Re-implementation of the whole system using smart dust is the next step [19].

4.6   Smart Materials for Military Applications

Smart dust particles can be injected in the uniform of soldiers making the uniform more durable as well as helping them from various environmental factors. In the military, there is a high chance for soldiers to be wounded. Relay information of this can be sent for medical help. The speed of the troop can also be tracked. Any unauthorized presence of people can also be monitored. Equipment failures and hazardous chemicals can be detected. In military soldiers do not get pure water always. For this smart dust can target contaminants in drinking water [20–23]. Border surveillance can also be done with the help of smart dust against terrorism [24, 25].

4.7   Smart Dust to Monitor Human Thought

Smart dust particles are spread in the brain of a paralyzed patient Cathy Hutchinson. She was connected with a robot arm through a rod-like pedestal driven into her skull. In one end of this pedestal, there is a bundle of gold wires is being attached to a very small sized array of microelectrodes implanted in primary motor cortex of her brain. Neural activities are recorded by these sensors. In another end, the external cables transmit neural data to a nearby computer for decoding (Fig. 3) [26]. This was very successful. But still, people think that this is risky. After few years there will be a change in the thinking of people with more and more development in this area [27–30].

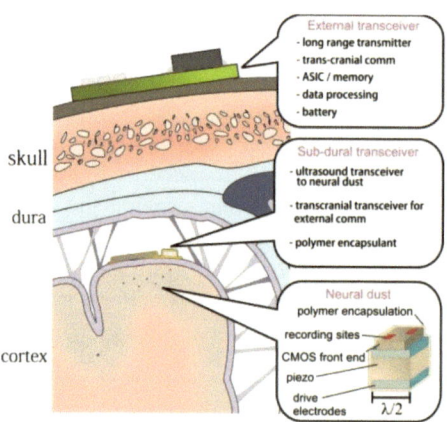

Fig. 3 Working of human thought monitoring working. *Credits* Michel Maharbiz, Neural Dust (August 2016)

4.8   Parking

A smart parking system [31] for cities of San Francisco and Los Angeles is being developed by dust networks customer Street line Inc. This system mainly focuses on helping people to find free space in busy and crowded parking lots. Sensors are embedded in the surface of parking spaces. In order to find the nearest area free to park drivers are given a password to activate the dashboard system [32]. 12,000 sensors are used in this system. Magnetometer helps this system to detect whether a car is located above the surface [33]. If this system becomes successful, the system can be used all over the world making the problem created by parking much simpler.

4.9   Temperature Prediction

Cambridge Consultants are working on smart dust. Their vision is to spread smart dust sensors all over the mountain so that the temperature can be continuously monitored. In our home, smart paint can be used to monitor the temperature of the room. Along with this humidity and noise can also be monitored [34]. The freshness of fruits and vegetables in a fridge and the quality of air can also be measured with the help of smart dust [35, 36].

4.10   Medical Field

The onset of diseases can be detected by smart dust. For example, cancer can be detected by injecting smart dust sensors into the blood. So as to cure brain disorders, Berkeley researchers are proposing to sprinkle smart dust sensors in the brain, so that they can monitor the neurons [3, 37]. Implanting of smart dust electrodes in the brain can make a paralyzed person to control a robotic arm. Chance of infection is less because there are no wired electrodes. This can also be a turning point in the area bladder control and appetite suppression [38]. Wireless nodes can be very helpful for elderly people to check their diabetes rather than going to the hospital each and every time [39].

4.11   Super-Sharp Images

3D printed lenses can take super sharp images. These are of the size of the grain. Also, these are 3D printed in one piece. There is very high scope in the area of medical imaging. Sharp and clear images can help doctors to be more precise about the area and intensity of fractures [27, 40]. Even in the manufacturing sector, this can be useful. Surface design restrictions and alignment difficulties can be solved easily [41]. This lens can be used for space telescopes also [42]. Micro cameras can be injected into the human body using a syringe, so that diseases, RBC rate, glucose rate can be monitored [43].

4.12   Agriculture

Damage to crops can be prevented by smart dust sensors, by deploying them in the land. Also, the purity of soil can be found with the help of smart dust sensors [42, 44]. The amount of irrigation and pesticides can be done in the right amount. Bacterial contamination can also be found out. Nowadays, people need farmers with higher skills about soil and other environmental factors. With the help of this unskilled labors can also be employed [45, 46].

4.13   Space and Planetary Exploration

In the planetary probes sent from Earth smart dust can be incorporated and then they can be injected into the atmosphere so that they float. The environmental conditions can be explored with the help of this. The existence of water and oxygen can also be detected. With this type of a system, other solar systems can also be monitored for research purpose [47, 48]. Micro-rockets are made using smart dust that can be sent to space [49].

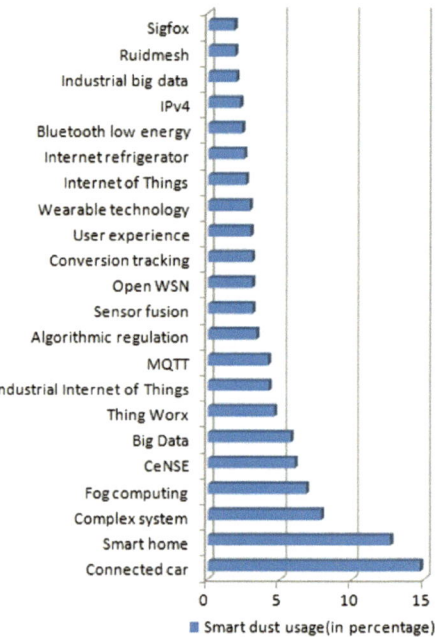

Fig. 4 Smart dust future applications. *Credits* Global Smart Dust may be the ultimate Internet of Things, September 14, 2016

Smart dust is not an isolated technology. For example, in the graph mentioned above (Fig. 4) there is collaboration with domains like the Internet of (IoT) or Big Data. Smart dust can be used widely for technologies like connected cars and smart home. Smart dust can also be used in the complex system, sigfox, ruidmesh, sensor fusion, open WSN, internet refrigerator, Bluetooth low energy, wearable technology, user experience, conversion tracking, MQTT, thingWorx, and IPv4. CeNSE is one of the projects in which smart dust is working on [50].

5   Conclusion and Scope

Smart dust is a never ending area of technologies. Much advancement is already done using smart dust. Due to the small size and low data transmission, smart dust meets with few limitations. All those limitations will be rectified in next few years. Iris scanning for the transaction in ATM, eco detection [51], oil drops in the ocean [52], and pollution checking [53] are some of the areas where developments are going on.

Smart tooth brush is another important smart dust application [54, 55]. Yet lots of applications are still to be done in the coming years.

In the coming years, we will be witnessing clouds made up of a large number of tiny sensors moving in the wind. This can be helpful for getting weather reports. Hidden clots will be found out with the sensors passing through the arteries. Brain disorders can be cured by sprinkling smart dust particles in the brain [32]. All these may sound impossible. But these are the technologies that are going to be the reality.

References

1. Gartner's 2016 hype cycle for emerging technologies identifies three key trends that organizations must track to gain competitive advantage. Stamford Connecticut 16 August 2016 [Online]. Available: http://www.gartner.com/newsroom/id/3412017. Accessed 10 May 2017
2. Panetta, K.: 3 trends appear in the gartner hype cycle for emerging technologies, 19 August 2016 [Online]. Available: http://www.gartner.com/smarterwithgartner/3-trends-appear-in-the-gartner-hype-cycle-for-emerging-technologies-2016/. Accessed 11 May 2017
3. Ryan, D.: Smart dust: communication systems and the future world, 10 December 2013 [Online]. Available: https://googleweblight.com/i?u=https://chaione.com/blog/smart-dust-communication-systems-and-the-future-world/&grqid=9N33obZC&hl=en-IN. Accessed 25 Apr 2017
4. Garmin.: Dimension of the New Bike Speed Sensor and Bike Cadence Sensor, 2017 [Online]. Available: https://support.garmin.com/faqSearch/en-US/faq/content/Z8GIB2rJz U67qkFzTK5by9. Accessed 11 May 2017
5. Crisp, S.: Camera Sensor Size: Why Does it Matter and Exactly How Big are They? News Atlas, 21 Match 2013 [Online]. Available: http://googleweblight.com/i?u=http://newatlas.com/camera-sensor-size-guide/26684/&grqid=HzRqMZDo&hl=en-IN. Accessed 21 Apr 2017
6. Electrochemical Sensors, SGX Sensor tech, 2017 [Online]. Available: https://google weblight.com/i?u=https://www.sgxsensortech.com/products-services/industrial-safety/elec trochemical-sensors/&grqid=RrZgY4f4&hl=en-IN. Accessed 21 Apr 2017
7. Mathas, C.: What You Need to Know About Vibration Sensor, Contributed By Electronic Products, 18 October 2012 [Online]. Available: https://www.digikey.com/en/articles/techzone/2012/oct/what-you-need-to-know-about vibration-sensors. Accessed 21 Apr 2017
8. Gravity: Analog Sound Sensor for Arduino, DFRobot, 2017 [Online]. Available: https://googleweblight.com/i?u=https://www.dfrobot.com/product-83.html&grqid=ONaMSmyI&hl=en-IN. Accessed 21 Apr 2017
9. Humidity Sensor, Digital Humidity Sensor SHTC1 (RH/T), Sensirion, 2017 [Online]. Available: https://www.sensirion.com/en/environmental-sensors/humidity-sensors/digital-humidity-sensor-for-consumer-electronics-and-iot/. Accessed 20 Apr 2017
10. Wikipedia
11. Sarangi, J., Pratikshya, P., Satpathy, P.K.: A case study smart dust and Major challenges towards ITS implementation. Int. J. Res. Adv. Eng. Technol. 1(1), 39–47 (2015)
12. Sabarish, B., Shanmugapriya, S.: Improved Data Discrimination in Wireless Sensor Networks. Wirel. Sens. Netw. 4 (4), 117–119 (2012). doi: 10.4236/wsn.2012.44016
13. Shah, M.D., Gala, S.N., Shekokar, N.M.: Lightweight authentication protocol used in wireless sensor network. In: International Conference on Circuits, Systems, Communication and Information Technology Applications (CSCITA), pp. 138–143, Mumbai

14. Cardei, M., Thai, M.T., Li, L., Wu, W.: Energy-efficient target coverage in wireless sensor networks. In: Proceedings IEEE 24th Annual Joint Conference of the IEEE Computer and Communications Societies, vol. 13, pp. 1976–1984 (2005)
15. Warneke, B., Last, M., Liebowitz, B., Pister, K.S.J.: Smart Dust: communicating with a cubic-millimeter computer. Computer 34(1), 44–51 (2001)
16. Fearing, R.S.: Challenges for Effective Milli robots. In: 2006 IEEE International Symposium on MicroNanoMechanical and Human Science, pp. 1–5, Nagoya (2006)
17. Scott, M.D., Boser, B.E., Pister, K.S.J.: An ultralow-energy ADC for Smart Dust. IEEE J. Solid State Circuits 38(7), 1123–1129 (2003)
18. Karakehayov, Z.: Zero-ppower design for smart dust networks. In: Proceedings First International IEEE Symposium Intelligent Systems, vol. 1, pp. 302–305 (2002)
19. Römer, K.: Tracking real-world phenomena with smart dust. In: Karl, H., Wolisz, A, Willig, A. (eds.) Wireless Sensor Networks. EWSN 2004. Lecture Notes in Computer Science, vol. 2920. Springer, Berlin, Heidelberg (2004)
20. Tiwari, A.: Smart materials for military applications. In: National Conference on Emerging Trends in Intelligent Computing and Communication 13–14 April 2012. Galgotias College of Engineering and Technology, Great Noida, 29 April 2012 [Online]. Available: https://www.slideshare.net/Anupam_Tiwari/smart-materials-for-military-applications. Accessed 3 May 2017
21. Schiller, B.: Forget the Internet of Thing. The Future Is Smart Dust, 12 February 2013 [Online]. Available: https://googleweblight.com/i?u=https://www.fastcompany.com/3022 114/forget-the-internet-of-things-the-future-is-smart-dust&grqid=jhVYgjZH&hl=en-IN. Accessed 21 Apr 2017
22. Weinz, S.: Future Military Sensors Could Be Tiny Specks of 'Smart Dust', 28 September 2014 [Online]. Available: https://medium.com/war-is-boring/smart-dust-is-getting-smarter-4b062abd7769. Accessed 26 Apr 2017
23. Bishop, R.: Smart Dust and Remote Sensing the Political Subject in Autonomous Systems, 2015 [Online]. Available: http://culturalpolitics.dukejournals.org/content/11/1/100.abstract. Accessed 27 Apr 2017
24. Mohan, S.C., Arulselvi, S.: Smart dust network for tactical border surveillance using multiple signatures. IOSR J. Electron. Commun. Eng. (IOSR-JECE) 5(5), 01–10. e-ISSN 2278-2834, p-ISSN 2278-8735
25. Dickson, S.A.: Enabling Battle Space Persistent Surveillance: The Form, Function, and Future of Smart Dust, Blue Horizons Paper Center for Strategy and Technology Air War College, 2007 April [Online]. Available: http://oai.dtic.mil/oai/oai?verb=getRecord&meta dataPrefix=html&identifier=ADA497553. Accessed 1 May 2017
26. Maharbiz, M.: Neural Dust. 3 August 2016 [Online]. Available: http://scienceatcal.berkeley. edu/aug-3-cafe-neural-dust-with-michel-m-marbiz/. Accessed 6 May 2017
27. Elise, A.: How Smart Dust Could be Used to Monitor Human Thought, Forbes, 7 July 2013 [Online]. Available: http://www.forbes.com/sites/eliseackerman/2013/07/19/how-smart-dust-could-be-used-to-monitor-human-thought/. Accessed 1 May 2017
28. Morgan, E.: Chemtrails: Neural Smart Dust Connects the Brain and Computer (Wireless mind control), 8 August 2016 [Online]. Available: http://googleweblight.com/i?u=http:// preparforchange.net/chemtrails-neural-smart-dust-connects-brain-and-computer-wireless-mind-control/amp/&grqid=en-IN. Accessed 24 Apr 2017
29. Farrell, J.P.: The Latest in Mind Manipulation Technology: Neural Smart Dust? Giza Death Star, 7 August 2016 [Online]. Available: https://gizadeathstar.com/2016/08/latest-mind-manipulation-technology-neural-smart-dust/. Accessed 27 Apr 2017

30. Whitman, R.: Smart neural dust could carry sensors deep into the human brain, send data back out, 17 July 2013 [Online]. Available: https://www.extremetech.com/extreme/161525-smart-neural-dust-could-carry-sensors-deep-into-the-human-brain-send-data-back-out. Accessed 28 Apr 2017
31. Sivaraman, R., Vasudevan, S.K., Kannegulla, A., Reddy, A.S.: Sensor based smart traffic regulatory/control system. Inf. Technol. J. **12**(9), 1863 (2013)
32. Smart Dust Future, Nanowerk, 7 December 2008 [Online]. Available: http://googleweblight.com/i?u=http://www.nanowerk.com/news/newsid%3D8535.php&grqid=aoXtEqOj&hl=en-IN. Accessed 23 Apr 2017
33. Sutter, J.D.: Smart Dust Aims to Monitor Everything, 3 May 2010 [Online]. Available: http://edition.cnn.com/2010/TECH/05/03/smart.dust.sensors/. Accessed 5 May 2017
34. Twentyman, J.: Future of IoT Will be 'Smart Dust', says Cambridge Consultants, 29 March 2017 [Online]. Available: https://googleweblight.com/i?u=https://internetofbusiness.com/future-of-iot-will-be-smart-dust-says-cambridge-consultants/&grqid=VV_Rc4Ps&hl=en-IN. Accessed 11 May 2017
35. Smart Dust & Ubiquitous Computing, 20 April 2015 [Online]. Available: http://googleweblight.com/i?u=http://www.nanotech-now.com/smartdust.htm&grqid=3mfAfe5C&hl=en-IN. Accessed 11 May 2017
36. Fell, J.: Electronica 2016: Smart Dust Sensor Designed for Indoor Air Quality Monitoring, 8 November 2016 [Online]. Available: https://googleweblight.com/i?u=https://eandt.theiet.org/content/articles/2016/11/electronica-2016-smart-dust-sensor-designed-for-indoor-air-quality-monitoring/&grqid=YcRuAV7-&hl=en-IN. Accessed 1 May 2017
37. How Smart Dust Could Spy on Your Brain, Emerging Technology from the arXiv, 16 July 2013 [Online]. Available: https://googleweblight.com/i?u=https://www.technologyreview.Com/s/517091/how-smart-dust-could-spy-on-your-brain/&grqid=UFSZGP4R&hl=en-IN. Accessed 21 Apr 2017
38. Lawerence, C.: Is Smart Dust the IoT Vector of the Future? Health, 20 Augest 2016 [Online]. Available: http://readwrite.com/2016/08/20/smart-dust-carrier-iot-future-dl4/. Accessed 1 May 2017
39. Darwish, A., Hassanien, A.E.: Wearable and Implantable Wireless Sensor Network Solutions for Healthcare Monitoring, 26 May 2011 [Online]. Available: https://www.ncbi.nlm.nih.gov/pubmed/22163914. Accessed 2 May 2017
40. Smart Dust Is Coming: New Camera Is the Size of a Grain of Salt, Trends FM-Trends and Future Megatrends, 29 June 2016 [Online]. Available: https://googleweblight.com/i?u=https://singularityhub.com/2016/06/28/smart-dust-is-coming-new-camera-is-the-size-of-a-grain-of-salt/&grqid=vqclH1i3&hl=en-IN. Accessed 29 Apr 2017
41. Dorrier, J.: Smart Dust Is Coming: New Camera Is the Size of a Grain of Salt, 28 June 2016 [Online]. Available: https://singularityhub.com/2016/06/28/smart-dust-is-coming-new-camera-is-the-size-of-a-grain-of-salt/. Accessed 20 Apr 2017
42. Gawlowicz, S.: Smart Dust Technology Could Reshape Space Telescopes 1 December 2014 [Online]. Available: https://www.rit.edu/news/story.php?Id=51127. Accessed 21 Apr 2017
43. Micro-camera Can be Injected with a Syringe, Phys.org, 27 June 2016 [Online]. Available: https://m.phys.org/news/2016-06-micro-camera-syringe.html. Accessed 29 Apr 2017
44. Dust That Changing The World, The World With 'Smart Dust'! LG CNS, 30 April 2015 [Online]. Available: http://www.lgcnsblog.com/features/dust-that-changing-the-world-the-world-with-smart-dust-2/. Accessed 20 Apr 2017
45. Rowinski, D.: Smart Dust Will Be The Future of The Internet Of Things, Connected World, 17 August 2016 [Online]. Available: https://arc.applause.com/2016/08/17/smart-dust-practical-applications/. Accessed 20 Apr 2017

46. Saracco, R.: The Future of Jobs: Smart Dust Programmers, IEEE FutureDirections, 26 December 2016 [Online]. Available: http://sites.ieee.org/futuredirections/2016/12/26/the-future-of-jobs-smart-dust-programmers/. Accessed 2 May 2017

47. Barker, J.: Smart Dust for Space Exploration, 26 January 2017 [Online]. Available: http://googleweblight.com/i?u=http://userweb.elec.gla.ac.uk/j/jbarker/sd.html&grqid=VBjcBm Xl&hl=en-IN. Accessed 24 Apr 2017

48. Smart Dust: Space Explorers of Future, The Daily Galaxy, 17 April 2007 [Online]. Available: http://googleweblight.com/i?u=http://www.dailygalaxy.com/my_weblog/2007/04/smart_dust_nano.html&grqid=84xvoGup&hl=en-IN. Accessed 25 Apr 2017

49. Teasdale, D., Milanovic, V., Chang, P., Pister, K.S.J.: Micro rockets for Smart Dust. Smart Mater. Struct. **10**(6) (2001)

50. Global Smart Dust May Be the Ultimate Internet of Things. Raw Science, 14 September 2016 [Online]. Available: https://www.rawscience.tv/global-smart-dust-may-be-the-ultimate-Internet-of-things/. Accessed 6 May 2017

51. Inman, M.: Smart Dust Sensors to Be Used for Eco Detection. National Geography News, 14 November 2006 [Online]. Available: http://news.nationalgeographic.com/news/2006/11/061114-smart-dust.html. Accessed 2 May 2017

52. Tao, A.: Smart Dust. Sailor Research Group, University of California, San Deigo, 17 February 2003 [Online]. Available: http://googleweblight.com/i?u=http://sailorgroup.ucsd.edu/research/smartdust.html&grqid=KT-dGdGc&hl=en-IN. Accessed 3 May 2017

53. Dickinson, B.: With 'Smart Dust,' A Trillion Sensors Scattered Around the Globe. ZDNet, 7 May 2010 [Online]. Available: http://www.zdnet.com/article/with-smart-dust-a-trillion-sensors-scattered-around-the-globe/. Accessed 3 May 2017

54. Low-cost IoT Breakthrough, Cambridge Consultants, 23 February 2017 [Online]. Available: https://www.cambridgeconsultants.com/media/press-releases/low-cost-iot-breakthrough. Accessed 11 May 2017

55. Twentyman, J.: Future of IoT Will be 'Smart Dust', says Cambridge Consultants, 29 March 2017 [Online]. Available: https://internetofbusiness.com/future-of-iot-will-be-smart-dust-says-cambridge-consultants/. Accessed 21 Apr 2017

Comparative Analysis of Biomedical Image Compression Using Oscillation Concept and Existing Methods

Satyawati S. Magar$^{(\boxtimes)}$ and Bhavani Sridharan

Department of ECE, Karpagam Academy of Higher Education,
Karpagam University, Coimbatore, Tamilnadu, India
magarss_123@redffmail.com, bhavanisns@yahoo.com

Abstract. Medical image compression has an important role in hospitals as they are moving towards filmless imaging and go completely digital. Medical imaging produces the great challenge of compression algorithms. While compressing the data to avoid diagnostic errors and yet have high compression rates for reduced storage and transmission time. In this paper, comparative analysis of biomedical image compression using oscillation concept & existing methods is done and we have achieved maximum compression ratio by using oscillation concept method. Other parameters like PSNR, MSE, MSSIM & Std. Deviation are also good as compared to existing methods.

Keywords: Oscillations concept · PACS · DCT · DWT · DFT
Fractal compression · CR · PSNR · MSE · MSSIM · Std. deviation

1 Introduction

In recent years, the development and demand of the medical image compression grows increasingly fast and contributing in the storage of diagnostic information. Therefore, the image compression becomes an essential in medical field and medical image compression has an important role in hospitals as they become completely digital. Biomedical images are essential and important data about patients. So hospitals should have to keep & maintain diagnostic information in the form of images, which requires a more space and transmission bandwidth to store the information. To reduce the file size for storage requirement purpose, Picture Archiving and Communication Systems (PACS) is used, which maintains relevant diagnostic information [1].

Medical imaging produces the great challenge of compression algorithms. The special care has to be taken, while compressing the data to avoid diagnostic errors and yet have high compression rates for reduced storage and transmission time. For this different image compression techniques are used. But to improve image quality and reducing errors, there is need of new techniques for compression of biomedical images.

© Springer International Publishing AG 2018
D. J. Hemanth and S. Smys (eds.), *Computational Vision and Bio Inspired Computing*,
Lecture Notes in Computational Vision and Biomechanics 28,
https://doi.org/10.1007/978-3-319-71767-8_17

In this paper, we propose an approach of oscillation concept for biomedical image compression, and we have achieved maximum compression ratio [2].

2   Image Compression

Compression is a very essential tool for archiving image data, image data transfer on the network etc. By compressing an image we can reduce the quantity of data which is used to represent a file, image content without exceptionally reducing the quality of the original data [3].

2.1   Techniques Available for Image Compression

- Lossy compressions.
- Lossless compressions.

2.2   Methods for Image Compression

Different methods are used for comparison are Discrete Cosine Transform (DCT), Discrete Fourier Transform (DFT), Discrete Wavelet Transform (DWT), Fractal Compression [2, 4–7].

2.3   The Objectives of an Image Compression Are

1. To reduce the redundancy of an image.
2. To store or transmit data in an efficient manner.
3. To find out oscillations in a biomedical image.
4. To extract principle component of biomedical image using oscillations.
5. To achieve maximum compression ratio of biomedical image [8].

3   Oscillation Concept Method

3.1   Image Compression Using Oscillation Concept

It outlines the new approach to biomedical image compression using oscillation concept. It introduces the theory of oscillations in images. Proposed theory states that in every image it has variations in pixels with respective to x and y axis of an image. These variations are nothing but oscillations at image. Appropriate oscillations can be utilized for image compression. Here we are applying this oscillation theory to a biomedical image. This method shows effective compression ratio [8–10].

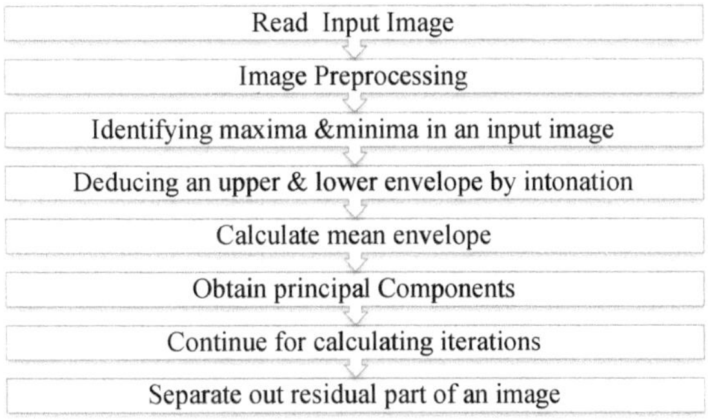

| Read Input Image |
| Image Preprocessing |
| Identifying maxima &minima in an input image |
| Deducing an upper & lower envelope by intonation |
| Calculate mean envelope |
| Obtain principal Components |
| Continue for calculating iterations |
| Separate out residual part of an image |

Fig. 1 Flow chart of oscillation concept Method

3.2 Steps for Image Compression Using Oscillations Concept

Following are the steps for the Image Compression using Oscillation Concept [8, 11].

4 Parameters to Be Analysed

4.1 Compression Ratio (CR)

It is the ratio of size of original image to the size of compressed image.

$$CR = \frac{\text{Size of Input Image}}{\text{Size of output Image}} \quad (1)$$

4.2 Mean-Square Error (MSE)

The MSE represents the cumulative squared error between the compressed and the original image. The lower the value of MSE, the lower the error. To compute the PSNR, the block first calculates the mean squared error using the following equation [12].

$$MSE = \sum_{i=1}^{x} \sum_{j=1}^{y} \frac{(Aij - Bij)^2}{x * y} \quad (2)$$

where,
x   Width of image, y: Height,
x * y   number of pixels (or quantities),
i   Column, j: Rows.

4.3 Peak Signal to Noise Ratio (PSNR)

The PSNR computes the peak signal to noise ratio, in dB between two images. This ratio is often used as a quality measurement between the original and a compressed image. When the PSNR value is more the quality of the compressed or reconstructed image is better. PSNR is generally used to analyze quality of image, sound and video files in dB (decibels) [4, 12].

$$PSNR = 10 * \log\left(\frac{255^2}{MSE}\right) \tag{3}$$

4.4 Structural Similarity Index (SSIM)

To compute structural similarity index between two images. It's value is between -1 and $+1$. When two images are nearby identical their SSIM is close to 1. Formula computing SSIM between two sequences seq 1 and seq 2 at a given pixel [12].

$$\mathbf{SSIM} = \frac{2 * \mathbf{mu1(P)} * \mathbf{mu}^2(\mathbf{P}) + \mathbf{C1}}{\mathbf{mu1(P)}^2 + \mathbf{mu2(P)}^2 + \mathbf{C1}} + \frac{2 * \mathbf{cov(P)} + \mathbf{C2}}{\mathbf{s1(P)}^2 + \mathbf{s2(P)}^2 + \mathbf{C2}} \tag{4}$$

where

mu1(P) and mu2(P)	Mean value of sequence 1 and sequence 2 computed over a small XY window located around P.
S1(P) and s2(P)	standard deviation of seq 1 and seq 2 computed over the same window
cov(P)	covariance between seq 1 and seq 2 computed over the same window
C1:(K1*L)^2	regularization constant(should be as small as possible)
C2:(K2*L)^2	regularization constant(should be as small as possible)
K1, K2	regularization parameters (must be > 0)
L	dynamic range of the pixel values (example: L = 255 if the sequence is 8 bit encoded) The default window is a Gaussian window with standard deviation 1.5 along both the X and the Y axis.

4.5 MSSIM (Mean Structural Similarity Index)

Mean SSIM index to evaluate the overall image quality [12, 13].

$$\mathbf{MSSIM} = \frac{1}{\mathbf{M}} \cdot \sum_{J=1}^{M} SSIM\,(Xj, Yj) \tag{5}$$

where

X	Reference Images Y: distorted images;
xj and yj	are the image contents at the jth local window;
M	The number of samples in the quality map.

5 Result

For experimental purpose, 8 bit brain images has been taken as test images as shown in Fig. 2. Matlab simulated results show that a high compression ratio and better PSNR was achieved. When Oscillation concept is used and executed using the more number of iteration of principle extraction while preserving the quality of an image. The Results of the simulation are shown in Figs. 2, 3, 4, 5, 6, 7 and 8; Tables 1 and 2.

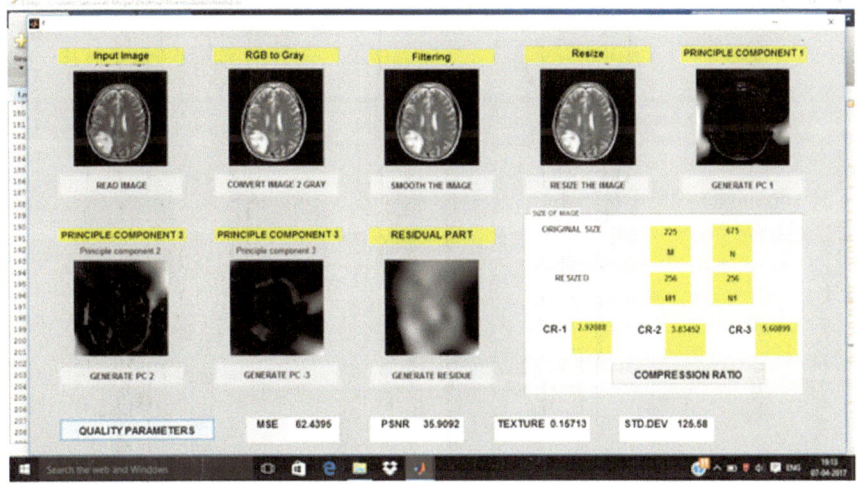

Fig. 2 Images and quality parameters of oscillation concept method

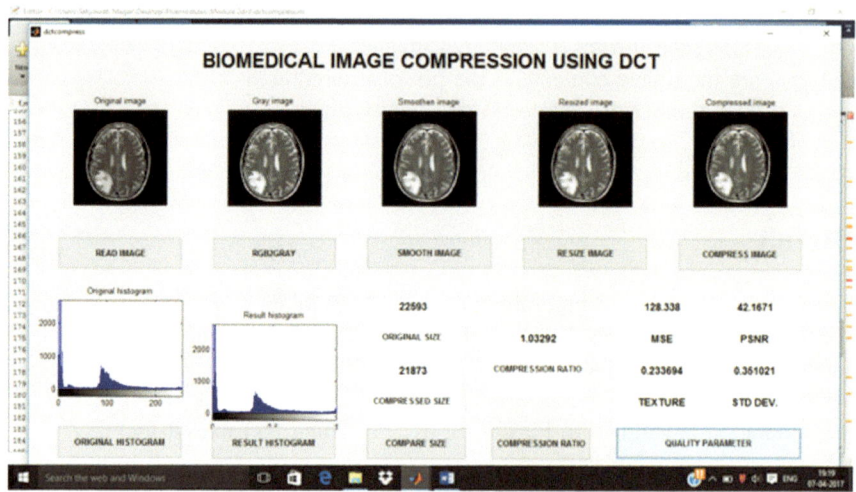

Fig. 3 Images and quality parameters of DCT

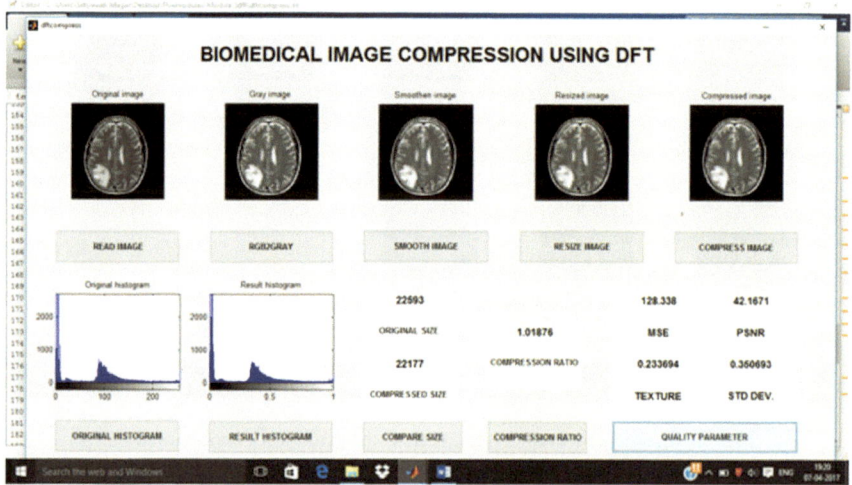

Fig. 4 Images and quality parameters of DFT

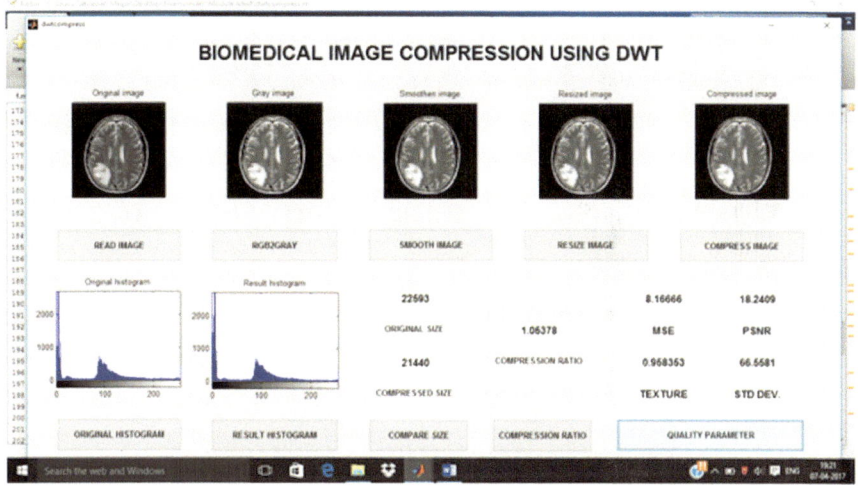

Fig. 5 Images and quality parameters of DWT

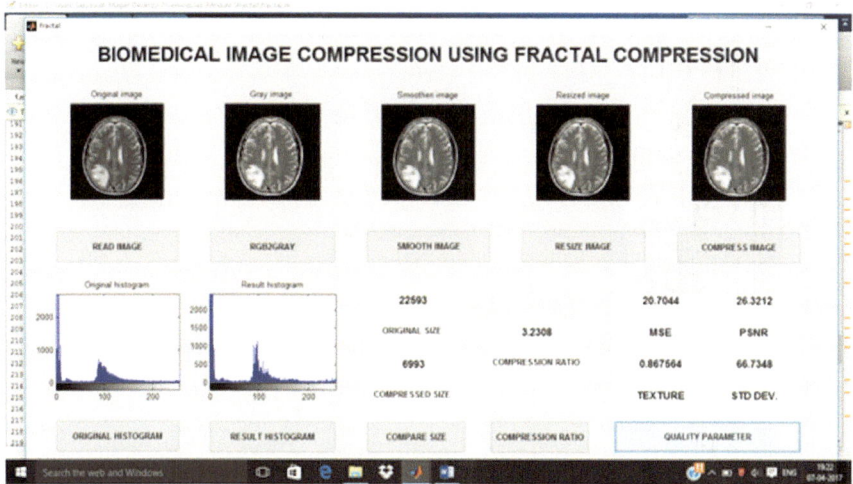

Fig. 6 Images and quality parameters of fractal compression

Table 1 Comparative analysis of existing image compression techniques

Sr. No.	Technique	PSNR	MSE	MSSIM	Std. Dev
1.	Oscillation concept method	35.9092	62.4396	0.15713	125.58
2.	DCT	42.1671	128.338	0.233694	0.351021
3.	DFT	42.1671	128.338	0.233694	0.350693
4.	DWT	18.2409	8.16666	0.9583	66.5581
5.	Fractal	26.3212	20.7044	0.86	66.7348

Table 2 Comparative analysis of CR for different image compression techniques

Sr. No.	Technique	Image size in bytes		Compression ratio (%)
		Input	Output	
1.	Oscillation concept method	22,593	4028	5.60899
2.	DCT	22,593	21,873	1.03292
3.	DFT	22,593	22,177	1.01876
4.	DWT	22,593	21,440	1.05378
5.	Fractal	22,593	6993	3.2308

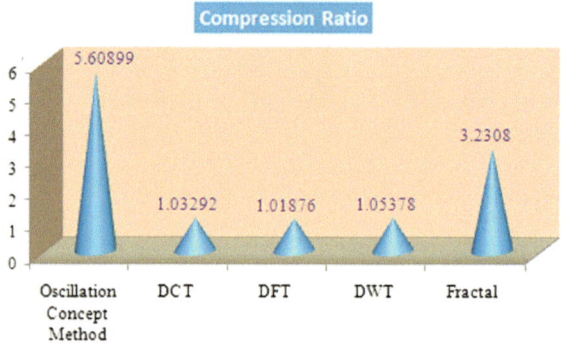

Fig. 7 Comparative analysis for Compression Ratio (CR)

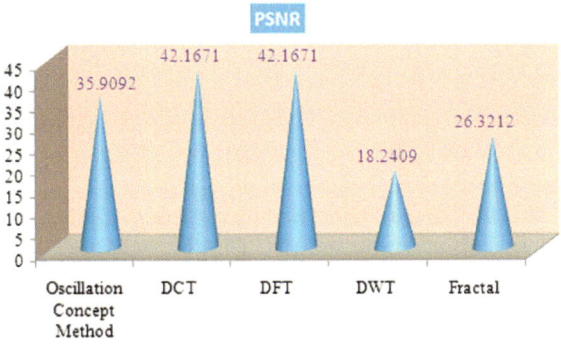

Fig. 8 Comparative analysis for Peak Signal to Noise Ratio (PSNR)

6 Conclusion

In this paper we have done comparative analysis of biomedical image compression using oscillation concept & existing methods. It is observed that with more number of iterations of principle extraction Oscillation concept method can achieve highest compression ratio as compared to other compression techniques. By using this technique Maximum Compression ratio is achieved i.e. 5.60%. It is also observed that other parameters like PSNR, MSE, MSSIM and Standard Deviation are also good as compared to other existing methods.

References

1. Dragan, D., Ivetic, D.: A comprehensive quality evaluation system for PACS. Special Issue on ICIT 2009 Conference—Bioinformatics and Image, **4**(3), 642–650 July 2009. ISSN 1992-8424
2. Shiwangi, S.K.: Analysis of image compression algorithm using DCT, DFT and DWT transform. Int. J. Adv. Res. Comput. Sci. Softw. Eng. (IJARCSSE), **6**(7) (2016). ISSN 2277-128x
3. Mathur, G., Mathur, R., Mathur M.R.: A comparative study of various lossy image compression techniques. In: ETRASCT, 14 Conference Proceedings on International Journal of Engineering Research and Technology (IJERT), pp. 165–169. ISSN: 2278-0181
4. Bhavani, S., Thanushkodi, K.: A novel fractal image coding for quasi lossless medical image compression. Eur. J. Sci. Res. **70**(1), 88–97 (2012)
5. Gupta, M., Amit Kumar Garg.: Analysis of image compression algorithm using DCT. Int. J. Eng. Res. Appl. (IJERA) **2**(1), 515–521 (2012). ISSN: 2248-9622
6. Beladgham, M.: Performance evaluation of DWT compared to DCT for compression of biomedical image. Int. J. Mod. Educ. Comput. Sci. (IJMECS) **4**, 9–15 (2014)
7. Kumar, T., Kumar, R.: Medical image compression using hybrid techniques of DWT, DCT and huffman coding. Int. J. Innovative Res. Electr. Electron. Instrum. Control Eng. (IJIREEICE) **3**(2) (2015). ISSN: 2321-2004

8. Magar, S.S., Sridharan, B.: Innovative approach to biomedical image compression using oscillation concept. International Conference on Automotive Control and Dynamic Optimization Techniques (ICACDOT), IEEE Conference Publications-IEEE Xplore, pp. 124–128
9. Pandey, P.K., Singh, Y., Tripathi, S.: Image processing using principle component analysis. Int. J. Comput. Appl. (0975 – 8887), **15**(4) (2011)
10. Chaudhary, R.N.: Waves and oscillations. New Edge International Publishers, India
11. Chaphman, S.: Matlab programming for engineers. Cengage Learning Publishers, India
12. Pinki, R.M.: Estimation of image quality under different distortions. Int. J. Eng. Comput. Sci. (IJECS) **5**(7), 17291–17296 (2016)
13. Saini, P.K., Singh, M.: Brain tumor detection in medical imaging using MATLAB. Int. Res. J. Eng. Technol. (IRJET) **2**(2), 191–196 (2015). e-ISSN: 2395-0056

Author Biographies

Mrs. Satyawati Magar was born at Maharashtra in the year 1971. She has completed her B.E E & TC from TPCT COE, Osmanabad in 1992, Marathwada University, Maharashtra and M.E in Electronics from JNEC, Aurangabad, BAMU, Maharashtra in 2006. She is having more than 20 years of experience in academics and currently working as Associate Professor in Dr. Vithalrao Vikhe Patil College of Engineering, Vilad Ghat, Ahmednagar, India. She has work experience of more than 7 years as Head of E & TC Department, Dr. Vithalrao Vikhe Patil College of Engineering, Vilad Ghat, Ahmednagar, Maharashtra, India. Her research interests are Digital Image Processing and Digital Communication. She has published more than 10 technical papers in National & International Journals & Conferences. She is also a research Scholar (Part-Time) in the Department of ECE at Karpagam University, Coimbatore, Tamilnadu, India. She is life member of ISTE & IEI.

Dr. S. Bhavani was born at Coimbatore in the year 1968. She has completed her B.E degree in Electronics and communication from V.L.B. Janaki Ammal College of Engineering and Technology, Coimbatore, Tamilnadu in 1990, M.E in Applied Electronics from Maharaja Engineering College, Coimbatore, Tamilnadu in 2006 and Ph.D. in Medical Image Fractal Compression from Department of EEE, Anna University, Coimbatore, Tamilnadu in 2013. She is having more than 25 years of experience in academics and currently working as Professor & Head of ECE Department, Faculty of Engineering, Karpagam University, Coimbatore, Tamilnadu, India. Her research interests are VLSI, Wireless Networks and Digital image Processing. She has published more than 25 technical papers in National & International Journals. She is also a life member of ISTE and National Scholarship holder.

Clustering of Trajectory Data Using Hierarchical Approaches

B. A. Sabarish[1(✉)], R. Karthi[1], and T. Gireeshkumar[2]

[1] Dept of Computer Science and Engineering, Amrita School of Engineering,
Amrita Vishwa Vidyapeetham, Coimbatore, India
ba_sabarish@cb.amrita.edu, r_karthi@ch.amrita.edu
[2] TIFAC CORE in Cyber Security, Amrita School of Engineering,
Amrita Vishwa Vidyapeetham, Coimbatore, India
t_gireeshkumar@cb.amrita.edu

Abstract. Large volume of spatiotemporal data as trajectories are generated from GPS enabled devices such as smartphones, cars, sensors, and social media. In this paper, we present a methodology for clustering of trajectories to identify patterns in vehicle movement. The trajectories are clustered using hierarchical method and similarity between trajectories are computed using Dynamic Time Warping (DTW) measure. We study the effects on clustering by varying the linkage methods used for clustering of trajectories. The clustering method generate clusters that are spatially similar and optimal results are obtained during the clustering process. The results are validated using Cophenetic correlation coefficient, Dunn, and Davies-Bouldin Index by varying the number of clusters. The results are tested for its efficiency using real world data sets. Experimental results demonstrate that hierarchical clustering using DTW measure can cluster trajectories efficiently.

Keywords: Trajectory · Dynamic time warping distance · Hierarchical clustering · Linkage methods

1 Introduction

With rapid development of global positioning systems, spatial trajectory data has been growing and spatial trajectory mining is increasing becoming an important area of research. A trajectory is the path a moving object follow in space as a function of time. Trajectories are generated by people, vehicle, animals, hurricane, ocean currents, robots etc. Extracting knowledgeable patterns from spatial datasets is difficult due to the complexity of spatial data. Attempts have been made to extend classical data mining algorithms such as association mining, classification, clustering, outlier analysis and time series analysis to spatial data [1, 2, 3]. Current trajectory clustering researchers focus on three aspects during algorithm development. They are finding a suitable feature representation for trajectories, similarity measure and an algorithm for

© Springer International Publishing AG 2018
D. J. Hemanth and S. Smys (eds.), *Computational Vision and Bio Inspired Computing*,
Lecture Notes in Computational Vision and Biomechanics 28,
https://doi.org/10.1007/978-3-319-71767-8_18

clustering spatial data. This paper focuses on trajectory clustering and analyses the effects of linkage measure and cluster size on generated clusters. Varying the methods of linkage measure and cluster size gives different cluster results for the same trajectory dataset. The efficiency of clustering is evaluated using clustering validity measures and by visualization. The paper is organized as follows: Sect. 2 discusses the related work and Sect. 3 discusses the problem of trajectory clustering and Sect. 4 give a general algorithm for trajectory clustering. Section 5 presents the results and discusses the effects of varying parameters of clustering algorithm. Section 6 summarizes the insights from the analysis and views on trajectory clustering.

2 Related Work

In this section, we focus on a survey of methods used for trajectory clustering. Zheng conducted a survey on trajectory data mining and discussed a framework for trajectory data mining. The framework reviews the techniques for pre-processing, indexing and retrieval, pattern mining and transformation for trajectory data [2]. Feng et al. proposed an architecture for trajectory data mining and investigated on the algorithms and applications in trajectory mining for path discovery, location prediction, movement behavior analysis, urban service [4]. Mazimpaka et al. aimed to represent trajectory algorithms into two categories primary and secondary. Primary methods categorize the trajectory based on properties of the data. Secondary methods analyze the data and aids in for movement pattern mining, outlier detection, and prediction [1]. Traclus algorithm was proposed by Lee et al. divides the clustering algorithm into two phases named partitioning and grouping. In partitioning phase, Minimum Description Length (MDL) principle is used and each trajectory is represented as group of sub-trajectories [5]. In the grouping phase, Density based clustering (DBSCAN) is done to group the similar trajectories based on density. Gafney et al. used probabilistic approach with regression model to represent cluster of trajectories. Expectation Maximization algorithm is used to find and solve the problem of hidden data including cluster membership for trajectories [6]. Chih et al. proposed a trajectory clustering approach using clues extracted from movement behavior of the user. The clues are extracted from the spatial and temporal points present in trajectory and clue based similarity and clustering algorithms are proposed [7]. Gudmundsson et al. proposed a method where trajectories are segmented based on the moving patterns and features of segments are represented using the geometric properties and motifs. These segments are clustered using density based approach and representative trajectories show the generalized movement of data in that cluster [8]. Kim et al. presented a method to cluster trajectory data using LCSS similarity measure and dbscan algorithm [9]. Vlachos et al. proposed a modified LCSS algorithm for finding similarity between two trajectories which are translated in space, noisy and moving at different speed. Hierarchical clustering is done to indexed trajectories and the nearest neighbors are identified using approximation algorithms [10]. Wang et al. conducted a comparative study of six commonly used trajectory similarity measures and studied the advantages and disadvantages of using these measures. The analysis is done by applying a series of transformations to trajectory such as adding noise, shifting and sampling and measured the similarity between the original and transformed trajectories [11]. Besse et al. proposed Symmetrized Segment-Path Distance algorithm based on distance between a point and trajectory. It is mean of all distances from a point in one trajectory to segments in another trajectory [12]. Zhao et al. proposed grid growing clustering algorithm for geo spatial data where a grid like structure is applied and every point will be identified a proper cell according to locations. Region starts growing on the grid by using some selected seed points from adjacent point and clusters are formed from the grid partitions [13]. In this paper, we focus on methods for clustering of trajectories and study the effects of linkage metrics on clustering. Hierarchical method is used for clustering and by varying the cluster size and linkage types analysis id done on the generated clusters. Analysis is done on the clusters based on validity metrics and by visualization.

3 Problem Statement

3.1 Trajectory

Trajectory TR is defined as a sequence of location points where the object has moved in space for a duration of time. $TR = [p_1, p_2, \ldots, p_n]$, where n is the number of points in the trajectory. Each point p is a 2-dimensional point representing the location information as latitude and longitude pair. The set T represents a list of trajectories $T = [TR_1, TR_2, \ldots, TR_m]$, where TR_i is a trajectory and m is the total number of trajectories that are used for analysis. Each trajectory TR_i is of varying length based on distance travelled and rate of sampling. The trajectories are sampled using Douglas–Peucker (DP) algorithm where the trajectories are approximated using line segments [14]. The points for the line segments are chosen based on the maximum error value, and point which has maximum error value is chosen as split point. The trajectory is constructed recursively by selecting split points, and the approximated trajectory is identified when the error value is below the specified threshold. The sampled trajectory is represented as TR_i^s, where s is the sample points in each trajectory and trajectory set is represented as T^s.

3.2 Distance Measure

Distance measure is used to find how close and related to two trajectories are from one another. There are many functions defined to measure similarity between trajectories. Existing distance measures are classified as global distance measures and local distance measures. Global distance measures capture the overall similarity between a pair of trajectories and local similarity work for short time intervals and sub trajectories [15]. Several global similarity measures include Euclidean distance, Dynamic Time Wrapping, Longest Common Sub-Sequence (LCSS), Edit Distance on Real Sequence, Edit Distance with Real Penalty, and Fréchet distance. Commonly used local similarity measures are Minimum Bounding Rectangles and Hausdorff distance [16]. In this work, we have used Dynamic Time Wrapping (DTW) as the distance measure which finds an optimal match between two sequences by alignment, where sequences are wrapped nonlinearly in the time dimension. The DTW algorithm finds a mapping path such that the cost is minimized [15]. A dynamic programming approach is used and DTW is defined by the following recurrence Eq. 1:

$$DTW(P_{1...n}, Q_{1...m}) = \|P_n - Q_m\| + \min \begin{bmatrix} DTW(P_{1...n-1}, Q_{1...m-1}) \\ DTW(P_{1...n-1}, Q_{1...m}) \\ DTW(P_{1...n}, Q_{1...m-1}) \end{bmatrix} \quad (1)$$

where $P_{1...n}$ and $Q_{1...m}$ are the two trajectories with length n and m.

3.3   Dissimilarity Matrix

A dissimilarity matrix captures the dissimilarity value between all pair of trajectories in the set $T = [TR_1, TR_2, \ldots, TR_m]$. The dissimilarity values are stored in a matrix $Sim[m, m]$ whose values are updated using DTW distance measure. $Sim[i, j] = DTW(TR_i, TR_j)$ where $i \in 1 \ldots m$ and $j \in 1 \ldots m$. The DTW measure maps a value of zero in the diagonal elements of $Sim[m, m]$ matrix.

3.4   Clusters

Clusters is a group where similar trajectories exist together and have similar properties. Let us consider m trajectories in $T = [TR_1, TR_2, \ldots, TR_m]$. The clustering algorithm group these trajectories into z clusters. The cluster groups are represented using $C = \{c_1, c_2 \ldots, c_z\}$, where c_i is cluster group i and each cluster group has n_i number of trajectories. The membership matrix is represented using $U = \{u_{ik}\}, i = 1, 2 \ldots, m, k = 1, 2, \ldots z$, where u_{ik} is either 0 or 1 based on trajectory membership to the cluster [17, 18].

3.5   Trajectory Clustering

Many researchers are attempting to development innovative trajectory clustering algorithms due to its wide range of applications and challenges involved in clustering. Trajectory data are of unequal length, so adapting partition based algorithms for clustering is a challenging task [12]. One solution is to convert the data using data reduction methods like sampling to make trajectories of equal length. The centroids of trajectories may not capture the relevant representative information of trajectories of the cluster. Density based algorithm like dbscan, optics require user to specify parameters for clustering that are hard to estimate for trajectory data. Hierarchical methods are most suitable for trajectory clustering as these algorithms require only the similarity metrics for clustering which can be calculated between trajectories. This work adapts the hierarchical method for clustering of trajectories.

4   Trajectory Clustering Algorithm

In section, we discuss the basic algorithm used for trajectories clustering using hierarchical approach. The algorithm consists of three phases. In the first phase the trajectories are generated and preprocessed to remove noise and missing values which are present in the data. Trajectories are transformed using aggregation functions to the required granularity. Trajectories are sampled and are represented using id, longitude, and latitude information. In the second phase, dissimilarity matrix is computed between the trajectory using the DTW measure. In the third phase, the trajectories are clustered using the hierarchical clustering algorithm and the clustering results are validated using clustering metrics. The algorithm for trajectory clustering is outlined below.

Algorithm for Trajectory clustering

```
Input: A Set of Trajectories T = [TR₁, TR₂,..., TRₘ] , where m is total num-
ber of trajectories.
Output: Trajectory clusters C = {C₁,C₂...C_z} where cᵢ represents clus-
ter i with nᵢnumber of trajectories
Step 1: Preprocessing
       For each TRᵢ ∈ T do
           Check for noise and process the trajectories to the required granu-
larity.
           Sampling is done to generate trajectories TRᵢˢ using Douglas-
Peucker (DP) algorithm
           Add TRᵢˢ to Tˢ
Step 2: Dissimilarity metric calculation
       For each trajectory pair (TRᵢˢ, TRⱼˢ) ∈ Tˢ do
           Calculate Sim[i,j] = DTW(TRᵢˢ, TRⱼˢ) wherei ∈ 1...m and j ∈ 1...m
Step 3: Cluster Trajectories
    Step 3.1: For each TRᵢˢ ∈ Tˢ do// create a cluster for each trajectory
           create a cluster cᵢ, i ∈ 1...m
    Step 3.2: For p in 1 to n // apply agglomerative clustering
           Find pairs of clusters cᵢ and cⱼ such that Sim[i,j] is minimum.
           Merge the selected clusters cᵢ and cⱼ to form a new cluster cᵢⱼ
           Update the matrix Sim of using selected linkage metrics.
    Step 3.3:
           Generate the dendrogram
           For l in 2 to k
           Cut dendrogram at level l to generate clusters. // generate clusters
```

4.1 Trajectory Clustering

Preprocessing of trajectories are done to remove noise and transform data as required for the application. Data are collected from various sources which are in different formats and are required to be converted into a coherent uniform format. Data reduction and normalization are done to reduce the effects of scaling and different level of granularity in data. Trajectories are generated from google maps, mapzen and open street map repositories. Data generated from these sources are cleaned and are extracted for the needed information. The selected information are id, latitude and longitude information which are stored as a sequence. The trajectories are sampled using Douglas–Peucker (DP) algorithm to reduce the size of original trajectories.

4.2 Dissimilarity Calculation

Dissimilarity is calculated between each pair of trajectory using DTW distance measure. The advantage of using DTW is that it allows to find similarity between trajectories of different length.

4.3 Clustering of Trajectories

Hierarchical clustering algorithms group data into a hierarchy or tree structure. Agglomerative method is used for clustering where initially each trajectory is assigned to a singleton cluster [18, 19]. The initial clusters are $C = \{c_1, c_2 \ldots, c_m\}$, where each trajectory is assigned to a cluster, m is the total number of trajectories. The clusters are iteratively merged to form larger clusters, until termination criteria are met or a single cluster is formed. The clusters for merging is selected based on the minimum distance between cluster pair (i, j), where (i, j) vary from 1 to L where L is the levels of cluster generated. The *dissimilarity* matrix is updated based on the distance between merged trajectories clusters generated a level L. Let I and J denote the set of indices of the trajectories of the two disjoint clusters, the distance between the clusters is $d(I, J) = f\left(\sum_{i \in I} \sum_{j \in J} DTW(i,j)\right)$, where f is based on the linkage metrics chosen. The linkage metrics are complete, ward, single, average, mcquitty, median and centroid method. Hierarchical clustering organizes the clustering results as a dendogram where root represent all trajectories as one cluster and leaf represents each trajectory in a single cluster. A dendrogram is constructed and is cut at level L which specify the numbers of clusters as needed by the user. The dendogram provide the visualization for clusters and captures the hierarchical pattern that exist between trajectories.

4.4 Cluster Validation

Clustering results are validated for their correctness using validity metrics where the metrics is independent of algorithm used for clustering. Different clustering algorithms lead to different clusters and the same algorithm run with different parameters may affect the clustering results. The clustering validation measures are classified into internal, external, and relative measures. Maria et al. presented the various clustering methods and a summary of various clustering validation methods used for evaluation [20]. Three validation metrics suitable for hierarchical clustering adapted for trajectory clustering are discussed below:

Cophenetic correlation coefficient

Cophenetic correlation coefficient is used to measure the linear correlation between the dissimilarity matrix between each pair of observation, and their corresponding cophenetic distance which measures the dissimilarity measure at which the trajectory objects merge together in the same cluster for the first time [21]. The dissimilarity matrix is represented as $Sim[i,j]$ where $i \in 1 \ldots m$ *and* $j \in 1 \ldots m$. The cophenetic matrix is represented using a matrix $CP[i,j]$ where $i \in 1 \ldots m$ *and* $j \in 1 \ldots m$.

The Cophenetic correlation coefficient (CC) is calculated as a correlation measure between the matrix *Sim* and *CP* and is defined by the Eq. 2

$$CC = (Sim * CP')/\sqrt{\text{var}(Sim)\text{Var}(CP)} \tag{2}$$

Dunn Index (DNI)

Dunn index characterize how intra-cluster distance and inter-cluster exist for the given clustering scheme. Distance between clusters should be large and diameter of each cluster should be small [20]. Larger value of Dunn index indicates well separated clusters and compact intra-clusters. The Eq. 3 defines Dunn index

$$D_k = \min_{i=1,\dots,k} \left\{ \min_{j=i+1,\dots k} \left(\frac{d(c_i, c_j)}{\max_{r=1,\dots k} diam(c_r)} \right) \right\} \tag{3}$$

where $d(c_i, c_j)$ is the dissimilarity function between two clusters c_i and c_j defined as $d(c_i, c_j) = \min_{x \in c_i, y \in c_j} (d(x, y))$ and $diam(c)$ is the diameter of the cluster. The diameter of the cluster is defined as $diam(c) = \max_{x,y \in C} d(x, y)$.

Davies-Bouldin Index

Davies Bouldin index (DBI) is a measure that takes ratio between dispersion within clusters and separation between clusters and a lower value indicate better clustering [20]. Smaller value of DBI means that there is a better separation of clusters and tightness inside the clusters. DBI is defined by Eq. 4

$$DB_k = \frac{1}{k} \sum_{i=1}^{k} R_i \tag{4}$$

where $R_i = max_{i=1,\dots k, i \neq j} R_{ij}$, $i = 1, \dots, k$. R_{ij} is measured based on $d_{ij} = d(c_i, c_j)$ the separation between clusters and s_i the within cluster scatter for cluster i. $R_{ij} = (s_i + s_j)/d_{ij}$.

5   Experimental Evaluation

To evaluate the performance of the clustering algorithms we have conducted experimentation using trajectory data set. In this study, trajectory data sets called Traj153 are generated by the authors and is for experimentation.

5.1   Trajectory Dataset: Traj153

This dataset was created by the authors and contains of 153 trajectories. These trajectories are the path traversed by faculty and students of University and these trajectories all fall within arrange of longitude and latitude of 50 km. The users are requested to draw their commonly used travelling routes in Google map and trajectories are generated from them.

5.2   Results and Discussion

Hierarchical clustering algorithm are used for the study and linkage metric are varied to study its effects on clustering. Six linkage methods are analyzed during clustering and they are complete (CL), ward (WL), average (AL), single (SL), mcquitty (MCL), median (ML) and centroid methods (CPL). The Cophenetic correlation coefficient values for the different linkage metrics considering Traj153 is shown in Table 1. From Table 1 we infer that average based hierarchical clustering produces good correlation between the dissimilarity matrix and dendrogram generated after clustering.

Table 1 Cophenetic correlation coefficient for various linkage methods

Bouldin index for various values	Bouldin Index for various values
Complete	0.5613
Ward	0.5346
Single	0.5512
Average	**0.7044**
Mcquitty	0.5594
Median	0.5958
Centroid	0.6276

The number of sampling points in the trajectory is set to 10 for DP algorithm. By varying the number of clusters k, the cluster formed are analyzed. The DNI and DBI metrics are analyzed by varying value k, the numbers of clusters from 2 to 10. Table 2 shows the Dunn value obtained for the trajectory clusters by varying for different value of k. The columns are sorted based on the values obtained for Dunn Index for each value of k and LM represent the linkage strategy used for clustering. From the results, we infer that best values are obtained for CPL method for different values of k. The table shows that centroid based linkage method has obtained Dunn index gradually decreasing as the k values increases from 2 to 10. From the table, we can infer that centroid based linkage metrics can be used for trajectory clustering.

Table 2 Dunn Index for various values of k = 2–10 for various linkage metrics

Dunn index (DNI)									
LM	K = 2	K = 3	K = 4	K = 5	K = 6	K = 7	K = 8	K = 9	K = 10
SL	0.783	0.545	0.211	0.211	0.473	0.211	0.211	0.211	0.211
WL	0.956	0.653	0.388	0.388	0.388	0.298	0.309	0.219	0.219
AL	1.178	0.653	0.545	0.473	1.211	0.506	0.327	0.327	0.327
CL	1.178	0.653	0.677	0.596	0.975	0.665	0.665	0.65	0.587
ML	1.221	0.973	0.889	0.717	0.593	0.94	0.833	0.846	0.65
MCL	1.294	1.038	0.989	0.844	0.211	1.023	0.992	0.861	0.861
CPL	1.3	1.186	1.124	1.129	0.737	1.082	1.067	1.101	1.014

Table 3 shows the Davies Bouldin index obtained for the trajectory clusters by varying for different value of k. The columns are sorted based on the values obtained for Davies Bouldin index for each value of k. From the results, we infer the best value is obtained for complete linkage method, From the table we see that Davies Bouldin index gradually is decreasing as the k values increases from 2 to 10 using complete linkage method. From the table, we can infer that complete based linkage metrics can be used for trajectory clustering.

Table 3 Davies Bouldin Index for various values of k = 2–10 for various linkage metrics

Davies Bouldin index (DBI)									
LM	K = 2	K = 3	K = 4	K = 5	K = 6	K = 7	K = 8	K = 9	K = 10
CL	1.028	1.065	0.988	1.054	1.072	1.072	1.109	1.103	1.102
AL	1.236	1.065	1.073	1.12	1.127	1.127	1.175	1.135	1.146
ML	1.265	1.185	1.328	1.252	1.222	1.222	1.225	1.146	1.157
SL	1.276	1.22	1.363	1.283	1.257	1.257	1.268	1.226	1.175
CPL	1.426	1.497	1.364	1.327	1.276	1.276	1.271	1.229	1.195
MCL	1.524	1.545	1.384	1.345	1.386	1.386	1.299	1.229	1.266
WL	**1.719**	1.58	**1.559**	**1.486**	**1.49**	**1.49**	**1.36**	**1.331**	**1.373**

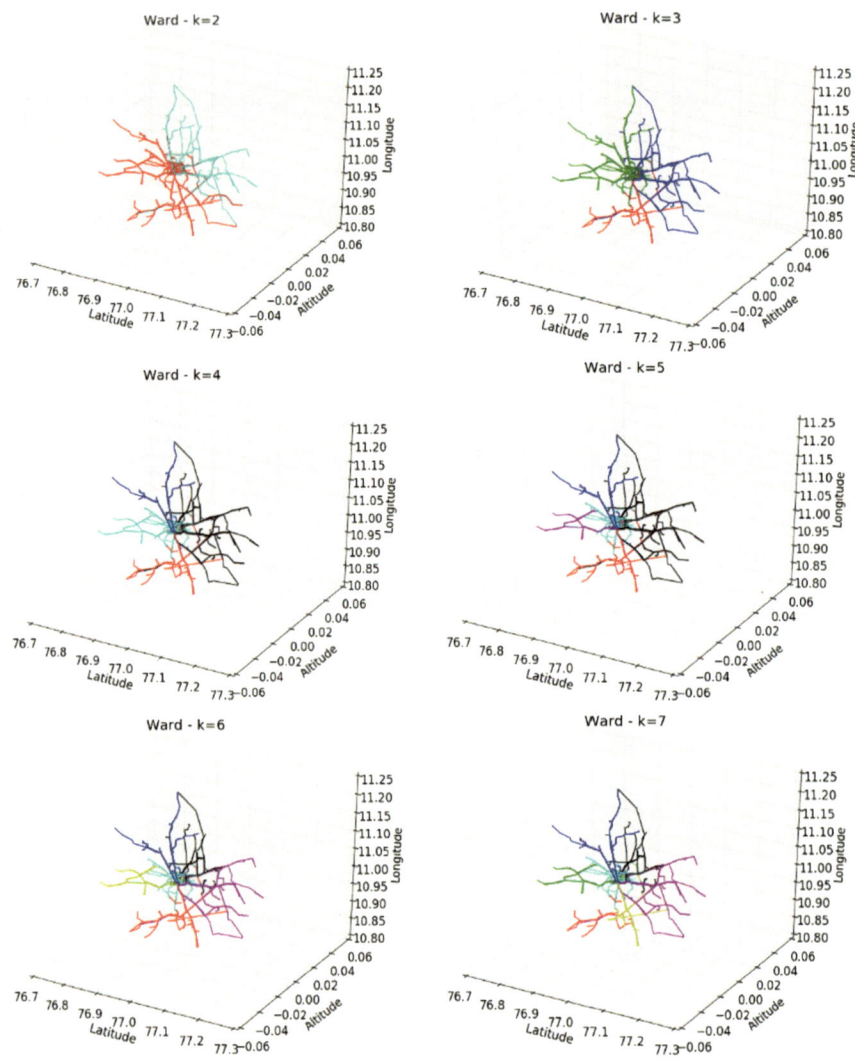

Fig. 1 Clustering of trajectories

We also observe the visual results of clustering by plotting for each of the trajectories of traj153 dataset for each value of k. The altitude axis value is zero, and is set to get a 3d visualization. Figure 1 shows the results of trajectory clustering using various values of k from k = 2 to 7. Ward method is used for clustering and each cluster is represented in varying colors. In all the figures, we see that algorithm has identified clusters that are spatially close to each other. From the Fig. 1, we clearly see the property of hierarchical agglomerative clustering where the initially formed clusters are split as the number of clusters are increased.

6 Conclusion

Clustering is an efficient way to find hidden and interesting patterns from large databases. In this paper, the focus is to analyses trajectories generated by movement of vehicles on road network using clustering strategies. Hierarchical clustering method using agglomerative principle is used to generate the trajectory clusters. The effects of linkage methods on clusters are validated using three clustering metrics. In the analysis, average linkage metric gives the optimal CC value. Centroid linkage method gives optimal value for DNI and complete method gives optimal value for DBI measure. The advantage of using hierarchical algorithm for clustering is due to its simplicity. Visual plots of clusters show that trajectories with similar spatial mapping are captured into same clusters by the algorithm. The algorithm can be used as a preprocessing step to identify hotspot, user similarity computation and compression of trajectories.

References

1. Mazimpaka, J.D., Timpf, S.: Trajectory data mining: a review of methods and applications. J. Spat. Inf. Sci. 61–99 (2016)
2. Zheng, Y.: Trajectory data mining: an overview. In: ACM Transactions on Intelligent Systems and Technology (TIST) (2015)
3. Praveen, V., Sivakumar, P.B.: Design of IoT systems and analytics in the context of smart city initiatives in India. Procedia Comput. Sci. **92**, 583–588 (2016)
4. Feng, Z., Zhu, Y.: A survey on trajectory data mining: techniques and applications. IEEE Access 2056–2067 (2016)
5. Lee, J.G., Han, J., Li, X., Gonzalez, H.: TraClass: trajectory classification using hierarchical region-based and trajectory-based clustering. In: Proceedings of the VLDB Endowment, pp. 1081–1094 (2008)
6. Gaffney, S., Smyth, P.: Trajectory clustering with mixtures of regression models. In: Proceedings of 50th ACM SIGKDD International Conference on Knowledge Discovery and Data Mining, pp. 63–72 (1999)
7. Chih, H.C., Peng, W.C., Lee, W.C.: Clustering and aggregating clues of trajectories for mining trajectory patterns and routes. VLDB J. **24**, 169–192 (2015)
8. Gudmundsson, J., Andreas, T., Jan, V.: Of Motifs and Goals: Mining Trajectory Data, ACM SIGSPATIAL GIS '12, pp. 129–138 (2012)
9. Kim, J., Mahmassani, H.S.: Spatial and temporal characterization of travel patterns in a traffic network using vehicle trajectories. Transp. Res. C Emerg. Technol. 375–390 (2015)
10. Vlachos, M., Kollios, G., Gunopulos, D.: Discovering similar multidimensional trajectories. 18th International Conference on Data Engineering, pp. 673–684 (2002)
11. Wang, H., Su, H., Zheng, K., Sadiq, S., Zhou, X.: An effectiveness study on trajectory similarity measures. In: Proceedings of the Twenty-Fourth Australasian Database Conference, pp. 13–22 (2013)
12. Besse, P., Guillouet, B., Loubes, J.M., François, R.: Review and perspective for distance based trajectory clustering of vehicle trajectories. IEEE Trans. Intell. Transp. Syst. **17**(11), 3306–3317 (2016)
13. Zhao, Q., Shi, Y., Liu, Q., Fränti, P.: A grid-growing clustering algorithm for geo-spatial data. Pattern Recognit. Lett. **53**, 77–84 (2015)

14. Douglas, D., Peucker, T.: Algorithms for the reduction of the number of points required to represent a digitized line or its caricature. Can. Cartographer **10**(2), 112–122 (1973)
15. Zheng, Y., Zhou, X.: Computing with spatial trajectories. Springer, New York (2011)
16. Yuan, G., Sun, P., Zhao, J., Li, D., Wang, C.: A review of moving object trajectory clustering algorithms. Artif. Intell. Rev. **77**, 123–144 (2017)
17. Xu, R., Wunsch, D.: Survey of clustering algorithms. IEEE Trans. Neural Netw. **16**(3), 645–678 (2005)
18. Han, J., Kamber, M., Pei, J.: Data Mining: Concepts and Techniques. Elsevier, Amsterdam (2012)
19. Sabarish, B.A., Karthi, R., Gireeshkumar, T.: A survey of location prediction using trajectory mining. Artificial Intelligence and Evolutionary Algorithms in Engineering Systems Springer India, pp. 119–127 (2015)
20. Halkidi, M., Batistakis, Y., Vazirgiannis, M.: On clustering validation techniques. J. Intell. Inf. Syst. **17**(2), 107–145 (2011)
21. Sinan, S., Nurhan, D., Ismet, D.: Comparison of hierarchical cluster analysis methods by cophenetic correlation. J. Inequalities Appl. **2013**(203), 1–8 (2013)

Sentiment Analysis of Twitter Data on Demonetization Using Machine Learning Techniques

N. M. Dhanya[1(✉)] and U. C. Harish[2]

[1] Department of Computer Science and Engineering, Amrita School of Engineering, Amrita Vishwa Vidyapeetham, Coimbatore, India
nm_dhanya@cb.amrita.edu
[2] Guruvayurappan Institute of Management, Coimbatore, India
harishuc@gmail.com

Abstract. Social media like twitter and Facebook is seen as a space where public opinions are formed in today's world. The data from these tweets and posts can provide valuable insights for policy makers and other agencies to propose and implement policies better. An attempt is made in this paper to understand the public opinion on the recently implemented demonetization policy in India. A sentiment analysis is carried out on twitter data set using machine learning approaches. Twitter data from November 9th to December 3rd is considered for analysis. The data set is pre-processed for cleaning the data and making it possible for analysis. A final set of 5000 tweets are analysed using machine learning techniques like SVM, Naïve Bayes classifier and Decision tree and the results are compared.

Keywords: Twitter data · Sentiment analysis · Machine learning
Demonetization

1 Introduction

Micro blogging sites such as twitter and Facebook are very powerful communication tools in the current highly interconnected world. Millions of messages are appearing daily in these media. The users can write about their life events, share their opinion and discussion on current issues can be done through this media. In the current scenario where the internet speed is very promising, these media act as an easy communication tool for spreading ideas and developing opinions. Any information such as new product launches, opinion polls [1] and current issues can be posted in these media to get responses of the public. The data collated from social media can be successfully utilized for analyzing people's opinion and sentiments. This analysis can be effectively used for product marketing.

© Springer International Publishing AG 2018
D. J. Hemanth and S. Smys (eds.), *Computational Vision and Bio Inspired Computing*,
Lecture Notes in Computational Vision and Biomechanics 28,
https://doi.org/10.1007/978-3-319-71767-8_19

1.1   Applications of Sentiment Analysis

Sentiment analysis has so many applications in real life across various domains.

E-Commerce: The most popular analysis that can be done using social media sentiment analysis is the e-commerce application. An e-commerce website allows users to submit reviews of the product which they have purchased. These reviews can be based on product quality, mode of delivery and overall customer satisfaction. Other customers can see the reviews and based on the review they can decide on the purchase. For the company management these reviews can be analyzed to make their products better. Based on the negative comments the products can be improved.

Reputation analysis: Reputation analysis can be done by using micro blogging sentiment analysis. Similar to customer analysis, reputation management is also equally important for management. The analysis helps them to manage and strengthen the brand reputation. Sentiment analysis helps companies to analyze how their products are being perceived by the online community.

Election exit polls: Sentiment analysis can also be used by government to analyze the strength and weakness of various government policies by analyzing the opinions of the public. The sentences like "The minister himself is corrupted, then how do you expect the truth will come out" clearly shows the negative sentiment and "What a nice policy, which will enhance the country's growth" specifies a positive sentiment towards the government. The polls like the one shown in Fig. 1 can be used by candidates to predict the votes. The opinion and the place from where the opinion is posted can be further analysed and used in campaigning.

Fig. 1 Sample Twitter poll

In India, a very significant policy initiative of demonetizing Rs. 500 and Rs. 1000 notes was announced on November 8, 2016 by the Prime Minister, Shri Narendra Modi. In this paper, the sentiment analysis of this policy is carried out using the most popular micro blogging site, twitter. In twitter, users post small lines of comments called tweets to mark their opinion. The following are the characteristic features of tweets.

1.2 Tweet Features

Length of a Tweet: The maximum length of text that is possible in twitter is 140 characters. Even though it cannot be a meaningful grammatically correct sentence, it can be considered as a single entity or a sentence. Normally in product reviews the sentences will be more structured and analysis will be much easier than a tweet.

Language: Tweets can be given through a variety of languages. Tweets will mostly be in colloquial languages filled with slangs and misspellings.

Hashtags: A Twitter hashtag is simply a keyword phrase, spelled out without spaces, with a hash sign (#) in front of it. For example, #inboundchat and #ILoveChocolate are both hashtags. The use of hashtag is gaining popularity in tweets. Analysis shows that almost all tweets contain hashtags. Hashtag campaigns are conducted for increasing the popularity of a product. Figure 2 shows an example.

Data availability: Another feature of tweets is the availability. With the Twitter API, in several languages it is possible to collect millions of tweets for training. Here we used python API for data collection. There are also some datasets [2, 3] available that have automatically and manually labelled tweets.

Fig. 2 Hashtags in Twitter

2 Related Work

Sentiment analysis can be considered as a natural language processing task, classifying the sentences into different clusters called positive, negative and neutral. It has many granularities right from the document level to the sentence level [4]. Micro blogging sites such as twitter and Facebook is updated with real time reactions and opinions. There has been large amount of prior research in sentiment analysis mostly in the domain of product reviews, movie reviews and election exit polls.

A distant supervision technique is introduced by Pak and Paroubek (2010) which automatically collects the dataset from the web [5]. They used happy and sad emoticons for positive and negative sentiments. The tweets from "New York Times" and "Washington Post" and other popular news paper accounts are collected. These tweets are used as training sets of subjective tweets. The techniques such as Unigrams and filtered n-grams are used for their sentiment classification. The negative sentiments are considered by attaching negation words like "no", "not" to the words before and after them. These negation words along with positive word can be considered as negative sentiments.

The sentiment analysis of tweets is considered to be different task as the tweets consists of informal language which are also sometimes new and creative [6]. Koulompis, Wilson and Moore (2011) use the earlier work done in hashtags and sentiment analysis to build their classifier. The most frequent hashtags are identified using Edinburgh Twitter corpus and they manually classify these hashtags. These are in turn used to classify tweets. They have used n-grams and part-of speech features. In addition, they have built a feature set from Internet lingo dictionary and MPQA subjectivity lexicon. Their analysis identifies that usage of Part-of-Speech causes a reduction in accuracy whereas n-gram features with lexicon features gives better results.

Another approach is discussed by Saif et al. [7]. They introduce a semantic based approach which identifies the items in a tweet, like a person, organization etc. [7]. The effect of stop words in the sentence is studied and they concluded that removal of such stop words may cause undesirable effect in the meaning of the sentence.

Identifying negation words is a technique usually studied in sentiment analysis. The words like "not", "no" and "never" can change the meaning of the sentence to the opposite direction. That is positive sentences can be changed to negative and negatives can be changed to positives. Because of such words the nearby word meanings may become opposite. For example "not good" change the meaning of a positive word "good" to negative. Thus negation words play an important role in sentiment analysis, which may change the entire meaning of a sentence.

The scope of negation words can be considered to be from the point of appearance to the next punctuation. This scope is analyzed in [8]. The scope is analyzed for the sentiment of that sentence. They identified explicit negation cues for each word in the sentence. The distance between the word and the nearest negative word from left and right is calculated and the sentiment score is calculated.

The adverse effect of drug [9] is analyzed in this work based on the drug related tweets. A pipelined algorithm is proposed which uses a simple drug-related classification and sentiment analysis to extract Adverse Drug Events on Twitter. The SVM and Naïve bayes classification is used for comparing the results and SVM is getting higher accuracy.

Bao et al. [10] compared the performance of different pre-processing approaches for twitter data. They compared the impact of pre processing on the accuracy of sentiment analysis. The effects of URLs on the tweets, negation words, repeated letters, stemming [11] and lemmatization were evaluated and conclude that removal of all these improves the accuracy significantly.

All the above stated techniques relay on n-gram [12] features. To improve the accuracy significantly some authors used feature selection also. In this paper we are creating a sample dataset from the tweets. The training data is created with the analysis on the tweets. By doing pre processing and feature extraction the important features are extracted and using machine learning approaches a classifier model is created.

3   Demonetization

In a dramatic announcement on the evening of 8th November, 2016, Mr. Narendra Modi, the Prime minister of India announced that currency notes of Rs. 500 and Rs. 1000 will no longer be a valid tender from 12.00 midnight. He announced that citizens could exchange or deposit the invalid notes at the banks till 30th December 2016. Also a number of restrictions on withdrawal of money from ATMs and banks were implemented. In his announcement he made it clear that this is an unavoidable step to eradicate fake notes and to curb black money which was in circulation in the country.

During the days that followed huge queues were seen in front of banks and ATMs and media reported that huge difficulties were faced by citizens due to the restrictions imposed. The opposition parties raised the issue in the parliament and parliament functions got disrupted many times.

There were also reports that the common citizens of the nation supported the move as they believed that this is an important step towards bringing out black money. The government also projected this move as a significant step towards making India largely a cashless economy where most of the transactions are made in electronic/digital form. The social media is also abuzz with discussions about the benefits and difficulties faced from the decision.

The decision and implementation of demonetization is seen as a historic decision with long term ramification in political and economic future of India. This analysis using tweets is an attempt to understand the general sentiment among the public about the demonetization policy implemented by the Government of India.

4   Dataset

Data set for the analysis is collected from twitter API methods for a period of 25 days. Around 12,500 tweets were collected over a period from 09-11-2016 to 03-12-2016. The data under the hashtag of demonetization and black money are collected and a live streaming of data on the day of 03rd December is also considered.

This data is pre processed and unwanted tweets are removed and a collection of 5000 tweets with two different topics, namely, demonitization and black money is collected and classified. The corpus which we collected contains the following information. Each entry contains a tweet id, the tweet text and the sentiment label. Python libraries for twitter can be used to collect the information about the tweets like creation date, creator name, etc. Each tweet is classified into three categories such as positive, negative and neutral. The irrelevant tweets are also classified as neutral. Positive is used for showing positive sentiment or positive opinion towards demonetization, Negative for showing negative sentiment towards the movement and Neutral for non English text or off-topic comments. The following table shows sample tweets with label (Table 1).

Table 1 Tweets in corpus

Label	Example
Positive	Very good way to handle current situation strategy demonitization
Negative	This demonetization has become such a mess that even Bhakts in my family have stopped defending it
Neutral	Narendramodi will respond in both the houses of parliament

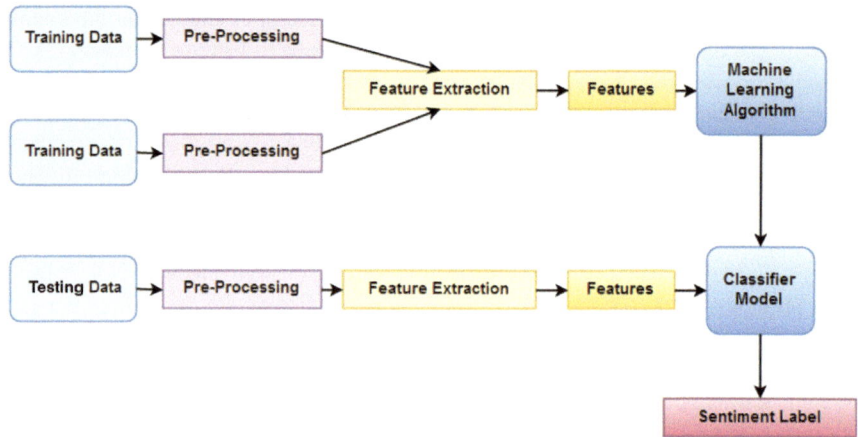

Fig. 3 Sentiment analyzer architecture

5 Methodology

A sample dataset with the labels of positive, negative and neutral is created as training dataset. The data is pre-processed to eliminate some of the unwanted features from the tweets. Different machine learning classifiers are used to find the best sentiment analysis (Fig. 3). Various pre-processing steps such as removing punctuations, emoticons, twitter specific terms and stemming are done and the data is cleaned for analysis. The machine learning techniques such as Naïve Bayers classifier, Decision tree and SVM are used for classification and the results are compared.

Pre Processing

The data generated from twitter is usually not suitable for learning or analysis directly. Hence the data should be normalized to make it in a better format before applying any techniques. Here we are using pre-processing techniques which will eliminate unwanted text from the tweets and hence reduce the features. Thus the data is made suitable for all the learning algorithms. Stemming is also used for pre-processing the data. The porter stemmer is used as stemming algorithm.

Figure 4 illustrates various features seen in micro-blogging. There can be Re tweet symbol, Special characters such as !,.. Emoticons, punctuations and handles. Our pre

Fig. 4 Sample tweet

processing module removes all these unimportant data from the tweet texts and makes it ready for feature extraction (Tables 2, 3, 4, 5 and 6).

Features

There can be a large variety of features based on which a tweet classification can be done. The most popularly used and basic feature set is word n-grams [12]. However, domain specific information present in tweets can also be used for classifying them. A sample dataset with sentiment classification is created. This is done by preparing a positive word corpus, which can appear as positive or words that are agreeing with the demonetization movement like awesome, good, hats off (awesome movement, good decision, and hats off to modiji). The negative word corpus is also created in a similar way like shocking, mess and slows down (Shocking decision, such a mess and slow down the economy rate). For each tweet a sentiment score is calculated from the positive and negative word corpus (Fig. 5; Table 7).

The sentiment scores are calculated on the basis of words used in the tweets. The calculation is shown in Table 7. Each tweet is evaluated and a numeric score is calculated. Based on this score, sentiment labels are attached according to the following rules.

If the score is less than 0 (negative values) a sentiment of negative is attached to the tweets and if the score is greater than 0 (positive) a positive sentiment is attached and finally if the score is 0 it is classified as neutral.

Table 2 Bernoulli naive Bayes

Tweets	Precision	Recall	F1 score
Negative	0.90	0.24	0.38
Neutral	0.64	0.99	0.78
Positive	0.83	0.26	0.39
Avg/total	0.74	0.67	0.61

The accuracy score is 66.75%

Table 3 Multinomial naive Bayes

Tweets	Precision	Recall	F1 Score
Negative	1.00	0.19	0.31
Neutral	0.61	1.00	0.76
Positive	1.00	0.12	0.22
Avg/total	0.78	0.63	0.54

The accuracy score is 63.25%

Table 4 SVM

Tweets	Precision	Recall	F1 Score
Negative	1.00	0.59	0.74
Neutral	0.79	1.00	0.88
Positive	0.95	0.84	0.83
Avg/total	0.87	0.84	0.83

The accuracy score is 84.00%

Table 5 LDA

Tweets	Precision	Recall	F1 score
Negative	0.32	0.68	0.44
Neutral	0.81	0.70	0.75
Positive	0.89	0.41	0.56
Avg/total	0.74	0.62	0.64

The accuracy score is 62.25%

Table 6 Decision tree

Tweets	Precision	Recall	F1 score
Negative	1.00	0.19	0.31
Neutral	0.61	1.00	0.76
Positive	1.00	0.12	0.22
Avg/total	0.78	0.63	0.54

The accuracy score is 63.25%

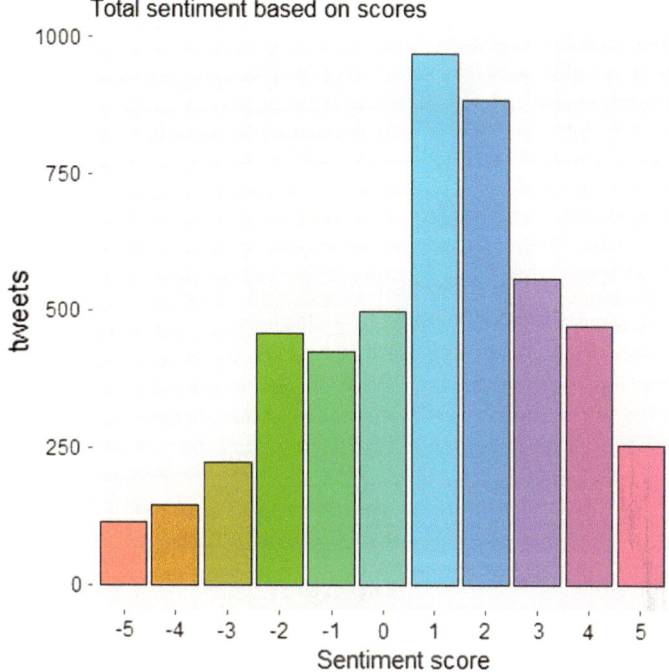

Fig. 5 Sentiment score

Table 7 Sentiment score

Tweets	Score
"Demonetization is brilliant. But people are suffering a bit"	0 (+brilliant, −suffering)
"Kejriwal posts pic of dead robber and claims it's #demonetization related death? How shameless has this man become?"	−3 (−dead, −death, −shameless)
"Putting Nation over Party Politics nitishkumar supports PM narendramodi on Demonetization"	+1 (+Supports)
"A man shaved his head at Jantar Mantar in protest against Demonetization"	−1 (−against)

6 Results

The result proves that positive sentiment score is more compared to the other sentiment scores, concluding that support for demonetization is high. There are also people who feel negative about it.

The following tables show the accuracy measures for different classification methods which are used in this paper. All the machine learning techniques are written in Python code with the help of NLTK [13] and sklearn packages. The data preprocessing and labeling is also done with Python [14] packages and a predefined positive and negative corpus.

The reduction in the accuracy rate is due to the Hindi and Tamil tweets which are present in the dataset and majority of these are classified as neutral. This can be improved by adding Hindi and Tamil words to the corpus with some technologies [15, 18] (Fig. 5).

7 Conclusion

Sentiment analysis on twitter dataset is done on the recent issue of demonetization. The analysis shows largely a positive opinion on the move though negative opinion is also present. Various machine learning algorithms [16] for classification are used for the results comparison. For our dataset, SVM shows the maximum accuracy. The major disadvantage of our approach is that many tweets are written in local languages like Hindi. The Hindi words are written in English language and we didn't include any such words in the positive and negative corpus. Because of this all tweets in Hindi, written in English are classified as neutral. In future those words can also be included and accuracy can be increased for analysis. As future work similar techniques like [17] can be used to reduce the neutral count.

References

1. Gokulakrishnan, B., Priyanthan, P., Ragavan, T., Prasath, N,, Perera, A.: Opinion mining and sentiment analysis on a Twitter data stream. In: 2012 International Conference on IEEE, Advances in ICT for Emerging Regions (ICTer) (2012)
2. Go, A., Bhayani, R., Huang, L.: Twitter Sentiment Classification Using Distant Supervision. Technical Report (2009)
3. Sanders, N.: Twitter Sentiment Corpus. http://www.sananalytics.com/lab/twitter-sentiment/. Sanders Analytics
4. Sun, S., Luo, C., Chen, J.: A review of natural language processing techniques for opinion mining systems. Inf. Fusion **36**, 10–25 (2017)
5. Pak, A., Paroubek, P.: Twitter as a corpus for sentiment analysis and opinion mining. European Language Resources Association(ELRA), Valletta Malta (2010)
6. Kouloumpis, E., Wilson, T., Moore, J.D.: Twitter sentiment analysis: the good the bad and the Omg! ICWSM 538–541 (2011)
7. Saif, H., He. Y., Alani, H.: Semantic Sentiment Analysis of Twitter. In: The Semantic Web-ISWC 2012, pp. 508–524. Springer, Berlin (2012)
8. Councill, I.G., McDonald, R., Velikovich. L: What's great and what's not: learning to classify the scope of negation for improved sentiment analysis. In: Proceedings of the Workshop on Negation and Speculation in Natural Language Processing, pp. 51–59. Association for Computational Linguistics (2010)
9. Peng, Y., Moh, M., Moh, T.S.: Efficient adverse drug event extraction using twitter sentiment analysis. IEEE/ACM International Conference on Advances in Social Networks Analysis and Mining (2016)
10. Bao, Y., Quan, C., Wang, L., Ren, F.: The role of preprocessing in Twitter sentiment analysis. Intell. Comput. Methodologies 615–634 (2014)
11. Smirnov, I.: Overview of Stemming Algorithms. Mechanical Translation (2008)

12. Pang, B., Lee, L.: Opinion Mining and Sentiment Analysis. Found. Trends Inf. Retrieval **2** (1–2), 1–135 (2008)
13. NLTK (Nature Language Tool Kit).: Last Retrieved on 21 March 2015 from http://www.nltk.org/
14. Bird, S., Klein, E., Loper, E.: Natural language processing with Python. O'Reilly Media, Inc. (2009)
15. Nivedhitha, E., Sanjay, S.P., Anand Kumar, M., Soman, K.P.: Unsupervised word embedding based polarity detection for tamil tweets. Int. J. Control Theor. Appl. **9**, 4631–4638 (2016)
16. Quinlan, J.R..: C4. 5: Programs for Machine Learning, vol. 1. Morgan Kaufmann, Burlington (1993)
17. Reshma, U., Barathi Ganesh, H.B., Anand Kumar, M., Soman, K.P.: Supervised methods for domain classification of tamil documents. ARPN J. Eng. Appl. Sci. **10**(8), 3702–3707 (2015)
18. Seshadri, S., Madasamy, A.K., Padannayil, S.K., Anand Kumar, M.: Analyzing sentiment in indian languages micro text using recurrent neural network. IIOAB J. **7**, 313–318 (2016)

Curvelet Based ECG Steganography for Protection of Data

Vishakha Patil[(✉)] and Mangal Patil

Deptartment of Electronics, Bharti Vidyapeeth University College
of Engineering, Pune, India
vishakhapatil569@gmail.com, mvpatil@bvucoep.edu.in

Abstract. A ECG steganography provides safe communication over internet such as transformation of patient information embed on ECG signals. The data hiding technique are used here in which steganography is hiding the patient information in image, video, audio and signal etc. The very important challenge to ECG steganography is extract data and signal without loss. In this technique, we use the curvelet transform for evaluating coefficients, converts information into binary format by using quantization and adaptive LSB algorithm for embedding. The ECG signals for steganography are taken from database of MIT-BIH. From outcome we have observed that the signal loss is less after embedding when the coefficients near to zero. The overlapping of embed watermark is avoided by using sequence. The performance is observed by using parameter such as peak signal to noise ratio, percentage residual difference, correlation and time elapsed at embedding. The extraction performance is measured by using correlation, BER and time elapsed parameters. At extraction bit error rate is used to calculate extract capacity. The size of patient information is raises then BER is zero but the original signal deteroiates. Curvelet transform removes the limitation of wavelet transform and watermarking technique. Hence the curvelet transform is very efficient technique.

Keywords: ECG steganography · Adaptive LSB algorithm · Watermark
position selection · Curvelet transform · Chaos encryption for embed
Key encryption

1 Introduction

Everywhere health concern monitoring is enabled by recent wearable biomedical equipment. The bio-physiological parameters are calculated by these devices and that data should be capable of sending via internet. The patients' information with medical data is transmitted safely by using steganography. The steganography is divided into different types which depends on embedding methods such as image, video, audio, etc. The personal information such as patient name, date of birth, age, address and details of previous treatment are store in database under personal information. This information is hidden in ECG signal which is known as ECG steganography. The ECG steganography

© Springer International Publishing AG 2018
D. J. Hemanth and S. Smys (eds.), *Computational Vision and Bio Inspired Computing*,
Lecture Notes in Computational Vision and Biomechanics 28,
https://doi.org/10.1007/978-3-319-71767-8_20

reduces the information loss produced at the extraction time. Most of the times frequency domain method is used in steganography. In this digital water-mark embedding, extraction and original ECG signal decomposition and also any selected transform method is used for signal de-composition. Generally the discrete wavelets transform (DWT) and fast discrete curvelet transform (FDCT) are used as transformation methods [1–4]. Chen et al. [5] have generalized an idea of Watermarking. For data copyright and biological protection this technique is mostly used. They apply watermarking technique by using quantization on ECG for protection of secret data. The quantization based watermarking method is implemented by using three transform domains such as DCT, DWT and DFT. While the watermarks embedding technique is not invertible, the very small change occurred between the amplitude and PQRST complex of ECG signal? Therefore the watermarked information can get together the necessities of physiological diagnostics.

By using wavelet transform, the representation the curves are in limitation. Curves are where main data such as characteristic points are there. Hence, the steganography is developed by subsequent features such as (i) for analysis of the coefficients which are evaluated and which store the secret data or information (ii) remove loss during extraction of patient information and (iii) identify and preserve original information and decline the redundant part of the signal.

In this way, ECG steganography, by using FDCT method, is available for protection of data. FDCT technique calculates the coefficients so as to re-present curves. The signal weakening is reduced by applying adaptive thresholding algorithm. For processing signal taken from MIT BIH database [6], the proposed technique decreases the loss which occurs during extraction of information and improves the limitation of the watermark method. The performance of steganography algorithm is calculated by using peak signal to noise ratio (PSNR), (MSE), percentage residual difference (PRD), correlation and time elapsed parameter at embedding side. Also at the receiver side the performance is measured by using correlation, bit error rate (BER) and time elapsed correspondingly. We present the performance of the proposed method for dissimilar sizes of watermark.

2 Literature Review

Chen et al. [7] have generalized a technique is hiding Patients Confidential data in the ECG Signal via Transform-Domain Quantization. The most of time the watermarking used for data hiding and provides security to patient information. This paper they proposes the watermarking encryption technique by using quantization and this technique applied on ECG signal and provide more security. They optimize the result by implementing DCT, DWT and DFT transform which is watermarking depends on quantization. The inverting of embed is not required in this process and this also called the blind watermarking. The extraction of hiding data with no understanding information of ECG signal. They verify result efficiency of this technique for result discussion. They proved Transform Domain Quantization technique is very useful and efficient.

The watermark algorithm by using curvelet transform for image this idea is generalized by author Xu et al. [8] In first step curvelet transform is used to decompose the carrier image and Arnold transform is used to scramble the watermark image. In second steps the embedding of binary image which is watermarked into coefficients of medium frequency. These coefficients are embed according to characteristics of human visualization. This method shows the performance is very good and security of system is better than other techniques. Noise robust is less with filtering, cropping and compression of jpeg is good than another attacker. The performance of scale and rotary motion is not good for this proposed technique. They gives future scope of proposed technique is watermark scheme adjacent to geometric hacker using curve let transform.

Degadwala et al. [9] has proposed High Capacity Image Steganography Using Curvelet Transform and Bit Plane Slicing. The representation of non-adaptive spare with edges and object by using multi dimensional curvelet transform. The image is dividing into eight planes with data compressed by using bit plane slicing method. For hiding information they used steam code RC4 method and result calculated by using parameter mean square error and peak signal to noise ratio. They estimate in general security is high and less distortion of extracted image by using cuvelet algorithm. It is also recognizing that robust and security is more. With the help of curvelet transform capacity of storing information is more and without hacking information, image is extracted by using BPS. They found that the future addition of proposed work is depends on the images size and also taken image without text. The Stream Cipher is not used in this technique they uses Block Cipher. They calculate robustness of Blurred, Rotation and other types of operation.

Jero et al. [10] they proposed ECG steganography using curvelet transform technique. They uses the curvelet transform for transformation of data and quantization used for embedding of information. The signal is divided into frequency sub bands by help curvelet and information is embedding in high frequency components. This proposed technique uses high frequency component for embedding for removing of the deterioration and loss from signal. The extraction performance is calculated by using bit error rate. The PSNR is increased in this technique by decreasing data size and increasing signal size. Also they calculate the effect on α due to curvelet transform.

Ramu et al. [11] they proposed a Discrete Wavelet Transform and Singular Value Decomposition Based ECG Steganography idea. For Secured Patient Information Transmission The signals are decomposed by using DWT and secrete information is embedding on ECG decomposed signal by using SVD. They embed information on 2D ECG image with the help of SVD. Then of information and ECG signal which is sub-band format extracted by taking invert secrete data values of singular decomposition. They calculate the results by applying PRD, PSNR, KL and BER parameter. The degradation of signal of proposed method is fewer than 0.6% and this signal collected from database of MIT BIH. Remember that the size of signal is more than the size of secretes information, so the transmission of patient data is reliable and lossless. The decomposition of 2D image is done by using four wavelets on transmitter and receiver side. They recognize the watermarking technique is by removing the SVD for data and signal.

Arvind Kourav et al. [12] has proposed review on Curvelet Transform and Its Applications. The Video surveillance, Investigation of criminal, Credit card security, Electronics banking and Forensic application used for face recognition. The face recognized by using curvelet transform is very correctly without distortion. This transform is multistate transform used for Image Processing Traditional Applications. They estimate the how curvelet transform is better than other transform and different applications are used for this estimation. The wavelet transform is not good for image information represented in singularities above Geometric Structures than curvelet transform. Curves on singularities for age information are designed by using curvelet and singularities of point are effective in wavelet.

3 Methodology of Proposed Technique

The fundamental structural design of steganography by using ECG signal consists is divided into two parts: Embedding (Transmitter) and Extraction (Receiver) which is shown in below Figs. 1 and 2. The ECG signal gives coefficients after any transformation. The embedding side of steganography hides the secrete information in coefficients of signals. Remember that the size of information is less than the size of signal. After embedding we have to apply inverse transform for extraction of original ECG signal. By using health care supplier we can receive the embedding signal. Now patient information is extracted by using stego signal taken at transformation and also key is used for extraction of embedding data. The watermarked method such as LSB algorithm coefficient matrix indicates the position of patient information in the array. Array is used as key which is used for security.

A. Preprocessing: (i) The one dimensional (1D) ECG signal is converted into two dimensional (2D) image.
B. Curvelet transform: With the help of curvelet transform the coefficients of image is calculated. The curvelet transform is divided into two types such as wrapping based fast Curvelet Transform and spaced Fast Fourier Transform. In proposed approach we uses the wrapping based fast curvelet transform.
C. Chaos encryption: The patient data taken as secrete information that data is more secured by using secrete key. The chaos encryption method is for key embedding and provides more security.
D. Data Concealment: The binary stream of patient data is embed in coefficients of ECG image by using adaptive LSB algorithm.

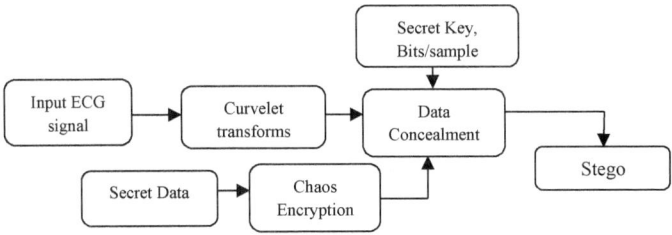

Fig. 1 Data embedding (transmitter)

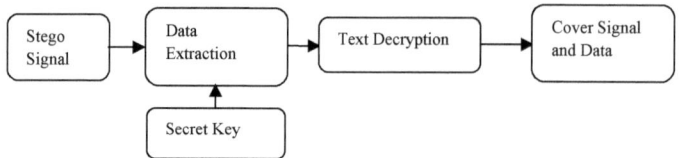

Fig. 2 Data extraction (receiver)

3.1   Preprocessing

In proposed approach the biomedical signals is taken from database of MIT BIH and choosing signals has 128 Hz sampling rate. The Tompkins algorithm is to convert 1D ECG signal into 2D ECG image. After converting signal into 2D ECG image we are applying curvelet transform which is explained below.

3.2   Curvelet Transform

In this technique we uses the curvelet transform for transformation of 2D image. By using curvelet transform the coefficients of image is calculated and these are represent as curved edges. The another method of fast discrete curvelet transform is unequally spaced fast fourier transform. This method is used to calculate the curvelet coefficients of 2D ECG image with the help of below Eq. 1.

$$c(j, l, k) = \int \hat{f}(S_{\theta l}\omega)\tilde{U}j(\omega)e^{i(b,w)}dw \tag{1}$$

With the help of FFT, the FDCT converts the time domain of image into frequency domain. The concentric square of scales that are achieved from frequency domain. The numbers of scales are produced from frequency space and this is measured by using logn. where n: size of image.

The wedges and scaled space of frequency are constructed by applying Cartesian window Eq. 2 is given below.

$$\tilde{U}_j(\omega) = W_j(\omega)V_j(S_{\theta l}\omega) \tag{2}$$

where $S_\theta = \begin{pmatrix} 1 & 0 \\ -\tan\theta & 1 \end{pmatrix}$, $W_J(\cdot)$ is radial windowpane and angular window is $V_j(\cdot)$.

The frequency components of image are very sensitive with coefficients. These frequency components are divided into four group such as scale 1, 2, 3, 4 etc. The scales have low frequency components, scale 2, 3 contains the components coefficients of intermediate frequency and a high frequency coefficient is in scale 4. The binary format of patient information is embedding in evaluated coefficients. Hence, the coefficient of scale 4 is modified for preservations the patient data. The scale 4 group is very good for embedding information because these remove the deterioration of data and 2D image.

3.3   Select Adaptive LSB Position

In this LSB position, the overlapping of LSB distribution and coefficient values modification is avoided. The number of adaptive LSB position represented in Fig. 6. The LSB location and its neighbors are embedded in watermark bit. However, the capacity of watermarking is reduced in this technique. Generally, the work out of sequence n*n is in sightless way.

For embedding the binary format of patient information is needed and that format is hides in C^D by using quantization Eq. 3 shown below

$$C^*(i,j) = \alpha |C^D(i,j)| w. \tag{3}$$

3.4   Watermark Extractions

For extraction of patient information ECG image is put in FDCT. The watermarked coefficients are finding by using key. Apply inverse watermark embedding and patient data are extracted with the help of Eq. 4 where

$$Wr = \begin{matrix} 1, & \text{if } \hat{c}(i,j) > 0 \\ 0 & \text{if } \hat{c}(i,j) < 0 \end{matrix} \tag{4}$$

Here we apply the adaptive LSB algorithm and achieves less signal deterioration. This outcome inside no concession going on diagnosibility with good information extraction.

4   Results and Discussion

We taken four different ECG signals and four different secrete biometric images for observation of parameters. This embedding done with text data (Patient Confidential information: Name: X.Y.Z, Date of Birth: 15/6/1994 Address: Pune Medicare Number: 9890124814 Telephone Number: 1234567 Patient Diagnoses information 1. Blood Pressure. 2. Temperature. 3. Glucose Level) is same for all signals. The PSNR, MSE, correlation, PRD and time elapsed is calculated for each signal. The data hiding capacity is depends on ECG signal size and number of bits stored per ECG sample. Below Table 1 shows the input signal secrete image and embed signal such as stego signal. Table 2 shows parametric evaluation of these signals after embedding.

Table 1 Difference between input and embed signal

Bits	Input signals	Secrete image	Stego signal
3	ECG signal 1	Sec img1 Finger Print Image	
3	ECG signal 2	Sec img2 Finger Print Image	
3	ECG signal 3	Sec img3 Finger Print Image	
3	ECG signal 4	Sec img4 Finger Print Image	

Table 2 Parameter evaluation of above table at embedding side

Parameters	ECG signal1, img1	ECG signal2, img2	ECG signal3, img3	ECG signal4, img4
PSNR(db)	60.28	59.9862	61.2277	61.3718
MSE	8.2925	9.3565	9.10725	8.6285
Correlation	0.999887	0.999825	0.999962	0.999862
PRD	0.00113096	0.00119855	0.00118666	0.00115305
Time elapsed	1.16463	0.490931	0.485996	0.484288

4.1 After Varying the Bits Per Sample Effect on Stego Signal and Parameters Is Given Below

The ECG signal 1 is taken for hiding secrete data such as secrete image1 and text (Patient Confidential information: Name: X.Y.Z, Date of Birth: 15/6/1994 Address: Pune Medicare Number: 9890124814 Telephone Number: 1234567 Patient Diagnoses information (1) Blood Pressure. (2) Temperature. (3) Glucose levels). The 1 and 2 bits hiding per ECG sample is not possible for this steganography and the 3bits per sample is very good than others. The number of bits per sample is increases then the PSNR is reduced, MSE is increased, correlation is decreases and PRD is increases which are shown in table and graph. The effect of increased bits per sample on stego signal is given below table a. After extraction process the secrete data is extracted completely and this proven by parameter correlation and BER. The correction value is '1' for all bits' per sample hence the correlation value 1 is indicates the all data are extracted completely. The BER value at extraction side is zero. The number of increasing bits per sample verses PSNR and correlation graphs has shown in Figs. 3 and 4; (Tables 3 and 4).

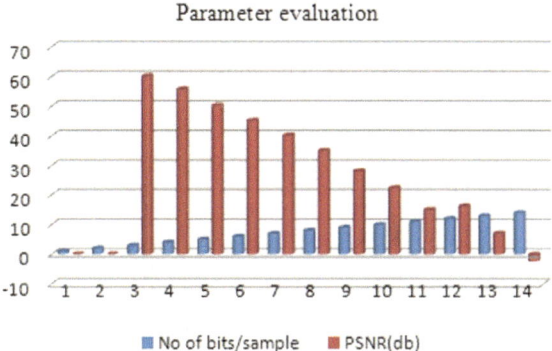

Fig. 3 No of bits/sample versus PSNR

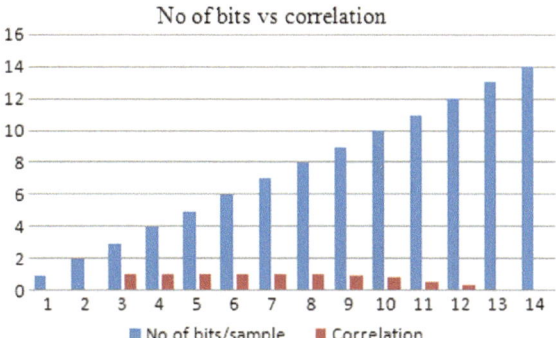

Fig. 4 No of bits/sample versus correlation

Table 3 Effect of rising bits/sample on signal

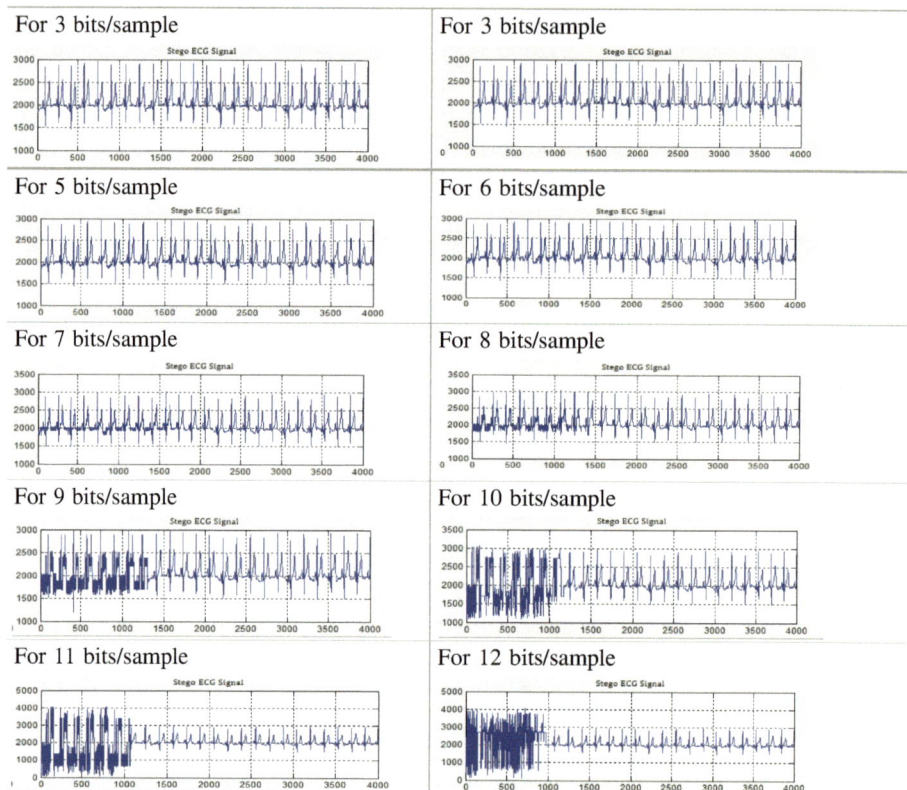

Table 4 After varying bits effect on parameter

No. of bits/sample	PSNR (db)	MSE	Correlation	PRD (%)	Time elapsed
3	60.2662	8.31875	0.999886	0.00113772	0.461194
4	55.8591	22.9492	0.999698	0.00163382	0.304611
5	50.3435	81.7205	0.998885	0.00279256	0.484966
6	45.2571	263.614	0.996595	0.00451624	0.483019
7	40.2338	838.104	0.988711	0.00750812	0.468957
8	35.1577	2697.16	0.967	0.0120463	0.489143
9	28.1867	13428	0.87288	0.0297179	0.457701
10	22.4296	50550.1	0.74025	0.567817	0.464007
11	15.0723	275077	0.53394	0.3226	0.479246
12	16.3363	205612	0.294792	0.143462	0.493095

4.2   Parameter Evaluation After Extraction Data

The correlation for all variable bits is one and BER is zero but signal deterioration is rises. The data extraction is completely done by using curvelet transform than wavelet transform is proven from this parameter. The time elapsed is varying by increasing number of bit's per sample this shown in below graph c (Fig. 5).

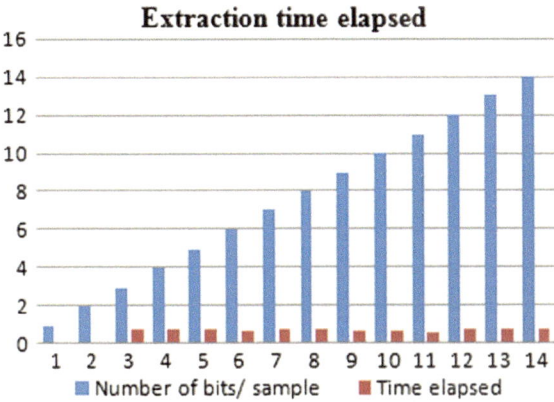

Fig. 5 No of bits/sample versus time elapsed extraction

5   Conclusions

Execution of curvelet based ECG steganography calculation utilizing versatile LSB system was considered in this work. At the point when coefficients are adjusted, the cover flag falls apart and influence diagnosability. It is watched that for around 1.3 times increment in mystery information measure is expanded, the flag decay is around 10%. We contemplate impact of number of bits per test on all parameters. The 0% BER for all watermark sizes demonstrates that proposed approach recovers the mystery information with no misfortune. PRD of proposed approach is right around zero for all watermark sizes. It is watched that the PSNR esteem is great all through which guarantees the indistinctness of watermark. For increment in watermark estimate, there is diminish in the PSNR esteem. Along these lines proposed philosophy can be utilized for dependable steganography.

Acknowledgements. I am greatly pleased to proof: M. V. Patil who guided me through my project curvelet based ECG steganography for protection of data.

References

1. Ibaida, A., Khalil, I.: Wavelet-based ECG steganography for protecting patient confidential information in point-of-care systems. In: IEEE Trans. Biomed. Eng, pp. 3322–3330. (2013)

2. Miyara, K., Chen, F., Nakao, Z.: Digital watermarking based on curvelet transform. Ninth Int. Symp. Signal Process. (2007)
3. Taby, A., Naima Fiete, F.: High capacity image steganography based on curvelet transform. In: Proc. Fourth Int. Conf. Dev. Systems Eng., pp. 191–196 (2011)
4. Ji, F., Huang, D., Deng, C., Zhang, Y., Miao, W.: Robust curvelet-domain image water-marking based on feature matching. Int. J. Comp. Math. **88**, 3931–3941 (2011)
5. Chen, S.T., Guo, Y.J., Huang, H.N., Kung, W.M., Tseng, K.K., Tu, S.Y.: Hiding patients confidential data in the ECG signal via transform-domain quantization scheme. J. Med. Syst. (2014)
6. Moody, G.B., Mark, R.: MIT-BIH arrhythmia database directory. In: MIT-BIH Database Distribution, Harvard-MIT Division of Health Sciences and Technology, Massachusetts Institute of Technology. Available from http://www.physionet.org/physiobank/database/html/mitdbdir/mitdbdir.htm (1992)
7. Chen, S.-T., Guo, Y.-J., Huang, H.-N., Kung, W.-M., Tseng, K.-K., Tu, S.-Y.: Hiding patients confidential data in the ECG signal via transform-domain quantization scheme. J. Med. Syst. 38–54. doi:https://doi.org/10.1007/s10916-014-0054-9 (2014)
8. Xu, J., Pang, H., Zhao, J.: Digital image watermarking algorithm based on fast curvelet transform. Software Engineering & Applications, 939–943. doi:https://doi.org/10.4236/jsea.2010.310111 (2010)
9. Degadwala, S., Thakkar, A., Nayak, R.: High capacity image steganography using curvelet transform and bit plane slicing. Int. J. Adv. Res. Comp. Sci. ISSN No. 0976–5697, **4** (2013)
10. Jero, E., Ramu, P., Ramakrishnan, S.: ECG steganography using curvelet transforms. Biomed. Sig. Process. Control **22**, 161–169. doi:https://doi.org/10.1049/el.2015.3218
11. Jero, S.E., Ramu, P., Ramakrishnan, S.: Discrete wavelet transform and singular value decomposition based ECG steganography for secured patient information transmission. J. Med. Syst. 38–132. doi:https://doi.org/10.1007/s10916-014-0132-z (2014)
12. Kourav, A., Singh, P.: Review on curvelet transform and its applications. Asian. J. Electr. Sci. ISSN 2249–6297. 2(1) pp. 9–13, (2013)

Kernel Based Approaches for Context Based Image Annotation

L. Swati Nair$^{(\boxtimes)}$, R. Manjusha, and Latha Parameswaran

Department of Computer Science Engineering, Amrita School of Engineering,
Amrita Vishwa Vidyapeetham, Coimbatore, India
swati.nov@gmail.com, r_manjusha@cb.amrita.edu,
p_latha@cb.amrita.edu

Abstract. The Exploration of contextual information is very important for any automatic image annotation system. In this work a method based on kernels and keyword propagation technique is proposed. Automatic annotation with a set of keywords for each image is carried out by learning the image semantics. The similarity between the images is calculated by Hellinger's kernel and Radial Bias Function kernel(RBF)kernel. The images are labelled with multiple keywords using contextual keyword propagation. The results of using the two kernels on the set of features extracted are analysed. The annotation results obtained were validated based on confusion matrix and were found to have a good accuracy. The main advantage of this method is that it can propagate multiple keywords and no definite structure for the annotation keywords has to be considered

Keywords: Automatic image annotation · Hellinger's kernel · RBF kernel
Semantics · Contextual keyword propagation · Gabor features · Haralick
features

1 Introduction

Due to the rapid growth of archiving of images, the need for indexing and searching images effectively has increased significantly today. In spite of the fact that many Content Based Image Retrieval methods are prevalent, searching based on image feature is rather difficult for the users. Most of the users prefer searching with textual queries. This can be achieved by annotating the images manually and then searching the annotated images using textual queries. But it is a known fact that manual annotation of a large number of images is very much time consuming, expensive and also involves considerable efforts. Hence, automatic image annotation methods are preferred over manual annotation for efficient retrieval of images. Thus, automatic image annotation with keywords is widely used which involves the learning of semantics of images. Automatic image annotation cannot be accurate if the context of the scene is not taken into account for any object in the scene. So context based image annotation becomes a very important aspect.

© Springer International Publishing AG 2018
D. J. Hemanth and S. Smys (eds.), *Computational Vision and Bio Inspired Computing*,
Lecture Notes in Computational Vision and Biomechanics 28,
https://doi.org/10.1007/978-3-319-71767-8_21

The most common methods used for object detection [1] and image annotation considers each keyword used for annotation as a separate class and then training a classifier correspondingly for the identification of images from each class. This method has been used in indexing of images linguistically as in [2, 3] and in annotation of images using SVM (support vector machine) as in [4]. The main drawback of these methods is that it cannot be incorporated for large data or large annotation. For any context based image annotation method the number of keywords increases enormously with the increasing number of images involved. In such a case considering each keyword as a separate class becomes practically very difficult which in turn will affect the overall performance of the system itself. Another type of approach of image annotation technique is by learning the correlation between annotation keywords and images [5]. These methods are based on probabilistic generative type models as in [6]. The joint probability between the keywords and region of images is estimated. Semi-supervised graph based learning as in [7–9] which does image annotation automatically by keyword propagation methods as in [10]. But the main drawback of these methods is that either the contextual details of the image or the correlation between the keywords is ignored. In the method proposed in [11], in order to extract the correlation between the keywords, a definite structure for the annotation keywords was considered. This limits the efficiency of this approach as it becomes more complicated with more number of keywords.

In order to overcome the drawbacks described above, a string kernel based method is proposed in this work which enables us to compute the similarity between images and also use the visual words context for contextual keyword propagation within each image. The method proposed here can propagate many keywords simultaneously from the training set of images to the images used for testing. The main advantage is that, the order or structure of the keywords is not an issue at all for annotation as the semantic context between the keywords can be extracted for any order. The gap between the semantics of annotations and visual features is learnt by creating the vocabulary of visual words. This method is very much similar to Latent Drichlet Allocation(LDA) [12] and PLSA(Probabilistic Latent Semantic analysis) [13] which are based on Bag of Words [14, 15] method.

In this work an attempt is made to capture the context within the images using linear kernel (Helinger's kernel) and a Gaussian kernel (RBF kernel) and then to compare their performances. In this work for kernel value computation the whole image is first converted into a two dimensional sequence of visual words as in [16]. It is then converted into a one dimensional sequence after which the normalized histogram of each combination of sequences is computed. The similarity between the test and the training images is calculated by determining the confidence scores between the images. The main advantage of this method is that it can be incorporated for any large data set and for any number of annotation keywords and it gives very good annotation results. This method also differs from others in a way that multiple keywords can be propagated to each image. In this method both the visual and the semantic contexts are considered which makes this work more robust. Here the global context is captured by computing the histogram for combination of classes followed by kernel computation.

2 Proposed Method

2.1 Overview

The proposed system consists of four modules namely Feature extraction, creating vocabulary of visual words, finding the Histogram for combination of classes, Context identification, Kernel computation and Contextual keyword propagation. The block diagram of the proposed method is shown in Fig. 1.

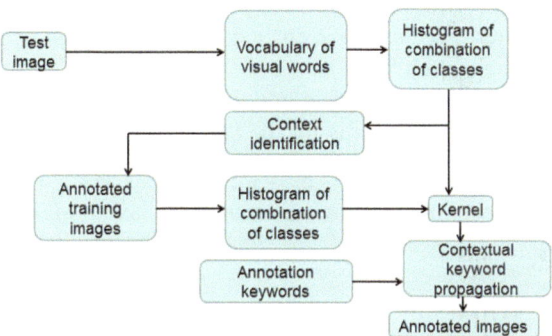

Fig. 1 Artifacts empowered by artificial intelligence

2.2 Feature Extraction

Feature extraction is done in the following contexts: color sub-band statistical features, features from Gabor filters and Haralick features. Totally **fifty one** features have been extracted: six color features, which is the mean and standard deviation of the three color channels (HSV color channel is used), twenty four Gabor features and twenty one Haralick features are extracted. The Gabor features are extracted by taking the average and standard deviation of Gabor filter outputs for three scales and four orientations. The Haralick features are extracted from the Gray Level Co-Occurrence Matrix (GLCM) with an orientation of zero and ninety degrees and displacement of one. The kernels are evaluated using combinations of color, Gabor and Haralick features. These features are extracted from each block and vocabulary of words is formed based on these features.

2.3 Vocabulary of Visual Words

A vocabulary of visual words is a set of visual words (image patches) which when combined together gives the semantic information of the image.

The steps involved are

Training phase:

1. The images in the training set are divided into blocks of $m \times m$ for all 'r' contexts.
2. The features are extracted and the blocks in each context are clustered into 'n' clusters (classes) and thus getting the $n*r$ (product of 'n' and 'r') $= t$ classes for the considered 'r' contexts.

Testing phase:

1. The test image is divided into blocks of size $m \times m$.
2. The color and texture features are extracted from each block.
3. The feature vector for each block of the image is formed.
4. Each block of the image is classified into any one of the 't' classes using KNN.
5. The 2-D sequence of visual words for the test image is formed.

A sample vocabulary of visual words thus formed for a test image is shown in the Fig. 2. where each number in the image represents the class of the corresponding block of image.

1	17	12	16	16	16
6	12	8	16	16	20
13	6	16	16	16	10
16	6	16	6	16	16

Fig. 2 Image and its visual word representation

2.4 Finding the Histogram of Combination of Classes

1. The 2-D sequence of visual words of the image is converted into 1-D sequence of visual words as shown in Fig. 3.
2. The number of times each combination of classes occur adjacently is counted and the count is put in an array (of length $t*t$).
3. This array is normalized and this forms the histogram for each combination of classes (Fig. 4).

```
Columns 1 through 12
   1 | 17    12    16    16    16     6    12     8    16    16    20
Columns 13 through 24
  13     6    16    16    16    10    16     6    16     6    16    16
```

Fig. 3 1-D sequence of Fig. 2

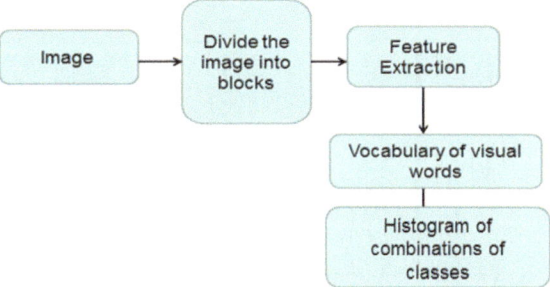

Fig. 4 Block diagram to find the histogram of classes

2.5 Identifying the Context

The normalized histogram for combination of classes in test and the training image is calculated. The mean of the histograms of the training images in each context is calculated. The kernel value is computed between the test image histogram and mean histograms of each context using Hellinger's kernel. In this work an attempt is made to capture and to identify the context using the most basic kernels in the most efficient way. Hence Hellinger's and RBF kernels are selected. This value gives the similarity of the test image to each context. The context which gives the maximum kernel value is determined and the test image is classified into this context. Once the context is identified the training images and descriptions of that particular context is taken for kernel computation. Figure 5 shows image and its contexts.

Fig. 5 Image and its context

2.6 Kernel Computation

Specific kernels are used to compute the similarity between the test and the query image. In this step the histogram of the classes is taken and normalized. The number of times each word occurring in an image is counted. The kernels are computed as given below:

$$\text{Helinger'kernel } k(h1, h2) = \sum_i \sqrt{h1(i)h2(i)} \tag{1}$$

$$\text{RBFkernel } k(h1, h2) = e^{\sum_i \frac{-0.5}{h1(i)h2(i)}} \tag{2}$$

where K is the kernel between the test and training image, h1 and h2 are the normalized histogram of combination of classes of test and training images.

2.7   Contextual Keyword Propagation

The "k" annotation keywords are assigned to each query image by calculating the confidence score. For this first a binary vector for each keyword is calculated: Each keyword is compared with the annotation keywords of each training image in the dataset, if the keyword is present then the binary vector element is set to one, zero otherwise. Then the confidence score is calculated as in Eq. (3).

$$Z_j = \sum_{i=1}^{N} [k(Q, Q_I)] t_{i,j} \tag{3}$$

where "k" is the kernel between test and query image and t is the binary vector element, Q is the query image, Q_I is the training image and Z is the confidence score value. Once the confidence score is calculated for each keyword, they are then sorted to obtain the top five values of confidence scores and the keywords corresponding to the top five confidence scores are assigned to the query image.

Confidence score is the amount of association of each keyword with an image. This value is calculated to determine how much a particular keyword will be suitable for an image. Hence the confidence score plays a very important role in assigning a keyword to an image.

3   Experimental Results

The experiments were conducted on datasets of University of Washington and MATLAB 2013(a) was used for implementation. Both the test and the training images are from the same dataset. The size of each image is 756×504. The training images are annotated images and the annotation keywords are in a text file. The number of annotation keywords used is 88. The images are resized to one fourth of its original size to increase the speed of computations. The test image and the training images are divided into size 16×16 blocks and the feature vectors are extracted. The vocabulary of words is formed for each of the images based on the clustering of the feature vectors. Then the kernel value is computed based on the Histogram for combination of classes formed. Then the confidence score is computed for each annotation keywords. Two types of kernels have been used (Hellinger's kernel and RBF kernel) and their results are analysed. The vocabulary of words of the test image formed after feature extraction from each block and classification is shown in the Figs. 6 and 7.

Fig. 6 Dataset and the results

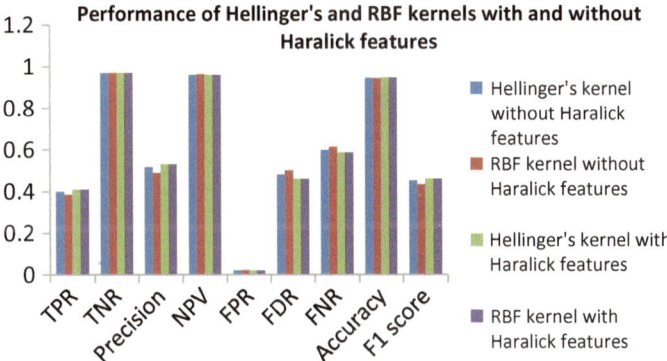

Fig. 7 Bar chart showing the performance of Hellinger's and RBF Kernel with and without using Haralick features

The annotated keywords of the test datasets are compared with the ground truth keywords and its precision is calculated for both the kernels and their performance is analysed using a Confusion matrix. The Confusion matrix is computed and various outcomes are calculated for the two kernels using the feature combinations.

Table 1 Derived values from the confusion matrix

Metrics	Helinger's kernel (Using Color + Gabor features)	RBF Kernel (Using Color + Gabor features)	Helinger's kernel (Using Color + Gabor Features + Haralick features)	RBF Kernel (Using Color + Gabor Features + Haralick features)
True positive	208	200	214	214
True negative	8605	8597	8615	8615
False positive	193	201	183	183
False negative	311	319	305	305
Sensitivity (true positive rate)	0.40	0.385	0.41	0.41
Specificity (true negative rate)	0.97	0.97	0.97	0.97
Precision (positive predicted values)	0.518	0.49	0.53	0.53
Negative predicted values	0.96	0.964	0.96	0.96
False positive rate	0.0219	0.0228	0.0208	0.0208
False discovery rate	0.4813	0.5012	0.46	0.46
False negative rate	0.599	0.614	0.587	0.587
Accuracy	0.945	0.944	0.947	0.947
F1 score	0.4522	0.4348	0.46	0.46

Table 1 shows the results obtained from the confusion matrix using the two kernels with and without using Haralick features along with color and Gabor features. The values of True Positive, True Negative, False Positive, and False Negative are calculated by comparing the annotation keywords of the test image with the ground truth image and the remaining values are derived from them. The results shows that Gabor features gave good accuracy and precision compared to Haralick features. Hellinger's kernel gives better results compared to RBF kernels while using any combination of Gabor and Haralick features. The performance of both the kernels was better by using Haralick features along with color and Gabor features.

4 Conclusion

In this work a new method for object categorization and image annotation using kernels (Helinger's kernel, RBF kernel) is presented. The features extracted are color, Gabor and Haralick. It was found that the performance of Hellinger's kernel with Haralick, color and Gabor features was the best. Experiments conducted on various datasets gives very convincing results for this method. The accuracy and efficiency of this work is better than the relevance based models [6, 17, 18] and graph based models [10]. This work can be further enhanced and extended to work for real time videos [19, 20] so that it can be made applicable for any real time scenario.

References

1. Divvala, S.K., Hoiem, D., Hays, J.H., Efros, A.A., Hebert, M.: An empirical study of context in object detection. In: Proceedings in IEEE Conference on Computer Vision and Pattern Recognition, pp. 1271–1278 (June 2009)
2. Li, J., Wang, J.: Automatic linguistic indexing of pictures by a statistical modelling approach. IEEE Trans. Pattern Anal. Mach. Intell. September 2003; 25(9), 1075–1088 (2003)
3. Li, L.J., Socher, R., Fei-Fei, L.: Towards total scene understanding: classification, annotation and segmentation in an automatic framework. In: IEEE Conference Computer Vision and Pattern Recognition, pp. 2036–2043 (2009)
4. Gao, Y., Fan, J., Xue, X., Jain, R.: Automatic image annotation by incorporating feature hierarchy and boosting to scale up SVM classifiers. In: ACM Multimedia Proceedings, pp. 901–910 (2006)
5. Okabe, T., Kondo, Y., Kitani, K.M., Sato, Y.: Recognizing multiple objects based on co-occurrence of categories. In: The Pacific–Rim Symposium on Image and Video Technology (PSIVT), pp. 497–508 (2009)
6. Jeon, J., Lavrenko, V., Manmatha, R.: Automatic image annotation and retrieval using cross-media relevance models. In: Special Interest Group on Information Retrieval (SIGIR) Proceedings, pp. 119–126 (2003)
7. Zhou, D., Weston, J., Gretton, A., Bousquet,O., Schölkopf, B.: Ranking on data manifolds. Adv. Neural Info. Process Sys. 16, 169–176 (2004)
8. Ladicky, L., Russell, C., Kohli, P., Torr, P.H.S.: Graph cut based inference with co-occurrence statistics. 11th European Conf. Comp. Vision. 6315(5), 239–253 (2010)
9. Yong Jae Lee: Kristen Grauman. Object-graphs for context-aware visual category discovery. IEEE Trans. Pattern Anal. Mach. Intell. 34(2), 346–358 (2012)
10. Liu, J., Li, M., Ma, W., Liu, Q., Lu, H.: An adaptive graph model for automatic image annotation. In: Proceedings of ACM International Workshop on Multimedia Information Retrieval, pp. 61–70 (2006)
11. Belongie, S., Malik, J., Puzicha, J.: Shape matching and object recognition using shape contexts. IEEE Trans. Pattern Anal. Mach. Intell. 24(4), 509–522 (2002)
12. Blei, D., Ng., A., Jordan M.: Latent dirichlet allocation. J. Mach. Learn. Res. 3(5), 993–1022 (2003)
13. Hofmann, T.: Unsupervised learning by probabilistic latent semantic analysis. Mach. Learn. 41(2), (2011)

14. Li, J., Zhang, H., Liao, Y.: Image annotation based on bag of visual words and optimized semi-supervised learning method. ICTACT J. Image Video Process. Spec. Iss. Video Process. Multimedia Sys. **5**(01), 223–226 (2014)
15. Wu, L., Hoi, S.C.H., Yu, N.: Semantics-preserving bag-of-words models and applications. IEEE Trans. Image Process. **19**(7), 1908–1920 (2010)
16. Lu, Z., Horace, H.S.Ip, Peng, Y.: Contextual Kernel and spectral methods for learning the semantics of images. IEEE Trans. Image Process. **20**(6), 1739–1750 (2011)
17. Lavrenko, V., Manmatha, R., Jeon, J.: A model for learning the semantics of pictures. Adv. Neural Info. Process. Sys. **16**, 553–560 (2004)
18. Feng, S., Manmatha, R., Lavrenko, V.: Multiple bernoulli relevance models for image and video annotation. Proc. IEEE Comp. Soc. Conf. Comp. Vision Pattern Recognit. **2**, 1002–1009 (2004)
19. Naphade, M.R., Basu, S., Smith, J.R., Lin, C.Y., Tseng, B.: Statistical modeling approach to content-based video retrieval. IEEE Proc. 16th Int. Conf. **2**, 953–956 (2002)
20. Hoffmann, M., Tuytelaars, T., Antanas, L., Frasconi, P., De Raedt, L.: 13 Proceedings of the 2013 IEEE Workshop on Applications of Computer Vision (WACV), pp. 133–139 (2013)

Analysis of Trabecular Structure in Radiographic Bone Images Using Empirical Mode Decomposition and Extreme Learning Machine

G. Udhayakumar$^{(\boxtimes)}$

Deptartment of EEE, Valliammai Engineering College, Anna University,
Chennai, India
udhayakumarg.eee@valliammai.co.in

Abstract. Biomechanical function is a primary concern in musculo-skeletal supporting system in our body. The functional adaptation of femur bone to the mechanical environment is an important component of bonemechanics research. The strength and architecture of these structures in femur bones are routinely analyzed by varied image based methods for diagnosis and monitoring of osteoporosis like metabolic disorders. In this work, an attempt has been made for investiagtion of trabecular femur bone architecture using spatial frequency decomposition and neural networks. Conventional radiographic femur bone images are recorded using standard protocols in this analysis. The compressive and tensile regions in the femur bone images are delineated using preprocessing procedures. The delineated images are analyzed using Fast and Adaptive Bi-dimensional Empirical Mode Decomposition (FABEMD) to quantify pattern heterogeneity and anisotropy of trabecular bone structure. The characteristic feature vectors are extracted and further subjected to classification using Extreme Learning Machine (ELM). Results show that FABEMD analysis combined with ELM could differentiate normal and abnormal images.

Keywords: Osteoporosis · Femur bone · Fast and adaptive empirical mode decomposition · Extreme learning machine

1 Introduction

Predicting the strength of the bone is an important goal of current research for the diagnosis of osteoporosis and femoral fractures. The bone strength is to estimate the loss of bone mass by Bone Mineral Density (BMD). As BMD correlates strongly with bone strength, fracture risk prediction in the individual patient relies chiefly on bone BMD measurements. The prediction of bone strength can be improved when BMD combined with measures of trabecular microarchitecture [1]. It is important to realize that bone strength depends not only on the amount of mineral measured but also on the architecture, geometric and mechanical properties of the bone [2]. Many attempts have

© Springer International Publishing AG 2018
D. J. Hemanth and S. Smys (eds.), *Computational Vision and Bio Inspired Computing*,
Lecture Notes in Computational Vision and Biomechanics 28,
https://doi.org/10.1007/978-3-319-71767-8_22

been made to quantify the quality of the trabecular structure captured using radiographs and assess its relationship to osteoporosis and BMD [3]. The relationship between plain radiographic patterns and 3D trabecular architecture shows that the plain radiograph contains architectural information directly related to the underlying 3D structures such as porosity and connectivity [4, 5]. However, two-dimensional projection-based images do not directly portray a material's 3D microarchitecture. Hence texture based analysis has been proposed as an indirect measurement to assess the architecture of the bone [6].

Huang proposed Empirical mode decomposition for extracting texture features at varying different scales or spatial frequencies [7]. This decomposition method is based on the local characteristic time scale of the data called adaptive; hence it is applicable to nonlinear and non-stationary processes. This empirical mode decomposition used to analyze the two-dimensional signals is called Bi-dimensional Empirical Mode Decomposition (BEMD) [8]. BEMD is better than Fourier, wavelet and other decomposition algorithms in extracting intrinsic components of textures because of its data driven property. The performance of this method depends on detection of maxima and minima points called extrema and the interpolation of the scattered extrema points.

In EMD or BEMD, extraction of each Intrinsic Mode Function (IMF) requires several iterations. Hence, extrema detection and interpolation at each iteration make the process complicated and time consuming. Fast and adaptive BEMD method enables the decomposition of images with any dimension in a very short period of time. In addition to eliminating the poor interpolation effects of each iteration, this process facilitates performing only one iteration for each IMF [9].

The automated classification of images is an important step towards clinical decision making. For past decades, gradient descent-based methods have mainly been used in various learning algorithms of feed forward neural networks. However, it is found that gradient descent-based learning methods are generally very slow due to improper learning steps and may easily converge to local minima. In order to acquire better learning performance need many iterative learning steps might be required by such learning algorithms [10]. The ELM is an efficient learning algorithm for Single-hidden Layer feed Forward Neural network. It avoids many difficulties faced by gradient-based learning methods such as stopping criteria, learning rate, learning epochs, and local minima.

In this work, Extreme learning machine and Fast and adaptive BEMD are employed to analyze compressive and tensile strength region of normal and abnormal femur bone using radiographic images.

2 Materials and Methods

Sixty pelvis images recorded using a standard clinical X-ray unit were considered for the analysis study. The anterior-posterior view was used to image all subjects and the recorded radiographs are digitized using an AGFA digitizer. Adaptive histogram equalization is employed to enhance the contrast of the radiographic images. A delineation method proposed by Singh et al. [11, 12] is used to select the compressive and tensile Regions of Interest (ROI). The quantitative analyses are also performed on these ROI to derive apparent porosity [12–14]. The percentage of apparent porosity is taken as the ratio of void area to the total area.

Spatial domain order-statistics filters, namely, MAX and MIN filters, are employed to get the running maxima and running minima of the data, which is followed by smoothing operation to get the upper envelope and lower envelope, respectively [15]. The local mean is estimated by averaging the upper and lower envelopes. The difference between the data and the local mean is designated as the first IMF component of the data. This process separates the finest local mode from the image based only on the characteristic multi-scale. This sifting procedure is repeated again to all the subsequent IMFs. The first IMF contains the highest local frequencies of oscillation, the second IMF corresponds to medium pattern and the residue to the lowest. This decomposition is suitable for feature extraction and classification in image processing.

Huang et al. [10] proposed randomly assigning values for parameters w_i and b_i, and thus the system becomes linear so that the output weights can be estimated as $\beta = H^+T$ using training set, various activation functions include sigmoidal, sine, hardlim, triangular and RBF function, and hidden node number N, where + is the Moore–Penrose generalized inverse [16] of the hidden layer output matrix H.

3 Results and Discussion

Representative planar radiographic images of normal and abnormal femur trabecular bones are shown in Fig. 1a, b, respectively. The trabecular patterns are distinct and are closely arranged in normal images and abnormal samples are seen with large trabecular spacing and high discontinuities. The overlap between trabeculae is found to be more in normal when compared to abnormal images. These images are processed with Singh index delineation method to identify the (1) compressive region and (2) tensile region as shown in Fig. 1c.

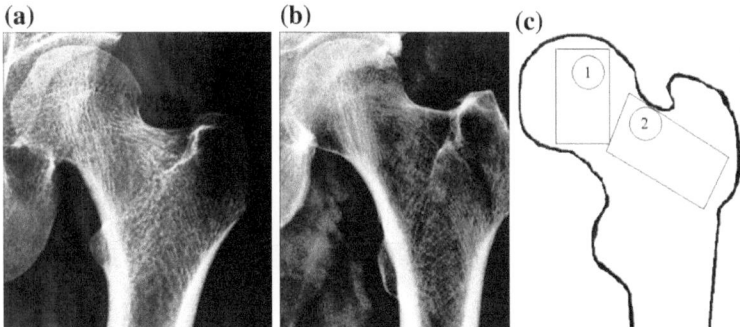

Fig. 1 Normal bone (**a**), abnormal bone (**b**), representative strength region (**c**)

In this study, regions of interest corresponding to the compressive and tensile regions are cropped using windows of constant size of 300 × 150. The apparent porosity has a higher value for abnormal images in the tensile region when compared to the compressive region and these values show better differentiation between normal and abnormal subjects. This may be due to the fact that structural variations due to pathology are well exhibited in tensile region. Two modes of decompositions are obtained for both compressive and tensile regions using fast and adaptive BEMD.

The representative compressive region of interest and their decomposed intrinsic mode functions of normal and abnormal images are shown in Fig. 2a–f. It is found that first IMF contains largest spatial variations compared to the second IMF. Distinct variations are observed among the different modes of decomposition of normal and abnormal images. The spatial frequency component of these modes decreases with increase of intrinsic mode function. Similar observation is found for tensile regions also. Statistical texture parameters such as kurtosis and energy are extracted from IMFs of normal and abnormal images. The variations in values of features extracted from IMF1 with the apparent porosity are shown in the following figures. The scattergram showing variations of kurtosis and energy with apparent porosity derived from compressive and tensile regions are shown in Figs. 3 and 4 respectively.

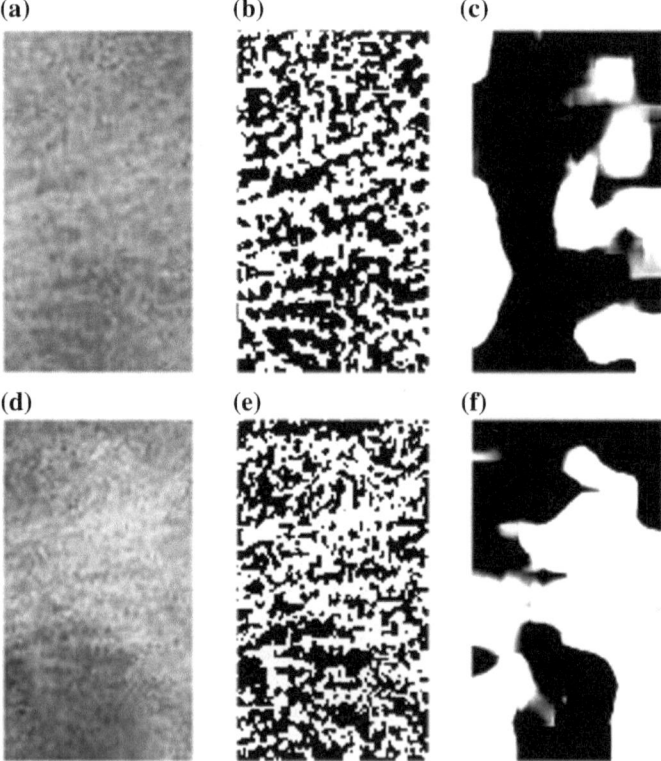

Fig. 2 Representative normal compressive region: Original (**a**), IMF1 (**b**), IMF2 (**c**) Representative abnormal Compressive region: Original (**d**), IMF1 (**e**), IMF2 (**f**)

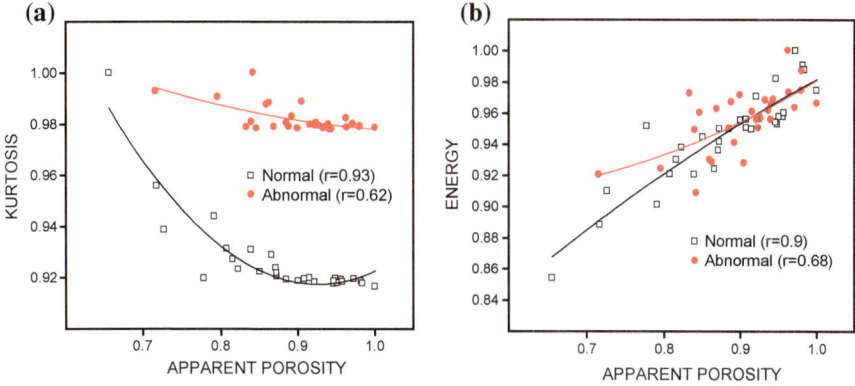

Fig. 3 Variations of values of features with porosity for IMF1 of compressive regions

The kurtosis values have distinct variation between normal and abnormal subjects in compressive region. It is found that these values derived from normal images are low due to homogenous distribution of pixels in compressive region. In normal subjects, feature values of kurtosis are decreases with increase in porosity and found to have high correlation. The difference in values of kurtosis for normal and abnormal images of compressive region is found to be statistically significant ($p < 0.000001$). The energy values derived from IMF1 of compressive region are increases with increase in porosity and found to have high correlation. These values derived from compressive region overlap of both normal and abnormal subjects. This may due to heterogeneous and complex biomechanical behavior of bones.

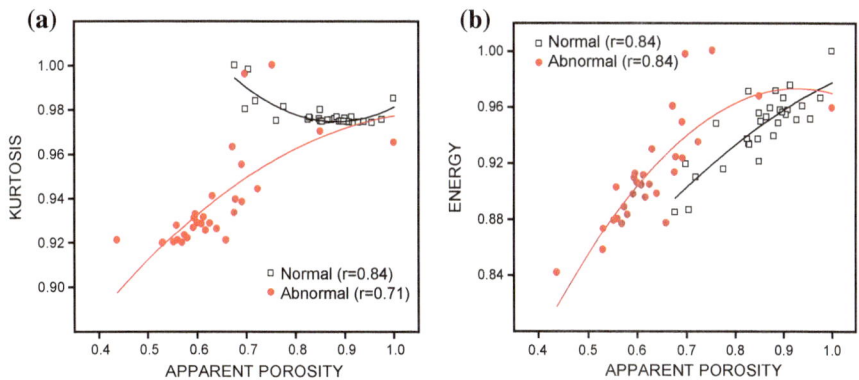

Fig. 4 Variations of values of features with porosity for IMF1 of tensile regions

In normal subjects, kurtosis values are high indicating the peakedness of probability distribution of gray levels in tensile regions due to the heterogeneity of femur bone. The energy values derived from IMF1 of tensile region are increases with increase in porosity and found to have high correlation. In normal subjects, feature values of energy are high due to homogeneous structure of femur bone. The kurtosis and energy values ($p < 0.0000001$) derived from IMF1 of tensile region are found to be distinct in normal and abnormal samples. Hence it appears that these index values could be used as discriminative measure.

The statistical texture features extracted from intrinsic mode functions of strength regions are subjected to classification using ELM. The various activation functions include sigmoidal, sine, hardlim, triangular and RBF are employed and the epochs are varied from 10 to 500. Figure 5 shows that classification accuracy of the ELM classifier of RBF for varying epochs. In ELM it is observed that the classification accuracy increases with increase in the number of epochs and becomes saturated beyond 70. Features derived from tensile region is found to have high accuracy, less training time with minimum number of hidden nodes. This reveals that the tensile region reflects exact architectural information in both the normal and abnormal trabecular femur bone images.

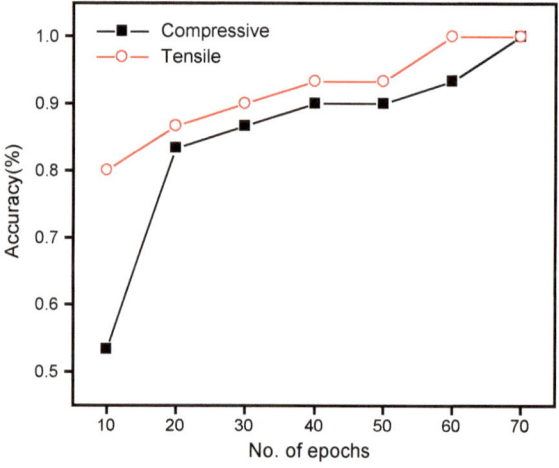

Fig. 5 Classification using ELM classifier

4   Conclusion

Analysis of trabecular bone is an essential component in the assessment of bone strength. Recent studies have indicated that BMD alone may be insufficient to determine the strength of trabecular bone and characterization of their architecture could be an important adjunct to the measurement of bone mass. FABEMD analysis is able to extract trabecular architectural variations in the strength regions of normal and abnormal femur bone images. The variations in values of texture features such as energy and kurtosis extracted from IMF1 using FABEMD show high correlation with the apparent porosity. The ELM of features derived from strength regions is found to have high accuracy, less training time with minimum number of hidden nodes. Thus it appears that FABEMD based spatial frequency decomposition method and ELM classification seems to be useful for trabecular architectural analysis.

References

1. Donnelly, E.: Methods for assessing bone quality a review. Clin. Orthop. Relat. Res. **469**(8), 2128–2138 (2011)
2. Stauber, M., Muller, R.: Age-related changes in trabecular bone microstructures: global and local morphometry. Osteoporos. Int. **17**, 616–626 (2006)
3. Gregory, J.S., Stewart, A., Undrill, P.E., Reid, D.M., Aspden, R.M.: Identification of hip fracture patients from radiographs using Fourier analysis of the trabecular structure: a cross-sectional study. BMC Med. Imaging **4**(4), 1–11 (2004)
4. Pothuaud, L., Carceller, P., Hans, D.: Correlations between grey-level variations in 2D projection images (TBS) and 3D microarchitecture: applications in the study of human trabecular bone microarchitecture. Bone **42**, 775–787 (2008)
5. Luo, G., Kinney, J., Kaufman, J., Haupt, D., Chiabrera, A., Siffert, R.S.: Relationship between plain radiographic patterns and three dimensional trabecular architecture in the human calcaneus. Osteoporosis Int. **9**, 339–345 (1999)
6. Corroller, T.L., Halgrin, J., Pithioux, M., Guenoun, D., Chabrand, P., Champsaur, P.: Combination of texture analysis and bone mineral density improves the prediction of fracture load in human femurs. Osteoporosis Int. **23**, 163–169 (2012)
7. Huang, N.E., Shen, Z., Long, S.R., Wu, M.C., Shih, H.H., Zheng, Q.N., Yen, N.C., Tung, C. C.: The empirical mode decomposition and the Hilbert spectrum for nonlinear and non-stationary time series analysis. In: Proceedings of the Royal Society A, Mathematical, Physical & Engineering Sciences, vol. 454, pp. 903–995, London (1998)
8. Nunes, J.C., Bouaoune, Y., Delechelle, E., Niang, O., Bunel, P.: Image analysis by bidimensional empirical mode decomposition. Image Vis. Comput. **21**, 1019–1026 (2003)
9. Bhuiyan, S.M.A., Reza, R.A., Adhami, R.R., Khan, J.F.: Fast and adaptive bidimensional empirical mode decomposition using order-statistics filter based envelope estimation. EURASIP J. Adv. Signal Process. **2008**, 1–18 (2008)
10. Huang, G.B., Qin-Yu, Z., Chee-Kheong, S.: Extreme learning machine: theory and applications. Neurocomputing **70**(3), 489–501 (2006)
11. Singh, M., Nagrath, A.R., Maini, P.S.: Changes in trabecular pattern of the upper end of the femur as an index of osteoporosis. J. Bone Joint Surg. **52**, 457–467 (1970)
12. Christopher, J.J., Ramakrishnan, S.: Assessment and classification of mechanical strength components of human femur trabecular bone using digital image processing and neural networks. J. Mech. Med. Biol. **7**(3), 315–324 (2007)
13. Udhayakumar, G., Sujatha, C.M., Ramakrishnan, S.: Trabecular architecture analysis in femur radiographic images using fractals. Proc. Inst. Mech. Eng. H, J. Eng. Med. **227**(4), 448–453 (2013)
14. Udhayakumar, G., Sujatha, C.M., Ramakrishnan, S.: Comparison of two interpolation methods for empirical mode decomposition based evaluation of radiographic femur bone images. Acta Bioeng. Biomech. **15**(2), 73–80 (2013)
15. Bhuiyan, S.M.A., Reza, R.A., Adhami, R.R., Khan, J.F.: Fast and adaptive bidimensional empirical mode decomposition using order-statistics filter based envelope estimation. EURASIP J. Adv. Signal Process. **2008**, 1–18 (2008)
16. Serre, D.: Matrices: Theory and Applications. Springer, New York (2002)

Design of an Optimization Routing Model for Real Time Emergency Medical Service System in Chennai Using Fuzzy Techniques

C. Vijayalakshmi[1($\boxtimes$)] and N. Anitha[2]

[1] SAS, Mathematics Division, VIT University, Chennai, Tamilnadu, India
vijusesha2002@yahoo.co.in
[2] Department of Mathematics, Kumaraguru College of Technology, Coimbatore, Tamilnadu, India
anitha.n23@gmail.com

Abstract. This paper mainly deals with the design of an Optimization Routing model for Emergency Medical Service System (EMSS) in Chennai. The effectiveness of Emergency Medical Service System (EMSS) plays a vital role towards society protection. The major idea of this article is to examine the real time flexible dispatching strategy so that crucial response time can be saved for EMSS. An optimization routing model is designed for developing flexible dispatching strategies with the help of duration information. This mathematical model and the is expressed as an IPP technique in which diminished path is assigned. Based on the numerical calculations and graphical representation it reveals to the fact that the different parameters are being analyzed such as duration prediction, incident/vehicles tracking, and consign Optimization and it is validated for road networks.

Keywords: Emergency Medical Service System · Response time
Fuzzy linear programming · EMS vehicle

1 Introduction

Emergency Medical Service is provided with a diversity of persons, using various of methods. In some extension, these will be determined by country and locale, with each entity country having its own way to how EMS should be equipped and by whom. The initial utilization of the ambulance was a particular vehicle, in combat reached through the ambulances volantes accomplished by Dominique Jean Larrey (1766–1842), Napoleon Bonaparte's chief surgeon. An especial actual confront within this proposal to develop transport scheme organization is to develop the synchronization of actions of authorities which plays a main task in transportation management. Specifically it is necessary to attain unified governing between state-owned and small town transport companies and state-owned and small town community security companies (generally police force, flames and EMS). In EMSS the reply moment shows a vital task in

© Springer International Publishing AG 2018
D. J. Hemanth and S. Smys (eds.), *Computational Vision and Bio Inspired Computing*,
Lecture Notes in Computational Vision and Biomechanics 28,
https://doi.org/10.1007/978-3-319-71767-8_23

diminish difficult contacts. Casualties and the loss of equity will be vastly abridged by improving the time of response to occurrences. The main EMS system is integrated the chronological with EMS consign. Real-time traffic and time of travel data are obtainable in EMSS consign center.

The designed EMSS implements detour under a sequence of restrictions. The Mathematical Formulation will help the consigned to make the conclusion. The hyper diminished path algorithm is designed using Optimization Routing technique. The diminished path is calculated among every connected O/D pair. A Mathematical model is constructed which will integrate the mathematical programming formulation and an algorithm for hyper diminished path into the proposed EMSS. The Mathematical model is resolved by the Fuzzy Multi Objective Linear Programming.

2 Literature Survey

Position of ambulance stations or individual ambulances within a region are analyzed with respect to response time [1]. Larson conducted significant research on EMS systems. Larson [2, 3] used a hypercube queuing model as a tool for facility Position and redistricting in urban emergency services. Haghani et al. [4] designed a Simulation Model for Real-Time Emergency Vehicle Consigning and Routing. Zhang et al. [5] proposed a novel probabilistic formulation for locating and sizing emergency medical service stations. Anitha and Vijayalakshmi [6] used a Fuzzy Linear programming to solve the Multi Objective Functions. Qin [7] used a solution method for the Emergency Vehicle Routing model based on the genetic algorithm. Uplenchwar et al. [8], used ASTUTE Routing System for Emergency Vehicle. Aboueljinane [9] studied the reviews on Simulation models applied to Emergency medical service operations. Ben Akiva [10] designed an on-line hyper traffic prediction, model for an inter-urban motorway network. Deo [11] formulated the shortest-path algorithms. Cooke [12] proposed the shortest route through a network with time dependency. Ziliaskopoulos [13] constructed the shortest-path algorithm for real-time intelligent vehicle highway system applications. Glodberg [14] validated and applied a model for locating emergency medical vehicles. Lubicz [15] simulated a model of emergency medical services. Savas [16] analysed the simulation and cost-effectiveness of New York's emergency ambulance service. Majzoubi [17] designed a mathematical programming for dispatching and relocating EMS vehicles. Glodberg [18] investigated about the emergency vehicle location based on service time calls. Ünlüyurt [19] estimated the performance of emergency medical service location models via discrete event simulation.

3 Emergency Vehicle Service System Model Design

The existing consigning models or simulation models do not use real-time traffic information in EMS consigning. Real-time traffic information will be used in the proposed EMS consigning model to make dispatch decisions. Flexible assignment strategies (route diversion and reassignment) are not considered in existing models.

3.1   Optimization Design Plan for Real-time Consigning

Characteristics of Emergency Occurrences

- Arrival Rates
- Time Intervals
- Position
- The Detection Time
- Required Response Time
- Treatment Time
- Hospital Treatment & Distribution.

3.2   Characteristics of the Emergency Medical Service System Vehicles

The characteristics of the EMS vehicles are its Position and Target along with Route.

The classification of vehicles according to their current status based on Optimization Model are as follows:

Type 1: Waiting for instruction at home station;
Type 2: Moving to an incident point;
Type 3: Staying on the scene to treat an incident;
Type 4: Transferring patient to a hospital;
Type 5: Hospital Stay;
Type 6: Moving back to home station.

3.3   Consign Strategy Rules

Rule 1: Change in target from home station to an incident without any constraints.

Rule 2: Vehicles could change to a new target only if there are significant improvements in the system. Significant improvement is defined as (The new overall response time) − (The old overall response time) > some pre-specified amount of time.

Rule 3: Change of target is not allowed unless significant improvements are achieved.

These rules define the condition of reassignment and route change.

An Optimization model is designed with respect to change of targets in EMS vehicles.

Based on numerical calculations it reveals the fact that response time is minimized.

The design of the Optimization Model is depicted in Figs. 1 and 2.

The objective of the Optimization Model is in Fig. 3.

The particulars about the Vehicles, Occurrences, Hospitals, Vehicle Status, Time and Constraints are given in Figs. 4, 5 and 6.

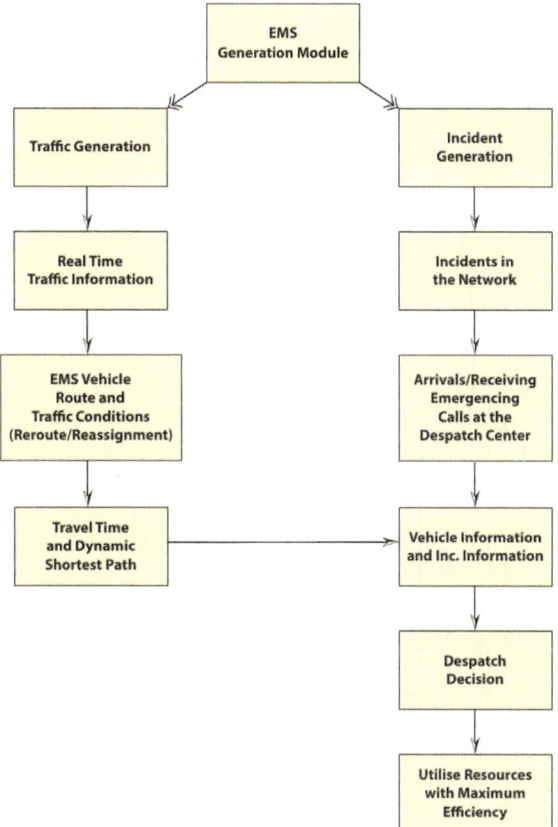

Fig. 1 EMS Generation Model

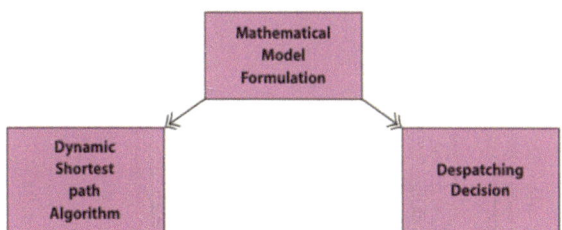

Fig. 2 Mathematical Model

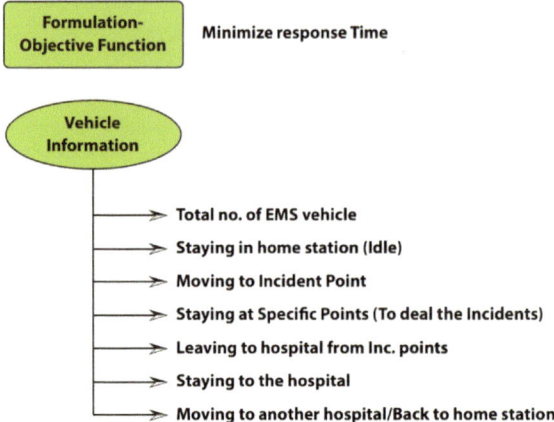

Fig. 3 Optimization Model

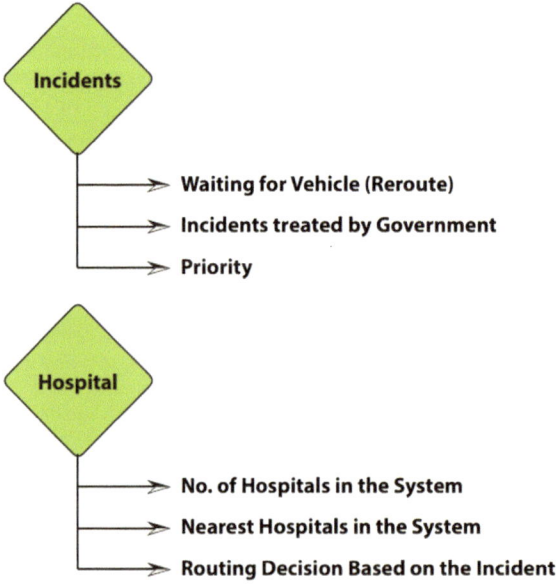

Fig. 4 Constraints Graph 1

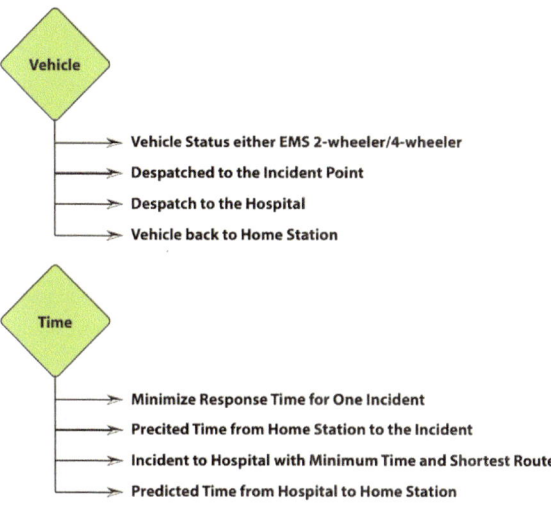

Fig. 5 Constraints Graph 2

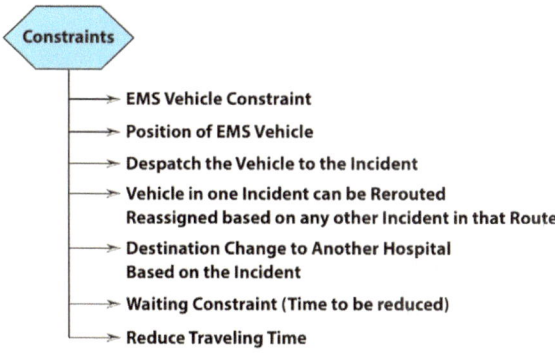

Fig. 6 Constraints Graph 3

3.4 Mathematical Model Formulation

The Mathematical Formulation for the real time emergency response vehicle consigning problem is designed which minimizes the response time subject to the duration of the EMS vehicle constraints is as follows:

The objective is minimizing total response time (where $x_{ij}t_{ij}(t)$ is the duration to the occurrences waiting for treatment, $y_{ik}t_{ik}(t)$ is the duration to the hospitals, and $z_{is}t_{is}(t)$ is the duration to home station) with the constraints (i) duration of EMS vehicle whose target could be either incident or hospital or its home station (ii) duration of EMS Vehicle whose target could be incident and hospital (iii) duration of EMS vehicle whose target could be hospital and its home station (iv) duration of EMS vehicle whose target could be from its home station to occurrences waiting for treatment. (v) duration of each vehicle those should go to occurrences for treatment (vi) duration of each vehicle those should go to hospital for treatment (vii) duration of each vehicle those should reach its home stations after finishing the tasks.

$$\sum_{i=1}^{N_v}\sum_{j=1}^{N_w} x_{ij} t_{ij}(t) + \sum_{i=1}^{N_v}\sum_{k=1}^{N_h} y_{ik} t_{ik}(t) + \sum_{i=1}^{N_v}\sum_{s=1}^{N_s} z_{is} t_{is}(t)$$

Subject to the constraints

$$\sum_{j=1}^{N_w} x_{ij} t_{ij}(t) + \sum_{k=1}^{N_h} y_{ik} t_{ik}(t) + \sum_{s=1}^{N_s} z_{is} t_{is}(t) = 1, \forall i \tag{A}$$

$$\sum_{j=1}^{N_w} x_{ij} t_{ij}(t) + \sum_{k=1}^{N_h} y_{ik} t_{ik}(t) \geq 0, \forall i \tag{B}$$

$$\sum_{k=1}^{N_h} y_{ik} t_{ik}(t) + \sum_{s=1}^{N_s} z_{is} t_{is}(t) \geq 0, \forall i \tag{C}$$

$$\sum_{j=1}^{N_w} x_{ij} t_{ij}(t) + \sum_{s=1}^{N_s} z_{is} t_{is}(t) \geq 0, \forall i \tag{D}$$

$$\sum_{j=1}^{N_w} x_{ij} t_{ij}(t) \geq 0, \forall i \tag{E}$$

$$\sum_{k=1}^{N_h} y_{ik} t_{ik}(t) \geq 0, \forall i \tag{F}$$

$$\sum_{j=1}^{N_s} z_{is} t_{is}(t) \geq 0, \forall i \tag{G}$$

Where
   V denotes the set of Emergency Medical Service vehicles in the station.
   N_V denotes the total number of vehicles in station
   i denotes the number of vehicles in set V, ie, i = 1, ..., N_V
   W denotes the set of occurrences that are waiting for Emergency Medical Service vehicles
   N_W denotes the total number of occurrences waiting for Emergency Medical Service vehicles
   j denotes the index of vehicles in set W, i.e., i = 1, ..., N_W.
   S denotes the set of Emergency Medical Service home stations
   N_s denotes the total number of home stations
   s denotes the number of home station in set S, s = 1,..., N_s
   H denotes the number of hospitals in the system

N_h denotes the total number of hospitals.

$t_{ij}(t)$ denotes the predicted duration for vehicle i to the incident j while departing at time t.

$t_{ik}(t)$ denotes the predicted duration for vehicle i to the hospital k while departing at time t.

$t_{is}(t)$ denotes the predicted duration for vehicle i to arrive at its home station s while departing at time t (Time duration between EMS Vehicles and incidents, hospitals, home stations are depicted in Tables 1, 2 and 3).

Table 1 Time (in minutes) among EMS vehicles and occurrences (4 EMS vehicles and 3 Occurrences)

	1	2	3
1.	30	40	20
2.	10	20	25
3.	12	18	10
4.	15	50	30

Table 2 Time (in minutes) between EMS vehicles and hospitals (4 EMS vehicles and 2 Hospitals)

	1	2
1.	20	25
2.	30	20
3.	40	30
4.	25	40

Table 3 Time (in Minutes) between EMS vehicles and Its Home station (4 EMS vehicles and 1 Home Station)

	1
1.	150
2.	115
3.	130
4.	170

3.5 Fuzzy Linear Programming Method

Consider $x_{ij}t_{ij}(t)$ be the duration to the occurrences waiting for treatment from i to j and z_1^0 is the respective objective desired level for objective function in reduction of duration to the occurrences & t_1 is the tolerance then $Z_1 : \sum_{i=1}^{N_v} \sum_{j=1}^{N_w} x_{ij}t_{ij}(t) \lesssim Z_1^0$.

Consider $y_{ik} t_{ik}(t)$ be the duration to the hospitals from i to k and z_2^0 is the respective objective desired level for objective function in reduction of duration to the hospitals & t_2 is the tolerance then $Z_2: \sum_{i=1}^{N_v} \sum_{k=1}^{N_h} y_{ikt_{ik}(t)} \overset{\sim}{\leq} Z_2^0$.

Consider $z_{is} t_{is}(t)$ be the duration to its home station from i to s and z_3^0 is the respective objective desired level for objective function in reduction of duration to the hospitals & t_3 is the tolerance then $Z_3: \sum_{i=1}^{N_v} \sum_{s=1}^{N_s} z_{ist_{is}(t)} \overset{\sim}{\leq} Z_3^0$.

The objective functions Z_1, Z_2, Z_3 are formulated for (the duration to the occurrences waiting for treatment, the duration to the hospitals, and the duration to home station) respectively. Their aspirations level are set as 280, 230, 565 by solving each objective function subject to given constraints is TSP and their tolerances are decided as 60, 60, 80. The Corresponding objective function are as follows:

$$\text{Min } Z_1 = 30x_{11}t_{11} + 40x_{12}t_{12} + 20x_{13}t_{13} + 10x_{21}t_{21} + 20x_{22}t_{22} + 25x_{23}t_{23}$$
$$+ 12x_{31}t_{31} + 18x_{32}t_{32} + 10x_{23}t_{23} + 15x_{41}t_{41} + 50x_{42}t_{42} + 30x_{43}t_{43} \leq 280 \tag{1}$$

tolerance $= t_1 = 60$

$$\text{Min } Z_2 = 20y_{11}t_{11} + 25y_{12}t_{12} + 30y_{21}t_{21} + 20y_{22}t_{22} + 40y_{31}t_{31} + 30y_{32}t_{32}$$
$$+ 25y_{41}t_{41} + 40y_{42}t_{42} \leq 230 \tag{2}$$

tolerance $= t_2 = 60$

$$\text{Min } Z_3 = 150z_{11}t_{11} + 115z_{21}t_{21} + 130z_{31}t_{31} + 170z_{41}t_{41} \leq 565 \tag{3}$$

tolerance $= t_3 = 80$.

The Fuzzy Membership function for the duration to the occurrences waiting for treatment, the duration to the hospitals, and the duration to home station are given below which are based on above equations.

$$\mu(Z_1) = \begin{cases} 0 & \text{if} & Z_1 \geq 340 \\ 1 - (Z_1 - 280)/60 & \text{if} & 280 \leq Z_1 \leq 340 \\ 1 & \text{if} & Z_1 \leq 280 \end{cases} \tag{4}$$

$$\mu(Z_2) = \begin{cases} 0 & \text{if} & Z_2 \geq 290 \\ 1 - (Z_2 - 230)/60 & \text{if} & 230 \leq Z_2 \leq 290 \\ 1 & \text{if} & Z_2 \leq 230 \end{cases} \tag{5}$$

$$\mu(Z_3) = \begin{cases} 0 & \text{if} & Z_3 \geq 645 \\ 1 - (Z_3 - 565)/80 & \text{if} & 565 \leq Z_3 \leq 645 \\ 1 & \text{if} & Z_3 \leq 565 \end{cases} \tag{6}$$

The FMOLP with max-min approach is given as follows:

$$\text{Maximize } CX \tilde{\geq} \alpha \tag{7}$$

Subject to:

$$\tilde{\alpha} \tilde{\leq} 1 - (z_1 - 280)/60 \tag{8}$$

$$\tilde{\alpha} \tilde{\leq} 1 - (z_2 - 230)/60 \tag{9}$$

$$\tilde{\alpha} \tilde{\leq} 1 - (z_3 - 565)/80 \tag{10}$$

$$x_{11}t_{11} + y_{11}t_{11} + z_{11}t_{11} = 1 \tag{11}$$

$$x_{12}t_{12} + y_{11}t_{11} + z_{11}t_{11} = 1 \tag{12}$$

$$x_{13}t_{13} + y_{11}t_{11} + z_{11}t_{11} = 1 \tag{13}$$

$$x_{11}t_{11} + y_{12}t_{12} + z_{11}t_{11} = 1 \tag{14}$$

$$x_{12}t_{12} + y_{12}t_{12} + z_{11}t_{11} = 1 \tag{15}$$

$$x_{13}t_{13} + y_{12}t_{12} + z_{11}t_{11} = 1 \tag{16}$$

$$x_{21}t_{21} + y_{21}t_{21} + z_{21}t_{21} = 1 \tag{17}$$

$$x_{22}t_{22} + y_{21}t_{21} + z_{21}t_{21} = 1 \tag{18}$$

$$x_{23}t_{23} + y_{21}t_{21} + z_{21}t_{21} = 1 \tag{19}$$

$$x_{21}t_{21} + y_{22}t_{22} + z_{21}t_{21} = 1 \tag{20}$$

$$x_{22}t_{22} + y_{22}t_{22} + z_{21}t_{21} = 1 \tag{21}$$

$$x_{23}t_{23} + y_{22}t_{22} + z_{21}t_{21} = 1 \tag{22}$$

$$x_{31}t_{31} + y_{31}t_{31} + z_{31}t_{31} = 1 \tag{23}$$

$$x_{32}t_{32} + y_{31}t_{31} + z_{31}t_{31} = 1 \tag{24}$$

$$x_{33}t_{33} + y_{31}t_{31} + z_{31}t_{31} = 1 \tag{25}$$

$$x_{31}t_{31} + y_{32}t_{32} + z_{31}t_{31} = 1 \tag{26}$$

$$x_{32}t_{32} + y_{32}t_{32} + z_{31}t_{31} = 1 \tag{27}$$

$$x_{33}t_{33} + y_{32}t_{32} + z_{31}t_{31} = 1 \tag{28}$$

$$x_{41}t_{41} + y_{41}t_{41} + z_{41}t_{41} = 1 \tag{29}$$

$$x_{42}t_{42} + y_{41}t_{41} + z_{41}t_{41} = 1 \tag{30}$$

$$x_{43}t_{43} + y_{41}t_{41} + z_{41}t_{41} = 1 \tag{31}$$

$$x_{41}t_{41} + y_{42}t_{42} + z_{41}t_{41} = 1 \tag{32}$$

$$x_{42}t_{42} + y_{42}t_{42} + z_{41}t_{41} = 1 \tag{33}$$

$$x_{43}t_{43} + y_{42}t_{42} + z_{41}t_{41} = 1 \tag{34}$$

$$x_{11}t_{11} + y_{11}t_{11} \geq 0 \tag{35}$$

$$x_{12}t_{11} + y_{11}t_{11} \geq 0 \tag{36}$$

$$x_{13}t_{11} + y_{11}t_{11} \geq 0 \tag{37}$$

$$x_{11}t_{11} + y_{12}t_{12} \geq 0 \tag{38}$$

$$x_{12}t_{12} + y_{12}t_{12} \geq 0 \tag{39}$$

$$x_{13}t_{13} + y_{12}t_{12} \geq 0 \tag{40}$$

$$x_{21}t_{21} + y_{21}t_{21} \geq 0 \tag{41}$$

$$x_{22}t_{22} + y_{21}t_{21} \geq 0 \tag{42}$$

$$x_{23}t_{23} + y_{21}t_{21} \geq 0 \tag{43}$$

$$x_{21}t_{21} + y_{22}t_{22} \geq 0 \tag{44}$$

$$x_{22}t_{22} + y_{22}t_{22} \geq 0 \tag{45}$$

$$x_{23}t_{23} + y_{22}t_{22} \geq 0 \tag{46}$$

$$x_{31}t_{31} + y_{31}t_{31} \geq 0 \tag{47}$$

$$x_{32}t_{32} + y_{31}t_{31} \geq 0 \tag{48}$$

$$x_{33}t_{33} + y_{31}t_{31} \geq 0 \tag{49}$$

$$x_{31}t_{31} + y_{32}t_{32} \geq 0 \tag{50}$$

$$x_{32}t_{32} + y_{32}t_{32} \geq 0 \tag{51}$$

$$x_{33}t_{33} + y_{32}t_{32} \geq 0 \tag{52}$$

$$x_{41}t_{41} + y_{41}t_{41} \geq 0 \tag{53}$$

$$x_{42}t_{42} + y_{41}t_{41} \geq 0 \tag{54}$$

$$x_{43}t_{43} + y_{41}t_{41} \geq 0 \tag{55}$$

$$x_{41}t_{41} + y_{42}t_{42} \geq 0 \tag{56}$$

$$x_{42}t_{42} + y_{42}t_{42} \geq 0 \tag{57}$$

$$x_{43}t_{43} + y_{42}t_{42} \geq 0 \tag{58}$$

$$y_{11}t_{11} + z_{11}t_{11} \geq 0 \tag{59}$$

$$y_{12}t_{12} + z_{11}t_{11} \geq 0 \tag{60}$$

$$y_{21}t_{21} + z_{21}t_{21} \geq 0 \tag{61}$$

$$y_{22}t_{22} + z_{21}t_{21} \geq 0 \tag{62}$$

$$y_{31}t_{31} + z_{31}t_{31} \geq 0 \tag{63}$$

$$y_{32}t_{32} + z_{31}t_{31} \geq 0 \tag{64}$$

$$y_{41}t_{41} + z_{41}t_{41} \geq 0 \tag{65}$$

$$y_{42}t_{42} + z_{41}t_{41} \geq 0 \tag{66}$$

$$x_{11}t_{11} + z_{11}t_{11} \geq 0 \tag{67}$$

$$x_{12}t_{12} + z_{11}t_{11} \geq 0 \tag{68}$$

$$x_{13}t_{13} + z_{11}t_{11} \geq 0 \tag{69}$$

$$x_{21}t_{21} + z_{21}t_{21} \geq 0 \tag{70}$$

$$x_{22}t_{22} + z_{21}t_{21} \geq 0 \tag{71}$$

$$x_{23}t_{23} + z_{21}t_{21} \geq 0 \tag{72}$$

$$x_{31}t_{31} + z_{31}t_{31} \geq 0 \tag{73}$$

$$x_{32}t_{32} + z_{31}t_{31} \geq 0 \tag{74}$$

$$x_{33}t_{33} + z_{31}t_{31} \geq 0 \tag{75}$$

$$x_{41}t_{41} + z_{41}t_{41} \geq 0 \tag{76}$$

$$x_{42}t_{42} + z_{41}t_{41} \geq 0 \tag{77}$$

$$x_{43}t_{43} + z_{41}t_{41} \geq 0$$

$$\begin{aligned}
&x_{ij}t_{ij} \geq 0 && \text{for } i = 1,2,3,4 \,\&\, j = 1,2,3 \\
&y_{ik}t_{ik} \geq 0 && \text{for } i = 1,2,3,4 \,\&\, k = 1,2 \\
&z_{is}t_{is} \geq 0 && \text{for } i = 1,2,3,4 \,\&\, s = 1 \\
&\tilde{\alpha} \geq 0
\end{aligned} \tag{78}$$

3.6   Results

By solving the above objectives with aspiration levels 280, 230, 565 and tolerance levels 60, 60, 80 the obtained solution is $Z_1 = 292.14$, $Z_2 = 245.50$, $Z_3 = 496.67$. Hence overall reduced response time is 1034.31 min.

4   Conclusion

A Fuzzy Optimization model for Emergency medical service system is designed with respect to various constraints. An optimization routing model is designed for developing flexible consigning strategies with the help of duration information. Based on the comparative analysis of the existing techniques, proposed IPP algorithm ensures that the proposed EMS system reduces the overall response time and the system can handle the targets in timely approach and reduce the important response times. By reducing the overall crucial response time, Emergency Vehicles will be utilized for further occurrences and hence it will help the Society.

References

1. Toregas, C., Swain, R., ReVelle, C., Bergman, L.: The location of emergency service facilities. Operat. Res. **19**, 1363–1373 (1971)
2. Larson, R.: A hypercube queuing model for facility location and redistricting in urban emergency services. Comput. Ops. Res. **1**, 67–95 (1974)
3. Larson, R.: Approximating the performance of urban emergency service systems. Operat. Res. **23**(5), 845–868 (1975)
4. Haghani, A., Tian, Q., Hu, H.: A simulation model for real-time emergency vehicle dispatching and routing. Submitted for presentation at the 82 nd annual meeting of the Transportation Research Board, Washington DC (2002)
5. Zhang, Z.-H., Li, K.: A novel probabilistic formulation for locating and sizing emergency medical service stations. Ann. Operat. Res. **229**(1), 813–835 (2014)

6. Anitha, N., Vijayalakshmi, C.: Traveling salesman problem model with fuzzy multi-objective linear programming. Glob. J. Pure Appl. Math. **12**(3), 538–543 (2016)
7. Qin, J., Ye, Y., Cheng, B., Zhao, X., Ni, L.: The emergency vehicle routing problem with uncertain demand under sustainability environments. Sustain. Supply Chain Manage. **9**(2), 288 (2017)
8. Uplenchwar, R., Deshmukh, P.R., Pawtekar, A., Bhange, G.: ASTUTE routing system for emergency vehicle. Int. J. Innovative Res. Comp. Commun. Eng. **5**(2), 1625–1628 (2017)
9. Aboueljinane, L., Sahin, E., Jemaiv, Z.: A review on simulation models applied to emergency medical service operations. Comp. Ind. Eng. **66**, 734–750 (2013)
10. Ben Akiva, M., Cascetta, E., Gunn, H.: An on-line hyper traffic prediction, model for an inter-urban motorway network. Urban Traffic Networks: Hyper Flow Modeling and Control, pp. 83–122 (1995)
11. Deo, N., Pang, C.: Shortest-path algorithms: taxonomy and annotation. Networks **14**, 275–323 (1984)
12. Cooke, K., Halsey, E.: The shortest route through a network with time dependent internodal transit times. J. Math. Anal. Appl. **14**, 493–498 (1996)
13. Ziliaskopoulos, A., Mahmassani, H.: Time dependent, shortest-path algorithm for real-time intelligent vehicle highway system applications. Transp. Res. Rec. **1408**, 94–100 (1993)
14. Glodberg, J., Dietrich, R., Chen, J., Mitwasi, M.: Validating and applying a model for locating emergency medical vehicles in Tucson, AZ. Eur. J. Oper. Res. **49**, 308–324 (1990)
15. Lubicz, M., Mielczarek, B.: Simulation modelling of emergency medical services. Eur. J. Oper. Res. **29**, 178–185 (1987)
16. Savas, E.: Simulation and cost-effectiveness analysis of New York's emergency ambulance service. Manage. Sci. **15**(12), B608–B627 (1969)
17. Majzoubi, F. 1983: A mathematical programming approach for dispatching and relocating EMS vehicles. Electronic Theses and Dissertations, 891 (2014)
18. Glodberg, J., Paz, L.: Locating emergency vehicle bases when service time depends on call location. Transp. Sci. **25**(4), 264–280 (1991)
19. Ünlüyurt, T., Tunçer, Y.: Estimating the performance of emergency medical service location models via discrete event simulation. Comp. Ind. Eng. **102**, 467–475 (2016)

Simulation of Cortical Epileptic Discharge Using Freeman's KIII Model

Pooja Vijaykumar[1], R. Sunitha[1(✉)], N. Pradhan[2], and A. Sreedevi[3]

[1] Department of Electronics and Communication Engineering, Amrita School of Engineering, Bengaluru, Amrita Vishwa Vidyapeetham, Bengaluru, India
poojavkl993@gmail.com, r_sunitha@blr.amrita.edu
[2] Former Head of Department of Psychopharmacology, National Institute of Mental Health and Neurosciences (NIMHANS), Bengaluru, India
nprnimhans@gmail.com
[3] Department of Electrical and Electronics Engineering, R. V. College of Engineering, Bengaluru, India
sreedevia@rvce.edu.in

Abstract. Advancements in Neuroscience have put forth many networks that are able to mimic cortical activity. One such biologically motivated network is the Freeman KIII Model. The Freeman KIII model is based on the mammalian olfactory system dynamics. It consists of a collection of second-order non-linear differential equations. This paper attempts to solve the equations using MATLAB in order to simulate cortical electroencephalographic (EEG) signals. We also attempt to simulate cortical epileptic discharge using this model. The degenerate state of epileptic seizure is analyzed by obtaining its frequency using the power spectrum and by plotting the phase plots.

Keywords: Freeman KIII model · Olfactory system · Chaotic EEG
Epileptic seizure EEG

1 Introduction

The Brain is the most complex system in the universe. Information processing in the brain takes place in the nerve cells or neurons and the information transfer occurs via synaptic connections, which in turn generates the 'brain signals' or the electroencephalographic signals (EEG). If an imbalance occurs in this communication, it leads to abnormal electrical discharge, known as seizures. A seizure is a sudden surge of electrical activity in the brain that usually affects how a person feels or acts for a short time. Seizures due to epilepsy are called epileptic seizures. Epilepsy is the tendency to have repeated seizures that begin in the brain, due to the abnormal communication among the neurons.

With recent advances in the field of Neuroscience, there have come about many networks that mimic the activity pattern of the brain. Such Neural Networks further help us in understanding the complex dynamics of the brain, while at the same time, are extensively used in multiple real-time applications. Out of these networks, the most biologically realistic is the cognitively motivated Freeman KIII model. The Freeman KIII model [1] is modeled based upon the anatomical and physiological characteristics of the olfactory system dynamics. The olfactory system was chosen by Freeman since it was the simplest and the most stable out of all sensory systems. Freeman KIII model is a hierarchical model containing neuron population dynamics that mediate between small neural networks and the overall brain activity.

The Freeman KIII model comprises of a collection of second-order non-linear differential equations [2] which on solving, simulates the chaotic EEG of the brain. The solutions are also used to generate background EEG as well as the degenerate state of epileptic seizures. Unlike the traditional neural networks, Freeman KIII model is a non-linear network which can represent complex brain dynamics and is an efficient classifier [3–5] that is used in many real-time applications, such as Bionic nose for odour classification [6, 7], Handwriting character classification [8], Face recognition [9], Motor Imagery classification [10], Mandarin digital speech recognition [11], etc.

With the Freeman KIII model we generate cortical epileptic discharge using MATLAB, and we analyze the output using power spectrum and phase plots, while verifying it using practical seizure EEG recordings obtained by Freeman [2]. By setting the required parameters, this model can be used to generate EEG signals of various cognitive activities, thus signifying the biological plausibility of the Freeman KIII model.

2 Olfactory System and Freeman KIII Model

2.1 The Mammalian Olfactory System

The olfactory system is made up of the receptor layer (R), olfactory bulb (OB), anterior olfactory nucleus (AON), and prepyriform cortex (PC) as shown in Fig. 1. Different receptors are tuned to different odour molecules, so a particular odorant causes the receptor neurons to depolarize and fire. The impulses are then transmitted by the primary olfactory nerve (PON) to the glomerular layer (GL), which consists of the periglomerular cells (PG), and the OB. The OB consists of two main cells, mitral cells (M) and granule cells (G), which are connected in negative feedback [12, 13]. The M and G cells generate oscillations which are coherent across the entire bulbar surface. These oscillations are said to be in the gamma range [14, 15]. The axons of M cells combine to form the lateral olfactory tract (LOT) which feeds into AON and PC. The AON and PC also consists of excitatory (E) cells and inhibitory (I) cells, and the negative feedback between them generates oscillations as in OB, but with difference in frequencies [16]. The PC and AON feeds back into OB via the medial olfactory tract (MOT). The oscillatory patterns that arise in these layers are encoded in the spatio-temporal patterns of the EEG.

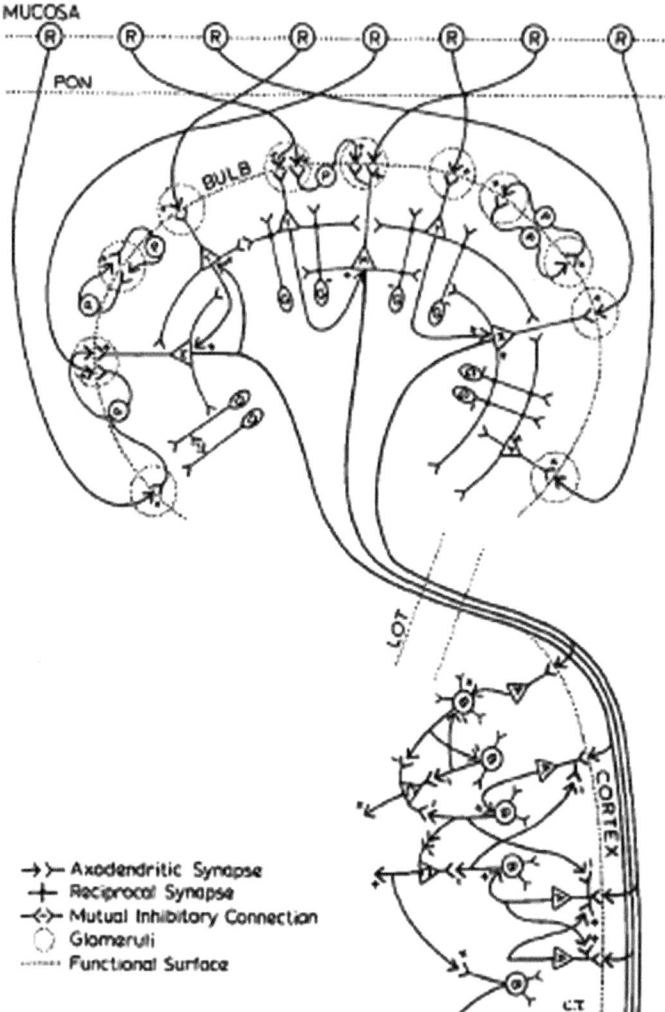

Fig. 1 Schematic diagram of the main cells of the olfactory system: R-receptor, PON-primary olfactory nerve, LOT-lateral olfactory tract, M-mitral cell, G-granule cell, A-superficial pyramidal cell, B-granule cell, C-deep pyramidal cell [1]

2.2   The Freeman KIII Model

The Freeman KIII model [2, 17, 18] is a hierarchical model containing neuron population dynamics that mediate between small neural networks and the entire brain activity (Fig. 2). The fundamental unit of the network is an ensemble of non-interactive neurons (excitatory or inhibitory) and is called K0 unit. When two K0 subsets (e-excitatory or i-inhibitory) are connected by positive feedback, it forms KIe or KIi subset. Negative feedback connection between KIe and KIi forms the KIIei. An array of KII subsets, complete with both feedforward and delayed feedback connections between the layers, forms the KIII network as shown in Fig. 2. The KII set itself is enough to simulate the response of all three layers of the KIII model [19–21].

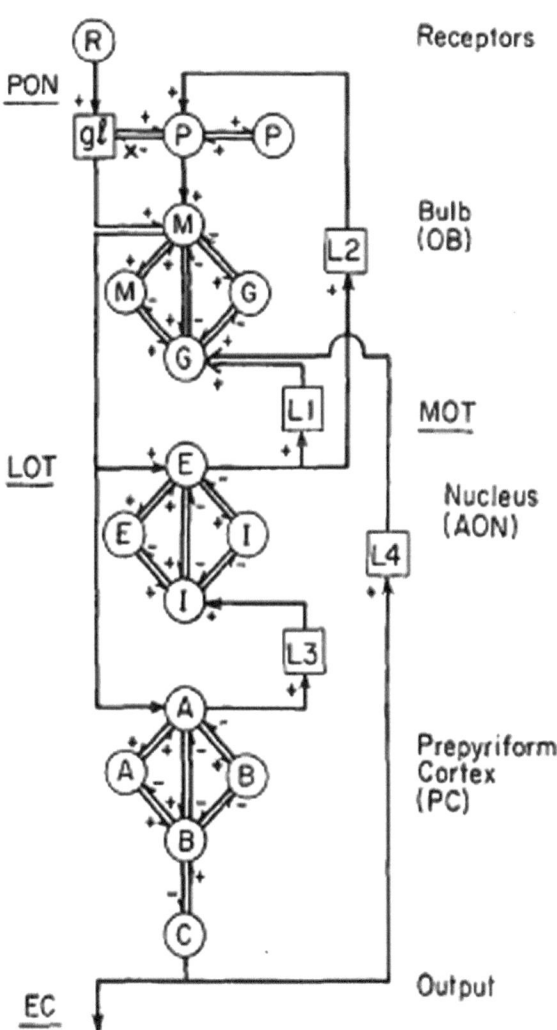

Fig. 2 Structure of the Freeman KIII model [2]

Each K0 node's dynamics is given by a second order differential equation [2]. Linear part of the equation is the left-hand side of (1), where voltage (v) with time (t), is given by

$$abF(v_n) = \ddot{v}_n + (a+b)\dot{v}_n + abv_n, \tag{1}$$

where a = 220/s and b = 720/s are the fixed rate constants obtained empirically [19].

The right-hand side of (1) is the non-linear part with input v and output p [2], given by

$$Q = Q_m\{1 - \exp[-(e^v - 1/Q_m)]\}, v > -u_0, \tag{2a}$$

$$Q = -1, v \le -u_0, \tag{2b}$$

$$u_0 = -\ln[1 - Q_m \ln(1 + 1/Q_m)], \tag{2c}$$

$$p = u_0(Q+1), \tag{2d}$$

where Q_m = 5.0 is the maximum of the sigmoid curve reached asymptotically [22]. The outputs of each node is multiplied by a gain coefficient k_{ij},

$$v_j = k_{ij}p_i \tag{3}$$

where v_j is the current density of neuron dendrite in A/sqcm cortical surface area and p_j is the pulse density of axon in pulse/s/sq cm [23]. The complete KII set with N nodes are given by,

$$F(v_{e1,j}) = k_{ee}p_{e2,j} - k_{ie}(p_{i1,j} + p_{i2,j}) + \sum_{k \ne j}^{N} k_{ee}p_{e1,k} + I_j, \tag{4a}$$

$$F(v_{e2,j}) = k_{ee}p_{e1,j} - k_{ie}p_{i1,j}, \tag{4b}$$

$$F(v_{i2,j}) = k_{ei}p_{e1,j} - k_{ii}p_{i1,j}, \tag{4c}$$

$$F(v_{i1,j}) = k_{ei}(p_{e1,j} + p_{e2,j}) + k_{ii}p_{i2,j} + k_{ii} \sum_{k \ne j}^{N} p_{i1,k}, \tag{4d}$$

where I_j is the external input as well as inputs from the neighbouring KII subsets. Equations (4a–4d) shows four internal weight parameters, which are $k_{ee}, k_{ei}, k_{ie}, k_{ii}$. These weights are optimized and selected based on the desired time series to be generated [24, 25].

When these set of equations are coupled together with feedforward and distributed feedback connections, we get the complete KIII model [26, 27]. The KIII model is built for generating chaotic activity with different parameter modifications [16]. KII-OB by itself is capable of only steady state or "DC" activity. When coupled with KII-AON, it results in multi-frequency oscillations. On coupling KII-OB with KII-AON, and giving a feedback through KI-P results in segments of chaotic activity. To this, if we introduce KII-PC, we get the complete KIII model which deregularizes the chaotic activity [2].

To obtain the degenerate state of epileptic seizure, modifications are performed on the feedforward and the feedback links connecting the layers [2]. The changes done in this paper are limited to the gain co-efficients connecting the KII sets. Other modifications that can also be performed are made in the internal weights within KII set and the height (Qm) of the sigmoid curves of KO sets. The effective changes for obtaining the degenerate seizure state is focused towards reducing the activity of excitatory elements while increasing the activity of inhibitory elements.

In this paper, we attempt to simulate chaotic EEG by solving (1–4d), as well as modify the parameter settings in these equations to generate the seizure state using MATLAB software. The output series is then analysed using phase plots and power spectrum. Power spectrum helps us in observing the range of frequencies that are present in the chaotic or seizure EEG series. Phase plots give us an insight into the

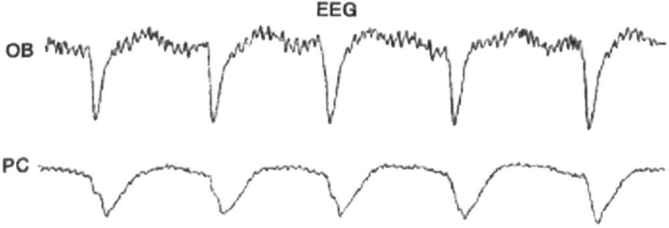

Fig. 3 An example of a 2 s seizure activity recorded from a rat amid seizure [2]

stability of the attractor in phase space. The different attractors are stable, limit cycle and chaotic attractors, and plotting the phase plot provides us with an insight into determining which of these attractors are dominant based on their geometric structure.

3 Results and Analysis

The Freeman KIII network is modeled in MATLAB using Eqs. (1–4d) and solved using Forward Euler method to generate chaotic EEG and seizure EEG. The parameters are set depending upon the desired series to be generated. The KIII network is sensitive even to the smallest parameter changes, hence variations in the output patterns are observed for the smallest changes in the network.

3.1 Modifications for Seizure Activity

As per Freeman [2], three changes suffice to obtain the degenerate state that mimics seizure activity. One is reduction of LOT input to the PC. Second is decreasing the feedback from KII-PC to KII-AON. Third is increasing the feedback from KII-AON to KII-OB. While these modifications suffice, other changes are also effective, as long as they are restricted to the feedforward and the feedback links between the KII sets, as described in Fig. 2. Figure 3 shows the short-time segments of activity recorded from a rat amidst a seizure [2] and Fig. 4 shows the seizure state as generated by the KIII model.

A change in the structure is required in the model for the simulation of seizure activity. This change indicates that seizure is due to degrading transition in state within the feedback mechanism that, on a normal basis, accounts for chaotic background activity. This change is because of an imbalance occurring between an abundance of inhibitory activity over excitatory activity, which goes against one of the rules of stability for KII sets [2].

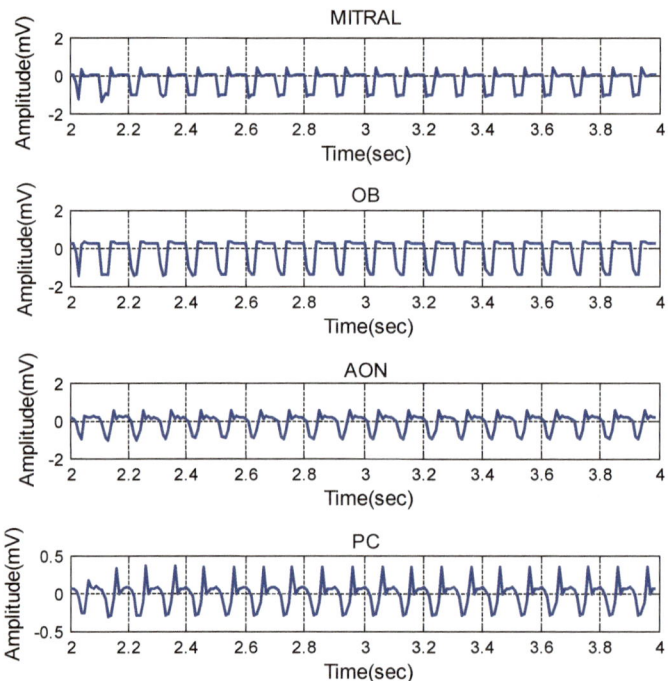

Fig. 4 The seizure activity of the Mitral cell, Olfactory Bulb (OB), Anterior Olfactory nucleus (AON) and Prepyriform Cortex (PC) layers generated from the Freeman KIII model in MATLAB. The activity is generated by modifying the gain coefficients of the links connecting the layers in the model. The plot shows activity between 2 and 4 s, in steps of 0.01

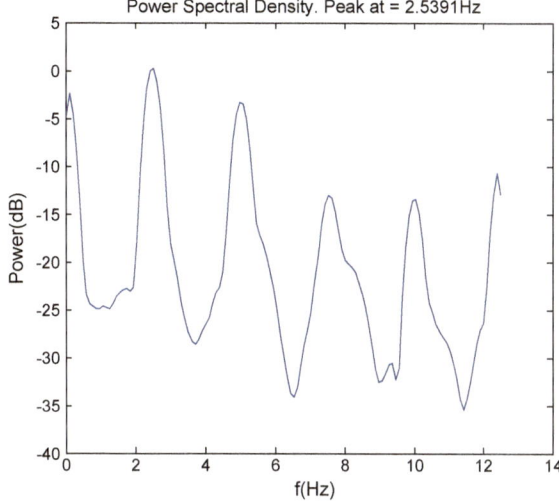

Fig. 5 Power spectral density plot of the seizure generated in Fig. 4

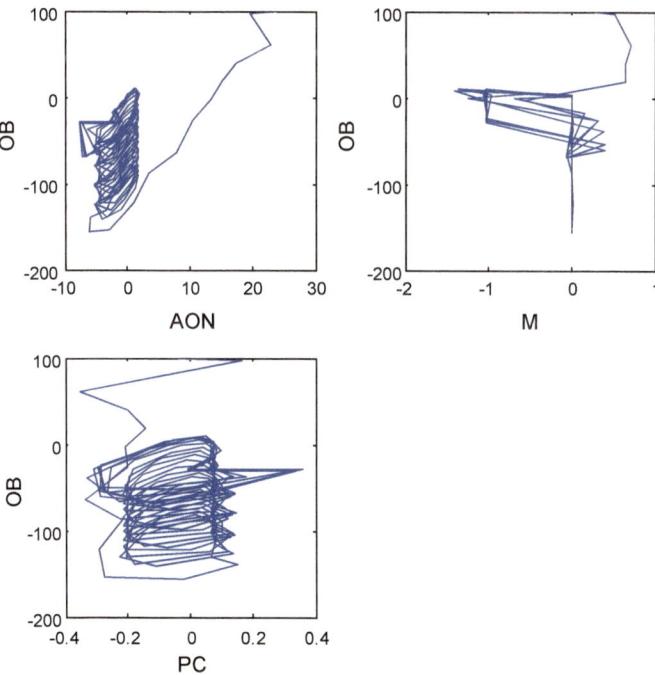

Fig. 6 The traces of phase plots of the seizure activity generated by the Freeman KIII model as in Fig. 4. The phase plots show a chaotic attractor with a 2-torus structure

3.2 Power Spectrum and Phase Plot

The seizure activity is, usually, characterized by low frequency oscillations (~ 3–4 Hz). The power spectrum of the seizure in Fig. 4 is shown in Fig. 5, with the peak at around ~ 2.5 Hz.

To better understand the behaviour of EEG series, an attempt is made at obtaining the phase plots. A phase plot is a geometric representation of the trajectories of a dynamical system in the phase plane. Since the EEG signals are chaotic, dynamical signals, phase plots are an invaluable tool in studying them, as they give us an insight into the stability and the complexity of the system. Construction of the attractor of chaotic EEG in 3 dimensions shows irregular activity with unknown geometric structure [1]. Meanwhile the phase plot of seizure, obtained in Fig. 6, shows a chaotic attractor whose geometric structure is a 2-torus or 'donut-shaped' structure [2].

The olfactory forebrain is neither more nor less unpredictable when contrasted with different parts of the nervous systems, however it is better known over most. Freeman [1] chose the olfactory system to base his model on with the reason being that, compared to other sensory systems, it is simple. The KIII model is a simplified version of the system, emphasizing features, especially the capability of generating chaotic background activity [28, 29]. The model was based on the anatomical properties, deleting all except the crucial features. Keeping this in mind, the model parameters are set accordingly to obtain seizure activity as shown in Fig. 4. The power spectrum and phase plots, Figs. 5 and 6 respectively, further help in analysing the output generated by the model.

4 Conclusion and Future Scope

The Freeman KIII model was modeled based on the olfactory system dynamics. The model generated chaotic background EEG, and with the modifications in the model parameters, we were also able to generate the degenerate state of seizure activity using MATLAB. The seizure activity was analysed using power spectrum, which gave the frequency of the simulated series, and the phase plots, which showed the geometric structure of the attractor of the seizure in phase plane. Furthermore, the seizure activity was verified with the help of practical seizure EEG recorded from a rat during seizure by Freeman. This KIII model gives us an insight into the dynamics of cortical activity as well as some of its properties. Although this model is sensitive in terms of its parameters, it can be used as a starting point to build and simulate other neural mechanisms which pertain to various other activities of cognition and perception.

References

1. Skarda, C.A., Freeman, W.J.: How brains make chaos in order to make sense of the world. Behav. Brain Sci. **10**, 161–195 (1987)
2. Freeman, W.J.: Simulation of Chaotic EEG patterns with a dynamic model of the olfactory system. Biol. Cybern. **56**, 139–150 (1987)

3. Rosa, J.L., Piazentin, D.R.: A new cognitive filtering approach based on Freeman K3 neural networks. Appl. Intell. **45**, 363–382 (2016)
4. Piazenti, D.R., Rosa, J.L.: A simulator for freeman K-sets in JAVA. In: International Joint Conference on Neural Networks (IJCNN), pp. 1–8 (2015)
5. Wang, L., Li, G., Liu, X., Wang, B., Freeman, W.J.: Study of a chaotic olfactory neural network model and its applications on pattern classification. In: IEEE Engineering in Medicine and Biology 27th Annual Conference, pp. 3640–3643, China (2005)
6. Fu, J., Li, G., Qin, Y., Freeman, W.J.: A pattern recognition method for electronic noses based on an olfactory neural network. Sens. Actuators, B **125**, 489–497 (2007)
7. Fu, J., Yang, X., Yang, X., Li, G., Freeman, W.J: Application of biologically modeled chaotic neural network to pattern recognition in artificial olfaction. In: Proceedings of the 2005 IEEE, Engineering in Medicine and Biology 27th Annual Conference, vol. 16, pp. 4666–4669, Shanghai, China (2005)
8. Obayashi, M., Koga, S., Feng, L., Kuremoto, T., Kobayashi, K.: Handwriting character classification using Freeman's olfactory KIII model. Artif. Life Robot. **17**(2), 227–232 (2012)
9. Zhang, J., Lou, Z., Li, G., Freeman, W.J.: Application of a novel neural network to face recognition based on DWT. In: International conference on Biomedical Robotics and Biomechatronics, Italy (2006)
10. Piazentin, D.R., Rosa, J.L: Motor imagery classification for brain-computer interfaces through a chaotic neural network. In: International Joint Conference on Neural Networks (IJCNN), vol. 11, pp. 4103–4108, Beijing, China (2014)
11. Zhang, J., Li, G. and Freeman, W.J.: Application of novel chaotic neural networks to mandarin digital speech recognition. In: International Joint Conference on Neural Networks, vol. 8, pp. 653–658, Vancouver, Canada (2006)
12. Freeman, W.J.: A model for mutual excitation in a neuron population in olfactory bulb. In: IEEE Transactions on Biomedical Engineering, vol. BME-21, No. 5, pp. 350–358 (1974)
13. Li, Z., Hopfield, J.J.: Modeling the olfactory bulb and its neural oscillatory processing. Biol. Cybern. **61**, 379–392 (1989)
14. Kay, L.M., et al.: Olfactory oscillations: the what, how and what for. Trends Neurosci. **32**(4), 207–214 (2009)
15. Kay, L.M., Freeman, W.J.: Bidirectional processing in the olfactory-limbic axis during olfactory behavior. Behav. Neurosci. **112**(3), 541–553 (1998)
16. Eisenberg, J., Freeman, W.J., Burke, B.: Hardware architecture of a neural network model simulating pattern recognition by the olfactory bulb. Neural Networks **2**, 315–325 (1989)
17. Barrie, J.M., Freeman, W.J., Lenhart, M.D.: Spatiotemporal analysis of prepyriform, visual, auditory, and somesthetic surface EEGs in trained rabbits. J. Neurophysiol. **76**(1), 520–539 (1996)
18. Kozma, R., Freeman, W.J.: Encoding and recall of noisy data as chaotic spatio-temporal memory patterns in the style of the brains. Int. Joint Conf. Neural Networks **5**, 33–38 (2000)
19. Chang, H., Freeman, W.J.: Parameter optimization in models of the olfactory neural system. Neural Networks **9**(1), 1–14 (1996)
20. Yao, Y., Freeman, W.J.: Model of biological pattern recognition with spatially chaotic dynamics. Neural Networks **3**, 153–170 (1990)
21. Yao, Y., Freeman, W.J.: Pattern recognition in olfactory systems: modeling and simulation. Int. Joint Conf. Neural Networks **1**, 699–704 (1989)
22. Hodgkin, A.L., Huxley, A.F.: A quantitative description of membrane current and it's application to conduction and excitation in nerve. J. Physiol. **117**, 500–544, Cambridge (1952)

23. Eledath, D., Ramachandran, S., Pradhan, N., Asundi, S.: Modelling and implementation of two coupled Hodgkin-Huxley neuron model. In: International Conference on Computing and Network Communications, Dec. 16–19, Trivandrum, India (2015)
24. Chang, H., Freeman, W.J., Burke, B.C.: Optimization of olfactory model in software to give 1/f power spectra reveals numerical instabilities in solutions governed by aperiodic (chaotic) attractors. Neural Networks **11**, 449–466 (1998)
25. Ilin, R. and Kozma, R: Stability conditions of the full KII model of excitatory and inhibitory Neural populations. International Joint Conference on Neural Networks, pp. 3162–3167, Canada (2005)
26. Freeman, W.J.: Understanding perception through Neural 'Codes'. IEEE Trans. Biomed. Eng. **58**(7), 1884–1890 (2011)
27. Obayashi, M., Suda, R., Kuremoto, T., Mabu, S.: A class identification method using Freeman's olfactory KIII Model. J. Image Graph. **4**(2) (2016)
28. Beliave, I., Kozma, R.: Studies on the memory capacity and robustness of chaotic dynamic neural networks. In: International Joint Conference on Neural Networks, July, Vancouver, Canada (2006)
29. Freeman, W.J.: Modelling olfactory pattern recognition by an adaptive spatial filter. In: IEEE Conference on Decision and Control including the 16th Symposium on Adaptive Processes and A Special Symposium on Fuzzy Set Theory and Applications, New Orleans, USA (1977)

Perceptual Features Based Rapid and Robust Language Identification System for Various Indian Classical Languages

A. Revathi[1], C. Jeyalakshmi[2(⊠)], and T. Muruganantham[2]

[1] Department of Electronics and Communication Engineering, School of EEE,
Sastra University, Thanjavur, Tamilnadu, India
revathidhanabal@rediffmail.com
[2] Department of Electronics and Communication Engineering, K. Ramakrishnan
College of Engineering, Samayapuram, Trichy, Tamilnadu, India
{lakshmikrce.2016,ananthusivam}@gmail.com

Abstract. In this paper, we have investigated the development of the robust recognition system to identify the language of the spoken utterance in several classical Indian languages. The ability of the machines to identify the language in which communication takes place globally is a paramount in the current scenario. In this work on VQ clustering based language identification, feature vectors extracted from training speeches are converted into set of language specific clustering models. During testing, features extracted from test speeches are applied to the language specific clustering models and hypothesized language is identified based on minimum of averages corresponding to the model using minimum distance classifier. Clustering technique is evaluated with variations in cluster size and length of the test utterances. Better performance is observed for the cluster size of 64 and 900 test utterances of 2–3 s duration. Noteworthy point to be mentioned is that though the clustering technique is an old technique, it provides 97% as average recognition accuracy, 3.13% as average false acceptance rate and 3.13% as false rejection rate for the testing done on the language specific clustering model with 64 as cluster size. To improve the accuracy, formants are also used as a feature and it is observed that formant frequencies do provide complementary evidence in ensuring better accuracy for some of the languages. Performance of the system is assessed by calculating the identification rate as the one corresponding to the correct identification with respect to either MFPLPC or Formants and the overall average accuracy obtained is 99.4% for clustering model with 32 as cluster size. Experiments are conducted on the database containing speech utterances in seven classical and phonetically rich speaker specific Indian languages such as Bengali, Hindi, Kannada, Malayalam, Marathi, Tamil and Telugu.

Keywords: Clustering · Mel frequency perceptual linear predictive cepstrum (MFPLPC) · Language identification (LID) · False acceptance rate (FAR)
False rejection rate (FRR) · Standard deviation (STD) · Vector quantization (VQ)

© Springer International Publishing AG 2018
D. J. Hemanth and S. Smys (eds.), *Computational Vision and Bio Inspired Computing*,
Lecture Notes in Computational Vision and Biomechanics 28,
https://doi.org/10.1007/978-3-319-71767-8_25

1   Introduction

Automatic language identification is the process of determining the identity of the spoken language from the set of speech utterances. Some of the notable applications of LID include global communications, call routing systems, multilingual dialog systems and multilingual translation systems etc. LID is also a topic of great interest and importance in areas of intelligence and security; it is necessary to establish the identity of recorded messages and archived materials before extracting any information. LID technology also makes massive, online language routing possible for enabling voice surveillance over telephone network.

With advancements in speech technology, humans can converse with computer backed systems in the language in which system has been trained to accomplish the required tasks such as air tickets and railway tickets reservation. In fact, systems would assume that all humans are speaking in an official language. But, it is always feasible and convenient for the people to converse in their native language with the system. So, it is necessary to develop the system which would understand the spoken utterances in any language and more reliable communication would be established between people and computer. And, the interface between multilingual system and human users would be a device that can quickly and accurately identify a language in which the user is speaking. All LIDs are classified into two types such as text dependent and text independent systems.

Text independent systems are implemented using some set of speech samples pertaining to the specific language for training and some other set of speech samples would be considered for testing. Four approaches [1] such as GMM, single language phone recognition followed by language modeling, multiple single language phone recognition and language dependent parallel phone recognition are implemented to assess the performance of LID system. The paper [2] provides compendious review about explicit and implicit LID system present in the literature. Text independent LID system [3] is developed for six Indic languages. The use of parallel phone recognition approach to develop LID system and the use of parallel phone recognition language modeling approach for LID system is discussed [4].

MFCC is used as a feature & language modeling [5] by PRLM (Phoneme recognition and language modeling) and classification is done by GMM. The paper [6] discusses the use of MFCC and DHMM for developing language models and evaluation is done on OGI database. Besides this, MFCC and DHMM [7] are utilized for developing LID system and evaluation is done on six Indic languages. ASM frame work and text independent phone models for LID systems is described [8] and evaluation is done on NIST 1996 and 2003 databases. The use of cluster based computation to derive new feature from MFCC and Formants and GMM is used for classification [9] and Evaluation is done on ODI database. The development of four approaches for LID system using OGI database is done [10]. The use of SVM for identifying languages has been described [11–15]. The development of probabilistic graphical models for identifying languages for five languages is considered [16]. LID system for two languages English and French are designed [17] using support vector machine with radial basis function kernel.

2   Description About Database Used

The IIIT-H Indic speech databases were developed at Speech and Vision Lab, IIIT-H (speech@iiit.ac.in) for the purpose of building speech synthesis systems in Indian languages. Currently the IIIT-H Indic speech databases consist of text and speech data in Bengali, Hindi, Kannada, Malayalam, Marathi, Tamil and Telugu. These languages were chosen, as the total number of Wikipedia articles in each of these languages was more than 10,000 and native speakers of these languages were available in the campus. Each of these languages does have several dialects. As an initial approximation, speeches are recorded in the dialect in which the native speaker was comfortable with. Wikipedia articles in Indian languages were used as text corpus. A set of 1000 sentences was selected for each language. These sentences were selected to cover 5000 most frequent words in text corpus of the corresponding language. The text data is made available in IT3 (a transliteration scheme) as well as in Unicode (UTF-8 format). The speech data was recorded by a native speaker of the language. The recording was done in a studio environment using a standard headset microphone connected to a Zoom handy recorder. Handy recorder was used for recording purpose as it was highly mobile and easy to operate. By using a headset the distance from the microphone to a mouth and recording level was kept constant.

3   Speech Analysis-Languages

Speeches in different languages are analyzed in time-domain and frequency domain. Figure 1 illustrates the Tamil speech signal and its spectrogram.

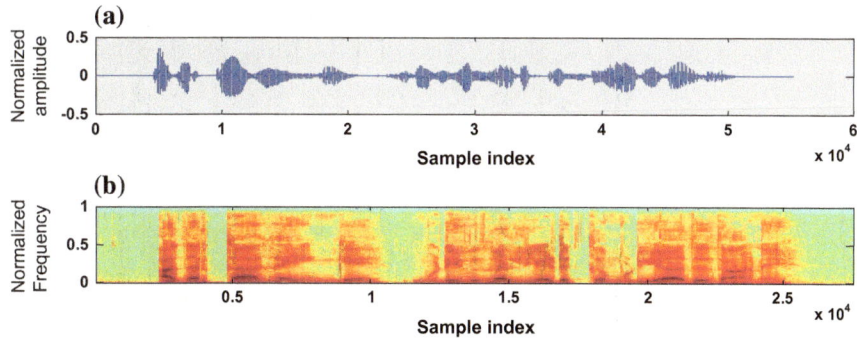

Fig. 1 Analysis on speech in Tamil language **a** Speech signal **b** Its spectrogram

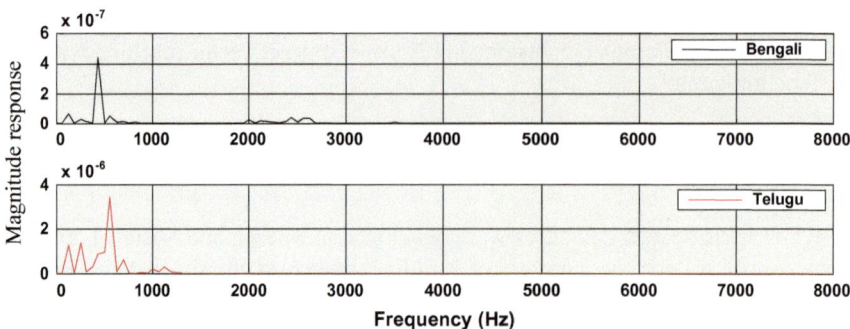

Fig. 2 Frequency response curves for utterances in Bengali and Telugu

Figure 2 depicts the frequency response of the speech signal in Bengali and Telugu languages. It indicates that the spectral peaks are occurring between 500 and 700 Hz for the speeches spoken in these languages. Frequency response plots in Fig. 3 indicate the Spectral variation between the speeches spoken in languages Marathi and Tamil. From these frequency response plots, it is understood that the speech frequencies are limited to 1000 Hz irrespective of the utterances spoken in different languages.

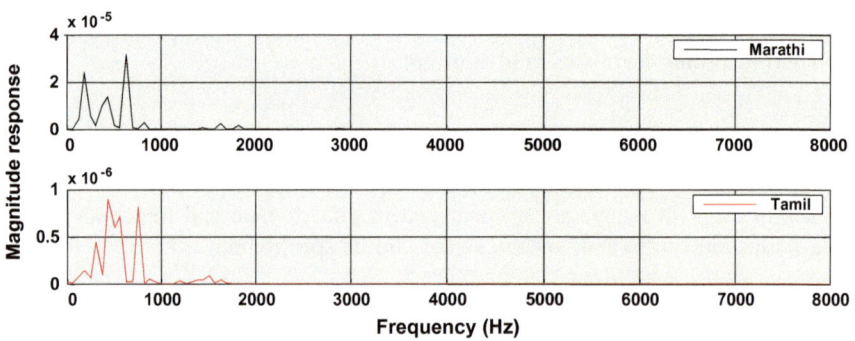

Fig. 3 Frequency response curves for utterances in Marathi and Tamil

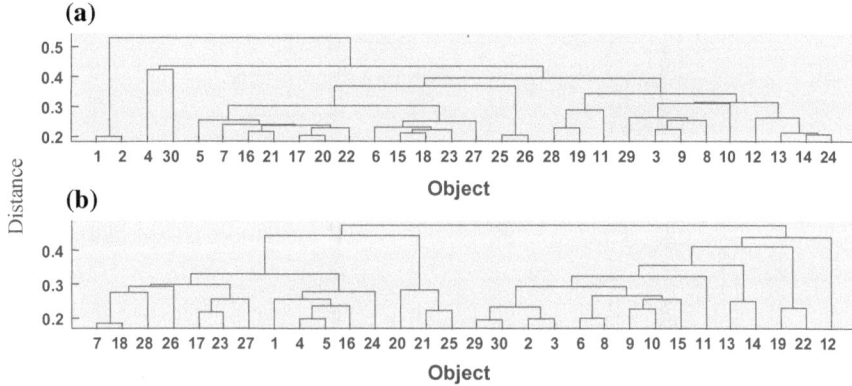

Fig. 4 Dendrogram plot **a** Hierarchical binary cluster tree for 30 training vectors (Bengali), **b** Hierarchical binary cluster tree for 30 training vectors (Hindi)

4 LID System Using MF-PLP and K Means Clustering Technique

MFCC features are widely used for speech/speaker/language identification [18–20]. LID system is developed using MF-PLPC [21–25] as a feature and K means clustering for developing training models.

4.1 Training and Testing Based on Clustering Technique

K-means clustering technique [26] is used to convert the training vectors into clusters and cluster centroids are representing the language specific training data. Figure 4 indicates the hierarchical binary cluster tree for the feature vectors of the training data taken in Bengali and Hindi languages.

Dendrogram plots reveal the production of distinct set of clusters for speeches in different languages. LID system involves extraction of features from the training data, building VQ codebook models for the speeches in languages and testing each utterance against a certain number of speech models to detect the identity of the speech of that utterance from among the speech models. Text independent LID system is evaluated using the data for utterances in speaker specific languages. For creating a training model, 60 utterances spoken in particular language are concatenated and the resultant signal is preprocessed to eliminate silence and low energy frames and pre-emphasized using a difference operator. Hamming window is applied on differenced speech frames of 16 ms duration with overlapping of 8 ms. Then, MF-PLPC features are extracted [21–25] and formants are extracted using LPC analysis. For each language, VQ codebook models are developed separately based on K-means clustering procedure with perceptual features and formants as input with variations in cluster size. In this algorithm, there is a mapping from L training vectors into M clusters. Each block is normalized to unit magnitude before giving as input to the model. Testing procedure used for testing the utterances spoken in languages is shown in Fig. 5. Distance is

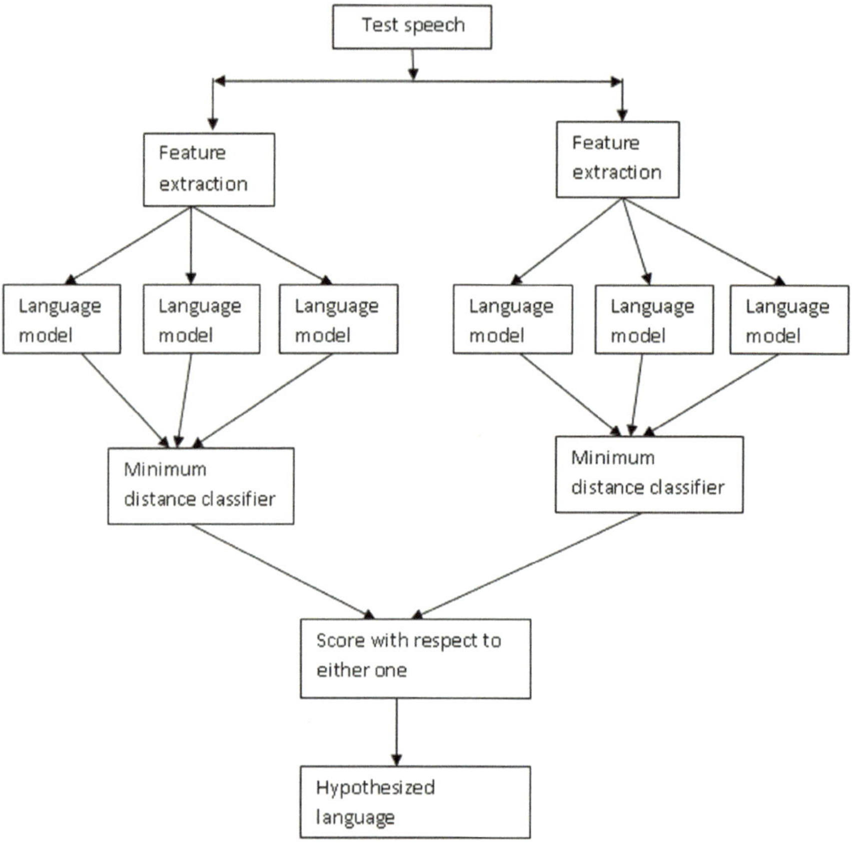

```
            ┌─────────────┐
            │ Test speech │
            └─────────────┘
```

```
┌──────────────┐                    ┌──────────────┐
│  Feature     │                    │  Feature     │
│  extraction  │                    │  extraction  │
└──────────────┘                    └──────────────┘
```

```
┌──────────┐ ┌──────────┐ ┌──────────┐   ┌──────────┐ ┌──────────┐ ┌──────────┐
│ Language │ │ Language │ │ Language │   │ Language │ │ Language │ │ Language │
│ model    │ │ model    │ │ model    │   │ model    │ │ model    │ │ model    │
└──────────┘ └──────────┘ └──────────┘   └──────────┘ └──────────┘ └──────────┘
```

```
    ┌──────────────────┐               ┌──────────────────┐
    │ Minimum          │               │ Minimum          │
    │ distance classifier │            │ distance classifier │
    └──────────────────┘               └──────────────────┘
```

```
              ┌──────────────────────┐
              │ Score with respect to │
              │ either one            │
              └──────────────────────┘
```

```
              ┌──────────────┐
              │ Hypothesized │
              │ language     │
              └──────────────┘
```

Fig. 5 Hypothesized language Classifier

calculated between feature vectors and clusters of the respective language models. Minimum distances are averaged for each model and hypothesized language is chosen based on minimum of averages [25]. Identification rate is computed for the test speech which reveals correctly identified index with respect to either MFPLPC or Formants. The testing system used for classifying different languages is shown in Fig. 5.

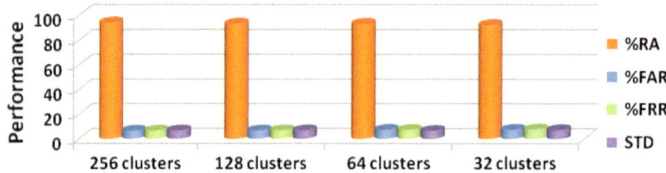

Fig. 6 Overall Performance analysis—various clustering models for rapid LID system

5 Results and Discussions

Language recognition rate is the number of correct choices over the total number of test speeches. Performance of the system for LID for the utterances in seven different indic languages using K means clustering technique is assessed by varying the cluster size as 32, 64, 128 and 256. Figure 6 elucidates the average performance of the LID system using MFPLPC as a feature with variations in cluster size in terms of the parameters such as RA, FAR, FRR and STD for first 100 vectors of test speeches being given to the language specific models. Though, clustering model of size 256 has provided better performance in terms of RA, FRR and FAR, but clustering model of size 64 has ensured low value of standard deviation. For this system, 60 speeches are taken for training.

Table 1 Individual performance analysis for rapid LID system

Clusters	Language	Confusion matrix							%RA	%FRR	% RA
		Be.	Hi.	Ka.	Mal.	Ma.	Ta.	Te.			
256	Bengali	854	0	5	0	34	0	7	95	5	**94**
	Hindi	0	900	0	0	0	0	0	100	0	
	Kannada	8	0	867	0	25	0	0	96.4	3.6	
	Mala.	5	2	2	866	24	0	1	96.2	3.8	
	Marathi	10	1	2	0	881	1	5	98	2	
	Tamil	29	35	0	0	49	783	4	87	13	
	Telugu	168	0	0	0	4	0	730	81	19	
	%FAR	**24.4**	**4.2**	**1**	**0**	**15**	**0.1**	**1.9**	% Frr = 6.66	% Far = 6.63	
128	Bengali	840	1	9	0	43	0	7	93.4	6.6	**93.2**
	Hindi	0	900	0	0	0	0	0	100	0	
	Kannada	8	1	867	0	24	0	0	96.4	3.6	
	Malayalam	3	0	0	871	25	0	1	97	3	
	Marathi	12	1	3	0	878	1	5	98	2	
	Tamil	41	35	0	0	51	769	4	86	14	
	Telugu	163	0	0	0	5	0	732	81.4	18.6	
	%FAR	**25**	**4.2**	**0.3**	**0**	**15.9**	**0.1**	**1.9**	% Frr = 6.8	% Far = 6.83	
64	Bengali	845	0	8	0	40	0	7	94	6	**93**
	Hindi	0	900	0	0	0	0	0	100	0	
	Kannada	12	1	864	0	22	0	1	96	4	
	Malayalam	6	0	2	860	31	0	1	95.6	4.4	
	Marathi	26	1	4	0	863	1	5	96	4	
	Tamil	33	35	0	0	60	764	8	85	15	
	Telugu	161	0	0	0	4	0	735	82	18	
	%FAR	**26.4**	**4.1**	**1.6**	**0**	**17.4**	**0.1**	**2.4**	% Frr = 7.43	% Far = 7.34	

(*continued*)

Table 1 (*continued*)

Clusters	Language	Confusion matrix							%RA	%FRR	% RA
		Be.	Hi.	Ka.	Mal.	Ma.	Ta.	Te.			
32	Bengali	849	0	10	0	18	0	3	94.4	5.6	**92.2**
	Hindi	0	900	0	0	0	0	0	100	0	
	Kannada	8	0	871	0	20	0	1	97	3	
	Malayalam	5	0	2	865	27	0	1	96	4	
	Marathi	27	1	14	0	848	5	5	94.2	5.8	
	Tamil	58	43	0	0	42	750	7	83.4	16.6	
	Telugu	162	0	2	0	12	0	724	80.5	19.5	
	%FAR	**28.9**	**4.9**	**3.1**	**0**	**13.2**	**0.6**	**1.9**	% Frr = 7.52	% Far = 7.79	

Table 2 Individual performance analysis—various clustering models for all test vectors of test utterances

Clusters	Language	Confusion matrix							%RA	%FRR	% RA
		Be.	Hi.	Ka.	Mal.	Mar.	Ta.	Te.			
256	Bengali	897	0	0	0	3	0	0	99.7	0.3	**96.8**
	Hindi	0	900	0	0	0	0	0	100	0	
	Kannada	0	0	886	0	14	0	0	98.5	1.5	
	Mala.	0	0	0	893	3	4	0	99.2	0.8	
	Marathi	1	0	0	0	893	6	0	99.2	0.8	
	Tamil	4	30	0	0	62	801	3	89	11	
	Telugu	65	0	0	0	0	11	824	92	8	
	%FAR	**7.8**	**3.3**	**0**	**0**	**9.1**	**2.3**	**0.3**	% Frr = 3.26	% FAR = 3.2	
128	Bengali	897	0	0	0	3	0	0	99.7	0.3	**96.6**
	Hindi	0	900	0	0	0	0	0	100	0	
	Kannada	0	0	876	1	23	0	0	97.4	2.6	
	Malayala	0	0	0	893	1	6	0	99.2	0.8	
	Marathi	0	0	0	0	893	7	0	99.2	0.8	
	Tamil	13	29	0	0	58	797	3	88.6	11.4	
	Telugu	56	0	0	0	0	15	829	92.1	7.9	
	%FAR	**7.7**	**3.2**	**0**	**0.1**	**9.4**	**3.1**	**0.3**	% FRR = 3.37	% FAR = 3.4	
64	Bengali	898	0	1	0	1	0	0	99.8	0.2	**97**
	Hindi	0	900	0	0	0	0	0	100	0	
	Kannada	0	0	884	0	16	0	0	98.2	1.8	
	Malayala	0	0	0	891	2	7	0	99	1	
	Marathi	1	1	0	0	890	8	0	99	1	
	Tamil	1	39	0	0	55	802	3	89.1	10.9	
	Telugu	49	0	0	0	0	14	837	93	7	
	%FA R	**5.7**	**4.4**	**0.1**	**0**	**8.22**	**3.2**	**0.3**	% FRR = 3.13	%FAR = 3.13	

(*continued*)

Table 2 (*continued*)

Clusters	Language	Confusion matrix							%RA	%FRR	% RA
		Be.	Hi.	Ka.	Mal.	Mar.	Ta.	Te.			
32	Bengali	897	0	1	0	2	0	0	99.7	0.3	**96.4**
	Hindi	0	900	0	0	0	0	0	100	0	
	Kannada	0	0	873	0	27	0	0	97	3	
	Malayala	0	0	0	894	3	3	0	99.4	0.6	
	Marathi	1	0	0	0	888	11	0	98.7	1.3	
	Tamil	14	55	0	0	28	800	3	89	11	
	Telugu	60	0	0	0	0	20	820	91.1	8.9	
	%FAR	**8.3**	**6.1**	**0.1**	**0**	**6.7**	**3.8**	**0.3**	% **FRR = 3.61**	% **FAR = 3.59**	

Table 3 Correlation matrix for clusters of speaker specific language models—256 cluster size

Language	Bengali	Hindi	Kannada	Malayalam	Marathi	Tamil	Telugu
Bengali	1	0.9845	0.9853	0.9638	**0.9863**	0.9857	0.9846
Hindi	0.9845	1	0.9817	0.9822	0.9849	**0.9853**	0.9813
Kannada	**0.9853**	0.9817	1	0.9819	0.9853	0.9826	0.9827
Malayalam	0.9838	0.9822	0.9819	1	**0.9840**	0.9824	0.9787
Marathi	**0.9863**	0.9849	0.9853	0.984	1	0.9845	0.9832
Tamil	**0.9857**	0.9853	0.9826	0.9824	0.9845	1	0.9842
Telugu	**0.9846**	0.9813	0.9827	0.9787	0.9832	0.9842	1

Table 1 indicates the performance of the LID system in detail for the first 100 test vectors being given to the language specific models with variations in cluster size. From Tables 1, 2, 3 and 4, bold letters shows the average recognition accuracy. It is understood that this rapid LID system has provided marginally the better accuracy for the clustering model of size 256 as compared to all other clustering models and number of test utterances considered is 900.

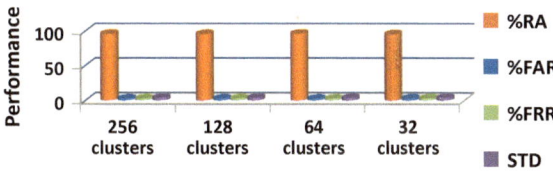

Fig. 7 Overall performance analysis of LID system for all test vectors of test utterances

Figure 7 elucidates the average performance of the LID system using MFPLPC as a feature with variations in cluster size in terms of the parameters such as RA, FAR, FRR and STD for all the test vectors of test speeches being given to the language specific models. Though, clustering model of size 64 has provided better performance in terms of RA, FRR, FAR and STD. Table 2 indicates the performance of the LID system in detail for all the test vectors being given to the language specific models with variations in cluster size.

It is understood that this LID system has provided marginally the better accuracy for the clustering model of size 64 and number of test utterances considered is 900. Table 3 indicates the analysis of results with correlation matrix between clustering models of 256 clusters for all languages derived from feature vectors of training data. For an example, in the first row of correlation matrix, second highest value is 0.9863. It actually illustrates some of the test speeches corresponding to Bengali would be misclassified as one corresponding to Marathi language.

Table 4 Performance of the system with MFPLC and Formants

Clusters	Languages	Classification of test speeches—individual accuracy (Total no. of test utterances taken for each language = 900)					
		MFPLPC		Formants		Either one of the features	
		Concerned language	Other languages	Concerned language	Other languages	Concerned language	Other languages
256	Bengali	897	3	674	226	900	0
	Hindi	900	0	886	14	900	0
	Kannada	886	14	739	161	888	12
	Malayalam	897	3	549	351	892	8
	Marathi	893	7	266	634	894	6
	Tamil	801	99	553	347	887	13
	Telugu	825	75	880	20	897	3
	Average % RA	**96.8%**		**72.2%**		**99.33%**	
128	Bengali	897	3	699	201	898	2
	Hindi	900	0	887	13	900	0
	Kannada	876	24	746	154	878	22
	Malayalam	899	1	521	379	899	1
	Marathi	893	7	254	646	892	8
	Tamil	797	103	554	346	887	13
	Telugu	841	59	869	31	890	10
	Average % RA	**96.9%**		**71.9%**		**99.11%**	
64	Bengali	898	2	662	238	899	1
	Hindi	900	0	877	23	900	0
	Kannada	884	16	736	164	886	14
	Mala.	896	4	557	343	896	4
	Marathi	891	9	239	661	891	9
	Tamil	802	98	549	351	889	11
	Telugu	839	61	885	15	897	3
	Average % RA	**97%**		**71.5%**		**99.33%**	

(*continued*)

Table 4 (*continued*)

Clusters	Languages	Classification of test speeches—individual accuracy (Total no. of test utterances taken for each language = 900)					
		MFPLPC		Formants		Either one of the features	
		Concerned language	Other languages	Concerned language	Other languages	Concerned language	Other languages
32	Bengali	897	3	662	238	898	2
	Hindi	900	0	875	25	900	0
	Kannada	886	14	747	153	888	12
	Mala.	897	3	552	348	897	3
	Marathi	891	9	223	677	893	7
	Tamil	804	96	514	386	890	10
	Telugu	825	75	880	20	895	5
	Average % RA	**96.82%**		**70.68%**		**99.38%**	

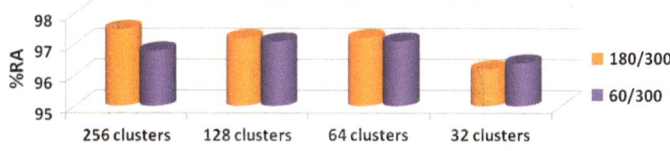

Fig. 8 Comparative performance with variations in no. utterances considered for training

Figure 8 indicates the comparative analysis between performance of the LID system by considering 180 utterances and 60 utterances for training. System shows very marginal improvement in accuracy when more utterances are being considered for training and the system is evaluated for testing on 300 utterances for each language.

Table 4 indicates the performance of the system with MFPLPC and Formants. LP analysis is done and poles corresponding to the highest magnitude are candidates of formant frequencies. System has shown considerable improvement in accuracy when correct score is considered with respect to either MFPLPC or Formants.

From the Table 4, it is clear that clustering model of size 32 provides better accuracy of 99.4% as compared to 64, 128 and 256 clusters for calculating the token of correctly identified hypothesized language with respect to either one feature among MFPLPC and Formants. Figure 9 depicts the distribution of cluster centroids for two languages Bengali and Telugu and it is evident that there is more correlation between the clusters of these two languages.

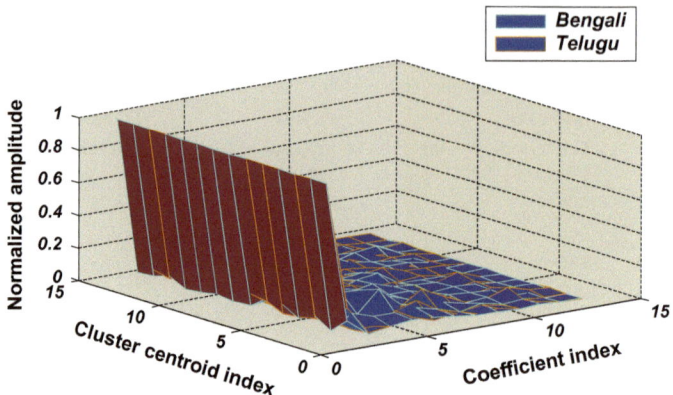

Fig. 9 Distribution of cluster centroids—Bengali and Telugu

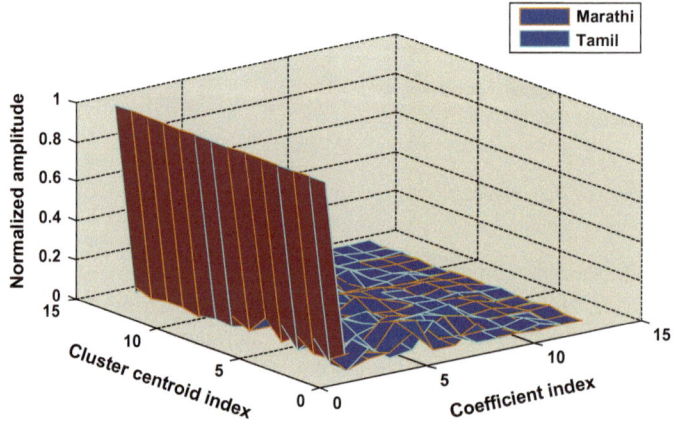

Fig. 10 Distribution of cluster centroids—Marathi and Tamil

Figure 10 depicts the distribution of cluster centroids for two languages Marathi and Tamil and clusters are more correlated between these languages. This fact is reflected in the results of the system developed using 60 speeches for training and all feature vectors of 900 test utterances applied. Figure 11 indicates the efficiency of minimum distance classifier used for classification with respect to the selection based on minimum of average of minimum distances computed between the features of the test data and the models of cluster size 256 for the test utterances in Hindi.

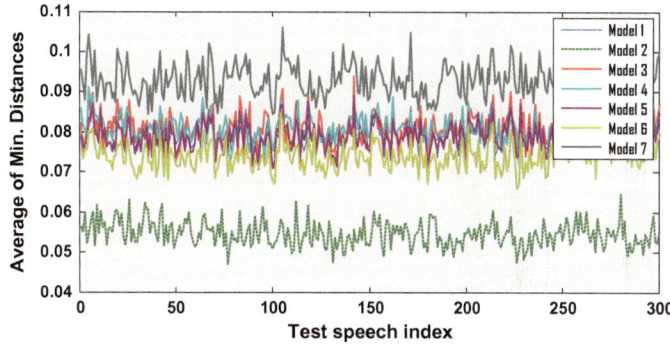

Fig. 11 Classification based on minimum of averages for Hindi Language

Figure 12 indicates the index of the correctly identified Hindi language for all the four cluster models of size 256, 128, 64 and 32 clusters and the testing is done on 300 test utterances. Results in Table 4 indicate that the individual accuracy for the languages Tamil and Telugu are relatively less as compared to other languages such as Bengali, Hindi, Kannada, Malayalam and Marathi. But, Formants do provide complementary evidence in ensuring better accuracy for Tamil and Telugu. As a whole, combination of MFPLPC and Formants has ensured better accuracy for these two languages and there is considerable increase in overall accuracy. Only, 37 test utterances out of 6300 test utterances are misclassified for clustering model of size 32 clusters. For testing one speech, this algorithm takes a few milliseconds to display the text information of the speech as visual output or the speech to be heard as the audio output in a language for the clustering model of 32 clusters.

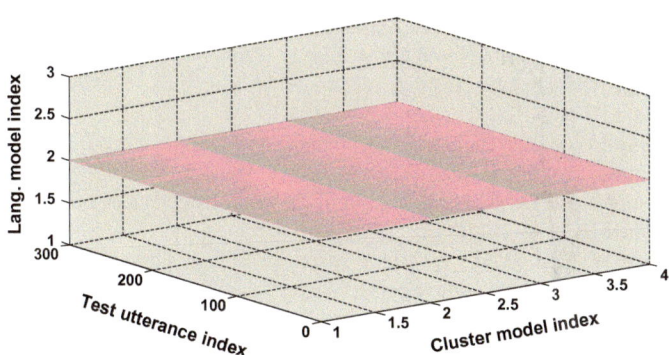

Fig. 12 Efficiency of the minimum distance classifier 300 utterances in Hindi—256 clusters

6 Conclusion

In this work on LID, system is developed to analyze the performance of the LID system to recognize seven Indian languages using speech utterances. LID system is implemented using two sets of features such as MFPLPC and Formants. Clustering is the procedure used for developing models using these features as input. MFPLPC has provided better accuracy for all languages except Tamil and Telugu. Formants ensured better accuracy for Telugu. But, the combination of MFPLPC and Formants has ensured better accuracy for all languages by calculating the score of correctly identified hypothesized language with respect to either MFPLPC or Formants. Though, the clustering technique is considered as an old technique, it is proved that it has provided good results on a par with the state-of-the art techniques for testing done on 900 test utterances. Results indicate that this LID system would be helpful for multilingual speech recognition and spoken document retrieval. Multilingual speech processing systems are likely to find applications in information kiosks in airports and hotels as well as in multilingual telephone based systems such as tele-shopping and travel reservations etc.... So, speech can be used as an alternative medium by web-based systems to provide reliable communication between users and systems irrespective of the language in which the users are interacting with the systems.

References

1. Zissman, M.A.: Automatic identification of telephone speech. Lincolns Lab. J. **8**(2), 115–143 (1995)
2. Rao, K.S., Nandi, D.: Language identification—a brief review. Springer briefs in speech technology, pp. 11–30
3. Sadanandam, M., Kamakshiprasad, V., Janaki, V.: Automatic language identification using new features and their weightage. Int. J. Adv. Comput. **35**(7), 380–385 (2012)
4. Suo, H., Li, M., Lu, P., Yan, Y.: Automatic language identification with discriminative language characterization based on SVM. IEICE Trans. Info. Sys. **91**(3), 567–575 (2008)
5. Rajasekar, S.: An automatic language identification using audio features. IJETAE, Special issues (January 2013)
6. Sadanandam, M., Kamakshiprasad, V., Janaki, V.: DHMM based automatic language identification system. Int. J. Info. Technol. Knowl. Manage. **6**(1), 93–97 (2012)
7. Sadanandam, M., Kamakshiprasad, V., Janaki, V.: Text independent language identification using DHMM. IJCA **48**(1) (2012)
8. Li, H., Bin, M. Lee, C.-H.: A vector space modelling approach to spoken language identification. IEEE Trans. Audio Speech Lang. Process. **15**(8), 271–284 (2007)
9. Sadanandam, M., Kamakshiprasad, V., Janaki, V.: GMM based language identification system using robust features. Int. J. Speech Technol. **17**, 99–105 (2014)
10. Zissman, Mark A.: Comparison of four approaches to automatic language identification of telephone speech. IEEE Trans. Speech Audio Process. **4**(1), 31–44 (1996)
11. Zhang, W., Li, B.,Qu, D., Wang, B.: Automatic language identification using support vector machines. In: Proceedings of 8th International Conference on signal processing, vol. 1 (2006)

12. Lee, K.A., You, C., Li, H.: Spoken language identification using support vector machine with generative front end. In: Acoustics, IEEE International Conference on Acoustics, Speech and Signal Processing, pp. 4153–4156 (2008)
13. Deng, Y., Liu, J.: Automatic language identification using support vector machine and phonetic N gram. In: International Conference on Audio, Language and Image Processing ICALIP, pp. 71–74 (2008)
14. Ziaei, A., Ahadi, S.M., Mirrezaie, S.M., Yeganeh, H.: Spoken language identification using a new sequence kernel-based SVM back-end classifier. In: IEEE International Symposium on Signal Processing and Information Technology, ISSPIT, pp. 324–329 (2008)
15. Verma, V.K., Khanna, N.: Indian language identification using k means clustering and support vector machine. Students Conference on Engineering and Systems (SCES), pp. 1–5 (2013)
16. Nicolai, G., Islam, M.A., Greiner, R.: Native language identification using probabilistic graphical models. In: International Conference on Electrical Information and Communication Technology (EICT), pp. 1–6 (2013)
17. Vyas, G., Dutta, M.K.: An integrated spoken language recognition system using support vector machine. In: Seventh International Conference on Contemporary Computing (IC3), pp. 105–108 (2014)
18. Murty, K.S.R., Yegnanarayana, B.: Combining evidence from residual phase and MFCC features for speaker recognition. IEEE Signal Process. Lett. 13(1), 52–55 (2006)
19. Han, Z.-Y., Wang, X., Wang, J.: Speech recognition system based on visual features and neural network for persons with speech-impairments. Int. J. Modell. Ident. Control 8(3), 240–247 (2009)
20. Jeyalakshmi, C., Revathi, A., Krishnamurthi, V.: Alphabet model-based short vocabulary speech recognition for the assessment of profoundly deaf and hard of hearing speeches. Int. J. Modell. Ident. Control 23(3), 278–286 (2015)
21. Revathi, A., Venkataramani, Y.: Speaker independent continuous speech and isolated digit recognition using VQ and HMM. In: International Conference on Communications and Signal Processing (ICCSP), IEEE, pp. 198–202 (2011)
22. Hermansky, H., Tsuga, K., Makino, S., Wakita, H.: Perceptually based processing in automatic speech recognition. In: Proceedings on IEEE International Conference on Acoustics, speech and signal processing, Tokyo, vol. 11, pp. 1971–1974 (1986)
23. Hermansky, H., Margon, N., Bayya, A., Kohn, P.: The challenge of Inverse E: the RASTA PLP method. In: Proceedings on Twenty fifth IEEE Asilomar conf. on signals, systems and computers, Pacific Grove, CA, USA, November, vol. 2, pp. 800–804 (1991)
24. Hermansky, Hynek, Morgan, Nelson: RASTA processing of speech. IEEE Trans. Speech Audio Process. 2(4), 578–589 (1994)
25. Revathi, A., Venkataramani, Y.: Perceptual features based isolated digit and continuous speech recognition using iterative clustering approach. In: First International Conference on Networks and Communications, IEEE, pp. 155–160 (2009)
26. Rabiner, L.R., Juang, B.H.: Fundamentals of Speech Recognition. Prentice Hall, New Jersey (1993)

Premature Cardiac Verdict Plus Classification of Arrhythmias and Myocardial Ischemia with *k*-NN Classifier

M. Inbalatha[1,2(✉)] and S. Kalaivani[3]

[1] SENSE, VIT, Vellore, Tamil Nadu, India
inbalathaphd@gmail.com
[2] Department of ECE, Dr.T.T.I.T, KGF, Kolar Gold Fields, Karnataka, India
[3] SENSE, DSP Division, VIT, Vellore, Tamil Nadu, India

Abstract. The exploration force cultivates a distinctive outline for feature extraction procedure based on Discrete Wavelet Transform with ECG Signal for Arrhythmias detection in addition to Stenosis detection of Myocardial Ischemia with Real Cardiac Computed Tomography Angiogram images of both the gender with different age groups. Succeeding both the cardiac diseases ingests been classified with the k-NN classifier separately based on their levels of severity as normal and abnormal for arrhythmia disease patients and moreover as standard, early, minor and serious for Myocardial Ischemia patients. The objective of this work is to pigeonhole perfectly and flourish a creative arrhythmia finding taxonomy and One hundredth stenosis detection that can sign to great exciting premature Myocardial Ischemia (Lack of oxygenated blood supply to the heart) analysis. In the virtual reality result, DWT features works commendable for the classifier with the utmost 94% truthfulness and with RCCTA images clearly detects the blockage area and classification also done more closely to the groundtruth values.

Keywords: Arrhythmias · ECG · K-NN classifier · Myocardial ischemia
RCCTA · Stenosis

1 Introduction

Coronary heart Disease is the captain disease all over the world with no warning sign which hints to demise. The situation possibly will be due to the stumbling block of life blood, decline in blood pour to the heart, too much impelling of blood hooked on the heart. The Cardiac analysis can be done first and foremost by traditional ECG. Auxiliary meticulous conclusion can be completed with 2D CT scan, 3D MRI Scan, etc. In this research work, two types of early cardiac diagnosis, Myocardial Ischemia using RCCTA imageries and Cardiac Arrhythmias by means of ECG waveforms shadowed by categorization with *k*-NN classifier resolve the accomplishment.

© Springer International Publishing AG 2018
D. J. Hemanth and S. Smys (eds.), *Computational Vision and Bio Inspired Computing*,
Lecture Notes in Computational Vision and Biomechanics 28,
https://doi.org/10.1007/978-3-319-71767-8_26

1.1 Arrhythmia

Arrhythmia is a disorder of the heart measure [1]. The heart can beat extremely wild (Tachycardia), very relaxed (Bradycardia), or irregular. Arrhythmia is triggered owing to the heart's electrical transmission structure during the existence of asymmetrical slimming down and easing of the heart, altered nerve cultures signals to the heart, obstructive of signals and also when signals transportable in new paths over the heart. Signals chronicled from the human body provide appreciated information about the actions of its body portion [2]. Explicit of the clues in the course of arrhythmia are life-threatening freaking, feebleness, chest- throbbing, vertigo, and anemia. As a consequence for estimating arrhythmia pointers such as II, III, AVR and V5 investigations. The attained morphology and the time durations of these bits are being used to produce optimal features of an ECG bit. Every pattern is produced by extracting these features [3]. The dataset, on which feature extraction and selection are applied, is MIT-BIH Arrhythmia. Thus for arrhythmia analysis, it starts with the abnormalities of the p wave. Time and voltage evaluation of P, Q, R, S, and T waves are the utmost important phases of an ECG signal [4]. The QRS complex is the utmost prognostic waveform in the electrocardiographic signal, with normal epoch from 0.06 to 0.1 s [5]. Its figure, period and phase of occurrence provide appreciated data about the existing state of the heart. In the meantime of its definite shape, the QRS compound aids as an access point for nearly all programmed ECG analysis algorithms and discovery of the QRS complex is the most important task in automatic ECG signal analysis [6]. Further the T wave is a succeeding wave which is generated owed to Ventricular repolarization.

1.2 Myocardial Ischemia

Myocardial Ischemia designates deterioration of lifeblood creek in the humanoid heart. This arises owing to the persistent revelation which little by little contracts the blood pour into the arteries. If the state of affairs rests, at big hand get up a starvation of oxygenated blood into the heart which clues to dead cells which slowly starts disturbing the Cardiac Sequence followed by sudden Heart Attack. The objective of this effort is to immediate detection of the threat features of heart attack with non-intrusive manner by radiologists for fast analysis [7]. By mining data from medicinal images with advances of digital image processing to home important facts from genuine RCCTA images for forecasting the ailment. Primarily pre-processing is executed with the median filter to exclude noise from the picked up input CCTA image. Then Segmentation of pre-processed left coronary arteries is achieved by smearing the Global thresholding procedure shadowed by hessian built Frangi Vesselness Filter which identifies the curl resembling blob assemblies [8]. The erosion and dilation morphological tasks for dissection of coronary branches [9]. The final vessel resemblance function [10] V(x) is defined as,

$$V = \left\{ exp\left(\frac{-R_b^2}{2\beta_1^2}\right) a^2 + b^2 = c^2 \right\} \tag{1}$$

where R_b = indicates tubular structure,

    V = normalized Voxel ratio,

    β_1, β_2 = weights representation,

    S = background region

Stenosis is a condition in which blood vessel arteries turn out to be tinny [11] Extent of stenosis condition in the individuals are measured by calculating the sectional area of blocked and unblocked coronary arteries [12]. Finally categorization with k-Nearest Neighbour Classifier for category of severity of the illness as Standard, Early, Minor and Serious.

2 Methodology

2.1 Arrhythmia Approach

In the anticipated plan, the input reserved are twenty individual's ECG signals are collected from MIT-BIH Arrhythmia Database. The signals are set aside for thirty minutes proportions from round the watch chronicles of twenty individualities and group filtered at one-hundred Hertz and digitized at three hundred and sixty Hertz. The waterway graph designates the evolvement of plan as shown in Fig. 1.

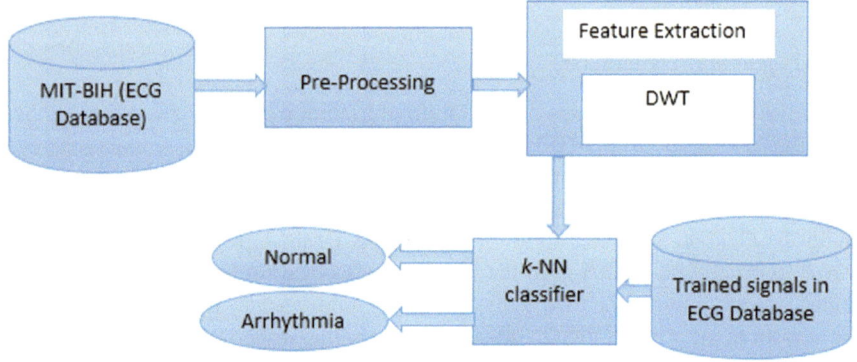

Fig. 1 Arrhythmia stream plan

The ECG gestures in habit for the dataset is seven signals for the test set and 13 signals for the trained set. From the preprocessed signal is at that point second hand to abstract primarily the six features such as QRS complex extent, PP interim, PR interim, RR interim, QT interim, and Heart Beat rhythm rate correspondingly for every distinct signals in the dataset. The six landscapes are indexed from one to six and it is systematized with two advancements separately through DWT. Then the extracted six features are then trained for thirteen signals in the trained dataset. Then the seven test signals relates each signal in the trained signals and pigeon-holes them with k-NN classifier into normal and anomalous for different k-values from one to five neighbors and the enactment valuation is acknowledged.

2.2 Myocardial Ischemia Approach

This approach aims the consequent sequential pecking order for suitable appraisal of heart outbreak and Stenosis discovery. The picked up 2D RCCTA datasets are from the Radiology division of Sri Devaraj Urs hospital, Tamaka in Karnataka are with different age groups of males and females. It consists of 9 male patients and 6 female patients with different levels of blockage of arteries and one normal patient of different age groups of 30–80 ages.

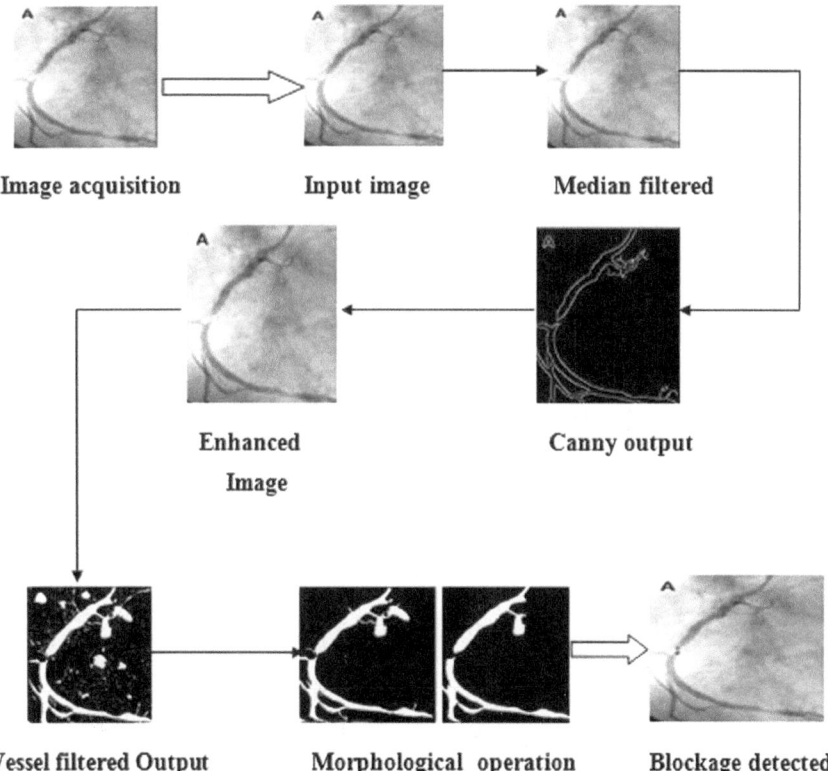

Fig. 2 Myocardial Ischemia stream plan

Primarily perception of Coronary pathways of the left ventricle from the RCCTA image. Followed by enlightening of Stenosis from the vessel strained segmented blood vessel to work out the CSA of healthy and diseased cardiac arteries. Then and there building of feature table for length, breadth, area, perimeter, pressure, blood stream, and

Stenosis for poles apart age collections of both males and females. To conclude, validating the largeness of stenosis with the virtual reality grades and categorized into diverse intensities of Stenosis like Standard, Early, Minor and Serious. The stream diagram designates the evolvement of plan as shown in Fig. 2.

3 Results and Discussion

3.1 Arrhythmia Outcomes

The six features such as QRS complex extent, PP interim, PR interim, RR interim, and QT interim, Heart Beat rhythm respectively for all distinct gesture in the dataset are mined. Discrete Wavelet algorithm is used to abstract the revealed six features and the virtual reality results are shown in Table 1. Also the Fig. 3 signifies the ECG sample.

The Fig. 4 represents the computer-generated outcomes with abstraction of six features of a distinct third individual's ECG Signal from the test database. Then the features are given to the k-NN Classifier and this checks with the features extracted from the each trained signals in the trained database and classifies the test signal category. More precisely, a k-NN Classifier structure for ECG arrhythmias nomenclature as normal and abnormal arrhythmias was anticipated. Planned method constitutes two phases-Feature extraction and Classification. The Fig. 5 gives a detailed comparison of extracted features of the selected test signal with the features of the all trained signal and classifies as normal and abnormal. Also it computes Accuracy, Sensitivity, Selectivity and Elapsed Time for performance evaluation of the k-NN Classifier. Openly 13 signals from the MIT-BIH arrhythmias ECG Database has been recycled for the training and testing our k-NN classifier.

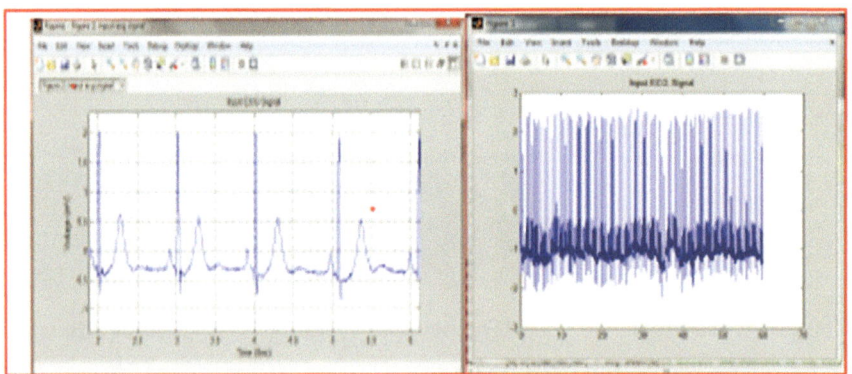

Fig. 3 ECG sample signal

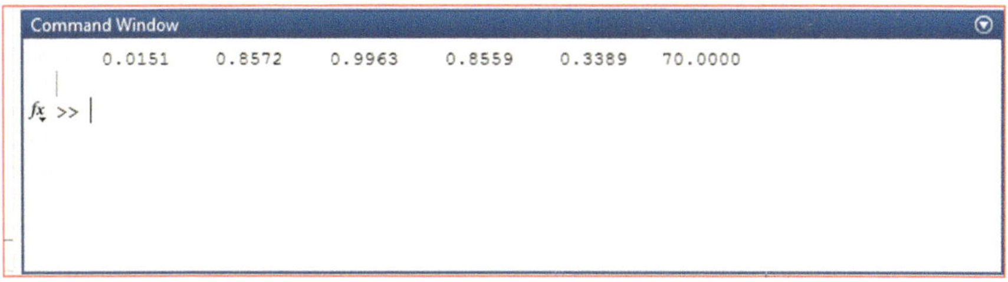

Fig. 4 Extracted six features from an ECG signal

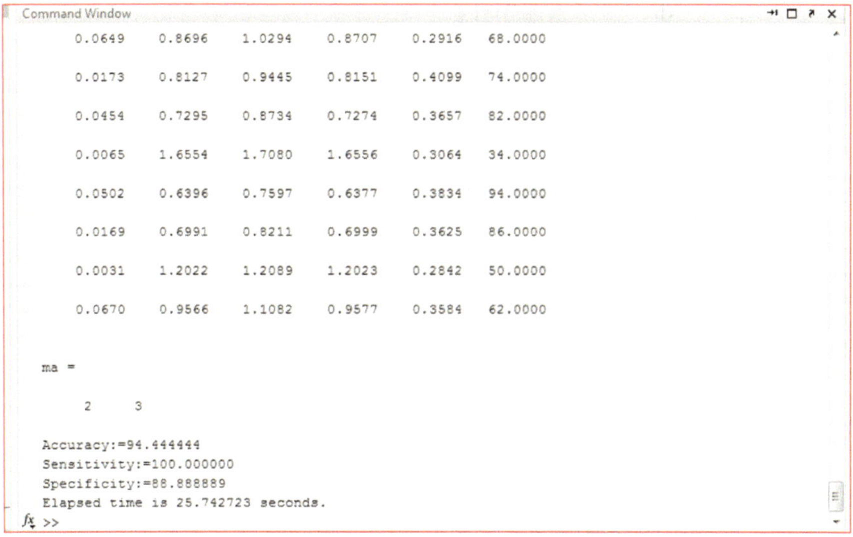

Fig. 5 *k*-NN classifier yield

In the simulation result, DWT features works worthy for the classifier for K = 1 and K = 2 with the utmost accuracy of 94.4% as shown in Table 1. Likewise it is witnessed that the sensitivity and specificity of the DWT time-outs honorable and enhanced for the *k*-NN classifier. In future, surplus extent of datasets can be puffed-up to construct unobtrusive admirable faithfulness conclusions. Further precise irregular ECG waveform due to noise or experimental investigation encounters and firm specific unpredictable cardiac inconsistencies are not seasoned due to scarcity of user-friendliness of databank.

In imminent, additional amount of datasets can be enlarged to create quiet

Table 1 DWT Process outcomes for some signals

K value	Accuracy (%)	Intervened time (milliseconds)
1	94.4	19.4
2	94.4	19.4
3	88.0	19.5
4	77.0	19.7
5	72.0	19.4

praiseworthy accuracy conclusions.

3.2   Myocardial Ischemia Outcomes

The stream diagram of the projected effort stays planned for the perception of stumbling block of the arteries. Pre-processing comprises of pass through a filter, advancement and edge innovation. The sieves castoff the Median filter to exclude the noise in the image. Canny edge detector stays castoff to recognize the marginal bounds of the image tailed by Frangi vesselness filter to splinter the exciting left coronary blood vessel. The meticulous splitting up of coronary arterial hierarchy leftovers a perilous factor in image processing because cardiac arteries are in hollow form thereby requiring computer systems. In this effort, coronary arteries are reserved as ROI to abstract from RCCTA images, Hessian Matrix based Frangi Vesselness filter remains castoff, which offers the Eigen values and Eigen Vectors. Then k-NN classifier classifies the tallness of heart outbreak as Standard, Early, Minor and Serious as shown in Table 2.

The below Table 3, illustrates the feature extraction of Six features of both gender

Table 2 Heights of heart outbreak

Heart blockage levels	Stenosis (in %)
Standard	0–0.9
Early	1–19
Minor	20–39
Serious	≥ 40

with different age groups like length, width, area, Pressure, Blood Flow and % Stenosis with the status of the patient condition is tabulated. If the % Stenosis varies between the range, 0–0.9, 1–19, 20–39 and ≥ 40, indicates the patient's disease status as Standard, Early, Minor and Serious respectively. Also the concluding simulated graph for blockage of area and pressure of blood flow are shown in both Fig. 6a, b prominently denote to Hagen Poiseuilles commandment, i.e., if the blockage area upturns, pressure of blood flow declines irrespective of gender and age groups.

Table 3 Extracted features of both gender with Status

S.No.	M/F	Length (mm)	Width (mm)	Area (mm²)	Pressure (Pascal)	Blood flow (ml/s)	Stenosis (%)	Status
1.	M	4.00	35.0	140.00	0.0714	0.0116	8.3732	Early
2.	M	14.0	64.0	840.00	0.0119	0.4030	348.5477	Serious
3.	F	9.00	6.00	54.00	0.1851	0.8423	8.424	Early
4.	F	32.0624	11.0	352.6868	0.0284	0.3434	131.1103	Serious
5.	F	14.5602	42.0	611.5292	0.0164	0.0419	35.6161	Minor

(a)　　　　　　　　　　　　　　　**(b)**

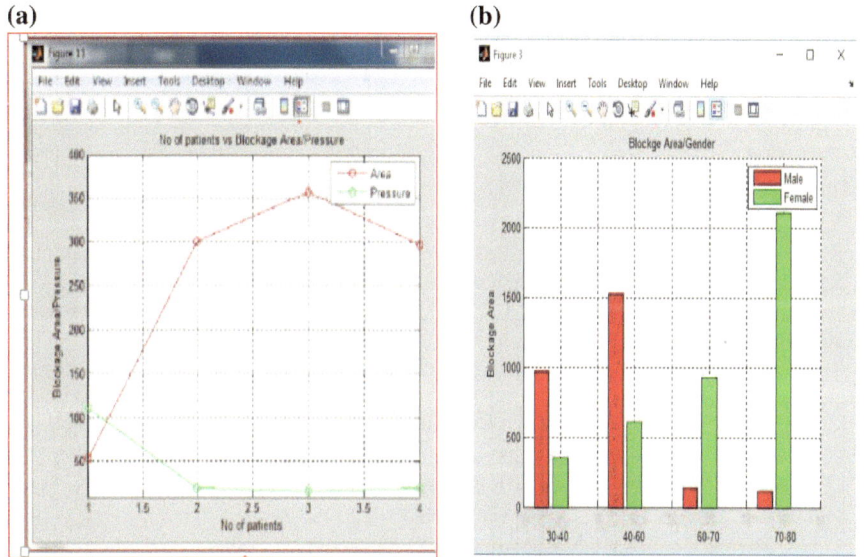

Fig. 6 **a** Blockage area and Pressure Graph **b** Comparison of area and Pressure (bar graph)

4 Conclusion

To conclude, early diagnosis of cardiovascular diseases like, Arrhythmia and Myocardial Ischemia are done separately with ECG signals from MIT-BIH Database for Arrhythmia detection and 2D CTA images for Myocardial Ischemia detection from Kolar Hospital for both men and women of different age groups. Simulation results for both the diagnosis are up and around to the acknowledged glassy. Finally for classification of both the diseases, we use k_NN Classifier to indicate the severity levels of cardiac Outbreak as normal and abnormal in universal and further in penetration as Standard, Early, Minor and Serious by indicating the levels of severity for the further treatment to be finished at a premature stage. In the simulation result, DWT features works worthy for the classifier with the utmost accuracy of 94.4%. To conclude, to detect Cardiac diseases mainly for Myocardial Ischemia detection and analysis RCCTA, EchoCardiac and MRI images are mostly suggested by the experts to detect the area of the blockage, which cannot be diagnosed by ECG signal of the patient. In forthcoming, additional amount of datasets can be enlarged to create quiet praiseworthy exactness outcomes. Moreover, different classifiers can be used to diagnose the disease in depth with different stages.

5   Conflict of Retreats

Altogether authors certainly not have any conflict of retreats to report at all times.

Acknowledgements. Real data used in this work were provided by the Radiology division of Sri Devaraj Urs hospital, Tamaka in Karnataka through the support of Sri Devaraj Urs Academy of Higher Education And Research. The authors would like to thank Dr. Jothilingam, from Bharat Earth Movers Limited (BEML) Hospital, Beml Nagar, Kolar Gold Field for medicinal justification for the outcomes attained and cherished suggestions delivered to proceed the work.

References

1. Rita ban Kirtania, Mali, K.: Cardiac Arrhythmia Classification using Optimal Feature Selection and k-Nearest Neighbor Classifier. IJARCSSE **5**(1) (2015)
2. Pal, S., Mitra, M.: Detection of ECG characteristic points using multiresolution wavelet analysis based selective coefficient method. Measurements **43** (2010)
3. Sharma, A., Sharma, T.: ECG beat recognition using principal component analysis and artificial neural network. IJEE **3**(1) (2011)
4. Sarkaleh, M.K., Shahbahrami, A.: Classification of ECG Arrhythmias using discrete wavelet transform and neural networks. IJCSEA **2** (2012)
5. Goldman, M. (ed.): Principle of Clinical Electrocardiography, 11th edn. Lange Medical Publication, Drawer L., Los Altos, California-94022 (1982)
6. Li, C., Zheng, C., Tai, C.: Detection of ECG characteristic points by wavelet transforms. IEEE Trans. Biomed. Eng. **42**(1) (1995)
7. Agrawal, S., et. al.: Early detection and segmentation of the coronary artery blockage using advanced imaging modalities for predicting risk factors of heart attack. Int. J. Sci. Eng. Res. **4**(4) (2013)
8. Sasidharan, G., George, A.: Frangi's vessel detection approach for coronary angiogram segmentation. Int. J. Eng. Trends Technol. **13**(5) (2014)
9. Agarwal, S.: Automated segmentation of cardiac stenosis and mathematical modeling of myocardial blood flow for early detection of heart attack using advanced imaging techniques. Int. J. Comp. Eng. **16**(6) (2014)
10. Lara, D.S.D., Faria, A.W.C., de A Araújo, A., Menotti, D.: A novel hybrid method for the segmentation of the coronary artery tree in 2-D angiograms. Int. J. Comp. Sci. Info. Technol. (IJCSIT) **5**(3) (2013)
11. Shahzad, R., Kiris, H.: Automatic segmentation, detection and quantification of coronary artery Stenosis on CTA. Int. J. Cardiovasc. Imag. **29**, 1847–1859 (2013). https://doi.org/10. 1007/s10554-013-0271-1
12. Inbalatha, M., Kalaivani, S.: Early detection of myocardial ischemia for predicting stenosis using frangi vesselness filter. Int. J. Adv. Res. Comp. Eng. Technol. (IJARCET) **5**(6), 1843–1848 (2016)

A Novel Algorithm for Triangulating 2d Point Clouds Using Multi-Dimensional Data Structures and Nearest Neighbour Approach

Sundeep Joshi and Shriram K. Vasudevan$^{(\boxtimes)}$

Department of Computer Science and Engineering, Amrita School
of Engineering, Amrita Vishwa Vidyapeetham, Amrita University, Coimbatore,
India
cb.en.p2cvil5014@cb.students.amrita.edu,
kv_shriram@cb.amrita.edu

Abstract. Generating a mesh from a point cloud has always been a complex and tedious task. It requires many guidelines to be followed and geometry specific boundary condition to be taken care of. Mostly we end up writing a new algorithm for each scenario. The existing algorithms like Delaunay triangulation end up taking a lot of time without taking care of the internal holes properly. Also the convergence is not guaranteed. In this paper, we present a recursion free algorithm for triangulating a structured surface with n-vertices with or without holes using Ball trees. The algorithm takes the input in the form of an image, creates a mask for the object to be triangulated and then creates a mesh for it with respect to computed vertices.

Keywords: Image segmentation · Point cloud · Triangulation
Ball trees

1 Introduction

Triangulating a 2d point cloud is a fundamental problem appearing in computer graphics, computational geometry, and fluid simulations and most importantly in geographical information systems. Several heuristic methods like delaunay triangulation [1] and its versions like the Bowyer-watson [2] algorithm are available and are recursive in nature and also computationally very expensive. As it happens, meshes are also very important when creating a 3d model. While working on an AR application, the only choice left was to use models available online among the free ones or the ridiculously high priced ones. Other options usually is to get it created with customisation. Either way the time taken is high. A skilled 3d artist takes a significant amount of time to create a 3d model from scratch. This can be a very time consuming and resource intensive process. This research carried out here is an attempt to automate the process of 3d model creation by taking image as input from various poses,

© Springer International Publishing AG 2018
D. J. Hemanth and S. Smys (eds.), *Computational Vision and Bio Inspired Computing*,
Lecture Notes in Computational Vision and Biomechanics 28,
https://doi.org/10.1007/978-3-319-71767-8_27

processing it by using images processing techniques, extracting the point cloud and triangulating it so that it can be used as a 3d model without any further alteration and modification. The already available algorithms for mesh generation are good at triangulation but a lot of care needs to be taken for different boundary conditions and geometrical complexities.

To be specific, this paper presents an algorithm, which allows the user to get effective triangulation of a 2d point cloud by means of nearest neighbour approach. The algorithm computes nearest neighbour for each vertex and then assigns faces in a way such that the overlapping faces if happen will be overlapping completely and not sharing any vertex with other face. We use python and blender to implement the algorithm and create visualisations.

2   Existing Solutions: State-of-the-Art

Mesh creation has been a topic of interest for a considerable time [1]. We emphasise on structured mesh generation. The already available existing methods for mesh generation were so helpful in building the foundation of the concept while getting a clear picture of their advantages and disadvantages for our application.

Woerdenweber [3, 4] discusses an approach of creating 2d mesh by topology decomposition approach. In this approach, he uses two elements as the fundamental element for the mesh. These two elements are the vertex and the edge. The vertex and the edge are represented by two attribute namely topology and position and shape. In case of edge, the geometrical information is represented by the curve on which it lies and the topological information is represented by the two endpoints. This approach creates mesh by connecting vertices to form triangles. The size of the triangle is not considered because of the object topology. The cons of this approach are that the hole cutting method is not proper and uses an operator to remove connected triangles doing making sure that the object boundary is not disturbed. The operator used here is similar to euler operator by Baumgart [5].

Cavendish [6], in his paper uses an approach to select the vertex in two ways. First time the vertex are selected by choosing the vertex lying on the boundary. Once this is done, it selects the points lying inside the object boundary depending on the density of mesh required. Based on the density parameter, the object is divided into zones. One point is added in the zone and the distance of that point from the boundary point is checked. If the distance is more than the set threshold, another point is not added. If the distance is less that the set threshold, point is added. A set number of attempt is made for each zone. This process is continued for all the points. Once the vertices are ready, the object is superimposed by a square grid. The size of the square grid is decided by the size of the size of the face required. The points inside the boundary are taken. For selecting boundary points, the point on the boundary closest to the nearest face is considered and the mesh is generated.

Delaunay triangulation [1] is perhaps the most popular and most discussed algorithm for mesh generation. This algorithm takes the vertex as input and gives triangulation results based on the in circle condition. This condition states that all the vertices of the triangle formed by delaunay triangulation should fall on the circumference of a circle. The triangulation starts by first creating a bigger triangle that contains all the vertices. After that points are introduced one by one and the circumcircle condition is checked. If it is satisfied, other existing points are checked for any possible violation of circumcircle condition. If any other point's circumcircle condition is violated, all the triangles are shuffled to fulfil that condition. This is a big disadvantage as reshuffling whole mesh on one vertex addition is a costly process.

Consolidating above point we can safely conclude that:

- The algorithm should take care of concave geometries, convex geometries and holes into account.
- It should create uniform mesh at the boundary and at the center.
- The time taken to complete the mesh generation process should also be reasonable.

3 Problem Statement

To design an algorithm that can take an image and input and provide a mesh as output by computing vertices and faces from the image. The generated mesh should also be usable by any 3D modelling software.

4 Evolution of Proposed Architecture

(a) **Segmenting object of interest from the image**: The first step of the algorithm is to segment the object of interest from the image. This is very crucial because our algorithm will take the mask image as input. For segmenting the object of interest from image, we have several methodologies and algorithms that are already available.

- Grab cut [7]
- Interactive segmentation tool by matlab.

But these algorithms fail in some complex scenarios such as different colored images. And it is not very intuitive to take some decision like number of iterations and selecting points from object of interest every time. So, to avoid these difficult decisions and automate the end to end to end flow, we decided to use deep learning for segmentation.

In deep learning, there are some state of the art research on image segmentation like deep lab and segnet. Deep lab uses atrous convolutions and fully connected Conditional Random Fields for segmentation. In this paper, they will use a VGG16 [8] or resnet deployed in a fully connected way and atrously convolved to reduce signal down sampling. Now a bilinear interpolation is applied to the image to convert the feature map to the full image resolution. Now CRF is used to refine the result of segmentation.

Other paper that claimed state of the art accuracy was segnet. This paper also claimed to be better that Deeplab [9] in many ways. In this paper, the author uses a encoder network topologically similar to VGG16 followed by a decoder network and a pixel wise classification layer. Since the feature map from the encoder is low in resolution, the decode maps it to same resolution feature map in a unique way by using pooling indices from the max-pooling step to upsample the image in a nonlinear fashion. This means that there will be no need to earn upsampling. They also claim better result that deep lab in terms of memory, number of trainable parameters, and computational time. So we decided to fine tune segnet for our application (Fig. 1).

(a)   **(b)**

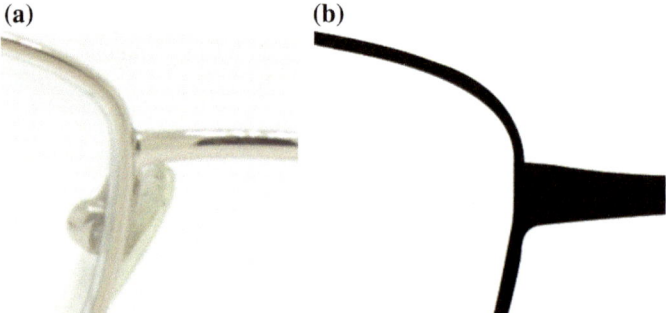

Fig. 1 **a** Input image, **b** Output of segmentation

(b) **Creating vertices from segmented image**: Once the segmentation results are obtained in the form of mask image, we clean the it and remove any noise present in the form of small speckles by doing contour analysis and removing any smaller contours and noise from the image.

Once the image is clean, we store all the vertices of the object in the form of a text file. Since the size of the input image is large, we scale down all the points by a factor of 10. This not only reduces the size of the model to a viewable scale, but it also makes it easy to modify. The vertices for input to ball tree are displayed in Fig. 2.

Algorithm:

- Select all the vertices from the foreground.
- Divide each of them by a scaling factor to bring down the total size.
- Store them in the form of tuples in a text file.

Fig. 2 Vertices for input to ball tree

(c) **Creating faces using ball tree**: Once we have the vertices stored in the form of text file, we start by creating a ball tree for a leaf size of 3 and compute five nearest neighbours for each vertex. The five nearest neighbours are then arranged into three faces in such a way that they all together combine all the five vertices to form a bigger face and no space is left vacant. In the similar way we compute for all the vertices. All the faces are stored in the form of vertex index and not the vertex itself. Figure 3 shows the Faces created by ball tree.

Algorithm:

- Compute a ball tree with default leaf size.
- Loop through all the vertices and make a query to the tree.
- Store five nearest neighbour, a, b, c, d, e for each vertices.
- Create three faces b arranging these vertices in a predefined order i.e., f1 = (a, b, c), f2 = (b, c, d), f3 = (c, d, e).
- Store them in the form of tuples in a text file.

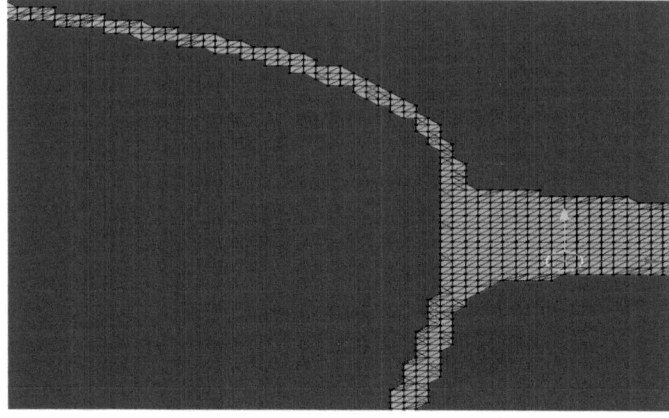

Fig. 3 Faces created by ball tree

(d) **Removing redundant faces**: Once all the faces are computed, we remove the repeating or overlapping faces from the list and write the information to a text file. This reduces the memory it takes to save the file. Figure 4 displays the Reduction of Face count.

Fig. 4 Reduction of face count

(e) **Merging adjacent faces to reduce memory**: Since there is still a lot of redundant information which can be removed. This information is in the form of number of small faces that represent an area which can also be represented by a larger face without distorting the geometry and uniformity of the model. This will reduce the number of vertices, edges, faces and will in turn result in total memory reduction of the model and a cleaner mesh.

5 Testing the Code

The code gives decent results in most of the use cases provided as input. Different kinds of images were given as input after segmentation and quality meshes was generated from them. Few of the mesh images have been put here for reference. The time taken by the algorithm to create meshes increases with increase in the number of vertices. The code also supports meshes with multiple holes also and the sample images can be seen below. Figure 5a, b shows the different images mesh and concave/ convex geometries respectively (Figs. 6 and 7).

(a)

(b)

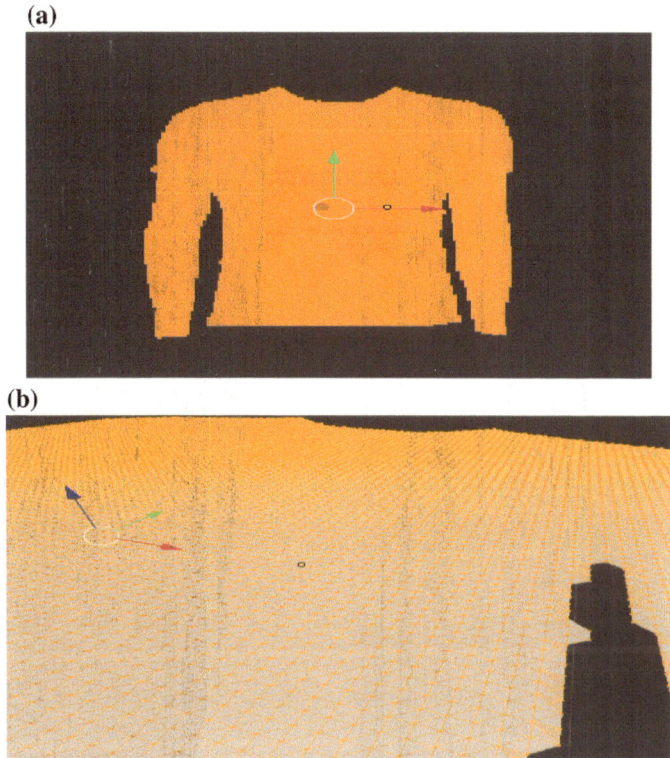

Fig. 5 **a** Different images mesh, **b** Different concave/convex geometries

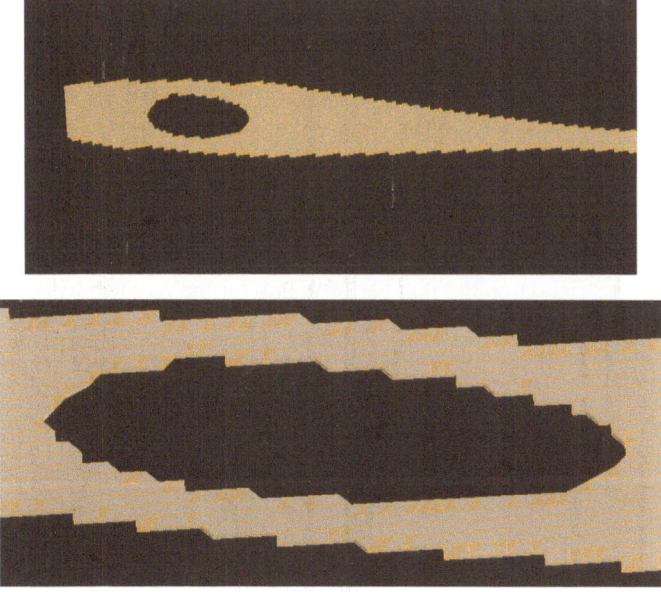

Fig. 6 Considers holes in a proper way

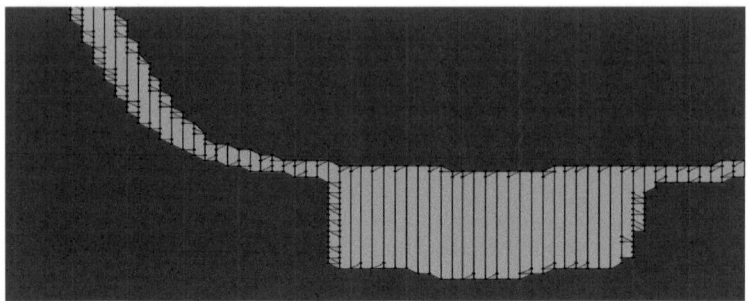

Fig. 7 After removing redundant faces

The code is also tested on different system configurations. The code seems to work fine with a minimum of dual core system with 2 GBs of ram although the time taken for mesh creation is more. The system on which the algorithm is developed and will be deployed is a core-i5 machine with 8 GB RAM and a dedicated 2 GB NVidia graphics card.

6 Conclusion and Future Scope

The meshes generated right now is are having faces parallel to the x axis. This means that bending the mesh along x axis will not be a problem. But when we bend the mesh along y axis, some small artefacts are seen on the mesh on both side. In future we would like to remove these artefacts. This can be done by adding some subdivisions along the y axis at the point of bending so that mesh becomes smooth at that point. Other way to achieve this is to add these subdivisions uniformly along the mesh along y axis. Other than this, there were few points that were noticed during the development:

a. This algorithm takes care of the boundary conditions on its own and no extra code has been written for it. Which makes it versatile for different geometrical shapes like concentric circles, concave shapes, convex shapes and others.
b. The way we choose faces is very different and we can firmly say from the examples that it works for challenging geometries also.
c. The algorithm works on nearest neighbour approach and is very easily understandable.
d. There is no restriction on the number of points for the mesh. It can take a mesh of any length and any number of points.
e. With the integration of various sensory input device like leap and Android [10, 11], this algorithm can really speed-up the development time.
f. Since 3D model plays a role in augmented reality and virtual reality applications and creating a 3d model is a time consuming job, this algorithm can reduce that time by almost 80% and that also be helpful in medical applications also [12].

References

1. Delaunay, B.: Sur la sphere vide. A lamemoirede Georges Voronoi, Bulletin de Academie des Sciences de l'URSS. Classe des sciences math´ematiques et na **6**, 793–800 (1934)
2. Bowyer, A.: Computing Dirichlet tessellations. Comput. J. **24**(2), 162–166 (1981)
3. Woerdenweber, B.: Finite element mesh generation. Comput. Aided Des. **16**(5), 285–291 (1984)
4. Ho-Le, K.: Finite element mesh generation methods. Rev. Classif. **20**(1), 27–38 (1988)
5. Baumgart, B.G.: Geometric modeling for computer vision. Report No. C5-463 Stanford Artificial Intelligence Laboratory, Computer Science Dept, Stanford, USA (October 1974)
6. Cavendish, J.C.: Automatic triangulation of arbitrary planar domains for the finite element method. Int. J. Numer. Meth. Eng. **8**, pp. 679–696 (1974)
7. Rother, C., Kolmogorov, V., Blake, A.: "GrabCut": interactive foreground extraction using iterated graph cuts, SIGGRAPH '04 ACM SIGGRAPH, pp. 309–314 (2004)
8. Simonyan, K., Zisserman, A.: Very deep convolutional networks for large-scale image recognition. Computer Vision and Pattern Recognition, arXiv:1409.1556
9. Chen, L.C., Papandreou, G., Kokkinos, I., Murphy, K., Yuille, A.L.: Semantic image segmentation with deep convolutional nets and fully connected CRFs. Computer Vision and Pattern Recognition. arXiv:1412.7062
10. Sundaram, V.M., Vasudevan, S.K., Santhosh, C., Barath Kumar, R.G.K., Kumar, G.D.: An augmented reality application with leap and android. Indian J. Sci. Technol. **8**(7), 678–682 (2015)
11. Vasudevan, S.K., Ritesh, A., Santhosh, C.: An innovative app with for location finding with augmented reality using CLOUD. Proc. Comp. Sci. **50**, 585–589 (2015)
12. Geethan, P., Jithin, P., Naveen, T., Padminy, K.V., Krithika, J.S., Vasudevan, S.K.: Augmented reality x-ray vision with gesture interaction. Indian J. Sci. Technol. **8**(S7), 43–47 (2015)

Cane Free Obstacle Detection Using Sensors and Navigational Guidance Using Augmented Reality for Visually Challenged People

Sriram Ramachandran$^{(\boxtimes)}$, Prashant R. Nair,
and Shriram K. Vasudevan

Department of Computer Science and Engineering, Amrita School of
Engineering, Amrita Vishwa Vidyapeetham, Coimbatore, Tamilnadu 641112,
India
cb.en.p2csel5022@cb.students.amrita.edu,
prashant@amrita.edu, kv_shriram@cb.amrita.edu

Abstract. The project concentrates on obstacle detection for visually impaired people using IR, Ultrasonic sensors and TSOP receiver which is of less cost, fast and efficient. Shortest distance algorithm is the method used to find the obstacles at each region sent from the TSOP receiver mounted on each angles. This works in environments such as indoor and outdoor. Experience comes from learning and the important requirements for learning will be the sense of vision of what normal human being sees and observes in the real world. But in the case of VI people, sense of vision will not help them in learning where other senses like ears, touch, smell helps them in various situations. The whole setup is wearable and lightweight mounted on a belt to make the VI people to walk into any situation like pits, steps by providing haptic feedback as the output. The project uses Arduino mega which helps us to program ease and also helps in testing various parameters with the available open source library routines. Since the targeted user is VI people, navigational guidance can be done through Augmented Reality domain as the technology is growing wide and can give more reliable to the user.

Keywords: Visually impaired (VI) · Arduino mega · Obstacle detection
Haptic feedback · Augmented reality

1 Introduction

According to the total human population worldwide as of August 2016, there are about 7.4 billion people and the United Nations estimates that it will further increase to 11.2 billion in the next century. Among this total population, there are 15% of physically disabled people and 285 million [1] people are visually impaired. 39 million are blind and 246 have low vision. 65% of visually impaired, and 82% of blind people are over 50 years of age, although this age group comprises only 20% of the world population. This project concentrates on providing one kind of solution to visually impaired people

© Springer International Publishing AG 2018
D. J. Hemanth and S. Smys (eds.), *Computational Vision and Bio Inspired Computing*,
Lecture Notes in Computational Vision and Biomechanics 28,
https://doi.org/10.1007/978-3-319-71767-8_28

where obstacle detection is the system which detects obstacles like fast moving vehicles approaching towards the person, steps, pits, animals in streets, side walls, etc. This can be achieved using IR blasters which calibrates in wide-angle and the whole system is mounted on wearable belt.

The important part is considering the environment which is indoor and outdoor. By using IR blasters and other electronic technologies, outdoor environment is challenging and it gives accurate results. Indoor environments can be predicted where the visually impaired people usually remembers the obstacles they experienced earlier and this system also detects nearby doors, walls, tables etc. In the reminder of the paper, Sect. 2 will have existing systems where obstacle detection are achieved in different manner, Sect. 3 contains system design which has both hardware and software implementations with architecture and process diagrams, Sect. 4 contains the tested results with screenshots and Sect. 5 details the conclusion and future enhancements in a brief.

2 Related Works

There are many projects on obstacle detection which emphasize on detecting obstacles linearly but not in a wide angle manner. Many project uses Ultrasonic sensor to calculate the distance of the obstacle approaching towards the person. But Ultrasonic sensor calibrates only linear and even in specific straight angle. It does not detect obstacles in left and right sides and not even backwards. Detecting obstacles on visually impaired person's back is not useful because even the normal human being never knew the obstacles back. Most projects uses Cane which have its own advantages and disadvantages and some of the projects have been discussed below.

Dimitrios Dakopoulos and Nikolaos G. Bourbakis members of IEEE written a survey on devices targeting on assistive technology domain in the title as "**Wearable Obstacle Avoidance Electronic Travel Aids for Blind: A Survey** [2]". This survey mainly concentrates on devices which assists visually impaired people in any environments and in any situations. This paper focuses on three main aspects for the "visual substitution".

1. Electronic travel aids (ETAs) captures environment through camera or other vision capturing device and output is in the form of auditory or tactile feedback.
2. Electronic Orientation Aids (EOAs) provides proximity or orientation where the user walks in. E.g.: Infrared transmitter.
3. Position Locator Devices: GPS tracking device to get current latitude and longitude.

These three aspects will fully contributes in developing projects for visually impaired people. Among the many, few projects were listed below. **GuideCane** [3] is one of the best device which is an update of NavBelt device. It uses ultrasonic sensors placed in the bottom of the Cane and with additional functionalities. It detects sideway walls and small obstacles at the ground. Output: Audio feedback. Limitations: Small area scanning which cannot detect tables and huge thing to hold. **CyARM** [4] developed by researchers in Japan, used ultrasonic sensors to detect the distance of the obstacles and gets a vibration feedback using a tension of wire attached to the user's belt. The results were good with static obstacles but not with moving object. It has got another disadvantage that the user should hold the device in hand and should continuously scan the environment. **Tactile Vision System** [5] developed a device which consisted of tactile belt having an array of 14 vibrator motors and a camera belt with two web cameras and microcontroller. The process of obstacle detection is complex by capturing an image using one of the cameras and checking the region of obstacle from the captured image and output is given to specify vibrator motor based on the obstacle detected in the region. It has not yet been performed in real environments.

Lasercane is a cane with three laser range sensors. It detects head-height, straight ahead and drop offs obstacles where warnings/alerts are given through sound and vibration. The device is costly with $3000 approximately. **Ultracane** does the same operation which costs approximately $900.

Lisa Ran, Sumi Helal and Steve Moore designed "**Drishti: An Integrated Indoor/Outdoor Blind Navigation System and Service** [6]" which provides navigational and obstacle detection service in indoor and outdoor environments. They uses voice command to switch from indoor to outdoor with the command "Indoor"/"outdoor" from the user. The indoor facility has some predefined and repetitive routes and when the user is lost in indoor it directs to nearest safest place or the device can direct the user to the destination. In outdoor environment, using DGPS (Differential Global Positioning System) they are getting current user latitude and longitude and direct the user to their destination. The hardware components are wearable computer, Differential GPS receiver, wireless network and ultrasound positioning device and location algorithms. In this paper, they used 4 HE900 M pilots and 2 HE900T beacon. The 2 beacons are attached to the shoulders to receive the ultrasound signals sent from the pilots in order to calculate the current orientation and location of user. The software components includes spatial database engine which acts as a gateway for the stored datasets (doors, tables, fans etc. with their position) on relational database management like Oracle 8i, route store for predefined routes to all places in indoor, IBM ViaVoice for voice commands. With all these components the system is effective but the results are not based on dynamic inputs. The indoor environments are completely predefined and doesn't know what if the user shifts to new indoor. The device can't be placed again for shifting of indoor.

Marina Rey, Inatan Hertzog, Nicolas Kagami, Luciana Nedel from University in Brazil developed a device which is free of using Cane. **Blind Guardian** [7] is independent of Cane consists of a headpiece(cap) which holds all hardware components like Arduino UNO and ultrasonic sensor and output feedback would be mobile vibrator. One side of the cap has ultrasonic sensor to capture the obstacle and the hood of the cap has all wirings connected to Arduino. This project is similar to other projects but the placing of sensor is different. The disadvantage of this project is the ultrasonic sensor is linear and can detect only in the angle of 30°. So to detect sideways, the user should always move their heads sideways which is not ease of use. Advantage is simple and cost effective but does not provide promising results for ground and hanging obstacles.

Multisensor guided walker for visually impaired elderly people [8] designed by Student Members of IEEE was one of the best prototype so far. They uses Cane which consists of multiple sensors like ultrasonic, IR, Pulse, Oxygen and Inertial Sensor. This project concentrates on visually impaired elderly people. Outputs are in the form of audible notifications. This system takes also into account of unstable grounds and potential fall or wall hits. Low Height obstacles like stairs, pits are detected by optical sensor like IR. The operating range of IR sensor are fixed for 25 cm and can detect objects on the near right, front and left. Higher obstacles are detected by ultrasound sensor. The system has also got servo motor to rotate the ultrasound sensor to 180° to detect in all directions. Accelerometer sensor is used to stop the cane not to get stuck or helpless in holes and pits. This sensor constantly measures the obstacles fall within three axis. This system was tested in real environments and results were promising.

Obstacle Detection from Still Images using Improved Background Subtraction Method [9] by Nishchala Thakur and Vikas Wasson published in Indian Journal of Science and Technology proposed an idea on finding the obstacle detection using still images. The captured image is then extracted by using improved Background Substitution Method which detects obstacles from video sequence where two consecutive frames are taken from the video backgrounds and arithmetic difference is calculated which gives the result of obstacles by the change of area between two video frames. This method was proposed to Mobile robot and not for the visually impaired people.

Blockage Detection in Seeder [10] **by** Shefali Malhotra, Surabhi Bhosale, Prachi Joshi and B. Karthikeyan published in Indian Journal of Science and Technology designed a prototype which detects blockage in seeder machines which is mostly used in the field of agriculture. The system contains an ultrasonic sensor placed inside the nozzle of seeder machines and keeping track of the flow of seeds by detecting obstacles using the distance calculated from the sensor. Since it's used in agriculture, there would be many seeder machines and each seeder has got a microcontroller and pair of ultrasonic sensor which is of more cost but efficient.

Camera based Text to Speech Conversion, Obstacle and Currency Detection for Blind Persons [11] by D. B. K. Kamesh, S. Nazma, J. K. R. Sastry and S. Venkateswarlu proposed a solution for detecting and handling currency for visually impaired people. This project is irrelevant from our project but can be used for recognizing the area of interest where the obstacles are present which is similar.

Augmented reality x-ray vision with gesture interaction [12] by Geethan P, Jithin P, Naveen T, Padminy KV, Krithika JS, Vasudevan SK developed a gesture movement control to move between the captured scenes. The algorithm used behind this project would be helpful because using gesture the visually impaired can move their hands to find the obstacles or objects in front of them by capturing video feeds in specific time. The algorithm is one kind of solution to our project.

An Augmented Reality Application with Leap and Android [13] by V. Meenakshi Sundaram, Shriram K. Vasudevan, C. Santhosh, R. G. K. Barath Kumar and G. Deepak Kumar proposed an idea on gesture movement control using leap sensor. The algorithm used in this project scans the surface area which will be useful for finding pits, steps or objects lying in our project. Applying Gaussian derivative, they find the difference between the objects located in different regions. Refining the surface area scanning process would bring our project more reliable and accurate results for finding obstacles.

3 Proposed Solution

Figure 1 shows the schematic view of the system and its functional component. The system contains hardware components like IR LED Sensor as a transmitter and TSOP 1838 as a receiver which receives transmitted IR rays. There are 2 transistors, ultrasonic sensor, and resistor, LED, Vibrators and Jumpers. All the mentioned hardware are mounted on a belt and given to the visually impaired user.

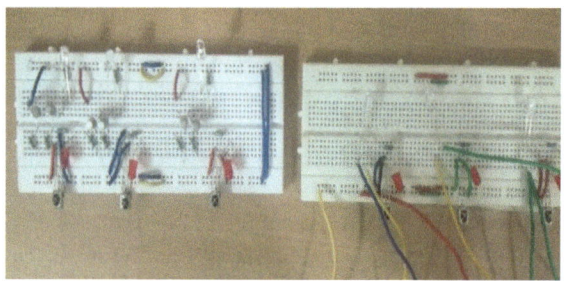

Fig. 1 IR LED's and TSOP 3 pairs in bread board connection

a. **Arduino Microcontroller**

This project uses Arduino UNO which is a microcontroller board based on ATmega328. It has got 14 digital input/output pins from 0 to 13. Out of 14 pins, 6 pins are PWM pins. (PWM—Pulse Width Modulation). 6 analog pins from A0 to A5. It contains everything much needed to support the microcontroller; simply connect it to a computer with a USB cable or power it with an AC-to-DC adapter or battery to get started.

b. **IR LED**

Infrared Emitting Diode (IR333-A) is a high intensity diode, mounted in a blue transparent plastic package. The device is spectrally matched with phototransistor, photodiode and infrared receiver module.

c. **TSOP 1838**

TSOP 1838 is a Thin Small Outline Package. Used in normal TV Remote Controller which receives the emitted IR rays. It usually works on 38 kHz and the main benefit is the reliable function even in disturbed and ambient lights and the protection against uncontrolled output pulses.

d. **Ultrasonic Sensor**

Ultrasonic sensor HC-SR04 which provides the non-contact measurement function of 2 to 400 cm. The sensing range reaches to 3 mm. The modules includes ultrasonic transmitters, receiver and control circuit. The speed of sound is 340 m/s or 29 μs/cm. This module has been used which is mounted on a belt facing downwards.

e. **Process Diagram**

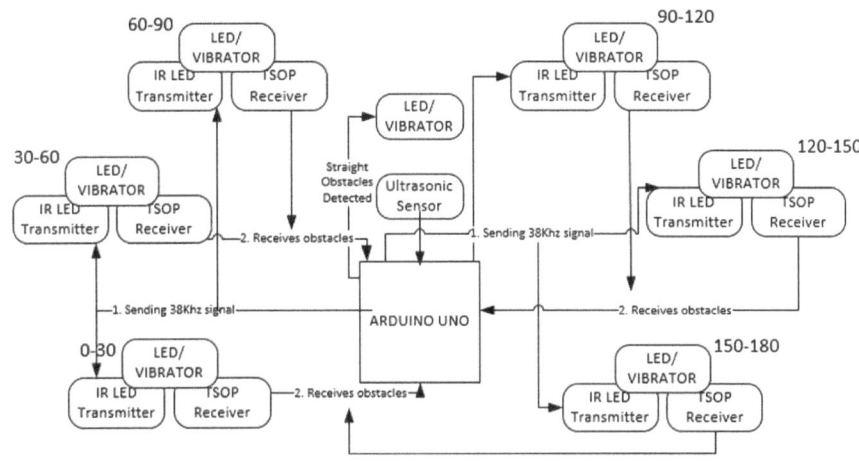

Fig. 2 Process diagram

The process diagram shows what are the functionalities carried out by each of the sensor at given time. The Fig. 2 contains the Ultrasonic sensors, 6 or 3 pairs of IR blasters which is IR transmitter and TSOP receiver. To control, Arduino UNO has been used. Vibrator Motors are used as the output.

For any embedded system to work properly and efficiently, input and output should be processed carefully. Inputs are generating 38 kHz in order to make the IR LED to send pulse and it will send pulse for every 38,000 times per second. The inputs are generated by writing a small timer program in ARDUINO IDE which then fed into ARDUINO UNO where this tasks would run in background. Arduino UNO has the capacity to run 16,000 tasks per second and running this small timer program works efficiently. The generated 38 kHz are given to all the 3 pairs of IR LED and waiting for the response. The TSOP will respond by receiving the reflected IR rays and will give output to ARDUINO. The output would be distance and based on the logic, for the respective distance received the vibrator motors will vibrate denoting the alertness to the user.

The IR LED and TSOP will work in straight, left and right sides with angles of 0–30, 30–60, 60–90, 90–120, 120–150, 150–180°. The below hip level obstacles are detected by Ultrasonic sensor which returns the distance directly to Arduino Uno and output is given based on the returned distance.

f. **Circuit Diagram**

To make the IR and TSOP module work, certain components needed to be added in case to avoid risks from misleading results. 220 Ω resistor and 2 BC 107 transistors are added to the circuit. The explanation of the circuit diagram is as follows.

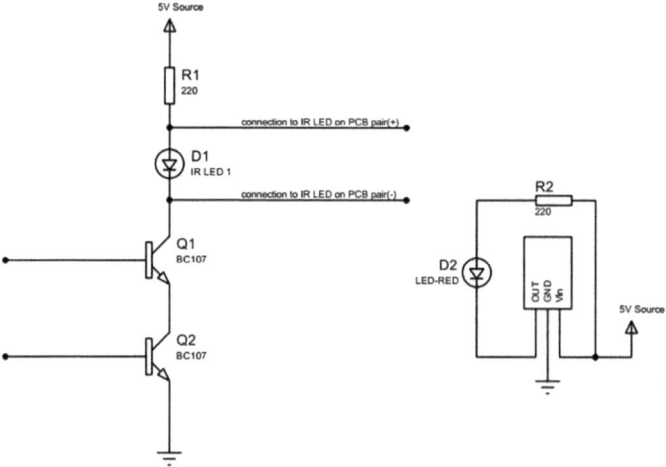

Fig. 3 Supporting modules

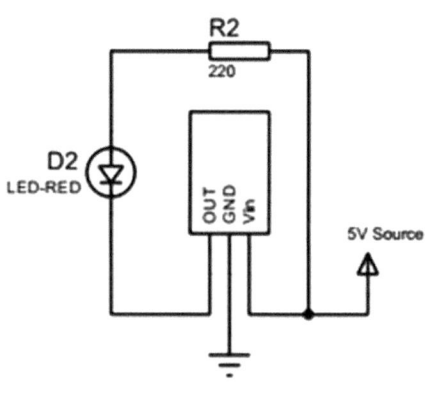

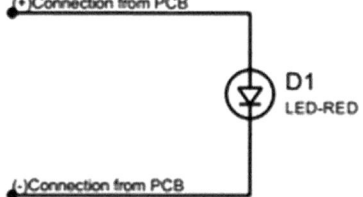

Fig. 4 Circuit diagram for main module

IR and TSOP should be together as a transmitter and receiver and to work that properly ground and power source should be supplied either from 9 V battery or 5 V from Arduino. From Arduino, one pin is taken as output which generates 38 kHz pulses and supplied to IR as an input. The circuit is made as simple because 3 circuits are constructed which is Fig. 3 and with those 3 circuits, the process is supplied to its pair circuits which is Fig. 4. Module 1, 2 and 3 are main circuits and 4, 5 and 6 are supporting circuits to main but all the 6 modules should work simultaneously. Given 38 kHz to all the 3 pairs of IR module, getting output will be tedious task. 2 BC 107 transistors are used because to get the output in specific amount of time which is the delay. One transistor is used to send 38 kHz and other transistor is used as a switching circuit which turns ON and OFF the module in given amount of time. The power and ground are supplied from main circuits to its paired/supporting circuits and also the transistor and output connections are also given accordingly.

g. **PCB Design**

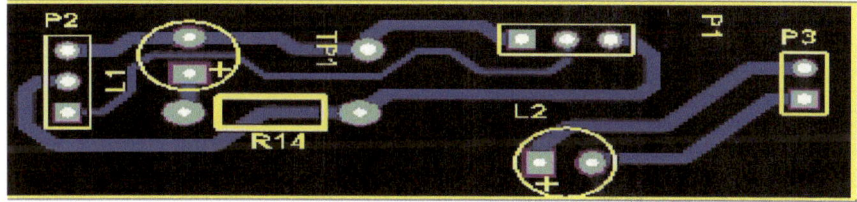

Fig. 5 PCB design for modules 4, 5 and 6

PCB—Printed Circuit Boards. The designed circuit diagram is converted to PCB because the whole setup was connected in bread board for testing purpose. Once the testing is done, the whole device is bought to small chip like module which will be helpful in placement of those modules on a wearable. The process would be same like given in circuit diagram. Figures 5 and 6 are the PCB designs of the circuit.

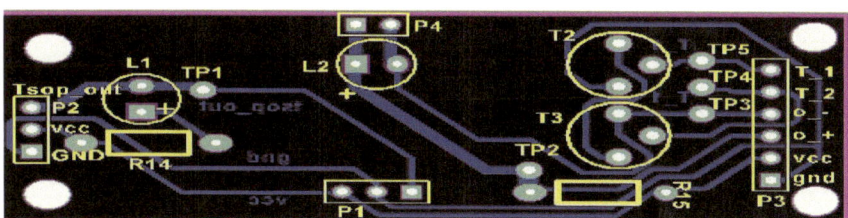

Fig. 6 PCB design for modules 1, 2 and 3

h. **Working**

The working principle of this project depends on calibrating the sensors at specific delays and getting the output and processing it accordingly to give alert to the user. There are several situations to detect obstacles precisely. Above Hip Level and Below Hip Level. In addition to this, Indoor and Outdoor environments too considered with this above and below hip levels. Indoor environments is where the user are residing at home which includes tables, chairs, wardrobes, walls etc. Outdoor environments would be traffic which covers all situations. By using the components like IR, TSOP and Ultrasonic sensors we can able to detect obstacles both in Indoor and Outdoor, above and below Hip level. The working procedure is as follows.

For Below Hip Level using Ultrasonic sensor

The maximum distance would be within 100 cm and there are certain conditions like Normal Floor—Distance is constant, Steps/Slope Downwards—Distance increases, Steps/Slope Upwards—Distance decreases, Animals/Parapets—Distance differ from normal floor distances.

The results from the Ultrasonic sensors would be distance in centimetre (cm) and the results will be received for every milliseconds. The output is received and processed programmatically to alert the user using haptic vibration. 3 samples are taken in an array for consideration. 1st distance received will be the 1st value in the array. 2nd distance detected –> checked if same as 1st value –> if yes: ignore to store, if no: add to the array which is 2nd index. 3rd distance detected –> checked if same as 2nd value –> if yes: ignore to store, if no: add to the array. 4th distance detected –> checked if same as 3rd value –> if yes: ignore to store, if no: update the 3rd to 2nd and last to 3rd and the loop continues.

The result would be distance but the output will be given in a pattern based on order to know the kind of obstacles the user is going to hit. Pattern differs by kind of obstacles, Slope Up—Moderate Vibration with delay 5 ms, Slope Down—Moderate Vibration with delay 10 ms, Straight Objects—Heavy Vibration, Distant Objects— Less Vibration. The vibrator motors are fixed in the abdomen on the belt such that the motor vibrates in the specific region denoting the direction the obstacles are present.

For Above Hip Level using IR and TSOP

The maximum distance would be 30 ft. 3 pairs of module are placed on a belt in different angles. To detect obstacles on various angles like 0–30, 30–60, 60–90, 90–120, 120–150, 150–180 which are 6 modules. Each module is switched ON and OFF for particular delay in order to receive the reflected IR rays which will be received by TSOP. The 1st and 6th pair are given 38 kHz first to detect obstacles on extreme left and right sides and received programmatically which is stored in an array. The 2nd and 5th pair next and received value stored in an array with corresponding module number like a Key Value Pair. 3rd and 4th pair is detected next. Within a millisecond, there will be 6 values in an array. Finding minimum value from the array denotes the obstacles closeness and giving alert to particular module is taken care of by the key stored. The input will be IR LED and output would be from TSOP. The output produced by TSOP will not be direct distance but it will be voltage which is received from normal blinking LED connected to each value. 0 V—LED blink—Short Distance and 5 V—LED OFF —Long Distance.

4 Result Analysis

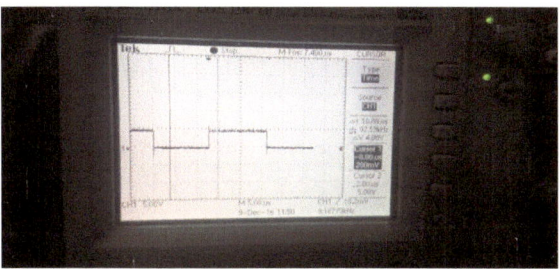

Fig. 7 Generating 38 kHz from Arduino

This project has been tested for detecting obstacles at below hip level. Figure 7 shows the 38 kHz generation using the Oscilloscope. Figure 8 shows that the obstacles are detected with certain distances and based on the returned distance, motors are vibrated in a specific pattern. Considering above Hip level, the LED's will be blown if there is obstacles in any angles/directions.

Fig. 8 Output from ultrasonic sensor receiving distances

5 Future Scope and Conclusion

This prototype would be of helpful to visually impaired people where it gives obstacle detection on the sides, front and down such that user can walk freely in any directions without any guidance from others Pedometer sensors + GPS—get current location and count the number of steps the user walks. The values are stored in cloud as KeyValuePair <stepcount, Location> so that if the user is walking again in same location, 2 comparisons results are given to reach the destination which is of one more step accuracy. Gyroscope can be added if the user walks in slopes/up or down roads which is one more additional accuracy to Obstacle Detection algorithm. With this prototype, the visually impaired people can gain more confidence to walk freely in roads since the results were promising.

References

1. Statistical Report—Total Disabled Population in World from World Health Organization
2. Dakopoulos, D., Bourbakis, N.G.: Wearable obstacle avoidance electronic travel aids for blind: a survey. IEEE Trans. Syst. Man Cybern. Part C (Applications and Reviews) **40**(1), 25–35 (2010)
3. Ulrich, I., Borenstein, J.: The GuideCane-applying mobile robot technologies to assist the visually impaired. IEEE Trans. Syst. Man Cybern. Part A Syst. Humans **31**(2), 131–136 (2001)
4. Ito, K., Okamoto, M., Akita, J., Ono, T., Gyobu, I., Takagi, T., Hoshi, T., Mishima, Y.: CyARM: an alternative aid device for blind persons. In: CHI'05 Extended Abstracts on Human Factors in Computing Systems 2005 Apr 2. ACM, pp. 1483–1488 (2005)
5. Johnson, L.A., Higgins, C.M.: A navigation aid for the blind using tactile-visual sensory substitution. In: Engineering in Medicine and Biology Society, 2006. EMBS'06. 28th Annual International Conference of the IEEE 2006 Aug 30, pp. 6289–6292. IEEE (2006)
6. Ran, L., Helal, S., Moore, S.: Drishti: an integrated indoor/outdoor blind navigation system and service. In: Pervasive Computing and Communications, 2004. PerCom 2004. Proceedings of the Second IEEE Annual Conference on 2004 Mar 14, pp. 23–30. IEEE (2004)
7. Rey, M., Hertzog, I., Kagami, N., Nedel, L.: Blind guardian: a sonar-based solution for avoiding collisions with the real world. In: Virtual and Augmented Reality (SVR), 2015 XVII Symposium on 2015 May 25, pp. 237–244. IEEE (2015)
8. Chaccour, K., Eid, J., Darazi, R., el Hassani, A.H., Andres, E.: Multisensor guided walker for visually impaired elderly people. In: 2015 International Conference on Advances in Biomedical Engineering (ICABME) 2015 Sep 16, pp. 158–161. IEEE (2015)
9. Thakur, N., Wasson, V.: Obstacle detection from still images using improved background subtraction method. Indian J. Sci. Technol. **9**(31) (Aug 24, 2016)
10. Malhotra, S., Bhosale, S., Joshi, P., Karthikeyan, B.: Blockage detection in seeder. In: Futuristic Trends in Research and Innovation for Social Welfare (Startup Conclave), World Conference on 2016 Feb 29, pp. 1–4. IEEE (2016)
11. Nazma, S., Kamesh, D.B., Sastry, J.K., Venkateswarlu, S.: Camera based text to speech conversion, obstacle and currency detection for blind persons. Indian J. Sci. Technol. **9**(30) (Aug 17, 2016)
12. Geethan, P., Jithin, P., Naveen, T., Padminy, K.V., Krithika, J.S., Vasudevan, S.K.: Augmented reality x-ray vision with gesture interaction. Indian J. Sci. Technol. **8**(S7), 43–47 (2015)
13. Sundaram, V.M., Vasudevan, S.K., Santhosh, C., Kumar, R.B., Kumar, G.D.: An augmented reality application with leap and android. Indian J. Sci. Technol. **8**(7), 678–682 (2015)

Preprocessing of Lung Images with a Novel Image Denoising Technique for Enhancing the Quality and Performance

C. Rangaswamy[1,2(✉)], G. T. Raju[3], and G. Seshikala[4]

[1] REVA University, Bengaluru, India
crsecesait@gmail.com
[2] Dept. of ECE, Sambhram Institute of Technology, Bengaluru, India
[3] Department of CSE, RNSIT, Bengaluru, India
gtraju1990@yahoo.com
[4] School of Electronics and Communication Engineering, REVA University, Bengaluru, India
seshikala.g@reva.edu.in

Abstract. Recently, digital image processing has attracted many researchers due to its significant performance in real-time applications such as bio-medical systems, security systems and automated computerized diagnosis systems. Lung cancer detection and diagnosis system is one such real-time health care application which requires automated processing, where in, the images are captured and processed through computer. Although various automated systems are already in place for this application, precise automatic lung cancer detection still remains a challenging task for researchers because of the unwanted signals get added into original signal during image capturing process which may degrade the image quality that intern resulting in degraded performance. In order to avoid this, image preprocessing has become an important stage with the key components as edge detection, image resampling, image enhancement and image denoising for enhancing the quality of input image. This paper aims to improve the quality of lung image with image denoising technique for enhancing the overall performance of the automated diagnosis system. A novel approach for image denoising by applying pixel classification using Multinomial Logistic Regression (MLR) and Gaussian Conditional Random Field (GCRF) is proposed in this paper. Proposed approach comprising of two major steps such as parameters generation by considering MLR and designing an inference network whose layer perform the computations which are tangled in GCRF formulation. Experimental analysis is done on LIDC, an open source benchmark database and the performance is compared with the state-of-the-art filtering schemes. Results show that proposed approach performs better in respect of PSNR values by factor 2.7 when compared to existing approaches.

Keywords: PSNR · MSE ANN · Fuzzy Inference System · Support Vector denoise · Image enhancement or preprocessing · Gaussian · Gradient Pixel

© Springer International Publishing AG 2018
D. J. Hemanth and S. Smys (eds.), *Computational Vision and Bio Inspired Computing*,
Lecture Notes in Computational Vision and Biomechanics 28,
https://doi.org/10.1007/978-3-319-71767-8_29

1   Introduction

In recent era of technology, medical imaging has grown rapidly and dramatic improvements have been noticed especially in clinical diagnostic systems with speed and accuracy. In clinical systems, MRI, X-Ray and CT scans are used widely for better diagnosis. These techniques provide efficient visualization of human anatomy for analysis [1]. These days doctors use various tools for acquisition of MRI, CT or X-Ray where in, during image acquisition, some unwanted signals get added into original image that may intern degrade the diagnostic performance and anatomy analysis [2]. In order to resolve this issue, medical imaging systems require image preprocessing techniques which can help to reduce the error and provide better quality image preserving the original information in the image. Image denoising techniques have attracted researchers due to their challenging nature in bio-medical applications. Conventionally, two basic models such as non-linear [3] and linear [4] image denoising models are widely used for image denoising. In general, linear models are accepted in medical application scenario due to its edge preserving nature. Recently Hua et al. [5] developed a model for medical image denoising with Gaussian Mixture Model (GMM) using edge information. Here, an extended GMM is presented where minimum mean square and Bayesian network frame is combined resulting in non-linear image mapping. Furthermore, a kernel density function is computed by considering edge information.

Chen et al. [6] discussed a similar approach, where markov random field is applied for image denoising model. Due to multivariate nature of medical imaging, images cannot be modeled based on their pixel information. However, for image analysis, pixel value and intensity value are correlated. In order to avoid this, Principle Component Analysis (PCA) scheme is utilized which helps to generate 5×5 patches of different filters. Based on the responses of these filters, noise distribution is estimated. Furthermore, Bayesian model is implemented to obtain the prior knowledge about image and finally a gradient computation model is implemented which helps to obtain the denoised image. Ouahabi et al. [7] studied about MRI imaging systems and its application in medical field. According to this work, a conventional model also can be implemented to obtain the denoised image where image signals are transformed into transform domain such as wavelet and counterlet domains. Further, coefficient comparing also can be implanted where fixed and adaptive threshold are considered.

Although enormous amount of work has been carried out in this field of medical image denoising to improve the quality of diagnostic system, due to the complex nature of medical imaging, still there is a need for improving the performance of image denoising systems which intern can provide noise free image for diagnostic analysis and can help to develop a better recommender system.

In this work, a novel approach for image denoising is proposed and implemented for lung image which can be used for detection and classification of lung cancer. This work mainly focuses on improving the quality of input lung image through image denoising technique. State of art models for image denoising suffers from performance issue for various raw datasets. To overcome this, a combined approach for denoising and classification is presented here by using Gaussian Conditional Random Field (GCRF) which is applied for pixel classification and denoising of image data. Main objective of this work is to remove the noise from the original source image and obtain image with better quality preserving the original features. Since, the lung images are multivariate in nature, here a multivariate learning technique is presented which improves the learning process.

Main contributions of this work are:

(a) Gaussian Random Field model is developed for learning process
(b) Pixel classification scheme is being used for image denoising
(c) A combined simultaneous learning and denoising model is presented.

Rest of the paper is organized as follows. Section 2 presents a related work in this field of image preprocessing and image denoising. Section 3 proposes a novel methodology for image denoising. Experimental setup and performance analysis is presented in Sect. 4 and paper concludes at Sect. 5.

2 Related Work

This section presents recent studies about image preprocessing, image filtering, image denoising and lung cancer detection systems. Image preprocessing combines various stages, image binarization, image transformation, color enhancement and image denoising. Image enhancement is a technique to improve the quality of image for further processing. Several researchers in their work on image denoising concluded that optimization of F-score can improve image denoising performance. Based on this assumption, Martin et al. [8] presented logistic regression method for optimizing the binary classification performance. According to this approach, a logistics training regression model is considered for problem formulation, employed for task retrieval and F-measurement is computed as measurement of utility function [9]. Joachims et al. [10] presented support vector machine (SVM) for learning process. This technique is also implemented for optimizing the multivariate nonlinear optimization problem to address the issue of F-Score. Similarly, Suzuki et al. [11] presented conditional random field (CRF) based modeling for solving the optimization problem for multivariate prediction module.

Image capturing in low-light conditions causes visibility degradation in original images resulting in performance degradation. This issue can be resolved by applying illumination improvement techniques. For illumination estimation, color components R, G and B and their maximum value considered for processing. With the help of this process, an initial illumination map is generated and later refined by imposing the structure resulting in final map [11]. Niu et al. [12] presented a new approach for image enhancement by concentrating on the contrast of input image. This technique is dependent on entropy maximization and tone-preserving of given input data. Vehicle detection in day-time video sequence is comparatively easier to night time sequences because night time sequences have low contrast and luminosity which affects detection process. To address this issue of night time vehicle detection, Kuang et al. [13] presented an improved approach by combining region or interest extraction and image enhancement technique. This technique uses image preprocessing model to obtain this outcome. For image enhancement and ROI extraction, improved multiscale retinex algorithm is used. However, this process provides better results but requires more computation time which makes increases implementation cost for real-time application scenario. For night time image acquisition system or low-contrast image systems, image dehazing is also considered as a promising technique. Recently, Mi et al. [14] discussed about image dehazing process where images are acquired under bad conditions such as illumination, low visibility etc. In this work, contrast and color of degraded image are considered as a key point for research work. In order to address this issue of contrast enhancement, multi-scale gradient domain approach is developed. According to this approach, an image is divided in multiple parts in the form of residual unlike conventional approach where entire image is processed and given to the system.

Image enhancement techniques are widely used for improving the quality of any given image. Afore mentioned techniques for image enhancement can work for a simple configuration of images whereas in real-time scenario, some unwanted signals get added in the original data and it may degrade the performance of application. To overcome this issue, image denoising technique is presented. Rizkinia et al. [15] presented an approach for image denoising by computing local spectral component and local distribution. The complete modelling is applied on multi-channel image system where linear combination among various images is computed in the spectral domain. For each channel of given image, linear feature vector is computed and later image is decomposed into three components as: single channel decomposition and two grayscale images. Image denoising is widely accepted and considered as a complex problem in medical image field application. Cong-Hua et al. [16] presented a generalized approach for medical image denoising. For image denoising application, information preservation is an important task which helps to improve performance analysis. This complete process is carried out by using Generalized Gaussian Mixture Model (GGMM) for medical imaging systems. This model is divided into two main stages. According to 1st stage, noisy image is modeled using extended Gaussian mixture model along with Bayesian framework. In second stage of implementation, performance is improved by using kernel density function. Xu et al. [17] presented a study for image denoising and discussed that wavelet domain approach also can be implemented for image denoising applications. Here, a threshold based approach is combined with artificial learning model and developed an adaptive approach for image denoising. Conventional models of neural network denoising uses zero threshold whereas these models computes coefficients with the help of polynomial function model. Furthermore, a sub band adaptive model is also used for threshold selection by considering each detail sub band.

Generally, image denoising is divided into two categories: (a) spatial domain and (b) transform domain. Yu et al. presented a model for Gaussian noise removal b using wavelet and bilateral filtering scheme. Wavelet based filtering models uses trivariate shrinkage modeling whereas bilateral filters are considered as spatial domain based filters. According to wavelet domain filtering process, Gaussian distribution is used for modeling the wavelet coefficients and a *posteriori* (MAP) estimator helps to formulate trivariate shrinkage filtering model. Authors concluded that wavelet-based approach shows a significant performance for image denoising application but it is not capable to generate salient features of given data and edge preserving remains unaddressed issue. On other hand, spatial approaches for image filtering produces significant results for image denoising and preserve image quality. Generally, these approaches are considered more complex and require more time to process, this issue is addressed by developing a joint bilateral filtering scheme for image denoising. Han et al. presented a model for image denoising by considering sparse and redundant representation modeling. Sparse based techniques for denoising fails to produce better performance when applied for low-light-level acquired image scenario due to complexity of noise. Hence this approach uses texture model for embedding noise invariant feature and finally Local Structure Preserving Sparse Coding (LSPSC) model is formulated for extracting the sparse and local structure of image data. Although various researchers have contributed a lot by proposing several algorithms for denoising applications, but still there is a scope for improving the performance of image denoising. In this paper, we propose a novel image denoising algorithm for denoising the lung image which is multivariate in nature and is implemented with reduced computation time and improved performance.

3 Proposed Denoising Technique

In order to develop a robust approach for image preprocessing with denoising, the input image needs to be modeled first and then analyzed using Gaussian random field model. Estimation of potential functions and hybrid model development processes are to be carried out next. These processes are discussed on the following sections.

3.1 Lung Image Modeling

Input lung images are multivariate in nature which consists of varied structure configuration. Let *Img* be the input image with varied intensity values represented by

$$f(x) : D \rightarrow S = \{s_{min}, s_{min+1}, \ldots, s_{max}\} \tag{1}$$

where D denotes computation space in a given functional domain f and S represent various channels for input image. if the input image is two dimensional in nature i.e., gray image then minimum and maximum values of s can be represented as $s_{min} = 0\ and\ s_{max} = 65535\ or\ 2^{16} - 1$ for any given 16 bit grayscale image. This modeling helps us to obtain the input image intensity given as $x = (x', y')$. On the other hand, if input image is color image, then this has to be converted into gray scale resulting in two-dimensional image. Hence, the intensity variation of any image can vary from s_{min} to s_{max}.

For experimental purpose, noise is expressed as:

$$f(x) = f_b(x) + f_p(x) \tag{2}$$

where $f_b(x)$ represents background pixel values and the amount of noise considered is denoted by $f_p(x)$ at point x. Now the effect of noise in respect of quality image is modeled as

$$f_p(x) = f_s(x) + f_{noise}(x) \tag{3}$$

where $f_s(x)$ is the structured model of input image and $f_{noise}(x)$ is the added noise to the original image.

3.2   Gaussian Random Field Modeling for Image Analysis

Once the modeling of two-dimensional multivariate image is done, we need to focus on reducing the noise to get a noise free image which intern used in further steps of diagnosis. This noise may be random in nature and is distributed identically in the input image. In order to address this, we apply Gaussian Random Field (GRF) which helps to generate a potential function for image analysis. Moreover, conditional random fields are also used for extracting the spectral, contextual and spatial information from the input image.

Let *Img* be an input noisy image and Y be the denoised output image inferred using random field modeling. The pixels in *Img* and Y are denoted as $Img(i,j)$ and $Y(i,j)$ respectively. In order to perform denoising, a probabilistic model is formulated for noise estimation and the complete model is expressed as follows:

$$p(Y/X) \propto \exp\{-\mathcal{F}(Y/X)\} \tag{4}$$

where

$$\mathcal{F}(Y|X) = \frac{1}{2\sigma^2} \sum_{ij} [Y(i,j) - X(i,j)]^2\} := \mathcal{F}_d(Y|X)$$
$$+ \frac{1}{2} vec(Y)^{tr} A(X) vec(\mathcal{Y})\} := \mathcal{F}_p(Y|X) \tag{5}$$

Generally, for any given probabilistic model, Gaussian random field provides a potential function for solving the probabilistic problem. In Eq. (5), noise variance for input image is denoted by σ, overall potential function parameters are denoted by A, for each data term, and data modeling is denoted by $\mathcal{F}_d$. Similarly, prior term modeling is also used which helps to obtain the generative mode of conditional random field.

3.3   Potential Function Solution

In previous section, potential function parameters are obtained using Gaussian conditional random field model for any given noisy image. Since the noise is random in nature, this has become more challenging task for researchers to estimate the exact amount of noise. In order to address this issue, input image is partitioned into multiple image *patches* with same pixel size for all patches. This patch combination information can be obtained by using $\mathcal{F}_p$. The partitioned *patch* for input image is denoted by $p \times p$ where x_{ij} represents pixel information. These *patches* are in the form of patch vector and are initiated from center location of pixel for noise consideration. Similarly, output patch vector is denoted as $p^2 \times 1$. Here, potential function estimation is an important step which can be inferred by computing mean values of noisy input image and inferred output image which can be expressed as $\bar{x}_{ij} = M_{x_{ij}}$ and $\bar{y}_{ij} = \mathcal{M}_{y_{ij}}$ respectively. For computation of mean of input image, mean subtraction matrix is used which is denoted by M and is expressed as:

$$\mathcal{M} = \mathcal{I} - \frac{1}{p^2} o^{2tr} \tag{6}$$

Here, o denotes $p^2 \times 1$ vector of one's and $\mathcal{I}$ is the identity matrix of size $p^2 \times p^2$. This process is based on the patch processing where potential function is defined for each path by considering each pixel value using a quadratic function which is given as:

$$Q\left(\frac{\bar{y}_{ij}}{\bar{x}_{ij}}\right) = \frac{1}{2} (\bar{y}_{ij})^{tr} \left(\frac{1}{\sum_{ij}(\bar{x}_{ij})}\right) \bar{y}_{ij}, \sum_{ij}(\bar{x}_{ij}) \geq 0 \tag{7}$$

In the next stage, all the *patches* at all pixels are combined together which results in potential function of complete image defined as

$$\mathcal{F}_p(\mathbb{D}|\mathbb{N}) = \sum_{ij} Q\left(\frac{\bar{y}_{ij}}{\bar{x}_{ij}}\right) \tag{8}$$

Alternatively, by considering subtraction matrix, potential function can be written as

$$\mathcal{F}_p(\mathbb{D}|\mathbb{N}) = \frac{1}{2} \sum_{ij} y_{ij}^{tr} \mathcal{M}^{tr} \left(\frac{1}{\sum_{ij}(\bar{x}_{ij})}\right) \mathcal{M} y_{ij} \tag{9}$$

In this work, image is divided into $p \times p$ *patches* which are centered on neighboring pixel. In each patch of the image, every pixel can interact with p^2 pixels. With this assumption, graphical model of neighborhood can be on image $\mathbb{D}$ given as $(2p - 1) \times (2p \times 1)$. Further, interference system is developed where input data is modeled using inference system. Let us consider the entire input data which is denoted as $\sum_{ij}(\bar{x}_{ij})$ obtained using potential function such as $\mathcal{F}_p(\mathbb{D}|\mathbb{N})$. This potential function helps to solve the Gaussian CRF problem as given in Eq. (10).

$$\mathbb{D}^* = \frac{argmin}{\mathbb{D}} \sum_{ij} \{\frac{1}{\sigma^2} [\mathbb{D}(i,j) - \mathbb{N}(i,j)]^2 + y_{ij}^{tr} \mathcal{M}^{tr} \left(\frac{1}{\sum_{ij}(\bar{x}_{ij})}\right) \mathcal{M} y_{ij} \tag{10}$$

Above expression is an optimization problem which can be resolved in a closed form, however this problem has unconstrained quadratic nature which requires an efficient solution. In order to address this issue, optimization problem is transformed into linear equation and is given by:

$$\mathcal{S}(\{a_{ij}\}) = \frac{argmin}{\mathbb{D}(i,j)} \{\frac{1}{\sigma^2} [\mathbb{D}(i,j) - \mathbb{N}(i,j)]^2 + \alpha \sum_{k,l=-\frac{p-1}{2}}^{\frac{p-1}{2}} [\mathbb{D}(i,j) - a_{kl}(i,j)]^2 \tag{11}$$

where S is optimal solution matrix, a_{ij} is an auxiliary variable which corresponds to each pixel from each patch, a_{kl} is intensity value of the pixel, *floor* and *ceil* operations are denoted as $\lfloor . \rfloor$ and $\lceil . \rceil$ respectively.

3.4   Multinomial Logistic Regression (MLR) for Pixel Classification

According to MLR modeling, probability of column vector x_{ij} belongs to class c can be denoted as

$$
P_c\left(x_i = c | x_{ij*}, \mathcal{A}\right) = \begin{cases} \dfrac{\exp\left(A_c^{tr} x_{i*}\right)}{1 + \sum_{m=1}^{M} \exp\left(A_m^{tr} x_i^*\right)} & if c < M \\[4mm] \dfrac{1}{1 + \sum_{m=1}^{M} \exp\left(A_m^{tr} x_i^*\right)} & if c = M \end{cases} \tag{12}
$$

Similarly for column vector y_{ij}, the probability is given as

$$
P_c\left(y_i = c | y_{ij*}, \mathcal{A}\right) = \begin{cases} \dfrac{\exp\left(A_c^{tr} y_{i*}\right)}{1 + \sum_{m=1}^{M} \exp\left(A_m^{tr} y_i^*\right)} & if\ c < M \\[4mm] \dfrac{1}{1 + \sum_{m=1}^{M} \exp\left(A_m^{tr} y_i^*\right)} & if\ c = M \end{cases} \tag{13}
$$

where $\mathcal{A}$ the potential parameter vector as discussed before which is denoted as $[a_{c1}, \ldots, a_{cl}]^{tr}$ for class c. $\mathcal{A}$ can be generated by concatenating $\{a_c c = 1, \ldots, M - 1\}$. If the classification problem is a binary problem, then value of M remains 2 and this model is known as a logistic regression model and if the value of $M > 2$ then the model is called as soft-max network. This complete process is summarized as an algorithm for inferring image using GCRF.

Algorithm 1: Image infereing using gaussian conditional random field

Input: input image (*Img*), labeled training samples for input image $\{x^{tr}, y^{tr}\}$, Band-pass filters, number of iterations, image scaling parameters

Output: GCRF model parameters

Step 1: Modeling of input image.

Step 2: Generate Noisy image with random noise

Step 3: Select noisy bands from input noisy image

Step 4: Initialize GCRF computation

Step 5: Apply probabilistic potential function computation

$p(Y/X) \propto \exp\{-\mathcal{F}(Y/X)\}$

Step 6: Input image patch generation $p \times p$

Step 7: Mean subtraction computation using $\mathcal{M} = \mathcal{I} - \frac{1}{p^2} o^{2tr}$ trained vector parameters

Step 8: Compute potential function parameters

$\mathcal{F}_p(\mathbb{D}|\mathbb{N}) = \frac{1}{2} \sum_{ij} y_{ij}^{tr} \mathcal{M}^{tr} \left(\frac{1}{\sum_{ij} (\bar{x}_{ij})} \right) \mathcal{M} y_{ij}$

In the next stage, we apply L_0 minimization scheme for image smoothing. In this work, we have incorporated 1D-image smoothing where highest contrast edges are further enhanced by applying gradient method. This process helps to obtain global smoothing for considered input image. Let us denote input discrete signal as d and smoothen resultant image as S. The global smoothing is now expressed as:

$$c(S) = \#d\{d | \|S_d - S_{d+1}\| \neq 0\} \tag{14}$$

In above expression, d and $d+1$ are index values for neighbouring pixels in the Gaussian distribution. Here gradient is denoted as $|f_d + - f_{d+1}|$ which is computed w. r.t. d. $\#\{\ \}$ denotes the counting operator where L_0 gradient norms are assumed if it satisfies the condition as $|f_d - f_{d+1}| \neq 0$. In order to obtain the smoothing, an objective function is defined by:

$$\min_{f} \sum_{p} (f_{d-g_d})^2 s.t.c(S) = k \tag{15}$$

Algorithm for L_0-grained minimization is presented as follows:

Algorithm 2: Lo Gradient minimization algorithm

Input: input image for smoothing I, weight vector for smoothing (λ), smoothing parameters $(\beta_0 \text{ and } \beta_{max})$ and smoothing rate K
Output: Smoothen image
Step 1: Initialize: $S \leftarrow I(input), \beta \leftarrow \beta_0$ and $i \leftarrow 0$
Step 2: Iterate With $S^{(i)}$ solve problem $h_p^{(i)}$ and $v_p^{(i)}$ which gives the binary solvers
With $h^{(i)}$ and $v^{(i)}$, solver for $S^{(i+1)}$
Step 3: Find relation such as $\beta \leftarrow K\beta, i++$
Step 4: Iterate until $\beta \geq \beta_{max}$

4 Experimental Results and Discussions

Proposed approach for denoising is implemented using MATLAB 2016b tool. For experimental analysis, lung images from the Lung Image Database Consortium image collection (LIDC) consists of 1018 cases have been considered. Performance of the proposed approach is compared with the state-of-art image denoising models such as median filtering, Weiner filter, wavelet transform based filtering schemes. Gaussian noise is considered for analyzing image denoising performance. Two sample images from LIDC are taken as test cases for image denoising demonstration and are given in Fig. 1.

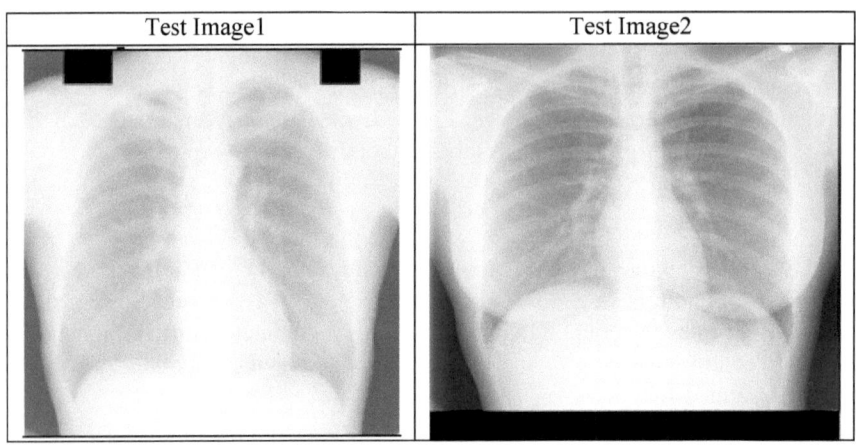

Fig. 1 Sample test images from LIDC for experimental purpose

The performance evaluation metrics such as Peak Signal to Noise Ratio (PSNR) and Mean Squared Error (MSE) and Structural Similarity (SSIM) index are considered for analyzing the performance of proposed denoising algorithm.

PSNR and MSE are computed as follows:

$$PSNR = 10\log_{10}\left(\frac{255^2}{MSE}\right)$$

$$MSE = \frac{\sum_i \sum_j \left(r_{ij} - x_{ij}\right)}{M \times N} \tag{17}$$

For experimental purpose, resampled lung image with 256×256 pixels is considered. Gaussian random noise with noise variation $\sigma = 20$ is added for each image. Input image, noisy image and the denoised image for test image1 and test image2 are shown in Figs. 2 and 3 respectively.

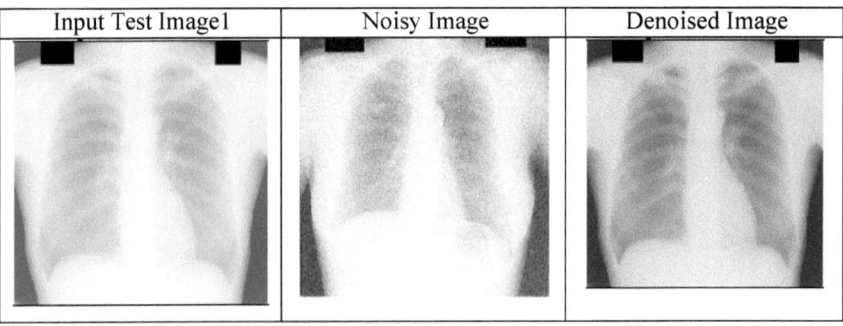

Fig. 2 Results of Image denoising for test image1 from LIDC

Input Test Image2	Noisy Image	Denoised Image
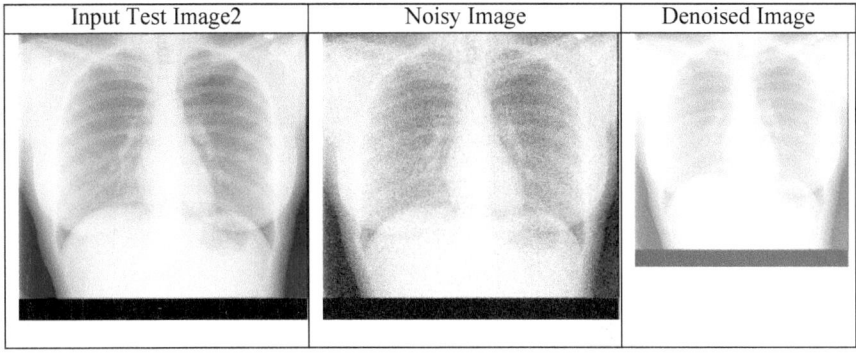		

Fig. 3 Results of Image denoising for test image2 from LIDC

To evaluate the performance of proposed denoising algorithm, we compute the PSNR values for denoising algorithms and are tabulated in the Tables 1 and 2 for test image1 and test image2 respectively.

Table 1 PSNR values for test image1 with $\sigma = 50$

Algorithm	PSNR(dB)
Median	14.1854
Vector Median	15.3966
Weiner	18.6224
SWT	22.1442
DWT	22.5897
Proposed	25.3

Table 2 PSNR values for test image1 with $\sigma = 40$

Algorithm	PSNR(dB)
Median	15.24
Vector Median	16.88
Weiner	18.702
SWT	22.037
DWT	25.142
Proposed	27.89

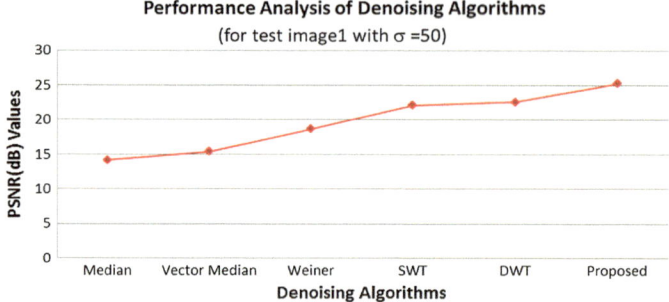

Fig. 4 Performance analysis of denoising algorithms for test image1 with $\sigma = 50$

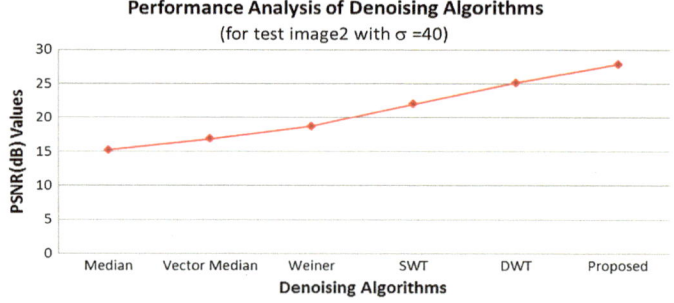

Fig. 5 Performance analysis of denoising algorithms for test image2 with $\sigma = 40$

It is observed from Figs. 4 and 5 that, the proposed approach for denoising outperforms in respect of PSNR values by a factor **2.7** when compared to the existing approaches. Hence the images (either gray scale or color) with the proposed denoising approach will have better quality preserving the features and is ready for further stage in diagnosis process which may yield better diagnostic results in terms of accuracy.

5   Conclusion

A novel approach for image denoising has been presented in this paper which helps to improve the quality of the input image for further processing steps in the diagnostic process. Proposed approach is a combination of two well-known models such as GCRF and MLR. Proposed approach is capable of handling high range of noise in gray scale or color images. Experiments have been conducted on sample test images drawn from LIDC benchmark data set and the results show that the proposed approach for denoising outperforms the existing approaches in terms of PSNR by a factor 2.7. The resultant denoised image preserving the features may be used in further stages of diagnosis process. Future work in this regard is to segment and extract the relevant features from preprocessed image for classification of lung image.

References

1. Nguyen, H.M., Peng, X., Do, M.N., Liang, Z.P.: Denoising MR spectroscopic imaging data with low-rank approximations. IEEE Trans. Biomed. Eng. **60**(1), 78–89 (2013)
2. Luo, J., Zhu, Y., Magnin, I.E.: Denoising by averaging reconstructed images: application to magnetic resonance images. IEEE Trans. Biomed. Eng. **56**(3), 666–674 (2009)
3. Nercessian, S.C., Panetta, K.A., Agaian, S.S.: Non-linear direct multi-scale image enhancement based on the luminance and contrast masking characteristics of the human visual system. IEEE Trans. Image Process. **22**(9), 3549–3561 (2013)
4. Chan, K.G., Streichan, S.J., Trinh, L.A., Liebling, M.: Simultaneous temporal super resolution and denoising for cardiac fluorescence microscopy. IEEE Trans. Comput. Imag. **2** (3), 348–358 (2016)
5. Cong-Hua, X., Jin-Yi, C., Wen-Bin, X.: Medical image denoising by generalised Gaussian mixture modelling with edge information. IET Image Proc. **8**(8), 464–476 (2014)
6. Chen, T.L.: A Markov random field model for medical image denoising. In: 2009 2nd International Conference on Biomedical Engineering and Informatics, Tianjin (2009)
7. Ouahabi, A.: Image Denoising Using Wavelets: Application in Medical Imaging, pp. 287–313 (2013)
8. Musicant, D., Kumar, V., Ozgur, A.: Optimizing F-measure with support vector machines. In: Proc. FLAIRS, pp. 356–360 (2003)
9. Jansche, M.: Maximum expected F-measure training of logistic regression models. In: Proceedings of the conference on Human Language Technology and Empirical Methods in Natural Language Processing (HLT '05). Association for Computational Linguistics, Stroudsburg, PA, USA, pp. 692–699 (2005)
10. Joachims, T.: A support vector method for multivariate performance measures. In: Proc. 22nd ICML, pp. 377–384 (2005)
11. Suzuki, J., McDermott, E., Isozaki, H.: Training conditional random fields with multivariate evaluation measures. In: Proc. 21st ICCL/44th AMACL, pp. 217–224 (2006)
12. Niu, Y., Wu, X., Shi, G.: Image enhancement by entropy maximization and quantization resolution upconversion. IEEE Trans. Image Process. **25**(10), 4815–4828 (2016)
13. Kuang, H., Chen, L., Gu, F., Chen, J., Chan, L., Yan, H.: Combining region-of-interest extraction and image enhancement for nighttime vehicle detection. In: IEEE Intelligent Systems, vol. 31(3), pp. 57–65 (May–June 2016)
14. Mi, Z., Zhou,H., Zheng, Y., Wang, M.: Single image dehazing via multi-scale gradient domain contrast enhancement. In: IET Image Processing, vol. 10(3), pp. 206–214 (2016)
15. Rizkinia, M., Baba, T., Shirai, K., Okuda, M.: Local spectral component decomposition for multi-channel image denoising. IEEE Trans. Image Process. **25**(7), 3208–3218 (2016)
16. Cong-Hua, X., Jin-Yi, C., Wen-Bin, X.: Medical image denoising by generalized Gaussian mixture modeling with edge information. IET Image Proc. **8**(8), 464–476 (2014)
17. Xu, J., Zhang, K., Xu, M., Zhou, Z.: An adaptive threshold method for image denoising based on wavelet domain. Proc. SPIE Int. Soc. Opt. Eng. **7495**, 165 (2009)
18. Suzuki, J., McDermott, E., Isozaki, H.: Training conditional random fields with multivariate evaluation measures. In: Proc. 21st ICCL/44th AMACL, pp. 217–224 (2006)

Author Biographies

C. Rangaswamy has received M.Tech., Degree from Visvesvaraya Technological University, Belagavi, Karnataka in 2010. Currently working as an Assistant Professor in the Department of Electronics and Communication Engineering Sambhram Institute of Technology, Bengaluru. He has 18 years of experience in teaching. His area of research interests include Image Processing, Communication. He is IETE life member. He has published 5 papers in leading reputed International Conferences/Journals/National conferences.

Dr. G. T. Raju has received M.E. Degree from Bangalore University, in 1995 and Ph.D. from Visvesvaraya Technological University, Belagavi, Karnataka in 2008. Currently working as Dean of Engineering, Professor and Head, in the Department of Computer Science & Engineering, RNS Institute of Technology, Bengaluru. He has 24 years of experience in teaching and research. His area of research interests include Web Mining, KDD, Image Processing, Pattern Recognition. He has published more than 90 papers in leading reputed International Journals/Conference proceedings. He has authored five Technical books. He has completed two funded research projects. Ten Research Scholars have been awarded Ph.D. Degree under his supervision.

Dr. G. Seshikala has received Ph.D. degree in Biomedical signal processing from JNTU-A, B.E. in ECE and M.E. in Digital Electronics. Currently working as a Processor in School of Electronics and communication Engineering, Reva University, Bengaluru. She has 21 years of teaching and 2 years of industrial experience, Her areas of specialization are image processing, pattern recognition, communication engineering and signal processing. She has published 16 research papers in international journals and conferences. She is ISTE life member, Presently guiding 5 Ph.D. students.

Computer Aided Diagnosis of Skin Tumours from Dermal Images

T. R. Thamizhvani[(⊠)], Suganthi Lakshmanan,
and R. Sivaramakrishnan

Department of Biomedical Engineering, SSN College of Engineering,
Kalavakkam, Chennai, Tamilnadu, India
thamizhvani15011@bme.ssn.edu.in, suganthil@ssn.edu.
in, raaju.shiv1@gmail.com

Abstract. Skin tumour is uncontrolled growth of skin cells which may be cancerous. The aim is to develop computer aided diagnosis for skin tumours. The dermal images of three types such as benign tumour, malignant melanoma and normal moles obtained from the authorised PH2 database. Pre-processing performed to remove hair cells. Contour based level set technique for segmentation of the lesion from which clinical and morphological features are extracted. The significant features are obtained using Random Subset Feature Selection technique. Classification is performed using three classifiers such as back propagation, pattern recognition and support vector machine. Classifier Efficiency of three classifiers is determined to be 94, 96 and 98% respectively with the Classifier performance parameters. One way ANOVA test is performed to analyse the efficiency of the three classifiers. With these results, Support vector machine is configured as accurate classifier for classification.

Keywords: Contour based level set · Random subset feature selection

1 Introduction

Skin is the outermost layer of the human body. Skin is formed of two layers epidermis and dermis. Melanin and keratin are the two pigment proteins produced by the special types of cells called melanocytes and keratinocytes respectively. Damage to any layer of the skin occurs in different ways and they are of various types. Skin tumours are developed due to the uncontrolled growth of the skin cells due to some metabolic changes. Majority of the skin tumours are benign. Malignancy of skin tumours defined as Skin cancers. Cancerous cells have the ability to spread all over the body. Skin tumours are mainly caused due to overexposure of ultraviolet radiation from the sun. Most common types of skin cancer are basal carcinoma, squamous carcinoma and melanoma. Melanoma is a common type of skin cancer that is dangerous and critical even leads to death of the human beings. The survival rate of the patients depends on the stages of cancer if diagnosed early it is completely curable. Mole or nevus is

© Springer International Publishing AG 2018
D. J. Hemanth and S. Smys (eds.), *Computational Vision and Bio Inspired Computing*,
Lecture Notes in Computational Vision and Biomechanics 28,
https://doi.org/10.1007/978-3-319-71767-8_30

chronic skin or mucosa lesions. Moles are benign but 25% of malignant melanomas arise from the existing nevi [1].

The signs and symptoms of melanoma include changes in the shape and colour of the moles or nodular melanoma. Melanoma also causes itching, ulcerate and bleeding. Early stages of melanoma are defined using Asymmetry, Borders, Colour variations and Diameter [2]. Melanoma is caused by DNA damage resulting from exposure to ultraviolet radiations from the sun. Genetics also plays an important role in the melanoma formation. Ultraviolet UVB and UVA radiations of wavelengths between 400 and 280 nm from the sun is absorbed by the skin cells and causes DNA damage leading to uncontrolled growth of the melanocytes. Common diagnostic technique is to analyse the signs and symptoms of skin tumours using Dermoscopy and Biopsy. Dermoscopy method is mainly for the examination of the lesion and outgrowths. Other technique used for diagnosis is skin biopsy in which a skin lesion is removed and sent to the pathologist for study.

Survey study has been carried out in India particularly in certain hospitals with the skin cancer patients [3, 4]. This study reveals the truth that skin cancer is becoming predominant and early diagnosis of this is very much necessary. Basal cell carcinoma, squamous carcinoma and malignant melanoma are common types of skin cancer seen in India. Melanoma rate increased in India [4]. Diagnosis and treatment is required for these skin cancers immediately for easy recovery. The purpose is to develop a computer aided diagnostic system for early detection and diagnosis of skin tumours. In this three different types of dermal images are categorised and classified. Malignant melanoma is very dangerous and critical form of skin cancer. Early diagnosis and detection is very much required in such cases. Many researches have been carried out related to the computer aided diagnosis [5].

Abuzaghleh et al. [6] explains that database images of melanoma can be analysed and classified into different types of pigmented skin diseases [2]. Real time alert system developed for the diagnosis and early detection of skin burns. Asymmetry, Border, Colour and Diameter (ABCD) are the features described for the diagnosis of melanoma skin cancer. Database images are used for segmentation, feature extraction by ABCD features and classification using SVM. Automatic image analysis model developed for acquisition, lesion segmentation, feature extraction and classification of skin images with great accuracy.

Fosu and Jouny [7] describes the development of a mobile application for melanoma detection in android smart phones. Applications are developed with the Android Developer Tools and also processed in MATLAB to classify the images as benign or malignant using support vector machines (SVM) model. The classification in MATLAB is based on the colour and symmetry analysis of the melanoma images. The efficiency of the SVM classifier model is determined with these database images. Karargyris et al. [8] describes that using machine language an automatic real time acquisition module can be developed and used for early detection of the skin cancer images. Application in real time is developed in iPhone for the early detection of skin cancer in general. Two class classifiers are used to determine and monitor the different types of skin cancer. With the review of the literature and study performed about the analysis of dermal images, the aim is to develop a computer diagnostic system for the skin tumours. Processing of the dermal images is defined to obtain an accurate and efficient system for diagnosis.

2 Proposed Methodology

This paper proposes a computer aided diagnosis technique for skin tumours in human body. Skin tumours are formed mainly due to over exposure of Ultraviolet Radiation that may also lead to cancer [3]. The proposed system defines that for development of the diagnosis system, dermal images from a standardised database as PH2 dermoscopic database is used. Three different categories of images (benign tumours, normal moles and malignant melanoma) are selected and processed. Hair removal process is carried out to define a clear lesion from the skin images for feature extraction. Level set segmentation for lesion separation and feature extraction to classify the images into three different categories. Feature selection to configure the dominant and significant features. The dominant features are considered for classification with three different types of classifier such as Back propagation network, Pattern recognition network and Support vector machines. Performance parameters of the classifiers are obtained to determine an accurate classifier for these database dermal images.

Dermoscopic images are obtained from the database of skin cancer for the diagnosis system. PH2 database is a standardised and authorised database accepted by many dermatologists all over the world. Dermal images from these databases are used in the process of diagnosis. Images in the databases are obtained with the help of the dermoscope system with magnification power of $20\times$ and resolution of 768×560 pixels. The images are acquired with a very high resolution and standard. Databases possess more than 2000 images of skin tumours which include melanomas, common nevi and benign tumours [2]. Out of which 1500 images are collected for three different types of skin tumours such as 500 from atypical nevus (benign tumours), 500 from malignant melanoma and 500 from common nevus (moles). Images in this database are not processed and normalised, it consists of all the features that the image possesses during a real time acquisition. Many Researches used these databases for the detailed study and analysis of the skin cancer images [2, 9, 10]. The database image is shown in Fig. 1.

(a)   **(b)**

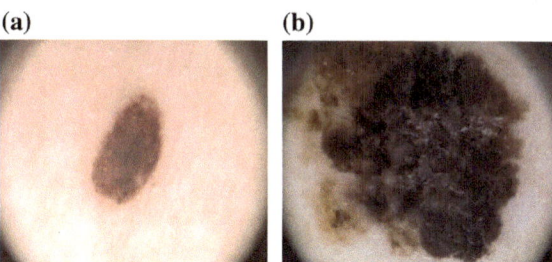

Fig. 1 Sample images of **a** benign tumour and **b** malignant melanoma. *Source* PH2 database

Melanoma lesions are irregular and asymmetrical. In the proposed methodology, dermal images from the PH2 database of three categories are processed. Dermal images are pre-processed through dilation. Segmentation is carried out with contour level set method and the features are extracted from the lesions. Random subset feature selection technique is defined to obtain significant features. With these significant features, classification is carried out. Three different classifiers (BPN, PNN & SVM) are used to categorise three different types of dermal images. Therefore the proposed methodology is shown in Fig. 2.

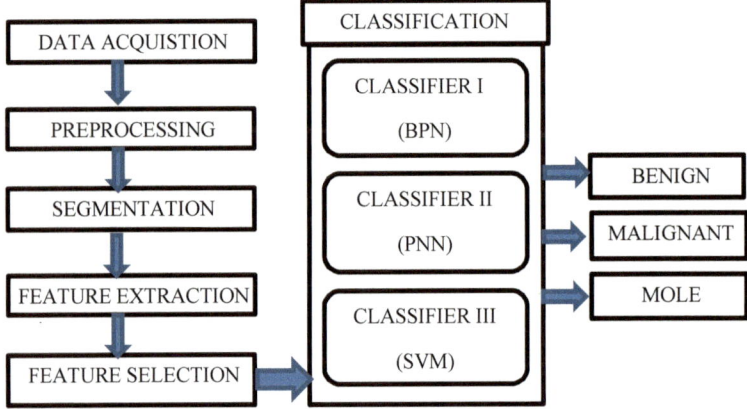

Fig. 2 Proposed methodology for diagnosis of dermal images

Pre-processing technique such as Dilation is carried out to remove the hair cells because the lesions are covered by the hair cells. Hair cells obstruct the clear vision of the lesion in the dermal image. In this process the original RGB image converted into grayscale image and hair cells are removed by morphological operation called grayscale dilation in which structural function is defined over the image. This enables better segmentation of lesion and extraction of features from the dermal images. Thus classification is more effective. Grayscale dilation is done with the help of the dilation operation as shown in Eq. 1 with $f(x)$ as image function and $b(x)$ as structuring function in Euclidean space (E).

$$(f \oplus b)(x) = \sup[f(y) + b(x - y)] \quad \text{for all } y \in E \tag{1}$$

Segmentation is mainly to separate the lesion from the background to extract features. In this process, contour based level set segmentation method is defined for separation of lesion from the skin surface. In this technique, Level set function determined for the curvature of the lesions in the grayscale dilated images and defines the lesion curvature in a binary image. Superimposition burning of the binary image over the original image derives the lesion without any background. The lesion image is used for feature extraction. Contour level set method is described below in Eq. 2 with Ψ as level set function.

$$\frac{\partial \Psi}{\partial t} = v |\nabla \Psi| \tag{2}$$

In feature extraction process, totally fifteen features are derived. They are Morphological features (2 parameters), Diagnostic clinical features (Asymmetry, Border irregularity, Colour variation and Diameter) of seven parameters, Statistical features and GLCM (Gray Level Co-occurrence Matrix) features (5 parameters). Clinical features described by the dermatologists as ABCD features which helps in the identification and determination of the different types of skin tumours. These features involve asymmetry index, border irregularity, colour variance and diameter that exactly and specifically explain the nature of the lesions or tumours.

Asymmetry index. Asymmetry Index defines that the shape of one half does not match the rest half. The severity of the skin cancer depends on asymmetry degree of a skin lesion. Biaxial asymmetry of the lesion helps more in diagnosis. So, asymmetry around both minor and major axes needs to be calculated. Asymmetry Index (AI) is calculated as shown in Eq. (3) with $\Delta Amax$ and $\Delta Amin$ correspond to the non-overlapping areas along major and minor axis respectively. A is the total area of the lesion.

$$AI = \frac{\Delta Amax + \Delta Amin}{2 * A} \tag{3}$$

Border irregularity. Irregularity of the lesion boundaries termed as border irregularity. The edges with pigmentation may extend into surrounding skin. Malignancy can be catalogued from benign one by analysing lesions boundary. Malignant lesions are more irregular than benign. Two parameters are used to define the border irregularity of the lesions. Compactness Index (CI) measures roundness in a two dimensional object. Index is sensitive to the noise defined around the boundary of the lesion. It is defined in Eq. (4) with P as perimeter of lesion and A as area of the lesion

$$CI = \frac{P^2}{4\pi A} \tag{4}$$

Circularity Index measures the circular nature of the lesion or defines the outline based on the curvature of the lesion. It is defined by Eq. (5)

$$\text{Circularity index} = \frac{4\pi A}{P^2} \tag{5}$$

Colour variation. Colour variance parameter analyses the colour distribution of lesion. It checks variation of colour from the centre to its boundary. Colour distribution throughout the lesion in the skin is defined with mean and standard deviation that is the statistical parameters.

Diameter. Diameter is mainly to determine the type of lesion which is defined by the major axis and minor axis of the lesion structure. Mostly malignant has more diameter than benign or nevus samples (larger than six millimetres or about a quarter inch for malignant melanoma).

Gray level Co-occurrence matrix used to define certain features corresponding to the second order statistical probability $P(i, j)$. Contrast, correlation, entropy, energy and homogeneity are GLCM features described below in Eqs. (6)–(10) with i and j as gray levels.

Contrast is a measure of local contrast or intensity variations between the gray levels i and j.

$$\text{Contrast} = \sum_{n=0}^{G-1} n^2 \left\{ G \sum_{i=1}^{G} \sum_{j=1}^{G} P(i,j) \right\}, |i - j| = n \tag{6}$$

The measure of linear dependency among the pixels at relative position specified to each other is defined as correlation.

$$\text{Correlation} = \sum_{i=1}^{G} \sum_{j=1}^{G} (\{i \times j\} P(i,j) - \{\mu x \times \mu y\}) / \sigma x \times \sigma y \tag{7}$$

Entropy is a statistical form of randomness to characterize the texture of the image.

$$\text{Entropy} = -\sum_{i=0}^{G-1} \sum_{j=0}^{G-1} P(i,j) \times \log(P(i,j)) \tag{8}$$

Maximum or periodic uniform values in the gray level distribution describe the maximum energy of texture.

$$\text{Energy} = \sum_{i=0}^{G-1} \sum_{j=0}^{G-1} (P(i,j))^2 \tag{9}$$

Homogeneity among the gray level pixels of the image are described in Eq. (10).

$$\text{Homogeneity} = \sum_{i=0}^{G-1} \sum_{j=0}^{G-1} \left(1/1 + (i - j)^2 \right) [P(i,j)] \tag{10}$$

The fifteen features undergo feature selection process to obtain significant and dominant features for effective classification. Feature selection technique is mainly to obtain dominant features which completely define the characteristics of the lesion in the image. Dominant features for classification process are determined using Random Subset Feature Selection (RSFC) technique. Random Subset Feature Selection (RSFS) is a feature selection method in which random subset of features is used to determine the significant features. These features are obtained by repetitively classifying the data with a k-NN classifier while using randomly chosen subsets of all possible features and adjusting the relevance value of each feature according to the classification performance of the subset. Each feature is evaluated in RSFS in terms of its average usefulness in the context of many other feature combinations. Random subsets of fifteen features are obtained and the probability of all subsets is compared using k-NN algorithm. Seven features such as Asymmetry Index, Colour Variation, Compact Index,

Mean, Energy, Contrast and Entropy with the 99% of probability are described as the dominant features. With these features, classification of images is carried out.

Three types of classifiers such as back propagation network, pattern recognition network and support vector machines are used to differentiate the different categories of images. Efficiency of the classifiers are studied and analysed with the help of the performance parameters of classification.

Back propagation neural network (BPN) is useful for complex pattern recognition and mapping functions. In BPN, a predefined set of input and output pair is used for learning with the help of two phase propagate adapt cycle [11]. The input layer is applied with stimulus which propagates to the upper layers to define a output unit. Backward flow of the error signals computed for each output unit to the input unit.

Pattern recognition neural network mainly focuses on the recognition of patterns and data regularity. Pattern recognition is supervised learning that are probabilistic in nature. Statistical inference in the pattern recognition provides a best result. The aim is to define the probability of each target of the training sample with the particular input. The probability recognises the complex patterns.

Support vector machines (SVM) are supervised machine learning algorithms or models. SVM is the representation of points that maps to form separate divisions and a clear boundary is defined called decision boundary [12]. In this training sample forms a hyper plane in D-dimensional space in such a way that the margin separates the positive and negative samples. Multiclass support vector machine in which different classes is optimised with the kernels. In multiclass SVM decision boundary between all the classes in comparison with other classes ids formed.

Dominant features are considered for classification using three different classifiers. Classification in which both training and testing phase occurs. Confusion matrix is obtained for each classifier in order to learn the performance of the classification algorithm. Performance of the three classifiers is defined using certain parameters such as accuracy, precision, specificity and sensitivity. These parameters are calculated based on the true positive (TP), false positive (FP), true negative (TN) and false negative (FN) values obtained from the confusion matrix.

Accuracy is defined as the observational error that describes the trueness of the classification function. Precision defines the statistical variability of random errors and related to the reproducibility and repeatability state of classification. Accuracy and precision are defined in terms of TP, TN, FN & FP is shown in Eqs. (11) and (12)

$$\text{Accuracy} = \frac{TP + TN}{P + N}, \quad P = TP + FN, \ N = FP + TN \tag{11}$$

$$\text{Precision} = \frac{TP}{(TP + FP)} \tag{12}$$

Sensitivity and specificity are classification performance tests which involves statistical measures. Sensitivity and specificity measures the positive and negative proportions that are correctly defined respectively. They are described in Eq. (13).

$$\text{Sensitivity} = \frac{TP}{(TP + FN)} \qquad (13)$$

$$\text{Specificity} = \frac{FP}{(FP + TP)} \qquad (14)$$

Performance parameters are used to define the effective and efficient classification nature of the three different classifiers. Analysis of variance (ANOVA) is a collection of statistical models that is used for analysis of difference between the group means and other associated process like variations between or among the groups. One way ANOVA test is performed for the performance parameters of the classification process. In this process each classifier is accessed for 10 trials and performance parameters are determined for each trial. With the overall values of performance parameters of three classifiers for all trials, the P value and honestly significant difference is obtained. This defines the efficient classification process. P value describes the significance of variations between the two samples. Honestly significant difference is defined using Tukey test which is a multiple statistical comparison procedure. Thus the best classifier for the diagnosis of skin images is obtained.

3   Results and Discussions

In the proposed method, the PH2 database dermoscopic images are used for the computer aided diagnosis system development process. The dermoscopic images of three categories benign tumours, common moles and malignant melanoma are processed and lesions are segmented separately. In which the RGB images are converted into grayscale images as shown in Fig. 3.

Grayscale dilation process is carried out to remove the hair cells for accurate feature extraction and classification. In this morphological operation, disk structuring element is used as a function over the image to obtain a pre-processed image. Dilation processed images are shown in Fig. 4.

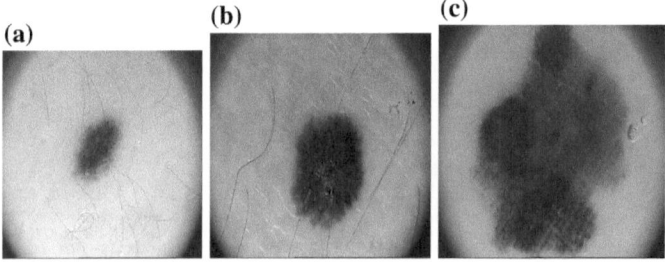

Fig. 3 Original images **a** normal mole, **b** benign tumour, **c** malignant melanoma

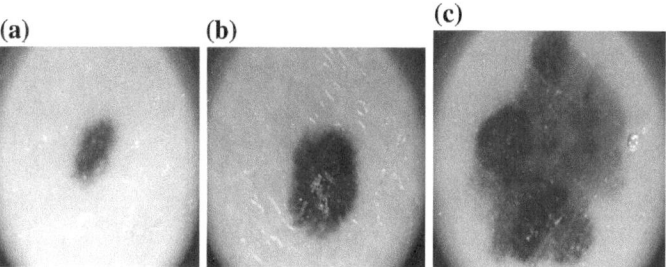

Fig. 4 Dilation processed images **a** normal mole, **b** benign tumour, **c** malignant melanoma

Contour based level set method is used in segmentation of the lesion from the images. Level set function defined for the curvature and a curvature image is obtained as a mask. This mask is used to obtain the lesion separately. Segmented images are defined in Fig. 5. Superimposition of the mask over original image occurs which is defined in Fig. 6.

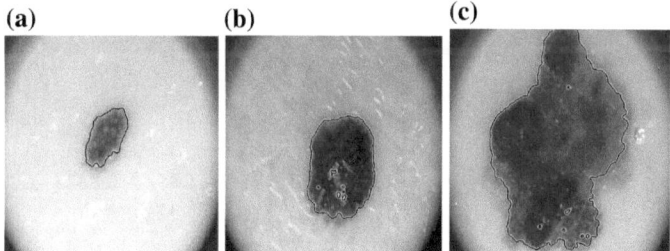

Fig. 5 Contour based level set segmented images **a** normal mole, **b** benign tumour, **c** malignant melanoma

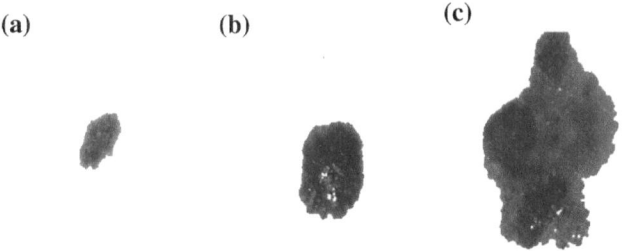

Fig. 6 Superimposed segmented images **a** normal mole, **b** benign tumour, **c** malignant melanoma

Feature extraction is carried out in the segmented lesion image. Fifteen features are extracted including the diagnostic clinical features, morphological and statistical features. Asymmetry Index, border irregularity, colour variation and diameter are the clinical features obtained from the segmented lesion. These features undergo selection process from which dominant features are derived for classification. Feature selection is carried out based on Random subset feature selection process. Seven features are described as dominant features to classify the three categories of dermal images. Random subset of features is determined and the probability of each random subset is calculated. Based on the probability a subset of random features is obtained as significant features. In this process from the random subset of fifteen features, seven significant features are determined with 99% of probability using k-NN algorithm. Significant nature of features are further analysed with the statistical values such as mean and standard deviation (SD). These statistical values of the seven significant features obtained as a result of random subset feature selection process is described in Table 1.

Table 1 RSFS based selected features (seven features)

Random subset feature selection (RSFS) based significant features							
Classes		Normal		Benign		Malignant	
		Mean	SD	Mean	SD	Mean	SD
Seven dominant features	Colour mean	0.444	0.020	0.431	0.054	0.426	0.066
	Mean	122.4	19.75	106.7	20.20	84.21	14.49
	Energy	0.727	0.159	0.672	0.143	0.588	0.145
	Contrast	0.081	0.057	0.108	0.043	0.183	0.086
	Entropy	1.505	0.878	1.830	0.831	2.297	0.917
	Compact index	1.790	0.608	2.032	0.788	2.705	1.268
	Asymmetry index	47.16	2.448	46.56	2.934	40.48	8.785

The significant features selected are defined for the classification process. Three different classifiers are used in which one type of classifier is based on machine learning and two are neural network based classifiers. Classifier 1 that is the back propagation network in which the both training and testing occurs. Output unit is formed corresponding to the stimulus applied to each input unit. Error signal formed along with the output unit has a backward flow. Thus seven significant features are framed as the input samples with defined output. Confusion matrix is defined with true negative (TN), true positive (TP), false negative (FN) and false positive (FP) for all the three categories of dermal images which is shown in Table 2.

Table 2 Confusion matrix for classifier 1 (BPN)

		Predicted class		
		Normal	Benign	Malignant
Actual class	Normal	**446**	54	0
	Benign	16	**477**	7
	Malignant	0	8	**492**

Classifier 2 is the pattern recognition mainly defined to recognise complex patterns and mappings. Pattern recognition is a supervised learning in which output unit to be obtained is predefined along with the input unit. Based the predefined units the recognition process occurs and also depends on the probability of the complex patterns. In this seven features are subjected for training and testing process with hidden layer to obtain a confusion matrix as shown in the Table 3.

Table 3 Confusion matrix for classifier 2 (PNN)

		Predicted class		
		Normal	Benign	Malignant
Actual class	Normal	**476**	24	0
	Benign	22	**468**	10
	Malignant	0	0	**500**

Classifier 3 is Support vector machine (SVM) in which a margin of separation so called decision boundary separates the positive and negative classes of the input samples with predefined values. SVM is a supervised learning process. Multi-class SVM is used in which all the three categories of dermal images are classified based on the predefined values set as a target for the input unit. Confusion matrix is shown in Table 4 that describes the classification performance of SVM.

Table 4 Confusion matrix for classifier 3 (SVM)

		Predicted Class		
		Normal	Benign	Malignant
Actual class	Normal	**483**	17	0
	Benign	4	**494**	2
	Malignant	0	4	**496**

Performance of classification of each classifier is determined using certain parameters such as Accuracy, Precision, Sensitivity and Specificity. These parameters are derived with the confusion matrix of each classifier that categorises the three different classes of dermal images. Percentage of each parameter describes the accurate and efficient classifier for dermal image diagnosis. Efficiency of classification can be increased by increasing the number of hidden units in the neural network and by changing the kernel function in the machine learning classifier such as support vector machine. With the increase in the hidden units or kernels, the classification performance increases which is shown in Table 5. The hidden units or kernels are increased from 10, 20, 30 & 40 which show changes in the efficiency of the classification process.

Table 5 Efficiency results with increasing hidden units or kernels

Classifiers	Number of hidden units or kernels	Performance parameters			
		Accuracy	Precision	Specificity	Sensitivity
BPN	10	93.8	0.938	0.969	0.9399
	20	94.33	0.9433	0.9717	0.9454
	30	94.6	0.946	0.973	0.9482
	40	94.4	0.944	0.9720	0.9465
PNN	10	94.5333	0.9453	0.9728	0.9451
	20	96.2667	0.9627	0.9814	0.9625
	30	96.5333	0.9753	0.9877	0.9756
	40	96.7667	0.9767	0.9883	0.9767
SVM	10	98.2	0.982	0.9910	0.9823
	20	98.4667	0.9847	0.9923	0.9851
	30	98.6667	0.9867	0.9933	0.9867
	40	98.7333	0.9873	0.9937	0.9875

Performance parameters of the classification process are described in percentage for the hidden units or kernels of value 40. These parameters are obtained with higher efficiency value is given below in Table 6.

Table 6 Percentage of performance parameters of classification

	Percentage of performance parameters			
	Accuracy (%)	Precision (%)	Sensitivity (%)	Specificity (%)
BPN	94.45	0.94	0.97	0.95
PNN	96.30	0.97	0.99	0.96
SVM	98.53	0.99	0.99	0.99

In this process, three different classifiers are compared with accuracy and efficiency of classifier 1 (BPN) is 94%, classifier 2 (PNN) is 96% and classifier 3 (SVM) is 98%. Thus the classifier 3 (SVM) which is based on machine learning defined to be more efficient compared to other classifiers. Performance parameters of the multiclass SVM classifier prove to be more efficient with the help of results.

One way ANOVA test is performed for the classification performance parameters which are obtained for each trial of classification by all the three classifiers. Maximum of ten trials are carried with the three classifiers for which accuracy, precision, specificity and sensitivity are calculated. With these values obtained for the ten trials, the ANOVA test is performed. One way ANOVA test reveals that the P value <0.01 or <0.05 then it is 99 or 95% of confidence level respectively for the performance parameter measures for ten trials. Consolidated results of one way ANOVA test is described in Table 7.

Table 7 Consolidated results of one way ANOVA test

Parameters	BPN	PNN	SVM	P
Accuracy	93.8	95.9333	98.2	<0.0001
	94.3333	96.2667	98.4667	
	94.6	97.5333	98.6667	
	94.4	96.6667	98.7333	
	94.6133	96.6	98.1333	
	94.6	97.2	98.4667	
	94.8	96.4667	98.6667	
	94.4	96.6667	98.7333	
	94.3333	97.6	98.5333	
	94.6	97.0667	98.7333	
Mean stats	94.4480	96.30001	98.53333	
Precision	0.9399	0.9451	0.9823	<0.0001
	0.9454	0.9625	0.9851	
	0.9482	0.9756	0.9867	
	0.9465	0.9767	0.9875	
	0.9454	0.9765	0.9814	
	0.9482	0.9727	0.9851	
	0.9495	0.9849	0.9867	
	0.9465	0.9767	0.9875	
	0.9454	0.9765	0.9854	
	0.9482	0.9806	0.9875	
Mean stats	0.9463	0.97278	0.98552	
Specificity	0.969	0.9728	0.9910	<0.0001
	0.9717	0.9814	0.9923	
	0.973	0.9877	0.9933	
	0.9720	0.9883	0.9937	
	0.9717	0.988	0.9907	
	0.973	0.986	0.9923	
	0.974	0.9924	0.9933	
	0.9720	0.9883	0.9937	
	0.9717	0.988	0.9927	
	0.973	0.9903	0.9937	
Mean stats	0.9721	0.98632	0.9927	

(*continued*)

Table 7 (*continued*)

Parameters	BPN	PNN	SVM	P
Sensitivity	0.938	0.9453	0.982	<0.0001
	0.9433	0.9627	0.9847	
	0.946	0.9753	0.9867	
	0.944	0.9767	0.9873	
	0.9433	0.976	0.9813	
	0.946	0.972	0.9847	
	0.948	0.9847	0.9867	
	0.944	0.9767	0.9873	
	0.9433	0.976	0.9853	
	0.946	0.9807	0.9873	
Mean stats	0.9442	0.97261	0.9823	

Table 8 Consolidated results of Tukey's test

Parameters	BPN	PNN	SVM	Tukey's test
Accuracy	93.8	95.9333	98.2	HSD[0.05] = 0.01
	94.3333	96.2667	98.4667	HSD[0.01] = 0.01
	94.6	97.5333	98.6667	M1 vs M2 $P < 0.01$
	94.4	96.6667	98.7333	M1 vs M3 $P < 0.01$
	94.6133	96.6	98.1333	M2 vs M3 $P < 0.01$
	94.6	97.2	98.4667	
	94.8	96.4667	98.6667	
	94.4	96.6667	98.7333	
	94.3333	97.6	98.5333	
	94.6	97.0667	98.7333	
Mean stats	94.4480	96.30001	98.53333	
Precision	0.9399	0.9451	0.9823	HSD[0.05] = 0
	0.9454	0.9625	0.9851	HSD[0.01] = 0.01
	0.9482	0.9756	0.9867	M1 vs M2 $P < 0.01$
	0.9465	0.9767	0.9875	M1 vs M3 $P < 0.01$
	0.9454	0.9765	0.9814	M2 vs M3 $P < 0.05$
	0.9482	0.9727	0.9851	
	0.9495	0.9849	0.9867	
	0.9465	0.9767	0.9875	
	0.9454	0.9765	0.9854	
	0.9482	0.9806	0.9875	
Mean stats	0.9463	0.97278	0.98552	

(*continued*)

Table 8 (*continued*)

Parameters	BPN	PNN	SVM	Tukey's test
Specificity	0.969	0.9728	0.9910	HSD[0.05] = 0.01
	0.9717	0.9814	0.9923	HSD[0.01] = 0.01
	0.973	0.9877	0.9933	M1 vs M2 $P < 0.01$
	0.9720	0.9883	0.9937	M1 vs M3 $P < 0.01$
	0.9717	0.988	0.9907	M2 vs M3 $P < 0.01$
	0.973	0.986	0.9923	
	0.974	0.9924	0.9933	
	0.9720	0.9883	0.9937	
	0.9717	0.988	0.9927	
	0.973	0.9903	0.9937	
Mean stats	0.9721	0.98632	0.9927	
Sensitivity	0.938	0.9453	0.982	HSD[0.05] = 0.49
	0.9433	0.9627	0.9847	HSD[0.01] = 0.62
	0.946	0.9753	0.9867	M1 vs M2 $P < 0.01$
	0.944	0.9767	0.9873	M1 vs M3 $P < 0.01$
	0.9433	0.976	0.9813	M2 vs M3 $P < 0.01$
	0.946	0.972	0.9847	
	0.948	0.9847	0.9867	
	0.944	0.9767	0.9873	
	0.9433	0.976	0.9853	
	0.946	0.9807	0.9873	
Mean stats	0.9442	0.97261	0.9823	

Tukey's test defines the honestly significant difference (HSD) between the group means and comparison of group variations. HSD also used to enumerate the efficiency of classification. The results of Tukey's test are shown in Table 8. Thus one way ANOVA and Tukey's test explains a best conclusion for the efficient classification process to be support vector machine.

4 Conclusion and Future Work

The incidence of skin cancer is increasing among the individuals. Early detection of skin cancer is necessary to treat patients effectively. Since surgical excision is the only lifesaving treatment method for skin cancers. Therefore early detection and diagnosis is necessary. In this methodology, database images of three different categories such as Common mole, benign tumour and malignant melanoma undergo image analysis process such as hair detection and removal, segmentation of the lesion, feature extraction, feature selection and classification. Classification is carried out with three different classifiers such as back propagation network, pattern recognition network and support vector machine. Efficiency and performance of the classification process of each classifier is determined using the performance parameters from the confusion matrix. One way ANOVA test is performed to define the significant classification process. Support vector machine (SVM) of 98% is described to be efficient for classification of dermal images. From the results, a computer aided diagnosis module is developed which classifies the three types of skin tumours.

Future work focuses on the development of a real time diagnostic tool is required which can help the dermatologists more effectively in early detection and diagnosis. Further development in the work includes formation of Indian database for skin tumours and extending the diagnosis for real time skin tumour images of different categories.

References

1. Saleh, F.S., Azmi, R.: Automatic multiple regions segmentation of dermoscopy images. In: Aritificial Intelligence and Signal Processing (AISP), pp. 24–29 (2015)
2. Rasanen, O., Pohjalainen, J.: Random subset feature selection in automatic recognition of developmental disorders, affective states and level of conflict from Speech. J. Health Med. **3**, 210–214 (2013)
3. Brar, B.K., Sethi, N., Khanna, E.: An epidemiological review of skin cancers in Malwa belt of Punjab India: a 3-year clinicopathological study. Sch. J. Appl. Med. Sci. **3**, 3405–3408 (2015)
4. Sharma, K., Mohanti, B.K., Rath, G.K.: Malignant melanoma: a retrospective series from a regional cancer center in India. Cancer J. **5**, 173–180 (2009)
5. Masood, A., Al-Jumaily, A., Anam, K.: Self-surpervised learning model for skin cancer diagnosis. In: IEEE EMBS on Neural Engineering, pp. 1012–1015 (2015)
6. Abuzaghleh, O., Bakara, B.D., Faezipour, M.: Automated skin lesion analysis based on colour and shape geometry feature set for melanoma early detection and prevention. Int. J. Comput. Vis. **9**, 203–208 (2015)
7. Fosu, K.P.O., Jouny, I.: Mobile melanoma detection application for android smart phones. J. NEBEC 1109–1112 (2015)
8. Karagyris, A., Karagyris, O., Pantelopoulos, A.: DERMA/care: an advanced image processing mobile application for monitoring skin cancer: J. Artif. Intell. **12**, 1–7 (2012)
9. Pirnog, I., Preda, R.O., Oprea, C., Paleologu, C.: Automatic lesion segmentation for melanoma diagnostics in macroscopic images. In: European Signal Processing Conference, vol. 3, pp. 659–663 (2015)
10. Abuzaghleh, O., Bakara, B.D., Faezipour, M.: Noninvasive real-time automated skin lesion analysis system for melanoma early detection and prevention. J. Health Med. **3**, 2168–2372 (2015)
11. Gonazalez-Castro,V., Debayle, J., Wazaefi, Y., Rahim, M., Gaudy, C., Grob, J.J., Fertil, B.: Automatic classification of skin lesions using geometrical measurements of adaptive neighborhoods and local binary patterns. In: International Conference on Image Processing, pp. 1722–1726 (2015)
12. Gautam, D., Ahmed, M.: Melanoma detection and classification using SVM based decision support system. INDICON **15**, 6541–6546 (2015)

13. Lau, H.T., Al-Jumaily, A.: Automatically early detection of skin cancer study based on neural network classification. In: Conference of Soft Computing and Pattern Recognition, pp. 375–380 (2009)
14. Pirnog, I., Oprea, C.: Cutaneous melanoma risk evaluation through digital image processing. In: IEEE Conference on Image Processing, pp. 7264–7268 (2014)
15. Nowak, L.A., Ogorzalek, M.J., Pawlowski, M.P.: Texture analysis for dermoscopic image processing. In: Biomedical Circuits and Systems Conference, pp. 292–295 (2012)

Segmentation of MRI Brain Images and Creation of Structural and Functional Brain Atlas

Hema P. Menon[1]([✉]), Reshma Hiralal[1], A. Anand Kumar[2], and Davidson Devasia[2]

[1] Department of Computer Science and Engineering, Amrita School of Engineering, Coimbatore, Amrita Vishwa Vidyapeetham, Amrita University, Coimbatore, India
p_hema@cb.amrita.edu, cb.en.p2cvil5010@cb.students.amrita.edu
[2] Department of Neurology, School of Medicine, Amrita Institute of Medical Sciences, Kochi, Amrita Vishwa Vidyapeetham, Amrita University, Coimbatore, India
anandkumar@aims.amrita.edu,
drdavidsondevasia@gmail.com

Abstract. In this work an analysis of the different segmentation algorithms has been done, in order to select the most appropriate method for segmenting the brain into its corresponding regions. This selection is done by comparing the segmented regions with the ground truth images using measures like Dice, Precision, Recall and Score. The segmented regions are then labelled by their respective part names and used for creation of structural and functional brain atlas. This atlas would be of great use to doctors in providing assistance in disease analysis and can also be used as a teaching/learning kit.

Keywords: Segmentation · Magnetic resonance imaging (MRI) images
Brain atlas · Level sets · Thresholding · Blob segmentation

1 Introduction

There are different modalities by which medical images can be obtained. This work concentrates on images obtained using MRI. An MRI brain image consists of more than 300 parts that are useful for disease analysis. In-order to create a full-fledged atlas all these parts needs to be segmented and labelled automatically. The automation of

© Springer International Publishing AG 2018
D. J. Hemanth and S. Smys (eds.), *Computational Vision and Bio Inspired Computing*,
Lecture Notes in Computational Vision and Biomechanics 28,
https://doi.org/10.1007/978-3-319-71767-8_31

segmentation process is the major problem faced in this work due the similarity between the regions to be segmented. This is because the complexity of the segmentation depends on the application for which the segmentation is done. For the detection of disease only a few (one or two) parts need to be segmented, whereas for the creation of brain atlas every part need to be segmented There is no single method using which all parts can be segmented. Complexity of the segmentation method varies for the segmentation purpose. In this work different segmentation algorithms are analysed for the segmentation purpose they are thresholding, multilevel thresholding, adaptive thresholding, region growing, level sets, watershed algorithm, k-means and manual segmentation. Since the application is to create a brain atlas the survey conducted is based on atlas based segmentation.

Development of brain and human body Atlases had gained lot of prominence of late. Among some of the atlases available Lancaster et al. [1] and Allan [2] are the mostly commonly used ones. They give a user friendly interface by which one can learn the parts of human brain. But they cannot be customized by doctors for their use. To customize the atlas creation automatic segmentation of parts is necessary. An automatic segmentation of MRI images was addressed by Govindan and Nair [3]. Renjini and Bhagavathi Sivakumar [4] compare the techniques available for segmenting images interactively and automatically. The thalamus region has been automatically segmented using fuzzy method by Amini et al. [5]. For Alzheimer's disease analyses hippocampus was segmented using an atlas based approach by Carmichael et al. [6]. Segmentation has been done by many researchers' using watershed transform [7] and level sets [8, 9] more commonly than the other methods like thresholding and region growing. Karim [10] has used an atlas as a template for segmenting human brain MRI images and Rohlfing et al. [11] has used a similar approach for bee brain microscopic image segmentation. The parts have to label for atlas creation and such an approach has been discussed by Lancaster et al. [12]. To access the accuracy of any segmentation algorithm different methods are needed like Dice, precision recall, score etc. these measures have been discussed by Taha and Hanbury [13] and Polak et al. [14]. An anaysis on the basic segmentation algorithms was done by the authors [15] based on which the algorithms for segmentation were selected for this work.

2 Proposed System Architecture

MRI image segmentation is the major task in MRI brain Atlas creation, several algorithms where employed in the segmentation of individual brain parts, the major segmentation algorithms used are thresholding, (global thresholding and local thresholding), region growing, clustering based segmentation (k-Means), level sets, blob segmentation, watershed and manual segmentation (Fig. 1).

The proposed system accepts MRI slices as input, one slice at a time i.e., axial, sagittal

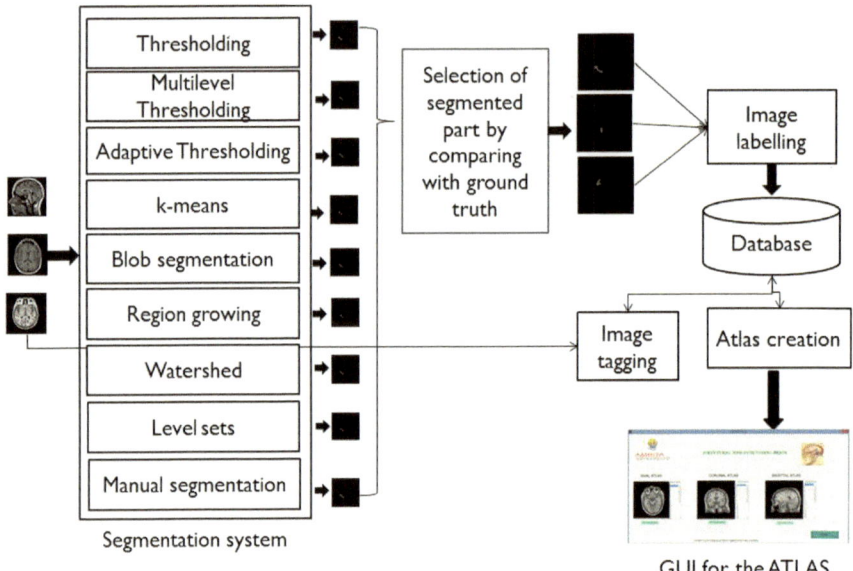

Fig. 1 Proposed system architecture

or coronal slices, segmentation toolbox consist of different segmentation algorithms namely thresholding, multilevel thresholding, adaptive thresholding, region growing, blob segmentation, level sets, watershed algorithm and manual segmentation. Each segmentation algorithm will segment the input image into its corresponding parts. These segmented parts are given for a subjective analysis in with best part is selected by quantitative analysis by comparing with the ground truth images. Those parts which are similar to the ground truth images are labeled and saved to their part name and slice number. Finally the atlas is created by linking the initial input images to the labeled image.

2.1 Data Set Details

Data set consist of 60 MRI images, which were collected from Amrita Institute of Medical Science (AIMS), Cochin. Initial file formats where DIACOM images which were converted into jpg images. Images from all three views, namely Axial, Coronal and Sagittal are considered. The dataset contains 20 images from each view.

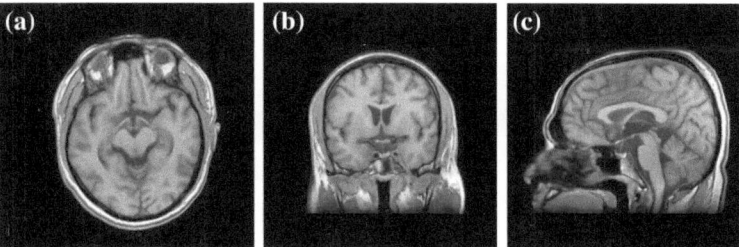

Fig. 2 **a** Axial, **b** coronal, **c** sagittal

The ethics committee comprised of the following doctors from AIMS, Kochi, namely,

- Dr. Anand Kumar A, Professor and Head, Department of Neurology, School of Medicine, Amrita Institute of Medical Sciences, Kochi
- Dr. Davidson Devasia, DM Neurology Resident, Department of Neurology, School of Medicine, Amrita Institute of Medical Sciences, Kochi
- Dr. Rajesh Kannan, Associate Professor, Department of Radiology, School of Medicine, Amrita Institute of Medical Sciences, Kochi.

The authentication certificate for the dataset and work done was issued by Dr. Davidson who was a part of the project, and is attached in Appendix.

3 Methodologies

For segmenting the images into its corresponding parts eight basic algorithms have been analyzed and the region which matches the best with the ground truth has been selected for atlas creation. This section discusses the results obtained on applying the different segmentation algorithms for the dataset shown in Fig. 2.

3.1 Thresholding

The accuracy of this method depends on the threshold value used. This has been set for each brain part by trial and error method and is found to range between 130 and 160 (Fig. 3).

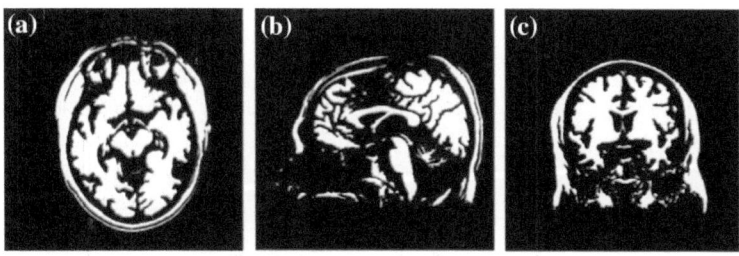

Fig. 3 Threshold based segmentation **a** axial, **b** sagittal, **c** coronal

Setting a single threshold is found to be insufficient in identifying multiple regions from the image.

3.2   Multilevel Thresholding

Multilevel thresholding is the process of using multiple threshold values in the segmentation process. Here we have used two thresholds in the implementation (Fig. 4).

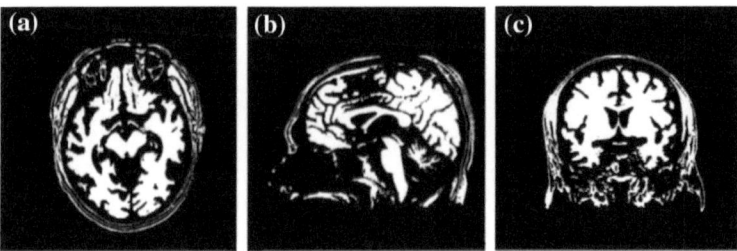

Fig. 4 Multilevel thresholding segmentation **a** axial, **b** sagittal, **c** coronal

3.3   Adaptive Thresholding

To get maximum regions a local adaptive threshold is selected for segmenting the image. Adaptive thresholding uses properties such as median or mean of the image is employed (Fig. 5).

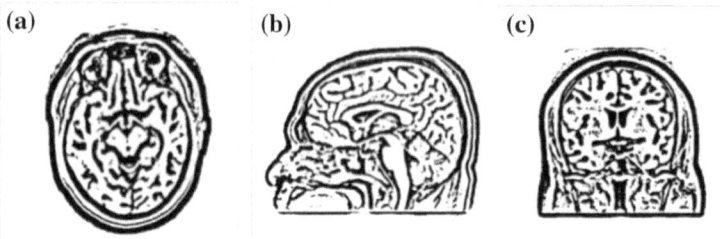

Fig. 5 Adaptive thresholding segmentation **a** axial, **b** sagittal, **c** coronal

3.4   Blob Segmentation

This is also a threshold base method. The regions greater than the threshold are detected as blobs (objects). A connected component analysis on these regions gives the required part (Fig. 6).

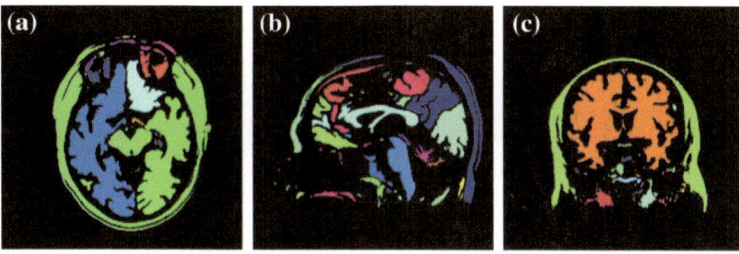

Fig. 6 Blob segmentation **a** axial, **b** sagittal, **c** coronal

It was found that Blob segmentation give the best result when the cut-off value is 135–140. For all other values the segmented outputs accuracy is low.

3.5 Region Growing

Region growing works with the principle of selecting a initial pixel which is called as a seed pixel. Here 8-connected adjacency relationship is employed for the initial seed pixel to grow as a region. The growing process will continue until it finds pixels which have a greater or lesser intensity value than the adjacent 8 connected neighbourhood (Fig. 7).

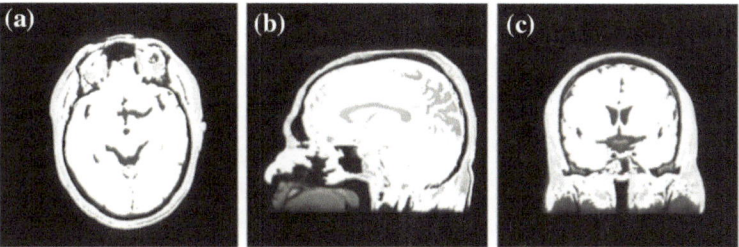

Fig. 7 Region growing segmentation **a** axial, **b** sagittal, **c** coronal

Region growing is applicable when the edge of the required part is well defined and distinct to all its adjacent pixels outside the region.

3.6 Level Sets Method

Level sets is an iterative approach for segmentation and in this work the number of iteration is found to be between 30 and 70 depending on the brain slice considered (Fig. 8).

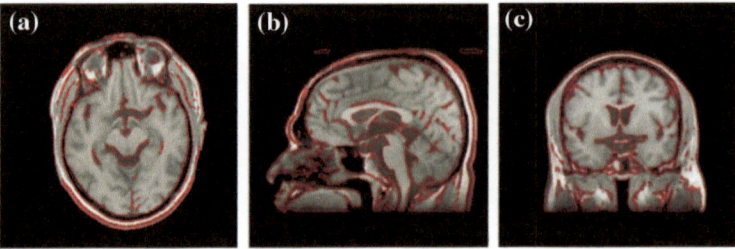

Fig. 8 Level set segmentation **a** axial, **b** sagittal, **c** coronal

3.7 K-Means

K-means algorithm works on the principal of selecting K seed pixel, these seed pixels are considered as the centroid of each clusters and based on the distance of every other pixel from this seed the clustering is done. The number of clusters as determined by the K value is decided for each slice based on the doctor's requirement of the number of regions from that slice (Fig. 9).

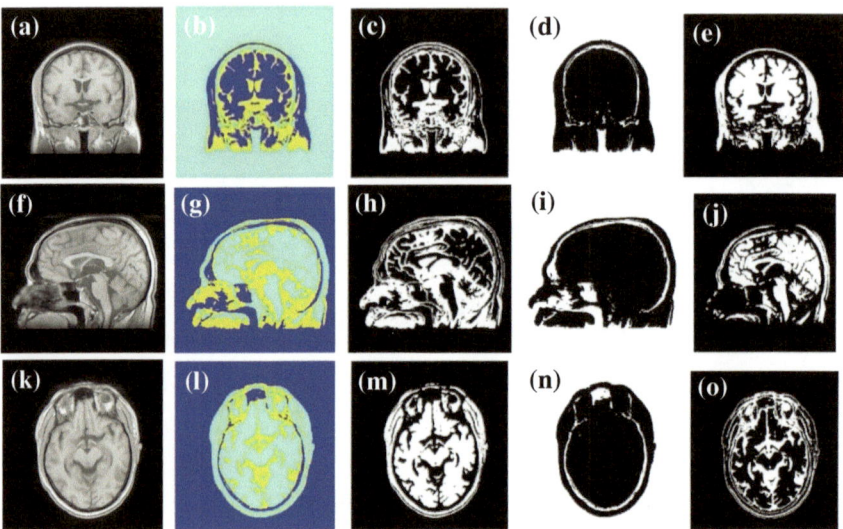

Fig. 9 k-means segmentation illustration: figure (**a, f, k**) are the input axial, sagittal and coronal respectively, (**b, g, l**) are the colour coded outputs of the segmented clusters, (**c, h, m**) are the cluster one of each orientation similarly (**d, i, n**) and (**e, j, o**) are the 2nd and 3rd clusters of corresponding orientations

3.8   Watershed

Segmentation using watershed algorithm gives to regions which have values equal to zero or greater than zero. Each distinctive label (i.e., 1, 2, 3…) represents a region. Here the algorithm used 8-connected region growing method (Fig. 10).

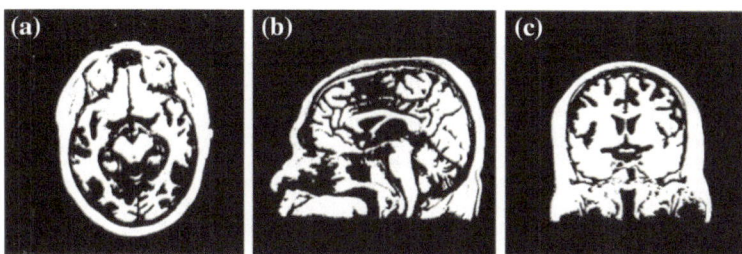

Fig. 10 Watershed segmentation **a** axial, **b** sagittal, **c** coronal

3.9   Manual Segmentation

In a MRI image the intensity value between the consecutive pixels are very low therefore employing automatic segmentation is very difficult. Automatic segmentation will fail if the region which need to be segmented does not have any unique properties, i.e. edge, texture or intensity value. Therefore manual segmentation is done on those parts which comes into those category which cannot be segmented buy any of the automatic segmentation algorithms. Figure 11 shows segmentation process and the steps performed is as given below:

1. Select two initial seed point from the image.
2. Connect the seed pints using a line drawing algorithm.
3. Select more points to make the selected region a closed region.
4. Copy the pints enclosed in the region.

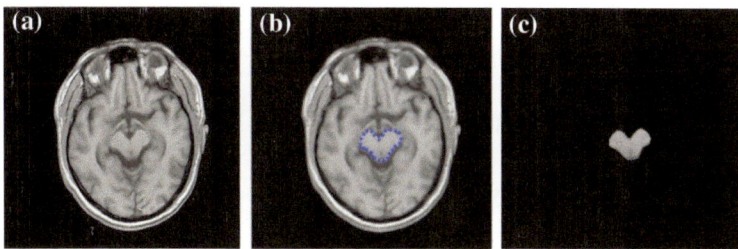

Fig. 11 Manual segmentation **a** input image, **b** seed point selection, **c** segmented region

3.10   Segmented Outputs

Figure 12 shows the segmented parts from the brain MRI for the creation of atlas. These outputs are obtained from the segmentation system which contains different algorithms.

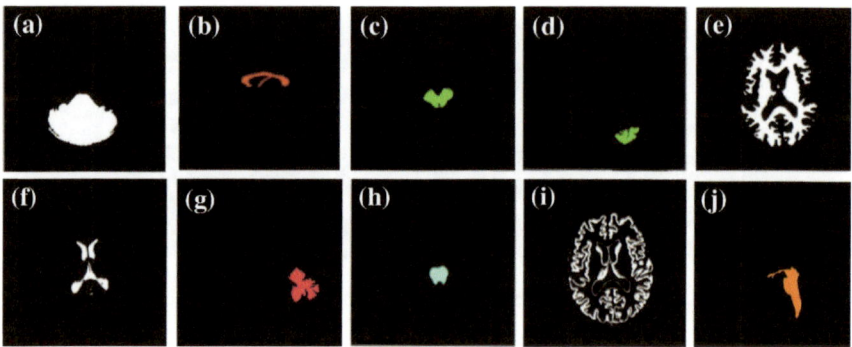

Fig. 12 a Cerebral hemispheres, **b** corpus callosum, **c** pons in axial slice, **d** occipital lobe, **e** white matter, **f** lateral ventricles and third ventricles, **g** cerebellum, **h** pons, **i** CSF, **j** brain stem

4   Accuracy Measurements

The accuracy of the methods used is computed by comparing the segmented parts with the ground truth available. For MRI images the ground truth is given by the doctors by manually marking the required regions. The measures that are used for accuracy check are given below.

Jaccard: $$Jaccard = \frac{(segmented\ image\ AND\ ground\ truth\ image)}{(segmented\ image\ OR\ ground\ truth\ image)}$$

Dice coefficient: $$DICE = \frac{2(segmented\ image\ AND\ ground\ truth\ image)}{segmented\ image + ground\ truth\ image}$$

Sensitivity: $$Recall = sensitivity = TPR = \frac{TP}{(TP+FN)}$$

Precision: $$Precision = PPV = \frac{TP}{(TP+FP)}$$

Score: $$Score = \frac{2*Precision*Recall}{Precision*Recall}$$

The analysis done on 2 of the parts is given below for exemplification.

4.1   Segmentation and Analysis of Brain Stem

See Fig. 13 and Table 1.

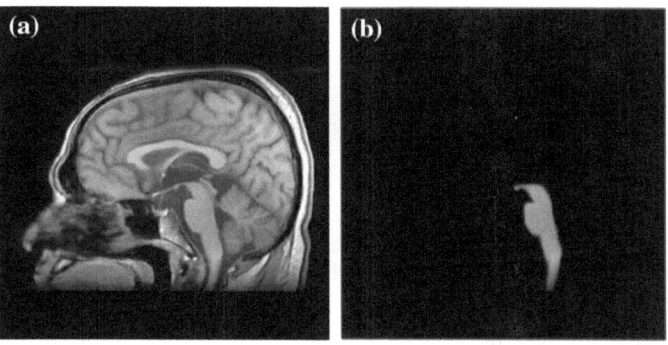

Fig. 13 **a** Input image, **b** ground truth image

Table 1 Comparison of accuracy measures for brain stem segmentation

Segmentation method	Segmented region	Accuracy measures				
		Dice	Recall	Precision	Score	Jaccard
Thresholding		0.0053	0.0182	0.0199	0.0190	0.1082
Multilevel thresholding		0.0219	0.0200	0.0220	0.0215	0.2981
Adaptive thresholding		0.0265	0.0098	0.0040	0.0057	0.4327
k-means		0.0206	0.0234	0.0249	0.0241	0.3000
Levelsets		0.0220	0.0521	0.0249	0.0337	0.2877

(*continued*)

Table 1 (*continued*)

Segmentation method	Segmented region	Accuracy measures				
		Dice	Recall	Precision	Score	Jaccard
Watershed	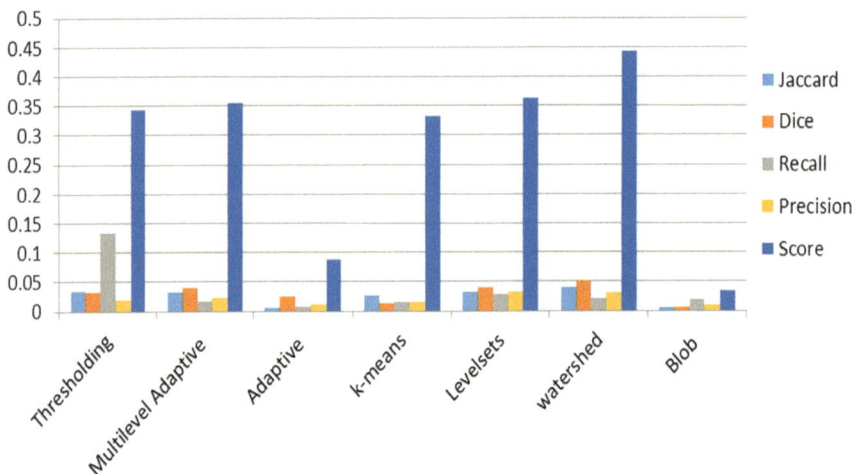	0.0193	0.0180	0.0199	0.0189	0.2862
Blob segmentation		0.0053	0.0138	0.0155	0.0164	0.0524

A comparison of the precision, recall and score is given in Fig. 14. On observing the values it can be inferred that the best method for brain stem segmentation from the considered MRI slice would be the use of thresholding.

Fig. 14 A comparison of the precision, recall, jaccard and score for brain stem segmentation

4.2   Segmentation and Analysis of Pons

See Fig. 15 and Table 2.

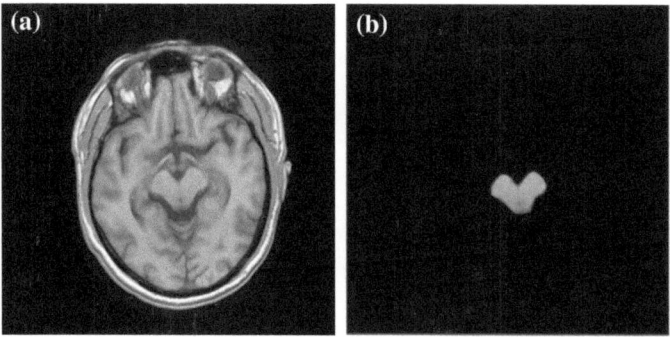

Fig. 15 **a** Input image, **b** ground truth image

Table 2 Comparison of accuracy measures for brain stem segmentation

Segmentation method	Segmented region	Accuracy measures				Jaccard
		Dice	Recall	Precision	Score	
Thresholding		0.0352	0.0330	0.133	0.0189	0.3440
Multilevel thresholding		0.0331	0.0400	0.0169	0.0238	0.3551
Adaptive thresholding		0.0053	0.0244	0.0083	0.0123	0.0871
k-means		0.0275	0.0138	0.0155	0.0146	0.3314
Level sets		0.0330	0.0408	0.0280	0.0332	0.3637

(continued)

Table 2 (*continued*)

Segmentation method	Segmented region	Accuracy measures				Jaccard
		Dice	Recall	Precision	Score	
Watershed	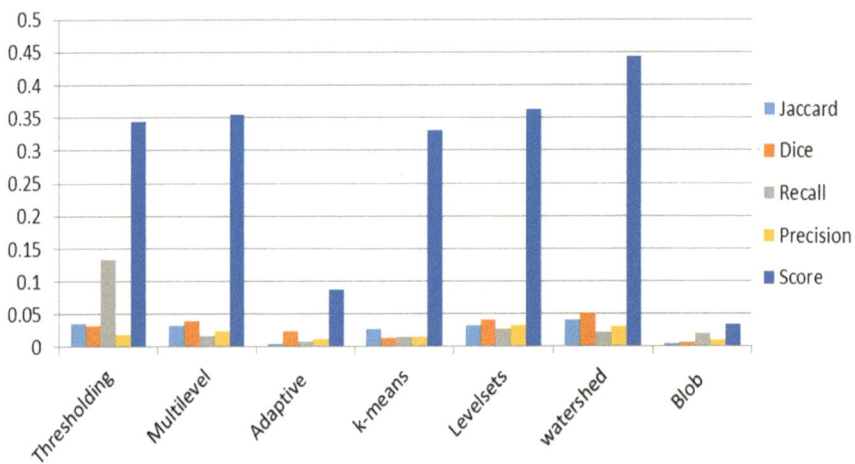	0.0409	0.0515	0.0218	0.0307	0.4432
Blob segmentation		0.0051	0.0068	0.0200	0.0102	0.0337

A comparison of the precision, recall, jaccard and score is given in Fig. 16. On observing the values it can be inferred that the best method for segmentation of pons from the considered MRI slice would be through thresholding.

For any method to be tagged as good, if the precision is high the recall has to be low, and hence a higher score. For a better algorithm the DICE value also has to be high compared to the other methods. On analysing these measures it has been inferred that for Brain Stem and Pons, segmentation can be got through simple thresholding itself. In similar way segmentation can be done for all the other parts. It was also found that there are certain parts like caudate nucleus which cannot be segmented by any of the segmentation algorithm therefore manual segmentation can be used in such parts.

Fig. 16 A comparison of the precision, recall, jaccard and score for pons segmentation

5   Atlas Creation

The sample screen shots of the Structural and Functional Brain Atlas is given in Figs. 17, 18, 19 and 20. Separate atlas has been created for Axial, Coronal and Sagittal views. The initial GUI for selecting the required view of MRI is shown in Fig. 17. The initial working starts with the selection of a slice from the required view of the MRI, which is given in GUI. Selecting a slice will transfer the control to the corresponding atlas with the selected slice number.

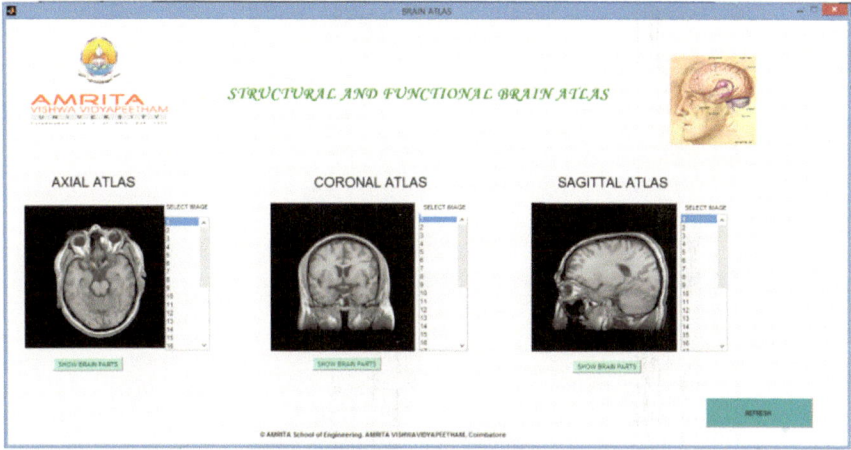

Fig. 17 Main atlas

Figure 18 shows the axial atlas in which we can select the required brain part of the corresponding slice number. By selecting the part name the corresponding segmented part will be displayed along with the functional details. Navigation across the slices of that particular view also has been provided. Similarly Figs. 19 and 20 shows the sagittal and coronal atlas.

The developed Brain Atlas was assessed by doctor at AIMS, Kochi for its appli-

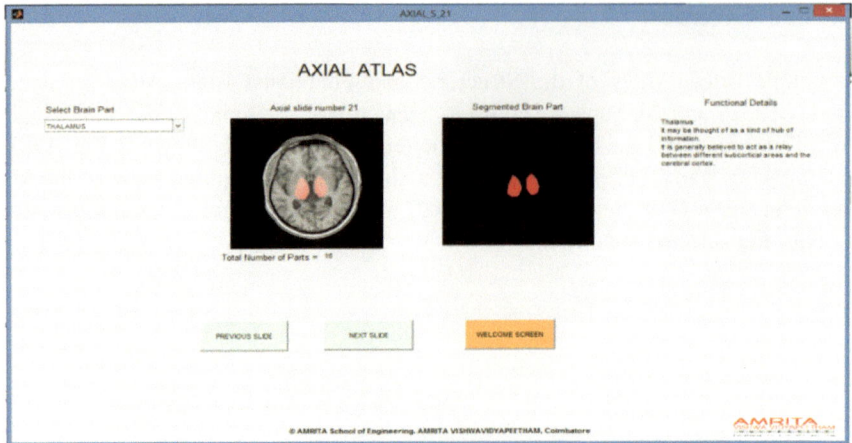

Fig. 18 Axial atlas

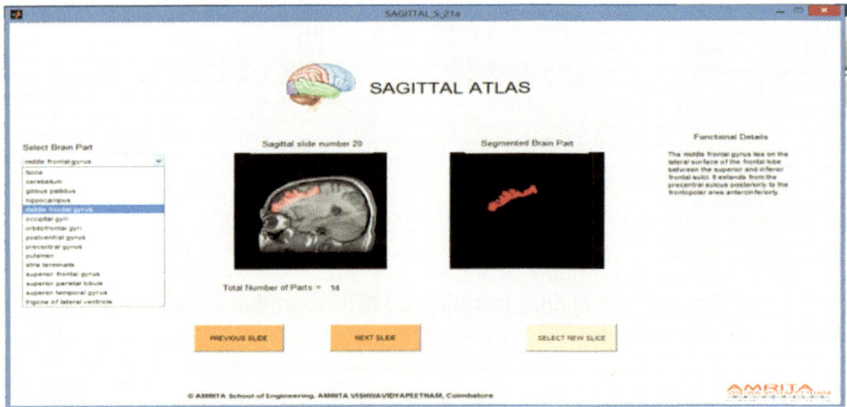

Fig. 19 Sagittal atlas

cability as a teaching and learning kit. The use of this could be further extended for analysing the effect of diseases on brain parts by mapping the patient MRI onto the Atlas for structural and functional details. This work provides a framework for creating a customized Brain Atlas as per the doctors requirements and are flexible enough for the user to make changes as and when required.

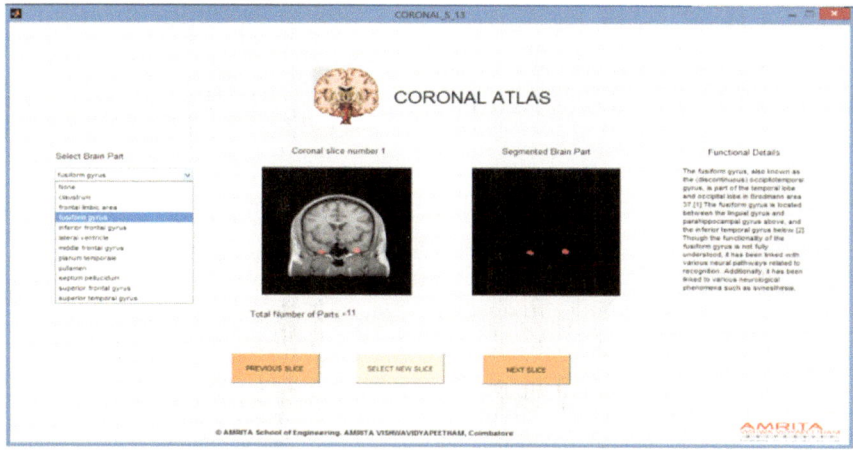

Fig. 20 Coronal atlas

6 Conclusion

In this work eight segmentation algorithms were implemented and tested for assessing their applicability on segmentation of brain MRI image slices, into its various homogenous regions. The work also creates a frame work for developing a structural and functional atlas customized to the user's requirements. Segmented parts are compared with the ground truth image and based on their accuracy measures, most appropriate segmented region is used for labelling and atlas creation.

Appendix (Certificate for Dataset and Work Authentication)

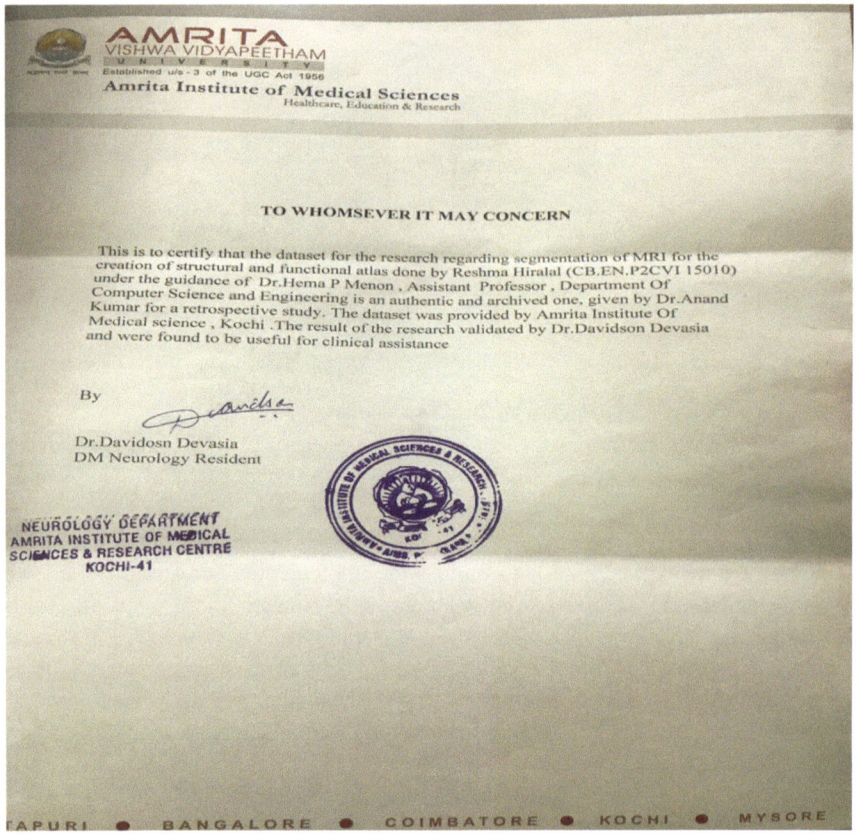

References

1. Lancaster, J.L., Rainey, L.H., Summerlin, J.L., Freitas, C.S., Fox, P.T., Evans, A.C., Toga, A.W., Mazziotta, J.C.: Automated labeling of the human brain: a preliminary report on the development and evaluation of a forward-transform method. Hum. Brain Map **5**, 238–242 (1997)
2. Jones, A.J., Overly, C.O., Sunkin, S.M.: INNOVATION: the allen brain atlas: 5 years and beyond. Nat. Rev. Neurosci. **10**, 821–828 (2009)
3. Govindan, V.K., Nair, J.J.: Automatic segmentation of MR brain images. In: Proceedings of the ICCCS International Conference on Communication, Computing and Security (ACM), NIT Rourkela (2011)
4. Renjini, H., Bhagavathi Sivakumar, P.: Comparison of automatic and interactive image segmentation methods. Int. J. Eng. Res. Technol. (IJERT) **2**(6), 3162–3170 (2013)

5. Amini, L., Zadeh, S.H., Lucas, C., Gity, M.: Automatic segmentation of thalamus from brain MRI integrating fuzzy clustering and dynamic contours. IEEE Trans. Biomed. Eng. **51**(5), 800–811 (2004)
6. Carmichael, O., Aizenstein, H., Davis, S., Becker, J., Thompson, P., Meltzer, C., Liu, Y.: Atlas-based hippocampus segmentation in Alzheimer' disease and mild cognitive impairment. NeuroImage **27**(4), 979–990 (2005)
7. Goshal, D., Acharjya, P.P.: MRI image segmentation using watershed transform. Int. J. Emerg. Technol. Adv. Eng. **2**(4), 373–376 (2012)
8. Kumaravel, M., Karthik, S.K.S., Sivraj, P., Soman, K.P.: Human face image segmentation using level set methodology. Int. J. Comput. Appl. **44**, 16–22 (2012)
9. Duy, N.H.M., Tuan, T.A., Duong, N.H.., Tuan, T.A., Dao, N.K., Yoshitaka, A., Kim, J.Y., Choi, S.H., Bao, P.T.: 3D-brain MRI segmentation based on improved level set by AI rules and medical knowledge combining 3 classes-EM and bayesian method. J. KIIT. **14**(5), 75–88 (2016)
10. Karim, B.M.: Atlas and snakes based segmentation of organs at risk in radiotherapy in head MRIs. In: Third IEEE International Conference in Information Science and Technology, pp. 356–363 (2014)
11. Rohlfing, T., Brandt, R., Menzel, R., Maurer, C.R.: Evaluation of atlas selection strategies for atlas-based image segmentation with application to confocal microscopy images of bee brains. NeuroImage **21**(4), 1428–1442 (2004)
12. Lancaster, J.L., Woldorff, M.G., Parsons, L.M., Liotti, M., Freitas, C.S., Rainey, L., Kochunov, P.V., Nickerson, D., Mikiten, S.A., Fox, P.T.: Automated Talairach Atlas labels for functional brain mapping. Hum. Brain Mapp. **10**, 120–131 (2000)
13. Taha, A.A., Hanbury, A.: Metrics for evaluating 3D medical image segmentation: analysis, selection, and tool. In: Taha and Hanbury BMC Medical Imaging (2015)
14. Polak, M., Zhang, H., Pi, M.: An evaluation metric for image segmentation of multiple objects. In: Image and Vision Computing (2009)
15. Hiralal, R., Menon, H.P.: A survey of brain MRI image segmentation methods and the issues involved. In: Corchado Rodriguez, J.M. et al. (eds.) Intelligent Systems Technologies and Applications 2016. Advances in Intelligent Systems and Computing, vol. 530, pp. 245–259 (2016)

Noninvasive Assistive Method to Diagnose Arterial Disease-Takayasu's Arteritis

Suganthi Lakshmanan[1(✉)], Dipanjan Chatterjee[2(✉)],
and Manivannan Muniyandi[2(✉)]

[1] Department of Biomedical Engineering, SSN College of Engineering,
Chennai, India
suganthi.lakshmanan@gmail.com
[2] Department of Applied Mechanics, Indian Institute of Technology Madras,
Chennai, India
dipanjan.c5@gmail.com, mani@iitm.ac.in

Abstract. Takayasu's arteritis (TA) is a rarely studied primary systemic vasculitis involving the aorta and other major arteries of the body. This work proposes a novel noninvasive assessment method, combining both time domain and frequency domain analysis of peripheral signals such as photo plethysmography (PPG) in normal and TA patients providing information about the severity of the TA disease. The novelty of the proposed method is twofold: one, a novel signal processing technique Auto-correlated Spectrum for analyzing PPG signals, and two, use of noninvasive techniques from multiple-site PPG for quantifying the severity. PPG from twenty TA patients and twenty normal subjects have been acquired from five different peripheral sites in the body and compared. The Auto-correlated Spectrums of multiple-site PPG signals are calculated. A novel parameter called P-measure is derived using the relation between the number of peaks and the average distance of the peaks from origin of the spectrum. P-measure is used for classifying normal and diseased using a binary classification method, when greater than or equal to 0.32 the subject is considered as normal and otherwise diseased. The sensitivity and specificity values of this classification method are 96 and 83% respectively. This method is also compared with other frequency domain analysis and this technique can be a simple cost-effective assessment tool to reduce cardiovascular morbidity and mortality in the rarely studied TA, and perhaps other arterial diseases. The small group of TA population due to the rarity of TA disease is a major problem in acquiring data.

Keywords: Power spectrum density (PSD) · Fast fourier transform (FFT)
Autocorrelation · Takayasu's arteritis (TA) · Photoplethysmography (PPG)

1 Introduction

Aorto-arteritis or Takayasu's arteritis (TA) is a rare, chronic inflammatory disease involving the aorta and other major arteries of the body. TA leads to vessel wall thickening, fibrosis, stenosis, and aneurysm. In approximately ninety percentages of

© Springer International Publishing AG 2018
D. J. Hemanth and S. Smys (eds.), *Computational Vision and Bio Inspired Computing*,
Lecture Notes in Computational Vision and Biomechanics 28,
https://doi.org/10.1007/978-3-319-71767-8_32

cases, TA disease appears in subjects younger than 30 years with a female: male ratio of 8:1, specifically among Asian. The TA is mostly diagnosed only after 80% of severity. The diagnosis of Takayasu's arteritis is confirmed by a radiographic procedure such as an angiogram or a magnetic resonance imaging study. Angiogram is the gold standard procedure for diagnosing TA [1–4].

In this paper, photo plethysmography (PPG) is proposed to be an assessment tool for the diagnosis of TA. PPG is a noninvasive technique to study the blood volume changes in the microvascular bed [5]. The motivation for using PPG as the source of biomedical signals comes from its unique advantages of being noninvasive, convenient and real-time availability [6–8]. Though the heart is the key source of the PPG signal, from a physiological perspective, blood has to travel a long distance along the artery before reaching the microvascular bed. Therefore, PPG signal not only has an influence on the function of the heart, but also can be affected by other factors including arterial vessel characteristics and blood parameters. The rhythm of any one of the internal organs, including the heart, lungs, liver, kidneys, intestines and stomach may impose direct or indirect influence on the PPG and induce changes in it [9]. PPG signal consists of a pulsatile AC waveform which corresponds to the arterial blood volume changes and in turn the cardiovascular functions, whereas the DC waveform corresponds to venous blood, the fixed quantity of arterial blood, and other stationary components like skin pigmentation.

For the analysis of beat to beat PPG amplitude variations, a wide variety of pulse wave analysis techniques have been used such as power spectral and correlation analysis [10–14]. Frequency domain analysis of PPG signals in healthy subjects and subjects with arterial disease has been well studied [10, 15, 16], but not used for TA analysis so far. Fast Fourier Transform (FFT) of PPG signals have been used to distinguish the healthy and subjects with vascular disease such as arteriosclerosis [15]. Frequency analysis shows that the higher harmonic frequencies reduced with age and consistent with the loss of the dicrotic notch feature in older subjects [16]. The existing FFT based analysis compares only the amplitude variations of peaks which represent fundamental frequency and its harmonics in control and subjects with arterial disease, the locations of these peaks are not analyzed. Other peaks introduced due to any change in circulatory system property, and in turn the location of these peaks, may be left out in such analysis.

Autocorrelation function is a time domain tool for finding changes in the circulatory system property as repeating patterns, such as the presence of a periodic signal which has been buried under noise, or for identifying the missing fundamental frequency. Autocorrelation of PPG signal has been used to measure differences in the signals due to the existence of the cycle-to-cycle variations [17]. The study referred in [18] compares only the shape difference between the cross correlated PPG signals of the control and subject with arterial diseases, and the frequency details are left out. Generally the Fourier transform of the auto correlated signal gives power spectral density (PSD) of the signal [19]. Autocorrelation function and the PSD can be combined together to explore the cycle to cycle variations and the repeating low frequency patterns present in the PPG signal of TA subjects for finding the severity of the disease.

Simultaneous measurements of PPG signals, from many anatomical locations on the body, can be used for detecting right–left body characteristics differences, especially if there are changes in cardiovascular parameters. Multi-site PPG measurements have been proposed for the detection of peripheral artery occlusive disease using time domain analysis [20–22]. The potential for using bilateral (right and left side) differences to detect lower limb stenosis has also been described using the time domain parameter, pulse wave velocity [23, 24]. Studies based on frequency analysis of PPG signal consider only the single site signal acquisition in control and subject with arterial disease [15, 16].

In this paper, an assessment technique for diagnosing TA using multi-site PPG signal analysis is described. The auto correlated power spectral density (APSD) spectrum is calculated and those plots of the healthy and TA subjects show clear difference with respect to the number of peaks and the location of the peaks in the spectrum. A new parameter (P-measure) has been proposed to classify the PPG signal of control and TA subjects. The proposed method has been compared with three other simple methods which analyze RR intervals of ECG signal in control and TA subjects.

The novelty in this work is, the use of multi-body site PPG signal of and TA subjects by combining both time domain and frequency domain signal processing methods such as correlogram (time domain) and PSD (frequency domain). A new parameter (P-measure) is introduced for accessing the severity of disease TA. The P-measure considers both the number of significant peaks and locations of the peaks in the autocorrelated spectrum. The proposed technique has been used for the analysis of multi-bilateral site PPG signals in the rarely studied disease TA to reliably locate disease site in the arterial tree.

2 Materials and Methods

This section details the signal acquisition method and signal processing method of PPG signal from control and TA subjects.

2.1 Acquisition and Processing of PPG Signals from TA Subject

Participants who are affected by Takayasu's arteritis with different disease severity level as well as affected in different parts of the body are chosen for this study. PPG signals are acquired from five different sites of the body, instead of single site, in order to study as comprehensive as possible. PPG signals are acquired from eighteen female and two male subjects (mean, standard deviation of age 33.50 ± 11.17 yrs, BMI 22.81 ± 3.98) with disease duration between 2 months and over 20 year. For validating the results of our proposed method disease duration and artery involvement were obtained from case notes. As a part of their routine clinical management, the cardiovascular analysis of the subjects was performed. From the participants written informed consent was obtained. According to the guidelines set down by the Christian medical college Hospital Ethics Committee and the study protocol confirms to the declaration of Helsinki as reflected in a priori approval by the Institution's Human Research and Ethic Committee. Eighteen healthy female and 2 male subjects (mean, standard deviation of age 29.6 ± 10.89 yrs., BMI 22.6 ± 3.66), are selected as control group without known cardiovascular diseases and any medication. Before the experiment, height and weight of the control subjects are noted.

PPG sensors are used to obtain the PPG signal from five different site of the participant. Four clip type and one band type PPG sensors are used to acquire the PPG signal from thumb fingers (left and right hand), toes (left and right leg), and neck (left carotid) and the amplification ratio of the signal acquisition is kept constant. The subjects are instructed to be in supine position. Through custom made data acquisition (CMC-DAC) card with eight channels, PPG and ECG signals are acquired at a sampling rate of 500 Hz per channel. The PPG sensor gain of 2000 and ECG sensor gain of 1000. Ambient temperature is maintained at $24 \pm 0.7\,°C$ to reduce the effect of temperature.

For the pre-processing of the PPG signals acquired from participants, Biopac Student Lab Pro was used. Analog high-pass filter with a cut-off frequency of 0.15 Hz was used to remove the DC component of the PPG signal and the AC component of the PPG signal is extracted using a low-pass filter with a cutoff frequency of 20 Hz. Sixty to hundred consecutive cardiac cycles are manually selected from the preprocessed data in order to decrease the influence of the respiratory system and other movement artifacts.

2.2   Description of the Proposed Method

In this section, we describe the proposed method for the robust classification of control and diseased subjects.

1. PPG signals of 60 s samples are auto correlated and the resultant samples are further processed as per the following steps. Autocorrelation (R_{xx}) of a real signal $x(m)$, is defined as in Eq. (1) which is existing only between 0 and M.

$$R_{xx}(n) = \sum_{m=0}^{M} x(m)x(m-n); \quad -M \le n \le M \tag{1}$$

   where M is number of samples in the real signal x, m and n are index term in time domain.

2. The auto correlated samples $R_{xx}(n)$ are partitioned into K overlapping sequences $R_{xxj}(n)$ with length L. The successive sequences are offset by D points and each sequence is L points long than jth sequence is given by Eq. (2).

$$R_{xxj}(n) = R_{xx}(n+jD); \quad n = 0, 1, \dots L-1; \ j = 0, 1, \dots K-1 \tag{2}$$

   In this analysis the auto correlated samples are divided into eight equal sections with 50% overlap (K = 8; D = L/2).

3. Each section is windowed with a Hamming window which defined as in Eq. (3) [25–27].

$$W(n) = 0.54 - 0.46\left(\frac{2\pi n}{L-1}\right); \quad 0 \le n \le L-1 \tag{3}$$

4. Auto correlated spectrum is defined as in Eq. (4). Discrete Fourier transform (DFT) is applied to the windowed data.

$$Sj(k) = \sum_{n=0}^{L-1} W(n)R_{xxj}(n+jD)e^{-i2\pi nk/L}; \quad k = 0, 1, \dots L-1; \; j = 0, 1, \dots K-1$$

(4)

where W(n) is Hamming window function, $S_j(k)$ is Fourier transform of windowed autocorrelated signal $(R_{xxi}(n))$, k is index term in frequency domain.

5. The (modified) periodogram of each windowed segment is computed which defined as in Eq. (5).

$$S_{Mj}(k) = \frac{1}{LU}\left|S_j(k)\right|^2; U = \frac{1}{L}\sum_{n=0}^{L-1}|W(n)|^2; \quad k = 0, 1, \dots L-1; \; j = 0, 1, \dots K-1$$

(5)

6. The K modified periodograms are averaged to form the spectrum estimate as in Eq. (6).

$$\hat{s}_{av}(k) = \frac{1}{k}\sum_{j=0}^{K-1} S_{Mj}(k); \quad k = 0, 1, \dots L-1$$

(6)

7. The resulting spectrum estimate is scaled to compute the power spectral density as $\hat{S}_{av}/F_s$. F_s is sampling frequency (500 Hz).
8. In the resulting two sided spectrum, the peaks are normalized with respect to maximum peak. The differences are observed in terms of number of peaks (many more for control) and locations of these peaks (farther for TA subjects) in the spectrum.
9. A new parameter called P-measure has been derived as an analytical expression based on these two parameters: number of peaks, locations of these peaks. The P-measure value is defined in Eq. (7) based on which control and TA subjects can be classified.

$$P - measure = 2\log_{10}(NP) - MD$$

(7)

where NP is number of significant peaks and MD is mean distance of the peaks from origin which is unit less. In order to find unit-less MD, mean distance is calculated in Hz and then it is divided by unit Hz. Even though the auto correlated spectrum is symmetric around 0 Hz, doubling the number of peaks and MD, by considering negative frequency axis, emphasizes the data points in the P-measure plot and hence the classification becomes accurate. The peaks with height greater than or equal to 5% of the max peak height are selected as significant peaks.

2.3   Spectral and Nonlinear Analysis of R-R Intervals of ECG Signal

In order to explore the variability in R-R interval of TA subject's ECG signal, frequency domain and nonlinear analysis of RR intervals have been performed. R-R intervals are extracted from the 1 min ECG data of control and TA subjects [28]. Kubios HRV (version 2.0) has been used for performing frequency domain and nonlinear analysis of RR intervals. The RR interval series is converted into equidistantly sampled series by interpolation (cubic spline interpolation) methods.

In the frequency-domain methods, the PSD estimation is carried out using FFT spectrum analysis and parametric Autoregressive (AR) spectrum analysis. FFT spectrum is calculated based on Welch's periodogram method with 256 window size and 50% overlap. In AR spectrum method, no factorization is used and the order of the system is 16. For both spectral analysis methods, Power ratio which is the ratio of low frequency power (LF) to high frequency (HF) power is calculated for all the controls.

A commonly used nonlinear method to find correlation between successive RR intervals, (plot of RR_{j+1} as a function of RR_j) is Poincaré plot [29]. The parameters SD1 and SD2 are derived from Poincaré plots and the ratio of these two parameters (SD1/SD2) is calculated for all the controls.

2.4   Sensitivity and Specificity Analysis

The statistical measures of the classifiers such as sensitivity (precision or true positive rate), specificity (recall), positive predictive value, negative predictive value, false positive rate (1—sensitivity), F-measure and area under curve (accuracy) are calculated based on which the performance of classification methods are compared [30].

3   Results

A rudimentary naked eye analysis revealed that a set of arbitrary shapes and patterns repeated for every 3 to 15 cycles in the PPG signal of TA subjects (Fig. 1b). However PPG of control subjects shows the cycles or oscillations repeated with almost every cycle (Fig. 1a). Figure 2a, b show the power spectrum of a control and TA subject. Under stationary conditions, the heartbeat frequency in humans is around 1 Hz, ranging from 0.6 Hz in sportsmen to 1.6 Hz in subjects with an impaired cardiovascular system and the smaller peaks round 0.15–0.4 Hz are from gastro-intestinal and respiratory processes which can be observed in Fig. 2a. FFT of a TA subject shows that induction of some low and high frequency component along with the heart beat harmonic components which are observed in Fig. 2b.

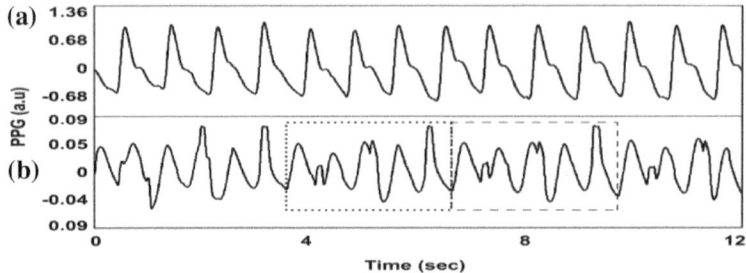

Fig. 1 Photoplethysmography (PPG). **a** Control subject. **b** TA subject. Dotted and dashed box indicates the repetition of pattern

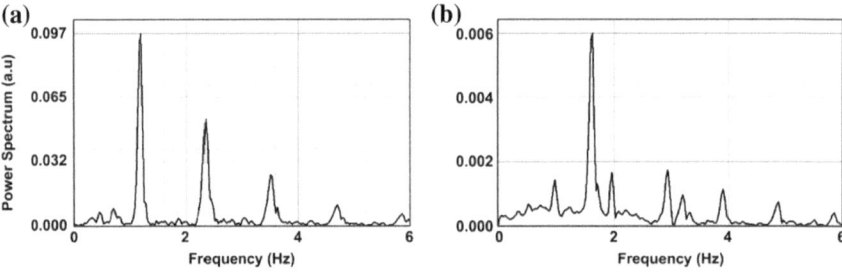

Fig. 2 Power spectrum of a PPG signal. **a** Control subject. **b** TA subject

The autocorrelation of the PPG signals which are obtained from the five peripheral sites of healthy subjects are similar to each other, less cluttered and evenly distributed along the horizontal axis (Fig. 3). Modified spectrum does not suppress the lower frequencies since they are relatively stronger and contain more energy, and this indicates the stronger correlation among the PPG cycles which continues for many cycles. In TA subject, the autocorrelations of PPG signals in limbs (both hands and toes) are not similar to each other (Fig. 4). The autocorrelation of PPG signals from the severely affected limbs of TA subjects indicates a behavior close to randomness and hence absence of any kind of repeating patterns to a large extent. The autocorrelation of TA subject signal (with less-severity), looks like an amplitude modulated waveform with a high frequency (generally sinusoidal) signal modulated by a low frequency signal.

ROC analysis provides important information about diagnostic test performance: the closer the apex of the curve toward the upper left corner, the greater the discriminatory ability of the test (the true-positive rate is high and the false-positive rate is low). This is measured quantitatively by the area under the curve (AUC), and AUC is greater than 0.9 for P-measure method (0.94) indicates excellent discriminatory ability (Fig. 5). AUC of FFT spectrum method, AR spectrum method and nonlinear analysis are 0.67, 0.68 and 0.68 respectively which indicate the poor discriminatory ability of

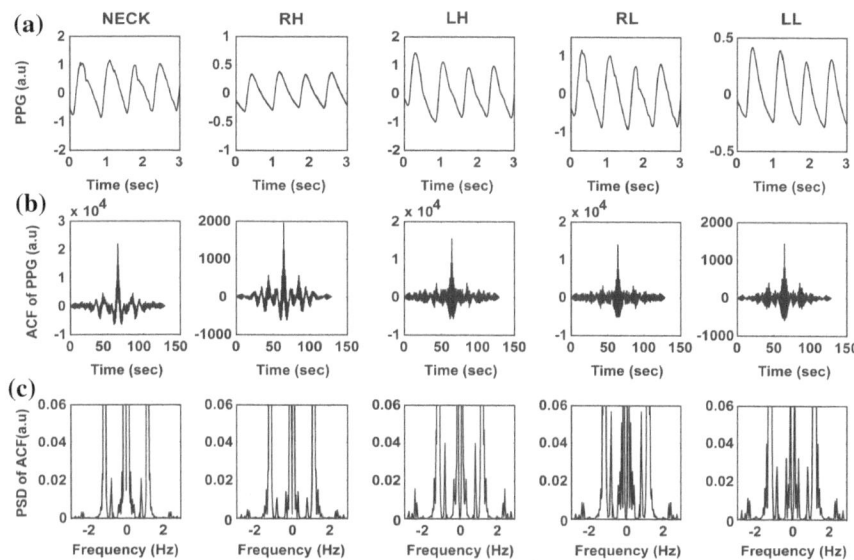

Fig. 3 Processing of PPG signals acquired from the neck, right hand, left hand, right leg and left leg for a control subject. **a** The original PPG signal. **b** The auto-correlation of PPG signal. **c** Power spectral density of the auto correlated PPG

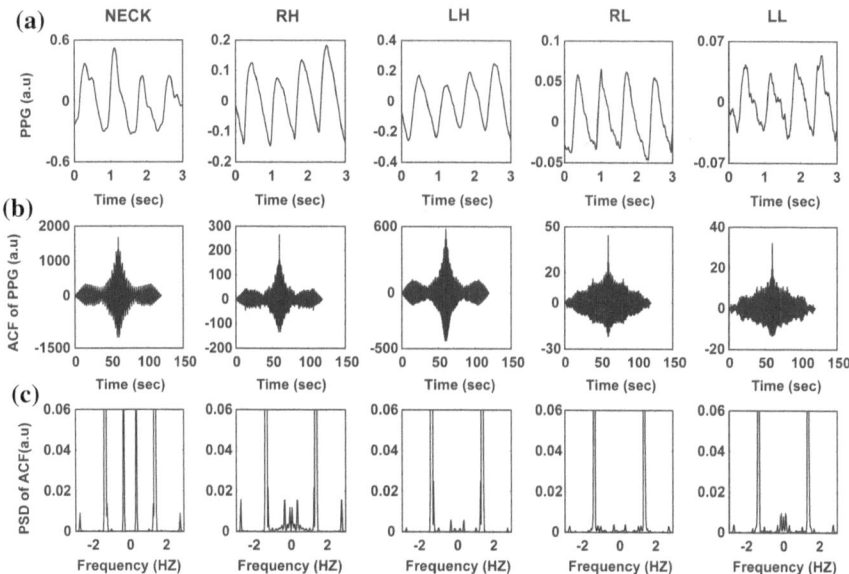

Fig. 4 Processing of PPG signals acquired from the neck, right hand, left hand, right leg and left leg for a TA subject. **a** The original PPG signal. **b** The auto-correlation of PPG signal. **c** Power spectral density of the auto correlated PPG

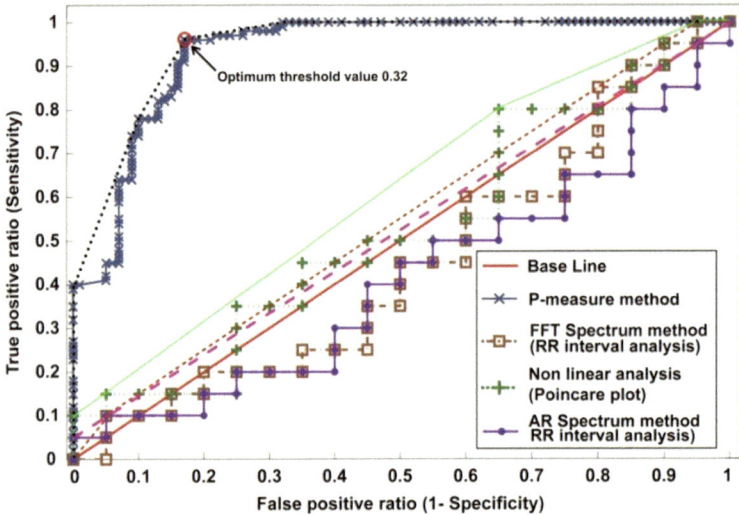

Fig. 5 Receiver operating characteristic ROC curve and convex hull of various classification methods

those classifiers. From the ROC curve, the optimum value of threshold with maximum sensitivity and selectivity has been found.

The P-measure value calculated for control and TA subject groups in five different locations are plotted in Fig. 6 and then a simple binary classification has been carried out using the plot. P-measure value greater than or equal to 0.32 (optimum threshold found from ROC) refers control subject, less than 0.32 refers TA subject and corresponding performance measures are calculated which is shown in Table 1 and it indicates the P-measure based classification method performs better than other three methods.

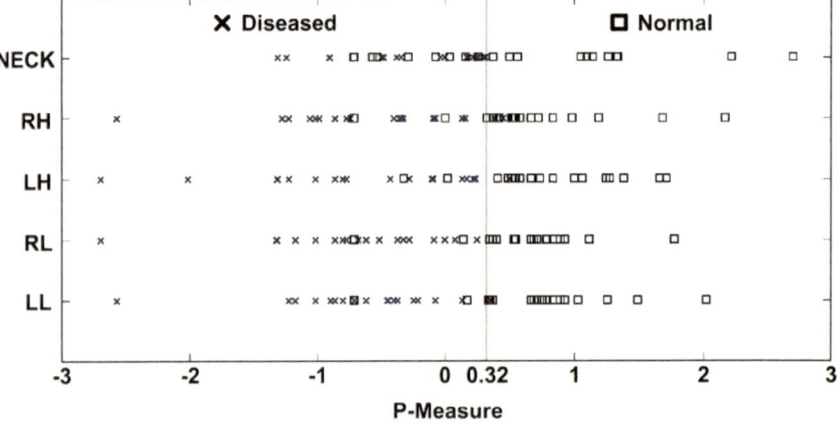

Fig. 6 Binary classification plot of P-measure value

Table 1 Performance measurements of different classifiers

Method	Site	Threshold	TP	FP	TN	FN	Sensitivity (precision) (%)	Specificity (recall) (%)	Positive predictive value (%)	Negative predictive value (%)	F-measure (%)	Accuracy (%)
P-measure method with one threshold		0.32	96	17	83	4	96	83	85	95	89	90
	Neck		20	9	11	0	100	55	69	100	71	78
	RH		18	2	18	2	90	90	90	90	90	90
	LH		19	2	18	1	95	90	90	95	92	93
	RL		20	2	18	0	100	90	91	100	95	95
	LL		19	2	18	1	95	90	90	95	92	93
P-measure method with site threshold	Neck	−0.71	20	0	20	0	100	100	100	100	100	100
	RH	0.001	19	1	19	1	95	95	95	95	95	95
	LH	0.41	20	2	18	0	100	90	91	100	95	95
	RL	0.14	20	1	19	0	100	95	95	100	97	98
	LL	−0.71	20	0	20	0	100	100	100	100	100	100
P-measure method based on angiogram	Neck [28] + (12)	−0.72	12	3	25	1	100	89	80	100	94	93
	RH [33] + (7)	−1.23	6	1	32	1	86	96	86	97	91	95
	LH [29] + (11)	−1.23	10	2	27	1	91	93	83	96	92	93
	RL[30] + (10)	−1.02	9	3	27	1	90	90	75	96	90	90
	LL [30] + (10)	−1.02	9	2	28	1	90	93	82	97	91	93

(continued)

Table 1 (*continued*)

Method	Site	Threshold	TP	FP	TN	FN	Sensitivity (precision) (%)	Specificity (recall) (%)	Positive predictive value (%)	Negative predictive value (%)	F-measure (%)	Accuracy (%)
FFT spectrum method (RR interval analysis)	ECG-RR interval	0.67	11	10	10	9	45	50	52	53	47	48
AR spectrum method (RR interval analysis)	ECG-RR interval	0.68	11	10	10	9	45	50	52	53	47	48
Nonlinear analysis (poincare plot)	ECG-RR interval	0.68	10	10	10	10	50	50	50	50	50	50

The performance measurements of different methods corresponding to the optimum threshold value are shown in Table 1. To check the classifier performance in each location separately, instead of considering single threshold value (0.32), we have found the optimum threshold value in each location separately to classify the diseased and control. If the value of a parameter is greater than its optimum threshold value, then it is considered as control else it is considered as TA participants. Performance measures for all parameters in five locations are given in Table 1. Number of subjects/limbs show normal patterns is indicated inside the square parenthesis.

4 Discussion

One of the reasons to acquire multisite PPG measurements is to identify the bilateral dissimilarity in peripheral pulses in TA subjects [20–22]. In control subjects, the right and left sides of the body are expected to have the similarity in pulse timing, amplitude and shape of the pulses since the anatomical structures are similar [31] and the peripheral vascular beds are innervated symmetrically by the autonomic nervous system [32]. Detection of TA disease with PPG is also possible because TA induces abrupt changes in geometry and elastic properties of arterial tree and hence the bilateral dissimilarity in peripheral pulses is induced. These findings were also observed in peripheral artery occlusive disease [21]. The changes in peripheral signals have significant diagnostic value and it can be used for the quantitative assessment of the severity of the TA disease.

The autocorrelation of PPG signals from the severely affected limbs of TA subjects indicates a behavior close to randomness and hence absence of any kind of repeating patterns to a large extent. The high frequency component of amplitude modulated wave form (Fig. 4) in the autocorrelation of TA subject signal (with less-severity) is the fundamental frequency representing the sampling frequency of the signal, and hence is carried forward through every autocorrelation operation. The amplitudes and the energy of this higher fundamental frequency signal are quite large as compared to the lower frequency signals. These small oscillations are the result of a lower extent of cycle-to-cycle to matching that exists in the signal; they are stronger than the small oscillations present in severely affected TA subject signal, but definitely weaker than the small oscillations present in signals from healthy subjects.

The number of significant peaks present in auto correlated spectrum requires a minimum height (5%) for peak detection, as a real-time signal contains lot of noise and artifacts which shows as low amplitude peaks in the signal. Also the spectrum is to be normalized first, a minimum threshold is necessary. Although NP and MD in auto correlated spectrum without any amplifying factors could be used to distinguish between control and TA subjects, suitable amplifying factor to each of these parameters improves the sensitivity. The log function in P-measure is used to amplify the smaller values than the larger values of these two parameters. The idea of P-measure mainly originated from the observation that certain features of the PPG wave form repeat every few cycles, and this low frequency repetitions are captured and amplified in P-measure.

From auto correlated spectrum in control subjects, it is absorbed that the minimum and maximum is 1.4 and minimum MD in TA subjects is 1.3. Using these values, the minimum expected P-measure value for control and maximum expected P-measure for TA subjects are calculated. $P_{min,control} = 2 \log_{10}(8) - 1.4 = 0.406$ and $P_{max,TA \ subject} = 2\log_{10}(6) - 1.3 = 0.256$. Therefore chosen threshold for P-measure is average of $P_{min,control}$ and $P_{max,TA \ subject}$ $[(0.406 + 0.256)/2 \sim = 0.32]$. This threshold value lies on the ROC curve with optimum sensitivity and selectivity value and hence it can be used for classification to distinguish control and TA subject. P-measure value greater than or equal to 0.32 refers control subject less than 0.32 refers TA subject.

5 Conclusion

In this work, a detailed analysis of PPG proves that PPG offers much information about cardiovascular system and also it can be used in noninvasive circulatory monitoring of TA. A new parameter (P-measure) is introduced by combining both time domain and frequency domain signal processing methods such as autocorrelation (time domain) and PSD (frequency domain) for accessing the severity of disease TA. The proposed technique has been used for the analysis of multi-bilateral site PPG signals in the rarely studied disease TA to reliably locate disease site in the arterial tree. The sensitivity and specificity values of the P-measure based classification of control and TA subjects with single threshold (0.32) are 96 and 83% respectively. Instead of considering single threshold value to classify the control and TA subjects in this P-measure method, with different optimal threshold for each site, we have cross verified the performance of this method based on angiogram report. It shows improved performance of the classification (sensitivity 86 to 100, Specificity 89 to 96). The small group of TA population due to the rarity of TA disease is a major problem in acquiring data. This is a challenge in validating the technique for early diagnosis, as the subjects are not diagnosed until after the 80% severity. Small population, the motion artifacts due to the inherent PPG measurements and absence of cross validation with other time and frequency domain analysis of PPG signals are the limitation of this study.

Acknowledgements. We thank Dr. George Joseph of Department of Cardiology, Dr. Debashish Danda of Department of Clinical Immunology and Rheumatology, Nisan kunju and Dr. Suresh Devasahayam of Department of Bioengineering, Christian Medical College, Vellore, India. No conflict of interest has been declared by the authors.

References

1. Subramanyan, R., Joy, J., Balakrishnan, K.G.: Natural history of aortoarteritis (Takayasu9s disease). Circulation **80**(3), 429–437 (1989)
2. Ishikawa, K., Maetani, S.: Long-term outcome for 120 Japanese patients with Takayasu9s disease. Clinical and statistical analyses of related prognostic factors. Circulation **90**(4), 1855–1860 (1994)
3. Schmidt, W.A., Nerenheim, A., Seipelt, E., Poehls, C., Gromnica-Ihle, E.: Diagnosis of early Takayasu arteritis with sonography. Rheumatology **41**(5), 496–502 (2002)
4. Rein, O.R., Markku, M.K.: Carotid and femoral artery stiffness in Takayasu's arteritis. Scand. J. Rheumatol. **31**(2), 85–88 (2002)

5. Hertzman, A.B.: Photoelectric plethysmography of the fingers and toes in man. Proc. Soc. Exp. Biol. Med. **37**(3), 529–534 (1937)
6. Allen, J.: Photoplethysmography and its application in clinical physiological measurement. Physiol. Meas. **28**(3), R1 (2007)
7. Eldrup-Jorgensen, S.V., Schwartz, S.I., Wallace, J.D.: A method for clinical evaluation of peripheral circulation: photoelectric hemodensitometry. Surgery **59**(4), 505–513 (1966)
8. Simonson, E.: Photoelectric plethysmography; methods, normal standards, and clinical application. Geriatrics **11**(10), 425 (1956)
9. Zheng, D., Allen, J., Murray, A.: Development of a method for determining arterial pulse propagation times and influence of arterial compliance. In: IEEE Computers in Cardiology, pp. 289–292 (2006)
10. Malvezzi, L., Castronuovo, J.J., Swayne, L.C., Cone, D., Trivino, J.Z.: The correlation between three methods of skin perfusion pressure measurement: radionuclide washout, laser doppler flow, and photoplethysmography. J. Vasc. Surg. **15**(5), 823–830 (1992)
11. Nitzan, M., Babchenko, A., Milston, A., Turivnenko, S., Khanokh, B., Mahler, Y.: Measurement of the variability of the skin blood volume using dynamic spectroscopy. Appl. Surf. Sci. **106**, 478–482 (1996)
12. Nitzan, M., Babchenko, A., Shemesh, D., Alberton, J.: Influence of thoracic sympathectomy on cardiac induced oscillations in tissue blood volume. Med. Biol. Eng. Comput. **39**(5), 579–583 (2001)
13. Nitzan, M., de Boer, H., Turivnenko, S., Babchenko, A., Sapoznikov, D.: Power spectrum analysis of spontaneous fluctuations in the photoplethysmographic signal. J. Basic Clin. Physiol. Pharmacol. **5**(3–4), 269–276 (1994)
14. Nitzan, M., Turivnenko, S., Milston, A., Babchenko, A., Mahler, Y.: Low-frequency variability in the blood volume and in the blood volume pulse measured by photoplethysmography. J. Biomed. Opt. **1**(2), 223–229 (1996)
15. Oliva, I., Ipser, J., Roztocĭl, K., Guttenbergerova, K.: Fourier analysis of the pulse wave in obliterating arteriosclerosis. VASA. Zeitschrift für Gefässkrankheiten **5**(2), 95 (1976)
16. Sherebrin, M.H., Sherebrin, R.Z.: Frequency analysis of the peripheral pulse wave detected in the finger with a photoplethysmograph. IEEE Trans. Biomed. Eng. **37**(3), 313–317 (1990)
17. Ingle, V.K., Proakis, J.G.: A Self-Study Guide for Digital Signal Processing, 3rd ed. Prentice Hall (2003)
18. Nitzan, M., Vatine, J.J., Babchenko, A., Khanokh, B., Tsenter, J., Stessman, J.: Simultaneous measurement of the photoplethysmographic signal variability in the right and left hands. Lasers Med. Sci. **13**(3), 189–195 (1998)
19. Proakis, J.G., Manolakis, D.G.: Digital Signal Processing (1996)
20. Allen, J.: Measurement and analysis of multi-site photoplethysmographic pulse waveforms in health and arterial disease. Doctoral Dissertation, University of Newcastle upon Tyne (2002)
21. Allen, J., Murray, A.: Similarity in bilateral photoplethysmographic peripheral pulse wave characteristics at the ears, thumbs and toes. Physiol. Meas. **21**(3), 369 (2000)
22. Allen, J., Murray, A.: Variability of photoplethysmography peripheral pulse measurements at the ears, thumbs and toes. IEE Proc. Sci. Meas. Technol. **147**(6), 403–407 (2000)
23. Erts, R., Spigulis, J., Kukulis, I., Ozols, M.: Bilateral photoplethysmography studies of the leg arterial stenosis. Physiol. Meas. **26**(5), 865 (2005)
24. Spigulis, J.: Optical noninvasive monitoring of skin blood pulsations. Appl. Opt. **44**(10), 1850–1857 (2005)
25. Hayes, M.H.: Statistical Digital Signal Processing and Modeling. Wiley, New York (1996)
26. Stoica, P., Moses, R.L.: Introduction to Spectral Analysis, vol. 1, pp. 3–4. Prentice hall, Upper Saddle River (1997)

27. Welch, P.: The use of fast Fourier transform for the estimation of power spectra: a method based on time averaging over short, modified periodograms. IEEE Trans. Audio Electroacoust. **15**(2), 70–73 (1967)
28. Salahuddin, L., Jeong, M.G., Kim, D.: Ultra short term analysis of heart rate variability using normal sinus rhythm and atrial fibrillation ECG data. In: IEEE 9th International Conference on e-Health Networking, Application and Services, pp. 240–243 (2007)
29. Brennan, M., Palaniswami, M., Kamen, P.: Do existing measures of poincare plot geometry reflect nonlinear features of heart rate variability? IEEE Trans. Biomed. Eng. **48**(11), 1342–1347 (2001)
30. Fletcher, R.H., Fletcher, S.W.: Clinical epidemiology: the essentials, 4th edn, pp. 156–199. Lippincott Williams & Wilkins, Baltimore (2005)
31. Anne, M.R.A., Arthur, F.D.: Grant's Atlas of Anatomy, 12th ed., pp 355–607. Lippincott Williams and Wilkins (2009)
32. Rowell, L.B.: Human cardiovascular Control. Oxford University Press, New York (1993)

An Automated Vision Based Change Detection Method for Planogram Compliance in Retail Stores

Muthugnanambika M.[1], Bagyammal T.[1], Latha Parameswaran[1], and Karthikeyan Vaiapury[2(✉)]

[1] Department of Computer Science and Engineering, Amrita School of
Engineering, Amrita Vishwa Vidyapeetham, Coimbatore, India
{cb.en.p2cvil5007, t_bagyammal, p_latha}@cb.amrita.edu
[2] TCS Innovation Labs, Chennai, India
karthikeyan.vaiapury@tcs.com

Abstract. Planogram are visual representations of a store's products and services designed to help retailers ensure that the right merchandise is consistently on display, and that inventory is controlled at a level that guarantees that the right number of products are on each and every shelf. The main objective of this work is to propose an algorithm using image processing and machine learning as its base to find and detect the changes in the arrangement of objects present in the retail stores. The proposed algorithm is capable of identifying void space, count objects of similar type and thus helps in tracking the changes. Blob detection superseded by classification using a discriminative machine learning approach with the extracted statistical features of the objects has been used in this proposed algorithm. Experimental results are quite promising and hence this algorithm can be used to detect any changes occurring in a scene.

Keywords: Blob detection · Multiclass discriminative machine learning approach · Change detection · Planogram compliance

1 Introduction

With advent of the revolutionized internet technology and cheap cameras Internet and cameras, more of digitization is happening in all commercial places. One major area where the camera and its technology named as computer vision can be much useful is machine vision. Automated change detection using cameras is more challenging due to factors such as illumination, geometric transformations while capturing the image, acquisition conditions etc. The biggest problem posed is "If a given scene can be read by a camera and interpreted as we human do, most businesses will improve". A planogram gives the visual representation of the products present in the retail stores [1]. While constructing a planogram, it is important to consider the placement of an item at each and every unique point of time. Without compliance, the effectiveness of the planogram cannot be accurately

D. J. Hemanth and S. Smys (eds.), *Computational Vision and Bio Inspired Computing*,
Lecture Notes in Computational Vision and Biomechanics 28,
https://doi.org/10.1007/978-3-319-71767-8_33

measured or traced out. To find the significant changes in an image, change detection plays a major role. Change detection finds its applications in remote sensing, medical diagnosis and treatment, underwater sensing, civil infrastructure, video surveillance etc.

Image of the same scene is shot at a different time interval to identify the change between the current and previous image. In departmental stores, the product arrangement in every rack will be stored as a catalog that could be the text or images. Planogram often evolves to be dynamic, since rearrangement happens each time when new products arrive or sales go down so as to attract the customers and maximize the sales. It is necessary to make sure that all the objects in the rack should satisfy the planogram and notify the position displacement and void space to increase the profit of the retailers. With this as the underlying problem, this proposed work is an attempt to read the items stacked on a shelf in a store, identify and track changes happening in order to improve sales and satisfy customer requirement.

2   Related Work

In the literature, it is observed that researchers have taken this problem of change detection in images and have proposed algorithms to identify the changes between a set of images. A few of them are discussed in this section. In [2], the authors have proposed an algorithm where the entire image in the stack was captured and the features extracted using Speeded Up Robust Features (SURF). Feature matching has been used to check compliance of placement of items. The main limitation of this algorithm is in estimating the void space. In [3–5] different types of change detection techniques have been compared based on the accuracy measure. The various techniques include Image Ratioing (IR), Image Differencing (ID), Principal Component Analysis (PCA), Change Vector Analysis (CVA), and Kauth-Thomas transformation or Tasseled Cap Transformation (KT). It has been reported that out of all these, CVA [3], gives the best accuracy value of 84.29% for detecting the changes. In [6], Support Vector Machine (SVM) was used for change detection in satellite images like vegetation area, urban area and bare land where images belonging to the period 1986 to 1990 were considered for discussion. In [7], Geospatial change detection and exploitation (GeoCDX) extracts the high-level features of the satellite images and identifies the changes. In [7, 8] Scale Invariant Feature Transform (SIFT) was used to extract all the features. For detecting the edges, Canny Edge Detection (CED) has been used and then it detects the blobs and allocates each blob as a key point. Laplacian of Gaussian (LOG) [9], Difference of Gaussian (DOG), has been discussed [10]. In [11], Multiclass Support Vector Machine (SVM) gives an outstanding performance for prediction and classification [12]. In [13], when the input is given as query shape the feature vector will be calculated and it will be compared with the bag of features using distance measures and the matching bins are formed respectively. The threshold value will be fixed and it will be compared with every distance value in the hash table. If the distance is greater than the threshold then the image will be discarded. Once it matches it will be taken to the next level where the histogram is found for every matched query. So dynamic programming similarity will be calculated for every query and it will be ranked accordingly. Finally based on ranks the query will be reordered which gives the top matches.

It can be observed from the literature survey, that still detecting changes in a given image is a difficult and challenging problem. It also poses additional difficulty to compute the area of void space. In this article, we have proposed an algorithm that meets the following major objectives: (i) identify items in a shelf with reference to object library (ii) identify and estimate the size of void space (iii) count the number of items of the same type and (iv) identify the changes between a pair of images taken at two different time.

3   Proposed Method

The proposed method aims to meet the objectives mentioned in Sect. 2. Object recognition is done by using the multiclass discriminative classifier. The statistical features used to train the classifier include area, mean intensity and diameter of the objects. Change detection is done by comparing the placement pattern of the objects stored in the database with the placement pattern of the objects in the current image.

3.1   Object Recognition

Object recognition includes training the discriminative classifier using the statistical features and then performing the testing.

3.1.1   Training Phase

In this phase, images in the database with its corresponding labels are used to train the classifier in order to identify the items present in a shelf with reference to the object library. Given color image is converted to a grayscale image and by applying threshold it has been converted to a binary image. Blob (Binary Large Object) detection is done which refers to a group of connected pixels in a binary image. Steps for training the classifier are depicted in Fig. 1. Statistical features extracted from the blobs are explained below.

Area (A):

The area A is defined as the count of the number of pixels present inside the boundary of every object and it is given as

$$A = \sum a_i \tag{1}$$

where a_i represents the intensity value of the pixels inside the object boundary.

Mean intensity (MI):

MI specifies the mean of all the intensity value in the object boundary.

Equivalent Circular Diameter (ECD):

Once the area is determined, ECD [14] has been calculated based on the pixel count.

$$ECD = \sqrt{\frac{4}{\pi} \cdot Area} \tag{2}$$

Multiclass Discriminative Classifier:

The multiclass discriminative classifier [15] is a supervised machine learning algorithm. It is based on the idea of finding a hyper plane that defines decision boundaries. A decision plane is the one that separates between a set of objects having different class memberships. Support vectors are the data points of a dataset nearest to the hyper plane if removed would alter the position of the hyper plane defining the decision boundary and hence considered as the critical elements of a data set. Data points in the input space are mapped to the feature space using a set of mathematical function known as kernels. This transformation would help to identify a feature space that is linearly separable which would otherwise require complex curve in the input space. The following are the steps carried out to find the optimal hyper plane.

Hyper plane is defined as:

$$f(x) = \beta_0 + \beta^T x \tag{3}$$

where β is known as the *weight vector* and β_0 as the *bias*.

Among all the possible hyper plane represented by different ways of scaling the bias and weight vector, the one chosen as per the following is the optimal hyper plane. This optimal hyper plane is known as the **canonical hyper plane**.

$$|\beta_0 + \beta^T x| = 1 \tag{4}$$

where x represents the training data point. The distance between a point x and a hyper plane (β, β_0) is

$$distance = \frac{|\beta_0 + \beta^T x|}{\|\beta\|} \tag{5}$$

For the canonical hyper plane, the numerator is equal to one and the distance to the support vectors is

$$distance_{support\ vector} = \frac{|\beta_0 + \beta^T x|}{\|\beta\|} = \frac{1}{\|\beta\|} \tag{6}$$

Margin here denoted as M, is twice the distance to the closest data point.

$$m = \frac{2}{\|\beta\|} \tag{7}$$

Maximizing M is equivalent to the problem of minimizing a function $L(\beta)$ subject to the constraints. The constraints model the requirement for the hyper plane to classify correctly all the training examples x_i. Formally represented by,

$$\min_{\beta,\beta_0} L(\beta) = \frac{1}{2} \|\beta\|^2 \; subject\; to\; y_i\left(\beta^T x_i + \beta_o\right) \geq 1 \forall i \tag{8}$$

where y_i represents each of the labels of the training examples.

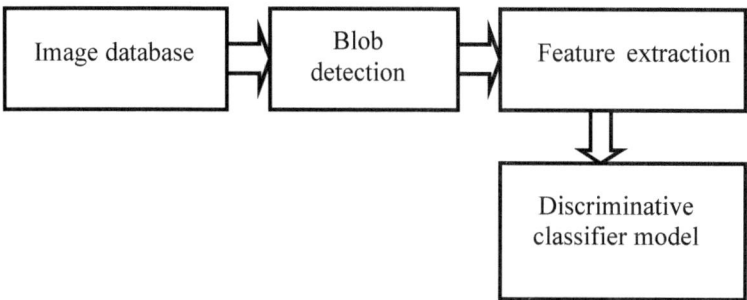

Fig. 1 Proposed architecture during training phase

3.1.2 Testing Phase

For the multi-class problem, a set of the hyper plane is constructed. If the point lies above the hyper plane then it belongs to class 1 and if it lies below the hyper plane, it belongs to class 2 and respectively for further classes. Steps for testing the classifier are depicted in Fig. 2.

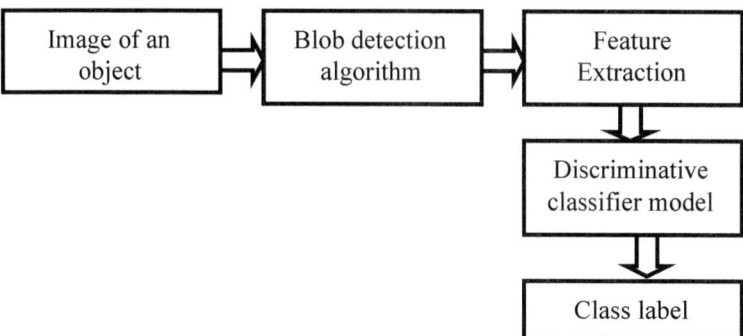

Fig. 2 Proposed architecture for the testing phase

3.2   Counting of Objects of Similar Type and Void Space Estimation

The image of the scene with multiple objects is given as input and all the objects present in the image are classified using the discriminative classifier and then the count of an object of each class has been calculated. Void space size is estimated by using the following algorithm.

Algorithm: Void Space Estimation

Steps for void space estimation are given below:

Step1: Input the image.
Step2: Extract the boundaries of all the objects in the image.
Step3: Repeat step4 and step5 for each pair of adjacent boundaries in the image.
Step4: Minimum distance = overall minimum distance between every pair of points.
Step5: If minimum distance > approximate size of the object then it is detected as "void space".
Step6: Stop.

Euclidean Distance (ED) [16] measure has been used in the above algorithm. P and Q are the two pixels with co-ordinates (p1, p2) and (q1, q2). The ED between P and Q is the square root of the sum of squares of the distance between the corresponding co-ordinates.

$$ED(P, Q) = \sqrt{(q1 - p1)^2 + (q2 - p2)^2} \tag{9}$$

Steps to estimate the size of the void space and to count the objects of similar type is depicted in Fig. 3.

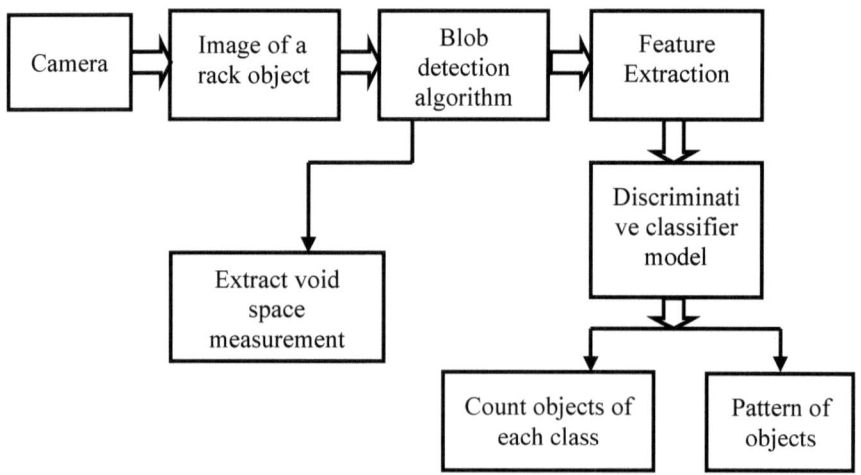

Fig. 3 Steps to estimate the size of void space and count the number of items of the same type

3.3 Change Detection

'From-to' Change is identified by comparing the placement pattern of objects of the reference image available in the database with the placement pattern of objects in the current image. At every period of time, the placement pattern and the void space will be monitored and notified to the retailer. The proposed algorithm for change detection has been shown in Fig. 4.

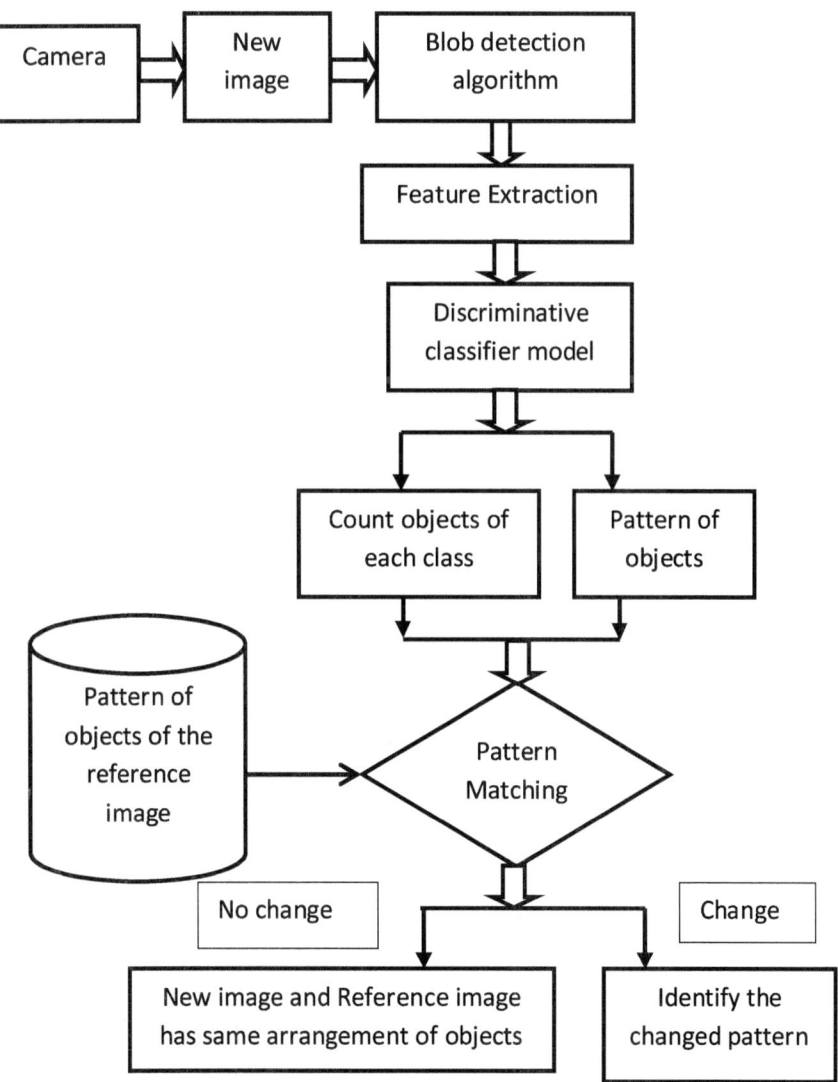

Fig. 4 Proposed algorithm for change detection

4 Dataset

The dataset used to experiment the proposed work is COIL (Columbia Object Image Library)-100 dataset [17] and the user defined dataset. Proposed change detection algorithm is experimented with three object categories consisting of 43 images. 50 samples from each object category have been considered for training the discriminative classifier. Objects that have been considered for classification are Container, Glue and Cetaphil lotion from COIL-100 dataset and Pantene, Yardley and Amway bottle from user defined dataset.

5 Experimental Results and Discussion

COIL dataset has the objects which have sharp edges as well as mixing of smooth and sharp edges. For cylindrical objects with a sharp edge, threshold value 75 have been used and for the objects with smooth edges as well as sharp edges, threshold value 45 has been used to binarize the image. Threshold value 100 gives good results for the user defined dataset.

Input Image	Boundary detection	Output
		The object belongs to 2 The object belongs to 2 The object belongs to 2 The object belongs to 2 out = 2 4 class two belongs to Glue

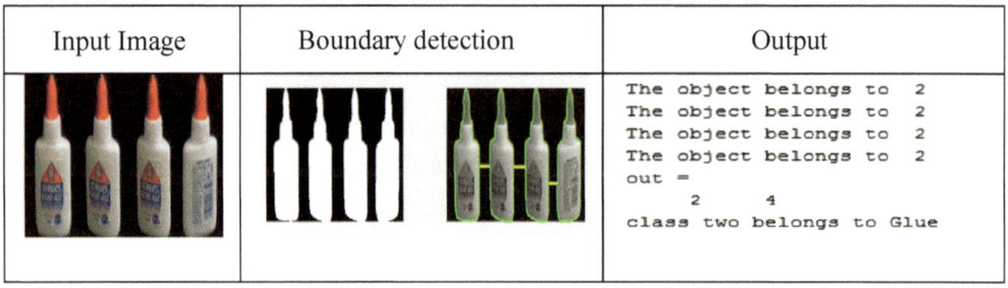

Fig. 5 Image with objects at different orientation (Left); Blob detection (Center); Object recognition and counting (Right)

In Fig. 5 the input image contains the images of a glue bottle placed at different orientations. The boundary has been identified for every product which corresponds to the blobs. Each blob has been labeled to its respective class using the discriminative classifier. The count of each product calculated is also shown. The performance of the classifier when tested using 25 objects for each class from the COIL dataset is measured using the confusion matrix shown in Table 1.

Table 1 Confusion matrix

Accuracy: 86.6%		Actual			
		Container	Glue	Cetaphil lotion	Class precision
Predicted	Container	25	0	0	**100%**
	Glue	0	21	4	**77.8%**
	Cetaphil lotion	0	6	19	**82.6%**
	Class recall	**100%**	**84%**	**76%**	

The overall accuracy rate obtained from the confusion matrix is 86.6%. Precision is calculated by taking the ratio of the number of objects classified correctly as class n to the total number of objects classified as class n where n denotes the class number.

$$PRECISION = \frac{number\ of\ objects\ classified\ correctly\ as\ class\ n}{total\ number\ of\ objects\ classified\ as\ class\ n} \quad (10)$$

Recall is calculated by taking the ratio of the number of objects classified correctly as class n to the total number of objects present in class n where n denotes the class number.

$$RECALL = \frac{number\ of\ objects\ classified\ correctly\ as\ class\ n}{total\ number\ of\ objects\ present\ in\ class\ n} \quad (11)$$

Classification fails when the products are oriented towards 90°. In Fig. 6 all the products are Cetaphil lotion. The middle product displayed at 90° has been misclassified as the product Glue.

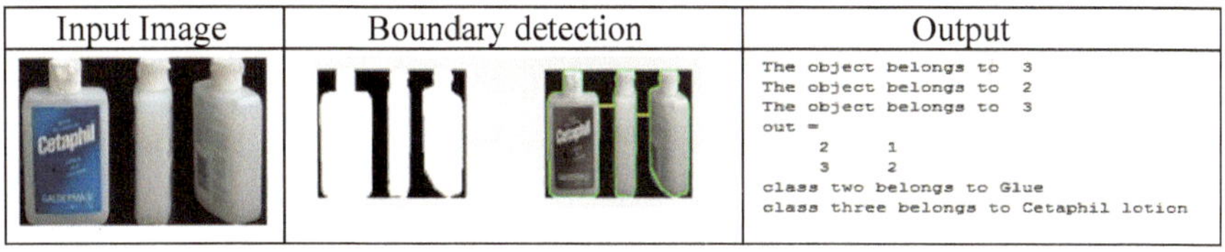

Input Image	Boundary detection	Output

		The object belongs to 3
		The object belongs to 2
		The object belongs to 3
		out =
		    2 1
		    3 2
		class two belongs to Glue
		class three belongs to Cetaphil lotion

Fig. 6 Image with objects at different orientation (Left); Blob detection (Center); Object recognition and counting (Right)

Void space detection result is shown in Fig. 7. Input image has three container bottles. There is a void space between the first object and the second object. This has been identified and the measurement is given in pixels. In this image the void space size is 127 pixels, which is approximately equal to the size of the object and hence it has been detected as void space. The minimum distance between object 2 and object 3 is small therefore not detected as void space.

Input Image	Boundary detection	Output
	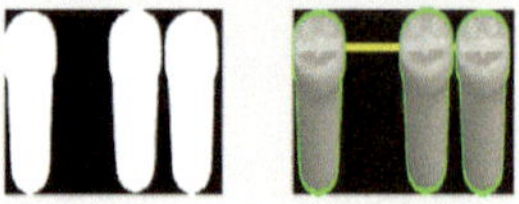	this object is 1 this object is 2 this object is 3 The minimum distance from region 1 to region 2 is 127.000 pixels The minimum distance from region 2 to region 3 is 8.000 pixels

Fig. 7 Given image (Left); Blob detection (Center); Object recognition and void space identification (Right)

Deviation of planogram compliance in the new image shown in Fig. 9 with respect to the reference image shown in Fig. 8 are done by comparing its corresponding object pattern and the result is depicted in Fig. 10. Result shows that the products at position 2 and position 3 have been misplaced.

Reference Image	Boundary detection	Output
	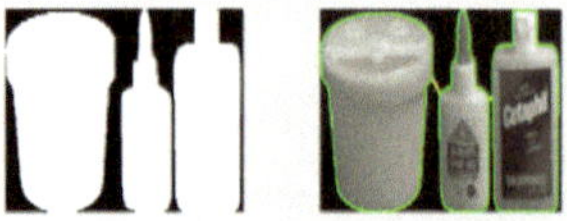	The object belongs to 1 The object belongs to 2 The object belongs to 3 out = 1 1 2 1 3 1 class one belongs to Container class two belongs to Glue class three belongs to Cetaphil

Fig. 8 Reference image (Left); Blob detection (Center); Object recognition and counting (Right)

New Image	Boundary detection	Output
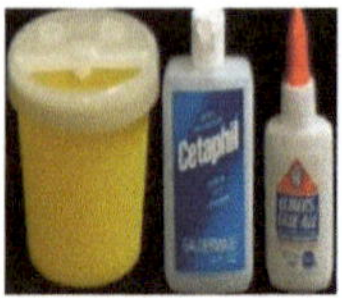	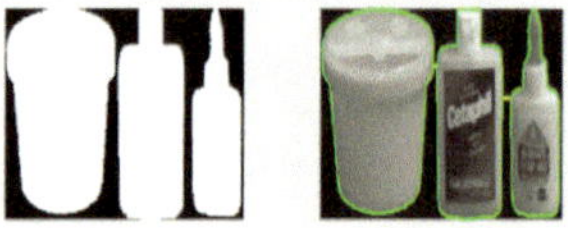	The object belongs to 1 The object belongs to 3 The object belongs to 2 out = 1 1 2 1 3 1 class one belongs to Container class two belongs to Glue class three belongs to Cetaphil lotion

Fig. 9 Given image (Left); Blob detection (Center); Object recognition and counting (Right)

Change has not occured in position 1
Change has occured in position 2
Change has occured in position 3

Fig. 10 Changes detected at the object level between the image shown in Figs. 8 and 9

For images with misplaced objects, when the testing is done for around 40 samples the changes occurred on the position of the objects have been detected. So we infer that if the objects are oriented in 90° then it doesn't show proper results for classification which affects the change detection also. Object recognition, counting the objects of similar type and void space detection on the images of user defined dataset is shown in Figs. 11 and 12 respectively. Figure 11 have three Pantene shampoo bottle and one Amway bottle which has been recognized correctly and its individual count is given. Void space between first and second object in the image which is portrayed in Fig. 12 (Left) is approximately equal to the size of the object thus it has been detected as void space.

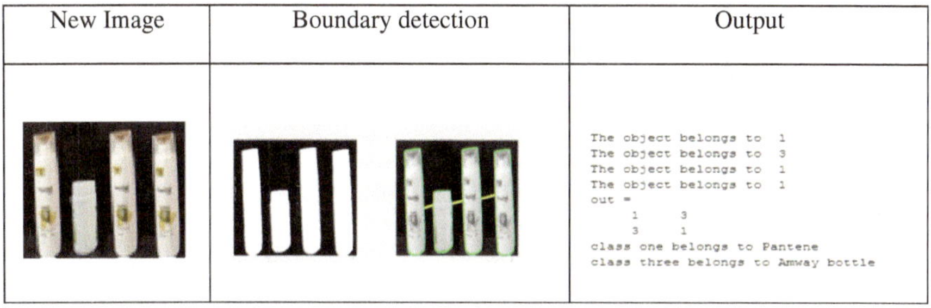

New Image	Boundary detection	Output
		The object belongs to 1 The object belongs to 3 The object belongs to 1 The object belongs to 1 out = 　　1　　3 　　3　　1 class one belongs to Pantene class three belongs to Amway bottle

Fig. 11 Given image (Left); Blob detection (Center); object recognition and counting (Right)

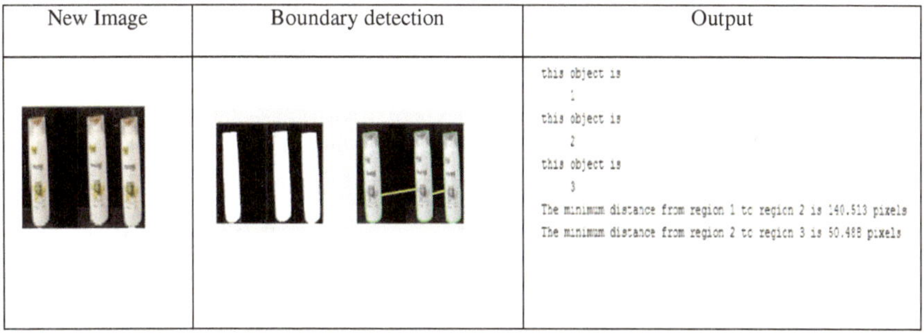

New Image	Boundary detection	Output
		this object is 　1 this object is 　2 this object is 　3 The minimum distance from region 1 to region 2 is 140.513 pixels The minimum distance from region 2 to region 3 is 50.489 pixels

Fig. 12 Given image (Left); Blob detection (Center); Object recognition and void space identification (Right)

6 Conclusion

Change detection algorithm has been efficiently implemented and we have arrived at better ways to detect the void space, count of each class and the occurrence of change patterns based on the experimental result that we have got. The accuracy of all these depends on the performance of the multiclass discriminative classifier. Statistical features are useful for effective classification. The proposed work can be extended to address major variation in the size of the object and also the orientation variant case where the current system shows less efficient results for the objects placed at 90°.

Acknowledgements. We thank Amrita Vishwa Vidyapeetham for having provided the required resources in the Amrita-Cognizant Innovation Lab for carrying out the research work.

References

1. https://www.cognizant.com/InsightsWhitepapers/Planogram-Compliance-Making-It-Work.pdf
2. Moorthy, R., Behera, S., Verma, S., Bhargave, S., Ramanathan, P.: Applying image processing for detecting on-shelf availability and product positioning in retail stores. In: Proceedings of the Third International Symposium on Women in Computing and Informatics. ACM (2015)
3. Minu S., Shetty, A.: A comparative study of image change detection algorithms in MATLAB. Aquat. Procedia **4**, 1366–1373 (2015)
4. Radke, R.J., Andra, S., Al-Kofahi, O., Roysam B.: Image change detection algorithms: A systematic survey. IEEE Trans. Image Process. **14**(3), 294–307 (2005)
5. Al-doski, J., Mansor, S.B., Shafri, H.Z.M.: Support vector machine classification to detect land cover changes in Halabja City, Iraq. Business Engineering and Industrial Applications Colloquium (BEIAC), IEEE (2013)
6. Klaric, M.N., Claywell, B.C., Scott, G.J., Hudson, N.J., Sjahputera, O., Li, Y., Barratt, S.T., Keller, J.M., Davis, C.H.: GeoCDX: An automated change detection and exploitation system for high-resolution satellite imagery. IEEE Trans. Geosci. Remote Sens. **51**(4), 2067–2086 (2013)
7. Grycuk, R., Gabryel, M., Korytkowski, M., Scherer, R., Voloshynovskiy, S.: From single image to list of objects based on edge and blob detection. In: International Conference on Artificial Intelligence and Soft Computing, Springer International Publishing (2014)
8. Kong, H., Akakin, H. C., Sarma, S. E.: A generalized laplacian of gaussian filter for blob detection and its applications. IEEE Trans. Cybern. **43**(6):1719–1733 (2013)
9. Han, K. T. M., Uyyanonvara, B.: A survey of blob detection algorithms for biomedical images. Inform. Commun. In: 7th International Conference of IEEE, Technol. Embedded. Syst. (IC-ICTES) (2016)
10. Huang, M. L., Hung, Y. H., Lee, W. M., Li, R. K., Jiang, B. R.: SVM-RFE based feature selection and Taguchi parameters optimization for multiclass SVM classifier. Sci. World J. **2014**:1–10 (2014)
11. Reddy, B. V., Reddy, A. S., Reddy, P. B.: BITSMSSC: Brain image tomography using SOM with multi SVM sigmoid classifier. Comput. Intell. Data Min. 2:497–505. Springer India (2016)
12. Biswas, S., Aggarwal, G., Chellappa, R.: An efficient and robust algorithm for shape indexing and retrieval. IEEE Trans. Multimedia **12**(5):372–385 (2010)
13. Venkateswaran, K., Kasthuri, N., Jeni, D. D.: A survey on unsupervised change detection algorithms. In: International Conference on IEEE, Circuits, Power Comput. Technol. (ICCPCT) (2013)
14. Ramanathan, R., Soman, K.P., Rohini,, P.A., Dharshana, G.: Investigation and development of methods to solve multi-class classification problems. In: International Conference on IEEE, Adv. Recent Technol. Commun. Comput. (ARTCom'09) (2009)

15. Sampath, A., Sivaramakrishnan, A., Narayan, K., Aarthi, R.: A study of household object recognition using SIFT-based bag-of-words dictionary and SVMs. In: Proceedings of the International Conference on Soft Computing Systems, Springer India (2016)
16. Bagyammal, T., Parameswaran, L.: Context based image retrieval using image features. Int. J. Adv. Inform. Sci. Technol. 29 (2014)
17. Nene, S.A., Nayar, S.K., Murase, H.: Columbia object image library (COIL-100). Tech. Rep CUCS-006-96, February (1996)

A Study on Various Quantification Algorithms for Diabetic Retinopathy and Diabetic Maculopathy Grading

Parvathy Ram and T. R. Swapna[(✉)]

Department of Computer Science and Engineering, Amrita School
of Engineering, Amrita Vishwa Vidyapeetham, Coimbatore, India
cb.en.p2cvil5009@cb.students.amrita.edu,
tr_swapna@cb.amrita.edu

Abstract. Diabetes also known as diabetes mellitus (DM) is a prominent disease all over the world. It is a metabolic disorder occurring due to high blood sugar levels over a prolonged period. Prolonged diabetes will cause diabetic retinopathy affects retina. Diabetes affecting macular area is called diabetic maculopathy. Developing automated systems for identification, grading and quantification of the retinal pathologies associated with DM is on the rise. There are four popular modalities that are useful for clinical diagnosis and treatment of diabetic maculopathy. They are slit-lamp biomicroscopy, color fundus images, fundus fluorescein angiograms (FFA) and optical coherence tomography (OCT). It is observed that FFA plays an vital role in the treatment of diabetic macular edema (DME). There are two major types of diabetic retinopathy: non-proliferative diabetic retinopathy (NPDR) and proliferative diabetic retinopathy (PDR). NPDR shows up as retinal exudates or cotton wool spots or microvascular abnormalities or as superficial retinal hemorrhages or as microaneurysms. PDR is characterized by severe small retinal vessel damage and reduced oxygenization of retina. Here a survey on the quantification of the macular edema, retinal exudates, microaneurysms and other retinal pathologies in diabetic maculopathy and diabetic retinopathy are elaborated.

Keywords: Fundus Fluorescein Angiogram · Optical Coherence Tomography
Diabetic Macular Edema · Non-prolifertive diabetic retinopathy · Proliferative
diabetic retinopathy

1 Introduction

Diabetic retinopathy (DR) and Diabetic Maculopathy (DM) is an issue of diabetes. It is the major cause of harm among people in many countries. By identifying the lesions at its earliest will reduce the severity of the disease. The main objective By detecting the exudates, microaneurysms and lesions will help the clinicians to identify the disease and make a correct follow up of it.

© Springer International Publishing AG 2018
D. J. Hemanth and S. Smys (eds.), *Computational Vision and Bio Inspired Computing*,
Lecture Notes in Computational Vision and Biomechanics 28,
https://doi.org/10.1007/978-3-319-71767-8_34

Diabetics is a major problem which is prevailing all over the world. India is one among the top three countries which are having diabetics. In 2013 it was estimated that over 382 million people throughout the world had diabetes (Williams textbook of endocrinology). As per the 2016 WHO report [1], 422 million adults had diabetics in 2014 i.e., the percentage of adults affected with diabetics is rising from 4.7 to 8.5%. It is due to the pancreas not producing proper count of insulin or the cells of the body not responding properly to the count of insulin produced. Diabetics can also affect different body organs such as: heart, liver, eyes etc... Even it can cause serious health issues which may or may not lead to the death of a person.

The other reason for vision loss is Diabetic Macular Edema (DME). DME is an accumulation of the fluid in the macula-a part of the retina that controls our most detailed vision abilities. This could be mainly due to the leakage of blood vessels. DME usually takes two forms: Focal DME and Diffuse DME. Focal DME occurs because of abnormalities in the blood vessels in the eye and Diffuse DME occurs due to the widening/swelling retinal capillaries.

There are mainly two types of diabetic retinopathy: Non-proliferative diabetic retinopathy (NPDR) and proliferative diabetic retinopathy (PDR). NPDR is the starting stage of diabetic retinopathy. PDR mainly occurs when the disease is at it high level. There are different modalities by which these diabetics can be identified. They are: color fundus images, Fundus Fluorescein Angiogram (FFA) images, Optical Coherence Tomography (OCT) images. There are many papers regarding the survey of these modalities.

2   Review of Literature Survey

Computer based quantification of diabetic images using image processing techniques generally involves the phases of preprocessing, converting to gray scale images, detection of the disease, quantification and counting. Quantification is mainly done to check the progress of the disease. This survey presents the various quantification methods adapted in the recent past to perform the above tasks in diabetic retinopathy and diabetic maculopathy.

2.1   Quantification in Diabetic Maculopathy

Quantification is major steps in the identification of the disease as this method will helps to know the intensity of the disease. Various methods are involved in this quantification.

Creel et al. [2], mainly focuses on quantifying the microaneurysms in digitized FFA. For the training set, the microaneurysm achieved sensitivity of 82%. Against the test images, it achieved sensitivity of 82%. Automated system reduced the time and increased the rate of success than the conventional methods.

Phillips et al. [3], quantification of the retinal pathologies is performed so that the progress of the retinal disease can be recorded easily. This will help the clinicians to provide a better treatment. A combination of shade correction, matched filtering and shape operators are used for identifying the microaneurysms. The sensitivity of the electronic technique was between 82 and 93% and specificity 93% or more. By the repeated analysis of the same region of the same angiogram frame, it detected almost 75% microaneurysms.

Ravishankar et al. [4], proposes a system which helps in localization of different features of red lesions in a color fundus image. The optic disk is located and their intersections are found out. It yielded a 97.1% of success rate for optic disk localization, 95.1 and 90.5% for microaneurysm detection.

Tariq et al. [5], using the vascular structure and optic disc location, it extracts the macula from digital retinal image. It has high sensitivity, specificity, positive predictive value and accuracy comparing to other methods.

Swapna and Chakraborty [6], here a review of the developed algorithms for FFA images on diabetic patients is proposed. This paper have insisted on the advancement of FFA database as it is one of the vital needs for mounting a system for analysis and cure of diabetic maculopathy.

2.2 Quantification of Diabetic Retinopathy

Different methods are comprised for the computerized detection of disease. Some results show a good result in the detection while some are not detecting all the prominent parts.

Esmaeili et al. [7], a new method for improving non uniform background has been proposed. The method Digital Curvlet Transform (DCUT) is also introduced here. The red lesions are detected from the digital color retinal images using this method. This proposed method was accurate and applicable for clinical works. Sekhar et al. [8], it presents a different method for the detection and localization of optic disk, fovea and macula in retinal fundus images. Hough transform is applied on the image to extract the shape of the image. It was capable of localizing 34 of those images successfully in the database. Marin et al. [9], here an algorithm is proposed which helps in the detection of red lesions in digital color fundus photographs. This algorithm is robust and accurate, but when a large set of images appears a further validation is required.

El Abbadi and Al-Saadi [10], here a novel method is proposed which helps to localize and isolate the exudates. It gives good results for the input given and hence is a promising algorithm comparing to the existing system. Welfer et al. [11], it explains in the detection of exudates in color eye fundus images using a new method based on mathematical morphology. Osareh et al. [12], the retinal exudates are detected automatically from the color images using fuzzy C-means clustering technique. Fuzzy C-means (FCM) clustering allows pixels to belong to multiple classes with varying degrees of membership. It showed 93.0% sensitivity and 94.1% specificity evidence of retinopathy.

Reza et al. [13], here a recent algorithm using watershed algorithm is used for the detection of bright lessions is proposed. Sopharak et al. [14], here a new technology known as fine-tuned segmentation with a morphological reconstruction is explained. Compared with FCM clustering, the results indicate better in accuracy. Walter et al. [15], a new algorithm for the detection of the optic disc and the vascular tree in noisy low contrast color fundus photographs is proposed based on mathematical morphology. These results are accurate for well contrasted images.

Eswaran et al. [16], this presents a new method for segmentation of automated exudates from color fundus images. Finally the optic disk is removed for the exact location of the exudates and for decreasing the negative rate. Chen et al. [17], here a method which helps in automatically detecting the hard exudates from the color retinal images is explained. Eadgahi and Pourreza [18], here a morphological approach is used here. By using the mixture of techniques like top-hat transform, bottom-hat transform and reconstruction operations the hard exudates are segmented out. It can obtain good results in the detection of exudates.

Dattaa et al. [19], it mainly deals with the hardware issues on the early detection of the disease. Later on, the techniques such as pre processing, segmentation and classification stages are applied on the images for knowing the progress of the disease. Akram et al. [20], here it uses a Gaussian mixture model for classifying whether it is an exudate or non-exudate. Also a new method known as hybrid classifier is used which can detect the exudates even in bright lesions. Zhang et al. [21], it helps in detecting hard exudates by using a novel method. This method gain good results than the existing methods.

Kaur and Kaur [22], it uses an unsharp mask for the pre processing which is applied on the green channel of the image. A hybrid approach based on minimum area detection and region based segmentation is used for exudate detection. This method is applied on Diartedb1 database, which shows good accuracy and improved results.

Kumar et al. [23], in this paper a segment based feature extraction is used for the extraction of exudates. Here the extraction is done on the basis of intensity values. This method is applied in the DRIVE database which shows an improved performance for automatic detection of exudates. Giancardo et al. [24], here a new method known as color wavelet decomposition and automatic lesion segmentation is introduced which helps in considering and diagnosing of DME. It was tested on the publically available database Messidor. Akram et al. [25], the exudates in the color retinal images is detected using a computer aided diagnosis system for automatic detection of optic disk. The results are good comparing to the existing systems. Niemeijer et al. [26], a hybrid approach for the identification of the red lesions is described in this paper. A wide spread estimation is performed in this paper on the images taken.

García et al. [27], here red lesions, hemorrhages, microaneurysms are automatically detected using their properties. A multi-layer perceptron classifier is used for final segmentation results. Liu et al. [28], a method using hough transform is used here. A Gaussian filter is used here to detect the exudates using dynamic thresholding. It will help the doctors to identify the disease easily and provide a better treatment. Siddalingaswamy and Gopalakrishna [29], here it proposes an automatic approach for the localization and boundary detection of the optic disc. For recognizing the optic disk, a thresholding and a connected component method is used. So it has good accuracy while comparing with other methods. Li and Chutatape [30], the novel methods for the detection and extraction of the main features in the retinal images are proposed here. The achievement rates of disc localization, disc boundary detection, and fovea localization are 99, 94, and 100%, respectively.

Júnior and Welfer [31], an automatic detection of microaneurysms and haemorrhages is proposed. A mathematical morphology is used, which gains mean sensitivity of 87.69% and specificity of 92.44%, when tested on a public database of fundus images. Mendels et al. [32], it describes the identification of the boundary of the optic disk using active contours. A new method Gradient Vector Flow is introduced which helped in the better detection of the contours in the given retinal images. This helps in the automatic detection of the optic disk in an easy manner.

Jadhav and Shaikh [33], this paper has described about the review of various methods for detecting and grading diabetic retinopathy using fundus images. It has given a good support for the future researchers to get a summary of the algorithms present. Prakash et al. [34], the exudates from the retinal images are detected and classified. A region of interest k-means clustering technique is used for detecting the optic disk and svm classifier is used for classification.

2.3 Judgements and Observations

I. To summarize the various quantification algorithms present in the literature for diagnosis of diabetic retinopathy, thereby providing researchers and academicians a detailed resource of the main algorithms employed in color fundus image quantification.

II. From the literature survey it is clear that though many techniques are available for quantification of diabetic retinopathy, there is no standard grading system for the diabetic retinopathy based on quantification of the various retinal pathologies. Developing new techniques for grading various types of diabetic retinopathy based on quantification algorithms will aid the clinicians to understand the progress of the disease, which in turn provides us with scope for further research.

Table 1 shows the summary of the surveys conducted.

3 Conclusion

In this paper, review of existing literature of quantifying the disease based on the different methods of diabetic retinopathy and diabetic maculopathy has been made. For proper quantification of the disease a thorough understanding of image is necessary. All

Table 1 Summary of the accuracies and technologies used in various techniques

Year	Database	Technology used	Accuracy	Authors
2016	E-ophtha-ex database	SVM classification	Average accuracy: 94.17%	Prakash et al. [34]
2015	DIARETDB1	Hybrid approach	Specificity: 98.12% Sensitivity: 90.83%	Kaur and Karur [22]
2014	DIARETDB1	Projection based algorithm	Accuracy: 99.7%	Eswaran et al. [16]
2014	E-ophtha EX database	Candidates segmentation method	AUC: between 0.93 and 0.95	Zhang et al. [21]
2014	HEI-MED, MESSIDOR	Gaussian mixture model and support vector machine	Sensitivity: 7.3%, Specificity: 95.9% Accuracy: 96.8%	Akram et al. [20]
2014	Standard database	A review	A review	Swapna et al. [6]
2013	MESSIDOR and STARE	Vascular structure and optic disc location + Gaussian mixture model	Accuracy of 97.53%	Tariq et al. [5]
2013	Standard database	Automated detection of exudates	Promising results	El Abbadi and Al-Saadi [10]
2013	DIARETDB1	Automatic detection of microaneurysms and hemorrhages in fundus images	Mean sensitivity: 87.69% Specificity: 92.44%	Júnior and Welfer [31]
2012	DIARETDB1	Histogram segmentation	Sensitivity: 94.7% Positive prediction value: 90.0%	Chen et al. [17]
2012	DIARETDB1	Mathemathical morphology operations	Sensitivity: 78.28%	Eadgahi and Pourreza [18]
2012	Fundus images of retina	Hardware based analysis	Good results	Dattaa et al. [19]
2012	DRIVE and STARE	Segment based technique	Sensitivity: 67.1%	Kumar et al. [23]

(*continued*)

Table 1 (*continued*)

Year	Database	Technology used	Accuracy	Authors
2012	MESSIDOR	Exudate-based diabetic macular edema detection	AUC between 0.88 and 0.94	Giancardo et al. [24]
2012	STARE, DIARETDB0, DIARETDB1	Candidate exudate detection, feature extraction and classification	Specificity: 98.25% Accuracy: 97.57%	Akram et al. [25]
2010	STAREAND DRIVE	Using marker controlled watershed transform	Sensitivity: 95%	Reza et al. [13]
2010	Digital retinal images	Using morphological reconstruction enhancement	Sensitivity: 88.1% Specificity: 99.2% Accuracy: 99%	Sopharak et al. [14]
2010	148 images	Automatic localization and accurate boundary detection of the optic disc	Sensitivity: 90.67 ± 5 Specificity: 94.06 ± 5	Siddalingaswamy and Gopalakrishna [29]
2009	DIARETDB1	Using morphological reconstruction and thresholding	Sensitivity: 70.48% Specificity: 98.84%	Welfer et al. [11]
2009	516 IMAGES	Automated feature extraction technique	Sensitivity: 95.7% Specificity: 90.5%	Ravishankar et al. [4]
2008	DRIVE database	Computer aided technique	Success rate of 90.25%	Sekhar et al. [8]
2008	Standard images	Automated three stage method	Sensitivity of 0.785%	Marin et al. [9]
2008	100 images	Multilayer Perceptron Neural Network	Mean sensitivity: 86.1%	García et al. [27]
2005	40 images	Automated detection of red lesions and kNN classifier	Sensitivity: 100% Specificity of 87%	Niemeijer et al. [26]
2004	89 retinal images	Model based approach	Sensitivity: 100% Specificity: 71%	Li and Chutatape [30]

(*continued*)

Table 1 (*continued*)

Year	Database	Technology used	Accuracy	Authors
2004	Standard database	Using active contours	Good results are obtained	Mendels et al. [32]
2003	Color retinal images	Fuzzy c means clustering	Sensitivity: 93% Specificity: 94%	Osareh et al. [12]
2001	30 color images	Morphological technique	Good results	Walter et al. [15]
1997	Standard database	Automated computer generated system	Sensitivity of 82%	Creel et al. [2]
1997	Fundus images	Automatic retinal image analysis system	Good results are obtained	Liu et al. [28]
1991	Standard database	Digital imaging of fundus	Good results	Phillips et al. [3]
1991	Standard database	Automatic detection	Good results	Esmaeili et al. [7]

the methods reviewed show a computer aided system for the detection and quantification of the diseases. Only a few methods are able to show a good accuracy in the result, while other methods are not able to show a proper quantification of the disease. The drawback of not identifying proper quantification may be due to the lack of image quality or due to the overlapping of the exudates or microaneurysms. By overcoming these drawbacks we will be able to quantify the disease with good accuracy. Automatic detection of the disease is needed to get more accuracy in the detection so that the severity of the disease can be understood easily. We hope that this survey will help researchers to have a proper understanding of the techniques for quantifying the disease. This learning shows that there is a huge scope for further research to quantify the disease in diabetic retinopathy and diabetic maculopathy.

References

1. WHO report: Global report on diabetics 2016. http://apps.who.int/iris/bitstream/10665/204871/1/9789241565257_eng.pdf
2. Creel, M.J., Olson, J.A., Mchardy, K.C., Sharp, P.F., Forrester, J.V.: A Fully Automated Comparative Microaneurysm Digital Detection System. Department of Bio-medical Physics and Bio-engineering, University of Aberdeen, Foresterhill, Aberdeen
3. Phillips, R.P., Spencer, T., Ross, P.G.B., Sharp, P.F., Forrester, J.V.: Quantification of Diabetic Maculopathy by Digital Imaging of the Fundus. Department of Ophthalmology, Department of Bio-medical Physics, Medical School, University of Aberdeen, Foresterhill, Aberdeen

 4. Ravishankar, S., Jain, A., Mittal, A.: Automated Feature Extraction for Early Detection of Diabetic Retinopathy in Fundus Images. University of Illinois at Urbana-Champaign, University of Maryland College Park, Indian Institute of Technology, Madras
 5. Tariq, A., Akram, M.U., Shaukat, A., Khan, S.A.: Automated detection and grading of diabetic maculopathy in digital retinal images
 6. Swapna, T.R., Chakraborty, C.: Diabetic maculopathy detection using fundus fluorescein angiogram images—a review. IJRET: Int. J. Res. Eng. Technol. **03**(15) (2014)
 7. Esmaeili, M., Rabbani, H., Dehnavi, A.M., Dehghani, A.: A New Curvelet Transform Based Method for Extraction of Red Lesions in Digital Color Retinal Images. Department of Biomedical Engineering, Department of Ophthalmology, Isfahan University of Medical Sciences
 8. Sekhar, S., Al-Nuaimy, W., Nandi, A.K.: Automated localization of optic disk and fovea in retinal fundus images. In: 16th European Signal Processing Conference (EUSIPCO 2008), Lausanne, Switzerland, 25–29 August 2008, Copyright by EURASIP
 9. Marin, O.,C., Ares, E., Penedo, M.G., Ortega, M., Barreira, N., Gomez-Ulla, F.: Automated Three Stage Red Lesions Detection in Digital Color Fundus Images. Grupo de Visión Artificial y Reconocimiento de Patrones University of A Coruña Campus de Elviña s/n, A Coruña, 15071, Spain
10. El Abbadi, N.K., Al-Saadi, E.H.: Automatic Detection of Exudates in Retinal Images. University of Kufa, Najaf, Iraq, IJCSI Int. J. Comput. Sci. Issues **10**(2), No 1 (2013)
11. Welfer, D., Scharcanski, J., Marinho, D.R.: A morphological three stage approach for detecting exudates in color eye Fundus images. In: Proceedings of the 2010 ACM Symposium on Applied Computing, pp. 964–968 (2010)
12. Osareh, A., Mirmehdi, M., Thomas, B., Markham, R.: Automated identification of diabetic retinal exudates in digital colour images. Published by group.bmj.com
13. Reza, A.W., Eswaran, C., Dimyati, K.: Diagnosis of diabetic retinopathy: automatic extraction of optic disc and exudates from retinal images using marker-controlled watershed transformation. Springer Science+Business Media, LLC (2010). Received: 9 Sept 2009/Accepted: 27 Dec 2009/Published online: 29 Jan 2010
14. Sopharak, A., Uyyanonvara, B., Barman, S.: Automatic exudate detection from non-dilated diabetic retinopathy retinal images using fuzzy c-means clustering. Sensors **9**, 2148–2161 (2009). doi:https://doi.org/10.3390/s90302148
15. Walter, T., Erginay, A., Ordoñez, R., Klein, J.: Automatic detection of microaneurysms in color fundus images. Med. Image Anal. (2008)
16. Eswaran, C., Saleh, M.D., Abdullah, J.: Projection based algorithm for detecting exudates in color fundus images. In: Proceedings of the 19th International Conference on Digital Signal Processing, 20–23 August 2014
17. Chen, X., Bu, W., Wu, X., Dai, B., Teng, Y.: A novel method for automatic hard exudates detection in color retina L images. In: Proceedings of the 2012 International Conference on Machine Learning and Cybernetics, Xian, 15–17 July 2012
18. Eadgahi, M.G.F., Pourreza, H.: Localization of hard exudates in retinal fundus image by mathematical morphology operations. J. Theor. Phys. Crypt. JTPC **1** (2012)
19. Dattaa, N.S., Banerjeeb, R., Duttac, H.S., Mukhopadhyayd, S.: Hardware based analysis on automated early detection of diabetic-retinopathy. Procedia Technol. **4**, 256–260 (2012)
20. Akram et al.: Retinal images: optic disk localization and detection. In: International Conference Image Analysis and Recognition, ICIAR, Image Analysis and Recognition, pp. 40–49 (2010)
21. Zhang, X., Thibault, G., Decencière, E., Marcotegui, B., Laÿ, B., Danno, R., Cazuguel, G., Quellec, G., Lamard, M., Massin, P., Chabouis, A., Victor, Z., Erginay, A.: Exudate

detection in color retinal images for mass screening of diabetic retinopathy. Med. Image Anal. (2014)

22. Kaur, M., Kaur, M.: A hybrid approach for automatic exudates detection in eye fundus image. Int. J. Adv. Res. Comput. Sci. Softw. Eng. **5**(6) (2015)

23. Kumar, A., Gaur, A.K., Srivastava, M.: A segment based technique for detecting exudate from retinal fundus image. Procedia Technol. **6**, 1–9 (2012)

24. Giancardo, L., Meriaudeau, F., Karnowski, T.P., Li, Y., Garg, S., Tobin, K.W., Chaum, E.: Exudate-based diabetic macular edema detection in fundus images using publicly available datasets. Med. Image Anal. Elsevier (2012)

25. Akram, M.U., Khan, A., Iqbal, K., Butt, W.H.: Retinal images: optic disk localization and detection. In: International Conference on Image Analysis and Recognition (2010)

26. Niemeijer, M., van Ginneken, B., Staal, J., Suttorp-Schulten, M.S.A., Abràmoff, M.D.: Automatic detection of red lesions in digital color fundus photographs. IEEE Trans. Med. Imaging **24**(5) (2005)

27. García, M., Sánchez, C.I., López, M.I., Díez, A., Hornero, R.: Automatic detection of red lesions in retinal images using a multilayer perceptron neural network. In: 30th Annual International IEEE EMBS Conference Vancouver, British Columbia, Canada, 20–24 Aug 2008

28. Liu, Z., Opas, C., Krishnan, S.M.: Automatic image analysis of fundus photograph. In: Proceedings—19th International Conference of the IEEE/EMBS, Chicago, IL, USA, 30 Oct–2 Nov 1997

29. Siddalingaswamy, P.C, Gopalakrishna, P.K.: Automatic localization and boundary detection of optic disc using implicit active contours. In: Int. J. Comput. Appl. (0975 – 8887), Volume 1 – No. 7 (2010)

30. Li, H., Chutatape, O.: Automated feature extraction in color retinal images by a model based approach. IEEE Trans. Biomed. Eng. **51**(2) (2004)

31. Júnior, S.B., Welfer, D.: Automatic detection of microaneurysms and hemorrhages in color eye fundus images. Int. J. Comput. Sci. Inform. Technol. (IJCSIT) **5**(5) (2013)

32. Mendels, F., Heneghan, C., Thiran, J.: Identification of the optic disk boundary in retinal images using active contours. Signal Processing Laboratory (LTS), Swiss Federal Institute of Technology (EPFL), semantic Scholar (2004)

33. Jadhav, M.L., Shaikh, M.Z.: Different methods for detecting & grading diabetic retinopathy using fundus images—a review. Int. J. Innovative Res. Electr. Electron. Instrum. Control Eng. **4**(3) (2016)

34. Prakash, N.B., Hemalakshmi, G.R., Mary, S.I.M.: Automated grading of diabetic retinopathy stages in fundus images using SVM classifer. J. Chem. Pharm. Res. ISSN: 0975-7384

Mango Leaf Unhealthy Region Detection and Classification

K. Srunitha[(⊠)] and D. Bharathi

Department of Computer Science and Engineering, Amrita School of
Engineering, Coimbatore, Amrita Vishwa Vidyapeetham, Amrita University,
Coimbatore, India
cb.en.p2cvil5013@cb.students.amrita.edu,
d_bharathi@cb.amrita.edu

Abstract. Diseases in any plant decrease the productivity and quality of product. Identification of plant leaf diseases by naked human eye is very difficult. Image processing techniques can identify the diseased leaf by preprocessing and classifying leaf unhealthy regions. This paper delivers an implementation on Mango leaf unhealthy region detection and classification. In the Proposed work Multiclass SVM is used for diseases classification and segmentation through k-means. The experimental results show the effectiveness of the proposed method in recognizing the diseases affected mango leaf.

Keywords: Multi class SVM · Image processing · k-means clustering

1 Introduction

India is famous for its plant cultivation and productivity. 3/4th of our nation's population depended on it. Individuals of here have wide variety of choice, for selecting plants for their cultivation. Many research works are going behind plant cultivation for improving the quality of production, reducing the expenditure, increasing the profit rate and decreasing the pesticide usage. The outcome of all the cultivation is a mixed combination of perfect soil, water, weather, chemicals…etc.

The major difficulty facing here is the infections found in plants, which cause reduction of product count and quality degradation. Diagnosis of diseases and timely cure is very important task in reducing substandard products. Each plant diseases depends on many characteristics, behavior and also various parameters like environment, nutrient, organism etc., up on which some of the diseases are not easily distinguishable. The visual symptom for each disease varies with respect to how deeply it is affected. Hence diagnosis is very difficult task to carry out manually. This makes development in image processing and pattern recognition techniques. With this it is possible to create a system which can recognize, classify and thus cure the diseases. Usage of pesticides actually damages the field and produce health issues in human.

© Springer International Publishing AG 2018
D. J. Hemanth and S. Smys (eds.), *Computational Vision and Bio Inspired Computing*,
Lecture Notes in Computational Vision and Biomechanics 28,
https://doi.org/10.1007/978-3-319-71767-8_35

In plants, leaf is considered as the area which is commonly prone to diseases. Majority of this are caused due to fungi, bacteria and viruses. This paper handles mainly four types of mango leaf diseases. They are Anthracnose, Red rust, Sooty mould, and Scab. Each disease differs from other with respect to their color, shape type of virus etc.

Anthracnose affected mango leaf will have dark brown spot covered with yellow patches. This will cause serious losses of young shoots. Red rust is caused by alga and it leads to serious reduction in leaf green regions white patches will cover all the green regions in leaf and it get damaged. Sooty mould shows black velvety patches all over the leaf. It spreads easily and damages the whole leaf. Scab is a fungal disease. This is similar to that of Anthracnose. Spots of this will be circular, slightly angular with dark black color. Identification of these diseases with naked human is impossible. Thus image processing methods are introduced (Figs. 1, 2, 3 and 4).

Fig. 1 Anthracnose

Fig. 2 Red rust

Fig. 3 Sooty mould

Fig. 4 Scab

For diagnosing the diseases images of mango leaf is captured using digital camera and mobiles, then the diseases spot is used for classification. The technique includes pattern recognition and image processing. The purpose of using image processing in plant leaf diseases identification is that spotting the diseases, recognizing the texture and color of the diseases spot area, etc.

Here, paper holds mainly following sections. Section 1 is a brief introduction about the importance of identifying leaf diseases and explains types of diseases that is handling in this paper. Section 2 narrates in short of literature survey and brief review on basic image processing technique for diseases identification. Section 3 detailed explanation of the proposed work and Sect. 4 delivers the experimental results. Finally Conclusions are summarized in Sect. 5.

2 Literature Survey

The various methods for finding the plant leaf diseases using different image processing technique mainly segmentation and classification phase is described here.

Anand et al. [1] used the machine learning method for identifying the diseases on the rice plant. The software prototype system uses the HIS model, boundary and spot detection for image segmentation and the Self Organizing Map neural network for the output classification. Pooja et al. [2] using hybrid intelligence system finds out the grape leaf infections identification from color imagery. There were mainly three phases involved. The very first step was leaf color segmentation. A self-forming feature map along with a back-propagation neural network is organized for identify the color diseases of the leaf. Following was segmenting the affected diseases and last stride was to find out and classify the diseases. Modified self-forming feature map together with optimization using genetic algorithm and for classification support vector machine is used, thus identification of the leaf diseases has done. To identify the color diseases gabor wavelet is used. For classifying the diseases support vector machine is used.

Krishnan and Sumithra [3] identification and recognition of diseases based on morphological operator which defines a system that contains four phases; the very prior phase was image enrichment, that includes, analysis of histogram, HSI detraction and intensity tuning. Segmentation of taken image was carried through fuzzy c-means algorithm. The features used to extract from leaf were color, size of effected diseases and the shape. Back propagation centered neural network was used for the classification of the diseases.

Table 1 Summary of methods in literature survey

Name of the algorithm	Advantages	Disadvantages
K-nearest neighbor (KNN)	Simpler classifier as exclusion of any training process. Applicable in case of a small dataset which is not trained	Speed of computing distance increases according to numbers available in training samples
Radial basis function (RBF)	Training phase is faster. Hidden layer is easier to interpret	It is slower in execution when speed is a factor
Probabilistic neural networks (PNN)	Tolerant of noisy inputs. Instances classified by more than one output	Long training time. Large complexity of network structure. Need lot of memory for training data
Back propagation neural network	Easy to implement. Applicable to wide range of problems. Able to form arbitrarily complex nonlinear mappings	Learning can be slow. It is hard to know how many neurons as well as layers are required
Support vector machine (SVM)	Simple geometric interpretation and a sparse solution. Can be robust, even when training sample has some bias	Slow training. Difficult to understand structure of algorithm. Large no. support vectors are needed from training set to perform classification task

Arivazhagan et al. [4] proposed a method for diagnosing the diseases found in citrus plant. Citrus canker, Anthracnose, Overwatering and Citrus greening disease are the major diseases that affect the citrus plant. Diseases identification and recognition play a vital role in improving the cultivation. For resolving the problem the system has mainly four sections. After the acquisition of leaf image the RGB image of citrus plant is translated to different color spaces. K-means was used as image segmentation technique for recognizing the diseases regions. Segmentation extracts the interest region. The extraction of feature for texture is carried through statistical GLCM and color through mean values. Finally SVM is used for classification.

Naikwadi and Amoda [5] the author found out the diseases in sugarcane. The input color image is transferred to L * a * b color space and thresholding a* element gives the spot information. The diseases regions are pulled out with maximum standard deviation. For identifying the severity of the diseases the white pixel area is subtracted from the total segmented area. The system has two features color and texture. GLCM is used for extracting texture feature and L * a * b is used for color. For classifying the diseases support vector machine uses both the feature information.

Arivazhagan et al. [6] there are certain diseases which are not easily identified with naked human eye like black leaf spot and sun scorch seen in orchid leaf. Paper gives a classification of diseases using morphological image segmentation. The area of white pixels in the inputted leaf image is calculated for recognizing the diseases. In segmentation phase some preprocessing steps like intensity adjustment and histogram equalization and filtering techniques like Gaussian filter, disc filter and median filter are used. Segmentation does the split-up between infected and uninfected areas of leaf. Morphological operations like opening, closing and filled holes are used. Interest region are chosen through binary mask. Complementary Binary mask will change zeros to one and vice- versa. Thus when this complement mask is subtracted from original the region of interest is founded and thresholding help to identify the edge. Classification is done based on the white pixel. White pixel for both the diseases were calculated earlier compares this with the query leaf image and it classifies the diseases (Table 1).

3   Proposed Method

The proposed work is feasible for all size of images. Before classification and segmentation some preprocessing is carried out. The pre-processing consists of image acquisition phase followed by image enhancement then segmentation in which the region of interest is segmented. After segmentation, extraction of feature and finally classification is carried out. For segmentation Color image segmentation using k-means clustering is used and GLCM (gray level color co-occurrence metrics) is used for feature extraction and Support vector machine (SVM) is used as classifier. The architecture diagram of the proposed method is explained in Fig. 5.

3.1 Image Acquisition

The first step is to collect dataset for input purpose. The dataset are collected from agricultural university. Dataset contains 5 classes. Below table shows the dataset details for each class. Number of image per class is 61 images in Anthracnose, 50 in Red rust, 55 images in Sooty mould, 75 in Scab and 45 healthy mango leaf image. The collected images are given for preprocessing to separate the diseases infected region.

3.2 Preprocessing of Leaf Image

All captured input images are preprocessed because all are corrupted by illumination, shadow, and noise. This can cause the loss of information that can be used for diseases diagnosis. Preprocessing can improve the image feature for further process. Preprocessing mainly includes contrast enhancement and color space transformation. The original images are of RGB (Red, Green, Blue) color space. They are converted to HSI (Hue, Saturation, and Intensity) color space.

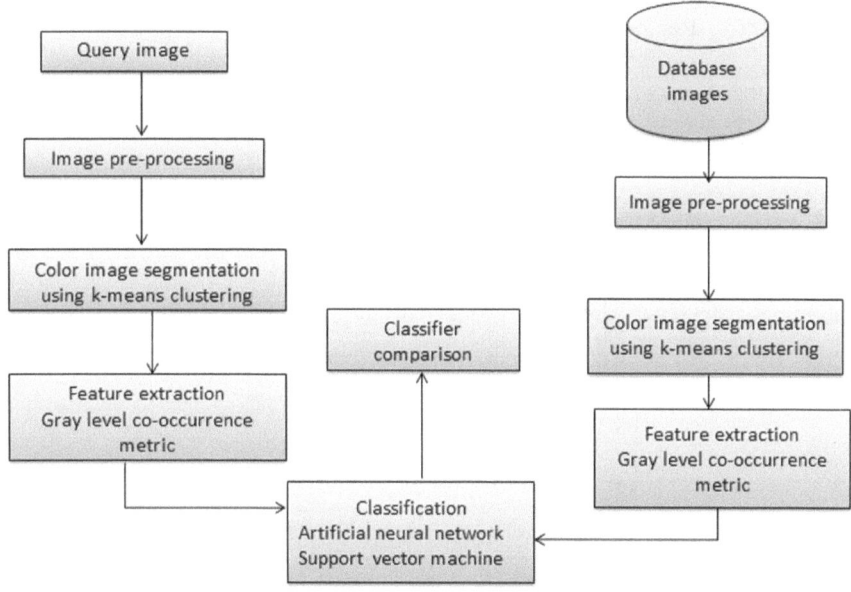

Fig. 5 Architecture of proposed work

The HSV represent the points in the RGB Space. Hue from HIS color space explains the color perceived by the viewer. Saturation is the white light added to hue. Intensity is the value of light amplitude. For each RGB individual color band histogram stretching is applied and pixel count is calculated. As a result color image are enhanced. Algorithm 1 explains the Image extraction and pre-processing [10].

Step 1: Collected Diseases infected mango leaf image are store in a folder on pc.
Step 2: Each image from the folder are read through MATLAB R2016a.
Step 3: Original images are converted into hue image, Saturation image and value image (HSV).
Step 4: Image histograms are calculated for individual RGB color bands i.e., pixel count (x, y).
Step 5: Apply each color band threshold range to its respective color band.
Step 6: Next regions which are smaller than 100 pixels are eliminated.
Step 7: Regions filled size filtered mask is created for the image.
Step 8: Regions which are smaller than 100 pixels are eliminated.
Step 9: The size filtered mask is applied to the original RGB image.
Step 10: Finally, we are displaying the masked original image of the specified color.

3.3   Segmentation

Mango leaf image are divided into multiple set of pixels. Different types of segmentation are there: Region based, Edge based, Threshold, Feature based, Clustering, Model based. Color image segmentation using K-Means Clustering is used in our work. The main aim of this algorithm is to find the mean value between points and group them depending on their minimum distance [11].

Color Image Segmentation using K-Means Clustering

Step 1: First, an image is taken as an input. The input image is in the form of pixels and is transformed into a feature space.
Step 2: Next similar data points are grouped together using k-means clustering. The distances are calculated using and Euclidean distance. Where d = distance, (p,q) = two data points.

$$d(p,q) = \sqrt{(q1 - p1)^2 + (q2 - p2)^2 + \cdots + (qn - pn)^2}$$
$$= \sqrt{\sum_{i=1}^{n}(qi - pi)^2}$$

The data points with minimum distance are grouped together to form the clusters.
Step 3: After clustering is done, the mean of the clusters is taken. The mean color in each cluster is calculated to be remapped onto the image.

The output of this clustering will have k = 3 clusters of Red, Green and Blue segments. These clusters are passed to feature extraction phase.

3.4   Feature Extraction and Classification

Feature extraction is a method for reducing the original dataset collection. Features are mainly considered as interesting points. It is done by measuring certain features like texture, color and shape. In this proposed method we uses GLCM (Gray level co-occurrence metrices) values such as Contrast, Correlation, Energy, Homogeneity, Mean, Standard-Deviation, Entropy, RMS, Variance, Smoothness, Kurtosis, and skewness. GLCM is a statistical method of extracting feature. Below is the formula for calculating GLCM features. Here N_g number of gray level and p_d is normalized metrics dimension of GLCM [12].

Contrast is a measure of gray level variations between the reference pixel and its neighbor. If the contrast is large it shows large intensity difference in GLCM metrics.

$$Contrast = \sum_i \sum_j (-j)^2 p_d(i,j) \tag{1}$$

Homogeneity shows how close the distribution of element to diagonal of GLCM is and when homogeneity increases the contrast decreases.

$$Homogeneity = \sum_i \sum_j \frac{1}{1 + (i-j)^2} p_{d(i,j)} \tag{2}$$

Entropy is the degree of randomness present in an image. When the co-occurrence metrics is having same element the value of entropy becomes high. It becomes small when elements become unequal.

$$Entropy = -\sum_i \sum_j p_{d(i,j)} \ln p_{d(i,j)} \tag{3}$$

Energy is derived from Angular Second Moment (ASM). Local uniformity of pixels is measured using ASM. The value of ASM becomes high when pixels are similar.

$$Energy = \sqrt{ASM}$$

$$ASM = \sum_i \sum_j p_d^2(i,j) \tag{4}$$

Correlation values are the linear dependency of the co-occurrence metrics.

$$Correlation = \sum_i \sum_j P_{d(i,j)} \frac{(i-\mu_x)(j-\mu_y)}{\sigma_x \sigma_y} \tag{5}$$

Mean is the average of pixel values present in an image.

$$Mean = \sum_i \sum_j (i-j) p_d(i,j) \qquad (6)$$

Standard deviation is noted as:

$$Standard\ deviation = \sum_i \sum_j (i-j)^2 p_d(i,j) \qquad (7)$$

Moment is measured as the asymmetry of the image measured as:

$$Moment = \sum_i \sum_j (i-j)^3 p_d(i,j) \qquad (8)$$

Kurtosis measures the relative peak of flatness in an image.

$$Kurtosis = \sum_i \sum_j (i-j)^4 p_d(i,j) \qquad (9)$$

Skewness gives the textural intensity values. It can be low or high.

$$Skewness = \sum_{i=0}^{G-1} (i-\mu)^3 p(i) \qquad (10)$$

Homogeneity of the diseased region is calculated through inverse difference metrics (IDM).

$$IDM = \sum_{i=0}^{G-1} \sum_{j=0}^{G-1} \frac{1}{1+(i-j)^2} p(i,j) \qquad (11)$$

$$Smoothness(R) = 1 - \frac{1}{1+\sigma^2} \qquad (12)$$

$$Angular\ second\ moment = \sum_{i=0}^{G-1} \sum_{j=0}^{G-1} \left\{ p(i,j)^2 \right\} \qquad (13)$$

The 13 features are extracted from the diseased mango leaf image and given to classification algorithm for classification. A comparison of two classifiers is carried out here. Multi class-SVM (Support vector machine) and ANN (Artificial neural network). Support vector machine (SVM) is a non-linear classifier, and is a latest method in machine learning algorithm. SVM is popularly used in many pattern recognition problems including texture classification [14]. SVM is designed to work with only two classes. This is done by maximizing the margin from the hyper plane. Support vectors are the samples that are closest to the margin, which helps in determining the hyperplane. Multiclass classification is applicable and basically built up by various two class SVMs to solve the problem, by using one-versus-all. The SVM uses RBF (Renal basis function) as kernel here. For a K-class classification One versus Rest creates K separate binary classifier. The nth binary classifier is trained by taking data from the nth class. This is considered as a positive example and the remaining classes (K − 1) considered as negative. When a test image came, depending on the maximum output value the class label is determined. Renal basis function used as a kernel for drawing hyperplanes. The RBF kernel on two feature point x and x^l is defined as

$$k(x, x^l) = \exp\left(-\frac{||x - x^l||^2}{\sigma^2}\right)$$

$$k(x, x^l) = \exp\left(-\gamma||x - x^l||^2\right) \quad \text{when } \gamma = \frac{1}{\sigma^2}$$

RBF kernel has two parameters gamma and C. Gamma parameter tells us how far a single training parameter can reach. The value of gamma is low means the sample is far and high means close. C gives the misclassification against the training surface. It makes the hyperplane smooth. A high C value means the misclassification is low. When gamma value is high C will prevent over fitting of data. Weaka 3.8 software is used for finding the accuracy and confusion metrics.

4   Experimental Results

This section holds the results obtained from the above experiment. In the preprocessing step collected diseased leaf image are taken as input to enhance the image quality by resizing, noise removal, contrast enhancement. The preprocessing original image is converted to binary image and individual RGB color band calculation is performed using histogram. Finally it displays the color threshold range (Fig. 6).

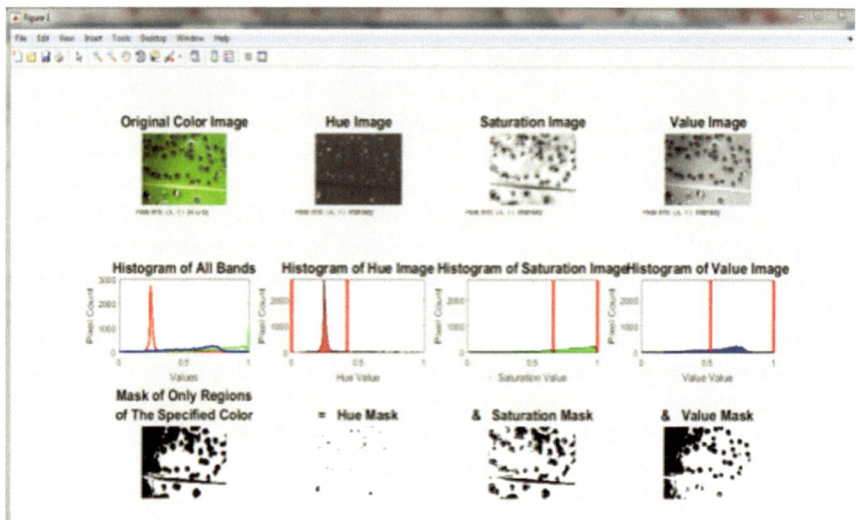

Fig. 6 Preprocessing for converting the original image into a binary image

A size filtered mask is applied to the original RGB image and display only the specified color (Fig. 7).

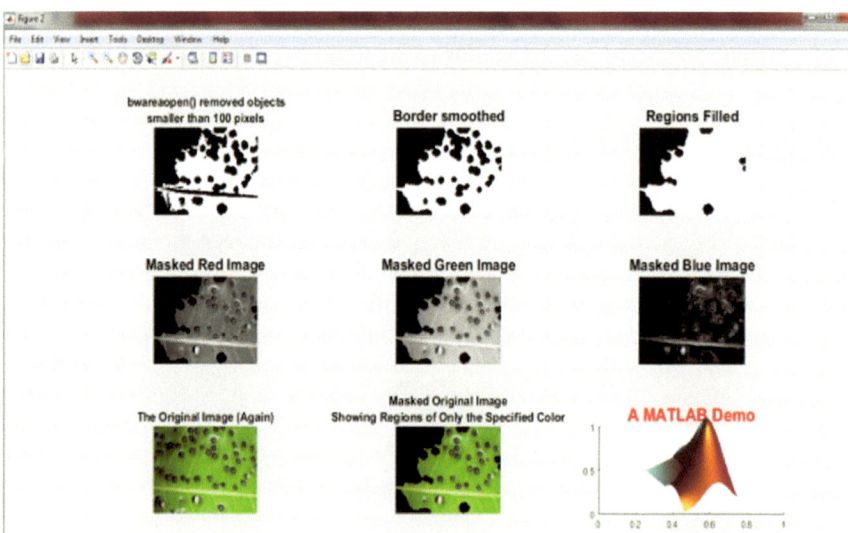

Fig. 7 Filtered mask for removing the objects smaller than 100 pixels and apply the filtering mask

The preprocessed diseased images are then passed to segmentation using k-means clustering. Here k = 3 (Red, Green, and Blue) (Fig. 8).

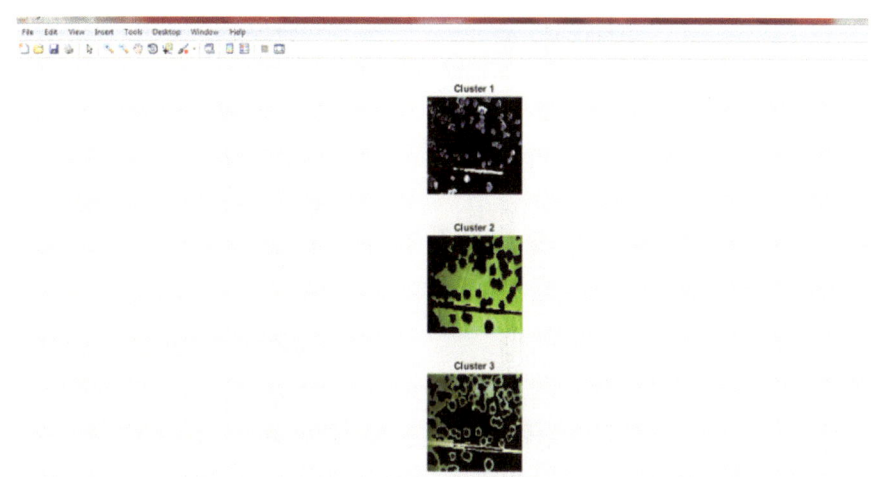

Fig. 8 Segmented mango leaf image with K = 3

After segmentation from each of this color band features are extracted through GLCM method. Thirteen harlick features are extracted for each band. So total 3 × 13 feature vector. Below is the feature vector for RED color band (Tables 2, 3 and 4).

Table 2 GLCM matrix

41781	264	383	442	567	212	46	30
371	1871	453	8	0	0	0	0
429	445	4679	550	18	0	0	0
389	28	534	2737	540	1	0	0
552	0	23	488	3114	337	0	0
209	0	0	6	316	1929	78	0
36	0	0	6	2	79	604	30
24	0	0	0	0	0	34	635

434 K. Srunitha and D. Bharathi

Table 3 Feature extraction using GLCM

Feature							
Contrast	0.07887	0.46683	0.36758	0.54123	0.51277	0.69762	0.07887
Correlation	0.97832	0.86570	0.91019	0.75103	0.71032	0.87389	0.97832
Energy	0.76258	0.79672	0.75731	0.53823	0.89470	0.48725	0.76258
Homogeneity	0.97487	0.95919	0.96254	0.9	0.97168	0.91041	0.97487
Mean	14.8438	14.150115	16.4441	17.9716	17.1185	31.5603	14.843851
Standard deviation	47.8116	48.1395	51.4194	37.6635	35.52045	56.45960733417924	47.8116
Entropy	1.70987	1.36584	1.66789	2.58288	2.84317	2.98298	1.70987
RMS	5.57477	4.31361	5.34037	7.40369	10.4504	8.11404	5.57477
Variance	2150.69625	1632.21	2305.04	1306.81	1162.22	2844.32	2150.69
Smoothness	0.99999	0.99999	0.99999	0.99999	0.99999	0.99999	0.99999
Kurtosis	15.5977	15.7654	13.7926	10.4951	27.6032	4.40083	15.5977
Skewness	3.63201	3.67442	3.40252	2.58833	4.68201	1.61292	3.63201
IDM	255	255	255	255	255	255	255

Classification gives result of classifier multi class-SVM (Support vector machine). The accuracy of svm is found out with the help of WEKA 3.8 software. Below is the result of SVM with weka.

Table 4 Confusion metrics plot using WEKA

Dataset	TP	FP	TN	FN
Anthracnose	55	6	75	150
Red rust	48	2	88	148
Sooty mould	50	5	91	140
Scab	72	3	81	130
Healthy leaf	44	1	96	145

5 Conclusion

The work contains mainly two zones the segmentation followed by classification using SVM. Over and done with the experimental analysis it is resolved that the proposed method works efficiently on leaf diseases recognition and classification. Presence of multiple diseases in one region of the leaf and the variation in their color, texture, and shape characteristics made difficult in segmentation and feature selection phase. SVM in classification phase performed well and get accuracy up to 96%. Upon implementing the process we get a conclusion that main requirements for any diseases detection are speed and accuracy. The work can also be extended by specifying suited organic timely diseases curing technique for each disease. Further need to compute the amount of disease area present on leaf.

References

1. Anand, R., Veni, S., Aravinth, J.: An application of image processing techniques for detection of diseases on brinjal leaves using K-means clustering method. In: IEEE International Conference on Circuit, Power and Computing Technologies, ICCPCT (2016)
2. Pooja, A., Mamtha, R., Sowmya, V., Soman, K.P.: X-ray image classification based on tumor using GURLS and LIBSVM. In: International Conference on Communications and Signal Processing (ICCSP'16) (2016)
3. Krishnan, M., Sumithra, M.G.: A novel algorithm for detecting bacterial leaf scorch (BLS) of shade trees using image processing. In: IEEE 11th Malaysia International Conference on Communications (2013)
4. Arivazhagan, S., NewlinShebiah, R., Ananthi, S., Vishnu Varthini, S.: Detection of unhealthy region of plant leaves and classification of plant leaf diseases using texture features. AgricEngInt CIGR J. **15**, 211–217 (2013)
5. Naikwadi, S., Amoda, N.: Advances in image processing for detection of plant diseases. Int. J. Appl. Innov. Eng. Manag. **2**(11) (2013)
6. Arivazhagan, S., NewlinShebiah, R., Ananthi, S., Vishnu Varthini, S.: Detection of unhealthy region of plant leaves and classification of plant leaf diseases using texture feature. CIGR **15**(1), 211–217 (2013)

7. Amoda, N., Naikwadi, S.: Advances in image processing for detection of plant diseases. Int. J. Appl. Innov. Eng. Manag. (IJAIEM) 2(11). ISSN: 2319-4847 (2013)

8. Jagtap, S.B., Hambarde, S.M.: Agricultural plant leaf disease detection and diagnosis using image processing based on morphological feature extraction. IOSR J. VLSI Signal Process. (IOSR-JVSP) 4(5), 24–30, Ver. I. e-ISSN: 2319-4200, p-ISSN: 2319-4197 (2014)

9. Gavhale, K.R., Gawande, U.: An overview of the research on plant leaves disease detection using image processing techniques. IOSR J. Comput. Eng. (IOSR-JCE) 16(1), 10–16, Ver. V. ISSN: 2278–8727 (2014)

10. Ratnasari, E.K., Mentari, M., Dewi, R.K., Hari Ginardi, R.V.: Sugarcane leaf disease detection and severity estimation based on segmented spots image. In: IEEE. ICTS 978-1-4799-6858-9/14/$31.00 © 2014

11. Fadzil, W.M.N.W.M., Rizam M.S.B.S., Jailani, R., Nooritawati, M.T.: Orchid leaf disease detection using border segmentation techniques. In: 2014 IEEE Conference on Systems, Process and Control (ICSPC 2014), Kuala Lumpur, Malaysia, 12–14 December 2014

12. Warne, P.P., Ganorkar, S. R.: Detection of diseases on cotton leaves using K-mean clustering method (IRJET) 02(04). e-ISSN: 2395 -0056 (2015)

13. Kaur, R., Kang, S.S.: An enhancement in classifier support vector machine to improve plant disease detection. In: IEEE 3rd International Conference on MOOCs, Innovation and Technology in Education (MITE), pp. 135–140 (2015)

14. Khirade, S.D., Patil, A.B.: Plant disease detection using image processing. Int. Conf. Comput. Commun. Control Autom. 978-1-4799-6892-3/15 $31.00 © 2015 IEEE

15. Padmavathi, S., Saipreethy, M.S., Valliammai, V.: Indian sign language character recognition using neural networks. In: IJCA Special Issue on Recent Trends in Pattern Recognition and Image Analysis, vol. RTPRIA, pp. 40–45 (2013)

16. Dinesh Kumar, C.K., Manjusha, R., Latha, P.: Comparision of image classification methods on event data. Int. J. Applied Eng. Res. 10, 29631–29640 (2015)

Personalized Research Paper Recommender System

Thota Sripadh and Gowtham Ramesh[✉]

Department of Computer Science and Engineering, Amrita School of
Engineering, Amrita Vishwa Vidyapeetham, Coimbatore, India
cb.en.p2csel5024@cb.students.amrita.edu,
r_gowtham@cb.amrita.edu

Abstract. Personalization is an emerging topic in the field of Research paper recommender systems and academic research. It is a technique to creative and efficient user profiles to achieve improved recommendations. Our work proposes a new user model to understand user behavior for personalization. This model initially extracts keywords based on the online behaviour of the user. The subsequent steps include concept extraction and user profile ontology construction to derive inferences and define relationships. The suggested model clearly depicts hierarchical ordering of the user's long-term and current research interests. Furthermore, the adoption of our model contributes to improvement of recommendations.

Keywords: Personalization · Ontology · User profile

1 Introduction

Personalization is a technique of meeting the customer's needs more effective and efficient way thus making interactions faster and easier, and increase the customer satisfaction and the likelihood of repeat visits. Personalization includes interaction with the customer, representing and analyzing customer data, tailoring based on customer profile, targeting each user separately. Some of the techniques used for personalization are 'Collaborative filtering' [1], 'Learning agent filtering' [2], 'RuleBased Filtering' [3]. Collaborative filtering means constructing user profile based on many other similar users' behaviour [1]. Learning agent filtering keeps track of user's behaviour by collecting user feedback and data implicitly. It is also known as 'Learning from the shoulder of the user' [2]. Rule-Based Filtering means constructing user profile based on the user's browsing history and refining the collected data by applying some data mining techniques [3].

Information Filtering is a process of constructing user profile which is also called as user modelling. The user profile is constructed based on the user preferences and needs of the user [4]. In this paper, user profile is constructed based on the collection of implicit or explicit user information. In our system user profile is mainly constructed

© Springer International Publishing AG 2018
D. J. Hemanth and S. Smys (eds.), *Computational Vision and Bio Inspired Computing*,
Lecture Notes in Computational Vision and Biomechanics 28,
https://doi.org/10.1007/978-3-319-71767-8_36

for identifying the user's long-term interests and current interests. The main objective of constructing user profile is finding users long term and current interest. The long-term interest is constructed based on the user's publications and areas of interest. Current interest is constructed based on the user's browsing history. These activities represent the user's interest as possible. By constructing user profile and using it in recommendation process will increase the performance of the recommender system relative to time and accuracy. Since there is more interaction with the user, accurate user's behaviour and user profile can be constructed for a specific user.

Ontologies are defined to relate knowledge to a particular domain in terms of relevant concepts, relationships between these concepts and the instances of these concepts. In the field of Information Retrieval (IR), when a user searches for a query, using traditional methods like keyword based or natural language query. A large amount of unsorted data is retrieved which is having poor semantic relations among the retrieved documents, which is irrelevant to the user. To avoid the above problems, we started constructing user profile for the user, using ontology which is specifically known as Ontology profile.

2   Related Work

This section summarizes the works that has been done in the same direction.

2.1   User Profile

Generally there are two methods to construct user profile, namely implicit construction and explicit construction. Under explicit construction the profile is constructed based on user's personal information. The drawback of the method is that it requires user's willingness and time to participate in the construction. For example, while a profile is created in MyYahoo! The web content is automatically organized based on the user information. Syskill & Webert! [5] Recommends web pages based on the explicit feedback given by the user. They recommend links based on the ratings given by the user. Apart from that system can construct a Lycos query and retrieve pages that might match a user's interest.

Implicit user information means collecting user data without the knowledge of user, where human intervention is not needed for constructing profile. Some of the data collection techniques are Browser cache, Proxy servers, Browser agents, Desktop agents, Web logs and Search logs. Browser cache is the process of collecting browser history, where the user needs to upload their browser cache in periodical time intervals. Problem with browser cache is that some users may not be interested in uploading browser cache. OBIWAN [6], Kleinberg et al. [7] and Trajkova [8] explained that proxy server works same as browser cache. Proxy servers will work only if user enables the proxy server. Problem with proxy servers is that they are used to enable user privacy and anonymous surfing. Letizia [9], WebMate [10] and Vistabar [11] explained how browser agents are added as plug-ins to the browser or standalone applications. Apart from collecting URLs and visit count, browser agents will also collect bookmarking and downloaded data. The problem with browser agents is that users have to install a separate application or plug-in and the main drawback is that the application is installed on personal computer. We can build user profiles only if the user is using that particular computer. Haystack [12] and Stuff I've Seen [13] explained desktop agents. These are the same as browser agents. Desktop agents collects all the user accessed files and activities. The drawback with desktop agents is that users need to install an extra application.

The above four approaches imposes burden on the user for collecting information. The last two approaches will not have any burden on the user as they collect data only using session ids, cookies, and/or logins.

Mobasher [14], Misearch [15], Liu et al. [16] explained web logs and search logs as capturing browsing history of the user at a given website. Drawback with weblogs is that very less information is collected since it collects information from only a single website.

All the works related to user profile, explored so far, generally give the details of extraction of user details, both implicitly and explicitly. Since these techniques are helpful for a strong foundation for our work, they have been used for progress.

2.2   Ontology

In this paper we considered open IE (OIE) for constructing ontology. Banko et al. [17] used OIE for extracting tuples from a large corpus without human input. They also implemented TEXTRUNNER which is used to arrange the tuples using probability and indexed for user queries. Wu et al. [18] proposed model for extraction of tuples from Wikipedia. To improve precision and recall, they also implemented new model called WOE, which gives better results than TEXTRUNNER. Soderland et al. [19, 20] proposed a model for constructing existing ontologies using open IE.

3   Architecture

In this section, we describe the proposed architecture. In the initial stage, we collect the browser history and publications implicitly and user's area of interests explicitly as shown in Fig. 1. From publications, we extract the abstract and is fed as input for vector converter. The similarity of the vector formats of the abstract and the browser history is found out for further proceedings in the whole architecture. When the similarity value exceeds 25% those files are recognized to be matched history files. The parallel operation to feed the ACM CCS taxonomy is done with two inputs; the user's area of interests and the keywords or indexed words collected from the user. The concept is extracted with the matched history files and the hierarchy obtained from the taxonomy. Once the concept is completely extracted, it can be developed to a user profile ontology as in Fig. 1.

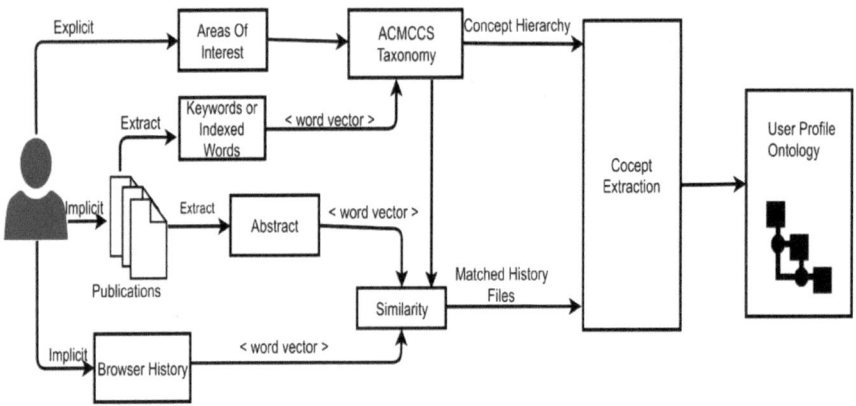

Fig. 1 Proposed architecture for building user profile ontology

3.1 Browser History

Browser history is collected implicitly from the user. It is collected in the form of URLs. From each URL, the content in the body of webpage is extracted. While extracting the content, content word count should be more than 200. While considering URLs like google.in, Gmail.in, youtube.in, Facebook.in, etc., the URLs consists only of content such as privacy, terms, settings, advertising, about, sign in and language, which is not useful. The webpage consists of a website, blog name, logo, or company name, search box, navigation bar, or menu, advertisement banners, social networking share links, breadcrumbs, feedback options, additional buttons like print, footers for visitors to continue to other web pages, hyperlinks. While collecting data, we consider only the body of the website. Each URL content is stored in a text file. Text files are pre-processed. In pre-processing, numbers, words which are a combination of alphabets and numbers, languages apart from English, special symbols are removed.

3.2 Areas of Interest

Areas of interest are collected explicitly from the user. Every research user will have his own areas of interest. Some examples are Web Security, Information Retrieval, Semantic Web, Artificial Intelligence, etc.

3.3 Publications

Publication documents are collected implicitly. Publications which we are considering are users' published research papers. From user publications, keywords or indexed words and abstracts are extracted. All the keywords are stored in a single file and abstracts are stored in separate files.

3.4 ACM CCS Taxonomy

ACM Computing Classification System (CCS) poly-hierarchical ontology is utilized in semantic web applications [21]. Collected keywords and areas of interest are matched with ccs taxonomy. The matched keywords' parent and child nodes are extracted from ACM ccs taxonomy and are considered as long term interest which are used for constructing the ontology.

3.5 Similarity

The similarity is used to find whether the user is having any current interest matched with the publications. Here cosine similarity is being used. Compared to any other similarities cosine similarity give accurate results. The input to the cosine similarity is given in the form of vectors. All abstract and web browser history files are converted into tf-idf vectors [22, 23]. Each file from browser is compared with all the abstract files. Before comparison, stop words are removed from browser history file. If the similarity measure is greater than or equal to 0.25 then the corresponding matched history file is considered.

3.6 New User

If a new user is not having any publications, then the user's areas of interest and browser history is collected. Area of interest is matched with ACM ccs taxonomy. Extracted keywords are used for user profile construction.

3.7 Matched History Files

Matched history files are considered if the cosine similarity value is greater than or equal to a threshold value of 0.25. Ontology is constructed for these matched files, using open information extraction (open IE) [24]. We used Stanford open IE to extract tuples in a sentence. Large corpus is divided into sentence, and each sentence is maximally shortened. These shortened sentence are passed to open IE, where each sentence is in triplet format.

3.8 User Profile Ontology

In user profile ontology output from the open IE and ACM ccs taxonomy are considered for constructing ontology. We considered is-a and related-to relations for constructing ontology. All the parent and child nodes are related using is-a relation. If any keyword matches in the ontology. Those are related using related-to. Output from open IE triples are considered directly for constructing ontology.

4 Experiment

This section contains experiment conducted for extracting the user interest. We have tried with four approaches and found which approach works better in extracting user interest. The experiment was conducted for small set of users and the data set includes browser history for duration of one month and corresponding user publication documents.

4.1 Approach-1

The first step to construct user profile, we consider TF-IDF (term frequency–inverse document frequency) for extracting trust key concepts. User's browser history is collected implicitly. In the next step, from each URL from history, body of the data is collected. On the collected data, TF-IDF is applied for weight calculation [25]. Words with the highest weight will be considered for trust key concepts. The main reason for using this method is, more frequently the word occurs, the higher weight is.

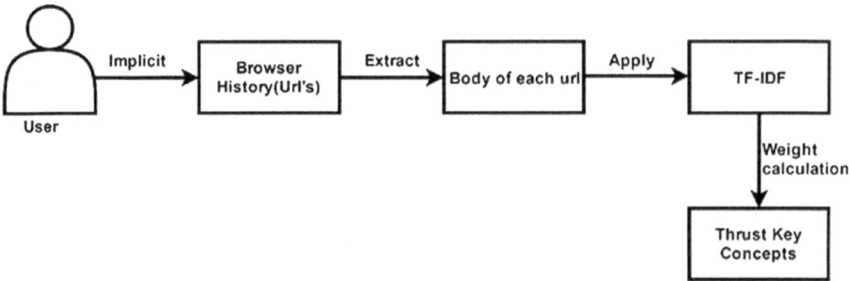

Fig. 2 Finding trust key concepts using TF-IDF

4.2 Approach-2

TF-IDF method failed to identify many of the user interested concepts as it is not considering the meaning of the words. POS tagger is able to differentiate class and its syntactic function of words.

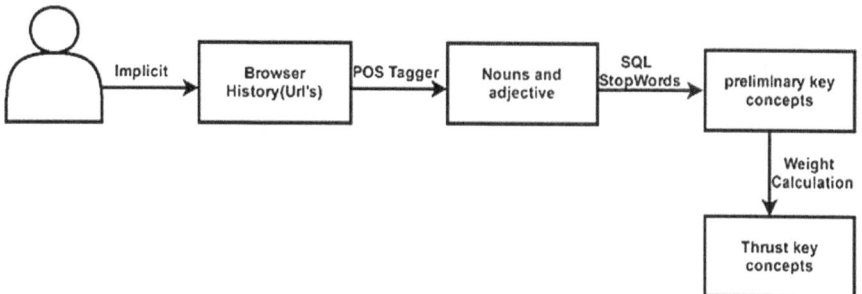

Fig. 3 Finding trust key concepts using pos tagger

The above architecture is same as Fig. 2, here we replaced tf-idf with POS (Parts of Speech) tagger. POS tagger is the process of tagging each word with parts of speech for which we used the Stanford POS tagger. To apply POS tagger, large corpus is divided into sentences and each sentence into words. Each word is tagged with parts of speech. Here we are considering only NOUNS (NN) and ADJECTIVES (JJ) since, important keywords in a sentence are either nouns or adjectives. This model is also interpreted as (Fig. 3):

$$(JJ) + (NN)*$$

where '*' indicates Zero or more occurrences and '+' indicates one or more occurrences. The above notation considers: (1) one or more nouns (2) one or more adjectives followed by one or more nouns. SQL stop words are also removed from the collected corpus.

4.3 Approach-3

Word2Vec proposed by Mikolov et al. is used to build word projections in high dimension. The author propose two architectures for learning word embedding's that are computationally less expensive than previous models. The float values in the vector represents the coordinates of the words in this high dimensional space [26].

Instead of TF-IDF, we used word2vec for getting word vectors. Word vectors of user publication document is compared against browser history documents and similar documents from browser history is retrieved. Word2vec algorithm doesn't gave better result as it was trained using a small corpus (Fig. 4).

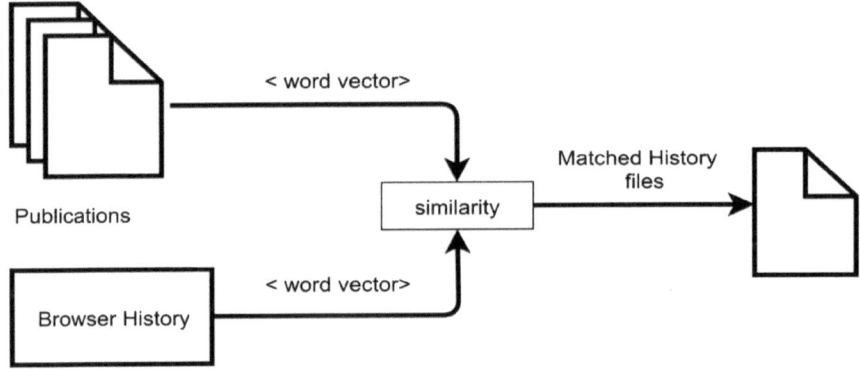

Fig. 4 Extracting matched history files using word2vec

4.4   Approach-4

Doc2vec algorithm is an extension of word2vec algorithm, used to generate high dimensional vectors which preserves the semantics of documents. Doc2vec algorithm learns fixed length representation for documents from a large corpus at much lower computational cost [27]. Document vectors of user publication as well as browser history files are generated using doc2vec. Documents in browser history which are more similar to user publication are retrieved by employing cosine similarity on doc2vec vectors. Doc2vec algorithm doesn't gave better result as it was trained using a small corpus (Fig. 5).

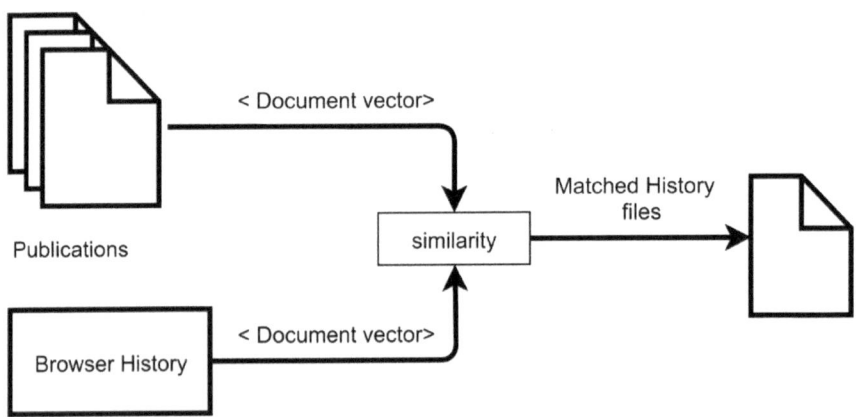

Fig. 5 Extracting matched history files using Doc2vec

Compared to word2vec and doc2vec representation TF-IDF vector gave better results.

5   Conclusion

Ontology is an efficient method for representing user profiles. A well-defined ontology could provide rich semantic information in a specific area, such as semantic equivalents of concepts, semantic types of concepts, and semantic relations between concepts, and thus has considerable potential to improve semantic understanding. In research paper recommender domain semantic understanding plays a major role in document retrieval. The personal content that captured a user's interests and computational activities are used for constructing user ontology. The proposed method outperforms the existing methods as it uses better pre-processing on user related data.

References

1. Linden, G., Smith, B., York, J.: Amazon.com recommendations: Item-to-item collaborative filtering. IEEE Internet Comput. **7**(1), 76–80 (2003)
2. Seo, Y.-W., Zhang, B.-T.: A reinforcement learning agent for personalized information filtering. In: Proceedings of the 5th International Conference on Intelligent User Interfaces. ACM (2000)
3. Adomavicius, G., Tuzhilin, A.: Expert-Driven Validation of Rule-Based User Models in Personalization Application. Applications of Data Mining to Electronic Commerce, pp. 33–58. Springer, US (2001)
4. Gauch, S., et al.: User profiles for personalized information access. Adapt Web 54–89 (2007) (Appendix: Springer-Author Discount)
5. Pazzani, M.J., Muramatsu, J., Billsus, D.: Syskill & Webert: Identifying interesting web sites. In: AAAI/IAAI, vol. 1 (1996)
6. Pretschner, A., Gauch, S.: Ontology based personalized search. In: Proceedingsof the 11th IEEE International Conference on Tools with Artificial Intelligence. IEEE (1999)
7. Kleinberg, J.M.: Authoritative sources in a hyperlinked environment. J ACM (JACM) **46**(5), 604–632 (1999)
8. Trajkova, J., Gauch, S.: Improving ontology-based user profiles. Coupling approaches, coupling media and coupling languages for information retrieval. LE CENTRE DE HAUTES ETUDES INTERNATIONALES D'INFORMATIQUE DOCUMENTAIRE, 2004
9. Lieberman, H.: Letizia: an agent that assists web browsing. IJCAI (1) **1995**, 924–929 (1995)
10. Chen, L., Sycara, K.: WebMate: A personal agent for browsing and searching. In: Proceedings of the Second International Conference on Autonomous Agents. ACM (1998)
11. Marais, H., Bharat, K.: Supporting cooperative and personal surfing with a desktop assistant. In: Proceedings of the 10th Annual ACM Symposium on User Interface Software and Technology. ACM (1997)
12. Adar, E., Karger, D., Stein, L.A.: Haystack: Per-user information environments. In: Proceedings of the Eighth International Conference on Information and Knowledge Management. ACM (1999)
13. Dumais, S., Cutrell, E., Cadiz, J.J., Jancke, G., Sarin, R., Robbins, D.C.: Stuff I've seen: A system for personal information retrieval and re-use. In: ACM SIGIR Forum, 49(2), 28–35. ACM (2016)
14. Mobasher, B.: Data Mining for Web Personalization. The Adaptive Web, pp. 90–135. Springer, Berlin, Heidelberg (2007)

15. Sieg, A., Mobasher, B., Burke, R.: Inferring user's information context from user profiles and concept hierarchies. In: Classification, Clustering, and Data Mining Applications, pp. 563–573. Springer, Berlin, Heidelberg (2004)
16. Liu, F., Yu, C., Meng, W.: Personalized web search by mapping user queries to categories. In: Proceedings of the Eleventh International Conference on Information and Knowledge Management. ACM (2002)
17. Banko, M., Cafarella, M.J., Soderland, S., Broadhead, M., Etzioni, O.: Open information extraction from the web. In: IJCAI vol. 7, pp. 2670–2676 (2007)
18. Wu, F., Weld, D.S.: Open information extraction using Wikipedia. In: Proceedings of the 48th Annual Meeting of the Association for Computational Linguistics. Association for Computational Linguistics (2010)
19. Soderland, S., Roof, B., Qin, B., Xu, S., Etzioni, O.: Adapting open information extraction to domain-specific relations. AI Mag. **31**(3), 93–102 (2010)
20. Venugopal, A., Ramesh, G.: A study on verbalization of OWL axioms using controlled natural language. Int. J. Appl. Eng. Res. (2015)
21. The 2012 ACM computing classification system. Retrieved November 22, 2017, from https://www.acm.org/publications/class-2012.
22. Gensim: models.tfidfmodel – TF-IDF model. Retrieved November 22, 2017, from https://radimrehurek.com/gensim/models/tfidfmodel.html
23. Sklearn.feature_extraction.text.TfidfVectorizer — scikit-learn 0.19.1 documentation. Retrieved November 22, 2017, from http://scikit-learn.org/stable/modules/generated/sklearn.feature_extraction.text.TfidfVectorizer.html
24. Angeli, G., Premkumar, M.J., Manning, C.D.: Leveraging linguistic structure for open domain information extraction. In: Proceedings of the 53rd Annual Meeting of the Association for Computational Linguistics. ACL (2015)
25. Gowtham, R., Krishnamurthi, I.: PhishTackle—a web services architecture for anti-phishing. Cluster Comput. **17**(3), 1051–1068 (2014)
26. Mikolov, T., Chen, K., Corrado, G., Dean, J.: Efficient estimation of word representations in vector space. arXiv preprint arXiv:1301.3781 (2013)
27. Dai, A.M., Olah, C., Le, Q.V.: Document embedding with paragraph vectors. arXiv preprint arXiv:1507.07998 (2015)

A Vision Based DCNN for Identify Bottle Object in Indoor Environment

Lolith Gopan and R. Aarthi[(✉)]

Department of Computer Science and Engineering, Amrita School of
Engineering, Coimbatore, Amrita Vishwa Vidyapeetham, Amrita University,
Coimbatore, India
cb.en.p2cvil5003@cb.students.amrita.edu, r_aarthi@cb.
amrita.edu

Abstract. Vision based detection and classification is an emerging area of
research in the field of automation. Due to the demand in automation different
fields artificial intelligent architectures plays vital role to address the issues.
Conventional architectures used for dealing computer vision problems are
heavily under control on user features. But the new deep learning techniques
have provided a substitute of automatically learning problem related features.
The classification problem can be designed based on feature learned from
DCNN. The performance of the DCNN algorithm vary based on the training. In
this paper the performance of Deep Convolutional Neural Network (DCNN) is
analyzed in classifying categories of bottle object.

Keywords: Deep convolutional neural network (DCNN) · Maxpooling
Classification

1 Introduction

One of the major challenges in developed and developing countries is to keep our
environment safe and clean. Million tons of garbage are accumulated in every year. The
survey says the electronic and plastic waste are high risk to environment safety and
human living organism. In order to solve this we are in need of more man power and
cost to segregating the waste as degradable and non-degradable items. The willingness
of people coming forward to perform segregation job is less because of less pay and
health issues. About 13 tons of hazardous wastes are producing in a single second all
over the world. And the numbers are increasing every generation. There are so many
projects are established to manage toxic wastes.

Still a waste management is an open and difficult task to solve all over the world.
The development in the field of science such as IOT, Machine learning and robotic
technologies have used to solve some of the issues. Our interest is to develop a vision
based artificial intelligent system to replace human activity of collecting and segregating the garbage. The following steps to have to address in order to build an independent decision making system.

© Springer International Publishing AG 2018
D. J. Hemanth and S. Smys (eds.), *Computational Vision and Bio Inspired Computing*,
Lecture Notes in Computational Vision and Biomechanics 28,
https://doi.org/10.1007/978-3-319-71767-8_37

1. Locating the objects such as paper, plastic and other in the given indoor environment.
2. Navigating to the object in the given environment.
3. Manipulating the object objects.

The process needs to be addressed as a combination of artificial intelligent and mechanical movement. The intelligent system plays a vital role in object learning and detection. Hence, our main objective is to detect the objects based on vision technology and current state of art methodology Deep convolution neural network (DCNN).

Vision based algorithms extracts the geometric and texture features and use these features to learn and later identify the types accordingly. In this paper the we have studied the performance of identifying the plastic bottles using the DCNN [1]. The performance of DCNN in classifying the objects is analyzed for various conditions are summarized in this paper.

2 Related Works

The greatest advantage of convolutional neural networks is that it can learn relevant features by itself. DCNN reduces the burden of feature selection in Computer vision algorithms. The extracted features from DCNN depicts the property of 'spatial invariance', where it can learn to recognize image features in any context. The machine learning algorithms [2–4] are used to train the machine to map features to labels automatically. The system is capable of learning various features like HOG [5], SIFT [6], SURF [7] and bag of features [8]. Designing subset of good features for each object class is a difficult task. But deep learning methods will extract the appropriate features from the raw image automatically. DCNN have the property of stacking analyzing the features across multiple scales and layers [9]. The layers are dedicated to accumulate the information one by one. For example, first few layers will detect edges and curves. Other layers will combines these for detecting geometric features until the detecting section is reached.

Kim et al. [10] proposed a method for detecting and tracking the underwater objects using template matching. The main focus is to navigate a robot through under water to identify various objects using template matching. It is an iterative method for finding small parts of an image that match with a template image. It is a straightforward process. In this technique template images for different objects are stored. When an image is given as input to the system, it is matched with the stored template images to determine the object in the input image.

As an extension of template based method Feris et al. [11] explains about object detection based on appearance based approach on a case study of vehicle detection in urban surveillance environments. In appearance-based models uses various two-dimensional perspectives of the object of-intrigue. Object recognition frameworks is an approach for managing 3D recognition of self-assertive objects within the sight of clutter and occlusion. The method improves the performance significantly when in scenarios with shadows, specularities, headlights, and occlusions are present.

Guodong et al. [12] introduces a model based object recognition system which uses shape-part features. The system is learned out object shape model from a small training set. Also it is made out of shape pieces of the objects in multi-scales. The major contribute of their work is to utilize the learned shape-parts based object detection in complex scene conditions.

Rhinehart et al. [13] explained a method by using arbitrarily shaped candidate regions of the images are iteratively used to detect the object from an image. They generate a simple class-specific algorithm to develop a candidate region instance in linear time in the number of low-level super pixels. This algorithm converts the input image into a directed graph which constructed through defined rules. The characteristics of the graph represents the shape information of the object globally. This strategy avoid the post processing traversing to the graph. So the computational time can be reduced.

Fan et al. [14] propose a robust color object identification and confinement algorithm. This technique can recognize every objects of similar color without any of the prior knowledge regarding the number of objects in template. To find the object candidate regions, an improved histogram back propagation algorithm is used. The intersection of the weighted histogram the will verify the presence of objects. With the color feature, their strategy can distinguish and find the item, calculate the number of objects with their scales and orientations. It works indoor as well as outdoor objects. The method can be extended to real time environments.

3 Object Detection Using DCNN

The object detection starts with the data acquisition process. The raw data (images) are directly feed into the network. The steps in DCNN depicted in Fig. 1. The preprocessed data is given as input into the defined neural network. The trained system can be extend to implement in real time environment.

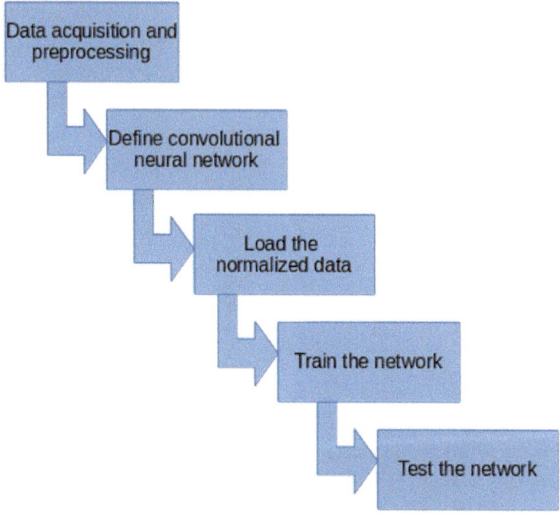

Fig. 1 Over all architecture of the bottle detection system

3.1 Data Acquisition and Preprocessing

Collect the images of bottles of different types like photographic, photo realistic and synthetic images. Images from the training set is allowed to pass through different preprocessing steps including noise removal, contrast stretching, resizing. The system will powerful if training set has images with different orientation, illumination and scale. After getting datasets the next step is to annotate the data and label it. Training image labeler is used for label object in image. It can be done manually and automatically. After performing the preprocessing operations, apply normalization on every instant of data. In normalization [15], our goal is to rescale the data along each data dimension so that the final data vectors lie in the range [0,1] or [−1,1]. For categorical data per sample mean subtraction is preferred. The mean and standard deviation of every data point is set to 0 and 1 in feature standardization method.

3.2 Define Neural Networks

A DCNN contains of a sequences of convolutional [16] and max-pooling layers, activation layer and each layer has connected with its previous layer. It is a general, hierarchical feature extractor [17] which will map input image pixel intensities into a feature vector. This will be classified by several fully connected layers in the next step. All adjustable parameters are optimized by minimizing the misclassification by reducing the error over the training set.

Each convolutional layer [18] performs a 2D convolution of its input maps with a filter of different size 3×3, 5×5, 7×7. The subsequent activations of the output maps are given by the total of the past convolutional responses which are gone through a nonlinear activation function.

Max pooling layer will perform the dimensionality reduction. The output of a thin layer is given by the most extreme activation over non-covering rectangular areas. Max-pooling makes location invariance and down-samples the image along every direction over bigger neighborhood.

Filter size of convolutional and max pooling layers are selected in such a way that a fully connected layer can combine the output into a one dimensional vector. The last layer will always be a fully connected layer which contains one output unit for all class. Here rectification linear unit is used as the activation function. Furthermore, it will be deciphered as the likelihood of a specific input image having a place with that class. Stochastic gradient is used to train the data along with negative likelihood criterion [19] as loss function.

3.3 Training the Networks

After the network has been structured for object detection application with all parameters, then it is ready for training. The classical stochastic gradient descent algorithm [20] is used for training the network. After each iteration, the network converges by reducing the error rate. The loop will terminate when it reaches a minimum error rate. Here it is 0.02. The network weight are adjusted subsequently in each iteration from initial value based on result until it converges to a value. Weight will decide the convergence. The weight value for each object is recorded in a backup file. The weight is further used to detect the object.

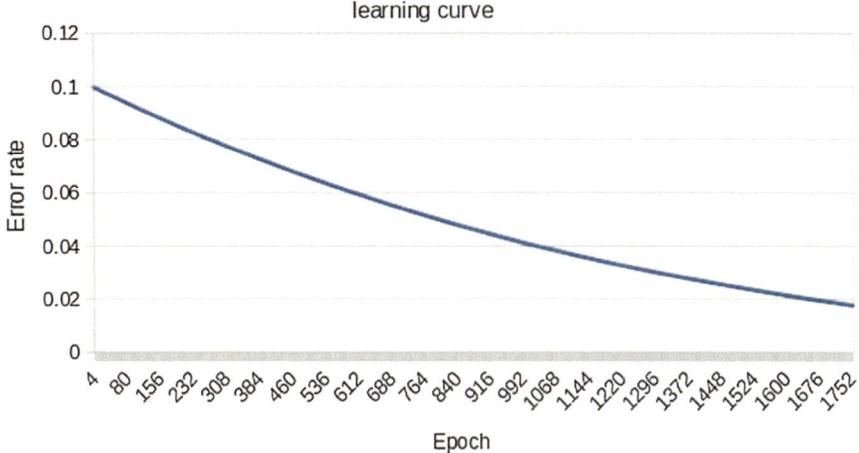

Fig. 2 Learning curve (between epoch and error rate)

Figure 2 shows the plot of number of epoch against error rate. During learning phase we can observe that at each iteration the error rate subsequently reduces. Error rate is high when the training process starts. After 300 epochs it reduces 0.02. The aim of learning process is to reduce the error.

3.4   Testing the Networks

The pre-trained weight which is obtained from the training phase is used in the testing phase. The input image is allow to pass through all layers of the neural network and parameters [21] are obtained. This values are crosschecked with the pretrained weight and identify the one which gives maximum matching with the classes. The system will consider the label to which it is closely matched.

4   Result Analysis

The system is developed using C language in darknet. The dataset consist of different images of bottle as explained in Sect. 3.1. The performance is analyzed by using the images that captured under different angles and illumination etc. Initial experiments are done on datasets with varying the parameters. The dataset consist of images are collected by considering of different size, illumination, and view point conditions. Changing the number of iteration with default parameters will not affect accuracy. Filters of different size are used in each convolutional layer. Different types of features like vertical components, horizontal components edges, corners etc. are extracted by changing the filter. Fine information of image are extracted by varying the dimension of filter. All possible lower dimensional features are extracted. In convolution layer minimum of 32 filters of size 3 × 3 used and maximum number is 1024 of same size (Table 1). Pooling layer will act as dimensionality reduction part. It will represent the entire feature vector in lower dimension. Linear rectification unit [22] (ReLu) is used as activation function. Activation function will replace the values less than zero with zero and greater than zero with one.

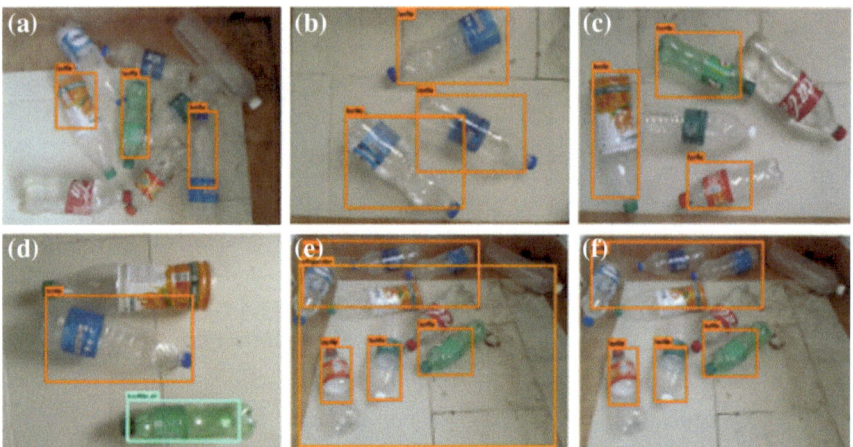

Fig. 3 Output of the system with object mask and label

Figure 3 shows the output of detecting the object at different scene. Prior detection systems will localizers to perform detection. That means it will divide the whole image into different sub regions. Then apply the DCNN model to an image at multiple locations and scales. High scoring regions of the image are considered detections. The detections are marked with a bounding boxes along with label. The result contains the correct classification an false positive. The first three images (Fig. 3a–c) are detected correctly with testing accuracy of 82%. But last three images a group of bottles are detected as a single bottle. Reducing the number of filters and filter size reduces the accuracy.

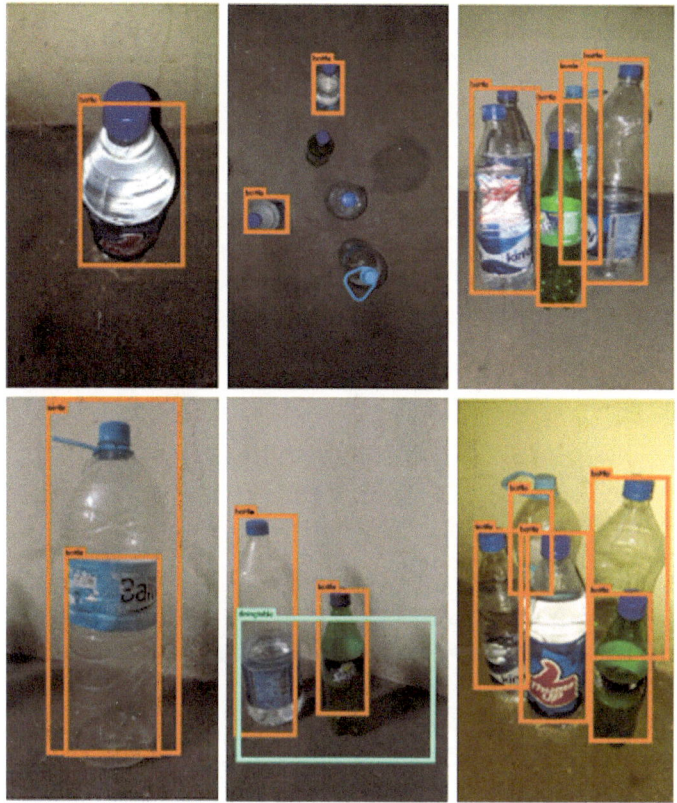

Fig. 4 DCNN output in different set of object

Figure 4 shows the result of the image with different angles. It is clear that top view detect the object at 120° with 63 and 44% testing accuracy. Also able to reduce the overlapping problem. But we can see that two bottles are detected as single and one single bottle is detected as two.

Table 1 Architecture of DCNN

	Layer	Filters	Size	Input	Output
0	conv	32	3 × 3/1	416 × 416 × 3	416 × 416 × 32
1	max		2 × 2/2	416 × 416 × 32	208 × 208 × 32
2	conv	64	3 × 3/1	208 × 208 × 32	208 × 208 × 64
3	max		2 × 2/2	208 × 208 × 64	104 × 104 × 64
4	conv	128	3 × 3/1	104 × 104 × 64	104 × 104 × 128
5	conv	64	1 × 1/1	104 × 104 × 128	104 × 104 × 64
6	conv	128	3 × 3/1	104 × 104 × 64	104 × 104 × 128
7	max		2 × 2/2	104 × 104 × 128	52 × 52 × 128
8	conv	256	3 × 3/1	52 × 52 × 128	52 × 52 × 256
9	conv	128	1 × 1/1	52 × 52 × 256	52 × 52 × 128
10	conv	256	3 × 3/1	52 × 52 × 128	52 × 52 × 256
11	max		2 × 2/2	52 × 52 × 256	26 × 26 × 256
12	conv	512	3 × 3/1	26 × 26 × 256	26 × 26 × 512
13	conv	256	1 × 1/1	26 × 26 × 512	26 × 26 × 256
14	conv	512	3 × 3/1	26 × 26 × 256	26 × 26 × 512
15	conv	256	1 × 1/1	26 × 26 × 512	26 × 26 × 256
16	conv	512	3 × 3/1	26 × 26 × 256	26 × 26 × 512
17	max		2 × 2/2	26 × 26 × 512	13 × 13 × 512
18	conv	1024	3 × 3/1	13 × 13 × 512	13 × 13 × 1024
19	conv	512	1 × 1/1	13 × 13 × 1024	13 × 13 × 512
20	conv	1024	3 × 3/1	13 × 13 × 512	13 × 13 × 1024
21	conv	512	1 × 1/1	13 × 13 × 1024	13 × 13 × 512
22	conv	1024	3 × 3/1	13 × 13 × 512	13 × 13 × 1024
23	conv	1024	3 × 3/1	13 × 13 × 1024	13 × 13 × 1024
24	conv	1024	3 × 3/1	13 × 13 × 1024	13 × 13 × 1024
25	conv	425	1 × 1/1	13 × 13 × 1024	13 × 13 × 425
26	Fully connected layer				

5 Conclusion and Future Work

In this paper we have analyzed the performance of existing DCNN in classifying the bottle in the indoor environment. The DCNN Architecture built with 22 convolutional layers and 5 max pooling layers for detecting the bottle. This work can be improved by increasing the number of training dataset to improve the accuracy. Further the investigation on difference DCNN network under varying parameter condition as needed for extending to multiclass classification problem. We have been stepping into the utilizing the latest technology to deal with the effective waste management.

Acknowledgements. We would like to extend the heartfelt gratitude to the faculty-in-charge of Amrita-Cognizant Innovation Lab, Department of Computer science and Engineering, Amrita school of Engineering, Coimbatore for the support extended in carrying out this work.

References

1. Ciresan, D.C., et al. Flexible, high performance convolutional neural networks for image classification. In: IJCAI Proceedings-International Joint Conference on Artificial Intelligence, vol. 22, no. 1 (2011)
2. Jain, A., Srivastava, S., Soman, S.: Transfer learning using adaptive SVM for image classification. In: 2013 IEEE Second International Conference on Image Information Processing (ICIIP), IEEE (2013)
3. Harisinghaney, A., et al.: Text and image based spam email classification using KNN. Naïve Bayes and Reverse DBSCAN algorithm. In: 2014 International Conference on Optimization, Reliability, and Information Technology (ICROIT), IEEE (2014)
4. Mottalib, M.M., et al.: Fabric defect classification with geometric features using Bayesian classifier. In: International Conference on Advances in Electrical Engineering (ICAEE). IEEE (2015)
5. Yamauchi, Y., et al.: Relational hog feature with wild-card for object detection. In: 2011 IEEE International Conference on Computer Vision Workshops (ICCV Workshops), IEEE (2011)
6. Alhwarin, F., et al.: Improved SIFT-features matching for object recognition. In: BCS International Academic Conference (2008)
7. Luo, Y., Chen, Y.: Robust matching algorithm based on SURF. In: 2015 12th International Computer Conference on Wavelet Active Media Technology and Information Processing (ICCWAMTIP), IEEE (2015)
8. Sampath, A., et al.: A study of household object recognition using SIFT-based bag-of-words dictionary and SVMs. In: Proceedings of the International Conference on Soft Computing Systems. Springer, India (2016)
9. Jose, J.T., Amudha, J., Sanjay, G.: A Survey on spiking neural networks in image processing. In: Advances in Intelligent Informatics, pp. 107–115. Springer International Publishing (2015)
10. Kim, D., et al.: Object detection and tracking for autonomous underwater robots using weighted template matching. In: 2012 Oceans-Yeosu, IEEE (2012)
11. Feris, R., et al.: Appearance-based object detection under varying environmental conditions. In: 2014 22nd International Conference on Pattern Recognition (ICPR), IEEE (2014)
12. Guodong, C., et al.: A learning algorithm for model-based object detection. Sens. Rev. **33**(1), 25–39 (2013)
13. Rhinehart, N., et al.: Visual chunking: a list prediction framework for region-based object detection. In: 2015 IEEE International Conference on Robotics and Automation (ICRA), IEEE (2015)
14. Fan, B., et al.: A novel color based object detection and localization algorithm. In: 2010 3rd International Congress on Image and Signal Processing (CISP), vol. 3, IEEE (2010)
15. Zhang, H., Lin, H., Li, Y.: Impacts of feature normalization on optical and SAR data fusion for land use/land cover classification. IEEE Geosci. Remote Sens. Lett. **12**(5), 1061–1065 (2015)
16. Liu, L., Shen, C., van den Hengel, A.: The treasure beneath convolutional layers: Cross-convolutional-layer pooling for image classification. In: Proceedings of the IEEE Conference on Computer Vision and Pattern Recognition (2015)
17. Akintayo, A., Sarkar, S.: A symbolic dynamic filtering approach to unsupervised hierarchical feature extraction from time-series data. In: 2015 American Control Conference (ACC), IEEE (2015)

18. Bayar, B., Stamm, M.C.: A deep learning approach to universal image manipulation detection using a new convolutional layer. In: Proceedings of the 4th ACM Workshop on Information Hiding and Multimedia Security. ACM (2016)
19. Lee, D.D., Seung, H.S.: Algorithms for non-negative matrix factorization. In: Advances in Neural Information Processing Systems (2001)
20. Mandic, D.P.: A generalized normalized gradient descent algorithm. IEEE Signal Process. Lett. **11**(2), 115–118 (2004)
21. Martinez, J.I., Nakano, M.K., Higuchi, K.: Parameter estimation in neural networks by improved version of simultaneous perturbation stochastic approximation algorithm. In: 2009 ICCAS-SICE, IEEE (2009)
22. Zhang, C., Woodland, P.C.: DNN speaker adaptation using parameterised sigmoid and ReLU hidden activation functions. In: 2016 IEEE International Conference on Acoustics, Speech and Signal Processing (ICASSP), IEEE (2016)

Segmentation of Brain Parts from MRI Image Slices Using Genetic Algorithm

K. Vikram, Hema P. Menon[(⊠)], and Dhanya M. Dhanalakshmy

Department of Computer Science and Engineering, Amrita School
of Engineering, Coimbatore, Amrita Vishwa Vidyapeetham, Amrita University,
Coimbatore, India
cb.en.p2cvil5015@cb.students.amrita.edu,
{p_hema, md_dhanya}@cb.amrita.edu

Abstract. In this work a genetic algorithm based approach for segmenting the parts of brain MRI (Magnetic Resonance Imaging) image slices has been presented. Segmentation of the brain MRI image has been a challenging task and an open area for research off late due to reason that, the intensity differences between the different regions present in the image is very less. Hence a complete automation of segmentation process is difficult. In this work the various parameters of the genetic algorithm has been analyzed and an oprimized threshold value has been determined based on the slice type. The complexities in the segmentation algorithm and the challenges have also been reported.

Keywords: Brain image · Evolutionary computing · Genetic algorithm
Magnetic resonance imaging (MRI) · Segmentation

1 Introduction

Medical Imaging is the process of producing a visual impression of the internal portions of a human for further analysis of clinical roles and medical purposes. Also, it can be viewed for the functions of the organs and tissues in a human body. The brain image could be acquired by various methods like (i) Magnetic Resonance Image (MRI), (ii) X-radiation (X-ray), (iii) Polyethylene terephthalate (PET), (iv) Functional MRI (fMRI), (v) Computed Tomography (CT), (vi) Single-photon Emission Computed Tomography (SPECT). In this paper, the technique of MRI has been addressed. The MRI is widely used for brain imaging, as it provides three views of a brain image, namely: Axial View, Coronial View and Sagittal View as shown in Fig. 1. The Axial View MRI will be a series of images from below of the chin to the top of the head, Coronial View will be of from the back of the head to the nose and the Sagittal View MRI took from the side of the ear to another ear.

The brain of a human contains different types of regions in it. The three types of the major areas that are present in the brain images are namely: (i) White Matter (WM), (ii) Grey Matter (GM) and (iii) Cerebro-spinal Fluid (CSF) is as shown in Fig. 1. The

© Springer International Publishing AG 2018
D. J. Hemanth and S. Smys (eds.), *Computational Vision and Bio Inspired Computing*,
Lecture Notes in Computational Vision and Biomechanics 28,
https://doi.org/10.1007/978-3-319-71767-8_38

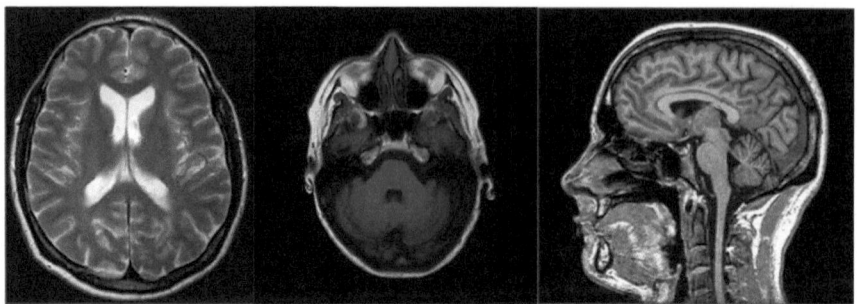

Fig. 1 Axial, coronial and sagittal view of MRI brain image

main objective in most of the segmentation process in medical imaging will be on extracting these regions and highlighting them accordingly.

Image Segmentation is the technique of partitioning an image into several regions based on the set of pixels and intensities in the digital image. Each of the regions that have been differentiated from one to another is based on its characteristics (i.e. color, texture or intensity). The White Matter, Grey Matter and CSF have been extracted using Image Segmentation techniques [1–4] and also the tumor detection in MRI Brain Image can be segmented by various studies [5–8]. El-Melegy et al. [2] proposed a non-parametric technique in such a way that it segments the 4 types of tissues on the MRI brain images, namely WM tissue, GM tissue, CSF and the remaining tissue as Non-Brain tissue. This paper concludes that it gives better performance in segmentation even at a higher degree of noise and bias. Shanthi and Kumar [9] has proposed a system of segmenting the DICOM images of brain using seed growth and the thresholding technique. This paper shows that the high frequency speckles in the images have been removed and gives good results. Unsupervised segmentation methods based on random walks proposed by Desrosiers [10] has been involved in segmenting the MRI of a brain image. The computed results on 3D Brain MRI from the IBSP (i.e., Internet Brain Segmentation Repository) shows that the computation is really efficient and performs well. Siddique et al. [3] used a method of region growing technique and seed pixel for the automatic segmentation of an brain MRI image. This paper implements in exact categorization of the brain parts such as WM, GM and CSF and the ventricular regions. Duth et al. [11] proposed a system of RSKFCM and level set method for the segmentation of brain MRI image which gives a promising results with a reduced time complexity. Arunkumar et al. [12] had proposed a system of optimzation algorithm to detect the ventricle region or the eye ball region of the MRI Brain Image. Song et al. [13] proposed a method of segmentation in MRI brain image into WM, GM, CSF using hierarchical tissue segmentation method. The neural disorders such as epilepsy, Alzheimer's disease can be detected at the stage of early by the proposed method of Shanthi et al. [14]. Hussain et al. [15] proposed a system that provides a result based on segmenting the normal tissues from the MRI images such as WM, GM, CSF and also it extracts the tumor from abnormal images to identify the pathological tissues like Edema.

1.1 Genetic Algorithm

Ghassabeh et al. [16] proposed a method of improver FCM algorithm that has been used to segment the brain MRI image using the optimization method in genetic algorithm. This paper concludes that the noisy images have been segmented using the efficient segmentation method which results in a good rate to get the desired values. Jansi and Subashini [17] proposed a method of clustering the brain image with the help of clustering methods such as K-Means and FCM using the Genetic algorithm. The Genetic algorithm is integrated with FCM to determine the global centroid value.

Balafar et al. [18] had proposed a system of segmenting the brain image using a combination of genetic algorithm and FCM. This proposed system is used in initializing the center of the clusters and this initialized center is founded out by the genetic algorithm. Saha and Bandyopadhyay [19] presented a method of fuzzy-VGAPS which automatically segment the brain tissue classes and also provides better results than the other 2 methods. Kumar et al. [20] has proposed a system of segmenting the MRI of a brain image using evolutionary computational technique. This paper thus concludes that the Robust Spatial Kerneled FCM (RSKFCM) with genetic algorithm provides better results than the other FCM methods.

2 Materials and Methods

2.1 Methods of Segmentation

There are various types of segmentation methods in partitioning the MRI images. Basically, they are of two types: (i) Basic Segmentation Methods and (ii) Advanced Segmentation Methods. The advanced segmentation methods have some more accuracy than the basic segmentation methods. The segmentation methods may be divided as many types such as (i) Manual Segmentation, (ii) Hybrid Segmentation Methods, (iii) Thresholding, (iv) Atlas Based Methods, (v) Intensity Based Segmentation, (vi) Surface Based Methods, (vii) Region Growing, (viii) Classification, (ix) Clustering, (x) Neural Network Methods.

2.2 Process of Segmentation

The MRI of a brain image is partitioned into several parts of regions based on its intensity. The main aim of the segmentation is to differentiate the different regions in an image and then label each region into its respective characteristics. The objects/boundaries in the images like lines, curves, etc. are located using the Image Segmentation. Each region that contains of different portions which varies from the other can share the certain characteristics. A collective set of areas that are pulled from the full picture or from a lot of contour regions, will render the resolution of image segmentation. The regions that are separated from the other region varies with different properties in each pixel of the region such as color, intensity or texture.

2.3   Methodology

(i) Initialization: Initialize the parameters of population, selection, mutation and crossover (n_population, n_iterations, n_thresholds, p_selection, p_mutation, p_crossover). Single chromosome is implemented as vector of binary numbers.

(ii) Population: Population is represented as a matrix, in which each row is denoted as a single chromosome and the number of rows corresponds to the size of population that was generated randomly. The algorithm works only with grayscale images, so we always convert each image at the beginning before we start to work with it.

(iii) Evaluation of fitness: The fitness is evaluated here by the ratio of inter and intra variance object. The evaluation made here is based on sum of intra-object variance. The solution with the lowest sum is the most accurate.

(iv) Selection: The best solution is selected as a current solution which passes as the new generation. The number of solution to be progressed here depends on the proportion of selection.

(v) Crossover: Chromosomes needed to crossover, are selected from two vectors randomly containing permutated indexes of chromosomes. The number of chromosomes needed to be altered depends on the ratio of crossover. One-point crossover is used here where the point of crossover is randomly generated with an uniform distribution.

(vi) Mutation: In the mutation process, the first chromosome is likely to be chosen from a randomly permutated indexes of chromosomes. The number of chromosomes to be chosen for transformation depends on the ratio of mutation. Here, only one gene of a chromosome is used to be mutated.

2.4   Algorithm

1. Input the MRI Image.
2. Assign the initialization factors such as n_population: defines the size of population; which contains different solutions n_iterations: number of iterations; decides how many times the loop should process n_thresholds: number of desired thresholds, i.e. the number of segmented regions we can get in the image. p_selection, p_mutation, p_crossover: probability of selection, mutation and crossover (The sum of the probability is set to be 1, so that the size of population does not change with the iterations).
3. The fitness function here works with the vectorized image (i.e. each pixel intensities of the input MRI image is assigned into a single column) multiplied with the population matrix, number of thresholds and the best solution is chosen by sorting these multiplied values and the thresholds. Here, it is the minimization problem so the solution with lowest sum is the most appropriate one.

4. A new population is then created by iterating the subsequent phases: (i) Two parent chromosomes are selected from the population, according to their fitness value. Here, the selection is based on the Roulette-Wheel selection method. (ii) The parents will form a new offspring (children) with the probability of crossover. If there's no type of crossover, the offspring copy will be the same as of the parents. (iii) A new offspring will be altered at each position in the chromosome, with a probability of mutation.
5. Thus the created new population is used for the continuous running of algorithm.
6. If the condition is satisfied it automatically stops the process and returns the best solution that's available in the current population, else go to step 3.

3 Results and Discussions

3.1 Varying Thresholds

Here (population, iterations, no_of_bins) = (20, 10, 256) and then the threshold is varied as on Table 1. Threshold 2 gives 3 segments of the image, threshold 3 gives 4 segments of the image and so on.

Table 1 Varying thresholds

Input MRI image	Output clustered image				
	Varying thresholds				
	2	4	6	8	10

3.2 Varying Population

Here (threshold, iterations, no_of_bins) = (10, 10, 256) and then the population is varied. Population = 20 fixed from the below results in Table 2, since there's no much variations. The inference from these results are changing population doesn't give much variations.

Table 2 Varying population

Input image	Output clustered image				
	Varying population				
	20	40	80	160	300

3.3　Varying Number of Bins

Here (population, iterations, threshold) = (20, 10, 10). The no of bins = 256 is fixed, since the grey thersholded image has the intensity variations of 0 to 255. So, maximum no of bins may be 256 (Table 3).

Table 3 Varying number of bins

Input MRI image	Output clustered image				
	Varying number of bins				
	50	100	150	200	256

3.4　Varying Number of Iterations

Here (population, no of bins, threshold) = (20, 256, 10) and the iterations can be varied based on the image intensity variations. If there's only slight intensity variations in the image, the iterations can be increased to give the good results as shown in Table 4.

Table 4 Varying number of iterations

Input MRI image	Output clustered image				
	Number of iterations				
	5	10	50	100	200

3.5 Genetic Algorithm Based Multi-thresholding

The results of the genetic algorithm based on thresholding with the image size and connected components is as shown in Table 5.

Table 5 Genetic algorithm based multi-thresholding

Input MRI image	Output clustered image					
	threshold—2		threshold—5		threshold—10	
Size conn_objs	[260 × 320] 40		[260 × 320] 30		[260 × 320] 21	
Size conn_objs	[1365 × 1365] 347		[1365 × 1365] 323		[1365 × 1365] 314	
Size conn_objs	[353 × 353] 45		[353 × 353] 32		[353 × 353] 21	

4   Conclusion

The applicability of genetic algorithms for optimizing the threshold in case of segmenting Axial MRI Brain Image slices has been discussed. Although there are many researches based on the MRI image segmentation, there is still scope for further research due to the challenges present in the characteristics of the MRI Brain Images. The parameters of the genetic algorithm has been varied and the results has been obtained with respective to the number of population, number of bins and thresholds. The proposed algorithm here employs the image segmentation using genetic algorithm with the multi-thresholding based techniques through which the number of thresholds is changed to get the desired segments that is needed. In future work, the genetic algorithm can be applied with the other clustering methods.

Acknowledgements. The authors would like to extend the heartfelt gratitude to the faculty-in-charge of Amrita-Cognizant Innovation Lab, Department of Computer Science and Engineering, Amrita School of Engineering, Coimbatore for the support extended in carrying out this work.

References

1. He, S.J., Weng, X., Yang, Y., Yan, W.: MRI brain images segmentation. In: The 2000 IEEE Asia-Pacific Conference on Circuits and Systems, IEEE APCCAS, Tianjin, pp. 113–116 (2000)
2. El-Melegy, M., Hasan, Y., Mokhtar, H.: MRI brain tissues segmentation using non-parametric technique. In: ICCES International Conference on Computer Engineering and Systems, Cairo, pp. 185–190 (2008)
3. Siddique, I., Bajwa, I.S., Naveed M.S., Choudhary, M.A.: Automatic functional brain MR image segmentation using region growing and seed pixel. In: ITI 4th International Conference on Information and Communications Technology, Cairo, pp. 1–2 (2006)
4. Ortiz, A., Górriz, J.M., Ramírez, J., Salas-Gonzalez, D.: MR brain image segmentation by growing hierarchical SOM and probability clustering. Electron. Lett. **47**(10), 585–586 (2011)
5. Dawngliana, M., Deb D., Handique, M., Roy, S.: Automatic brain tumor segmentation in MRI: hybridized multilevel thresholding and level set. In: International Symposium on Advanced Computing and Communication (ISACC), Silchar, pp. 219–223 (2015)
6. Vaishnavee, K.B., Amshakala, K.: An automated MRI brain image segmentation and tumor detection using SOM-clustering and proximal support vector machine classifier. In: IEEE International Conference on Engineering and Technology (ICETECH), Coimbatore, pp. 1–6 (2015)
7. Birgani, P.M., Ashtiyani, M., Asadi, S.: MRI segmentation using fuzzy C-means clustering algorithm basis neural network. In: ICTTA 23rd International Conference on Information and Communication Technologies: From Theory to Applications, Damascus, pp. 1–5 (2008)
8. Deepa, S.N., Devi, B.A.: Artificial neural networks design for classification of brain tumour. In: International Conference on Computer Communication and Informatics (ICCCI), Coimbatore, pp. 1–6 (2012)

9. Shanthi, K.J., Kumar, M.S.: Skull stripping and automatic segmentation of brain MRI using seed growth and threshold techniques. In: International Conference on Intelligent and Advanced Systems, ICIAS, Kuala Lumpur, pp. 422–426 (2007)
10. Desrosiers, C.: An unsupervised random walk approach for the segmentation of brain MRI. In: IEEE 11th International Symposium on Biomedical Imaging (ISBI), Beijing, pp. 337–340 (2014)
11. Duth, P.S., Viswanath, V., Sreekumar, P.: Robust MRI brain image segmentation method: a hybrid approach using level set and fuzzy C-means clustering. Int. J. Eng. Technol (IJET) **8** (2) (2016)
12. Arunkumar, C., Husshine, S.R., Giriprasanth, V.P., Prasath, A.B.: Automated classification and segregation of brain MRI images into images captured with respect to ventricular region and eye-ball region. ICTACT J. Image Video Process. 831–834 (2014)
13. Song, T., Huang, M., Lee, R.R., Gasparovic, C., Jamshidi, M.A.: A hierarchical tissue segmentation approach in brain MRI images. In: Proceedings of the World Automation Congress, Seville, pp. 1–8 (2004)
14. Shanthi, K.J., Kumar, M.S., Kesavadas, C.: Neural network model for automatic segmentation of brain MRI. In: Asia Simulation Conference—7th International Conference on System Simulation and Scientific Computing, ICSC, Beijing, pp. 1125–1128 (2008)
15. Hussain, S.J., Savithri, A.S., Devi, P.V.: Segmentation of brain MRI with statistical and 2D wavelet features by using neural networks. In: 3rd International Conference on Trendz in Information Sciences and Computing (TISC2011), Chennai, pp. 154–159 (2011)
16. Ghassabeh, Y.A., Forghani, N., Forouzanfar, M., Teshnehlab, M.: MRI fuzzy segmentation of brain tissue using IFCM algorithm with genetic algorithm optimization. In: IEEE/ACS International Conference on Computer Systems and Applications, Amman, pp. 665–668 (2007)
17. Jansi, S., Subashini, P.: Modified FCM using genetic algorithm for segmentation of MRI brain images. In: IEEE International Conference on Computational Intelligence and Computing Research (ICCIC), Coimbatore, pp. 1–5 (2014)
18. Balafar, M.A., Ramli, A.R., Saripan, M.I., Mahmud, R., Mashohor, S., Balafar, H.: MRI segmentation of medical images using FCM with initialized class centers via genetic algorithm. In: International Symposium on Information Technology, Kuala Lumpur, Malaysia, pp. 1–4 (2008)
19. Saha, S., Bandyopadhyay, S.: A fuzzy genetic clustering technique using a new symmetry based distance for automatic evolution of clusters, ICCTA'07, Computing: theory and application, pp. 309–314 (2007)
20. Kumar, S.V.A., Harish, B.S., Guru, D.S.: Segmenting MRI brain images using evolutionary computation technique. In: International Conference on Cognitive Computing and Information Processing (CCIP), Noida, pp. 1–6 (2015)

Air Pollution Modelling from Meteorological Parameters Using Artificial Neural Network

Sateesh N. Hosamane[1](✉) and G. P. Desai[2]

[1] Department of Chemical Engineering, K. L. E. DR. M. S. Sheshgiri College
of Engineering and Technology, Belagavi 590 008, Karnataka, India
satishosamane@gmail.com
[2] Department of Chemical Engineering, Bapuji Institute of Technology,
Davanagere 577005, Karnataka, India
desai_gp@yahoo.com

Abstract. The aim of this study is to develop neural network air quality prediction model for PM_{10} (particle whose diameter is less the 10 μm), NO_2 and SO_2. A multilayer neural network model with a hidden recurrent layer is used to predict pollutant concentrations at four monitoring sites in Belagavi city of Karnataka State, India. The Levenberg Marquardt algorithm is used to train the network. A combination of input variables were investigated taking into the predictability of meteorological input variables and the study of model performance. The meteorological variables air temperature, wind speed, wind direction, rainfall and relative humidity were considered as input variables for this study. The results show very good agreement between measured and predicted pollutant concentrations. The performance of the developed model was assessed through performance index. The models developed have good prediction performance (>85%) for all the pollutants. The proposed models were predicted pollutant concentration with relatively good accuracy and outputs were proven to be satisfactory by measuring of the goodness of fit and by mean absolute percentage error.

Keywords: ANN · Air pollution · Modelling · Prediction

1 Introduction

Air pollution is a serious environmental problem across major cities in India. The problem of air pollution has significant health impacts on human and the environment [1, 2]. Atmospheric pollution sourced by industrial activities and congestion of roads and traffic is of principal concern. Air pollution concentration is mainly due to various combination of pollutant and their physiochemical interactions or processes with other components in the atmosphere, properties of earth surface and geometry [3]. PM_{10}, SO_2 and NO_2 are the three main pollutants responsible for the degradation of the ambient air quality. These pollutants exert a wide range of impacts on biological, physical, and economic systems, especially, effects on plant and human health are of particular

© Springer International Publishing AG 2018
D. J. Hemanth and S. Smys (eds.), *Computational Vision and Bio Inspired Computing*,
Lecture Notes in Computational Vision and Biomechanics 28,
https://doi.org/10.1007/978-3-319-71767-8_39

concern. Worldwide there are more deaths due to poor air quality than from automobile accidents [4]. Air pollution is a serious problem in urban areas, expose to higher air pollutant concentrations for longer period may increase the risk of asthma, respiratory and cardiovascular systems, cancer and mortality. Therefore, there is an urgent need to development an efficient forecasting system to provide air quality information to the general public. In recent years statistical methods have been used for air pollution predictive models which are not capable of predicting short term pollution levels. Regression modelling is most popular statistical approach has been used to develop air quality predictive models in number of studies [5]. Linear regression models have good predictability with linear process. They will underperform when we try to model nonlinear processes. Therefore, artificial neural networks (ANN) approach is the best as compared with statistical linear methods, especially where the problem being analyzed includes nonlinear behavior [6]. ANN models have been used for the forecasting a wide range of pollutants and their concentrations at various time scales, with very good results [7, 8]. The modelling tools are widely used in many scientific fields, especially in environmental sciences and have been widely applied for modelling air pollutant concentrations with the aim to forecast them. ANN modelling is the best tool for nonlinear relationships. Their performance capability is superior when compared to other statistical methods [9, 10]. The literature showed that fairly good estimates can be achieved by different models developed using ANN. The most common structure of the neural network is the "feed forward" where the data flow from input to output units is strictly feed forward and are able to find and identify complex patterns in datasets which may not be well described by a simple mathematical formula or a set of known processes [11]. The neural network approach is applied for highly complex pattern recognition and to solve the problem in presence of noisy dataset [12, 13]. ANN models performed better when they combined with traditional deterministic modelling approach [14]. In this paper, the back propagation algorithms are used to model and predict the daily average concentrations of PM_{10}, NO_2 and SO_2 using meteorological variables (Input).

1.1 Artificial Neural Network (ANN)

ANNs are computing systems motivated by biological models, and made up of a number of easy and highly interconnected processing components, which process information by its dynamic state response to external inputs. The processing components called neurons are organized into interconnected layers [14–16]. Multilayer perceptron (MLP) was introduced by Rumelhart in 1986. It is most widely used and composed of three layers of neurons. Several neurons are organised into input layer, hidden layer and the output layer. The input layers take the value from the model input and serves to pass the values to the hidden layer as shown in Fig. 1. In this study, feed forward ANN was used to predict air pollutant concentrations based on different meteorological variables. The number of hidden layers were selected by using trial and error procedure varying for 8–12 hidden layers in the network structure was examined in order to check the effectiveness of network predictability.

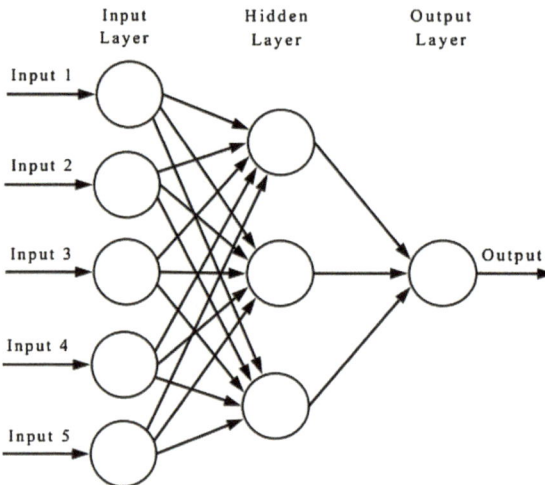

Fig. 1 Architecture of multilayer neural network

2 Study Area

Figure 2 shows satellite view of Belagavi city highlighted with four monitoring stations, is located at 15.87 °N 74.5 °E, with an average elevation of 751 m. The city is in the north western parts of Karnataka state and is along the border of two states, Goa and Maharashtra. Belagavi is a district head quarter and officially known as second capital of Karnataka State, India.

Fig. 2 Satellite picture of Belagavi city showing four monitoring sites

Belagavi is known for its foundry Hub. More than 200 foundries are producing automotive and industrial castings of ferrous base and supporting ancillaries and one of the largest alumina producing company is located at Belagavi city. Vehicle population and traffic, industrial emissions are the major sources of pollution in the Belagavi atmosphere, a problem that has been annoyed by the drastic increase in the number of mobile sources from last five years. Therefore, there is an urgent need for the assessment and evaluation of air quality in Belagavi city. It has been found that air pollutant concentrations of PM_{10}, SO_2, and NO_2 distribution is strongly affected by meteorological factors.

2.1 Data Sets

The meteorological data temperature (in °C), wind speed (in mps), wind direction (in degrees), relative humidity (in %) and rain (in mm) were collected from meteorological station located at Sambra, Belagavi city. Table 1 shows the mean, standard deviation, minimum and maximum values of meteorological parameters and pollutant concentrations for the year 2011–2013. The mean annual temperature is around 24.65 °C, annual relative humidity is between 17.15 and 99.07% and prevailing wind direction NNE to NNW and annual wind speed is between 2 and 13.73 mps and heavy rain occurs in the period of June to September is between 0 and 9.23 mm.

Table 1 Daily average means, standard deviation, minimum and maximum values of Meteorological parameters and pollutant concentrations of PM_{10}, SO_2 and NO_2

Variable	Mean	Standard deviation	Minimum	Maximum
Temperature, °C	24.65	2.57	20.12	34.09
WS, mps	6.58	2.42	2.00	13.73
WD, degrees	195.09	82.52	22.05	337.51
Rain, mm	0.42	0.88	0.00	9.23
Humidity, %	66.04	20.93	17.51	99.07

Autonagar				_Railway Station_			
	Mean	SD	Min	Max	Mean	SD	Min	Max
PM_{10}, µg/m^3	63.07	34.88	12.35	245.85	88.71	34.46	15.03	208.95
SO_2, µg/m^3	5.98	2.36	0.56	14.87	7.71	2.44	1.98	17.57
NO_2, µg/m^3	20.98	11.98	2.43	77.56	26.58	9.65	7.16	61.26

Vadgaon				_Udyambag_			
	Mean	SD	Min	Max	Mean	SD	Min	Max
PM_{10}, µg/m^3	46.75	17.41	10.15	101.23	53.78	18.25	9.87	152.36
SO_2, µg/m^3	3.97	1.58	0.98	10.56	10.11	4.87	1.06	28.36
NO_2, µg/m^3	12.16	5.33	1.66	36.91	15.32	7.02	3.02	53.30

The site characteristics are different resulting industrial (Autonagar and Udyam-bag), traffic (Railway station) and residential (Vadgaon). Air quality monitoring sites were classified as industrial, traffic and residential sites and the air pollution monitoring was carried for the period of 2011–2013. Respirable dust sampler RDS (Envirotech APM 460 NL) was used to monitor and measure the concentration of PM_{10} ($\mu g/m^3$) and Gaseous pollutant sampler (Environtech APM 433) was used to measure NO_2 ($\mu g/m^3$) and SO_2 ($\mu g/m^3$) concentrations using suitable reagents. The data sets of meteorological parameters and pollutant concentrations are used on their daily average for the period of 2011–2013.

We had problems in the data sets mainly outlier and missing data. The outlier was due to the malfunction of instrument or incorrect measurement of the pollutant. The outliers are maximum or minimum values of data. They are analysed with care, because they cause more deviation in the model development and prediction. Missing data was due to instrument calibration or malfunctions and this problem was very limited (2%), these gaps are filled by linear interpolation method [16]. In order to support the neural network to efficiently handle a data, all the input variables are normalised to the range (0, 1) by Eq. 1.

$$X_n = \frac{(X_i - X_{min})}{(X_{max} - X_{min})} \tag{1}$$

X_n is the normalised data, X_i actual measured data, X_{min} minimum value of the measured data, X_{max} maximum value of the measured data.

2.2   ANN Modelling

The ANN models were developed by using the MATLAB R2015a software from Mathworks group Inc. The feed forward Back propagation (BP) multilayer perceptron (MLP) network model was used for the present study. The BP algorithm is used to train a given feed-forward multilayer neural network for a known set of input patterns with known classifications. The BP algorithm is based on Widrow-Hoff delta learning rule which is based on weight adjustment through mean square error of the output to the sample input [17]. Therefore, BP networks are the simplest and most widely used network models [18]. The meteorological variables (inputs) and pollutant concentrations (outputs) are divided into training, validation and testing subsets. One third data was used for validation, one third data was used for testing and two third of the data was used for the training set. The neuron number in hidden layer has the significant importance for the model development and its accuracy and performance. Another important step in model development is determination of activation function. The most widely used activation functions are liner, sigmoid and hyperbolic tangent. The optimization of one hidden layer was conducting by several tests with various network structures, optimised networks have selected based on lower prediction error and smaller convergence times. Present study applied back propagation network with three layers to predict air pollutant concentrations. Selected network structure was used and trained after definition of subset with *Levenberg-Marquardt* optimization (*trainlm*) in hidden layer (nodes of 10, 12, 18), log-sigmoid (*logsig*) transfer function is used in output layer [19, 20].

2.3   Evaluation of Model Performance

Evaluation of the performance of developed model is very important and is important to evaluate forecast accuracy. We have selected statistical indicator to describe goodness of the estimates. The accuracy of model was determined by considering how well a model performed with new data which are not used in model fitting. The model performance was checked using Mean Absolute Error (MAE), and Root Mean-Square Error (RMSE) and correlation Coefficient (R) are given by Eqs. (2)–(6).

$$MSE = \frac{\sum_{i=1}^{n} (Y_i - X_i)^2}{n} \tag{2}$$

$$MAE = \frac{\sum_{i=1}^{n} |Y_i - X_i|}{n} \tag{3}$$

$$RMSE = \sqrt{\frac{\sum_{i=1}^{n} (Y_i - X_i)^2}{n}} \tag{4}$$

$$MAPE = \sqrt{\frac{\sum_{i=1}^{n} (Y_i - X_i)^2}{n}} \tag{5}$$

$$R = \frac{\sum_{i=1}^{n} (Y_i - \overline{Y_i})(X_i - \overline{X_i})}{\left\{ \left[\sum_{i=1}^{n} (X_i - \overline{X_i})^2 \right] \left[\sum_{i=1}^{n} (Y_i - \overline{Y_i})^2 \right] \right\}^{1/2}} \tag{6}$$

where, n is the number of data, Y_i is the modeled pollutant concentration, X_i is the observed concentration. Zero Error indicates that all the modeled concentrations of various pollutants computed by ANN models were perfectly match the observed concentrations.

3   Results and Discussions

Overall, the annual average concentrations of PM_{10} were above the national ambient air quality standard (NAAQS) at Autonagar and Railway station and is almost at alarming stage at Udyambag and Vadgaon. PM_{10} concentrations are mainly due to construction activity, industrial activity and bad road conditions. The annual average NO_2 and SO_2 concentrations were below NAAQS standards. The concentrations of NO_2 are higher on traffic sites (Railway station) and industrial area (Autonagar and Udyambag) and it confirms industries and traffic as important sources of NO_2 in Belagavi city. SO_2 concentration is very low (negligible) and is mainly due to diesel vehicles and old vehicles, industrial burning and commercial burning of various fuel oils.

The optimisation of a neural network is most important objective to developed ANN based models [21]. The process of optimization plays an important role in the selection and performance of the network. Hence, an optimisation was carried out with number of neurons and MSE [19]. Then, the multilayer layer neural network was evaluated using BP algorithm with 10, 12 and 18 nodes in the hidden layer. With increasing in neuron number, the network gave several local minimum values and different MSE values were obtained for the training set. Increasing neuron number to more than 20 gave unrealistic results for all the pollutants.

The various performance indicators were used to determine to measure the goodness of the fit and the results of ANN model are summarized in Table 2. The best performing ANN network was *trainlm* in the hidden layer. The results shows excellent performance for the developed models for PM_{10} for all four monitoring sites are optimised with nodes of 10, 10, 18 and 12 according to values of MAPE was found to be 7.2, 6.2, 10.8, and 7.7%. Good results were obtained for NO_2 with nodes of 18, 12, 10, 10 with values of MAPE are 12.1, 3.5, 8.6, 8.7%. Similarly, good performance by the developed network of SO_2 for all monitoring sites using nodes of 12, 12, 18 and 10 with MAPE of 2.0, 7.1, 3.6 and 8.6%. The smallest MSE was obtained for *trainlm* function and the mean absolute percentage error was used to select the models. The following models have performed best based on the change in number of hidden nodes and keeping the *logsig* transfer function. The training stopped after 8, 9 and 9 iterations for PM_{10}, The training stopped after 10, 11 and 9 iterations for SO_2, for NO_2, training stopped after 6, 7 and 8 iterations.

Table 2 Performance indices for the testing data for four monitoring sites

PM_{10}—Autonagar					PM_{10}—Railway station				
Nodes	MAD	MSE	RMSE	MAPE	R	MAD	MSE	RMSE	MAPE	R
10	47.56	71.88	6.757	0.072	0.9592	88.18	28.71	4.212	0.061	0.8309
12	64.56	74.18	7.199	0.200	0.9072	88.70	94.39	7.691	0.114	0.8467
18	65.71	41.91	5.587	0.163	0.9447	90.40	141.6	9.613	0.149	0.8550

PM_{10}—Udyambag					PM_{10}—Vadgaon				
Nodes	MAD	MSE	RMSE	MAPE	R	MAD	MSE	RMSR	MAPE	R
10	48.19	27.90	4.384	0.130	0.8420	52.11	29.15	4.409	0.093	0.8863
12	46.22	28.93	4.367	0.115	0.8420	50.79	26.60	4.152	0.077	0.9015
18	47.09	21.92	3.851	0.108	0.8511	58.02	29.09	4.594	0.123	0.8982

NO_2—Autonagar					NO_2—Railway station				
Nodes	MAD	MSE	RMSE	MAPE	R	MAD	MSE	RMSE	MAPE	R
10	3.911	5.06	2.075	0.341	0.8389	5.462	5.101	2.155	0.280	0.8785
12	4.240	3.58	1.745	0.287	0.8632	5.681	4.125	1.937	0.252	0.8899
18	5.186	0.981	0.819	0.121	0.8982	7.716	0.061	0.191	0.035	0.8925

NO_2—Udyambag					NO_2—Vadgaon				
Nodes	MAD	MSE	RMSE	MAPE	R	MAD	MSE	RMSE	MAPE	R
10	4.576	0.3824	0.603	0.223	0.8721	9.589	1.221	0.847	0.087	0.8781
12	4.087	0.0534	0.1978	0.086	0.9249	10.68	1.92	1.194	0.190	0.8927
18	4.517	0.3412	0.549	0.216	0.8828	10.54	1.084	0.896	0.144	0.9320

(continued)

Table 2 *(continued)*

SO2—Autonagar					SO2—Railway station				
Nodes	MAD	MSE	RMSE	MAPE	R	MAD	MSE	RMSE	MAPE	R
10	20.14	5.05	1.432	0.063	0.9063	29.58	14.84	3.434	0.174	0.8673
12	20.59	0.59	0.485	0.020	0.9690	25.15	6.15	1.939	0.071	0.9014
18	22.10	1.44	1.200	0.072	0.9877	22.00	6.00	1.501	0.080	0.8847
SO2—Udyambag					SO2—Vadgaon				
Nodes	MAD	MSE	RMSE	MAPE	R	MAD	MSE	RMSE	MAPE	R
10	13.39	1.021	0.912	0.100	0.9348	14.58	4.12	1.363	0.086	0.8531
12	12.70	0.204	0.354	0.036	0.9568	15.18	3.60	1.340	0.099	0.8884
18	13.60	0.411	0.324	0.078	0.9435	15.78	3.296	1.399	0.116	0.8916

Figure 3 shows observed and predicted PM_{10}, NO_2 and SO_2 concentrations using developed ANN models. It demonstrates that these models can estimate air pollutant concentration for the given set with an accuracy of approximately 90%. As discussed earlier the used data set was divided into two subsets, the first set was included data from the period of 2011–2012 and 2013 data was totally unknown data for the model was used to evaluate the forecasting ability of the developed model is shown in Fig. 4. According to Fig. 4, it seems that the prediction of all developed models is in a very satisfactory level ($p < 0.05$). The performance of developed models for training data set and performance in terms R are shown in Fig. 4. The results indicate that developed ANN models predicated the PM10 with good accuracy of 95.92, 85.50, 85.11 and 90.15%, NO_2 with 89.82, 89.27, 92.49 and 93.20%, for SO_2 with excellent accuracy of 98.77, 88.47, 95.68 and 89.16% for the monitoring sites Autonagar, Railway station, Udyambag and Vadgaon.

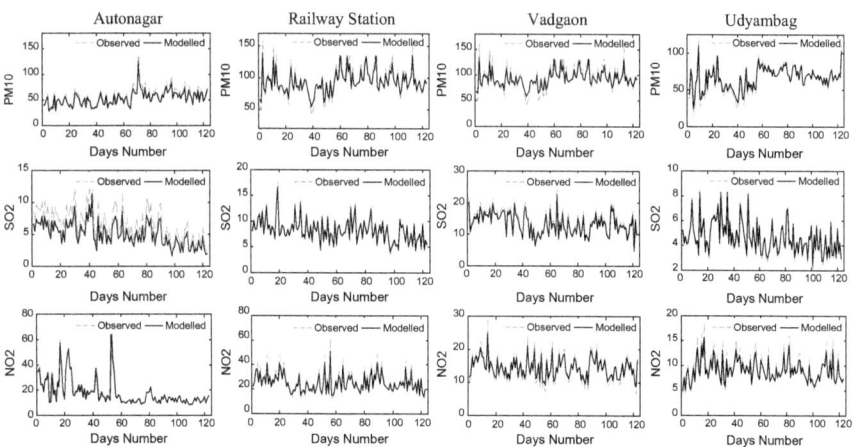

Fig. 3 Measured and predicted concentrations of PM_{10}, SO_2 and NO_2 for training data

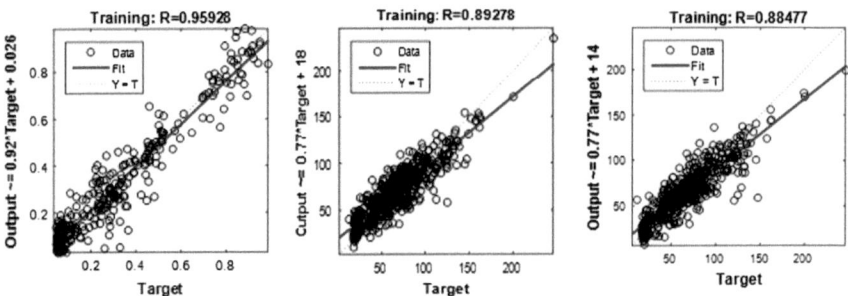

Fig. 4 ANN outputof training data set for PM$_{10}$ (Autonagar), NO$_2$ (Vadgaon) and SO$_2$ (Railway station)

4   Conclusion

The optimization study was done for better ANN mode with different structures in terms of hidden layer and number of nodes. Three layer model used for the present study showed a precise and effective prediction of air pollutant concentrations. The *trainlm* has given good prediction with relatively good accuracy. For the air pollution predictive the ANN models have faster predictive power and are viable as compared to other statistical models. The performance of the model in all four locations was satisfactory and is able to efficiently simulate the atmospheric time-series pollutant concentrations. ANN models can be considered as appropriate for operational usage in urban air quality management. The models discussed in this study are easily implemented and are helpful to the local authorities in providing the information to the general public and to protect the health of people by implementing appropriate controlling measures.

References

1. Dimitriou, K., Paschalidou, A.K., Kassomenos, P.A.: Assessing air quality with regards to its effect on human health in the European Union through air quality indices. Ecol. Ind. **27**, 108–115 (2013)
2. Pope III, C.A., Burnett, R.T., Thun, M.J., Calle, E.E., Krewski, D., Ito, K., Thurston, G.D.: Lung cancer, cardiopulmonary mortality, and long-term exposure to fine particulate air pollution. JAMA **287**(9), 1132–1141 (2002)
3. Karatzas, K.D., Kaltsatos, S.: Air pollution modelling with the aid of computational intelligence methods in Thessaloniki. Greece. Simul. Model. Pract. Theor. **15**(10), 1310–1319 (2007)
4. Deleawe, S., Kusznir, J., Lamb, B., Cook, D.J.: Predicting air quality in smart environments. J. Ambient Intell. Smart Environ. **2**(2), 145–154 (2010)
5. Gardner, M.W., Dorling, S.R.: Neural network modelling and prediction of hourly NO$_x$ and NO$_2$ concentrations in urban air in London. Atmos. Environ. **33**(5), 709–719 (1999)
6. Viotti, P., Liuti, G., Di Genova, P.: Atmospheric urban pollution: applications of an artificial neural network (ANN) to the city of Perugia. Ecol. Model. **148**(1), 27–46 (2002)

7. Karaca, F., Alagha, O., Ertürk, F.: Application of inductive learning: air pollution forecast in Istanbul. Turkey. Intell. Autom. Soft Comput. **11**(4), 207–216 (2005)
8. Athanasiadis, I.N., Karatzas, K., Mitkas, P.: Contemporary air quality forecasting methods: a comparative analysis between classification algorithms and statistical methods. In: Fifth International Conference on Urban Air Quality Measurement, Modelling and Management, Valencia, Spain (2005)
9. Kolehmainen, M., Martikainen, H., Ruuskanen, J.: Neural neworks and periodic components used in air quality forecasting. Atmos. Environ. **35**(5), 815–825 (2001)
10. Kukkonen, J., Partanen, L., Karppinen, A., Ruuskanen, J., Junninen, H., Kolehmainen, M., Cawley, G.: Extensive evaluation of neural network models for the prediction of NO_2 and PM_{10} concentrations, compared with a deterministic modelling system and measurements in central Helsinki. Atmos. Environ. **37**(32), 4539–4550 (2003)
11. Rumelhart, E., Hinton, J., Williams, R.: Learning internal representations by error propagation, in parallel distributed processing: exploration in the microstructure of cognition, vol. 1. MIT press, Cambridge (1986)
12. Hertz, J.A., Krogh, A.S., Palmer, R.G.: Introduction to the theory of neural computation. Addison Wesley, Canada (1995)
13. Bishop, A.: Neural networks for pattern recognition. Oxford University Press, UK (1995)
14. Fausett, L.: Neural Networks: Architectures, Algorithms, and Applications. Prentice-Hall Inc., New Jersey (1994)
15. Gardner, M.W., Dorling, S.R.: Artificial neural networks (the multilayer perceptron)—a review of applications in the atmospheric sciences. Atmos. Environ. **32**(14), 2627–2636 (1998)
16. Kandasamy, S., Baret, F., Verger, A., Neveux, P., Weiss, M.: A comparison of methods for smoothing and gap filling time series of remote sensing observations application to MODIS LAI products. Biogeosciences **10**(6), 4055–4071 (2013)
17. http://wwwold.ece.utep.edu/research/webfuzzy/docs/kk-thesis/kk-thesis-html/node22.html
18. Niska, H., Hiltunen, T., Karppinen, A., Ruuskanen, J., Kolehmainen, M.: Evolving the neural network model for forecasting air pollution time series. Eng. Appl. Artif. Intell. **17**(2), 159–167 (2004)
19. Velasquez, G.: A Distributed approach to a neural network simulation program. Master's thesis, The University of Texas at El Paso, El Paso (1998)
20. Cai, M., Yin, Y., Xie, M.: Prediction of hourly air pollutant concentrations near urban arterials using artificial neural network approach. Transp. Res. Part D Transp. Environ. **14**(1), 32–41 (2009)
21. Akkoyunlu, A., Yetilmezsoy, K., Erturk, F., Oztemel, E.: A neural network-based approach for the prediction of urban SO_2 concentrations in the Istanbul metropolitan area. Int. J. Environ. Pollut. **40**(4), 301–321 (2010)

Grayscale Image Encryption Based on Symmetric-Key Latin Square Image Cipher (LSIC)

V. Pawan Kumar$^{(\boxtimes)}$, A. R. Aswatha, and Smitha Sasi

Department of Telecommunication Engineering, Dayananda Sagar College
of Engineering, Bangalore, India
{pawan13995,aswath.ar,smitha.sasi24}@gmail.com

Abstract. The increased computing power and improving technologies demand the necessity for stronger encryption algorithms. Here we are introducing a new encryption method, Latin Square Image Cipher (LSIC) for grayscale image. This includes Latin square whitening, S-box, P-box and also LSB noise embedding for probabilistic encryption. As a result, using all the above primitives, LSIC is constructed as a Substitution-Permutation Network (SPN) consisting of eight stages of whitening, substitution and permutation using different Latin squares of order 256 at each stage. The proposed method has a good resistance against brute-force attacks, ciphertext attacks and plaintext attacks.

Keywords: LSIC · Whitening · S-box · P-box · SPN

1 Introduction

The necessity for image encryption has increased over the years due to the increase in the digital images and digital devices all around the world [1]. There is variety of images captured from digital devices such as smart phones, tablets, digital cameras. There is also a large amount of images and scanned documents on the internet, which causes severe problems to the owners such as invasion of personal privacy, loss for a company due to leakage of confidential information, leakage of medical reports and many more.

The conventional method for encrypting digital images were using block and stream ciphers [2–5]. There are two types of block ciphers; they are AES (Advanced Encryption Standard) [5, 6] and DES (Data Encryption Standard) [4, 6]. The main disadvantage of stream cipher was that it required the generation of very large key sizes, for example, to encrypt a file of 2MB, a 2MB length of key was necessary. So this raised the concerns regarding the storage and key exchange issues.

The stream and block ciphers are not ideal for encrypting digital images for the following reasons.

© Springer International Publishing AG 2018
D. J. Hemanth and S. Smys (eds.), *Computational Vision and Bio Inspired Computing*,
Lecture Notes in Computational Vision and Biomechanics 28,
https://doi.org/10.1007/978-3-319-71767-8_40

- Insufficient large block size: Digital images are usually of the size in terms of several Kb's or Mb's, while the block and stream ciphers usually have a block size less than 256 bits.
- Neglect the nature of the image: Digital images are two-dimensional data, while the conventional ciphers encrypt a one-dimensional stream, i.e., indirectly they are encrypting the pixel sequence extracted from the image.

The Latin Square image cipher proposed in this paper is a non-chaotic image cipher that can be effectively implemented on the hardware and software. In this method a 256 bit encryption key is used to generate Latin squares, that are then used in the various staged of the Substitution Permutation network along with whitening, S-box and P-box to provide good confusion and diffusion properties [7].

The proposed algorithm is mainly designed for medical purposes. As most of the patient information is stored on the computers such as MRI scans, X-rays and so on, it is necessary to prevent the unauthorized access to these images.

In case of military purposes, this algorithm can be used to embed secret messages by integrating steganography rather than using the conventional encryption methods.

This paper is organized as follows: In the second section we have the background related to the Latin squares, the third section contains the proposed Latin square image cipher, the fourth section contains the results related to the proposed method (including the plaintext, cipher text, histograms and the comparison tables) and then it finally ends with a general conclusion.

2 Background

2.1 Latin Squares

A Latin square is an $N \times N$ array consisting of N unique symbols, with each symbol appearing only once in each row and column. The term Latin square was derived from the famous mathematician Leonhard Euler who used Latin characters as symbols.

Mathematically, Latin square of order N can be defined as follows

$$fL(r,c,i) = \begin{cases} 1, & L(r, c) = Si \\ 0, & Otherwise \end{cases} \tag{1}$$

where r is the row index, c is the column index, i is the symbol index and S_i is the ith symbol of the set $S = \{S_0, S1, \dots, S_{N-1}\}$.

For a Latin square of order n, we have, (Fig. 1)

$$\sum_{r=0}^{N-1} fL(r,c,i) = 1 \tag{2}$$

$$\sum_{c=0}^{N-1} fL(r,c,i) = 1 \tag{3}$$

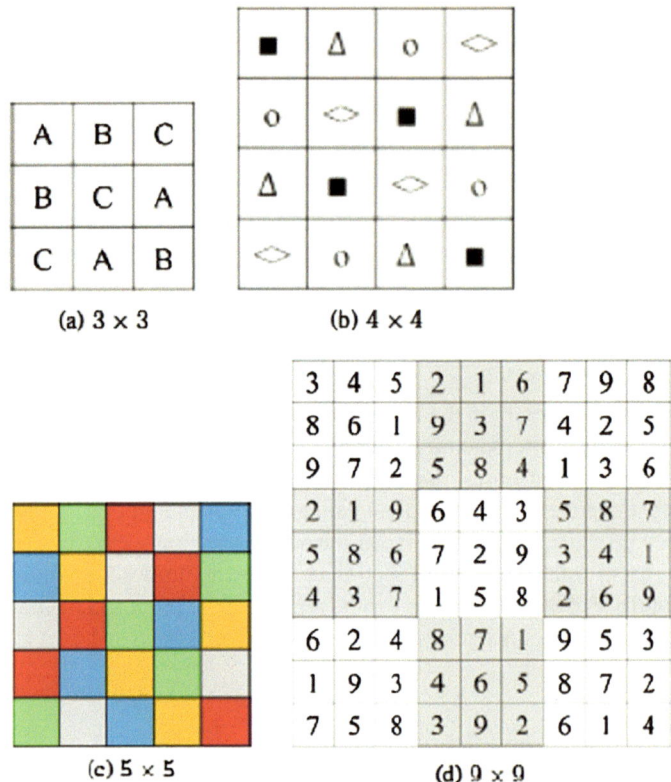

A	B	C
B	C	A
C	A	B

(a) 3 × 3

(b) 4 × 4

(c) 5 × 5

3	4	5	2	1	6	7	9	8
8	6	1	9	3	7	4	2	5
9	7	2	5	8	4	1	3	6
2	1	9	6	4	3	5	8	7
5	8	6	7	2	9	3	4	1
4	3	7	1	5	8	2	6	9
6	2	4	8	7	1	9	5	3
1	9	3	4	6	5	8	7	2
7	5	8	3	9	2	6	1	4

(d) 9 × 9

Fig. 1 Figure shows the examples for Latin squares of different order [8]

2.2   Latin Square Generator

There are a variety of methods to generate a Latin square, Algorithm 1 is one of the methods,

Algorithm 1. **A Latin Square Generator** L = LSG (Q₁, Q₂)

Require: *Q_1 and Q_2 are two length-N sequences*
Ensure: *L is a Latin square of order N*
Q_{seed} = *SortMap* (Q_1)
Q_{shift} = *SortMap* (Q_2)
for *r = 0 : 1 : N − 1* ***do***
$L(r, :)$ = *RowShift* ($Q_{seed,\ Q_{shift}}(r)$)
end for

Here, Q1 and Q2 are generated using a linear Congruential generator (LCG) which is a pseudo-random number generator.

For better understanding, let us take an example of a 4 × 4 Latin Square L, Q1 = [1, 2, 5, 3] and Q2 = [2, 5, 3, 9].

The function SortMap (.) sorts Q1 and Q2 in ascending order. Hence we get, Q_1^* = [1, 2, 3, 5] and Q_2^* = [2, 3, 5, 9].

Comparing Q_1 with Q_1^* and Q_2 with Q_2^*, we get,

$$Q_{\text{seed}} = [0, 1, 3, 2]$$

$$Q_{\text{shift}} = [0, 2, 1, 3]$$

Here Q_{seed} is the seed value and Q_{shift} represents the number of times the seed value has to be shifted left.

Hence we get the Latin square as follows

$$L = \begin{bmatrix} 0 & 1 & 3 & 2 \\ 3 & 2 & 0 & 1 \\ 1 & 3 & 2 & 0 \\ 2 & 0 & 1 & 3 \end{bmatrix}$$

2.3   Substitution-Permutation Network

The Substitution-Permutation network is a structure consisting of a number of stages of substitution and permutation that is used to provide good diffusion and confusion properties (Fig. 2).

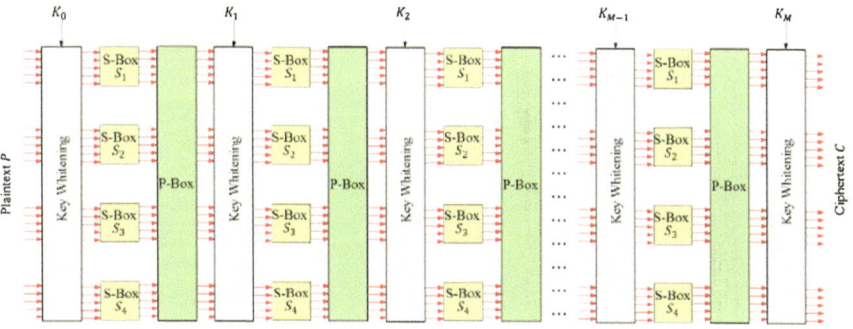

Fig. 2 M-round Substitution-Permutation Network [8]

3   Latin Square Image Cipher

3.1   Overview

The plain text is encrypted in the block size of 256×256 grayscale block. Each pixel intensity here is denoted by an 8-bit byte. In this paper the 256×256 plain text is denoted by P, cipher text by C, the 256-bit encryption key as K, and the keyed Latin as L. The structure of the Latin square image cipher is of a substitution-permutation network with eight rounds of whitening, substitution and permutation. It also provides the probabilistic encryption by using LSB noise embedding. The overview of LSIC is shown in the Fig. 3.

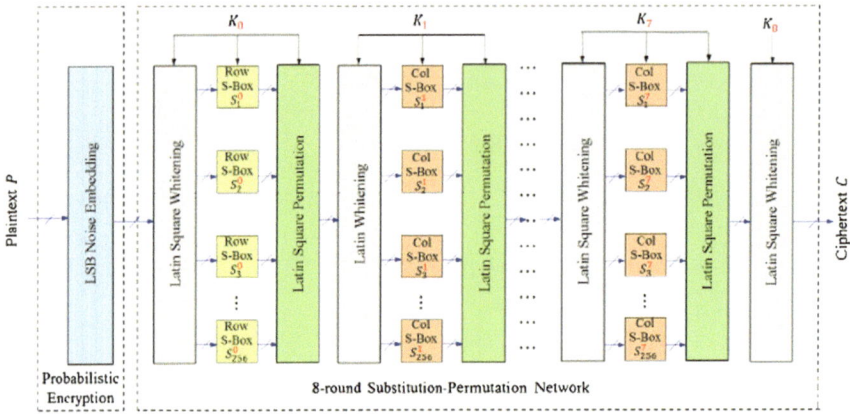

Fig. 3 Overview of Latin square image cipher [8]

3.2 LSB Noise Embedding

In this paper randomness is introduced by embedding noise in the least significant bit of the image. This is done by XOR operation between a randomly generated 256×256 bit plane with LSB of the plaintext image (Fig. 4).

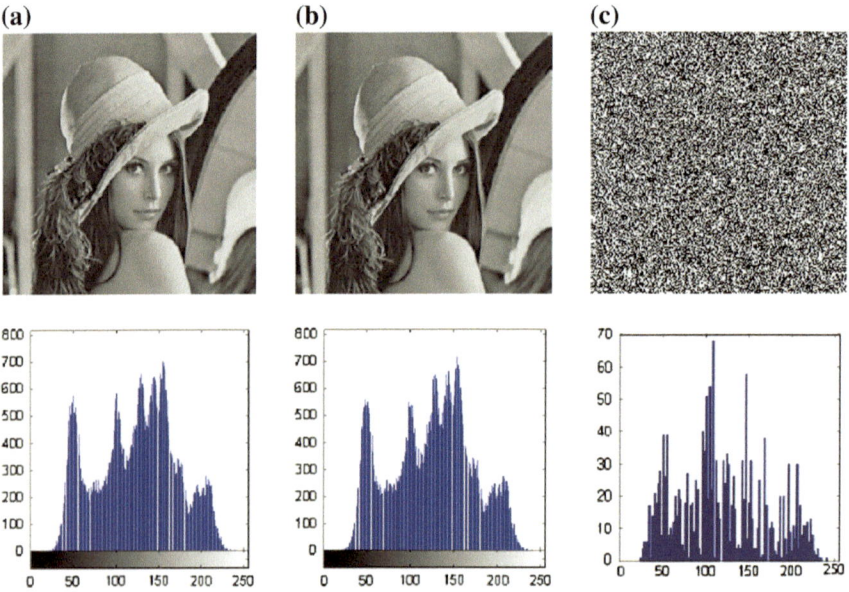

Fig. 4 Noise embedding in LSB—**a** plaintext Lena P with histogram **b** noise embedded plaintext P0 with histogram, and **c** P − P_0 with the difference of histograms

3.3 Key Translation

In the proposed method a 256-bit encryption key is used along with key translation to get eight Latin squares used in different stages of the SPN. This is done as follows.

- The 256-bit encryption key is divided into eight 32-bit subkeys
- Then a pair of pseudorandom numbers Q_1 and Q_2 is generated using a linear Congruential generator which is a PRNG
- These pseudorandom sequences are then used to generate the Latin squares of order 256.

Algorithm 2. Key Dependent Sequence Generator $(Q1, Q2) = KDSG (Key, M)$

Require: *K is a 256-bit key*
Require: *n is a nonnegative integer*
Ensure: $Q1 = \{ Q_1^0, \_\_\_\_, Q_2^M \}$ and $Q2 = \{ Q_2^0, \_\_\_\_, Q_2^M \}$ are n-element set of random sequences, each of a length 256.
$K_0 = K$
for $n = 0 : 1 : M$ *do*
     $[k_0, k_1, \_\_\_, k_7] = SubKeyDiv (Kn)$
     for $i = 0 : 1 : 8$ *do*
         $q^i(0) = PRNG(k_i)$
         for $j = 1 : 1 : 63$ *do*
             $q^i(j) = PRNG(q_i (j - 1))$
         end for
     end for
     $Q_1^n = q^0(0 : 31), q^1(0 : 31), \_\_\_\_, q^7(0 : 31)$
     $Q_2^n = q^0(32 : 63), q^1(32 : 63), \_\_\_\_, q^7(32 : 63)$
     $K_{n+1} = q^0(63), q^1(63), \_\_\_\_, q^7(63)$
end for

3.4 Latin Square Whitening

Whitening is a process of mixing the plain text with the round key usually using XOR operation. But in case of an image each pixel is represented by several binary bits. Hence it becomes inefficient to use XOR operation. Here the whitening is done as a transposition cipher.

$$Y = SR(X, d) = \begin{cases} X, & if\ d = 0 \\ FLIPXup \rightarrow down, & if\ d = 1 \\ FLIPXleft \rightarrow right, & if d = 2 \end{cases} \qquad (4)$$

Here SR denotes the spatial rotating function; SR(X, d) rotates the image X according to the different values of the direction d.

3.5   Latin Square Row and Column Bijections

The Bijections can be defined for the rows and columns of Latin square L of order N by mapping the integer number sequence [0, 1…N−1] to either row or a column. Hence forward row mapping and inverse row mapping can be constructed for a given r in L. Similarly forward column mapping and inverse column mapping can be constructed for a given c in L (Fig. 5).

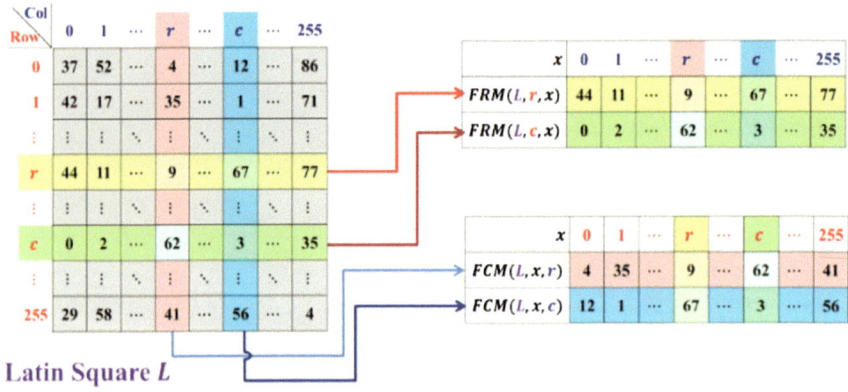

Fig. 5 Examples of forward row mappings and forward column mappings [8]

$$\begin{cases} \text{IRM(L,r,FRM(L,r,x))} = x \\ \text{FRM(L,r,IRM(L,r,y))} = y \end{cases} \tag{5}$$

$$\begin{cases} \text{ICM(L, FCM(L, x, c), c)} = x \\ \text{FCM(L,ICM(L,y,c), c)} = y \end{cases} \tag{6}$$

3.6   Latin Square Substitution

The byte substitution is performed using an S-box. S-box is a fancier form of a look up table. The substitution is done to provide diffusion property. An image pixel is represented as a sequence of bits. For example an 8-bit grayscale image has 256 pixel intensities each represented by a sequence of 8 bits. The substitution in this cipher is performed using bijection from rows and columns in a Latin square. The substitution with respect to a row is called a Latin square Row S-box (LSRS) and with respect to the column is called as Latin square Column S-box (LSCS).

$$\text{LSRS} = \begin{cases} C = Ecr_s^{row}(L, P) \\ P = Dcr_s^{row}(L, C) \end{cases} \tag{7}$$

$$\text{LSCS} = \begin{cases} C = Ecr_s^{col}(L, P) \\ P = Dcr_s^{col}(L, C) \end{cases} \qquad (8)$$

3.7   Latin Square Permutation

Permutation is used to perform byte shuffling or scrambling. It is used to provide confusion property. The p-box can also be defined as a bijection.

$$\text{LSRP} = \begin{cases} C(r, cy) = P(r, \text{FRM}(L, r, cx)) \\ P(r, cx) = C(r, \text{IRM}(L, r, cy)) \end{cases} \qquad (9)$$

$$\text{LSCP} = \begin{cases} C(ry, c) = P(\text{FCM}(L, rx, c), c)) \\ P(rx, c) = C(\text{ICM}(L, ry, c), c)) \end{cases} \qquad (10)$$

3.8   Encryption and Decryption Algorithms

Algorithm 3. **Latin Square Image Cipher—Encryption** $C = E(P, K)$

Require: *K is a 256-bit key*
Require: *P is a 256 X 256 8-bit grayscale image block*
Ensure: *C is a 256 X 256 8-bit grayscale image block*
$(Q_1, Q_2) = KDSG\ (K, 8)$
for $n = 0 : 1 : 7$ ***do***

    if $n == 0$ ***then***
        $C_{LSP} = LSBNoiseEmbedding(P)$
    end if
    $L_n = LSG(\ Q_1^n, Q_2^n)$
    $D_n = L_n(0, 0)$
    $C_{LSW} = Ecr_w(L_n, C_{LSP}, D_n)$
    if $mod(n, 2)\ != 0$ ***then***
        $C_{LSS} = Ec\ r_s^{col}\ (L_n, C_{LSW})$
    else
        $C_{LSS} = Ec\ r_s^{row}\ (L_n, C_{LSW})$
    end if
    $C_{LSP} = Ecr_p(L_n, C_{LSS})$
end for
$L_8 = LSG(\ Q_1^8 , Q_2^8)$
$D_8 = L8(0, 0)$
$C = Ecr_w(L_8, C_{LSP}, D_8)$

Algorithm 4. **Latin Square Image Cipher—Encryption** $P = D(C, K)$

Require: *K is a 256-bit key*
Require: *P is a 256 X 256 8-bit grayscale image block*
Ensure: *C is a 256 X 256 8-bit grayscale image block*
$(Q_1, Q_2) = KDSG\ (K, 8)$
for $n = 7 : -1 : 0$ ***do***

if n == 7 then
  $L_8 = LSG(\ Q_1^8\ ,\ Q_2^8)$
  $D_8 = L_8(0,\ 0)$
  $P_{LSW} = Dcr_w(L_8,\ C,\ D_8)$
end if
$L_n = LSG(\ Q_1^n,\ Q_2^n)$
$D_n = L_n(0,\ 0)$
$P_{LSP} = Dcr_w(Ls,\ P_{LSW})$
if mod(n, 2) != 0 then
  $P_{LSS} = Dc\ r_s^{col}\ (L_n,\ P_{LSP})$
else
  $P_{LSS} = Dc\ r_s^{row}\ (L_n,\ P_{LSP})$
end if
  $P_{LSW} = Dcr_w\ (L_n,\ P_{LSS},\ D_n)$
end for
$P = P_{LSW}$

4 Results

A good encryption algorithm must be able to resist all types of attacks. To provide this resistance, the algorithm must have both good confusion and diffusion properties. These diffusion and confusion properties are provided by the substitution and permutation respectively.

The proposed algorithm has good resistance against the following attacks.

- **Brute-Force Attack**: Brute-force attack is the kind of attack in which all the possible keys are searched until the right key is found. In this attack the strength is directly proportional to the length of the key. The proposed system uses a 256-bit key which is larger than or equal to most of the standard encryption algorithm.
  Also the theoretical key size is even larger than 256-bits. The proposed algorithm uses eight 256×256 Latin squares, where each is uniquely different from the other. Therefore the number of Latin Squares that can be generated is $256! \times 256! = 2^{3368}$. As there are 8 Latin squares, the theoretical key size is about $(2^{3368})^8 = 2^{26944}$, i.e. 26944 bits. Thus the proposed scheme has extremely large key length and hence provides a very good resistance against the brute-force attacks.
- **Ciphertext and Plaintext attacks**: The proposed LSIC has a good resistance against ciphertext and plaintext attacks because of the following reasons:
  → Nonlinear Key Translation: The use of key translation of 256-bit encryption key into eight 256×256 Latin squares instead of directly encrypting the plaintext provides a good level of security.
  → Probabilistic Encryption: The proposed algorithm uses probabilistic encryption by embedding random noise in the LSB of the intensity pixels of the image.
  → Dynamic S-boxes and P-boxes: All the s-boxes and P-boxed employed in this algorithm are dependent on the Latin squares of the respective rounds. Hence even

if the attacker manages to get hold of the encryption key, all the other keys are safe as the S-boxes and P-boxes are dependent on different keyed Latin (Fig. 6 and Tables 1, 2, 3 and 4).

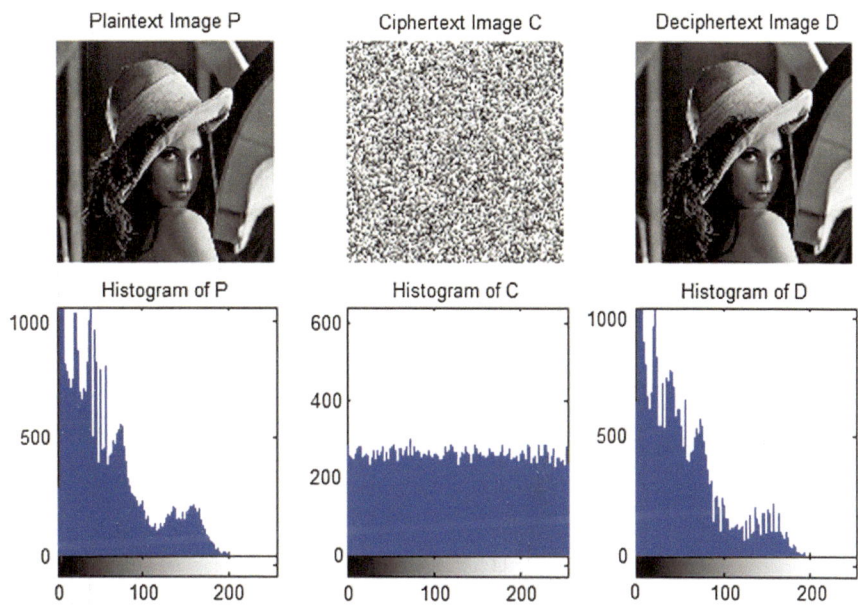

Fig. 6 Output of encryption and decryption. Encryption execution time = 0.2782 s. Decryption execution time = 0.1939 s

Table 1 Results of information entropy analysis

File	Plain text	Ciphertext			
		Proposed Method	BmpPacker-AES[2]	Chaotic 3D mapping [9]	Non-chaotic LSIC [8]
Airplane	6.4522	7.9969	7.9294	7.9965	7.9967
Clock	6.7056	7.9976	7.3635	7.9967	7.9967
Moon	6.7093	7.9969	7.9067	7.9970	7.9972
Airplane	6.4522	7.9969	7.9294	7.9965	7.9967
Areal	7.3118	7.9974	7.9418	7.9971	7.9976
Chemical plant	7.3424	7.9975	7.9925	7.9972	7.9970
Resolution chart	1.5483	7.9978	7.9035	7.9971	7.9972

Table 2 Results of information entropy analysis

File	Plaintext	Ciphertext	
		Proposed method	Latin square and CNN [10]
Lena	6.9095	7.9972	7.9968

Table 3 Results of PSNR, MSE and RMSE

File	PSNR	MSE	RMSE
Lena	40.1690	6.3035	2.5107

Table 4 Comparison of NPCR and UACI scores of image Lena

Encryption method	NPCR (%)	UACI (%)
Self adaptive wave transmission, 2010 [11]	99.65	33.48
Chaos-based encryption, 2011 [12]	99.63	33.48
Non-chaotic LSIC, 2013 [8]	99.66	33.49
Chaotic 3D mapping, 2004 [9]	99.25	33.14
Proposed method	99.65	33.53

5   Conclusion

A good encryption algorithm must be able to resist against all known attacks. Along with the stronger resistance it must provide good confusion and diffusion properties. Since the proposed system is dependent only on integers, it can be easily implemented on hardware and software. Also this algorithm encrypts the data in terms of bytes instead of bits which is more efficient for encrypting images. The use of probabilistic encryption by embedding noise in LSB of the image bit plane provides different cipher text every time even with the use of same key.

The security analysis against different types of attack are performed and it is determined that it has a good resistance against brute-force, cipher text and plaintext attacks. The proposed system has a large key space compared to any other algorithm. The results also show that the proposed algorithm provides a higher security level in terms of NPCR, UACI and entropy of the cipher images.

Due to all these aforesaid advantages, the proposed cipher has wide applications in medical, military and any other confidential transmission.

References

1. Yang, M., Bourbakis, N., Li, S.: Data-image-video encryption. Potentials, IEEE **23**(3), 28–34 (2004)
2. Anderson, R., Schneier, B.: Description of a new variable-length key, 64-bit block cipher (Blowfish). ser. Lecture Notes in Computer Science, vol. 809, pp. 191–204. Springer Berlin, Heidelberg (1994)

3. Schneier, B.: The two fish encryption algorithm: a 128-bit block cipher. John Wiley (1999)
4. Data encryption standard: Federal Information Processing Standards Publication 46, (1977)
5. Advanced encryption standard: Federal Information Processing Standards Publication 197, (2001)
6. Stallings, W.: Cryptography and network security: Principles and practice. Upper Saddle River, N.J: Prentice Hall, Print (1999)
7. Shannon, C.E.: Communication theory of secrecy systems. Bell Syst. Tech. J. **28**(4), 656–715 (2010) (compilation and indexing terms, Copyright 2010 Elsevier Inc. 19490005984)
8. Wu, Y., Zhou, Y., J.P. Noonan, Agaian, S., Philip Chen, C.L.: A Novel Latin Square Image Cipher. (2013)
9. Chen, G., Mao, Y., Chui, C.K.: A symmetric image encryption scheme based on 3d chaotic cat maps. Chaos, Solitons Fractals **21**(3), 749–761 (2004). https://doi.org/10.1016/j.chaos.2003.12.022
10. Lin, M., Long, F., Guo, L.: Grayscale image encryption based on Latin square and cellular neural network. IEEE (2016). doi:https://doi.org/10.1109/CCDC.2016.7531456
11. Liao, X., Lai, S., Zhou, Q.: A novel image encryption algorithm based on self-adaptive wave transmission. Signal Process. **90**(9), 2714–2722 (2010)
12. Zhu, Z., Zhang, W., Wong, K., Yu, H.: A chaos-based symmetric image encryption scheme using a bit-level permutation. Inf. Sci. **181**(6), 1171–1186 (2011)

Enhanced Defogging System on Foggy Digital Color Images

Sarath Krishnan, B. A. Sabarish$^{(\boxtimes)}$, V. Gayathri, and S. Padmavathi

Department of Computer Science and Engineering, Amrita School
of Engineering, Amrita Vishwa Vidyapeetham, Amrita University,
Coimbatore, India
ba_sabarish@cb.amrita.edu

Abstract. Images which are captured using camera can cause degradation in images by the effect of climatic conditions such as haze and fog. Image restoration makes a notable change in performing different application of computer vision and pattern recognition. The main aim of this paper is to improve the effect of fog and hazy images compared to the existing methods. The enhanced defogging system [EDS] consists of different image improvement techniques with a Dark Channel Prior [DCP] Algorithm to estimate the amount of fog is there in the images and transmission as well. Fusion based fog removal will reduce the amount of haze remained in those images. Experiments were done more than 100 images and the results are discussed below.

Keywords: Enhanced defogging system · Computer vision · DCP technique
Fog

1 Introduction

The Presence of fog is a serious issue in current scenario. The main occurrence of fog is due to climatic conditions and pollution. Surveillance and poor visibility can make several difficulties in surveillance, Flight landing and Vehicle accidents. Fog is also considered as major reason for accidents in land and water as well. A fog can be highly useful in satellite images and military applications. In this paper the fog removal is considered as an image recognition problem and could be able to remove the haze and fog as well. The above algorithm performs better than the existing techniques and can be used in real time as well [1, 2].

Foggy images are similar to noisy images, but the change is that the fog is continuous in a manner and hence filters cannot be used in those scenarios to get an efficient output [3]. The fog will not uniform in all cases. Hence some part is visible and other parts will not be visible. Recognizing the foggy area and improving that region is vital to get an improved output. This requires an efficient defogging algorithm. The foggy areas in an image will create difficulties in feature recognition and extraction [4–6].

© Springer International Publishing AG 2018
D. J. Hemanth and S. Smys (eds.), *Computational Vision and Bio Inspired Computing*,
Lecture Notes in Computational Vision and Biomechanics 28,
https://doi.org/10.1007/978-3-319-71767-8_41

The fog will not uniform in all cases. Hence some part is visible and other parts will not be visible. Recognizing the foggy area and improving that region is vital to get an improved output. This requires an efficient defogging algorithm [7]. The foggy areas in an image will create difficulties in feature recognition and extraction.

Many such defogging techniques are mostly based on the outside scene design, atmospheric light estimation models. Narasimhan et al. analyzed different image degradation conditions such as cloud and smoke. Based on the given factors a hazy model has been designed in his paper [8–10]. This model uses the evidence present in the non foggy area and assumes a smaller presence of haze. Hence in this paper an EDS is used to enhance these regions distinctly and a fusion at multi-scale is done to recover the overall performance and establish an Enhanced Defogging System.

The paper is organized with Sect. 2 on literature Survey Sect. 3 on proposed work, Sect. 4 on result analysis and Sect. 5 with conclusion.

2   Related Works

A. *Dark Channel Prior Algorithm*

A Hazy image model is designed and analysed with dissimilar atmospheric models and estimated the atmospheric light and transmission light image which is free from fog. In these images the hazy regions is represented by light color blocks and the fog free regions by dark color blocks [11, 12]. For a haze free image the dark channel becomes zero. The air light is calculated from the block values using DCP.

B. *Dehazing using Multi-Scale Fusion Technique*

In the reference, for an input image, white balancing and contrast enhancement are performed separately. On each inputs, performed calculation of weightmaps on luminance, chrominance and saliency.

(i) *Luminance Weightmap*

The luminance weight map is calculated as given in Eq. (1). The luminance weightmap acts as an identifier for the amount of degradation occurred on this particular region of the image.

$$W_L^k = \sqrt{1/3\left[\left(R^k - L^k\right)^2 + \left(G^k - L^k\right)^2 + \left(B^k - L^k\right)^2\right]} \qquad (1)$$

WL is the weightmap and the L value is luminance value which is the average of RGB channels in an image.

(ii) *Chrominance Weightmap*

The chromatic weightmap computes the amount of saturation by converting it from RGB to HSV. The S values as given in Eq. (2) with Smax = 1 and sigma is given as 0.3.

$$W_C^K(x) = \exp\left(-\frac{\left(S^{k(x)} - S\ max^k\right)^2}{2\sigma^2}\right) \qquad (2)$$

(iii) *Saliency Weightmap*

The weightmap helps to amount of conspicuousness in human beings visual quality and it is represented by a kernel with particular values and produced a kronickular tensor product of a matrix with 5 × 5 in size and it is multiplied with the image to get the particular weightmap.

Gaussian pyramids are computed for these weight maps of the two input image [13, 14]. Laplacian pyramid is computed for the input images. A multi-scale fusion of these Gaussian and Laplacian pyramids is done to get the restored output.

3 Proposed System

Multi-scale fusion technique performs very poor when the image is completely covered by haze. Some kind of haze will be there in the output image as well. DCP algorithm improves the lighter and darker regions. Enhanced Defogging system proposed in this paper uses the multi-scale fusion of weightmaps after Enhancing the darker regions and lighter regions separately. This algorithm improves the details of an image in an efficient manner as compared to the methods discussed above. The architecture diagram of the proposed system is given in Fig. 1.

Input image for the algorithm is the foggy degraded image which is improved using gamma correction. The lighter and darker are made clear using DCP technique that is followed by multi-scale fusion based on weightmaps. The edges of the result are also used by unsharp masking and given as output.

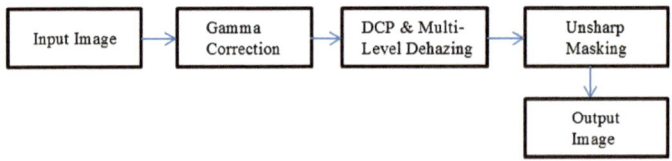

Fig. 1 Proposed EDS system

The details of each block is discussed in detail below,

Gamma Correction

Gamma correction can be used to improve a fully whitened image/fully darkened image so as to reduce the amount of light by selecting the gamma value anywhere between 0 to 1 as represented in Eq. (3) [14, 15].

$$V_{out} = AV_{in}^{\gamma} \tag{3}$$

A is the magnitude value and V_{in} is the input hazy image.

DCP and Multi-level Fusion method

Dark Channel Prior method and Multi-level fusion methods are discussed previous and hence any of the systems never combined the techniques together [11, 14]. The DCP algorithm is most important in terms of identifying the foggy and non-foggy regions by using lighter and darker blocks [6, 7]. From that we could able to get a fog free image by using the formula represented in Eq. (4),

$$I(x) = u(x) * t(x) + A_{ir}(1 - t(x)) \tag{4}$$

u(x) is the fog free image and t(x) is the transmission coefficient were Air is the atmospheric light which is the main degradation condition for any foggy image. So getting u(x) from the equation helps to get a haze free image model.

The transmission is computed and represented in Eq. (5),

$$t(x) = 1 - w(\min(\min(Ic/Air))) \tag{5}$$

Here the w value lies between 0 to 1 and depending on the amount of haze that need to be kept in those images. The transmission and atmospheric light is computed to get an efficient output as given in Eq. (6).

$$u(x) = \frac{I(x) - A_{ir}}{\max(t(x), t_0)} + A_{ir} \tag{6}$$

The t_0 value is typically assigned as 0.1 and the improved output is retrieved and it is a fog free image. Hence, we also discussed that the major disadvantage that we cannot able to recover information from image densely covered with fog [16, 17]. Dehazing by multi-scale fusion is discussed (Fig. 2).

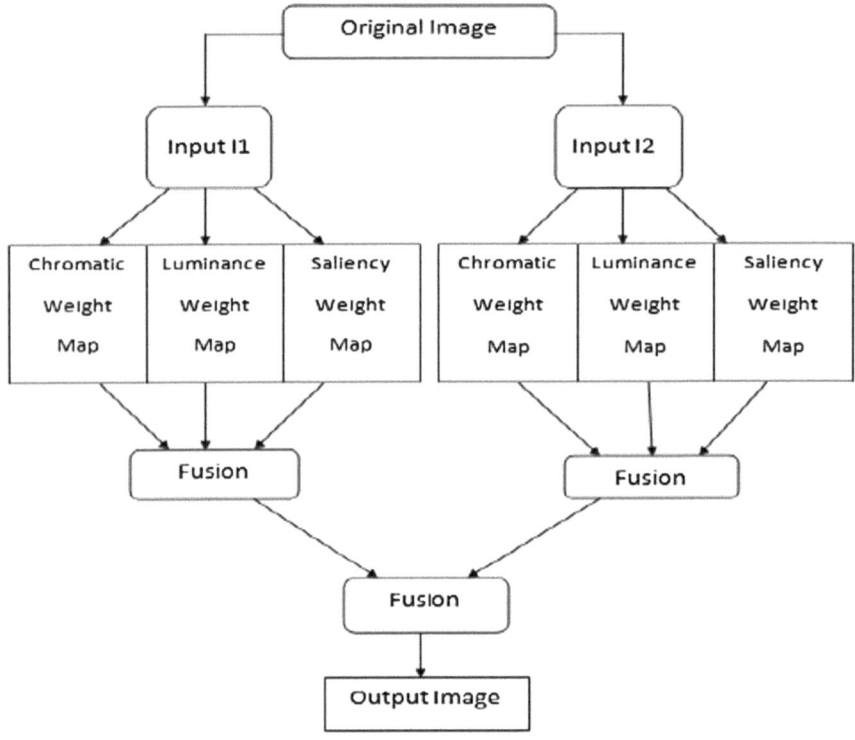

Fig. 2 Multi-level fusion method

Multi-scale fusion has been done on the output of the DCP technique and it is separated into 2 parts [13, 14]. White balancing is performed on the first part and contrast enhancement on the other side. Compute the chrominance, luminance and saliency weightmaps and obtain a required output. Gaussian pyramid is applied on the equivalent weight mapped image and laplacian pyramid on the white balanced, contrast enhancement inputs [18]. Both pyramids are fused together to get a restored output, pyramid level is chosen as 5.

Unsharp Masking

Gaussian filter is applied on the output image to get a filtered result by using suitable sigma value and it is absolutely depends on the images that is going to be tested [15, 18]. Edges have been extracted by subtracting the original image with the Gaussian filtered output [5, 9]. Add the edges with the enhanced original image to get a improved output and hence we get a non-hazy image and most of the information will be restored by using the new EDS.

4 Result Analysis

Images taken under light and dense haze conditions with city and natural background are acquired from different land views. The dataset consists of 100 such images on which the EDS is tested. The output of a dense fog image, Fig. 3a is in Fig. 3b.

(a)   Original Image   (b)   EDS Method

Fig. 3 Comparison between input image and EDS method

Figure 3a shows the input image and main portion of the image is fully covered with heavy fog and the details will not be visible. The robustness of EDS method is proven in Fig. 3a as the bison is not visible in the input image. In Fig. 3b, all the information is preserved and we can see that the details are retained too. This gives you a better picture of the EDS system.

(a) Input Image   **(b) DCP method**   **(c) Multi-level Dehazing**   **(d) EDS Method**

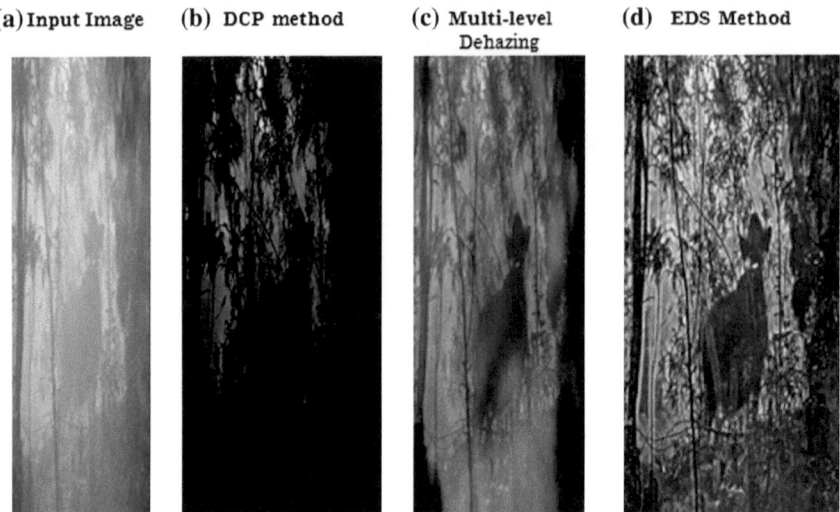

Fig. 4 Comparison with all the discussed methods

Figure 4 shows the contrast between the existing methods and the proposed method for land view of a forest region in Fig. 4a. In Fig. 4b gives the corresponding output of DCP as discussed in Sect. 2. As we can realise that most of the area is very much dark and the details are not visible. The output of Multi-scale dehazing is given in Fig. 4c. Multi-scale Dehazing method shows a better picture when compared to the DCP method, but the amount of fog removed is not complete and still some areas of image will be hidden by the fog. The output of the EDS system is shown in Fig. 4d. It is observed that more details can be retrieved by the EDS. The EDS has an advantage of enhanced robustness in case of heavily degraded(hazy) images and it is discussed above.

5 Conclusions

In this proposed EDS initially gamma correction is used to get a better improved view of the hazy image. DCP method helps to approximate the amount of fog present in those regions. The darker objects and shadows and the lighter region will be the fog and it will be in whitish colour. The important significant in DCP is that it helps in calculating the atmospheric light and the haze free image can be obtained using the formulas [18]. In Multi-scale dehazing, the weightmaps are obtained for luminance, chrominance and saliency on increased brightness and enhanced contrast images [3, 11, 12]. Laplacian and Gaussian pyramids are used to restore the images by reconstruction. Levels in pyramids is fused to get the improved result [11, 18]. Unsharp masking is applied on the output to get further efficient results.

Fog removal by EDS is efficient when compared to other methods as it improves in retrieving the foggy regions. When a dense fog image is considered, the objects which are near is clearly visible. The far off objects may introduce some kind of colouring artifacts and blur can also be viewed in the output.

References

1. Xu, Y., Wen, J., Fei, L., Zhang, Z.: Review of video and image defogging algorithms and related studies on image restoration and enhancement. IEEE Access **4**, 165–188 (2016)
2. Latha, M., Poojith, A., Reddy, B.V.A., Kumar, G.V.: Image processing in agriculture. Int. J. Innovative Res. Electr. Electron. Instrum. Control Eng. **2**(6) (2014)
3. Tripathi, K., Mukhopadhyay, S.: Removal of fog from images: a review. IETE Tech. Rev. **29**(2), 148–156 (2012)
4. Yu, X., Xiao, C., Deng, M., Peng, L.: A classification algorithm to distinguish image as haze or non-haze. In: Proceeding IEEE International Conference Image Graph. pp. 286–289 (2011)
5. Fang, S., Zhan, J., Cao, Y., Rao, R.: Improved single image de-hazing using segmentation. In: IEEE International Conference on Image Processing (ICIP), pp. 3589–92 (2010)
6. Tarel, J.P., Hautiere, N., Caraffa, L., Cord, A., Halmaoui, H., Gruyer, D.: Vision enhancement in homogeneous and heterogeneous fog. IEEE Intell. Transport. Syst. Mag. **4**(2), 6–10 (2010)
7. Ding, M., Ruo, F.T.: Efficient dark channel based image dehazing using quadtrees. Sci. China Inf. Sci. 56(9), 1–9 (2013)
8. Xu, Z., Liu, X., Ji, N.: Fog removal from color images using contrast limited adaptive histogram equalization. In: Image and Signal Processing, 2009. CISP'09. 2nd International Congress on. IEEE (2009)
9. Vasudevan, S.K, Venkatachalam, K., Anandaram, S., Menon, A.J.: A novel method for circuit recognition through image processing techniques. Asian J. Inf. Tech. **15**, 1146–1150 (2016)
10. Nayar; S.K., Narasimhan, S.G.: Vision in bad weather. In: IEEE International Conference on Computer Vision (ICCV), vol. 2, pp. 820–827 (1999)
11. Mishra, S., Sharma, T.: Image restoration technique for fog degraded image. Int. J. Comput. Trends Tech. **18**(5), 208–213 (2014)
12. He, K., Sun, J., Tang, X.: Single image haze removal using dark channel prior. In: Proceeding IEEE Conference Computer Vision Pattern Recognition, pp. 1956–1963 (2009)
13. Jiang, J., Hou, T., Qi, M.: Improved algorithm on image haze removal using dark channel prior. Chinese J. Circuit Syst. **16**(2), 7–12 (2011)
14. Sharma, R., Chopra, V.: A review on different image dehazing methods. Int. J. Comput. Eng. Appl. **6**(3), 77–87 (2014)
15. Narasimhan, S.G., Nayar, S.K: Removing weather effects from monochrome images. In: Proceeding of the IEEE Computer Society Conference, Computer Vision Pattern Recognition (CVPR), vol. 2, pp. II-186–II-193 (2001)
16. Chen, Z., Shen, J., Roth, P.: Single image defogging algorithm based on dark channel priority. J. Multimedia **8**(4) (2013)
17. Lan, X., Zhang, L., Shen, H., Yuan, Q., Li, H.: Single image haze removal considering sensor blur and noise. EURASIP J. Adv. Signal Process. **2013**(1), 86 (2013)
18. Sabarish, B.A., Mohan, S.B, MamthaShri, D.P.B., Ajit, R.C.B., Arun, A.V.R.B.: Automating runout decisions in cricket using image processing. Int. J. Appl. Eng. Res. **10**, 25493–25500 (2015)

Identification of the Risk Factors of Type II Diabetic Data Based Support Vector Machine Classifiers upon Varied Kernel Functions

A. Sheik Abdullah[1(✉)], N. Gayathri[1], S. Selvakumar[2],
and S. Rakesh Kumar[3]

[1] Department of Information Technology, Thiagarajar College of Engineering,
Madurai, India
asait@tce.edu
[2] Department of Computer Science and Engineering, G.K.M College
of Engineering, Chennai, India
[3] Department of Computer Science and Engineering, Bannari Amman Institute
of Technology, Sathyamangalam, India

Abstract. In the innovation in making a data driven decision model, the technological impacts across medical data processing has been increasing rapidly. Type II diabetes is one among the majorly increasing non-communicable disease across India. Advancements in the field of medicine its therapeutic procedures have to be considered in determining the risk factor related to disease prediction and its major cause. This research work, focus towards the applicability of Support Vector Machine (SVM) classifiers for predicting the risk related to type II diabetes. The model involves the deployment of various kernel functions upon the building of SVM models. The experimental results shows that data classification using SVM upon varied kernel function has improvement over the accuracy with 86.65%, precision 76.21% and recall with 81.11% respectively. Among the kernels experimented, polynomial kernel performed better than the other kernels with increased correlation results. The variation in the kernel with model enhancements can be deployed for risk factor prediction in type II diabetes.

Keywords: Data classification · Decision model · Risk factors · Kernel functions

1 Introduction

As more data is gathered, with the amount of data doubling every three years, we go for data mining. Data mining is the process of extraction of hidden predictive information from large databases. It is a powerful technology with great potential to help companies focus on the most important information in their data warehouses. Data mining systems can be classified according to the kinds of databases mined, the kinds of knowledge mined, the techniques used or the applications. Nowadays, Health care has gained more attention as diseases grow exponentially with years. Type 2 diabetes which is of non-insulin form is found to be 95% predominant in most people. There are no proper signs or techniques to detect early onset of diabetes. Diabetes can be predicted using predictive analysis on the basis of various factors like age, sex, BMI index etc. Various techniques emerged to detect diabetes at various stages of human life based on the human body parameters.

Classification is a supervised learning technique. It predicts categorical labels. It is used to classify each item in a set of data into one of predefined set of classes or groups. Classification method makes use of mathematical techniques such as decision trees, linear programming, neural network and statistics. The classification model is developed to classify the data items into groups. The prediction as it name implied is one of a data mining techniques that discovers relationship between independent variables and relationship between dependent and independent variables. It predicts continuous valued functions. For instance, prediction analysis technique can be used in sale to predict profit for the future if we consider sale is an independent variable, profit could be a dependent variable. Then based on the historical sale and profit data, we can draw a fitted regression curve that is used for profit prediction.

The literature presents that gene of the human plays a vital role in the examination and determination of diabetes. Various machine learning methods like regression, neural networks and support vector technique plays a vital role in identifying the correlation between T2D and nucleotide polymorphism. This proposed work helps in classification of people on the basis of Type 2 diabetes using support vector machine. The various kernel functions in SVM are used for further exploration and training that helps in determining the efficient technique to increase the accuracy and infer other parameters like specificity and sensitivity.

2   Literature Review

The paper by Nathan et al. [1] proposes therapy which helps to overcome the complications of Type 1 diabetes such as nervous issues. Intensive therapy for diabetes involves more than three injections of insulin daily to maintain the glucose level. Maintaining glucose level was not a goal of conventional therapy. The proposed therapy aims in lowering glycemic levels which minimizes risk of diabetic neuropathy and retinal problems. The proposed theory reduces the risk of cardiovascular problems by 57%. Thus the intensive therapy achieves risk reduction of cardiovascular problems improving health condition of the patients.

This paper [2] uses a feature selection via supervised model construction (FSSMC) for selecting appropriate attributes relating to diabetic control. On selecting the attributes, various algorithms like Naïve Bayes, C4.5 and IB1 were applied to predict how well it controls the patients. By using the proposed technique, a predictive accuracy of 95% is achieved.

This paper [3] proposes diagnosis based on Quantum particle swarm optimization (QPSO) algorithm and also based on weighted least squares SVM rather than single kernel SVM function. The accuracy and performance obtained through the proposed technique is more than the regular neural networks. This paper [4] focuses on predicting the type 2 diabetes disease using six biomarkers for people above the age of 39 and with certain BMI index. The various factors like age, sex etc. are used as a parameter for risk factor identification of Type 2 diabetes mellitus. Result shows that using six biomarkers could reveal the disease to the greater extent than the routine risk predictions.

Paper [5] focuses on the prediction of diabetes using SVM and rule based technique which further aids in giving deep insights and also in the detection of undiagnosed people. SVM based rule set produced an accuracy of 94%, sensitivity of 93%. The extracted rules also proves to be an efficient tool in diagnosis and predictions. The future extensions are proposed to use various parameters like waist circumference to predict the disease accurately with the rules so that diagnosis could be done at the earlier stages to avoid medical expenses. This paper [6] proposes feature selection based on SVM technique which relates gene-gene interaction in Type 2 diabetes. With the proposed system, a prediction rate of 65.3% has been achieved and individual values of 70.9 and 70.6% prediction rates have been obtained for datasets of men and women respectively. Further if throughput for single nucleotide polymorphism improves, prediction rate would be much higher.

The proposed system uses SVM technique and a web based tool [7] namely diabetes classifier was developed to facilitate easy classification of pupil with and without diabetes [8]. Various set of variables have been taken into consideration and receiver operating characteristics (ROC) curve was found to be 83.5%. This paper [9] proposes SVM technique of classifying the diabetes data using RBF kernel function. The accuracy of the proposed system is found to be 78% and further the receiver operating characteristic for the dataset is also being studied.

This paper [10] proposes a breadth analysis system which helps in diagnosing diabetes and monitoring blood glucose level. A subject-specific prediction model [11, 12] is built to diagnose accurate blood glucose level thereby reducing the error to 27%. The proposed method has shown better results with specificity 90.77% and sensitivity 91.51%. A food classification tool [13] has been proposed by using predictive modeling which helps diabetes patients to select their diet easily. There are facilities to know whether a particular food falls under one of the three classifications namely "choose more often", "choose less often", "moderate". The proposed technique could achieve an accuracy of 93% for diabetes patients, managing diet effectively.

The relation between the hypertriglyceridemic waist (HW) phenotype and type 2 diabetes is being elaborated in [14]. On the basis of triglyceride level, the predictive power of the phenotype is assessed. Logistic regression technique is being applied to differentiate normal people from those who have type 2 diabetes. Further machine learning algorithms [15], Naïve bayes and LR helps to validate the predictive power of various phenotypes. The technique proposed helps in initial screening of type 2 diabetes. This paper [16] proposes a hybrid prediction model for diabetes prediction by using MATLAB in the initial stage to select the significant predictors which reflects the presence of the disease. In the next stage, classification is applied on the filtered data by using a two-layered approach of support vector machine and neural networks to increase the recognition of the proposed technique. The proposed technique achieves 96.09% accuracy using Pima diabetic dataset which is more when compared to the previous values of accuracy. Then various parameters like mean absolute error and recognition rate were used to evaluate the efficiency.

The proposed technique [17] uses three methods of medication adherence namely met for-min (M), M plus rosiglitazone (M+R) and M plus an intensive lifestyle program(M+L).Then the three methods were compare and found that M+R was performing well than M and M+L was intermediate. Major shortcoming is pill count, which is taken as a measure of medication adherence, as it creates issue when visits are missed. The paper [18] discusses about type 2 diabetes and the risk factors which promote to the disease. The polymorphic genotype namely AA and GG were found to one of the risk factor promoting for Type 2 diabetes mellitus. A SVM based two staged classifier [19] is used which makes use of a simple linear iterative clustering method for super-pixel segmentation. A two-staged classifier is used in which a set of k binary classifier is used for training subset of images and in the second stage incorrectly classified set is used to train the SVM classifier. The proposed system provides high performance rates in terms of 94.6% specificity than the existing comparable systems. This paper [20, 21] proposes a phenol-typing approach using machine learning which enables various phenol-typing algorithms used for identifying the Type 2 diabetes patients and also for initial screening process. To evaluate the system (Fig. 1), parameters like high sensitivity and high positive predictive value are used.

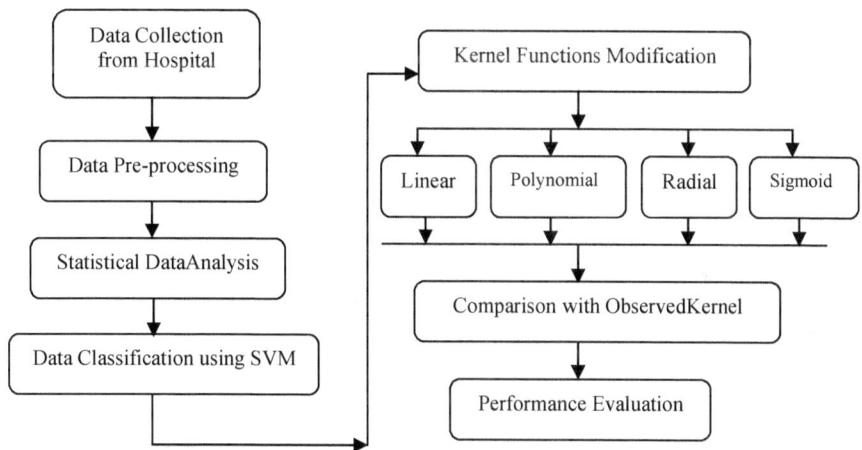

Fig. 1 Workflow of the proposed model

Materials and Methods

Diabetes dataset is collected from UCI repository. It consists of 12 attributes including 1 class label and 400 patient records. The attributes in Table 1 are Patient Number, patient race, age, sex, weight, type of admission of the patient, Period of patient in the hospital, diagnosis number, significance of glucose serum test, significance of A1c test, medication and medication change

Table 1 Dataset description

S. no	Attribute	Description
1.	Age	Age in years
2.	Sex	Sex (1 = male; 0 = female)
3.	Patient No	Each patient's Unique identifier
4.	Patient's Race	Nominal value (Asian, African, American)
5.	Weight	In pounds
6.	Type of Admission of the patient	Integer notations (1–9) [newborn, emergency etc.]
7.	Period of the patient in hospital	Interval between admission and discharge of the patients(In days)
8.	Diagnosis Number	Count of diagnosis in the system
9.	Significance of Glucose serum test	Specifies the outcome of the test if taken or not (">200", " > 300", "normal", "none")
10.	Significance of A1c test	Specifies the outcome of the test if taken or not (">7", " > 8"),
11.	Medications	Specifies if any medicine for diabetes have been suggested
12.	Medication Change	Specifies if there is any medication change for diabetes.

In this approach, the classification accuracy rates for the datasets were measured. For example, in the classification problem with two-classes, positive and negative, an single prediction has four possibility. The True Positive rate (TP) and True Negative rate (TN) are correct classifications. A False Positive (FP) occurs when the outcome is incorrectly predicted as positive when it is actually negative. A False Negative (FN) occurs when the outcome is incorrectly predicted as negative when it is actually positive. The measures are

1. Sensitivity: The ability to correctly detect disease.

$$Sensitivity = \frac{TP}{TP + FN} \qquad (1)$$

2. Specificity: The ability to avoid calling normal as disease.

$$Specificity = \frac{TN}{FP + TN} \qquad (2)$$

$$Accuracy = (TP + TN)/(TP + FP + FN + TN) \tag{3}$$

4. Precision—is the fraction of retrieved instances that are relevant.

$$Precision = \frac{TP}{TP + FP} \tag{4}$$

5. Recall—is the fraction of relevant instances that are retrieved.

$$Recall = \frac{TP}{TP + FN} \tag{5}$$

6. Root-Mean-Squared-error—It is a statistical measure of the magnitude of a varying quantity. It can be calculated for a series of discrete values or for a continuously varying function.

Experimental Results and Discussion:

1. **Results Obtained**
   SVM technique is used for classification of the medical data. This technique is more efficient than the linear regression technique as it plots value near to the real values. SVM technique uses various kernel functions to determine the performance of the data in SVM. Basically kernel types are used for training the system and also for future predictions. The following kernel types are available namely linear, polynomials, radial and sigmoid.

(a) linear: $u'*v$
(b) polynomial: $(gamma^* u'^* v + coef0)^\wedge degree$
(c) radial: $exp(-gamma^*|u-v|^\wedge 2)$
(d) sigmoid: $tanh(gamma*u'^* v + coef0)$

where
  coef0—coef0 for polynomial and sigmoid kernel. (default 0)
  degree—degree for polynomial kernel. (default 3)
  gamma value—gamma for polynomial, radial and sigmoid kernel. (default 1)
  The dataset is evaluated for the three kernel functions under SVM namely linear polynomial and radial and the estimated parameters are shown below:

Table 2 Comparison of the medical data based on various Kernel functions in SVM

Parameter	Linear	Polynomial	Radial
Correlation coefficient	0.6685	0.7115	0.7099
Mean absolute error	0.6613	0.6339	0.6355
Root mean squared error	0.913	0.8706	0.8889
Relative absolute error (%)	65.0692	62.37	62.5356
Root relative squared error (%)	74.2658	70.8215	72.3026

The Table 2 shows that the polynomial function performs better than the other two giving high correlation coefficient value of 0.71 which is an indication of positive correlation. Further root relative squared error percentage is also relatively minimized in polynomial kernel function compared to other kernel functions. The Figure 2 shows the plot of linear and polynomial kernel functions. The plot shows that a linear kernel function prediction does not exactly fit into the real values. Whereas, the polynomial prediction matches the real values and thus error is minimized by fine-tuning the parameters of the kernel functions further.

2. Support Vector Machine Characteristic with the Observed Kernel

This algorithm mainly maximizes the margin between the classes and to minimizes the distance between the hyper plane points. SVM split the dataset in two vector sets under n dimensional space vector. It constructs a hyper plane environment so that each element is been compared respective to the separated linear line. Error ratio can be reduced by classifying the largest margin classifier. This work also includes the analysis based on margin vector along with support vector analysis [5]. The data is clearly differentiated by using hyper plane at the margins. If there is a large margin value, then it depicts that data is clearly separated by the hyper plane. The purpose of an SVM for a training set $\{(x_i, y_i)\}$, where x_i is the observed data and y_i is the labels, where SVM helps in identifying the maximum margin value for the hyper plane.

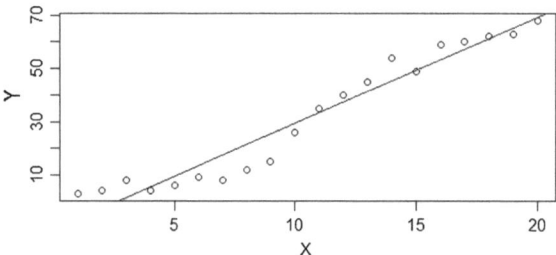

Fig. 2 Linear-function

Linear kernel function (Fig. 2) is best suited for text classification process. Training SVM with linear kernel is much faster when compared to other kernels. C regularization parameter needs to be optimized when using SVM with linear kernel. Linear functions give better results for test classification process with large amount of files and words in each document.

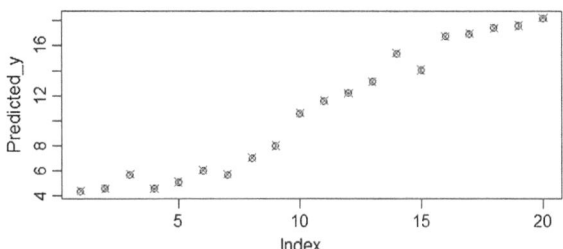

Fig. 3 Polynomial-function

Polynomial function (Fig. 3) not only takes the input samples with given features as such to compute similarity for classification, but also include various combination of the features for classification. This makes classification more effective. Polynomial kernel is used for mapping of φ similar to polynomial regression. The performance of polynomial function is comparatively better than other kernel in terms of accuracy and provides a higher degree of correlation. The polynomial kernel aids Support vector machine in better classification thus reducing the classification error percentage to 10.

3. **Comparing the Efficiency of the Proposed Work with Other Techniques**
   The proposed technique of SVM performs better than the other classification techniques like decision trees etc. The varied kernel functions used in this paper also provides varied parameters (Table 3) based on the function used for each kernel.

Table 3 Performance Measure

Performance Measures	Linear	Polynomial	Radial
Accuracy (%)	85.55	86.65	85.00
Precision (%)	74.04	76.21	75.11
Recall (%)	72.06	81.11	80.10
Classification error (%)	14.44	10.37	11.21
Correlation	0.782	0.867	0.794

The results shows that the polynomial kernel function of SVM gives higher accuracy and precision than the other techniques giving 86.65% accuracy, precision of 76.21% and recall of 81.11%. The observation shows that using SVM with polynomial kernel function provides better results in terms of correlation and accuracy thereby reducing the classification errors to a greater extent.

Conclusion

Type II diabetes has become a major disease affecting people across various ages. To predict the T2D, various classification algorithms were used namely C4.5, ID3, Naïve Bayes and SVM. This paper dealt with the SVM technique of classification of various people as with or without diabetes. This classification is based on various kernel functions which help in identifying the category with better precision. The various kernel functions used include linear, radial and polynomial function of which polynomial gave better accuracy of 86.65% and correlation when compared to other kernels. The further studies could be made on the complexity of the techniques proposed in the paper to optimize the classification process.

References

1. Nathan, D.M., Patricia A., Backlund, J.Y.C.: Intensive diabetes treatment and cardiovascular disease in patients with type 1 diabetes. N. Engl. J. Med. **353**(25) (2005)
2. Huang, Y., McCullagh, P., Black, N., Harper, R.: Feature selection and classification model construction on type 2 diabetic patient's data. Artif. Intell. Med. **41**, 251—262 (2007)
3. Yue, C., Xin, L., Kewen, X., Chang, S.: An intelligent diagnosis to type 2 diabetes based on QPSO algorithm and WLS-SVM. In: International Symposium on Intelligent Information Technology Application Workshops. IEEE computer society (2008)
4. Kolberg, J.A., Rowe, M.W., Jorgensen, T.: Development of a type 2 diabetes risk model from a panel of serum biomarkers from the Inter99 cohort. Diabetes Care **32**(7) (2009)
5. Barakat, Nahla H., Bradley, Andrew P., Mohamed Barakat, Nabil H.: Intelligent support vector machines for diagnosis of diabetes mellitus. IEEE Trans. Inf. Technol. Biomed. **14**(4), 1114–1120 (2010)
6. Ban, H.J, Heo, J.Y., Oh, K.J., Park, K.J.: Research article Identification of type 2 diabetes—associated combination of SNPs using support vector machine. BMC Genet. **11**(26) (2010)
7. Yu, W., Liu, T., Valdez, R., Gwinn, M., Khoury, M.J.: Application of support vector machine modeling for prediction of common diseases: the case of diabetes and pre-diabetes. BMC Med. Inf. Decis. Making **10**(16) (2010)
8. Liu, X., Chen, H.: Identifying adverse drug effects from patient social media-a case study for diabetes. IEEE Intell. Syst. 1541–1672 (2015)
9. Anuja Kumari, V., Chitra, R.: Classification of diabetes disease using support vector machine. Int. J. Eng. Res. Appl. **3**(2), 1797–1801 (2013)
10. Yan, K., Zhang, D., Wu, D., Wei, H., Lu, G.: Design of a breath analysis system for diabetes screening and blood glucose level prediction. IEEE Trans. Biomed. Eng. **61**(11), 2787–2795 (2014)
11. Zarkogianni, K., Litsa, E., Mitsis, K., Wu, P.Y., Kaddi, C.D., Cheng, C.W., Wang, M.D., Nikita, K.S.A.: Review of emerging technologies for the management of diabetes mellitus. IEEE Trans. Biomed. Eng. **62**(12), 2735–2749 (2015)
12. Georga, E.I., Protopappas, V.C., Ardig, D., Marina, M., Zavaroni, I., Polyzos, D., Fotiadis, D. I., Member, S.: Multivariate prediction of subcutaneous glucose concentration in type 1 diabetes patients based on support vector regression, IEEE J. Biomed. Health. Inform. **17**(1), 71–81 (2013)
13. Luo, Y., Ling, C., Ao, S.: Mobile-based food classification for type-2 diabetes using nutrient and textual features. In: International Conference on Data Science and Advanced analytics (DSAA) (2014)

14. Lee, B.J., Kim, J.Y.: Identification of type 2 diabetes risk factors using phenotypes consisting of anthropometry and triglycerides based on machine learning. IEEE J. Biomed. Health Inf. **20**(1), 39–46 (2016)
15. Sheik Abdullah, A., Selvakumar, S., Karthikeyan, P., Mahesh, M., Deepchand, P.K.: An efficient prediction model using multi swarm optimization empowered by data classification for type II diabetes. In: Third International Conference on Business Analytics and Intelligence, ICBAI. Indian Institute of management, Bangalore (2015)
16. Gill, N.S., Mittal, P.: Computational hybrid model with two level classification using SVM and Neural network for predicting the diabetes disease, J. Theor. Appl. Inf. Technol. **87**(1), (2016)
17. Katz, L.L, Anderson, B.J., Mckay, S.V.: Correlates of medication adherence in the TODAY cohort of youth with TYPE 2 diabetes. pp. 1–7 (2016)
18. Luna, G.I., da Silva I.C.R, Sanchez, M.N.: Association between 308G/A TNFA polymorphism and susceptibility to TYPE 2 Diabetes mellitus: A systematic review. J. Diab. Res. **16**, 1–6 (2016)
19. Wang, L., Pederson, P.C., Agu, I., Strong, D., Tulu, B.: Area determination of Diabetic food ulcer images using a cascaded two-stage SVM based classification. IEEE Trans. Biomed. Eng. (2016) (In press)
20. Kagawa, R., Kawazoe, Y., Ida, Y., Shinohara, E., Tanaka, K., Imai, T., Ohe, K.: Development of type 2 diabetes mellitus phenotyping framework using expert knowledge and machine learning approach. J. Diab. Sci. Technol. Jul. **11**(4):791–799 (2017)
21. Sheik Abdullah, A.: A data mining model to predict and analyze the events related to coronary heart disease using decision trees with particle swarm optimization for feature selection. Int. J. Comput. Appl. **55**(8), 973-93-80870-77-4 (2012). https://doi.org/10.5120/8779-2736

Role of Data Analytics in Blended Learning Models of Educational Technology

Ananthi Sheshasaayee and S. Malathi[(✉)]

PG & Research Department of Computer Science, Quaid-E-Millath Govt College
for Women (Auto), Anna Salai, Chennai 600 002, Tamil Nadu, India
{ananthi.research,malathi.research}@gmail.com

Abstract. Blended models the new buzz word" blended" is making its impact on the society, even though not introduced in recent times their existence are dated from age old days. With several models in use their role based for learners and teachers vary according to their accessibilities. This paper discusses the models and need for study on blended models with a role based difference, its benefit's (from teacher's and student's perception) and challenges. It also emphasises the need for the introduction to data analytics of models and the impact on learners.

Keywords: Blended models · Major components · Role based
Analytics

1 Introduction to Blended Models

Blended learning models are the combination of teacher instruction along with online technology to enable student oriented learning. Technological integration used by teachers in a lesson or when students create own curriculum standards is called as blended or hybrid model [1–3]. Most way to integrate varied technology into a curriculum is to use the technology, pedagogy and content. There are different levels of technology integration for a teacher that helps to see where the issue lies in delivering the knowledge. Blended/hybrid learning gives student more control over time, area/location, path and learning pace [4].

2 Models and the Components

There are several models based on blended learning. The major shift in the traditional model is the pedagogy of blended models. These models are based on the teacher's role to help student by providing instruction and to mastery goals [5].

The major components based on blended models:

1. The personal classroom activities which are supervised by a trained instructor/faculty

D. J. Hemanth and S. Smys (eds.), *Computational Vision and Bio Inspired Computing*,
Lecture Notes in Computational Vision and Biomechanics 28,
https://doi.org/10.1007/978-3-319-71767-8_43

2. The online learning materials which includes pre-recorded lectures given by the instructor/faculty
3. Both structured and independent study based on time. It involves guidance by the online material during the lectures and varied skills developed during the classroom [3] experience.

Blended models use the classroom time for activities. All the activities are mostly benefited from direct interaction between the student and the instructor. Traditional/classical education tends to place an impact on delivering the source material through lectures [3]. While in a blended learning model the lectures pre-recorded well ahead of time. By repeating the recorded material the students get benefited [4], by learning at individual own space and time. The classroom time is utilized for application and to solve problems or work through tasks. The classroom time tended to be more than the practical work during the initial [4] stages and gradually the amount of work increases than time. In case of pre-recorded lecturer the students can watch the session any number of times at their own pace. The "flipped" classroom [5] also referred as structured classroom contains a combination of lectures and assignment. In flipped method students are given topic, pre-recorded lectures and online lecturers to watch at off-campus and do assignment. There are other various models present to blend the learning as well to benefit the student's community.

3 Role Base Differences Among Blended Models

Blended models can be differentiated based on roles:

The Learner's Role

3.1 The Rotation Model

Either a course or particular subject in which students rotate based on a time/schedule or for a particular period [6, 7].

1. Station Rotation—Students belonging to either an entire course or particular subject within a classroom or belonging to different group of different classrooms tend to rotate based on courseware [6]
2. Lab Rotation—Either an entire course or particular subject in which students rotate to a lab environment for the online-learning sessions [6] or for any hands on session provided by the supervisor
3. Flipped Classroom—Either an entire course or particular subject in which students take part in online learning, and then for a face-to-face/interactive sessions where the teacher physical presence for guiding and for practice [6]
4. Individual Rotation—Either an entire course or particular subject in which each student is provided with separate set of individual activities [6].

3.2 The Flex Model

Either a course or a particular subject forms the learning basic for students. Occasionally it even provides directions to the learners towards offline activities. In this model the learners are placed as groups and move on an individually customized, schedule among same learning modalities [6].

3.3   The a La Carte Model

A course to be undergone by the student is provided with the option of taking the course entirely online or try accompany experiences that other student is having to at learning centre. In this model students/learner enrol within a particular subject area and will be rolled into the same traditional/classroom by an instructor. During the same duration the learner can enrol in multiple online courses [7]. Through this model approach, technologies can be integrated with the use of enhanced learning activities. Students can access the courses at anytime and anywhere and can also go for extra practice or to review material several times. Even when the learners are not able to attend the lecture sessions the learners can supplement their lost sessions through online material [7].

3.4   The Enriched Virtual Model

An Enriched Virtual Model allows students to divide their time between attending—campus instruction and off-learning remotely. Students interact with instructors for scheduled interactive sessions only on requirement bases. This form of blended learning can begin as a fully online educational experience and then add the on-campus experience along with the supplement classes taken over the Internet. Student's time is spent to do more of online learning than to attend traditional classes [6, 7].

The Teacher's Role

The main aim of instructors/tutor/teacher's is focused to learner's and render to help in order to target the right instruction, to provide all the ongoing system support and to adapt to the current technologies. The role of the teacher is to identify the learning goals and provide the right blend in technology. The teacher's custom blend is to support the learner [8].

3.5   The Rotation Model

Most of this model [9] is based on the teacher's discretion. The mode involves learning modalities in which at least one of the module being online learning.

1. Station Rotation—Either a course or an interested area of subject in which students experiences the model rotation within students in a classroom or group of students belonging to different classrooms [6]
2. Lab Rotation—Either a course or an interested area of subject in which students rotates within a lab environment for online-learning sessions [6]
3. Flipped Classroom—Either a course or an interested area of subject in which students participate in both online learning being away from campus compared to the in—place of traditional assignment [6]
4. Individual Rotation—Either a course or an interested area of subject in which each student has personalized time schedule and rotate to their need bases [6].

3.6 The Flex Model

Model [9] is based only on similar field of interest among students. Flexible grouping is more than moving the student's position. It is the most effective practical way to differentiate as learning needs. The presence of teacher/instructor of record is onsite [1]. Teachers use different instruction strategies for each learner. Each technique is meant to be implemented according to each learner's learning style and method to adaptable to learn new ideas, field of interest as well as individual involvement using a variety of different instructional methods. Based on individual field of interest, teachers group students irrespective of their age or ability. Based on flex models [10] teachers are discovering that grouping and regrouping students helps in a variety of ways throughout the course as the students become more productive [6].

3.7 The a La Carte Model

The a la carte model [9] of blended learning is based on online courseware. This approach integrates technology with the use of enhanced learning activities, allowing the teacher to use student's data to target instruction based on small groups [11]. The model supports in creating individual learning paths for each student. Based on the students work grading can be done to access the student's capability and to improve the performance accordingly [7].

3.8 The Enriched Virtual Model

Entire course or particular subject in which [12] students have required face-to-face interactive learning sessions with the teacher. The sessions include partially individual interaction with the teacher and the remaining sessions are completed remotely. Model [9] represents classroom experience in which each course is divided between student's time and online content along with instructions [6].

4 Benefits and Challenges of Models and Need for Data Analytics

Benefits of models [13]

1. Provides independent support for the student as different needs of the students are catered through personalized instructions [3, 14]
2. The availability of online material [15] to cater the knowledge of students by providing any time and any where material facilities [3, 14]
3. As modern technological devices permits instructors to reach out students across the classroom, in-campus and off-campus [3, 14]
4. Certain models provide more interactive experience like the lab rotation [3, 14]
5. Provides better support [13], better communication [3, 14]and better control over the process
6. Certain models provides ample time [3, 14] for learning and practicing.

Challenges of models

1. Each technology usage can be more challenging than being useful for both students as well as for the teacher. Issues [3, 14] such as availability of technological literacy
2. Certain models need more blended work which is treated as over worked by teachers, the over work [3, 14] as got the same impact with students as increase in cognitive load can be experienced
3. Plagiarism [3, 14] and credibility of sources forms a bigger issue as many digital resources forms to be from unreliable sources.

Learning data analytics and impact on learners

Data analysis in education is not a new trend. Several, previous information supports data driven decision making. There have been several useful reports from disciplines with respect to academic analytics and educational analytics. Nature of learning analytics, the availability of big data in education offers many opportunities and answers to all unsolvable solutions about education, and also offers more to pressing challenges in the education category [16, 17].

Since data is present abundantly studying the trends and severing the learners are important, learning analytics is as important for education [18]. A well reliable and robust architecture is needed to manage any dynamic learning environments thereby, preserving anywhere, anytime access to students and teachers as well as maintaining the integrity of data [18]. Learning data analytics provides a formal process of measuring details about learning contents, usefulness at different levels as well, predictive levels and reporting of the outcomes data [18]. Blended learning data analytics refers to science of organizing, sequencing and monitoring instruction [18]. Advancement in digital age has lead to the enormous growth of data supporting the pattern based through analyses of these data. With patterns assist all possible factors with respect to learner's success can be improved [19].

With the availability of online data tools and applications about student's performance, learning abilities and the availabilities of teaching resources [20] data can be analysed to study the pattern of trend. The analytic tools such as educational data mining and learning analytical tools support the analysis of educational perspective [19]. Big data and learning analytics converge towards quantitative traditional method. Data are collected through various sources to learn the pattern of learner's trend. With perfect skill and access analyst transform the data to a better understanding form known as knowledge [21]. Learning Analytics prepares students to meet all the competitive requirements. Through various analytical skills based on theories, concept, exploration and discovery, students learn to select the needed skills, prepare for competency, implement various technologies, interpret the data observed, and evaluate using quantitative tools to utilise appropriate learning analytic models [21].

5   Conclusion

Blended models form an inspiration to redefine traditional roles. Models have brought different focus to education. The instructor places students with varied skill set and knowledge. Most of the online and relevant material, independent and individual study time, knowledge sharing, and various methods of guiding students toward the most meaningful and needed experience have become possible due to blended method of teaching. Learning through supervised activities and with blended models has proven to be adaptable to most of blended learners and teachers. In spite of challenges and issues present with each model blended method in education has its own place in institutions. By providing personal analytics to students all the common errors in understanding can be provided. These analytics identify less appropriate techniques used by the students and can provide the individual with better context which reflect on individual development thereby addressing timely improvements for betterment.

References

1. Christensen, C., Staker, H., Horn, M.B.: Blended-learning (2013)
2. Hobgood, B.: Blended learning (2013)
3. Winstead, S.: 6 disadvantages of blended learning you have to cope with (2016) (Blog)
4. Bharti, P.: 6 great tools for blended learning (2014)
5. Jasmine, G., Michaela, P., Suyin, D., Scott, M.: Children's learning environments (2015)
6. Acrobatiq Guest: Seven blended learning models used today in higher Ed(2016)
7. Wylie, J.: Mobile learning technologies for 21st century classrooms (2011)
8. Horn, M.: The benefits of blended learning for students and educators. dreambox.com
9. Hunt, V.: Pros and cons of blended learning at college (2016)
10. John, A.: 4 Best blended learning resources to drive engagement (2015)
11. Edudemic Staff: 16 of the best blended learning resources (2015)
12. Edudemic Staff: 6 of the Best blended learning resources (2015)
13. Blog: Raise your hand texas: Understanding the different blended learning models, blended learning (2016)
14. Kaplanis, D.: Benefits of the blended learning approach (2013)
15. Waitzkin, J.: Technology integration and blended learning—there is a difference (2014)
16. Pardo, A., Dawson, S., Gašević, D., Steigler-Peter, S.: The role of learning analytics in future education models. Telstra
17. Corbett, K.: Blended learning: 4 blended learning models, EdTech (2017)
18. Sweeney, S.: Victoria University taps analytics to provide blended learning (2014)
19. Charlton, P., Mavrikis, M., Katsifli, M.: The emerging trend of learning analytics and big data can support and empower learning and teaching (2013)
20. Giarla, A.: The benefits of blended learning (2016)
21. Dean, M.: The growing impact analytics is having on education (2016)
22. Blog Advancement courses: Which blended learning model is best for your classroom?: (2016)
23. Cox, J.: Flexible grouping as a differentiated instruction strategy. Teachhub
24. Smith, S.: Blended learning moving to centre stage in higher education (2015)
25. John, A.: 4 Best blended learning resources to drive engagement (2015)
26. Kaplanis, D.: Benefits of the blended learning approach (2013)

Query by Example—Retrieval of Images Using Object Segmentation and Distance Measure

S. Sathya$^{(\boxtimes)}$, Latha Parameswaran, and R. Karthika

Department of Computer Science and Engineering, Department of Electronics
and Communication Engineering, Amrita School of Engineering, Amrita Vishwa
Vidyapeetham, Coimbatore, India
cb.en.p2cvil5012@cb.students.amrita.edu, {p_latha,
r_karthika}@cb.amrita.edu

Abstract. Image segmentation is the process of extracting object and regions of interest which have applications in computer vision, object detection, classification of retrieval. Many researchers have developed algorithms for segmenting objects from digital images. This paper is an attempt to retrieve images based on existing segmentation algorithms and distance measures. A diverse dataset consisting of images of various categories has been used for experimentation and this work suggests the best segmentation algorithm for retrieving the best match for given query images using distance measures.

Keywords: Global threshold · Multilevel threshold · Active contour model
Super pixels · CCA · Eucledian distance

1 Introduction

Image segmentation is a classical area of research which has more challenges in the field of computer vision. Segmenting an image is normally done to enhance the quality of a complex image thereby reducing the burden on feature extraction, object boundary extraction and for labeling. Most of the segmentation algorithms [1] have similarity with edge detection for clustering and boundary encoding of objects. In [2–4] the authors have suggested hybrid methods for block based segmentation, based on the survey done, it is evident that retrieval of matching images using segmentation techniques have not been much explored. Thus this proposed work aims at tapping, the potential of segmentation algorithms for retrieval of matching images using distance metric as criteria to find the best match.

2 Literature Survey

Image segmentation and image retrieval are two major challenges in any computer vision and image processing application. In [5, 6] the authors have reported their study on content based image retrieval system using global features, region based retrieval,

© Springer International Publishing AG 2018
D. J. Hemanth and S. Smys (eds.), *Computational Vision and Bio Inspired Computing*,
Lecture Notes in Computational Vision and Biomechanics 28,
https://doi.org/10.1007/978-3-319-71767-8_44

color, texture and shape based retrievals. In [7] hybrid techniques using texture based features in wavelet domain have been explored where Eucledian distance and Manhattan distance metric have been used for retrieval. In [8] the authors have proposed an algorithm to extract matching images using SIFT key points as they are good in detecting corners of a given image. Image retrieval based on HOG, SIFT, SURF, and Color histogram has been developed by the authors. In [9] the authors have concentrated on image retrieval system, RGB color combination, and retrieval of images with a large image database. In [10] the authors discussed primitive feature extraction technique using color, shape, texture, similarity measures and matching images. In [11] the authors have addressed the different segmentation algorithms like thresholding, edge, and region based segmentation [12]. Though, in the literature many algorithms have been proposed by the researchers 'No single Technique' has been found to segment objects on all types of images. In [13] developed a algorithm for maximizing the between class variance of pixel intensity to perform thresholding. In [14] facial features are extracted using gabor wavelet and classification is done using SVM.

Based on the algorithms and methods reported in the literature it is inferred that segmentation algorithms have not been exploring for query by example based retrieval of images. This paper is an exploration for retrieving the matching images for a given query image using the existing segmentation algorithms and the available distance metrics.

3 Proposed Algorithm

In this proposed system two stages of work has been carried out. The first stage concentrates on extracting the best features from the given image by performing segmentation. In the second stage given an example image, this proposed algorithm will extract best matching images from the available dataset in a similar manner. The distance metric chosen for the match is Eucledian distance.

In Fig. 1 the architecture of the proposed algorithm is shown with various intermediate steps such as Filtering, choice of segmentation algorithms, connected component analysis and feature extraction for feature database generation.

The second stage of the proposed system which includes feature comparison and retrieval of matching images is shown in Fig. 2

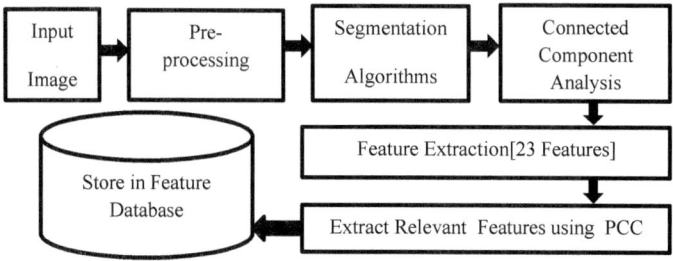

Fig. 1 Stage1 architecture of the proposed system

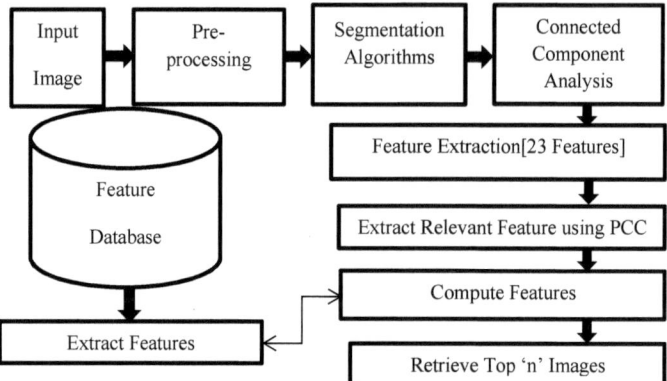

Fig. 2 Stage2 architecture of the proposed system

4   Proposed Work

The three major steps involved in this algorithm: Segmentation, Feature Extraction, and Distance Measure computation are explained below.

4.1   Segmentation

In this work, four segmentation algorithms namely Global thresholding using OTSU's method, Multilevel thresholding, Active Contour Model and Super pixels based segmentation have been chosen after exhausted study and experimentation. A detailed study on OTSU's method is discussed [2, 15–18] where the main concept is in choosing a dynamic threshold value to increase the inter-class variance and minimize the intra-class variance. An Algorithm for Multilevel thresholding for segmentation has been discussed in [13, 16, 19] where the perception is based on the histogram of the original image and the selection of threshold has been expressed as an objective function for optimization. Active Contour Model-based segmentation has been discussed in [15, 20, 21] where the contours of the image have been extracted using a snake function and final points have been extracted using energy minimization. Super pixels [3] based segmentation algorithm produced better gradient image and serves as the key for segmentation.

4.2   Feature Extraction

After performing segmentation using the algorithms mentioned above a set of features based on the segmented output of the image has been extracted as shown in Table 1. These 9 features have been carefully chosen out of the 23 features extracted from the segmented image, after analyze using PCC.

Table 1 Features extracted from segmented image

Features	Definition
Area	The aggregate number of 'on' pixels in the image
Major axis length	The length is measured in pixels and indicates the longest diameter of the object
Minor axis length	The length is measured in pixels and indicates the shortest diameter of the object
Orientation	Orientation of edge pixels that indicate their direction. Pixels on a vertical line, horizontal line or slanted. It is represented degrees or radians. Range of orientation: $-90°$ to $+90°$
Convex area	The area which is covered by convex is greater than or equal to the region area
Filled area	Number of pixels in filled image
Euler number	Difference between Number of objects and number of holes
Equivalent diameter	The equivalent diameter is identical to the diameter encountered in stoke's law. Equiv diameter = sqrt(4 * Area/pi)
Perimeter	Boundary of the labelled component

4.3 Distance Measure

In this proposed algorithm the distance measure used for retrieving matching images has been chosen as Eucledian distance Absolute difference, Chessboard distance and Bray-cutis distance. It was empirically found that Eucledian distance measure provides better retrieval of matching images due to its simple line distance based computation, of straight line distance between two points in eucledian space. If a and b are two points on the original line then,

$$\mathbf{D_{Euc}(b, a)} = \mathbf{D_{Euc}(a, b)} = \sqrt{(a - b)^2} \qquad (1)$$

where, $\mathbf{D_{Euc}(b, a)}$, is the distance between the points a and b.

This proposed work has explored major segmentation algorithms, extracted various features from the segmented output and retrieval has been performed using Eucledian distance measure. Suitable pre-processing algorithms have been used for preparing the image towards the same.

5 Experimental Results

Exhaustive experimentation has been done on the proposed algorithm on COIL-100 dataset. COIL 100 dataset [22] has been taken which has images of single objects such as bottles, glue…etc. out of which 70 images belonging to objects of ten classes has been considered for feature extraction and storing in a database in the first stage. Nearly 30 images have been used for retrieval in stage2. Though many researchers have used this dataset but very few used if for retrieval. After applying smoothing filter 23 features namely Area, Major Axis Length, Minor Axis Length, Orientation, Convex Area, Filled Area, Euler Number, Equiv Diameter, Perimeter, Centroid, Bounding Box, Sub array Idx, Eccentricity, Convex Hull, Convex Image, Image, Filled Image, Extrema, Solidity, Extent, Pixel Idx List, Pixel List, and Perimeter old have been extracted and first 9 features have been chosen by removing redundancy using PCC and stored in the database in stage1.

During the second stage, any random image from the dataset has been taken and similar steps are executed and the same set of features extracted. Using the Eucledian distance measure as the criteria for the retrieval, matching images are retrieved. Results pertaining to the various segmentation techniques are shown in Fig. 3.

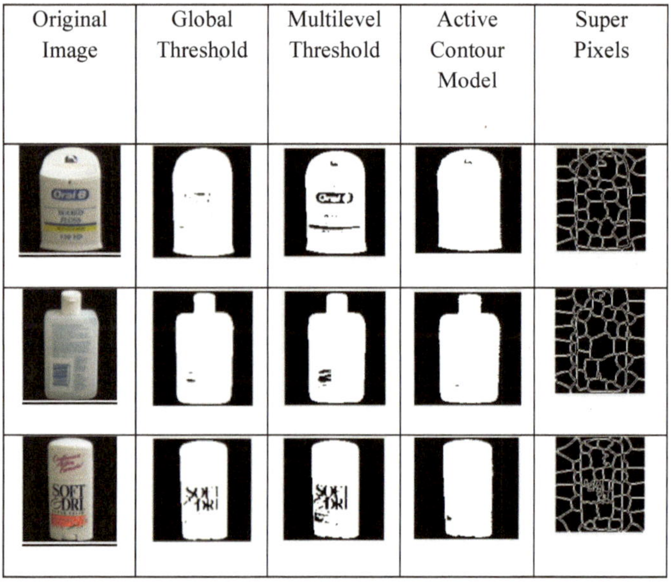

Fig. 3 Coil-100[single object]

It has been observed that 'Active Contour Model' provides best segmentation results visually. Relevant 9 features have been extracted from these images and the results pertaining to 3 images are shown in Table 2.

Table 2 Feature extraction [single object]

Features	Area	Major Axis Length	Minor Axis Length	Orientation	Convex Area	Filled Area	Euler Number	Equiv Diameter	Perimeter
Global threshold	7464	143	88	−87	9041	3319	−18	98	529
	7081	127	79	−85	7975	7543	0	95	354
	5228	150	63	−84	7269	5492	−15	82	559
Multilevel threshold	16,900	150	150	0	16,900	16,900	1	147	505
	16,900	150	150	0	16,900	16,900	1	147	505
	16,900	150	150	0	16,900	16,900	1	147	505
Active contour	8958	137	36	−89	9273	8979	0	107	377
	7911	130	80	89	8353	7917	−1	100	380
	7443	142	69	−39	7652	7457	0	97	368
Super pixels	4020	153	131	87	16,130	10,876	−35	72	972
	3847	152	137	90	16,220	10,674	−32	70	109
	4295	143	125	−75	16,291	10,686	−47	74	105

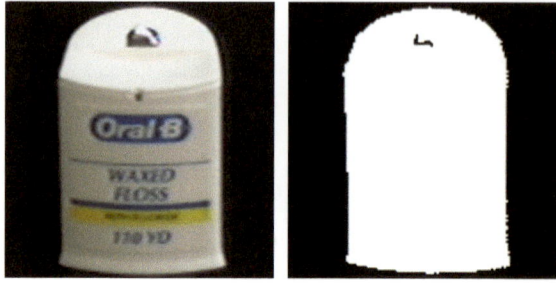

Fig. 4 Segmented Image

Results pertaining to a specific image after segmentation are shown in Fig. 4.

The Eucledian distance between the features stored in the database during stage1 and the graph is shown in Fig. 5

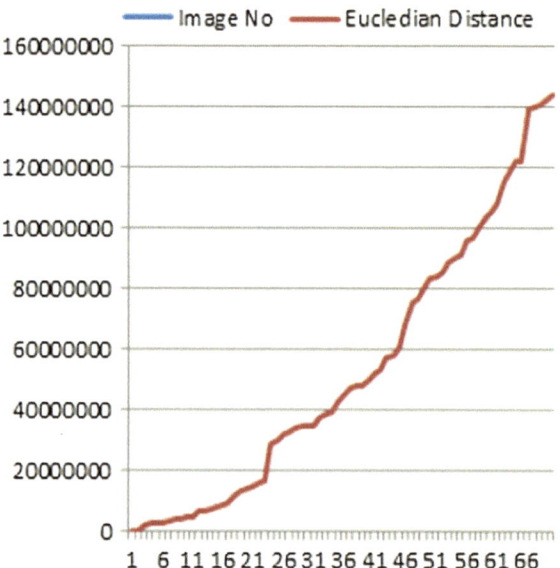

Fig. 5 Eucledian distance [single object]

The images which have minimal distance are considered to be relevant to the specified image. The top 5 retrieved images are shown in Fig. 6.

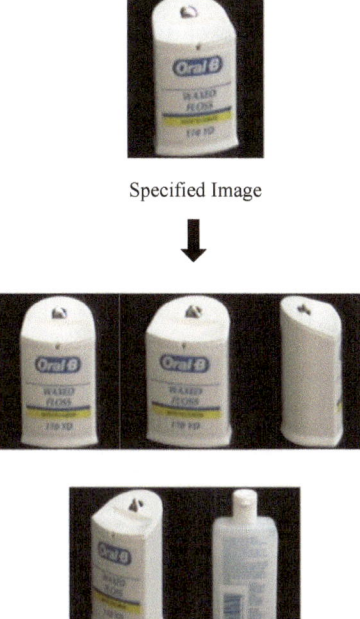

Specified Image

Fig. 6 Top 5 retrieved images

Similar experimentation has been considered for two objects from the Coil dataset and results pertaining to segmentation algorithms are shown in Fig. 7.

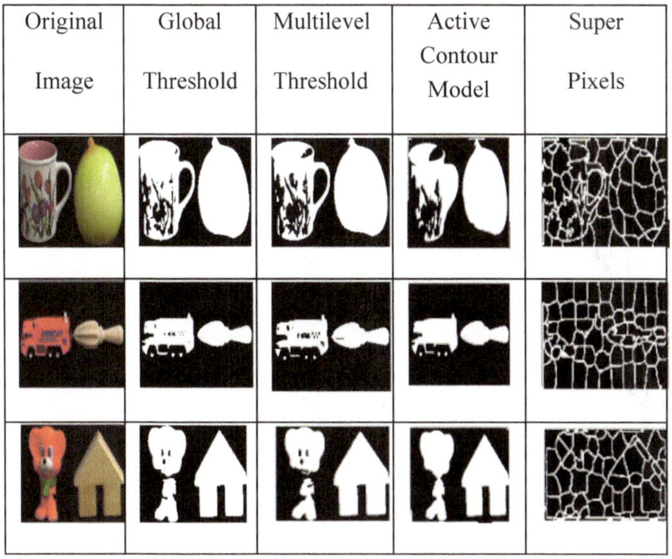

Original Image	Global Threshold	Multilevel Threshold	Active Contour Model	Super Pixels

Fig. 7 Coil-100[two objects]

It can be observed that 'Multilevel Thresholding' provides better segmentation in this case. Results pertaining to features of 3 images are shown in Table 3.

Table 3 Feature extraction [two objects]

Features	Area	Major axis length	Minor axis length	Orientation	Convex area	Filled area	Euler number	Equiv diameter	Perimeter
Global threshold	20376	133	122	−88	12,743	12,262	1	125	416
	3775	131	45	2	5193	3858	−10	69	430
	18,040	297	145	−7	34,834	34,834		152	765
Multilevel threshold	34,965	299	156	0	34,965	34,965	1	211	768
	34,534	303	152	0	34,584	34,584	1	210	768
	34,846	303	154	0	34,846	34,846	1	211	786
Active contour	9740	124	102	−38	10,593	9740	1	111	453
	3794	130	44	2	5135	3794	1	70	448
	8124	136	106	−17	13,750	8124	1	102	776
Super pixels	7664	295	152	1	34,577	26,127	−58	99	1623
	6236	302	123	0	33,014	13,794	−48	89	1864
	6349	302	161	0	34,159	30,117	−38	90	1220

Figure 8 shows the results of the Eucledian distance of the given query image and the images in the dataset.

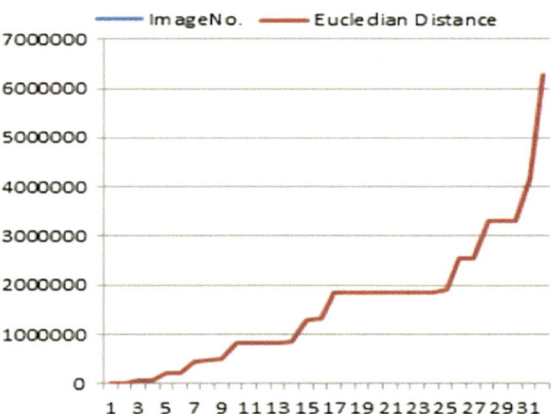

Fig. 8 Eucledian distance [two objects]

As a measure for retrieval, the top 5 retrieved images are shown in Fig. 9.

Fig. 9 Top 5 retrieved images

Hence, it may be inferred that retrieval of images based on segmentation is much powerful as it extracts most of the shape features of the given image.

6 Conclusion

A query by example based image retrieval system by using existing segmentation algorithms and Eucledian distance measure has been explored in this work. Experimentation results help to conclude that images with a single object produce best retrieval results when 'Active Contour Model' is used as the base algorithm for segmentation. In the case of, multiple objects it is observed that 'Multilevel Thresholding' algorithm is best suitable for features extraction to retrieve similar images. This work can be extended to images with a complex background and occluded objects.

References

1. Nilima, S., Dhanesh, P., Anjali, J.: Review on image segmentation clustering and boundary encoding. Int. J. Innovative Res. Sci. Eng. Technol. **2**(11), 6309–6314 (2013). ISSN 2319-8753
2. Zaitoun, N.M., Aqel, M.J.: Survey on image segmentation techniques. In: International Conference on Communication, Management and Information Technology (ICCMIT), pp. 797–896. (2015)
3. Radhakrishna, A., et al.: SLIC superpixels compared to state-of-the-art superpixel methods. IEEE Trans. Pattern Anal. Mach. Intell. **34**(11), 2274–2282 (2012)
4. Khan, W.: Image segmentation techniques: a survey. J. Image Graph. **1**(4), 166–170 (2013)
5. Tunga, S., Jayadevappa, D., Gururaj, C.: A comparative study of content based image retrieval trends. Int. J. Image Process. (IJIP) **9**(3), 127–155 (2015)
6. da Silva Junior, J.A., Marcal, R.E., Batista, M.A.: Image retrieval: importance and applications. In: X Workshop de Vis-ao Computacional, pp. 311–315 (2014)
7. Abood, Z.I, Mushin, I.J, Tawfiq, N.J.: Content-based image retrieval (CBIR) using hybrid technique. Int. J. Comput. Appl. **83**(12), 17–24 (2013)

8. Bagyammal, T., Parameshwaran, L.: Context based image retrieval using Image features. Int. J. Adv. Inf. Eng. Technol. (IJAIET) **9**(9), 27–37 (2015)

9. Chary, R., Rajya Lakshmi, D., Sunitha, K.V.N.: Feature extraction methods for color image similarity. arXiv preprint arXiv:1204.2336(2012)

10. Kondekar, V.H., Kolkure, V.S., Kore S.N.: Image retrieval techniques based on image features: a state of art approach for CBIR. Int. J. Comput. Sci. Inf. Secur. (IJCSIS) **7**(1), 69–76 (2010)

11. Jogendra kumar, M., Raj Kumar, G.V.S., Vijay Kumar, R.: Review on image segmentation technique. Int. J. Sci. Res. Eng. Technol. (IJSRET) **3**(6) (2014). ISSN 2278-0882

12. Ramadevi, Y., Sridevi, T., Poornima, B., Kalyani, B.: Segmentation and object recognition using edge detection techniques. Int. J. Comput. Sci. Inf. Technol. (IJCSIT) **2**(6), 153–161 (2010)

13. Liao, Ping-Sung, Chen, Tse-Sheng, Chung, Pau-Choo: A fast algorithm for multilevel thresholding. J. Inf. Sci. Eng. **17**(5), 713–727 (2001)

14. Karthika, R., Parameswaran, L.: Study of Gabor wavelet for face recognition invariant to pose and orientation. In: Proceedings of the International Conference on Soft Computing Systems, Springer India (2016)

15. Arbela, P., Maire, M., Fowlkes, C., Malik, J.: Contour detection and hierarchial image segmentation. IEEE Trans. Pattern Anal. Mach. Intell. **33**(5), 898–916 (2011)

16. Arora, S., Acharya, J., Verma, A., Panigrahi, P.K.: Multilevel thresholding for image segmentation through a fast statistical recursive algorithms. Pattern Recognit. Lett. **29**, 119–125 (2007)

17. Kondekar, Vipul, et al.: Image retrieval techniques based on image features: a state of art approach for cbir. Proceedings of the International Conference and Workshop on Emerging Trends in Technology. ACM, (2010)

18. Otsu, N.: A threshold selection method from gray—level histograms. IEEE Trans. Syst. Man Cybern. **9**(1), 62–66 (1979)

19. Luessi, M., Eichmann, M., Katsaggelos, A.K.: Framework for efficient optimal multilevel image thresholding. J. Electron. Imaging **18**, 1–10 (2009)

20. Kass, M., Witkin, A., Terzopoulos, D.: Snakes: active contour models. Int. J. Comput. Vision **1**, 321–331 (1987)

21. Xu, X., et al.: Characteristic analysis of Otsu threshold and its applications. Pattern Recognit. Lett. **32**(7), 956–961 (2011)

22. Columbia University Image Library (COIL-100). http://www.cs.columbia.edu/CAVE/software/softlib/coil-100.php

Adaptive Weighted Median Filter for Motion Estimation

Prashant Bhalge$^{(\boxtimes)}$ and Salim Amdani

Computer Science and Engineering Department, Babasaheb Naik College
of Engineering, Pusad, Pusad, Maharashtra, India
prashbhalge@gmail.com, salimamdani@yahoo.com

Abstract. Smoothing techniques can be constructive to the motion vectors that can detect defective vectors and put forward alternatives. The substitute motion vectors can be used in place of those recommended by the block match algorithm. If frames are going to be interpreted by the receiver then motion vector amendment is expected to be precious. Design of an adaptive weighted median filter whose weights alters according to the confined characteristics is possible which can be used for smoothing intention.

Keywords: Block distortion measure · Motion compensation · Motion estimation · Video compression

1 Introduction

Video compression has always been an imperative area of study due to realistic precincts on the extent of information that can be stored, processed, or transmitted. To make use of the spatial and temporal redundancies expected in all video sequences predictive coding is favored. In this format, a predicted value is anticipated for each pixel of a video frame, and the variation between the predicted and actual values of each pixel is the only pixel information mandatory to be encoded.

Motion Estimation

The same set of motion parameters is shared by cluster of closest pixels that cover a moving object. Assuming that the stirring objects can be approximated logically well by standard shaped blocks, the image is alienated into regular, non-overlapped blocks. Then under the belief that all the pels in the block share the same motion (displacement) vector, a single displacement vector is anticipated for the entire image block.

D. J. Hemanth and S. Smys (eds.), *Computational Vision and Bio Inspired Computing*,
Lecture Notes in Computational Vision and Biomechanics 28,
https://doi.org/10.1007/978-3-319-71767-8_45

2 Overview of Existing Motion Estimation Algorithms Using Filters

L. Alparone, M. Barni, F. Bartolini and R. Caldelli suggested the use of weighted median filter in 1999 to regularize optic flow field across motion boundaries. Vector median filter have inadequacy of lack of tuning. Weights were defined taking into consideration reliability factors. The proposed method was based on vector median filtering adaptively weighted by a confidence estimate of motion vectors. It was observed that proposed method had better outlier elimination capability and preservation of motion edges [1].

L. Alparone, M. Barni, F. Bartolini and L. Santurri analyzed the impact of an effective regularization of MV on the performance of H.263 video coder in 2001. The proposed vector filtering was found much more effective in terms of bit saving when coder was operated at extremely low rates. The use of filter weight definition and vector median filtering of motion vectors allowed a consistent saving of computational effort. For 'mother_daughter' sequence experiment had observed PSNR of 31 dB [2].

Z. Liu, Y. Song, T. Ikenaga and S. Gota proposed algorithm in 2006 for motion estimation. The proposed algorithm used three approaches. In the first approach low pass filter based subsampling scheme was used. In the second approach SMVP algorithm avoids the data dependency among sub partition within one macro block. The third search window approach effectively reduced the search positions. The experiment had observed 7.15% absolute difference operations of simplified full search [3].

A multistage motion vector processing method was proposed in 2008 by A. Huang and T. Nguyen considering the reliability of each received motion vector. By analyzing the distribution of residual energies and effectively merging blocks that have unreliable motion vectors, the proposed method had the capability of preserving the structure information. To refine motion vector using a constrained vector median filter, the motion vector reliability information was used. Smoothness measure for all directions were defined. The smoothness measurement were aimed to lessen the difference among motion vector. By using proper motion vector, blockiness was effectively decreased. The proposed method had observed PSNR value greater than 1.39 dB compared to other method [4].

An adaptive motion vector smoothing scheme based on weighted vector median filter was proposed by J. Guo, J. Kim in 2011. The algorithm eliminated motion outliers more effectively to improve the quality of side information. A simple motion vector outlier reliability measure was used for each block in motion compensated interpolated frame. The proposed algorithm adaptively selected the candidate motion vectors according to the sum of absolute difference value of each corresponding block. If SAD value was less than a predefined threshold, motion vector was unchanged otherwise it was needed to be smoothed. For filtering purpose weights were defined taking into consideration predicted errors of blocks. The weighted vector median filtering was applied only to blocks with unreliable motion vectors. The proposed method had produced PSNR of 29.70 dB for '*Foreman*' sequence [5].

M. Stengel, P. Bauszat, M. Eisemann, E. Eisemann and M. Magnor suggested a post process in 2015 that can be used to simulate different shutter effect or can be used for other artistic purpose [6].

2.1   Adaptive Weighted Median Filter for Smoothing

To eradicate noise and for smoothing purpose many state of the art filters have been accessible. Bulk of them are based on median filter [7]. A median filter which is non-linear filter, commonly used as spine filter in noise exclusion and smoothing techniques.

Motion Smoothing is extra step in video stabilization. In motion smoothing, unnecessary global displacement vectors are cleaned. There are diverse types of filters used to smooth the motion. Some of the past techniques worn were low pass filter, IIR, FIR filters. While recently Gaussian filters, Adaptive IIR filter, Kalman filters are used. Motion smoothing is obligatory to smooth undesired camera motion after motion estimation and to remove buildup error prior to motion compensation.

Smoothing is post process that take place after getting motion vectors from the motion estimation process. In this work motion vector smoothing is completed to assist conservative block matching algorithm in finding true motion [8].

In case of block based motion assessment, the distortion criteria often consequences in motion vectors which are untrustworthy. Some factors such as noise, deformable motion, motion edges existing with in block. Object occlusion challenges distortion criteria, so the block preferred with least DFD may not correspond to real motion.

The size of descending window as matter of fact is only adaptable parameter; this can be trounce by weighted median filter. Estimator selectivity can be controlled by a set of weights, thin details are conserved and thus flexible to use. Bigger weights are linked with those producing lesser displaced frame difference values. The computational complexity of adaptive weighted median filter is quite fair.

The performance of adaptive filters is much enhanced than mean and order-statistics filters [9].

Binary Matrix of search:

The Binary matrix of search MS can be defined, which elements are

$$MS(k, 1) = 0; \text{ if } MAD_B(x, y, 0, 0) < T_1.$$
$$MS(k, 1) = 1; \text{ if } MAD_B(x, y, 0, 0) > T_1.$$

$$MAD_B(x, y, 0, 0) = \frac{1}{MN} \sum_{m=1}^{M} \sum_{n=1}^{N} abs\left[X_q(m, n) - X_{q-1}(m, n) \right]$$

where X_q and X_{q-1} are picture elements of matching block of frame q and (q−1).The size of matching block is M by N. The threshold value T_1 set by using variance and T_2 set by correlation.

$$W(i,j) = \frac{MAD_B(k,l,u_B,v_B).MS(k-2+i,l-2+j)+\eta}{MAD(k-2+i,l-2+j)+\eta}$$ (1)

$$i, j \in \{1, 2, 3\}$$

where the number η prevents dividing by zero ($\eta = 2^{-10}$). The weights of neighbor-hoods depend on the magnitude of their MAD's.

Modification

The yield value of AWM filter can be some of close vectors considering the actual vector, which provides slighter value MAD in its position, however it needn't provide a fewer value of $MAD_{AWM}(x, y, u_{AWM}, v_{AWM})$ in the position of the actual vector. In this modification, after the practice of AWM filtering in the position with a value MS $(k, l) = 1$ this value of MV (u_{new}, v_{new}) is accepted, for which

New Vector $V = \arg(\min(MAD_B(X, Y, U_B, V_B), MAD_{AWM}(X, Y, U_{AWM}, V_{AWM})))$.

In this case the modified hexagonal search is used for motion estimation [10].

Three step Hexagon-based Search

Algorithm:

Step 1:	Initial bristly pattern is centered at the origin of the search window. Now, experiment each points in the search pattern. If the least block distortion point is the center point go to step3. If not go to step2
Step 2:	Arrange a new LHSP with the MBD point as the center point. If the new MBD point is at the center location, go to step3
Step 3:	Form the fine pattern with preceding MBD point as the center point. The new MBD point acquired in this step becomes the final solution i.e., the motion vector (x, y). The number of search points depends on the site of MBD point also regulates the search direction

Result:

See Table 1.

Table 1 Average PSNR for different algorithms and different video sequences

Video sequence	BMA(PSNR in dB)		
	FSA	MHS	Median (AWM)
Claire	22.42	17.84	18.10
Missa	35.61	34.85	34.90
Susie	35.34	32.23	32.60
Football	22.59	17.98	18.75

Conclusion:

This paper presents the smoothing of motion vectors. Many fast ME/MC algorithms to diminish the complexity of compression practice have been developed by researchers. The smoothing using adaptive weighted median filter produced enhanced compensation than three step hexagonal search. Smoothing motion vectors, conversely, can add noteworthy intricacy to a video compression algorithm and should only be used where the reimbursement are more important than these costs. For the '*Susie*' sequence the PSNR obtained using AWM is 32.60 dB which is better than modified Hexagonal search. For other fast, medium and slow video sequences the PSNR obtained using smoothing AWM manner is found superior than modified Hexagonal search.

References

1. Alparone, L., Barni, M., Bartolini, F., Caldelli, R.: Regularization of optic flow estimates by means of weighted vector median filtering. IEEE Trans. Image Process. **8**, 1462–1467 (1999)
2. Alparone, L., Barni, M., Bartolini, F., Santurri, L.: An improved H.263 decoder relying on weighted median filtering of motion vectors. IEEE Trans. Circuits Syst. Video Technol. **11**, 235–240 (2001)
3. Liu, Z., Song, Y., Ikenaga, T., Gota, S.: Low pass filter based variable block size motion estimation algorithm for H.264. In: IEEE International Conference ICASSP, pp. 253–256 (2006)
4. Huang, A., Nguyen, T.: A multistage motion vector processing method for motion compensated frame interpolation. IEEE Trans. Image Process. **17**, 694–708 (2008)
5. Guo, J., Kim, J.: Adaptive motion vector smoothing for improving side information in distributed video coding. J. Inf. Process. Syst. **7**,103–110 (2011)
6. Stengel, M., Bauszat, P., Eisemann, M., Eisemann, E., Magnor, M.: Temporal video filtering and exposure control for perceptual motion blur. IEEE Trans. Visual. Comput. Graphics **21**, 663–671 (2015)
7. Jain, A.: Fundamentals of Digital Image Processing, 1st edn. PHI Learning (2010)
8. Haylin, S., Kailath, T.: Adaptive filter theory. LPE, Pearson (2002)
9. Gamcova, M., Marchevsky, S., Gamec, J.: Higher efficiency of motion estimation methods. J. Radioengineering **13**, 58–64 (2004)
10. Bhalge, P., Amdani, S.: Modified Hexagonal search for motion estimation. In: Proceeding of IEEE international conference ICICCS-2017, Madurai (2017)
11. Wai, L.: Efficient Block Matching Motion Estimation Algorithms for Video Coding. Thesis, City University of Hong Kong (2004)
12. Halsall, F.: Multimedia Communication. Pearson Education (2001)

Detection of Menisci Tears in Sports Injured and Pathological Knee Joint Using Image Processing Techniques

Mallikarjunaswamy M. S.[1](✉), Rajesh Raman[2], Mallikarjun S. Holi[3], and Sujana Theja J. S.[4]

[1] Department of Instrumentation Technology, Sri Jayachamarajendra College of Engineering, Mysuru 570006, Karnataka, India
msm@sjce.ac.in
[2] Department of Radio-Diagnosis, J. S. S. Medical College and Hospital, J. S. S. University, Mysuru 570015, Karnataka, India
[3] Department of Electronics and Instrumentation Engineering, University B.D.T. College of Engineering, Constituent College of VTU, Belagavi, Davangere 577004, Karnataka, India
[4] Department of Orthopedics, J. S. S. Medical College and Hospital, J. S. S. University, Mysuru 570015, Karnataka, India

Abstract. In many physical and sports activity, knee joint encounters extreme level of stresses leading to traumatic injury. The menisci are located between the tibial plateau and femoral condyles. These menisci act to distribute body weight evenly across the knee joint. Any damage to menisci may lead to uneven weight distribution and cause the development of abnormal excessive forces leading to early damage of the knee joint. Menisci act as 'shock absorbers' between femur and tibia, and prevent the lateral movement of the joint when the knee is fully extended. It also helps to provide a lubricating effect on the knee joint, provides some degree of stability, and is essential for the normal biomechanics of the knee joint. The menisci are often injured, particularly during athletics. Magnetic resonance imaging (MRI) is the widely used imaging modality in detection of menisci injuries. There is a chance of missing visibility of menisci tears in MRI during diagnosis. In this work, image processing method based on seeded region growing is developed to detect and visualize menisci tears from MRI. The segmentation accuracy is evaluated using dice similarity coefficient (DSC). The developed method segments the menisci from knee joint MRI with good accuracy and visualizes the tears in different compartments of menisci. The developed method is helpful in treatment and surgery planning of knee joint injured patients.

Keywords: Menisci tears · Seeded region growing algorithm · MRI
Biomechanics of knee

© Springer International Publishing AG 2018
D. J. Hemanth and S. Smys (eds.), *Computational Vision and Bio Inspired Computing*,
Lecture Notes in Computational Vision and Biomechanics 28,
https://doi.org/10.1007/978-3-319-71767-8_46

1 Biomechanics of Knee Joint

Knee joint is a major weight bearing and most often injured joint of the human body. The joint consists of three bones namely the femur which is located on superior side, tibia on inferior side and patella on anterior part. The capsular structure of knee joint includes articular cartilage, menisci, cruciate ligaments, synovial membrane with synovial cavity filled with synovial fluid, and burse. The motion of the knee joint is controlled by several muscles and ligaments and they protect it from damage due to overturn. The medial and lateral collateral ligaments stabilize the knee on either side. The surface of femur, tibia and patella are covered with thin layer of tissue called cartilage. The medial meniscus and lateral meniscus are the two tissues between the cartilage surfaces of femur and tibia act as shock absorbers. The horseshoe shaped menisci reduce the stress on the cartilage surfaces during activity of knee joint. The articular cartilage and menisci in combination provide friction less movement of the knee [1]. Figure 1 shows the anatomy of knee joint.

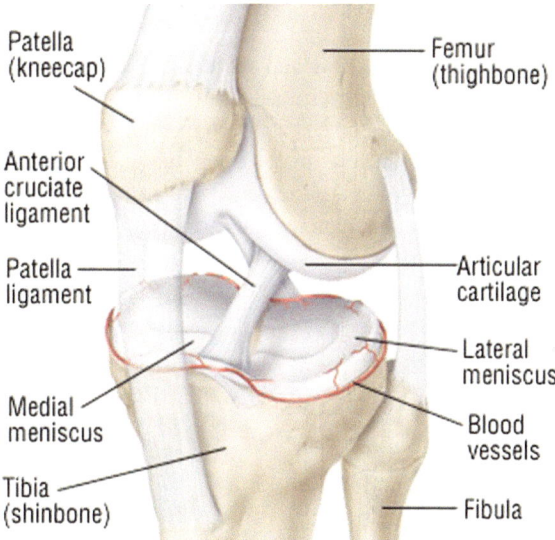

Fig. 1 Knee joint showing lateral and medial meniscus [2]

Movements of knee joint: Knee joint is a typical hinge joint with flexion (bending) and extension movements. It also rotates slightly when it is flexed. The knee joint when extended to straight, it is considered as 0°. The flexion of the knee joint is possible up to 135° during normal activity. The range of motion in normal knee joint is 0–135°. The extension of knee joints to 0° is an indication of healthy condition. In daily routine activity the healthy knee joint rarely moves to 95° in flexion. During an activity such as squatting, knee flexion may reach as much as 160° as the hip and knee are both flexed and the body weight is superimposed on the joint. The normal gait on plane ground requires approximately 60–70° knee flexion, whereas ascending stairs requires about

80°, and sitting down into and arising from a chair requires 90° of flexion or more. The menisci are fibro cartilage structures composed of thick collagen fibers. Forces across the knee joint are as high as two to four times body weight during walking and up to six to eight times the body weight during running. Menisci play a major role in cushioning and distributing the stress due to body weight over the cartilage surface of femur and tibia. They also contribute in joint lubrication and provide stability for the joint with the support of ligaments [3].

The meniscus performs important functions such as stress distribution over the articular cartilage surface and shock absorption during axial loading of knee joint. In addition meniscus plays vital role in stabilization of knee joint during flexion and extension. Menisci also contribute in joint lubrication. They provide minor contribution in secondary stabilization in cruciate ligament injured knee [4].

Menisci tears: Meniscal tears are among the most common knee injuries. Athletes, particularly those who play sports such as football, hockey and basketball are at the risk of meniscal tears. The menisci are easily injured by the force leading to rotation of the knee while bearing weight. A partial or total tear of a meniscus may occur when a person quickly twists or rotates the upper leg while the foot stays still. When a meniscus is torn, it often does not heal and there are chances that the pieces of the cartilage may become trapped which results in pain with certain twisting activities of the knee. If the tear is small, the meniscus stays connected to the front and back of the knee. If the tear is large, the meniscus may be left hanging by a thread of cartilage. Menisci tears are named based on their appearance and location. Common tears include longitudinal, transverse, flap, bucket handle, and torn horn. Figure 2 shows menisci and different types of tears on menisci.

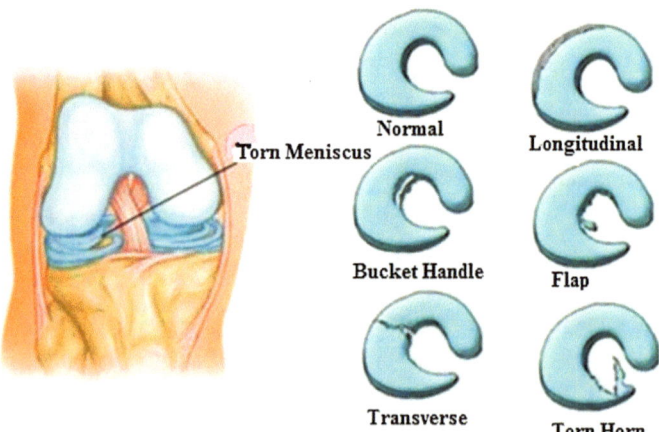

Fig. 2 Knee joint menisci and different types of menisci tears [5]

The longitudinal and radial tears occur as a result of increased force on meniscus in normal knee joints. Horizontal tears occur because of normal forces on degenerative meniscus on the posterior half region of meniscus. The menisci are not repairable when tears are too long [6]. The menisci reduce the stress on the tibia by uniform distribution of load on cartilage surface. When menisci are torn in knee joints cartilages are overstressed due to load and results in osteoarthritis (OA). Detection of menisci tear at early stage is important for its repair and prevention of OA. Early detection and treatment helps in betterment of quality of life in sports personal and elderly people.

2 Diagnosis and Treatment of Menisci Tears

The knee joint is one of the most often injured joints in the human body. Menisci are torn and dislocated during injuries to the knee joint. Meniscal injuries occur during sports activities due to over extension, twisting forces at the joint and stretching of legs beyond the limitations. Usually MRI is recommended to confirm the diagnosis. Occasionally, doctor may use arthroscopy to diagnose and treat the meniscal tear.

Meniscectomy: The complete meniscectomy or partial meniscectomy are the invasive procedures performed on the knee joint to diagnose and repair the menisci. This procedure may further damage the cartilage and its load distribution, later this may result in the faster degradation of cartilage in such knee joints [7]. Meniscectomy is a surgical procedure to remove torn meniscus partially or completely. The possibility of repair of menisci tears is decided based on the pattern of tears in injured knee joint. Usually part of meniscus is removed surgically in cases of horizontal and flap tears of meniscus. The age, activity level of the injured person also considered in addition to size and location of tears to decide the surgical repair of meniscus. Meniscal surgeries are carried out by orthopedic surgeon with arthroscopy. Arthroscopic based surgeries helps in localized repair of meniscus and limits the damage to other parts of the knee joint especially cartilage. In certain meniscus injuries, total meniscectomy is inevitable because of type and pattern of the meniscal tear. In this surgical procedure the entire meniscus is removed. In partial meniscectomy a small portion of the menisci is removed and remaining edge are smoothened. The decision on type meniscectomy procedure is based on location of the tear (red zone or white zone) and also on healing conditions of the knee injured person. In certain OA affected knee joints, along with cartilage, the menisci are also get affected and degenerated leading to severe joint problems. The knee joint arthroscopy procedure conducted on such patients may result in further damage to the cartilage. During sports injuries, when the menisci are not repairable the meniscus transplantation is performed. It is essential to detect menisci tear and its location before the surgery [8]. Therefore, there is a need of noninvasive procedure for visualization of menisci for diagnosis and surgical planning.

MRI: MRI is a useful imaging technique in diagnosis of meniscal injuries and pathologies of knee joint. MR imaging is useful in assessing the presence as well as the size and location of the meniscal tear. Along with the symptoms and clinical examination it is useful in the determination of the timing of surgery. In OA affected knee joints the cartilage loss results in menisci alterations over a period of time. The sensitivity of menisci alterations and tears are high in MRI findings. The sensitivity, specificity and accuracy of detection of meniscal tears are more in MRI than with arthroscopy. MR images give valuable information regarding the tear pattern and configuration which helps in preoperative planning and assessment of the reparability of the tear. Even though MRI can visualize tears, it is difficult to notice because of the complex structure of knee joint. There are chances of missing visibility and detection of menisci tears during diagnosis of knee injuries. Figure 3 shows the knee joint with different types of meniscal tears.

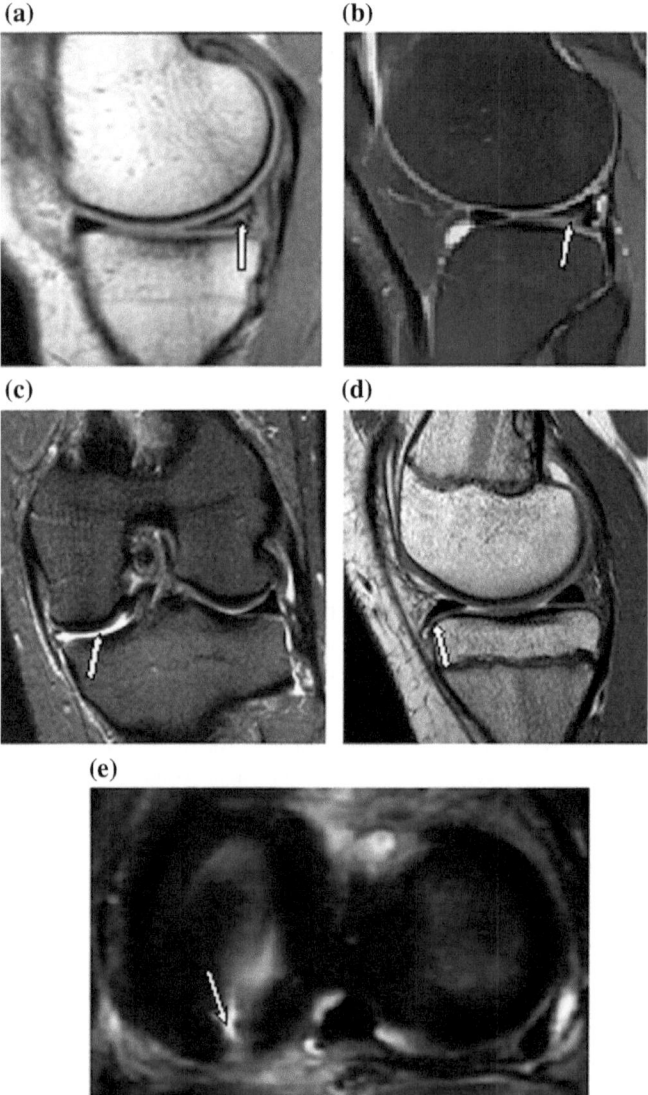

Fig. 3 Knee joint with menisci abnormalities **a** posterior horn longitudinal tear **b** radial and horizontal tear **c** bucket handle tear **d** discoid meniscus **e** meniscus with radial tear [8]

The discoid meniscus, meniscal ossicles and meniscal flounce are the abnormalities which can be observed by MRI. Different types of discoid meniscus are named as complete, incomplete and ring shaped. Meniscal ossicles usually appear in MRI with menisci tears. The wavy appearance of meniscus along the edges is called as meniscal flounce and observed in sagittal plane of knee joint MRI. The menisci extrusion is usually more than 3 mm and is another abnormality observed using MRI. The observation, preservation, detection of menisci tear and repair are possible using MRI. Suphaneewan et al. [9] shown in their study that the 3-Tesla MRI diagnosis shown higher sensitivity in detection of menisci tears compared to 1.5-Tesla. Stone et al. [10] measured meniscal sizes using 1.0T MRI for population of interest with different parameters for their study. In all these studies by medical community, the usefulness of MRI for diagnosis and treatment of menisci is well established. For detection of menisci tears, there is a need of image processing method to confirm the menisci tears and help the doctors in locating the tear for further treatment. Therefore, there is a need of non-invasive visualization of the menisci, which helps medical practitioners in diagnosis and treatment of different menisci pathologies.

3 Segmentation of Menisci

Dataset: The knee joint MRI dataset is obtained from JSS Medical College and Hospital, Mysuru. The images are acquired using SIEMENS 1.5T and PHILIPS 3.0T MRI systems in different image protocol. The imaging parameters for the sequence are: TR/TE: 16.3/4.7 ms, matrix: 384×384, FOV: 140 mm, slice thickness: 0.7 mm, x/y resolution: 0.365/0.365 mm. For this study, approval is obtained from Ethical committee of medical research, J. S. S. University, Mysuru to review the medical records and images of the patients of J. S. S. Hospital who had been clinically diagnosed for knee joint problems and undergone MR imaging. Informed consent for participation was obtained from all the subjects of this study. The data set includes knee joint MR images of normal and pathological subjects. The knee joint images are processed for segmentation and visualization using MATLAB software. The images are preprocessed for noise removal using median filter of size (3×3). Median filter removes noise without affecting edges and boundary information in an image. Figure 4 shows input knee joint MRI image and resultant image after preprocessing.

(a)   (b)

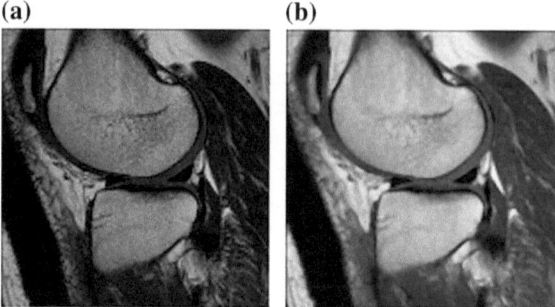

Fig. 4 Preprocessing of knee joint MRI **a** input image **b** image after preprocessing

SRG algorithm: The algorithm developed by Adams and Bischof [11] is used to segment the menisci from the surrounding cartilage, synovial fluid and other tissues. Region growing methods can correctly detect the region of pixels of similar properties and segment the regions which have clear edges. In SRG method the seed points are initialized manually on the region of interest. From the seed point a set of connected neighboring pixels are selected based on the connectivity criteria. The different connectivity possible is shown in Fig. 5.

(a)   (b)   (c)

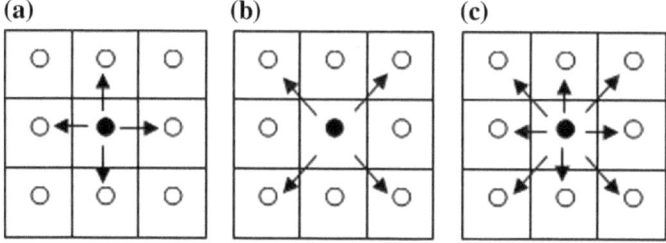

Fig. 5 Connectivity of neighboring pixels **a** 4 connected **b** 4 diagonally connected **c** 8 connected

Figure 5a shows 4 adjacent connected neighbors, Fig. 5b shows it's diagonally connected neighbors and Fig. 5c shows 8 connected surrounding neighbors. For menisci the segmentation region is very small and narrow, therefore 4 adjacent neighboring pixels are considered for region growing algorithm.

The algorithm to segment menisci from knee joint MRI [12] works as follows

1. Initialization: The size of the region is initialized to zero and mean intensity of the region is initialized to zero. The maximum size of the region to be grown is limited to the size of the input image. The distance of the regions newest pixel to the region mean is also initialized to zero. Initialize the region max intensity distance to a threshold value t. Initialize t value to 0.2.

2. Seed point marking: The seeded region growing is started by selecting one seed point on each of the lateral and medial menisci region of the knee MRI. In the initial slices of MRI data set menisci is visible as one region and one seed point is visualized. In the later slices menisci is visible as two regions and segmentation is started with two seed points in each portion of the menisci. The intensity of the seed point pixel is assigned to the mean intensity of the region. The region growing is started with seed point pixel in each region and the region size is assigned with a value one.

3. Calculate the neighboring pixels of seed point with four connectivity. If the connected pixels are not already part of segmented area retain them and make a list of neighbor pixel coordinates with pixel intensity values.

4. Compare the region mean intensity value with pixel intensity values of each neighboring pixels and find the distance. The distance $\delta(x)$ is simple measure of the intensity of the pixel x from the intensity mean value of the region under consideration.

$$\delta(x) = \left| g(x) - \mathrm{mean}_{y \in A(i)}[g(y)] \right| \tag{1}$$

where $g(x)$ is the gray value of the image pixel x and $g(y)$ is the mean intensity of pixels y which are already in the region. A_i is the ith set of initialized seed point.

5. Add the pixel which is nearest to region mean intensity to the region. The region size is incremented by one since new pixel is added to the region. Update the region mean intensity considering the newly added pixel to the region.

$$\delta(z) = \min_{x \in T}\{\delta(x)\} \tag{2}$$

where T is the set of unallocated pixels in the border of the region. The z is unallocated pixel which gives minimum intensity distance according to Eq. (1) is appended to the region. The coordinates of appended pixel is initialized as new seed point $A_i(z)$ for the next search.

6. Consider the coordinates of newly added pixel as new seed point. Remove the pixel from the list of neighboring pixels.

7. If the pixel distance is less than the maximum intensity distance of 0.2 or region size is less than the image size repeat procedure from step 2.

8. Display the binary image of the segmented region using the pixels of the region.

Segmentation of menisci using SRG: The menisci were segmented from MRI of normal knee joint using SRG algorithm the results are shown in Fig. 6. The Fig. 6a shows segmented images of menisci of lateral side Fig. 6b–d shows segmented menisci from different slices of MRI from lateral to medial.

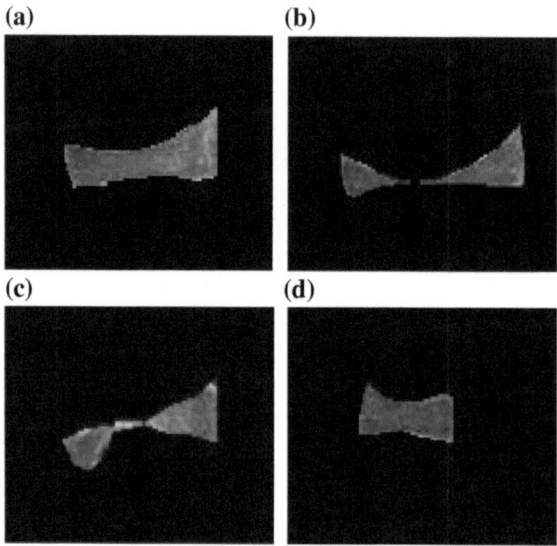

Fig. 6 Segmented images from normal knee joint MRI using SRG **a–d** segmented images from different MRI slices from lateral to medial

4 Results and Discussion

The developed SRG algorithm is used to detect the menisci tears in different knee joints MRI of the dataset. Figure 7 shows the detection of menisci tears using SRG algorithm.

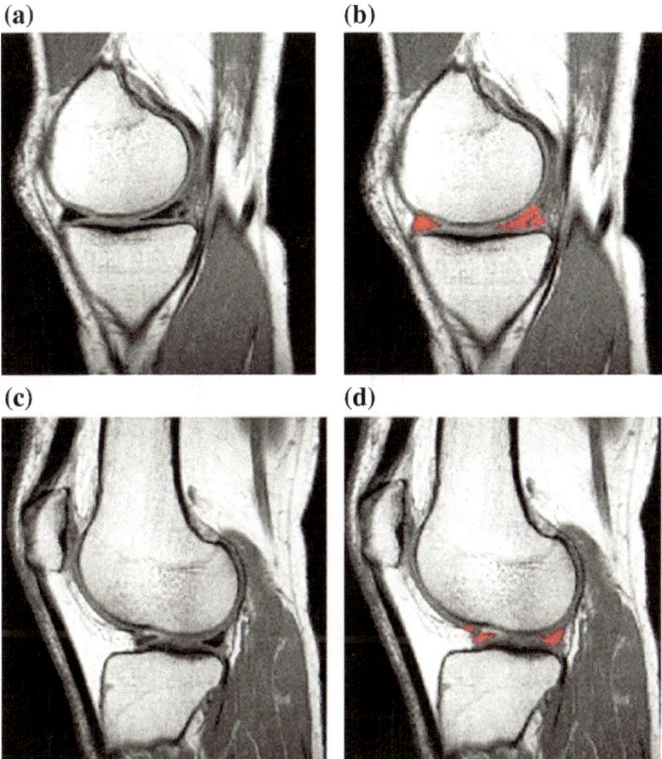

Fig. 7. Detection of menisci tears **a** input knee joint MRI with tear in posterior horn **b** segmented menisci (in color) with detected tear **c** input knee joint MRI with menisci tear in the anterior horn **d** segmented menisci (in color) with detected tear

The developed algorithm of menisci segmentation using SRG is used to segment and visualize the tears in few more cases of menisci tears and Fig. 8 shows the results of visualized tears of menisci.

Accuracy of menisci segmentation: The knee joint menisci are segmented using developed interactive segmentation technique based on SRG. For comparison of segmentation accuracy the similarity index is computed and described as dice similarity coefficient [13] as

$$DSC = 2\frac{|M_1 \cap M_2|}{|M_1 + M_2|} \tag{3}$$

where M_1 and M_2 are the segmented images of menisci, M_1 is the segmented image of menisci using SRG method and M_2 is the segmented image of menisci using manual segmentation method (ground truth). M_1 and M_2 are obtained by computing number of 1's in segmented images of menisci (binary). The manual segmentation was carried out by radiology students who are expertise in identifying menisci abnormalities.

The MRI slices knee joints are randomly selected from data set and menisci are segmented using manual segmentation method and SRG method for comparison. Figure 9 shows the images of menisci segmented and overlapped on the input image. Figure 9a shows input image, Fig. 9b shows segmented of menisci using manual method and overlapped on input image (white color). Figure 9c shows segmented menisci using SRG method overlapped on input image with yellow color. Figure 9d shows segmented menisci images of both methods to show the overlap of segmented regions. The tiny white color traces around the menisci shows the minute difference in segmentation.

(a)   (b)

(c)   (d)

(e)   (f)

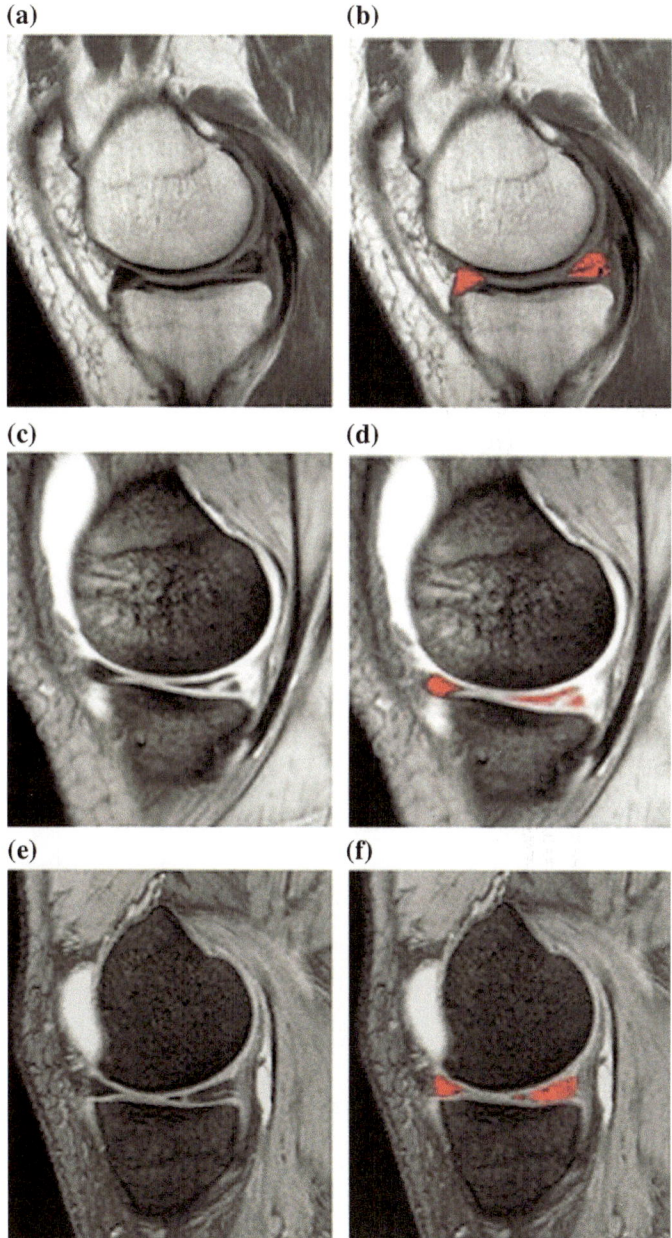

Fig. 8 Input MRI with menisci tears and visualization of menisci tears; **a**, **c** and **e** are input MRI images, **b**, **d** and **f** are segmented menisci overlapped on respective input images

The DSC computed for segmented images of normal and menisci tear knee joint MRI of dataset. The mean DSC index computed shows the value 0.85 for normal knee joint menisci subjects. The value is slightly lesser in segmented menisci images with tears due to sports injuries and OA. The mean of DSC index values and standard deviation (SD) from the mean value were computed using segmented images of 49 normal and 52 teared menisci knee joint subjects are shown in Table 1. Segmentation results of SRG method are in acceptable limits (greater than 0.75) according to DSC index obtained.

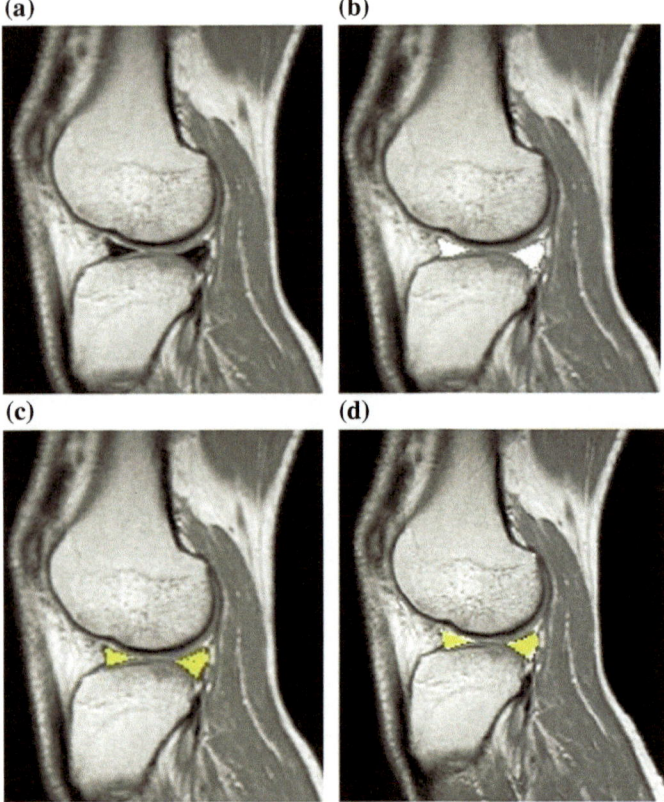

Fig. 9 Comparison of menisci segmentation using SRG with manual segmentation method **a** input MRI **b** segmented image of manual method **c** segmented image of SRG method **d** overlapped segmented regions of menisci

Table 1 DSC index of menisci segmentation

Case Description	Mean of DSC	SD
Normal	0.85	0.06
Menisci tear	0.82	0.01

5   Conclusion

The knee joint menisci were segmented from MRI of the obtained data set from Hospital. A method based on seeded region growing was developed to segment the menisci from MRI. An interactive segmentation approach is adopted for segmentation of menisci from knee joint MRI using SRG. The method is interactive to locate seed points in the region of menisci and other steps are automated. The method segments menisci and visualizes the menisci tears. The DSC was computed to compare the segmentation results with the manual method. The method shows good segmentation accuracy with DSC value of greater than 0.75. Detection of menisci tears is helpful in treatment and surgery planning of knee due to sports injury and pathological conditions.

Acknowledgements. J. S. S. Medical College and Hospital, J. S. S. University, Mysuru, INDIA for providing MRI data set.

References

1. Levangie, P.K., Norkin, C.C.: Joint structure and function a comprehensive analysis, 4th edn. F.A. Davis Company, Philadephia (2006)
2. Torn Meniscus Guide: Causes, symptoms and treatment options. Available, at https://www.drugs.com/health-guide/torn-meniscus.html. (2000)
3. Bronzino, J.D.: The biomedical engineering handbook, 3rd edn. CRC Press (2006)
4. McCarty, E.C., Marx, R.G., Wickiewicz, T.L.: Meniscal tears in the athlete operative and nonoperative management. Phys. Med. Rehabil. Clin. N. Am. **11**(4), 867–880 (2000)
5. Skinner, H.B.: Current diagnosis & treatment in orthopedics. Available at http://www.accessmedicine.com/. (2006)
6. De Smet, A.A.: How I diagnose meniscal tears on knee MRI. Am. J. Roentgenol. **199**, 481–499 (2012)
7. De Smet, A.A., Mukherjee, R.: Clinical, MRI, and arthroscopic findings associated with failure to diagnose a lateral meniscal tear on knee MRI. Am. J. Roentgenol. **190**, 22–26 (2008)
8. Wang, Y., Yu J., Luo, H., Yu, C., Ao, Y., Xie, X., Jiang, D., Zhang, J.: An anatomical and histological study of human meniscal horn bony insertions and peri-meniscal attachments as a basis for meniscal transplantation. Chinese Med. J. **122**(5), 536–540 (2009)
9. Jaovisidha, S., Virayavanich, W., Woratanarat, P., Siriwongpairat, P.: Three-Tesla MRI diagnosis of meniscal tears of the knee. J. Med. Assoc. Thai **92**(12), 1662–1668 (2009)
10. Stone, K.R., Abhi Freyer, B.A.S., Thomas Turek, B.S., Ann W Walgenbach, R.N, Wadhwa, S., Crues, J.: Meniscal sizing based on gender, height and weight. Arthrosc. J. Arthroscopic Relat. Surg. **23**(5), 503–508 (2007)
11. Adams, Rolf, Bischof, Leanne: Seeded region growing. IEEE Trans. Pattern Anal. Mach. Intell. **16**, 641–647 (1994)
12. Mallikarjunaswamy, M.S., Holi, M.S., Raman, R.: Knee joint menisci segmentation, visualization and quantification using seeded region growing algorithm. J. Med. Imaging Health Inform. **5**(3), 552–560 (2015)
13. Yoshioka, H., Schlechtweg, P.M., Kose, K.: Imaging of arthritis and metabolic bone diseases in clinical magnetic resonance imaging (MRI). Elsevier Health Sciences, pp. 34–48 (2009)

An Optimized Adaptive Random Partition Software Testing by Using Bacterial Foraging Algorithm

K. Devika Rani Dhivya[1(✉)] and V. S. Meenakshi[2]

[1] Department of BCA & M.Sc. (SS), Sri Krishna Arts and Science College,
Bharathiar University, Coimbatore 641008, Tamilnadu, India
Devika58@gmail.com
[2] Department of Computer Science, Chikkanna Government Arts College,
Bharathiar University, Tirupur 641602, Tamilnadu, India
meenasri70@yahoo.com

Abstract. Software testing is a procedure of investigating a software product to find errors. Optimized Adaptive Random Partition Testing (OARPT) is a combined approach which comprises of Adaptive Testing (AT) and Random Partition Testing (RPT) in an alternative manner depending on test case. ARPT consists of two strategies are ARPT 1 and ARPT 2. The parameters of ARPT 1 and ARPT 2 need to be assessed for different software with different number of test cases and programs. The process of assessing the parameters of ARPT consumes high computational overhead and therefore it is more necessary to optimize parameters of ARPT. In this paper, the parameters of ARPT 1 and ARPT 2 are optimized by using Bacterial Foraging Algorithm (BFA) which improves the performance of ARPT software testing strategies. The experiments are conducted in different software and the proposed method improves defect detection efficiency and high code coverage, reduces time consumption and reduces memory utilization.

Keywords: Adaptive random partition testing · Adaptive testing
Random partition testing · Bacterial foraging algorithm

1 Introduction

Software testing [1] has been broadly recognized as a mainstream technique for improving and assessing the quality of software. Software testing and retesting occur incessantly throughout the software development lifespan. The objective of the software testing is to assess the quality of an application and to progress it. A primary purpose of testing is to spot software failures so that flaws may be exposed and corrected. There are many software testing strategies used for testing software such as Random Testing, Adaptive Random Testing [2], Partition Testing [3], Adaptive Testing and Random Partition Testing. Random testing is a software testing technique where programs are tested by generating random, independent inputs. Randomized

© Springer International Publishing AG 2018
D. J. Hemanth and S. Smys (eds.), *Computational Vision and Bio Inspired Computing*,
Lecture Notes in Computational Vision and Biomechanics 28,
https://doi.org/10.1007/978-3-319-71767-8_47

testing [14] uses randomization for some characteristics of test input data choice. The vital value of randomized testing is the capability to generate numerous distinct test inputs in a petite time, including test inputs that may not be selected by test engineers but, which may, nevertheless force failures. Randomization for some aspect of test input is used by the randomized testing process [13]. Partition testing is an another software testing technique that splits the input data of a software unit into partitions of equal data from which test cases can be derived. Random Partition testing combines Random and Partition Testing techniques. A specific testing technique recognized as the adaptive testing (AT) approach is using the testing history collected online to improve the testing effectiveness of RPT.

Adaptive Random Partition Testing (ARPT) [4] is a hybrid approach which is used for software testing. ARPT used AT and RPT testing in an alternative manner. ARPT 1 and ARPT 2 are the two variants of ARPT. In ARPT 1 parameters determines the stages of AT and RPT are initiated. The prime idea of ARPT 1 is an alternative mean it's depend upon the test case, if Complexity of test case is less than it adapts AT else if the test case is Complexity than it adapts RPT. Then test cases are partitioned and each partition switch between AT and RPT testing strategies. In ARPT 2 test cases are divided into two different lengths. ARPT 2 adapts AT and RPT in a random manner. AT is applied for first length and RPT is applied for second length of test cases. The computational complexity problem of ARPT was resolved by improving the random partitioning through clustering algorithms [5]. In addition to that, the APRT 1 and ARPT 2 testing strategies consist of different parameters. Those parameters values are varied based on the software. It needs to estimate over and over for different software. Thus the parameter estimation for different software leads to computation overhead and time consumption.

JUnit [17] is a straightforward and practical testing Structure for Java classes. It empowers the nearby incorporation of testing with advancement by permitting a test suite be made incrementally. Isolate testing code must be composed and kept up in synchrony with the code being worked on, as the test class must acquire from the JUnit system. This test class must be investigated when the code under test changes, and, if vital, additionally updated to reflect the progressions. The trouble and cost of composing the test class are exacerbated during improvement, when the code being tested changes frequently.

In this paper, the parameters of ARPT 1 and ARPT 2 are optimized based on different optimization techniques. In this proposed work, Bacterial Foraging Algorithm (BFA) algorithm is used to optimize both the ARPT 1 and ARPT 2 parameters. This optimization algorithm optimizes the parameters effectively, which improves the ARPT based software testing process.

2 Literature Survey

Schwartz and Do [6] proposed two extra Adaptive Test Prioritization (ATP) systems using Weighted Sum Model (WSM) and fluffy Analytical Hierarchy Process (AHP) to test programming. Three systems were produced in this paper. To begin with methodology used the fluffy AHP which address the issue of the outcomes from AHP and the second procedure used a fluffy master framework to get the advantages of a system which does not require a couple shrewd correlations. The last system utilized WSM which research the viability of technique for ATP.

Yong et al. [7] An approach, mechanized test era for EFSM models. Configuration by contract approach is connected to formalize detail necessities. Expanded Finite State Machines (EFSMs) are frequently utilized as a part of model-based improvement and for displaying VHDL details. Hereditary calculation is proposed to discover set of qualities that triggers given way in the EFSM and uncovers irregularities with the determination. This enhances the proficiency of the model.

Shenga et al. [18] A hybrid algorithm (GA-PSO) which consolidates Genetic Algorithm and Particle Swarm Optimization (PSO) the new calculation is demonstrated successful by an agent trial of the "triangle sort of segregation". The analysis demonstrates that the new calculation has higher execution estimation of 20%.

Lv et al. [8] investigated asymptotic behavior of adaptive testing and proposed Adaptive Testing with Gradient Descent (AT-GD) method to enhance the global performance of adaptive testing without losing the local optimality. In this paper gradient descent method was introduced into the adaptive testing framework from this the asymptotic performance of AT-GD was examined and the upper certain for the global performance of adaptive testing was explored. The flexibility of the testing framework was not effective in the proposed Adaptive Testing with Gradient Descent method.

Huang et al. [9] proposed an improved Mirror Adaptive Random Testing called as Dynamic Mirror Adaptive Random Testing (DMART) to lessen the calculation over-head of Adaptive Random Testing (ART) procedure. Mirror Adaptive Random Testing can't diminish the request of size for computational overhead of ART technique while keeping up comparable disappointment identification adequacy. The proposed system parts the information space incrementally alongside the testing procedure by utilizing new reflecting plan. The reflecting plan of DMART is autonomous of solid ART methodology. This approach still sets aside more opportunity for creating the experiments.

Chen et al. [10] presented a new adaptive random testing method using the notion of iterative partitioning for software testing. In this proposed strategies the info area was spitted into similarly measured cells by frameworks. In perspective of the relative range of grids, the systems were masterminded into unique gatherings to effective experiments. This strategy easily chooses the grid cells which are far isolated from every single viable test for try time. This method diminished the time complexity of testing methodology and has high fault recognition ability.

Shahbazi et al. [11] proposed Random Border Centroidal Voronoi Tessellations (RBCVT) for effective software testing process through effective test case generation. RBCVT enhanced the random testing by improving the coverage of the input space. The created test cases by the other methods such as Random Testing, Adaptive Random Testing and Centroidal Voronoi Tessellations were act as input to RBCVT method. Then the RBCVT generated an improved set of test cases. The major drawback of this method is it dependent on other testing strategies to generate improved set of test cases.

Chen et al. [12] introduced DividE-and-conquer methodology for identifying categorieS, choiceS, and choicE Relations for software testing process proposed methodology is abbreviated as DESSERT. In this paper, extends the choice relation framework, abbreviated as chocolate, which helps to program analyzers in the utilization of class/decision method to testing. Chocolate accepts that the analyzer. Chocolate assumes that the tester is able to construct a single choice relation table from the entire specification; this table then forms the basis for test case generation using the associated algorithms. This assumption, however, may not hold true when the requirement is complex and contains many specification components.

Singh [15] Multi-objective test suite minimization problem is to select a set of test cases from the available test suite while optimizing the multi objectives like code coverage, cost and fault history. Regression Test suite optimization is an effective technique to reduce time and cost of testing. Many researchers have used computational intelligence techniques to enhance the effectiveness of test suite. These approaches optimize test suite for a single objective. Introduction of nature inspired algorithms like GA, PSO and BFO may be used to optimize test suite for multi-objective selection criteria. Main focus of our approach is to find a test suite that is optimal for multi-objective regression testing.

Multi-target test suite minimization issue is to choose an arrangement of experiments from the accessible test suite while enhancing the multi goals like code scope, cost and blame history. Relapse Test suite enhancement is a viable strategy to decrease time and cost of testing. Numerous specialists have utilized computational insight procedures to upgrade the adequacy of test suite. These methodologies enhance test suite for a solitary goal. Presentation of nature roused calculations like GA, PSO and BFO might be utilized to improve test suite for multi-target choice criteria. Fundamental concentration of our approach is to discover a test suite that is ideal for multi-target relapse testing.

3 Methodology

ARPT consists of two variants are ARPT 1 and ARPT 2. In ARPT 1, parameters that determines stage of AT and RPT are initiated. The test cases are partitioned and then each partition is switched between the AT and RPT testing techniques. Thus it considerably reduces the computational overhead for test case selection. In ARPT 2, the test case is divided into two different lengths. First AT is used and then for the other length RPT is used. The parameter is set only to alter the length of the test case because there is only one switching between the two AT and RPT strategy.

The parameters of ARPT 1 and ARPT 2 are optimized using Bacterial Foraging Algorithm (BFA). Thus the optimal parameters are used for any subject programs which reduce time consumption and improves the detect detection efficiency.

Table 1 The parameters of ARPT

Parameters of ARPT				
ARPT 1			ARPT 2	
S. No.	Parameters	Denotes	Parameters	Denotes
1	S	The state of the present testing process, The state characterizes an AT segment and RPT segment	X	A Parameter utilized To govern when to alter the testing strategy
2	$Sig_R(S)$	Signal reserved to record whether any defect is detected	Y	The testing length for the AT Process.
3	X	A parameter utilized to govern when to alter the testing strategy		
4	k(0)	The test case of AT testing strategy		
5	l(0)	The test case of RPT testing strategy		

The Table 1 shows the parameters of ARPT. ARPT 1 consists of five parameters are S, x, $Sig_R(S)$, k(0) and l(0). ARPT 2 consists of two parameters are x and y. S signifies the state of the present testing process, the state represents an AT segment and RPT segment, $Sig_R(S)$ represents signal retained to record whether any defect is detected. k(0) is the test case of AT testing strategy and l(0) is the test case of RPT testing strategy, x is a parameter utilized to govern when to alter the testing strategy and y is the testing length for the AT process. These parameters need to be optimized using optimization algorithm. The work of ARPT is derived and it shows in the following form [4] (Table 2):

Table 2 Two ARPT Strategies [4]

ARPT-1: ARPT Approach #1	ARPT-2: ARPT
initialize(S=0, n=0, $Sig_R(S)=0$ and $k(0)=k_0, l(0)=l_0$); while(the *stopping criterion is not satisfied)* { x = x+ 1; If *(x≤k(S))* { T_n= adaptive( ); } *If(k(S)0<n≤k(S)+l(S))* { T_n = random partition( ); } execute(T_n); If (failure == TRUE) { removeDefect( ); $Sig_R(S)=1$; } If *(x == k(S)+l(S))* { *(k(S+1),l(S+1))=CalcSteps(k(S),l(S),* *$Sig_R(S)$;S = S+1; x = 0; $Sig_R(S) = 0$;* } }	initialize(x=0 and y); while(the *stopping criterion is not satisfied)* { If *(x < y)* { T_n = adaptive( ); } else{ T_n = random partitions; } execution(); If (failure == TRUE) { removeDefect(); } x = x + 1; }

The fitness function is more important factor in the optimization algorithm. In this proposed work, the fitness function is calculated based on weighted sum of time consumption t, defect detection efficiency E_d, memory consumption M, number of test cases n and code coverage C. It is mathematically represented as follows:

$$Fitness = \sum w_1 \frac{1}{t} + w_2 E_d + w_3 M + w_4 \frac{1}{n} + w_5 C \tag{1}$$

where $w_1 + w_2 + w_3 + w_4 + w_5 = 1$.

3.1 Optimal Parameter Setting Using Bacterial Foraging Algorithm

Bacterial Foraging algorithm is an optimization algorithm where a Microscopic organism's searches for supplements are a way to expand vitality acquired per unit time. The individual bacterium additionally communicates with others by sending signals. A bacterium takes foraging choices in the wake of considering two previous factors. The procedure, in which a bacterium moves by making little strides while looking for supplements, is called chemo taxis. The key thought of BFA is impersonating chemotactic development of virtual bacteria in the problem search space. The BFA algorithm is based on four steps, they are the Chemotaxis, swarming, reproduction and elimination dispersal.

Chemotaxis: Here, swimming and tumbling are the two prime ways which define the manner in which bacteria search for food [15]. In the Chemotaxis procedure simulates the crusade of an *E. coli* cell through swimming and plummeting through flagella. Biologically an *E. coli* bacterium can move in two different ways. It can swim for a timeframe a similar way or it might tumble and alternate between these two modes of operation for the entire lifetime.

Suppose $\theta^i(j, k, l)$ represents the i-th bacterium at the j-the chemotactic, k-th reproductive and l-th elimination-dispersal step. $C(i)$ is the size of the step taken in the random direction specified by the tumble (run length unit). Then in computational chemotaxis the movement of the bacterium may be represented by

$$\theta^i(j+1, k, l) = \theta^i(j, k, l) + C(i) \frac{\Delta(i)}{\sqrt{\Delta^T(i)\Delta(i)}} \tag{2}$$

In the above Eq. 2, Δ represents a vector in the random direction whose elements lie in $[-1, 1]$.

In swarming the cell to cell signaling in E. coli is defined and it is represented as follows:

$$
\begin{aligned}
Fitness_{cc}(\theta, P(j, k, l)) &= \sum_{i=1}^{S} Fitness_{cc}(\theta, \theta^i(j, k, l)) \\
&= \sum_{i=1}^{S} \left[-d_{attractant} \exp\left(-w_{attractant} \sum_{m=1}^{P} (\theta_m - \theta_m^i)^2 \right) \right. \\
&\quad \left. + \sum_{i=1}^{S} h_{repellant} \exp\left(-w_{repellant} \sum_{m=1}^{P} (\theta_m - \theta_m^i)^2 \right) \right]
\end{aligned}
\tag{3}
$$

In Eq. 3, $Fitness_{cc}(\theta, P(j,k,l))$ is the impartial function worth to be summed to the real impartial function to present a time changing objective function, S is the aggregate number of bacteria, p is the number of variables to be optimized, which are present in each bacterium and $\theta = [\theta_1, \theta_2, \ldots, \theta_p]^T$ is a fact in the p-dimensional pursuit domain, $d_{attractant}, w_{attractant}, w_{repellant}$ are different coefficients.

Then the minimum healthy bacteria ultimately die while each of the healthier bacteria (those yielding lower value of the objective function) asexually split into two bacteria, which are then placed in the same location [19].

This keeps the swarm estimate consistent. At long last in the disposal and dispersal process, Gradual or sudden changes in the neighborhood condition where a bacterium populace lives may happen because of different reasons e.g. a huge nearby ascent of temperature may slaughter a gathering of bacteria that are right now in an area with a high centralization of energy gradients. Events can happen in such a fashion, to the point that every one of the microbes in an area is executed or a gathering is scattered into another area. To recreate this wonder in BFA a few microscopic organisms are sold aimlessly with a little likelihood while the new substitutions are arbitrarily introduced over the search space.

BFA based Optimal Parameter setting algorithm

$\oplus$ *if (ARPT_strategy == ARPT 1)*
$\oplus$ *Initialize S bacteria with* $S_1, \ldots S_h$, $x_1, \ldots x_h, Sig_R(S)_1, \ldots Sig_R(S)_h, k(0)_1, \ldots$
    $k(0)_h$ *and* $l(0)_1, \ldots l(0)_h$.
$\oplus$ *Call BFA algorithm*
$\oplus$ *else*
$\oplus$ *Initialize S bacteria with* $m_1, \ldots m_h$ *and* $n_1, \ldots n_h$
$\oplus$ *Call BFA algorithm*

BFA Algorithm

$\oplus$ *Initialize p, S,* $N_c, N_s, N_{re}, N_{ed}, P_{ed}$, *C(i) (i = 1,2,....S) and* θ^i
    //p denotes dimension of search space, S denotes total number of bacteria in the palpation, N_c *denotes number of chemotactic steps,* N_s *denotes swimming length,* N_{re} *denotes number of reproduction steps,* N_{ed} *denotes number of elimination-dispersal events,* P_{ed} *denotes elimination-dispersal probability and C(i) denotes the variation amount of parameters of either ARPT 1 and ARPT 2 during chemotactic steps.*
$\oplus$ *Dismissal dispersal loop: l = l+1*
$\oplus$ *Reproduction loop: k = k+1*
$\oplus$ *Chemotaxis loop: j = j+1*
$\oplus$ *for (i = 1; i < S; i ++)*
$\oplus$ *Compute the fitness function Fitness(i, j, k, l)*
$\oplus$ *Fitness(i, j, k, l) = Fitness(i, j, k, l) + Fitness_{cc}$ $(\theta^i(j, k, l), P(j, k, l))$*

⊕ *Fitness$_{cc}$ is calculated using* **Eq.** *3*
⊕ *Let Fitness$_{last}$ = Fitness(i, j, k, l) to save this value since a better fitness may find*
⊕ *Generate a random number vector $\Delta(i)$*
⊕ *Move using* **Eq.** *2*
⊕ *Compute Fitness $(i, j + 1, k, l)$ using following equation*
⊕ *Fitness$(i, j + 1, k, l) = Fitness(i, j, k, l)Fitness_{cc}\left(\theta^i(j, k, l), P(j, k, l)\right)$*
⊕ *Swim: for($m = 0$; $m < N_s$; m ++)*
⊕ *if $(Fitness(i, j + 1, k, l) < Fitness_{last})$*
⊕ *Assign Fitness$_{last}$ = Fitness$(i, j + 1, k, l)$ and let*

$$\theta^i(j + 1, k, l) = \theta^i(j, k, l) + C(i)\frac{\Delta(i)}{\sqrt{\Delta^T(i)\Delta(i)}}$$

⊕ *else $m = N_s$*
⊕ *end if*
⊕ *end for*
⊕ *Go to next bacterium $(i + 1)$ if $i \neq S$*
⊕ *if $j < N_c$, go to Chemotaxis loop*
⊕ *reproduction: Calculate health of the bacterium i using following equation*

$$Fitness^i_{health} = \sum_{j=1}^{N_c + 1} Fitness(i, j, k, l)$$

⊕ *Sort bacteria and chemotactic parameters $C(i)$ in order of ascending cost Fitness$_{health}$*
⊕ *The S_r bacteria with the lowest Fitness$_{health}$ die and the remaining S_r bacteria with the best values split*
⊕ *if $(k < N_{re})$*
⊕ *go to reproduction loop*
⊕ *Dismissal dispersal: for($i = 1$; $i < S$; i ++)*
⊕ *eliminate and disperse each bacterium*
⊕ *if $(l < N_{ed})$*
⊕ *go to Elimination-dispersal loop*
⊕ *else*
⊕ *end*

The above BFA algorithm optimizes the parameters of ARPT 1and ARPT 2 which is used to select the optimized test suite for software testing process.

4 Results and Discussion

In this section the experiments results are showed in order to prove the effectiveness of the proposed optimization algorithm in ARPT testing strategies. The comparison is based on the time consumption, defect detection efficiency, memory utilization, number of test cases and code coverage of both existing and proposed techniques. For the experimental purpose, three software namely univocityparser, marc4j and jsoup are used. The univocity-parser is a suite of extremely fast and reliable parsers for Java. It provides a consistent interface for handling different file formats, and a solid framework for the development of new parsers. The marc4j is software which provides an easy to use Application Programming Interface (API) for working with MARC and MARCXML in Java. jsoup is a Java library for working with real-world HTML. It provides a very convenient API for extracting and manipulating data, using the best of DOM, CSS, and jquery-like methods. Each software is described in the following Table 3, simple to utilize Application Programming Interface (API) for working with MARC and MARCXML in Java. Jsoup is a Java library for working with real-world HTML. It gives an exceptionally helpful API to separating and controlling information, utilizing the best of DOM, CSS, and jquery-like techniques. Each software is described in the following Table 3.

Table 3 Software description

Software test cases	No. of test cases	No. of programs
univocityparser	1000	137
marc4j	1000	93
jsoup	1000	76

4.1 Time Consumption

Time consumption is a measure of amount of time taken to test software based on optimized ARPT testing strategy.

Figure 1, shows the comparison of time consumption between proposed optimization techniques Bacterial Foraging Algorithm (BFA) algorithm with different software's. X axis represents the univocityparser, marc4j and jsoup software and Y axis represents time consumption in seconds. From the Fig. 1, it is proved that the proposed BFA optimization algorithm consumes less time in both ARPT testing strategies.

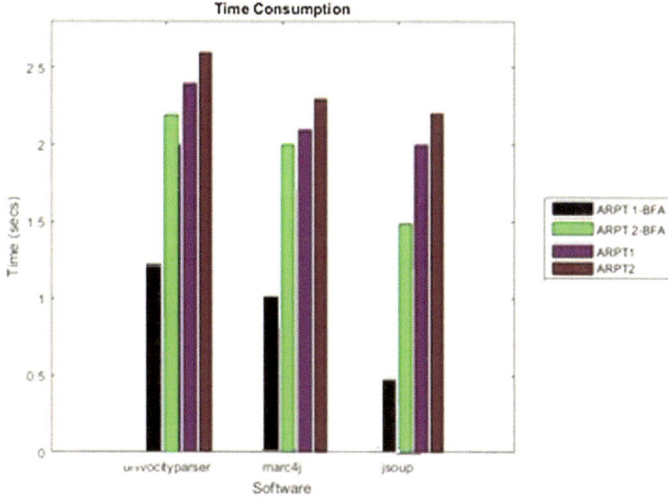

Fig. 1 Comparison of time consumption

4.2 Defect Detection Efficiency

Defect detection efficiency (DDE) is the quantity of deformities identified amid a stage/arrange that are infused during that same stage separated by the aggregate number of defects injected during that stage. It can be computed by utilizing following formula [16]:

$$DDE = \frac{No.of\ defects\ injected\ AND\ Detected\ in\ a\ phase}{Total\ No.of\ Defects\ injected\ in\ that\ phase} \times 100\%$$

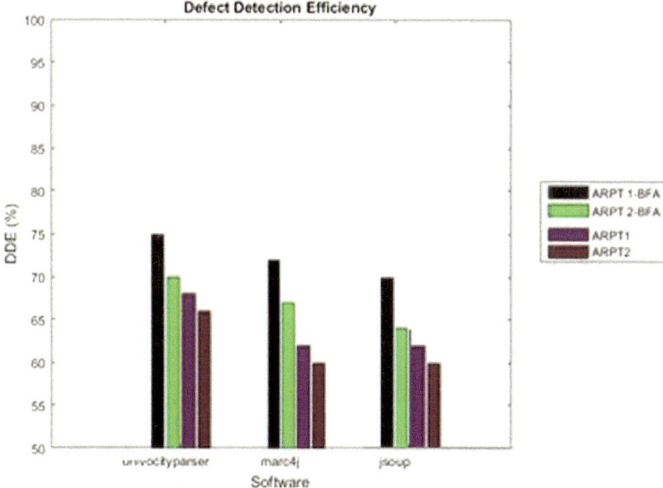

Fig. 2 Comparison of defect detection efficiency

Figure 2, shows the comparison of Defect Detection Efficiency between proposed optimization techniques are Bacterial Foraging Algorithm (BFA) with ARPT and the existing APRT with different software. X axis represents the univocityparser, marc4j and jsoup software and Y axis represents Defect Detection Efficiency in %. From the Fig. 2, it is proved that the proposed BFA optimization algorithm has high defect detection efficiency in both ARPT testing strategies.

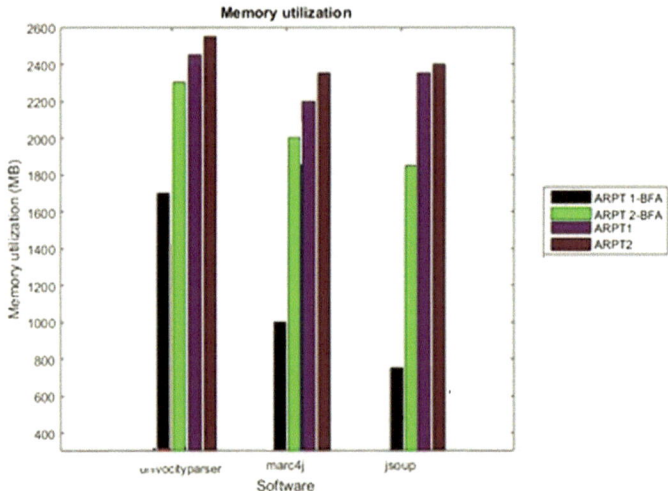

Fig. 3 Comparison of memory utilization

4.3 Memory Utilization

Memory utilization is the amount of memory utilized by optimized ARPT testing strategies. The Number of memory allocated during minus Total memory is the Memory taken for each strategy.

$$Memory = Total\,Memory - Memory\,allocated\,during\,Testing$$

Figure 3, shows the comparison of Memory Utilization between proposed optimization techniques are Bacterial Foraging Algorithm (BFA) and ARPT. X axis represents the univocityparser, marc4j and jsoup software and Y axis represents Memory Utilization in MB. From the Fig. 3, it is proved that the proposed BFA optimization algorithm utilized less memory in both ARPT testing strategies.

4.4 Code Coverage

Code coverage is a measure utilized to define the degree to which the source code of a program is executed when a particular test suite runs. The process of assigning optimization algorithm for parameters of ARPT testing strategies of an application shows the effective test coverage [14, 20].

Code Coverage = Number of code exercised − Total number of code*100%.

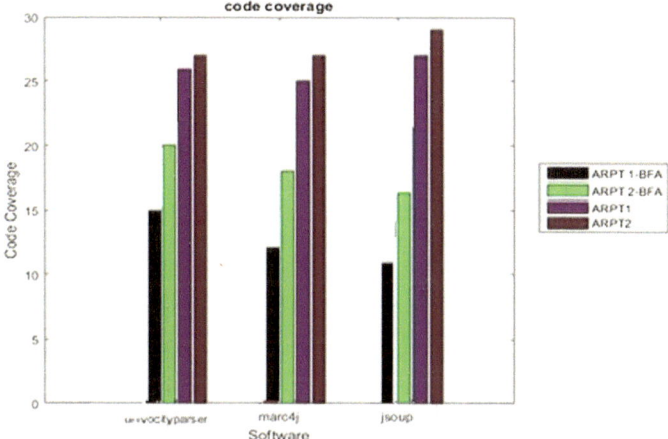

Fig. 4 Comparison of code coverage

Figure 4, shows the comparison of code coverage between proposed optimization techniques are BFA and ARPT with different software. X axis represents the univocityparser, marc4j and jsoup software and Y axis represents code coverage. From the Fig. 4, it is proved that the proposed BFA optimization algorithm has better code coverage in both ARPT testing strategies.

5   Conclusion

In this paper, optimization algorithm is utilized to improve the ARPT testing strategies. The ARPT has two testing options ARPT 1 and ARPT 2 and these strategies have different parameters. These different parameters are optimized used BFA. Initially the testing strategies are checked whether the testing strategy is ARPT 1 or ARPT 2. If the testing strategy is ARPT 1, the five parameters of ARPT 1 are optimized by BFA algorithm otherwise two parameters of ARPT 2 are optimized. The optimized parameters improve the accuracy of software testing process. The experiments are conducted in univocityparser, marc4j and jsoup test cases which prove that proposed BFA optimization algorithm is better in terms of time consumption, defect detection efficiency, memory utilization and code coverage.

References

1. Chi, Z., Xuan, J., Ren, Z., Xie, X., Guo, H.: Multi-level random walk for software test suite reduction. IEEE Comput. Intell. Mag. **12**(2), 24–33 (2017)
2. Nie, C., Wu, H., Niu, X., Kuo, F.C., Leung, H., Colbourn, C.J.: Combinatorial testing, random testing, and adaptive random testing for detecting interaction triggered failures. Inf. Softw. Technol. **62**, 198–213 (2015)
3. Zachariah, B.: Analysis of software testing strategies through attained failure size. IEEE Trans. Reliab. **61**(2), 569–579 (2012)
4. Lv, J., Hu, H., Cai, K.Y., Chen, T.Y.: Adaptive and random partition software testing. IEEE Trans. Syst. Man Cybern. Syst. **44**(12), 1649–1664 (2014)
5. Devika Rani Dhivya K.: Improved time performance of adaptive random partition software testing by applying clustering algorithm. In: International Conference on Interdisciplinary Research Innovations in Computer Science, Bioscience (2016, on 29th and 30th)
6. Schwartz, A., Do, H.: Cost-effective regression testing through Adaptive Test Prioritization strategies. J. Syst. Softw. **115**, 61–81 (2016); Bashir, M.B., Nadeem, A.: Improved genetic algorithm to reduce mutation testing cost. IEEE Access (2017)
7. Yong, C., Yong, Z., Tingting, S., Jingyong, L.: Comparison of two fitness functions for ga-based path-oriented test data generation. ICNC'09. Fifth International Conference on Natural computation, 2009, 14–16 Aug 2009, vol. 4, pp. 177–181
8. Lv, J., Yin, B.B., Cai, K.Y.: On the asymptotic behavior of adaptive testing strategy for software reliability assessment. IEEE Trans. Software Eng. **40**(4), 396–412 (2014)
9. Huang, R., Liu, H., Xie, X., Chen, J.: Enhancing mirror adaptive random testing through dynamic partitioning. Inf. Softw. Technol. **67**, 13–29 (2015)
10. Chen, T.Y., Huang, D.H., Zhou, Z.Q.: On adaptive random testing through iterative partitioning. J. Inf. Sci. Eng. **27**(4), 1449–1472 (2011)
11. Shahbazi, A., Tappenden, A.F., Miller, J.: Centroidal voronoi tessellations-a new approach to random testing. IEEE Trans. Software Eng. **39**(2), 163–183 (2013)
12. Chen, T.Y., Poon, P.L., Tang, S.F., Tse, T.H.: DESSERT: a DividE-and-conquer methodology for identifying categorieS, choiceS, and choicE Relations for Test case generation. IEEE Trans. Software Eng. **38**(4), 794–809 (2012)
13. Devika Rani Dhivya K.: Analysis on generating test case for random testing using optimization technique. In: Second International Conference on Information Technology & Society pp. 200–207. Proceeding of IC-ITS 2015, Meliá Hotel Kuala Lumpur, Malaysia (2015) e-ISBN: 978-967-0850-07-8
14. Devika Rani Dhivya K., Meenakshi, V.S.: Weighted particle swarm optimization algorithm for randomized unit testing. In: Electrical, Computer and Communication Technologies (ICECCT), 2015 IEEE International Conference. ICECCT 2015, vol 2, pp. 0828–0834. IEEE Xplore: 2, Coimbatore, 5–7 Mar 2015
15. Singh, R.R.: Test suite minimization using evolutionary optimization algorithms: review. Int. J. Eng. Res. Technol. (IJERT) **3**(6) (2014, June)
16. Zhang, X., Teng, X., Pham, H.: Considering fault removal efficiency in software reliability assessment. IEEE Trans. Syst. Man Cybern. **33**(1), 114–120 (2003, Jan)
17. Cheon, Y., Leavens, G.T.: A simple and practical approach to unit testing: the JML and JUnit way. Computer Science Technical Reports. 181 (2001)

18. Shenga, Z., Zhang, Y., Zhou, H., He, Q.: Automatic path test data generation based on GA-PSO. In: 2010 IEEE International Conference on Intelligent Computing and Intelligent Systems (ICIS), 29–31 Oct 2010, vol. 1, pp. 142–146
19. Mai, X., Li, L.: Bacterial foraging algorithm based on gradient particle swarm optimization algorithm. In: 2012 8th International Conference on Natural Computation (2012)
20. http://m.softwaretestinggenius.com/?page=details&url=know-the-basic-white-box-testing-techniques-based-upon-code-coverage

Global Skew Detection and Correction Using Morphological and Statistical Methods

Sharfuddin Waseem Mohammed$^{(\boxtimes)}$ and Narasimha Reddy Soora

Department of Computer Science and Engineering, Kakatiya Institute of
Technology and Sciences, Warangal, Telangana, India
waseem7602@gmail.com, snreddy75@yahoo.co.uk

Abstract. In this paper we have proposed a technique for skew detection and correction for printed documents, and have used an existing Optical Character Recognition (OCR) to recognize the characters. The proposed algorithm has the following steps (a) Applying the morphological dilations by defining the various structure elements (SE) (b) extracting the longest connected components (CC) (c) finding the global skew angle by statistical analysis of connected component (d) reference text line estimation and regression line fit to rotate the individual line by estimated angle of rotation. We have conducted experiment using printed images having different languages i.e. English, Devanagari, and Arabic (custom dataset) and have achieved significant performance.

Keywords: Morphological dilations · Statistical analysis · Regression line fit
Connected components analysis

1 Introduction

Most of the historical documents are preserved in digital format by scanning the document with proper dpi (dots per inch). To retrieve the information from these digital images, processing needs an OCR to recognize the characters with the help of techniques such as [1, 2].

In document recognition systems, the quality of the input image is essential to the output performance. During the scanning process, adverse effects of tilting the document produces noise or skew, which are unavoidable. Many techniques have been proposed in the literature to overcome the noise. Many OCR systems need a preprocessing of document to improve the efficiency; it involves noise removal, skew correction.

2 Related Work

Skew detection and correction can be performed on local text regions in images, which is referred to as local skew detection, based on the distance between the characters, words and lines proposed by Saragiotis and Papamarkos [3], and if analysis is performed on whole

© Springer International Publishing AG 2018
D. J. Hemanth and S. Smys (eds.), *Computational Vision and Bio Inspired Computing*,
Lecture Notes in Computational Vision and Biomechanics 28,
https://doi.org/10.1007/978-3-319-71767-8_48

document then is referred to as global skew detection, most of the related skew estimation algorithms are proposed to work for global skew. Many different approaches have been proposed for skew estimation such as projection profiles [4–8], Hough transform [9], nearest neighbor clustering [10], and interline cross correlation [11].

Most traditional method is projection profile approach which is simple to detect the skew angle of document image. It is proposed by Postl [4], which is based on horizontal projection profiles are calculated, profiles with maximum variation refers the best alignment to the text lines, with this projection angle the document is rotated to correct the skew. In order to reduce the computational effort many different algorithms are proposed. Baird [5] proposed a technique for selecting the midpoint of each CC the bottom side of the bounding box is projected, the objective function is to compute the sum of the squares of the profiles, Ciardiello et al. [6] projected selected sub-region, and the objective is to maximize the mean square deviation of the profile. Ishitani [7] used a different approach a cluster of parallel line on the image is selected and it stores the number of black/white transitions along the lines. Bloomberg et al. [8] proceeded with extraction of projection profile from a sample image and skew is estimated for sample image rather than the whole document hence this method is faster skew estimation method. All the mentioned projection profiles are horizontal approaches even vertical projections Papandreou and Gatos [12] method is also proposed for skew estimation. Projection profile methods are limited to estimate skew angle within $\pm 10°$ to $\pm 15°$, [5].

Hough Transform method is generally used to find the shapes in binary digital image. Srihari and Govindaraju [9] proposed a method in which an image is transformed to Hough domain and peak is calculated to detect the skew and validate in image domain, if more text is scattered, then it's difficult to find maximum peak which is major limitation of this approach. Nearest neighbor clustering method is based on page layout analysis. Gorman [10] proposed a method which clusters the nearest neighbor CCs, which exhibits a poor text line segmentation which is limited to certain languages. In interline cross-correlational method, the cross correlation between two lines with a fixed distance is calculated. Yan [11] proposed correlation functions for all pairs of lines to find the shift of interline cross correlation to determine the skew rate, this method is suitable for small skew angles up to $10°$.

Our approach is based on statistical method for estimating the skew based on accumulating the 10 longest CCs where we consider the mean and standard deviation, and a reference text line is estimated from the selected longest CCs, and skew rate is estimated to rotate the document. Experiment is conducted on different printed images which belong to different languages i.e. English, Devanagari, and Arabic (custom dataset). Results shows a robustness support for the proposed algorithm.

3 Proposed Work

Historical documents are scanned using an electronic scanner to convert it into a digital image either a color or grayscale image, which consist of R rows and C columns of matrix M and contains a value of intensity depending on color or grayscale image, if the digital image is color then value of intensity is combination of RBG (i.e. Red, Blue and Green) and if the digital image is grayscale it consist of values between a range of

{0,1,2... 255}, then M(i, j) ∈ {0, 1, 2... 255} where i = 1,... R and j = 1... C. After performing the binarization procedure the image M is converted into binary image B(i, j) whose value is either 0 or 1.

Proposed algorithm consists of the following steps:

Step 1: Preprocessing of the image, remove the CCs which are <5 pixel in dimension and apply image binarization, fix a bounding box.
Step 2: Apply the morphological dilations by defining the structuring Elements (SE).
Step 3: Extract the Longest CC.
Step 4: Find the Global Skew angle by statistical method.
Step 5: Reference Text line Estimation by regression line fit.
Step 6: Document image rotation with the defined skew angle.

Step 1: Preprocessing of the image.

In preprocessing stage, the CCs which are <5 pixels in dimension are removed such as punctuation marks (comma and dot), then the image binarization technique [13] is applied to convert the image into robust binary version. We consider the a threshold value of 5 pixel with a heuristic approach, after conducting a number of experiments on the scanned images where non-text content such as comma, dots, and noise components are removed.

Step 2: Applying the morphological dilations by defining the SE.

We have applied dilation method to the input image by considering 1×3 SE. We found the longest connected component and found the skew of the line by fitting a regression line to find the angle that it makes with x-axis. Based on this angle, we have defined three different SE's for morphological dilation of the input image as shown below.

```
se = [1, 1, 1];
if Angle > 5
   se = [1,0;0,1];
end
if Angle < -5
   se = [0,1;1,0];
end
```

After SE's are selected based on the angle, we have applied again the morphological dilation process using the selected SE's on the input document. In Eq. (1) C is the input document, SE is the selected structuring element based on the angle of the longest CC.

$$X_n = C \, \varphi \, SE \tag{1}$$

X_n is resultant matrix after applying of dilation.

Step 3: Extracting the Longest CC's.

From the resultant X_n image, we extract all the longest CC by analyzing the subsequent connectivity of component from the top left most bounding box to the next bounding box.

$$LCC = max\left(\bigcap_{n=1}^{3} Xn(i,j)\right) \qquad (2)$$

The longest CC LCC_{ALL} is shown in Fig. 1.

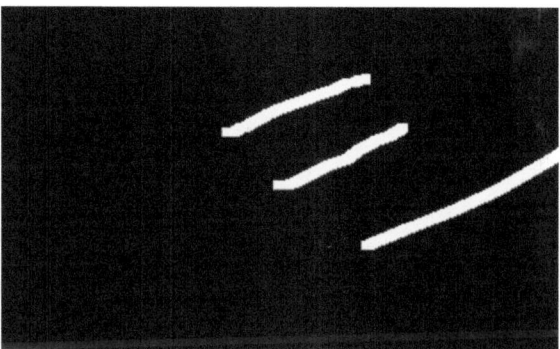

Fig. 1 The extraction of all longest connected components LCC_{ALL}

Step 4: Finding the Global Skew angle by statistical method.

Here, we have considered 10 longest CCs whose aspect ratio (AR) is more when compared with all the CCs (LCC_{ALL}) to find the average slope to compute the skew angle. We have applied the mean and standard deviation on these longest 10 CCs height and width.

Figure 2 shows the centroid of individual bounding box of a longest CC where X_{min}, Y_{min} are the bottom left most coordinates of the bounding box, similarly X_{min}, Y_{max} are top left coordinates, X_{max}, Y_{min} are the bottom right coordinates and X_{max}, Y_{max} are the top right coordinates.

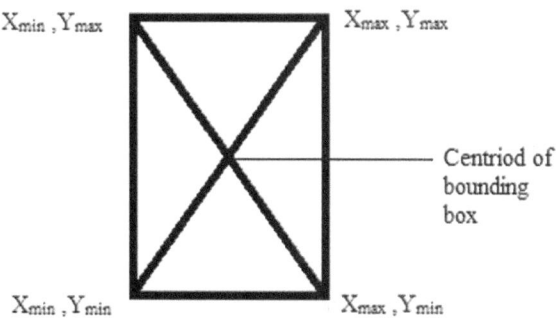

Fig. 2 Shows centroid of bounding box

Centroid of CCs is considered as follow:

$$CCC = \left(\frac{X_{min} + X_{max}}{2}, \frac{Y_{min} + Y_{max}}{2} \right) \tag{3}$$

We have considered the centroid of individual CC i.e. CCC (centroid of CC) to fit a reference line in the component.

Step 5: Reference Text line Estimation by Regression line fit.

Reference text line for printed document is almost linear hence we can use first-degree polynomial equation for fitting a line.

$$y = mx + c \tag{4}$$

where m is the slope of line and c is y-intercept these can be calculated as follows

$$m = \frac{Y_{max} - Y_{min}}{X_{max} - X_{min}} \tag{5}$$

The above equation is used to calculate the slope of line with the corresponding bounding box which is applied to the longest CC.

$$C = \frac{Y_{max} - m.X_{max}}{P} \tag{6}$$

where P is the number of bounding boxes in individual CC's, we can calculate the average slope for the top 10 longest CCs as follows:

$$\theta = \arctan(m) \tag{7}$$

where m is the slope of the regression fitted line from the centroid of the CCs.

Step 6: Document image rotation with the defined skew angle.

Estimated skew angle θ is computed for all the top 10 longest CCs and mean angle ($\theta mean$) is computed by using which the skewed image document is rotated. We have considered $\theta 1, \theta 2, \theta 3 \ldots \theta 10$ as the angle of the longest CCs with x-axis and computed mean angle as $\theta mean$ from the angles mentioned.

Figure 3 illustrates the computation of θ from mean angles of 3 longest CCs, similarly we compute the angles for all 10 longest CCs.

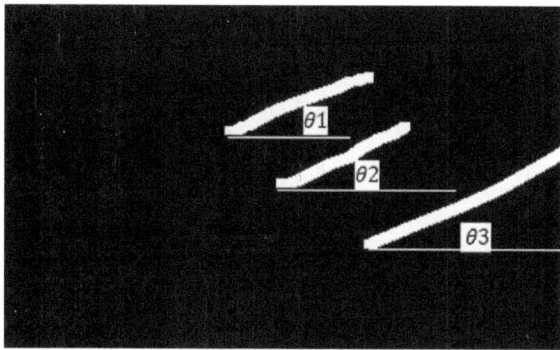

Fig. 3 Computing of θ1, θ2 and θ3 from longest connected components

4 Experiment Results

Tables 1, 2, 3, and 4 demonstrate the result of the proposed algorithm where the input images and output images (without skew) are listed. Experiment is conducted on different text document images from languages like English, Telugu, Devanagari and Arabic. We have considered both printed and hand written documents having multi-lingual scripts and also have multi skewed text lines to test the performance of the proposed algorithm.

Table 1 Result showing English printed and hand written documents

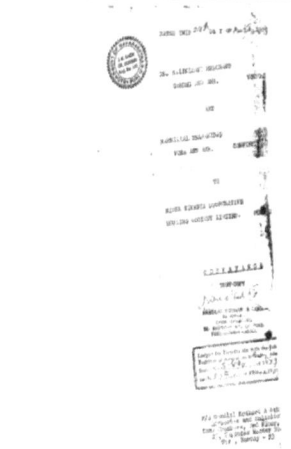

| (a1) Original Image with stamps and skew | (a2) a1 Image after de-skew |

(*continued*)

Table 1 (*continued*)

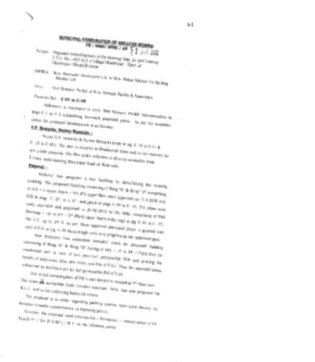

(b1) Original Image with English characters

(b2) b1 Image after de-skew

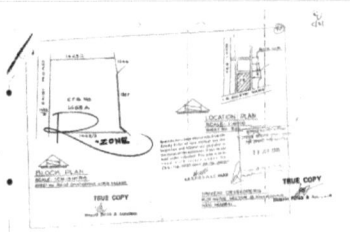

(c1) Original Image with layout and tables

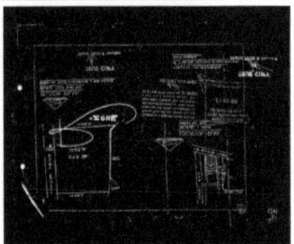

(c2) c1 Image after de-skew

(d1) Original Image with rectangular boxes

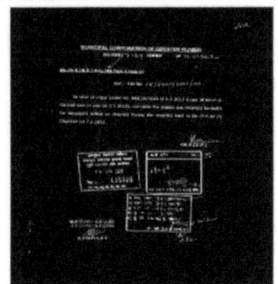

(d2) d1 Image after de-skew

Table 2 Result Showing Devanagari printed and hand written documents

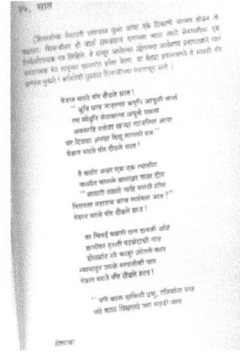	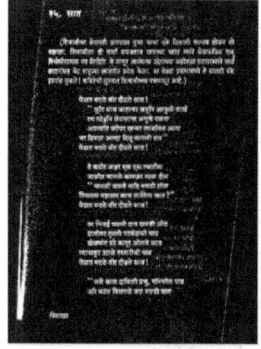
(e1) Original Image with Devanagari	(e2) e1 Image after de-skew

(f1) Original Image with Devanagari Max skew	(f2) f1 Image after de-skew

Table 3 Results showing Telugu printed and hand written documents

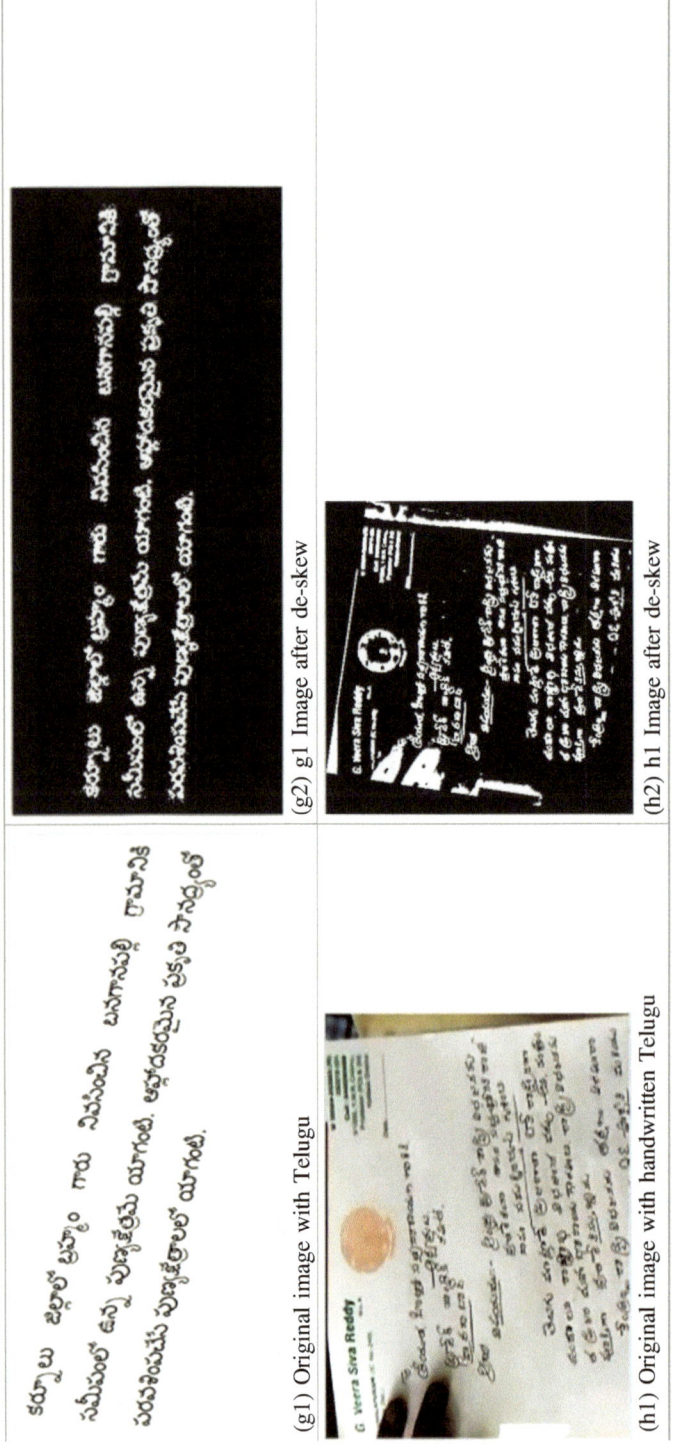

(g1) Original image with Telugu

(g2) g1 Image after de-skew

(h1) Original image with handwritten Telugu

(h2) h1 Image after de-skew

(continued)

Table 3 (*continued*)

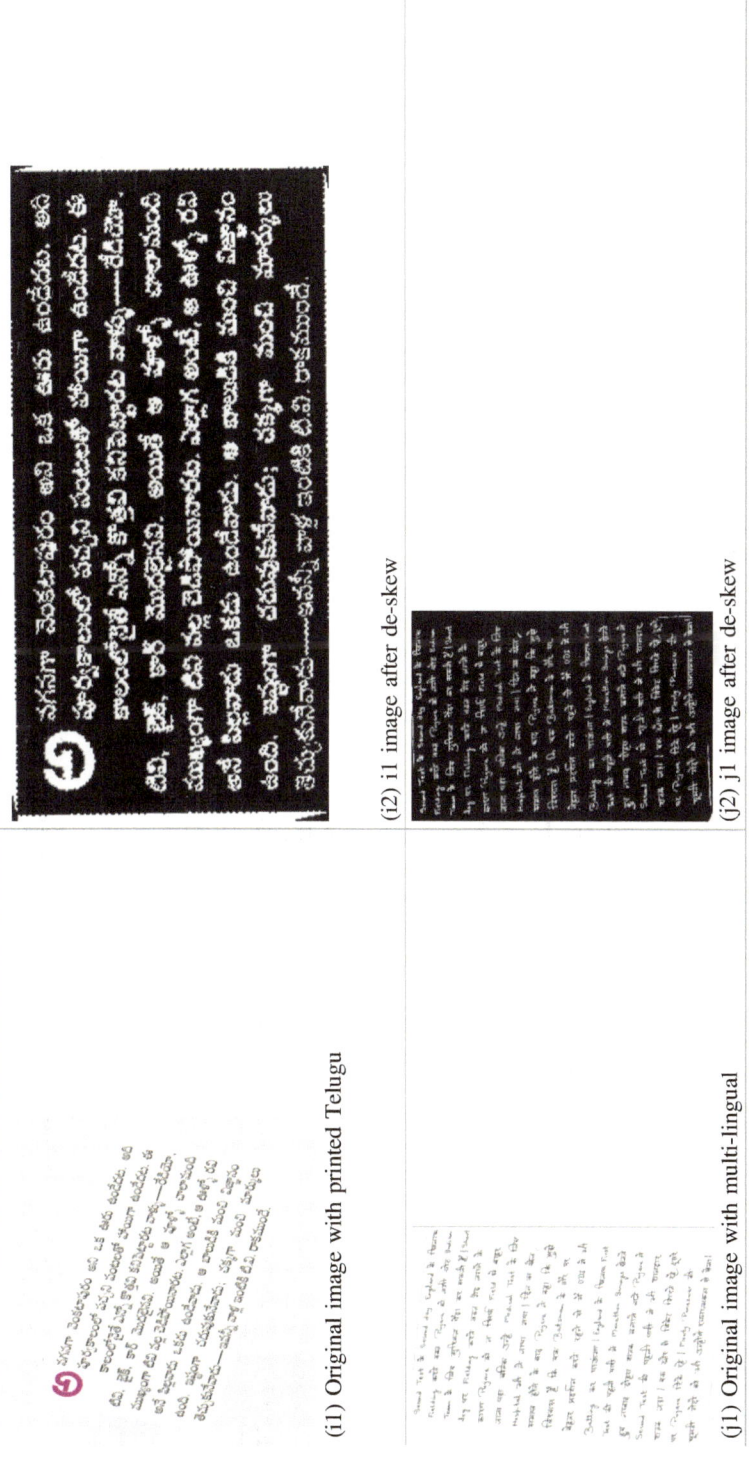

(i1) Original image with printed Telugu

(i2) i1 image after de-skew

(j1) Original image with multi-lingual

(j2) j1 image after de-skew

Table 4 Results showing Arabic printed and hand written documents

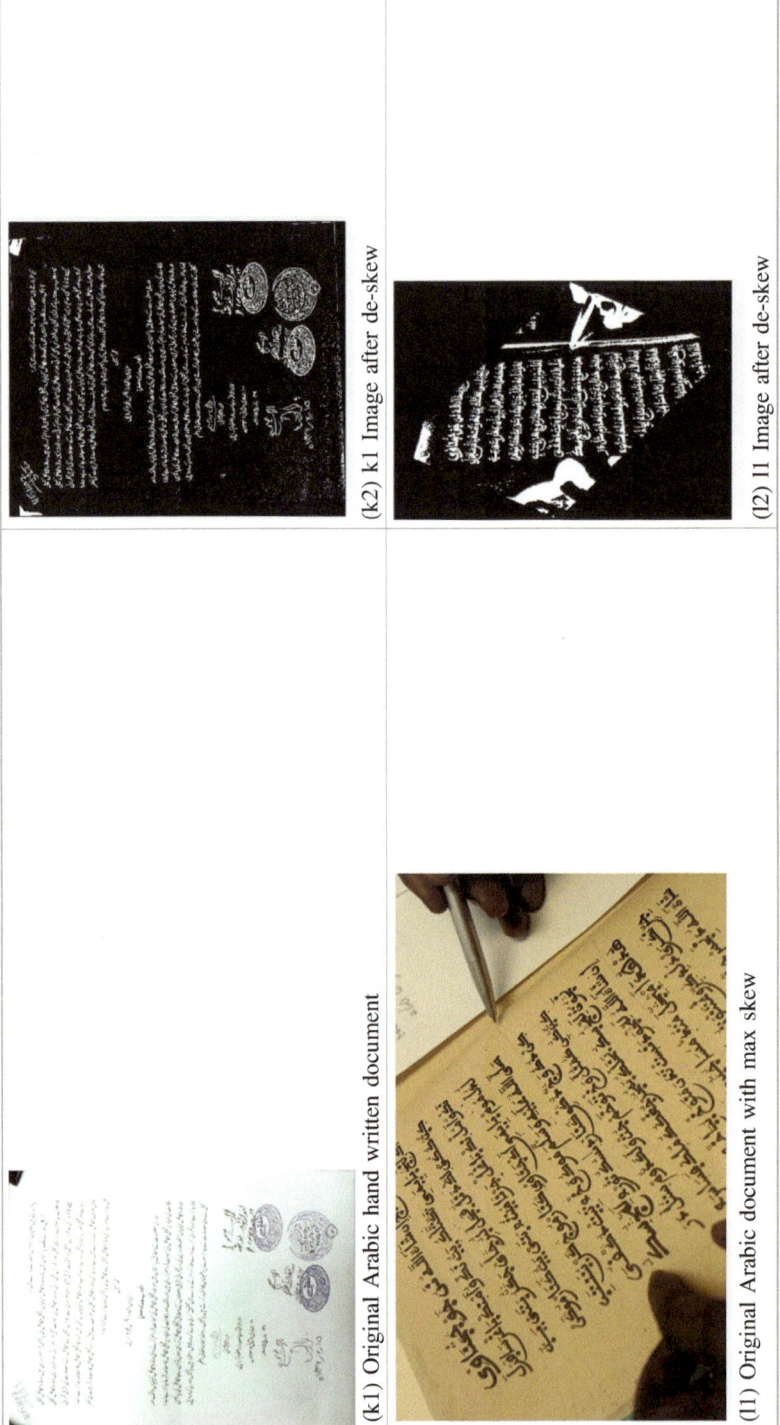

(k1) Original Arabic hand written document

(k2) k1 Image after de-skew

(l1) Original Arabic document with max skew

(l2) l1 Image after de-skew

(continued)

Table 4 (*continued*)

(m1) Original Arabic printed document

(m2) m1 image after de-skew

(n1) Original Arabic image with blur and skew

(n2) n1 image after de-skew

5 Conclusion

This paper proposes a technique to detect and correct the global skew in printed and handwritten documents by considering the top 10 longest CC's by applying the morphological operations and statistical methods. We have tested the proposed algorithm on multi-lingual languages like English, Telugu, Devanagari and Arabic and have observed encouraging results as shown in Tables 1, 2, 3 and 4.

References

1. Soora, N.R., Deshpande, P.S.: Novel geometrical shape feature extraction techniques for multi-lingual characters recognition. IETE Tech. Rev. (2016). https://doi.org/10.1080/02564602.2016.1229583
2. Soora, N.R., Deshpande, P.S.: Robust feature extraction technique for license plate characters recognition. IETE J. Res. **61**(01), 73–80 (2015)
3. Saragiotis, P., Papamarkos, N.: Local skew correction in documents. Int. J. Pattern Recognit. Artif. Intell. **22**, 691–710 (2008)
4. Postl, W.: Detection of linear oblique structures and skew scan in digitized documents. In: 8th International Conference On Pattern Recognition, pp. 687–689 (1986)
5. Baird, H.S.: The skew angle of printed documents. In: 40th Symposium Hybrid Imaging Systems, Rochester, NY, pp. 739–743 (1987)
6. Ciardiello, G., Scafuro, G., Degrandi, M.T., Spada, M.R., Roccotelli, M.P.: An experimental system for office document handling and text recognition. In: 9th international conference on pattern recognition, pp. 739–743 (1988)
7. Ishitani, Y.: Document skew detection based on local region complexity. In: 2nd International Conference On Document Analysis And Recognition, Tsukuba, Japan, pp. 49–52 (1993)
8. Bloomberg, D.S., Kopec, G.E., Dasari, L.: Measuring document image skew and orientation. Doc. Recognit. **2422**, 302–316 (1995)
9. Srihari, S.N., Govindaraju, V.: Analysis of textual images using the Hough transform. Mach. Vis. Appl. **2**, 141–153 (1989)
10. Gorman, L.: The document spectrum for page layout analysis. IEEE Trans. Pattern Anal. Mach. Intell. **15**(11), 1162–1173 (1993)
11. Yan, H.: Skew correction of document images using interline cross-correlation. In: CVGIP: Graphical Models and Image Processing, vol. 55, no. 6, pp. 538–543 (1993)
12. Papandreou, A., Gatos, G.E.: A novel skew detection technique based on vertical projections. In: International Conference on Document Analysis and Recognition, pp. 1384–388 (2011)
13. Sauvola, J., PietikaKinen, M.: Adaptive document image binarization. Pattern Recogn. **33**, 225–236 (2000)

Image Pyramid for Automatic Segmentation of Fabric Defects

Ankita Sarkar[(✉)] and S. Padmavathi

Department of Computer Science and Engineering, Amrita School of
Engineering, Amrita VishwaVidyapeetham, Amrita University, Coimbatore,
India
cb.en.p2cvil5001@cb.students.amrita.edu,
s_padmavathi@cb.amrita.edu

Abstract. Automatic fabric detection is required by the textile industries to improve their quality. For extraction of defective fabric areas, process of segmentation is needed to distinguish the defective region from the background. This paper investigates a method to construct image pyramid by Gaussian method wherein the images are decomposed into multiple levels. Noises are removed and features are extracted for fifteen different defects. Various levels were analyzed and the best level required for proper segmentation is identified for each defect. Region based watershed segmentation and edge based Sobel edge segmentation were experimented on multiple levels. The base level and best level of all decomposed images were compared for all fabric defects investigated.

Keywords: Image pyramid · Gaussian pyramid · Sobel edge detection
Watershed segmentation

1 Introduction

These days fabric defect detection plays vital role for controlling the quality of the fabric in textile industries. Automated fabric defect detection technique helps in detecting as well as locating the defect in the fabric. If the process of detection is accurate we can reduce the wastage. Often, defects are being detected by eyes, but the efficiency for manual defect detection is too low due to fatigue and not suitable for high speed processing. It also requires huge scale of man power and time consumed is much more for such process. Therefore, various computer based vision methods can help in reducing the cost of production. For automatic detection of defects, segmentation method is needed, aiming for separating the fabric defected region from non-defective area and also it helps in distinguishing the background from the foreground. In vision system, segmenting is considered as greatest challenge. In this paper Gaussian Pyramid is generated by decomposing the original image into various levels. Extraction of the defective area at all these levels is tried for segmentation. The two major categories of segmentation namely region based and edge based segmentation are experimented on these pyramid generated. The segmentation of

© Springer International Publishing AG 2018
D. J. Hemanth and S. Smys (eds.), *Computational Vision and Bio Inspired Computing*,
Lecture Notes in Computational Vision and Biomechanics 28,
https://doi.org/10.1007/978-3-319-71767-8_49

these defective area gives better result at certain levels. These levels are analyzed for various defect. The entire process is summarized in block diagram as shown in Fig. 1.

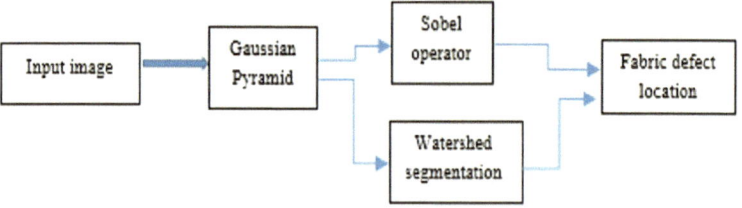

Fig. 1 Block diagram

The three major steps involved in this paper are Image pyramid construction by Gaussian decomposition, Watershed segmentation and edge detection using Sobel operator. In Sect. 2, related work on fabric defect detection and segmentation techniques have been discussed. In Sect. 3, proposed method of the paper have been discussed. The results of various defects are analyzed in Sect. 4. In Sect. 5, conclusion and future work needed is presented.

2 Related Work

For controlling the quality of the fabric in textile industries inspection of the defects in the fabric is needed. For this locating and detecting the defect in the fabric is very important [1]. GLCM approach was used like entropy and measurement of brightness [2]. Based on mathematical operation morphology has been done [3]. Fourier transformation, Gabor filters approach has also been performed [4, 5]. But the main drawback of all these methods are that it has the capability of detecting but only for detecting specific kinds. Because of enormous amount of computation, the above methods limits its use. It has been found out that by using segmentation method, the rate of detecting defective image is very high and the accuracy is good. The main challenge in image processing is segmentation [6, 7]. In this paper, author have used morphological operator for detecting the edges but the drawback is due to noise effect, the edges are hugely effected [8]. Common defect detection methods have been analyzed by two methods that is accuracy and rate of computing accuracy. It included GLCM approach and blob detection but if large size defect is present then computing may cause difficulty. For segmenting the text and binarization [9] in order to inspect the detects in the fabric images have used.

3 Proposed Work

In this proposed technique initially fabric defect image is captured, then image pyramid is developed using Gaussian decomposition. The decomposed image is generated to twelve levels and the best of all is identified. Region based watershed segmentation and edge based sobel edge segmentation are applied at each level of the pyramid. The level in which the location of fabric defect is best achieved is selected as the best level for that defect. The details of these steps are discussed in the following sections.

3.1   Construction of Image Pyramid

It has been seen that many of the image processing techniques, better results can be obtained with the multiresolution decomposition of the original image [10, 11]. The simplest form of multiresolution decomposition of an image can be performed using image pyramid. More precisely, the image pyramid is the collection of decreasing resolution images, from a single original image [12]. The base of such a pyramid always contains the original image and is called as the level 0 image in the pyramid. Then the successive levels of the pyramid can be obtained by down-sampling or sub-sampling the original image by a factor of 2i, where I is the level of the pyramid. So, here the area of the resulting image is exactly one-quarter of the preceding level image. Successively on repeating the procedure an image of size 1×1, i.e. having only one pixel is obtained. This will form the apex of the pyramid. The process can be stopped at any size of the image depending on the need. Let the original image be of size $M \times N$ (for convenience of description, square image is considered, but the same will be followed for a rectangular image as well). Then the level 1 image will become of $M/2 * N/2$ size and so on. Each layer of pyramid signifies a different band of image's frequency, we may use many filters, to each band we can apply filters followed by convolution of the image. Another alternative is, instead of using many filters to the image, we can use a various image for convoluting with a single filter. The multiple images can be obtained by size reduction of the original image (Figs. 2, 3).

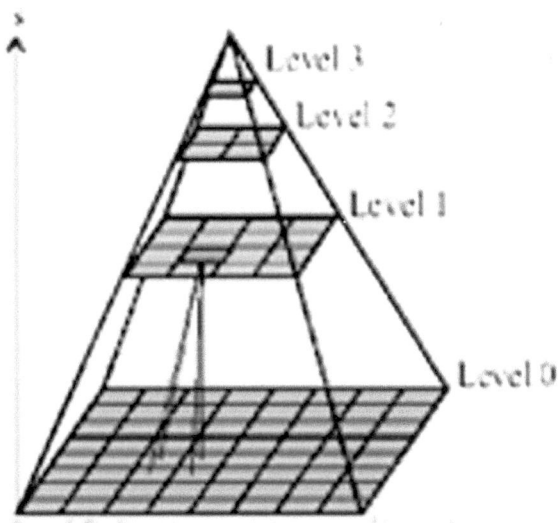

Fig. 2 Showing different levels of pyramid

3.2   Construction of Gaussian Pyramid

The Gaussian pyramid is constructed by using the low-pass filter. This low pass filter is applied to the preceding layer and then the size is being reduced by one fourth (down sampling). This process is mostly termed as reduced operation. As discussed previously that the base level is the original image. The images are made to blur by a Gaussian convolution kernel and to form the image of resolution that is lower, sampling is performed. So, Gaussian pyramid, for all reduced image, it is useful for estimating those images which are present at higher resolution.

Let there be an array f0 containing "M" columns and "W" rows which represents the image. Here each and every pixel will represent the intensity of light by some integer value, ranging between 0 and L-1. Eventually, in the Gaussian Pyramid, this same image will be the zero or the base level. Now, the reduced or low pass filtered of f0 is f1 which will be in level 1 of the pyramid. All values in level 1 are the weighted average of those in level 0 in a 5-by-5 neighborhood. Similarly, level 2 is represented by f2, is obtained from level 1. The average method is followed on a level to level decomposition process, by using the function called REDUCE.

$$fk = REDUCE(fk - 1) \tag{1}$$

This implies, for levels $0 < 1 < N$ and nodes u, s $0 \leq u < M_l$, $0 < j < W$

$$F_l(u, s) = \sum_{a=-2}^{2} \sum_{b=-2}^{2} w(a, b) f_{l-1} \tag{2}$$

where, N = Number of levels, M_l = Dimension of column in lth level W_L = Dimension of row in lth level.

Eventually, decomposed the fabric defective image into information at multiple scales. So, that the noise can also be removed and extract the features as well.

3.3   Watershed Segmentation

The watershed algorithm, proposed by L. Vincent and P. Soille, which is an iterative labeling process [13]. It is an approach for segmenting the objects. Here the input image is changed into different images, hence the catchment basins are identified. After doing this transformation, the distance transform from the complimented image is computed. Now after performing distance matrix, the negative of the distance matrix is taken and is converted to two areas which are bright into the catchment basins.

So, it is important that before doing the watershed transform there is need to do appropriately transform on the fabric defected image. The background image is being subtracted from the image by removing all the features smaller than the structuring elements. Hence noise is removed from the image. After this the level of grey threshold has been taken for finding the correct threshold and therefore the image has been divided into two nodules. Then the image has been converted into black and white image, such that background becomes white and object turns into black. Nearest non zero values are found out by computing the distance transform. So, basically the watershed transformation such that the edge lines are given by 0 and the catchment basins are non-zeros. And finally, convert the image into RGB, hence distinguishing the edges of the defected fabric is easy.

3.4 Sobel Edge Detection Operator

The Sobel operator is used to perform measurement of gradient change in 2-D spatial on image. Usually in an input grayscale image, Sobel operator is used to find the approximate absolute gradient magnitude for each point [14, 15]. Gradient is calculated by sobel edge operator in the x-direction and the other in y-direction, using 3*3 convolution masks.

2 and −2 values is consisted in sobel operator, firsty in the centre,third column of horizontal mask and first, and in vertical mask, third rows. This gives more weightage to the pixel values around the edge region, hence increases the edge intensity.

For finding a square of pixels, the mask is being slide on image's top. It works likes a row pattern, sliding over an area where it is found original image changes with that value of pixel and then shifting of one pixel at a time to the right happens, continuing to the right until reaches the end of row, and then starts again from the starting position of the next row. Although 3×3 masks has a limitation that it cannot manipulate pixels in the first and last row, as well as the first and last column. As the name suggests, Gm mask works on highlighting the edges in the horizontal direction while the Gn mask works in vertical direction, where m represents $\times$ direction and n represents y direction. The resulting output detects edges in both directions, after taking the magnitude of both Gm and Gn. It is simpler to implement than the other operators. So with the help of Sobel operator we can find the edges of the defect present in the fabric.

The change or the gradient of the magnitude is given by;

$$|G| = \sqrt{g_m^2 + g_n^2} \qquad (3)$$

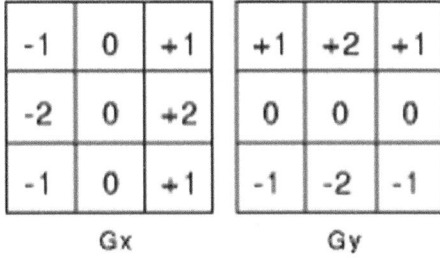

G_x G_y

Fig. 3 Showing vertical and horizontal Sobel edge detection

4   Experimental Results

For experimentation fifteen defective images has been considered. These are extracted from twenty four different fabric material, few of them are listed in Fig. 4.

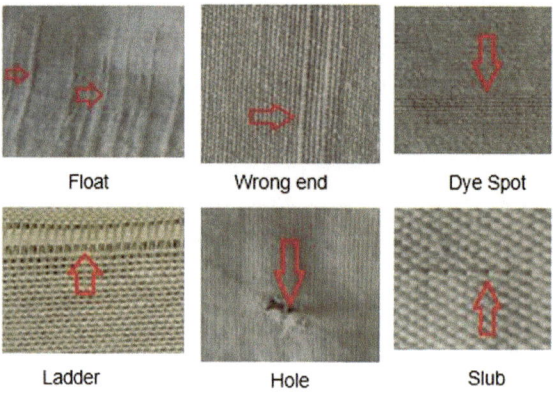

Fig. 4 Datasets

An image pyramid is constructed using Gaussian decomposition. Twelve levels of image pyramid is generated out of the original image. The segmentation algorithms are applied on those level of the pyramid. The level showing the defective area clearly is considered as the best level for that defect.

In this multiresolution decomposition process, base level and best level are compared. The Figs. 5 and 6 shows the pyramid base and the best level of a Hole defective image.

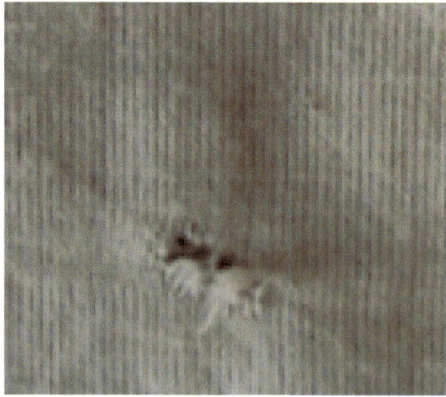

Fig. 5 Base level

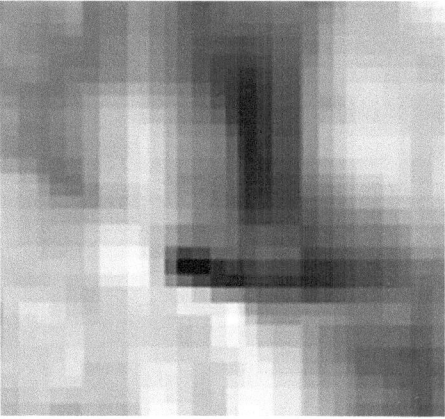

Fig. 6 Best level

Sobel edge detector was used in finding the edges of the defect in image by detecting the vertical and horizontal edges (Figs. 7, 8).

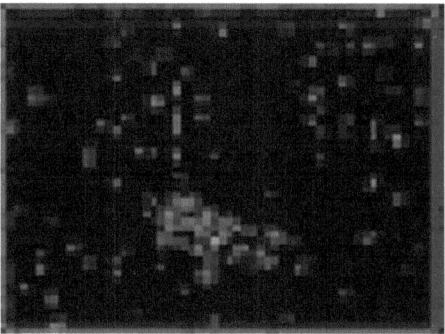

Fig. 7 Level 1

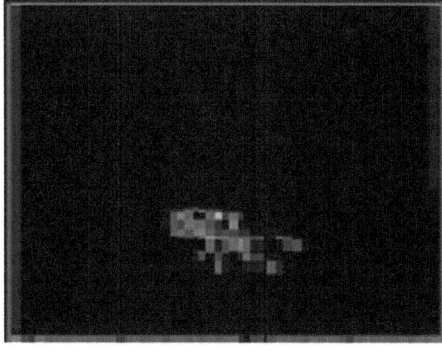

Fig. 8 Best level

Sobel edge detection was performed in all the levels of a defective image and the best level is identified as explained above.

In watershed segmentation, the regions of the defective area is segmented from non-defective area in the given image.

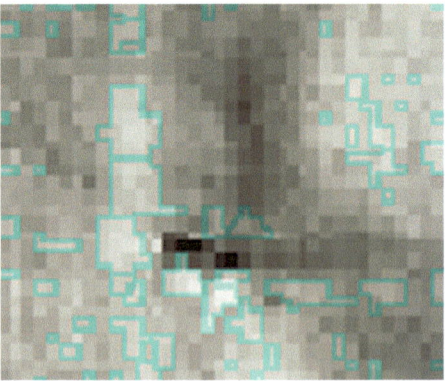

Fig. 9 Level 1

Fig. 10 Best level

Similar to edge segmentation, watershed segmentation is performed in all the levels of a defective image. The level showing the best region extracted,is assigned the best level. Figures 9 and 10, Shows the watershed segmentation result for the level 1 and best level for Hole defective image. The best level for sobel and watershed algorithm are identified as level 9 and 11 respectively for Hole defective image. Similarly, for Mispick defective image, level eleven and ten was best level for sobel and watershed algorithm respectively. For Starting place defect, level 8 and 9, Slough off weft image, level 11 and 12, for Wrong end defective image level 9 and level 10. For Float defective image level 9 and 10. Missing loop level 12 and 11. Stain best levels achieved are level 11 and 12. For Bar defective image level 11 and 10. In Bar defective image level 10 and 12. In Ladder defective image level 7 and 8 achieved the best level. For Color out decfective image level 11 and 12. In Broken color pattern defective image level 9 and 10 were the best. Crease Mark defect attained the best output in level 10 and 11. Taking Sobel and watershed segmentation respectively for all the defects mentioned above.

5 Conclusion

At present, segmentation techniques have been applied for detection of the fabric defects. The Gaussian image pyramid have been evaluated to decompose the fabric defective image into information at multiple scales. This is done to reduce the effect of thread and other fabric details affecting the detection process. After Gaussian decomposition, the Sobel edge segmentation and Watershed segmentation was applied to segment the defective area from the image. The various levels of decomposed image were analysed to find better segmentation and location of the defect. The best level for each category is specified as a result of experimentation. The levels were found to be different for various defects and fabric. Hence this paper suggests appropriate levels for the defects. The automatic identification of the levels thereby reducing the natural intervention is the future scope of the paper.

References

1. Karlekar, V.V., Biradar, M.S., Bhangale, K.B.: Fabric defect detection using wavelet filter. In: International Conference on Computing Communication Control and Automation (ICCUBEA), 2015. IEEE (2015)
2. Rebhi, A., Abid, S., Fnaiech, F.: Fabric defect detection using local homogeneity and morphological image processing. In: International Conference on Image Processing, Applications and Systems (IPAS), 2016. IEEE (2016)
3. Wang, D., Liu, H.: Edge detection of cord fabric defects image based on an improved morphological erosion detection methods. In: Sixth International Conference on Natural Computation (ICNC), 2010, vol. 8. IEEE (2010)
4. Arnia, F., Munadi, K.: Real time textile defect detection using GLCM in DCT-based compressed images. In: 6th International Conference on Modeling, Simulation, and Applied Optimization (ICMSAO), 2015. IEEE (2015)
5. Zuo, H. et al.: Fabric defect detection based on texture enhancement. In: 5th International Congress on Image and Signal Processing (CISP), 2012. IEEE (2012)
6. Benjelil, M. et al.: Steerable pyramid based complex documents images segmentation. In: 10th International Conference on Document Analysis and Recognition (ICDAR'09), 2009. IEEE (2009)
7. Loo, P.-K., Tan, C.-L.: Using irregular pyramid for text segmentation and binarization of gray scale images. In: Seventh International Conference on Proceedings of Document Analysis and Recognition, 2003. IEEE (2003)
8. Wang, H., Huang, L.-L., Zhao, X.-J.: Automated detection of masses in digital mammograms based on pyramid. In: International Conference on Wavelet Analysis and Pattern Recognition (ICWAPR'07), 2007, vol. 1. IEEE (2007)

9. Subudhi, P., Mukhopadhyay, S.: A pyramidal approach to active contours implementation for 2D gray scale image segmentation. In: International Conference on Wireless Communications, Signal Processing and Networking (WiSPNET). IEEE (2016)

10. Okamoto, T. et al.: Image segmentation of pyramid style identifier based on Support Vector Machine for colorectal endoscopic images. In: 37th Annual International Conference of the IEEE, Engineering in Medicine and Biology Society (EMBC), 2015. IEEE (2015)

11. Miller, E.H.: A note on reflector arrays (Periodical style—Accepted for publication). IEEE Trans. Antennas Propag., to be published

12. Wang, J.: Fundamentals of erbium-doped fiber amplifiers arrays (Periodical style submitted for publication). IEEE J. Quantum Electron., submitted for publication

13. Karthika, R., Parameswaran, L.: Study of Gabor wavelet for face recognition invariant to pose and orientation. In: Proceedings of the International Conference on Soft Computing Systems. Springer, New Delhi (2016)

14. Padmavathi, S., Soman, K.P.: Comparative analysis of structure and texture based image inpainting techniques. In: International Journal of Electronics and Computer Science Engineering (IJECSE), vol. 1. (2012)

15. Ishu, G., Kaur, B.: Color based segmentation using K-mean clustering and watershed segmentation. In: 3rd International Conference on Computing for Sustainable Global Development (INDIACom), 2016. IEEE (2016)

Multi Focus Region-Based Image Fusion Using Differential Evolution Algorithm Variants

K. C. Haritha and S. Thangavelu[(⊠)]

Department of Computer Science and Engineering, Amrita School
of Engineering, Amrita Vishwa Vidyapeetham, Coimbatore, India
cb.en.p2csel5007@cb.students.amrita.edu,
s_thangavel@cb.amrita.edu

Abstract. This work focuses on an optimum process of image fusion on multiple focus images using an optimization algorithm viz., Differential Evolution (DE) algorithm. The input image is divided into regions and sharper regions are selected from these two images. The selected clear blocks are used for constructing final resultant image. The main purpose of using differential evolution algorithm is to find out optimum block size, which is more useful during division of image rather than fixed block size. And also, this work compares different variants of differential evolution algorithm based image fusion to find out which one will be suitable for getting more focused image. The major focus of the research is finding out which type of differential evolution algorithm is best suitable for almost all type of images. Block based and pixel based method are used together to achieve a better resultant image. Performance of fused image is calculated using image quality measures and found out better fusion method, which can be used in almost all situations.

Keywords: Image fusion · Image processing · Differential evolution algorithm
Criterion function

1 Introduction

Fusion of image is the method of associating valid details from more than two input images and forming one final fused image of good quality. During image fusion, all the redundant details within the input image needs to be removed. First and foremost, pre-processing stage in image fusion is the registration of both the input images. Image registration is the process of making point-to-point matching between a numbers of input images, which are related to same scene [1].

Spatial domain, transform domain/frequency domain are the two types of image fusion methods [2]. Most general frequency domain methods are laplacian pyramid, morphological pyramid, discrete wavelet transform etc. In all these methods, fusion coefficients are selected either by pixel-based or region-based fusion rules. The problem of transform domain fusion is that, during inverse transform operation some information gets lost. As in spatial domain method, fused image is obtained by using

© Springer International Publishing AG 2018
D. J. Hemanth and S. Smys (eds.), *Computational Vision and Bio Inspired Computing*,
Lecture Notes in Computational Vision and Biomechanics 28,
https://doi.org/10.1007/978-3-319-71767-8_50

maximum/average method. Fused image can also be obtained by selection of arbitrary pixels among input images. But there may be some problems in both the methods, such as blurredness and pixel-discontinuity in some parts of image [3].

Some block based segmentation has been developed in spatial domain for avoiding the problems caused by transform domain methods. Image segmentation and fusion are the main steps in spatial domain based method [4]. Spatial domain operations are simple, as it doesn't involve the use of transforms or inverse transforms. Fusion rules are employed to find out proper block in fused image. This fusion rule can be either pixel-based or region-based according to the area of application of fused image. Registration is the first and foremost step in both transform and spatial fusion methods [5]. Through this process, we will be getting point-by point matching between two or more input images.

Multi-focus image fusion is the effect of joining two or more images of a same scene to form one final fused all in focus image. Using this type of fusion, image with better focus can be formed [6]. Many varieties of multi focus fusion type are available; among this empirical mode decomposition is the best known method. Multi-view fusion is used for fusing images taken from different cameras [7]. In this method format of image can be same or different, but can be of same modality. Multi modal takes different styles of the same scene as input and fuse them together to form the output image. It is a method of combining information from various image formats.

Based on the level of image fusion, there are three type, pixel-based fusion, region-based fusion, and decision based fusion [8]. In pixel-based method, fused images are formed after selecting sharpest pixels [9]. Most common pixel based sharpness measures are spatial frequency and variance. Spatial characteristics are computed for doing pixel based fusion method. In region-based fusion, it helps to overcome some drawbacks of pixel-based fusion, like blurring, susceptibility to noise and mis-registration [10]. In decision-based fusion, only sharper regions or sharper pixels are taking into consideration. Unnecessary comparison can be reduced to a great extent.

Structural information of an image can't be fully expressed by employing arbitrary pixel selection method. Corner portion of an image is also taken into consideration while dividing image into blocks [11]. In region based fusion also group of pixels are taken for doing various pixel operations. Region based fusion is high level compared to pixel based operation, as it is dealing with structural information of an image rather than the basic features like colour, pixel etc. in pixel based fusion method [12].

2   Literature Survey

An approach of fusion of image by DE algorithm is proposed in [13]. V. Aslantas, R. Kurban have used block based fusion method. Criterion functions like sum-modified laplacian, variance and SF are used for sharper block selection. Fused image is formed using the selected blocks, which contains only focus regions. The authors concluded that this fusion method works better than other primitive fusion methods and GA based method. Quantitative performance evaluation is carried out with reference image and without using reference image (Blind image fusion method) using MSE, PSNR and MI. In [1], Feng et al., divided image into small overlapping blocks and found that if the sharpness values of blocks from two images are equal, extends block mechanism can be applied to find the highest sharpness valued block. Extends block mechanism used includes criterion functions like sum-modified laplacian, variance and SF. The two main advantages of this method are, block size is purely determined using differential evolution algorithm and the proposed algorithm ignores size of the image. Calculating block size of large images become more complex and time consuming as algorithm doesn't consider image size. Improved DE is similar to the original DE algorithm, but the former initializes population at different intervals and consumes less time than the latter. MSE and PSNR were used for measuring the performance of resultant image [14].

In [15], Aslantas et al., used spatial domain image fusion on multiple focus images. Differential evolution algorithm is used for estimating optimal point spread function (PSF) of source images (blurred and sharp images). Each image is blurred artificially by convolving it with PSF which is used to find sharpest pixel in the input images and are selected to form the resultant image. Qualities of Edges (QE) and Fusion Factor (FF) were used for the performance evaluation in [15]. They have proved quantitatively and visually that this method outperforms DWT and DCT (Discrete Cosine Transform) methods [16].

From the above listed papers, dealing with fusion process using differential evolution algorithm, it can be concluded that DE is mainly used for selecting the optimum block size in almost image fusion process.

2.1 Image Quality Metrics

Mainly criterion functions comprise of spatial frequency, energy of gradient, sum modified laplacian, and variance. A criterion function is used for choosing the sharper blocks from input images so that to construct the final sharp end image [13, 15]. Highly focused images contain sharp edges which is having high frequency as compared with the out of focus image [7]. Using the frequency, sharpness value of an image is measured. The details of these functions are given below.

2.1.1 Spatial Frequency (SF)

SF computes the sharpness value of an image. As spatial frequency value increases, image becomes clearer. The row and column gradients of an image pixel are squared and added, so as to get maximum addition from large gradients. Image gradients are used for deriving information from images [9].

$$SF = \sqrt{C^2 + R^2} \tag{1}$$

Column (C) and row (R) value of an image can be formulated as following [15],

$$R = \sqrt{\frac{1}{MN} \sum_i \sum_j (F(i,j) - f(i,j-1))^2} \tag{2}$$

$$C = \sqrt{\frac{1}{MN} \sum_i \sum_j (F(i,j) - f(i-1,j))^2} \tag{3}$$

2.1.2   Variance (VAR)

Variance is the difference between mean grey level intensity value and the individual intensity value [13]. Variance and focus measure of an image can be calculated using the below equations, where focus measure (F) is mainly used to measure relative focus of the image.

$$VAR = \frac{1}{MN} \sum_{i=1} \sum_{j=1} \left(F(i,j) - \overline{F}\right)^2 \tag{4}$$

$$F = \frac{1}{MN} \sum_{i=1} \sum_{j=1} F(i,j) \tag{5}$$

2.2   Image Quality Metrics for Performance Evaluation

Image quality metrics measures the deviation of fused image from the ideal image. Image quality metrics are used for evaluating the performance of various image fusion methods used in an experiment.

2.2.1   Peak Signal to Noise Ratio (PSNR)

PSNR gives the ratio between the powers of a signal to the power of pestiferous noise which affects the reliability of image delegation. Higher PSNR indicates that the rejuvenated image is of high quality. As larger the PSNR value, the fusion value also become pronounced [8, 9].

$$PSNR = 10 \log 10 \frac{L^2}{\frac{1}{mxn} \sum_{i=1}^{m} \sum_{j=1}^{n} (R(i,j) - F(i,j))^2} \tag{6}$$

2.2.2   Mean Square Error (MSE)

MSE is a reference based quality judgement which is commonly used when the reference image is present. R(i, j) is the pixel value of reference image and F(i, j) is the pixel value of fused image. Here in this equation, m and n are the dimensions present in an image. As minor the value of MSE, the fused image becomes prominent [9, 15].

$$MSE = \frac{1}{mxn} \sum_{i=1}^{m} \sum_{j=1}^{n} (R(i,j) - F(i,j))^2 \tag{7}$$

2.2.3   Normalized Cross Correlation (NCC)

This metric is used to measure the degree of affinity or dissimilarity between two correlated images [8]. NCC gives the normalized value of correlation value between the fused image and reference image.

$$NCC = \left(\sum_{i=1}^{M} \sum_{j=1}^{N} R_{ij} * F_{ij}\right) \div \left(\sum_{i=1}^{M} \sum_{j=1}^{N} R^2\right) \qquad (8)$$

2.3   Differential Evolution (DE) Algorithm

Differential evolution algorithm is a heuristic algorithm which follows optimization concepts. It is a stochastic and population based evolutionary algorithm introduced by Storn and Price in 1996 [13]. DE is developed for optimizing real time parameters and real valued functions. Genetic algorithm is also used for optimizing parameters, but global optimization is necessary in image processing fields [6, 12]. In image processing applications, statistics, and finance DE suits well than other evolutionary algorithms. DE can be used to find approximate solution to many practical problems, which have objective functions that are non-differentiable, non-linear, noisy, or multi-dimensional.

DE is defined by the four-step process, which are initialization, mutation, crossover and selection.

2.4   State of the Art in Image Fusion Using Differential Evolution Algorithm

Emerging field in fusion scenario is the method of fusion which is with respect to regions or blocks. Pixel based image fusion is the most primitive form of fusion method, which only deals with pixels; hence it is a lower version of image fusion method. Region based and decision based image fusion methods are the most emerging and advanced type of fusion method. Region based came into account with the assumption that single region carry more information than a single pixel. Actually, regions are again a group of pixels, but doing various operations on region needs lesser time compared with individual pixels. And also, pixels taken collaboratively as a region can be handled efficiently rather than separate individuals. Region based overcomes several disadvantages of pixel based method, like blurring, mis-registration effect, or effect of noise.

Multi-focus image fusion can be done in two ways, spatial domain based fusion and transform domain based fusion. Spatial domain fusion method deals with directly fusing the source images due to the spatial characteristics of image. Spatial features of an image can be represented either in object mode or in image mode. Object model is the basic level of representation as a point or polygon. In image mode, individual objects are represented as contiguous cells or regions. Most of the multi focus fusion methods deals with spatial domain based fusion process. Spatial domain method can be pixel based or region based, according to the need for an image analyst. Most of the applications need region based fusion, as it is more meaningful than pixel-based. Region-based fusion is actually dealing with pixels only, but it is a more meaningful way of representing pixels for further use.

3 Proposed Method

An efficient way of doing fusion process on images with optimum selection of block size with the help of a heuristic algorithm, viz., DE algorithm is proposed. How optimization plays an important role in fusion process is explored. How DE/best and DE/rand acts in image fusion process has been compared and found out results. Idea of using various evolutionary algorithms for doing image processing applications came into existence due to the fact that optimum block size is better than some random block size. DE is one among the evolutionary algorithm, which produces better results compared to genetic algorithm (GA) for some real-time applications. GA and Particle Swarm Optimization [17] fall into local convergence, means it will assume a solution after some run. But DE do not fall in local convergence, it will evolve to find global convergence. So, for getting better fused image in image processing applications, DE will be suitable.

Region-based fusion using differential evolution is the main focus of this work. Region-based method deals with blocks rather than pixels as in primitive fusion types. Selecting suitable block size is the most tedious task in block-based fusion process. Blocks should be somewhat large to include most of the focused portions, and small enough to get sufficient number of blocks for comparison. Selecting a suitable block size is the main challenge in block based fusion process. Heuristic choices must be needed for efficient selection of block size in a more meaningful way. Different size of block selection is possible, so it is hectic to find any optimum solution using some primitive methods. Hence, heuristic algorithms such as differential evolution algorithm came into picture. The main aim of image fusion is to select all the sharper regions from source images and hence produces final combined image of high quality.

3.1 Motivation of Image Fusion Using DE

The main motivation for designing the current system for doing block based image fusion is the fact that region-based fusion procedures are most effective and reliable than primitive pixel based methods. Pixel based method just consider individual pixels or nearby pixels for fusion process, not consider the whole portion of image for comparing sharpness.

Fusion of image is mainly consisting of image registration, which is the primitive step for dealing with noisy regions. DE will handle noisy, irregular, unstructured data, so there is no need to do image registration. First step in image fusion process is segmentation. Region-based fusion processes blocks than small pixel regions. This helps to avoid problems like noisy region in image, issues due to mis-registration.

3.2 Multi-focus Image Dataset

Two input images are used for doing image fusion, in first image Fig. 1, time piece is focused and man in front of computer is not focused, in Fig. 2, it is vice versa. In Fig. 1, focus is on the clock other portions are blurred, including human. So, this focused area is the point of attraction for our method. All sharper/clear portions of image need to consider for doing image fusion.

3.3 Optimization of Parameters Using DE

DE algorithm starts with forming initial members for the population. Each member in the first fusion result consists of the solution for the DE algorithm, which will be used for forming further individuals within the population size. For using DE for image fusion, a set of solution must be found out first. In this case, row and column size will be the candidates inside solution set. In every stage of optimization, sharpness measure is used to find out the fitness value of the current population in a generation.

Fitness value is being assigned with each individual solution in a population. The members which produced highest fitness value is transferred to the immediate hierarchy (generation). The members for next generation are formed from the members using DE operations. DE operations include primitive mutation, crossover and selection processes.

3.4 Algorithmic Steps

The basic steps of the region-based image fusion process can be described as follows:

Step 1   Initialize the population
Step 2   For each candidate of the population, carry out fusion process (divide image
          as block, selection of best block, fuse the best blocks to form fused image)

Fig. 1 Human and clock image, focus on clock

Fig. 2 Human and clock image, focused on human

Step 3   Evaluate the fused image using fitness function, spatial frequency is used because of its effectiveness in comparing the blocks

Step 4   For each generation, with in the population size, perform mutation, crossover as in standard DE process

Step 5   Find fitness using newly formed individuals; perform selection operation to find out individuals for next generation

Step 6   Make next population as current population

Step 7   Do the process specified in above steps, till we get an optimum solution. That is until getting a fully focused fused image.

3.5   Mutation Process

Select population from the initial population, and do the following steps,

Step 1   Select three individuals randomly from the initial population

Step 2   Find mutated vector by using the following equation. Finding difference between to random numbers and multiplying this result with mutation constant, result is added with third random individual

Step 3 $v(i,:) = x(r1,:) + F*(x(r2,:) - x(r3,:))$, where r1, r2, r3 are the random individuals and F is the mutation rate within the range [0, 1].

3.6 Crossover Process

Crossover process is carried out after mutation, i.e., from the mutated vector do the following steps:

Step 1   If (rp <= CR)||(j == rj), where CR is in the range [0, 1] and u(i, j) = v(i, j))

Step 2   Else u(i, j) = x(i, j), rp is random number, rj = ceil(rand*D).

3.7 Selection

Step 1   Selection for next generation by comparing trial and target vector

Step 2   Calculate fitness (maximizing sharpness) using spatial frequency measure. Select whether the trial vector or target vector survive into next generation by comparing sharpness of fused image.

3.8 Fitness Function

Fused image is directly dependent on the clarity or sharpness of regions of image. Hence, higher the brightness, solution or (member) becomes better than the input image. So, for calculating the fitness of each member, following sharpness measure can be used:

$$Fitness = Global\ sharpness\ value\ (fused\ image)$$

As global sharpness value, any criterion functions which will provide better sharper fused image can be used. Spatial frequency is the best known global sharpness value measure employed in fitness measure.

3.9 Steps for Combined Region and Pixel Based Fusion Process

Combined block based and pixel based fusion method is done for improving the final fused image quality. Each divided block is compared according to the intensity of pixels. Blocks are nothing but a group of pixels itself. Hence block comparison indirectly means comparison of pixels. Pixel is the basic way of representing an image. Pixel level intensity is measured using either of the sharpness measures listed in literature survey portion.

Step 1   Divide image into blocks per the size obtained after running DE algorithm.

Step 2   Compare the intensity value of the corresponding pixels of the input pair of blocks.

Step 3   Form final fused image by taking all the high intensity pixel points and leaving low intensity.

Step 4   Run Differential Evolution algorithm (Mutation, Crossover, and Selection) using spatial frequency as the fitness function (Fig. 3).

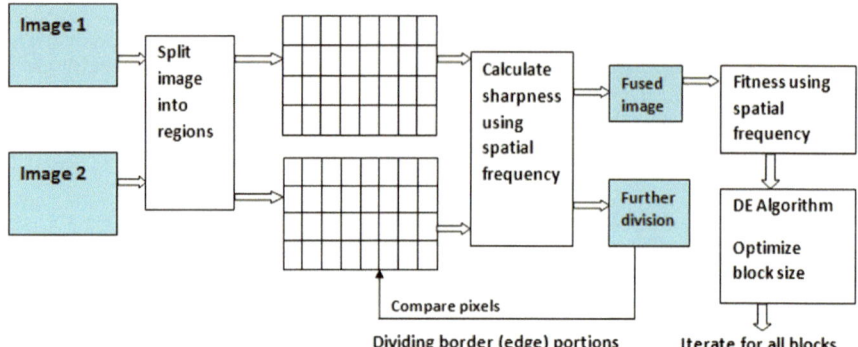

Fig. 3 Architecture of proposed work. *Courtesy* [13]

4 Experimental Results

This section holds the results obtained from the above experiment. In the preprocessing step collected multi focus image dataset and resize both the input images (similar size image can be fused). DE algorithm is selected for optimizing the block size for dividing the image. Initial population size is fixed which is of 10. Number of generation is fixed as 50 only single run is carried out. Crossover constant and mutation rate is varied from 0 to 1 and evaluated the final fused image (Fig. 4).

Fig. 4. Pixel based image fusion on human and clock image

Fig. 5 Combined pixel and block based fusion using DE/rand/1/bin

After doing pixel based image fusion, the resultant fused image looks as above image, with all areas focused, but still some blurred portion exists in some areas.

Combination of pixel and block based fusion process with all-in-focus image as shown in Fig. 5. More clarity is achieved in this type of fusion (Fig. 6).

Fig. 6 Region based image fusion using DE/rand/1/bin

Pixel based fusion is the primitive type of fusion used in almost all fusion algorithms. Three types of image fusion are carried out and result is evaluated using spatial frequency, PSNR, MSE, NCC and SSIM measures [5]. Spatial frequency is the most reliable measure that has been used in most of the image fusion processes (Table 1).

Spatial frequency is used as sharpness criterion function for measuring clarity of a portion (individual block) of an image. DE variant such as DE/rand/1/bin, DE/best/1/bin and DE/best/2/bin were used with image fusion for selecting best block size and to divide the whole image into small region. Pixel-based fusion is carried out with three DE variants and like that region-based, combined region and pixel based were carried out (Table 2).

Table 1 Image fusion types and quality measures

Image fusion types	Spatial frequency	PSNR	MSE	NCC	SSIM
Pixel-based fusion	3.9791	21.8021	429.4079	1.019	0.3928
Region-based fusion	5.1272	21.1885	494.5691	1.0075	0.3880
Combined pixel and block based fusion	8.645	20.9676	520.3828	0.9845	0.3294

Table 2 Three types of image fusion using DE variants using spatial frequency as sharpness measure

DE variant	Pixel-based	Region-based	Region and pixel-based
DE/rand/1/bin	11.645	12.565	14.6945
DE/best/1/bin	11.675	11.99	13.695
DE/best/2/bin	11.680	12.655	15.295

Similarly, variance is used as sharpness criterion function for measuring pixel intensity or clarity in image. Three DE variants are combined with pixel-based, region-based together with a combination of block based and pixel based fusion method (Table 3).

Table 3 Three types of image fusion with variance as the sharpness measure

DE variant	Pixel based	Region based	Region and pixel-based
DE/rand/1/bin	120.2	122.45	123.46
DE/best/1/bin	120	122.5	124.5
DE/best/2/bin	120.3	122.5	124.8

5   Conclusion

This work divided input images into small block, and then will select the sharper image block out of two corresponding blocks from input images. The selection is based on sharpness measure which can be spatial frequency or variance. The blocks selected after calculating clarity of blocks is used to form final fused image of good quality. Performance of the fused image is evaluated between reference image and resultant image. Fused image is evaluated using visual evaluation and by using quantitative or qualitative evaluation. The source images used here is the black and white multi focus image, taken using digital camera; size of the two input images is fixed. Out of the DE variants, DE/best/2/bin produces better fusion result compared to DE/rand/1/bin and DE/best/1/bin.

As an extension, methods that can be used for doing image fusion can also be suggested to user using DE optimization. It will be a better extension, as there is no optimum algorithm available for fusing all type of images. DE algorithm can suggest the fusion types together with the optimum block size to the user.

References

1. Feng, Y., Li, T., Zhou, S.: Enhanced differential evolution algorithm and extends block selection mechanism-based multi-focus images fusion. J. Inf. Comput. Sci. **8**, 2637–2644 (2011)
2. Shah, S.K., Shah, D.U.: Comparative study of image fusion techniques based on spatial and transform domain. Int. J. Innovative Res. Sci. Eng. Technol. (IJIRSET) **3**(6) (2014)
3. Li, S., Yang, B.: Multifocus image fusion using region segmentation and spatial frequency. Image and Vision Comput. **26**(7), 971–979 (2008)
4. Li, Q. et al.: Region-based multi-focus image fusion using the local spatial frequency. In: 2013 25th Chinese Control and Decision Conference (CCDC). IEEE (2013)
5. Geetha, G., Raja Mohammad, S., Murthy, Y.S.S.R.: Multifocus image fusion using multiresolution approach with bilateral gradient based sharpness criterion. J. Comput. Sci. Inf. Technol. **10**, 103–115 (2012)
6. Kong, J., Zheng, K., Zhang, J., Feng, X.: Multi-focus image fusion using spatial frequency and genetic algorithm. Int. J. Comput. Sci. and Netw. Secur. **8**(2), 220–224 (2008)
7. Gupta, R., Awasthi, D.: Wave-packet image fusion technique based on genetic algorithm. In: 2014 5th International Conference on Confluence the Next Generation Information Technology Summit (Confluence), September 2014, pp. 280–285. IEEE (2014)
8. Krishnamoorthy, S., Soman, K.P.: Implementation and comparative study of image fusion algorithms. Int. J. Comput. Appl. (0975–8887) Volume (2010)
9. Sruthy, S., Parameswaran, L., Sasi, A.P.: Image fusion technique using DT-CWT. In: Proceedings—2013 IEEE International Multi Conference on Automation, Computing, Control, Communication and Compressed Sensing, iMac4s 2013, Kerala, pp. 160–164. (2013)
10. Erkanli, S., Oguslu, E., Li, J.: Fusion of visual and thermal images using genetic algorithms. INTECH Open Access Publisher (2012)
11. Zhang, J., Feng, X., Song, B., Li, M., Lu, Y.: Multi-focus image fusion using quality assessment of spatial domain and genetic algorithm. In: 2008 Conference on Human System Interactions, May 2008, pp. 71–75. IEEE (2008)

12. Jyothi, V., Kumar, B.R., Rao, P.K., Reddy, D.R.K.: Image fusion using evolutionary algorithm (GA). Int. J. Comp. Tech. Appl. **2**(2), 322–326 (2011)
13. Aslantas, V., Kurban, R.: Fusion of multi-focus images using differential evolution algorithm. Expert Syst. Appl. **37**(12), 8861–8870 (2010)
14. Bedi, S.S., Khandelwal, R.: Comprehensive and comparative study of image fusion techniques. Int. J. Soft Comput. Eng. (IJSCE) ISSN, **3**, 2231–2307, 2013
15. Aslantas, V., Toprak, A. N.: Multi focus image fusion by differential evolution algorithm. In: 11th International Conference on Informatics in Control, Automation and Robotics (ICINCO), September 2014, vol. 1, pp. 312–317 (2014)
16. Anish, A., Jebaseeli, T.J.: A survey on multi-focus image fusion methods. Int. J. Adv. Res. Comput. Eng. Technol. (IJARCET), **1**(8) (2012)
17. Zhang, X., Sun, L., Han, J., Chen, G.: An application of swarm intelligence binary particle swarm optimization (BPSO) algorithm to multi-focus image fusion. Optica Applicata **40**(4), 949–964 (2010)

A Hybrid Classification Model Using Artificial Bee Colony with Particle Swarm Optimization and Minimum Relative Entropy as Post Classifier for Epilepsy Classification

Harikumar Rajaguru[✉] and Sunil Kumar Prabhakar

Department of ECE, Bannari Amman Institute of Technology, Coimbatore, India
harikumarrajaguru@gmail.com

Abstract. One of the most striking features of the human brain is that it has an amazing spatio temporal dynamics. Due to the excessive and irregular electrical action flowing in the human brain, the seizure activity happens. Because of epileptic seizures, it gives rise to unusual and strange sensations thereby affecting the behaviour and quality of life of the person. The electrical activities of the human brain can be easily measured with the help of Electroencephalogram (EEG) signals. In this paper, the hybrid model of Particle Swarm Optimization (PSO) and Artificial Bee Colony (ABC) algorithms is implemented to get a first level optimization in the epilepsy risk level of the EEG signals. Further Minimum Relative Entropy (MRE) is used as a second level post classifier to optimize it further for the perfect classification of epilepsy risk levels from EEG signals. The results show that an average post classification accuracy with MRE classifier is 97.90% along with an average time delay of 2.06 s.

Keywords: Epilepsy · EEG · PSO · ABC · MRE

1 Introduction

The human brain is most complex in nature [1]. The disorders of the human brain include stroke, Alzheimer's disease, severe brain injury, seizure disorders, learning disabilities etc. One of the severe disorders of the human brain is epileptic seizures. If there is a seizure attack, then the neurons start flowing in an abrupt manner, thereby affecting the quality of life of the epileptic patients [2]. To observes the seizures, EEG recordings are quite important. The electric potential in the brain is generated by the electrochemical process occurring in the human brain. The variations in the electrical activities of the brain are recorded with the help of EEG signals by placing the electrodes at various positions on the scalp of the patient [3]. For the diagnosis of main diseases like seizure disorders and problems related to sleep disorders and alcoholic consumption, EEG is used. Automatic seizure detection systems have become the need of the hour as it can assist the clinicians and visual encephalographers to a great extent [4]. A lot of

© Springer International Publishing AG 2018
D. J. Hemanth and S. Smys (eds.), *Computational Vision and Bio Inspired Computing*,
Lecture Notes in Computational Vision and Biomechanics 28,
https://doi.org/10.1007/978-3-319-71767-8_51

interesting and variety of works has been reported in literature regarding the classification of epilepsy risk levels from EEG signals. Some of the important works are discussed here as follows.

For the classification of epileptic EEG signals, the local pattern transformation based feature extraction techniques were done by Jaiswal and Banka [5]. A new framework based on the ideas of recurrence quantification analysis for the detection of epileptic seizures was done by Niknazar et al. [6]. Using complex-valued classifiers, a novel method was designed for the automated diagnosis of epilepsy by Peker et al. [7]. With the help of a novel variation of Empirical Mode Decomposition (EMD), the seizure detection from EEG and the diagnosis of epilepsy was done by Kaleem et al. [8]. With the help of statistical pattern recognition and wavelets, the EEG signal was classified for the epileptic seizure detection by Gajic et al. [9]. For epileptic brain signal classifications, the wavelet entropy and the relative wavelet energy was used by Kumar et al. [10]. With the aid of wavelets, approximate entropy and Support Vector Machine (SVM), a high-performance seizure detection system with a clinical validation was found by Shen et al. [11]. Based on the non-linear analysis and the time-frequency analysis, the epileptiform activity in the EEG signals was detected by Gajic [12]. With the help of phase-space reconstruction, wavelet transform and Euclidean distance, the classification of epileptic seizure and normal EEG signals was found by Lee et al. [13].

In this work, the raw EEG signals are obtained. It is sampled and then with the help of hybridizing Artificial Bee Colony with Particle Swarm Optimization (ABC-PSO), it is further optimized. Finally it is classified with Minimum Relative Entropy (MRE) as the post classifier for classification of epilepsy risk levels from EEG signals. The organization of the paper is as follows. In Sect. 2, the materials and methods are discussed, followed by the Hybridization of ABC with PSO in Sect. 3 for first level classification. In Sect. 4, the further classification with the help of MRE is done and the results are discussed in Sect. 5. Section 6 gives the consolidated conclusion of the work.

2 Materials and Methods

In this paper, for nearly 20 patients who are suffering from epilepsy, the EEG dataset has been acquired through clinical EEG monitoring system from Sri Ramakrishna Hospital, Coimbatore. According to the standard 10–20 International System, the 16 channel electrodes are kept on the patient's scalp and the recordings are obtained for different stages like awake stage, sleeping stage, eyes open and closed stage, resting stage and random muscle movement stage. The data is obtained in European Data Format (EDF). With the help of neurologist, the recordings of EEG which had the most distinct features were selected. Certain types of artifacts like eye blinks and motion artifacts too are present in the recorded EEG signals. For the sake of the amplification of the EEG signals, versatile EEG machines were deployed. For more than 55 min, the recordings are done and each recording is split into epochs for computational ease. The block diagram of the work is shown in Fig. 1.

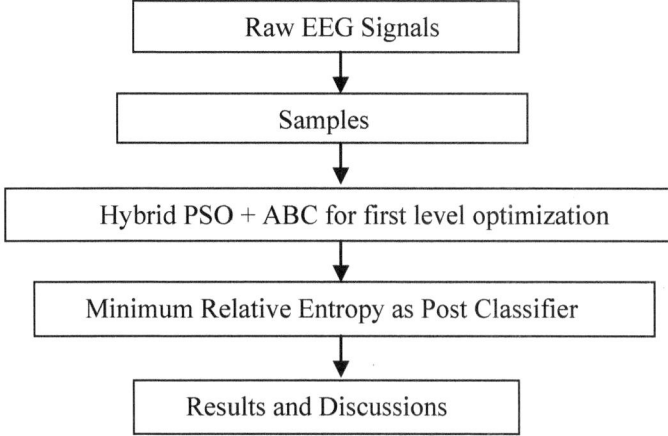

Fig. 1 Block diagram of the work

3 Hybridization of PSO with ABC

A hybrid algorithm of PSO and ABC is constructed here. The algorithm is initiated here by the building of random and initial personal best positions for both bees and particles. Using the PSO_PHASE stage, the personal best can be improved in the main loop [14]. Once the termination of the main loop is done, the recent personal best position is scanned by the algorithm and returns the best value among them as the global best. Initially the description about PSO_PHASE technique and ABC_PHASE technique is discussed followed by the Hybridization of the PSO with ABC [14]. Then the actual hybrid model of PSO_PHASE (M) and ABC_PHASE (Z) is discussed.

3.1 PSO_PHASE Technique

The following steps are involved in the PSO_PHASE technique:

(a) A list of best positions of particles is considered initially and then it is updated with better positions and finally returns back.
(b) For each particle, a random and new current position is assigned which is totally in contrast to the personal best.
(c) A fixed *niterpso* times is used to run the main loop of the method.
(d) The most of the technique behaves like the normal PSO algorithm [15], except the output.
(e) PSO algorithm always returns a single global best position when PSO_PHASE gives the modified version of the personal bests back.

3.2   ABC_PHASE Technique

The following steps are involved in ABC_PHASE technique:

(a) As a parameter, an initial list of employed bee position is considered, then it is improved and finally returns them back.
(b) In the PSO and ABC hybridization, this particle list of positions is obtained from the personal best of the previous PSO_PHASE.
(c) Such a task is performed to make sure that the incremental improvement of the same list of personal best positions is done by PSO_PHASE and ABC_PHASE.
(d) A fixed *niterabc* times is used to run the main loop of the method.
(e) The rest of the technique behaves like the normal ABC algorithm [16] except for the outputs.
(f) While the list of present food source positions is being returned by ABC_PHASE, the global best which represents the best of the food sources is returned by the ABC algorithm.

3.3   Hybridization of PSO with ABC

The hybridization of PSO with ABC [14] is done as follows:

PSOABC (*h*, **assessmentcount**, *l*)

(1) //inputs are *h*, function to be maximized, *maxassessment*; It represents the total number of assessments which is allowed by the algorithm before it breaks out the main loop, *l* = no of particles/number of food sources
(2) //output is represented by '*q*' which represents the coordinate which has the best fitness value in the entire search space
(3) To the variable assessment count, the value of 0 is assigned
(4) By calling the function *h* and where the fitness is obtained, incrementing the *assessmentcount* by one is done
(5) A random initial position is assigned for each employed bee or particle
(6) The initial position is assigned as the personal best for each particle or employed bee
(7) A set of personal best value is represented as *P*
(8) The following condition is implemented;
     Repeat while
     Assessmentcount < maxassessment
     (a) The personal best is improved using PSO_PHASE as shown; P <- PSO_PHASE (P)
     (b) Using ABC_PHASE, the personal best is improved as shown in; P <- ABC_PHASE (P)
(9) The set of personal bests is scanned and the one with the best fitness value which represents the global best '*q*' is returned.

3.4   PSO_PHASE (M)

The PSO_PHASE (M) [14] is explained as follows:

(a) //input: $M = \{m_1, m_2, \ldots\}$ represented as initial best positions of the particle
(b) //output: $M = \{m_1, m_2, \ldots\}$ represented as final best position of the particle
(c) For each and every particle
    (i)   A random position y is assigned
    (ii) If y has a better fitness value than m, then the personal best is updated
    (iii) A random velocity v is also assigned
(d) The position with the best fitness is selected as global best q
(e) It is repeated while the iteration is done less than niterpso. For every particle
    (i)   The new velocity v is calculated
    (ii) The new position y is calculated using $y = y + v$
    (iii) If the new position 'y' has a better fitness than the personal best m, then y is assigned to P
    (iv) If the new position y is found to have a better fitness value than the global best q, then y is assigned to q
    (v)   A set of particle best positions M is returned finally.

3.5   ABC_PHASE (Z)

The ABC_PHASE (Z) [14] is explained as follows:

(a) //input: $Z = \{z_1, z_2, \ldots\}$ (the initial employed bee positions)
(b) //output: $Z = \{z_1, z_2, \ldots\}$ (the final employed bee positions)
(c) While the iteration is done less than *niterabc*:
//Employed Bee Phase
For each and every food source
    (i)   The position x of the employed bee is tweaked to get a new position
    (ii) The new position is accepted if it has a better fitness value
    (iii) Or else the trial value of the food source is increased by one.

    //Dance Phase:

(i)   A probability 'r' of being selected by the onlooker bees is assigned to every employed bee here.
(ii) The value of 'r' should be in a manner such that the bee which is employed with a particular position has a better fitness value than the probability value.
//Onlooker Bee Phase:

(i)   For each bee which is employed and returned to the hive and it is performed a dance.
(ii) With a certain probability r, onlooker bee visits the food source.
(iii) The fitness of a particular position is checked by the onlooker bee in the neighborhood premises of the employed bee positions.

(iv) The new positions is accepted if it is better and therefore onlooker bee is turned into an employed bee
(v) Or else the trial no. of the food source is increased by one.

//Scout bee phase:

(i) For every position of the food source, the trial number always exceeds the limit parameters. A new position is randomly determined by a scout bee and the neighborhood is chosen as the food source.
(ii) Therefore the scout bee is turned easily into and employed bee
(iii) Return Z.

4 Minimum Relative Entropy as the Post Classifier

The values optimized through the hybrid model of PSO and ABC is then classified with the help of Minimum Relative Entropy. Due to the non Gaussian, non linear and non stationary nature of the EEG signals, they are highly complicated. In the recent years, an information theoretic approach to pattern recognition has been widely used. For the recognition criteria, two major concepts have been widely used known as relative entropy and Shannon's entropy [17]. The discrepancy between similar patterns can be explored with the help of relative entropy. Minimizing the relative entropy for a similar group of patterns is important so that an optimum criterion can be easily obtained.

The representation of MRE optimization is explained below. For the sources A and B, let c_a and d_b represents the probability measures. The relative entropy distance denoted as $H(B\|A)$ is defined as

$$H(B\|A) = \sum_b d_b(y) \log \frac{d_b(y)}{c_a(y)}$$

where $H(B\|A)$ represents the non-negative continuous function. It will be equal to zero if c_a and d_b coincide together. Thus $H(B\|A)$ is seen as a distance in between the measures c_a and d_b. However, $H(.\|.)$ is not considered as a metric as it neither satisfies the triangle inequality nor symmetric. $H(B\|A)$ can be equal to zero and it is quite an easy condition to observe. The conditional entropy rate $G(B|A)$ should be large always. Therefore it serves as an optimizer for clinical decisions where the information distance is dependent on relative entropy. Let $J = [P_{tw}]$ be considered as the co-occurrence matrix with (t, w) elements which depicts the ABC-PSO based epilepsy risk levels for a single epoch. As there are $(16 \times 3 = 48$ epochs) available, Minimum Relative Entropy (MRE) can be optimized using three stage process.

(a) The 16×3 matrix epilepsy risk levels are deduced into 16×1 by means of row wise optimization through MRE.
(b) The 16×1 matrix is deduced into 4×1 matrix by means of column wise optimization.
(c) The 4×1 matrix is deduced into a single and optimum epilepsy risk level.

Stage 1:

(1) To avoid, the padding of the 16×3 matrix elements is done with the same elements

(2) Finding out $C(t, w)$ relative entropy of (t, w)th element in the $J(t, w)$ matrix through 4 neighborhoods.

$$C_{t,1}(t, w) = C(t - 1, w) + C(t + 1, w) + C(t, w + 1) + C(t, w - 1)$$

where

$$C(t - 1, w) = V_{t-1} \ln(V_{t-1}/V_t)$$

(3) Likewise the following are found out $C_{t,2}(t, w + 1), C_{t,3}(t, w - 1)$ and find $\min\left(C_{t,1}(t, w), C_{t,2}(t, w + 2), C_{t,3}(t, w - 1)\right)$

Now the conversion of row of three elements to a single element is done and the value of $\min(C(t, w))$ is replaced with original probability values. For all the 16 rows, it is repeated and the nature is reduced into 16×1 matrix.

Stage 2:

(1) The 16×1 matrix is grouped into 4 co-occurrence matrices of 4×1.

(2) The relative entropy is found using adjacent neighborhood of the $(t, 1)$ element as

$$C(t) = C(t + 1) + C(t - 1)$$
$$C(t + 1) = C(t) + C(t + 2)$$
$$C(t - 1) = C(t) + C(t - 2)$$

(3) A minimum $\{C(t), C(t + 1), C(t - 1)\}$ for a member in that specific group is found out.

(4) A minimum relative entropy is found out in a similar fashion for other members in that group.

(5) Hence the minimum points will be four and so the least min in the group is found out and so likely 4×1 matrices are obtained.

Stage 3:

The stage 3 is repeated and so the reduction of the 4×1 matrix into a single optimum value is done which depicts the optimum and best epilepsy risk level.

5 Results and Discussion

With a first level optimization performed with the help of hybrid ABC + PSO technique it is further classified with Minimum Relative Entropy Classifier, on the basis of parameters like Accuracy, Sensitivity, Specificity, Performance Index, Time Delay and

Quality Values the average results are computed in Table 1 respectively. The Figs. 2, 3 and 4 shows the specificity versus sensitivity analysis, time delay versus quality value analysis and performance index versus accuracy analysis respectively. The mathematical formulae for the parameters like Performance Index (PI), Sensitivity, Specificity and Accuracy are expressed as follows

$$PI = \left(\frac{PC - MC - FA}{PC} \right) \times 100$$

where PC denotes the Perfect Classification, MC indicates the Missed Classification and FA explains the False Alarm. The Sensitivity, Specificity and Accuracy measures are mathematically formulated by the following

$$Sensitivity = \frac{PC}{PC + FA} \times 100$$
$$Specificity = \frac{PC}{PC + MC} \times 100$$
$$Accuracy = \frac{Sensitivity + Specificity}{2}$$

The Quality Value Q_V is defined as

$$Q_v = \frac{C}{(R_{fa} + 0.2) * (T_{dly} * P_{dct} + 6 * P_{msd})}$$

where C denotes the scaling constant,
R_{fa} specifies the number of false alarm per set,
T_{dly} indicates the average delay of the onset classification in seconds
P_{dct} shows the percentage of perfect classification and
P_{msd} expresses the percentage of perfect risk level missed.

The time delay is given as follows

$$Time \quad Delay = \left[2 \times \frac{PC}{100} + 6 \times \frac{MC}{100} \right]$$

Table 1 Consolidated final result analysis with MRE classifier

Name of the parameter	Average
PC (%)	95.21
MC (%)	2.70
FA (%)	1.87
PI (%)	95.25
Specificity (%)	97.29
Sensitivity (%)	98.12
Time delay (sec)	2.06
Quality values	22.24
Accuracy (%)	97.70

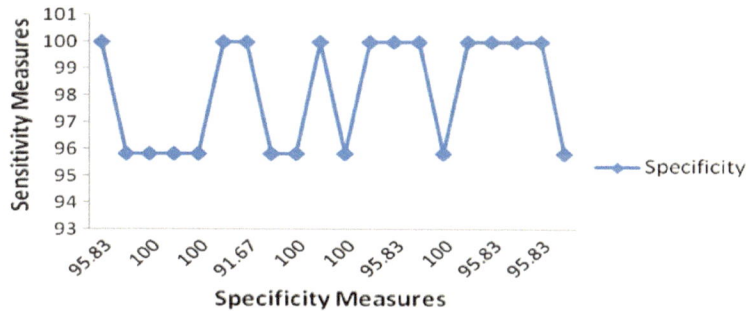

Fig. 2 Specificity and sensitivity measures

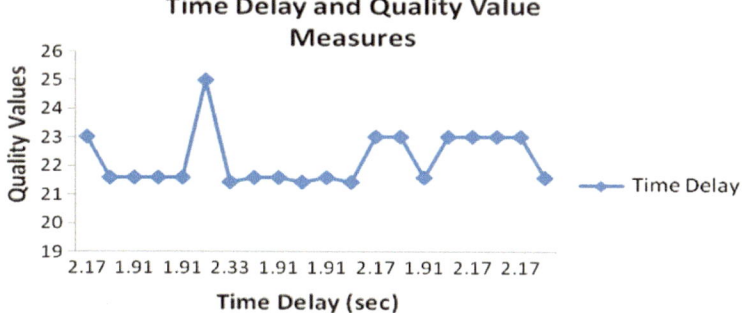

Fig. 3 Time delay and quality value measures

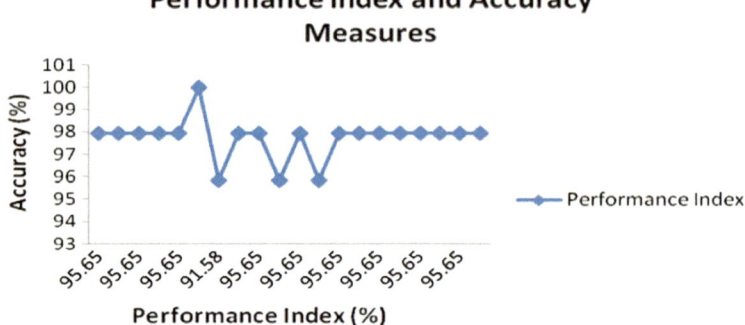

Fig. 4 Performance index and accuracy measures

6   Conclusion

In this work, the raw EEG signals are considered and then sampled. It is initially optimized with the help of PSO and ABC optimization technique and then it is further classified with the help of Minimum Relative Entropy as a post classifier. Results show that an average perfect classification of about 95.41% is obtained, average missed classification of about 2.70% is obtained, average false alarm of about 1.87% is obtained, average accuracy of about 97.70% is obtained along with an average quality value of about 22.24. Future works deal with the modification in the Minimum Relative Entropy post classifier for the better classification of epilepsy risk levels from EEG signals.

References

1. Prabhakar, S.K, Rajaguru, H.: Development of patient remote monitoring system for epilepsy classification. In: 16th International Conference on Biomedical Engineering (ICBME), Singapore, 7–10 December 2016 (2016)
2. Prabhakar, S.K, Rajaguru, H.: Efficient wireless system for telemedicine application with Reduced PAPR using QMF based PTS technique for epilepsy classification from EEG signals. In: IFBME Proceedings (Springer), International Conference on Advancements of Medicine and Health Care through Technology (MEDITECH), 12–15 October 2016, Romania (2016)
3. Prabhakar, S.K, Rajaguru, H.: Entropy based PAPR reduction for STTC system utilized for classification of epilepsy from EEG signals using PSD and SVM. In: IFBME Proceedings (Springer), 3rd International Conference on Movement, Health and Exercise (MoHE), 28–30 September 2016, Malaysia (2016)
4. Prabhakar, S.K, Rajaguru, H.: Factorization and particle swarm based sparse representation classifier for epilepsy classification implemented for wireless telemedicine applications. In: IFBME Proceedings (Springer), 6th International Conference on the Development of Biomedical Engineering, Ho Chi Minh City, Vietnam, pp. 474–478 (2016)

5. Jaiswal, A.K., Banka, H.: Local pattern transformation based feature extraction techniques for classification of epileptic EEG signals. Biomed. Signal Process. Control **34**, 81–92 (2017)
6. Niknazar, M., Mousavi, S.R., Vahdat, B.V., Sayyah, M.: A new framework based on recurrence quantification analysis for epileptic seizure detection. IEEE J. Biomed. Health Inf. **17**, 572–578 (2013)
7. Peker, M., Sen, B., Delen, D.: A novel method for automated diagnosis of epilepsy using complex-valued classifiers. IEEE J. Biomed. Health Inf. **20**, 108–118 (2016)
8. Kaleem, M., Guergachi, A., Krishnan, S.: EEG seizure detection and epilepsy diagnosis using a novel variation of empirical mode decomposition. In: Proceedings of the 2013 35th Annual International Conference of the IEEE Engineering in Medicine and Biology Society (EMBC), Osaka, Japan, 3–7 July 2013, pp. 4314–4317
9. Gajic, D., Djurovic, Z., Gennaro, S.D., Gustafsson, F.: Classification of eeg signals for detection of epileptic seizures based on wavelets and statistical pattern recognition. Biomed. Eng. Appl. Basis Commun. **26**, 1450021 (2014)
10. Kumar, S.P., Sriraam, N., Benakop, P.G., Jinaga, B.C.: Entropies based detection of epileptic seizures with artificial neural network classifiers. Expert Syst. Appl. **37**, 3284–3291 (2010)
11. Shen, C.P., Chen, C.C., Hsieh, S.L., Chen, W.H., Chen, J.M., Chen, C.M., Lai, F., Chiu, M. J.: High-performance seizure detection system using a wavelet-approximate entropy-fsvm cascade with clinical validation. Clin. EEG Neurosci. **44**, 247–256 (2013)
12. Gajic, D., Djurovic, Z., Gligonjevic, J., Gennaro, S.D., Gajic, I.S.: Detection of epileptiform activity in eeg signals based on time-frequency and non-linear analysis. Front. Comput. Neurosci. **9**, 38 (2015). https://doi.org/10.3389/fncom.2015.00038
13. Lee, S.H., Lim, J.S., Kim, J.K., Yang, J., Lee, Y.: Classification of normal and epileptic seizure EEG signals using wavelet transform, phase-space reconstruction, and euclidean distance. Comput. Methods Progr. Biomed. **116**, 10–25 (2014)
14. Altun, O., Korkmaz, T.: Particle Swarm Optimization-Artificial Bee Colony Chain (PSOABCC): a hybrid metaheuristic algorithm. In: Scientific Cooperation International Workshop on Electrical and Computer Engineering Subfields, pp. 42–49 (2014)
15. James, E., Russell, K.: Particle swarm optimization. In: 1995 IEEE International Conference on Neural Networks, vol. 4, pp. 1942–1948 (1995)
16. Yan, X., Yunlong, Z., Wenping, Z.: A hybrid artificial bee colony algorithm for numerical function optimization. In: 11th International Conference on Hybrid Intelligent Systems (HIS) (2011)
17. Sukanesh, R., Harikumar, R.: Minimum Relative Entropy (MRE) method for fuzzy based classification of epilepsy risk levels from EEG signals. In: ICBPE 2006, pp. 93–98 (2006)

Fuzzy Optimization with Modified Adaboost Classifier for Epilepsy Classification from EEG Signals

Harikumar Rajaguru$^{(\boxtimes)}$ and Sunil Kumar Prabhakar

Department of ECE, Bannari Amman Institute of Technology,
Sathyamangalam, India
harikumarrajaguru@gmail.com

Abstract. Epilepsy is recognized as a chronic neurological condition with the occurrence of recurrent seizures that alters the normal electrical activities in the neurons of the brain. Epilepsy can occur to any person irrespective of time and age as the exact cause of epilepsy is very difficult to know. For the diagnosis of epilepsy, Electroencephalograph (EEG) signals are widely used. The EEG signals are non-stationary and non-linear sequences of data which can be easily traced and detected when the electrodes are placed on the scalp of the patient. In this work, fuzzy optimization is used as a first level classifier to classify the epilepsy risk levels and then for second level classification, Adaboost Classifier and the Modified Adaboost Classifier are used for classification of epilepsy risk levels from EEG signals. Modified Adaboost Classifier is done through the implementation of Linear Discriminant Analysis (LDA) to Adaboost Classifier thereby enhancing the performance of the classifier. Results show that an average accuracy of 96.68% and an average quality value of 21.77 are obtained when the Modified Adaboost Classifier is used and an average accuracy of 97.20% and an average quality value of 22.51 is obtained when the Adaboost Classifier is used for the classification of epilepsy risk levels from EEG signals.

Keywords: Epilepsy · EEG · Adaboost · LDA

1 Introduction

One of the serious neurological disorders affecting the human brain is epilepsy which is recognized by recurrent seizures [1]. Due to the abnormal electrical activities happening in the brain, seizures arise. Therefore seizures are a type of neurological dysfunction which terribly affects the life of the patient. Seizures are usually followed by a high level of unconsciousness and muscle cramps, which may lead to traumatic sequences [2]. The greatest boon to detect the epileptic seizures is done through EEG [3]. It is the most commonly used technique for detecting the epileptic seizures. An EEG recording unit comprises of a set of electrodes which helps to take the potential to the amplifier from the scalp, a basic storage unit to have all the EEG recordings stored for analysis in future, and a

© Springer International Publishing AG 2018
D. J. Hemanth and S. Smys (eds.), *Computational Vision and Bio Inspired Computing*,
Lecture Notes in Computational Vision and Biomechanics 28,
https://doi.org/10.1007/978-3-319-71767-8_52

display unit from which one can visualize things easily [4]. During a seizure, the EEG signal readings from epileptic patients are quite different from the normal brain signals and it varies drastically in both frequency and time. The channel electrodes which are located on the scalp records the abnormal electrical activities of the cortical regions of the brain. As EEG has a high resolution, which produces result in real time, it is one of the easiest techniques to detect the brain disorders [5]. The potential difference between an active electrode which is located at the place of neural activity with a reference electrode can be easily measured with the help of EEG. Some of the most important works in EEG signal processing and epilepsy classification are discussed here. By using an automated multi-channel algorithm, the high inter-reviewer variability of spike detection on intracranial EEG was addressed by Barkmeier et al. [6]. The epileptiform discharges that were mimicked by EEG patterns but do not have any association with seizures was done by Pedley [7]. The spatiotemporal scales and the micro seizures of human partial epilepsy was done by Stead et al. [8]. For the laterization of epileptogenetic hippocampus in Magnetic Resonance images, the Support Vector Machines (SVM) with non linear kernel optimization was used by Nazem-Zadeh et al. [9]. Independent Component Analysis (ICA) algorithm was used by Arunkumar et al. for the automatic detection of epileptic seizures [10]. With the help of Discrete Wavelet Transform (DWT) and ICA along with neural networks, the detection of the epileptic seizures was done by Mercy [11]. An adaptive thresholding technique was used by Hopfeng et al. for the automatic detection of seizures in long term scalp EEG recordings [12]. The epileptic seizures were detected automatically in long-term EEG records by Correa et al. [13].

In this work, the fuzzy optimization is used as a level one classifier and it is further classified with the help of Adaboost and Modified Adaboost Classifier. The organization of the paper is as follows. In Sect. 2, the materials and methods are discussed followed by the fuzzy optimization of epilepsy risk levels in Sect. 3. Section 4 gives the details about the Adaboost and Modified Adaboost Classifier and it is followed by results and discussion in Sect. 5 and ended with conclusion in Sect. 6.

2 Materials and Methods

The standard 10–20 International system was used for the acquisition of EEG data from the Department of Neurology, Sri Ramakrishna Hospital, Coimbatore, India. The 16 channel electrodes were placed on the scalp of the patient and for all the 16 channels of EEG data the recordings were done simultaneously and it was for more than 55 min. The recordings of the EEG signals were done for various stages like resting period, deep sleep period, eyes open and closed period, awake period, random muscle movement period and hyperventilation stage. The EEG signals were amplified with the help of EEG machines. For the sake of easy computation, the entire EEG recordings are broken into sections or epochs. Then with the help of neurologist, the recordings of the EEG which had the best and distinct features were selected. Approximately 1% of the entire EEG data had artifacts like eye blink, chewing and motion artifacts. To find out the important changes happening in the activities of the signal, a two second epoch is found out. With the help of graphics programming, each epoch is sampled at a frequency range of 200 Hz. For the quantification of EEG signals, different parameters are used and it is computed using the amplitude

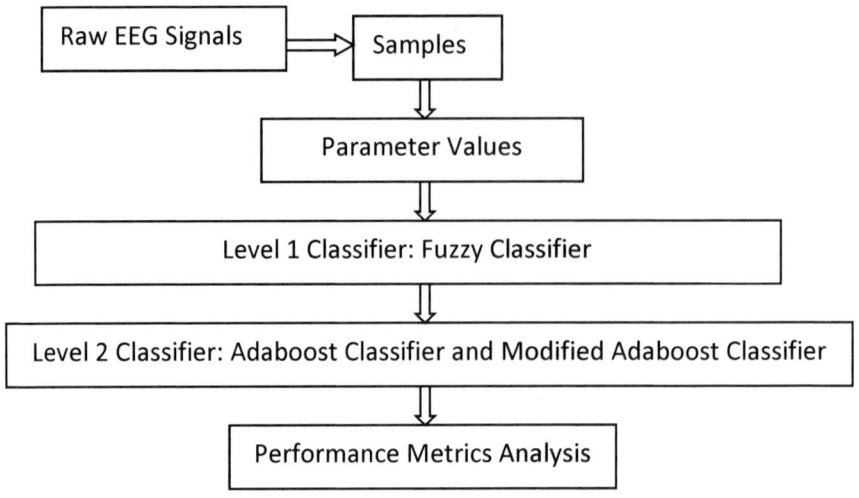

Fig. 1 Block diagram of the work

values of the EEG signals with the help of appropriate programming codes. The Fig. 1
shows the block diagram of the work.

3   Fuzzy Optimization as a Pre Classifier

For handling noisy and imprecise data, fuzzy set theory is very effective [14]. Moreover, a linguistic approach can be used in fuzzy set theory, thereby a good approximation to medical contexts can be sorted out [15]. The fuzzy system designed here
accomplishes the following tasks:

(a) At each channel, the fuzzy classification for epilepsy risk levels is done
(b) The optimization is done for each channel result because they are at different risk levels
(c) After the analysis of optimization, the performance of fuzzy classification is determined.

The following parameters are derived from EEG signals

(1) Energy: The energy is represented as $E = \sum_{i=1}^{n} q_i^2$, where q_i is the sample value
     of the signal and the number of samples is denoted as n. Only the normalized
     energy should be considered here.
(2) Spikes and Sharp waves: The former is detected when the zero crossing duration of
     very high amplitude peaks lies between 20 to 70 ms and the latter is detected when
     the zero crossing duration of very high amplitude peaks lie between 70 and 200 ms.
(3) The number of peaks (both positive and negative) which exceeds a threshold is
     found out.
(4) The events are recorded as the total number of sharp and spike waves.
(5) The variance is mathematically computed as σ and is given by $\sigma^2 = \sum_{i=1}^{n} \dfrac{(q_i - \mu)^2}{n}$,

where $\mu = \frac{\sum_{i=1}^{n} q_i}{n}$ represent the amplitude (average) of the epoch.

(6) The average duration is represented as $D = \frac{\sum_{i=1}^{l} m_i}{l}$, where m_i represents one peak to peak duration and l represents the total number of such durations.

(7) The covariance duration is mathematically expressed as $CD = \sum_{i=1}^{l} \frac{(D-m_i)^2}{lD^2}$.

3.1   Membership Functions of Fuzzy

The comparison of energy with six other input features is done in order to give six outputs. Every input feature is classified totally into 5 different fuzzy linguistic levels such as very low, low, medium, high and very high. The type of membership function used here is triangular for every input variable [16]. The classification of the output risk levels is done into 5 linguistic levels as normal, low, medium, high and very high.

3.2   Rule Set of Fuzzy

In the following format the rules are framed.
IF the variance is low AND the Average Duration is low, THEN the output risk level is low. Totally there are five linguistic levels of variance and five other linguistic levels of 6 features such as average duration, peaks, events, spike and sharp waves, covariance duration and energy. Depending on six sets which has 25 rules totally, a rule base of 150 rules is obtained. The type of fuzzy rule based system employed here is exhaustive. The training of the fuzzy system is done with the simulation of both input and output variables. Under a single category, nine inputs are present for every fuzzy set variable. As there are five sets such as very low, low, medium, high and very high, totally 45 test inputs are being simulated for all the inputs like average duration, energy etc.... The fuzzy system's sample output which contains the original patient readings shows that a low risk level is found in channel 1 and a high risk level is found in channel 7. In between the adjacent epochs, there is a huge variation in the risk level classification. Therefore Adaboost algorithm and Modified Adaboost algorithm are used to optimize the fuzzy result to get a best epilepsy risk level for each and every patient.

4   Adaboost as a Post Classifier for Epilepsy Risk Level Classification

The fuzzy optimized values are fed inside the Adaboost Classifier for further optimization and classification. Adaboost is a kind of ensemble learner where an overall strong classifier is made up of the joint decision rule of multiple weak classifiers [17]. An initial uniform weight $W_1(i) = 1/n$ is assigned by Adaboost to each and every training sample x_i, where the number of training samples is represented as n. At a particular iteration 'f', the Adaboost is used to find the classifier g_k trained with the help of samples x_i so that the weighted error E_f is minimized according to $W_f(i)$. The updation of the weights is done with the help of the following equation

$$W_{f+1}(i) = \frac{W_f(i).\exp\left(-\alpha_f z_i g_k(x_i)\right)}{Q_f}$$

where

$$\alpha_f = \frac{1}{2}\log\left[\frac{1-E_f}{E_f}\right],$$

z_i is nothing but the ground truth tables and the normalization factor is denoted as Q_f. Based on these formulae, the weights for the perfectly classified samples are increased and there is a decrease in the weights for incorrectly classified samples. This channel permits the Adaboost to focus more on the hectic and information training samples. The classifier which results because of the Adaboost algorithm is expressed as

$$G(x) = sign\left(\sum_{t=1}^{T}\alpha_t g_t(x)\right)$$

where g_t represents the hypothesis of the tth weak classifier and $G(x)$ represents the hypothesis of the storage classifier. The combination of Adaboost with a very simple threshold classifier is done so that the weighted error on a particular feature is minimized. At a particular iteration f, the features are selected by Adaboost so that the weighted error is minimized. $W_i(f+1)$ is increased when x_i is perfectly classified by g_k and vice versa.

4.1   Application of Linear Discriminant Analysis to Adaboost Classifier

LDA helps to maximize the ratio of 'between-class scatter' to 'within-class scatter'. For every individual class, the variance of the projections are measured by the between-class scatter. From all the data samples, the variances of the projections are measured with the help of within-class measure. The feature combination which gives the highest separations in between the classes is being employed by LDA [18]. The weights are not taken into consideration by the traditional LDA implementation. Assuming there are totally 'J' training samples which are equally split into C classes, $P_1, \ldots, P_c$. The standard LDA formulae for the between class scatter P_B and within class scatter P_W is expressed as

$$P_B = \sum_{d=1}^{C} J_d(\mu_d - \bar{x})(\mu_d - \bar{x})$$

$$P_W = \sum_{d=1}^{C}\sum_{i \in P_d}(x_i - \mu_d)(x_i - \mu_d)^T$$

$$\mu_d = \frac{1}{J_d} \sum_{i \in P_d} x_i$$

$$\bar{x} = \frac{1}{J} \sum_{i=1}^{J} x_i = \frac{1}{J} \sum_{d=1}^{C} J_d \mu_d$$

where J_d represents the total number of cases in class d. The LDA criterion function can be expressed as

$$L(w) = \frac{w^T P_B w}{w^T P_w w}$$

The vector w is found out by LDA where the $L(w)$ is maximized. A linear transformation to the data is applied by this vector. The separation in between the classes is maximized after this specific transformation. A threshold w_0 is mandatory in order to have a complete classifier.

4.2 Weighted LDA

To use the LDA with Adaboost, the formulae of LDA is modified to get the best weights. With this combination, the selection of the best feature combination at each boosting iteration can be obtained easily. The scatter matrices are reformulated for LDA with sample weights as follows

$$P'_B = \sum_{d=1}^{C} \left(\sum_{i \in P_d} S(i) \right) (\mu'_d - \bar{x}') (\mu'_d - \bar{x}')^T$$

$$P'_W = \sum_{d=1}^{C} \sum_{i \in P_d} S(i)(x_i - \mu'd)(x_i - \mu'_d)^T$$

$$\mu'_d = \frac{1}{\sum_{i \in P_d} S(i)} \sum_{i \in P_d} S(i)x_i$$

$$\bar{x}' = \sum_{i \in P_d} S(i)x_i$$

where $S(i)$ is the parameter weight which is assigned to each of the ith training sample with $\sum_{i=1}^{J} S(i) = 1$ in the Adaboost Algorithm [19]. The feature correlation is made effective use in this case as with Weighted LDA, combinations can be done easily. A WLDA classifier is trained with all possible combination with many fuzzy features at each iterations. The selection is done in such a manner that the weighted error is totally minimized in the entire feature combination. At each iteration, the updation of the sample weights are done.

5   Results and Discussion

The fuzzy optimization is used as a pre classifier and the post classifier used here is Adaboost and Modified Adaboost Classifier which is dependent on Linear Discriminant Analysis and on the basis of parameters like Time Delay, Accuracy, Sensitivity, Specificity, Quality Values and Performance Index the average results are compared and shown in Table 1 respectively. The mathematical expression for the parameters like Performance Index (PI), Sensitivity, Specificity and Accuracy are expressed as follows

$$PI = \left(\frac{PC - MC - FA}{PC}\right) \times 100$$

where PC means the Perfect Classification, MC denotes the Missed Classification and FA abbreviates the False Alarm. The Sensitivity, Specificity and Accuracy measures are mathematically expressed by the following

$$Sensitivity = \frac{PC}{PC + FA} \times 100$$

$$Specificity = \frac{PC}{PC + MC} \times 100$$

$$Accuracy = \frac{Sensitivity + Specificity}{2}$$

The Quality Value Q_V is defined as

$$Q_v = \frac{C}{(R_{fa} + 0.2) * (T_{dly} * P_{dct} + 6 * P_{msd})}$$

where C denotes the scaling constant,
R_{fa} shows the number of false alarm per set,
T_{dly} specifies the average delay of the onset classification in seconds
P_{dct} indicates the percentage of perfect classification and
P_{msd} denotes the percentage of perfect risk level missed
The time delay is expressed mathematically as follows

$$Time\ Delay = \left[2 \times \frac{PC}{100} + 6 \times \frac{MC}{100}\right]$$

Table 1 Performance Comparison of Adaboost and Modified Adaboost Classifier

Parameters	Adaboost Classifier	Modified Adaboost Classifier
PC (%)	94.99	94.16
MC (%)	2.91	2.91
FA (%)	2.08	2.91
PI (%)	94.57	93.67
Sensitivity (%)	97.31	96.28
Specificity (%)	97.08	97.08
Time delay (sec)	2.07	2.06
Quality value	22.51	21.77
Accuracy (%)	97.20	96.68

6 Conclusion

The temporary impairment of the electrical activity of functions of brain causes epileptic seizures. Therefore for the automatic epileptic seizure detection, EEG signal recordings are quite important. In this paper, fuzzy optimization is used as a pre classifier for epilepsy risk level classification. The fuzzy optimized values are further optimized using the post classifiers such as Adaboost Classifier and Modified Adaboost Classifier. The Modification of the Adaboost Classifier is done by means of the application of Linear Discriminant Analysis. Results show that the traditional Adaboost Classifier performed slightly better than the Modified Adaboost Classifier. A high classification accuracy of 97.20% is obtained along with a quality value of 22.51 when the combination of fuzzy optimization with Adaboost Classifier is used. A slightly less classification accuracy of 96.68% is obtained along with a quality value of 21.77 when the combination of fuzzy optimization with Modified Adaboost Classifier is used. Future works is to make a lot of modifications in the fuzzy optimization concept and the Adaboost Algorithm to obtain a very high classification accuracy and quality value.

References

1. Prabhakar, S.K., Rajaguru, H.: GMM better than SRC for classifying epilepsy risk levels from EEG signals. In: Proceedings of the International Conference on Intelligent Informatics and BioMedical Sciences (ICIIBMS), Okinawa, Japan, 28–30 Nov (2015)
2. Prabhakar, S.K., Rajaguru, H.: Cascaded feed forward neural networks and generalized regression for epilepsy risk level classification—a study. In: Proceedings of the 3rd MEC International Conference on Big Data and Smart City (ICBDSC), Muscat, Oman, 15–16 Mar 2016
3. Prabhakar, S.K., Rajaguru, H.: Factorization and particle swarm based sparse representation classifier for epilepsy classification implemented for wireless telemedicine applications. In: IFBME Proceedings, 6th International Conference on the Development of Biomedical Engineering, Springer, pp. 474–478, Ho Chi Minh City, Vietnam (2016)

4. Prabhakar, S.K., Rajaguru, H.: Entropy based PAPR reduction for STTC system utilized for classification of epilepsy from EEG signals using PSD and SVM. In: IFBME Proceedings, 3rd International Conference on Movement, Health and Exercise (MoHE), Springer, Malaysia, 28–30 Sept 2016

5. Prabhakar, S.K., Rajaguru, H.: Wireless systems with reduced PAPR using K-means modified PTS implemented for epilepsy classification from EEG signals. In: IFBME Proceedings, International Conference on Advancements of Medicine and Health Care through Technology (MEDITECH), Springer, Romania, 12–15 Oct 2016

6. Barkmeier, D.T., Shah, A.K., Flanagan, D., Atkinson, M.D., Agarwal, R., Fuerst, D.R., Jafari-Khouzani, K., Loeb, J.A.: High inter-reviewer variability of spike detection on intracranial eeg addressed by an automated multi-channel algorithm. Clin. Neurophysiol. **123**(6), 1088–1095 (2012). Elsevier

7. Pedley, T.A.: EEG patterns that mimic epileptiform discharges but have no association with seizures. In: Current Clinical Neurophysiology: Update on EEG and Evoked potentials. Elsevier/North Holland, New York (1980)

8. Stead, M., Bower, M., Brinkmann, B.H., Lee, K., Marsh, W., Meyer, F.B., Litt, B., Van, G., Worrell, G.: Microseizures and the spatiotemporal scales of human partial epilepsy. Brain p. awq190 (2010)

9. Nazem-Zadeh, M.-R., Schwalb, J.M., Bagher-Ebadian, H., Mahmoudi, F., Hosseini, M.-P., Jafari-Khouzani, K., Elisevich, K.V., Soltanian-Zadeh, H.: Lateralization of temporal lobe epilepsy by imaging-based response-driven multinomial multivariate models. In: Proceeding of IEEE International Conference of Engineering in Medicine and Biology Society (EMBC), pp. 5595–5598. (2014)

10. Arunkumar, N., Balaji, V.S., Ramesh, S., Natarajan, S., Likhita, V.R., Sundari, S.: Automatic detection of epileptic seizures using independent component analysis algorithm. In: Proceeding of IEEE International Conference of Advances in Engineering, Science and Management (ICAESM), pp. 542–544. (2012)

11. Mercy, S.: Performance analysis of epileptic seizure detection using dwt & ica with neural networks. Int. J. Comput. Eng. Res. **2**(4), 1109–1113 (2012)

12. Hopfeng, R., Kasper, B.S., Graf, W., Gollwitzer, S., Kreiselmeyer, G., Stefan, H., Hamer, H.: Automatic seizure detection in long-term scalp eeg using an adaptive thresholding technique: a validation study for clinical routine. Clin. Neurophysiol. **125**(7), 1346–1352 (2014). Elsevier

13. Correa, A.G., Orosco, L., Diez, P., Laciar, E.: Automatic detection of epileptic seizures in long-term EEG records. Comput. Biol. Med. (2014) (Elsevier)

14. Harikumar, R., Sunil Kumar, P.: Fuzzy techniques and aggregation operators in classification of epilepsy risk levels for diabetic patients using EEG signals and cerebral blood flow. J. Biomater. Tissue Eng. **5**(4), 316–322 (2015)

15. Sukanesh, R., Harikumar, R.: A comprehensive analysis on post processing mathematical models (MRE, Aggregation operators & Soft Decision Trees) for patient specific fuzzy based epilepsy risk level classifier from EEG Signals. I.E. India J. Interdisc. panels **89**(2), 3–12 (2008)

16. Sukanesh, R., Harikumar, R.: Diagnosis and classification of epilepsy risk levels from EEG signals using fuzzy aggregation techniques. J. Eng. Lett. **14**(1), 90–95 (2007)

17. Yang, G., Wang, Z., Ren, J.: Facial expression recognition based on adaboost algorithm. Chinese J. Comput. Appl. **25**(4), 946–948 (2005)

18. Rajaguru, H., Prabhakar, S.K.: LDA, GA and SVM's for classification of epilepsy from EEG signals. Res. J. Pharm. Biol. Chem. Sci. **7**(3), 2044–2049 (2016)

19. Okada, K., Flores, A., Linguraru, M.G.: Boosting weighted linear discriminant analysis. Int. J. Adv. Stat. IT & C Econ. Life Sci. **2**, 1–10 (2010)

Application of Morphological Filtering with Modifications in Linear Discriminant Analysis Classifier for Epilepsy Classification from EEG Signals

Harikumar Rajaguru$^{(\boxtimes)}$ and Sunil Kumar Prabhakar

Department of ECE, Bannari Amman Institute of Technology, Coimbatore, India
harikumarrajaguru@gmail.com

Abstract. One of the most prominent neurological disorders posing a huge peril to the human community is epilepsy. Because of certain electrical disturbances happening in the function of brain, epilepsy occurs and it is characterized by recurrent seizures. Because of these epileptic seizures, both the physical and mental condition of the patient deteriorates thereby the patient is prone to more physical attacks and injury. Only if the seizures are detected and classified properly, then a good health care can be provided to the patients. For detection of the seizure activities, Electroencephalograph (EEG) signals are used. In this paper, morphological filtering concept is applied to the code converters which is obtained from processing EEG signals and it is employed as a preclassifier, and later it is post classified with Linear Discriminant Analysis (LDA), Log LDA (L-LDA) and Kernel LDA (K-LDA) classifiers. Results show that when LDA is used as a post classifier, an average classification accuracy of 97.39% along with an average quality value of 21.3 is obtained. Similarly if L-LDA is used as a post classifier, then an average classification accuracy of 96.87% along with an average quality value of 21.4 is obtained and when K-LDA is used as a post classifier, then an average classification accuracy of 96.45% along with an average quality value of 20.7 is obtained.

Keywords: Epilepsy · EEG · LDA · L-LDA · K-LDA

1 Introduction

Epilepsy is regarded as the second most common neurological disorder which affects about 3–5% of the whole world's population [1]. It is a neurological anarchy represented by continuous and impulsive versions of chaos in sensory parts of the brain [2]. The natures of the seizures are sometimes uncontrollable and are highly persistent. As a result of some unwanted firing in the cerebral neurons, the function of the brain get disturbed and altered heavily and so epilepsy occurs. In medical literature, there are

© Springer International Publishing AG 2018
D. J. Hemanth and S. Smys (eds.), *Computational Vision and Bio Inspired Computing*,
Lecture Notes in Computational Vision and Biomechanics 28,
https://doi.org/10.1007/978-3-319-71767-8_53

various techniques available for the detection of seizure related activities in the human brain. The widely used technique for the detection of electrical activities is EEG [3]. To measure the electrical impulses which travel through the scalp or the cortical regions of the brain, EEG is used as it is a non-invasive technique. Conventional techniques like Electrocardiograph (ECG), electro dermal machines and Epilepsy Monitoring Unit (EMU) are used to detect these seizures continuously but it is very expensive and consumes a lot of time and so EEG is preferred. Anti-epileptic drugs are not always helpful and for some patients the effect of it is highly unresponsive. To evaluate the activities of the human brain, EEG seems to be acting as an easy access clinical system tool and is highly informative. With the advancements and applications of programming and computers in medicine, automated seizure detection and classification systems have come into existence. A lot of algorithms have been developed for the thorough analysis and classification of the significant electrical activities of the brain. Various complicated steps such as signal acquisition, pre-processing of it, decomposition, filtering and finally the classification of those extracted features have to be used. A lot of works has been proposed in the literature for EEG signal processing and classification of epilepsy from EEG signals. Wavelet transforms was used by Garg and Narvey for the denoising and feature extraction of EEG signals [4]. A comprehensive review on various techniques for the extraction of brain signals for human machine interface was done by Bhatia and Sharma [5]. The Extreme Learning Machines and the non linear features were used for epileptic EEG signal classification by Yuan et al. [6]. The Hilbert marginal spectrum analysis was used for the automatic detection of seizures in EEG signals by Fu et al. [7]. For the non-stationary EEG signal classification, a noise robust analysis of sparse representation based classification technique was done by Shin et al. [8]. Empirical Mode Decomposition (EMD) and Support Vector Machines (SVM) are used for feature extraction and recognition of ictal EEG by Li et al. [9]. For the computer aided detection of seizures and diagnosis of epilepsy, a wavelet based EEG processing was done by Faust et al. [10]. The application of variable weight neural networks on the epilepsy seizure phase classification was done by Lam et al. [11].

In this paper, the morphological filtering concept is used and then it is classified with the help of three classifiers like LDA, L-LDA and K-LDA. The organization of the paper is as follows: In Sect. 2, the methods and materials are discussed with the application of Morphological Filtering and in Sect. 3, the post classifiers such as LDA, L-LDA and K-LDA are discussed. It is followed by results and discussion in Sect. 4 and ended with a conclusion in Sect. 5. The block diagram of the paper is shown in Fig. 1.

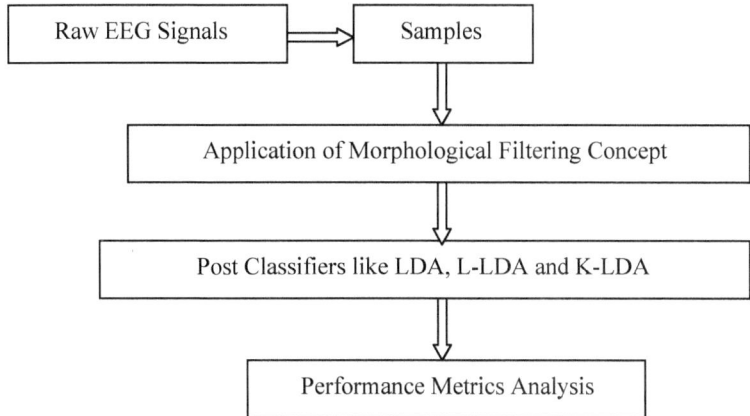

Fig. 1 Block diagram of the paper

2 Materials and Methods

For 20 epileptic patients who were under the evaluation and treatment for epilepsy at Sri Ramakrishna Hospital, Coimbatore, India, the recordings of EEG are obtained. The readings are obtained in European Data Format (EDF). Based on the standard 10–20 International System, the 16 channel electrodes were placed on the scalp of the patient and the EEG recordings are obtained. As the EEG recordings are continuous and have a duration of about 45 min each, it is divided into epochs which has two second duration. To detect the most important changes happening in the signal and to avoid redundancy, a 2 s epoch is used. As the maximum frequency of EEG is 50 Hz, each epoch of it is sampled at a frequency of 200 Hz. The instantaneous amplitude value of the signal represents each sample value and so totally for an epoch, 400 values are present.

The main aim of this work is to accomplish the following steps:

(1) With the help of morphological filtering, four parameters out of seven are extracted from EEG signals.
(2) For each channel the code converter risk level classification methodology is applied.
(3) Optimization of the code converter results are done with the help of classifiers like LDA, L-LDA and K-LDA.
(4) The performance analysis of code converter classification results along the further post classification results with the help of LDA, L-LDA and K-LDA classifiers are analyzed thoroughly.

The parameters such as Energy, positive and negative peaks, sharp waves, variance, spikes, average duration and covariance of duration are calculated accordingly [2].

2.1 Morphological Filtering of EEG Signals

To diagnose the epileptic disorder in a clinical manner, detection of spikes is quite significant. Generally, an experienced neurologist always detects the spikes in EEG recordings either visually or manually in a hospital. For long term EEG recordings, the process is difficult and time consuming and therefore spike detection technique in an automated sense is required. One of the highly useful techniques for the detection of spikes automatically in epileptic EEG signals is with the help of morphological filters. Detecting the spikes in EEG automatically is quite a hectic task because there is a huge variety in the morphology of spikes, its similarity to noise and artifacts or other background noise, and its different amplitude values. As there is a huge necessity to process an enormously large quantity of EEG data in real time scenarios, expressing their features like appearance, duration, frequency, shape and amplitude is important. A relatively large number of spike detection techniques was proposed in literature such as temporal approach, wavelet based approach and so on [12]. Since Morphological Filtering has quite a high percentage of detection, this method is chosen for detection of spikes.

2.2 Basic Operators of Mathematical Morphology

For a Euclidean space (G), assume that $b(t)$ and $d(t)$ be the subsets and $b(t)$ is considered as the one dimensional time series data and $d(t)$ is considered as the pre-defined structuring element. The domains of $b(t)$ and $d(t)$ are denoted as B and D respectively. $d^s(t) = d(-t)$ is assumed as the reflection of d which has a 180 degree rotation with respect to the origin. For one-dimensional time series data $b(t)$, the Minkowski addition and subtraction with functional structuring element $d(t)$ is defined as follows:

Addition:

$$(b \oplus d)(t) = \max_{\substack{t-u \in B \\ u \in B}} \{b(t-u) + d(u)\}$$

Subtraction:

$$(b \ominus d)(t) = \min_{\substack{t=u \in B \\ u \in D}} \{b(t-u) - d(u)\}$$

The opening and closing morphological filters are defined as follows:
Opening: $(b \circ d)(t) = [(b \ominus d^s) \oplus d](t)$
Closing: $(b \bullet d)(t) = [(b \oplus d^s) \ominus d](t)$ For the original signal $b(t)$, the smoothing of the convex peaks is done by the opening operation and for the signal $d(t)$, the smoothing of concave peaks is done by closing operator [13]. Thus for the detection

of peaks and valleys in the particular signal, the opening and closing operators can be applied.

2.3   Morphological Filtering for Spike Detection

Depending on the shape of the structuring element, the smooth extraction of the different parts of the signal can be done with the help of different operators. An important task here is to select the structuring element so that the spiky regions of the signal can be separated. A filter can be easily produced by the combination of morphological operators so that an original signal is separated into two signals, first signal is assessed by the structuring element and the second signal is assessed as the residue of the signal. Therefore selecting a good morphological filter is an important task. In an epileptic EEG signal, the spikes has both positive and negative phase. The morphological filters can be easily implemented with opening operators and then it is followed by closing operator or vice versa for the detection of bi-directional spikes.

Open-Closing Operator:

$$OC(b(t)) = b(t) \circ d_1(t) \bullet d_2(t)$$

Close-Opening Operator:

$$CO(b(t)) = b(t) \bullet d_1(t) \circ d_2(t)$$

The structuring elements are denoted as $d_1(t)$ and $d_2(t)$ respectively. The open-closing filter has a similar influence to that of the low pass filter. By measuring the differences between the input signal and its open-closing, a high pass filter can also be constructed. A lower amplitude is found when compared to the original signal as a result of open-closing operation and a higher amplitude is found when compared to the original signal where the close-opening operation is implemented. The amplitude's distortion can cause missing detection or pseudo positive detection if the detection of the spikes is selected by some threshold of the amplitude. Therefore it is quite essential to consider an average combination of both open-closing and close-opening operators so that exact and precise extraction of bi-directional spikes can be obtained.

$$OCCO(b(t)) = \frac{[OC(b(t)) + CO(b(t))]}{2}$$

The presentation of the original EEG signal $b(t)$ is expressed as follows:

$$b(t) = q(t) + OCCO(b(t))$$

Here the activities of the background are represented as $OCCO$. $q(t)$ denotes the spiky part or the transient activity of the signal. Spike is quite a transient and highly distinguishable pointed peak from background activity at a conventional paper speed and a duration ranging from 20 to below 70 ms approximately. As there are various spike waves with different frequency and amplitude, the adjustment of the structuring elements should be made to the proper size so that the extraction of the spike component can be at its best form. To clearly separate the components of the spike from the activity of the background, the structuring element pair $d_1(t)$ and $d_2(t)$ is understood as 2 parabolas and it is represented as

$$d_i(t) = r_i t^2 + s_i, i = 1, 2$$

The above mentioned elements will not fit into the spike waveform and it will fit into the shape of the background EEG waves.

Depending on the selection of the amplitude and width of $d_1(t)$ and $d_2(t)$, the background parts of the original signal and the transient parts can be separated, (i.e.), the selection of s_i and r_i, $i = 1, 2$

On the same signal and for all the signals of EEG recordings, the width and the amplitude of spikes can vary. Based on the spike wave characteristics, the selection of these parameters is done.

2.4   Code Converters as a Preclassifier

The sampled output values are processed through an encoding method and the net result is obtained as individual code. The outputs are encoded as a string of alphabets as working with definite alphabets is more much easier than working with numbers having large decimal accuracy [14]. The five different classifications of the outputs is represented alphabetically as shown in Table 1. The processing of characteristic representation eases the operation. For the five states, the risk level encoding is done and as a result a string of seven characters is obtained for all the 16 channels of every epoch. A sample output which has all the actual readings of the patient is depicted in Table 2 for 8 channels over 3 different epochs. The Performance Index of the code converters is mathematically expressed as

$$PI = \frac{PC - MI - FA}{PC} \times 100$$

Table 1 Alphabetic representation of the risk levels of epilepsy

Risk level classification	Alphabetic representation
Normal risk	U
Low risk	W
Medium risk	X
High risk	Y
Very high risk	Z

Table 2 The output of the code converters with morphological operators dependent feature extraction

Epoch-1	Epoch-2	Epoch-3
XYYWYXX	WXXWYYX	WZYXWWW
YZZYXXY	XYYYXXY	YYYXXYX
YYZXYYZ	YXYYYXY	YXXYXXY
YZZYXXY	YZXXYYX	YYYYYYY

For the code converters classifier, the Performance Index is as low as 33.26%. A highly non periodic pattern is emulated in the code converter output as so for the purpose of optimization, closed form of solutions will not work. As a result, in this paper, LDA, L-LDA and K-LDA are used for further optimization and for getting a better classification.

3 LDA, L-LDA and K-LDA as a Post Classifier

The morphologically filtered values are then fed inside the three post classifiers utilized here. In this section, LDA, L-LDA and K-LDA are utilized as post classifiers for epilepsy risk level classification from EEG signals.

3.1 LDA and Log LDA

Those vectors in the underlying space which best discriminates among classes is being explored by LDA [15]. If a number of independent features related to data is explained, then a linear combination of these data is created by LDA so that the largest mean differences is achieved between the desired classes. An optimal discrimination projection matrix Q_{opt} is solved by LDA and is represented as

$$Q_{opt} = \arg\max_q \frac{|Q^T R_b Q|}{|Q^T R_q Q|}$$

(or)

$$Q_{opt} = \arg\max_q \frac{|Q^T R_b Q|}{|Q^T R_t Q|}$$

where the scatter matrices are represented as R_b, R_q, and R_t respectively.

$$R_b = \sum_{i=1}^{K} n_i (\mu_i - \mu)(\mu_i - \mu)^T$$

$$R_q = \sum_i (y_i - \mu_{k_i})(y_i - \mu_{k_i})^T$$

where R_b denotes the between class scatter matrix and R_q denotes the within class scatter matrix. Here, $R_t = R_b + R_q$ is denoted as the total scatter matrix.

The mean feature vector of class i is represented as μ_i and n_i represents the number of samples in class i. K denotes the total number of samples in the whole signal set. For a sample, its feature vector is y_i and μ_{k_i} denotes the vector of the signal class that y_i belongs to.

For log LDA, the logarithm of the optimal discrimination projection matrix Q_{opt} is computed and then the total scatter matrix is computed.

3.2   Kernel LDA

Kernel LDA is simply a non linear extension of LDA. In the feature space ϕ, the performance of the kernel LDA is evaluated [16]. With the help of non linear mapping as stated below;

$$\phi : C \rightarrow G | y \rightarrow \phi(y)$$

where the compact subset of $\Re^n$ is represented by C. In the feature space ϕ, the linearly non-separable configuration becomes separable easily. The quantities in the new feature space is denoted by ϕ and the objective function can be rewritten as

$$Q_{opt} = \arg \max_{q} \frac{|Q^T R_b^\phi Q|}{|Q^T R_q^\phi Q|}$$

(or)

$$Q_{opt} = \arg \max_{q} \frac{|Q^T R_b^\phi Q|}{|Q^T R_t^\phi Q|}$$

where

$$R_b^\phi = \sum_{i=1}^{K} n_i(\mu_i^\phi - \mu^\phi)(\mu_i^\phi - \mu^\phi)^T$$

$$R_q^\phi = \sum_{i} (\phi(y_i) - \mu_{k_i}^\phi)(\phi(y_i) - \mu_{k_i}^\phi)^T$$

4   Results and Discussion

The application of morphological filtering concept was applied to code converters and the classification results were obtained. As it was not satisfactory, it was further optimized with LDA, L-LDA and K-LDA and on the basis of parameters like

Accuracy, Time Delay, Sensitivity, Specificity, Performance Index and Quality Values, the average results are compared and shown in Table 3 respectively. The performance comparison of quality value analysis is shown in Fig. 2 and the performance comparison of time delay analysis is shown in Fig. 3 respectively. The mathematical expression for the parameters like Performance Index (PI), Sensitivity, Specificity and Accuracy are expressed as follows

$$PI = \left(\frac{PC - MC - FA}{PC}\right) \times 100$$

where PC denotes the Perfect Classification, MC indicates the Missed Classification and FA denotes the False Alarm. The Sensitivity, Specificity and Accuracy measures are mathematically expressed by the following

$$Sensitivity = \frac{PC}{PC + FA} \times 100$$

$$Specificity = \frac{PC}{PC + MC} \times 100$$

$$Accuracy = \frac{Sensitivity + Specificity}{2}$$

The Quality Value Q_V is defined as

$$Q_v = \frac{C}{(R_{fa} + 0.2) * (T_{dly} * P_{dct} + 6 * P_{msd})}$$

where
C   explains the scaling constant,
R_{fa} denotes the number of false alarm per set,
T_{dly} indicates the average delay of the onset classification in seconds,
P_{dct} specifies the percentage of perfect classification and
P_{msd} explains the percentage of perfect risk level missed.

The time delay is mathematically expressed as follows.

$$Time\ Delay = \left[2 \times \frac{PC}{100} + 6 \times \frac{MC}{100}\right]$$

Table 3 Performance comparison of LDA, log LDA and kernel LDA classifier

Parameters	LDA classifier	Log LDA classifier	Kernel LDA classifier
PC (%)	94.8	93.8	92.97
MC (%)	1.0	3.1	1.0
FA (%)	4.2	3.1	6.0
PI (%)	94.6	93.4	92.6
Sensitivity (%)	95.8	96.9	94.0
Specificity (%)	99.0	96.9	99.0
Time delay (s)	2.0	2.1	1.9
Quality value	21.3	21.4	20.7
Accuracy (%)	97.39	96.87	96.45

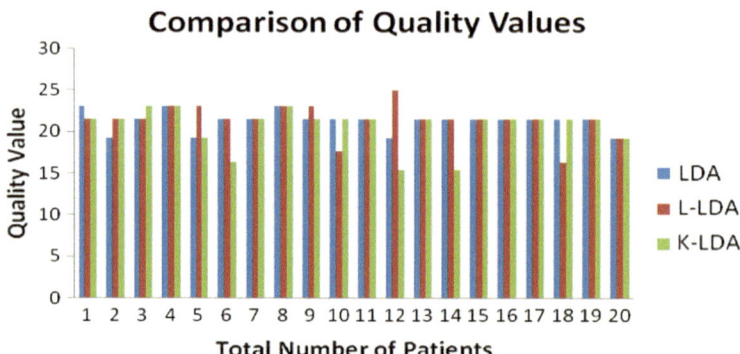

Fig. 2 Comparison of quality values for LDA, L-LDA and K-LDA classifiers

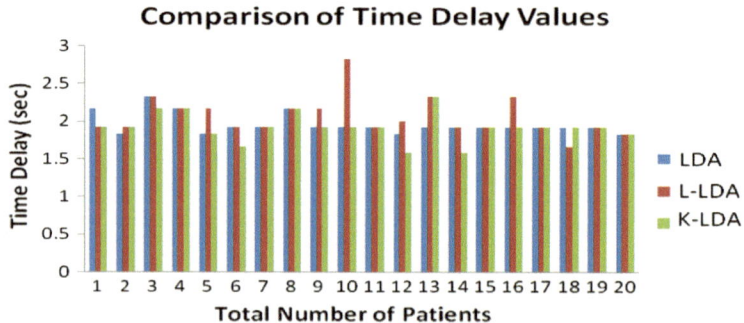

Fig. 3 Comparison of time delay for LDA, L-LDA and K-LDA classifiers

5 Conclusion

Thus the human brain is quite a complex system which exhibits a variety of spatio-temporal dynamics. As seizures are manifestations of epilepsy, it can be assessed and analyzed with the help of EEG signals. In this paper, the concept of morphological filtering is applied to code converters and it acts as a pre classifier for epilepsy risk level classification. The post classifiers used here are LDA, L-LDA and K-LDA for the best epilepsy risk level classification. Results show that an average classification accuracy of 97.39% is obtained when the concept of morphological filtering applied to code converters is classified with LDA classifier. A comparatively less classification accuracy of 96.87% is obtained when classified with L-LDA classifier and it is followed by K-LDA classifier as it gives a classification accuracy of about 96.45%. In terms of quality value analysis, L-LDA classifier surpasses the other two classifiers with a value of about 21.4. Future works is to incorporate the modifications in morphological filtering concept and post classification schemes for the perfect epilepsy risk level classification.

References

1. Prabhakar, S.K., Rajaguru, H.: Entropy based PAPR reduction for STTC system utilized for classification of epilepsy from EEG signals using PSD and SVM. In: IFBME Proceedings (Springer), 3rd International Conference on Movement, Health and Exercise (MoHE), Malaysia, 28–30 September 2016 (2016)
2. Prabhakar, S.K., Rajaguru, H.: Comparison of fuzzy output optimization with expectation maximization algorithm and its modification for epilepsy classification. In: International Conference on Cognition and Recognition (ICCR 2016), Mysore, India, 30–31 December 2016 (2016)
3. Prabhakar, S.K., Rajaguru, H.: Performance analysis of ApEn as a feature extraction technique and time delay neural networks, multi layer perceptron as post classifiers for the classification of epilepsy risk levels from EEG signals. In: Computational Intelligence, Cyber Security and Computational Models, Advances in Intelligent Systems and Computing, Coimbatore, India, Series vol. 412, pp. 89–97. Springer Verlag (2015)
4. Garg, S., Narvey, R.: Denoising and feature extraction of eeg signal using wavelet transform. Int. J. Eng. Sci. Technol. **5**, 1249–1253 (2013)
5. Bhatia, P., Sharma, A.: Different techniques for extracting brain signals for human machine interface, a review. Aust. J. Inf. Technol. Commun. **2**(2), 31–34 (2015)
6. Yuan, Q., Zhou, W.D., Li, S.F., Cai, D.M.: Epileptic EEG classification based on extreme learning machine and nonlinear features. Epilepsy Res. **96**(1–2), 29–38 (2011)
7. Fu, K., Qu, J., Chai, Y., Zou, T.: Hilbert marginal spectrum analysis for automaticseizure detection in EEG signals. Biomed. Signal Process. Control **18**, 179–185 (2015)
8. Shin, Y., Lee, S., Ahn, M., Cho, H., Jun, S.C., Lee, H.N.: Noise robustness analysis of sparse representation based classification method for non-stationary EEG signal classification. Biomed. Signal Process. Control **21**, 8–18 (2015)
9. Li, S.F., Zhou, W.D., Qi, Y., Geng, S.J., Cai, D.M.: Feature extraction and recognitionof ictal EEG using EMD and SVM. Comput. Biol. Med. **43**(7), 807–816 (2013)
10. Faust, O., Acharya, U.R., Adeli, H., Adeli, A.: Wavelet-based EEG processing forcomputer-aided seizure detection and epilepsy diagnosis. Seizure **26**, 56–64 (2015)

11. Lam, H.K., Ekong, U., Xiao, B., Ouyang, G., Liu, H.B., Chan, K.Y., Ling, S.H.: Variable weight neural networks and their applications on material surface and epilepsy seizure phase classifications. Neurocomput **149**, 1177–1187 (2015)
12. Rajaguru, H., Thangavel, V.: Performance analysis of wavelet transforms and morphological operator-based classification of epilepsy risk levels. EURASIP J. Adv. Signal Process. **2014**, 59 (2014). https://doi.org/10.1186/1687-6180-2014-59
13. Prabhakar, S.K, Rajaguru, H.: Morphological operator based feature extraction technique along with suitable post classifiers for epilepsy risk level classification. In: Proceedings of the International Conference on Intelligent Informatics and Biomedical Sciences (ICIIBMS), Okinawa Japan, 28–30 November (2015)
14. Prabhakar, S.K, Rajaguru, H.: Code converters with city block distance measures for classifying epilepsy from EEG signals. In: Fourth International Conference on Recent Trends in Computer Science & Engineering, Proceedings bought out in Procedia Computer Science, Chennai, India, 29–30 April 2016, vol. 87, pp. 5–11 (2016)
15. Prabhakar, S.K, Rajaguru, H.: LDA, GA and SVM's for classification of epilepsy from EEG signals. Res. J. Pharm. Biol. Chem. Sci. **7**(3), 2044–2049 (2016)
16. Roth, V., Steinhage, V.: Nonlinear discriminant analysis using kernel functions. In: Solla, S. A., Leen, T.K., Mueller, K.R. (eds.) Advances in Neural Information Processing Systems, vol. 12, pp. 568–574. MIT Press (2000)

Analysis of Dimensionality Reduction Techniques with ABC-PSO Classifier for Classification of Epilepsy from EEG Signals

Harikumar Rajaguru$^{(\boxtimes)}$ and Sunil Kumar Prabhakar

Department of ECE, Bannari Amman Institute of Technology,
Sathyamangalam, India
harikumarrajaguru@gmail.com

Abstract. Epilepsy is a commonly occurring neurological disorder which is characterized by recurrent seizures and it is of different types. The seizures can occur at various times irrespective of any symptoms. The central nervous systems are severely disturbed by the seizures activity. The abnormal electrical behaviour of a collection of cells in the brain leads to seizures and it can be detected by clinical symptoms. For the detailed study and diagnosis of epilepsy, Electroencephalography (EEG) signals are most commonly used. The electrical activity representation which results due to the generation by the cerebral cortex neurons are shown by EEG recordings. Because of this reason, it forms an integral component in the brain activity evaluation, epilepsy diagnosis and perception of epileptic attack. As the EEG recordings are quite long in nature, processing it is quite difficult. Therefore in this paper, the dimensions of the original EEG recordings are reduced with the help of dimensionality reduction techniques such as Singular Value Decomposition (SVD), Principal Component Analysis (PCA), Independent Component Analysis (ICA), Fast ICA and Linear Discriminant Analysis (LDA). The dimensionally reduced values are then fed inside the hybrid classifier called as Artificial Bee Colony-Particle Swarm Optimization (ABC-PSO) Classifier and the epilepsy risk level classification from EEG signals are analyzed. The results show that the highest classification accuracy of 97.42% along with a highest quality value of 22.76 is obtained when Fast ICA is used as a dimensionality reduction technique and classified with ABC-PSO Classifier.

Keywords: Epilepsy · EEG · SVD · PCA · Fast ICA · LDA · ABC-PSO

1 Introduction

Characterized by unprovoked and recurrent seizures, epilepsy is a common neurological disorder and a huge threat to the mankind [1]. The firing rate of neurons during the normal state varies greatly with the neuronal firing rate during the occurrence of a seizure. The epileptic people suffer from problems like depression, strange sensation, loss of consciousness, mental illness and physical injury [2]. Therefore, it is important to concentrate more on the diagnosis and classification of epilepsy. The electrical activity of the brain can

© Springer International Publishing AG 2018
D. J. Hemanth and S. Smys (eds.), *Computational Vision and Bio Inspired Computing*,
Lecture Notes in Computational Vision and Biomechanics 28,
https://doi.org/10.1007/978-3-319-71767-8_54

be easily recorded with the help of EEG [3]. Placing the electrodes on the head for recording EEG signals is of 2 types, namely, intracranial and scalp. As scalp EEG recordings are non-invasive in nature, it is widely preferred as the electrodes placed on the scalp of the patient possess good electrical and mechanical properties. The essential details about the pathology of the brain affected by epilepsy can be bought out with the help of EEG. The spectral and temporal characteristics of the EEG signals varies incessantly with respect to time, so these major variations make the analysis and detection procedure quite difficult. In the literature, several works have been reported in EEG signal processing for epilepsy classification, time based measures, frequency based measures and so on for analysis and classification of epilepsy from EEG signals. Some of the pioneering works done in this field are as follows. The detailed analysis of EEG signals under flash stimulation for epileptic and migraine patients was done by Akben et al. [4]. The advantages and applications of mobile phones in epilepsy care were detailed by Ranganathan et al. [5]. A comparative study analysis on the epileptic seizure predictors based on various computational intelligence techniques was done by Teixeira et al. [6]. The epileptic seizure detection was done in an efficient manner by differentiating the spectral analysis of EEG's by Tawfik et al. [7]. A hybrid methodology of automatic detection of epileptic seizures in EEG recordings was done by Kang et al. [8]. An improved methodology for epileptic EEG signal classification based on spectral factors using K-NN classifier was done by Anu and Thomas [9]. For detecting epileptic seizures in long term EEG signal, an unsupervised methodology is used by Tsiouris et al. [10]. In this paper, the dimensionality reduction techniques have been utilized effectively and it has been classified with ABC-PSO classifier for obtaining the best epilepsy risk level classification. Figure 1 shows the block diagram of the work.

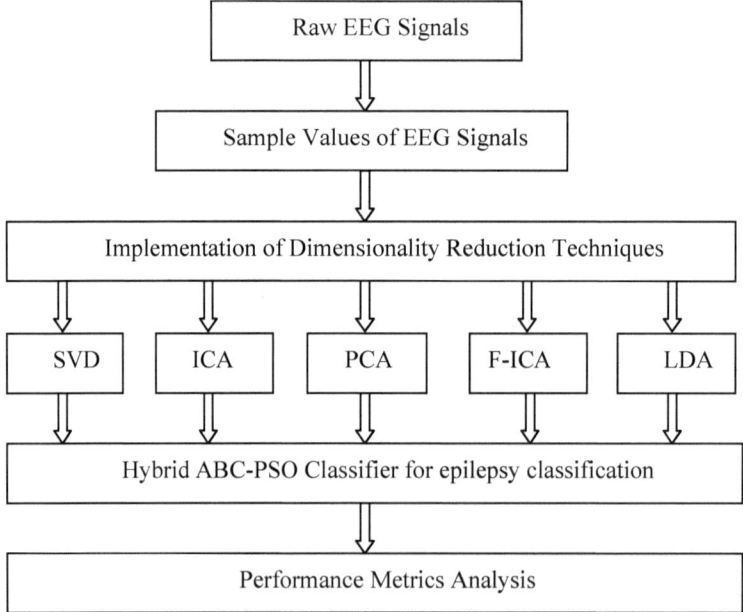

Fig. 1 Block diagram of the work

2 Materials and Methods

From 20 epileptic patients who were under the treatment for epilepsy at Sri Ramakrishna Hospital, Coimbatore, India, the acquisition of EEG data was done according to the standard 10–20 International system. By placing the 16 channel electrodes on the scalp of the patient and for different periods like fully sleep, partially awake, active stages etc. the recordings of the EEG signals are done. The EEG amplification is done appropriately after the acquisition of signals is done. The recordings are done for more than 55 min and it is split into epochs for easy mathematical analysis and computational interpretation. As there are 16 channel electrodes to get the EEG data from the epileptic patient, the data obtained is very large to process and so the dimensions of the data has to be reduced for obtaining a good classification accuracy. Therefore, in this paper, 5 different dimensionality reduction techniques are analyzed when it is classified with ABC-PSO classifier for epilepsy classification from EEG signals.

3 Dimensionality Reduction Techniques

The dimensionality reduction techniques employed here in this paper are SVD, PCA, ICA, Fast ICA and LDA.

3.1 Singular Value Decomposition

Factorizing a real or complex matrix can be done with the help of SVD. SVD has been utilized in a plethora of applications ranging from computational biology, applied electronics and mathematics, signal processing algorithms, data mining etc. For computing the pseudo inverse, to perform matrix approximation and to obtain the least squares fitting of data, SVD is widely used [11]. For applications like determination of rank, null space and range of a particular matrix, SVD is used. In this case, SVD is utilized as a dimensionality reduction technique. When schemes such as Gaussian elimination and various other decomposition techniques don't provide an efficient satisfactory answer, SVD is used often. With the help of SVD, a real or complex rectangular matrix is decomposed into a product of 3 matrices, out of which one is diagonal and 2 are orthogonal. For a particular matrix Q which has a size $C \times D$, the SVD is observed as the factorization of the given form and represented as $Q = U\Sigma V^T$, where U denotes an unitary matrix which is column orthogonal of size $C \times C$, Σ denotes the diagonal matrix of size $C \times D$ with elements $p_{ij} = 0$ if $i \neq j$ and $p_{ij} \geq 0$ in a descending order along the diagonal, and V^T is a $D \times D$ unitary matrix. The singular values of Q represent the elements on the diagonal of Σ.

3.2 Principal Component Analysis

PCA is a widely used and powerful linear technique in signal processing, image processing, statistics etc. A multidimensional dataset can be easily simplified to lower dimensions for the sake of visualization analysis or data compression [12]. The data is represented in a new coordinate system by PCA and in that system the basis vector follows modes which have the highest variance in the entire data. So for the specific data set, the calculations of new basis vectors are done. If a dataset is having P observations each of S variables represented as dimensions, so that $P \gg S$. The main aim is to reduce the dimensionality of the data so that each observation is represented with only H variables, $1 \leq H < S$. The data is arranged as a set comprising of P column data vector, each column data vector represent a single observation comprising of S variables. The pth observation is a column vector represented as $y_p = (y_1, \ldots, y_S)^P$, $p = 1, \ldots, P$. Thus a $S \times P$ data matrix Y is found out. The data normalization is required initially and so this procedure is not applied to the raw data J but instead to the normalized data Y as follows. In a matrix J, the raw observed data is arranged and the calculation of empirical mean along each row of J is done. The storage of the result is done in a vector w where elements are scalars and it is represented as

$$w(s) = \frac{1}{P} \sum_{p=1}^{P} J(s,p), \quad \text{where } s = 1, \ldots, S$$

From each column of J, the empirical mean is subtracted. If 'e' is represented as a unitary vector which has a size P, then it rewritten as

$$Y = J - we$$

If the approximation of Y in a lower dimensional space S is done by the lower dimensional matrix Z (of dimensionality H), then the MSE ε^2 is written as follows:

$$\varepsilon^2 = \frac{1}{P} \sum_{p=1}^{P} |y_p|^2 - \sum_{i=1}^{H} b_i^T \left(\frac{1}{P} \sum_{p=1}^{P} y_p y_p^T \right) b_i$$

where $b_i, i = 1, \ldots, H$ forms the basis vectors of the linear space of dimensions H. The following terms has to be made maximum if the error ε^2 is to be minimized. Therefore, $\sum_{i=1}^{H} b_i^T \text{cov}(y) b_i$, where $\text{cov}(y) = \sum_{p=1}^{P} y_p y_p^T$ is the covariance matrix. Various special properties are possessed by this covariance matrix such as symmetric, real and positive semi-definite. Using this technique, the dimensionally reduced values are obtained.

3.3   Independent Component Analysis (ICA) and Fast ICA

In two versions, the fast ICA was proposed, (a) one-unit approach (b) symmetric one [13]. The preliminary step is common for both the kinds and other similar ICA algorithms, where the sample mean and the decorrelation of the data J is removed (i.e.)

$$H = \hat{C}^{-1/2}(J - \bar{J})$$

Here $\hat{C}$ represents the sample covariance matrix, $\hat{C} = (J - \bar{J})(J - \bar{J})^T / N$ and $\bar{J}$ represents the sample mean of the measured data. The one unit ICA is dependent on minimization/maximization of the criterion $c(v) = E[Q(v^T H) - Q_o]^2$, where v denotes the to be found vector of coefficients which aims to separate a desired independent components from the mixture. E represents the sample mean, $Q(.)$ represents the approximate non linear function which is also called as contrast function. Q_0 denotes the expected value of $Q(\eta)$, where η denotes the standard normal random variable. All the signals are estimated in parallel fashion by the symmetric Fast ICA and each step is completed by a symmetric orthogonalization as follows

$$V^+ \leftarrow q(VH)H^T - diag[q'(VH)1_N]V$$
$$V \leftarrow (V^+ V^{+T})^{-1/2}V^+$$

where $q(\bullet)$ denotes the first derivation and $q'(\bullet)$ denotes the second derivation of $Q(\bullet)$ respectively. With the help of gain matrix, $Q = \hat{V}A$, the treatment of the separation quality can be done, which identifies the relative presence of the mth original signal component in the estimated nth component $m, n = 1, \ldots, d$.

3.4   Linear Discriminant Analysis

One of the most widely used dimension reduction methods is LDA. The basic assumption here is that the classes are normally distributed with the help of different means but the same covariance matrix is used. For the original dimensions, the optimal orthogonal linear combinations are computed with the help of LDA [14]. The separability of the classes are maximized by the new base vectors instead of the variance of the full dataset. It is often considered as an optimization problem and represented as

$$\max_v \frac{v^T Q_b v}{v^T Q_v v}$$

$$Q_b = \sum_{i=1}^{C} (\mu_i - \mu)(\mu_i - \mu)^T$$

$$Q_v = \sum_{i=1}^{C} \sum_{j=1}^{n_i} (\mu_i - d_{i,j})(\mu_i - d_{i,j})^T$$

where Q_b denotes the between-class scatter matrix, Q_v denotes the within class scatter matrix, Q_v denotes the within class scatter matrix, μ represents the mean of the dataset, μ_i represents the mean of the ith class, $d_{i,j}$ is the jth vector of the ith class, n_i denotes the total number of vectors in the ith class, C represents the total number of classes. It is worth observing that Q_v is replaced with the total scatter matrix Q_t because of the following equation

$$Q_t = Q_b + Q_v, \quad \text{where } Q_t = \sum_{i=1}^{n} (d_i - \mu)(d_i - \mu)^T$$

where the total number of vectors is denoted by n. A generalized Eigen value–Eigen vector problem is obtained because of this optimization criterion. The extraction of more discriminant dimensions may be done by obtaining the Eigen vectors which corresponds to the largest Eigen value.

4   ABC-PSO as a Post Classifier

The dimensionally reduced values are fed inside the ABC-PSO Classifier for the classification of epilepsy from EEG signals. A hybridization of ABC with PSO is done which aims to enhance the accuracy of the classification rule [15]. The classification rules are discovered from the training data set. To solve the problem, the hybrid algorithm is used to mix the component from both ABC ad PSO algorithms respectively. The particle moving in an n-dimensional space is made use by the PSO algorithm in order to find a solution for the nth variable function optimization problem. Based on its own experience and the current prevailing situation, a particle decides where to move next. This enables to find the global best position in the neighborhood by PSO. With the help of ABC, the global position velocity (V_{ij}) related with every dimension is updated as

$$V_{ij} = X_{ij} + \phi_{ij}(X_{ij} - X_{kj})$$

The above equation helps to produce a new solution V_{ij} in the neighborhood particles X_{ij} (the current position of particle i on j) for the employee bee. k denotes the solution in the neighborhood of i, ϕ denotes the random number in the range of $(-1,1)$. The evaluation is done using the greedy solution process in between them.

Hybrid ABC-PSO algorithm:

(i)   Initially the Rule Set is made empty
(ii)   **For** each Class C, the training samples belonging to all classes is considered
(iii) **While** (number of uncovered training sample of class C > max uncovered
           example per class)
(iv)   **For** any particle i, do
(v)   The velocity is updated as

$$V_{ij} = wV_{ij} + C_1 r_1 (pbest_{ij} - X_{ij}) + C_2 r_2 (gbest_{ij} - X_{ij})$$

where w is the inertia weight, C_1 is the determination of effect of the cognitive component (self confidence factor) and C_2 is the determination of effect of the social component (swarm confidence factor), r_1 and r_2 are the random variables which are uniformly distributed between 0 and 1.

(vi)   The position is updated as $X_{ij} = X_{ij} + V_{ij}$
(vii) **If** $f(pbest) \leq f(X_i)$ then
           $pbest = X_i$
(viii) End **If**
(ix)   End **For**

(x)   The *gbest* value is updated
(xi)   **For** every particle *i* do
    (a) The random problem variable is chosen
    (b) To the *pbest*, apply ABC update rule
    (c) Update $pbest_i$ and *gbest*
(xii) End **For**
        Iteration_num = Iteration_num + 1
(xiii) End **while**
(xiv) End **For**
(xv)   Return *gbest*

5 Results and Discussion

For SVD, PCA, ICA, Fast ICA and LDA as dimensionality reduction technique followed by ABC-PSO as Post Classifier, based on the parameters like Classification Accuracy, Performance Index, Time Delay, Quality Values, Specificity and Sensitivity, the results are calculated and shown in Table 1. The mathematically expression for the performance metrics like Performance Index (PI), Sensitivity, Specificity and Accuracy are given as follows

$$PI = \left(\frac{PC - MC - FA}{PC} \right) \times 100$$

Perfect Classification is indicated as PC, Missed Classification is denoted as MC and the False Alarm is specified as FA. The Sensitivity, Specificity and Accuracy measures are expressed by the following equations

$$Sensitivity = \frac{PC}{PC + FA} \times 100$$

$$Specificity = \frac{PC}{PC + MC} \times 100$$

$$Accuracy = \frac{Sensitivity + Specificity}{2} \times 100$$

The definition of Quality Value Q_V is expressed as follows

$$Q_v = \frac{C}{(R_{fa} + 0.2) * (T_{dly} * P_{dct} + 6 * P_{msd})}$$

where C means the scaling constant,
R_{fa} implies the total number of false alarm per set
T_{dly} explains the average delay expressed in seconds
P_{dct} specifies the perfect classification expressed in terms of percentage and
P_{msd} means the risk level missed expressed in terms of percentage

The time delay is expressed as follows

$$Time\, Delay = \left[2 \times \frac{PC}{100} + 6 \times \frac{MC}{100} \right]$$

Table 1 Analysis of dimensionality reduction techniques with ABC-PSO classifier

Parameters	SVD	PCA	ICA	Fast ICA	LDA
PC (%)	87.29	92.36	87.77	94.85	89.79
MC (%)	12.63	7.63	11.94	4.99	9.92
FA (%)	0.069	0	0.27	0.13	0.27
PI (%)	84.6	91.98	84.94	94.31	88.09
Sensitivity (%)	99.93	100	99.72	99.86	99.72
Specificity (%)	87.36	92.36	88.05	94.79	90.07
Time delay (sec)	2.50	2.30	2.47	2.19	2.39
Quality value	20.22	21.77	20.37	22.76	20.89
Accuracy (%)	93.64	96.18	93.88	97.42	94.89

6 Conclusion

On the analysis of the dimensionality reduction techniques with the ABC-PSO Classifier, an average classification accuracy of 97.42% is obtained when Fast ICA is utilized as a dimensionality reduction technique. When PCA is classified along with ABC-PSO classifier an average classification accuracy of 96.18% is obtained and when LDA is classified with ABC-PSO classifier an average classification accuracy of about 94.89% is obtained. A comparatively less time delay of about 2.9 s along with an average quality value of 22.76 is obtained when the Fast ICA is utilized as a dimensionality reduction technique and when classified with ABC-PSO classifier. SVD with ABC-PSO classifier produces a comparatively less result in terms of accuracy, time delay, performance index and quality values when compared to the other dimensionality reduction techniques. Future works is to analyze various other dimensionality reduction techniques with ABC-PSO Classifier for getting the best epilepsy classification accuracy from EEG signals.

References

1. Prabhakar, S.K., Rajaguru, H.: A novel combination of code converters and sparse representation classifier for an efficient epilepsy risk level classification. In: Proceedings of the 8th Biomedical Engineering International Conference (BMEiCON), Pattaya, Thailand, 25–27 Nov 2015

2. Prabhakar, S.K., Rajaguru, H.: Morphological operator based feature extraction technique along with suitable post classifiers for epilepsy risk level. In: Proceedings of the International Conference on Intelligent Informatics and BioMedical Sciences (ICIIBMS), Okinawa, Japan, 28–30 Nov 2015
3. Prabhakar, S.K., Rajaguru, H.: Development of patient remote monitoring system for epilepsy classification. In: 16th International Conference on Biomedical Engineering (ICBME), Singapore, 7–10 Dec 2016
4. Akben, S., Subasi, A., Tuncel, D.: Analysis of EEG signals under flash stimulation for migraine and epileptic patients. Med. Syst. **35**, 437–443 (2011)
5. Ranganathan, L.N., Chinnadurai, S.A., Samivel, B., Kesavamurthy, B., Mehndiratta, M.M.: Application of mobile phones in epilepsy care. Int. J. Epilepsy **2**(1), 28–37 (2015). ISSN: 2213-6320
6. Teixeira, C.A., Direito, B., Bandarabadi, M., Quyen, M.L.V., Valderrama, M., Schelter, B., Schulze-Bonhage, A., Navarro, V., Sales, F., Dourado, A.: Epileptic seizure predictors based on computational intelligence techniques: a comparative study with 278 patients. Comput. Methods Programs Biomed. **114**(3), 324–336 (2014). ISSN: 0169-2607
7. Tawfik, N.S., Youssef, S.M., Kholief, M.: A hybrid automated detection of epileptic seizures in EEG records. Comput. Electr. Eng. Available online 28 Sept 2015, ISSN: 0045-7906
8. Kang, J.H., Chung, Y.G., Kim, S.P., An efficient detection of epileptic seizure by differentiation and spectral analysis of electroencephalograms. Comput. Biol. Med. **66**(1), 352–356 (2015). ISSN: 0010-4825
9. Anu, N.S., Thomas, P.: An improved method for classification of epileptic EEG signals based on spectral features using k-NN. SSRG Int. J. Electron. Commun. Eng. **2**, 35–38 (2015)
10. Tsiouris, K.M., Konitsiotis, S., Markoula, S., Koutsouris, D.D., Sakellarios, A.I., Fotiadis, D.I.: An unsupervised methodology for the detection of epileptic seizures in long-term EEG signals. IEEE 1–4 (2015)
11. Forsythe, G.E., Malcolm, M.A., Moler, C.B.: Computer Methods for Mathematical Computations. Prentice-Hall (1977)
12. Rajaguru, H., Prabhakar, S.K: An exhaustive analysis of code converters as pre-classifiers and K means, SVD, PCA, EM, MEM, PSO, HPSO and MRE as post classifiers for classification of epilepsy from EEG signals. J. Chem. Pharm. Sci. **9**(2), 818–822 (2016)
13. Oja, E.: Convergence of the symmetrical FastICA algorithm. In: 9th International Conference on Neural Information Processing (ICONIP), Singapore, 18–22 Nov. 2002
14. Rajaguru, H., Prabhakar, S.K.: LDA, GA and SVM's for classification of epilepsy from EEG Signals. Res. J. Pharm. Biol. Chem. Sci. **7**(3), 2044–2049 (2016)
15. Altun, O., Korkmaz, T.: Particle swarm optimization-artificial bee colony chain (PSOABCC): a hybrid meteahuristic algorithm. In: Scientific Cooperation International Workshop on Electrical and Computer Engineering Subfields, pp. 42–49 (2014)

E-health Design with Spectral Analysis, Linear Layer Neural Networks and Adaboost Classifier for Epilepsy Classification from EEG Signals

Harikumar Rajaguru$^{(\boxtimes)}$ and Sunil Kumar Prabhakar

Department of ECE, Bannari Amman Institute of Technology,
Sathyamangalam, India
harikumarrajaguru@gmail.com

Abstract. About 1–2% of the population in the whole world is suffering from a serious neurological disorder called epilepsy which is characterized by spontaneous seizures. A lot of temporary disruptions occur in the ongoing electrical activities of the brain if the seizure attack is present. Antiepileptic drugs may be favourable for some patients while for other patients it may not respond well. To explore the electrical behaviour of the human brain, the measurement and the recordings of the electrical brain activity is done. By analyzing the Electroencephalography (EEG) signals and extracting all its features including both univariate and multivariate, various algorithms for seizure prediction, detection, classification have been developed. In this paper, an e-health design for epilepsy classification with the help of spectral analysis, Linear Layer Neural Networks (LLNN) and Adaboost Classifier has been proposed. The LLNN has been used as the preliminary level classifier and as the results obtained through it are not satisfactory, further optimization and classification is done with the help of Adaboost Classifier. Results show that when classified with Adaboost Classifier an average classification accuracy of about 99.43%, an average quality value of 24.38, an average less time delay of 1.99 s along with an average performance index of 99.13% is obtained.

Keywords: Epilepsy · EEG · LLNN · Adaboost

1 Introduction

The abnormalities in the functionality of the brain are caused due to many neurological disorders and one such prominent disorder is epilepsy [1]. In the case of epilepsy, there is excessive synchronization in the neuronal activities of the brain and so seizures occur. The patterns of the brain activity in epileptic patients are quite different from the patterns of the brain activity in the normal patient [2]. The asymmetries caused due to the irregular connectivity in the brain regions in the case of seizure disorder are helpful to distinguish the epileptic patients from the other healthy subjects. To assess the

© Springer International Publishing AG 2018
D. J. Hemanth and S. Smys (eds.), *Computational Vision and Bio Inspired Computing*,
Lecture Notes in Computational Vision and Biomechanics 28,
https://doi.org/10.1007/978-3-319-71767-8_55

activities of the brain, a valuable measurement known as EEG is used [3]. To analyze the electrical behaviour of the brain, EEG signals are highly useful. Due to the action of chemical transmitters on the post synaptic cortical neurons, the recordings can be done easily. For the processing of diagnostic disorders like epilepsy, EEG signals are highly useful. Using classical mathematical techniques along with artificial intelligence and soft computing techniques, a lot of works has been proposed [4]. Some of the most relevant and useful works in processing of EEG signals and classification of epilepsy from EEG signals are discussed as follows.

For the epileptic EEG signal classification, a feature extraction technique based on the local binary patterns was done by Kaya et al. [5]. A sparse functional linear model based on wavelets for the application of seizure detection from EEG signals was done by Xie and Krishnan [6]. Higher Order Statistics were utilized in the Empirical Mode Decomposition (EMD) domain by Alam and Bhuiyan for seizure detection in EEG signals [7]. The cross-correlation analysis based on multi fractional detrends for epileptic seizure detection was done by Ghosh et al. [8]. Hilbert Huang Transform was used by Fu et al. for the seizure classification from EEG signals [9]. A variational Bayesian GMM which has zero-crossing intervals are used for prediction of the seizures in EEG signals by Zandi et al. [10]. For detection of epileptic seizures, a genetic algorithm based expert model was developed by Dhiman et al. for epilepsy detection [11]. For the detection of neonatal seizures, the EEG signal description with spectral envelope was used by Temko et al. [12]. With the advent of Neural Fuzzy networks, Sadati et al. tried to detect the epileptic seizures [13]. For the neonatal seizure detection, the atomic decomposition with a new dictionary was used by Nagaraj et al. [14]. In this paper, Power Spectrum Density (PSD) is analyzed as a dimensionality reduction technique and then it is classified with LLNN. As the performance of it is not satisfactory, it is further optimized and classified with the help of Adaboost Classifier. The illustration of the work is shown in Fig. 1.

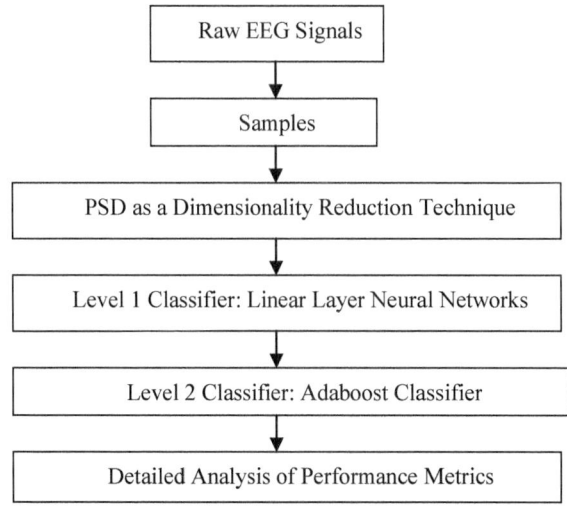

Fig. 1 Illustration of the work

The paper is organized as follows. In Sect. 2, the materials and methods are discussed followed by the application of PSD as a dimensionality reduction technique. The preliminary classification using the Linear Layer Neural Networks is discussed in Sect. 3 and the final classification using Adaboost Classifier is done in Sect. 4. The results and discussion are given in Sect. 5 and the paper is concluded in Sect. 6.

2   Materials and Methods

The experiment performed here is for 20 patients suffering from epilepsy. The EEG recordings of 20 epileptic patients were obtained from Sri Ramakrishna Hospital, Coimbatore, India with the help of a qualified neurologist. For the acquisition of EEG data, the standard 10–20 International system is employed. The EEG data is obtained in the standard European Data Format (EDF). The 16-channel electrodes were placed on the scalp of the epileptic patient and the recordings were done for more than 55 min. The signals were amplified well and the artifacts were removed using suitable filtering techniques. For the easy interpretation of the huge EEG recordings, it is split into sections or epochs. As the EEG recordings are too large to process, Power Spectral Density is employed to reduce the dimensions of the original data.

2.1   PSD as a Dimensionality Reduction Technique

The power spectrum of a particular time series explains the power distribution into the frequency components which composes that signal. The spectral energy distribution per unit time is referred as PSD. Over the entire range of frequency, the distribution of power is done with the help of spectral analysis. For clinical diagnosis, spectral analysis of EEG signals can provide the most relevant information. A finite average power is usually possessed by a random signal and so it is characterized by the average PSD as follows.

$$PSD(i) = \sum_{k=0}^{N} |X(k)|^2$$

where $X(k)$ is the output of Fast Fourier Transform (FFT) and k denotes the number of FFT components.

3   Linear Layer Neural Networks (LLNN)

As a first level classifier, LLNN is utilized effectively. It is more similar to Perceptron and the transfer function is linear rather than hard-limiting [15]. As the transfer function is linear, it allows its output to take on any value and can solve the linearly separable problems. A linear layer neural network can be designed with a particular set of given input vector and it produces the outputs of the corresponding target vectors. For every input vectors, the calculation of the output vectors of the network can be done. The error is calculated as the difference between an output vector and its target vector.

The values for the biases and the network weights are found out so that the sum of the squares of the errors is minimized. As a single error minimum is found in linear systems, this problem can be easily managed. In most cases, a linear network can be calculated directly so that there is minimum error for the given input vector and target vector. With the help of least mean squares algorithm, the training of the network is done in order to have a minimum error. Therefore linear layers are single layers comprising of linear neurons and can be static or dynamic. For simple linear time series problem, it can be easily trained. The adjustments in the relationship between inputs and outputs can be done adaptively in order to ensure the continuous learning. The results obtained after employed LLNN were not satisfactory and so it is further optimized with the help of Adaboost Classifier.

4   Adaboost Classifier for Classification of Epilepsy

As the classification results with Linear Layer Neural Networks are not satisfactorily obtained, it is further optimized and classified with the help of a Adaboost Classifier. For improving the performance of a weak classifier, boosting technique is used. The most important and prominent member in the boosting family is Adaboost [16]. For the objects in the dataset, a particular set of weights is maintained in boosting techniques, so that more weights are acquired to the objects that have been hectic to classify. By repeatedly running a learning algorithm, these methods can work well on different distributions on the training data. Let it can be combined to form a single composite classifier. The input in the Boosting classifier is taken as a training set of n examples $Q = \langle (a_1, b_1), \ldots (a_n, b_n) \rangle$, where a_i is an instance obtained from some space A, and $b_i \in B$ is the class label associated with a_i. The assumption here is that the set of labels B is of finite cardinality k. The weak learning algorithm is called repeatedly in series of rounds by the Boosting algorithm. A hypothesis is computed by the weak learning algorithm as *hypothesis* $h_t : A \rightarrow B$, which should definitely misclassify a non trivial fraction of the training examples, related to the distribution D_t. The main intention of the weak learning is to seek a hypothesis h_t so that the training error $\varepsilon_t = \text{Pr}_{i \approx D_t}[h_t(a_i) \neq b_i]$ is minimized. With respect to the distribution D_t, the measurement of training errors is done that has been provided to the weak learner. For T rounds, the process is incessant and finally, the combination of the weak hypothesis $h_1, \ldots, h_T$ into a final single hypothesis h_{fin} is done. The algorithm is shown below where it utilizes a simple rule where the initial distribution D_t is uniform over Q so that $D_1(i) = 1/n$ for all i. To update the distribution, if h_t classifies a_i correctly then the weights of the example i is multiplied by some number $\beta_t \in [0, 1]$ or else the weight is left unchanged and so it is divided by the normalization constant W_t. Therefore examples which are 'hard' and is prone to be highly misclassified gets higher weights and the examples which are 'easy' and is correctly classified by the previous weak hypothesis gets lower weight. Therefore Adaboost mainly focuses the highest weight on the examples that seem to be hardest for the algorithm which is weak learning. The algorithm is given below:

The function of ε_t is the number β_t. The weighted representation or vote of the

---

Input: sequence of n examples $\langle (a_1, b_1), \ldots, (a_n, b_n) \rangle$ with labels $b_i \in B = \{1, \ldots k\}$ weak learning algorithm

The number of iterations is specified as integer T

Initialize $D_1(i) = 1/n$ for all i.

Do for $t = 1, 2, \ldots, T$

    (a) The weak learning algorithm is called for and provided it with the distribution D_t

    (b) The hypothesis $h_t : A \to B$ is got back

    (c) The error of h_t is calculated as $\varepsilon_t = \Sigma_{i:h_t(a_i) \neq b_i} D_t(i)$

    (d) If $\varepsilon_t > 1/2$, then set $T = t - 1$ and then the loop is aborted

    (e) The $\beta_t = \varepsilon_t/(1 - e_t)$ is set

    (f) The distribution is updated as $D_t : D_{t+1}(i) = \frac{D_t(i)}{W_t} \times \begin{cases} \beta_t & if \quad h_t(a_i) = b_i \\ 0 & otherwise \end{cases}$, where W_t is

    the normalization constant

    (g) Output: The final hypothesis is given as $h_{fin}(a) = \arg\max \Sigma_{t:h_t(a)=b} \log \frac{1}{\beta_t}$.

---

hypothesis which is weak is given as the final hypothesis h_{fin}. For the hypothesis h_t, the weight is defined to be $\ln 1/\beta_t$ so that for the hypothesis with lower error, greater weight is given. The main success of Adaboost algorithm is the diversity creating ability of it. This algorithm makes sure that more concentration is provided to inaccurate classifiers by forcing them to focus on toughest objects and to ignore the rest of the information, thereby leading to a large diversity boosting the ensemble performance.

5   Results and Discussion

For the PSD as dimensionality reduction technique followed by Linear Layer Neural Network as an initial post classifier and Adaboost Classifier as the secondary post classifier, based on the different parameters like Classification Accuracy, Performance Index, Quality Values, Time Delay, Specificity and Sensitivity, the results are calculated and shown in Table 1. The expression for the different performance metrics like Performance Index (PI), Sensitivity, Specificity and Accuracy are given mathematically as follows.

$$PI = \left(\frac{PC - MC - FA}{PC} \right) \times 100$$

Perfect Classification = PC, Missed Classification = MC and the False Alarm = FA. The Sensitivity, Specificity and Accuracy measures are expressed by the following equations.

$$Sensitivity = \frac{PC}{PC + FA} \times 100$$

$$Specificity = \frac{PC}{PC + MC} \times 100$$

$$Accuracy = \frac{Sensitivity + Specificity}{2} \times 100$$

The mathematical definition of Quality Value Q_V is given as follows

$$Q_v = \frac{C}{(R_{fa} + 0.2) * (T_{dly} * P_{dct} + 6 * P_{msd})}$$

where C expresses the scaling constant,

R_{fa} tells the total number of false alarm per set,

T_{dly} indicates the average delay expressed in seconds

P_{dct} mentions the perfect classification expressed in terms of percentage and

P_{msd} identifies the risk level missed expressed in terms of percentage.

The time delay is expressed as follows

$$Time\ Delay = \left[2 \times \frac{PC}{100} + 6 \times \frac{MC}{100} \right]$$

Table 1 Analysis of results with spectral analysis classified with linear layer neural networks and optimized further with Adaboost classifier

Parameters	Average
PC (%)	99.16
MC (%)	0.208
FA (%)	0.624
PI (%)	99.13
Sensitivity (%)	99.07
Specificity (%)	99.79
Time delay (sec)	1.99
Quality value	24.38
Accuracy (%)	99.43

6 Conclusion

Thus the brain generates erratic signals in a chronic condition called epilepsy. The epileptic seizure analysis in an automated sense refers to techniques such as seizure prediction, detection and localization and the EEG recordings are highly useful for the analysis of epilepsy. With PSD utilized as a dimensionality reduction technique and when it is classified with LLNN as the initial classifier and Adaboost as the final classifier, an average perfect classification of about 99.16%, an average false alarm rate of 0.624%, average time delay of 1.99 s and average accuracy of 99.43% is obtained.

Future works aims to replace the Adaboost classifier with various machine learning and soft computing techniques for obtaining the best epilepsy classification rate from EEG signals.

References

1. Prabhakar, S.K., Rajaguru, H.: Analysis of centre tendency mode chaotic modeling for electroencephalography signals obtained from an epileptic patient. Adv. Stud. Theor. Phys. **9** (4), 171–177 (HIKARI Ltd) (2015). http://dx.doi.org/10.12988/astp.2015.5117
2. Harikumar, R., Kumar, P.S.: Frequency behaviors of electroencephalography signals in epileptic patients from a wavelet Thresholding perspective. Appl. Math. Sci. **9**(50), 2451–2457 (HIKARI Ltd) (2015). http://dx.doi.org/10.12988/ams.2015.52135
3. Harikumar, R., Kumar, P.S.: Dimensionality reduction techniques for processing epileptic encephalographic signals. Biomed. Pharmacol. J. **8**(1), 103–106 (2015)
4. Harikumar, R., Kumar, P.S.: Dimensionality reduction with linear graph embedding technique for electroencephalography signals of an epileptic patient. Res. J. Pharm. Technol **8**(5), 554–556 (2015)
5. Kaya, Y., Uyar, M., Tekin, R., Yıldırım, S.: 1D-local binary pattern based feature extraction for classification of epileptic EEG signals. Appl. Math. Comput. **243**, 209–219 (2014)
6. Xie, S., Krishnan, S.: Wavelet-based sparse functional linear model with applications to EEGs seizure detection and epilepsy diagnosis. Med. Biol. Eng. Compu. **51**, 49–60 (2013)
7. Shafiul Alam, S.M., Bhuiyan, M.I.H.: Detection of seizure and epilepsy using higher order statistics in the EMD domain. IEEE J. Biomed. Health Inform. **17**(2), 312–318 (2013)
8. Ghosh, D., Dutta, S., Chakraborty, S.: Multifractal detrended cross-correlation analysis for epileptic patient in seizure and seizure free status. Chaos, Solitons Fractals **67**, 1–10 (2014)
9. Fu, K., Qu, J., Chai, Y., Dong, Y.: Classification of seizure based on the time-frequency image of EEG signals using HHT and SVM. Biomed. Signal Process. Control **13**, 15–22 (2014)
10. Zandi, A.S., Tafreshi, R., Javidan, M., Dumont, G.A.: Predicting epileptic seizures in scalp eeg based on a variational bayesian gaussian mixture model of zero-crossing intervals. IEEE Trans. Biomed. Eng. **60**(5), 1401–1413 (2013)
11. Dhiman, R., Saini Priyanka, J. S.: Genetic algorithms tuned expert model for detection of epileptic seizures from EEG signatures. Appl. Soft Comput. **19**, 8–17 (2014)
12. Temko, A., Nadeu, C., Marnane, W., Boylan, G.B., Lightbody, G.: EEG signal description with spectral-envelope-based speech recognition features for detection of neonatal seizures. IEEE Trans. Inf. Technol. Biomed. **15**(6), 839–847 (2011)
13. Sadati, N., Mohseni, H. R., Magshoudi, A.: Epileptic seizure detection using neural fuzzy networks. In: Proceeding of IEEE International Conference on Fuzzy Systems, 16–21 July 2006, pp. 596–600 (2006)
14. Nagaraj, S.B., Stevenson, N.J., Marnane, W.P., Boylan, G.B., Lightbody, G.: Neonatal seizure detection using atomic decomposition with a novel dictionary. IEEE Trans. Biomed. Eng. **61**(11), 2724–2732 (2014)
15. Prabhakar, S.K., Rajaguru, H.: Performance analysis of linear layer neural networks for oral cancer classification. In: 6th IEEE ICT International Student Project Conference 2017 (ICT-ISPC), Universiti Teknologi Malaysia, Johor Bahru, Malaysia, 23–24 May (2017)
16. Rajaguru, H., Prabhakar, S.K.: Power spectral density and KNN based Adaboost classifier for epilepsy classification. In: IEEE Proceedings of the International Conference on Electronics, Communication and Aerospace Technology (ICECA 2017), Coimbatore, India, pp. 441–445

Biomedical Voice Based Parkinson Disorder Identification for Homosapiens

B. Anusha[(⊠)] and P. Geetha

Department of Information Science and Technology, College of Engineering
Guindy, Anna University, Chennai, Tamilnadu, India
bsanusharaj@gmail.com, geethap@annauniv.edu

Abstract. A long term neuro degenerative genetic and sporadic based Parkinson disorder affects the central nervous system of homosapiens. Most of the existing approaches have detected the Parkinson Disorder using more number of voice features such as MDVP:Fo(Hz), jitter, shimmer, and so on, which causes complexity O (mxn) because of inclusion of unrequired extra features in the field n. In order to reduce the processing time, the feature subset is identified using Correlation based Feature Selector, Principal Component Analysis and Genetic Algorithm. In this paper supervised approach named Neural Network based Genetic Algorithm is used to construct the probabilistic model for existing data set. This model is compared with existing approach. The experiment reached a performance improvement of 95.38%.

Keywords: Parkinsonism disorder · Tele-monitoring data set · Correlation based feature selector · Principal component analysis · Genetic algorithm
Neural network based genetic algorithm

1 Introduction

Parkinson disorder is one of the chronic neuro degenerative disorders which happen because of genetic or environmental conditions. 15% of Parkinson Disorder (PD) occurrences are caused by genetic reasons and remaining was occurred because of environmental condition such as smoking, radiations, etc. Parkinson's incorporates the breakdown and end of basic nerve cells in the cerebrum, called neurons. It primarily impacts neurons in a region of the cerebrum known as substantia nigra. Ahead of schedule in the illness, the most obvious symptoms are shaking, unbending nature, gradualness of development, discourse, and trouble with walking, speaking, thinking and behavioural issues.

PD investigation by using voice signal of speaker has been expanded in a decade ago. Here there is a tele-monitoring data set with huge number of voice signal associated with speech is collected by clinical studies from healthy and PD affected speakers. In order to decrease the incorporation of additional features which are not necessary were removed by using feature sub selection algorithms. Consideration of extra features causing the increased complexity of O(mxn) for PD's tele-monitoring data set, where m is number of healthy and PD affected speakers and n is number of attributes. Such above challenge was addresses by feature sub selection algorithms,

© Springer International Publishing AG 2018
D. J. Hemanth and S. Smys (eds.), *Computational Vision and Bio Inspired Computing*,
Lecture Notes in Computational Vision and Biomechanics 28,
https://doi.org/10.1007/978-3-319-71767-8_56

which selects a subset of typical voice derived features adequate to differentiate healthy speaker from PD affected speakers.

In this paper tele-monitoring voice data set were considered for PD identification. In order to decrease the time for extracting large set of features in the data set, feature sub selections were carried out using Correlation based Feature Selection (CFS), Principal Component Analysis (PCA) and Genetic Algorithm (GA). It recognizes most required features for reducing the number of measures. Finally the voice from healthy speaker people and people with Parkinson's disease speaker is classified using probabilistic Neural Network based Genetic Algorithm (NN-GA) model to quantify the improvement in classification performance.

Beginning with the computational model of Parkinson disorder prediction system, the rest of the paper is structured as follows. In Sect. 2, we discuss about related works. In Sect. 3, gives our proposed system framework for Parkinson Disorder identification. In Sect. 4, give a discussion about Feature sub selection algorithms and Parkinson disorder prediction algorithm using NN-GA. In Sect. 5, carry a discussion about implementation setup and experimental results. In Sect. 6, we conclude the Parkinson prediction system and provide a way of doing future works.

2   Related Works

Hazan et al. [1] developed a recognition system on the basis of dysphonia. It has been recommended that discourse anomalies may be available at early phases of the illness. Karnebäck [2] low frequency modulation and Jennifer C. et al. Cepstral analysis were two speech waveform analysis methods which involve two steps while prediction such as extraction of features and classification. Difficulty in the part of this methods to decide which voice feature are non redundant and provide accurate prediction while classification needs feature sub selection.

Revett et al. [3] proposed Decision class based rough set approach were used as a feature sub selection method, which uses correlation measurements. It does not require any additional information's about features, which simply takes too low and too high correlated values based on decision classes as the required features. Particularly PPE is represented as a highly required feature.

Hall [4] defined Correlation based feature sub selection algorithm uses heuristic function to select the required feature subset, which is highly correlated with class and uncorrelated with each other features. CFS carried out filtration in large data set at both continuous and discrete intervals and also reduces dimensionality of data drastically to improve the performance of learning algorithms. Doshi [5] proved that the correlation based feature sub selection improves the classification accuracy through the removal of redundancy in attributes in student performance prediction project.

Malhi et al. [6] provided that PCA have a capacity to discriminate directions with the largest fluctuation in a data set, the appropriateness was contributed as an input to a well defined defect classification scheme to distinguishing the most required representative features from set of features. Yu et al. [7] use PCA to avoid the complicated as well as high dimensional financial ratios features, PCA extricate the low-dimensional and proficient features for enhancing the accuracy and efficiency while classification.

Sun et al. [8] defined a fitness function based high dimensional data clustering approach for feature subset selection uses genetic algorithm, which improved the performance through the redundant dimensions to some extent. Gu et al. [9] proved that GA could be effective for large scale feature sub selection.

Biologically inspired ANN was organized as human brain composed of neurons, which works as a decision making process. Lin et al. [10] prefer ANN for pattern recognition and classification. It has strong fault-tolerant and adaptive ability. Endo et al. [11] came up with GA for flexible learning and Arifovic et al. [12] suggested GA for MSE reduction.

For some PD, it is hard to visit the centre for checking and treatment; in any case, the getting to the Internet and the enhanced band-width of media telecommunication frameworks offer the opportunity to remotely screen the patients, that is, telemedicine. At that point so as to use these open doors, there is the requirement for dependable clinical observing system.

Nevertheless huge number of measures of dysphonia is computationally infeasible to test every single attributes combination. Besides, hypothetical examination demonstrates that as the list of capabilities size increments, dependable order is weakened by revile of dimensionality. Therefore there is a need for attribute determination to pick a subset that contains the most favourable amount of information for effective classification. A rigorous system model is required to evaluate an existing prediction system using quantifiable measurement from voice, and give intends to improve the accuracy of prediction. Vagueness in voice measurements need to be identified, hence accurate prediction of PD depends upon the erroneous observations. Redundant or irrelevant measurements not only prompt long analysis time, as well as increment finding unpredictability.

In this paper, we propose a probabilistic model based on supervised NN-GA is applied to identify the PD. The main contribution is selecting required features by comparing the various approaches such as CFS, PCA and GA and the results were analysed.

3 System Framework of Proposed Work

PD classification model utilizes tele-monitoring data set acquired from UCI repository. Data set contains biomedical voice features of 22 attributes and 4 patients related information such as name, age, sex and status of disorder. For limiting redundancy caused because of unrequired feature consideration in voice, feature sub selection utilizes rankers search based methods such as CFS, PCA and GA. For predicting PD affected patients from healthy individuals a newly developed NN-GA based classification model were used in whole data set and also data sets with subset features selected by using CFS, PCA and GA. Accuracy in the PD prediction system was based on the performance of NN-GA. The proposed work based on biomedical voice based PD classification system framework is represented in Fig. 1.

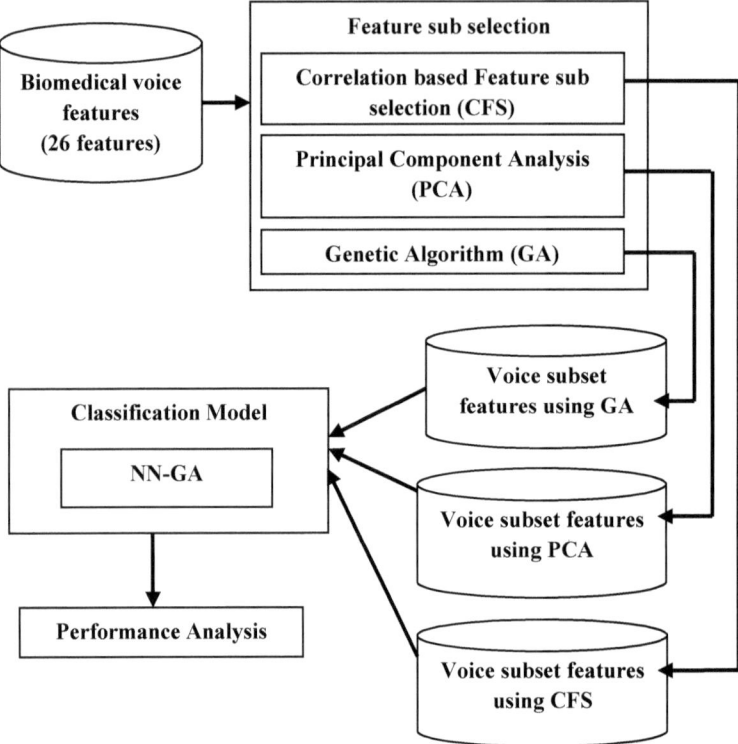

Fig. 1 System framework of biomedical voice based PD identification

4 Feature Sub Selection and Prediction

Feature sub selection methods selects the subset of features for the purpose of model simplification, reducing training time, reduce over fitting and curse of dimensionality [13].

4.1 Feature Sub Selection Methods

Feature sub selection methods carried out the subset selection by using CFS, PCA and GA and perform the comparison analysis to get accurate PD prediction system without redundancy and over fitting of feature values.

4.1.1 Correlation Based Feature Sub Selection

Rank search based filtering algorithm [4, 14] uses correlation based heuristic estimation function to detect the relationship among existing features. Heuristic function screened out the subset feature $(S_1, S_2 \ldots S_{nvf}$ where $S_{nvf} < 22)$ which was highly correlated with class. Redundant features were identified using high correlation with one feature to another. Feature subset evaluation function of CFS is given in Eq. (1) which was called as merit value (*Merit$_{val}$*). In *Merit$_{val}$* neumerator provide the value of how set of voice features predict the class and denominator offer the how much redundancy occurs among voice features.

$$Merit_{val} = \frac{nvf * m_{fc}}{\sqrt{nvf + nvf(nvf - 1)a_{ff}}} \qquad (1)$$

Subset features $S_1, S_2 \ldots S_{nvf}$ were calculated from heuristic merit function in which nvf be the total number of voice features which was $S_{nvf} < 22$. The mean of voice feature with class is m_{fc} where feature f is the subset of class c and the average correlation exists between features with other features. The diagramatic representation of working of CFS is represented in Fig. 2.

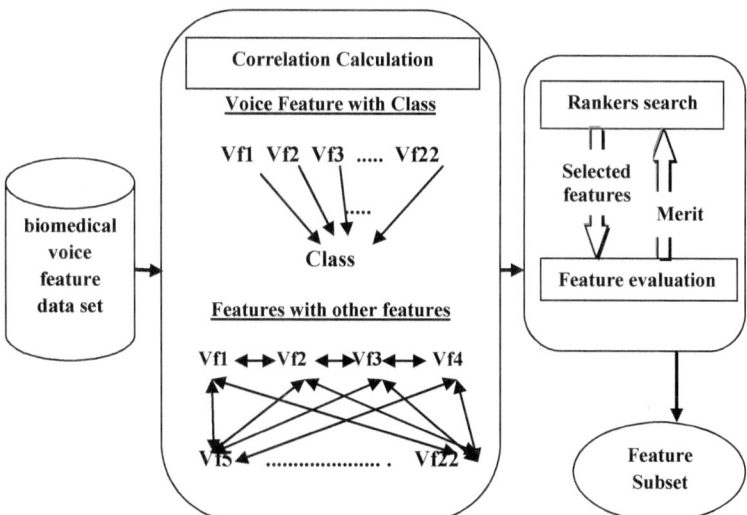

Fig. 2 Working of CFS algorithm based on heuristic function

4.1.2   Principle Component Analysis Based Feature Sub Selection

PCA is the leading linear dimension reduction method. Boutsidis et al. [15] extensively applied PCA on data sets in all domains, which seeks the finest required subset of exactly n columns from $m \times n$ data matrix. It extensively deliberates in Linear Numerical Algebra community. The stpes for selecting subsets attributes subset using PCA is given as follows.

Steps for selecting attribute subset using PCA

Step 1: *Input*: Provides the voice features in the biomedical data set as data vectors having N dimensions.

Step 2: Normalizes the voice data set in which each and every voice features were comes under certain range.

Step 3: Computes principal components as K orthogonal vectors where $K \leq N$. Each voice input features are the linear combination of K principal components.

Step 4: Sorts the principal component based on strength or significance.

Step 5: The input features are reduced by removing weak components, where weak components have low variance.

Step 6: *Output*: Strong components are taken as a selected voice features for avoid redundancy exists in the biomedical voice data set.

4.1.3   GA Based Feature Sub Selection

GA based feature sub selection algorithm derived by Sun et al. [8] quoted in Sect. 2, discriminate the targets from the natural clutter in specific application domains and also used to automatically identify the required relative significance of various features. The stpes for selecting subsets attributes subset using GA is given as follows.

Steps for feature sub selection using GA

Step 1: Create voice features in the biomedical tele-monitoring data set as initial population.

POPFUNCTION()

pop ← Population Size in binary matrix * Genome Length

Return pop

Step 2: Evaluate the fitness value by calculating the KNN error rate.

Step 3: Selection of parent chromosome until offspring < parent by performing elicit, crossover and mutate children.

Step 4: *Output*: Fittest chromosome (unredundant voice features).

4.1.4   Classification Model (NN-GA)

In this part of classification, training NN to determining the best possible set of chromosome called as connection weights. Minimization of NN error function over training data can be measured by iteratively adjusting the link weights. The mean square error between the target and the actual output averaged overall output nodes serve as a superior estimate of the fitness of the network configuration for the current input [16]. The stpes for NN-GA to identify PD is given as follows.

NN-GA algorithmic steps

Input: N1, N2 … NN be the number of populations (voice features).

Step 1: NN weights in a network were combined to form a chromosome.

Step 2: Performance Criterion for model selection.

- Divide the biomedical voice feature data set into three groups before and after selecting features using CFS and PCA
  - ➤ Training data set (70%)
  - ➤ Testing data set (20%)
  - ➤ Unseen data set (10%)
- Calculate the error in training data set $Error_{training}$ using Eq. (2)

$$Error_{training}^2 = \frac{1}{N_{TR}} \sum_{i=1}^{N_{TR}} DO_i^{TR} - \left(M_{TR}\left(O_i^{TR}\right)\right)^2 \qquad (2)$$

- Calculate the error in testing data set $Error_{testing}$ using Eq. (3)

$$Error_{testing}^2 = \frac{1}{N_{TE}} \sum_{i=1}^{N_{TE}} DO_i^{TE} - \left(M_{TE}\left(O_i^{TE}\right)\right)^2 \qquad (3)$$

- Calculate the networks performance criterion P_C using Eq. (4)

$$P_C = Error_{training}^2 + Error_{testing}^2 + wf\left(Error_{training}^2 - Error_{testing}^2\right)^2 \qquad (4)$$

- Calculating the error in unseen biomedical voice data set $Error_{Unseen}$ using Eq. (5)

$$Error_{Unseen}^2 = \frac{1}{N_U} \sum_{i=1}^{N_U} DO_i^U - \left(M_U\left(O_i^U\right)\right)^2 \qquad (5)$$

where N_{TR} is a number of training voice data, N_{TE} is a number of testing voice data, and N_U be the number of unseen voice data taken from Parkinson affected and healthy individuals. Desired outcome of training, testing and unseen data to be represented as DO_i^{TR}, DO_i^{TE} and DO_i^U were used for calculating Error in data. Training, testing and unseen models outcomes are represented as $M_{TR}\left(O_i^{TR}\right)$, $M_{TE}\left(O_i^{TE}\right)$ and $M_U\left(O_i^U\right)$ which are also involved in error calculation. Weight factor wf were used for calculating performance criterian P_C.

Step 3: Using genetic algorithm to select the parent by using performance criterion $Perfor_C$ which met all $Error_{training}$, $Error_{testing}$, and $Error_{Unseen}$ data set and also avoid overfitting which is represented in Fig. 3.

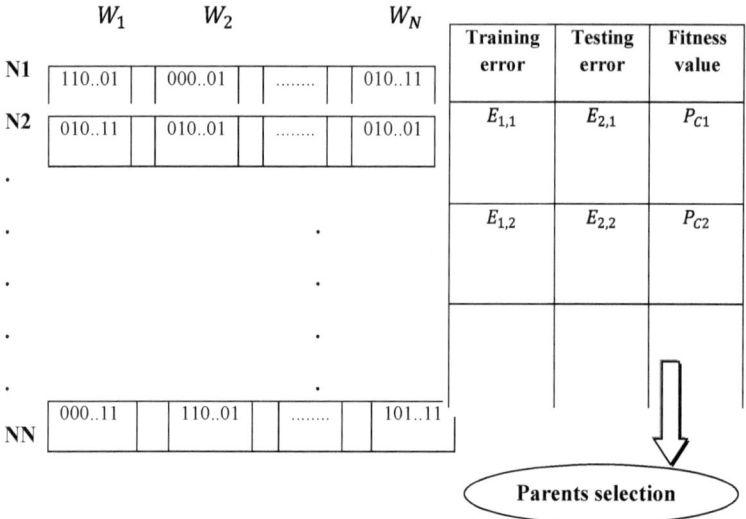

Fig. 3 Working of NN-GA

Step 4: For each iteration fittest Parent is selected and new generation is created by performing crossover with their weights and mutating its parameter.
Step 5: Until offsprings are less fit than their parent, repeatedly estimate a new fitness using Eq. (3 and 4).

Output: Correct prediction of PD class.

5   Implementation

5.1   Implementation Set-up

Implementation of PD prediction system was done on software MATLAB version 7.14 and released name 2012a. This software uses tele-monitoring PD disorder data set of size 890 KB from UCI repository, which have biomedical voice features such as MDVP:F0(Hz), Shimmer, Jitter, PPE and so on.. are used for accurate prediction of PD.

5.2   Experimental Analysis

In this session offer various feature sub selection techniques selects various attributes based on their perspective explained in session 4. After predicting the results made a comparison with existing Rough set (RS) approach based feature sub selection proposed by Revett et al. [3].

The metrices used to evaluate the proposed system are true positive (TP rate), false positive (FP rate), Precision, Recall, F-measure, ROC area as illustrated in Table 1.

Table 1 Performance evaluation of PD identification system

Techniques	TP rate	FP rate	Precision	Recall	F-measure	ROC area
NN-GA	0.821	0.193	0.844	0.821	0.827	0.907
CFS + NN-GA	0.872	0.306	0.870	0.872	0.864	0.824
PCA + NN-GA	0.846	0.184	0.859	0.846	0.850	0.831
GA + NN-GA	0.954	0.078	0.954	0.954	0.953	0.992

Figure 4 shows comparison of different error metrics such as mean absolute error, root mean square error, relative absolute error, and root relative squared error. Different error measurements analysis have their own perspective, which offer the same result of low error rate such as 6.45%, 17.77%, 15.24%, and 38.67% when GA was used for feature sub selection.

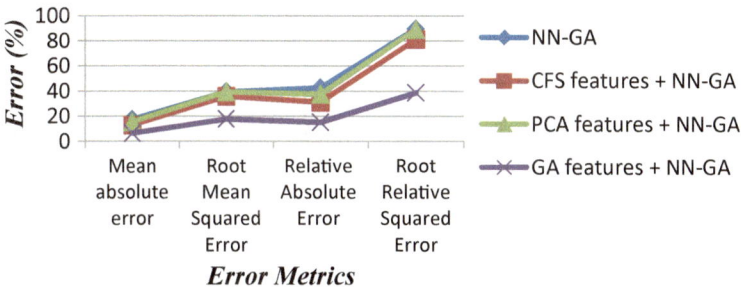

Fig. 4 Comparison of different error metrics

Based on proposed system GA based feature sub selection offer more prediction accuracy. To classify the GA based subset features using NN-GA produces the highest accuracy of 95.38%. Our system get much better when compared with exising rough set based feature sub selection using decision classes also involved in NN-GA classifier offer 84.32%. The direct involvement of entire datset classification have low prediction accuracy because of redanancy in voice attributes are represented in Fig. 5.

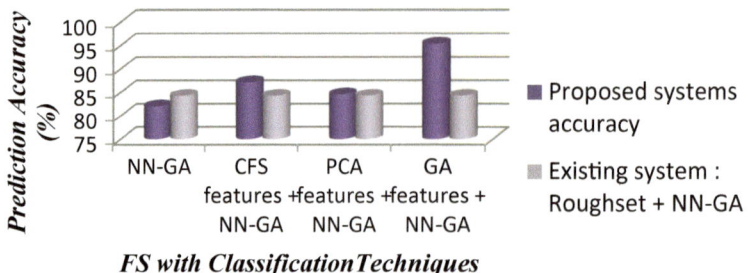

Fig. 5 Prediction accuracy compared with existing rough set based feature sub selection with NN-GA

6 Conclusion and Future Work

This work proposed an efficient PD identification system using NN-GA to classify the bio medical voice in the tele-monitoring PD database. Before that the most significant number of voice samples features is identified with GA as feature sub selection. Initial findings attained that GA as a training algorithm for the neural network for the recognition of PD. Experimental results of Parkinson's disorder identification system offers the entire data set classification accuracy using NN-GA be 82.05%. While feature sub selection carried out through GA reached the peak 95.38% accuracy in prediction. These findings shows the proposed method has higher prediction power compared to existing roughest based feature subset selection method. Further studies can be carried out for getting better classification power of this technique at medical fields.

References

1. Hazan, H., Hilu, D., Manevitz, L., Ramig, L.O., Sapir, S.: Early diagnosis of Parkinson's disease via machine learning on speech data. In: 2012 IEEE 27th Convention of Electrical and Electronics Engineers in Israel (IEEEI), pp. 1–4. IEEE 2012
2. Karnebäck, S.: Discrimination between speech and music based on a low frequency modulation feature. In: Seventh European Conference on Speech Communication and Technology 2001
3. Revett, K., Gorunescu, F., Salem, A.B.M.: Feature selection in Parkinson's disease: a rough sets approach. In: 2009 International Multiconference on Computer Science and Information Technology IMCSIT'09, pp. 425–428. IEEE 2009
4. Hall, M.A.: Correlation-based feature selection of discrete and numeric class machine learning (2000)
5. Doshi, M.: Correlation based feature selection (Cfs) technique to predict student perfromance. Int. J. Comput. Netw. Commun. **6**(3), 197 (2014)
6. Malhi, A., Gao, R.X.: PCA-based feature selection scheme for machine defect classification. IEEE Trans. Instrum. Meas. **53**(6), 1517–1525 (2004)
7. Yu, H., Chen, R., Zhang, G.: A SVM stock selection model within PCA. Procedia comput. sci. **31**, 406–412 (2014)

8. Sun, M., Xiong, L., Sun, H., Jiang, D.: A GA-based feature selection for high-dimensional data clustering. In: 2009 3rd International Conference on Genetic and Evolutionary Computing WGEC'09, pp. 769–772. IEEE 2009

9. Gu, Y., Tan, S.L., Wong, K.J., Ho, M.H.R., Qu, L.: Using GA-based feature selection for emotion recognition from physiological signals. In: 2008 International Symposium onIntelligent Signal Processing and Communications Systems, pp. 1–4. IEEE 2009

10. Lin, Y., Zhang., Y.: Prediction of number of depression patients based on neural network. In: 2012 IEEE 3rd International Conference on Software Engineering and Service Science (ICSESS), pp. 599–602. IEEE 2012

11. Endo, S., Ohuchi, A.: Genetic based concept learning from positive examples. In: Proceedings of the 34th SICE Annual Conference. International Session Papers SICE'95, pp. 1619–1622. IEEE 1995

12. Arifovic, J., Gencay, R.: Using genetic algorithms to select architecture of a feedforward artificial neural network. Phys. A Stat. Mech. Appl. **289**(3), 574–594 (2001)

13. Wikipedia [Online]. Available: https//En.Wikipedia.Org/Wiki/Feature_Selection

14. Cui, Y., Jin, J., Zhang, S., Luo, S., Tian, Q.: Correlation-based feature selection and regression. Adv. Multimed. Inf. Process. PCM **2010**, 25–35 (2010)

15. Boutsidis, C., Mahoney, M.W., Drineas, P.: Unsupervised feature selection for principal components analysis. In: Proceedings of the 14th ACM SIGKDD International Conference on Knowledge Discovery and Data Mining, pp. 61–69. ACM (2008)

16. Karimi, H., Dastranj, J.: Artificial neural network-based genetic algorithm to predict natural gas consumption. Energ. Syst. **5**(3), 571–581 (2014)

Cancelable Biometrics Using Geometric Transformations and Bio Hashing

R. Parkavi[⊠], K. R. Chandeesh Babu, T. Neelambika, and P. Shilpa

Department Information Technology, Thiagarajar College of Engineering,
Madurai, Tamil Nadu, India
parkaviravi@gmail.com

Abstract. Biometrics is a security system which includes the physical traits of a person such as facial features, fingerprints, iris detection etc for the verification and authentication of a system. These features are stored as a template also known as a biometric template in database or server. Once the Biometric template is compromised it cannot be changed like the passwords and therefore to ensure cancelability of several biometric templates several papers were introduced. Cancelability is a method of generating a new biometric template from the original template using several approaches and algorithms. In this paper cancelability in biometrics is achieved by applying algorithms such as geometric transformations and bio hashing algorithms. First, at the time of registering user, the traits of a user such as a fingerprint and a face is scanned using a scanner and the biometric template thus obtained is encrypted first using geometric transformation algorithms such as Cartesian transform, polar transform and functional transforms. Then bio hashing is applied to the user specified tokenized random number to the transformed template to obtain the encrypted template. The encrypted template is stored in the database which can be further used for verification. In the case of security attacks which may compromise the biometric template a new set of a template is obtained from the same user by using the different set of tokenized random numbers.

Keywords: Fingerprint · Minutiae · Biometric template · Cancelability
Projections · Features

1 Introduction

Biometrics refers to the physical characteristics of a user. The characteristics of a user involve facial features, fingerprints, iris, palm etc. These features can be used for authentication which is known as a biometric security system. Biometric is an authentication system which includes identification and process control using several human traits such as fingerprints, palm, iris recognition, face features etc [1]. These features are recorded and stored as a template which can be further used for verification. Several security issues are associated with biometric such as maintaining its secrecy. The other security systems which include passwords may be memorized by the owner and can maintain secrecy but biometric is exposed to all making it difficult to

© Springer International Publishing AG 2018
D. J. Hemanth and S. Smys (eds.), *Computational Vision and Bio Inspired Computing*,
Lecture Notes in Computational Vision and Biomechanics 28,
https://doi.org/10.1007/978-3-319-71767-8_57

maintain secrecy. Some other issues are that if application involving biometrics is compromised then other applications involving the same biometrics may also be compromised. Also, it is difficult to generate the same biometric template as the user may correspond to some disturbance during authentication. Due to security issues such as several attacks which may compromise the biometric template. Unlike other security systems which involve generation of new passwords, we cannot generate a new fingerprint. So in order to ensure cancelability in biometric template several algorithms are used which transforms the original biometric template to maintain cancelability. If the biometric template is compromised several new unique biometric templates can be produced from several algorithms. Providing cancelability in biometrics provides enhanced security and privacy [2].

2 Literature Survey

In this paper, detailed survey papers on cancellable biometrics are discussed and problems on cancelability in biometrics are addressed here.

2.1 On Mixing Fingerprints

The objective of protecting the biometric template is achieved by mixing two different fingerprints and generating a new fingerprint. Each fingerprint is divided into two component, Continuous component, and spiral component [3]. The continuous component of one fingerprint is mixed with a spiral component of another fingerprint. Fingerprint mixing can be used to generate a large set of virtual identities. These virtual identities can be used to conceal the original identities. The Main steps used in fingerprint mixing are Fingerprint decomposition, Fingerprint pre-alignment, Mixing Fingerprints. In fingerprint decomposition, the fingerprint is decomposed into continuous fingerprint and spiral fingerprint. Fingerprint pre-alignment involves locating a reference point and finding the alignment line. After pre-alignment, the fingerprints are combined into one by using the formula.

$$MF_1 = \cos(\Psi_{v2} + \Psi_{v1}) \tag{1}$$

Similarly, MF2 is calculated. The continuous phase of F2 (F1) is combined with the spiral phase of F1 (F2). Advantages of the proposed approach prevent identity linking by preventing the possibility of successfully matching the original part with the mixed part. It is computationally infeasible to obtain original fingerprint features from a mixed fingerprint. In case if a stored fingerprint is compromised the newly mixed fingerprint can be generated by mixing original with a new fingerprint. Disadvantages are Generation of new several sets of different unique fingerprints is not possible since the template is created from the original fingerprint only. Also, variations in orientations and frequencies of ridges between fingerprints can result in visually unrealistic fingerprint images.

2.2   Biometric Cryptosystem Involving Two Traits and Palm Vein as Key

The proposed approach involves multiple traits of a person are used where if anyone of the trait fails the other could be used for verification [4]. The biometric traits are a face, fingerprint palm which are fused together with anyone of the feature is used as a key for verification such that it cannot be verified unless it belongs to the same person. The methodologies involved are Fingerprint feature extraction: Each person has different fingerprints and these fingerprints are unique. Fingerprints consist of many ridges and furrows and are distinguished by minutia. The fingerprint feature vector is created by image pre-processing, minutiae extraction, minutiae pre-processing, minutia matching. Face Feature Extraction: Each individual has a unique face feature. These features are extracted from the face and are represented as objects. These features are represented in a matrix which is the feature vector of the face. Key generation from palm vein: The features of the palm are identified and extracted. These features are represented in the matrix. The key vector is generated from the matrix which is more secure and irrevocable. This key is used to encrypt the template obtained from fingerprint or face feature using RSA algorithm. Advantages of the proposed algorithm differentiate imposture minutiae pairs from genuine minutia pairs in a certain confidence level. The algorithm is capable of differentiating fingerprints at a good correct rate by setting an appropriate threshold value. If the biometric template is compromised new template can be generated by using the different key. Disadvantages are Cancelability cannot be achieved since only a single biometric template can be created for a particular person. Also, the biometric template is invertible i.e. if the biometric template is inverted the original features of a person are obtained.

2.3   Cancelable Biometrics Realization with Multispace Random Projections

One of the main disadvantages of biometrics is that if the biometric template is compromised then it cannot be changed. To meet this disadvantage the author proposed an approach which is a two-factor biometric formulation [5]. The biometric data is transformed into a revocable but irreversible by transforming the biometric data into a fixed-length feature vector and then projecting the feature vector onto a sequence of random subspaces that were derived from a user-specific pseudorandom number (PRN). The multi-space random projection is carried out with the help of user specified a pseudorandom number (PRN). Since different users hold their own set of PRNs, the formulation can be extended to include multiple users or applications to produce multiple random subspaces. Multispace random projection consists of two stages namely feature extraction and multi-space random projections. In feature extraction, the biometric image is transformed into a fixed length feature vector by using a feature extractor. The biometric feature vector x is further projected onto a random subspace, as determined from an externally derived pseudorandom sequence.

Advantages are Multispace random projections provides a high level of security where access to a system requires a valid token and individual's face features whereas the traditional system has single factor authentication system. Also, cancelability is achieved. If the biometric template is compromised the distortion characteristics are changed and the same biometrics is mapped to a new template. Disadvantages are the PRN overpowers the biometrics in the MRP formulation, and the resulting biometrics role is nullified. Also each time the pseudo random number varies thereby making it difficult for verification. Also, the verification performance depends on the quality of feature extractor.

2.4 Cancelable Templates for Sequence-Based Biometrics with Application to On-Line Signature Recognition

From the proposed approach which involves Bio Convolving, it is able to guarantee security and renewability to biometric templates [6]. A set of non-invertible transformations, which can be applied to any biometrics whose template can be represented by a set of sequences, in order to generate multiple transformed versions of the template. Once the transformation is performed, retrieving the original data from the transformed template is computationally as hard as random guessing. The methodologies used are Generating Cancelable sequence based biometric templates: The proposed approach which involves Bio convolving provides protection to templates characterized by a set of discrete finite sequences extracted from a given biometrics. Then transformations are applied which involves two methods namely Noninvertible transform: Baseline approach and Noninvertible transform extended approach. In baseline approach, transformations are applied to a set of finite sequences obtained from Bio convolving to obtain the transformed template. Each transformed sequence is obtained through the linear convolution of parts of the corresponding original sequence. Since the baseline approach does not possess a renewable capability two extended approaches were used they are Mixing approach in which the scrambled version of a vector is obtained as a column matrix. Shifting approach where an initial shift is applied to original sequences. Advantages are from the given paper the security and privacy issues are handled. Biometric template renewability is provided. Cancelability is achieved by generating multiple biometric templates from the given template each of which is non-invertible. Disadvantages are although with the proposed approaches, it is not possible to obtain an infinite number of discriminable templates. Also, the baseline protection approach presented produces only a slight loss of performance with respect to an unprotected system. If the key is obtained then it can easily be inverted which reveals the original features of the individual.

2.5 Generating Cancelable Fingerprint Templates

In this paper, we demonstrate several methods to generate multiple cancelable identifiers from fingerprint images to overcome these problems [7]. In essence, a user can be given as many biometric identifiers as needed by issuing a new transformation "key." The identifiers can be canceled and replaced when compromised. We empirically compare the performance of several algorithms such as Cartesian, polar, and surface folding transformations of the minutiae positions. It is demonstrated through multiple experiments that we can achieve revocability and prevent cross-matching of biometric databases. It is also shown that the transforms are noninvertible by demonstrating that it is computationally as hard to recover the original biometric identifier from a transformed version as by randomly guessing. The cancelable fingerprint template is generated using a one-way transformation [8]. Instead of storing the original minutiae features, the minutiae locations and orientations that are stored transformed irreversibly. The steps involved are Registration, Intra user variability tolerance, Entropy retention, Transformation function design. Also, several transformations such as Cartesian transformation, polar transformation, and functional transformations are used. The Cartesian transformations consist of changing the cell positions. The minutiae positions are obtained in rectangular coordinates with reference to the position of a singular point. Then the positions are changed. In polar transformation, the minutiae positions are measured in polar coordinates with reference to core positions. The transformation consists of changing the sector positions. In functional transform, the minutiae positions are identified and are passed to the functions to obtain the transformed template. Advantages are the obtained biometric template is distinct and unique i.e. the original fingerprint and its transformed version are not correlated. The local smoothness of biometric template is achieved by a surface folding algorithm. Also, the biometric template obtained is non-invertible i.e. if it is inverted the original data cannot be revealed. Disadvantages are the sharp boundaries in the Cartesian transform and polar schemes can sometimes move minutiae point to a totally different part of the image if its detected position changes just slightly. Also during authentication generating the same biometric template for matching is made complex here.

2.6   Alignment Free Cancelable Fingerprint Templates based on Local Minutiae Extraction

Several approaches used either the relative position of minutiae to the core point or absolute position of minutiae in a given fingerprint image. In this paper a new approach has been proposed which involves translation and rotation for each minutia and the invariant value is computed from the orientation information the invariant value is used as input to two changing functions that output two values for the translational and rotational movements of the original minutia respectively in the cancelable template. The methodologies involved in this approach are pre-processing, minutiae extraction, calculation of invariant value, changing functions, minutia movement, and generation of a cancelable template and minutia matching. Image pre-processing involves pre-processing of an input fingerprint image consisting of segmentation, orientation estimation, ridge enhancement and thinning [9]. The image is pre-processed and minutiae are extracted. The invariant value is calculated for each minutia. Two changing functions which include distance changing function and orientation changing function are applied and the cancelable template is obtained. While matching it checks the similarity of minutiae in two fingerprints with respect to their positions, orientations, and types.

Advantages are the verification accuracy and performance is good in matching. There is also a huge dissimilarity between the original fingerprint and obtained fingerprint. Also, cancelability and changeability are achieved. Disadvantages are the

Biometric template generated is invertible [10]. Because of the amount of transformation for each minutia which is determined by its invariant value and two changing functions, if the invariant key is obtained the original minutiae can be recovered from the transformed minutiae.

3   Generating Cancelable Biometric Template

In order to achieve both cancelability and non-invertibility in the biometric template a new type of transformation is applied and combined with bio-hashing [7]. The biometric image is transformed using any one of the non-invertible geometric transforms such as Cartesian transform, polar transform and functional transform. The biometric template is applied to functional transform in which each minutiae position is transformed using non-invertible function $y = f(x)$. Each minutia is identified using a feature extractor and using a user specified tokenized random number (TRN), n orthogonal pseudo-random vectors are generated. Then the dot product of the feature vector and all the random vectors is calculated. Finally, a binary discretization is applied to compute n bit Bio-Hash template. The Bio-hashing framework is a one-way process. The steps involved in generating cancellable biometric template are Face Detection: The user's face is detected using a camera in an authentication system at the time of authentication. Convert to gray scale and scaling: The obtained template is converted to gray scale and scaling is done to identify the distinctive features of the obtained template. Feature extraction: The features of the biometric template are extracted and the pixel position is identified. The obtained pixel position is then passed through the polar transform involving transformation. Key generation and encryption: The sixteen-bit hexadecimal key is then generated randomly and then the dot product of the user given tokenized random number and the key is computed. Using the resultant key and the obtained template after transformation bio hashing is applied to obtain the encrypted template. Face detection: First using the camera of the authentication system the face of the user is captured and it is stored as a template for further processing. The captured image size is 320×240. Scaling: If the obtained input image is a color image it is converted into a grayscale image and the template is converted to binary to perform the computation. The data i.e. the template is reshaped and padding is applied

$$\text{Data} = \text{reshape}(\text{Data}, [\text{size}(\text{Data}, 1) * \text{size}(\text{Data}, 2)1]);$$

$$\text{padding} = \text{mod}(\text{length}(\text{Data}));$$

Polar transformation: In this method, the minutiae positions are measured in polar coordinates with reference to the core position. The angles are measured with reference to the orientation of the core. The coordinate space is now divided into polar sectors (L levels and S angles) that are numbered in a sequence. The process of transformation now consists of changing the sector positions. The minutiae angles also change in accordance with the difference in the sector positions before and after transformation. The non-invertible transform consists of changing the polar wedge positions. The minutiae angles also change with differences in the wedge positions before and after transformation [11]. Unlike the Cartesian transformation, an unconstrained mapping is not feasible in polar coordinates since the angular shift is transformed to a positional shift at a distance r from the core. This leads to a situation where minutiae pairs that occur within a tolerance distance of each other before transformation no longer match after transformation due to the large divergence that occurs away from the core [12]. The positions of the sector before and after transformation are related as

$$C(n) = C + M \qquad (2)$$

where C(n) is the position of the cells after transform and C is the position of the cell before transform and M are the mapping matrix. Key generation and encryption: The sixteen-bit hexadecimal key is generated randomly and several computations are applied on the key. First, they obtained the sixteen-bit hexadecimal key is converted to a binary key. A four cross four matrixes each of 16 bit is generated. One round of circular shift is applied for each cell positions in a matrix. The column-wise transposition of the matrix is applied. The tokenized random number is obtained from the user and is converted into binary. Then the dot product of the resultant hexadecimal key and the tokenized random number obtained from the user is calculated by performing AND operation since it is converted into binary. The resultant key is 64 bit since each hexadecimal bit contains four binary digits and it is a sixteen-bit hexadecimal key. The 64-bit key is concatenated and a circular shift is performed and then converted. The resultant key is converted to a sixteen-bit hexadecimal key. This key is used for encryption. Now using the key bio hashing is applied using the formula, where sig(.) is defined as a signum function and x is an empirically determined threshold. The above equation only applies to a user who holds the user-specific random vectors.

$$c = \mathrm{Sig}\left(\sum_i x b_i - \tau\right) \quad b_i \in R^N, i = 1, \ldots, n \qquad (3)$$

4 Experiments and Observations

Figure1 shows the overall working principle of Bio Hashing and Fig. 2 shows encryption of a template. Performance is calculated on the basis of Histogram analysis.

Performance Analysis

Histogram analysis: We performed a histogram test on plotting both the original template and the obtained cancellable test and the analysis shows huge variance which means that the obtained template is different from the original template. From the Fig. 3, we can see that the top one is the histogram of the obtained cancellable template and the bottom one is the histogram of the original template. By comparing these two images we can see that there is a huge difference which indicates that the obtained cancellable template is distinct i.e. the original template differs from the obtained template. The test also indicates that both the original template and the obtained template are not correlated.

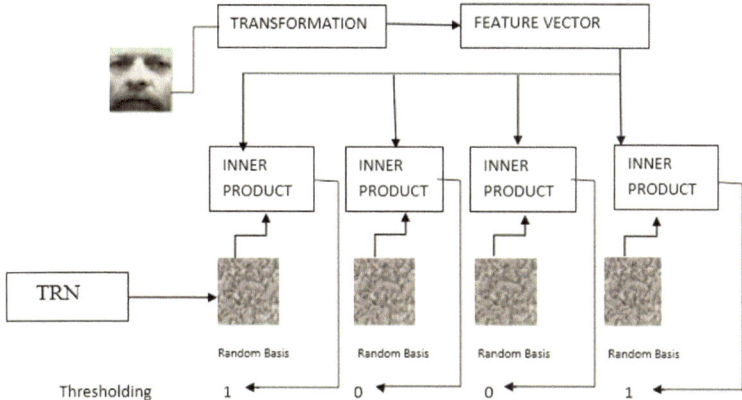

Fig. 1 An overview of bio hashing

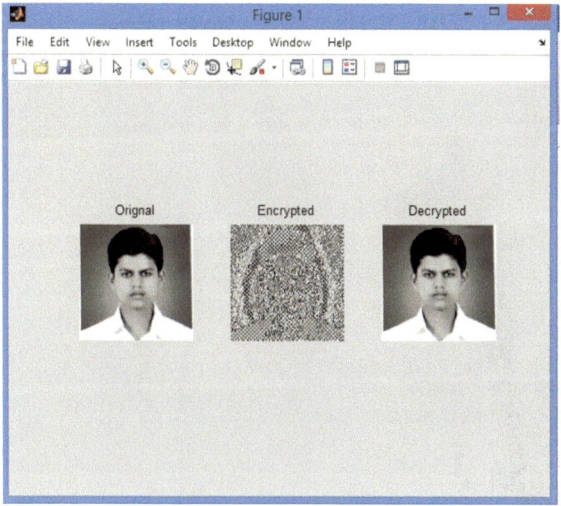

Fig. 2 Encryption of template

Non-Invertibility: The biometric template obtained also is non-invertible i.e. the generated cancelability cannot be converted or decrypted back to the original template since the key cannot be determined. Also if the randomly generated key is determined the user given random numbers cannot be hacked since both are needed for generating the key. Therefore we used brute force attack to test all possible combinations of key and the test failed which shows that the biometric template is noninvertible.

MATLAB benchmark test: We also did MATLAB benchmark test which showed how effective was that system in generating the template and it took no less than a second to generate the template. Also, the time taken to generate the template varies by hardware.

Cancelability: This is one of the main factors in this project. The template if in case hacked or attacked the new set of a biometric template can be generated for the same user using the different hexadecimal key. Also, the template is dynamic i.e. the template keeps changing due to changing hexadecimal key but the user given tokenized random number is constant as in Fig. 4. The obtained template, therefore, is cancelable.

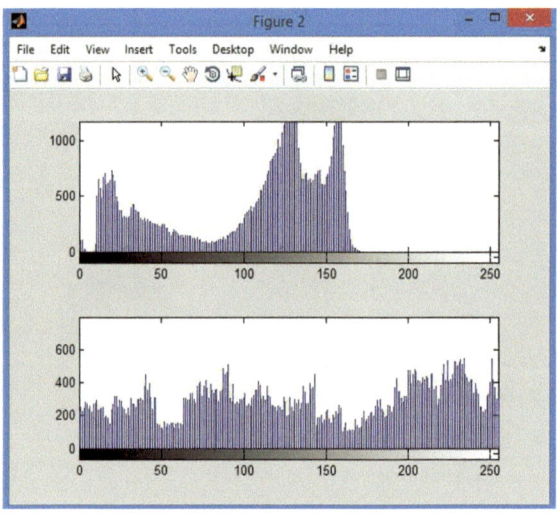

Fig. 3 Histogram analysis

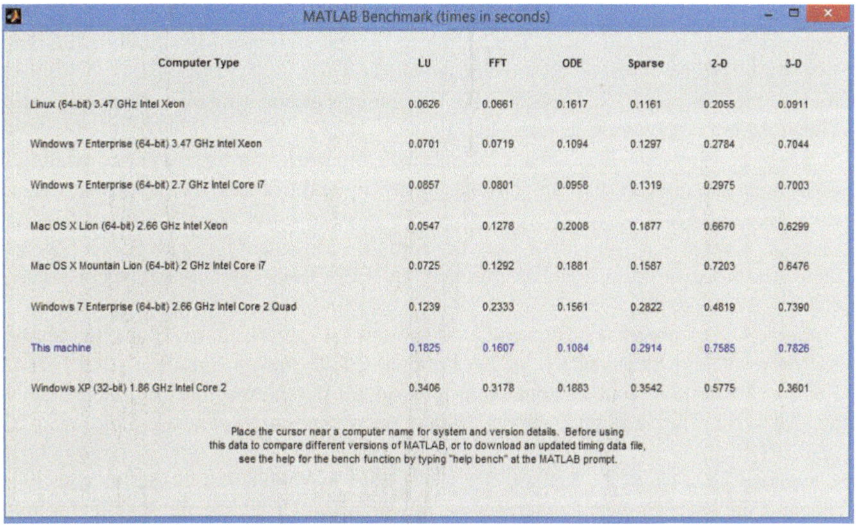

Computer Type	LU	FFT	ODE	Sparse	2-D	3-D
Linux (64-bit) 3.47 GHz Intel Xeon	0.0626	0.0661	0.1617	0.1161	0.2055	0.0911
Windows 7 Enterprise (64-bit) 3.47 GHz Intel Xeon	0.0701	0.0719	0.1094	0.1297	0.2784	0.7044
Windows 7 Enterprise (64-bit) 2.7 GHz Intel Core i7	0.0857	0.0801	0.0958	0.1319	0.2975	0.7003
Mac OS X Lion (64-bit) 2.66 GHz Intel Xeon	0.0547	0.1278	0.2008	0.1877	0.6670	0.6299
Mac OS X Mountain Lion (64-bit) 2 GHz Intel Core i7	0.0725	0.1292	0.1881	0.1587	0.7203	0.6476
Windows 7 Enterprise (64-bit) 2.66 GHz Intel Core 2 Quad	0.1239	0.2333	0.1561	0.2822	0.4819	0.7390
This machine	0.1825	0.1607	0.1084	0.2914	0.7585	0.7826
Windows XP (32-bit) 1.86 GHz Intel Core 2	0.3406	0.3178	0.1883	0.3542	0.5775	0.3601

MATLAB Benchmark (times in seconds)

Place the cursor near a computer name for system and version details. Before using this data to compare different versions of MATLAB, or to download an updated timing data file, see the help for the bench function by typing "help bench" at the MATLAB prompt.

Fig. 4 Time consumed

5 Conclusion

From the given paper we can infer that several algorithms ensure cancelability in a biometric template. From the proposed approaches several sets of new unique biometric templates can be generated if the biometric template is compromised. Both the security and privacy is enhanced by using cryptography and non-invertible property is achieved. Also in combining fingerprint approach a database of virtual identities is generated from a fixed dataset. In general overall, all these papers address the solution of cancelability in biometrics.

References

1. Toli, C.A. et al.: A survey on multimodal biometrics and the protection of their templates. International Federation for Information Processing (2015)
2. Hirano, T., Ito, T., Kawai, Y.: A practical attack to AINA2014's countermeasure for cancelable biometric authentication protocols. In: International Symposium on Information Theory and Its Applications (ISITA) (2016)
3. Othman, A., Ross, A.: On mixing fingerprints. IEEE Trans. Inf. Forensics and Secur. **8**(1) (2013)
4. Prassanalakhsmi, B., Kannammal, A., Gomathi, B., Deepa, K., Sridevi, R.: Biometric cryptosystem involving two traits and palm vein as key. In: International Conference on Communication Technology and System Design (2011)
5. Tech, A.B.J., Yuang, C.T.: Cancelable biometrics realization with multispace random projections. IEEE Trans. Syst. Man Cybern. (2007)

6. Maiorana, E., Campisi, P., Ortega-garcia, J., Neri, A.: Cancelable templates for sequence-based biometrics with application to on-line signature recognition. IEEE Trans. Syst. Man Cybern.—Part A: Syst. Humans **40**(3) (2010)
7. Ratha, N.K., Chikkerur, S., Connell, J.H.: Generating cancelable fingerprint templates. IEEE Trans. Pattern Anal. Mach. Learn. (2007)
8. Piciucco, E., Maiorana, E., Kauba, C.: Cancelable biometrics for finger vein recognition. In: First International Workshop on Sensing, Processing and Learning for Intelligent Machines (SPLINE) (2016)
9. Lee, C., Choi, J.Y., Toh, K.A., Lee, S., Kim, J.: Alignment-free cancelable fingerprint templates based on local minutiae information. IEEE Trans. Syst. Man Cybern.—Part B: Cybern. **37**(4) (2007)
10. Rathgeb, C., Breitinger, F., Busch, C.: Alignment-free cancelable iris biometric templates based on adaptive bloom filters. In: International Conference on Biometrics (ICB) (2013)
11. Kaur, K, Khanna, P.: Gaussian random projection based non-invertible cancelable biometric templates. In: Eleventh International Multi-Conference on Information Processing-2015 (IMCIP) (2015)
12. Choudhury, B., Then, P., Raman, V.: Cancelable iris biometrics based on data hiding schemes. In: IEEE Student Conference on Research and Development (SCOReD) (2016)

Automated Tool for Assessment of Satellite Image Quality and Characteristics

P. Muni Deepak[1,2], V. Kamalaveni[1(✉)], E. Venkateswarlu[2],
Thara Nair[2], G. P. Swamy[2], and B. Gopala Krishna[2]

[1] Department of Computer Science and Engineering, Amrita School of
Engineering, Amrita Vishwa Vidyapeetham, Amrita University, Coimbatore
641112, India
cb.en.p2cvil5006@cb.students.amrita.edu,
v_kamalaveni@cb.amrita.edu
[2] Data Processing, Products Archival and Web Applications Area, National
Remote Sensing Center, ISRO, Hyderabad, India
{venkateswarlu_e,thara_nair,swamy_gp,bgk}@nrsc.gov.in

Abstract. The functionality of automated tool developed for computing quality metrics to assess the satellite image quality and characteristics is presented in this paper. The tool facilitates to analyse quantitatively different sets of satellite images. This tool can also be used to compare the different sets of satellite images. In this paper, we present quality metrics based on single image and multiple images, thus aiding in relative characterization. The tool estimates the Modulation Transfer Function (MTF) and converts the gray image to reflectance image. GUI has been developed for operational convenience and a log file is maintained for recording the generated parameters.

Keywords: Quality · Image · Characteristics · Assessment · Automated tool
Entropy · Radiance · Reflectance · MTF

1 Introduction

In remote sensing the Satellite images are interpreted at pixel level using various analysis methods. One key requirement for almost all the applications is the accuracy of the images in terms of radiometric and geometric quality. Any kind of noise and perturbation in pixel value or geometric position will cause misinterpretation of the data resulting in inaccurate predictions. Data quality assessment is a critical process for satellite data as each sensor has its own quality standards to cater to various applications. This would lead to a major challenge for continuous assessment and control of the image quality.

The significance of Image Quality metrics (IQM) lies in its creating multidisciplinary focus that generally incorporates Computer vision, visual psychophysics, neural physiology, information theory, machine learning, and image acquisition. Image quality metrics (IQM) are figures of merit used for the evaluation of imaging systems or of coding/processing techniques.

© Springer International Publishing AG 2018
D. J. Hemanth and S. Smys (eds.), *Computational Vision and Bio Inspired Computing*,
Lecture Notes in Computational Vision and Biomechanics 28,
https://doi.org/10.1007/978-3-319-71767-8_58

2 Literature Review

Quality metrics in [1] is utilized for single entity of the image. If the input image is accessible, then the image combination the corrected and un-corrected images is assessed utilizing the measurements RMSE, PSNR, MSE, SSIM. If the input image is not accessible then the image will be assessed utilizing measurements like standard deviation, mean, median, entropy. In [2], the definition of image quality and evaluation based on corrected and un-corrected images are described. The quality of image can be progressively controlled and balanced utilizing these estimations [3, 4]. In [5], a review of subjective and objective image quality measures are mentioned and grouped according to the strategies and techniques used.

In [6] and [7], the assessment strategies (qualitative and quantitative) are investigated to asses the quality of corrected and un-corrected images. In qualitative assessment, if a visual correlation is not found under the same visual condition, then the correlation won't give solid outcomes and it likewise relies upon the experience of the viewer. For quantitative assessment, there are a few reference image based and non-reference image based measurements available to assess the image quality. The proposed approach examines the scope of legitimacy of PSNR in image and video. A better quality reproduced image has higher PSNR value. In spatial domain enhancement filters PSNR is utilized to evaluate the quality of image produced by the filter algorithm.

Mean value provides individual pixel value for whole image, whereas variance represents how every pixel differs from the neighbouring pixel and utilized to classify the image into various regions.

A robust method for estimating the MTF of high resolution remote sensing images is proposed in [8]. On-orbit estimation techniques are discussed to measure the along scan and across scan of MTF profiles. In [9], a detailed procedure is proposed to assess radiometric quality of high resolution satellite images.

3 Overview of the Work

This paper is structured in the following manner: Section 4, describes quality metrics of single image. In Sect. 5 we defined and computed the image quality metrics using two images that is uncorrected and corrected image. The objective of work in Sects. 4 and 5 is to develop automated quality metric assessment package for satellite images. In Sect. 6 we discussed Modulation Transfer Function estimation incorporating edge spread function and line spread function at most prominent edges. In Sect. 7 we discussed about conversion of gray image to reflectance image in two steps using spectral radiance scaling method and radiance to Top of Atmosphere (ToA) reflectance method. In Sect. 8, we discussed the implementation of GUI with single image metrics as well as corrected and uncorrected image metrics along with log file. Section 9 deals with data set details. Section 10 gives implementation results corresponding to different quality metrics of single image, relative characterization of two images and estimation of MTF. Sections 11 concludes the paper.

4   Characteristics of Single Image

The quality examination describes the substance of an image and its surface. Essentially, the assessment measurements can be grouped into 1st order, 2nd order and higher order. The 1st order metric provides essential information like mean, standard deviation and variance. So, it is working exclusively on individual pixels of an image and is not about the spatial connection between the pixels. Then again, the 2nd and higher order measurements measures the properties of two or a significant number of pixels in particular areas with respect to each other. Figure 1 demonstrates the architecture outline of Image Quality Metrics.

4.1   Mean

Arithmetic mean of any image corresponds to the mean brightness of the image [10].

$$\text{Mean} = \frac{1}{MN} \sum_{i=1}^{M} \sum_{j=1}^{N} |(x(i,j))| \tag{1}$$

where M is the No. of horizontal pixels, N is the No. of vertical pixels and x(i, j) is the input image pixel at location (i, j).

4.2   Median

Median is the middle value in the frequency distribution [11].

$$y[m,n] = median\{x[i,j]\}, (i,j) \in \omega \tag{2}$$

where ω represents a neighborhood defined by the user, centered around location [m, n] in the image.

4.3   Mode

Having considered the mean and the median of the local intensity distribution as candidate intensity values for image characterization, it also seems relevant to consider the mode of the distribution. Indeed, this is more important than the mean or the median, since the mode represents the most probable value of any distribution [11]. In the image band histogram, mode refers to the pixels with particular DN value which is having higher frequency of occurrence.

4.4   Standard Deviation (SD)

The standard deviation estimates the quantity of variations in an image. This can be optimum metric to assess the standard of reconstructed images and it can be utilized in many applications wherever the image is degraded, due to impulse noise [11].

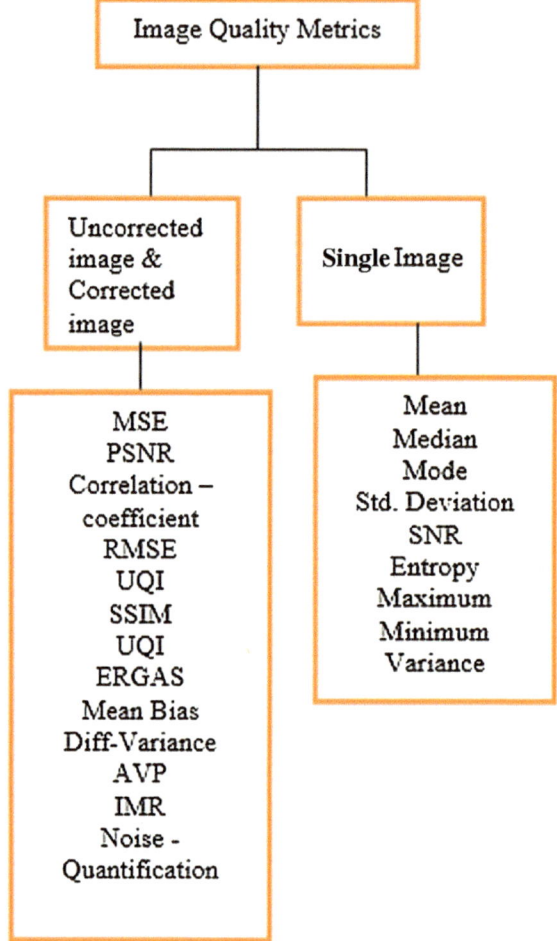

Fig. 1 Architecture diagram of IQM

$$SD = \sqrt{\frac{1}{Totol\,no.\,of\,pixels} \sum_{i=1}^{M} \sum_{j=1}^{N} (x_{i,j} - \mu)^2} \tag{3}$$

where M is the No. of horizontal pixels, N is the No. of vertical pixels and x(i, j) is the pixel in the input image and μ will be calculated mean of image.

4.5   Maximum

Maximum is defined as the maximum of all pixels within a local region of an image and is applied to an image to remove negative outlier noise [11].

$$f(x, y) = \max_{(s,t)\in S_{x,y}} \{g(s,t)\} \tag{4}$$

g(s, t) is the pixel of an input image.

4.6 Minimum

Minimum is defined as the minimum of all pixels within a local region of an image and it is typically applied to an image to eliminate positive outlier noise [11].

$$f(x, y) = \min_{(s,t)\in S_{x,y}} \{g(s,t)\} \tag{5}$$

g(s, t) is the pixel of an input image.

4.7 Variance

Variance defines every pixel value varies from neighbouring pixel. It additionally defines statistical information of an image. This is a necessary quality parameter which is employed to compute the standard of reconstructed images and it can be utilized in many applications wherever image is degraded due to different diffusions.

$$\text{Variance} = \frac{1}{Total\,no.\,of\,pixel} \sum_{i=1}^{M} \sum_{j=1}^{N} (x_{i,j} - \mu)^2 \tag{6}$$

where M is the No. of horizontal pixels, N is the No. of vertical pixels and x(i, j) is the input image and μ is the calculated mean of an image.

4.8 Entropy

Image entropy describes the quantity of information that should be coded for by a compression algorithm. Low entropy images, have little distortion and huge runs of pixels with identical or similar DN values. A picture that is utterly flat can have associate entropy of zero. Consequently, they will be compressed to a comparatively tiny size. On the other hand, high entropy pictures have an excellent amount of distinction from one pixel to next and consequently cannot be compressed to a minimum amount as low entropy images [12].

$$\text{Entropy} = -\sum_{i} p_i \log_2 p_i \tag{7}$$

where, p_i is the probability that the difference between two adjacent pixels is equal to i.

5   Relative Characterization of Images

5.1   Mean Square Error (MSE)

This metric can be used to measure the quality of image enhancement algorithm after removing the noise and blur. Image degradation increases along with increasing of MSE value. When MSE reaches zero, the corrected and expected images are similar. So, the output image will be denoised image [13].

$$MSE = \frac{1}{MN} \sum_{i=1}^{m} \sum_{j=1}^{n} \left((x(i,j) - y(i,j))^2 \right) \tag{8}$$

M is no. of pixels in horizontal, N is no. of pixels in vertical, x(i, j) is the value of the pixel in the uncorrected image and y(i, j) is the value of the pixel in the corrected image.

5.2   Peak Signal to Noise Ratio (PSNR)

PSNR is the ratio between maximum possible power of an image and the power of the corrupting noise (MSE) [14].

$$PSNR = 10 \log_2 \frac{\max I^2}{MSE} \tag{9}$$

Here highest possible value of image pixel is I. A good quality image has higher PSNR value.

5.3   Structural Similarity Index (SSIM)

SSIM quality metric which is used for estimating similarity between two images. It depends on comparison between image luminance, contrast and structure [14].

$$SSIM(x, y) = \frac{(2\mu_x\mu_y + c1)(2\sigma_{xy} + c2)}{(\mu_x^2 + \mu_y^2 + c1)(\sigma_x^2 + \sigma_y^2 + c2)} \tag{10}$$

where μ_x and μ_y are mean of uncorrected and corrected images respectively. Next σ_x and σ_y are standard deviation of uncorrected and corrected images respectively. σ_{xy} is co-variance between x and y. c1 and c2 are constants used to stabilize weak parameters.

5.4   Universal Quality Index (UQI)

UQI is defined by modeling any image distortion as a combination of three different factors: loss of correlation, luminance distortion and contrast distortion [14].

$$Q = \frac{4\sigma_{xy}\bar{x}\bar{y}}{(\sigma_x^2 + \sigma_y^2)[(\bar{x})^2 + (\bar{y})^2]} \quad (11)$$

5.5 Correlation Coefficient (CC)

Linear relationship between two measured quantities can be defined by correlation coefficient. Pearson's correlation coefficient (r) the primary formal correlation measure which is widely utilized in applied analysis, pattern recognition and image processing [15].

$$r = \frac{\sum_i (x_i - x_m)(y_i - y_m)}{\sqrt{\sum_i (x_i - x_m)^2}\sqrt{\sum_j (y_i - y_m)^2}} \quad (12)$$

where x_i is intensity of ith pixel in uncorrected image, y_i is the intensity of ith pixel in corrected image, x_m is mean intensity of uncorrected image, and y_m is mean intensity of corrected image.

5.6 Root Mean Square Error (RMSE)

It corresponds to difference in pixels between any two images. If the images are similar, then the RMSE value is equal to zero and it will increase when the dissimilarity between the two images increases [10].

$$RMSE = \sqrt{MSE = \frac{1}{MN}\sum\sum((x(i,j) - y(i,j))^2} \quad (13)$$

M is no. of pixels in horizontal, N is no. of pixels in vertical, x(i, j) is uncorrected image and y(i, j) is the corrected image.

5.7 Ergas

ERGAS is a commonly used parameter to calculate the amount of spectral distortions in the comparative assessment of multiple images [10].

$$ERGAS = 100\frac{h}{l}\sqrt{\frac{1}{N}\sum_{N=1}^{N}\left(\frac{RMSE(n)}{\mu(n)}\right)^{-2}} \quad (14)$$

where h/l is the ratio between pixel sizes of panchromatic images and multispectral images, $\mu(n)$ is mean of nth band, and N is no. of bands.

5.8   Mean Bias

It is the difference between the means of the two images considered. This parameter checks degree of biased intensity between low resolution original image and fused image.

5.9   Difference Variance

This metric measures quantity of information lost or added during estimation. A positive value indicates loss of information and negative value shows some added information.

$$DIV = \frac{\sigma^2_{uncorrected} - \sigma^2_{corrected}}{\sigma^2_{uncorrected}} \tag{15}$$

5.10   Average Percentage Variation (AVP)

It defines the percentage mean variation between two images. It takes average difference of two images and divides this average difference by average of original image.

$$AVP = \frac{Avg(orignal) - Avg(distorted)}{Avg(original)} \times 100 \tag{16}$$

5.11   Image Mean Ratio (IMR)

Image mean ratio is calculated as mean ratio between original image and distorted image.

$$IMR = \frac{Avg(original)}{Avg(distorted)} \tag{17}$$

5.12   Noise Quantification

The total estimation of the noise in a image is hard to evaluate because ground truth images are frequently inaccessible. The relative extent of noise is

$$Noise = \frac{\sqrt{\frac{1}{N}\sum(d1 - d2)^2}}{\max(d1, d2) - \min(d1, d2)} \times 100 \tag{18}$$

where N is the quantity of pseudo-invariant pixels present in the image pair, and d1 and d2 refers to the spectral values of leaf-on and leaf-off images respectively.

6   MTF Estimation

There are several ways to estimate the Modular Transfer Function (MTF) like pulse technique, edge technique, and comparison technique. The determination of MTF supported by the edge technique involves following three steps [16]:

- Edge detection and Edge Spread Function (ESF) g(x) computation
- Line Spread Function (LSF) *l*(x) computation

$$\frac{dg(x)}{dx} = l(x) \tag{19}$$

- MTF is estimated by finding fast fourier transform on LSF. This process is shown in Fig. 2.

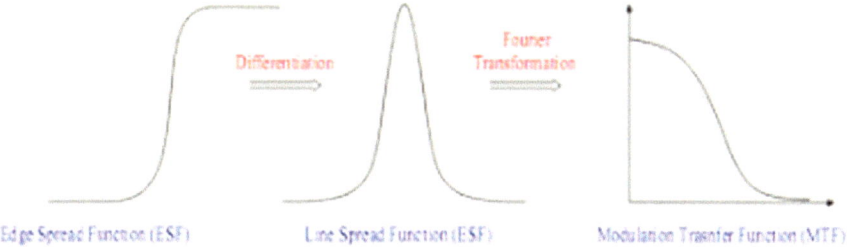

Fig. 2   Process of MTF estimation

7   Converting DN to Top of Atmosphere (ToA) Relectance

The satellite optical sensors captures reflected solar energy, converts to radiance, and afterwards rescale this information into Digital number (DN) with a range. These DNs can be converted to ToA reflectance using two-step process.

7.1   DN to Radiance

The first step is to convert the DNs to radiance using the LMAX and LMIN values specified in the Meta data.

$$L_\lambda = ((LMAX_\lambda - LMIN_\lambda)/(QCALMAX - QCALMIN))* \\ (QCAL - QCALMIN) + LMIN_\lambda \tag{20}$$

where:
$L\lambda$	= is the cell value as radiance
QCAL	= digital number
$LMIN_\lambda$	= spectral radiance scales to QCALMIN
$LMAX_\lambda$	= spectral radiance scales to QCALMAX

QCALMIN   = the minimum quantized calibrated pixel value
QCALMAX   = the maximum quantized calibrated pixel value.

7.2 Radiance to ToA Reflectance

The second step converts the radiance data to ToA reflectance using the formula (21):

$$\rho_\lambda = \pi * d^2 / ESUN_\lambda * \cos\theta_z \qquad (21)$$

where:

$\rho\lambda$	= unitless planetary reflectance
$L\lambda$	= spectral radiance
d	= Earth-Sun distance in astronmoical units
$ESUN\lambda$	= mean solar exoatmospheric irradiance
Θz	= solar zenith angle.

8   Graphical User Interface (GUI)

The above Sects. 4 and 5 are developed by using MATLAB version 2014Ra with GUI. The Fig. 3 shows how the interactive module will enable the user to load the image by clicking the load button. Image characteristics can be computed by clicking appropriate button on the screen. The corresponding metric value will be displayed. For the purpose of analysis, different quality metrics are computed and displayed. The user can also save the output image and quality metrics at user defined location. Figure 4 shows the log file of the quality metrics computed for images.

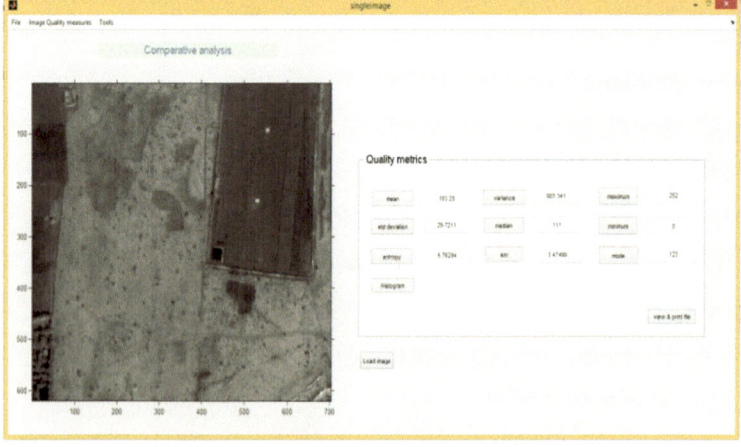

Fig. 3 GUI representation for single image

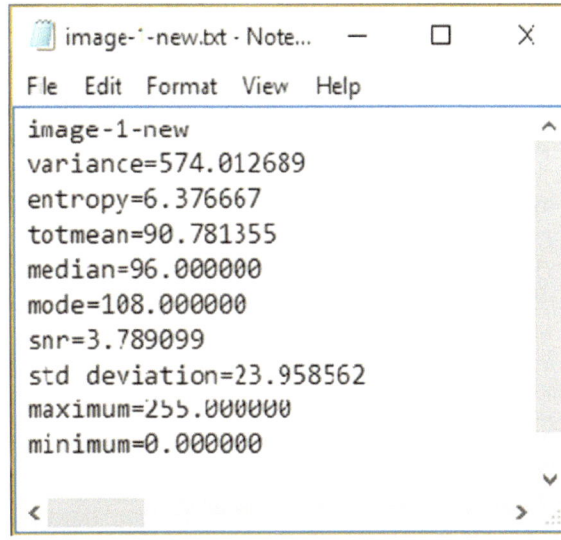

Fig. 4 Log file of computed IQMs for single image

The Fig. 5 shows the interactive module and describes how the user can load multiple images by clicking the load button. The user can also save the output image and quality metrics at user defined location. Figure 6 shows the log file of the quality metrics computed for multiple images.

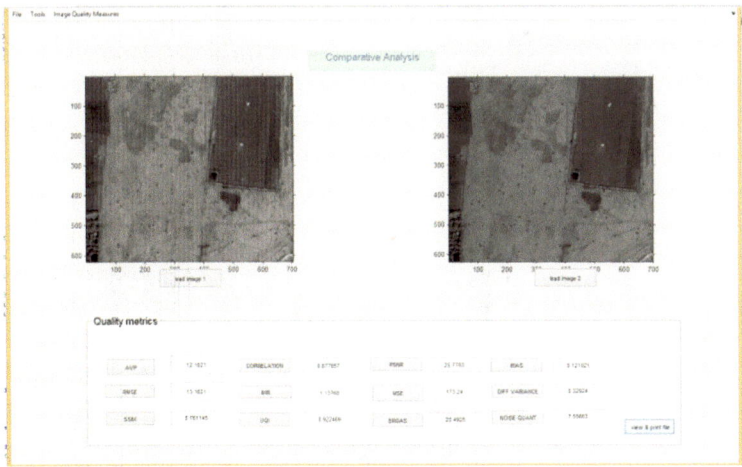

Fig. 5 GUI representation for multiple images

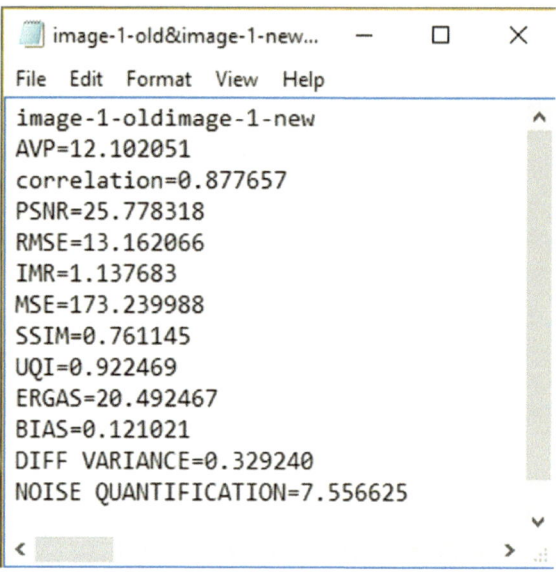

Fig. 6 Log file of uncorrected and corrected images

9   Datasets

See Table 1.

Table 1 Details of data sets

Satellite	Sensor	Date of pass
IRS-P6	AWiFS	08 Jan 2015
IRS-P6	LISS-3	03 Oct 2015
Cartosat-1	AFT	12 Feb 2015

10   Results and Analysis

10.1   Assessment of Image Characteristics

All the selected data sets are processed with above GUI and the quality metrics are computed. Table 2 gives the results of Single Image Metric. Table 3 gives quality metrics of various image pairs.

Table 2 Quality metrics of single image

Quality metrics	AWiFS BAND-2	LISS-3 BAND-2	Carto-1 AFT
Median	85	75	96
Mode	0	0	86
Entropy	0.13	4.69	0.17
Std.dev.	41.17	33.36	15.37
Maximum	308	245	517
Minimum	0	0	42
Variance	574.01	1113.12	236.51
Mean	90.78	60.49	96.77

Table 3 Quality metrics of multiple images

Quality metrics	Pair-1	Pair-2	Pair-3
MSE	173.23	301.95	326.05
PSNR	25.77	23.36	23.03
Correlation Coeff.	0.877	0.75	0.94
RMSE	13.16	17.38	18.06
SSIM	0.76	0.76	0.99
UQI	0.92	0.87	0.97
ERGAS	20.49	28.9	23.52
Mean Bias	0.12	0.12	0.086
Diff-Variance	0.32	0.22	−0.08
Noise	7.55	9.88	10.91
AVP	12.01	12.44	8.21
IMR	1.14	1.14	1.09

10.2 MTF Estimation

From the given image, suitable edge is detected then ESF and LSF are computed. Figure 7 shows the marked edge in the image for analysis.

Figure 8 shows the plot of edge spread function for the detected edges in the satellite image shown in Fig. 7. Here, we computed the Edge Spread Function by detecting five horizontal edges.

Fig. 7 Satellite image for edge analysis

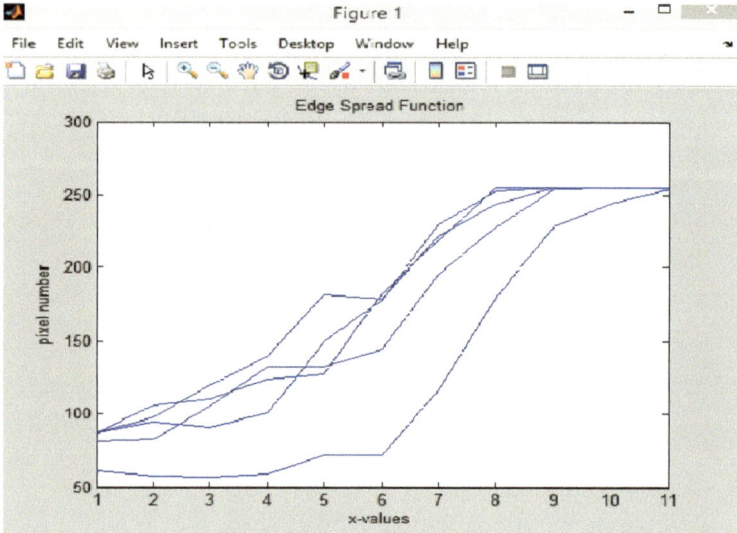

Fig. 8 Edge spread function

Figure 9 shows the plot of Normalized Edge Spread Function, Line Spread Function and Modulation Transfer Function. Normalized ESF is obtained by normalizing the ESF in the above Fig. 8 and LSF is obtained by differentiating the ESF. MTF is computed by applying Fourier transform to LSF and scaling it in frequency axis, in order to represent MTF at Nyquist frequency. MTF value at Nyquist frequency was estimated for selected edge region in the Fig. 7. MTF results achieved for Cartosat-2, Cartosat-2B and Cartosat-2C Panchromatic camera imagery are presented in Table 4.

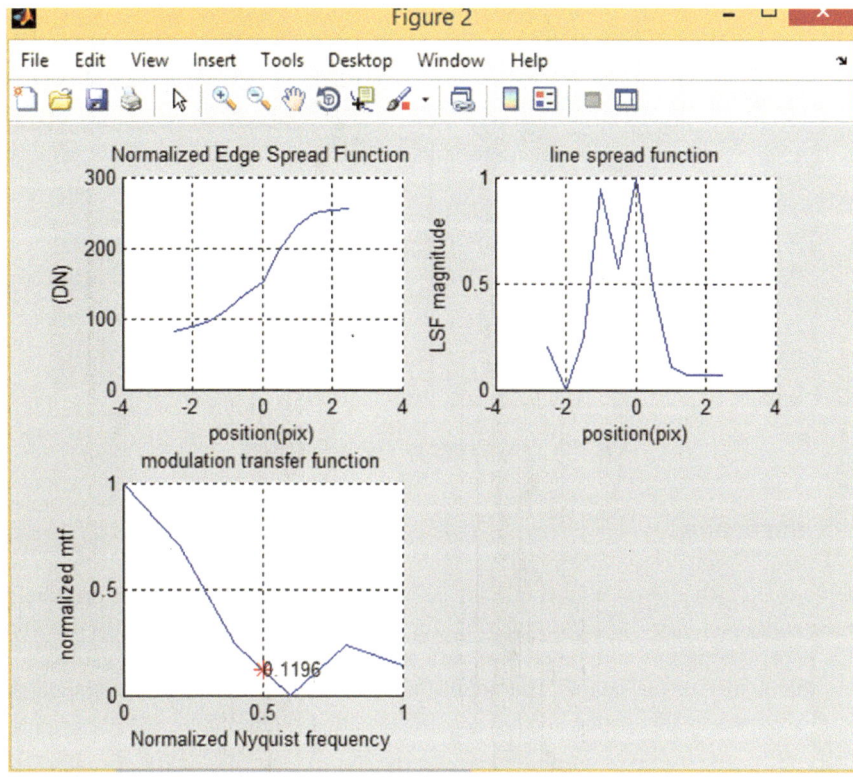

Fig. 9 Graphical representation of ESF, LSF and MTF

Table 4 MTF results

Imagery type	MTF (%)
Cartosat-2	16
Cartosat-2B	17
Cartosat-2C	24

10.3 DN to Top of Atmosphere (ToA) Reflectance

The Fig. 10 shows the output of LISS-3 band-2 image conversion from DN to ToA reflectance. The result is obtained using Eqs. (20) and (21).

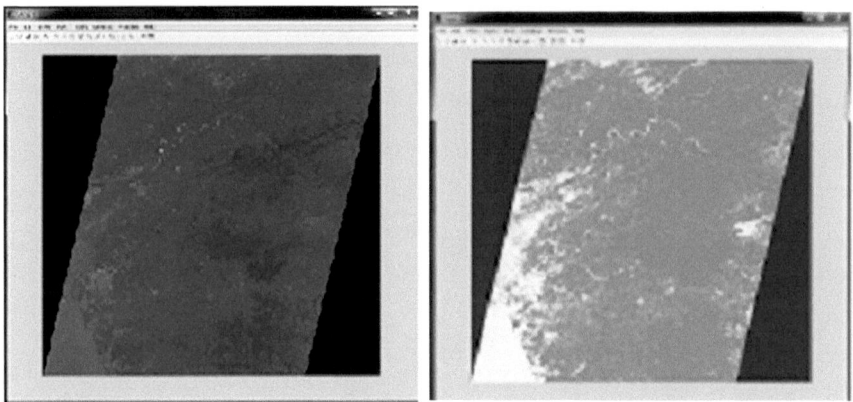

Fig. 10 Gray image to ToA reflectance

11   Conclusions

Different metrics based on both single image and multiple images to assess the satellite image quality and also estimation of MTF for the selected region are clearly presented in this paper. Using this automated tool, different image characteristics are estimated to assess the quality of the image. The quality metrics computed for satellite image data set are tabulated in this paper.

References

1. Jagalingam, P., Hegde, A.V.: A review of quality metrics for fused image. In: International Conference on Water Resources, Coastal and Ocean Engineering (ICWRCOE-2015), pp. 133–42 (2015)
2. Thung, K.H., Raveendran, P.: A survey of image quality measures. In: IEEE International Conference for Technical Post Graduates (TECHPOS), Kuala Lumpur, pp. 1–4 (2009)
3. Wang, Z., Bovik, A.C., Sheikh, H.R., Simoncelli, E.P.: Image quality assessment: from error visibility to structural similarity. IEEE Transactions of Image Processing. **13**(4), 600–612 (2004)
4. Silva, E.A., Panetta, K., Agaian, S.S.: Quantify similarity with measurement of enhancement by entropy. In: Proceedings: Mobile Multimedia/Image Processing for Security Applications, SPIE Security Symposium. (2007)
5. Thung, K.-H., Raveendran, P.: A survey of image quality measures (2010). https://doi.org/10.1109/TECHPOS.2009.5412098. Source: IEEE Xplore
6. Jagalingam, P, Hegde, A.V.: A review of quality metrics for fused image. In: International Conference on Water Resources, Coastal And Ocean Engineering (ICWRCOE 2015) (2015)
7. Pedersen, M., Bonnier, N., Hardeberg, J.Y., Albregtsen, F.: Image quality metrics for the evaluation of print quality. In: Farnand, S.P., Gaykema, F. (ed.) Image Quality and System Performance VIII, Proceeding of SPIE-IS&T Electronic Imaging, SPIE, vol. 7867, 786702 © 2011 SPIE-IS&T (2011)

8. Kohm, K.: Modulation transfer function measurement method and results for the orbview-3 high resolution imaging satellite. In: Proceedings of ISPRS 2004, Istanbul, Turkey, July 12–23, (2004).
9. Crespi, M., De Vendictis, L.: A procedure for high resolution satellite imagery quality assessment. J. MDPI/Sensors **9**, 3289–3313 (2009). https://doi.org/10.3390/s90503289
10. Soumya, B.S., Suresh, M.B.: Image fusion for improving spatial resolution of multispectral satellite images. Int. J. Sci. Eng. Res. **5**(6) (2014) (RMSE, ERGAS)
11. Kumar, V., Gupta, P.: Importance of statistical measures in digital image processing. Int. J. Emerg. Technol. Adv. Eng. **2**(8) (2012)
12. Yakhdani, M.F., Azizi, A.:Quality assessment of image fusion techniques for multisensor high resolution satellite images (Case study: Irs-P5 And Irs-P6 Satellite Images). IAPRS **XXXVIII** (2010)
13. Sruthy, S., Parameswaran, L., Sasi, A.P.: Image fusion technique using DT-CWT. In: Proceedings—2013 IEEE International Multi Conference on Automation, Computing, Control, Communication and Compressed Sensing, iMac4 s 2013, Kerala, pp. 160–164 (2013)
14. Al-Najjar, Y.A.Y., Soong, D.C.: Comparison of image quality assessment: PSNR, HVS, SSIM, UIQI. Int. J. Sci. Eng. Res. **3**(8) (2012)
15. Kaur, A., Kaur, L., Gupta, S.: Image recognition using coefficient of correlation and structural similarity index in uncontrolled environment. Int. J. Comput. Appl. **59**(5) (2012) ((0975-8887))
16. Xia, Y., Chen, Z.: Quality assessment for remote sensing images: approaches and applications. In: 2015 IEEE International Conference on Systems, Man, and Cybernetics

Reconstruction of Recaptured Images Using Dual Dictionaries of Edge Profiles

J. Hridhya$^{(\boxtimes)}$ and A. Shyna

Department of Computer Science and Engineering, TKM College of
Engineering, Kollam, India
hridhya672@gmail.com, shyna@tkmce.ac.in

Abstract. The recapture detection based on high-quality LCD screen is really challenging as the recaptured image from LCD screen seems to be like the original and very difficult to distinguish by human eye. An image recapture detection algorithm is used to classify single captured and recaptured image. This paper proposes a novel approach for the reconstruction of recaptured images for improving quality using a dual dictionary of which, one is for single captured and the other for recaptured images. Here K-SVD algorithm is used to train both dictionaries in which the orthogonal matching pursuit algorithm has been used to generate sparse approximation from the dictionary of recaptured image and Line Spread Profile matrix. With the help of sparse approximation of recaptured set and dictionary of captured set, captured image can be reconstructed from recaptured image. The Experimental results indicate that the proposed algorithm results better quality of captured image from recaptured set in terms of PSNR.

Keywords: Dual dictionary learning · Line spread profile · Orthogonal matching pursuit · K-SVD

1 Introduction

Now a day's several image editing tools are widely available so that the manipulation of images and changing their content is becoming a trivial task. An attacker can easily recapture the forged image from an LCD monitor which is very difficult to detect. But now detecting whether the image is originally captured or whether it is recaptured from an LCD monitor is existing [1]. Image recapture detection is to distinguish images of real scenes from the recaptured images, that is real-scene images such as printed pictures or LCD display. One of the important applications of recapture detection is in face authentication system. Image recapture detection is also useful for general object recognition to differentiate the objects on a poster from the real ones, which improves the intelligence of robot vision. Another important application for Image recapture detection is in composite image detection. One way to cover composition in an composite image is to recapture it. In many applications only recaptured image is available and single captured image reconstruction problem is particularly important in those situations.

© Springer International Publishing AG 2018
D. J. Hemanth and S. Smys (eds.), *Computational Vision and Bio Inspired Computing*,
Lecture Notes in Computational Vision and Biomechanics 28,
https://doi.org/10.1007/978-3-319-71767-8_59

Recaptured image is mainly characterized by the blurriness of the edges. So line spread profile of captured and recaptured images will be different. To characterize the differences between the line spread profiles of originally captured and recaptured images, two parameters related to edge, a sparse representation error and an average line spread width [2] can be considered. These parameters has been calculated using two over complete dictionaries of single captured and recaptured set which can be trained by using K-SVD and Orthogonal Matching Pursuit approach. Based on these parameters we can classify recaptured and single captured images. Dictionary training can be used as a tool to learn the characteristics of the distortion patterns present in edges since the descriptions in single capture and recaptured images are fundamentally different due to the sharpness degradation introduced by the recapture process.

The reconstruction of recaptured image is an enhancement of recapture detection using learning dictionaries of edge profiles [2]. The reconstruction of recaptured image using dual dictionary is mainly based on the inverse procedure of detection criteria. There are applications where only recaptured image will be available which possess low contrast and sharpness of edges. In such cases, recaptured images need to be reconstructed. So reconstruction of recaptured image to captured image will improve the quality of image. This paper introduces a novel idea that if we know the sparse representation of recaptured set and dictionary of captured set, captured image can be reconstructed from recaptured image.

The paper is organized as follows: Sect. 2 describes some related works and different image reconstruction techniques. Section 3 describes proposed methodology in detail. Experimental results are described in Sect. 4. In Sect. 5, Conclusion of this work is presented.

2 Related Works

Thongkamwitoon et al. [2] proposed an algorithm to detect recapture image based on learning dictionaries of edge profiles. They identified the fact that the edge profiles of single and recaptured images are different due to distortion and blurriness. The line spread functions of selected edges were used to train single capture and recapture dictionaries following the K-SVD approach. An SVM classifier was then built to detect recaptured images.

A lot of research and work has been done on image reconstruction techniques since the mid nineteen eighties. Gallo et al. [3] proposed the reconstruction of multi-view images by integrating images from multiple cameras. It combines the relevant information from two or more source images into a single resultant image that describes the scene better and retains useful information from the source image.

In Muammar and Dragotti [4], investigated one approach to detecting an image that has been recaptured from an LCD monitor is to search for the presence of aliasing due to the sampling of the monitor pixel grid. This method shows that aliasing can be completely eliminated in a recaptured image by setting the camera to monitor distance to a value determined by the camera lens focal length, the pixel pitch of the LCD monitor and the pixel pitch of the cameras image sensor. The main Advantages of this approach is very selective technique to eliminate aliasing detect from recaptured images. And the disadvantages are Paper investigation is in aliasing only, camera dependency and the user should have knowledge about technical details of captures of camera.

Flohr et al. [5] introduced Multi–detector row CT systems for CT image-reconstruction. The multi–detector row computed tomography (CT) was a milestone with regard to increased scan speed, improved z-axis spatial resolution, and better uti-lization of the available x-ray power. Xu et al. [6] proposed low-dose X-ray CT recon-struction via dictionary learning. Compared to the discrete gradient transform used in the TV method, dictionary learning is proven to be an effective way for sparse representation. They developed a dictionary learning based approach for low-dose X-ray CT. This approach is evaluated with low-dose X-ray projections collected in animal and human CT studies, and the improvement associated with dictionary learning is quantified relative to filtered back projection and TV-based reconstructions. Needell and Tropp [7] introduced basis pursuit algorithm for sparse signal recovery which falls under convex optimization algorithm category. The important advantages of the basis pursuit algorithm are prior sparsity knowledge of signal is not necessary compared to other algorithms, and the reconstruction problem can be formulated easily as linear programming problem and easy to solve and finally the performance of basis pursuit under noisy scenarios is good.

3   Proposed System

The proposed system mainly consists of image recapture detection [1] and image reconstruction phase as shown in Fig. 1. The image recapture detection phase is used to detect whether the image is recaptured from LCD monitor or single capture of a natural scene. By keeping dictionary learning phase as same as that of the existing works [2], we have focused a new approach for recaptured image reconstruction using dual dic-tionary method. The steps are shown below.

1. Generate line spread profile for single captured and recaptured images.
2. Using this line spread profiles obtain two over complete dictionaries (D_{SC}, D_{RC}) of single captured and recaptured set using K-SVD approach.
3. Based on the line spread profile and dictionaries two parameters can be determined. The parameters sparse representation error and average width are used for detection.
4. Determine the training feature matrices of recaptured image set using line spread profile matrix which is given as input to the OMP along with D_{RC} to obtain sparse representation of recaptured image.
5. Use the sparse representation of recaptured image and D_{SC} to obtain reconstructed captured image.

3.1   Line Spread Profile Extraction

Canny Edge Detection method can be used to detect all the edges in an image. The Canny edge detection algorithm is known to many as the optimal edge detector. The advantages of the canny edge detection are Improving signal to noise ratio, better detection especially in noise conditions. There is a block based computation for block detection. The image is divided into a number of non-overlapping square block $B(m, n)$ of size W × W with W = 16 pixels. Here m and n are the horizontal and vertical indices of the image block respectively.

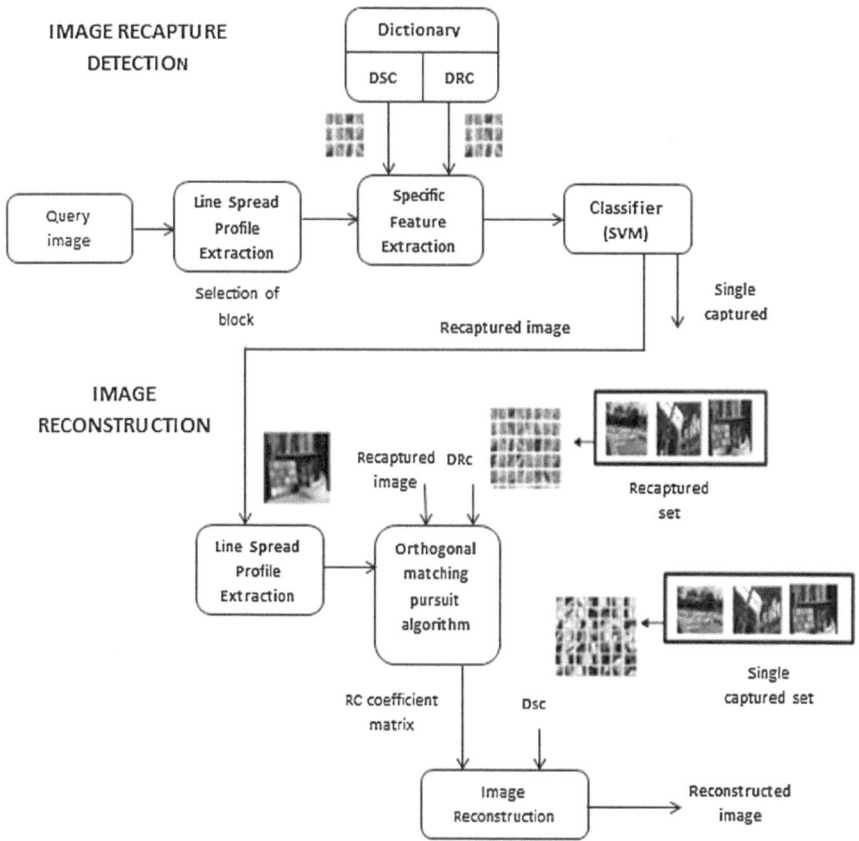

Fig. 1 Proposed System Design: Detection and Reconstruction of Recaptured image

For each block check whether it contains horizontal and near the horizontal sharp single edge. Then the block will be detected only when the condition, $\beta = 0.6$ will be satisfied [2]. Most significant features in the line spread profile extraction are its sharpness and contrast. The detected blocks can be ranked according to their sharpness and edge contrast. The line spread function can be easier to determine than edge spread function or point spread function. Figure 2 shows that how line spread profile can be extracted from an image.

(a) Create Feature matrix: the matrix which represents the gray scale values of a block. Each column of the matrix Y can be used to represent an edge profile of the image.

(b) Find the normalized line spread profile qi;

$$q_i = y_{i(1)} / \|y_{i(1)}\|_2^F \qquad (1)$$

Where $y_{i(1)}$ is the first derivative of y_i. The differentiated edge profile is normalized in order to standardize the feature [2]. The line spread profile qi can be cropped and centered before zero padding is applied in order to maintain the length matrix dimension. If the total number of line spread profile is M, so that the feature matrix dimension is to W × M. This feature matrix is used for reconstruction purposes.

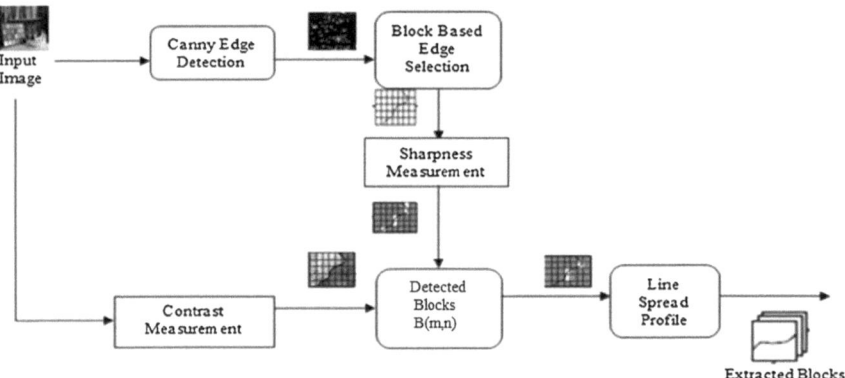

Fig. 2 Diagram of Line Spread Profile Extraction

3.2 Dictionary Learning Phase

Two over complete dictionaries, D_{SC} (for single captured) and D_{RC} (for recaptured) can be obtained by dictionary learning that provide an optimal sparse representation of line spread profiles from single captured and recaptured images, respectively. Dictionary training can be used as a tool to learn the characteristics of the distortion patterns present in edges [8]. Since the recapture process introduce a sharpness degradation, the description in the recaptured and captured image will be different. In this approach, K-SVD dictionary learning method [9] can be used for the learning process of the image set. The K-SVD method is an iterative learning scheme based on two important steps for each round of computation: sparse coding and dictionary update. In sparse coding, given an initial dictionary D, X is chosen such that each of its columns xi provides the best L-sparse representation of q_i.
Specifically:

$$\min_{xi} \|q_i - Dx_i\|_2^F \; subject \cdot to \|x_i\|_0 \leq L \qquad (2)$$

In practice, this is achieved using the orthogonal matching pursuit (OMP) algorithm [1] which is known to provide near optimal sparse coding. Next, given X, D is updated so as to achieve

$$\min_{D}\|S - DX\|_F^2 \qquad (3)$$

3.3 Feature Extraction and Classification

For the image recapture detection, two parameters related to edges, a sparse representation error and an average line spread width. These parameters were chosen because they provide a concise but informative description of the differences between the line spread profiles of original and recaptured images.

The first parameter, is the difference in the errors, E_{SC} and E_{RC}, between the extracted line spread profiles and their sparse representations determined using the dictionaries, D_{SC} and D_{RC}, respectively. The value of E_d is determined by taking the differences between E_{SC} and E_{RC}. If the image is considered as original then $E_{SC} < E_{RC}$ and $E_{SC} \geq E_{RC}$ if the image was a recaptured image. The second parameter, λ provides a description of the width of an extracted line spread profile. Higher values of λ correspond to blurry edges, while small values to sharp edges.

These parameter pairs are collected on an image by image basis and the set of parameter pairs is then used to train a 2-dimensional SVM classifier. When the training procedure is complete a hyperplane that optimally separates the two sets of images based on their values E_d and λ is determined.

3.4 Generation of Sparse Representation and Reconstruction of Recaptured Images

For the reconstruction of recaptured image, we need the coefficient matrix or sparse representation of recaptured image. The sparse representation of a pixel is expressed as a sparse vector whose nonzero entries correspond to the weights of the selected training samples. Using the line spread profile matrix of recaptured image and the corresponding dictionary for recaptured image set, sparse representation of recaptured image has been obtained using orthogonal matching pursuit algorithm [10]. Orthogonal matching pursuit algorithm generates sparse matrix or the coefficient matrix for the corresponding input. Using these sparse representations of recaptured image, captured image has been constructed if we have the trained dictionary of captured set.

4 Experimental Results

A database [11] of images recaptured from an LCD monitor and single capture of natural scene was developed for the purposes of testing and evaluating the reconstruction of recaptured images algorithm. The database comprised 9 set of single captured images captured using set of 9 cameras, Nikon D70s, Canon EOS 600D, Sony RX100, Nikon D40, Olympus E-PM2, Panasonic TZ7, Kodak V550 (Black), Kodak V550 (Silver) and Kodak 610 cameras. The set of 8 cameras used to perform recapture were Nikon D70s, Canon EOS 600D, Olympus E-PM2, Sony RX100, Nikon D3200, Panasonic TZ10, Canon A700 and Canon EOS 60D. The proposed reconstruction algorithm for recaptured image is implemented in MATLAB R2015 a using K-SVD and OMP algorithms, on Intel(R) Core(TM) i3–5005U CPU at 2.00 GHz with 4 GB of RAM.

The images used in the experiments are shown in Fig. 3. Figure 3a–d is a recaptured image captured by Canon EOS 600D cameras and Fig. 3e–h shows its reconstructed version.

Table 1 shows the comparison of SVM kernel functions in term of accuracy and execution time (second). On comparison it is found that SVM RBF kernel function shows better accuracy with least execution time (average) of 7.18 s.

Table 1 Comparison of different SVM Kernel functions

Kernel's	Original captured (50)	Accuracy (%)	Recaptured (50)	Accuracy (%)	Average execution time (s)
RBF	49	98	50	100	7.18
Linear	46	92	48	96	8.27
Polynomial	41	82	43	86	11.35
Quadratic	45	90	46	92	9.75

(a) Recaptured   (b) Recaptured   (c) Recaptured   (d) Recaptured

(e) Reconstructed image   (f) Reconstructed image   (g) Reconstructed image   (h) Reconstructed image

Fig. 3 Example of the images used in the experiments **a–d** Recaptured and **e–h** Reconstructed image

Table 2 shows the performance comparison of the same set of test images. Proposed approach uses the least number of features compared to the existing methods. By analyzing the results, it is obvious that proposed method shows better results in terms of accuracy. For single captured image set, accuracy of 95.56% and for the recaptured image set, accuracy of 98.19% is obtained.

Table 2 Dimensions of features and performance on datasets

Method	Number of features	Success rate (%)	
		Single captured	Recaptured
LBP + MSWS + Color features [12]	129	83.67	92.02
Higher-order wavelet statistics [13]	216	87.56	90.04
Markov statistics [14]	486	81.35	74.84
Geometry based [15]	192	86.31	80.12
Physics [16]	166	91.30	86.66
Proposed method	2	95.56	98.19

Table 3 shows the PSNR comparison of the reconstruction procedure. By analyzing the results it is obvious that proposed approach shows better result for the recaptured image reconstruction.

Table 3 Performance analysis of reconstruction procedure in terms of PSNR

Camera model	No: of images	PSNR
D70s	100	34.37
EOS 600D	100	35
RX 100	100	34.67
EOS 60D	100	36.88
E-PM 2	100	32.46
TZ 7	100	35.48
Average		34.81

5 Conclusion

In this paper, a novel approach has been proposed for reconstruction of recaptured image for improving the quality which can be used in applications where only recaptured image is available. Dual dictionaries concept has been used to obtain two over complete dictionaries of captured and recaptured set using K-SVD. The sparse representation of recaptured image has been obtained from line spread profile set which is provided as input to the OMP along with dictionary of captured set to construct captured image. Proposed approach has been applied on different sets of captured and recaptured images which is taken by using different camera model. It is found that proposed method shows better accuracy in terms of PSNR. In future we can extend this work for video recapture detection with the use of dual dictionaries.

References

1. Thongkamwitoon, T., Muammar, H., Dragotti, P.L.: Robust image recapture detection using a k-svd learning approach to train dictionaries of edge profiles. In: 2014 IEEE International Conference on Image Processing (ICIP), pp. 5317–5321. IEEE (2014)
2. Thongkamwitoon, T., Muammar, H., Dragotti, P.L.: An image recapture detection algorithm based on learning dictionaries of edge profiles. IEEE Trans. Inf. Forensics Secur. **10**(5); 953–968 (2015); Maxwell, J.C.: A Treatise on Electricity and Magnetism, vol. 2, 3rd edn, pp. 68–73. Clarendon, Oxford (1892)
3. Gallo, A., Muzzupappa, M., Bruno, F.: 3D reconstruction of small sized objects from a sequence of multi-focused images. J. Cult. Heritage **15**(2), 173–182 (2014)
4. Muammar, H., Dragotti, P.L.: An investigation into aliasing in images recaptured from an LCD monitor using a digital camera. In: 2013 IEEE International Conference on Acoustics, Speech and Signal Processing (ICASSP), pp. 2242–2246. IEEE (2013)
5. Flohr, T.G., Schaller, S., Stierstorfer, K., Bruder, H., Ohnesorge, B.M., Schoepf, U.J.: Multi–detector row CT systems and image-reconstruction techniques. Radiology **235**(3), 756–773 (2005)
6. Xu, Q., Yu, H., Mou, X., Zhang, L., Hsieh, J., Wang, G.: Low-dose X-ray CT reconstruction via dictionary learning. IEEE Trans. Med. Imaging **31**(9), 1682–1697 (2012)
7. Needell, D., Tropp, J.A.: CoSaMP: iterative signal recovery from incomplete and inaccurate samples. Appl. Comput. Harmonic Anal. **26**(3), 301–321 (2009)
8. Anitha, S., Nirmala, S.: Representation of Digital Images Using K-SVD Algorithm
9. Aharon, M., Elad, M., Bruckstein, A.: K-SVD: an algorithm for designing overcomplete dictionaries for sparse representation. IEEE Trans. Signal Process. **54**(11), 4311–4322 (2006)
10. Pati, Y.C., Rezaiifar, R., Krishnaprasad, P.S.: Orthogonal matching pursuit: Recursive function approximation with applications to wavelet decomposition. In: 1993 Conference Record of The Twenty-Seventh Asilomar Conference on Signals, Systems and Computers, pp. 40–44. IEEE (1993)
11. Recapture Image Database. [Online]. Available: http://www.commsp.ee.ic.ac.uk/~pld/research/Rewind/Recapture/. Accessed 24 Oct 2014
12. Sanas, P., Gupta, P.: Image detection and verification using local binary pattern with SVM. Int. J. Eng. Res. **5**(6), 489–493 (2016)
13. Faridy, H., Lyu, S.: Higher-order wavelet statistics and their application to digital forensics. In: Proceedings of the IEEE Workshop on Statistical Analysis in Computer Vision, pp. 1–8 (2003)
14. Jiang, X., Wang, W., Sun, T., Shi, Y.Q., Wang, S.: Detection of double compression in MPEG-4 videos based on Markov statistics. IEEE Signal Process. Lett. **20**(5), 447–450 (2013)
15. Ng, T.-T., Chang, S.-F., Hsu, J., Xie, L., Tsui, M.-P.: Physics-motivated features for distinguishing photographic images and computer graphics. In: Proceedings of the 13th Annual ACM International Conference on Multimedia, pp. 239–248 (2005)
16. Gao, X., Ng, T.-T., Qiu, B., Chang, S.-F.: Single-view recaptured image detection based on physics-based features. In: Proceedings of the IEEE International Conference on Multimedia and Expo (ICME), pp. 1469–1474 (2010)

Brain Tumor Segmentation Based on Clustering Using Pixel Intensity Variance of Pattern Recognition

M. Muthalakshmi$^{(\boxtimes)}$ and R. Dhanasekaran$^{(\boxtimes)}$

Department of Electrical and Electronics Engineering, Syed Ammal Engineering
College, Ramanathapuram, India
muthalakshmi.2010@gmail.com, rdhanashekar@syedengg.
ac.in

Abstract. In medical image processing of tumor segmentation process are very effective process of Magnetic Resonance Imaging (MRI) segmentation are using the radiology or clinical diagnosis. The proposed a novel technique of image clustering to segment the MRI brain tumor and segment the region of filtering process. The image filtering and image histogram analysis of the Circulation based Non-Linear Median Filtering (CNMF). The pixel intensity of labelling process of the neighboring structures are the pixel image intensity and quality recognize the Neighborhood Interior Edge detection (NIED) of segmentation technique. The neighboring pixel variation and detection of the image pattern recognition of the normal and abnormal image category level of processing to the Pixel Intensity Variance of Pattern Recognition (PIVPR) technique. The techniques are using the predict the abnormal category level of tumor spot and high accuracy level information.

Keywords: Edge detection · Clustering · Median filtering · Pattern analysis
Segmentation · MRI

1 Introduction

The convolutional neural network architecture is provided by the time and speeding process of subsequent brain neural information [1]. The supervised classification to segment the brain tumor. To implement the Convolutional Neural Network (CNN) to extract the Region of Interest (ROI). This perform patch based learning of image set to classify the abnormality region from trained dataset [2]. The adaptive image segmentation technique to extract the tumor region of brain MRI. The two-tier classifier to

© Springer International Publishing AG 2018
D. J. Hemanth and S. Smys (eds.), *Computational Vision and Bio Inspired Computing*,
Lecture Notes in Computational Vision and Biomechanics 28,
https://doi.org/10.1007/978-3-319-71767-8_60

retrieve the abnormality or normal class of image dataset. The segmentation and classification provide abnormality level of MRI image [3]. The brain image segmentation by using improved intuitionistic fuzzy c-means clustering algorithm. Image filtering by consider the noise applied to it this segment the image into Gray Matter (GM), White Matter (WM), and Cerebro Spinal Fluid (CSF) region of brain MRI image [4]. The singular value decomposition (SVD) are smoothening brain image segmentation. The intracranial tumor identification are diagnosing the problem of tumor identification and abnormal category of MRI brain tumor images [5]. In supervised learning technique to segment the tumor spot and abnormal tissue region of MRI image. The sparse information of each image to annotate the difference between normal and abnormal tissue region of MRI image [6]. The morphological process for brain tumor metastasis of early stage of primary tumor detection and helpful for diagnosis of the metastasis analysing the tumor infection [7]. Fuzzy k Means (FKM) algorithm and Self Organising Map (SOM) are combined to the multiple category of the tumor infection and analyze the benign and malignant category of brain tumor and time consisting detection of MRI brain images [8]. The Cuckoo Search (CS) algorithm are using the multiple formation of the size and shape depending upon the brain tumor diagnosing process are compared to the polar search methods and rectangular search methods of the rating process [9]. The random field framework segmentation process is the quality and time consuming. The brain tissues are the white matter, gray matter and cerebro spinal fluid and other tissues are the consisting by the quality of disease assessment [10]. 3-D image data set are using the patches are selection to be the framework registration and unsupervised learning process [11]. Model-based methods are using the detection of metastatic brain injury with cancer of MRI brain image to guide the radio surgery treatment planning process [12].

2 MRI Brain Tumor Segmentation

MRI Brain tumor generally consist of three components of segmentation process. These are the segmentation process are, the proposed system as shown in Fig. 1 to implement the T1-weighted image were affected by random speckle noise. To remove this additive noise of Circulation based Non-Linear Median Filtering (CNMF) technique used. It removes the noise component and also enhances the histogram image pixel intensity labelling process. In this research we remove the skull region and normalize background at the pre-processing stage which may affect the segmentation result. Above work is done by using Neighborhood Interior Edge Detection (NIED) method. The significant of correct classification through the Pixel Intensity Variance of Pattern Recognition (PIVPR). It can be clear segmentation of clustering pattern and predict the tumor spot pattern recognition.

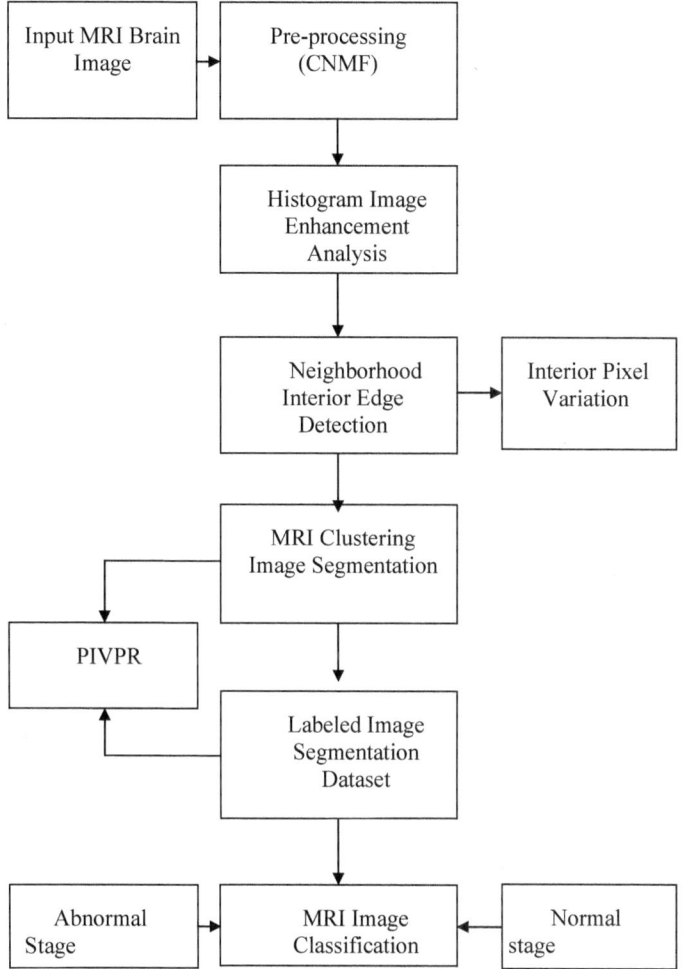

Fig. 1 Flow diagram of proposed system in MRI brain tumor segmentation

3 Segmentation Process

3.1 Image Enhancement

In this section to present a Circulation based Non-Linear Median Filtering (CNMF) of MRI image segmentation. Figure 2 shows the pre-processing steps are the median filtering process. The planning of filtering operation to apply the mask of median filter with the help of Gaussian distribution function. The depending upon the variation of center and boundary pixel, the noise pixel and eliminate by merging the median of boundary pixel. This provide smoothening effect to the image which decrease the noise in MRI brain. The filtered image to perform image Non-Linear function to apply the image normalization and equalized pixel to the image. The pixel image intensity of that image get normalization and enhancement of the pre-segmentation process.

(a)   **(b)**   **(c)**

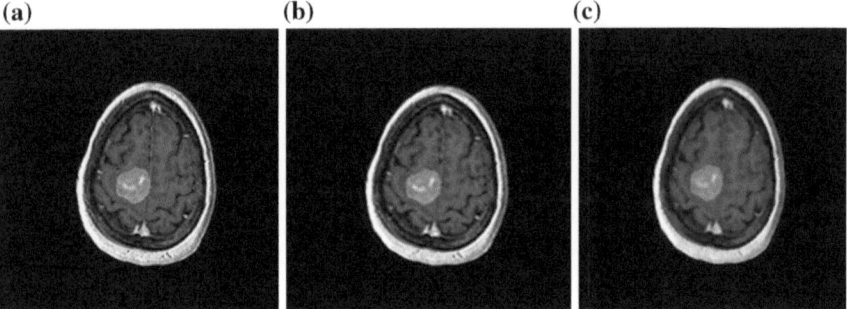

Fig. 2 Pre-Processing **a** Filtering enhancement of the MRI brain **b** Normalization image **c** Filtering image of the MRI brain

3.2 Skull Stripping

In this section to remove the skull boundary region of the segmentation MRI brain process as shown in Fig. 3 to eliminate the boundary region of skull and other tissues of the brain. The form of the segmentation brain of outer region is perfectly remove the skull and another tissues. From that skull stripping process of inner and outer layer of background skull region normalization process to suppress the highest pixel intensity of image. Figure 4 shows to Remove the skull and other bright region of the segmentation using the connected region to estimate the morphological operation. The brain segmentation to separate the brain region from skull. These are the Skull-Stripping process of MRI Brain background normalization.

(a)   **(b)**   **(c)**

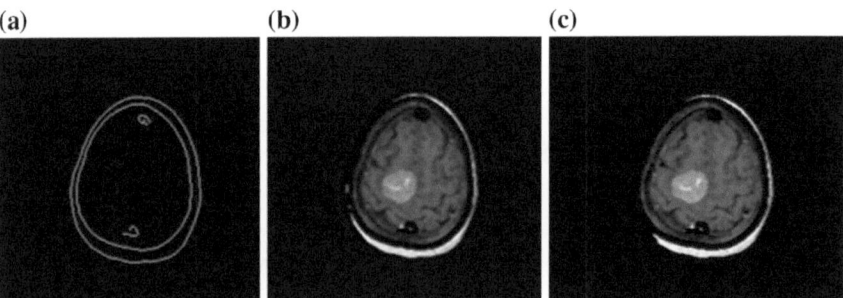

Fig. 3 Skull stripping **a** Skull removal of background region **b** Inner and outer layer of skull region **c** Suppress the maximum intensity level of MRI brain

(a) (b)

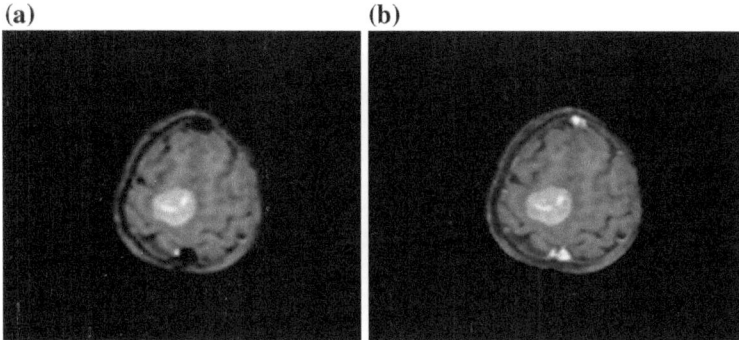

Fig. 4 Removal of another tissues **a** Extract the inner layer of MRI brain image **b** Remove the skull stripping process of background normalization

3.3 Clustering of MRI Brain Segmentation

In this section to perform image clustering pattern are using intensity pixel variance pattern analysis and pattern recognition of MRI brain. Shows in Fig. 5 clearly illustrate the minimum and maximum range of image pixel intensity to separate the clustering and apply the location of tumor infection. The cluster process steps through the initial cluster segmentation process. Another process to extract the bright intensity of binary output and labelling cluster formation of MRI brain to rectify the tumor and other tissues as shown in Fig. 6 the clustering output estimate the maximum level of image intensity to predict the tumor spot and also find the pixel intensity level of MRI brain image. From that clustering, to extract pattern recognition of that clustered data to exact the tumor spot of MRI brain segmentation. The pattern recognition of image histogram analysis the normal and abnormal category of brain. From that histogram analysis of that pattern recognition to identify the MRI brain image tumor infection of the accuracy level.

(a) (b)

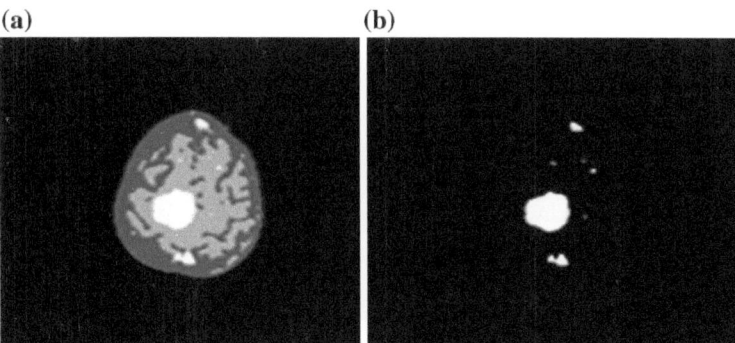

Fig. 5 Clustering process **a** Cluster result of the initial clustering MRI image **b** Binary extract initial clustering MRI brain image process

(a)   (b)

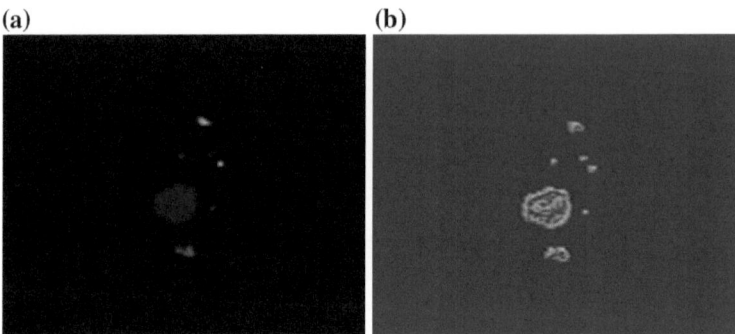

Fig. 6 Pattern recognition **a** Labeled clustering of MRI brain tumor segmentation **b** Intensity variance of texture pattern recognition

4   Result Analysis

The result analysis to perform segmentation and classification of MRI brain image by using machine learning of unsupervised learning methods. These unsupervised learning methods to verify the histogram of that pattern recognition and to generate the histogram image analysis. Table 1 shows the proposed image classification are compared to the Support Vector Machine (SVM) and Relevance Vector Machine (RVM). The classifier to rectify the image was normal and abnormal category of MRI brain image. The proposed technique to extract the pattern recognition process and analyze the binary level of image segmentation to redrive the tumor spot analysis. To predict the Region of Interest (ROI) and analyze the category of MRI brain tumor analysis. The analysis of ROI tumor spot from the Pixel Intensity Variance of Pattern Recognition (PIVPR) method to apply the accurate level of brain tumor infection.

Table 1 The Proposed System Comparsion of SVM and RVM

Parameters	Proposed	SVM	RVM
TP	66	55	57
TN	53	53	53
FP	0	0	0
FN	5	16	17
Sensitivity	92.95	77.46	80.28
Accuracy	95.96	87.09	88.70

Figure 7 Shows the rate analysis for the ratio parameters are compared to the analysis of proposed system to apply the Global Acceptance Ratio (GAR), False Acceptance Ratio, and False Rejection Ratio (FRR). The ratio are comparing to the global level information of the parameter analysis. The parameter analysis are useful

for the True Positive (TP), True Negative (TN), False Positive (FP), False Negative (FN), Sensitivity, and Accuracy level of highest information. The most level of pixel intensity of accuracy information to gets the Global Acceptance Ratio MRI brain tumor analysis.

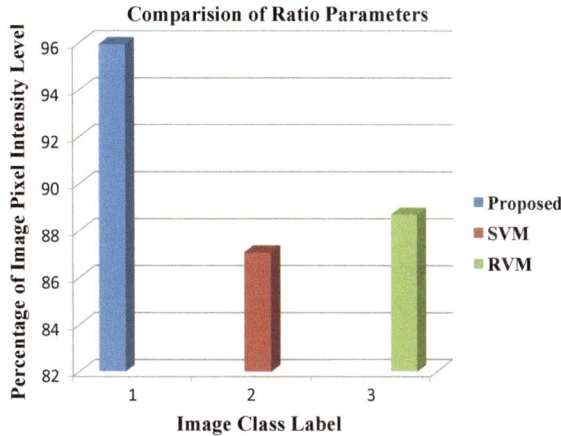

Fig. 7 Rate analysis for GAR, FAR, FRR

5 Conclusion

The techniques are used by the Image Enhancement, skull stripping, and clustering process of MRI brain image segmentation. The process are the normal and abnormal category of image pixel intensity level are using the three segmentation process are the Circulation based Non-Linear Median Filtering, Neighborhood Interior Edge Detection, and Pixel Intensity Variance of Pattern Recognition. The comparative analysis of the proposed system is Support Vector Machine (SVM) and Relevance Vector Machine (RVM). The histogram analysis are using Predict the MRI brain tumor spot and category of analysis and also effective solution of high accuracy pixel level information and also comparing the rate analysis for the Global Acceptance Ratio is high compare to the False Acceptance Ratio, False Rejection Ratio. It is useful for diagnosis for radiology clinical expert systems.

References

1. Havaei, M., Davy, A., Warde-Farley, D., Biard, A., Courville, A., Bengio, Y., Pal, C., Jodin, P.-M., Larochelle, H.: Brain tumor segmentation with deep neural networks. Med. Image Anal. **35**, 18–31 (2017)
2. Pereira, S., Pinto, A., Alves, V., Silva, C.A.: Brain tumor segmentation using convolutional neural networks in MRI images. IEEE Trans. Med. Imaging **35**(5), 1240–1251 (2016)

3. Anitha, V., Murugavalli, S.: Brain tumour classification using two-tier classifier with adaptive segmentation technique. IET Comput. Vision **10**(1), 9–7 (2016). https://doi.org/10.1049/iet-cvi.2014.0193
4. Verma, H., Agrawal, R.K., Sharan, A.: An improved intuitionistic fuzzy C-means clustering algorithm incorporating local information for brain image segmentation. Appl. Soft Comput. **46**, 543–557 (2016)
5. Subudhi, B.N., Thangaraj, V., Sankaralingam, E., Ghosh, A.: Tumor or abnormality identification from magnetic resonance images using statistical region fusion based segmentation. Magn. Reson. Imaging **34**, 1292–1304 (2016)
6. Goetz, M., Weber, C., Binczyk, F., Polanska, J., Tarnawski, R., Bobek-Billewicz, B., Koethe, U., Kleesiek, J., Stieltjes, B., Maier-Hein, K.H.: DALSA: domain adaptation for supervised learning from sparsely annotated MR images. IEEE Trans. Med. Imaging **35**(1), 184–196 (2016)
7. Perez, U., Arana, E., Mortal, D.: Brain metastases detection methods in magnetic resonance imaging. IEEE Lat. Am. Trans. **14**(3), 1109–1114 (2016)
8. Vishnuvarthanan, G., Pallikonda Rajasekaran, M., Subbaraj, P., Anitha, V.: An unsupervised learning method with a clustering approach for tumor identification and tissue segmentation in magnetic resonance brain images. Appl. Soft Comput. **38**, 190–212 (2016)
9. Ilunga-Mbuyamba, E., Cruz-Duarte, J.M., Avina-Cervantes, J.G., Correa-Cely, C.R., Linder, D., Chalopin, C.: Active contours driven by Cuckoo Search strategy for brain tumor images segmentation. Expert Syst. Appl. **56**, 59–68 (2016)
10. Pereira, S., Pinto, A., Oliveira, J., Mendrik, A.M., Correia, J.H., Silva, C.A.: Automatic brain tissue segmentation in MR images using random forests and conditional random fields. J. Neurosci. Methods **270**, 111–123 (2016)
11. Wu, G., Kim, M., Wang, Q., Munsell, B.C., Shen, D.: Scalable high-performance image registration framework by unsupervised deep feature representations learning. IEEE Trans. Biomed. Eng. **63**(6), 1505–1516 (2016)
12. Perez, U., Arana, E., Moratal, D.: Brain metastases detection methods in magnetic resonance imaging. IEEE Lat. Am. Trans. **14**(3), 1109–1114 (2015)

Outlier Detection in Time-Series Data: Specific to Nearly Uniform Signals from the Sensors

Sourabh Suman and B. Rajathilagam[(✉)]

Department of Computer Science and Engineering, Amrita School
of Engineering, Coimbatore, Amrita Vishwa Vidyapeetham, Amrita University,
Coimbatore, India
cb.en.p2csel5019@cb.students.amrita.edu,
b_rajathilagam@cb.amrita.edu

Abstract. In this paper, the complexity of detecting an outlier has been shown. The importance of an outlier has been presented with the need to interpret these outliers. The sensors collect data with certain sampling time period and these data are stored which contribute to the huge database. These sensors can be electrocardiogram sensor which monitors electrical and muscular functions of the heart, they can be pollution monitoring stations at the airports, they can be heat sensors in a building, and they can be flight data recorders (FDR) and so on. Sometimes, these sensors miss to detect the signal due to some technical fault and hence the output is "Not Available (NA)". These NA time stamps create unnecessary problems which lead to unwanted outputs when the data is processed. In this paper, an algorithm is presented which replaces these NA values with most probable values. When the data is ready with all NA values removed, the data is processed for detecting the outlier. In this paper, an outlier is being detected by analyzing the signal in the frequency domain along with the mean in the time domain. A large data set is divided into equal sized blocks. Each block is then converted to its frequency domain and mean is calculated in the time domain. These two parameters are considered to detect any outlier in the block. This approach removes the complexity in the algorithm without compromising in the efficiency of detecting an outlier. Hence, a large database of values is processed in relatively less time with appreciable accuracy.

Keywords: Outlier · Time-series data · Fast Fourier Transform (FFT)
Block mean

1 Introduction

An outlier is all about the abnormality that exists in a signal. Any unusual observance as compared to the natural behavior of the signal can be inferred as an outlier [1, 2].

Human beings have the natural capability to detect an outlier with the help of brain. The brain does the necessary processing of the signal received by the five senses in the human body namely eyes, ear, nose, tongue and skin. Ear detects the health of a person by hearing and observing the change in the frequency or loudness of the sound produced by that person. If there is any change than usual, the brain is easily able to detect

© Springer International Publishing AG 2018
D. J. Hemanth and S. Smys (eds.), *Computational Vision and Bio Inspired Computing*,
Lecture Notes in Computational Vision and Biomechanics 28,
https://doi.org/10.1007/978-3-319-71767-8_61

the change. The smell of the food is another example of the signal. It acts as an input for the nose and the processing takes place in the brain which generates the signal whether the food is good or foul. Similarly, other senses can detect any unusual phenomena with the help of brain which processes the signal and detects any outlier occurring in nature and easily makes the necessary conclusion.

Making machines to work according to humans is a complex job. All the signals are converted to the electrical signal and then the processing takes place in the processor of a computer. Algorithms are written to detect any outlier in the signal if it occurs. Detecting an outlier is a very important job as this contains very useful information. An outlier can inform about the quality of a product in a manufacturing industry [3]. Any change in the heart beat of a patient can warn the doctors about the health of that patient. Any outlier in the nuclear reactors can give the information about the amount of nuclear fission taking place and help the scientists to take appropriate measures to control the nuclear reactions. Traffic levels on the road can help to detect the amount of pollution generated. Traffic in the network helps to detect any unauthorized access in an organization. Outliers help to detect any breach in the physical security system with the help of sensors installed in a building. Hence, detecting an outlier is a very important and useful phenomenon.

But, there doesn't exist a reliable and robust algorithm which can detect an outlier in a signal generated from any source. Basu and Meckesheimer [3] put forward the method of one-sided and two-sided median methods. But, this method is inefficient when outliers spanning longer than window are consecutive as this makes it difficult to differentiate between same value and actual signal. Akouemo and Povinelli [4] detected an outlier by introducing ARIMAX model. But, each and every step has to be trained on cleaner data. If an algorithm detects an outlier with high accuracy, then the processing time is high. Tian et al. [5] gave a method that combined density-based clustering algorithm with soft sensor modeling. But, this method detects an outlier accurately only for soft sensor modeling in complex industrial processes. Nouira and Trabelsi [6] proposed the method of graphical approaches and Gibb's sampling approach. It has been promised that this information may be used to improve outlier detection. But, no experimentation has been performed using these two methods. Zwilling and Wang [7] detected an outlier using the information from covariance of time series data. It provides the user to choose any number and type of features and the algorithm will correctly identify the outlier. Then this method becomes accurate. Zwilling and Wang [8] proposed Multivariate Voronoi Outlier Detection (MVOD) method. Its performance is accurate, robust and sensitive for multivariate time series data. Ferdousi and Maeda [9] gave the method to find an outlier in time-series financial data using Peer Group Analysis (PGA). In this method, pattern of behavior is checked around the target and then compared with pattern of behavior of the target. Suh et al. [10] proposed the method of echo-state conditional variational autoencoder (ES-CVAE) for anomaly detection in multivariate time-series data which combines echo-state network with variational encoder. Martins et al. [11] detected an outlier online in non-stationary time-series data. The method is based on a Least Squares Support Vector Machine along with a sliding window-based learning algorithm. Lv [12] presented Autoregressive and Moving Average Model (ARMA). This model is a real-time outlier detection and replacement algorithm. If the processing time to detect an outlier is less, then one has to compromise with the accuracy of detecting an outlier. In this paper, an algorithm is proposed which can detect outlier in time-series data with less processing time and appreciable accuracy.

2 Method Proposed

The method proposed in this paper to detect an outlier gives more accuracy when the outlier occurs in a uniform signal or a nearly uniform signal. Sensors installed at the airports to check the pollution level due to sulphur dioxide, carbon dioxide, particulate matter, etc. generally produces an output which follows nearly a uniform distribution. Figure 1 shows the continuous ambient air quality due to nitric oxide (NO) at DTU station in Delhi. The value is available every after 4 h. Hence, the sampling time period is 4 h. Such, signals are the examples of time-series signals.

The data accumulated is very large in quantity. The stored data is later checked for detecting the outlier. In order to detect an outlier in a time-series data, following are the steps involved:

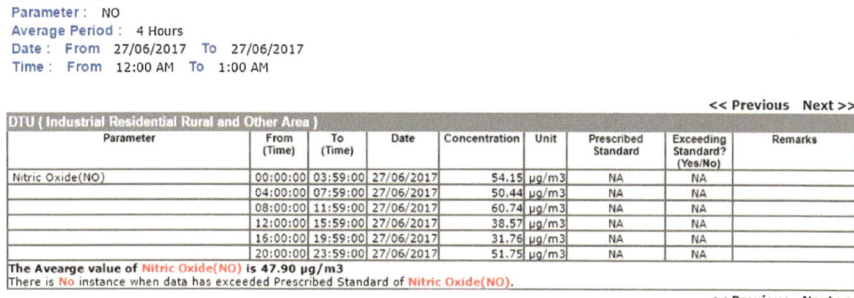

State : Delhi
City : Delhi
Station : DTU
Parameter : NO
Average Period : 4 Hours
Date : From 27/06/2017 To 27/06/2017
Time : From 12:00 AM To 1:00 AM

<< Previous Next >>

DTU (Industrial Residential Rural and Other Area)

Parameter	From (Time)	To (Time)	Date	Concentration	Unit	Prescribed Standard	Exceeding Standard? (Yes/No)	Remarks
Nitric Oxide(NO)	00:00:00	03:59:00	27/06/2017	54.15	µg/m3	NA	NA	
	04:00:00	07:59:00	27/06/2017	50.44	µg/m3	NA	NA	
	08:00:00	11:59:00	27/06/2017	60.74	µg/m3	NA	NA	
	12:00:00	15:59:00	27/06/2017	38.57	µg/m3	NA	NA	
	16:00:00	19:59:00	27/06/2017	31.76	µg/m3	NA	NA	
	20:00:00	23:59:00	27/06/2017	51.75	µg/m3	NA	NA	

The Avearge value of Nitric Oxide(NO) is 47.90 µg/m3
There is No instance when data has exceeded Prescribed Standard of Nitric Oxide(NO).

<< Previous Next >>

Fig. 1 CPCB data on continuous ambient air quality for DTU due to NO level

(1) A large time-series data is checked for the presence of NA. If any NA is found, it is removed appropriately (algorithm to remove NA is explained in the next section).
(2) Then the data is divided into equal sized blocks.
(3) Each block is converted into the frequency domain by performing FFT over the block.
(4) Mean is calculated for each and every block in the time domain.
(5) Then the block is checked against the threshold in the frequency domain as well as in the time domain.
(6) The combination of the two thresholds helps to increase the accuracy to detect an outlier.

After processing the signal, the output is in the form of block numbers in which outliers have been detected.

3 Experimentation and Result

While processing the data, if the data contains any NA value, then the output is undetermined. When a block containing NA values, is converted into its frequency domain, the output varies from negative infinity to positive infinity (−Inf, Inf). Hence, these NA points are replaced with most expected value. The procedure to remove NA is as follows:

(1) If NA occurs in the middle index, this value is replaced by the mean of the previous and the next value.
(2) If NA occurs in the first index, this value is replaced with next non-NA value.
(3) If NA occurs in the last index, this value is replaced with previous non-NA value.

For the demonstration of the algorithm to detect an outlier, Fig. 2 describes the data.

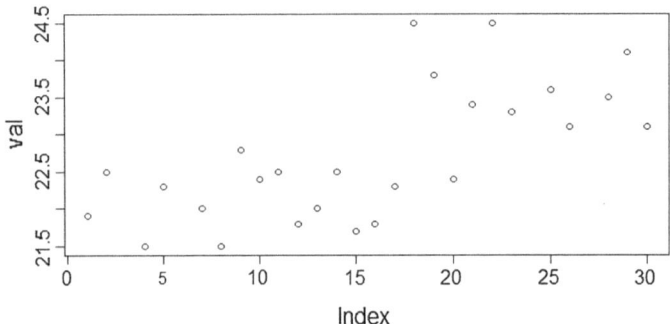

Fig. 2 True data obtained from the sensor

NA values are present at the index 3, 6, 24 and 27. After removing NA points, the data is shown in Fig. 3.

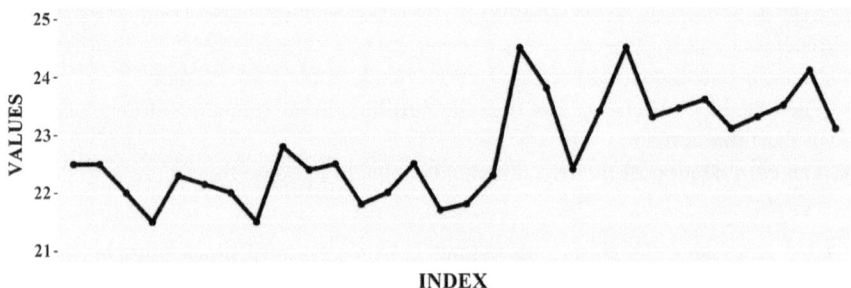

Fig. 3 The data after removing NA points

This data is divided into three equal sized blocks and analyzed for detecting the outlier. Figure 4a shows the block 1. The threshold value of the block means is 23 while in the frequency domain, the threshold value is 3. Figure 4b shows the frequency domain of the block 1. Value at index 1 is not taken to check for the threshold. This is because intuitively it can be understood that a large amount of information can lie in the lower frequency range. Figure 4c shows that no information lies above 3. Hence, it can be inferred that there is no outlier in this block.

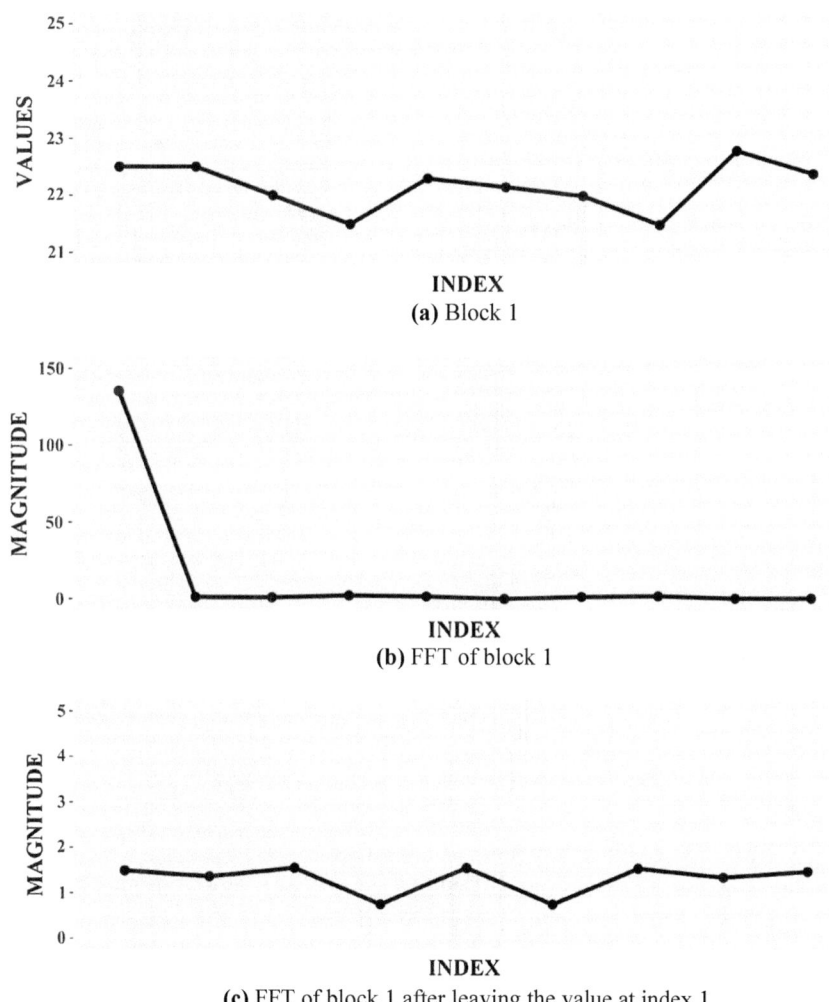

(a) Block 1

(b) FFT of block 1

(c) FFT of block 1 after leaving the value at index 1

Fig. 4 Block 1 in time and frequency domain

Figure 5a shows the block 2. From Fig. 5c, it can be observed that some amount of information lies above 3. So, it can be inferred that an outlier lies in this block.

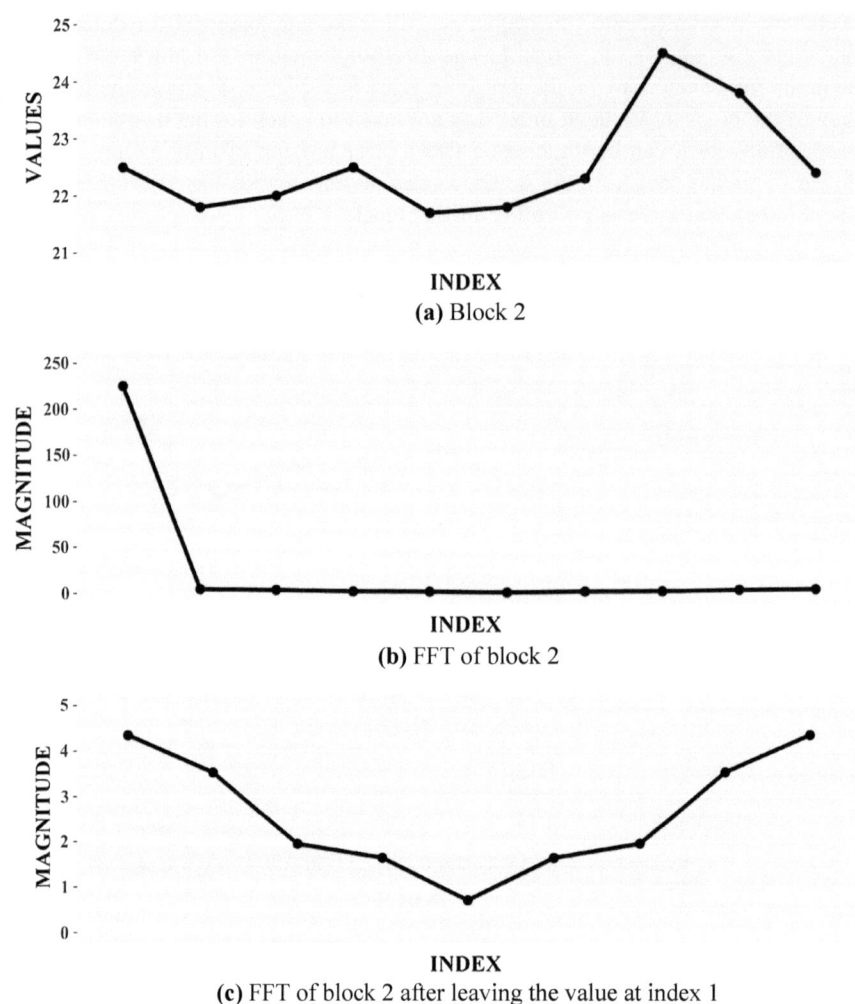

(a) Block 2

(b) FFT of block 2

(c) FFT of block 2 after leaving the value at index 1

Fig. 5 Block 2 in time and frequency domain

Figure 6a shows the block 3. All the values lie above 23. Hence, the whole block 3 is an outlier. But, Fig. 6c shows that all the information lies below 3. At such instances, block mean helps to detect the outlier. Mean of the block is 23.535 which is more than the threshold. Hence, this block contains an outlier.

This experiment can further be extended to detect a proper signal or a random signal. Also, the extent of randomness can be determined by setting proper threshold.

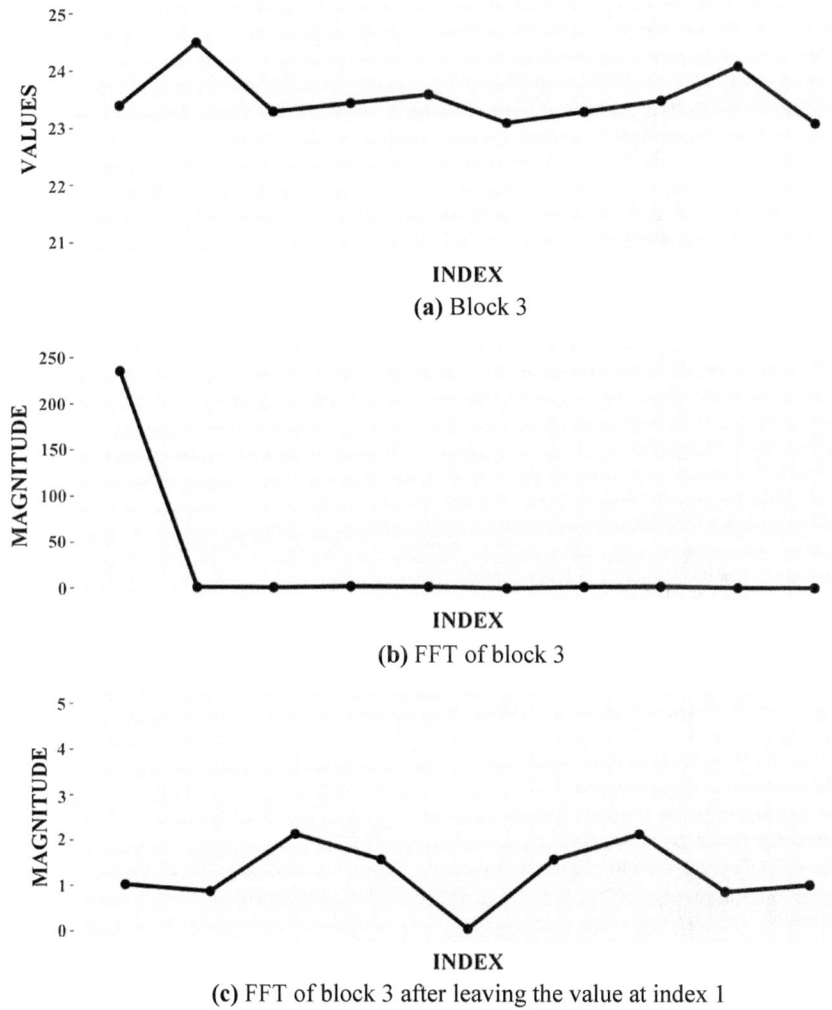

(a) Block 3

(b) FFT of block 3

(c) FFT of block 3 after leaving the value at index 1

Fig. 6 Block 3 in time and frequency domain

4 Conclusion

Many algorithms have been proposed to detect an outlier. But, there is a trade-off between processing time and accuracy. This paper introduced a generic approach to detect an outlier in relatively less time with appreciable accuracy. Block means in the time domain along with frequency domain analysis poses a simple and fast algorithm which detects an algorithm with good accuracy. In future, this project can be extended to detect a proper data and random data. Also, the degree of randomness can be obtained.

References

1. Ashok, A., Smitha, S., Kavya Krishna, M.H.: Attribute reduction based anomaly detection scheme by clustering dependent oversampling PCA. In: 2016 International Conference on Advances in Computing, Communications and Informatics (ICACCI). IEEE (2016)
2. Prathibhamol, C.P., Amala, G.S., Kapadia, M.: Anomaly detection based multi label classification using Association Rule Mining (ADMLCAR). In: 2016 International Conference on Advances in Computing, Communications and Informatics (ICACCI). IEEE (2016)
3. Basu, S., Meckesheimer, M.: Automatic outlier detection for time series: an application to sensor data. Knowl. Inf. Syst. 11(2), 137–154 (2007)
4. Akouemo, H.N., Povinelli, R.J.: Time series outlier detection and imputation. In: 2014 IEEE PES General Meeting Conference & Exposition. IEEE (2014)
5. Tian, H., Liu, X., Han, M.: An outliers detection method of time series data for soft sensor modeling. In: 2016 Chinese Control and Decision Conference (CCDC). IEEE (2016)
6. Nouira, K., Trabelsi, A.: Time series analysis and outlier detection in intensive care data. In: 2006 8th International Conference on Signal Processing, vol. 4. IEEE (2006)
7. Zwilling, C.E., Wang, M.Y.: Covariance based outlier detection with feature selection. In: 2016 IEEE 38th Annual International Conference of the Engineering in Medicine and Biology Society (EMBC). IEEE (2016)
8. Zwilling, C.E., Wang, M.Y.: Multivariate voronoi outlier detection for time series. In: 2014 IEEE Healthcare Innovation Conference (HIC). IEEE (2014)
9. Ferdousi, Z., Maeda, A.: Unsupervised outlier detection in time series data. In: Proceedings of the 22nd International Conference on Data Engineering Workshops, 2006. IEEE (2006)
10. Suh, S., et al.: Echo-state conditional variational autoencoder for anomaly detection. In: 2016 International Joint Conference on Neural Networks (IJCNN). IEEE (2016)
11. Martins, H., et al.: A machine learning technique in a multi-agent framework for online outliers detection in Wireless Sensor Networks. In: 41st Annual Conference of the IEEE on Industrial Electronics Society, IECON 2015. IEEE (2015)
12. Lv, Y.: An adaptive real-time outlier detection algorithm based on ARMA model for radar's health monitoring. In: IEEE AUTOTESTCON, 2015. IEEE (2015)

Smart Energy Management System Based on Image Analytics and Device Level Analysis

S. Birindha, V. Ananthanarayanan$^{(\boxtimes)}$, and P. Bagavathi Sivakumar

Department of Computer Science and Engineering, Amrita School of
Engineering, Amrita Vishwa Vidyapeetham, Coimbatore, India
cb.en.p2csel5003@cb.students.amrita.edu,
{v_ananthanarayanan, pbsk}@cb.amrita.edu

Abstract. Large educational institutes, organization, and industries face large challenges on energy utilities, consumption and its management strategies. But smart energy management technology solutions help the high energy consumption complexities during while putting the best and greenest foot forward. Smart Energy Management technology solutions, improve and respond quickly to power spikes at times of demand, expedite data gathering, reporting and regulatory compliance, automate services to control operating costs and enable to save energy. Connecting smart meters to data stores requires a reliable, intelligent network. The Smart Energy Management System introduced here deals with a device level analysis that gives information of the device that has caused the peak rise in the total power consumption of the organization. Predictive analysis technique is used on the database to predict the future maximum demand and load balancing technique is applied to reduce the consumption of power from generator source. Therefore the total power consumption from exceeding the maximum demand can be avoided and the maximum demand of the power supply for the organization can be maintained. Further, on application of AI techniques this system control becomes fully automated.

Keywords: Internet of things · Predictive analysis · Smart sensor · Threshold
Wi-Fi · Facial recognition system · Image analytics

1 Introduction

Electricity is one of the easiest resources to waste. So, every year, in order to effectively overcome energy crises, efficient energy management systems are introduced [1]. The paradigm of energy management shifted from being supply-centered to demand-centered and most up-to-date supplied and researched energy management systems include energy management through basic scheduling demand prediction/management through data analysis [2].

© Springer International Publishing AG 2018
D. J. Hemanth and S. Smys (eds.), *Computational Vision and Bio Inspired Computing*,
Lecture Notes in Computational Vision and Biomechanics 28,
https://doi.org/10.1007/978-3-319-71767-8_62

1.1   Role of IoT in Energy Management System

In data analysis of energy, IoT plays a major role. IoT is supported by communication technologies like Bluetooth low energy, ZigBee, Wi-Fi which enables devices to communicate [3]. In support of intelligent decision making, Internet of Things bridges diverse technologies to enable new applications by connecting physical objects together. Data from related objects within the network are gathered by sensors and sends it to a data warehouse, database, or cloud [4]. Typically, the IoT nodes should operate using low power in the presence of lossy and noisy communication links [5]. So, for a smart energy management, it must be interconnected to embedded systems and more specifically to the android applications to allow a remote management.

Electricity is typically metered at such a high level that it is difficult to be sure where it is being consumed, and by what. If the power consumption is measured at a discrete level, monitoring each device or process, it would be easy to estimate the energy cost of each item or the energy cost of a particular process. So to collect data from different locations and provide it to a software application that can make sense all of it is by using the Internet of Things technology (IoT) [6]. Using sensors a measure of current flow can be made. It's no longer necessary to hardwire each one to a network. Instead, wireless network can be used to transmit data from each device to the database in the system or cloud application on the Internet [7].

In this paper, a device level energy management system has been proposed. The rest of the paper is organized as follows: Sect. 2 describes about the proposed energy management system. Section 3 describes the case study on energy management system based on feeder level. Section 4 describes the system architecture for energy management systems based on different levels. Section 5 describes about the device level analysis. Implementation of smart energy management system based on device level analysis is given in Sect. 6. Section 7 presents the result analysis and future work is given in Sect. 8. Conclusion is given in Sect. 9.

2   A Device Level Energy Management System

When an organization as a whole is considered, the Energy Management hierarchy can be applied from bottom to top level. So this starts right from the end equipment placed in a laboratory, a fan within a faculty cabin, a printer in an administrative office etc., as such the complete detail of each end device which consumes power. The energy management technique applied, analyses each end load, the amount of power consumed with time factor. This can be termed as "Device level analysis".

Systems that consume large electrical energy for their effective functioning have significant impact on fuel economy while using generators. By adopting energy management algorithms considerable energy savings could be attained by managing the heavy power consumption by the Electrical and Electronic equipment not to exceed the maximum power usage allocated for an organization. This maximum power usage allocated for an organization can be otherwise termed as "The Maximum Demand".

Next a database is created for each device with the detail of its power consumption per second, cumulative power consumed as a function of time and maximum usage of it in a day, (i.e.) maximum usage. This data can be used to predict the power consumption factor of an end device. Based on this the predictive analysis and load balancing techniques are framed and applied which in turn as a whole supervises the total energy consumption. Each device is given a priority number. Thus the entire campus is networked, and using specialized sensors at device levels, the energy management is fully automated which could be termed as the "Smart Energy Management System applied at device level strategies—"**SEMS-d**".

3 Case Study on Energy Management System Based on Feeder Level

A case study on power consumption was done in the University campus. The university campus has two power houses and each power has about seven gensets. A unit of 825 kVA had been allotted to the university campus. But this is not sufficient to meet the maximum demand of the university campus. Figure 1 shows the number of units consumed from the power supplied by diesel generator and TNEB.

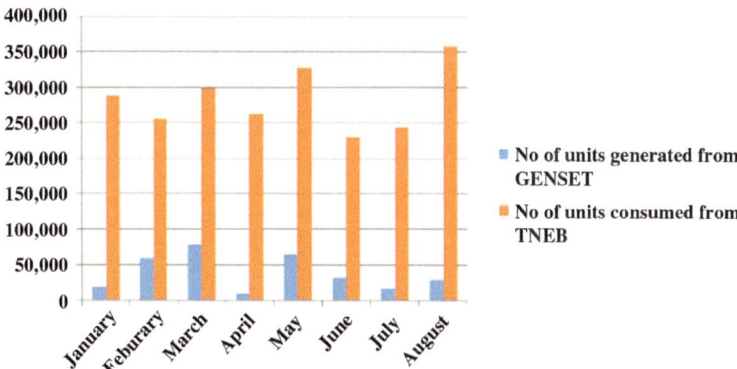

Fig. 1 Total no of units consumed using diesel and units consumed from TNEB (2014)

From the case study it was found that the cost incurred per unit using the diesel generator is higher than that of the cost incurred per unit consumption from TNEB. So, it is necessary to design an energy management system that exploits the power supply from TNEB and avoid the usage of generators. Figure 2 shows the cost incurred in generating one unit of power from diesel generator and TNEB.

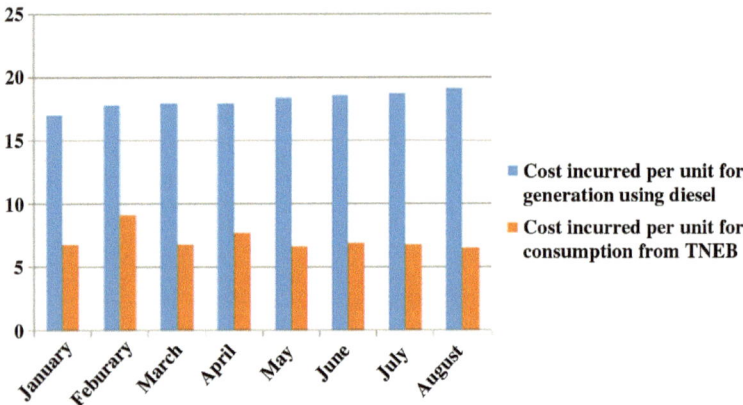

Fig. 2 Comparison of cost incurred per unit for generation using diesel and units consumed from TNEB (2014)

3.1 Smart Control of Devices

As a first step of smart energy management system, smart technologies have been used to efficiently control the electrical devices. A smart technology such as Bluetooth, ZigBee [8], Wi-Fi, and GSM has been implemented in the existing switch boxes. The choice of selecting the smart technologies depends on the distance between the user and the device. For example, Bluetooth could be used to control an electrical device at a distance of 10 m from the user. The user can use his/her mobile/desktop computer to control the electrical devices using the smart switch box.

3.1.1 Bluetooth Module

The Bluetooth module shown in the Fig. 3 has a step-down transformer to reduce the incoming current of 50 Hz, 230 V AC to 50 Hz, 12 V AC (only the peak to peak voltage was reduced, not the frequency). A LCD (Liquid Crystal Display) 16 × 2 is placed to display the operations performed by the microcontroller. A maximum of 32 characters can be displayed by the LCD. HC-05 Bluetooth is used in the module and it is a trans-receiver Bluetooth. The Bluetooth works with 3.3 V, and hence a regulator LM1117 is connected, which in turn reduces the voltage from 5 to 3 V. The Rx (receiver) of the Bluetooth is connected to the Rx of the microcontroller ATMega8. The data/signals from the android mobile are transmitted to the Bluetooth module (HC-05) and it receives those data. Thus here the Bluetooth module is used as a receiver. The ATMega8 microcontroller has got 28 pins totally. The pins 7 and 20 are connected to 5 V power supply. A regulator (7805) is used to reduce the voltage since the microcontroller and the LCD requires only 5v. So this regulator reduces 12–5 V. The pins 8 and 22 are connected to the ground. The pins 9, 10, 14–19 are used for the LCD (i.e. PB0–PB7) and pins 23–26 (from PC0 to PC3) are connected to the relay.

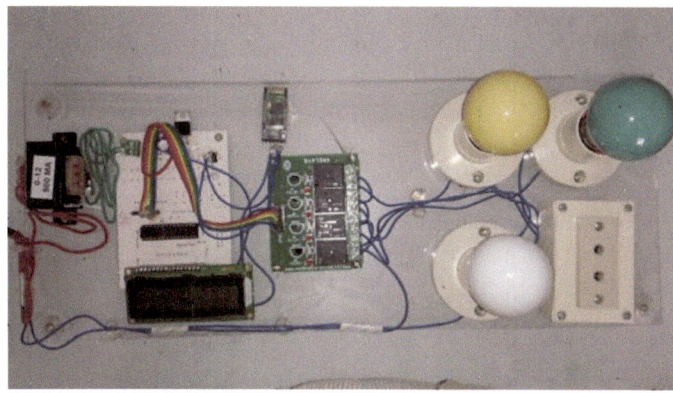

Fig. 3 Bluetooth Module

There are four relay, four LEDs and three bulb holders and a two pin plug socket. The microcontroller receives the data/signals from the Bluetooth (HC-05) and does the corresponding operations. For example, if the user wants to control the first bulb, then one could select the first option in the Bluetooth application and the signals are given to the microcontroller through the Bluetooth. There are four relays of 12 V each and four transistors (BC5478). The transistor acts as an electronic switch to control the device.

There are four LEDs and they have very less voltage of about 3 V, and so the resistors of 2.2 k is connected in series to reduce the flow of current. This Bluetooth module is used when the user is within the range of Bluetooth (<10 m). For example: The Bluetooth module works efficiently within a room of area 10 m². To use this smart switch box, user's mobile is connected to the Bluetooth inside the module.

3.1.2 ZigBee Module

The ZigBee module is used when the user is within the ZigBee range of 10–100 m. For example, a person in ground floor can control the electrical device in first floor of the same building.

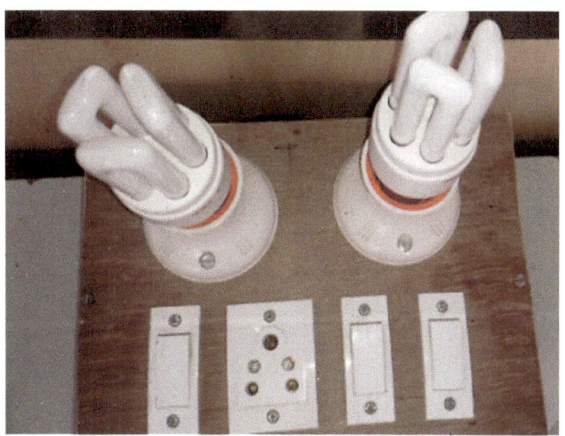

Fig. 4 ZigBee Module

It consists of an intelligent circuit that can control the device according to the command it receives. To make the device intelligent a microcontroller is added to each node and is connected with an RF transceiver. The test bed shown in the Fig. 4 consists of Smart device, Smart Server and the SEM intelligent sensor node.

3.1.3   Wi-Fi/GSM

The Wi-Fi module shown in the Fig. 5 is used when the user is within the Wi-Fi range of 20 m. For example, if a person in one building wants to control the electrical device in another building.

Fig. 5 Wi-Fi Module

3.2   Feeder Level Energy Management

All the devices have been connected to different feeders. There are about twenty eight feeders in the power house-1 and forty four feeders in the power house-2. Each feeder has a digital energy meter fixed to it. Using Wi-Fi all the digital energy meters are connected to a main server in the power house-1.

The power consumption of each feeder is fed into the energy management algorithm and it calculates the total power consumption. Also, the power consumption details are stored in a database for future reference. A case study was done on the energy management system in the university campus. Based on the case study, a list of critical and non-critical loads was identified. Critical loads are those loads to which the power supply has to be maintained under any circumstances. Power supply to these loads should never be interrupted.

3.3   Automated Energy Management System

The automated energy management system has a decision making algorithm that keeps the total power consumption from exceeding 825 kVA. The decision making algorithm takes action depending on various cases.

Case 1: When the total power consumption equals 800 kVA, it gives the message "warning, total power consumption exceeding 800 kVA".

Case 2: If the total power consumption is greater than or equal to 810 kVA and retains the same condition for thirty minutes then it is said to be in the critical condition. The source of power supply for the feeder with the highest priority is changed to generator. A message "feeder number: 5, is changed to generator power supply" is sent.

Case 3: If the total power consumption reaches below 800 kVA, then it is a safe state and the source of power supply of the feeder is again switched back to power supply from TNEB.

Case 4: If the total power consumption does not reach below 800 kVA and retains in the critical condition for the next thirty minutes, then the source of power supply of the feeder in the second priority is changed to generator.

This decision making algorithm was implemented in python and simulated using raspberry-pi. The results showed that for an efficient energy management system, it is necessary to go for a device level. The device that caused a sudden peak rise in the total power consumption cannot be found in a feeder level energy management since many devices may be connected to a feeder. To reduce or optimize the power consumption, it is necessary to know the power consumed by each and every device. Thus in order to reduce the total power consumption of an organization and of the most very effectively reduced usage of generators, a device level analysis must be introduced which gives the information of the device which caused hikes in the total power consumption to exceed the threshold.

4   System Architecture

The feeder level energy management system shown in Fig. 6 has devices from various buildings of the institution connected to different feeders. A feeder may have about fifteen devices connected to it. The power consumed by the devices is got from the feeder by fixing a digital energy meter to it. The power details are stored into a database through Wi-Fi. The data collected is fed into the automated system module.

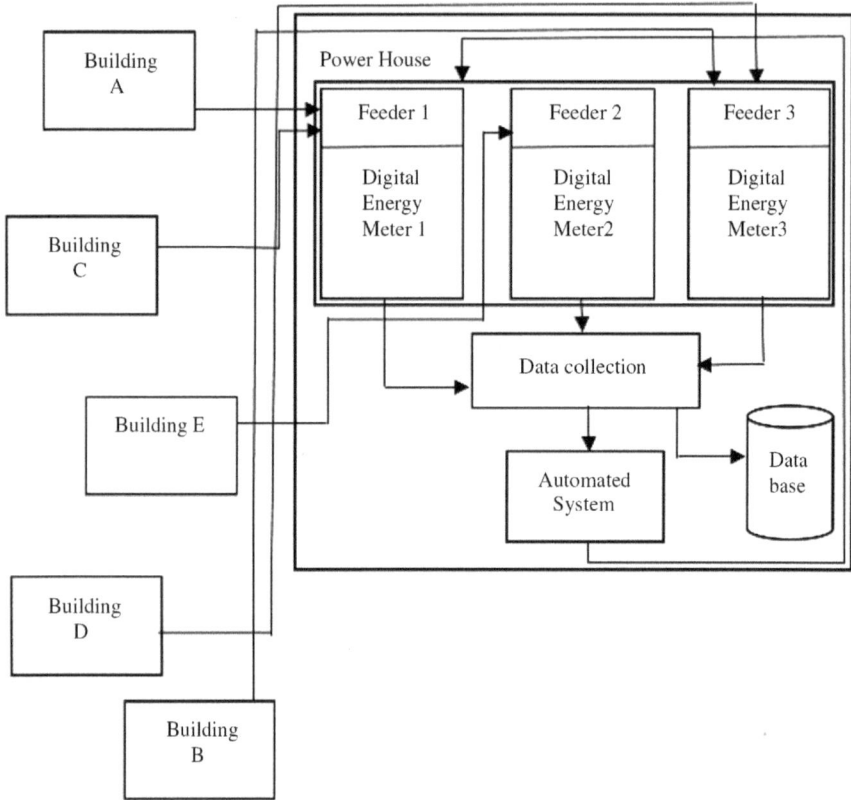

Fig. 6 System Architecture for Feeder Level Smart Energy Management System

The system architecture for device level smart energy management system shown in the Fig. 7 has three different modules. A sensor is mounted on each device to get the individual power consumption (DP). In the first module, details of each device are received from the smart senor [9] or smart energy meters [10]. Data preprocessing steps are done on the senor data to avoid noisy data, outliers and anomalies [11]. After the data preprocessing, total power consumption (TP) is calculated and threshold is determined for individual device. The threshold for each device (D-TH) is determined by taking its average power consumption. Also, the threshold for total power consumption is determined (T-TH). The second module is the decision making module. It has three different test cases. The third module is the predictive analytics which predicts the future maximum demand [12]. Finally the entire energy management system is made automated using artificial intelligence [13]. The main system (PC) is under surveillance for security purpose.

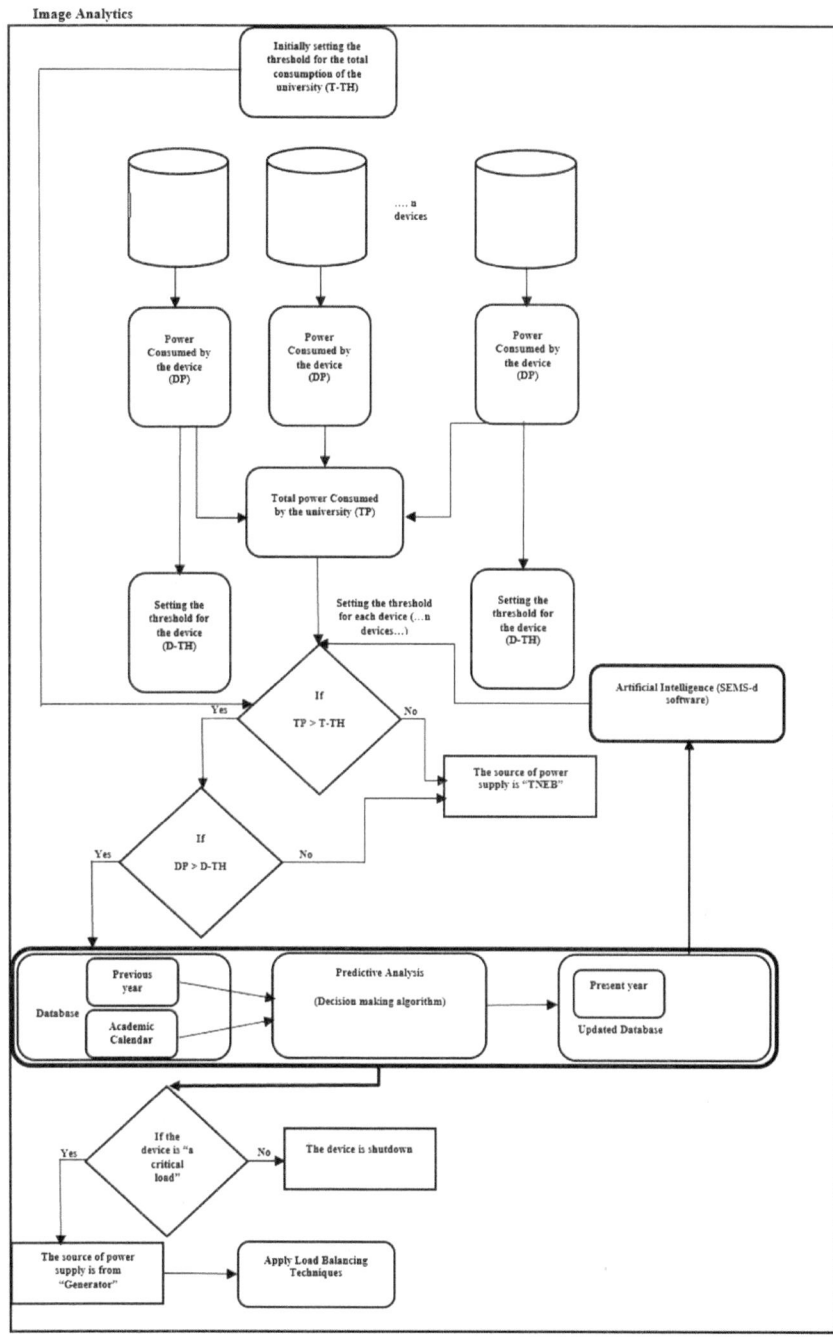

Fig. 7 System Architecture for Device Level Smart Energy Management System

5 Device Level Analysis

A device level analysis introduced here involves a smart sensor mounted on each device. This transfers the required data set of the various parameters under consideration from the device to the database. In addition to this, the database also contains the maximum usage of each end device per day. From this the period of maximum usage, minimum usage and period of shut down of each device can be determined. Next stage is to determine which of the device has to be given more importance as per the requirements. At this level the priority factor is fixed for each end device as such for all the devices coming under the network covering the entire campus. On a comparative relative study between the maximum powers limitation of the organization and the cumulative maximum power consumption made a threshold value is fixed for each end device.

When the total power consumption exceeds the maximum demand, each end device is also checked for its respective threshold value. If a device exceeds its threshold but has a higher priority, then its source of power is switched over to generator. If it is not a prioritized device then load balancing techniques are applied to it. This process continues until the maximum demand is maintained. More importance is given for very much reduced usage of generator source. Also for example special sensors are fixed to identify the wastage of power for fan, light etc., in cabins when no personalities are within, by which the total power consumption level could be further reduced.

5.1 Image Analytics

The system (PC) which has the entire power consumption details is under surveillance for security. The data can be tampered to trigger the peak value during power consumption. Therefore it becomes essential that system has to be accessed only by the authorized operators. Facial recognition system is being used for recognizing the authorized persons. This system scans the face of the active user and compares the facial details with the existing database in the system. An alarm is given when an unauthorized user is found.

The PCA (Principal Component Analysis) algorithm using Eigenface algorithm is used here. A training set of face images of authorized people is prepared. The original images of the training set are transformed into a set of eigenface. Then the weights are calculated for each image of the training set and stored in a set. When an authorized image is found, weights are calculated for that particular image and stored in a vector. Then the vector is compared with the weights of the existing images in the set (authorized). Based on the comparison, analysis is done as per the category listed in the Table 1.

Table 1 Data/Image Analytics

S.No.	Questions	Category
1	Who was the person?	Image Analytics
2	What was the time?	Image Analytics
3	At what condition the system was abnormal?	Data Analytics
4	What was the value during the abnormal condition?	Data Analytics
5	What time this abnormality happened?	Data Analytics

6 Implementation

A building was taken as the test bed with many end devices into consideration. A smart sensor fixed on each device updates the power consumption of the devices (DP) in a database, named as "power consumption database" with a predetermined time interval via Wi-Fi which is illustrated in the Fig. 8.

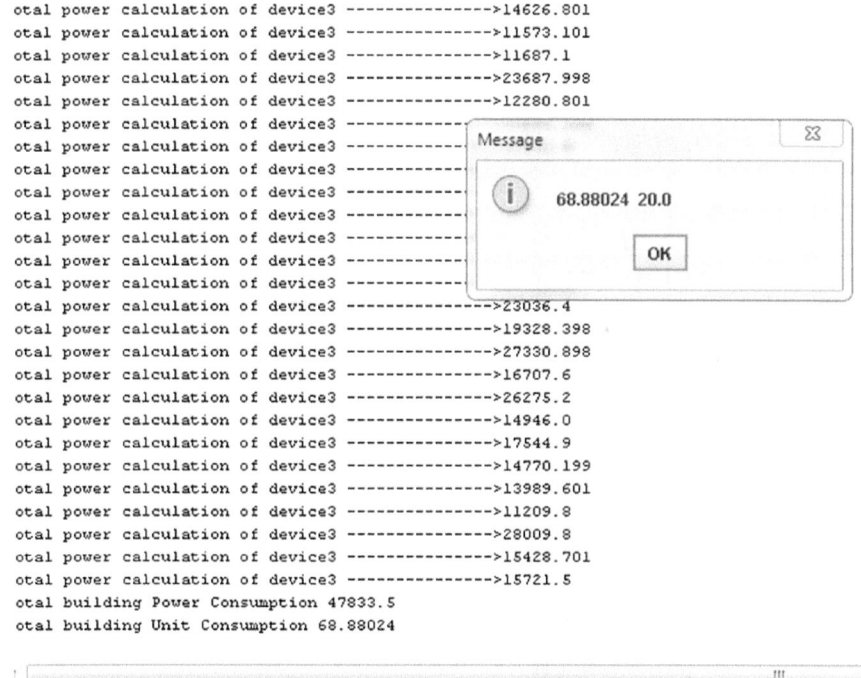

Fig. 8 Calculation of total building power consumption

The collected data is then pre-processed and analyzed. From the power consumption database, the units (consumed by each device) are calculated by analyzing the different parameters of the devices and a threshold for each device is set. Threshold of a device (-TH) is determined by taking the average of maximum and minimum power consumed. Figure 9 illustrates the priority list which contains the list of critical devices.

Step 1: Find the power consumption of each device

Step 2: Find the total power consumption of the building

Step 3: If the total power consumption > the total threshold

Individual device > device threshold

Device IsCritical load

"The source of power supply is GENERATOR"

Device Noncritical load then "the device is shutdown"

Else "the source of power supply is the same -TNEB"

Else "the source of power supply is the same -TNEB"

Step 4: End

Fig. 9 Algorithm for device level energy management system

The maximum demand of an organization is known by analyzing the continuous data of total power consumption (TP) and a threshold is determined for this maximum demand. When the total power consumption of the organization exceeds its threshold (T-TH), power consumption of all the devices (individually) are compared with its respective threshold. If the device is present in the priority list, then the source of power supply is changed to generator. If it is not present in the priority list, load balancing techniques are applied. If the total power consumption of the organization is below its threshold value, then the source of power supply for all the devices will be TNEB. The implementation is done using java programming language.

7   Result Analysis

The device level analysis was tested for three different conditions.

Case 1: When the total power consumed exceeded the total threshold of the institution, that is the determined threshold level set, and the device that was identified for this cause comes under the priority list, then this extra power requirement is being balanced by switching the control to generator source.

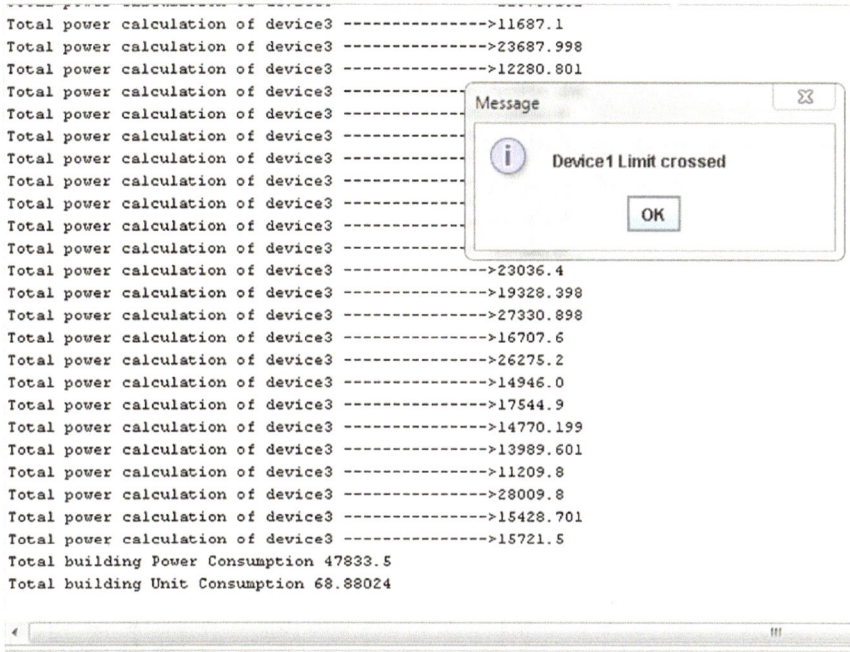

```
Total power calculation of device3 --------------->11687.1
Total power calculation of device3 --------------->23687.998
Total power calculation of device3 --------------->12280.801
Total power calculation of device3 ------------
Total power calculation of device3 ------------
Total power calculation of device3 ------------
Total power calculation of device3 ------------
Total power calculation of device3 ------------
Total power calculation of device3 ------------
Total power calculation of device3 ------------
Total power calculation of device3 ------------
Total power calculation of device3 --------------->23036.4
Total power calculation of device3 --------------->19328.398
Total power calculation of device3 --------------->27330.898
Total power calculation of device3 --------------->16707.6
Total power calculation of device3 --------------->26275.2
Total power calculation of device3 --------------->14946.0
Total power calculation of device3 --------------->17544.9
Total power calculation of device3 --------------->14770.199
Total power calculation of device3 --------------->13989.601
Total power calculation of device3 --------------->11209.8
Total power calculation of device3 --------------->28009.8
Total power calculation of device3 --------------->15428.701
Total power calculation of device3 --------------->15721.5
Total building Power Consumption 47833.5
Total building Unit Consumption 68.88024
```

Message

(i)   Device1 Limit crossed

OK

Fig. 10 Case 1: The total power consumed exceeded the total threshold of the institution

```
Total power calculation of device3 --------------->12319.199
Total power calculation of device3 --------------->23265.0
Total power calculation of device3 --------------->27448.2
Total power calculation of device3 --------------->23036.4
Total power calculation of device3 --------------->19328.398
Total power calculation of device3 --------------->27330.898
Total power calculation of device3 --------------->16707.6
Total power calculation of device3 --------------->26275.2
Total power calculation of device3 --------------->14946.0
Total power calculation of device3 --------------->17544.9
Total power calculation of device3 --------------->14770.199
Total power calculation of device3 --------------->13989.601
Total power calculation of device3 --------------->11209.8
Total power calculation of device3 --------------->28009.8
Total power calculation of device3 --------------->15428.701
Total power calculation of device3 --------------->15721.5
Total building Power Consumption 47833.5
Total building Unit Consumption 68.88024
Device1  is Connected to Generator
BUILD SUCCESSFUL (total time: 36 seconds)
```

Fig. 11 Case 1 result

Figures 10 and 11 illustrates the output screenshots under case 1. Case 2: When the total power consumed exceeded the total threshold of the institution and the device that was identified for this cause did not come under the priority list, then the extra power requirement to operate this device is identified to be not essential and hence the supply of power to this device is switched off. This is illustrated in Fig. 12 12.

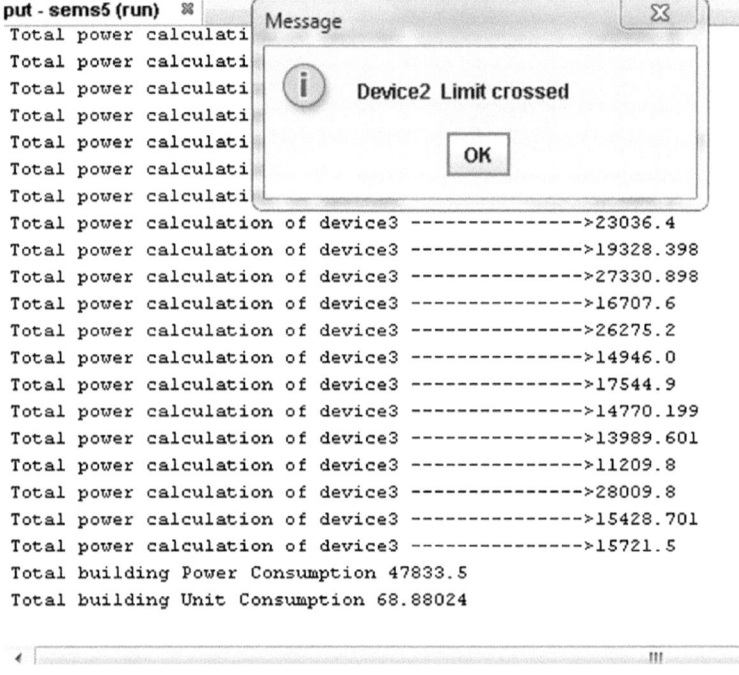

Fig. 12 Case 2: The total power consumed exceeded the total threshold of the institution and the device that was identified for this cause did not come under the priority list

Case 3: When the total power consumed is under the required total power demand of the university, that is the determined threshold level set, and then no devices would be identified. Under this condition the power requirement is identified to be normal and hence the supply of power would be utilized from the Electricity board.

The implementation part was tested for the above conditions. The results worked out very satisfactorily. The results show that the power from TNEB can be used efficiently as well as preventing the total power consumption from exceeding the maximum demand. This was made possible by analyzing the power consumed by each device, so as to know for which device optimization technique must be used. This also helped in determining the device threshold which is shown in Fig. 13.

```
Total power calculation of device3 ----------------->12319.199
Total power calculation of device3 ----------------->23265.0
Total power calculation of device3 ----------------->27448.2
Total power calculation of device3 ----------------->23036.4
Total power calculation of device3 ----------------->19328.398
Total power calculation of device3 ----------------->27330.898
Total power calculation of device3 ----------------->16707.6
Total power calculation of device3 ----------------->26275.2
Total power calculation of device3 ----------------->14946.0
Total power calculation of device3 ----------------->17544.9
Total power calculation of device3 ----------------->14770.199
Total power calculation of device3 ----------------->13989.601
Total power calculation of device3 ----------------->11209.8
Total power calculation of device3 ----------------->28009.8
Total power calculation of device3 ----------------->15428.701
Total power calculation of device3 ----------------->15721.5
Total building Power Consumption 47833.5
Total building Unit Consumption 68.88024
Shutdown the device2
BUILD SUCCESSFUL (total time: 22 seconds)
```

Fig. 13 Case 2 result

8 Future Work

The power consumption database contains past details of the power consumption of each device, device threshold. Predictive analytics techniques can be used on the database to predict the future maximum demand of power [14]. The results of the prediction techniques can also be updated to the database. The database is updated for every 10 min. The academic calendar is merged with the database for better analysis of power consumption. For example, the variation of power consumptions can be studied during the semester vacation for various years. The predictive analytics technique helps in enhancing the automated energy management system. To implement a more secure energy management system, Constrained Application Protocol (CoAP), an application protocol specially designed for IoT devices can be applied [15].

9 Conclusion

This device level analysis helps in finding which device has consumed more power and further load shifting and power reduction techniques can be implemented to reduce the total power consumption of the organization. Based on the case study of cost analysis of power consumption of the organization, the cost of power consumed from TNEB (Tamil Nadu Electricity Board) as a source of power supply is less when compared to the cost of power consumed using generator. Therefore load balancing techniques is used to reduce the power consumption, which in turn reduces the usage of generator. As an enhancement to this smart energy management system artificial intelligence introduced, makes decision making automated. This results with a very effective package of SEM-d system. Therefore implementation of this device level analysis reduces the total power consumption and also avoids it from exceeding the maximum demand of the organization. The image analytics module helps in avoiding data tampering. Thus this SEMS-d could find wide application at various industrial sectors in power saving.

References

1. Yang, T.-Y., Yang, C.-S., Sung, T.-W.: An intelligent energy management scheme with monitoring and scheduling approach for IoT applications in smart home. In: 3rd International Conference on Robot, Vision and Signal Processing (RVSP). Kaohsiung (2015)
2. Cho, K., Kim, S.H., Kang, B., Jang, S.M., Park, S.: Intelligent office energy management system by analysis in Hyper-connected-IoT environments. IEEE International Conference on Consumer Electronics (ICCE). Las Vegas (2017)
3. Ma, H., Liu, L., Zhou, A., Zhou, D.: On networking of internet of things: explorations and challenges. IEEE Internet Things J. 3(4), 441–452 (2015)
4. Sharad, S., Bagavathi Sivakumar, P., Anantha Narayanan, V.: A novel IoT-based energy management system for large scale data centers. In: ACM Sixth International Conference on Future Energy Systems, pp. 313–318. Bangalore (2015)
5. Al-Fuqaha, A., Guizani, M., Mohammadi, M., Aledhari, M., Ayyash, M.: Internet of things: a survey on enabling technologies, protocols, and applications. IEEE Commun. Surv. Tutorials 17(4), 2347–2376 (2015)
6. Lee, C.-H. Lai, Y.H.: Design and implementation of a universal smart energy management gateway based on the Internet of Things platform. In: IEEE International Conference on Consumer Electronics (ICCE). Las Vegas (2016)
7. Saha, A., Kuzlu, M., Khamphanchai, W., Pipattanasomporn, M., Rahman, S., Elma, O., Selamogullari, U.S., Uzunoglu, M., Yagcitekin, B.: A home energy management algorithm in a smart house integrated with renewable energy. In: IEEE PES Conference on Innovative Smart Grid Technologies. Istanbul (2015)
8. Anjana, M.S., Athira, K., Devidas, A.R., Ramesh, M.V.: A smart positioning system for personalized energy management in buildings. In: International Conference on Wireless Communications, Signal Processing and Networking (WiSPNET). Chennai (2016)
9. Farnham, T.: Performance optimisation for visitor information systems using smart sensors and analysis of trial data. IEIET Netw. 4(6), 329–337 (2015)
10. Sun, Q., Li, H., Ma, Z., Wang, C., Campillo, J., Zhang, Q., Wallin, F., Guo, J.: A comprehensive review of smart energy meters in intelligent energy networks. IEEE Internet Things J. 3(4), 464–479 (2016)
11. Ploennings, J., Chen, B., Palmes, P., Lloyd, R.: e2-Diagnoser: a system for monitoring, forecasting and diagnosing energy usage. In: IEEE International Conference on Data Mining Workshop (ICDMW). Shenzhen (2014)
12. Hock, K.P., Radjabli, K., McGuiness, D., Boddeti, M.: Predictive analysis in energy management system. In: 16th IEEE International Conference on Environment and Electrical Engineering (EEEIC). Florence (2016)
13. Dehghanpour, K., Nehrir, M.H., Sheppard, J.W., Kelly, N.C.: Agent-based modeling in electrical energy markets using dynamic bayesian networks. IEEE Trans. Power Syst. 31(6), 4744–4754 (2016)

14. Balac, N., Sipes, T., Wolter, N., Nunes, K., Sinkovits, B., Karimabadi, H.: Large scale predictive analytics for real-time energy management. In: IEEE International Conference on Big Data. Silicon Valley (2013)
15. Rahman, R.A., Shah, B.: Security analysis of IoT protocols: a focus in CoAP. In: 3rd MEC International Conference on Big Data and Smart City (ICBDSC). Muscat (2016)

Assisting Visually Challenged Person in the Library Environment

D. Kavin Kumar$^{(\boxtimes)}$ and Senthil Kumar Thangavel

Department of Computer Science and Engineering, Amrita Vishwa
Vidyapeetham, Coimbatore, India
cb.en.p2cvil5016@cb.students.amrita.edu,
t_senthilkumar@cb.amrita.edu

Abstract. Nowadays, there is lot of assistive technologies to support visually impaired people. Among them computer vision based methods provide a feasible solution. An indoor environment provides challenges like object recognition, character and scene recognition. It is important to understand that people need to know information about things; places and they may feel insecure in places like their working places, shopping mall because of their challenges in vision. It is essential that a technology based solution can be provided to support the people so that they can be guided along in their pathways, rooms, shopping malls and they can also access things in their living environment. In this paper a model is proposed for detecting text from library book shelf scene video and informing the user about book name through audio to assist visually impaired people in accessing their book which is kept on shelf in library. Key frames are extracted using PSNR and Edge Change Ratio method. Text on the key frame is detected and localized using MSER and Projection profiles. CNN is used to recognize characters from the localized text. This paper gives an outline of different techniques which are combined to extract key frames, localize and recognize text from natural library scenes.

Keywords: Edge Change Ratio · Keyframe Extraction · Maximally Stable Extremal Regions · Text Localization · Convolutional Neural Network

1 Introduction

Visual Impairment alludes to nonattendance of sight or visual keenness not more than 6/60 or 20/200 in the better eye even with focal points or Limitation of the field of vision sub tending an edge of 20° or more terrible. A man is said to have low vision on the off chance that he has lessening of fields less than 50°, heminaopia with macular contribution, altitudinal imperfection including lower fields.

The visual impedance is brought about because of uncorrected refractive blunders, waterfalls and glaucoma. In 2014, there were 285 million individuals outwardly disabled individuals. Among them 246 million had low vision and 39 million individuals were visually impaired. The greater part of individuals with low vision is in the creating nations.

© Springer International Publishing AG 2018
D. J. Hemanth and S. Smys (eds.), *Computational Vision and Bio Inspired Computing*,
Lecture Notes in Computational Vision and Biomechanics 28,
https://doi.org/10.1007/978-3-319-71767-8_63

The World Health Organization found that almost 80% of visual hindrance is either reparable or preventable with therapeutic treatment. Any gadget or administration which bolsters handicapped individuals to have a free life is considered as assistive innovation. Assistive innovation is utilized by impaired individuals to help with regular undertakings in both work and home life. The most regularly utilized assistive innovations are white stick, focal points, screen magnifiers, recognizing cautioning surfaces, advanced mobile phone based advances and PC vision based strategies like optical character   acknowledgment,   stereo   profundity,   question   discovery   and acknowledgment.

The following section will contain the previous related works, text localization using MSER and recognizing characters from those identified text using CNN to get information which helps in accessing books.

2   Related Works

The assistive technologies based on image and video processing is in its developing stage since it requires more computation, speed and accuracy. Video based assistive technology requires an efficient key frame extraction technique to acquire most important frames from video, to avoid redundant frames and to reduce computation time. Recognizing characters in an indoor environment like library to support visually impaired people has two major problems to be concentrated. Recognizing characters in document is an easier process compared to recognizing text in natural scenes because in document form the noise will be less. In natural scene text recognition process there are some factors like lighting conditions, atmospheric conditions, occlusions, different font color, orientations, …etc. which makes text recognition in natural scenes hard. Recognizing machine printed text is simpler when compared to handwritten text since it doesn't have various printing styles and different writing patterns. In handwritten text recognition process there is a problem of handling text with different writing styles and patterns. Some of recent technologies which had tried to solve these problems of key frame extraction, detecting text in natural scenes and unconstrained character recognition are discussed below.

2.1   Techniques for Extracting Key Frames from Video

In Sandhu and Agarwal [1], video rundowns are produced by extricating key casings of a video utilizing visual elements like RGB channel, shading histograms, surface, snapshots of idleness, SSIM. This method produces rundowns with low illumina repetition and furthermore functions admirably with recordings of low or high change in edge content or even with recordings with shifting conditions. In Raikwar et al. [2], key edges are picked in view of power of movement vitality. The movement vector is utilized to gauge the movement vitality of that point. The optical flow can be used to

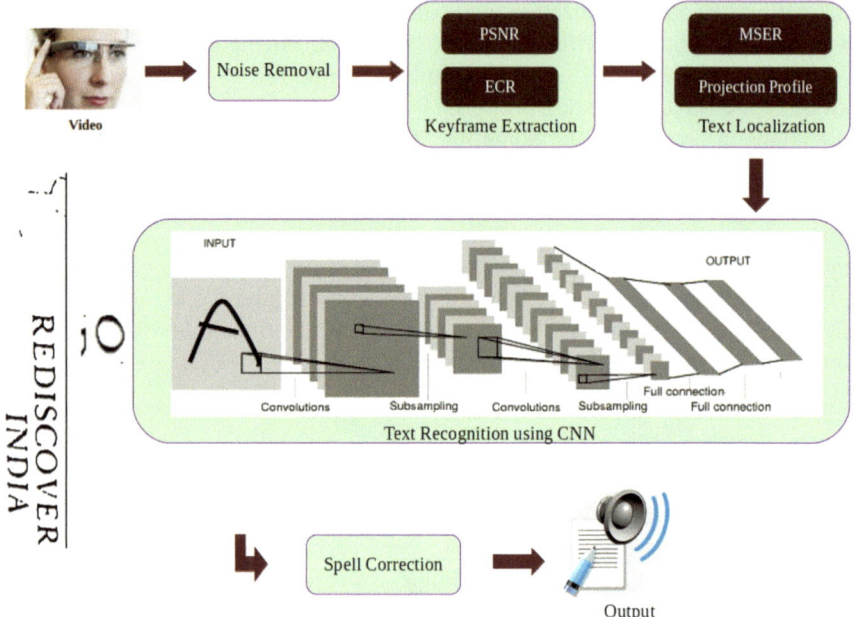

Fig. 1 The proposed system architecture

detect the motion vectors and to reduce time complexity motion vectors are detected at the corner of an edge. The corner points are detected by Harris Corner detector. By utilizing Hierarchical Clustering Algorithm the key edges are separated in [3]. The qualities of picture data entropy is utilized to quantify the similitude level of two casings, if the comparability achieves a specific esteem, they will be converted into a similar bunch. At that point we separate the grouping focus as the key casings. Video Summarization and key edge extraction is finished utilizing QR Decomposition in [4] (Fig. 1).

2.2 Technologies in Detecting Text in Natural Scenes

In Epshtein et al. [5], the text in natural scenes has been detected using stroke width transform of character. Stroke width is defined as the contiguous part of an image with constant width. Canny edge detector is used to obtain edges and then the edge pixel p's gradient direction (d_p) at one side is computed until an edge pixel q at another end is found. From this **q**, compute the gradient direction (d_q). If d_q is opposite to d_p then stroke width is computed by $\|p - q\|$ and if it's not opposite discard it. This method obtains a precision of 0.73, Recall of 0.60 and f-measure of 0.66 and this method can be applied to many languages and fonts. This method cannot handle curved text lines.

In paper Greenhalgh and Mirmehdi [6], a novel technique for identifying and perceiving activity signs in street has been proposed. Discovery of applicants is performed utilizing maximally stable extremal districts and acknowledgment is finished by the straight bolster vector machine with the histogram of situated angle highlights. The

Maximally stable extremal locales are associated parts which keep up their shapes through a few limit levels. This technique functions admirably with fluctuating lighting and climate conditions. The exactness of this technique is 86.8%, review of this strategy is 80.7% and F-measure is 0.84. Strong content discovery in Natural Scene Images [7] depicts an imperative strategy to distinguish message in regular scenes which is a substance based picture examination issue. Maximally Stable Extremal Region calculation is utilized to concentrate character hopefuls and non-characters are decreased by utilizing MSER pruning calculation with the procedure of limiting regularized varieties. Content competitors are made by bunching character applicants utilizing single-connection grouping calculation with the parameters got the hang of utilizing proposed metric learning calculation. Character classifier is utilized to quantify the posterior probabilities of text candidates which compares to text and non-text candidates with high probabilities for non-text are expelled. An Adaboost classifier is utilized to choose whether recognized text candidate is a genuine content or not. This method produces a recall rate of 68.26, precision rate of 86.29 and f-measure of 76.22.

2.3 Techniques for Unconstrained Character Recognition

The acknowledgment of manually written malayalam script which comprises of 36 consonants and 15 vowels utilizing wavelet change is talked about in [8]. In the first place, the character picture is changed over into dark scale picture and scale the picture with the goal that it fits precisely into 64×64 networks. 2D Haar wavelet change is performed on the scaled picture. The bolster vector machine is utilized to characterize manually written malayalam characters. The execution of SVM relies on upon the determination of portion, delicate edge parameter and bit's parameters. This technique gets grouping precision of 90.25%.

Tensor portrayal based picture fix examination is a framework proposed in [9] to recognize and see content. Picture patches are addressed as tensors and low dimensional portrayals of these tensors are discovered using focalized tensor portrayal learning computation. An arbitrary woodland classifier is set up with the informed tensor subspace to recognize content locales in new report pictures. Neural Network is utilized as a part of manually written character recognition [10]. The Intelligent K Means clustering algorithm is used to find the nearly optimal value of K and segment the object. This strategy is based in finding for which value of K the variance is maximum [11]. Multilayer bolster forward system with back engendering has been utilized as a part of this paper [12]. A sigmoid layer, systems with slants and a direct yield layer are fit for approximating any capacity with a predetermined number of discontinuities. This technique can't deal with varieties in pivot, interpretation, or scale.

3 Proposed Method

3.1 Noise Removal

Noise arises when there occurs departures from ideal signals which were generated by sensors. Noises are produced by unmodellable or unmodelled processes which occur in the production of real signal. There are different types of noises which may cause due to environmental conditions and sensor conditions. The input to this model is the video with quality of 1920 × 1088 taken using mobile phone. In this model the motion blur effect acts as a major noise in input video. There is a chance of Gaussian noise in input video due to electronic problems. So, in this proposed model there are two different types of filters which have been applied. First, the Non-Local Means denoising filter to reduce motion blurs effect in the input video and then the Gaussian filter to reduce Gaussian noise in the input video. Non-Local Means filter is one of the efficient filters in removal of blur in images which has been caused due to unfocused optics or linear motion. In perspective of signal processing, blurring due to linear motion in a photograph is the result of poor sampling. When the camera is in motion and shutter speed is slow, the output image produced would be blurred. In order to maintain effectiveness of the proposed model first a keyframe is checked whether it is blurred or not. The NL-Means denoising filter is only applied to blurred keyframes.

This blur detection method works on the principle that the blurred image will have low amount of high frequencies. The blurred image will contain less edge. The Laplacian operator with kernel k,

$$k = [0, 1, 0; 1, -4, 1; 0, 1, 0]$$

is applied to compute edges. Then, the variance of obtained edge image is computed. If the variance falls below threshold, then that image is considered to be blurry. If the variance is greater than threshold, then that image is considered to be not blurred. In [13], Non-Local Means denoising filter will replace the color of a pixel with an average of colors of similar pixels. In this method, a vast amount of pixels of the image have to be considered to identify pixels that resemble the pixel to be denoised. Denoising is performed by computing the average color of most similar pixels. The Non-Local Means filter is represented by,

$$NLM(P) = \frac{1}{Z(P)} \int f(d(B(p), B(q))u(q)dq$$

where Z(P) is the normalizing factor, $d(B(p), B(q))$ is an euclidean distance between image patches centered at p and q, f is a decreasing function.

There are two different types of Non-Local Means denoising filer implementations as follows,

1. **Pixel wise Implementation**—The denoising of color image u = (u$_1$, u$_2$, u$_3$) at certain pixel p is as follows,

$$u(p) = \frac{1}{Z(p)} \sum_{q \in B(p,r)} u(q)w(p,q), \quad Z(p) = \sum_{q \in B(p,r)} w(p,q),$$

where i = 1, 2, 3 and B(p, r) represents a neighborhood centered at p and with size (2r + 1) * (2r + 1) pixels. Each pixel value is the average of the most similar pixels. An exponential kernel is used to acquire weights w (p, q)

$$w(p,q) = e^{-\frac{\max\left(d^2 - 2\sigma^2, 0.0\right)}{h^2}}$$

where σ is the standard deviation of the noise and h is the filtering parameter which depends on the value of σ.

2. **Patch wise Implementation**—The denoising of a color image $u = (u_1, u_2, u_3)$ and a patch $B = B(p, f)$ which is centered at p and with size $(2f + 1) * (2f + 1)$.

$$B_i = \frac{1}{C} \sum_{Q=Q(q,f) \in B(p,r)} u_i(Q)w(B,Q), \quad C = \sum_{Q=Q(q,f) \in B(p,r)} w(B,Q)$$

where $i = 1, 2, 3$, $B(p, r)$ represents a window centered at p and with size $(2r + 1) * (2r + 1)$ pixels.

Gaussian noise is generated during the process of image acquisition. It can be reduced using spatial filters like median and Gaussian smoothing filters.

$$\mathbf{P(z)} = \left(\frac{1}{\sigma\sqrt{2\pi}}\right) e^{\frac{-(z-\mu)^2}{2\sigma^2}}$$

Z-gray level value, μ-the mean value, σ-the standard deviation.

3.2   Key Frame Extraction

The key frame extraction plays a vital role in video processing. A video is the collection of frames. A video may consists of varying numbers of frames based upon its length. A frame resolution decides the quality of video which depends upon the sensors. If every frame of input video is considered then it needs more computational time and it also means some amount of time is wasted unnecessarily by processing redundant frames. A key frame defines where a transition should start or stop. The key frames provide short representation of a video.

In this proposed method, two different techniques have been implemented in obtaining key frames. Peak Signal-to-Noise Ratio which determines the impact of noise on real image is used. PSNR value is defined using the Mean Squared Error as follows,

$$\text{MSE} = \frac{1}{mn} \sum_{i=0}^{m-1} \sum_{j=0}^{n-1} [I(i,j) - K(i,j)]^2$$

where I is the original image and K is the noisy approximation,

$$\text{PSNR} = 10.\log_{10}\left(\frac{\text{MAX}_I^2}{\text{MSE}}\right)$$

where MAX_I is the maximum pixel intensity value.

If two images are identical then MSE value will be 0 and PSNR value would be infinite. In this proposed model, the difference between PSNR values of each and every frame is computed. If difference is less than threshold value, then it will be allowed for further processing. Then Edge Change Ratio method is applied to the obtained frames. Consider two frames X and Y to which sobel edge detector is applied to detect the edges. To upgrade the edges expansion is connected. At that point X's expanded picture is XOR-ed with a transformed form of Y's edge picture. Essentially, Y's widened picture is XOR-ed with an altered variant of X's edge picture.

$$ECR_i = \max\left(\frac{\rho_{in}}{s_i}, \frac{\rho_{out}}{s_i + 1}\right)$$

In the got pictures add up to number of shaded pixels is checked. E_{in} is the approaching edge pixel and E_{out} is the developing edge pixels. At that point E_{out} is isolated by number of edge pixels in the edge picture of X. E_{in} is isolated by number of edge pixels in the edge picture of Y. Thoutcome is ρ_{in} edge leave proportion and ρ_{out} edge entrance proportion. Greatest of two qualities relate to the edge division change. While partitioning ECF values by s_i and s_{i+1} the edge change proportion esteem is acquired. S_i is the quantity of all edge pixels in first picture and s_{i+1} is the quantity of all edge pixels in second picture. The casings with most extreme contrast esteem are picked as key frames.

3.3 Text Localization

After extraction of key frames next step is to detect and localize the text regions. In proposed method, they are two methods involved in the process of detecting and localizing text as follows, first one is the Maximally Stable Extremal Region method and other method is based upon projection profiles. MSER is the connected components of gray level sets of image. In the gray level image various sets of threshold is applied. The connected components are computed. A region of image is said to be extremal only if its inner region pixel intensities is either brighter or darker than its outer boundaries.

$$\varphi_l(R_t) = \frac{A(R_t)}{\frac{d}{dt}A(R_t)}$$

where $A(R_t)$ represents the area of region R_t. A region is known as stable region only if its area changes slightly with the different thresholds. A region is considered as maximally stable only if it has local maximum at threshold t. MSER regions are affine invariant. MSER features are one of the best features for problems with view point changes and scale changes. But they are sensitive to lighting effects and shadows. From [14], the projection profile based character segmentation has been acquired. The Vertical Projection Profile is utilized as a part of the procedure of line division and flat projection is utilized as a part of the procedure of word division. The even projection profile of a line speaks to aggregate number of pixels which have power esteem more prominent than or lesser than particular edge in that column. Thus, the vertical projection profile of a section speaks to aggregate number of pixels which have power esteem more noteworthy than or lesser than particular limit in that segment.

3.4   Text Recognition

Deep learning has been used in this part of the proposed approach. Convolutional Neural Networks is used to recognize text from the key frame. The CNN model used is shown in Fig. 2.

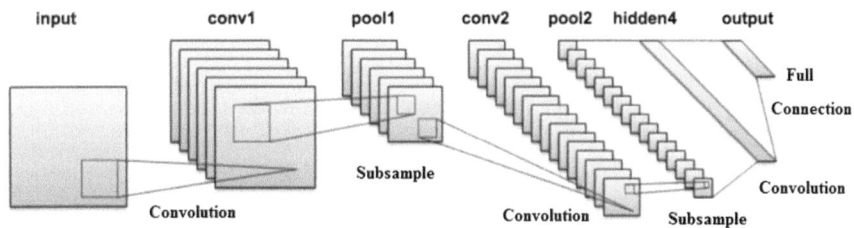

Fig. 2 LeNet CNN architecture

The CNN consists of three layers as follows,

3.4.1   Convolutional Layer
In this layer, given input image is convolved with specified number of kernels to get desired number of feature maps. CNN is working on the spatial information of image which makes it most effective. Kernel size is defined by the user.

3.4.2   Max-Pooling Layer
The arrangement of highlight maps acquired in the convolutional layers is given as contribution to this layer. This layer diminishes the measure of highlight guide. Consider a 2 × 2 window which moves along the element outline walk length of 2. The model will take most extreme estimation of these four values and place it in the new measurement diminished component outline.

3.4.3   Fully Connected Layer
The output of max-pooling is given as input to this layer. This layer is also known as dense layer. The feature map values are given as input to neural network and the class with maximum probability is chosen as output. Then, finally to ensure the meaning of recognized word an auto-spell correction tool is used. By using this tool natural language meaning of the recognized sentence is preserved.

The recognized text is produced as audio to user through a tool known as pyttsx.

4 Findings

From the study it is found that key frame extraction using PSNR and Edge Change Ratio is one of the best key frame extraction techniques. However the results on applying with dataset of library book shelf and other standard dataset shows PSNR and edge change ratio is better in terms of time. The result of PSNR and ECR Key frame Extraction method is shown in Table 1. The blend of Maximally Stable Extremal Regions and projection profiles functions admirably on the issue of content identification and restriction. The consequence of MSER and Projection profile based content identification and content confinement is appeared in Figs. 3 and 4.

Table 1 Results of PSNR and ECR method

Video source	Dimensions	No. of frames	No. of keyframes	Average time per keyframe
Self Video using 13 MP Lenovo K3 Note Mobile				
	1920 × 1088	690	23	9.2 min
Standard Video from Matlab Toolbox Dataset				
	320 × 240	120	4	11.5 s
Standard Video from Hong Kong University of Science and Technology Dataset				
	352 × 144	570	19	7.16 s
	352 × 240	660	22	23.83 s

The Convolutional Neural system which is one of the cutting edge strategies is connected to the issue of content acknowledgment. The CNN model is relied upon to create exactness around 80%. Since this method is concentrating on recognizing text on book side. Two major rotations 90° and 270° of each lowercase and uppercase character have been considered as separate class. Totally there are 104 classes based upon the training data used. The training data per class will be 2000 and totally the amount of data used is 208,000. The result of CNN model predicting characters based upon the segmented character images from text localization phase is shown in Table 2.

Fig. 3 Result of MSER based text detection

Fig. 4 Result of character segmentation

Table 2 From the above table it is known that fused images like A and I which was generated during text localization and character segmentation will not work here

Input image	Predicted class	Actual class
	28-I	28-I
	28-I	28-I
	155-Z	75-N
	16-D	16-D
	35-G	4-A

5 Conclusion

Computer vision based assistive advancements to bolster visually weakened individuals is an essential technology. The different strategies which help in proposed arrangement of unconstrained text recognition for visually disabled individuals in the library environment have been discussed. In the proposed system, key frame extraction using PSNR and edge change ratio is used which is comparatively better. The text detection and localization based on MSER and projection profiles is used. The state-of-the-art deep learning technique CNN is used in the process of text recognition to produce best result. There are certain drawbacks in the proposed system due to natural lighting, shadows and skewness which would be solved in future investigation.

References

1. Sandhu, S.K., Agarwal, A.: Summarizing Videos by Key frame extraction using SSIM and other Visual Features. ICCCT'15 (2015)
2. Raikwar, S.C., Bhatnagar, C., Jalal, A.S.: A framework for key frame extraction from surveillance video. In: 5th International Conference on Computer and Communication Technology (2014)
3. Liu, H., Hao, H.: Key frame extraction based on improved hierarchical clustering algorithm. In: 11th International Conference on Fuzzy Systems and knowledge discovery (2014)
4. Amiri, A., Fathy, M., Naseri, A.: Key-frame extraction and video summarization using QR-Decomposition. In: 6th International Conference on Digital Content, Multimedia Technology and its Applications (2010)
5. Epshtein, B., Ofek, E., Wexler, Y.: Detecting text in natural scenes with stroke width transform. Microsoft Corporation (2010)
6. Greenhalgh, J., Mirmehdi, M.: Real-Time detection and recognition of road traffic signs. In: IEEE Transactions on Intelligent Transportation Systems, 13(4), 1498–1506, Dec. 2012
7. Yin, X.C., Yin, X., Huang K., Hao, H.W.: Robust text detection in natural scene images. In: IEEE Transactions on Pattern Analysis and Machine Intelligence, 36(5), 970–983, May 2014

8. John, J., Pramod, K.V., Balakrishnan, K.: Unconstrained handwritten Malayalam character recognition using wavelet transform and support vector machine classifier. Procedia Engineer. **30**, 598–605 (2012)
9. Zhong, G., Cheriet, M.: Tensor representation learning based image patch analysis for text identification and recognition. Pattern Recogn. **48**(4), 1211–1224 (2015)
10. Patel, C.I., Patel, R., Patel, P.: Handwritten character recognition using neural network. Int. J. Sci. Eng. Res. **2**(5), 1–6 (2011)
11. Gautam, K.S., Thangavel, S.K.: Hidden object detection for classification of threat. In 2017 4th International Conference on Advanced Computing and Communication Systems (ICACCS), Coimbatore, pp. 1–7 (2017)
12. Gautam, K.S., Thangavel, S.K.: Discrimination and Detection of Face and Non-face Using Multilayer Feedforward Perceptron. Proceedings of the International Conference on Soft Computing Systems, **397**, 89–103. Springer (2016)
13. Buades, A., Coll, B., Morel, J.-M.: Non-local means denoising. Image Process. On Line **1** (2011), 208–212 (2013)
14. Mamatha H.R, Srikantamurthy, K.: Morphological operations and projection profiles based segmentation of handwritten Kannada document. Int. J. Appl. Inf. Syst. **4** (2012)

Tensor Flow Based Analysis and Classification of Liver Disorders from Ultrasonography Images

K. Raghesh Krishnan[1(✉)], M. Midhila[1], and R. Sudhakar[2]

[1] Department of Computer Science and Engineering, Amrita School of
Engineering, Coimbatore, Amrita Vishwa Vidyapeetham, Amrita University,
Coimbatore, India
k_raghesh@cb.amrita.edu, cb.en.p2cvil5005@cb.
students.amrita.edu
[2] Department of Electronics and Communication Engineering, Dr. Mahalingam
College of Engineering and Technology, Udumalai Road, Pollachi, Coimbatore
642 003, Tamilnadu, India
sudhakar.radhakrishnan@gmail.com

Abstract. In the field of medical imaging, Ultrasonography is a popular and most frequently used diagnostic tool owing to its hazard-free, non–invasive and the cost effective nature. Liver being the largest and vital organ in the human body, liver disorders are treated very important and initial detection of the disorder is made using ultrasound imaging by the radiologists that leads to additional biopsies for confirmation, if necessary. This work focusses on the automated classification of nine types of both focal and diffused liver disorders using ultrasound images. A deep convolutional neural network architecture codenamed Inception is used. The technique achieves a new state for classification and detection of liver disease. The disease is predicted based on the score obtained as a result of training. The classification is achieved using tensor flow and it outputs the predicted labels and the corresponding scores. The method achieves reasonable accuracy using the trained model.

Keywords: Tensor flow · Convolutional neural networks (CNN) · Inception model · Transfer learning · Bottle necks

1 Introduction

Liver disorders are typically differentiated into two categories namely focal disorders and diffused disorders. Focal liver diseases such as hemangioma, hepatitis and cyst are confined to small regions of the liver surface while the diffused liver disorders such as fatty liver and cirrhosis, affect major portions of the liver without any definite shape or confinement. Interpretation of liver disorders from ultrasound images is a tedious task and requires the knowledge and assistance of trained and experienced radiographer. The

© Springer International Publishing AG 2018
D. J. Hemanth and S. Smys (eds.), *Computational Vision and Bio Inspired Computing*,
Lecture Notes in Computational Vision and Biomechanics 28,
https://doi.org/10.1007/978-3-319-71767-8_64

disease diagnosis is based on human vision, the experimental and experiential knowledge of the observer. Digital image processing plays a key role in the above biomedical classification problem. Conventional segmentation techniques which mostly rely on definite shapes and intensity variations fail to segment the ultrasound images particularly in the case of liver diseases. No single segmentation technique fits all the diseases. Similarly shape based features would work for some diseases whereas texture based features would be the solution for others. The current approach combines the inception model and transfer learning process for classifying the liver ultrasound images. The classification mechanism is robust enough to handle the complex similarities and differences of medical images especially liver ultrasound images. This work aims to provide a second opinion tool for physicians for diagnosis.

The rest of the paper is organized as follows. Section 2 discusses related work and Sect. 3 provides an outline of the method. Sections 4 and 5 throw light on inception model and transfer learning. Sections 6 and 7 elaborate on bottlenecks and training. Sections 8 and 9 introduce classification and present the results and analysis respectively. Section 10 gives the conclusions and scope for future work.

2 Related Work

Literature presents the classification of liver disorders using conventional approaches such as preprocessing, segmentation, feature extraction and classification with various selective or collective techniques applied in every phase [1–6]. Of late deep learning and convolutional neural networks find their importance in biomedical image classification. Wu et al. [7] classify focal liver lesions with contrast enhanced ultrasound using Deep learning and report an accuracy of 86%. In [8], cirrhosis is diagnosed using liver capsules and patch triplets are classified using CNNs with transfer learning and achieve highest area under the curve as 97%. Hassan et al. in [9] apply level set contours and fuzzy c-means segmentation to isolate focal diseases from liver ultrasound images and represent the features using stacked sparse auto encoder and classify the features using softmax classifier. An accuracy of 97.2% is reported. The dataset consists of 110 images of cyst, hem and HCC. In [10], Transfer learning with the CNNs CaffeNet and VGGNet are used to classify 185 ultrasound images of the abdomen into 11 categories. The current work aims at classifying both focal and diffused liver disorders using tensor flow deep learning method.

3 Methodology Outline

Figure 1 illustrates the methodology outline. Images are input to the inception model which is a convolutional neural network model trained by Google on 1000 categories supplied by the ImageNet competition to near human accuracy. The model is trained by a process termed as transfer learning which results in the creation of a training set. The training set thus obtained is used for classification which is performed based on prediction scores. Prediction is achieved by presenting the image to a retrained model using tensor flow session. The following sections discuss each portion of the methodology in detail.

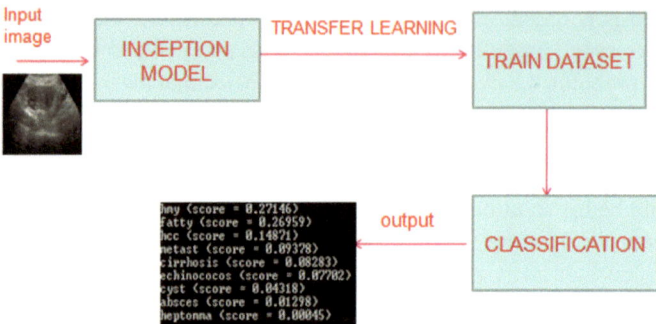

Fig. 1 Methodology outline

4   Inception Model

Inception model is a huge image classification model that consists of millions of parameters that can distinguish a large count of images. The nodes in the inception model represent mathematical operations and the edges represent multidimensional data arrays known as Tensors [11]. Figure 2 illustrates the inception model. Each layer of the model represents a different set of abstractions. The initial layers are for Cell edge detection and the middle layers are for shape detection. The final layers contain the highest level of detectors. Training is performed only for the last layer of a particular network. It is possible to retrain the final classification as CNN uses multiple layers to fine tune classification

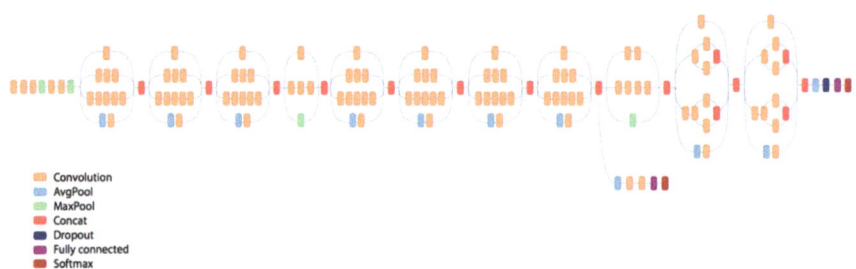

Fig. 2 Inception model

A specified number of convolutional filters are applied by the *Convolution* layer on each sub region of the image resulting in a single valued output feature map. The *Concat layer* concatenates all the tensor values along one dimension. *Fully Connected layer*, also known as *dense layer* performs classification on the features extracted from the previous layer. The results are down sampled by the *pooling* layers. *Maxpool* represents the statistics over neighboring features, to reduce the size of the feature maps and selects the maximum values from the map. Similarly *AvgPool* (average pool) selects the average value from the maps. *Softmax* function is used in the final layer of the neural network. The networks are commonly trained under cross-entropy, which gives a non-linear variant of multinomial logistic regression. Dropout is a regularization technique to improve the accuracy on the validation set by reducing overfitting issues.

Every node in the dense layer is connected with previous layer. The final layer in the neural network is the *logits* layer which returns the raw values for prediction. The classification results are analyzed based on these values [12].

5   Transfer Learning

Transfer learning refers to the process of applying the learning from a previous training session to a new training session. In transfer learning, the last layers are retrained for features. The two possibilities available here are fine-tuning the network with problem-specific data or extracting the features from the network and training a new classifier. A schematic representation of transfer learning is presented in Fig. 3.

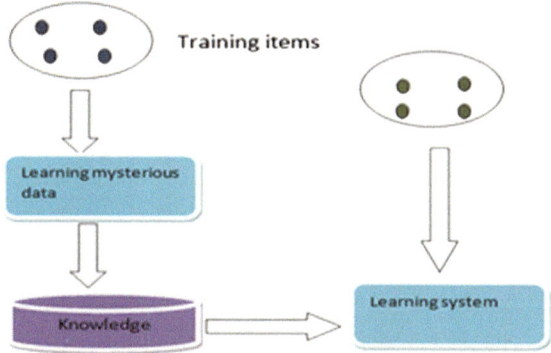

Fig. 3 Transfer learning

6   Bottlenecks

The term bottleneck is used for the layers prior to the final output layer, used for image classification. The output layer is trained to output a set of values that are fine enough for the classifier. The layer should have sufficient information about the image, for the classifier to make an appropriate choice on a very small set of values. The images are reused multiple times during the training phase and the bottleneck values are calculated. Bottleneck directory is used to cache the output of the lower layers (stored as *.tf* files) to avoid recalculation of values. Figure 4 shows the creation of bottleneck files for each disease.

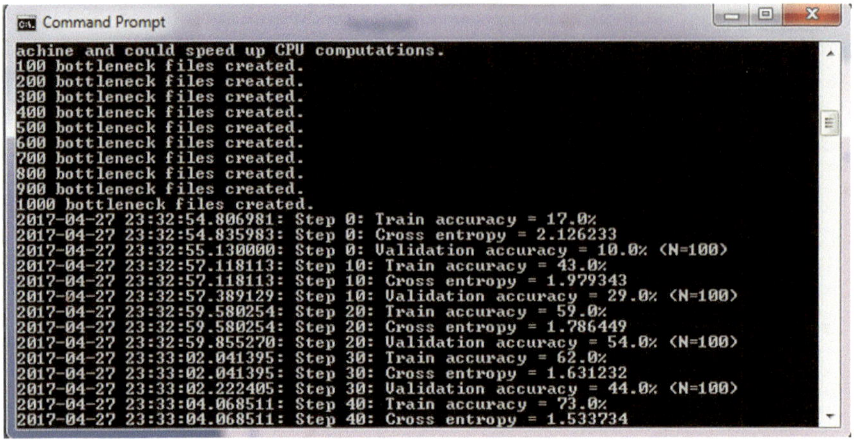

Fig. 4 Bottle neck creation

7 Training

Training operates competently by feeding the cached values of each image into the bottleneck layer. For every image, the true value of label is fed into the node labeled *GroundTruth*. The inputs thus obtained are adequate for the calculation of training updates, various performance metrics and classification probabilities. A series of outputs indicating the training accuracy, precision and cross entropy are obtained.

8 Classification

Classification is accomplished based on the prediction score obtained. Prediction is made by presenting the image to the retrained model. The session creates an environment with three variables: one to store the user input image path, second one to store the data from the image and a third variable to load the label file. The *softmax* function uses the final layer to map the input data probabilities to an expected output. The classification output consists of predicated label along with the predicated score. The label indicates the class name and the score represents the chance of classifying the image into a particular class. If the value of the score is high then the corresponding label will be the class name. The scores are displayed in descending order of relevance.

9 Results and Analysis

9.1 System Specifications

The work was carried out on an Intel core i3 2.7 GHz processor computer with 4 GB random access memory. The implementation was in *Python* supported by *Anaconda* which is a leading open data science platform bundling other data manipulation and machine learning packages like *Numpy*, *SciKit* etc. In this work, an Inception CNN from google for image classification is used.

9.2 Dataset

Classification problems require considerable amount of data sets for efficient training, testing and validation. The images were obtained from Sunitha Scan and Diagnostic Centre, Guntur, Sono Scan center, Coimbatore. The images were acquired using Philips HD15 PureWave Ultrasound System with 5 MHz convex probe and Voluson 730 Expert GE Ultrasound System with 3.5 MHz convex probe. As the images were of varying resolutions, they were trimmed and resized to a common resolution of 512 × 512. Figure 5 shows some sample images and Table 1 shows the data set count.

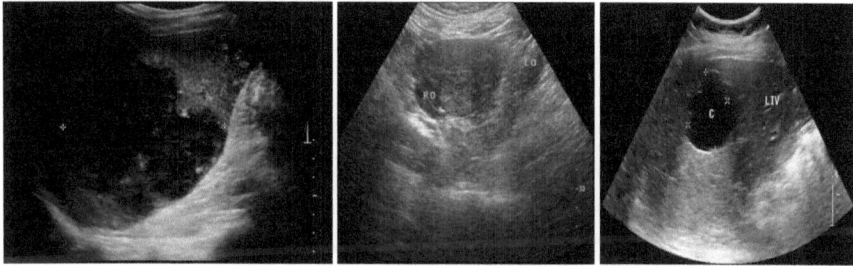

Fig. 5 Some sample images from the dataset

Table 1 Dataset count

S.No.	Disorder	No. of images
1	Abscess	55
2	Cirrhosis	65
3	Cyst	34
4	Echinococcosis	45
5	Fatty Liver	41
6	Haemangioma	46
7	Hepatocellular Carcinoma	214
8	Hepatomegaly	20
9	Hepatitis	10
10	Metastases	66

9.3 Analysis of the Predicted Scores

Figure 6 shows the predicted scores for abscess images against all the diseases exported into spreadsheet for analysis. The spreadsheet indicates the scores obtained against each class for a particular input image. For example, the first row shows the scores obtained for a particular abscess image labeled *rstr_abs019.jpg*. The score for the image getting classified as abscess is 0.18416, as metastasis is 0.146678 and so on which shows that the input image has a high score for getting classified into abscess.

image name	abscess	metast	echinococos	cirrhosis	hcc	cyst	hmy	fatty	heptomma
rstr_abs019.jpg	0.18416	0.146678	0.145187095	0.132495	0.108491	0.095314	0.093732	0.04871173	0.045230582
rstr_abs038.jpg	0.630279	0.205352	0.065156773	0.034317	0.026431	0.023696	0.007632	0.00442205	0.002713742
rstr_abs039.jpg	0.286303	0.18758	0.183437198	0.101723	0.096959	0.073011	0.02999	0.02225989	0.018737156
rstr_abs040.jpg	0.864085	0.065913	0.028656039	0.012233	0.010223	0.009959	0.007354	0.00095516	0.000621321
rstr_abs041.jpg	0.840699	0.036501	0.03289526	0.032029	0.018514	0.016484	0.010537	0.00653084	0.005810174
rstr_abs044.jpg	0.134484	0.297241	0.037831698	0.159276	0.038802	0.028295	0.035575	0.26565498	0.265654981
rstr_abs045.jpg	0.231946	0.187952	0.143614337	0.134549	0.114337	0.107283	0.040461	0.02968304	0.010173911
rstr_abs046.jpg	0.05107	0.335089	0.235970557	0.11982	0.117937	0.073543	0.041751	0.02164552	0.00317306
rstr_abs047.jpg	0.221999	0.17639	0.134143487	0.106305	0.106206	0.079764	0.07599	0.07263599	0.026566947
rstr_abs048.jpg	0.205963	0.488889	0.137562588	0.064391	0.037897	0.031482	0.016448	0.00935735	0.008008887
rstr_abs049.jpg	0.163915	0.158355	0.15817377	0.144415	0.108961	0.088623	0.073245	0.06247148	0.041840151
rstr_abs050.jpg	0.279072	0.180919	0.172252551	0.136708	0.118579	0.041981	0.027527	0.02304922	0.019911243
rstr_abs051.jpg	0.668471	0.09157	0.080975696	0.075879	0.046399	0.014649	0.010768	0.00809151	0.003196615
rstr_abs052.jpg	0.599581	0.132426	0.121237636	0.060811	0.030409	0.018778	0.017631	0.01437683	0.004748776
rstr_abs053.jpg	0.551156	0.265885	0.086453527	0.038932	0.022746	0.017035	0.013157	0.00244789	0.002187053
rstr_abs054.jpg	0.092639	0.116198	0.078410596	0.142337	0.068869	0.064543	0.094425	0.05249571	0.290081978
rstr_abs055.jpg	0.878669	0.046959	0.043429922	0.009702	0.008329	0.006216	0.005267	0.00106325	0.00036429

Fig. 6 Spreadsheet for abscess

Figure 7 shows the graphical analysis of the scores obtained for the abscess image inputs illustrated in Fig. 6 with the X axis indicating the input image label and the Y axis indicating the score obtained in the range 0–1. From the Fig. 7 it can be seen that 13 of the 17 abscess images obtain high scores against abscess label. Some of the abscess images get misclassified into metastases and fatty liver. Likewise the analysis is carried out for all the images and analysis is performed.

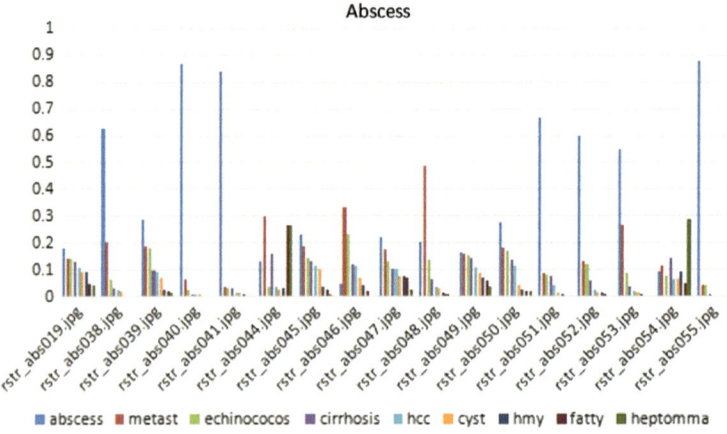

Fig. 7 Graphical representation of scores obtained for abscess

9.4   Accuracy Calculation

The predicted scores are converted into a *confusion matrix* and the four quantitative parameters namely *True Positive* (*TP*), indicating disease as disease, *True Negative* (*TN*), indicating no disease as no disease, *False Positive* (*FP*), indicating no disease as disease and *False Negative* (*FN*), indicating disease as no disease are obtained as shown in Table 2. From the above values, Precision, FN rate, recall and classification accuracy are calculated as shown in the Eqs. (1)–(4) respectively and shown in Table 3.

$$\text{Precision} = \frac{\text{TP}}{(\text{TP} + \text{TN})} \times 100 \tag{1}$$

$$\text{FN rate} = \frac{\text{FN}}{(\text{FN} + \text{TN})} \times 100 \tag{2}$$

$$\text{Recall} = \frac{\text{TP}}{(\text{TP} + \text{FN})} \times 100 \tag{3}$$

$$\text{Classification} = \frac{\text{TN} + \text{TP}}{(\text{TP} + \text{FP} + \text{FN} + \text{TN})} \times 100 \tag{4}$$

Table 2 TP, TN, FP, FN values obtained for the diseases

Disease name	TP	FP	FN	TN
Abscess	25	30	65	380
Cirrhosis	60	5	100	335
Cyst	30	4	66	400
Fatty Liver	39	2	45	360
HCC	63	2	40	395
Echinococcosis	11	4	113	352
Haemangioma	9	2	156	333
Hepatomegaly	3	2	110	385
Metastases	12	7	109	372

Table 3 Classification accuracy

Disease name	Specificity	Precision	Recall	Accuracy (%)
Abscess	0.92	0.454	0.277	81
Cirrhosis	0.176	0.923	0.375	79
Cyst	0.991	0.882	0.312	86
Fatty Liver	0.951	0.097	0.4642	78
HCC	0.994	0.962	0.3841	91.6
Echinococcosis	0.982	0.733	0.0887	72
Haemangioma	0.994	0.818	0.0545	68
Hepatomegaly	0.994	0.6	0.0265	75
Metastases	0.981	0.631	0.060	76.8

An overall accuracy of 79% is achieved. Highest accuracy is achieved for HCC and lowest accuracy is achieved for Hemangioma. Increasing the number of training images would increase the accuracy of the method. The classification accuracy is depicted graphically in Fig. 8.

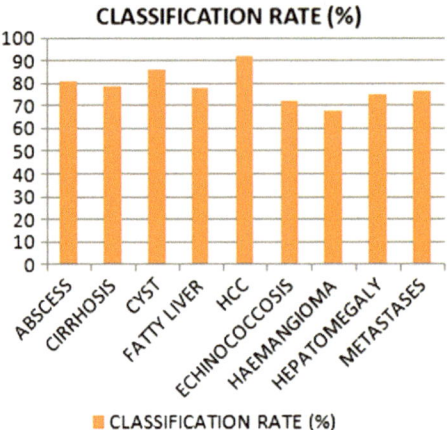

Fig. 8 Classification accuracy

10   Conclusion and Scope for Future Work

This work aims to apply deep learning networks for the classification of liver disorders from ultrasound images. Results show that increasing the number of images would increase the accuracy. Also preprocessing techniques such as isolation of the disease portion from the image by segmentation techniques and presenting the portion images as input to the CNN rather than the whole image may improve the accuracy. Further work could aim in these directions.

Acknowledgements. The authors thank *Sonoscan Center—Coimbatore, Scan Point—Pollachi, Sunita Scan and Diagnostic Center—Guntur, Orean Scans—Udumalpet* and *N.M. Hospital— Coimbatore* for kindly providing the images which has made this work possible.

References

1. Raghesh Krishnan, K., Radhakrishnan, S.: Focal and diffused liver disease classification from ultrasound images based on isocontour segmentation. IET Image Proc. **9**(4), 261–270 (2015). https://doi.org/10.1049/iet-ipr.2014.0202
2. Raghesh Krishnan, K., Radhakrishnan, S.: Hybrid approach to classification of focal and diffused liver disorders using ultrasound images with wavelets and texture features. IET Image Proc. **11**(7), 530–538 (2017). https://doi.org/10.1049/iet-ipr.2016.1072

3. Midhila, M., Raghesh Krishnan, K., Sudhakar, R.: A study of the phases of classification of liver diseases from ultrasound images and gray level difference weights based segmentation. In: Proceedings of the International Conference on Communication and Signal Processing, ICCSP'17 (2017)
4. Mittal, D., Kumar, V., Saxena, S.C., Khandelwal, N., Kalra, N.: Enhancement of the ultrasound images by modified anisotropic diffusion method. Med. Biol. Eng. Compu. **48** (12), 1281–1291 (2010)
5. Sakr, A.A., Fares, M.E., Ramadan, M.: Automated focal liver lesion staging classification based on Haralick texture features and multi-SVM. Int. J. Comput. Appl. **91**(8), 17–25 (2014)
6. Vijayarani, S., Dhayanand, S.: Liver disease prediction using SVM and Naive Bayes algorithms. Int. J. Sci. Eng. Technol. Res. (IJSETR) **4**(4), 816–820 (2015)
7. Wu, K., Chen, X., Ding, M.: Deep learning based classification of focal liver lesions with contrast-enhanced ultrasound. Optik - Int. J. Light Electron Opt. **125**(15), 4057–4063 (2014). https://doi.org/10.1016/j.ijleo.2014.01.114
8. Liu, X., Song, J.L., Wang, S.H., Zhao, J.W., Chen, Y.Q.: Learning to diagnose cirrhosis with liver capsule guided ultrasound image classification. Sensors **17**(1), 149 (2017). https://doi.org/10.3390/s17010149
9. Hassan, T.M., Elmogy, M., Sallam, E.S.: Diagnosis of focal liver diseases based on deep learning technique for ultrasound images. Arab. J. Sci. Eng. 1–14 (2017). https://doi.org/10.1007/s13369-016-2387-9
10. Cheng, P.M., Malhi, H.S.: Transfer learning with convolutional neural networks for classification of abdominal ultrasound images. J. Digit. Imaging **30**, 234–243 (2017). https://doi.org/10.1007/s10278-016-9929-2
11. https://www.tensorflow.org/
12. Zeiler, M.D., Fergus, R.: Visualizing and understanding convolutional networks. Lect. Notes Comput. Sci. **8689**, 818–833 (2014)

A Powerful and Lightweight 3D Video Retrieval Using 3D Images Over Hadoop MapReduce

Chandra Mohan Ranjith Kumar[1($\boxtimes$)] and Sangayah Suguna[2]

[1] Bharathiar University, Coimbatore, Tamil Nadu, India
ranjithnetm@gmail.com
[2] Sri Meenakshi Govt. Arts College, Madurai, Tamil Nadu, India
kt.suguna@gmail.com

Abstract. Content Based Video Retrieval (CBVR) is an approach for redeeming 3D videos from several video sources such as YouTube, Viki etc. Various modern video retrieval applications are revealed by using approaches like key frame selection, features extraction, similarity matching, etc. We propound a new framework that gathers key frame selection, feature extraction, denoising and similarity matching for effective retrieval of videos from 3D images. We initiate Relative Entropy based Fast Key Frame Selection (REFKFS) for nominating optimal key frames for a video. BM3D filter with Bayesian threshold (BB) is proposed for decreasing white Gaussian noises on 3D key frames. We extract texture features, shape features, motion features and color features from the key frames. For extricating shape features, we prefer moments of objects and regions. After extricating features, we propose similarity matching process which is established from Multi-Featured Light-weight (MFLW) matching scheme for successful retrieval of videos with 3D images. Here we qualify Hadoop environment for handling massive sized database on video retrieval. Our experimental outcome includes larger database and furnish efficacious result as 98% of accuracy obtained for overall proposed framework.

Key Terms: 3D video retrieval · Filtration · Denoising · Key frame selection Feature extraction · Hadoop · Similarity matching

1 Introduction

Content Based Video Retrieval was an efficacious approach for retrieval of videos on Multimedia domain. Usually video has certain characteristics such as richer content than individual images, Immensive amount of raw data and compact prior structures [1]. Image and video retrieval is a famous application that is based on artificial intelligence, digital signature processing, statistics, pattern recognition etc. [2]. Multimedia files are now emerged in the world due to expansion of different social networks such

© Springer International Publishing AG 2018
D. J. Hemanth and S. Smys (eds.), *Computational Vision and Bio Inspired Computing*,
Lecture Notes in Computational Vision and Biomechanics 28,
https://doi.org/10.1007/978-3-319-71767-8_65

as Whatsapp, Facebook, Twitter, etc. Many organizations prefer Hadoop which can handle larger data with volume, variety, veracity, value, velocity. Map reduce is a kind of programming model that consists of Map() and Reduce() method [3]. For retrieving Near Duplicate Videos (NDV), Graphics Processing Unit (GPU) based Map Reduce was proposed in [4]. In [5], sketch based video redeeming was presented for higher representation with spatiotemporal layouts and behaviors of semantic objects. With the help of user queries and database videos, sketch events are established. Instance learning [6] was focused on video retrieval based on considering Object of Interest (OOI) that includes a multiple instance learning approach [7, 8]. Novel data mining algorithms based machine learning approaches like classification and clustering [9, 10] were proposed for reclaiming videos with affordable features. A detailed view of classifying the videos was presented based on SVM is explained on [11]. In [12], a spatial bag of words were used for retrieving the 3D models of CAD with special hierarchal feature descriptors are concentrated that considers the bag of words, local feature vectors and sub-graphs. In [13], Parallel key frame extraction was proposed for surveillance video service applications. A novel key frame extraction and selection approach was proposed on [14] which centralized on properties like un-supervision, efficiency and scalability. In [15], authors proposed automatic extraction of video shots from web videos with specific constraints like actions, comedies, etc.Videos are also retrieved using Hasamard Matrix Discrete Wavelet Transform (HDWT) with sparse representation [16, 17]. In order to obtain fruitful matching result between two 3D frames, noises must be removed. In [18], for diminishing noises on the 3D videos, surfacelet transformation was used that contains directional decomposition and mul-tiscale decomposition. Fast bilateral filtering was presented for reducing the noises and it was applied based on computing Gaussian convolutions [19].

Section 1, we examine the related work on CBVR systems with different approa-ches. Section 2 we provide the detailed proposed work of our 3DCBVR framework. Section 3, we experimentally analyze the result of our proposed framework and we conclude the proposed framework with an outlook on future direction in Sect. 4.

2 Proposed Framework

In this section, we illustrate our overall propound work. Here we choose HDFS servers in order to manage videos. Here admin is the person who can add the 3D videos from various data owners on the HDFS server likewise YouTube, Viki, etc. This server has 3D videos. Our main approach is to retrieve 3D videos based on furnishing the 3D image so we have incorporate HDFS and MapReduce framework.

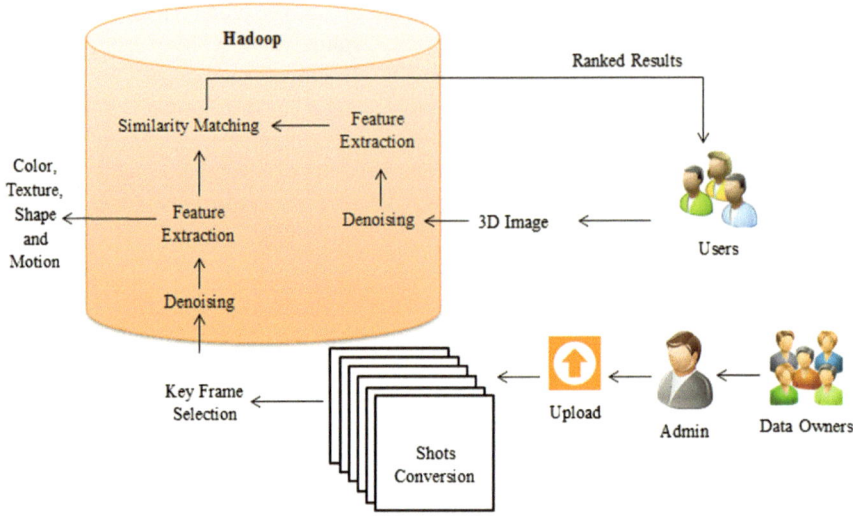

Fig. 1 Overall proposed framework

Here this MapReduce framework has mainly three operations such as Map(), Reduce() and Shuffle(). Figure 1 describes the representation of overall proposed framework with proposed methodologies. As an initial step of our framework is that we partite the 3D videos into shots (a series of frames) based on shot boundary detection mechanism. Hence the shots are appraised as the dataset. From the bulk of series, we select the optimal key frames. For that we use our proposed Relative Entropy based Fast Key Frame Selection methodology. This reveals the optimal key frame and it is followed by removal of white Gaussian noise on the key frame, because noises are conflicts for similarity matching. For denoising the nominated key frames, we prefer BM3D filter with Bayesian threshold technique and this issues a denoised key frame. Thus decreasing noises on key frames is a new significance for our proposed work. After elimination of noises, we extricate the features on key frame such as colors, textures, shapes and motions. In shape based feature we consider moments and regions for efficacious feature extraction. All above specified methods are done for stored database images and on the other hand, 3D image is arrived from user i.e. who are accessing from the website. For 3D query image we need to denoised and features are extricated. At last our propound framework involves an novel method for similarity verification which are called as multi-featured light-weight similarity matching. Here we appraised color feature and motion feature. This mechanism also involves a ranking mechanism which proceeds top rank to the video having top similarity value.

2.1 Key Frame Selection

Key frame selection methodology is heart of our proposed approach that plays a major for reclaiming the 3D videos because it provides the exact content of videos. Many approaches proposed several solutions for key frame selection that does not provide efficacious outcome. To solve these troubles, we prefer a novel approach called "Relative Entropy based Fast Key Frame Selection (REFKFS)". Here we use RE which is one of the best distance measures between probability distributions. SRRE is the additional highlighting distance measure for identifying the few changes or little contents recognition on the frames with their fewer differences. Our proposed key frame selection mechanism also eradicates the redundant frames on the frame series. We arrange the video frames from left to right direction with time series of each frame. Based on time direction of videos, we select the first frame and last frame. Then on remaining frames, we select the key frame based on classical approach calculating the Relative Entropy (RE) and Square Root of Relative Entropy (SRRE) with their distance between frames. They are calculated by,

$$D_{RE}(F_i, F_{i+1}) = \sum_{k=1}^{n} P_i(k) \log \frac{P_i(k)}{P_{i+1}(k)} \tag{1}$$

$$D_{SRRE}(F_i, F_{i+1}) = \sqrt{\sum_{k=1}^{n} P_i(k) \log \frac{P_i(k)}{P_{i+1}(k)}} \tag{2}$$

From the above equations, $P_i = \{P_i(1), P_i(2) \ldots P_i(n)\}$ is the Probability Distribution Function (PDF) of the frames F_i and F_{i+1} based on normalized intensity histogram with n bins where n = 256 and k is the total number of frames on a shot. After calculating the entropy value of frames with distance, we compare with time direction, RE and SRRE values. While comparing, we detect the changes on frames and we remove the redundant frame that has same RE values and time. Again we need to calculate the RE and SRRE for the remaining continuous frames,

$$\{D(F_j, F_{j+1}), D(F_{j+1}, F_{j+2}), D(F_{j+n-1}, F_{j+n})\}$$

and at last key frames are determined as the frame which has minimum differences between $D(F_{key}, F_{key+1})$ and average of all RE or SRRE frame distances:

$$F_{key} = F_{\substack{avg\ \min_k}} \left| D(F_j, F_{j+1}) - \frac{\sum_{l=j, l \neq k}^{j+n-1} D(F_l, F_{l+1})}{n-1} \right| \tag{3}$$

Based on Eq. 3, we get the final average of key frames on the remaining consecutive frames. Thus we get first frame, key frames and last frame. This frame sequence is appraised for efficacious video retrieval.

2.2   Denoising

Denoising is defined as cleaving of noises from 3D frames and that encouraging the visual quality on video retrievals. Our proposed work is redeeming the 3D videos from several video sources. In order to procure visual videos for many users, we concentrate on noise removal and we solve noise removal at high noises levels on every frame. So we proposed BM3D filter with Bayesian thresholding. This approach involves Taylor series for acquiring the high resolution wavelet subbands. We initially include Taylor series with variables i.e. 3 variables (x, y, z) which is a 3D image,

$$T(X) = (a) + (X - a)^T Df(a) + \frac{1}{2!}(X - a)^T \{D^2 f(a)\}(X - a) + \ldots (4)-)$$
$$= imageber\ variables,$$

This Taylor series is a mathematical series which represents infinite sum of terms that are calculated from function derivatives, so this identifies the pixels changes. So we analyze the pixel changes clearly. Based on this series, we calculate the noise on all sub bands and we appraise higher noise subband for calculating the Bayesian thresholding. Here we calculate the thresholding for all subbands,

$$T_B = \frac{\sigma_N^2}{\sigma_s}$$

where $\sigma_s \neq 0$ and it is given by,

$$\sigma_s = \sqrt{\max((\sigma_y^2 - \sigma_y^2), 0)} \tag{5}$$

$$\sigma_y = \frac{1}{N}\left(\sum SB_i\right) \tag{6}$$

where $SB_i = \{LLL, LLH, LHL, LHH, HLL, HLH, HHL, HHH\}$ is the total subbands and noise estimation (σ_N) is done by,

$$\sigma_N = \left[\frac{median(SB)}{0.6745}\right] \tag{7}$$

After calculating T_B, we average the values for obtaining Bayesian threshold. Then we acquire curve fitting method, which can be written as,

$$\Upsilon = \frac{p_1\beta^2 + p_2\beta + p_3}{\beta + q} \tag{8}$$

Here β is the noise standard deviation, $p_1 = 0.9592$, $p_2 = 3.648$, $p_3 = -0.138, q = 0.1345$. Then BM3D filtering algorithm is implemented as,

$$\hat{Y}S_{SB} = T_{3D}^{-1}\left(\propto \left(T_{3D}(Z_{SB}), T_B \Upsilon \sqrt{2\log(N^2)}\right)\right) \tag{9}$$

Here SB denotes subbands, T_{3D} specifies unitary 3D transform, $\hat{Y}S_{SB}$ is the stacked subbands, Z represents all subband sizes and $\propto$ describes threshold operator. Thus this Eq. 9 reconstructs the all subbands and results the denoised images.

2.3 Feature Extraction

Feature extraction is nothing but extricating information from a frame and it is one of the significant processes on proposed video redeeming framework which start its execution from measured data and it pick-outs the derived features that will be informative, non-redundant and considered as the subsequent learning. Usually frames are composed of shapes and edges of grey level. Our proposed framework includes extraction of visual low-level features such as color, texture and shapes whereas visual semantic features are motions. We discuss the detailed extraction of features in the following sections.

2.3.1 Extraction of Visual Low-Level Features

(i) Texture and Color Feature Extraction:

For extracting texture features, we concentrate classical methods which can be calculated from co-occurrence matrix. This has 14 statistics texture measurement which are supported from co-occurrence matrix. The co-occurrence matrix use orientation of angle ($\theta = 0, 45, 90, 135$) degrees that occurs between pair of grey level and axis. The grey level co-occurrence matrix is given by,

$$P(i,j|d,\theta) = \{(x_1,y_1),(x_2,y_2)\} \tag{10}$$

where, $i = I(x_1,y_1), j = I(x_2,y_2), |x_1 - x_2| = 0°, |y_1 - y_2| = d$. The grey level co-occurrence is defined as $p(i,j|d,0°)$ with respect to distance (d) and angle (θ). The probability value of grey level co-occurrence matric is given by,

$$p(i,j|d,\theta) = \frac{p(i,j|d,\theta)}{\sum_{i=1}^{256}\sum_{j=1}^{256}p(i,j|d,\theta)} \tag{11}$$

We list out a few textures in our proposed framework that are discussed below:

Angular Second Moment:

It is defined as the sum of squares of entries in the Grey Level Co-occurrence Matric (GLCM) which measures the image homogeneity. This is also known as the uniformity or energy. This measure has effective result like better homogeneity otherwise the pixels are similar. Here i, j are the spatial coordinates of function p (i, j) where Ng is grey tone

$$ASM = \sum_{i=0}^{Ng-1} \sum_{j=0}^{Ng-1} P_{ij}^2 \qquad (12)$$

Entropy:

Entropy specifies the total contents of the image that are needed for image compressions. This is a measure of loss of information or message in a transmitted signal and also measures the image information.

$$Entropy = \sum_{i=0}^{Ng-1} \sum_{j=0}^{Ng-1} -P_{ij} * \log P_{ij} \qquad (13)$$

Correlation:

This measures the grey level linear dependency on neighboring pixels. It returns the measure of how the pixel is correlated to its neighbors. Here μ and σ are the mean and standard deviation.

$$Correlation = \sum_{i,j}^{K} \frac{(i - \mu_i)(j - \mu_j)}{\sigma_i \sigma_j} \qquad (14)$$

Contrast:

This measures the local variations on GLCM where i and j are equal and k is the row or column dimension of the 3 * 3 square matrix, $P_{i,j}$ is the probability of pixel pairs which satisfy the offset. The contrast texture is also called as inertia.

$$Contrast = \sum_{i,j=0}^{K} P_{i,j}(i - j)^2 \qquad (15)$$

Mean:

This supports measuring the mean of grey level in an image and it is calculated by,

$$Mean = \frac{1}{2} \sum_{i}^{U} \sum_{j}^{V} (iP[i,j]) + (jP[i,j]) \qquad (16)$$

The above mentioned equations are applied for texture matrices which results in single output. We combine all above mentioned features for effective similarity matching between 3D frames and 3D input query image. We also extract color feature in our proposed process, so that we initially convert the images to grey scale and we also calculate the intensity values of each frames. RGB to grey scale conversion is based on following equation,

$$I_Y = 0.333 * F_R + 0.5 * F_G + 0.1666 * F_B \qquad (17)$$

Here, F_R, F_G, F_B are the intensity of R, G, B components respectively and I_Y is the intensity of equivalent grey level image of RGB. Initially we convert the RGB image to HSV and YCC color space. Then we extract the hue, saturation, luminance and intensity values (I_Y).

(ii) **Shape Feature Extraction**:

- *Region*

Shape features are one of the low level prime features for video retrievals. In general 3D video has different shapes based on regions and moments. In our proposed system we consider both moments and regions for extracting the shape features. The previous work of SA-DWT is done for 2D space regions extraction. Here they extracted the region on all subbands. We use SA-DWT for 3D key frames in order to extricate regions. According to 3D key frames, we get subbands such as,

$$3D\ KF_{(x,y,z)} = subbands(LLL, LLH, LHL, LHH, HLL, HLH, HHL\ and\ HHH)$$

We apply SA-DWT because it allows decomposition of regions on rows and columns and selecting the approximate subbands is not suitable region extractions. Resulting pixel values are related with pixel values inside the segment regions. We involve image mask information that specifies with edge. Finally we involve convolution high pass filters and low pass filters for both horizontal and vertical directions. As a result we acquire regions from the key frames.

- *Moments*

The 3D Zernike moments are used to detect the moments of the 3D frames, we initially specify the size of object such that N * N * N is a 3-dimensional array of voxels. Here the intensity function of the voxel points are,

$$x_i = \frac{2i - N - 1}{N\sqrt{3}}, y_j = \frac{2j - N - 1}{N\sqrt{3}}, z_k = \frac{2k - N - 1}{N\sqrt{3}} \qquad (18)$$

where the sampling intervals in x, y and z directions $\Delta x_i = x_{i+1} - x_i, \Delta y_j = y_{j+1} - y_j, \Delta z_k = z_{k+1} - z_k$ with i, j and k = 1, 2 ... N. The exact 3D geometric moments of the order (r + s + t) is defined by,

$$G_{rst} = \sum_{k=1}^{\lfloor N/2 \rfloor} I_t(z_k) R_{rsk} \qquad (19)$$

Where, G is the geometric moment of the order $(r + s + t)$, I specifies intensity, rsk represents the exponents at point z and R is the augmented intensity function. Based on the above equation, we compute the moments of the key frames. Thus moments and region extractions are effective for 3D video retrieval.

2.3.2   Extraction of Virtual Semantic Features

(iii)   Motion Feature Extraction:

Our proposed framework involves redeeming of 3D videos from server database, so we include the motion features. The motion information is a salient feature for video retrievals because the information is taken from foreground pixels leaving the background pixels. Let us consider width (W) and Height (H) of a frame and $P_{(i,j)}(k)$ is the binary value of a pixel in ith row and jth column on kth frame. It is '1' when the pixels are foreground pixel else it is '0'. For kth frame, the motion information is calculated by,

$$MI(k) = \sum P_{(i,j)}(k), \quad 0 < = i < W, 0 < = j < H \tag{20}$$

Based on the above equation, we calculate the motion information on each frame. Thus we extracted all features which are combined for 3D video retrievals.

2.4   Similarity Matching

Similarity matching is the salient process for our proposed framework that implicates a multi-featured light-weight matching scheme on Hadoop framework. This scheme involves more than one feature such as the texture, shape, color and motion features. Here f1 is the key frame and f2 is the query image.

This similarity value is calculated by,

$$\begin{aligned} \text{Similarity} = \text{Color-Similarity}(f1, f2) + \text{Motion-Similarity}(f1, f2) \\ + \text{Shape-Similarity}(f1, f2) + \text{Texture-Similarity}(A, B) \end{aligned} \tag{21}$$

Here higher value obtained for the video is considered as the exact video for the query image. We also provide ranking for accurate video retrieval. Here higher similarity video gets the '1' rank whereas the nearer similarity values has ranks according to ascending order. For example, if a video gets second similarity value for the query images then its rank is assigned to be 2, and goes on.

2.5   Hadoop Framework

Hadoop is a framework which has the MapReduce programming model with Map(), Reduce() and Shuffle() operations. According to our proposed framework, after key frame selection, all proposed methodologies are executed under Hadoop environment. In map operation, we calculate the similarity values for query image and frames as discussed on previous section. Here this operation has value pair based on (frames, similarity values).

This operation maps the similarity values to their corresponding frames which is done by,

Map operation:

$$[f1 \rightarrow sim(f1), f2 \rightarrow sim(f2), f3 \rightarrow sim(f3)...] \tag{22}$$

Then it is followed by the shuffle operation, according to proposed framework, we have a frame sequence that acquired on key frame selection i.e. first frame, middle frames and last frame. We sort these frames according to corresponding similarity values. The shuffle operation has key value pair based on (sequence of frames and their similarity values).

Shuffle operation:

$$(FFr \quad MFr \quad LFr) = [Similarity\ values] \tag{23}$$

At last, our reduce operation takes place, which has the videos with their corresponding ranking value. This has the value pair based on (videos, rank). Thus it behaves like an indexing operation for the videos.

Reduce operation:

$$(V(1), V(2), V(3), ...) = [R1, R2, R3, ...] \tag{24}$$

When a user provides a 3D query image to the Hadoop server, they get the accurate video as first ranking and they also has related videos results. These related videos have second ranking. Thus our proposed framework provides efficacious and effective results.

3 Experimental Evaluations

Our proposed framework is examined based on the various experimental datasets that are used in research proposals such as images and videos. But the sizes of the videos are larger than the images. In our demonstration, we collect videos and covert to shots based on shot boundary detection mechanism, and then shots are considered as dataset. Our demonstration considered dataset from different servers such as YouTube, Viki, etc. We show the hardware configurations for the demonstration on Table 1.

Table 1 Configuration of Hardware

Specifications	N_1	N_2	N_3	N_4	N_5
Processor	I3	I5	I7	AMD	Opteron
CPU Speed	3.4 GHz	2.4 GHz	2.9 GHz	3.5 GHz	3.0 GHz
CPU Core	2	4	4	6	16
RAM	4	4	4	3	3

3.1   Performance Analyses

Our performance analysis is done in two ways (1) overall MapReduce result and (2) comparative result. Our overall proposed framework results are broadly categorized into three different scenarios for each process which are takes place in proposed MapReduce framework in order to find processing time for proposed framework for execution of different nodes. We considered scenarios such as FIRST SCENARIO: Working of Node 1 and Node 2, SECOND SCENARIO: Working of Node 1, Node 2 and Node 3 and LAST SCENARIO: Working all nodes Node 1–Node 5. Our implementation work is done on software specifications of Ubuntu 14.04 and Hadoop 2.7.0 and JDK 1.7 version. We appraise parameters for evaluating 3D videos redeeming of our proposed system and by selecting the parameters like MapReduce time, Precision-Recall and E-measure.

3.2   Experimental Results

This section deals with experimental results such as Overall MapReduce results and comparative results. Our framework is composed of 12,890 videos and here we acquired 39,905 key frames for the videos. First we discuss about the comparative results of our proposed framework. This comparative result concentrates on frame selection and frame detection [14], Noise Level [19], accuracy [9], Precision-Recall [12].

3.2.1   Comparative Results

Figure 2 shows the comparison of key frame selection between proposed ERKFS methodology and existing work [14]. Here we detect the key frames from video shots with relative entropy and square root of relative entropy whereas in existing research they focus on change detection test with a threshold value

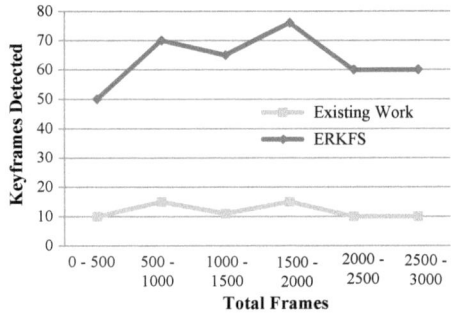

Fig. 2 Comparison of key frame detection

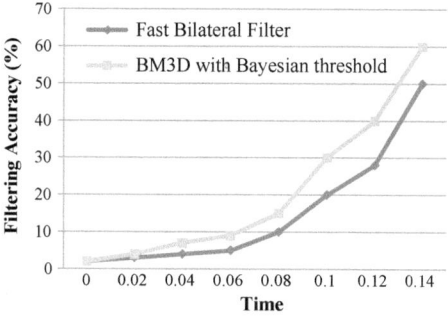

Fig. 3 Comparison of filtering accuracy

From this graph, we see the robustness of key frame detection for both proposed and existing work. Thus our proposed framework results 30% higher detection of key frames than the existing work. Figure 3 illustrates the comparison of filtering accuracy for the noise removal between proposed noise removal and existing fast bilateral filter [19]. Here, our proposed noise removal approach involves block matching 3D filter with Bayesian thresholding whereas in fast bilateral filter based on diagonal modeling, this filter acquires lower noise removal accuracy than our proposed BM3D filter because it leaves some spaces on noise removal. We calculate accuracy of the filter based on inverse of mean square error operation which is done by,

$$\text{MSE} = \frac{1}{n} \sum_{i=1}^{n} \left(\hat{Y} - Y_i \right)^2$$

where $\frac{1}{n} \sum_{i=1}^{n}$ is the mean, $\left(\hat{Y}_i - Y_i \right)^2$ is the square of errors.

$$Accuracy = \left(\frac{1}{MSE} \right) * 100$$

Thus our both filter acquires same accuracy up to 0.04 timing, after 0.04 timing there is a slight variation of accuracy on our proposed work.

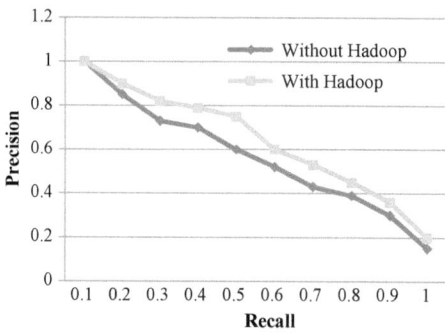

Fig. 4 Comparison for Precision-Recall with and without Hadoop

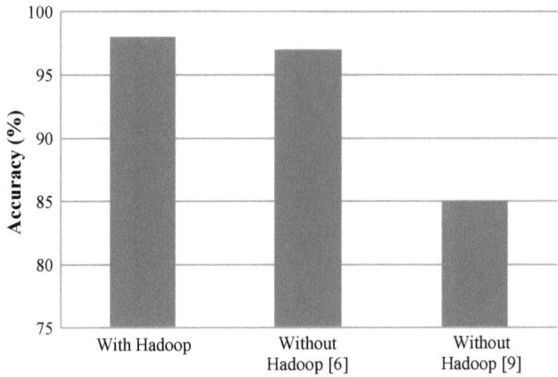

Fig. 5 Comparison for accuracy with and without Hadoop environment

Figure 4 describes the comparison of Precision-Recall values on Hadoop environment and without Hadoop environment [12]. In without Hadoop environment has effective ratio between the query image and retrieved videos in terms of precision values and recall values but using Hadoop environment, processing time is increased which results in slight variation of Precision-Recall ratio, whereas entire e-measure acquires 68% of video redeeming. Thus the graphical plot illustrates the effective performance of proposed framework using Hadoop. Figure 5 compares of accuracy of existing works without Hadoop environment and our proposed framework using Hadoop environment. Both existing works involves the classification based learning algorithms for redeeming the videos and they acquire less accuracy but our proposed framework acquires 98% of accuracy.

4 Conclusion

In this paper, we presented implied framework for redeeming 3D videos from larger dataset by supporting Hadoop environment. This framework predominantly concentrated on processing speed for dataset which has an advantage of scalability. Our framework has various methodologies such as key frame selection, denoising, feature extraction and similarity matching. Here all methodologies except key frames selection are executed under Hadoop framework. From our experimental result, we proved our framework is the best framework when comparing our methodologies with other existing works. Our framework has Relative Entropy based Fast Key Frame Selection is presented for nominating the key frames and BM3D filter with Bayesian threshold is presented for solving noises on 3D key frames, we also extract and combine features from the key frames such as texture features, shape features, motion features and color features. Then a light-weight Multi-Featured similarity matching scheme is presented for acquiring the fruitful result such as relevant videos for the queried 3D image. Thus our proposed framework furnishes efficacious result in terms of 98% of accuracy. In our future work, we concentrate on retrieving 3D videos with elimination of duplicate videos using Hadoop environment and enhanced operations.

References

1. Hu, W., Xie, N., Li, L., Zeng, X., Maybank, S.: A survey on visual content-based video indexing and retrieval. IEEE Trans. Syst. Man Cybern. **41**(6), 797–819 (2011)
2. Lew, M.S., Sebe, N., Eakins, J.P.: Challenges of Image and Video Retrieval, vol. 2383, pp. 1–6. Springer, Heidelberg (2002)
3. Jain, P., Gyanchandani, M., Khare, N., Singh, D.P., Rajesh, L.: A survey on big data privacy using Hadoop architecture. Int. J. Comput. Sci. Netw. Secur. **17**(2), 148–155 (2017)
4. Wang, H., Zhu, F., Xiao, B., Wang, L., Jiang, Y.-G.: GPU-based MapReduce for large-scale near-duplicate video retrieval. In: Multimedia Tools Applications, pp. 1–20. Springer, Heidelberg (2014)
5. Zhang, Y., Chen, X., Lin, L., Xia, C., Zou, D.: High-level representation sketch for video event retrieval. In: Science China Information Sciences, vol. 59, pp. 1–15. Springer, Heidelberg (2016)
6. Lin, T.-C., Yang, M.-C., Tsai, C.-Y., Wang, Y.-C.F.: Query-adaptive multiple instance learning for video instance retrieval. IEEE Trans. Image Process. **24**(4), 1330–1340 (2015)
7. Fridman, L., Reimer, R.: Semi-automated annotation of discrete states in large video datasets. Association for the Advancement of Artificial Intelligence, 1–6 (2016)
8. Sabokrou, M., Fayyaz, M., Fathy, M., Klette, R.: Deep-cascade: cascading 3D deep neural networks for fast anomaly detection and localization in crowded scenes. IEEE Trans. Image Process. 1–13 (2017)
9. Brindha, N., Visalakshi, P.: Bridging semantic gap between high-level and low-level features in content-based video retrieval using multi-stage ESN–SVM classifier. In: Indian Academy of Sciences, pp. 1–10. Springer, Heidelberg (2016)
10. Choi, M.-K., Wang, Z., Lee, H.-G., Lee, S.-C.: A bag-of-regions representation for video classification. In: Multimedia Tools Application, pp. 1–20. Springer (2015)
11. Nagaraja, G.S., Rajashekara Murthy, S., Deepak T.S.: Content based video retrieval using support vector machine classification. In: IEEE Conference on Applied and Theoretical Computing and Communication Technology, pp. 821–827 (2015)
12. Huangfu, Z.-M., Zhang, S.-S., Yan, L.-H.: A method of 3D CAD model retrieval based on spatial bag of words. In: Multimedia Tools Application, pp. 1–29. Springer, Heidelberg (2016)
13. Zheng, R., Yao, C., Jin, H., Zhu, L., Zhang, Q., Deng, W.: Parallel key frame extraction for surveillance video service in a smart city. PLoS ONE 1–8 (2015)
14. Lu, G., Zhou, Y., Li, X., Yan, P.: Unsupervised, efficient and scalable key-frame selection for automatic summarization of surveillance videos. In: Multimedia Tools Application, pp. 1–23. Springer, Heidelberg (2016)
15. Nga, D.H., Yanai, K.: Automatic collection of web video shots corresponding to specific actions using web images. In: IEEE Computer Society Conference on Computer Vision and Pattern Recognition Workshops, pp. 15–20 (2012)
16. Mohamadzadeh, S., Farsi, H.: Content based video retrieval based on HDWT and sparse representation. Image Anal. Stereol. **35**, 67– 81 (2016)
17. Sajjad, M., Mehmood, I., Baik, S.W.: Sparse representations-based super-resolution of key-frames extracted from frames-sequences generated by a visual sensor network. Sensors **14**, 3652–3674 (2014)
18. Khalid, M., P. Sajith Sethu, P.S., Sethunadh, R.: Video denoising using surfacelet transform. In: IEEE International Conference on Recent Trends in Electronics Information Communication Technology, pp. 1502–1506 (2016)
19. Papari, G., Idowu, N., Varslot, T.: Fast bilateral filtering for denoising large 3D images. IEEE Trans. Image Process. 1–12 (2016)

An Improvement of Iterative Algebraic Reconstruction Technique by Using Bisectors: An Illustration in Computerized Tomography

Mohamad Soubra$^{(\boxtimes)}$ and Ömer Özgür Tanrıöver

Department of Computer Engineering, Ankara University, Ankara, Turkey
mhd.m.soubra@gmail.com, ozgurtanriover@yahoo.com

Abstract. This paper proposes the utilization of bisectors for iterative algebraic reconstruction techniques to attain better image reconstruction. We use bisector hyperplanes and the average of the multiple solutions for beam equations. We also discussed why bisector hyperplanes may improve the approximate solution for pixel intensities and how they are obtained. Then, we defined the ART procedure with bisector hyperplanes and implemented the algorithm. Through an application, we compared the solution points obtained with other ART based methods and our ART procedure. As closer solution points may be obtained at each iteration; the images formed by pixels intensities based on these solutions may have higher quality.

Keywords: Algebraic reconstruction · Computerized tomography · Image accuracy

1 Introduction

The problem of obtaining images of the internal formations of objects occurs in different scientific domains ranging from engineering to medicine. Procedures based on magnetic resonance, emission tomography and X-ray computerized tomography have been used for this purpose. Among these, X-ray computerized tomography (CT) is one of the most common procedures for obtaining cross-sectional images of the human body [1, 2]. In CT, X-ray pictures are formed by X-rays that are projected perpendicular to the plane of the picture. After X-rays pass through the cross section in the body, the intensities of the X-ray beams are measured by an X-ray detector and these measurements are relayed to a computer for process. Each pixel of the image matrix is shaded a level of gray proportional to its X-ray density. Because different tissues within the human body have different X-ray densities, the image is obtained.

Although there are different tomography procedures, they do share similar conceptual foundations. The main idea is the reconstruction of the cross-section of the body by its layer-by-layer projections. Methods for image reconstruction are divided into two main categories. The first category includes methods based on the Radon transform. As an example, filtered back-projection (FBP) algorithm can be seen as a

© Springer International Publishing AG 2018
D. J. Hemanth and S. Smys (eds.), *Computational Vision and Bio Inspired Computing*,
Lecture Notes in Computational Vision and Biomechanics 28,
https://doi.org/10.1007/978-3-319-71767-8_66

computer implementation of Radon's inversion formula for reconstructing a 2D function from its 1D line integrals. This reconstruction requires many projections which may pose a concern for the patients [3, 4].

The second one is based on thin X-rays sent parallel to the projection axis. Instead of back projecting the acquired projection data, the data is interpreted as a system of linear equations. However, the size of the variable matrix is proportional to the square of the image size multiplied by the number of projections and in some applications may reach to the order of 10^{10}–10^{12} entries [5]. Such a huge system cannot be solved by standard methods; instead, feasible iterative methods are used. With the help of iterative algebraic reconstruction technique (ART), an approximate solution to the (generally) inconsistent system of linear equations can be obtained [6]. In ART procedure starting with an arbitrary initial point in n-dimensional space by calculating the orthogonal projections from one hyperplane to the other, some points on each one of the hyperplanes which are relatively close to other hyperplanes can be obtained [1–3]. One of these points is considered as an approximate solution to the system.

In the past, various iterative algorithms based on ART [6, 7] such as simultaneous iterative reconstruction technique (SIRT), simultaneous algebraic reconstruction technique (SART) [2, 3], expectation maximization (EM) and ordered subsets expectation maximization (OSEM) have been developed. In general, ART reconstructions suffer from salt and pepper noise produced by propagation of approximation errors through iterations. SIRT and EM algorithm alleviate this problem and produces higher fidelity images in expense of slower convergence. Furthermore, SART and OSEM converge rapidly and produce higher quality images [3, 7].

The main advantage of iterative ARTs in general is the need of little number of projections hence less radiation. When the projection data is incomplete, dynamic or noisy, ARTs are able to generate higher quality CT images when compared with FBP algorithm [1]. Hence, when large number and uniform projections are not acquired or attenuation or a certain amount of ray bending occurs, ARTs may be more amenable [8]. However, iterative methods require high computation time and capacity to reach acceptable high-quality images; they have not been employed by commercial CT scanners [9]. Meanwhile, with the increase in capacity and power of computer systems, given their potential, the algebraic methods are regaining attention [10].

One problem with ARTs is that, the final approximate solution point obtained is on one of the hyperplanes hence may be far from other planes in the system. This situation may result in inaccuracy especially when some hyperplanes are very far from each other. In this case, it is quite possible that these hyperplanes correspond to the image segment where there is high difference between pixels intensities. Hence, if a solution point far from these hyperplanes is chosen, for example, the shape of the object may not be represented accurately.

This paper proposes an improvement for iterative ARTs in computed tomography. Hyperplanes interpreted from beam equations are replaced with bisector hyperplanes. This approach may result in a more appropriate and fair solution without increasing the order of complexity of the algorithm. In addition, we propose that the average of the solution points obtained in the final iteration of the ART procedure also may improve the accuracy. It must be noted that the proposed extension is not a replacement but may be used in conjunction with ART based methods in general.

2 Related Work

In this section, in order to put this study into perspective we present a review of some of the recent related work especially aiming to improve ART.

Cengiz and Kamasak [11] utilized the pure (Karcmarz) ART and the Multiplicative Algebraic Reconstruction Technique (MART) as a sort of comparison with a 3D Shepp-Logan phantom to test the performance of each respectively. Two main values are used to expose the characteristics of the techniques are the Root Mean Square Error (RMSE) and the Mean Structural Similarity (MSSIM) as the index of the image quality. These two values aided also to bring out the Layer Of Interest (LOI) which was also essential for the comparison. As a conclusion, the author points out that MART gave a better result with respect to the RMSE value while, ART gave a better result regarding the MSSIM value. MART gave the same result as ART but converged faster.

Li and Song [12] compared the pure ART with SART and modified SART with the new improvement on the SART named as Self-correlative Algebraic Reconstruction Technique (SSART). The techniques were compared by the means of taking 50×50 rectangular elements and doing the projection of each. For each projection, the graph of the constructed image is calculated after 200 iterations of each technique. The comparison showed that the shape of the graph got worse when MSART was utilized. On the other hand, SSART showed a highly similar graph compared with the regular ART thus producing the most precise result. As a conclusion and with respect to the MSE which was chosen to analyze the images' accuracy, it was observed that SSART converged faster than all the above-mentioned techniques in which the MSE's value dropped 96.4% from that of ART while MSART was the most divergent thus proving that SSART was the ideal choice in this test.

Chetih and Messali [13] proposed the utilization of Filtered Back Propagation (FBP) with Hamming filter as a better way to construct images with respect to ART. For the comparison, 128×128 and 256×256 grey scale images were selected. The comparison was based to the relative norm error. That would take every pixel then compare it with the calculated value of the projected pixel respectively. The Normalized Cross-correlation (NCc) is basically used for template matching to find certain patterns within an image. Structural Content (SC) is used to measure the images' similarity base on small regions in the images, which would contain significant low-level structural information. The results based on 16, 32, 64, and 180 projections and using ART with 200 iterations showed that FBP with Hamming filter provided a better image quality with respect to low error values.

Wan et al. [14] compared ART, SART, and MART with Weighted Back Propagation (WBP) in the field of Cryo-electron tomography (Cryo-ET) which combines electron microscopy and the principles of tomography. The author states that SART suffers from short comings regarding how it starts in an arbitrary point then begins to converge which will make the number of iterations quite high. As a conclusion, the author states that MSART could be the solution intended to solve the high number of iteration problem. So an implementation was made to confirm that the proposed technique compared with the implementation SART with respect to a 171×171 sized phantom. MSART was found to be providing a more accurate and less time expensive.

Kim et al. [15] discusses the utilization of a fast innovative approach to reconstruct images using on-line directional algebraic reconstruction technique (OLDART) which is the improved version of DART. It produces the same quality of images such as DART but with less time consumption. The comparison was done with respect to Electric Capacitance Tomography (ECT) which is used to obtain the cross-sectional image about the distribution of a mixture of dielectric materials inside a domain of interest. DART was compared with both Linear Back Propagation (LPB) which is commonly used for image reconstruction in ECT. OLDART proved its efficiency by using the same proposed weighting matrix with DART by constructing 1948 triangular elements in which the on-line reconstruction. Essentially it differs from DART as a non-iterative construction technique, its characteristic gave a noticeable edge over DART and gave a better image quality than that of LBP.

On the other hand, Jiang and Wang [16] insisted that SART would converge much quicker if the coefficients of the linear image are located in the non-negative space and thus providing more efficiency than regular ART. The author had the opportunity to place a comparison between SART and ART through the providing mathematical proof about the discussed topic. The provided example included the utilization of Hilbert space as a starting point to then doing proofing and propositions. The author also discussed that with the help of the EM formula, SART is shown to be more accurate and to diverge quicker.

Saha and Tahtali [17] defined a technique composed of three main steps to reconstruct images from tomography (CT) scans. The author recommended the utilization of a compressed sensing mechanism which has been under development. At first, the scans would be treated by the back-propagation technique which will help the Kaczmarz method to better to detect the objects' contour i.e. determining the intensity of the pixels within the boundary of the objects. In the second step, the derived results would be then treated by ART that would be responsible for the reconstruction of the images' cross-section. In the final step, the compressed sensing technique (adaptive regularization) would be performed to help in quicker convergence. The technique was tested and as a conclusion, the technique helped in minimizing the number of iterations when compared with regular ART.

In their paper, Oliveira et al. [18] intended to utilize ART highlighting the importance of the number of iterations and the images' noise after the iterations. The author recommended a relaxation parameter as a factor to help in minimizing these. After the implementation and testing of the technique, it was noticed that with relaxation parameter between 0.05 and 0.30, ART converges within 20 iterations and reduces the value of noise in the constructed image thus, proving that efficiency of such a factor.

3   Bisector Hyperplanes and Improved Art

A hyper-plane equation is of the form $a_1x_1 + a_2x_2 + \ldots + a_nx_n + d = 0$. Here, $(a_1, a_2, \ldots, a_n) = N$ is the normal vector of the hyperplane, $(x_1, x_2, \ldots, x_n)$ is any point of the plane, d is a constant number. This plane can be compactly represented in the form $N \cdot X + d = 0$.

Let hyperplanes be given with equations $H_1: N_1 \cdot X + d_1 = 0$ and $H_2: N_2 \cdot X + d_2 = 0$. Let not be N_1 and N_2 parallel to each other. In this case these planes intersect. These intersecting planes have one inner bisector plane and one outer bisector planes as shown in Fig. 1. If the distance of a point Y in the bisector plane, is denoted by l_1 and l_2

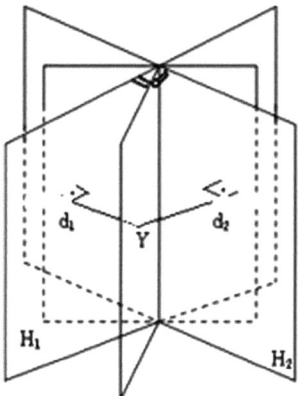

Fig. 1 Bisector hyperplanes

$$l_1 = l_2 \Rightarrow \frac{|N_1 \cdot Y + d_1|}{\|N_1\|} = \frac{|N_2 \cdot Y + d_2|}{\|N_2\|} \Rightarrow \frac{N_1 \cdot Y + d_1}{\|N_1\|} = \frac{N_2 \cdot Y + d_2}{\|N_2\|} \text{ or}$$

$$\frac{N_1 \cdot Y + d_1}{\|N_1\|} = -\frac{N_2 \cdot Y + d_2}{\|N_2\|}$$

is found. The first and second equations may be either for inner bisector hyperplane and outer bisector hyperplanes respectively. The first equation can be written as;

$$\underbrace{\left(\frac{N_1}{\|N_1\|} - \frac{N_2}{\|N_2\|}\right) \cdot Y}_{N_3 \cdot Y} + \underbrace{\left(\frac{d_1}{\|N_1\|} - \frac{d_2}{\|N_2\|}\right)}_{d_3} = 0 \tag{1}$$

In a special case when $N_1 \neq N_2$ and $\|N_1\| = \|N_2\|$, this equation becomes

$$\underbrace{(N_1 - N_2) \cdot Y}_{N_3 \cdot Y} + \underbrace{d_1 - d_2}_{d_3} = 0 \tag{2}$$

or

$$N_1 \cdot Y + d_1 = N_2 \cdot Y + d_2 \tag{3}$$

The orthogonal projection of any point A, denoted by A^* is calculated as;

$$A^* = A - \frac{(N \cdot A) + d}{(N \cdot N)} \cdot N \tag{4}$$

In the ART procedure used in CT [1], the system of linear equations obtained from X-ray beams sent through slice planes is a system of hyperplanes in the form of

$$\left. \begin{array}{ll} H1: & N_1 \cdot X + d_1 = 0 \\ H2: & N2 \cdot X + d_2 = 0 \\ \cdots & \cdots \cdots \cdots = 0 \\ H_m: & N_m \cdot X + d_m = 0 \end{array} \right\} \tag{5}$$

In general, this system has not an exact solution. In ART procedure, starting with an arbitrary initial point A, an orthogonal projection from this point to an hyper plane is calculated. The orthogonal projection point reached after high number of iterations approximates the exact solution of the systems. Although, the final point obtained, is on the plane it belongs, it is far from other planes in the system. This situation may result in inaccuracy especially when some hyperplanes are very far from each other. Hence if a solution point far from these hyperplanes is chosen, the shape of the object may not be represented accurately. This problem may be alleviated, if inner bisector hyperplanes obtained from (1) are used in system (5) instead, the solution on inner bisector hyperplanes calculated by (4) will be a more appropriate solution. This is depicted in Fig. 2.

Consider Fig. 2, the hyper-planes H_1, H_2, H_3, ..., H_m are interpreted from beam equations. The hyper-planes L_1, L_2, L_3, ..., L_m on points P, Q, R, S are the interior bisector hyperplanes of H_1, H_2, H_3, ..., H_m. In Fig. 2, the interior region PQRS of the bisector-hyperplanes is always within the region BCDE of the hyperplanes H_1, H_2, ..., H_m. Hence, if we consider the inner bisector hyperplanes L_1, L_2, ..., L_m instead of the hyperplanes H_1, H_2, ..., H_m, we can obtain an approximate solution point on to the edges of the region PQRS by the iteration. This new approximate solution is better than the approximate solution when original hyperplanes are used, because the new point is closer to the edges H_1, H_2, ..., H_m than the previous one. In the following steps, we define an ART procedure extended with bisector hyperplanes.

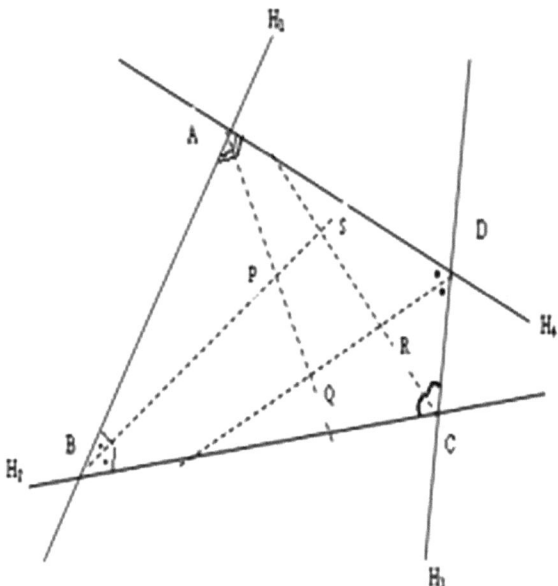

Fig. 2 Inner bisector hyperplanes and iteration region

Step 0 Find inner bisector hyper-planes L_1, L_2, ..., L_m of hyperplanes of H_1, H_2, ..., H_m based on the formula

$$\underbrace{\left(\frac{N_i}{\|N_i\|} - \frac{N_{i+1}}{\|N_{i+1}\|}\right) \cdot Y}_{N_i \cdot Y} + \underbrace{\left(\frac{d_i}{\|N_i\|} - \frac{d_{i+1}}{\|N_{i+1}\|}\right)}_{d_i} = 0$$

where i = 1,2, ..., m − 1.

The subscript i indicate the ith inner bisector hyperplane to be determined in ith iteration. Note that, for obtaining mth bisector hyperlane L_m, H_m and first hyperplanes H_1 is used.

Step 1 Choose any arbitrary point A in the space R^n.
Step 2 Project A orthogonally on to the first inner bisector hyperplane L_1 with the formula;

$$A_1^1 = A - \frac{(N \cdot A + d)}{(N \cdot N)} \cdot N$$

and call this projection A_1^1. The superscript 1 indicates that this is the first of iterations to be conducted. The subscript 1 indicates that the orthogonal projection is on to the

inner bisector hyper-planes L_1. N is a normal vector of the inner bisector hyperplanes
L_1: N · X + d = 0.

Step 3 Project A_1^1 orthogonally on to the second hyperplane L_2 and call this projection A_2^1.

Step 4 Repeat the process and find A_m^1, that is, orthogonally project A_{m-1}^1, found in the previous cycle, on to mth hyperplane L_m.

Step 5 Take the point A_m^1 as the new value of the point A and cycle from step 2 through step 5 again.

Step 6 Calculate the average of solution points A_1^n to A_m^n obtained in nth iteration.

4 Illustration of the Bisector Art Procedure

As an simple illustrative example, consider x1, x2 to be unknown pixel densities and H1: x + y − 2 = 0, H2: x − 2y + 2 = 0 and H3: 3x − y − 3 = 0 are three beam equations obtained from a scan. Note that in order to be able to see the results of proposed ART, we have written the procedure as a simple Matlab program given in Appendix. When, we use the conventional ART, we obtain an orthogonally projected point $A_3^{45} = (1.40909 \ 1.22727)$ in 45th iteration. It can be seen from the beam equations that although this point is on H3, it is far from H1 and H2. It may be possible to find a more appropriate approximate solution point. In order to do this, we have determined the three inner bisector hyperplanes as given in Table 1 by using (3).

Table 1 Beam equations and inner bisector hyperplanes

Beam equations	Obtained bisector hyperplanes
H_1: x + y − 2 = 0	L_1: 0.2598x + 1.6015y − 2.3086 = 0
H_2: x − 2y + 2 = 0	L_2: 0.5014x + 0.5781y − 1.8431 = 0
H_3: 3x − y − 3 = 0	L_3: 1.6550x + 0.3308y − 2.3628 = 0

Then, we use the ART procedure with the inner bisector hyperplanes, we obtain an orthogonally projected point $A_3^{45} = (1.06903 \ 1.79430)$ in 45th iteration. The average of the points from A_1^{45} to A_3^{45}, obtained in the final iteration is; (1.18708 1.65253). It can be seen from the beam equations that although this point is not exactly on H_3, it is closer to H_1 and H_2 then the one found by the ART with original hyperplanes. Therefore, this point may be considered as a fair or more accurate solution for obtaining the pixel densities of the beam equations.

Table 2 Example beam equations (hyperplanes) in conventional ART and obtained inner bisector hyperplanes in proposed ART

Beam equations	Obtained bisector hyperplanes
H_1: $x_7 + x_8 + x_9 = 13.00$	$L_1 \ldots (H_1, H_2) \Longrightarrow -x_4 - x_5 - x_6 + x_7 + x_8 + x_9 + 2 = 0$
H_2: $x_4 + x_5 + x_6 = 15.00$	$L_2 \ldots (H_2, H_3) \Longrightarrow -x_1 - x_2 - x_3 + x_4 + x_5 + x_6 - 7 = 0$
H_3: $x_1 + x_2 + x_3 = 8.00$	$L_3 \ldots (H_3, H_4) \Longrightarrow x_1 + x_2 - x_3 - x_6 - x_8 - x_9 + 7.79 = 0$
H_4: $x_6 + x_8 + x_9 = 14.79$	$L_4 \ldots (H_4, H_5) \Longrightarrow -x_3 - x_5 + x_6 - x_7 + x_8 + x_9 - 0.48 = 0$
H_5: $x_3 + x_5 + x_7 = 14.31$	$L_5 \ldots (H_5, H_6) \Longrightarrow -x_1 - x_2 + x_3 - x_4 + x_5 + x_7 - 10.40 = 0$
H_6: $x_1 + x_2 + x_4 = 3.81$	$L_6 \ldots (H_6, H_7) \Longrightarrow x_1 + x_2 - x_3 + x_4 - x_6 - x_9 + 14.19 = 0$
H_7: $x_3 + x_6 + x_9 = 18.00$	$L_7 \ldots (H_7, H_8) \Longrightarrow -x_2 + x_3 - x_5 + x_6 - x_8 + x_9 - 6 = 0$
H_8: $x_2 + x_5 + x_8 = 12.00$	$L_8 \ldots (H_8, H_9) \Longrightarrow -x_1 + x_2 - x_4 + x_5 - x_7 + x_8 - 6 = 0$
H_9: $x_1 + x_4 + x_7 = 6.00$	$L_9 \ldots (H_9, H_{10}) \Longrightarrow x_1 - x_2 - x_3 + x_4 - x_6 + x_7 + 4.51 = 0$
H_{10}: $x_2 + x_3 + x_6 = 10.51$	$L_{10} \ldots (H_{10}, H_{11}) \Longrightarrow$ $-x_1 + x_2 + x_3 - x_5 + x_6 - x_9 + 5.62 = 0$
H_{11}: $x_1 + x_5 + x_9 = 16.13$	$L_{11} \ldots (H_{11}, H_{12}) \Longrightarrow$ $x_1 - x_4 + x_5 - x_7 - x_8 + x_9 - 9.09 = 0$
H_{12}: $x_4 + x_7 + x_8 = 7.04$	$L_{12} \ldots (H_{12}, H_1) \Longrightarrow x_4 - x_9 + 5.06 = 0$

As a more elaborate example, let x_1, x_2, x_3, x_4, x_5, x_6, x_7, x_8, x_9 are unknown pixel densities, which are established by a X-ray beam on a row with 9 pixel of the image matrix. Suppose that from the ith X-ray beam (i = 1, 2, 3, 4, 5, 6, 7, 8, 9, 10, 11, 12) 12 beam equations are obtained by center of pixel method. These equations are presented in first column of Table 2. Matlab program in appendix implements the underlying procedure of the ARTs [1]. We call these shared procedural steps as the conventional ART. Even the recent improvements share these basic procedure steps [6].

It is known that large number of iteration gives the best approximate solution point in ART. An orthogonally projected point A_{12}^{45} is obtained in 45th iteration on to the hyperplane H_{12}. The coordinates $(x_1, x_2, x_3, x_4, x_5, x_6, x_7, x_8, x_9)$ of the projection point A_{12} found in [1] as $A_{12}^{45} = (1.32, 0.60, 5.32, 2.15, 7.49, 4.59; 1.76, 3.14, 7.32)$ where origin O (0, 0, 0, 0, 0, 0, 0, 0, 0) is chosen as the initial point. If we choose initial point A (1, 2, 3, 4, 5, 6, 7, 8, 9), our implementation of the conventional ART as a Matlab program gives; $A_{12}^{45} = (1.3191, 0.6008, 5.3169, 2.1444, 7.4900, 4.5922, 1.7597, 3.1958, 7.3209)$. The two points are almost the same. Therefore, we verified that conventional ART procedure is correctly defined converges and the program implements it correctly.

For the improved ART procedure, we defined the following procedure. This extension improves the shared steps of the ART procedures and may be used as the new bases. This procedure is;

Procedure BisectorART [initial point (A), planes (N_i), (d_i)]; where i = number of planes

1. For $i \leq k$ where k is number of iteration
   2.2. For every plane (i);

      2.2.1. If (iteration number = 1)
           2.2.1.1. $N_{i-1} = N_{i-1}/\text{norm}(N_{i-1}) + N_i/\text{norm}(N_i)$
           2.2.1.2. $d_{i-1} = d_{i-1}/\text{norm}(d_{i-1}) + d_i/\text{norm}(d_i)$
         2.2.2. $P_i = A^T - ((N_i * A + d_i)/(N_i * N_i^T)) * N_i$
         2.2.3. If (iteration number = k) sum_of_P = P + sum_of_P
   2.3. Average = sum_of_P/i.

As explained in Sect. 2, the approximate solution to pixel densities obtained by conventional ART is on to the hyperplane H_{12} but may be relatively far from the other hyperplanes H_1, H_2, ..., H_{11}. It is possible to calculate a more appropriate approximate solution point. To do this, we obtain inner bisector hyperplanes by using (3) as shown in second column of Table 1. For example, for L_{12} N = (0, 0, 0, 1, 0, 0, 0, 0, −1) and d = 5.06 as shown in Table 2. By using the extended Matlab implementation based on the ART procedure defined in Sect. 2, a new approximate solution point is found. This point is A_{12}^{45} = (0.61765, 1.84976, 4.21617, 3.05894, 7.64213, 5.70816, 1.13118, 3.99459, 6.72950). Furthermore, the average of the points from A_1^{45} to A_{12}^{45} obtained in the final iterations is; (0.58927, 1.83453, 4.19735, 3.01395, 7.60885, 5.67830, 1.08734, 3.94587, 6.68656). This point is within the region PQRS of Fig. 2 and is a fair and more appropriate solution.

5 Application of Bisector Art Method and Results

In order to evaluate the developed bisector ART, we have implemented a test application in Matlab available at this link https://drive.google.com/open?id=0Bx08mHLlL_J9OEVYN1RsUlZEZXc [19].

For a good illustration of the effect, we have used the old benchmark 64 × 64 simple Shepp-Logan Phantom image as this image is widely used many previously developed methods. We compared the results of our procedure with three previous procedures specifically based on ART approach namely Kaczmarz, SIRT, SART. As generally ART's real advantage is the need of little number of projections hence less radiation, we set the projection angle 5 degrees hence simulated and ran the algorithms with only 36 projection data. The Fig. 3 shows the obtained results.

In Fig. 3, image (6) is the original Shepp-Logan Phantom image and the first image

(1) is the result of the pure (Karczmarz)ART method. The forth and the fifth images represent the derived results of the SART and the SIRT Landweber algorithms respectively. The Karczmarz ART method and has comparable quality in comparison to bisector Art image (2) and bisector ART with averaging (3).

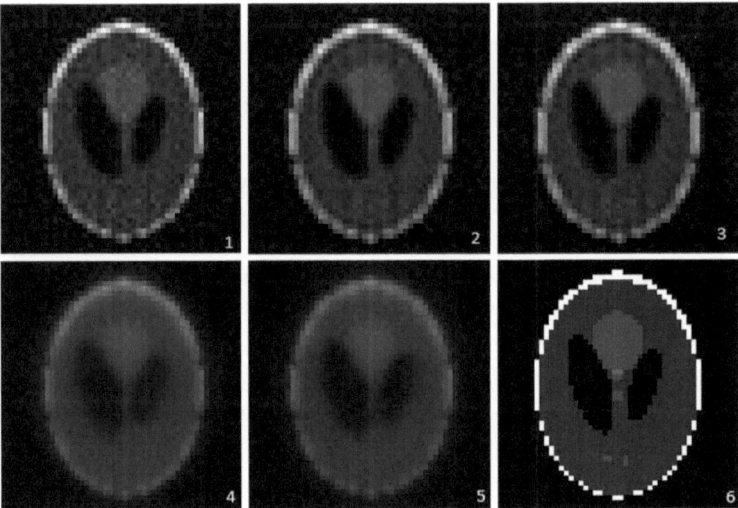

Fig. 3 Image (1) is the constructed image using Kaczmarz ART. Image (2) is the constructed image using Bisector ART. Image (3) is the constructed image using Bisector ART with average result. Image (4) is the constructed image using SART. Image (5) is the constructed image using SIRT Landweber. Image (6) is the original Shepp-Logan Phantom image utilized in reconstruction

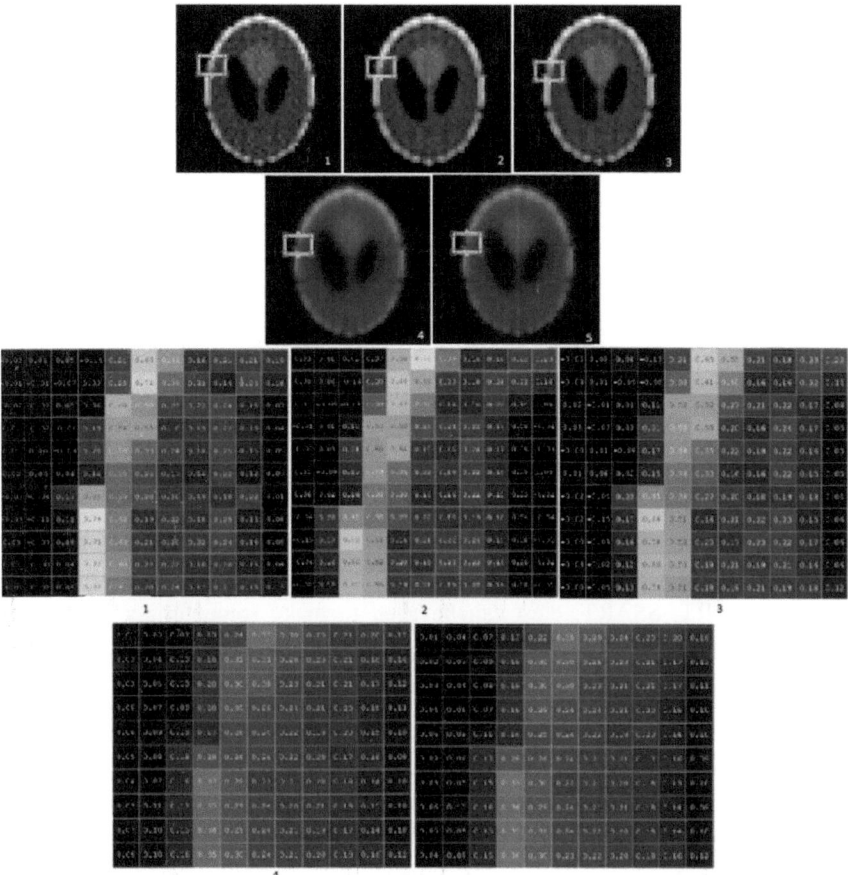

Fig. 4 Taken a 11 × 11 pixel snapshot to demonstrate the effect of pixilation and broken edges. From the top left to right, the pixels were taken ART, Bisector ART, Bisector ART with averaging, SIRT Landweber, and at the right bottom corner SART

However, in (1) the edges are sometimes broken due to pixilation and the details in between the edges were also suffering from the pixelation problem. As it can be observed on (2), (3) images represent slightly better shades of gray than (1) and lowers considerably the pixilation in non-edge regions i.e. considering surfaces in the middle. The (4) and the (5) images represent the derived results of the SART and the SIRT Landweber algorithms respectively. As it's can be observed in Fig. 4, the results of both algorithms were the worse as the edges were slightly blurry and the details between the edges were also suffering from the same problem. As for the round mean square error (RMSE), the value of the SART algorithm was 0.1739 and that of SIRT Landweber was 0.1744. Regarding the round mean square error, Karczmarz ART, had the highest round mean square error as 0.1913 somehow close to bisector ART but it is noticed that proposed bisector ART algorithms had a slightly less round mean square value. Specifically, RMSE of (3) was 0.1861 while the RMSE of (2) was 0.1865. However, the real difference is less pixilation and slightly more edge smoothness.

Regarding time consumption, similar to the methods discussed in the literature, bisector extended method also suffers time consumption as ART algorithms' time complexity is $O(n^2)$. The image reconstruction of 64×64 simple Shepp-Logan Phantom with the proposed bisector ART with averaging method converged in 207 ms. The result was obtained through the utilization of a Toshiba Qosmio X70-A series bearing i7 core and 16 GBs of RAM computer.

6   Conclusion

This study shows that, if the image matrix of the cross-section is determined with interior bisector hyperplanes and the average of the solutions are used; the new image of cross-section constructed may be more accurate than the one obtained with conventional ARTs. The higher proximity of the new solution points to hyperplanes and averaging these in the final iteration may improve the accuracy of pixel intensities and decrease the pixilation of the constructed image.

As for some limitations, other ART methods the methods/techniques discussed above (i.e. MART, SSART, etc.) can be used to make some further experiments. However, using bisectors based procedure improves the shared steps of the ARTs in general and may be used in conjunction with them.

Finally, as mentioned in the introduction, many studies related to CT image quality focused on filtering methods, helped to improve image quality. However, with less projection and radiation ART based procedures are more advantageous for human body images. Therefore, this study aimed at better image reconstruction as accuracy and fidelity of images are still outstanding issues.

Disclaimer: The authors declare that there is no conflict of interest regarding the publication of this paper.

Appendix

```
function PROPART()
clear;
fprintf('------------------------------------\n');
fprintf('Please Enter The Number of ');
NumberOfEquations = input('Beam Equations: ');
fprintf('Please Enter The Number of ');
column = input('Variables in Beam Equation: ');
fprintf('Please Enter The Initial Point \n(!You should enter %d values in [1x%
d] vector format!) ',column,column);
Point = input('\nPoint Values: ');
fprintf('------------------------------------\n');
for i=1:NumberOfEquations
```

```
    fprintf('Please Enter the ''%d''. beam equation coefficient ',i);
    A = input('Values: ');
    B{i} = A;
    fprintf('Please Enter the ''%d''. beam equation constant ',i);
    C = input('Values: ');
    D{i} = C;
    Dinitial{i} = C;
    Binitial{i} = A;
end
Dfirst = D{1};
Bfirst = B{1};
Boo = B{NumberOfEquations};
n=45;
P = Point;
sumpoint = 0;
initialpoint = P - ((( dot(P,B{1}) + D{1} ) / dot(B{1},B{1})) * B{1});
%first point on the first hyperplane
P = initialpoint;
previouspoint = P;
% Approximation calculation loop
for count = 1:n
    for j=0:NumberOfEquations
        % Calculation step for bisector hyperplanes
        if (count==1) & (j >0) &(j<NumberOfEquations)
        Bo = B{j};
        B{j} = (B{j}/(norm(B{j}))) + (B{j+1}/(norm(B{j+1})));
        D{i} = (D{j}/(norm(Bo)))  + (D{j+1}/(norm(B{j+1})));
        end;
        if (count==1) & (j == NumberOfEquations)
        B{NumberOfEquations} = ((Bfirst/norm(Bfirst)) + B{NumberOfEquations}/
norm(B{NumberOfEquations})));
        D{NumberOfEquations} = ((Dfirst/norm(Bfirst)) + (D{NumberOfEquations}/
norm(Boo)));
        end;
        % End of calculation step for bisector hyperplanes
        if j>0
        P = P - ((( dot(P,B{j}) + D{j} ) / dot(B{j},B{j})) * B{j});
        end
        PSize = size(P,2);
        if (count == n) & (j>0)
        sumpoint = sumpoint + P;
        fprintf('\n Solution point %d after %d iterations: [',j,n);
        for k=1:PSize
        fprintf('%.5f ',P(k));
        end
    end;
```

```
    end
  end
end
% End of Approximation loop
avgpoint = sumpoint / NumberOfEquations;
PSize = size(avgpoint,2);
fprintf('\n-------------------------------------\n');
fprintf('The avarage of solution the points:  [ ');
for k=1:PSize
   fprintf('%.5f ',avgpoint(k));
end
   fprintf(' ]\n———————————————————————\n');
```

References

1. Anton, H., Rorres, C.: Elementary Linear Algebra Applications, pp. 685–697. Wiley, New York (1994)
2. Natterer, F., Ritman, E.L.: Past and future directions in X-ray computed tomography. Int. J. Imaging Syst. Technol. 75–187 (2002). https://doi.org/10.1002/ima.10021
3. Kak, A.C., Slaney, M.: Principles of Computerized Tomographic Imaging, pp. 276–294. SIAM, New York, (2001)
4. Sakas, G.: Trends in medical imaging: from 2D to 3D. Comput. Graph. 577–587 (2002)
5. Jiri, J.: Medical Image Processing, Reconstruction an Restoration, pp. 374–400. Taylor and Francis, New York (2006)
6. Herman, G.T.: Algebraic reconstruction techniques. In: Fundamentals of Computerized Tomography, pp. 193–216. Springer (2009). https://doi.org/10.1007/978-1-84628-723-7_11
7. Kesidis, A.L., Papamarkos, N.: Exact image reconstruction from a limited number of projections. J. Vis. Commun. Image Represent. 285–298 (2008)
8. Dai, X.B., Shua, H.Z., Luoa, L.M., Hanb, G.N., Coatrieux, J.L.: Reconstruction of tomographic images from limited range projections using discrete Radon transform and Tchebichef moments. Pattern Recogn. 1152–1164 (2010)
9. Cierniak, R.: A new approach to image reconstruction from projections using a recurrent neural network. Int. J. Appl. Math. Comput. Sci. 147–157 (2008). https://doi.org/10.2478/v10006-008-0014-y
10. Bharkhada, D., Yu, H., Liu, H., Plemmons, R., Wang, G.: Line-source based X-ray tomography. Int. J. Biomed. Imaging (2009). https://doi.org/10.1155/2009/534516
11. Cengiz, K., Kamasak, M.: Comparison of algebraic reconstruction techniques for tomosynthesis. In: IWSSIP 2014 Proceedings, Dubrovnik, pp. 15–18 (2014)
12. Li, Z., Song, Y.: Improving algebraic reconstruction techniques with nonlinear iterating algorithms. In: 2009 Fifth International Conference on Natural Computation, Tianjin, pp. 387–391 (2009)
13. Chetih, N., Messali, Z.: Tomographic image reconstruction using filtered back projection (FBP) and algebraic reconstruction technique (ART). In: 2015 3rd International Conference on Control, Engineering & Information Technology (CEIT), Tlemcen, pp. 1–6 (2015)

14. Wan, X., Zhang, F., Liu, Z.: Modified simultaneous algebraic reconstruction technique and its parallelization in cryo-electron tomography. In: 2009 15th International Conference on Parallel and Distributed Systems (ICPADS), Shenzhen, pp. 384–390 (2009)
15. Kim, J.H., Kang, B.C., Choi, B.Y., Lee, S.H., Kim, K.Y.: On-line directional algebraic reconstruction technique for electrical capacitance tomography. In: 2006 5th IEEE Conference on Sensors, Daegu, pp. 923–926 (2006)
16. Jiang, M., Wang, G.: Convergence of the simultaneous algebraic reconstruction technique (SART). IEEE Trans. Image Process. **12**(8), 957–961 (2003)
17. Saha, S., Tahtali, M., Lambert, A., Pickering, M.: Compressed sensing inspired rapid algebraic reconstruction technique for computed tomography. In: IEEE International Symposium on Signal Processing and Information Technology, Athens, pp. 000398–000403 (2013)
18. Oliveira, N., Mota, A.M., Matela, N., Janeiro, L., Almeida, P.: Dynamic relaxation in algebraic reconstruction technique (ART) for breast tomosynthesis imaging. Comput. Methods Programs Biomed. **132**, 189–196 (2016)
19. Implementation files for experimentation of proposed Bisector ART. https://drive.google.com/open?id=0Bx08mHLIL_J9OEVYN1RsUlZEZXc (2017)

Use of Predictive Analytics Towards Better Management of Parking Lot Using Image Processing

K. A. Maheshwari$^{(\boxtimes)}$ and P. Bagavathi Sivakumar

Department of Computer Science and Engineering, Amrita School of
Engineering, Amrita Vishwa Vidyapeetham, Coimbatore, India
cb.en.p2csel5008@cb.students.amrita.edu, pbsk@cb.
amrita.edu

Abstract. As more and more smart cities are planned in India, there is a growing need for smart parking and smart transportation. Parking has been identified as a major challenge to traffic network and urban life quality. Already most of the cities are facing the problem of pollution. Due to drivers struggling for finding the parking area, 30% of traffic congestion occurs according to industry data. There is also a need for secure, efficient, intelligent and reliable systems that can be used for searching the unoccupied parking facilities, guide towards the parking facilities, and negotiate the parking fee. This would help in the proper management of the parking facility. There is no publically available data on parking in India. This work would be useful in creation of such datasets. Image based model has been proposed to identify the slot occupancy status. A prediction model has also been incorporated in the system to predict the occupancy rate and thereby help the management in better management of parking lots. One of the machine learning method, linear regression is used for predicting the number of car parked every hour. A slot based approach was used and the performances of prediction algorithms were compared.

Keywords: Smart parking · Occupancy rate · Prediction model · Payment rate
Image processing

1 Introduction

At present 30% of the Indian population are living in cities and expecting more by 2030. This is equal to US total population. To manage the services and infrastructures of cities more and more smart cities are required. Smart city integrates multiple information and communication technology, mainly concentrate on urban development [1]. The main goal of building a smart city using technology is to improve the quality of life and to improve the efficiency of services. The smart city is now emerging with the internet of things [2], which include network of interconnected objects with the

© Springer International Publishing AG 2018
D. J. Hemanth and S. Smys (eds.), *Computational Vision and Bio Inspired Computing*,
Lecture Notes in Computational Vision and Biomechanics 28,
https://doi.org/10.1007/978-3-319-71767-8_67

current internet into it. There are many automated sensor networks, parking meters, data Centers, energy measuring devices, and actuators with the technological platform involved. Machine learning is a data analysis method that automates building model for analytics. The process is similar to data mining. Machine learning algorithm find the hidden insights without being explicitly programmed. When expose to dataset, it focuses on the development of computer programs.

Smart city transportation is an important pillar for quality of life of citizens in a city. In most of the cities, public and private road transportation is the key mode of commuting and logistics. Some of the technologies relevant for smart cities includes geospatial—enabled efficient transportation system, dynamic carpooling or car sharing, GPS-based tracking [3] and route information of public transport, smart parking, smart toll etc. Image processing plays a major role in identifying the empty slots in the parking area [4, 5]. From the captured image the red circle extracted to count the empty slot.

One of the major issues in cities is car parking and traffic. Searching for the availability of the car parking slot is always troublesome for the drivers [6]. 30% of the world traffic is because of searching the unoccupied parking slot. If the driver is informed prior about the available parking slot, the traffic congestion can be efficiently controlled. This requires a sensor to monitor every slot to calculate the occupancy rate and thereby inform the users accordingly.

1.1 Need and Benefits of Smart Parking in India

1. Parking space usage is optimized.
2. Guides the user to park their vehicle in available parking slot and find the occupancy in real time [3].
3. Easy for the merchants and drivers to park their vehicles.
4. Help the city to be free from traffic.
5. Reducing the emission of CO_2 and pollutants.
6. The increase in the revenue by creating smart parking and real-time management.
7. Tools are provided to optimize the workspace.

In this work, the role of predictive analytics has been explored with the real-time data acquired from a parking lot. The rest of the paper is organized as follows: Sect. 2 presents related work. Section 3 describes the implementation of slot allocation and prediction. Section 4 describes the analysis and results of real-time data. Conclusion is given in Sect. 5.

2 Related Work

In manning and automating the parking lot, occupancy rate and payment rate are the two main crucial parameters.

2.1   Occupancy Rate

The occupancy rate [1] of the parking area can be calculated as the share of total spaces available which are also occupied at any snapshot in time. This system measures the occupancy rate of vehicles using parking sensor and continuous data and it is used to calculate the occupancy rate in units of seconds [7]. The parking information was filtered using database to calculate the general metered parking (GMP) occupancy rate. Metered time is only considered as the general metered parking to general public which is available for parking the cars.

Occupancy rate is calculated as [8],

$$= \frac{(\text{Total occupied time})}{(\text{Total occupied time}) + (\text{Total vacant time})} \tag{1}$$

2.2   Payment Rate

The meter payment rate of the parking area can be calculated as the share of total spaces available that are paid at any snapshot in time [1]. This system calculates meter payment rate from smart meters using real-time payment information in units of seconds. From the database of parking meter configuration, information is filtered and to calculate the general metered parking (GMP) meter payment rate. Same as the occupancy rate, payment rate also considers metered time that was available for parking to the general public [9].

Payment rate is calculated as [10],

$$= \frac{(\text{Total paid time})}{(\text{Total paid time}) + (\text{Total unpaid time})} \tag{2}$$

2.3   Communication in Parking Lot

Some of the exiting parking technologies use Lifting the balloon [11] when the parking space is empty, using sensors or using LED lights [12]. In sensor based technology [13], each car parking slot has one Arduino attached with sensor and ZigBee [14]. A gateway is present in each column of the parking slot. It has one Arduino attached to ZigBee and Wi-Fi. Through Wi-Fi, occupancy data is sent to the cloud. From the cloud, the data can be disturbed to the respective parking areas.

2.4   Image Based Slot Occupancy Detection

Image based model is used to process the car parking area image to identify the whether the parking slot is occupied or not. Initial the empty parking slot image is captured with the red color dot slot number indication as show in the figure. The images capture is converted into HSV image. To calculate the threshold HSV image [15] is converted into grey scale image. Conversion of grey scale image from HSV image make easy for the comparison of each pixel from the capture image.

$$\text{GRAY} = (0.299 * r + 0.587 * g + 0.114 * b) \tag{3}$$

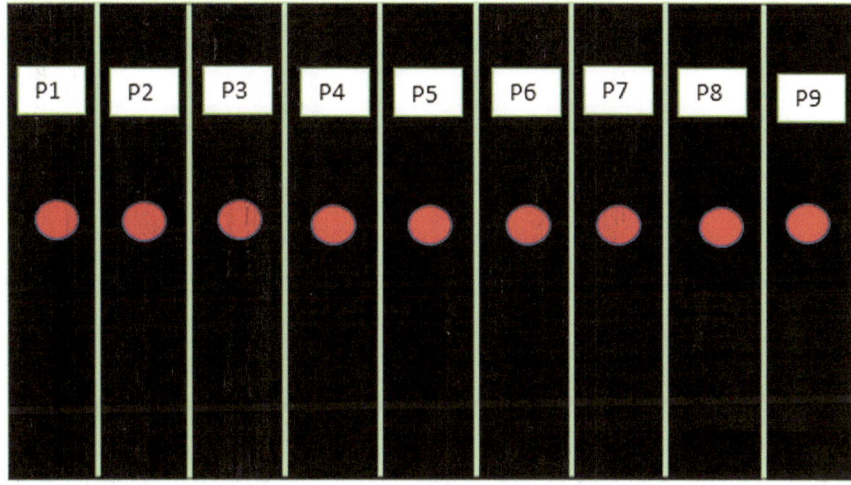

Fig. 1 Model of captured car parking image

Equation 3 is used for the conversion from RBB value image to gray scale value [16]. Signal processing is used in [17]. The Fig. 1 show the model of parking area captured image. The green color line indicates the slot separation in the parking area. Red dot denoted the empty parking area. When the image is captured for every 30 s, it will count the red dot in the capture image. From the counted red dots the occupancy status can be easily calculated. Slot number is indicated in each slot from the image the empty slot is separated and displayed in the board for the occupancy status in every parking area. Figure 2 shows the working process to find occupancy status.

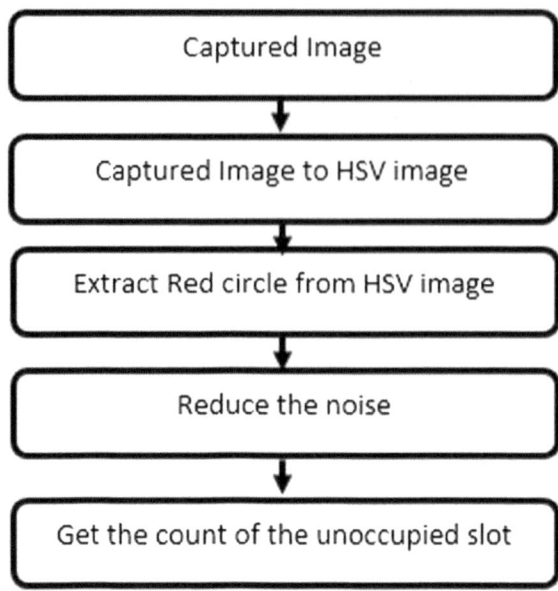

Fig. 2 Process step

3   Implementation

In this work, data has been collected from a parking lot. Slot allocation and prediction has been made. The slot allocation system includes modules for entry, exit, and payment. Occupancy prediction has been done on an hourly, daily, weekly and monthly basis. During the processes of analysis, it has been found that the pattern of parking varies a lot during weekdays and weekends. Outlier analysis is also carried out. Modules are described below.

3.1   Slot Allocation

In this working system user interface is created for monitoring the slots. When the car enters the parking lot, the entry ticket will be given to each customer. The entry ticket contains car number, entry time, basement number and the slot number allotted by the system. The existing system does not contain slot allocation mechanism at the entry point. Customer needs to search for the parking slot. Thus allocating a slot at the entry point using proposed slot allocation system avoids traffic inside the parking lot and saves considerable time.

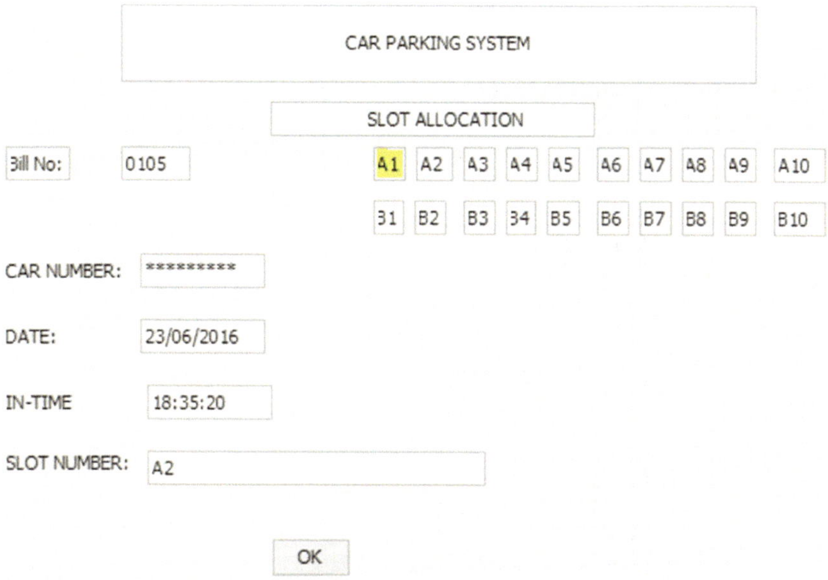

Fig. 3 Entry system

Figure 3 shows the interface of the entry system. A1 is yellow in color and this indicates that A1 slot is occupied and other slots are free. This information is stored in the cloud database and the bill status is shown in Fig. 4.

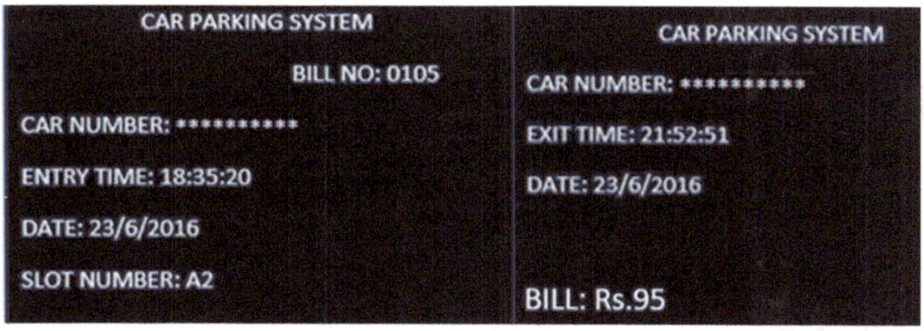

Fig. 4 Entry bill and exit bill

When the car reaches the exit point the sensor reads the car number and calculates the occupancy rate based on the exit time and number of occupied cars. Occupancy rate is calculated using Eq. 1. Payment rate has been calculated as per [1]. Customer charges are more when the occupancy rate is high and. Table 1 shows the payment rate corresponding to the occupancy rate. Payment rate depends on the occupancy rate.

Table 1 Payment rate

Occupancy rate (%)	Payment rate
80–100	+Rs. 10
50–80	Rs. 0
0–50	−Rs. 10

3.1.1 Parking Occupancy Status

Table 2 Parking occupancy status

Basement	No. of slots occupied	No. of slots empty	Occupancy rate (%)
B1	68	132	34
B2	120	80	60
B3	98	102	49

Table 2 shows the parking occupancy status. Parking management can check the occupancy status at any time for the better management of the lot. Here each basement has 200 car parking slots. Occupancy rate is calculated for each basement and number of free slot is computed. This information can be displayed on the LED display board if required.

3.2 Prediction

As discussed earlier, Car parking dataset has been collected from parking lot at Coimbatore. Data is stored in Microsoft SQL server. From the database number of vehicle parked for every hour duration (e.g. 1–2 pm) and "x" hours (x = 1, 2, 3, 12) parked vehicle have been calculated. Graphical output is provided to visualize the data. Weekdays and weekend analysis has been performed.

Based on the analysis, prediction has been done to find the number of vehicle parked for upcoming normal and special days, where a normal day corresponds to weekdays and special days correspond to holidays. Thus the prediction will help in effective management of slot, whenever the slot capacity is filled.

4 Result and Analysis

The result and analysis is described below. This includes hour-wise analysis, day-wise analysis, weekday and weekend analysis, number of vehicles parked for a particular duration, outlier analysis, etc.

4.1 Vehicles Parked for Different Duration

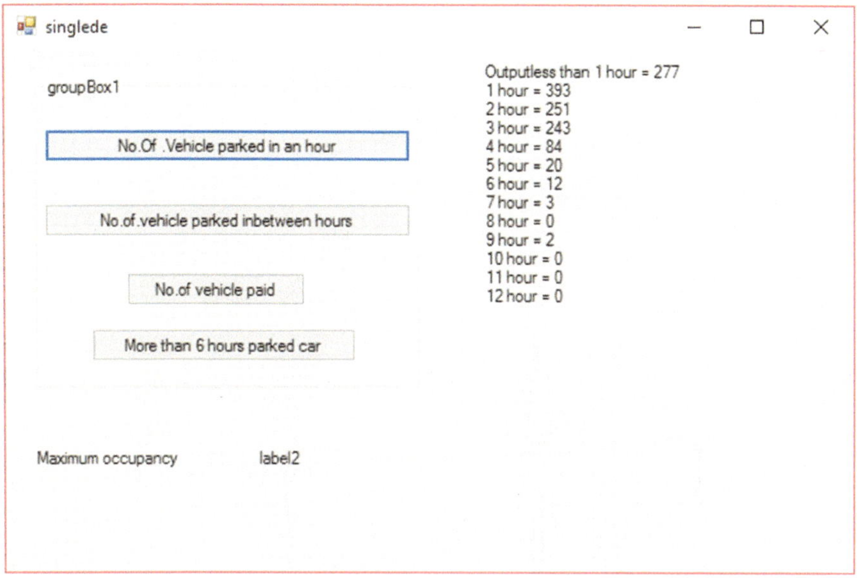

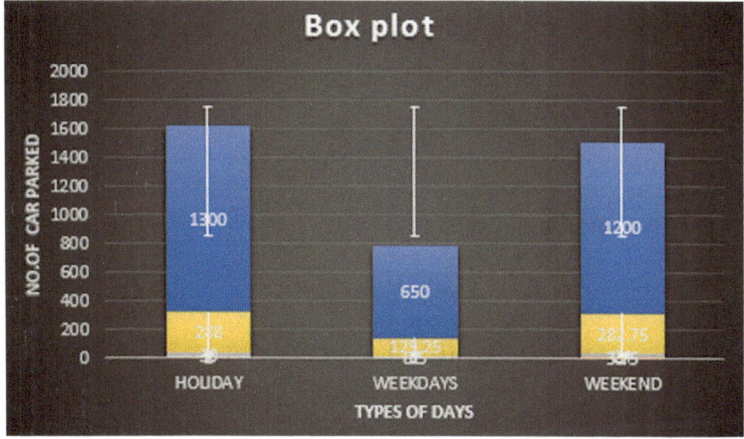

Fig. 5 Vehicles parked for different duration

Figure 5 shows how many vehicles have been parked for duration of 1, 2 h, etc., as an example the statistics on number of vehicle parked on 6-6-2016 is shown in the Fig. 5. The system can be used to retrieve the information corresponding to any particular day.

Fig. 6 Box plot

Figure 6 shows the box plot for holidays, weekends and weekdays. As an example, the statistics on number of vehicle parked on all weekdays in June, all weekends in June and special days in the dataset (April to August) is shown in the Fig. 6. For weekend and holidays the median range is almost same but differs for weekdays. From this graph, the analysis shows that the median value for holiday and weekend are in the range of 288. The range of weekdays is 129, that which is half of the holidays and weekends. The occupancy rate of the parking area is almost same during holidays and weekends.

4.2   Hour-Wise Parking Status

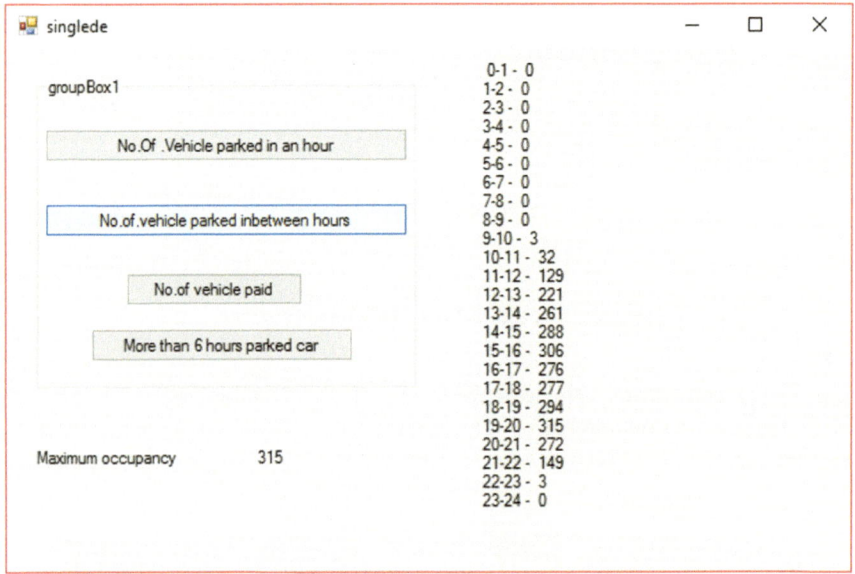

Fig. 7 Hour-wise parking status

The Fig. 7 shows the number of vehicle parked every hour. As an example the statistics on number of vehicle parked on 6-6-2016 is shown in the Fig. 7. The system can also be used to retrieve the value for any or all days.

4.3 Weekdays and Weekends Analysis

4.3.1 Hour

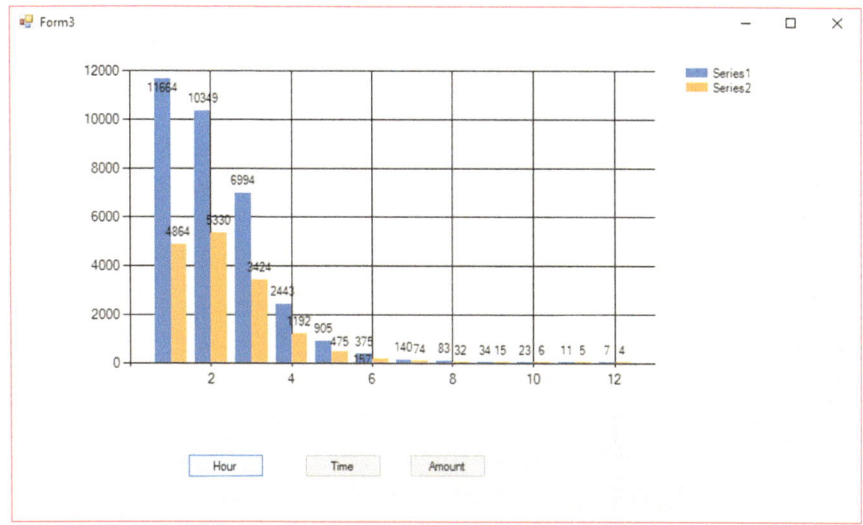

Fig. 8 Comparison of weekdays and weekends based on duration of parking

The Fig. 8 shows the graphical representation of weekdays and weekends for every hour duration (1, 2 h, etc.). Series-1 indicates from Monday to Friday. Series 2 indicates Sunday and Saturday. Figure 8 shows the comparative analysis of weekdays and weekends for a whole month from 6-6-2016 to 7-7-2016.

4.3.2 Time

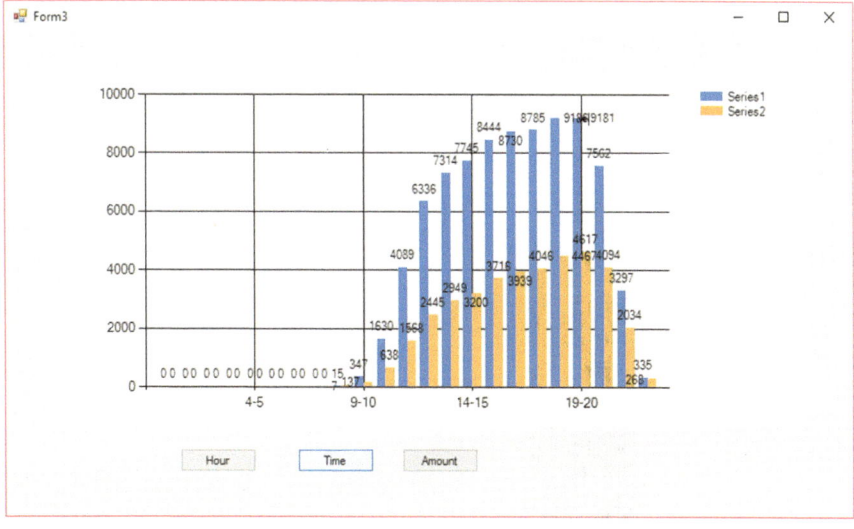

Fig. 9 Comparison of weekdays and weekends based on time

This Fig. 9 shows the graphical representation of weekdays and weekends for duration of time (1–2 am, 2–3 am, etc.). Series-1 indicates from Monday to Friday. Series 2 indicates Sunday and Saturday. Figure 9 shows the comparative analysis of weekdays and weekends for a whole month from 6-6-2016 to 7-7-2016.

4.4   Outlier Analysis

Sometimes management is concerned to find the vehicle which has been parked abnormally. So this analysis helps mainly to tackle threats. As part of the study, all the vehicle parked more than 6 h can be found and an alert message is given to the parking management.

4.5   Prediction

Prediction is carried out using linear Regression. For example the prediction result on 09-080-2016 (normal day), 14-08-2016(weekends), 15-06-2016(special day which falls on weekdays) is shown in the Table 3.

Table 3 Prediction result on number of car parked

Time	Normal (9-8-2016)		Weekend (14-6-2016)		Special (15-6-2016)	
	O	P	O	P	O	P
0–7	0	0	0	0	0	0
7–8	0	0	1	0	2	6
8–9	0	0	2	2	7	12
9–10	7	16	77	71	63	68
10–11	33	46	201	217	180	178
11–12	135	141	404	399	417	422
12–13	242	243	533	528	616	620
13–14	275	266	560	572	696	691
14–15	299	296	588	580	720	724
15–16	333	318	670	682	745	741
16–17	304	305	738	730	747	738
17–18	315	318	763	761	777	768
18–19	365	354	732	735	708	711
19–20	337	334	701	720	710	699
20–21	323	313	545	556	577	585
21–22	171	179	249	257	562	565
22–23	8	14	17	21	289	294
23–24	5	8	13	14	11	13

In Table 3 O indicates original number of car parked and P indicates the predicted number of car parked. Using Linear Regression, Mean R^2 value and mean MSE are calculated for the dataset (April to August). Table 4 shows the R^2 value and mean MSE for normal days, weekend, and special days. The analysis shows the prediction is more accurate for special days.

Linear regression equation for normal day,

$$Y = 8.37521 + 0.95773 * X \tag{4}$$

For weekend,

$$Y = -3.95417 + 1.00758 * X \tag{5}$$

For special day,

$$Y = 3.80289 + 0.99609 * X \tag{6}$$

Here x represents the average number of car parked in weekdays, weekends and holidays respective to their model.

Table 4 Performance measure of linear regression

	Normal	Weekend	Special
Mean R^2	0.9808	0.9969	0.9941
Mean MSE	0.045	0.027	0.010

4.6 Occupancy Rate Calculation

Table 5 Occupancy rate

Time	No. of vehicle parked (2016-07-01)	Occupancy rate (2016-07-01)
10–11	54	7.94118
11–12	165	24.2647
12–13	264	38.8235

Occupancy rate is calculated using the Eq. (1) mentioned before. The system calculated occupancy rate for the whole dataset is shown in Table 5 and the management can find the occupancy rate of the parking area at any time.

4.7   Weekly Analysis

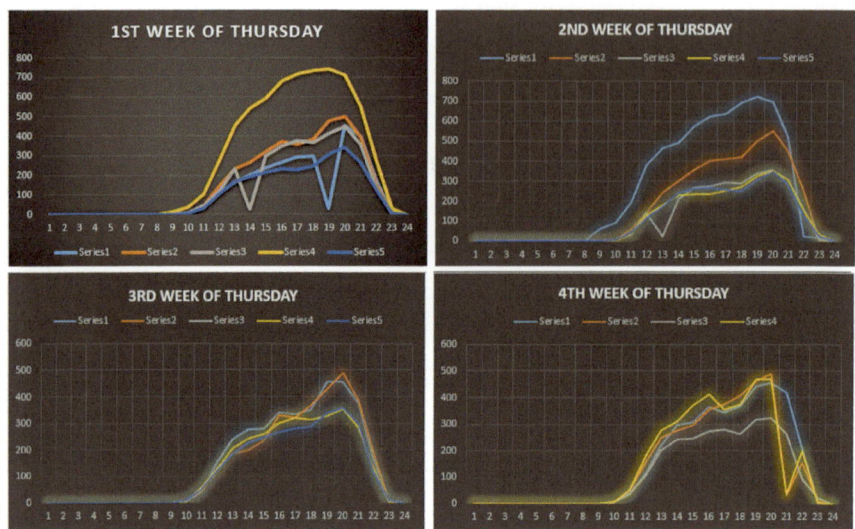

Fig. 10 Weekly analysis

Graphical representation is done for every month and every day. Graphical representation is shown in Fig. 10 for Thursday of 1st week, 2nd week, 3rd week and 4th week.

Series 1 shows the 1st week of Thursday's in all months, Series 2 shows 2nd week of Thursday's in all months and respective series for respective month.

In the 1st week of Thursday series 4 and 2nd week of Thursday series 1 has maximum occupancy rate because both these days are special days (Holidays).

5   Conclusion

The work has demonstrated the use of predictive analytics towards better management of parking slots in the context of smart city initiatives of India. Through this study a model has been proposed to calculate occupancy rate using image processing by counting the red circle in the parking slot and there by a means of a reasonable rate. For this purpose the real-time data acquired from a parking lot was used. Form the analysis, it was observed that there was a considerable differences in the pattern of vehicles parked during weekends, weekdays and holidays. The model takes care of this aspect and separate models were built. The performance was compared in terms of MAD, R^2 value and MSE. Thus this system can be adopted to any parking lot. This would also help in reducing manpower and meet all the mentioned objectives of smart parking.

Acknowledgements. We would like to extend our heartfelt gratitude to the Mobile and Wireless networks Lab, Dept. of CSE, Amrita School of Engineering, Coimbatore and Brookefields, Coimbatore for the support extended in carrying out this work.

References

1. Zheng, Y., Rajasegarar, S., Leckie, C.: Parking availability prediction for sensor-enabled car parks in smart cities. In: 2015 IEEE Tenth International Conference on Intelligent Sensors, Sensor Networks and Information Processing (ISSNIP). IEEE (2015)
2. Vijai, P., Bagavathi Sivakumar, P.: Design of IoT systems and analytics in the context of smart city initiatives in India. Proc. Comput. Sci. **92**, 583–588 (2016)
3. Bogoslavskyi, I., et al.: Where to park? Minimizing the expected time to find a parking space. In: 2015 IEEE International Conference on Robotics and Automation (ICRA). IEEE (2015)
4. Idris, M.Y.I., et al.: Smart parking system using image processing techniques in wireless sensor network environment. Inf. Technol. J. **8**(2), 114–127 (2009)
5. Al-Kharusi, H., Ibrahim Al-Bahadly, I.: Intelligent parking management system based on image processing. World J. Eng. Tech. **2**(2), 55–67 (2014) http://dx.doi.org/10.4236/wjet.2014.22006
6. Charette, R.: Smart parking systems make it easier to find a parking space. http://spectrum.ieee.org/green-tech/advancedcars/smart-parking-systems-make-it-easier-to-find-aparking-space/0 (2007)
7. IoT Deployment: IoT deployment in the City of Melbourne. http://issnip.unimelb.edu.au/research_program/Internet_of_Things/iot_deployment (2015)
8. SFpark: Sfpark. http://sfpark.org (2013)
9. Dias, G.M., Bellalta, B., Oechsner, S.: Predicting occupancy trends in Barcelona's bicycle service stations using open data. In: SAI Intelligent Systems Conference (IntelliSys), 2015. IEEE (2015)
10. SFPark: San Francisco parking sensor locations. http://sfpark.org/howit-works/the-sensors/, 7 (2012)
11. Bonde, D.J., et al.: Automated car parking system commanded by Android application. In: 2014 International Conference on Computer Communication and Informatics (ICCCI). IEEE (2014)
12. Zheng, Y., et al.: Smart car parking: temporal clustering and anomaly detection in urban car parking. In: 2014 IEEE Ninth International Conference on Intelligent Sensors, Sensor Networks and Information Processing (ISSNIP). IEEE (2014)
13. City of Melbourne: Melbourne cbd in-ground sensor implementation map. http://www.melbourne.vic.gov.au/ParkingTransportandRoads/Parking/Pages/InGroundSensors.aspx,   5 (2012)
14. Sulaiman, H.A., et al.: Wireless based smart parking system using zigbee. Int. J. Eng. Technol. 3282–3300 (2013)
15. Wang, L., Bai, J.: Threshold selection by clustering grey levels of boundary. Pattern Recogn. Lett. **24**, 1983–1999 (2003)
16. Liu, J., Mohandes, M., Deriche, M.: A multi-classifier image based vacant parking detection system. In: 2013 IEEE 20th International Conference on Electronics Circuits and Systems (ICECS), pp. 933–936 (2013)
17. Megalingam, R.K., et al.: Smart, public buses information system. In: 2014 International Conference on Communications and Signal Processing (ICCSP). IEEE (2014)

Enforcement of Automatic Penalty (e-Penalty) to Govern the Traffic Rule Violators in Digitized INDIA Using I.C.T.

Shiv Kumar Goel[1(✉)], Kavita[1], and Manoj Shukla[2]

[1] JV Women's University, Jaipur, Rajasthan, India
shiv.banty@gmail.com, drkavita@jvwu.ac.in
[2] Amity University, Noida, Uttar Pradesh, India
mkshukla001@gmail.com

Abstract. Government of INDIA (GOI) has a dream to make every organization and transaction digitized. Government has initiated so many steps to digitize the organizations and forced the citizens to use digitized mode in their day to day life. According to the W.H.O. Road accidents are not a small matter now a days, it is listed in their top health agendas now a days. According to the statistics road accidents are the leading cause of death in 15–29 years age group. In this digitized country there should be some solution which can enforced automatic penalty on the traffic rule violator so that they will stop themselves to do the road safety offences. Government should also go ahead to curb such road offenders by penalize them forcibly from their AADHAR linked accounts automatically. In this paper a solution is proposed which can integrate easily with the existing methods of penalty in INDIA to penalize the violators automatically.

Keywords: E-penalty · E-challan · E-Governance · Intelligent expert system (IES) · ICT · MVA

1 Introduction

Information and Communication technology (ICT) is playing a vital role in this digitized era. This technology with the help of other technologies can provide a solution to curb the traffic rule violators on road due to which so many unwanted deaths have been taken place. So many prominent people have lost their life in road accidents who can serve a lot to nation growth like our Transport minister Late Shree Gopi Nath Munde. There are so many accidents take place in metro cities like Mumbai, Delhi, Kolkata and many other places everyday in which sometimes survivals are minors only. Traffic police statistics shows that the 60% of the accidents in metro city like Delhi are because of the driver's fault. Late Shree Munde's death was due the jumping of the signal, this all statistics have given the idea to propose a system which can help in curb road offences caused to innocent human loss. GOI in RTO offices has taken an initiative to make all vehicles and Drivers data digitized with the help of VAHAN and SAARTHI

© Springer International Publishing AG 2018
D. J. Hemanth and S. Smys (eds.), *Computational Vision and Bio Inspired Computing*,
Lecture Notes in Computational Vision and Biomechanics 28,
https://doi.org/10.1007/978-3-319-71767-8_68

respectively. VAHAN store all information about the vehicle like, chasis number, engine number, model, make of the vehicle and the vehicle owner's address and contact number. VAHAN and SAARTHI provide a chip based smart card having all digitized data of vehicle and driver in the form of RC and DL respectively. Government is forcing the citizens to link their AADHAR number everywhere like, link AADHAR in Banks and giving subsidies, pension etc. automatically in the beneficiary account. Recently GOI has asked every citizen to link their mobile phone with AADHAR for some rule enforcement, linking of PAN card etc. Such enforcement can also be imposed to govern the traffic violator by linking the vehicle RC with AADHAR and Bank account so that penalty can be enforced automatically from the vehicle owner's accounts. Now days government is installing cameras in every city to surveillance the traffic and as well as terrorist activities. In some metropolitan cities these cameras are working well to detect the traffic rule violators and they give e-challan to the vehicle owner. This method is also not foolproof as the registered address of the owner does not match with the current address and hence e-challans are also not fulfilling the purpose of imposing the e-penalty to such offenders. This method of e-challan is very time consuming also and required much man power and efforts. Proposed system with the help of enforcement of government rules which is a type of G2C Governance can forced the citizen not to do the road offences in future by keeping their offence records through Intelligent Software System and penalize them with more amount next time or may be cancellation of their vehicle registration or DL according to the modified enforced rule.

2   Existing Systems

In India, existing systems to penalize the offenders are not foolproof even the traffic rules are not very stringent like the other countries. There is a need to make Motor Vehicle Act (MVA) more stringent and as well as an expert system is required which can deduct the penalty amount automatically. Following are the rules and methods of penalty on violator of the traffic rules in other countries and INDIA.

2.1   Penalty System in Italy

In Italy, very reliable, inexpensive road-side speed monitoring cameras were installed in the market. These cameras take the photograph of the driver and the car license plate, and record the speed and location of the car. The traffic ticket is issue to the offender and a fine will be imposed. The recipient of the ticket has 60 days to pay or appeal. If you don't pay, the amount is doubled [1]. They provide the photograph of the offence done by the offender. For this, offender has to enter the vehicle registration and the photograph is available for which the offender is punished. Fine range varies from 90 to 150 euro.

2.2   Penalty System in Canada

In Canada, A traffic ticket is a notice issued by a law enforcement official to a driver (motorist) or other road user, accusing violation (offence) of traffic laws. A traffic ticket constitutes a notice that a penalty, such as a fine or deduction of points, or both, has been or will be assessed against the driver or owner of a vehicle. In Canada, traffic laws are made at the provincial level. Some serious violations are considered criminal and are located under the Criminal Code of Canada [2]. Each province maintains a database of drivers, including their convicted traffic offence. Traffic rules violators come under the Ontario demerit point system, where if an offender violates the traffic rules then demerit point will be accumulated till 2 years. If this point exceeds the threshold limit, offender Driving Licence will be suspended and the person can be jailed up-to 6 months.

2.3 Penalty System in UAE

Dubai Police has installed advanced cameras to monitor motorists violating rules by activating a sophisticated technical program. These new devices will be able to monitor and seize offender's vehicles that do not follow lane markings, intersections, will detect hard shoulder drivers and any other road offences [3]. These devices, which look like cylindrical radars, perform different tasks highlighting colour and type of cars and these devices can also extract its data. Sophisticated cameras that feed Al Motabea'a (the supervisor) would monitor the violating vehicles and record the violations automatically in the system without the need of human intervention. The device 'Rasd' can detect any vehicle which has violated the traffic rules and will immediately issue alarm to the command and control center to pursue the car and pin-pointing its location by tracking devices and after that patrolling seize the such vehicles and penalize them.

2.4 Penalty System in USA

In the United States, most traffic laws are structured or arrange in a variety of state, county and municipal laws or ordinances, with most minor violations classified as infractions, civil charges or criminal charges. The monetary fine for a red-light running (RLR) traffic violation varies widely in the U.S., with a fine of $50 in North Carolina and as much as $490 in California. Each state's Motor Vehicles department or Bureau of Motor Vehicles maintains a database of motorists, including their offences related to traffic violations. When a ticket is issue to a motorist, he/she is given the option to mail into the local court in which the violation is alleged within the ten to fifteen days. Additionally, the motorist can request a mitigation hearing, In this matter both the motorist and the officer have to present the case in front of the judge, if violation is not proven then penalty will be dismissed otherwise he has to pay the penalty and the court fees also.

2.5 Penalty System in France

In France, speeding is punishable by law. Punishments sometimes look severe, like prison or licence suspensions. If offenders are arrested by the police for moderate speeding, fine to be paid on the spot and points lost if offender possesses a French driving licence [4]. Offender is arrested with speed more than 50 km/h, fine to be paid, licence confiscated, vehicle might be impounded. In France along with the hefty amount of punishment, citizens may also be liable to immediate licence disqualification and/or seizure of the vehicle.

2.6 Penalty System in UK

In UK, parliament sets the maximum penalties for road traffic crimes. It is for the courts to decide what sentence to impose according to crime. Offenders may lose their point which in turn suspension or disqualification for their driving Licence for ever according to the severeness of the offence. The penalty point system is proposed to alarm drivers and motorcyclists from various unsafe motoring [5]. A driver or motorcyclist who accrues 12 or more penalty points within 3-year period will be disqualified. This will be for a minimum period of 6 months, or longer if the driver or motorcyclist has previously been disqualified.

2.7 Penalty Systems in INDIA

In India there are so many ways to apply the penalty of traffic rule violator but none of the method is foolproof which can curb the offenders strictly.

2.7.1 Traditional Approach of Penalty

This approach is very old approach, in this kind of approach, offender commits the violation of traffic rules, if caught by the police officer, then he may or may not pay the actual penalty of offence and can give some amount of penalty as a bribe to cops and will go away. There is no record of repetitive offences by the same offender at all. Adamant people sometimes refuse to pay fine, which leads to fight. Sometimes, due to political influence, the offender might escape from paying fine. This kind of penalty system has a big hole in revenue and as well as it encourages the offender to commit the traffic violation again and again and also increases the corruption in system. In many small cities and town in INDIA, this approach is in use and could not stop the offenders to commit the violation of traffic rules.

2.7.2 On-Spot e-Penalty System

On spot penalty system is based on, if an offender is caught in violation of traffic rule by the traffic police then they will be given on spot electronic challan and impose a penalty on them. In this system Police officer is equipped with Android tablets and printers to issue spot challans. Police officer will collect the penalty in cash and the offender's vehicle number will be registered in the Central database for the future reference. This system, very first implemented in Bangalore, and was successful also; the rate of offences went down compare to manual penalty system. Now this system is partially implemented in Navi Mumbai, some metropolitan cities of Haryana and many other metropolitan cities in INDIA. In this system, it is difficult to curb all the offenders on road as police cannot be deployed at every place and all the time. Police officer may take bribe and leave the offender without actual penalty. So corruption cannot be completely removed by this system (Fig. 1).

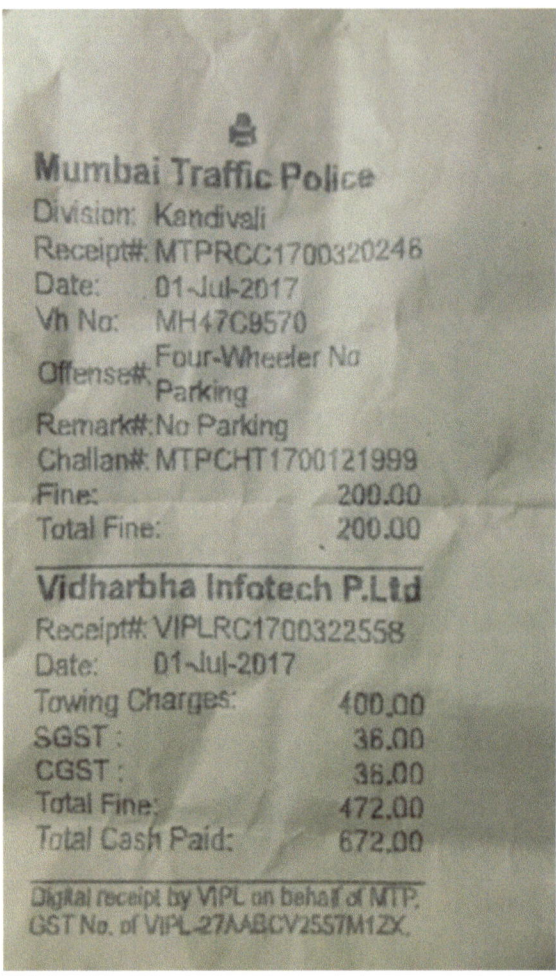

Fig. 1 Sample of on spot challan for offence on road

2.7.3 E-Challan Penalty System

In this system, e-challan will be issued on the basis of the offender's vehicle image captured in CCTV camera and sent to the owner's registered address in the records of VAHAN database. Offender can pay the penalty online or at consumer service centers, list of such centers are available on website. If offender does not pay the penalty within a certain period of time then the matter will go to court and he has to respond to court summons and has to face punishment decided by court. This e-challan penalty system is also implemented in Ahmedabad, Bengaluru and Hyderabad. There is a hitch in this penalty system, if the address is not correct then offender will not get the e-challan ever and hence repeat the offences even though this e-challan system is very well integrated with VAHAN records. Hence this system is required to be enhanced which can assure the penalty on the offender and stop them not to repeat such offences.

Inference of penalty systems in different countries throw a light that there should be the stringent traffic rules in INDIA and as well as the enforcement of government rules on the citizens should be more strict to curb the traffic road violator. There should be a change in Motor Vehicle Act (MVA), specially the amount of penalties should be increased so that no one will have the courage to repeat the offence. Government has made amendments in MVA in this regard. Traffic rule violation should be treated as a crime. Database of all registered vehicles and their owner should be available at every node of traffic police in every city across INDIA, so there should be an efficient mechanism to penalize the offenders automatically with the help of the images captured by the CCTV cameras installed at road side.

3 Proposed System

E-challan system can be further extend to this proposed system by using the image pre processing techniques to extract the Licence Plate number from the Registration plate and integrate this with VAHAN data to fetch all information about the vehicle owner and impose penalty automatically from their account which is linked with the AAD-HAR card and Vehicle registration with help of Intelligent Expert System (IES) and ICT.

There are many Literatures on Licence Plate Recognition by prominent authors. Authors Satadal, Subhadip have used Histogram Equalization method to enhance the contrast of each image and Median Filtering for reducing the noise within the image after Gray scale conversion, Sobel edge detection operator was used to extract the vertical edges created by the license plate characters [6]. L. Angeline, Choong have proposed a novel algorithm to tracking and localizing the moving vehicle number plate on the basis of signature analysis and connected component analysis [7]. Yu M., Kim in their research article, authors have suggested a vertical edge matching based algorithm to recognize a Korean license plate from an input gray-scale image [8]. Author Eric Groft had given a novel idea in their patent which was useful to getting the idea of LPR [9]. The techniques explained by the author Batuwita, Bandara and Timar, in their technical paper Fuzzy Recognition of offline Handwritten Numeric [10, 11] and authors kamlakannam, Fernado in their paper Fast Character recognition using Expert System [12], Suen in article Handwriting recognition [13] and Zhao P., in their paper On-Line cursive writing recognition system [14] are really useful in extracting the essential feature of the number and character both in order to recognize them. Many authors have suggested that features of character like horizontal line, vertical line, left oblique line, right oblique line, curve shapes are not sufficient to recognize a character as mapping between feature space and pattern space is difficult to establish. Presence of lines, curves and intersection themselves can not recognize a character but the temporal information i.e. the sequence in which they appeared makes them uniquely recognizable [15]. Some other advance techniques like structural approaches, neural network approaches, statistical approaches and vector machine (SVM) have all been well implemented on LPR. Each approach has its own pros and cons. Statistical approaches are very effortless but not very truthful. Neural network approach is good in learning, capturing features and their generalization. The neural network is also good at handling erroneous, deficient and inexact information [16]. Additionally, the innovative idea depicted on socket programming was really useful in order to understand the socket programming that will be used in designing tracking system [17, 18]. So keeping all pros and cons of these approaches suggested by different authors in mind, a new algorithm is proposed in this paper for LPR which gives prominent results in simplest way.

3.1   Extracting the Registration Plate (RP) from the Captured Image

The detailed process to localize the Registration Plate (RP) and recognition of the Registration Plate Registration Number (RPRN) is as follows. There are four steps to get the number from the RP.

(1) Capturing the vehicle's image (2) Extraction of registration plate from the image (3) Extracting the characters from the plate image (4) Recognizing registration plate characters (Fig. 2).

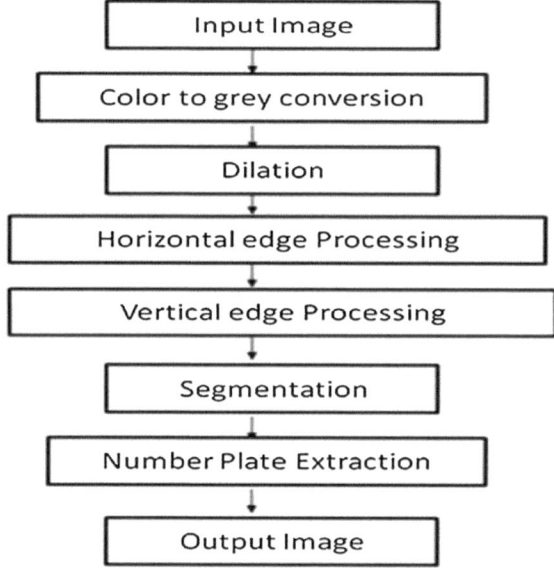

Fig. 2 Proposed method for Registration Plate (RP) localization

3.1.1   Convert a Color Image into Gray Image

The proposed algorithm is independent of the type of colours in image and relies mainly on the gray level of an image for processing and extracting the required information. Color components like Red, Green and Blue value are not used throughout this proposed algorithm. So, if the input image is a coloured image represented by 3-dimensional array in MATLAB, it is converted to a 2-dimensional gray image before further processing. The sample of original input image (Fig. 3a) and a gray image (Fig. 3b) are shown below.

(a)   (b)   (c)

Fig. 3 **a** Original color input image **b** Gray image **c** Dilated Image

3.1.2   Dilate Image

Registration plate image may contain different kinds of shades and brightness. During RGB to grey conversion difference in colours, lighter edges etc. may get lost. Dilation helps to abolish these kinds of losses. Figure 3c is the image after dilation.

3.1.3   Horizontal and Vertical Edge Processing of an Image

This phase uses horizontal and vertical histogram which is column-wise and row-wise histogram. These histogram shows row-wise and column-wise sum of differences of gray values between connected pixels. In this phase horizontal histogram is calculated first and the algorithm traverses through each column. In each column, the proposed algorithm starts from second pixel from top and difference is calculated between second and first pixel, after checking with certain threshold, if it exceeds the value, it is added to the total sum of differences. Then difference between third and second pixel is calculated and it goes on until it will move end of column and calculates total sum of differences between connected pixels, an array of column-wise sum is created at the end. Same process is used to find vertical histogram. In this process instead of processing columns, rows are processed. Horizontal and Vertical edge processing histogram is shown in Fig. 4a, b respectively.

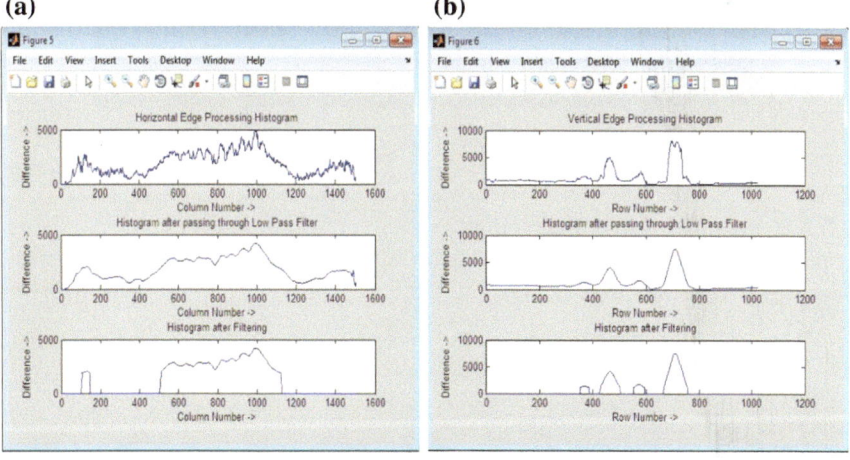

(a)   (b)

Fig. 4 **a** Horizontal edge processing histogram **b** Vertical edge processing histogram

3.1.4   Passing Histograms Through a Low Pass Digital Filter

In this performing phase, after considering the values on right-hand side and left-hand side, each histogram value is averaged out. This phase is performed on both the horizontal and the vertical histogram as well. Below are the figures showing the histogram before passing through a low-pass digital filter and after passing through a low-pass digital filter.

3.1.5   Filtering Out Unwanted Regions in an Image

In this phase, histogram is passed through low pass filter. A low pass histogram value indicates image contains less variation among connected pixels. As registration plate contains plain background with letters and numbers in it, the difference in connected pixels, edges of the letters and numbers is very high. Therefore registration plate may have high horizontal and vertical histogram values. These areas can be removed out by using dynamic threshold.

3.1.6   Segmentation

In this phase, find all the regions in an image having high probability of containing a registration plate. Co-ordinates of all such probable regions are stored in an array. The output image displaying the probable license plate regions is shown in Fig. 5.

Fig. 5 Output of segmentation

3.1.7   Region of Interest Extraction

Segmentation process output provides all the regions that have maximum probability of containing a registration plate. Out of these regions, the one with the maximum histogram value is considered as the most probable candidate for registration plate. All the regions are processed row-wise and column-wise to find a common region having maximum horizontal and vertical histogram value. Detected registration plate region is shown in Fig. 6.

Fig. 6 Detected registration plate

3.2 Registration Plate Recognition

There are several pre-processing steps to recognize the RP.

3.2.1 Segmentation
These are the steps involved in this

1. Filter the noise level present in the image.
2. Clip the plate area in such a way that only numbers of plate area extracted.
3. Separate each character from the plate.

3.2.2 Number Identification
Following steps are involved for identification of number

1. Create the template file from the stored template images.
2. Resize image obtained from segmentation to the size of template (24 * 42).
3. Compare each character with the templates.
4. Store the best matched character.

3.2.3 Display Registration Plate Number
To display the Plate Number following steps are involved

Templates Used for Character Recognition

Following templates are used to recognize registration plate number. 50 Templates of size 24 * 42 are used which are given below (Fig. 7).

Fig. 7 Templates used for number plate recognition

Recognized Characters

Recognized character after applying pattern matching algorithm (Figs. 8 and 9).

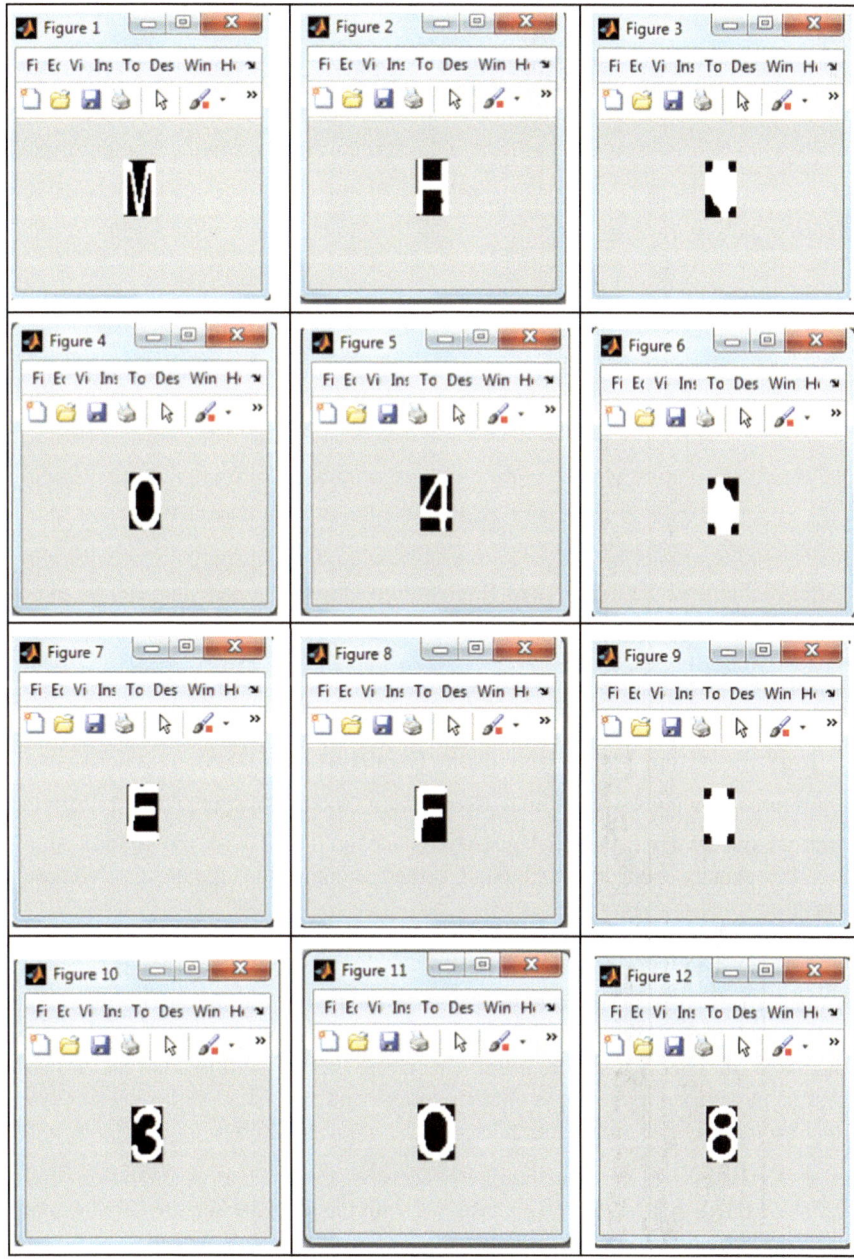

Fig. 8 Recognized license plate characters

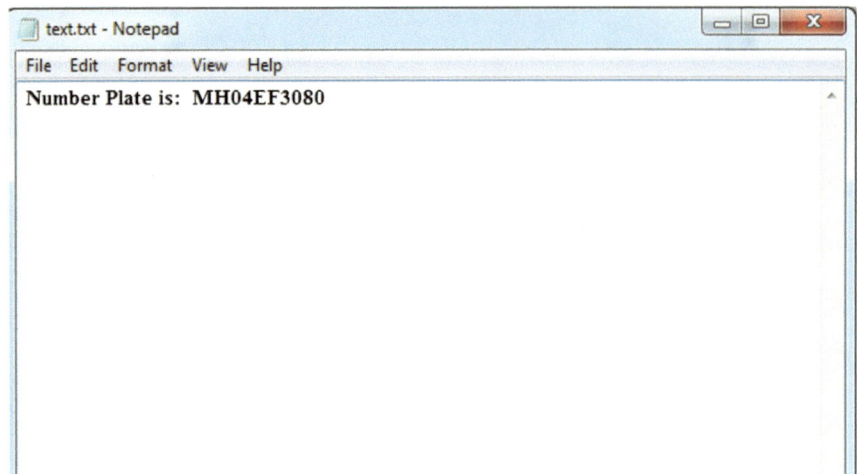

Fig. 9 Recognized number of the Registration Plate (RP)

After recognized the numbers of Registration Plate (RP), this data can be matched with the VAHAN data and IES can impose the penalty on the registered owner from their AADHAR linked account registered at the time of registration and stored in VAHAN.

3.3 Implementation Step of Intelligent Expert System (IES)

This Intelligent Expert System will work like the ATM system of any bank and extract the data of the owner from the VAHAN data base linked with registration number, impose the penalty from the AADHAR linked account and update the database for further use.

3.3.1 Implementation Step of IES

This system will work in the following manner to impose the penalty on the offender. This system will also keep the record for future use and impose more penalty amount on the same offender. This system will overcome the problem of incorrect registered address as in the case of current e-challan system and as well as in this current system, there is no record of the same offender.

Step 1 Extracted registration number received from Plate Recognition System (Example: MH 46 AB 1595) will be matched with the state data centre (SDC) in which the vehicle belongs to, for this purpose the first two character will identify the state. Here MH i.e. Maharashtra.

Step 2 After identifying the state, next step is identifying the RTO area, for this next two digits are used. Example MH 46 then 46 belongs to Kalamboli RTO in Navi Mumbai.

Step 3 After this the next numbers are Registration number, here in example AB1595 will be matched with the registration digitalized data of VAHAN with the help of Registration Certificate (RC) handled by RTO.

Step 4 After this the owner's detail will be identified and penalty will be deducted from his AADHAR card linked account and SMS will be sent to the vehicle owner and this data will be stored for future references (Fig. 10).

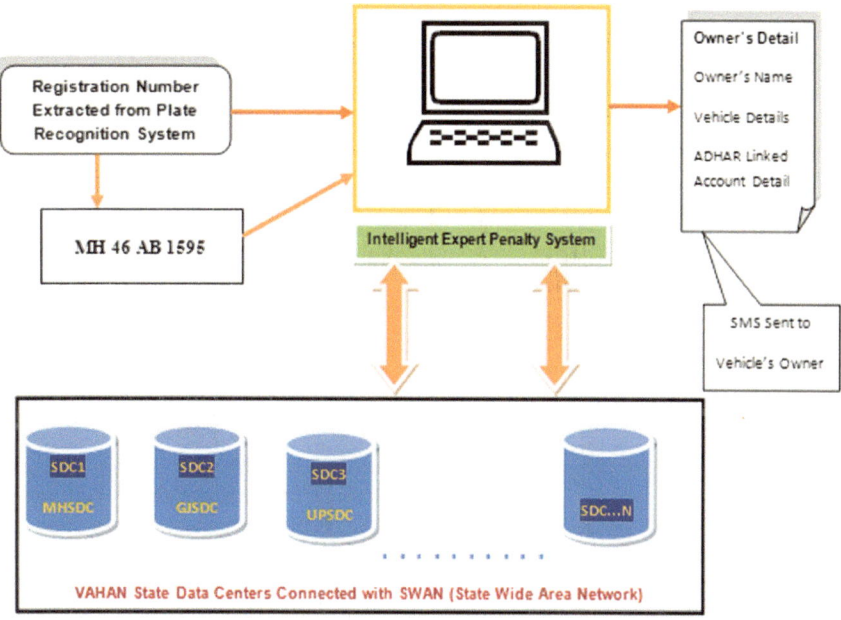

Fig. 10 Framework of Intelligent Expert System (IES)

4 Conclusion

We proposed a system which can penalize the offenders automatically from the Bank account of the vehicle owner linked with AADHAR card and overcome the problem of existing e-challan system where the penalty challan may or may not be reached to the owner due to incorrect address or even after receiving the e-challan they may not pay the penalty hence fail to impose e-penalty. This system reduces the bribe in traffic department and will also control the harassment of citizens by traffic police. Since the system keeps the record of offence offended by an owner hence the same owner can be penalized with more amount of penalty if he repeats the offence in future. This system will reduce the man power involved in the existing e-challan system and improve the efficiency of the current system by applying automatic penalty on the errant drivers with the help of ICT and enforcement of the Government rule on citizens and help the society by controlling the unwanted human loss.

References

1. The ultimate information portal on European speeding and traffic violation. http://www.speedingeurope.com/italy
2. Traffic Tickets in Canada: http://allontario.ca/2013/05/trafic-tickets
3. Khaleej Times in UAE: http://www.khaleejtimes.com/nation/transport/list-of-traffic-violations-and-fines-in-dubai
4. https://www.french-property.com/guides/france/driving-in-france/driving-offences/
5. UK government Portal: https://www.gov.uk/speeding-penalties
6. Saha, S., Basu, S., Nasipuri, M., Basu, D.K.: License Plate localization from vehicle images: an edge based multi-stage approach. Int. J. Recent Trends Eng. 1(1), 284–288 (2009)
7. Angeline, L., Choong, M.Y., Wong, F., Teo, K.T.K.: Tracking and localisation of moving vehicle license plate via Signature Analysis. In: 4th International Conference on Mechatronics (ICOM), 17–19 May 2011, Kuala Lumpur, Malaysia (2011)
8. Yu, M., Kim, Y.D.: An approach to Korean license plate recognition based on vertical edge matching. In: IEEE International Conference on Systems, Man, and Cybernetics, vol. 4, pp. 2975–2980 (2000); Foster, I., Kesselman, C.: The Grid: Blueprint for a New Computing Infrastructure. Morgan Kaufmann, San Francisco (2000)
9. Groft, E., Andrews, K., Kuff, H.: Smart meter parking system. US Patent 20070016539
10. Batuwita, K.B.M.R., Bandara, G.E.M.D.C.: Fuzzy recognition of offline handwritten numeric characters. In: IEEE Conference on Cybernetics and Intelligent Systems, 7–9 June (2006)
11. Timár, G., Karacs, K., Rekeczky, C.: Analogic preprocessing and segmentation algorithms for off-line handwriting recognition. In: Proceedings of the 2002 7th IEEE Conference on Cellular Neural Networks and Their Applications(CNNA) (2002)
12. Ganapathy, K., fernado, C.G., Davari, A.: Fast character recognition using expert system. In: Proceedings of the Thirty-Seventh Southeastern IEEE Symposium on System Theory (SSST-05) (2005)
13. Suen, C.Y., Kim, J., Kim, K., Xu, Q., Lam, L.: Handwriting recognition—the last frontiers. In: Proceedings of the 15th International Conference on Pattern Recognition (ICPR) (2000)
14. Zhao, P., Yasuda, T., Sato, Y.: Cursivewriter: on-line cursive writing recognition system. In: Proceedings of the Second International Conference on Document Analysis and Recognition (ICDAR), pp. 703–706, 20–22 October (1993)
15. Jameel, A.: Experiments with various recurrent neural network architectures for handwritten character recognition. In: Sixth International Conference on Tools with Artificial Intelligence (1994)
16. Liu, H., Ding. X.: Handwritten character recognition using gradient feature and quadratic classifier with multiple discrimination schemes. In: Proceedings of the Eighth International Conference on Document Analysis and Recognition (2005)
17. Ahmed, M.J., Kumar, D., Singh, P.P., Sarfraz, M., Zidouri, A., Alkhatib, W.G.: License plate recognition system. In: Proceedings of the 10th IEEE International Conference on Electronics, Circuits And Systems (ICECS), Sharjah, United Arab Emirates (UAE) (2003)
18. Fujisawa, H., Liu, C.-L.: Directional pattern matching for character recognition revisited. In: Proceedings of Seventh International Conference on Document Analysis and Recognition (ICDAR) (2003)

Optimization of Rules in Neuro-Fuzzy Inference Systems

J. Amudha and D. Radha$^{(\boxtimes)}$

Department of Computer Science and Engineering, Amrita School of
Engineering, Amrita Vishwa Vidyapeetham, Amrita University, Bengaluru, India
d-radha@blr.amrita.edu

Abstract. Optimization of rule based system with Neuro Fuzzy Inference system results better regarding accuracy and interpretability. Dynamic Evolving Neuro Fuzzy Systems (DENFIS) model is used to find out an optimized rule base using computational intelligence techniques for a target search application. The process of optimization starts at the beginning of the target search process by selecting an appropriate selection of rule based Fuzzy Inference System. Further, optimization has been addressed in the choice of the number of rules by reducing the number of attributes used in the input. The integrated approach of input selection and rule selection results in accurate target predictions. The ability of knowledge-representation, highly interpretable if…then rules and imprecision tolerance are the major features of the proposed model.

Keywords: Neuro fuzzy inference · Target · DENFIS · Rule selection
Computational intelligence

1 Introduction

The efficiency regarding memory and computational power [1] can be increased by Computational Intelligence (CI) paradigm. It is constituted by Artificial Neural Network (ANN), Swarm Intelligence (SI), Fuzzy Logic (FL), Evolutionary Algorithm (EA), etc. Hybrid systems help in optimizing the convenience and intelligence compared to other systems. Fuzzy Inference Systems (FIS) are widely used for process simulation or control in applications like recognition, detection, automation, etc. Fuzzy systems can broadly categorized into Mamdani model [2] and Sugeno Model [3]. Knowledge from expert or data is the source for their design. For complex systems, fuzzy rules inferring from data gives better accuracy than expert Knowledge. Fuzzy Logic systems do not possess any inherent method of learning. So the rule generation always requires the aid of a human expert.

The computational power of Artificial Neural Networks replicates the biological neural networks and bestows the systems with some of the (higher-level) cognitive abilities of biological organisms. Neural Network exhibits the ability to learn and generalize from training patterns with its interconnected nodes. Learning capability and distributed representation are two major features of neural networks. The combination

© Springer International Publishing AG 2018
D. J. Hemanth and S. Smys (eds.), *Computational Vision and Bio Inspired Computing*,
Lecture Notes in Computational Vision and Biomechanics 28,
https://doi.org/10.1007/978-3-319-71767-8_69

of learning ability of Neural networks and fuzzy inference system as Neuro-Fuzzy Inference System end up in self-learning ability.

The Neuro–fuzzy approach has merits of connectionist and fuzzy approach combination and forms an important component of soft computing. If required, a learning process can be used as a part of knowledge acquisition. Reinforcement learning is a better choice than supervised learning in the absence of sufficient time or data or an expert Knowledge which can be expressed as linguistic rules that can build a fuzzy system. On the other hand, Artificial Neural Network is a better choice for simulation or real time task. Integration of neural and fuzzy systems adds their merits in neuro-fuzzy approach [4]. A fuzzy system can lead to the better combination for numeric data and can be easily extended for higher order data. Neural networks, on the other hand, can blindly generate and refine fuzzy rules from training data [5]. Fuzzy sets are beneficial in the logical field, and in handling higher order processing efficiently. Learning leads to greater flexibility and performs better in data-driven processes [6]. Hayashi and Buckley [7] proved that there is an approximation method existing between the rule-based fuzzy system and neural network [8].

2 Literature Survey

The comparison of various rule based approaches [1] showed that Dynamic Evolving Neuro Fuzzy Systems (DENFIS) approach is better in learning ability, generation of the membership function, learning rate and in supporting multiple output models. It can be used in offline and online mode. As the rules depend on the number of input attributes [1], the integrated approach of selection of relevant input attributes and rule optimization in DENFIS enhances the performance. It is even compared with other models like ARX and HEC HMS and showed the out-performance [9].

In general, neuro fuzzy system has redundancy. The redundancy can be minimized by reducing the structure of the system consecutive learning of the neuro fuzzy system and by selecting appropriate fuzzy rules from the input data set. Then neuro-fuzzy networks can be implemented on a reduced set by learning the attributes of the membership functions.

The reduced neuro-fuzzy model [10] is obtained using dependency in Rough Set theory. Dependency between each rule and the output determine the selection of a rule. This reduces the complexity and redundancy of the structure without affecting the performance. Rough set theory is a mathematical tool to deal with vagueness and uncertainty.

The integrated method [11] proposes a novel neuro-fuzzy system that can do feature analysis and System Identification simultaneously. The five-layered feed-forward network is used in the system. The modulator function with fuzzification of the input enables online selection of essential features by the network. The system checks for non-negative characteristic of certainty factors of rules. The conflicting rules are eliminated by pruning nodes and links and then re-training the structure. Re-training the structure results in optimal network architecture. The performance of pruned network is maintained as the original one.

A four-layered feed-forward network [12] is used in simultaneous feature selection and fuzzy rule system. The network is trained in three phases in which the system learns the important features and the classification rules in first phase. In subsequent phases, the redundant nodes as detected by the feature attenuators are pruned, and the network is re-tuned in its reduced architecture, and then the incompatible rules, zero rules and less used rules are pruned to further reduce the design. The final set of rules achieves better performance by tuning their membership functions.

An interpretable three-layered feed forward network [13] for Linguistic rule selection and rule based classification. Fuzzification is the membership function layer along with input and output layers. Each input feature node from input layer are connected to small (S), medium (M) and large (L) membership value nodes in fuzzification layer represented by a Gaussian membership function. Output layer is used for feature selection by their weights. The selected relevant features based on weights comprehend its structure to logical rules. The small number of features gives an informative subset of the input dataset. This classifier can produce good classification results from the direct calculation or logical rule extraction.

Another work based on DENCLUE clustering method [14] is used for input space partitioning. This method is a new approach to neuro-fuzzy system modeling based on DENCLUE using a dynamic threshold and similar rules merging (DDTSRM). Results are good at the determination of some partitions, improvement in accuracy of the structure identification and reduce the rule redundancy by considering the similarity measures between fuzzy sets.

The integrated framework [15] is to optimize the number of inputs and the number of rules simultaneously. The most significant rules are selected along with the removal of unnecessary correlations. A trade-off is balanced between interpretability and accuracy by finding an appropriate number of inputs and rules to be included in the model. The optimal solution is obtained by adding a backward stage to refine and remove the redundancy. An improved Akaike Information criterion (AIC) is used in the model to avoid over fitting. The AIC criterion helps in the simultaneous selection of rules and input.

The integrated neuro fuzzy system with input selection may result in optimization of rules without any trade-off with accuracy. The following section proves the same.

3   System Overview

The aim is to design a structurally optimized neuro fuzzy model for a target search application. The concept of DENFIS model is used to design a neuro fuzzy model. The model so obtained is refined to develop an optimized structural model by implementing feature selection method which results in reduced number of fuzzy rules. The input to the model is an image which will follow a feature extraction procedure before providing it to the model. Feature extraction is done with the help of VOCUS [16] model. Structured Optimization is shown in the Fig. 1 which is required in the training phase.

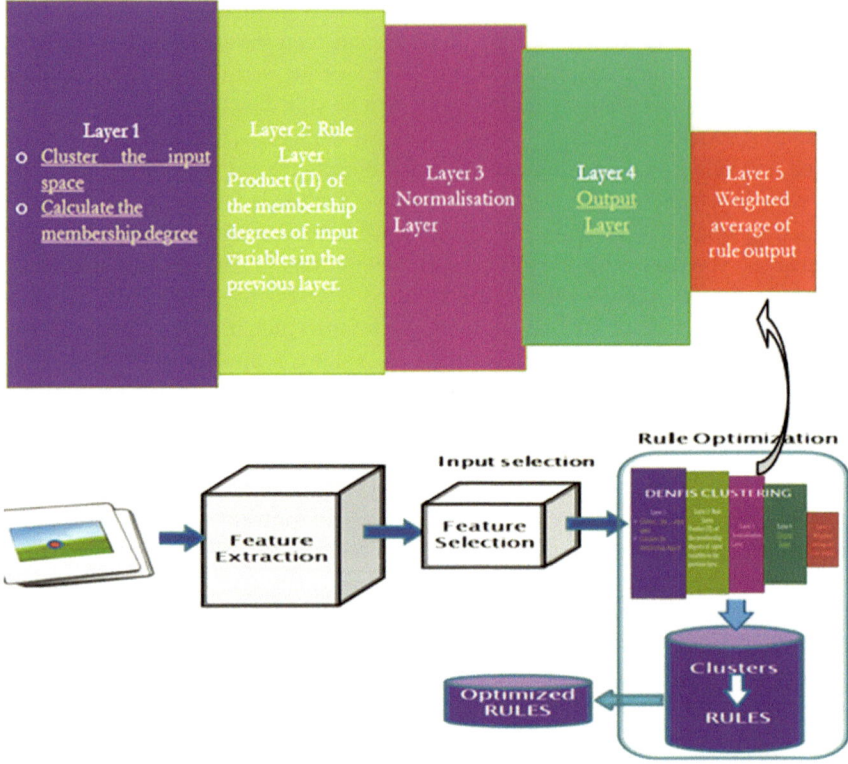

Fig. 1. Structure optimization: training phase

3.1 Feature Extraction

In learning mode, the system is provided with a region of interest containing the target. The computational attention systems help in extracting features of a target for generating the training set for DENFIS model. The features may include color features, intensity features and orientation features that contribute to identifying the target.

3.2 Feature Selection

Since the rules depend on the input, selecting the relevant input is also a concern. This stage explains about the feature selection which can eventually lead to an optimized rule base. Since the number of rules depends on the number of input attributes, attribute reduction will result in a reduced number of rules. Since input attributes are the properties of an instance that may be used to determine its classification, retaining the relevant attributes is a major concern in feature selection. DENFIS model is thus incorporated into a feature selection stage, and the user can select the required number of features that are to be deleted. The features are eliminated based on their rank obtained after ReliefF algorithm [17]. The requested number of features is removed before the training.

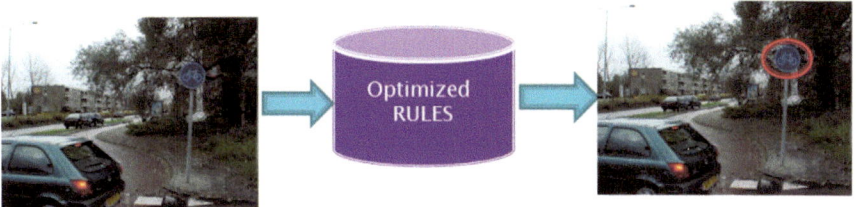

Fig. 2. Structure optimization: testing phase

Feature selection algorithms are advantageous [18] as it is independent on heuristics, time complexity, noise-tolerance and robust to feature interactions and also applicable for binary or continuous data. Using these algorithms, less relevant input gets deleted, and the model is trained using this reduced features. While performing the testing, these irrelevant attributes get rejected, and rest of the features are considered for testing to search the target as shown Fig. 2.

3.3   DENFIS Structure and Rule Optimization

DENFIS proposed by Kasabov [19], evolves by incremental, integrated, local element tuning. An evolving clustering method (ECM), is introduced in both online and offline models. DENFIS uses Takagi-Sugeno inference system for both online and offline model.

3.3.1   DENFIS Online Model for Learning

In DENFIS online model, Evolving Clustering Method (ECM) is used for input data partitioning. The steps in ECM are:

- Clusters are created by ECM.
- The distance between current input sample and already created cluster centers are calculated for all samples of the data stream have been processed.
- If the consequent functions are crisp values, then that systems are termed as zero-order Takagi-Sugeno systems (TSK). If the consequent functions are linear functions, then it is termed as First-order systems and if the output is a nonlinear function that systems are termed as higher order Takagi-Sugeno systems.

Generation of first m fuzzy rules in the following steps with n_0 learning data pairs

- First n_0 pairs are taken from the learning data set.
- m cluster centers are obtained using ECM method.
- Data points which are closest to the each cluster center C_i, are taken as p_i data points.
- The antecedents of the fuzzy rule are created using cluster center and data pairs which in turn used to obtain fuzzy rule corresponding to a cluster center. The weights are calculated as the distance between data points and the cluster center.

3.3.2   DENFIS Offline Model for Learning

The DENFIS offline model works efficiently for larger data sets. This is done applying batch-mode training on the offline model.

DENFIS offline learning process is implemented in following way:

- The input space is partitioned to make n cluster centers using the offline ECM.
- The antecedent part is created using the current position of the cluster center.
- n datasets corresponding to n cluster centers and p learning data pair which is closest to cluster centers are found. In the general case, one data pair may be closest to more than one cluster.

3.4   Target Selection

DENFIS model is reframed to act as a zero order TSK system where the consequent is a crisp output rather than a linear function. The constant consequent output varies depending on the training data. In this work of detecting the target region, which is designed as a classification problem, the crisp output is determined by the clustering technique. The training data is provided as input as well as output class. The cluster elements are compared with their original output class, and the frequent output class becomes the crisp output of that cluster.

The optimized DENFIS model is used as the learning model for target search application. The learning model is fed with training data obtained after feature extraction by computational models. The dataset used for Training and Testing are the features extracted from the image. The optimized rules are identified by running the model for different values of clustering parameter. The trained model is followed by a testing phase to analyze the performance.

4   Implementation Details

The optimized rule based model for target search application is implemented using MATLAB. The final optimized rule based model was developed through a series of 3 stages.

Stage 1: Incorporating Feature selection method to DENFIS for structural optimization.
Stage 2: Implementation of DENFIS model and validation of the model with a standard dataset (Iris Dataset).
Stage 3: Reframing the DENFIS model for a Target search application.

4.1   Incorporating Feature Selection Method to DENFIS for Structural Optimization

The computational attention system named as Visual Object Detection with a Computational attention System (VOCUS) consists of a learning mode and search mode. Learning mode of VOCUS has been used for feature extraction in this work. In learning mode, the system is provided with a region of interest containing the target. Thirteen features are extracted using this model for generating the training set for DENFIS model. Thirteen features include four color features (RGBY), 4 Intensity features (0°, 45°, 90° and 135°), on-off intensity and off-on intensity and three conspicuity maps of intensity, color, and orientation. An RGB color space representation is considered for color feature representation. Orientation maps are computed using oriented pyramids. Log Gabor filters are used for obtaining the different orientation maps and is implemented using the Greenspan's method.

RELIEF is a feature selection algorithm used in binary classification proposed by Kira and Rendell in 1992 [18]. The system is advantageous as it is independent on heuristics, less time complexity, noise-tolerance and robust to feature interactions and also applicable for binary or continuous data; But, redundant features are not differentiated and low numbers of training instances may mislead the algorithm.

RELIEF algorithm is repeated for m times by for a dataset having many instances with a specific number of features which are belonging to two different classes. The features are scaled in the range of 0 to 1. The weight vector is initialized to zeros for the length as the particular number of features.

A random instance is taken at each iteration, and its feature vector (X) is compared with other vectors from each class. The closest feature vector from the same-class instance is called 'near-hit,' and the closest feature vector from the different-class instance is called 'near-miss.' The weight vector is updated with the following equation.

$$W_i = W_i - (x_i - \text{nearHit}_i)^2 + (x_i - \text{nearMiss}_i)^2$$

Thus the weight of any given feature changes according to the near-hit instance and near-miss instance. The relevance vector is calculated by dividing each element of the weight vector by the number of iterations m. Selection of features is dependent on the threshold value τ of relevance vectors.

The comparison of a different number of features and its impact on true positive rate on different clustering parameter is shown in Fig. 3.

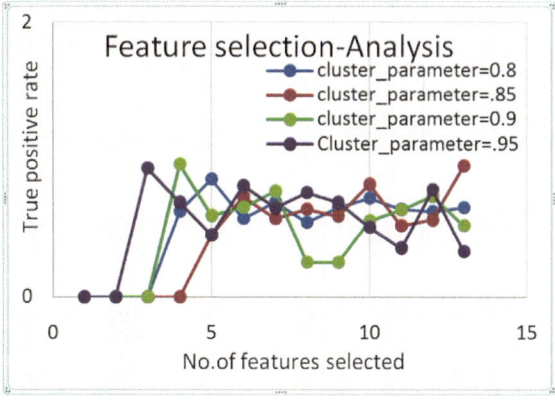

Fig. 3. Analysis of the impact of feature selection on modified DENFIS model

The performance analysis of different features concerning the true positive rate for different clustering parameter is plotted in the graph. The maximum true positive rate is obtained with 13 features at the clustering parameter 0.85 which corresponds to optimal 14 rules. Analysis has been done on all feature combinations starting from 13 to 1. High true positive rates have been observed when selecting four attributes at the clustering parameters 0.9 and 0.95. This corresponds to the optimal rule of 3 and two respectively. Without feature selection, the optimal rule was 14, and the feature selection succeeds in reducing the optimal rule to three and four without compromising on performance. Since the same performance can be delivered by reduced number of rules, the feature selection can be incorporated in the model there by reducing the complexity of the model. Thus a combined approach of rule selection and input selection leads to an optimized structural model for the target search applications.

4.2   Implementation and Validation of DENFIS Model with a Standard Dataset (Iris Dataset)

This stage follows feature selection of the system design. The development starts with reading the training dataset with target features and ends with the formation of a rule based model.

The step by step procedure of training a DENFIS model is explained below.

1. Read training data.
2. Cluster the input space using Evolving Clustering Method (ECM) algorithm.
3. Determine the antecedent part of each rule based on cluster centers.
4. Determine the consequent part of the rule.
5. Generate rules.

4.2.1   Evolving Clustering Method (ECM)

This method is used for input space partitioning. It takes one input vector at a time and updates the cluster details. The flow chart of Evolving clustering method is given in Fig. 4. Evolving clustering method cluster all the input data to one or more clusters, and cluster center and cluster radius are updated for all the clusters.

4.2.2   Rule Generation

Rule-based systems in the form IF...THEN rules are used as a way to store and manipulate knowledge to interpret information in a useful way.

The general form of a DENFIS rule is:

If x_1 is in cluster 1 and x_2 is in cluster 1 and............ x_q is R_{1q} then y is f $(x_1, x_2$ ...., $x_q)$
If x_1 is in cluster 2 and x_2 is in cluster 2 and .... x_q is R_{2q} then y is f $(x_1, x_2$ .... $x_q)$

where the parameters of the cluster (cluster center and radius) determine the antecedent part of the rule and is generated in membership generation phase and Y is the consequent part. The model obtained after training is subjected to testing. The testing

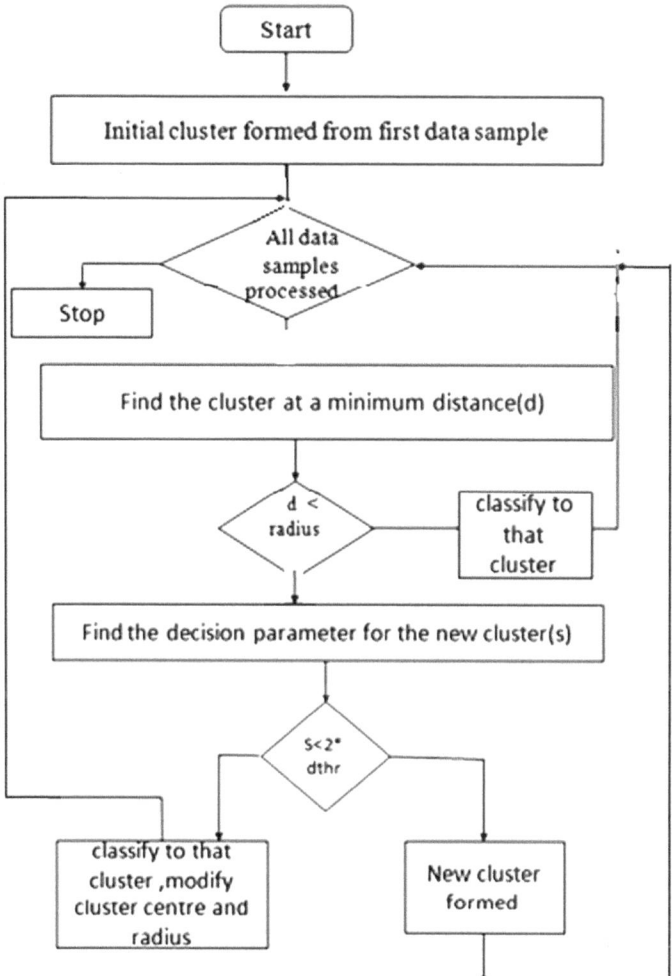

Fig. 4. ECM algorithm

dataset used in this phase to validate the model is Iris membership generation, antecedent part calculation, and consequent part calculation using least dataset.

Rule-based systems in the form of IF...THEN rules are used as a way to store and manipulate knowledge to interpret information in a useful way.

An optimal rule based model is generated by running the model with different values for the clustering threshold distance which determines the inter cluster distance. The optimal model generated by following this procedure is stored for further testing processes.

4.3   Reframing DENFIS Model for a Target Search Application

The following sections were framed for explaining the training and testing phases in the target search application.

4.3.1   Training Phase

In the training phase, an image with the specified target is given to the system. The GUI of the training phase is given in Fig. 5 with the target marked with the rectangle. The image undergoes a feature extraction procedure to make it equipped for the training procedure. Robust features [20] can be selected with different algorithms like chi-square etc. Feature extraction is done with the help of VOCUS [16] model. Using VOCUS model, 13 feature values required for identifying the target is extracted. The thirteen features are four color features (Red, Green, Blue, Yellow), four orientation features (0°, 45°, 90°, 135°), on-off intensity, off-on intensity, intensity conspicuity, color conspicuity and orientation conspicuity.

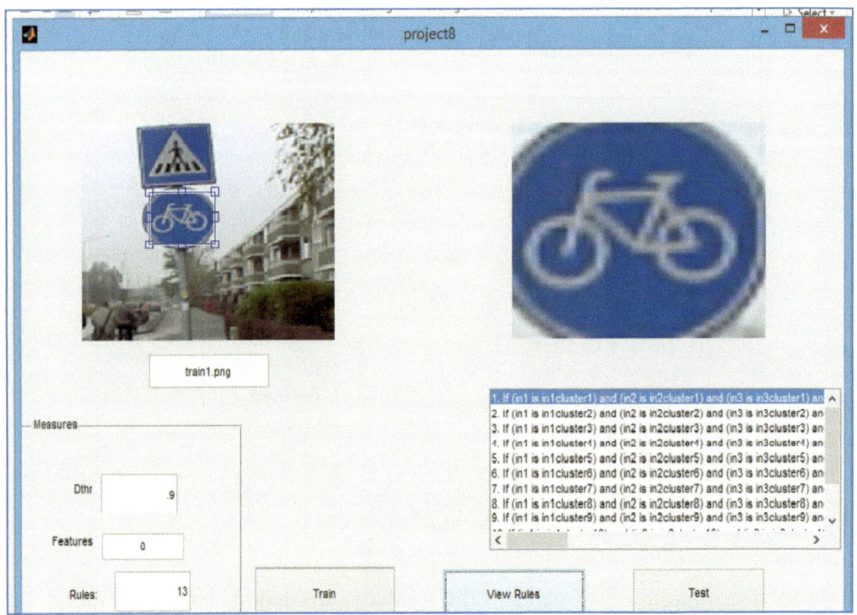

Fig. 5. GUI after training phase with target marked and rules displayed

These 13 features served as the input part of training data. The output part can be labeled as 1 or 2 depending on whether the particular pixel belongs to a non-target or target area respectively. The input image given to the VOCUS model is resized to 68×90. From these 13 feature maps, the corresponding (i, j)th locations of each matrix form the single row vector of training data. The training input data obtained is subjected to removal of redundant data pairs, and this unique data pair forms the training data.

Once the training data is obtained, next step is the implementation of DENFIS model. DENFIS process starts with clustering of data to obtain the membership functions and rules. The cluster center determines the antecedent part of the rule. The number of rules is determined by the number of clusters and the rule layer is obtained as the product of previous layer output. The consequent part of the rule is obtained by considering the most frequent class in that cluster. Next stage corresponds to the final output, it is calculated as the average of each rules output. A fuzzy inference system is formed from these stages through self-learning ability of neuro fuzzy systems.

The rule based system transforms to an optimized rule based system by clustering the training data at the optimal threshold clustering parameter. This clustering parameter corresponds to the inter cluster distance, greater the value more the size of the cluster. The optimized rule based model obtained is retained and stored for the target search application. Distance threshold for clustering and training image are the inputs given to the system.

The training phase ends with generating the set of rules for determining the target. The GUI of the model after training with the rules is shown in Fig. 4. The rules contain the information about the antecedent part and consequent part. The antecedent part of the rule is determined using the cluster center, and the consequent part of the rule is designed using the concept of zero order TSK systems.

4.3.2 Testing Phase

In testing phase also the input to the system is an image from which the target has to be detected using the optimized rule base formed after the training phase. The feature extraction from the image and collecting the testing input is similar to the steps of training data collection in the training phase. The GUI of testing phase is shown in Fig. 5. Each testing vector, when given to the rule base, follows a series of execution as follows.

1. Calculate the membership degree by using the antecedent part of rule—the rule antecedent contains the parameter for calculating the triangular membership function. Based on these values the membership degree of each input are calculated.
2. Calculate firing strength of rules—the rule firing strength is the product of previous layer membership functions.
3. Normalize the rule strength.
4. Identify each rule's consequent part—each rule's consequent part is already stored in the rule base.
5. Calculate overall output by calculating the weighted sum of each rule's output.

Every pixel of the testing input image passes through the execution steps described above. The results of each pixel can be mapped as a target or a non-target. The consecutive set of target class corresponds to the target image. Figure 6 shows the detected target as white portions in the black background.

The performance of the target search application is tested by comparing the results of the testing image with the ground truth. The performance metrics used to analyze the results are the true positive rate, True negative rate, false positive rate and false negative rate. The true positive rate determines the portion of target class correctly determined and true negative rate determines the proportion of correctly identified nontarget class.

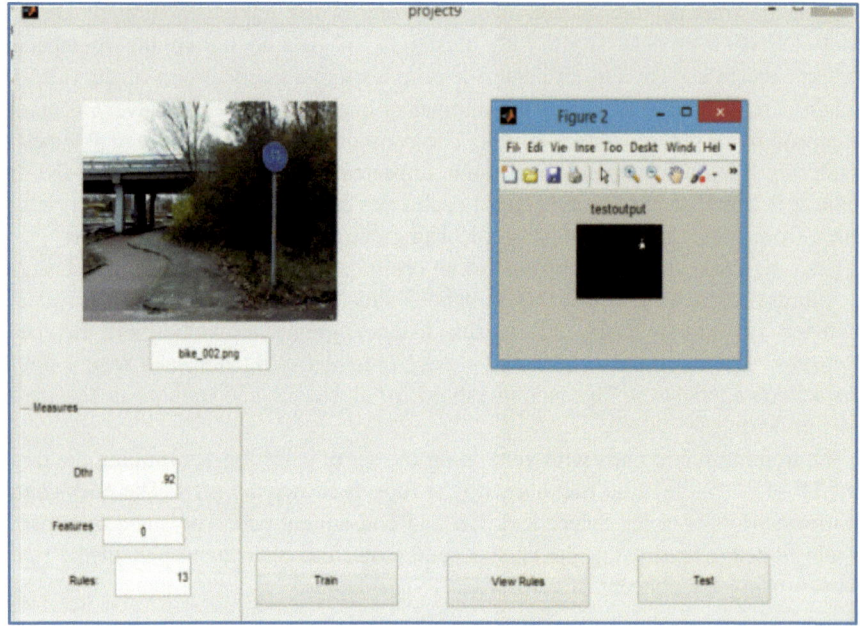

Fig. 6. GUI for testing

5   Analysis and Results

After the three stages of implementation as feature selection, testing of DENFIS model with iris dataset, modifying DENFIS model for a target search application, the testing results are as follows.

5.1   Testing of DENFIS Model

A first order TSK DENFIS model has been trained and validated with iris dataset. Iris dataset is a standard dataset obtained from UCI repository.

5.1.1   Dataset Used

The first stage of this model was to implement a DENFIS model and validate that model. Testing of that model is done with the help of Iris dataset. Classification problem of novel multivariate Iris flower data set is created by Sir Ronald Aylmer Fisher in 1936. There are 150 instances of IRIS dataset which is taken from the following species:

- *Iris setosa*
- *Iris virginica*
- *Iris versicolor*.

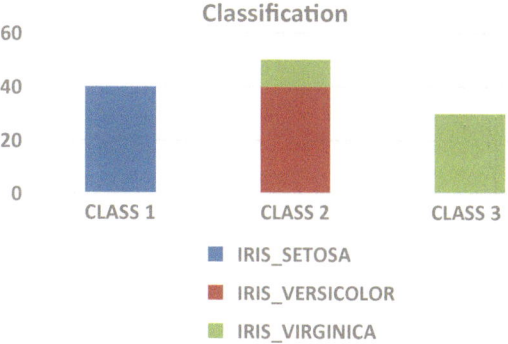

Fig. 7. Classification accuracy of iris dataset

Sepal Length, Sepal Width, Petal length and Petal width are the features derived from each sample of three selected species of Iris flower. The classification is based on the combination of the four features; The dataset split up for training and testing is 80–20%. Therefore the first 40 instances of each class are taken as the training dataset, i.e., a total of 120 instances. The next ten instances of each class have been taken as the testing instances, i.e., a total of 30 instances.

The DENFIS model has been tested and the classification accuracy of *I. Setosa* and *I. versicolor* is 100% giving an overall classification accuracy of 91.6% which is shown in Fig. 7.

5.2   Modified DENFIS Model for Target Search Application

The optimized DENFIS model is used as the learning model for target search application. The learning model is fed with training data obtained after feature extraction done by VOCUS model. The dataset used for Training and Testing are the features extracted from the image. Thirteen features are extracted using VOCUS model. Extracted features are four color features (Red, Green, Blue, Yellow), four orientation features (0°, 45°, 90°, 135°) and color, intensity and orientation conspicuity maps. The optimized rules are identified by running the model for different values of clustering parameter. The trained model is followed by a testing phase to analyze the performance. The testing phase analyses the model by comparing the parameters true positive rate, true negative rate, false positive rate and false negative rate. The performance of the model is good when the true positive rate is high.

True positive is the event which gives a positive result for the positive prediction, and a false positive is the event which gives a negative result for the positive prediction. Similarly true negative and false negative events give a negative result for negative prediction and a positive result for a negative prediction respectively. The performance is compared with true positive rate and true negative rates. The analysis results regarding true positive rate, true negative rate, false positive rate and false negative rate are shown in Fig. 8.

$$accuracy = \frac{true\,positive + true\,negative}{number\,of\,instances} \tag{1}$$

The optimal value corresponds to the true positive rate of 0.9524 for the clustering parameter 0.85. The optimized number of rules can be fixed at this clustering parameter, and it is 14 rules. A training data with 13 attributes with a minimum of 2 membership functions can make 2^{13} rule combinations and hence find out the best rules out of this combination is a tedious task. So this proves that the work done here has developed an optimal model of rule base which makes a reduction of rules from 2^{13} to 14. With these 14 rules, the model gives a high true positive rate and true negative rate, which shows the models ability to successfully identify the target and non-target areas. Based on the equation for accuracy given in Eq. (1), which depends on true positive and true negative values, higher the true positive and true negative higher the accuracy. The graph shown in Fig. 8. supports the higher accuracy of the model owing to its higher true positive and true negative rates.

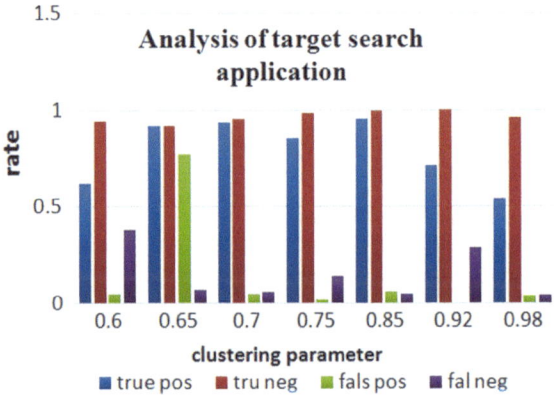

Fig. 8. Accuracy of target search application

The accuracy of this model lies in the success of identifying the correct threshold parameter for clustering the input space. It is observed that even a small change in the threshold value can result in varying performance based on the testing error. Based on the value of clustering parameter the number of the rule can vary from 1 to a maximum of any value. Selecting the appropriate value which gives a high level of accuracy determines the optimal number of rules. A set of experiments has to be conducted with the training and testing images with different clustering parameters to identify an optimized rule based model. The optimized model obtained after experimenting with different parameters can be retained as a standard model for the further target search identification procedures.

5.3 Observations

The model has been undergone a multiple sets of experimentation to arrive at an optimized rule base. Clustering parameter has been the important parameter which changes every experiment. Later on, it is observed that selecting the training images also have an important role in getting the optimized rule base. There should be a balance in the pixel areas belonging to target and non-target class. This will reduce the over fitting of rules to a particular class and hence improves performance. When the training image includes non-target areas as a significant portion of it, it will lead to the over fitting of rules to non-target class and vice versa. So this can be overcome by giving equal importance to target class and non-target class in the training image.

6 Conclusion and Future Work

Fuzzy rule based system is a renowned method to deal with imprecise data. Incorporating neural network to the fuzzy rule based systems makes the system an adaptive model [21]. Rule optimization is to a large extent influenced by the clustering techniques used in these neuro fuzzy models. The improved performance of DENFIS model for rule optimization owes to the evolving cluster method implemented. ECM enhances the performance by identifying broad visualizations of similar data points into the same cluster and hence capable of optimizing rules. The DENFIS model has been used for learning and testing of target search applications. The model has been modified to a zero order Takagi-Sugeno Inference system to get better performance results. A feature selection technique incorporated in the model leads to an optimized structural model. A reduced rule base in the form of IF…THEN is obtained by selected relevant attributes which can successfully identify the data element with accuracy and interpretability of information in a useful way [1]. This model helps in reducing the complexity of manipulating with many rules.

The model can be extended by considering more number of feature channels like shape, size, and hue and fine-tuned to overcome the over fitting and under fitting challenges faced while selecting the training input.

References

1. Amudha, J., Radha, D., Smitha, S.: Analysis of fuzzy rule optimization models. Int. J. Eng. Technol. 7(5), 1564–1570 (Nov 2015) ISSN: 0975-4024
2. Mamdani, E.H., Assilian, S.: An experiment in linguistic synthesis with a fuzzy logic controller. Int. J. Man-Mach. Stud. 7, 1–13 (1975)
3. Sugeno, M., Yasukawa, T.: A fuzzy-logic-based approach to qualitative modeling. IEEE Trans. Fuzzy Syst. 21(1), 7–31 (1993)
4. Pal, S.K., Mitra, S.: Neuro-fuzzy Pattern Recognition: Methods in Soft Computing. Wiley, New York (1999)
5. Kosko, B.: Neural Networks and Fuzzy Systems. Prentice-Hall, Englewood Cliffs, NJ (1991)

6. Takagi, H.: Fusion technology of fuzzy theory and neural network Survey and future directions. In: Proceedings of International Conference on Fuzzy Logic Neural Networks, Iizuka, Japan, pp. 13–26, (1990)
7. Hayashi, Y., Buckley, J.J.: Approximations between fuzzy expert systems and neural networks. Int. J. Approx. Reas. **10**, 63–73 (1994)
8. Julia, TJ., Amudha, J., Sanjay, G.: A Survey on Spiking Neural Networks in Image Processing, Advances in Intelligent Informatics, pp. 107–115. (2015)
9. Kwin, C.T., Talei, A., Alaghmand, S., Chua, L.H.: Rainfall-runoff modeling using dynamic evolving neural fuzzy inference system with online learning. In: 12th International Conference on Hydro Informatics, HIC 2016, Elsevier Procedia Engineering, pp. 1103–1109, (2016)
10. Yen, J.H., Yang S.M., Jeon, H.T: Structure optimization of fuzzy-neural network using rough set theory. In: IEEE International Fuzzy Systems Conference Proceedings, Seoul, Korea, 22–25 Aug. (1999)
11. Chakraborty, D., Pal, N.R.: Integrated feature analysis and fuzzy rule based system identification in a neuro fuzzy paradigm. IEEE Trans. Syst. Man Cybern. Part B Cybern. **31**(3), June (2001)
12. Chakraborty, D., Pal, NR.: A neuro fuzzy scheme for simultaneous feature selection and fuzzy rule based classification. IEEE Trans. Neural Netw. **15**(1), Jan. (2004)
13. Eiamkanitchat, N., Theera-Umpon, N., Auephanwiriyakul, S.: A novel neuro-fuzzy method for linguistic feature selection and rule-based classification. In: 2nd International Conference on Computer and Automation Engineering, vol. 2, p. 247–252. (2010)
14. He, J., Pan, W.: A DENCLUE based approach to neuro-fuzzy system modeling, 2010. In: 2nd International Conference on Advanced Computer Control (ICACC), vol. 4, pp. 42–46. (2010)
15. Pizzileo, B., Li, K., Irwin, GW., Zhao, W.: Improved structure optimization for fuzzy-neural networks. IEEE Trans. Fuzzy Syst. **20**(6), Dec (2012)
16. Frintrop, S.: VOCUS A Visual Attention System for Object Detection and Goal-Directed Search. Springer, Berlin (2006)
17. Robnik-Sikonja, M., Kononenko, I.: Theoretical and empirical analysis of ReliefF and RReliefF. Mach. Learn. **(53)**1–2, 23–69, Oct.–Nov. (2003)
18. Kira, K., Rendell, L.: A practical approach to feature selection. In: Sleeman, D., Edwards, P. (eds.) International Conference on Machine Learning, pp. 368–377. Aberdeen, July (1992)
19. Song, Q., Kasabov, N.: Dynamic evolving neuro-fuzzy inference system (DENFIS): on-line learning and application for time-series prediction. IEEE Trans. Fuzzy Syst. **10**(2), 144–154, Apr. (2000)
20. Anandaraj, G., Anupriyanka, P., Aaswin, K., Sri Saravanan, JB., Sumangala, V., Gopinath, R., Santhosh, K.: Robust features for load independent machine fault diagnosis using feature selection. Int. J. Mech. Prod. Eng. **2**(9), Sept. (2014). ISSN: 2320–2092
21. Kaur, R., Sangal, AL., Kumar,K.: Modeling and simulation of adaptive neuro-fuzzy based intelligent system for predictive stabilization in structured overlay networks. Eng. Sci. Technol. Int. J. **20**(1), 310–320 Feb. (2017)

DNA Based Image Steganography

R. E. Vinodhini and P. Malathi[(✉)]

Department of Computer Science and Engineering, Amrita School of
Engineering, Amrita Vishwa Vidyapeetham, Coimbatore, India
cb.en.p2csel5029@cb.students.amrita.edu,
p_malathy@cb.amrita.edu

Abstract. Providing security to all kinds of data is an essential task in the
world of digital data. The data can be secured using the several methods like
cryptography, steganography and watermarking. Steganography hides the data
behind the cover object to secure it. To increase the security of the data, dual
cover objects can be used. In this paper, image and DNA are the two covers,
which are used to secure the data. The DNA insertion algorithm is used to hide
the data in the DNA sequence and it results a fake DNA sequence. The capacity,
payload, BPN and Cracking Probability are calculated for the fake DNA
sequence to ensure the security. The fake DNA is hidden in a cover image using
LSB and F5 algorithm. The MSE and PSNR values are calculated and compared. To conform the security to the data, steganalysis method called histogram
attack is performed and compared. This works shows that F5 Algorithm is better
compared to LSB algorithm in the second layer of hiding the data.

Keywords: Insertion method · LSB algorithm · F5 algorithm

1 Introduction

Steganography is an art of hiding the data behind a cover object. It is one of the
traditional techniques used in olden days to communicate secret messages. The word
steganography originates from the Greek and it means "concealed writing". The area of
steganography is divided into different divisions based on the cover objects used. The
cover objects like Image, DNA, Video, Audio and Text files can be used to hide the
secret data. The image steganography uses image as a cover object and it has several
techniques to hide the secret data inside the image. The image steganography has two
broad domains known as spatial and frequency domain. In spatial domain the image
pixels are directly manipulated to hide the secret data inside the image. The frequency
domain uses the orthogonal transformations to hide the data inside the image. In image
steganography, there are several techniques available to hide the data inside the cover
object. The least significant bit technique is one of the most common technique used
for hiding the secret data inside the cover object. DNA steganography is an emerging
technique of hiding a secret message inside a DNA. The advanced technology, computing enables the simulation of biological structure of DNA (Fig. 1).

© Springer International Publishing AG 2018
D. J. Hemanth and S. Smys (eds.), *Computational Vision and Bio Inspired Computing*,
Lecture Notes in Computational Vision and Biomechanics 28,
https://doi.org/10.1007/978-3-319-71767-8_70

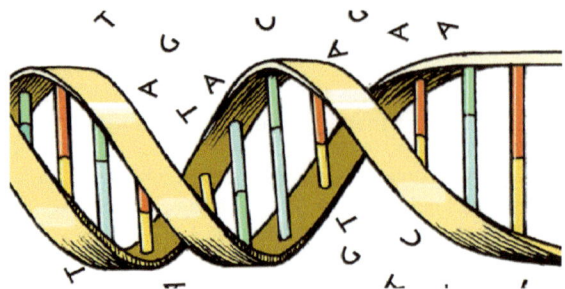

Fig. 1 Structure of DNA

Deoxyribonucleic acid is the base of all lifestyles of all living things. The nucleotide forms the biological macromolecule called DNA. A single base forms a nucleotide and as the whole there are four types of bases [Adenine (A), Thymine (T), Cytosine (C), and Guanine (G)]. From EBI (European Bio Informatics Institute) the first DNA sequence litmus with 154 nucleotides is retrieved [1].

In this paper, dual cover steganography is used to secure the data from the attackers. DNA insertion algorithm is used to hide the data in the DNA sequence. The capacity, payload, BPN values are calculated. The cracking probability of the proposed method is very low and it is impossible to the fake DNA is hidden in a cover image using LSB and F5 algorithm. The MSE and PSNR values are calculated and compared. To conform the security to the data, steganalysis technique called histogram attack is performed and compared.

The rest of this paper is organized as follows: Sect. 2 explains related works, problem formulation is done in Sects. 3 and 4 gives the brief of results and discussions and the conclusion with future enhancement of the research work is done in Sect. 5.

2 Related Works

Chan et al. [2] proposed a paper on hiding data in the cover image using the simple LSB substitution method. In this paper, the secret data is hidden inside the cover image by substituting the least significant bits of the image. The substitution method increases the image quality than LSB insertion method. The MSE and PSNR values are calculated for the stego and a cover image.

Das et al. [3] proposed a system on highly secured DNA using image steganography. The image used as a primary cover object and the secondary object is a DNA sequence which forms a dual cover. Before the embedding of secret data inside the cover object, it is encrypted to increase the security. The DNA with secret data is embedded inside an image which gives extra security to the data. During the extraction process the Meta data related to secret message is acquired to verify the integrity of the secret data steganalysis techniques like statistical attack and visual attack are performed.

An LSB algorithm using insertion method is developed by Sutone et al. [4]. This article uses LSB insertion technique to embed and extract the secret data from the cover image. The image files like BMP, JPEG, and PNG files are used as a cover object. To hide the data, the data bits are XOR-ed between each character of the bits of cover object. During the decoding process the message is extracted from the image by applying XOR operation again. The PSNR values are calculated for image files and they are compared.

A paper that relates the embedding efficiency of the LSB technique in the domain of spatial and transform in image steganography is proposed by Malathi et al. [5]. In this paper, F5 and matrix embedding are combined with the LSB method. This combination is applied to the transform and spatial domain of the images. The size of 256 × 256 images is taken as the cover image and the length of 3096 is taken as the message. The outcome of the techniques are compared by using histograms and the performance is measured by using the PSNR and MSE values.

3 Proposed Work

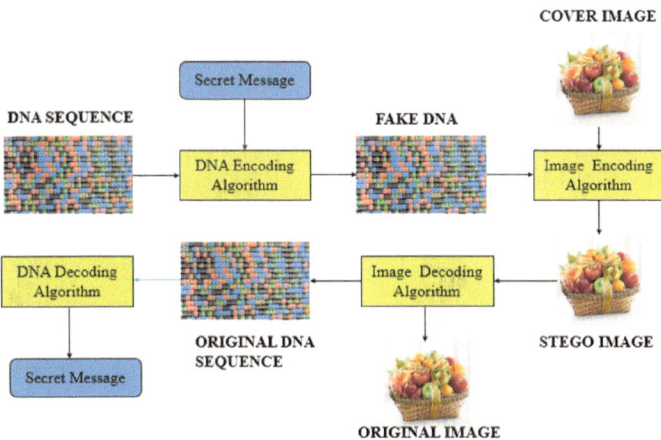

Fig. 2 Proposed architecture

The secret data like password, phone number, key values, and pin number are used in this paper to secure it behind the cover object. The secret data is converted into binary using the ASCII values of the characters. The binary value of the message is embedded into the DNA sequence using the insertion technique. The reference DNA sequence is binary converted and are segmented using the key value. The binary converted data is embedded at the end of the segment using DNA insertion technique. The segments are concatenated to form the binary sequence. The binary sequence in turn converted into dictionary method. This results the fake DNA sequence. The fake DNA sequence is hidden inside the cover image in the LSB technique and also the proposed technique uses the F5 algorithm to hide the fake DNA inside the cover image. Before the embedding process the binary converted and it uses LSB and F5 technique to hide inside the cover image. The stego image undergoes several steganalysis technique called histogram attack to ensure the security of the hidden data. The metrics like MSE, PSNR, during the implementation for conforming the output quality and the secrecy of the hidden message (Fig. 2).

3.1 DNA Insertion Algorithm

3.1.1 Embedding Algorithm

//Input: Secret data $S \in \{0, 1\}^S$, Reference DNA $d \in \{0, 1\}^s$
//Output: Fake DNA sequence with hidden data

Step 1   The secret data is converted into binary sequence.
Step 2   The reference DNA D is binary converted by using the dictionary method (A = 00, C = 01, G = 10, T = 11).
Step 3   The random sequence R_1, R_2, R_n are generated and the value t is found using the formula $E_i = \sum_{n=1}^{i} R_n > |S|$ and the key value is chosen from the random sequence R_1, R_2, R_{n-1}.
Step 4   Choose key value from the above random sequence R_1, R_2, R_{t-1}. (Here k = 3) and equally segment D with k value results d_1, d_2.
Step 5   The binary sequence of S_t in appended to front d_1, d_2.
Step 6   Transform the binary sequence to DNA base using dictionary method.

3.1.2 Decoding Algorithm

//Input: Fake DNA $f \in \{0, 1\}^s$
//Output: secret data, Reference DNA

Step 1   The fake DNA f is binary converted by using the dictionary method (A = 00, C = 01, G = 10, T = 11)
Step 2   Using random seed U and V, generate random sequence $U_1, U_2....U_n...$and $V_1, V_2...V_n...$
Step 3   The value p is found using the formula $\sum_{n=1}^{t}(U_n + V_n) \leq |f_1|$ and the sequence formed is $U_1 + V_1...U_P + V_P$, then the fake DNA f is segmented using the sequence.
Step 4   For each segment n, $1 \leq n, \leq p$, extract first U_n bits called $S_n.$ For segment n, $1 \leq n \leq p$, extract last V_n bits called f_n.
Step 5   Concatenate all f_n, $1 \leq n, \leq p$, results the reference DNA by applying inverse function of dictionary method.
Step 6   Concatenate all S_n, $1 \leq n \leq p$, results the secret data [6].

3.2 Least Significant Bit Algorithm

The LSB mechanism is used here to hide the fake DNA inside the RGB image. The binary conversion of fake DNA and the cover image is done. The fake DNA binary sequence's each bit is replaced with the LSB of the RGB color image. The obtained binary sequence with secret data is converted into a stego-image. This process is reversed by using the stego-image as input and the fake DNA is retrieved [7] (Fig. 3).

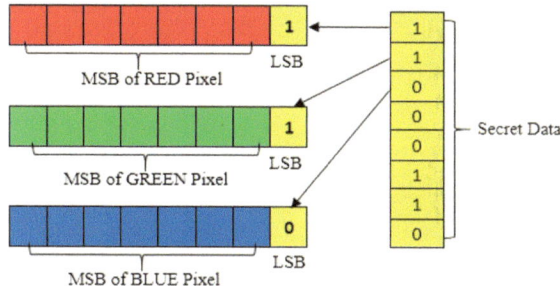

Fig. 3 Least significant bit technique

The algorithm for the LSB embedding and extraction is as follows.

3.2.1   LSB Embedding Algorithm

//Input: secret data (Fake DNA sequence) s € {0,1}m, cover image a € At
b = a;
s = min(s,t);
//If secret_data is longer than capacity available, truncate it
for x = 1 to s
{

$$B[x] = a[x] + s[x] - a[x] \; mod \; 2;$$

}
//b is the stego image with secret data bits

3.2.2   LSB Extraction Algorithm
LSB Extraction Algorithm

//Input: stego image b£At
for x = 1 to s
{

$$s[x] = b[x] \; mod \; 2;$$

}//s is the extracted secret data

3.3   F5 Algorithm

The F5 algorithm embeds the binary converted fake DNA sequence bits in the LSBs of DCT coefficients. So, embed instead of flipping the LSB the DCT coefficient value is decreased by value 1. The coefficient is equal to zero are skipped by the F5 algorithm to avoid easily detectable artefacts. Then the secret message is embedded the coefficient is equal to one is modified to zero. For receiver read the message bits from LSB of non-zero AC DCT coefficient. Shrinkage leads to loss of embedded data [8] (Fig. 4).

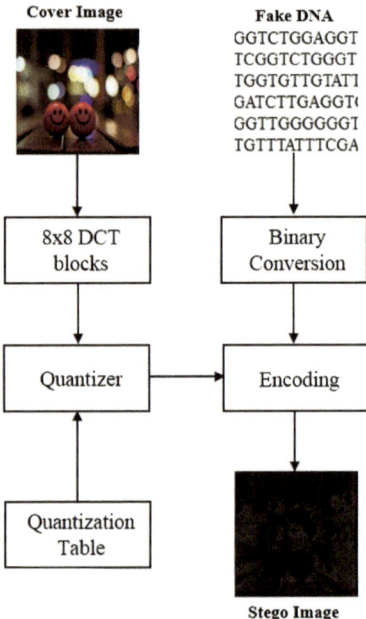

Fig. 4 Flow diagram for F5 algorithm

3.3.1   F5 Embedding Algorithm
//Input: secret data (Fake DNA sequence) s $\in \{0,1\}^s$, quantized JPEG DCT coefficients a $\in A^t$

```
B = a;
x = 1 ; y = 1;
while (x <=s) & (y <= t)
{
        If (a[x] ≠ 0) & (a[y] is not a DC term)
        {
                If LSB(a[y]) = s[x]
                {        x = x+1;
                }
                Else
```

```
                {
                B[y] = a[y] - sign(a[y])
                If b[y] ≠ 0
                {
                        x = x+1;
                }
        }
}
        y = y+1;
}
```

3.3.2 F5 Extracting Algorithm

```
//Input: quantized JPEG DCT coefficients b ∈ A_t
x = 1; y = 1;
while (y <= t)
{
        If(b[y] ≠ 0) & (b[y] is not DC term)
        {
                s[x] = LSB (b[y])
                {
                        x = x + 1;
}
        }
        y = y+1;
}
```

4 Performance Metrics

The BPN, Capacity, Payload and Cracking probability are calculated for measuring the quality of the fake DNA. The metrics like Mean Squared Error (MSE), Peak signal to noise ratio (PSNR) are calculated for confirming the quality of the stego image. The cracking probability is calculated by the following formula:

$$\mathbf{P(r)} = \frac{1}{1.63 \times 10^8} \times \frac{1}{\mathbf{n}-1} \times \frac{1}{2^{\mathbf{m}}-1} \times \frac{1}{2^{\mathbf{s}}-1} \times \frac{1}{24}$$

Here

- n is the number of bits in the Fake DNA sequence
- m is the number of bits in the secret data
- s is the number of bits in the reference DNA sequence.

5   Steganalysis

5.1   Histogram Analysis

Histogram analysis is a statistical test that is performed to detect the significant changes in brightness, frequency, and color between the cover image and the stego-image. Based on the variation between the cover and stego-images the histogram structure changes [9].

6   Results and Discussion

The implementation of the system is carried out using MATLAB 2015 and the results are compared. The techniques like DNA insertion, LSB and F5 are performed.

The following table shows the explanation for tested DNA (Table 1).

Table 1 Explanation for tested DNA

Locus	Nucleotide	Species definition	Common name
DP001064	120054	Oryctolagus cuniculus ENCODE region ENm011 genomic scaffold	Rabbit
DP001042	174523	Oryctolagus cuniculus ENCODE region ENr332 genomic scaffold	Rabbit
AC254805	147388	Canis lupus familiaris strain Doberman Pinscher clone rp81-5g12	Dog
AC254806	173111	Canis lupus familiaris strain Doberman Pinscher clone rp81-2a16	Dog
AC254802	144410	Canis lupus familiaris strain Doberman Pinscher clone rp81-181h12	Dog

The above table gives the different types of DNA sequence and its scientific names and the number of nucleotides in the DNA [10].

The following table shows the results of payload, capacity, BPN for the different DNA sequence for 10 k secret data (Table 2).

Table 2 Result and analysis using the cover "DNA"

Locus	Payload	Capacity	BPN
DP001064	40,000	160,054	0.5
DP001042	40,000	214,523	0.37
AC254805	40,000	187,388	0.42
AC254806	40,000	213,111	0.37
AC254802	40,000	184410	0.43

6.1   Using the Second Layer Cover Object

6.1.1   Image Results (LSB Technique)

The size of the cover image is 440 × 300 pixels and it is a high-quality RGB image. The stego image and cover image are of same size and the cover image with the secret data embedded within that image (Fig. 5).

Fig. 5 LSB Technique of cover image and stego image (440 × 300)

Steganalysis on LSB technique

The histogram of stego-image and cover image helps to find the variation in pixels. The change of resolution and pixels are compared using the histograms. The first histogram represents the cover image and the second represents the histogram of stego-image. By comparing those histograms, the variation in resolution of a pixel is easily identified because of the changes in the histogram. So, by using the histogram the LSB technique can be easily attacked to retrieve the hidden fake DNA sequence.

Fig. 6 F5 embedding algorithm cover image and stego image

6.1.2   Image Results (F5 Embedding Algorithm)

The cover image size is 440×300 pixels and it is a high-quality RGB image. The stego image is the DCT converted image of the size similar to that of the cover image with the secret message inserted within that image (Fig. 6).

Steganalysis on F5 technique

The first histogram represents the histogram of the cover image and the second represents of stego-image. These histograms help to attack the F5 technique then it is very hard for the attackers to detect using the histogram.

6.1.3   Tested Images

Figure 7 and Table 3.

Football.jpg   Flower.jpg   Tree.jpg

Baboon.png   Peppers.png

Fig. 7 Tested images

Table 3 Implementation and evaluation of above techniques in tested images for 800 characters

Techniques	Images	MSE	PSNR
LSB algorithm	Football.jpg	0.498406	51.15
	Flower.jpg	0.514668	51.05
	Tree.jpg	0.502551	51.12
	Peppers.png	0.484375	51.27
	Baboon.png	0.507015	51.08
F5 embedding algorithm	Football.jpg	0.194196	55.24
	Flower.jpg	0.201531	55.08
	Tree.jpg	0.197066	55.18
	Peppers.png	0.131059	56.95
	Baboon.png	0.138393	56.17

7 Conclusion

The proposed system with dual covers provides more security than the other existing techniques. The two covers used to hide the data in this approach are DNA sequence and Image in-order to provide higher security. The DNA insertion technique is used to hide the secret data inside DNA sequence. The resulting fake DNA sequence is hidden inside the image using LSB and F5 algorithms. The MSE, PSNR, values are calculated for the different techniques and they are compared. The histogram analysis is performed to check the security of the data stored in the cover image and the cracking probability is calculated for ensuring the security provided by the DNA. The insertion techniques have the lowest cracking probability in DNA steganography and LSB, F5 algorithms are used. The F5 provides better security compared to the LSB technique and it outperforms the existing systems.

References

1. Wang, Z., Zhao, X., Wang, H., Cui, G.: Information hiding based on DNA steganography. In: 4th IEEE International Conference on Software Engineering and Service Science, pp. 946–949. IEEE (2013)
2. Chan, C.K., Cheng, L.M.: Hiding data in images by simple LSB substitution. Pattern Recogn. **37**(3), 469–474 (2004)
3. Das, P., Nirmalya, Kar.: A highly secure DNA based image steganography. In: International Conference on Green Computing Communication and Electrical Engineering, pp. 1–5. IEEE (2014)
4. Sutaone, MS., Khandare, MV.: Image Based Steganography Using LSB Insertion, pp. 146–151 (2008)
5. Malathi, P., Gireeshkumar, T.: Relating the embedding efficiency of LSB steganography techniques in spatial and transform domains. Procedia Comput. Sci. **93**, 878–885 (2016)
6. Shiu, H.J., Ng, K.L., Fang, J.F., Lee, R.C., Huang, C.H.: Data hiding methods based upon DNA sequences. Inf. Sci. **180**(11), 2196–2208 (2010)
7. Karim SM, Rahman MS, Hossain MI.: A new approach for LSB based image steganography using secret key. In: 14th International Conference on Computer and Information Technology (ICCIT), pp. 286–291. IEEE (2011)
8. Hmood, A.K., Jalab, H.A., Kasirun, Z.M., Zaidan, A.A., Zaidan, B.B.: On the capacity and security of steganography approaches: An overview. J. Appl. Sci. **10**(16), 1825–1833 (2010)
9. Meera, M., Malathi, P.: An improved embedding scheme in compressed domain image steganography. Intern. J. Appl. Eng. Res. **10**(55), 1933–1937 (2015)
10. NCBI Homepage https://www.ncbi.nlm.nih.gov/
11. Fridrich, J.: Steganography in Digital Media: Principles, Algorithms, and Applications. Cambridge University Press, 12 Nov. (2009)
12. Lou, D.C., Hu, C.H.: LSB steganographic method based on reversible histogram transformation function for resisting statistical steganalysis. Inf. Sci. **188**, 346–358 (2012)

Performance Analysis of Combined k-mean and Fuzzy-c-mean Segmentation of MR Brain Images

K. V. Ahammed Muneer$^{(\boxtimes)}$ and K. Paul Joseph

National Institute of Technology, Calicut, India
ahammedcet@gmail.com, paul@nitc.ac.in

Abstract. Magnetic resonance imaging (MRI) plays a vital role among the advanced techniques for the imaging of internal organs. It is the least harmful method compared to other existing medical imaging techniques like computed tomography scan, X-ray etc. Image segmentation is the basic step to analyse images and hence to extract data from them. In this paper, we concentrate on brain MRI segmentation, where the performance of algorithms such as k-mean, fuzzy-c-mean (FCM) and their combination (k-FCM) is evaluated. In the proposed methodology, MR brain images of different tumor types like meningioma, sarcoma, glioma, etc. are preprocessed and separate segmentation are being performed using k-mean and FCM methods. Further, the k-mean segmented image is given to the FCM and their performance is compared. The hybrid segmentation scheme gives better results for extraction of tumor regions. The segmented image can be given to a good classifier to detect tumor types and hence the physicians can execute better treatment.

Keywords: MRI · Brain tumor segmentation · FCM · k-mean and k-FCM

1 Introduction

Magnetic resonance imaging (MRI) is a commonly used imaging method in medical field. Several diseases like tumor, alzheimer, epilepsy etc. are diagnosed from MR brain images. So, it is very important to segment out the affected portion in the brain images. The main objective of medical image segmentation is to divide it into various anatomical structures, thereby separating regions of interest such as tumors, blood vessels, etc. from their background.

A lot of works have been done in the field of image segmentation, especially in MR images. Basic image segmentation methods and their comparative study is done in [1]. A survey on MRI based tumor segmentation methods are given in [2, 3]. Though there exist many algorithms, FCM and its modified versions are widely used for MRI segmentation. Kernalized FCM and its application in MR image segmentation is described in [4, 5]. Y. K. Dubey et al. proposed rough set based intuitionistic fuzzy clustering in

© Springer International Publishing AG 2018
D. J. Hemanth and S. Smys (eds.), *Computational Vision and Bio Inspired Computing*,
Lecture Notes in Computational Vision and Biomechanics 28,
https://doi.org/10.1007/978-3-319-71767-8_71

[6]. A qualitative comparison of FCM and k-mean segmentation with histogram guided initialization, on tumor edema complex MR images is performed in [7]. Eman Mohammed et al. in [8] performed a hybrid model of k-mean and FCM in a complex way such that they applied level set thresholding, tumor contouring etc. Moreover, [9] and [10] utilized the spatial information of the image for segmentation purpose.

On the contrary, our proposed methodology is more simple where the raw images are preprocessed and then the hybrid segmentation so called k-FCM is being performed. The segmented results are accurate and better compared to the other state of the art. The methodology and results are discussed in the following sections.

2 Methodology

Figure 1. shows the block diagram of the proposed method for tumor segmentation. Various MR tumor images of 256×256 in plane resolution are collected from the Harvard medical school website [11]. The images are of different tumor categories like Meningioma, sarcoma, metastatic-bronchogenic carcinoma etc.

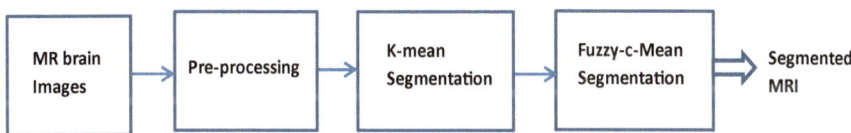

Fig. 1 Block diagram of the proposed scheme

2.1 Preprocessing Stage

Preprocessing involves image enhancement and skull stripping. Here contrast improvement is done for image enhancement where *imadjust* function in MATLAB is used. Skull stripping is done by double thresholding of grey levels in MRI. This technique converts the image to binary form. This is achieved by generating a mask and setting each pixel in the range of 0.1 * 255 to 0.88 * 255 as 1, that corresponds to white and the remaining pixels as 0, that corresponds to black. This is because, the intensity level of normal human brain varies from 10 to 88% [12]. Obviously non-brain tissue pixels get discarded by skull stripping.

$$f(x_{ij}) = \begin{cases} 0.5 * x_{ij}, & x_{ij} < 0.1 \\ 0.1 + 1.5 * (x_{ij} - 0.1), & 0.1 \leq x_{ij} \, and \, x_{ij} \leq 0.88 \\ 1 + 0.5 * (x_{ij} - 1), & x_{ij} > 0.88 \end{cases} \quad (1)$$

2.2   Segmentation

From the preprocessed images given in Fig. 2, the exact tumor portions get separated out using the following algorithms.

k-mean clustering

k-mean is a hard clustering method where a single point strictly belongs to only one cluster. Initially, we fix the cluster number as three, since a raw MRI contains normal brain tissue, tumor tissue and background. The algorithm always clusters the point nearest to the centroid or mean. The mean value of the clusters is again calculated where the resulting mean is fixed as new mean. If the error between the succeeding means is greater than a threshold, then the point to the clusters are assigned with the new mean and the process will be repeated until the error threshold is minimum. Here the objective function J is minimized as per the following equation.

$$J = \sum_{j=1}^{K} \sum_{i=1}^{N} \left\| x_{i,j} - c_j \right\|^2 \tag{2}$$

where K is the number of clusters and N is the number of data points.

FCM Clustering

FCM is a clustering method where each data point may belong to multiple groups where the degree of membership is governed by the probability distribution over the clusters. FCM is useful when the number of required clusters are known. In our application, the cluster number is fixed as three. Here the following objective function is minimized.

$$J = \sum_{i=1}^{N} \sum_{j=1}^{C} \delta_{ij} \left\| x_i - c_j \right\|^2 \tag{3}$$

where, N is the number of data points, C is the number of clusters, c_j is the center vector of cluster j and δ_{ij} is the degree of membership for the ith data point of x_i
in cluster j. With respect to the data point x_i, the fuzziness membership to jth cluster is determined as follows.

$$\delta_{ij} = \frac{1}{\sum_{k=1}^{C} \left(\frac{\left\| x_i - c_j \right\|}{\left\| x_i - c_k \right\|} \right)^{\frac{2}{m-1}}} \tag{4}$$

where m is the fuzziness coefficient. The centroid c_j for each clusters is calculated as per the following equation.

$$c_j = \frac{\sum_{i=1}^{N} \delta_{ij}^m * x_i}{\sum_{i=1}^{N} \delta_{ij}^m} \tag{5}$$

Hybrid Clustering

In the hybrid algorithm (k-FCM), the combined advantage of both the k-mean and FCM clustering is exploited. For that, k-mean segmented image is given as input to the FCM segmenting section. FCM results in three clusters as normal, abnormal tissue and the background. In the case of k-FCM, it results only two clusters, one with tumor portion and the other with zero intensity value. It reveals that the k-FCM segmented image is more useful for further analysis.

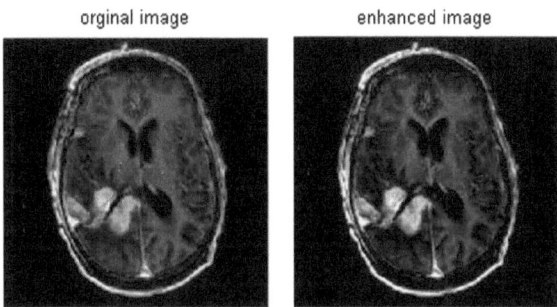

Fig. 2 Preprocessed output of a tumor image

3   Results and Discussions

The experiment is carried out in intel core i3 processor and windows 8.1 operating system. The simulations are done using MATLAB version R2013b. Figure 2 illustrates the preprocessed result of a tumor affected MR image. Here the contrast of the image is increased. The segmented results for the three algorithms are depicted in Fig. 3a–c. In the case of k-mean segmentation, the white matter content in tumor portion cannot be easily distinguished, but segmentation time required is very low. In FCM segmentation, accuracy is fine and the execution time required is higher compared to k-mean. FCM shows each of the clusters in different colours and has higher visual perception as well as an abrupt cutting in each cluster boundary. The k-FCM is comparatively faster and at the same time visual perception quality is excellent.

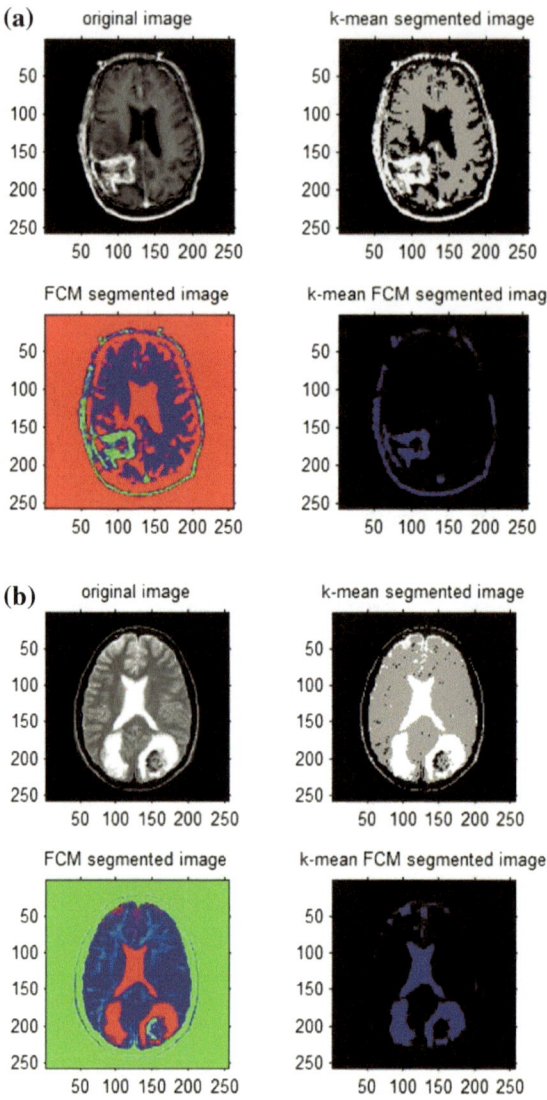

Fig. 3 **a** Segmented results of meningioma affected brain MRI, **b** Segmented results for sarcoma affected brain MRI, **c** Segmented results for carcinoma affected brain MRI

The speed of execution for each type of segmentation algorithms is compared and is executed for a particular image where the number of iterations is set as 100. The performance evaluation is summarized as given in Table 1.

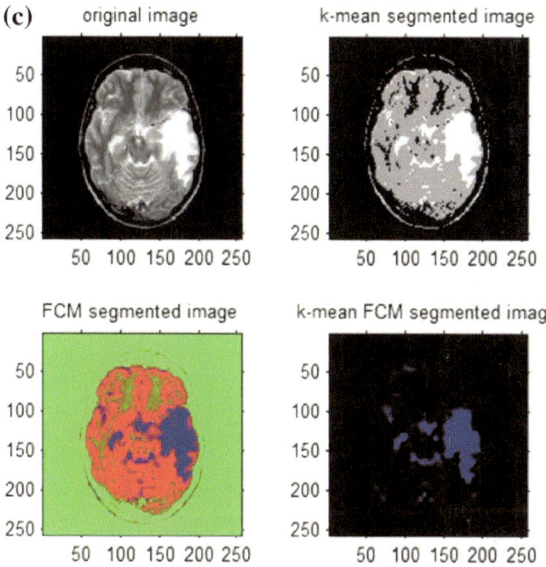

Fig. 3 (*continued*)

Hence it is obvious that k-FCM segmentation can be achieved with moderate speed and at the same time resulted image is found to have better quality.

Table 1 Performance of segmentation algorithms

Segmentation algorithm	Time elapsed in Seconds
FCM	41.59
k-FCM	27.81
k-mean	6.3

4 Conclusions

Various MR brain tumor images are segmented using k-mean, FCM and k-FCM clustering algorithms. As per the performance measures we have arrived at, FCM segmented image is found to have better visual perception quality. But in terms of computational time, k-mean algorithm is better. On an average, the hybrid method (k-FCM) performs better where it can extract the tumor portion efficiently. The performance of k-FCM can be considered to be better when compared to the other modified FCM algorithms mentioned in the state of the art, as it is fully automatic where it requires no manual intervention.

References

1. Khan, A.M., Ravi, S.: Image segmentation methods: a comparative study. Int. J. Soft Comput. Eng. **3**, 84–92 (2013)
2. Gordillo, N., Montseny, E., Sobrevilla, P.: State of the art survey on MRI brain tumor segmentation. Magn. Reson. Imaging **31**, 1426–1438 (2013)
3. Liu, Jin, Li, Min, et al.: A survey of MRI based brain tumor segmentation methods. TSINGHUA Sci. Technol. **19**, 578–595 (2014)
4. Zanaty, E.A.: Determining the number of clusters for kernelized fuzzy-c-mean algorithm for automatic medical segmentation. Egypt. Inf. J. **13**, 39–58 (2012)
5. Kannan, S.R., et al.: Effective fuzzy-c-means based kernel function in segmenting medical images. Comput. Biol. Med. **40**, 572–579 (2010)
6. Dubey, Y.K., et al.: Segmentation of brain MR images using rough set based intuitionistic fuzzy clustering. Bio-Cybern. Biomed. Eng. **114**, 1–14 (2016)
7. Madhukumar, S., Santhiyakumari, N.: Evaluation of k-maens and fuzzy-c-means segmentation on MR images of brain. Egypt. J. Radiol. Nucl. Med. **46**, 475–479 (2015)
8. Abdel-Maksoud, E., Elmogy, M., Al-Awadi, R.: Brain Tumor Segmentation Based on Hybrid Clustering Technique, Faculty of computers and information, Cairo University (2015)
9. Chuang, K.S., Tzeng, H.L et.al.: Fuzzy-c-means clustering with spatial information for image segmentation. Comput. Med. Imaging Graph. **30**, 9–15 (2006)
10. Li, B.N., Chui, C.K., et al.: Integrating spatial fuzzy clustering with level set methods for automated medical image segmentation. Comput. Biol. Med. **41**, 1–10 (2011)
11. The whole brain Atlas, http://www.med.harvard.edu/AANLIB/01/10/2012
12. Parveen, Singh, A.: Detection of brain tumor in MR images using combination of fuzzy-c-means and SVM. In: 2nd International Conference on Signal Processing and Integrated Networks, SPIN, 2015

Generation of Author Topic Models Using LDA

G. S. Mahalakshmi, G. Muthu Selvi[✉], and S. Sendhilkumar

Department of Computer Science and Engineering, Anna University, Chennai,
Tamilnadu, India
gsmaha@annauniv.edu, gmuthuselvi16@gmail.com,
ssk_pdy@yahoo.co.in

Abstract. Copyright and ownership of research ideas is questionable as to
which author the credit should be attached to. Mining author contributions has to
be approached more semantically to solve this issue. Representing the research
ideas using topic distributions substantiate the measuring of author contributions. Author Topic Models (ATM) are generally obtained by applying topic
modeling approaches over an author's research articles. ATMs form the blueprints of an author. Given a research paper and the blueprints of it's' authors,
identifying the contribution of every author in the article becomes easy. This
paper proposes the generation of ATMs by applying Latent Dirichlet Allocation
(LDA).

Keywords: Document topic model · Author-topic model · Author profile
model · Latent Dirichlet allocation · Hierarchical Dirichlet process

1 Introduction

Determination of authorship roles is quite interesting and complex [1]. Representing
the author semantically would motivate to finding the author contribution and part of
ownership in legal research issues. To resolve potential research controversies, a
semantic representation of the author is very much essential. To perform semantic
contribution analysis, analyzing the research article is necessary. Analyzing the contribution of an author [2] in a multi-author paper is a tedious task. As the writing of
research article involves high creative skills, exact words might not appear in the
manuscript as like the author's published articles. Every author has unique style and
idea which is reflected during manuscript writing. Therefore, representing the research
article as topics is quite convincing. This paper proposes the methodology behind
generation of author topic models (ATMs) which represent a given author. These
ATMs shall be applied for author attribution [3] or author contribution research [4, 5].

© Springer International Publishing AG 2018
D. J. Hemanth and S. Smys (eds.), *Computational Vision and Bio Inspired Computing*,
Lecture Notes in Computational Vision and Biomechanics 28,
https://doi.org/10.1007/978-3-319-71767-8_72

2 Related Work

Modeling the entire research article with topic models is the fundamental step for Author Contribution Mining [6]. It brings out the most frequently and most likely used words in the documents. Topic models are a measure to discover the themes in large text collections [7]. Topic Models work by defining a probabilistic representation of the latent factors of corpora called topics. Girgis et al. [8] discusses about the Authorship Attribution using different Topic Modeling Algorithms. Pratanwanich N et al. [9] proposed LDA implementation using the Collapsed Gibbs Sampling Method. Rubin TN et al. [10] models Flat-LDA, Prior-LDA and Dependency-LDA but the performance rapidly drops off as the total number of labels and the number of labels per document increase.

Author-Topic Model (ATM), works with the aim of identifying topic word distribution of anonymous documents and authors. ATM identifies the topics of Authors based on the publishing history Dataset. The author publishing history is generally obtained from valid bibliographic repositories. Authorship profiling is the problem of determining the characteristics of a set of authors based on the text they produce [3].

3 Generation of ATMs

Author Topic Models are a means to model an author's research interests. If the topics are modeled only from one research paper, we refer to it as Document Topic Model (DTM). When an author's entire research articles are collected, to form topic models, we refer to it as ATMs. If the author has published only one research article which forms the publishing history, ATM is same as the obtained DTM. However, document topics are far different from author topics. Author topics shall be obtained by aggregation of document topics of every research article of any author. Alternatively, all the research articles of that author shall be collectively fed as a single input document to form document topics, which might be the author topic as well. We refer to this approach as collective DTM. Author topics obtained by (i) DTM + aggregation and (ii) Collective DTM tend to differ in topic distributions. However, in this work, we assume obtaining author topics from aggregating the document topics of the author. Document topics are word distributions that arise from documents, while author topics are word distributions that characterize the authors. LDA uses only document topics, whereas ATM generates only author topics based on the publishing history Dataset to predict the Author's interest Fig. 1.

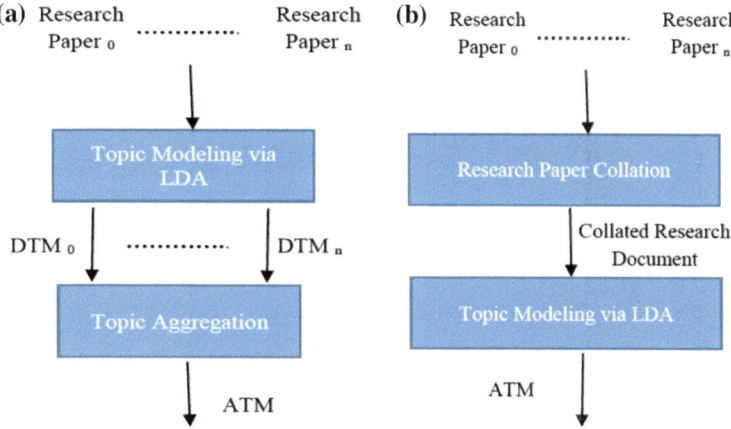

Fig. 1 Methods of ATM generation for an author **a** ATMs via topic aggregation **b** ATMs using collated research input

The research papers are from author's publishing history. The topics obtained are removed for duplicates and representative quantity of words per topic are chosen for record as ATMs. There are two things to note at this ATM generation: (1) LDAs obtain less words than the original research paper vocabulary. This is due to the very nature of LDA and which shall be rectified if appropriate state-of-art topic modeling algorithms are used. (2) The author might have publications across domains, and the topic models generated with mix of research domains will not be appealing enough. Therefore, for generation of ATMs we need to identify and categorize the research papers per domain [11] before feeding into the ATM generation system, which might improvise the quality and focus of the ATMs obtained.

Alternatively, there is another interesting challenge in ATM generation which is believed to be the modeling of a given author. A researcher might publish scientific articles either as single author [12], or in collaboration with other researchers [12], where, the researcher's name appears in one of the author positions. For analysis of ATMs and their quality, we assume scientific articles 'solely' published by the researcher. However, challenges might arise if a researcher has not published any articles exclusively and have always published in collaboration with others. Generation of ATMs to such authors is quite not possible if the sole author publishing history is assumed. However, feeding the research articles written in collaboration with other researchers might not be too convincing to model this author.

Let us assume, if researcher R_0 has only published with his co-authors ever, say R_1, R_2. Let us assume the research details as per Table 1. The two articles published by R_0 have different co-authors. Therefore, DTM shall be obtained for individual articles, and intersection of topics would provide the topics relevant to R_0. This is a mere assumption that the research concepts which are common to both the articles would belong to R_0 because the articles have R_0 as common author. This might not be true and the validity of the topics have to be examined by other intelligent machine learning methods to establish the claim. The issue shall also be solved if either R_1 or R_2 have sole author ATM. We refer to this as Subtraction. However, topics would still be different for R_0 when obtained via intersection or via subtraction, or via subtraction from different co-author (refer Table 1).

Table 1 Statistical approach for ATM generation

Article #	Authors	Topic modeling	Generation of ATM	Comments
P_0	R_0, R_1	DTM (P_0)	ATM	Recommended approach
P_1	R_0, R_2	DTM(P_1)	(R_0) = DTM $(P_0) \cap$ DTM(P_1)	
			ATM (R_0) = DTM $(P_0) - $ ATM(R_1)	Both ATM might be differing; the original issue remains recursive
			ATM (R_0) = DTM $(P_0) - $ ATM(R_2)	

4   Results and Discussion

The objective of generating ATMs is to obtain author blueprints which might facilitate research contribution analysis. Therefore, we have assumed the research articles from 'Journal of Informetrics'. The author for examination is 'Catherine Blake'. The author has 5 articles as sole author, and 15 articles as co-author (refer Tables 2, 3). Table 4 gives the top 10 words of top 5 topics for research articles in Table 2. Table 5 gives the top 10 words of top 5 topics for research articles in Tables 2, 3. It is evident that most words are common between Tables 4, 5, which indicates the author modeling. Table 5 also indicates (in bold case) the topic words which are new to the generated author topic model. As these words did not appear in sole author ATM (Table 4), we might conclude in either ways: (1) these new topics are the footprints of co-author(s) (2) These new topics are of the author 'Catherine Blake' where the author's research interest might have shifted while working in research collaboration. Challenges remain if the list of co-authors in an article is high in number and the co-author(s) do not have any sole publishing history. In such cases, the ATMs need to be generated for those new co-author(s) beforehand to be able to conclude the ATM of 'Catherine Blake'.

Table 2 Publications of 'Catherine Blake' as sole author

S. No.	Single-author publications
1	Blake, Catherine. "A Technique to resolve contradictory answers." New directions in question answering. 2003
2	Blake, Catherine. "Information synthesis: A new approach to explore secondary information in scientific literature." Proceedings of the 5th ACM/IEEE-CS joint conference on Digital libraries. ACM, 2005
3	Blake, Catherine. "A comparison of document, sentence, and term event spaces." Proceedings of the 21st International Conference on Computational Linguistics and the 44th annual meeting of the Association for Computational Linguistics. Association for Computational Linguistics, 2006
4	Blake, Catherine. "Beyond genes, proteins, and abstracts: Identifying scientific claims from full-text biomedical articles." Journal of biomedical informatics 43.2 (2010): 173–189
5	Catherine Blake. "Text mining." ARIST 45(1): 121–155 (2011)

Table 3 Publications of 'Catherine Blake' as Co-Author

S. No.	Multi-author publications
1	Blake, Catherine, and Wanda Pratt. "Multiple categorization of search results". Proceedings of the AMIA Symposium. American Medical Informatics Association, 2000
2	Blake, Catherine, and Wanda Pratt. "Better rules, fewer features: a semantic approach to selecting features from text." Data Mining, 2001. ICDM 2001, Proceedings IEEE International Conference on. IEEE, 2001
3	Blake, Catherine, Wanda Pratt, and Tammy Tengs. "Automated information extraction and analysis for information synthesis." Proceedings of the AMIA Symposium. American Medical Informatics Association, 2002
4	Blake, Catherine, et al. "A study of annotations for a consumer health portal." Digital Libraries, 2005. JCDL'05. Proceedings of the 5th ACM/IEEE-CS Joint Conference on. IEEE, 2005
5	Blake, Catherine, et al. "Cataloging on-line health information: a content analysis of the NC Health Info portal." AMIA Annual Symposium Proceedings. Vol. 2005. American Medical Informatics Association, 2005
6	Blake, Catherine, and Meredith Rendall. "Scientific Discovery: A view from the trenches." International Conference on Discovery Science. Springer Berlin Heidelberg, 2006
7	AbdelRahman, Samir, and Catherine Blake. "A rule-based human interpretation system for semantic textual similarity task." Proceedings of the First Joint Conference on Lexical and Computational Semantics-Volume 1: Proceedings of the main conference and the shared task, and Volume 2: Proceedings of the Sixth International Workshop on Semantic Evaluation. Association for Computational Linguistics, 2012

(continued)

Table 3 (*continued*)

S. No.	Multi-author publications
8	Guo, Jinlong, Yujie Lu, Tatsunori Mori, and Catherine Blake. "Expert-guided contrastive opinion summarization for controversial issues." In Proceedings of the 24th International Conference on World Wide Web, pp. 1105-1110. ACM, 2015
9	Blake, Catherine, et al. "The Role of Semantics in Recognizing Textual Entailment." TAC. 2010
10	Blake, Catherine, and Wanda Pratt. "Collaborative information synthesis." Proceedings of the American Society for Information Science and Technology 39.1 (2002): 44–56
11	Blake, Catherine, and Wanda Pratt. "Collaborative information synthesis I: A model of information behaviors of scientists in medicine and public health." Journal of the Association for Information Science and Technology 57.13 (2006): 1740–1749
12	West, Suzanne L., Catherine Blake, Zhiwen Liu, J. Nikki McKoy, Maryann D. Oertel, and Timothy S. Carey. "Reflections on the use of electronic health record data for clinical research." Health Informatics Journal 15, no. 2 (2009): 108–121
13	Lučić, Ana, and Catherine L. Blake. "A syntactic characterization of authorship style surrounding proper names." Digital Scholarship in the Humanities 30.1 (2015): 53–70
14	Zheng, Wu, and Catherine Blake. "Using distant supervised learning to identify protein subcellular localizations from full-text scientific articles." Journal of biomedical informatics 57 (2015): 134–144
15	Blake, Catherine, and Ana Lucic. "Automatic endpoint detection to support the systematic review process." Journal of biomedical informatics 56 (2015): 42–56

Table 4 Top-10 Topic words for 'Catherine Blake' sole author

Topic 0		Topic 1		Topic 2		Topic 3		Topic 4	
Information	0.15591	Claim	1.180595	Text	1.239711	Information	1.277354	Term	1.171213
Cancer	0.11703	Claims	1.177662	Literature	1.175049	Text	1.239538	idf	1.090641
Breast	0.078148	Change	1.164763	Extraction	1.08265	Mining	1.214833	Document	1.045225
Analysis	0.078145	Collection	1.163519	Journal	1.071279	Proceedings	1.202818	Documents	1.036315
Articles	0.039269	Development	1.157005	Based	1.066135	Conference	1.18805	Terms	1.034943
Alcohol	0.039267	Quality	1.133028	Document	0.481363	System	1.180953	Language	1.030342
Study	0.033878	Grammar	1.124858	Provide	0.429107	Knowledge	1.14205	Sentence	1.002258
Metis	0.032874	Framework	1.122141	Words	0.415522	Methods	1.037112	Corpus	0.511338
Synthesis	0.032178	Precision	1.095219	Terms	0.412051	Association	1.013535	Frequency	0.505411
Article	0.031075	Sentence	0.486901	Relationship	0.404105	Discovery	0.434391	Spaces	0.495512

Table 5 Top-10 Topic words for 'Catherine Blake' as author (sole and collaborative)

Topic 0		Topic 1		Topic 2		Topic 3		Topic 4	
Text	1.160422	Claim	0.293294	**Set**	**1.181214**	Information	0.412051	Documents	1.249711
System	1.109624	**Sentences**	0.195638	**Reviews**	**1.090641**	**Medical**	**0.11265**	Document	1.175049
Protein	**1.106813**	Sentence	0.06543	**Number**	**1.045225**	Analysis	0.039269	Sentence	1.08265
Discovery	1.067982	**Object**	**0.033878**	**Test**	**1.036315**	**Group**	**0.037799**	**Rules**	**1.071279**
Knowledge	1.047738	**agent**	**0.032874**	**Sentences**	**1.034943**	**process**	**0.036791**	**Model**	**1.066135**
Information	1.024398	Claims	0.032178	**Features**	**1.030342**	Articles	0.034799	Based	0.481363
Proceedings	1.014361	Precision	0.031075	**Training**	**1.002258**	Study	0.034099	**Word**	**0.429107**
Relation	**0.650263**	Change	0.030878	**Performance**	**0.511338**	Cancer	0.033878	Term	0.415522
Location	**0.596116**	**System**	**0.023876**	**Sentence**	**0.505411**	**Health**	**0.032178**	**Semantic**	**0.412051**
Instances	**0.56844**	**Recall**	**0.012878**	**Author**	**0.495512**	Breast	0.023876	Language	0.404105

The topics obtained for the author 'Catherine Blake' is the ATM of the author. For empirical assessment of establishing the ATM, we have attempted at calculating the distance between ATMs obtained for the author via single author history (Table 4) and total research history (Table 5) that includes single authored and co-authored publications. We have used various distance measures (Table 6) from Cha [13]. Ideally the distance measure has to be close to 1 which indicates that the author's topics from single author history (Table 4) are inclusive in that of total history. However, the empirical investigation projects differences in distance measures. This proves that the ATM out of single author history is only overlapping in the ATM obtained out of total research history (Fig. 2). This is supported with the evidences of 20 different similarity measures [13] in Table 6. According to Czekanowski, Ruzicka, Lorentzian and Cosine [13] similarity measures, the ATMs of the author in both the approaches are similar; which means, the author's blueprint or ATM has remained same irrespective of the co-author's topics, which are a sizeable proportion in co-authored research articles. In other words, there is no or very less influence of the co-authors (Fig. 3). However, other similarity measures compute very low support stating that both the ATMs of the author are totally different, thereby, indicating strong research influence from the co-authors (Fig. 4).

(a)   **(b)**

ATM – single author research history

ATM – total research history

Fig. 2 ATM **a** Ideal distance **b** Actual distance

Table 6 ATM Distance Measures—single author research history and total research history

Similarity measures	Distance between ATMs	Similarity measures	Distance between ATMs
Czekanowski	0.960	Inner product	0.156
Ruzicka	0.923	Euclidean	0.127
Lorentzian	0.902	Wave hedges	0.077
Cosine	0.835	Tanimoto	0.064
Minkowski	0.701	Gower	0.050
Kulczynski	0.641	Sørensen	0.040
Chebyshev	0.493	Kulczynski	0.040
Motyka	0.480	Canberra	0.040
City block	0.469	Intersection	0.020
Soergel	0.202	Harmonic mean	0.020

Though, ideally, ATMs will be pure (Fig. 3) if and only if they are generated with single authored research articles, the additional research disciplines which the author has initiated their research in collaboration, would not be measured. Therefore, it is evident that generation of ATMs including co-authored research articles is essential for generating the ATM. However, the overlap of research topics from co-authors (aka) Author footprints have to be avoided. Figure 4 showcases the level of co-author influence calculated using various distance metrics [13]. To keep it simple, co-author Influence is assumed as the inverse of ATM similarity. Deriving topics based only on probability is proving to be less sufficient for ATM generation. Since the writing of research manuscripts involves greater creativity, more intelligent measures are needed to generate ATMs. This implies that, unlike other authorship analysis research like authorship attribution, which totally build upon LDA [14], we have proved that LDA is unsuccessful for ATM generation since there are less number of topics recommended. LDA is probabilistic and generates only specified number of topics. Using other non-parametric [11] methods which could automatically generate the sufficient quantum of topics by utilising the rich vocabulary of the research article is the need of the hour.

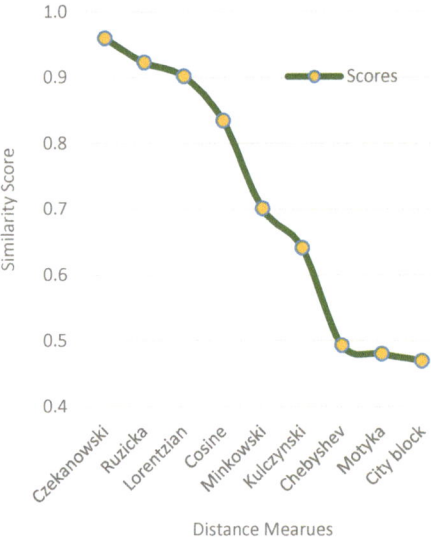

Fig. 3 Distance between ATMs to demonstrate author research dominance

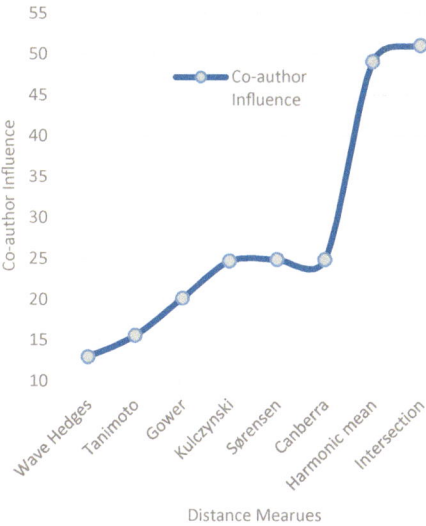

Fig. 4 Co-author influence as derived using various distance measures across ATM topics

5 Conclusion

In this paper, we have proposed methods for Author Topic Model generation and have discussed the challenges involved. Every ATM generated out of collaborated authorship would certainly involve footprints of co-authors. We attempt to explore employing deep machine learning algorithms to arrive into more intelligent solutions.

Acknowledgements. The authors would like to thank Department of Science and Technology for supporting this research by Inspire Fellowship No: DST/INSPIRE Fellowship/2013/505.

References

1. Mahalakshmi, G.S., Muthu Selvi, G., Sendhilkumar, S.: A bibliometric analysis of journal of informetrics—a decade study. Int. J. Control Theory Appl. (2017)
2. Mahalakshmi, G.S., MuthuSelvi, G., Sendhilkumar, S.: Gibbs sampled hierarchical Dirichlet mixture model based approach for clustering scientific articles In: Proceedings of Pattern Recognition and Machine Intelligence, India, Springer LNCS, (2017) (Under Review)
3. Muthu Selvi, G., Mahalakshmi, G.S., Sendhilkumar, S.: An investigation on collaboration behavior of highly cited authors in journal of informetrics (2007–2016). J. Comput. Theor. Nanosci., (2017)
4. Mahalakshmi, G.S., Muthu Selvi, G., Sendhilkumar, S.: Authorship analysis of JOI articles (2007–2016). Int. J. Control Theory Appl. **9**(10), 1–11 (2016). ISSN: 0974-5572
5. Mahalakshmi, G.S., Muthu Selvi, G., Sendhilkumar, S.: Measuring author contributions via LDA. In: 2nd International Conference on Advanced Computing and Intelligent Engineering, Central University of Rajasthan, Ajmer, India, Springer book series on Advances in Intelligent Systems and Computing. (Under Review)

6. Mahalakshmi, G.S., Muthu Selvi G., Sendhilkumar, S.: Hierarchical modeling approaches for generating author blueprints. In: International Conference on Smart Innovations in Communications and Computational Sciences (ICSICCS-2017) organizing by North West Group of Institutions, Moga, Punjab, India, Springer Book Series on Advances in Intelligent Systems and Computing, 23–24 June 2017

7. Blei, D.M., Ng, A.Y., Jordan, M.I.: Latent Dirichlet allocation. J. Mach. Learn. Res. **3**, 993–1022, Jan. (2003)

8. Girgis, M.R., Aly, A.A., Azzam, F.M.E.: Authorship attribution with topic models. Comput. Linguist. **40**(2), 269–310 (2014)

9. Pratanwanich, N., Lio, P.: Who wrote this? Textual modeling with authorship attribution in big data. In: Proceedings of the 2014 IEEE International Conference on Data Mining Workshop, **16**(4), 645–652, (2014)

10. Rubin, T.N., Chambers, A., Smyth, P., Steyvers, M.: Statistical topic models for multi-label document classification. Mach. Learn. **88**(1–2), 157–208 (2012)

11. Mahalakshmi, G.S., Muthu Selvi, G., Sendhilkumar, S.: Measuring authorial indices from the eye of co-author(s). In: International Conference on Smart Innovations in Communications and Computational Sciences (ICSICCS-2017) organizing by North West Group of Institutions, Moga, Punjab, India, Springer Book Series on Advances in Intelligent Systems and Computing, 23–24 June 2017

12. MuthuSelvi, G., Mahalakshmi, G.S., Sendhilkumar, S.: Author attribution using stylometry for multi-author scientific publications. Advan. Nat. Appl. Sci. **10**(8), 42–47 (2016)

13. Cha, S.H.: Comprehensive survey on distance/similarity measures between probability density functions. Int. J. Math. Models Methods Appl. Sci. **4**(1), 300–307 (2007)

14. Seroussi, Y., Bohnert, F., Zukerman, I.: Authorship attribution with author-aware topic models. In: Proceedings of the 50th Annual Meeting of the Association for Computational Linguistics: Short Papers vol. 2, Association for Computational Linguistics, 2012

Comparison Analysis of Biowatermarking Using DWT, DCT and LSB Algorithms

N. V. Brindha[1]([⊠]) and V. S. Meenakshi[2]

[1] Department of Computer Science, Bharathiar University, Coimbatore, India
brindhavijay25@yahoo.com
[2] Department of Computer Science, Chikkanna Goverment Arts and Science
College, Tirupur, India
meenasri70@yahoo.com

Abstract. Nowadays data exchange through computer networks has increased to a great extent. Though the advancement in internet technology has made our life more easy and convenient, this rapid growth in internet usage and data sharing has given many challenges to the researchers. Providing security to the data sent through the network which may be a normal text, image, audio or video has been a big challenge to the network community till now. Watermarking, a technique applied to hide an information, has seen a lot of research interest recently. Many E-Commerce and E-Governance applications needs some method to protect ownership and copyrights by which the unauthorized access to digital products can be detected and controlled. This paper presents an implementation and comparison of Biowatermarking with palmprint using DWT, DCT and LSB algorithms. The experimental results show that the quality of watermarking using LSB is higher than DCT and DWT.

Keywords: Text watermarking · Data hiding · Authentication
Biowatermarking · Biometrics

1 Introduction

Watermarking is a technique which is applied to hide a proprietary information using digital media like images, digital audio, digital music, or digital video. Watermark can be considered as a kind of a signature that reveals the owner of the multimedia object. Copyrighted material can be easily exchanged over peer-to-peer networks, and this has caused major concerns to those content providers who produce these digital contents.

The watermark system undergoes through three stages [1]: first one is the process of generation and embedding the watermark in the original media. Then, transmission of watermarked data in the network to the receiver. The process of detection of the embedded watermark is the final stage.

Most of watermarking techniques can be classified into several categories based on some criteria [2] as shown in Fig. 1. The watermarking technique can be visible or invisible. In visible watermarking technique, the watermark can be observed by the

© Springer International Publishing AG 2018
D. J. Hemanth and S. Smys (eds.), *Computational Vision and Bio Inspired Computing*,
Lecture Notes in Computational Vision and Biomechanics 28,
https://doi.org/10.1007/978-3-319-71767-8_73

human eye like logos. Invisible watermarking technique is used more in Steganography. They can also be classified into robust, fragile and semi-fragile techniques based on the modification could be occurring on the watermark. In a robust watermarking technique, the watermark cannot be affected from attacks when it detected or extracted. The watermarking technique is called a fragile technique, if the watermark is changed or destroyed. In semi-fragile watermarking technique, the watermark can be affected from some types of attacks. Watermarking techniques can be visual, semi blind or blind. A blind technique does not require the original data for detection or extraction. While in non-blind technique, the original data is needed for detection or extraction process.

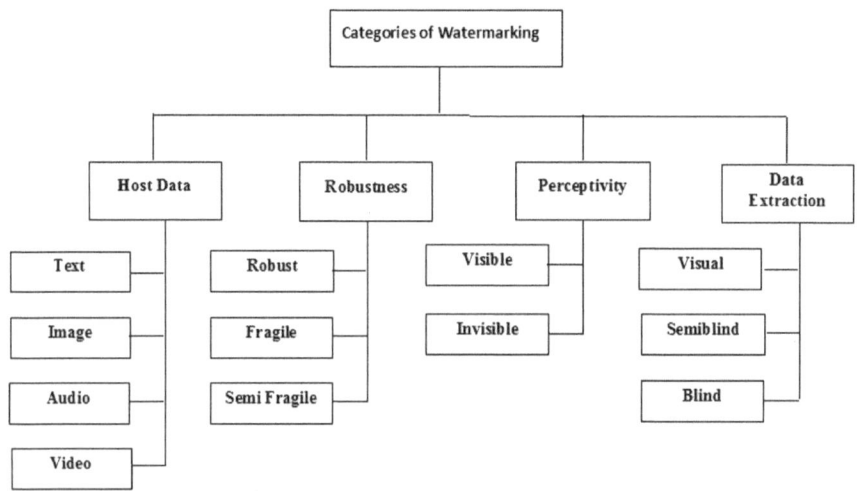

Fig. 1 Watermarking types based on some criteria

An effective watermarking technique should have several requirements [2, 3]. These requirements differ depending on the application and the type of data to be watermarked. The watermarking process has the following general requirements:

- Robustness: The method used to watermark should be able to resist any kind of distortion or any kind of attack. The watermark should still be detectable in the extraction process [4]. The watermark should be able to withstand any image processing or signal processing operations.
- Security: An unauthorized should not be able to detect the hidden watermark. Cryptographic keys are used to achieve this property. The watermark is embedded in such a way that an unauthorized person should not be able to extract the watermark.
- Capacity: It means the number of hidden watermark bits the technique can encode in a unit amount of time.

The capacity ratio in text watermarking could be computed as:

Capacity ratio (Bits/KB) = (Watermark data (Bits))/(Cover file (KB)) [5, 6].
Capacity ratio (%) = (File size (Bytes))/(Cover size (Bytes)) × 100 [6].

- Imperceptibility: The original text should not noticeably destroy after the watermark embedding process; the modifications are done by a small amount. There should not be no any difference visibly seen between the original and watermarked image.

A watermarking technique should be preferably easy to implement, imperceptible, robust, and adaptable to different text formats, have high information carrying capacity.

2   Literature Review

Kaur and Mahajan [7], presented various techniques for text watermarking. A text, being the simplest mode of communication and information exchange, brings various challenges when it comes to copyright protection. Any transformation on text should preserve the meaning, fluency, grammaticality, writing style and value of the text. Short documents have low capacity for watermark embedding and are relatively difficult to protect. Text watermarking algorithms are also dependent on text size, its language, rules, grammar, conventions and writing styles. Figure 2 shows the various text watermarking approaches.

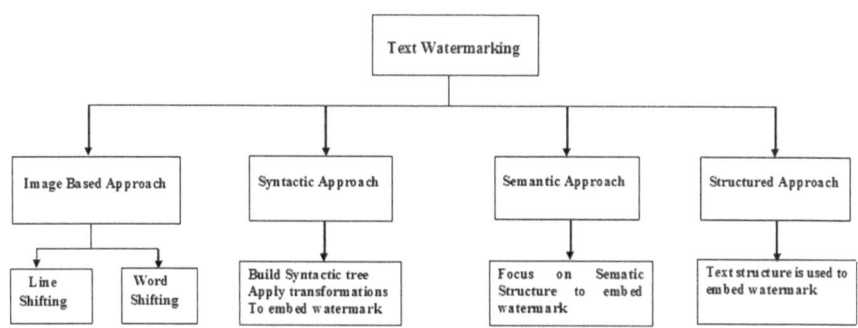

Fig. 2 Text watermarking techniques

In [8] the novel text watermarking scheme was analyzed by Zhang and Qin. The good robustness is provided for the word document. This paper discusses the technique of implanting the secret signals in the special properties of word object. Different groups get formed from the encrypted image by using the water marking techniques. Then it was packed into the message before embedding in the word document circularly. All these operations made the scheme performance excellent on robustness when encountered with attacks as compared to the methods based on the features of character font.

A public zero knowledge watermark detection protocol was proposed by Fu et al. [9] to prevent the owner from cheating by ambiguity attacks. Four steps were concluded in this proposed protocol to achieve the goal: robust feature extraction method, watermarks generation method, watermarking embedding method and zero knowledge watermark detection protocol. Any verifier can check that whether the medium contains a watermark claimed by proverb or not. The proposed method can satisfy the three requirements of zero knowledge proof of identity: completeness, soundness, zero knowledge. The method ensures the security for watermark verification in the watermarking detection process without revealing any secret information related to watermarking.

Du and Zhao [10] proposed a text digital watermarking algorithm based on human visual redundancy. According to this paper, the human eye is not sensitive to the slight change for text color; watermarks were embedded by changing the low-4 bits of RGB color components of characters. Experiment showed that the proposed method has good invisibility and robustness to resist deletion, modification attack etc. This algorithm can be used for secret communication of confidential information.

For integrity and confidentiality of the document, a technique was proposed by Goyal et al. [11]. In this technique, watermark was created based on the contents of the document and embedded without changing the contents of the document and also encrypted the text to provide confidentiality. To authenticate and prove the integrity of the document the watermark could be easily extracted and verified for tampering.

In [12] Ranganathan et al. presented the combined text watermarking technique. In this paper an effective approach is used which combines the advantage of image based text water marking and syntactic water marking technology to provide an robust security mechanism.

In [13] Vashistha et al. has proposed an image watermarking algorithm where fingerprint is used for authentication also. Watermarking is performed using integer DCT algorithm. Inverse algorithm is also proposed for watermark extraction. The three color components of RGB is used for embedding the watermark.

In [14] Patel et al. has proposed an image watermarking model based on discrete wavelet transform. Alpha blending technique is used to embed and extract biometric watermark. A general architecture and modified algorithm for DWT based digital watermarking is illustrated in this paper.

In [15] Verma has proposed a watermarking algorithm based on LSB. The data to be embedded is transformed into binary and depending upon the length of the watermark text, it is embedded in the third and fourth or second and fourth LSB. For color image the bits are embedded in the blue component of the original image.

In [16] Alenzi et al. has proposed a DWT based watermarking scheme for video authentication. Orthonormal filters are used to decompose the video frames into subbands. Polynomial Coefficients are generated randomly based on which filter banks are generated and used for both wavelet decomposition and reconstruction. The bands in the middle frequency is used for watermark embedding.

Qianli et al. in [17] has presents a digital watermarking algorithm that works on 2D discrete wavelet and cosine transforms. The image is transformed into discrete wavelet and three level decomposition is applied to split into subblocks. The subblocks are transformed into discrete cosine domain and the watermark is embedded in maximum coefficient. The extraction process is the reverse of the embedding process.

A semi blind watermarking technique was proposed by Inamdar and Rege in [18]. The principal component analysis is used to extract the features from face image. The face features are embedded in the fingerprint image. The original image is transformed into subblocks using singular vector decomposition and the face features are embedded in the resultant singular vectors. The revers process is applied to extract the watermark and correlated with the features stored in database to recognize the actual face image.

In recent days watermarking techniques are used to protect biometric data. Laadjel et al. in [19] have introduced a new blind watermarking technique for a palmprint recognition system to protect it from replay attacks. Redundant Discrete Wavelet Transform is used to embed the watermark in ROI of Palmprint image. For watermark detection Maximum likelihood ratio is used. The results are accessed based on the PSNR values and ROC curves.

In [20] Qadir and Ahmad proposed an encoding scheme which can be inserted on the plain text without changing the text format. The invisible security needs to be implemented to avoid the attacks. The document gets encoded by the existing text in the intelligent way without changing the content of the document. In this way the hidden information gets preserved. The invisible security gets provided by hiding the information without changing the text format.

3   Application of Biometrics in Watermarking

Biometrics is used to uniquely identify and authenticate a person based on the individual's physiological or behavioral traits. The physiological traits include fingerprint, palmprint, iris, retina, face, hand geometry, ear etc. The behavioral traits include gait, voice, speech, odor, and signature etc. Figure 3 shows the types of biometrics. As the biometric features are unique to a person, it can be acquires and stored as a template in a database and used for authentication at any time required. So the individual need not carry any tokens, cards or IDs and also need not remain passwords in mind to prove his identity and for doing any transactions. Biometric recognition systems are more advantageous than traditional authentication systems as they are concerned directly with the feature what an individual possess. The biometric systems are easy to use, fast and less expensive and the biometric features are also attack resistant. A biometric trait used for authentication purpose should possess the following characteristics universality, uniqueness, permanence, collectability, performance, acceptability, circumvention.

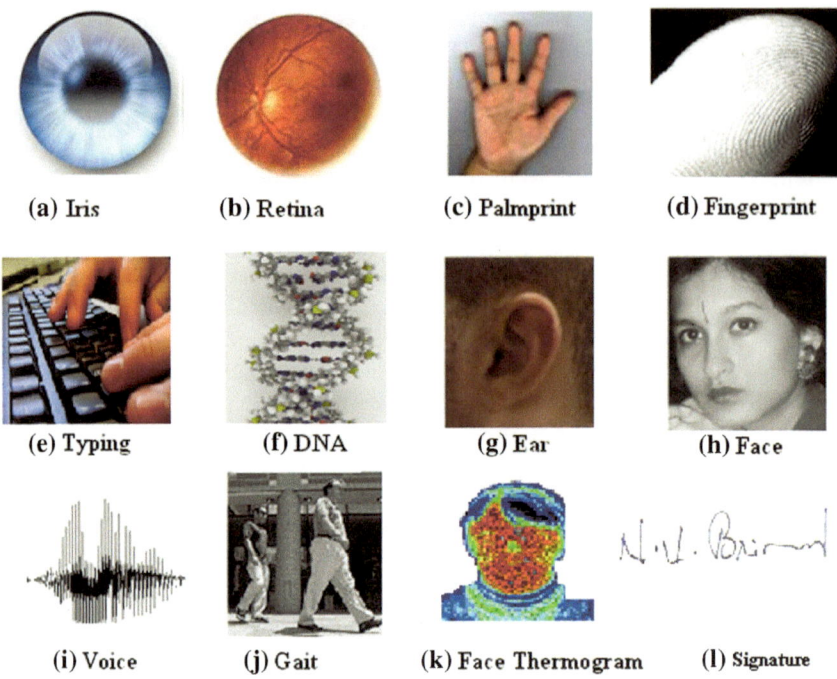

(a) Iris   (b) Retina   (c) Palmprint   (d) Fingerprint

(e) Typing   (f) DNA   (g) Ear   (h) Face

(i) Voice   (j) Gait   (k) Face Thermogram   (l) Signature

Fig. 3 Types of biometrics adapted from [22]

3.1   Palmprint

Palmprint is one of the biometric traits used in biometric systems and it has gained more importance recently and has emerged as an active research topic in the last few decades. The large area of Palmprint, number of features to extract, less cost of image acquiring devices has brought the attention of researchers to concentrate more upon this biometric feature. The features that can be extracted from palmprint are minutia, ridges, valleys, principal lines, wrinkles, creases. The line features in the palmprint are stable in nature. The features can be extracted from a low resolution image itself. Even elder people are able to produce their palmprint for authentication. Figure 4 shows Palmprint image.

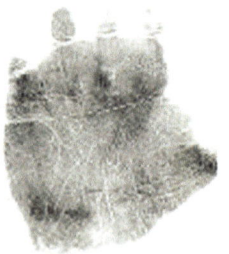

Fig. 4 Palmprint

4 Implementation of Biowatermarking Using DWT, DCT and LSB

This section explains the implementation of biowatermarking using DWT, DCT and LSB algorithms. Palmprint is used for embedding the data that can be a text or image.

4.1 Biowatermarking Using DWT

The Discrete Wavelet Transform is a watermarking technique used to embed watermark in the palmprint image. The watermark can be text, image, video or audio data. The Haar filter is used to decompose the input palmprint image into four subbands. A pair (h, g) of low pass filtering and high pass filtering is applied to the image decomposition. The low pass filter produces the components with low frequency as output that contains approximate part of the Palmprint image and the high pass filter produces the components with high frequency as output that contains the detailed part of the image. This 1D decomposition is applied for 2D DWT for each row and for each column. Figure 5 shows the image decomposition for Palmprint into four sub bands LL, LH, HL, and HH. Watermarking is usually done on the high frequency components because the man's eye is somewhat insensitive to high frequencies like edges [8]. The HH band contains high intensity pixels which is used to embed the watermark. The bits in the HH band is replaced by watermark bits.

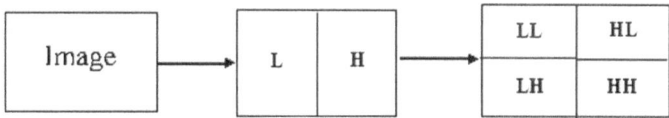

Fig. 5 Image decomposition using DWT

Algorithm:

* Use Haar filter to decompose the palmprint input image into four sub bands LL, HL, LH, and HH.
* The high frequency band HH is selected for watermark embedding.
* Replace the HH band with watermark bits.
* Figure 6 shows the DWT process of watermarking with Palmprint.

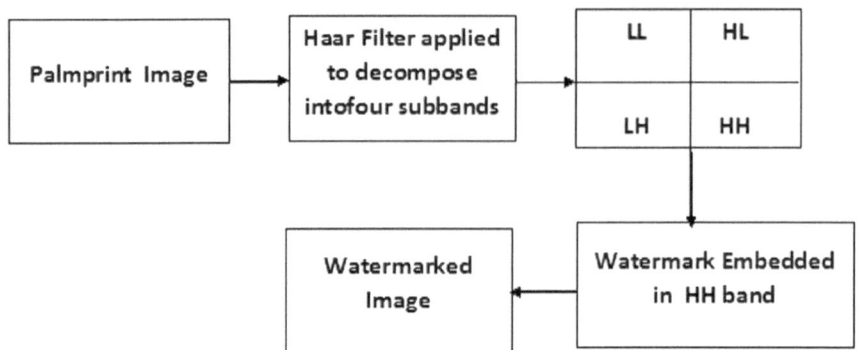

Fig. 6 Biowatermarking of palmprint using DWT

The image function for the 2D wavelet decomposition is given by:

$$LL = [(f(x,y)^{*}\Phi(-x)\Phi(-y))(2n,2m)]^{2}_{(n,m)\varepsilon z} \tag{1}$$

[21]

$$LH = [(f(x,y)^{*}\Phi(-x)\Psi(-y))(2n,2m)]^{2}_{(n,m)\varepsilon z} \tag{2}$$

[21]

$$HL = [(f(x,y)^{*}\Psi(-x)\Phi(-y))(2n,2m)]^{2}_{(n,m)\varepsilon z} \tag{3}$$

[21]

$$HH = [(f(x,y)^{*}\Psi(-x)\Psi(-y))(2n,2m)]^{2}_{(n,m)\varepsilon z} \tag{4}$$

[21] The watermark detection or extraction process is the reverse of embedding process is extracted.

4.2   Biowatermarking Using DCT

Discrete Cosine transform is a transformation function used to transform a signal that may be text or image from spatial domain to frequency domain. It is used in JPEG compression to reduce the redundant bits. Biowatermarking in Palmprint using DCT is done through the following two steps:

- Embedding of watermark in Palmprint image.
- Extraction of data from the watermarked Palmprint image.
    Embedding of Watermark in Palmprint image
- Divide the Palmprint image taken for embedding watermark into 8 × 8 non overlapping blocks.
- The watermark is converted into its binary equivalent.
- Find the DCT coefficients for each and every block.
- Calculate the average pixel intensity i.
- Select the block having DC coefficients that has the pixel intensity greater than i and also that has the number of bits equal to number of watermark bits. DC coefficient withstands many watermark attacks.
- JPEG zigzag scanning is applied which distributes the frequency from low to high and high to low and middle frequency domain is selected for inserting the watermark.
- The watermark bits are inserted one by one in the selected block using the following equation:

For watermark bit = 1

$$IM'(x, y) = |IM(x, y)| + s \qquad (5)$$

For watermark bit = 0

$$IM'(x, y) = |IM(x, y)| - s \qquad (6)$$

IM denotes image, x and y are the coordinates of the image and s denotes the strength of the watermark and its value is 64. Figure 7 depicts the process of watermarking using DCT.

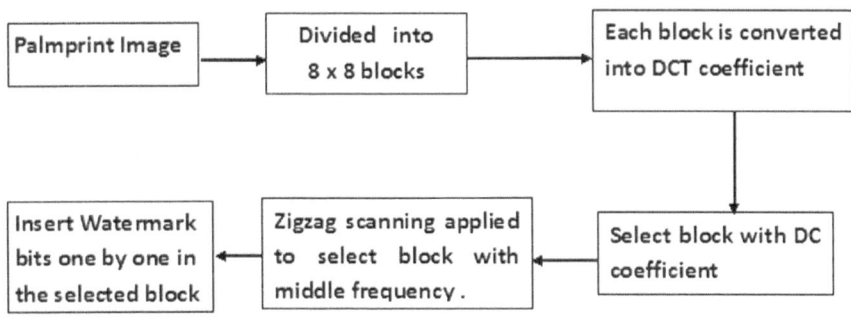

Fig. 7 Biowatermarking of palmprint using DCT

Extraction of data from the watermarked Palmprint image.

- Divide the watermarked palmprint image taken for extracting watermark into 8 × 8 non overlapping blocks.
- Find the DCT coefficients and blocks used for embedding.
- Apply zigzag scanning.
- Watermark bits are extracted using the equation.

$$\text{Watermark bit is } 1 \quad \text{if} \quad IM'(x, y) > 0 \tag{7}$$

$$\text{Watermark bit is } 0 \quad \text{if} \quad IM'(x, y) < 0 \tag{8}$$

4.3   Biowatermarking Using LSB

Least Significant Bit is an approach used to embed data in the image. The image is transformed into binary format and the least significant bits of all the bytes or some of the bytes are changed to a bit of Red, Blue, Green color components. Least Significant Bit approach is of two categories one is insertion and another one is substitution. In the present approach insertion method is used as it is more imperceptible when compared to the substitution method. The Palmprint image is divided into four quadrants and the fourth quadrant is taken for embedding the secret data. The secret data and the image quadrants are converted into binary format that is 0 and 1. The secret data bits are embedded in the third and fourth LSB of fourth quadrant image to get the resultant watermarked palmprint image.

Embedding of data in the Palmprint Image

- Divide the Palmprint image into four quadrants.
- Convert fourth quadrant into its equivalent binary format.
- Transform the secret data into binary digits.
- Embed the secret data bits in the third and fourth LSB of the fourth quadrant image.
- The resultant output is watermarked image.

Extraction of data from Watermarked Palmprint
    The extraction and detection of watermark is the same process followed for Embedding in LSB algorithm.

- Divide the Watermarked Palmprint image into four quadrants.
- Convert fourth quadrant into its equivalent binary format.
- Extract the secret data bits from the third and fourth LSB of fourth quadrant.
- The secret data and Palmprint image are separated.
- Figure 8 shows the Biowatermarking of Palmprint using LSB.

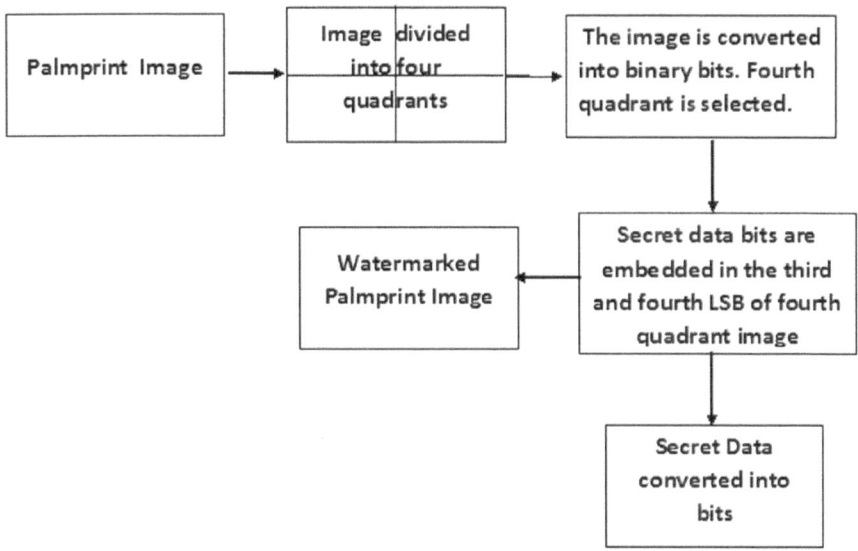

Fig. 8 Biowatermarking of palmprint using LSB

5 Experimental Results and Discussion

The present work is implemented using MATLAB tool. The palmprint image taken for watermarking is 256×256 color image. The database used for Palmprint image is Poly print database. The input secret data given is text and 64×64 image for all the three process and the results are tabulated. The performance measures taken for comparison are PSNR and MSE.

PSNR Peak Signal to Noise Ratio is a commonly used measure in watermarking to evaluate and find the quality of the watermarked image. Its unit is Decibels Db. R is the maximum fluctuation in the input data type. For example, if the input data has a double-precision floating-point data type, then R is 1. If it has an 8-bit unsigned integer data type, R is 255, etc.

$$PSNR = 10\log 10\left(\frac{R^2}{MSE}\right) \tag{9}$$

MSE Mean Square Error the MSE represents the cumulative squared error between the Clustered and the original data, whereas PSNR represents a measure of the peak error. The lower the value of MSE, the lower the error. To compute the PSNR, the block first calculates the mean-squared error using the following equation:

$$MSE = \frac{\sum_{MN}[I_1(m, n) - I_2(m, n)]^2}{M * N} \tag{8}$$

I1 = Input data
I2 = Clustered data

M and N are the number of rows and columns in the input data, respectively.

The comparison of MSE parameter and PSNR parameter of the three watermarking algorithms for the 5 input color images and 5 different Palmprint images are shown in Tables 1 and 2. The comparison of MSE parameter and PSNR parameter of the three watermarking algorithms for the 4 input text messages and 5 different Palmprints are shown in Tables 3 and 4. It shows that the mean square error is less and PSNR is high for LSB algorithm when compared to other two (Graphs 1, 2, 3, and 4).

Table 1 Comparison of MSE for image data

Images	DCT	DWT	LSB
Palmprint1.jpg	2.0010	0.2011	0.0446
Palmprint2.jpg	2.1835	0.2195	0.1747
Palmprintt3.jpg	1.6372	0.1647	0.1361
Palmprint4.jpg	1.5070	0.1516	0.1337
Palmprint5.jpg	1.6611	0.1672	0.0908

Table 2 Comparison of PSNR for image data

Images	DCT	DWT	LSB
Palmprint1.jpg	45.1183	55.0957	61.6407
Palmprint2.jpg	44.7393	54.7159	55.7089
Palmprintt3.jpg	45.9898	55.9644	56.7955
Palmprint4.jpg	46.3495	56.3226	56.8700
Palmprint5.jpg	45.9720	55.8996	59.6286

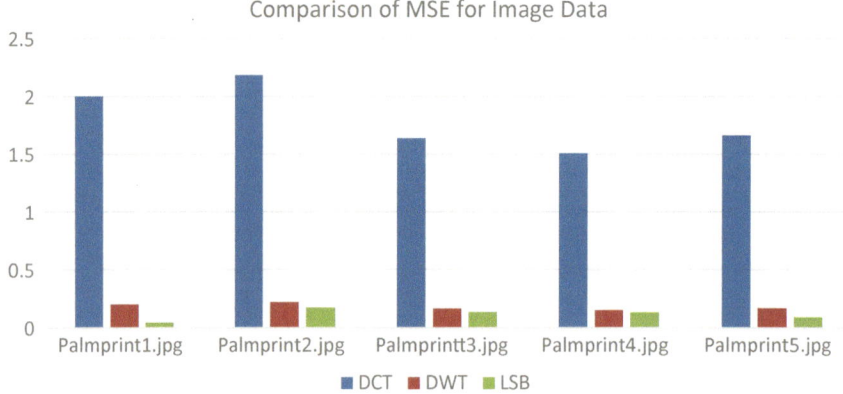

Graph 1 Comparison of MSE for image data

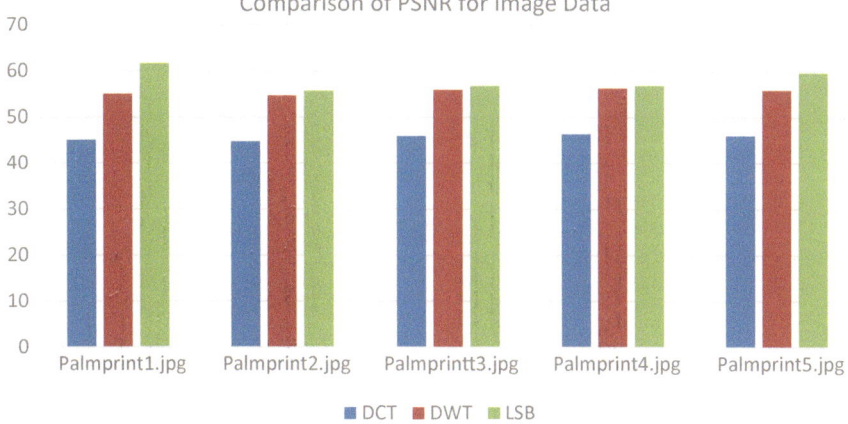

Graph 2 Comparison of PSNR for image data

Table 3 Comparison of MSE for text

Text	DCT	DWT	LSB
Hai	1.865	0.28	0.12
Hello	2.22	0.45	0.18
MATLAB	1.78	0.13	0.12
Watermarking	1.35	0.17	0.16

Table 4 Comparison of PSNR for text

Text	DCT	DWT	LSB
Hai	44.23	55.12	57.3
Hello	44.6	54.23	55.57
MATLAB	43.9	55.04	54.9
Watermarking	43.4	56.07	56.08

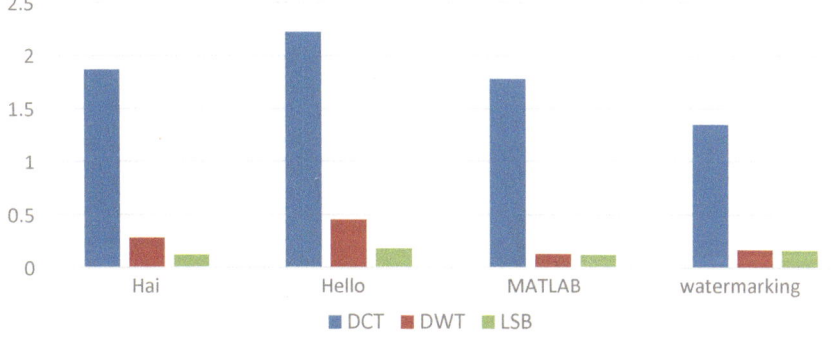

Graph 3 Comparison of MSE for text data

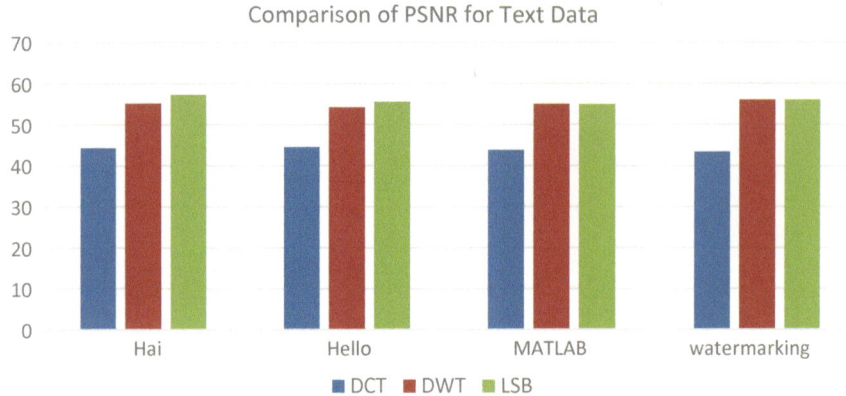

Graph 4 Comparison of PSNR for text data

6   Conclusion and Future Work

This paper presents a comparative analysis of DWT, DCT and LSB algorithm for watermarking text and image data in Palmprint. The input images and Palmprint images used are color images. The PSNR for LSB is higher than for DWT and DCT and MSE is very low compared to the other two algorithms which shows that LSB algorithm is better for color images among the three. The combination of biometrics with watermarking can be used for data hiding and authentication simultaneously. But the present work concentrates only on data hiding. This work can be enhanced by using the Palmprint for personal authentication.

References

1. Irena, O.: Multimedia Signals and Systems, Springer Science and Business Media, (2012)
2. Katzenbeisser, S., Petitcolas, F.: Information Hiding Techniques for Steganography and Digital Watermarking, Artech house, Massachusetts, (2000)
3. Robert, L., Shanmugapriya, T.: A study on digital watermarking techniques. Int. J. Recent Trends Eng. **1**(2), 223–225 (2009)
4. Hsieh, M.S.: Perceptual copyright protection using multiresolution wavelet-based watermarking and fuzzy logic. Int. J. Artif. Intell. Appl. (IJAIA) **1**(3), 45–57 (2010)
5. Aabed, M.A., Awaideh, S.M., Abdul-Rahman, M., Elshafei, A.R.M., Gutub, A.A.: Arabic diacritics based steganography. In: IEEE International Conference on Signal Processing and Communications, pp. 756–759, Dubai, UAE, 24–27 Nov. (2007)
6. Bensaad, M.L., Yagoubi, M.B.: High capacity diacritics-based method for information hiding in Arabic text. In: Innovations in Information Technology (IIT), International Conference on. IEEE, (2011)
7. Kaur, M., Mahajan, K.: An existential review on text watermarking techniques. Int. J. Comput. Appl. **120**(18), 0975–8887 (2015)
8. Zhang, Y., Qin, H.: A Novel Robust Text Watermarking for Word Document, IEEE, (2010)

9. Fu, Z., Sun, X., Shu, J., Zhou, L., Wang, J.: Verifiable Text Watermarking Detection to Improve Security, (2014)
10. Min, D., Zhao, Q.: Text watermarking algorithm based on human visual redundancy. Adv. Inf. Sci. Serv. Sci. **3**(5), (2011)
11. Goyal, L., Raman, M., Divan, P., Vijay, M.: A robust method for integrity protection of digital data in text document watermarking. Int. J. Innov. Res. Sci. Technol. **1**(6), (2014)
12. Ranganathan, S., Ali, A.J., Kathirvel, K., Mohan Kumar, M.: Combined text watermarking. Int. J. Comput. Sci. Inf. Technol. **1**(5), (2010)
13. Vashistha, A., Joshi, A.M.: Fingerprint based biometric watermarking architecture using integer DCT. In: IEEE Region 10 Conference (TENCON)—Proceedings of the International Conference, (2016)
14. Patel, M., Sajja, P.S.: The significant impact of biometric watermark for providing image security using DWT based alpha blending watermarking technique. Int, J. Innov. Res. Comput. Commun. Eng. **3**(5), May (2015)
15. Verma1, R., Tiwari, A.: Copyright protection for watermark image using LSB algorithm in colored image. Adv. Electron. Elect. Eng. **4**(5), 499–506 (2014)
16. Alenizi, F., Kurdahi, F., Eltawil, A.: DWT-Based Watermarking Technique for Video Authentication. IEEE, Cairo (2015)
17. Qianli, Y., Yanhong, C.: A digital image watermarking algorithm based on discrete wavelet transform and discrete cosine transform. In: International Symposium on Information Technology in Medicine and Education, (2012)
18. Inamdar, V.S., Rege, P.P.: Face features based biometric watermarking of digital image using singular value decomposition for fingerprinting. Int. J. Secur. Appl. **6**(2), April (2012)
19. Laadjel, M., Zebbiche, K., Kurugullu, F., Bouridane, A., Nibouche, O.: Watermarking for palmprint image protection. In: 13th International Machine Vision and Image Processing Conference, (2009)
20. Qadir, M.A., Ahmad, I.: Digital Text Watermarking: Secure Content Delivery and Data Hiding in Digital Documents, IEEE A&E Systems Magazine, Nov. (2006)
21. Sathik, M.M., Sujatha, S.: A Novel DWT based invisible watermarking technique for digital images. Int. Arab J. E-Technol. (2012)
22. Brindha, N.V., Meenakshi, V.S.: Biometric based secure architecture for mobile adhoc network. In: ICIRCBE Proceedings, Nov. (2016)
23. Zunera Jalil, M., Jaffar, A., Mirza, A.M.: A novel text watermarking algorithm using image watermark. Int. J. Innov. Comput. Inf. Control **7**(3), (2011)

Diagnosis of Carious Legions Using Digital Processing of Dental Radiographs

Harsh Vikram Singh[(✉)] and Raghav Agarwal

Electronics Engineering Department, Kamla Nehru Institute of Technology,
(an Autonomous Government Engineering Institute), Sultanpur, UP, India
harshvikram@gmail.com

Abstract. Diagnosis of caries is a difficult process in the clinical setting, because of the obvious reasons like edge line in the Approximal surfaces is not so clear and also the complicated setting of pits and fissures in the Occlusal surface. Due to this, dependency on the expert advice of doctor increases which changes, with difference in opinion of different doctors. This indirectly affects the Sensitivity and Specificity of the disease in the radiographs used. The goal of this work is to study digital radiographs in this case we are using IOPA for dental caries. Then find out which image is having caries, whether it is visible or not, if yes than what amount of caries it is comprised of, if no than what is the reason behind it. Now to find the visibility we have preprocessed the image using an algorithm which de noises the image and also enhances the image quality, maintaining its medical standards. Also, we have used artificial intelligence to initially diagnose the carious legions by detecting those areas where caries generally occurs like the approximal surface and occlusal surface of the tooth. This approach of diagnosing the carious legions has opened gates for diagnosis of some others diseases also.

Keywords: Digital radiography · Medical image processing · Approximal caries · Occlusal caries

1 Introduction

Medical Image processing is a fast growing field contributing tons of datasets for study and analysis of different diseases. Different types of techniques are being used to diagnose different diseases. CT scanner, Ultrasound, Magnetic Resonance Imaging, Radiographs, and Cone Beamed Computer Tomography (CBCT), fiber optic transillumination (FOTI), electrical conductance (EC), laser fluorescence are some of the techniques used to see inside out of the human body without cutting much of the parts [1–4].

Image Enhancement is the technique in which we try to amplify the visual information for deep analysis or for better visual display. Image Enhancement does not change the basic structure of image but find out some region of interests using processes like image segmentation, feature extraction, edge detection, denoising the specific areas and then apply these settings to amplify the characteristics of that particular area which changes the visual perception of the image [5–7].

© Springer International Publishing AG 2018
D. J. Hemanth and S. Smys (eds.), *Computational Vision and Bio Inspired Computing*,
Lecture Notes in Computational Vision and Biomechanics 28,
https://doi.org/10.1007/978-3-319-71767-8_74

In this work, we have taken dental radiographs into account to diagnose the problem of dental caries in the Human tooth. Dental Caries is the term given to tooth decay or cavities. There are certain specific types of bacteria's which are responsible for caries. An acid is being produced by those bacteria's which destroys the tooth Enamel and also the Dentin layer. Finally dissolving these two layers a cavity is created [8, 9]. Here, the main challenge was to bring out total amount of caries hidden inside the tooth. There should be less number of false positives once the image is processed and also it should retain its image quality since it is very sensitive data providing information regarding disease. Also, since we have used artificial intelligence to detect the carious legions, it was important to select correct pixel values which genuinely represent carious legions in the image [10, 11].

2 Methods and Materials

2.1 Datasets

The goal of this work is to diagnose carious legions. So for this reason database of DICOM standard has to be acquired. The images used in this work are IOPA (periapical view). A periapical X-ray is a specific type of intraoral X-ray that is used to investigate the structural integrity of an individual tooth. A periapical X-ray provides an image of a tooth from the tooth's crown to the tip of its root. Periapical X-rays provide a more highly focused, finely detailed image than common bitewing X-rays that survey three-to-four teeth at a time [12, 13].

In this work we have taken 23 IOPA's of dental problems from a verified source. These images are the mixture of all types of caries. This is done in order to completely diagnose the disease.

2.2 Software

The software used in this work is MATLAB [14]. The purpose of this study is to improve features and gain better characteristics of medical images for a right diagnosis. Using this software we first of all removes the noise in the image using Median Filter than to sharp the images we have used unsharp mask filters. Now the main problem with medical images is that their poor contrast so to overcome that we have used Contrast Limited Adaptive Histogram Equalization (CLAHE) which enhances the contrast of the image in appropriate manner. It is pretty obvious that after so much of processing the image needs to be smoothen so for that we have used average filter which smoothens the image and hence we get a very high quality image which is perfectly enhanced [15, 16].

2.3 Median Filtering

Image filtration is the essential part of image enhancement process which is used to maintain its edges. Edges procurement generally results in lowering of noise in the image. Now since we are using medical images which are assumed to have so much of salt and pepper noise due to the variations in the image resulting from minute gray particles, this technique is the most sorted one for removal of such types of noise. The advantage of this image is that image sharpness is not affected. Now how this filter works, we first of all rank all the values of odd number of samples and then the value lying in the median is used as the output of the filter [17]. The signal length is finite. Therefore

$$Samples = X(0) \, to \, X(L-1) \tag{1}$$

Length:

$$N = 2k+1 \, (Filter's \, window \, length) \tag{2}$$

The Filtering Procedure

$$Y(n) = med[X(n-k), ., X(n), .X(n \, k)] \tag{3}$$

X(n) is input; Y(n) is output

2.4 Unsharp Mask Filtering

The Unsharp Filter is a process in which the high frequency components in an image are enhanced which indirectly enhances the edges by subtracting the smoothed version of an image from the original image.

Now the working of this filter could be understood as follows:

1. Unsharp masking produces an edge image g(x, y) obtained from an input image f(x, y) from the Eq. 4, (Fig. 1):

$$g(x,y) = f(x,y) - f_{smooth}(x,y) \tag{4}$$

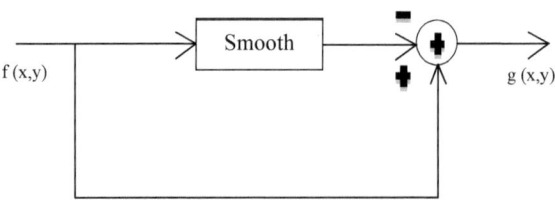

Fig. 1 Spatial sharpening

2. Now, If we have a signal and subtracting away the lowpass component of that signal, yields the highpass, or edge, representation.
3. This edge image can be used for sharpening if we add it back into the original signal.
4. Thus, the complete unsharp sharpening operator is shown in Fig. 2.

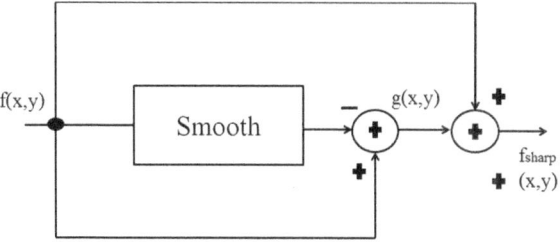

Fig. 2 The complete unsharp filtering operator

5. The Final equation will be formed after combining all forms will be

$$f_{sharp}(x, y) = f(x, y) + k \times g(x, y) \tag{5}$$

where k is a scaling constant. Reasonable values for k vary between 0.2 and 0.7, with the larger values providing increasing amounts of sharpening.

2.5 Adaptive Histogram Equalization

To improve the local contrast of images we use the technique called Contrast limited adaptive histogram. In this technique the generalization of ordinary Histogram Equalization has been done which has been followed by change in its operation i.e. unlike Histogram Equalization this technique does not work on whole image but operates on small areas termed as tiles. Now, there are problems which are related to both HE and AHE so to limit these problems the reduction in contrast enhancement has been used not in whole image but in some homogeneous areas [18, 19]. This particular algorithm allows maximum numbers of pixels over a local histogram comprised of bins. Now after clipping the local histograms the amount of pixels stored inside them are also get clipped which are then equally redistributed over the complete histogram so that we may get a new histogram with identical counts [20, 21]. Hence we can say that this algorithm has limited the slope associated with gray level assignment scheme to prevent saturation of the image.

Steps included in the working of CLAHE are:

1. The Medical image is divided into contextual regions of size (8 × 8)
2. The histograms of each contextual regions are calculated
3. The histograms of each contextual regions are clipped by limit of 0.01. The no of pixels in each region is equally distributed to each gray level. Therefore, average no of pixels in each gray level is defines as follows:

$$N_{av} = N_{cr-x} * N_{cr-y} N_g \tag{6}$$

where

N_{av} = Average number of pixels
N_g = Gray levels number in the contextual region

$N_{cr\text{-}x}$ = Pixels number in the x dimension of the contextual region
$N_{cr\text{-}y}$ = Pixels number in the y dimension of the contextual region

the N_{ac} can be calculated by the Eq. 7:

$$N_{ac} = N_{cx}N_{av} \qquad (7)$$

where

N_{ac} is actual clip-limit;
N_c is the maximum multiple of average pixels in each gray level of the contextual region.

The original and clipped histograms are shown in Fig. 3. In Fig. 3a if the number of pixels is greater than Nc, the pixels will be clipped. The total number of clipped pixels is defined as NΣc, and then the number of pixels distributed averagely into each gray level is given by Eq. 8:

$$N_{ac} = N_c N_g \qquad (8)$$

After the above distribution, the remaining number of clipped pixels is expressed as LP N and then the step of distributed pixels is given by

$$P_d = N_g N_{lp} \qquad (9)$$

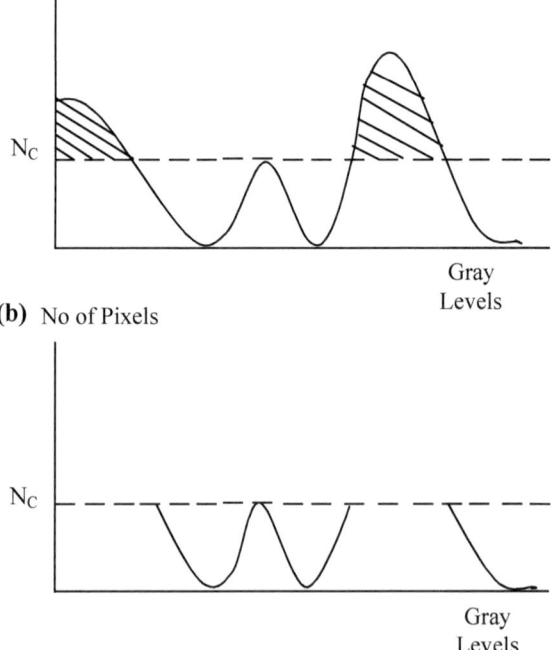

Fig. 3 Original and clipped histograms **a** original histogram **b** clipped histogram

2.6 Masking

Masking is the process in which we apply color mask to those areas which we want to classify or we can say to find area of interest. Here we have used this technique in the form of artificial intelligence which uses the color to mark the carious legions in the image or the areas which are doubted to be having dental caries.

In this step first of all we will check whether the image is color image or not. If it is a color image then we will extract the red, green and blue color channels individually. Then we have used a threshold which is checking the pixels darker than the given value and masking them. Since we are using here the green channel mask so the value will be 120 for green channel and 40 for other two channels. Now in the third step we will apply the mask by keeping all the channel value to 255.

This will mark those areas in the image which are subjected to caries hence will help in diagnosing the carious legions in any image by generalizing certain pixel values. Through this technique we will be able to make a generalized pixel cluster for medical image which will according to the disease will check whether the disease is present or not [22].

Finally, we will calculate the total number of pixels in which masking is done. After that we will divide that value of pixel with total number of pixels of the image which will give us the total percentage of caries or probable carious legions present in the image.

3 Results

3.1 Software Based Results Based on Clinical Reviews

In this work we have taken 23 samples of the dental IOPA having different types of caries. In them some of the images were easily identified with their type of caries. But in some cases it was very difficult to predict the type of caries in them. This was verified by the group of doctors to whom these images have been shown before using the software and they have reviewed the images with best of their knowledge and availability of sources.

Total no of Positive Cases————————16
Total no of Negative Cases————————2
Total no of cases unpredicted————————5

Using these values in software and calculating the values for sensitivity, specificity, P, and the amount disease prevalence.

1. *Sample size*
   First the program displays the number of observations in the two groups. Concerning sample size, it has been suggested that *meaningful qualitative conclusions* can be drawn from ROC experiments performed with a total of about 23 observations.
2. *Disease Prevalence*
   The term prevalence refers to the Particular time at which a particular group was found infected with a particular type of Disease.

3. *Sensitivity*

It is a probability that a test result will be positive when the disease is present (true positive rate, expressed as a percentage).

Sensitivity = a/(a + b)

Where a = True Positive; b = False Negative

4. *Specificity*

It is aprobability that a test result will be negative when the disease is not present (true negative rate, expressed as a percentage).

Specificity = d/(c + d)

Where c = False Positive; d = True Negative

Variable	TEST1	
Classification variable	DIAGNOSIS	
Sample size		23
Positive group [a]		16 (69.57%)
Negative group [b]		7 (30.43%)
[a] DIAGNOSIS = 1		
[b] DIAGNOSIS = 0		
Disease prevalence (%)		69.6
Area under the ROC curve (AUC)		
Area under the ROC curve (AUC)		0.643
Standard Error [a]		0.146
95% Confidence interval [b]		0.418 to 0.829
z statistic		0.980
Significance level P (Area=0.5)		0.3269
[a] DeLong et al., 1988		
[b] Binomial exact		
Youden index		
Youden index J		0.3661
Associated criterion		<19
Sensitivity		93.75
Specificity		42.86
Optimal criterion		
Optimal criterion [a]		<21
Sensitivity		100.00
Specificity		28.57

[a] Taking into account disease prevalence (69.6%) and estimated costs: cost False Positive: 0; cost False Negative: 0

Fig. 4 Diagnosis result for TEST 1 data

After applying the algorithm on the samples we get an enhanced set of images which were again showed to the same group of doctors with change in their order just to make sure the result may not get affected and this time they were able to identify those cases also which were unpredictable before. Therefore (Fig. 4)

Total no of Positive Cases————————21

Total no of Negative Cases————————2

Total no of cases unpredicted——————0

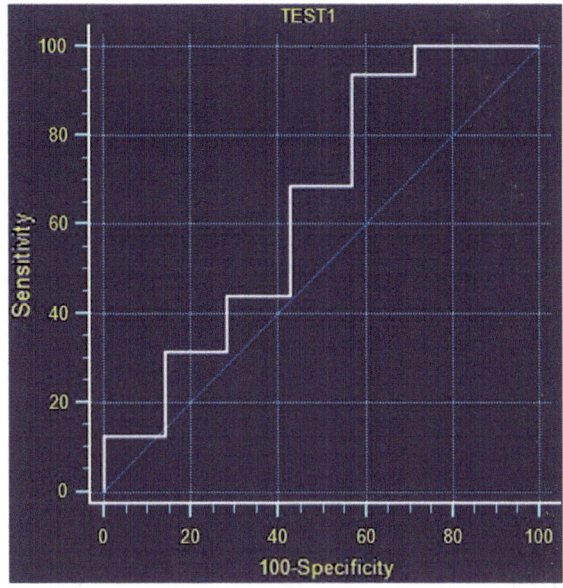

Fig. 5 ROC—curve for TEST 1 data

Variable	TEST2
Classification variable	DIAGNOSIS
Sample size [a]	23
Positive group [a]	21 (91.30%)
Negative group [b]	2 (8.70%)
[a] DIAGNOSIS = 1	
[b] DIAGNOSIS = 0	
Disease prevalence (%)	91.3
Area under the ROC curve (AUC)	
Area under the ROC curve (AUC)	1.000
Standard Error [	0.000
95% Confidence interval [	0.852 to 1.000
Significance level P (Area=0.5)	<0.0001
DeLong et al., 1988	
[b] Binomial exact	
Youden index	
Youden index J	1.0000
Associated criterion	≤21
Sensitivity	100.00
Specificity	100.00
Optimal criterion	
Optimal criterion [a]	≤21
Sensitivity	100.00
Specificity	100.00
[a] Taking into account disease prevalence (91.3%) and estimated costs:	
cost False Positive: 0; cost False Negative: 0	
cost True Positive: 21; cost True Negative: 2	

Fig. 6 Diagnosis result for TEST 2 data

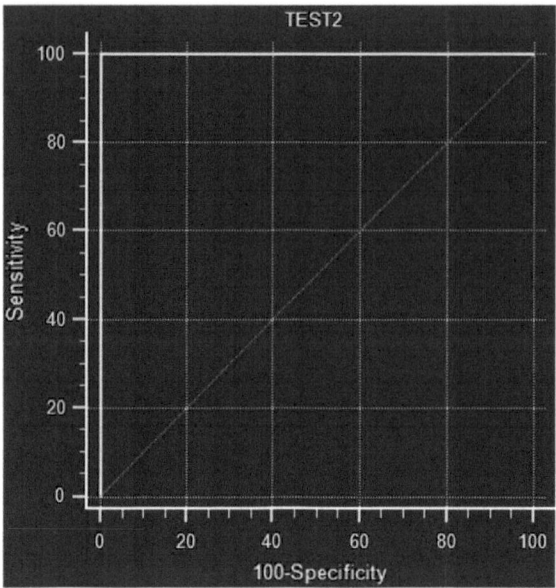

Fig. 7 ROC—curve for TEST 2 data

Using these values again in software and calculating the values for sensitivity, specificity, P, and the amount disease prevalence we have (Figs. 5, 6 and 7).

5. ***ROC curve***

In a ROC curve the true positive rate (Sensitivity) is plotted in function of the false
positive rate (100-Specificity) for different cut-off points. Each point on the ROC
curve represents a sensitivity/specificity pair corresponding to a particular decision
threshold. A test with perfect discrimination (no overlap in the two distributions)
has a ROC curve that passes through the upper left corner (100% sensitivity, 100%
specificity). Therefore the closer the ROC curve is to the upper left corner, the
higher the overall accuracy of the test.

6. ***P-level***

The Significance level or P-value is the probability that the observed sample Area
under the ROC curve is found when in fact, the true (population) Area under the
ROC curve is 0.5 (null hypothesis: Area = 0.5). If P is small (P < 0.05) then it can
be concluded that the Area under the ROC curve is significantly different from 0.5
and that therefore there is evidence that the laboratory test does have an ability to
distinguish between the two groups.

3.2 Images of Samples Before and After Applying Algorithm

Some of the images from the samples have been shown to prove the usefulness of the
algorithm in the diagnosis of the dental caries. (Figs. 8, 9, 10, 11 and 12)

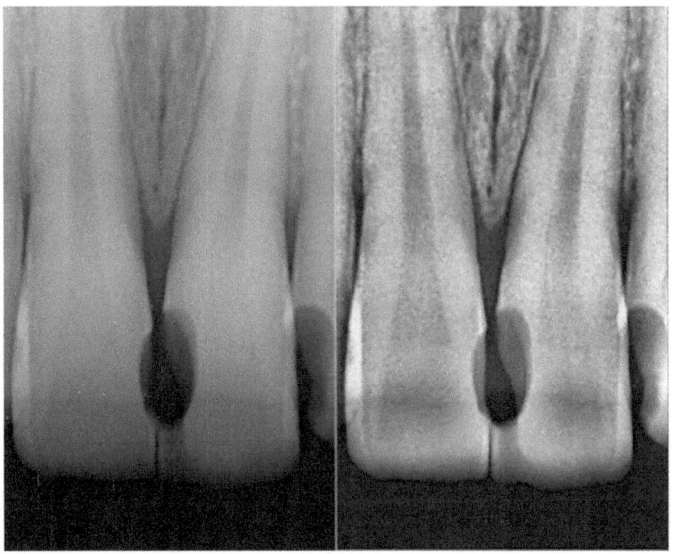

Fig. 8 Image sample 1 before and after processing

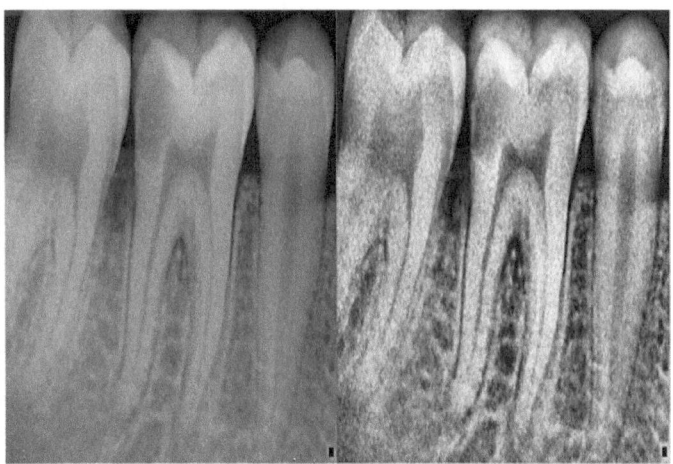

Fig. 9 Image sample 2 before and after processing

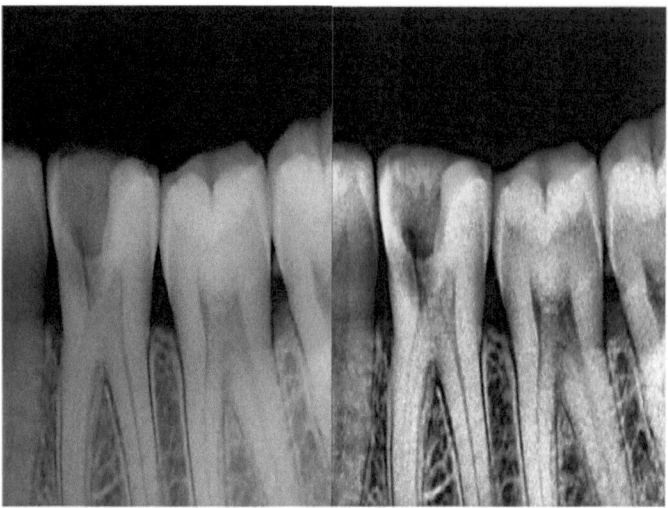

Fig. 10 Image sample 3 before and after processing

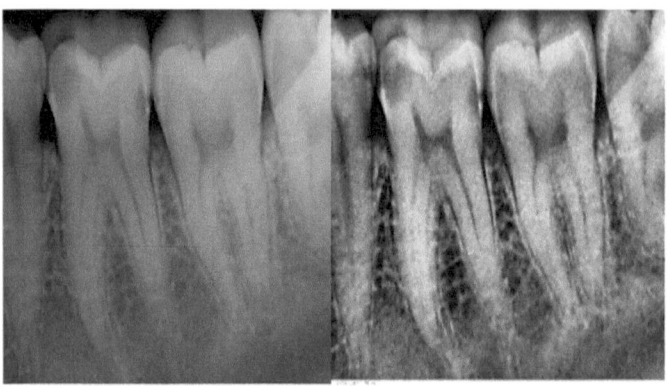

Fig. 11 Image sample 4 before and after processing

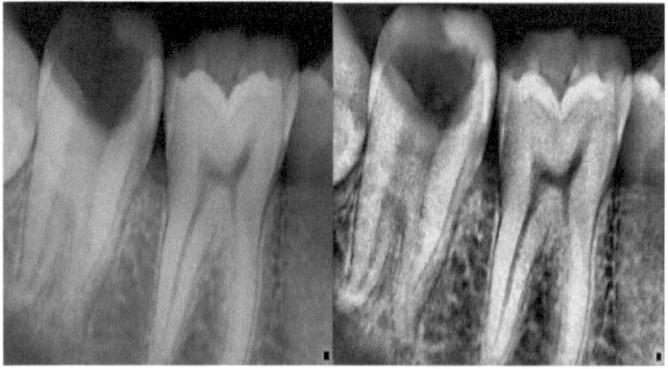

Fig. 12 Image sample 5 before and after processing

3.3   Images Obtained After Masking

These images samples are obtained after color masking and could be helpful in depicting the probable carious legions in the image along with the carious legions already present (Figs. 13 and 14).

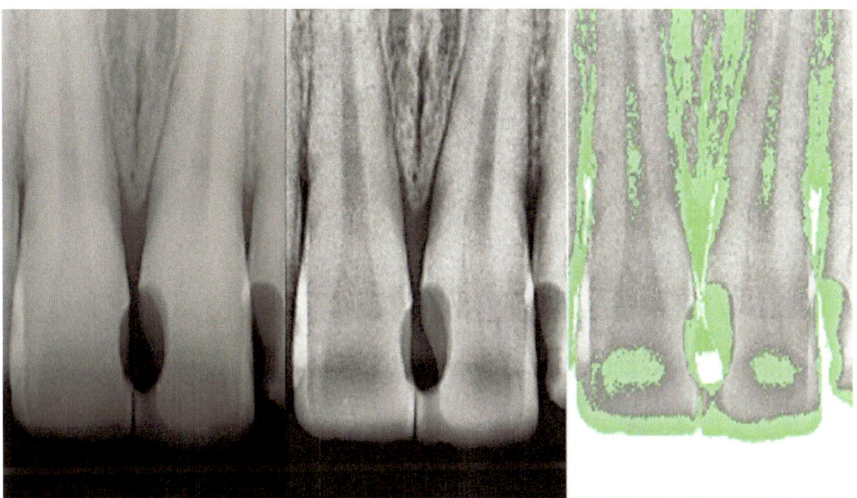

Fig. 13 Image sample 1 showing masking of the probable carious legions in dental IOPA

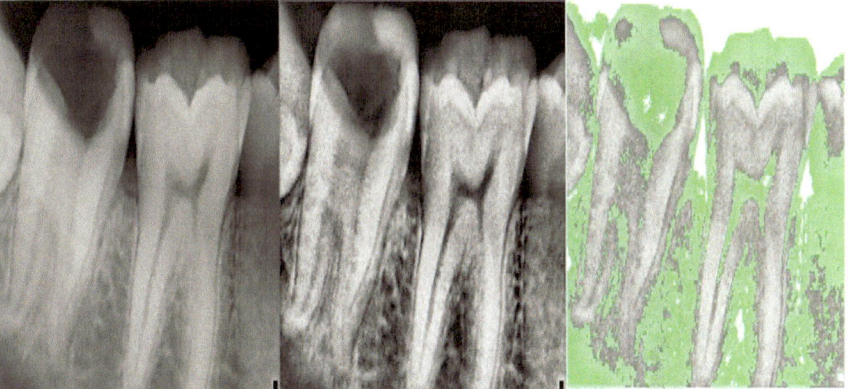

Fig. 14 Image sample 1 showing masking of the probable carious legions in dental IOPA

3.4　Images Obtained After Masking

1. **Mean Square Error between original image and median filtered image**
   In Table 1 we will see that there was certain amount of noise which was present in the original image which was removed to some extent by applying median filtering. This proves that it is important to use median filtering in the image so that we can diagnose the caries in a better way.

Table 1 Mean square error between original image and median filtered image

S. No.	Mean square error of original image	Mean square error of median filtered image
1	19.94	17.83
2	41.40	43.02
3	61.64	52.88
4	30.56	32.65
5	26.57	18.94
6	17.22	13.24
7	26.87	18.89
8	35.44	28.60
9	53.55	42.93
10	36.14	33.06
11	36.27	27.61
12	38.47	38.94
13	57.26	44.96
14	24.00	16.24
15	14.04	18.71
16	19.84	18.86
17	35.44	42.06
18	35.65	22.34
19	26.51	18.25
20	24.29	27.44
21	26.46	21.09

2. **Peak Signal to Noise Ratio between original image and median filtered image**
   In Table 2 we see that the PSNR is increased for most of the images which shows that the quality of image is enhanced.

Table 2 Peak signal to noise ratio between original image and median filtered image

S. No.	Peak signal to noise ratio of original image	Peak signal to noise ratio of median filtered image
1	35.17	35.65
2	31.99	31.82
3	30.26	30.93
4	33.31	33.02
5	33.92	35.39
6	35.80	36.94
7	33.87	35.40
8	32.67	33.60
9	30.87	31.83
10	32.58	32.97
11	32.56	33.75
12	32.31	32.26
13	30.58	31.63
14	34.36	36.05
15	36.69	35.44
16	35.18	35.40
17	32.66	31.92
18	32.64	34.67
19	33.93	35.55
20	34.31	33.78
21	33.93	34.92

3. **Mean Square Error between original images, Restored images and Masked Image**

In Tables 3 and 4 we compare the MSE for all the images before applying algorithm, after applying algorithm and after applying masking. We see that mean square error value increases but the PSNR value remains approximately same which shows that after processing of images the quality of image remains same.

Table 3 Mean square error between original images, restored images and masked image

S. No.	Mean square error of original image	Mean square error of restored images	Mean square error of masked images
1	19.94	52.69	18.49
2	41.40	73.05	25.78
3	61.64	88.48	10.27
4	30.56	63.45	25.85
5	26.57	56.93	22.11
6	17.22	35.16	28.50

(*continued*)

Table 3 (*continued*)

S. No.	Mean square error of original image	Mean square error of restored images	Mean square error of masked images
7	26.87	50.90	27.12
8	35.44	46.57	31.56
9	53.55	76.54	31.61
10	36.14	106.73	22.59
11	36.27	55.54	35.82
12	38.47	63.64	29.19
13	57.26	79.68	28.60
14	24.00	60.14	18.48
15	14.04	57.39	9.25
16	19.84	65.23	10.74
17	35.44	180.06	11.27
18	35.65	41.62	18.17
19	26.51	38.89	14.50
20	24.29	66.49	10.51
21	26.46	44.89	16.56

4. Peak Signal to Noise Ratio between original images, Restored images and Masked Image

Table 4 Peak signal to noise ratio between original images, restored images and masked image

S. No.	Peak signal to noise ratio of original images	Peak signal to noise ratio of restored images	Peak signal to noise ratio of masked images
1	35.17	30.95	35.49
2	31.99	29.52	34.05
3	30.26	28.84	38.04
4	33.31	30.14	34.04
5	33.92	30.61	34.71
6	35.80	32.70	33.61
7	33.87	31.09	33.83
8	32.67	31.48	33.17
9	30.87	29.32	33.16
10	32.58	27.88	34.62
11	32.56	30.71	32.62
12	32.31	30.12	33.51
13	30.58	29.16	33.60
14	34.36	30.37	35.49
15	36.69	30.57	38.50
16	35.18	30.02	37.85

(*continued*)

Table 4 (*continued*)

S. No.	Peak signal to noise ratio of original images	Peak signal to noise ratio of restored images	Peak signal to noise ratio of masked images
17	32.66	25.61	37.64
18	32.64	31.97	35.57
19	33.93	32.26	36.55
20	34.31	29.93	37.94
21	33.93	31.64	35.97

5. Structural Similarity Measure index for restored images with respect to original images

In Table 5 we see the structural similarity index between the original image and restored image. We were able to see that after restoring the images we have achieved approximately 60% of similarity between the original image and restored image. In this similarity the disease content i.e. dental caries which was not visible earlier was included and also quality of image has been enhanced.

Table 5 Structural similarity measure index for restored images with respect to original images

S. No.	Structural similarity measure index for restored image
1	0.6360
2	0.5835
3	0.5621
4	0.6164
5	0.6929
6	0.5632
7	0.6057
8	0.6632
9	0.6000
10	0.6506
11	0.6058
12	0.6132
13	0.5758
14	0.6251
15	0.7798
16	0.6799
17	0.7295
18	0.7074
19	0.7232
20	0.6826
21	0.6826

6 Percentage of Disease in Original Images, Restored Images and Masked Images found using computer based diagnosis

In Table 6 we have shown the percentage of disease in the original images, restored images and masked images. This percentage was calculated by finding region of interest in the image where the caries was detected. It was done by marking that particular area and then calculating the total number of pixels present in that particular area. After finding those pixels we divide them with total pixel value of the image and after multiplying it by 100 we get the percentage value of disease in that particular image. For masked images the percentage is more due to the areas which are caries probable have been included in the region of interest.

Table 6 Percentage of disease in original Images, restored images and masked images found using computer based diagnosis

S. No.	Percentage of disease in original images (%)	Percentage of disease in restored images (%)	Percentage of disease in masked images (%)
1	1.80	1.86	66.42
2	1.14	1.12	67.09
3	29.30	17.66	63.60
4	3.19	21.30	66.72
5	14.34	9.23	65.64
6	15.51	22.24	18.83
7	12.92	13.98	64.69
8	1.84	18.51	76.12
9	11.46	15.02	32.07
10	20.02	7.00	18.51
11	4.16	22.85	28.52
12	19.04	19.24	30.20
13	1.42	5.67	29.51
14	3.99	4.33	35.26
15	17.84	43.97	44.91
16	21.87	22.58	31.87
17	23.28	23.94	26.76
18	29.57	7.05	21.35
19	10.03	11.54	27.36
20	30.28	24.07	26.16
21	3.82	10.30	36.29

4 Conclusion

In this work, we have taken 23 samples of images having different type of caries present in them. As we can see in the results images acquired were having not so good brightness and contrast level also there was some noise present. With the help of our algorithm we were able to recover the lost information about the disease by enhancing the image sample in terms of brightness and contrast. Also we have been able to remove the noise up to a descent level. Secondly, with the help of doctors (specialized in dentistry) we were able to identify the carious legions in all the samples before and after processing and then were able to derive the software based analysis in which we have been able to increase the value of True positive cases from 16 to 21 using our algorithm and also the with that the disease prevalence has also been increased i.e. from 69.6 to 91.6%. Thirdly, using the color masking as a base for artificial intelligence we were able to derive certain carious probable areas in the image which would be able to help in diagnosing of diseases. In this case we were able to show the carious legions present in the samples and also those areas where caries could be present if diagnosed correctly.

References

1. Rafael, C.G., Richard, E.W.: Digital Image Processing, 3rd edn. Pearson Education, 2013
2. Rafael, C.G., Richard E.W.: Digital Image Processing Using Matlab, 2nd edn. Pearson Education, 2013
3. Wolberg, G.: Digital Image Wrapping, IEEE Computer Society Press, 1999
4. Singh, H.V.: Information Hiding Techniques for Image Covers, LAP LAMBERT Academic Publishing GmbH & Co. KG, Ddudweiler Landstr. 99, 66123 Saarbrücken, Germany, 2010
5. Pratt, W.K.: Digital Image Processing, 3rd edn. Wiley & Sons, Inclusive, New Jersey (2001)
6. Singh, H.V., Rai A.: SVM Based Robust Watermarking For Enhanced Medical Image Security, Computers and Electrical Engineering, Elsevier Publication, Amsterdam, 2009
7. Rasche, K., Geist, R., Westall, J.: Re-coloring images for gamuts of lower dimension. Comput. Graph. Forum (Proc. Euro Graph.) 24, 423–432 (2005)
8. Strickland, R.N., Kim, C.S., McDonnell, W.F.: Digital colour image enhancement based on the saturation component. Opt. Eng. 26, 609–616 (1987)
9. Cheng, S.-C. Hsia, S.-C.: Fast algorithms for colour image processing by principal component analysis. J. V. Commun. Image Represent. 24, 184–203 (2003)
10. Hanbury, A.G., Serra, J.: Morphological operators on the unit circle. IEEE Trans. Image Proc. 10, 842–1850, (2001)
11. Chellappa, R.: Digital Image Processing, 2nd edn, IEEE Computer Society Press, 1992
12. Uprichard, K.K., Potter, B.J., Russel, C.M., Schafer, TE., Adair, S., Weller, R.N.: Comparison of direct digital and conventional radiography for the detection of proximal surface caries in the mixed dentition. Am. Acad. of Padiatr. Dent. 22(1), 10–15 (2000)
13. Wenzel, A.: Digital radiography and caries diagnosis. Dentomaxillofacial Radiol. 27, 3–11 (1998)
14. MATLAB and Image Processing Toolbox Release.: The MathWorks, Inc., Natick, Massachusetts, United States (2015a)
15. White, S.C., Yoon, DC.: Comparative performance of digital and conventional images for detecting proximal surface caries. Dentomaxillofacial Radiol. 26, 32–38 (1997)
16. Tyndall, DA., Ludlow, JB., Platin, E., Nair, M.: A comparison of Kodak Ektaspeed Plus film and the Siemans Sidexis digital imaging system for caries detection using receiver operating characteristic analysis. Oral Surg. Oral Med. Oral Pathol. 85, 113–118 (1998)
17. Price, C., Ergul, N.: A comparison of a film-based and a direct digital dental radiographic system using a proximal caries model. Dentomaxillofac Radiol. 26, 45–52 (1997)

18. Naitoh, M., Yuasa, H., Yoyama, M., Shiozima, M., Nakemerra, M., Ushida, M., Iida, H., Hayashi, M., Ariji, E.: Observer agreement in the detection of proximal caries with direct digital intraoral radiography. Oral Surg. Oral Med. Oral Pathol. **85**, 107–112 (1998)
19. Selwitz, R.H., Ismail, A.I., Pitts, N.B.: Dental caries. The Lancet **369**(9555), 51–59 Jan. (2007)
20. Nielsen, L.L., Hoernoe, M., Wenzel, A.: Radiographic detection of cavitation in approximal surfaces of primary teeth using a digital storage phosphor system and conventional film, and the relationship between cavitation and radiographic lesion depth: an in vitro study. Int. J. Paediatr. Dent. **6**, 167–172 (1996)
21. Metz, C.E.: Basic principles of ROC analysis. Semin. Nucl. Med. **8**, 283–298 (1978)
22. Hanley, J.A., McNeil, B.J.: A method of comparing the areas under the receiver operating characteristics curves derived from the same cases. Dentomaxillofacial Radiol. **148**, 839–843 (1983)

Development of Glare Recognition
for Advanced Driver Assistance System

N. Madan$^{(\boxtimes)}$ and K. S. Geetha$^{(\boxtimes)}$

Department of Electronics and Communication, R V College of Engineering,
Bengaluru, India
`madan.nataraj@gmail.com`, `geethaks@rvce.edu.in`

Abstract. Glare poses a major threat in the implementation of Advanced Driver Assistance Systems. The video or image captured by vehicle bound camera can be affected by sun glare at daytime or any artificial light during night hindering the visibility and causing problems for object detection by ADAS. The paper proposes an algorithm that effectively and precisely detects the glare affected regions in video by the method of finding and drawing contours. The simulations are performed in Visual studio platform and the results show that the glare boundaries are detected accurately and detection speed of proposed method is 3 times faster than existing method of Circle Hough Transform.

Keywords: ADAS · Glare · Contours · Circle hough transform (CHT)
Intensity · Saturation

1 Introduction

There is an increased rate in the road accidents in current years urging the need for road safety. Advanced Driver Assistance System (ADAS) fall under the category of Active safety of the Automobiles where an accident can be avoided before it possibly can happen. ADAS system uses a camera bound on the outer surface of the automobile for monitoring purpose by capturing videos or images. Glare comes in the way of functioning of the ADAS system by creating very high intensity regions in frames of video being captured by camera hindering the visibility and decreasing the Performance of ADAS Systems. An overview of ADAS, its behavioral impacts, the benefits and functionalities the system provides is described in paper [1]. Maria Staubach describes the design of ADAS in [2] by the study and analysis of causes for accidents on roads. The paper comes up with an approach of examining number of accidents as close to 474 by different means of data sources. The reason for road accidents is found out to be the distraction that driver faces and reduction in the activity of the driver. In [3], an approach which analyses two mutual helping ADAS features which are warning the upcoming danger and assistance for merging is proposed. Simulations are performed under multiple driver conditions for the study of individual driver behavior or behavior of group of drivers. In [4], a brief description of different external environmental factors

© Springer International Publishing AG 2018
D. J. Hemanth and S. Smys (eds.), *Computational Vision and Bio Inspired Computing*,
Lecture Notes in Computational Vision and Biomechanics 28,
https://doi.org/10.1007/978-3-319-71767-8_75

that affect the performance of incident detection system is described. Varied environmental conditions like snowfall, rain, shadow and glare which causes the false detection of objects or makes the objects to be detected going unnoticed is discussed. The detection of the night time glare due to headlights of the oncoming traffic using various sensors have been discussed in papers [5, 6].

Tan and Feris in [7] proposes a method which uses multiple images for the removal of glare from a frame of a video or image. The camera set-up is slightly altered so that the source of light is positioned differently for different images captured by camera. The features of an image which are not visible in one image due to glare can be obtained from other image which is not affected by glare at that position due to difference in position of light source. In [8], a method which could decrease the wrong detection of automobile glare caused due to vehicle headlights is discussed. The method combines two techniques. One is to convert the image from RGB to HSV and the other is that, the brightest spot obtained from HSV is subjected to the calculation of magnitude of the vector at that point. Both together performed decreases the false detection of glare. In [9], a specular reflection detection and removal method is proposed for endoscopic images where a thresholding algorithm is used which transforms the input color image to gray scale and this gray scale image is binarized using a specific threshold for detection of specular reflection regions. The detection of glare in the endoscopic images by the use of image segmentation and its removal by the very popular Inpainting algorithm has been discussed in paper [10]. The detection of static glare at night caused by reflection of street light from the wet roads which was interfering in Automatic Incident detection system using the method of background generation and background subtraction has been discussed in paper [11]. Glare recognition for ADAS by using Photometric features like Intensity, Saturation and local contrast and geometric features of glare which assumes all the glare sources to be circular whether the glare at daytime by sun or the streetlights causing glare at night. Considering all these, the paper proposes the Photometric map which is the combination of Intensity, Saturation and local contrast maps forming input to Circle Hough transform (CHT) algorithm to detect circular glare sources has been discussed in paper [12]. Figure 1a shows the input image with glare and Fig. 1b shows the glare detected by the method of CHT.

The demand for making ADAS systems to work efficiently in adverse weather conditions like rainfall, snowfall, mud on camera and intense glare conditions is of prime importance. Our research goes on detecting the glare hindering the visibility of ADAS.

In this paper, we develop a Glare recognition system using the method of finding and drawing contours. The method uses certain image processing techniques and computer vision algorithms eliminating the need of sensors which was the case in researches [1, 2], and thus removing the need to synchronize the sensors for operation of ADAS system. The proposed method of finding and drawing contours outperforms the method of Circle Hough transform as described in [4] by following means:

- Time taken to detect the glare region using the proposed method is one third the time taken by the existing Circle Hough Transform method.
- Computationally simplistic when compared to the method of CHT which demands the calculation of mean and standard deviation for each block of Image.
- Glare boundaries are well defined and the false detection rate is less in the proposed method than the existing CHT method.

The rest of the paper is organized as follows. In Sect. 2, the methodology of the proposed glare recognition is described. In Sect. 3, the simulation results are discussed. Finally, the paper is concluded in Sect. 4.

(a)   **(b)**

Fig. 1 **a** Input frame of video containing sun glare at day-time **b** Sun glare detected by the method of circle hough transform

2 Methodology

Glare affected regions usually are the very high intensity regions with low color saturation. We take these properties of glare to our advantage to detect the glare which affects the performance of ADAS systems. From a 3-channel BGR video, we extract each of the frame and store it in separate buffers. The extracted 3-channel BGR image leads to the formation of the single channel intensity and the Saturation maps that are used for the glare detection. Figure 2 shows the block diagram of the proposed Glare detection method for ADAS system. The proposed method is simplistic enough by means of the computational complexity but reliably detects both the static and moving glare. static glare means that the position of glare does not vary between the frames of the video. Moving glare is the one in which the position of glare regions keeps changing in each frame of the video. The method is even proven to detect both daytime glare due to sun and night time glare due to artificial sources of light like headlights of vehicles and street lights reliably.

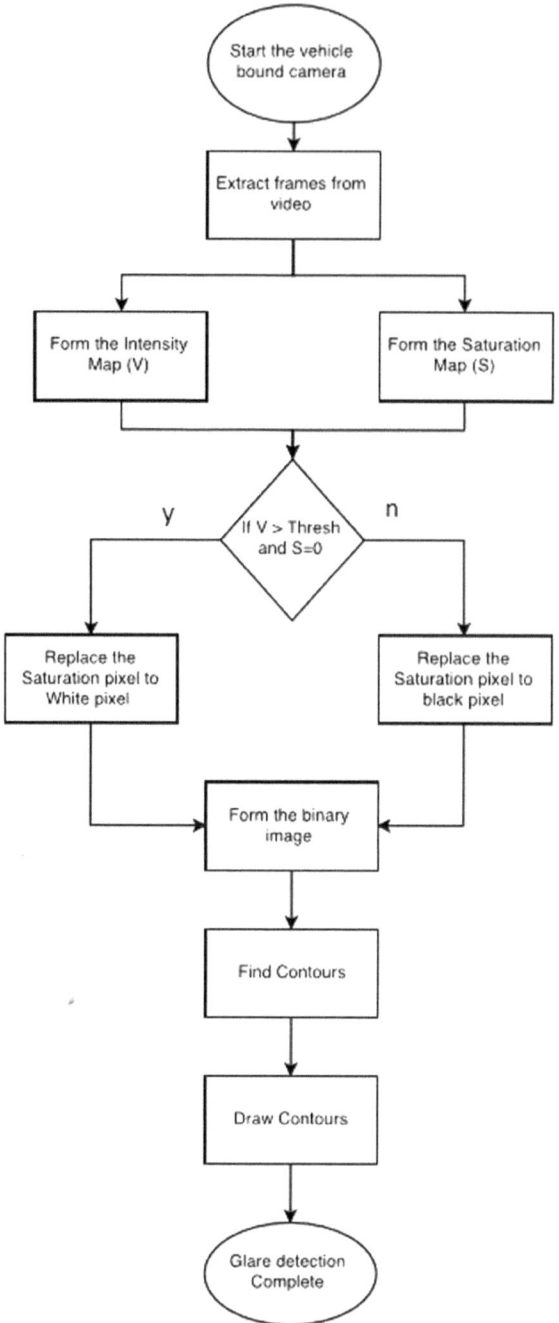

Fig. 2 Flow-chart for glare detection by the method of finding and drawing contours

The frames are extracted from video being captured by the vehicle bound camera. One such frame from the videos containing Sun glare at daytime and streetlight glare at night time are shown in Fig. 3a, b respectively. For each of the frame which is a 3-channel BGR image, we form single channel Intensity and saturation map by the use of following formulas respectively.

$$Int(x,y) = \max\{Image\ R(x,y), Image\ G(x,y), Image\ B(x,y)\} \tag{1}$$

$$Sat(x, y) = Int\,x, y - minImageRx, y,\ ImageGx, y,\ ImageBx, yInt\,x, y,$$
$$for\ Int\,x, y > 00, \tag{2}$$
$$otherwise$$

The Intensity Maps for two input images obtained by Eq. (1) are shown in Fig. 4a, b respectively.The Saturation Maps obtained by Eq. (2) are shown in Fig. 5a, b respectively.

Since Glare candidate pixels will usually have very high intensity and very low color saturation, the pixels with very high intensity value exceeding the threshold value in the intensity map and pixels with very low color saturation value with 0 in the Saturation Map is considered as Glare pixels. The threshold is selected based on trialing with few input videos. The glare pixels are given a value of 255 and the rest pixels are assigned a value of 0. This gives the binary image depicting the Glare in the image. The binary images for two input images are shown in the Fig. 6a, b respectively. Dilation and erosion operations on the binary image can be considered to make glare regions more precise.

The Input to finding and drawing contour algorithm is a binary image that is obtained. The algorithm of finding and drawing contours is an efficient one which finds each of the small segments with same intensity value in the image and stores it as a Vector of points. The algorithm will preserve the edges of each contour and also have a count on the number of contours found.

By the help of the number of contours found and preserved edges of each of the contour, a draw function can be used to mark the boundaries of the detected glare region. This detects the glare region in each frame of the input video of Sun glare and streetlight glare reliably and quickly than the existing method.

3   Results Discussion and Analysis

The proposed method of glare detection by finding and drawing contours was tested on a number of Sun-glare and street light glare videos. The images on the left are concerned with the day-time sun glare and the images on the right are concerned with the night time glare due to street lights. Simulations are performed on Visual studio platform

(a)   (b)

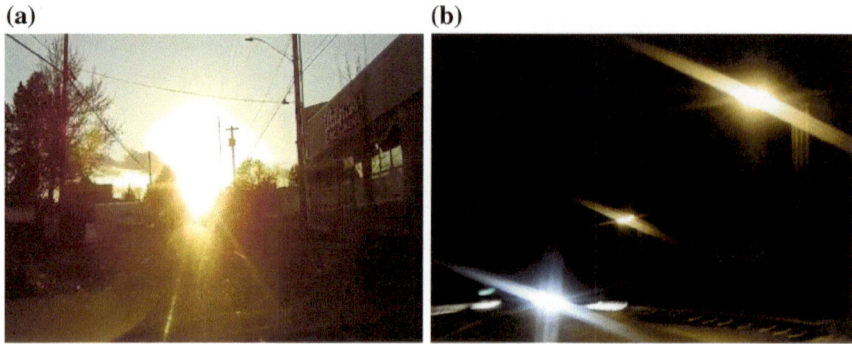

Fig. 3 a Input frame of a video containing sun glare at day-time **b** Input frame of a video containing night-time glare due to street lights

(a)   (b)

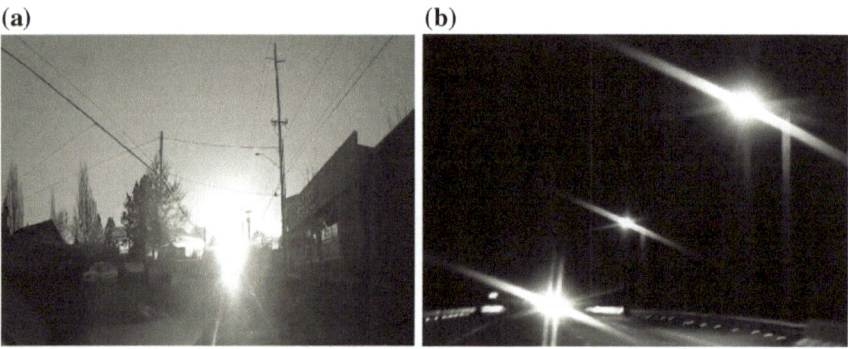

Fig. 4 a Intensity map for input frame containing Sun glare **b** Intensity map for input frame containing glare due to street lights

(a)   (b)

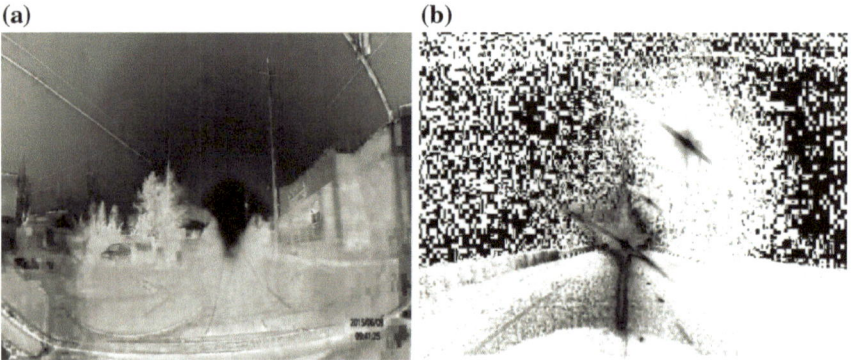

Fig. 5 a Saturation map for input frame containing sun glare **b** Saturation map for input frame containing glare due to street lights

(a)   (b)

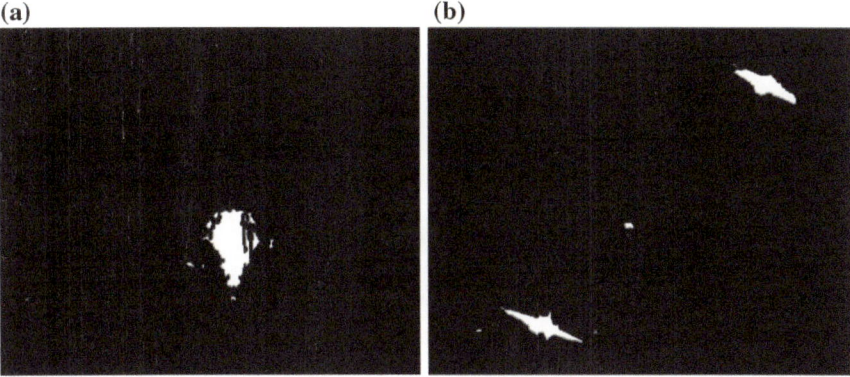

Fig. 6 a Binary Image showing the glare region caused due to sun **b** Binary image showing the glare region caused due to street lights

Figure 7 depicts the glare detected in both day-time and night-time glare videos. It can be seen that the glare boundaries are well defined unless the case of CHT method used in paper [4].

(a)   (b)

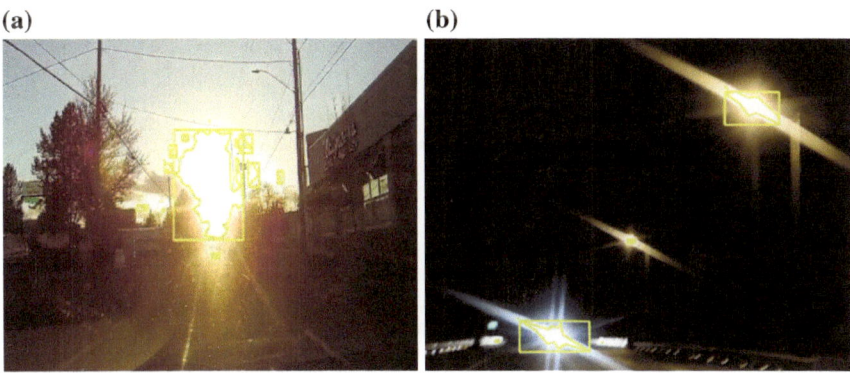

Fig. 7 a Sun glare detected using the method of finding and drawing contours **b** Glare at night due to street lights detected by the method of finding and drawing contours

Table 1 gives the comparison between the existing CHT method of glare recognition and the proposed method of finding and drawing contours. The table clearly shows that the detection speed of proposed algorithm is nearly 3 times faster than the CHT method.

Table 1 Comparison between the existing and proposed methodology for day-time and night-time glare videos

Video type	Method	Resolution	Frames	Video duration in seconds	Avg time taken/frame in secs
Day-time	CHT method	640 * 480	900	30	1.974
Day-time	Proposed method	640 * 480	900	30	0.652
Night-time	CHT method	640 * 480	360	12	1.621
Night-time	Proposed method	640 * 480	360	12	0.574

4 Conclusion

In this paper, development of glare recognition system by the method of finding and drawing contours is proposed. The proposed method performs significantly well over the existing method of CHT. The simulation was performed in Visual studio platform where both day-time and night-time videos were tested with the following parameters listed in the table. The PC used for the simulation had configuration of i5 processor 2.7 Ghz clock-speed. The detection speed was around 0.54 to 0.68 s which is almost 3 times faster than the CHT method. The proposed method is simplistic enough in terms of computational complexity and reliably detects both day-time and night-time glare. Porting this on to a hardware platform and test this on real-time environment is part of our future work.

References

1. Brookhuis, K.A., De Waard, D., Janssen, W.H.: Behavioural impacts of advanced driver assistance systems–an overview. EJTIR **1**(3), 245–253 (2001)
2. Staubach, M.: Factors correlated with traffic accidents as a basis for evaluating advanced driver assistance systems. Accid. Anal. Prev. **41**(1025–1033), Elsevier, (2009)
3. Maag, C., Mühlbacher, D., Mark, C., Kruger, H.-P.: Studying effects of Advanced Driver Assistance Systems (ADAS) on individual and group level using multi-driver simulation. IEEE Intell. Trans. Syst. Mag. (2012)
4. Shehata, M.S., Cai, J., Badawy, W.M., Burr, T.W., Pervez, M.S., Johannesson, R.J., Radmanesh, A.: Video-based automatic incident detection for smart roads: the outdoor environmental challenges regarding false alarms. IEEE Trans. Intell. Trans. Syst. (2008)
5. Menon, K.A.U., Ananyase, A.D., Pradeep, P.: Consensual and co-ordinated vehicular headlight attenuation using wireless sensor networks. In: Computing, Communication and Networking Technologies (ICCCNT), International Conference 2014, Hefei, China (2014)
6. Devassy, A., Gopinath, N., Narayanan, V., Ramachandran, A.: Coordinated, progressive vehicular headlight glare reduction for driver safety using wireless sensor networks. In: International Conference on Connected Vehicles and Expo (ICCVE) (2014)

7. Feris, R., Raskar, R., Tan, K.-H., Turk, M.: Specular reflection reduction with multi-flash imaging. In: Proceedings of Computer Graphics and Image Processing, 17th Brazilian Symposium (2004)
8. Pharadornpanitchakul, S., Duangchit, A., Chaisricharoen, R.: Enhanced danger detection of headlight through vision estimation and vector magnitude information and communication technology. In: 4th Joint International Conference on Electronic and Electrical Engineering (JICTEE), (2014)
9. Guo, J.J., Shen, D.F., Lin, G.S., Huang, J.C., Liu, K.C., Lie, W.N.: A specular reflection suppression method for endoscopic images. In: IEEE Second International Conference on Multimedia Big Data (BigMM), (2016)
10. Karapetyan, G., Sarukhanyan, H.: Automatic detection and concealment of specular reflections for endoscopic images. In: Ninth International Conference on Computer Science and Information Technologies (2013)
11. Cai1, J., Shehata, M., Badawy, W., Pervez, M.: An algorithm to compensate for road illumination changes for AID systems. In: Proceedings of the 2007 IEEE Intelligent Transportation Systems Conference Seat.tle, WA, USA, Sept. 30–Oct. 3, 2007
12. Andalibi, M., Chandler, D.M.: Automatic glare detection via photometric, geometric, and global positioning information. Soc. Imaging Sci. Technol. Feb. (2017)

High Resolution Remote Sensing Image Denoising Algorithm Based on Sparse Representation and Adaptive Dictionary Learning

Yuwei Qiu[1], Yuda Bi[2], Yang Li[3], and Haoxiang Wang[4(✉)]

[1] Tsinghua University, Beijing, China
[2] University of Georgia, Georgia, USA
[3] George Washington University, Washington, USA
[4] Cornell University, NY, USA
hw496@goperception.com

Abstract. Aiming at the denoising algorithm of high resolution remote sensing images, in this paper, we propose a novel method based on sparse representation and adaptive dictionary learning. The proposed algorithm uses the strong correlation between the bands of high resolution remote sensing images, which combines the non local self similarity of the image with the local sparsity to improve the denoising performance. By means of sparse representation of image noise, we extract texture information from image noise so as to improve the quality of image denoising. A learning based super-resolution algorithm learns a dictionary through a set of training examples, and combines the missing high-frequency information from the low resolution image, and finally obtains the corresponding high-resolution image. The traditional denoising algorithm still has noise residue after noise removal, and the image denoising effect is not obvious when the noise is large. Experimental results show that the peak signal-to-noise ratio of the proposed method is higher than the existing similar algorithms, and it can better preserve the details and texture information of the image, and improve the visual effect.

Keywords: Sparse representation · Adaptive dictionary · Learning method
High resolution · Remote sensing image · Denoising · Algorithm

1 Introduction

Digital images are often contaminated by various noise sources. The existence of noise seriously affects the validity and reliability of image feature extraction, target detection and recognition. In order to improve the image quality, we need to remove the noise from the image.

D. J. Hemanth and S. Smys (eds.), *Computational Vision and Bio Inspired Computing*,
Lecture Notes in Computational Vision and Biomechanics 28,
https://doi.org/10.1007/978-3-319-71767-8_76

In the process of digital image acquisition and transmission, the experimental results will be affected by noise interference, resulting in lower image quality. However, in many applications, we need clear, high-quality images. Therefore, image denoising is of great significance, and it is an important research topic in the field of image processing. The traditional methods of noise reduction include neighborhood filtering, median filtering and frequency domain filtering. These methods often result in the loss of detail and texture information while removing noise.

In recent years, the denoising method based on dictionary learning and the method based on non local self similarity have attracted extensive attention of scholars at home and abroad while the development of denoising algorithm based on dictionary learning benefits from the development of sparse representation theory. The basic idea is to make use of the local sparseness of image denoising, such as K-SVD algorithm and some improved algorithms. The K-SVD denoising algorithm divides the noisy image into small overlapping blocks, then the algorithm is trained by K-SVD dictionary algorithm to obtain adaptive redundant dictionary, which completes sparse representation, so as to achieve the purpose of image denoising. The LSSC algorithm and the CSR algorithm firstly classify the overlapping image blocks by using image similarity, and obtain redundant dictionary from various image blocks, finally implement image denoising. The above 3 algorithms have achieved great success in image denoising, but at the same time, the existing denoising algorithm will lead to the loss of a large number of image details and texture information, which affects the subsequent image processing. Therefore, it is necessary to study how to preserve image details and texture information while denoising [1–3].

Aiming at the shortage of the existing image denoising algorithm, and considering the kernel sparse representation model can capture nonlinear data structure, which fully describes the details of the image, we propose an image representation based on sparse denoising algorithm and sparse representation and adaptive dictionary learning (Fig. 1).

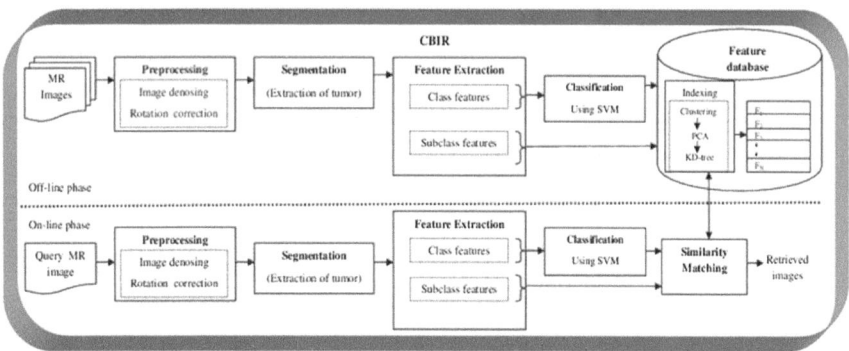

Fig. 1 The process of image denoising

2 Proposed Method

2.1 Guiding Filtering

Assume we have a guiding image I, input image is denoted as P, thus the output image satisfies the following equations:

$$q_i = a_k I_i + b_k \forall i \in \omega_k \tag{1}$$

$$a_k = \frac{\frac{1}{|\omega|} \sum_{i \in \omega_k} I_i p_i - u_k \overline{p_i}}{\sigma_k^2 + \varepsilon} \tag{2}$$

$$b_k = \overline{p_i} - a_k u_k \tag{3}$$

where ω_k represents 2-dimensional window of image center; p_i and q_i respectively represents the i-th pixel value of input image and output image; I_i represents the i-th pixel value of guiding image; u_k and σ_k^2 respectively represents mean and variance; $\overline{p_i}$ represents the mean of input image. i.e.: (Figs. 2, 3)

$$\overline{p_i} = \frac{1}{|\omega|} \sum_{i \in \omega_k} I_i p_i \tag{4}$$

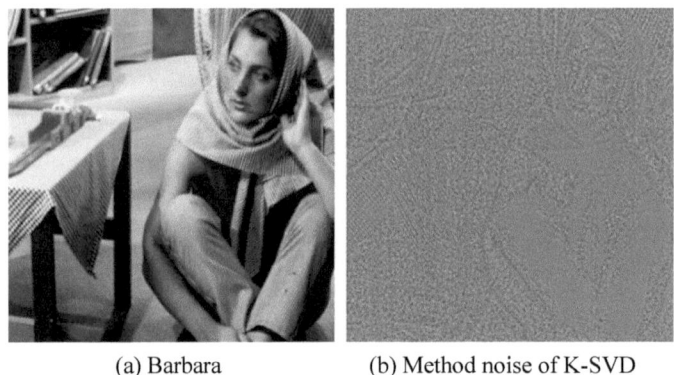

(a) Barbara   (b) Method noise of K-SVD

Fig. 2 Image Barbara and K-SVD algorithm method noise.

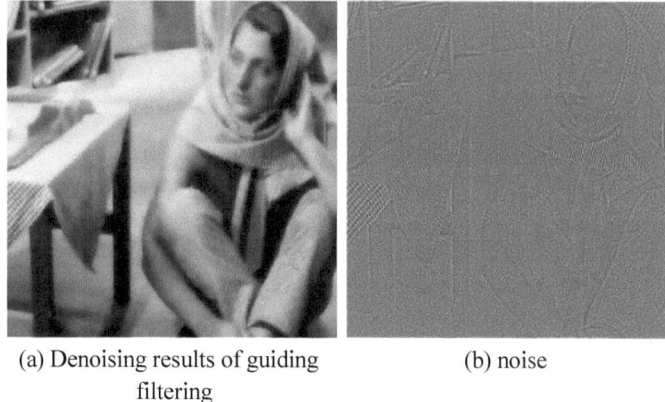

(a) Denoising results of guiding   (b) noise
filtering

Fig. 3 Denoising results of guiding filtering and method noise.

2.2 Kernel Sparse Representation Model

The sparse representation problem is to calculate the most sparse vector of given image vector in dictionary matrix. The given image vector is denoted as $y \in R^n$, and the redundant dictionary matrix is denoted as $D = (d_1, d_2, \ldots, d_m) \in R^{n \times m}$.

$$\begin{cases} \min_x ||y - Dx||_2^2 \\ s.t. ||x||_o \le T_0 \end{cases} \tag{5}$$

where d is vector, T_0 is sparsity. For above question, there are several solutions, such as orthogonal matching pursuit, regularized orthogonal matching pursuit, basis pursuit de-noising and smoothed l_0 norm [4, 5].

Assume the nonlinear mapping matrix is a Hilbert space in the kernel space, thus the kernel function corresponding to the nonlinear mapping matrix is expressed as:

$$\Phi : R^n \to F \subset R^N (n < < N) \tag{6}$$

$$k(x, x') = \left\langle \Phi(x), \Phi(x') \right\rangle = \Phi(x)^T \Phi(x') \tag{7}$$

The corresponding kernel sparse representation problem is described as:

$$\begin{cases} \min_x ||\Phi(y) - \Phi(D)x||_2^2 \\ s.t. ||x||_o \le T_0 \end{cases} \tag{8}$$

The algorithm proposed in this paper is divided into 3 main processes. First, the noise image is divided into several overlapping image blocks, and image blocks were randomly selected forming a sample set. The adaptive redundancy dictionary is obtained by using the sample set, and the coefficients and the learned dictionary are used to recover the image.

Considering that the image noise is additive noise, the observation model is:

$$Y = X + v \tag{9}$$

2.3   Remote Sensing Image Denoising Model

Any ideal remote sensing image can be represented as $y_0 \in R^N$, and we add into mean value, variance and Gauss white noise, thus the observed image can be modeled as:

$$y = y_0 + v \tag{10}$$

The purpose of image denoising is to recover y_0 from the observed image y. According to sparse and redundant representation, a dictionary matrix is defined as follows:

$$D \in R^{n \times k} \tag{11}$$

Each image block can be represented by a sparse dictionary:

$$\hat{a} = \arg \min ||a||_0 subject to ||Da - y||_2^2 \leq \varepsilon \tag{12}$$

According to the idea of the optimization algorithm, the problem can be transformed into a typical *triplet* model:

$$\hat{a} = \arg \min ||Da - y||_2^2 + u||a||_0 \tag{13}$$

First, assume that D is a known item, and $X = Y$, then t optimal solution for each image region is obtained:

$$\hat{a}_{ij} = \arg \min_a u_{ij} ||a||_0 + ||Da - y||_2^2 \tag{14}$$

In this paper we use the OMP algorithm to solve this problem. The iterative process is ended if $||Da - x||_2^2 \leq \varepsilon$. Each image is processed at the same time according to the sparse encoding of the sliding window, and X is updated. i.e.:

$$X = \arg \min_x ||X - Y||_2^2 + \sum ||Da_{ij} - R_{ij}X||_2^2 \tag{15}$$

The solution of binomial is generally translated into the following model:

$$\widehat{X} = \left(\lambda I + \sum_{ij} R_{ij}^T R_{ij}\right)^{-1} \bullet \left(\lambda Y + \sum_{ij} R_{ij}^T R_{ij} D \hat{a}_{ij}\right) \tag{16}$$

3 Adaptive Dictionary Learning

Image blocks are extracted from high-quality remote sensing images as training libraries to generate dictionaries. This method is relatively simple, but the efficiency is relatively low. In order to obtain the dictionary which can satisfy the remote sensing image characteristic, we use the noisy image itself to train the dictionary, and extract the image block from the noisy remote sensing image [6, 7].

3.1 Dictionary Learning

Given a set of samples, denoted as follows:

$$\beta = (\beta_1, \beta_2, \ldots, \beta_K) \in R^{n \times K} \tag{17}$$

Dictionary learning is to find redundant dictionaries so that each sample can be represented as a sparse matrix Γ.

$$\Gamma = (\Gamma_1, \Gamma_2, \ldots, \Gamma_K) \in R^{n \times K} \tag{18}$$

$$\begin{cases} \min\limits_{D,\Gamma} ||\beta - D\Gamma||_F^2 \\ s.t. ||\Gamma_k||_0 \le T_0 \end{cases} \tag{19}$$

Dictionary learning problems are commonly solved by MOD algorithm and K-SVD algorithm.

3.2 Adaptive Dictionary Learning Model

The data structure of high resolution remote sensing image is a cube. Assume that the image contains S bands. The image is divided into overlapping image blocks, thus dictionary learning is represented as:

$$\begin{cases} \min\limits_{D,\alpha_{n,s}} \sum\limits_{n=1}^{N} \sum\limits_{s=1}^{S} ||p_{n,s} - D\alpha_{n,s}||_2^2 \\ s.t. ||\alpha_{n,s}||_0 \le T_0 \end{cases} \tag{20}$$

where $p_{n,s}$ is the s-th image block vector of the n-th image block, D is redundant dictionary, and $\alpha_{n,s}$ coefficient of sparse representation.

Considering the use of non local self similarity, the N cube data are divided into K classes by using K-means clustering algorithm, and use the cube data in each class, we learn the sub dictionary, and use the sub dictionary to represent the data in the class. Assume cube data contained in each class is denoted as M_k, thus the adaptive dictionary learning model is represented as:

$$\begin{cases} \min\limits_{D,\alpha_{m,s}^{(k)}} \sum\limits_{m=1}^{M_k} \sum\limits_{s=1}^{S} ||p_{m,s}^{(k)} - D\alpha_{m,s}^{(k)}||_2^2 \\ s.t. ||\alpha_{m,s}^{(k)}||_0 \leq T_0 \end{cases} \tag{21}$$

where D_k is the sub dictionary of the k-th class, $\alpha_{m,s}^{(k)}$ is sparse representation coefficient, and $p_{m,s}^{(k)}$ can be represented by D_k, namely, $||\alpha_{m,s}^{(k)}||_0 \leq T$. Thus nonlocal dictionary learning model of the k-th class is as follows:

$$\min\limits_{D,\alpha_{m,s}^{(k)}} \sum\limits_{m=1}^{M_k} \sum\limits_{s=1}^{S} ||p_{m,s}^{(k)} - D\alpha_{m,s}^{(k)}||_2^2 \tag{22}$$

Since each band of high resolution images has strong correlation, the coefficient matrix for the full band data of the n-th cube block is a low rank matrix.

$$\alpha_m^{(k)} = \left[\alpha_{m,1}^{(k)}, \ldots, \alpha_{m,s}^{(k)} \right] \tag{23}$$

4   Experiment and Analysis

In this section, the performance of the proposed algorithm is verified by experimental analysis and compared with the K-SVD algorithm. Experiment 1 compares the denoising algorithm of different noise variance with the value of the PSNR; Experiment 2 compares the visual effect of denoising algorithms; Experiment 3 gives the effect of overlapping pixels on the performance of the algorithm; In Experiment 4, the influence of the number of atoms on the performance of the algorithm was investigated [8].

The block size of the noisy image is 5 * 5, and the overlapping pixel of adjacent images is 4, we randomly extract 20,000 image blocks for dictionary learning.

Experiment 1 Comparison of PSNR values of denoising algorithms

We test with Barbara, boat and house respectively. Table 1 compares the PSNR values of 3 algorithms when the mean square of variance is different. As can be seen from Table 1, in the same case, the PSNR value of the MNSR algorithm is higher than the other 2 algorithms.

Table 1 Comparison of PSNR values of denoising algorithms

α	Barbara		Boat		House	
	K-SVD	MNSR	K-SVD	MNSR	K-SVD	MNSR
20	30.87	31.81	30.25	31.12	32.88	33.22
25	29.33	30.66	29.16	30.26	32.15	32.90
30	28.51	29.75	28.43	29.41	31.06	31.68
50	25.48	26.49	25.89	26.77	27.99	28.46
100	21.99	23.12	22.79	23.59	24.38	25.99

Experiment 2 Comparison of Visual effect of denoising algorithm

In this experiment, we use three remote sensing images to compare the visual effects of different algorithms, and $\sigma = 30$. The results are shown as follows (Fig. 4):

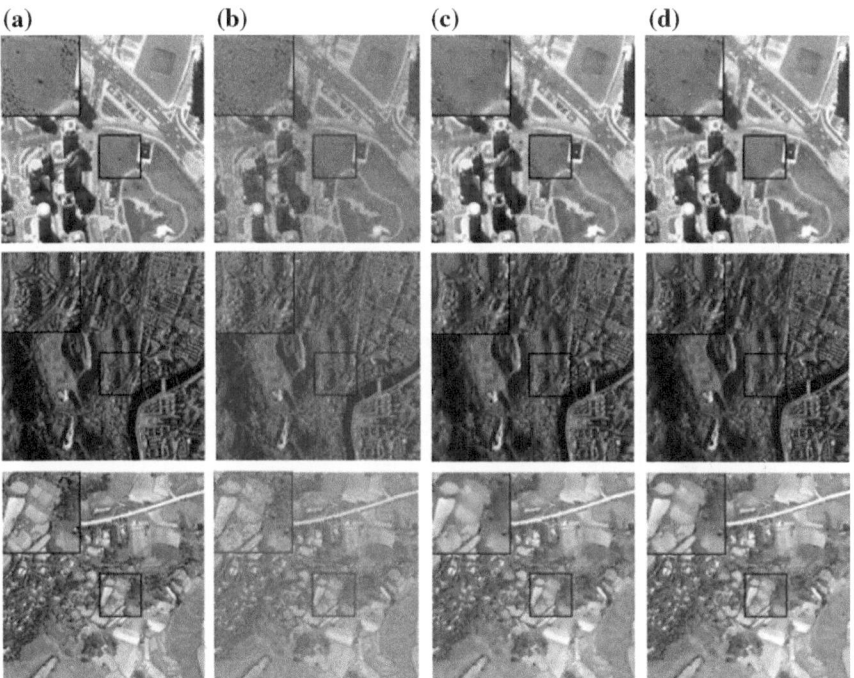

Fig. 4 a Original image; **b** Noise image; **c** K-SVD; **d** MNSR

Experiment 3 Influence of the number of overlapping pixels on the performance of the algorithm.

This experiment gives the influence of the number of overlapping pixels on the performance of the algorithm, and $\sigma = 30$. We can figure out that with the increase of the number of overlapping pixels, the performance of the algorithm is improved (Fig. 5).

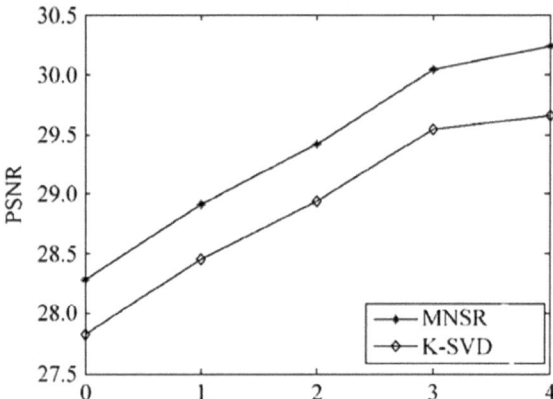

Fig. 5 The influence of the number of pixels overlap

Experiment 4 Influence of number of dictionaries on performance of algorithm

The experiment shows the influence of the number of dictionaries on the performance of the algorithm, and $\sigma = 50$. Other simulation conditions remain unchanged, change the number of atoms ranging from 100 to 1000 (Fig. 6).

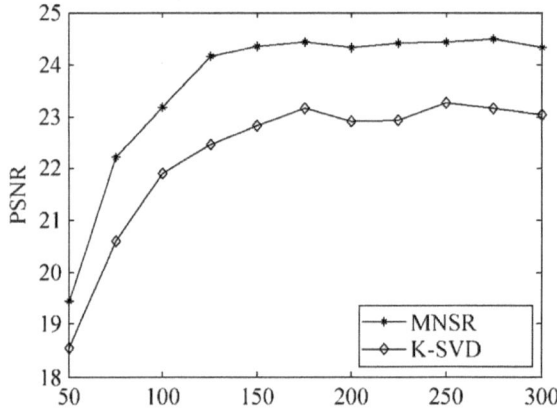

Fig. 6 The influence of the number of dictionaries

5   Conclusion

The traditional denoising algorithm is difficult to keep the details and texture information of the image. Therefore, this paper proposes an image denoising algorithm based on sparse representation and adaptive dictionary learning. The proposed algorithm uses guiding filtering to obtain image noise, and then uses the dictionary learning and sparse representation algorithm to extract image texture information contained in noise, so as to enhance the image quality after denoising. The experimental results show that the proposed method is better than the existing algorithms in the peak signal to noise ratio, and can better preserve the texture and detail information of the image, thus improving the visual effect.

References

1. Gu, S., et al.: Weighted nuclear norm minimization with application to image denoising.In: Proceedings of the IEEE Conference on Computer Vision and Pattern Recognition, (2014)
2. Zhao, Y.Q., Yang, J.: Hyperspectral image denoising via sparse representation and low-rank constraint. IEEE Trans. Geosci. Remote Sens. **53**(1), 296–308 (2015)
3. Zhang, S., Wang, H., Huang, W.: Two-stage plant species recognition by local mean clustering and weighted sparse representation classification. Clust. Comput. 1–9 (2017)
4. Ram, I., Elad, M., Cohen, I.: Image denoising using nl-means via smooth patch ordering. In: IEEE International Conference on Acoustics, Speech and Signal Processing (ICASSP), 2013. IEEE (2013)
5. Shi, Y., Yang, X., Guo, Y.: Translation invariant directional framelet transform combined with Gabor filters for image denoising. IEEE Trans. Image Process. **23**(1), 44–55 (2014)
6. Wang, H., Wang, J.: An effective image representation method using kernel classification. In: Tools with Artificial Intelligence (ICTAI), IEEE 26th International Conference on IEEE, pp. 853–858, Nov. (2014)
7. Om, H., Biswas, M.: A generalized image denoising method using neighbouring wavelet coefficients. SIViP **9**(1), 191–200 (2015)
8. Aguerrebere, C., et al.: A Bayesian hyperprior approach for joint image denoising and interpolation, with an application to HDR imaging. IEEE Trans. Comput. Imaging (2017)

4-Share VCS Based Image Watermarking for Dual RST Attacks

Sheshang D. Degadwala[(⊠)] and Sanjay Gaur

Madhav University, Rajasthan, India
{sheshang13, sanjay.since}@gmail.com

Abstract. The startling improvement about web need aggravated the transmission, conveyance Also right to advanced networking really useful. Therefore, networking makers need aid more usually managing illegal What's more unapproved utilization from claiming their productions. In our proposed approach, first enter the user name and password then generate QR-code using zxing library that will converted into the three shares using Binary Visual cryptography algorithm. Now share-4 is save in the database that is for future reference at receiver side. Remaining share-1, share-2 and share-3 are embedding into the Red, Green and Blue-component LL bit using of block DWT-SVD and Pseudo Zernike moment. After embedding add RGB Component. Now Color watermark image transfer from the network. As in network there are different attackers apply combination of Rotation, Scale and Translation attacks on the color watermark image. For recover the attacks first apply Pseudo Zernike moment, Surf feature on R, G and B-component they will extract the attacks pixel and recover the scale-angle using affine transformation. Now share-1, share-2, share-3 and another share-4 is in data base so we will apply EX-OR operation to get the QR-code. The final QR-code is decoded and we get the user name and password. This research work can give a way for providing authentication to all online Services.

Keywords: QR codes · Visual cryptography · RGB-Embedding
Block-DWT · Surf · Affine and dual RST attacks

1 Introduction

The enormous Growth in e-world which will be coupled for reality totally Web furthermore headway to machine execution encouraged the initial circulation of advanced information. Done globe totally Web because of rupture in security advanced picture camwood a chance to be undoubtedly duplicated and disseminated without straight reasonably. Those advanced watermarking schemes have been recommended will flexibility these sorts for unapproved right about advanced media information. Toward starting stage, encryption and control get systems are used to copyright protection, content verification Also proprietorship security. In any case presently days, the advanced watermarking strategies are utilized prominently on stay with advanced media secure [1, 2].

© Springer International Publishing AG 2018
D. J. Hemanth and S. Smys (eds.), *Computational Vision and Bio Inspired Computing*,
Lecture Notes in Computational Vision and Biomechanics 28,
https://doi.org/10.1007/978-3-319-71767-8_77

Watermarking may be an example about odds embedded under an advanced image, sound alternately feature record that identifies those files copyright majority of the data. The same advanced watermarking hails from the faintly noticeable watermark imprinted in stationary that identifies the maker of the stationary. The reason for the advanced watermarks may be will gatherings give copyright insurance for licensed innovation that's on advanced arrangement [3]. In this way watermark will be those concealed data inside the advanced indicator. There would a number requisitions for advanced watermarking Anyway Around the greater part copyright protection, substance authentication, duplicate and use control Furthermore content portrayal are imperative provision region of the advanced watermarking (Fig. 1).

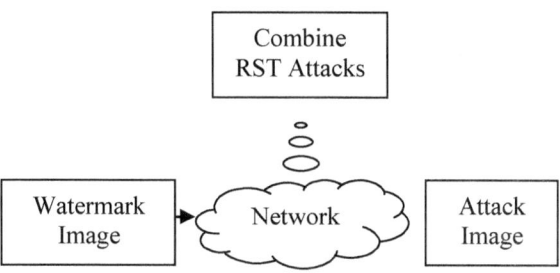

Fig. 1 Network scenario

Watermarking systems are arranged into spatial space techniques and change area strategies. Spatial area techniques are less unpredictable, however less strong against assaults. The watermarking plan in view of the change areas can be further divided into discrete wavelet transform (DWT) and discrete cosine transform (DCT), the discrete Fourier transform (DFT). Capacity of DWT-SVD based plan is more than DFT. We have made system to do secure transaction which is visual cryptography scheme and, for copyright protection and deal with geometrical attacks the watermarking scheme is used. It's absolutely impossible that anybody could decode the data contained inside some of shares. At the point when the shares are stack together, decoding is conceivable when the shares are set more than each other. Now, the data turns out to be in a flash accessible. No additional computational power is required keeping in mind the end goal to decode the data.

2 Related Works

2.1 Proposed Visual Cryptography

Share Generation: In this phase to generate white pixel share-1 will use 1 0 and in share-2 it will use 0 1. 1 indicates white pixel and 0 indicates black pixel. Share-1 will use 1 1 and share-2 uses 0 0 Repeating again to generate black pixel by 0 0 and white pixel by 1 1, Again Same to generate white pixel 0 1 is used and 1 0 to black in share-1 and in share-2. By using visual cryptography scheme, it creates a share from secret image. Firstly secret image is taken and converted into binary image. Then every pixel image is divided into eight sub groups and then into four pixel in each share. By selecting image randomly one can encode schemes out of three given figure.

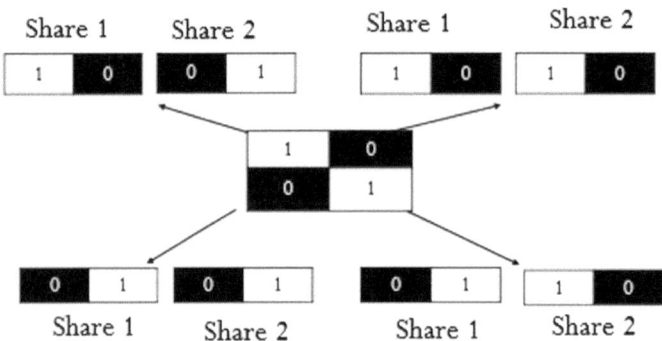

Fig. 2 Proposed visual cryptography

In the last phase, the process of Visual Cryptographic Combination is performed. Here by applying the binary XOR operation, on both shares, we are going to get back the original secret image.

2.2 Discrete Wave Late Transformation (DWT) [4, 5]

Wavelet transform disintegrates a picture under an course of action about band compelled segments which could make reassembled to redo the principal picture without shortcoming since those information transmission of the ensuing coefficient sets is more diminutive over that of the 1st picture, those coefficient sets might make down inspected without reduction for information. Propagation cost of the 1st banner may be master toward up sifting, inspecting also summing the individual sub gatherings. To 2-D pictures applying DWT compares to get ready those picture toward 2-D channels previously, each estimation.

Those channels disconnect the information picture under four non-covering multi-determination coefficient sets, an easier determination estimation picture (LL1) also even (HL1), Verthandi (LH1) Furthermore inclining (HH1) point of interest segments seemed for Fig. 5. The sub-band LL1 identifies with those coarse-scale DWT coefficients same time the coefficient sets LH1, HL1 What's more HH1 talk of the fine-size about DWT coefficients.

To get those accompanying coarser extent of wavelet coefficients, those sub-band LL1 is further took care of until a few completing up scale n will be come to. In the perspective At n may be attained we will need 3 N + 1 coefficient sets including of the multi-determination coefficient sets LLN What's more LHX, HLX and HHX the place x ranges starting with 1 until n.

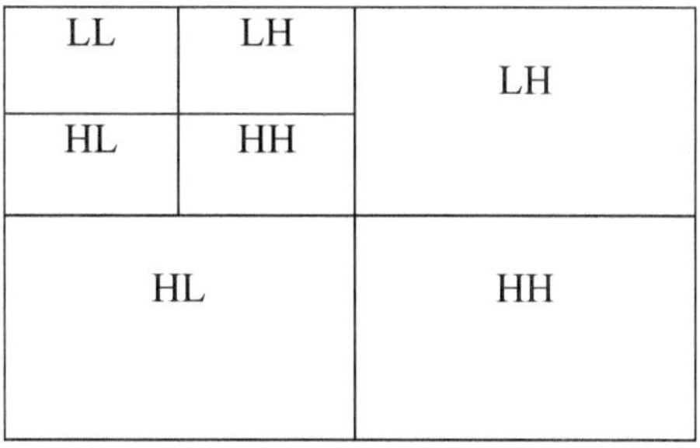

LL	LH	LH
HL	HH	
HL		HH

Fig. 3 Decomposition model of DWT at level 2

At that point again, the helter skater repeat coefficient sets HHX fuse those edges What's more surfaces of the picture and the mankind's eye will be not commonly fragile should progressions clinched alongside such coefficient sets. This permits the watermark with a chance to be embedded without being seen Eventually Tom's perusing the human eye.

2.3 SVD [6]

SVD is a standout amongst the most valuable apparatuses in direct variable based math with a few applications in picture pressure, watermarking, and other flag handling zones. In the event that an is a n × n network then SVD of lattice A can be characterized as:

$$A = U * S * VT, \tag{1}$$

where U and V are the orthogonal networks and S is a corner to corner lattice. Askew components of S are the solitary qualities and they fulfill the accompanying property.

$$S(1,1) > S(2,2) > S(3,3) > \ldots\ldots > S(n, n), \tag{2}$$

Singular Value Decomposition will be great known to those watermarking done light of the certainty that few about singular qualities might talk with considerable section about banner vitality, SVD could a chance to be associated with square What's more rectangular pictures, the SV's of a picture bring extraordinary upheaval invulnerability, i.e., SV's don't progress inside and out The point when An minimal inconvenience is included should An picture energy values, SV's talk will intrinsic scientific properties.

2.4 SURF [7]

SURF features will be a scale-invariant characteristic identifier In view of those hessian matrix, likewise is, the Hessian-Laplace identifier. However, as opposed utilizing an alternate measure to selecting the area and the scale, those determinant of the hessian may be utilized for both. The hessian grid will be harshly approximated, utilizing a situated for box sort filters, Furthermore no smoothing may be connected the point when setting off starting with you quit offering on that one scale of the next (Fig. 4).

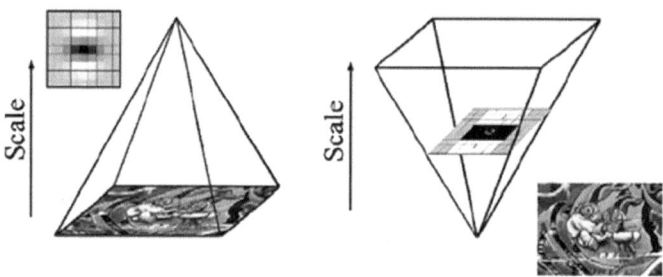

Fig. 4 SURF Feature

Gaussian would ideal for scale-space analysis, but that to act they must be ruined which introduces artifacts, specifically in little Gaussian Kernels. Surf pushes the close estimation indeed further, utilizing those box filters. These estimated second-order Gaussian derivatives, and camwood make assessed exceptionally utilizing essential analytics images, freely for their measure. Surprisingly, despite those harsh approximations, those execution of the characteristic identifier is tantamount to the outcomes got with those ruined Gaussian. Box filters could process addition close estimation of the Gaussian subsidiaries Likewise there are large number other sources about huge clamor in the preparing chain. Surf need been appeared for make more than five times speedier over difference of Gaussian.

2.5   Pseudo Zernike Moments [8]

Pseudo-Zernike polynomials are illustrious and broadly used in the exploration of optical schemes. Image analysis uses shape descriptors. PZM is geometric-based moment that uses the worldwide info in an image for extracting features. The orthogonal moments of PZM are shift, rotation, and scale invariants which are suitable for pattern recognition applications. Pseudo-Zernike contains several orthogonal sets of complex-valued polynomials defined as:

$$S_{rc}(X,Y) = R_{rc}(X,Y) \exp(jm \tan^{(-1)}(X/Y)), \tag{3}$$
where $X^2 + Y^2 \leq 1$, $r \geq 0$, $|c| \leq r$.

$$PZM_{rc} = (r+1)/\pi \sum X \sum Y f(X,Y) S_{rc}(X,Y) \tag{4}$$
A = absolute (Z)
Angle (Z) = $\tan^{-1}$ (imag(Z), real(Z));
Phi = angle (Z) * 180/pie

It should be noted that the PZM is computed for positive m because $(x, y) = vnm * (x, y)$. If an image is rotated, phase of moments in PZM will be varied and its absolute value remains constant. Thus, if the absolute value or value of PZM is considered as the feature, the feature f is independent of rotation [9]. Pseudo Zernike polynomials of order $\leq P$, contain $(+1)2$ linearly independent polynomial of degree $\leq P$. Pseudo Zernike moment is used in optical system, pattern recognition and in image analysis as shape descriptors.

3 Proposed Preserving Method

After studying various visual cryptography schemes and watermarking schemes, we propose new technique for secure bank transaction. In this scheme we provide authenticity and data integrity of the shares using watermark technique. In our scheme we take one QR-image as original image or host image and create shares using 2-out-of-2 VC scheme. When two shares will be created, server share is stored in bank database and client share is kept by user. The user will present with client share during all the transactions with bank. After that we apply the watermark technique on that client share image for providing the authentication and data integrity and send it on the open communication channel.

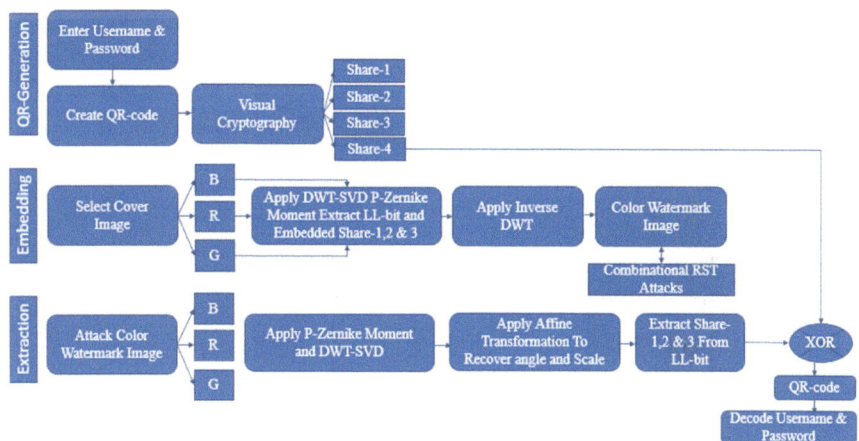

Fig. 5 Proposed block diagram

QR-Generation: As shown in the Fig. 2 first select the user name and password. Now using zxing library generating the QR-code. That QR-code is now in invisible form so now one can see the data inside. Further we have Apply VCS scheme to generate two shares of QR-Code.

Embedding: In this process as shown in the Fig. 3 select the color cover image. Extract the R, G and B component. Now Select R, G and B-component and Apply P-Zernike Moment and DWT-SVD transformation and Extract LL-bit. In the LL-bit embedding the Share-1 data. After Inverse DWT-SVD transformation to generate R, G and B-Embedded Image Now Add all to Create Color Water Mark Image. Color Watermark Image is transmitted over the Network Different Attackers Apply RST attacks on it.

Extraction: After RST attacks getting the Attack Color Image Which is now apply the P-Zernike Moment with Surf Feature Extraction to recover attacks. Now Extracting the share-1, share-2, share-3 and it will combine with another database share-4 to generate QR-image. QR decoder will decode the Username and Password.

The beauty of our system lies in the fact that, if any attacker makes a copy of any image share to forge it later, the watermark will be distorted so for such forged image share our system will not allow the generation of host image from the stack of 4 image shares. Thus, the attacker will not get the original image.

Here we use Singular Value Decomposition discrete wavelet transform based watermarking technique which is geometrically invariant. This type watermarking scheme is robust against the RST attacks, various JPEG and noise attacks.

4 Results and Discussion

4.1 Results

Figure 6.

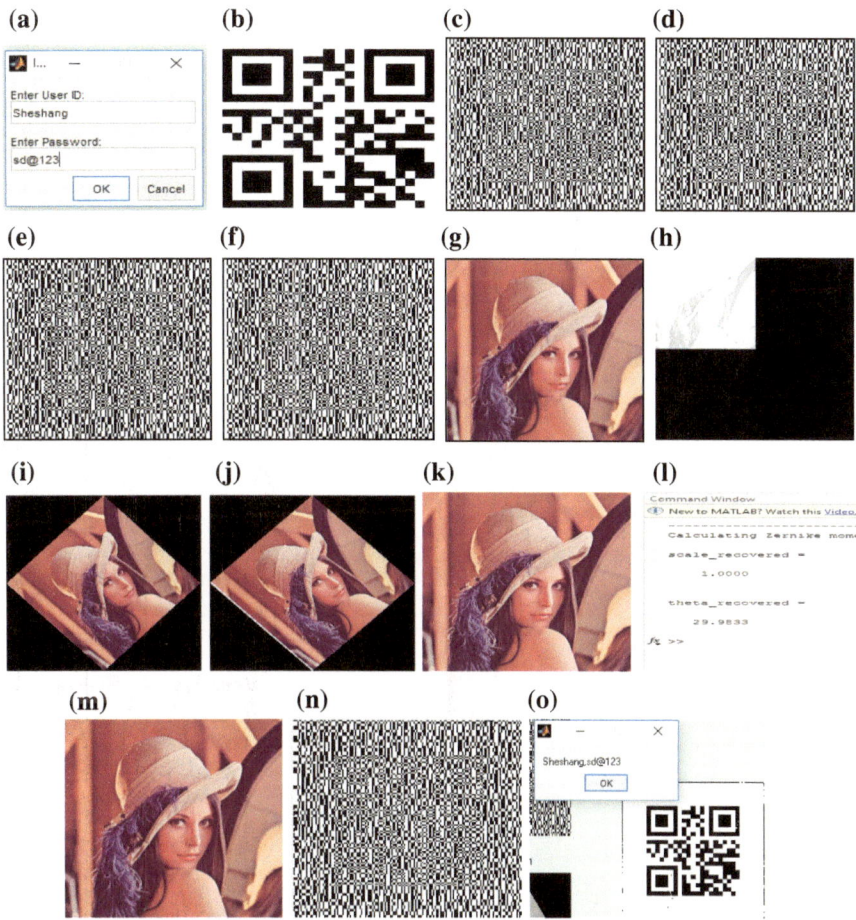

Fig. 6 **a** Enter USR and PSW. **b** QR-code. **c** Share-1. **d** Share-2. **e** Share-3. **f** Share-4. **g** Cover image. **h** DWT-SVD. **i** Rotation-scale attack. **j** Rotation-translation attack. **k** Scale-translation attack. **l** Recover theta and angle. **m** Recover image. **n** Recovered share-2. **o** QR-code recover

4.2 Analysis of Different Attacks

See Figs. 7, 8 and Tables 1, 2, 3.

Table 1 Rotation with scale

Rotation	Scale	PSNR	MSE
0	2	63.083	0.028
15		64.043	0.025
30		65.063	0.023
35		65.053	0.022
40		66.021	0.021
45		66.081	0.023
55		64.093	0.022
65		65.023	0.024
100		65.033	0.019
120		66.071	0.022
180		64.081	0.024

Table 2 Rotation with translation

Rotation	Translation	PSNR	MSE
0	10	62.051	0.032
15		63.072	0.033
30		64.022	0.031
35		65.064	0.028
40		66.014	0.025
45		66.032	0.023
55		64.086	0.025
65		65.068	0.026
100		65.029	0.012
120		65.032	0.021
180		64.055	0.023

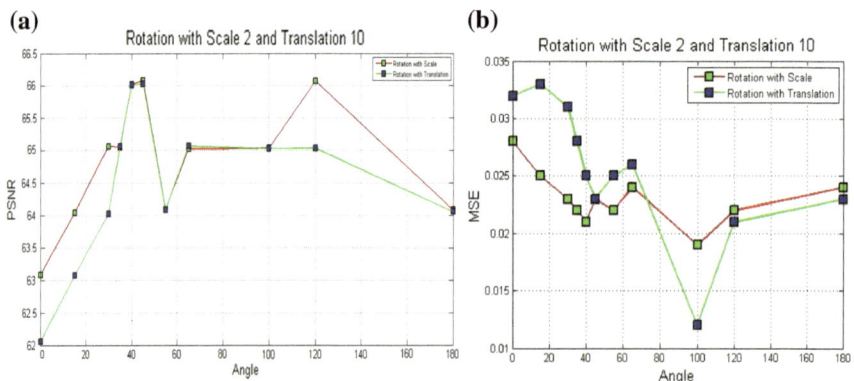

Fig. 7 Rotation with translation **a** PSNR and **b** MSE

Table 3 Translation and scale

Translation	Scale	PSNR	MSE
1	2	66.023	0.032
2		66.019	0.033
3		66.015	0.031
4		66.011	0.030
5		66.007	0.031
−1		66.003	0.023
−2		65.999	0.024
−3		65.995	0.028
−4		65.991	0.028
−5		65.987	0.027

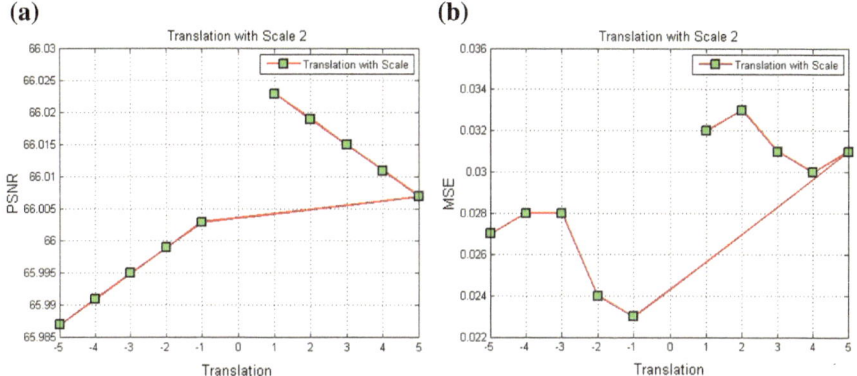

Fig. 8 Translation and scale **a** PSNR and **b** MSE

5 Conclusion

An arrangement for QR code produce with secondary security which comprises about image encoding, information embedding and data extraction/image-recovery stages. In the primary phase, generate QR-code at that point second stage it will a chance to be for offers four Shares. In those receiver side Decoder, it could extracted the information and recuperate the first content without loss. Dual RST attacks is apply on Watermark image in our Proposed Privacy Preserving System. For Recovery of Attacks here we have use Combine approach of Block DWT-SVD and Pseudo Zernike Moment with surf feature. Affine transformation is also apply for recover attack watermark image. So after extraction of all shares are combine to get QR-code image. Our Proposed system will increase PSNR value for Recovered QR-image.

References

1. Delphin Raj, K.M, Nancy, V.: Secure QR coding of images using the techniques of encoding and encryption. Int. J. Appl. Eng. Res. **9**(12), 2009–2017 (2014). ISSN 0973-4562
2. Ajish, S., Rajasree, R.: Secure Mail Using Visual Cryptography (SMVC), 5th ICCCNT, Hefei, China 11–13 July 2014
3. Benoraira, A., Benma hammed, K., Boucenna, N.: Blind Image Watermarking Technique Based on Differential Embedding in DWT and DCT Domains, Springer (2015)
4. Aparna, J.R., Ayyappan, S.: Comparison of digital watermarking techniques. In: International Conference for Convergence of Technology, (2014)
5. Singh, B., Dhaka, V.S., Saharan, R.: Blind Detection Attack Resistant Image Watermarking, IEEE, (2014)
6. Huang, C.H., Wu, J.L.: Attacking visible watermarking schemes. IEEE Trans. Multimed. **6**(1), Feb. (2004)
7. Bay, H., Ess, A., Tuytelaars, T., Van Gool, L.: Speeded-Up Robust Features (SURF), Elsevier, (2007)
8. Degadwala, S.D., Gaur, S.: Privacy preserving system using Pseudo Zernike moment with SURF and affine transformation on RST attacks. Int. J. Comput. Sci. Inf. Secur. **15**(4), April (2017)
9. Gupta, A.K., Raval, M.S.: A Robust and Secure Watermarking Scheme Based on Singular Values Replacement Sadhana, vol. 37, Part 4. Indian Academy of Sciences, pp. 425–440 Aug. (2012)

An Intelligent Framework for Road Safety and Driver Behavioral Change Detection System Using Machine Intelligence

Idhant Haldankar[1(✉)], Mohit Tiwari[2(✉)], G. Usha[3], and S. Aruna[3]

[1] Software Engineering Department, SRM University, 603, Wellington Hiranandani Estate, Thane West, Mumbai, India
idhant123@gmail.com
[2] Software Engineering Department, SRM University, N-571 Aashiyana Colony, Lucknow 226012, Uttar Pradesh, India
mohitt533@gmail.com
[3] Software Engineering Department, SRM University, Chennai, India

Abstract. Road accidents have been a major concern for every countries all over the world. According to the report submitted by WHO 3400 people die on roads every day and millions of people are getting injured or disabled every year. In India itself 1214 (according to the report by NCBI) people met in road accidents every day in which 377 people die, 20 of them are children below the age of 14 years. More than half of all road traffic deaths occur among young adults aged 15–44. These alarming rate of deaths call for immediate attention and finding a way to reduce this high rate of death toll on the world road. Keeping all these in our mind we propose to develop a safety system in a form of a "smart-car" which deduces the safety of the driver based on his past driving patterns or driving habits. The Driving patterns could be tracked by taking the vehicle GPS coordinates and usage in general. would be recorded, passing these features into Regression Analysis would help us in predicting a possible accident.

Keywords: Image processing · Drunk driving detection · Behavior monitoring Distraction detection · Experience learning · Object detection

1 Introduction

Improvement of public safety and reduction in the risk factor while driving is the major issue on which the traffic authorities of the world are focusing on at present. Various surveys have shown that most of the accident take place due to the mental, psychological and physical condition of the driver at the time of driving. It was noticed that most of the accidents take place because driver was distracted. This distraction could be the result of alcohol, stress or some other reason (Fig. 1).

© Springer International Publishing AG 2018
D. J. Hemanth and S. Smys (eds.), *Computational Vision and Bio Inspired Computing*,
Lecture Notes in Computational Vision and Biomechanics 28,
https://doi.org/10.1007/978-3-319-71767-8_78

MUARC study of 340 casualty crashes in VIC and NSW 2000-2011	
13.5%	Intoxication
11.8%	Fell asleep
10.9%	Fatigued
3.2%	Failed to look
3.2%	Passenger interaction
2.6%	Felt ill
2.6%	Blacked out
1.8%	Feeling stressed
1.5%	Looked but failed to see
1.4%	Animal or insect in vehicle
0.9%	**Using a mobile phone**
0.9%	Changing CD/cassette/radio
0.9%	Adjusting vehicle systems
0.9%	Looking at vehicle systems
0.3%	Searching for object

Fig. 1 The table describes the statistics of accidents and the respective reasons for the same. Results from the Monash University Accident Research Centre (MUARC)

Keeping all these factors in mind we propose to device a smart system that can be installed in the four wheelers and could monitor driver's behavior and keep in track the action of the driver and process it. This system will also keep in track distraction of all sorts, like alcohol [1] bottle and relate it with the change in behavior of the driver to get the final result. Furthermore it will track the road route taken by the driver to check if there is any change in the driving pattern of the driver or not.

To evaluate which parameters we need to look after to resolve driver's behavior we have discussed "Parameters in Consideration" in Sect. 2. The parameters which we would be considering will be enough to calibrate and fit in our Logistic Regression Model (LRM). The most important of them all is "Drowsiness Detection" which is further explained in Sect. 2.1.2. We also discuss Cognitive and Visual distraction in Sect. 2.2. In Sect. 2.3 we describe the "Driving Pattern Detection" where using Google Maps API we locate the driver's position and get supporting data which make it useful understand the surroundings of the road and possible risk of encountering an accident.

2 Proposed System Architecture

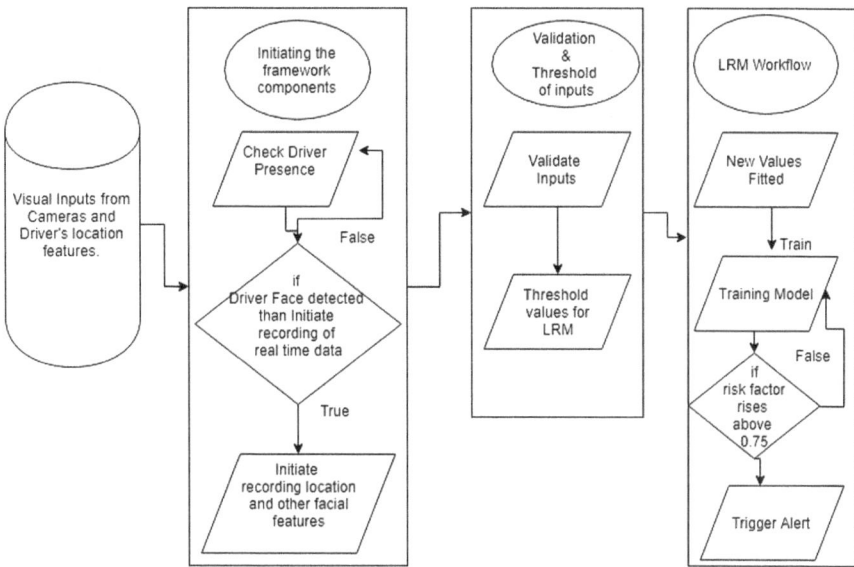

Fig. 2 Architecture diagram

- Visual inputs will be coming from two cameras, one which is fitted in front of the driver and the other is fitted facing from the passenger seat (Fig. 2).
- Front Camera inputs will be used for Face detection, Eye detection and Sentiment detection.
- The side Camera will track the body movements of the driver and check for any distractions which could cause a risk to his life.
- Driving pattern would involve data about the road information which the driver is currently using. Based on the location, traffic density, and population density of that place, we can estimate the risk level which is present for the driver's untoward actions. A GPS system is essential to track the Driving patterns.
- The data coming from these peripherals should be evaluated and ranked according [2] to the level of risk they indicate to the driver. The values will be fitted into the (LRM).
- The [3] pre-trained model will evaluate the inputs in real time and check of the risk level rises above 75% or 0.75.

We now explain the parameters, in the next section. These will be providing us the values which we can fit in out LRM.

3 Parameters in Consideration

The data which need to be interpreted are given below.

3.1 Facial Expression Tracking in OpenCV

OpenCV is used with a trainer as well as detector OpenCV already contains many pre-trained classifiers for face, eyes, smile etc. We will be using it as a preliminary test which detects the presence of driver's presence, before any of the other data processing takes place.

3.2 The Drowsiness Detection Algorithm

A Facial landmark detector needs to be implemented which is inside Dlib and produces 68 (x, y)-coordinates that map to specific facial structures. There will be 68 point mappings we obtained by training a shape predictor [4] (Adrian Rosebrock, Face Alignment with OpenCV and Python).

Below we can visualize what each of these 68 coordinates map to (Fig. 3).

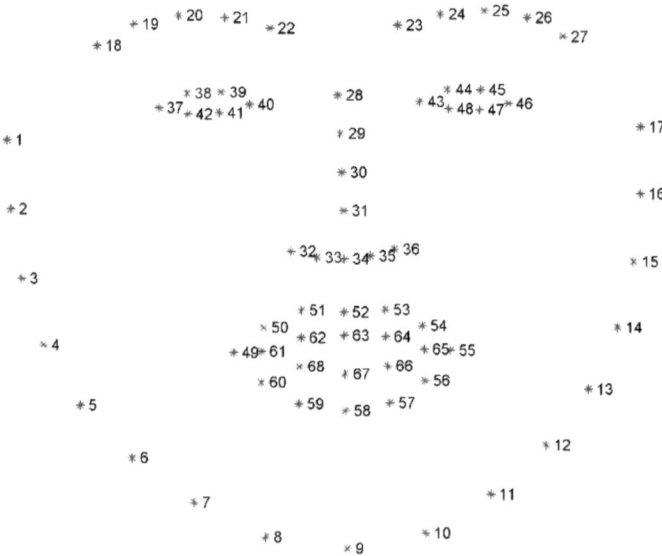

Fig. 3 68 coordinates map

We set up cameras which would give visual inputs for Image Processing on the facial features. When a face is detected, we apply FLD and extract the eye regions [5]. Now we have the eye regions, we can compute the eye aspect ratio (EAR) to determine if the eyes are closed [6]. If the EAR indicates that the eyes have been closed for significant amount of time, we'll trigger an alarm to alert the driver (Fig. 4).

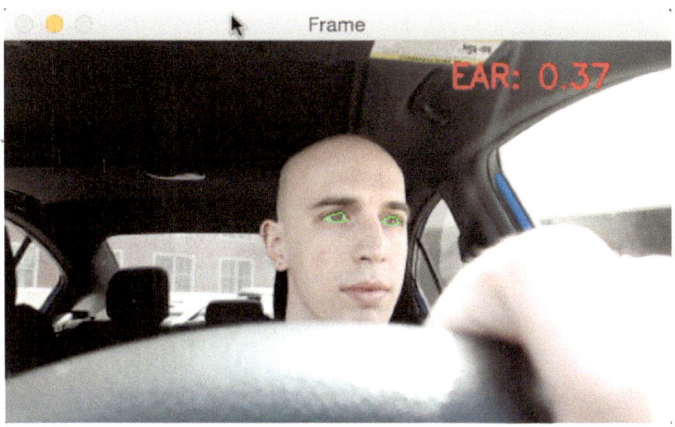

Fig. 4 Extract eye regions from the face and apply facial landmark localization

By using the above technique we can get data about whether the driver is too tired to drive and is falling asleep during driving. In a case when if the driver's eye is closed for a while, say more than 30 s then we can reduce the speed of car or even stop the car.

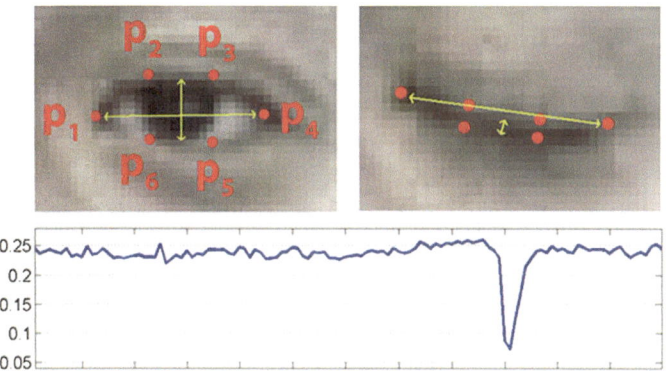

Fig. 5 EAR for different plots

The regions marked in the Fig. 5 show [7] the coordinates of P1, P2, P3, P4, P5, P6. The Euclidian distance between these coordinates contributes to the different EAR values which we would encounter in real time execution of the algorithm.

3.3 Detecting Distraction

Distraction is the main reason why most of the accidents take place. Distraction can be due to mental state of the driver or due to some other activity. Based on these distraction can be classified as:

- Visual distraction
- Cognitive distraction (Fig. 6).

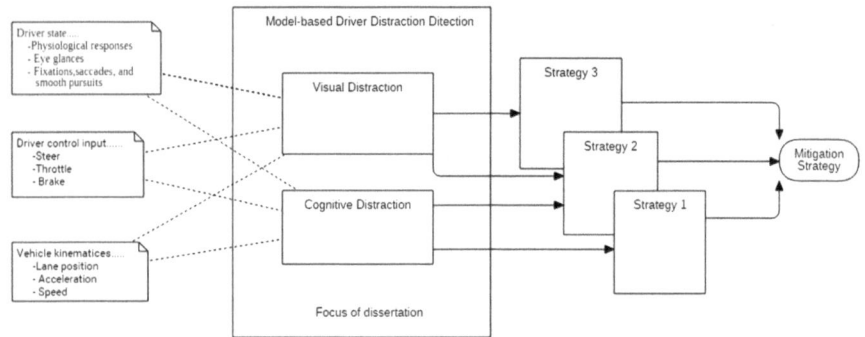

Fig. 6 Architecture of the distraction detection system

3.3.1   Estimating Visual Distraction

Detection of visual distraction involves prediction of continuously changing states of the distraction that corresponds to if the driver's attentive on the roadside or he is distracted and looking anywhere else [8]. These visual distraction pattern suggest that the algorithms used for the estimation of the behavior of the driver and his distraction level have to take account of the immediate changes of eye patterns. An algorithm can be created to find the percentage of visual distraction applying the top-down approach as the result of driver's off glances is a general case among drivers. Several scholars have devised models to calculate the amount of risks due to visual distraction using drivers glance pattern. The release time for acceleration release, the time taken for the lead car's brakes until the time the driver suspends the accelerator, has been estimated by applying the amount of driver's off glances. It can be represented by using a linear equation [9]: (reaction time for accelerator release) = 1.58 + 1.65 × (percentage of off-road glances) (Table 1).

Table 1 Summary of visual and cognitive distraction [12, 6]

Parameters	Visual distraction	Cognitive distraction	Combined distraction (hypothesized)
Eye movement	Frequent off glances, long period of distraction from road, and less amount of attention towards road	Low	Frequent off glances, long period of distraction from road, and less amount of attention towards road
Lane position	Huge change in lane	Low or no lane variation	Huge lane changes
Steering control	Discrete steering correction and large correction magnitude (<5, high steering error)	Small correction magnitude (<3, low steering error)	Discrete steering correction and both high and low correction magnitude (<5)

3.3.2   Detection of Cognitive Distraction

Generally detection of cognitive distraction is more convoluted as compared to visual distraction. Recent techniques find the quantized state of cognitive distraction. Detection of cognitive distraction would mostly require a coagulation of multiple performance measures which are spread over relatively long period of time, and personalized for different drivers. The task of analyzing cognitive distraction is to assemble a large dataset of performance records, such as eye gaze measures, in a logical order and intuituitively derive the driver's cognitive state. Although some theories, like ACT-R [10], have made predictions regarding driver distraction, these are efficient at reporting, rather than estimating performance and thus cannot be utilized as a singular method for detecting cognitive distraction.

To detect cognitive distraction data mining methods have been used. Cognitive workload can be estimated with the help of eye glances and driving performance by using a decision tree. SVMs and Bayesian Networks (BNs), in earlier works, were successfully accepted for cognitive distraction by eye activity and performance of driving [11].

3.4   Tracking Driving Patterns

In order to monitor and analyze the driving route pattern of the driver, we are planning to use pre-existing API from Google or some other GPS provider. Google provides a Google Road API that helps us track the path followed by the driver using GPS breadcrumbs (Fig. 7).

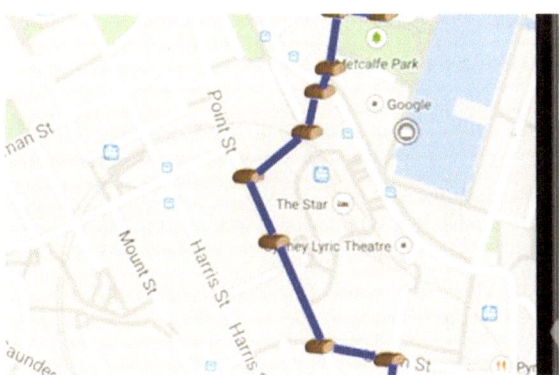

Fig. 7 Illustration of working of Google road API

Google Road API [3] regularly grabs GPS breadcrumbs as the driver drives and then using these breadcrumbs it can give the exact path followed by the user during his journey. Not only this using this API we can get data of the various speed limit of the various road traveled by the user and also the speed at which the user traveled these roads so we can easily compare these data to get the current state of the driver for our analysis [3].

3.4.1   Request

In order to get all these data one only need to send a request to the Google API via HTTPS, in the below given way:

"https://roads.googleapis.com/v1/snapToRoads?parameters&key=YOUR_API_KEY"key=YOUR_API_KEY" (Fig. 8).

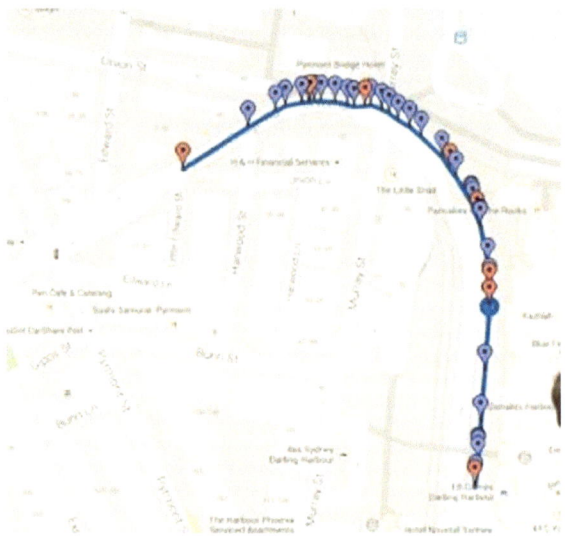

Fig. 8 Illustration of multiple breadcrum in a path

3.4.2   Required Parameters

- Path—This is the path that is snapped. This parameter takes a collection of latitude/longitude pairs. Latitude and longitude is seperated from each other by commas. And the Coordinates are seperated from each other by a pipe character "|".
- Key—This is the API KEY for the Google roads application. This is for the purpose as the application needs to identify itself everytime it sends a request to the GOOGLE ROAD API, this is done by attaching application key with each request.

3.4.3   Optional Parameters

Interpolate—Default value for this is false. In the case the value is true then in that case more interpolated points are returned which results in more smooth path which is according to the geometry of the road, tunnel corners, etc.

3.4.4   Responses

For every request received by the GOOGLE MAPS ROAD API, it will return a response in a format given within the URL. Elements of the response are:

SnappedPoints—A collection of snapped points. Each one of the point have following fields:

- Location—This consists of the latitudes and longitudes values.
- OriginalIndex—It is nothing but an integer that gives the corresponding value in the request. Each value in the request corresponds to the value in the response. But if you have the value of the interpolate as true, then the response can have more value then the request. Interpolated values do not have an original Index. Indexing of these values starts from 0, so if a point have an original Index as 4, then it will have a snapped value of the 5th longitude/latitude that has been passed to the parameter of the path.
- PlaceId—Each of the Ids that is sent back by the Google Maps Road API corresponds to a segment of a road. These Ids can be used with other Google APIs, that includes Google Places API and the Google Maps JavaScript API.

4 Logistic Regression Algorithm (Data Processing)

The certain estimated data is used to train the model using Logistic Regression. The following would be the key ideas behind choosing this Regression Algorithm over others:

- Develop a model that will provide a probability and the odds of being approved for any given Risk Score.
- Decide approximately what score is associated with a probability is 75% for being safe.
- Input the real-time score into the model to determine the probability and the odds of the driver putting himself at risk.
- Return the greatest possible Variable which caused an increase in risk probability.

Logistic Regression seeks to:

- Model the probability of an event occurring depending on the values of the independent variables, which can be categorical or numerical.
- Estimate the probability that an event occurs for a randomly selected observation versus the probability that the event does not occur.
- Predict the effect of a series of variables on a binary response variable.
- Classify observations by estimating the probability that an observation is a particular category (such as approved or not approved in our problem).

Table 2 Description of variables

Variables	Definition	Characteristics
T1	Eye aspect ratio	0 = NO 1 = Yes
T2	Face sentiments (mood)	0 = Angry 1 = Not Angry
T3	Distractions	0 = In control 1 = Distracted
T4	Location	0 = Safe zone 1 = Risk zone

4.1   Model with Data

See Table 2.

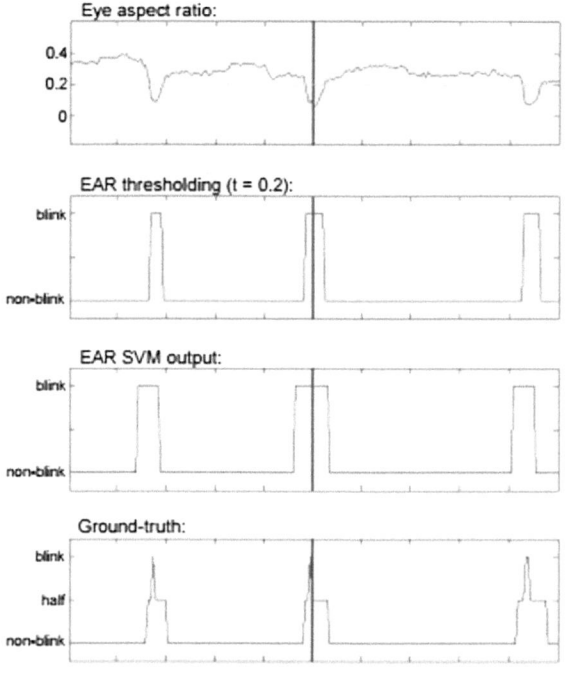

Fig. 9 EAR (blink detection)

4.2   Evaluating the Variables—EAR (Eye Aspect Ratio)

An EAR value above 0.4 would be treated under the non-risk category while the ones below it will be under risk category. The values would indicate and detect if the eyes are wide enough for drivers attention.

The following table is an excellent statistics proven in the paper "Real-Time Eye Blink Detection using Facial Landmarks" (Fig. 9).

During implementation, there would be a need to estimate an average time of blinking and adjust it with the low EAR values which would be generated. A low EAR value for a short amount of time may be due to blinking and this should not be used to indicate the drowsiness of the driver. Therefore these values prove to be false-positives and should be validated before evaluated against the model.

Driver's distraction value can be modeled by modeling the SVR and various BN values to get an accurate level of distraction and classify it whether it is harmful and could result in an accident.

Data from above factors can be mapped with the map route taken by the driver which we get from the Road API to give the final result and take necessary actions.

5   Conclusion

The major cause for a road mishap is driver's intoxications or negligence in general. This project intends to track those characteristics and make an alert to the driver or concerned people who should be made aware of the risk at which the driver is to himself and to others around. The alert may be in any form and completely depends on the implementation specifications.

An additional concept of getting a history of past driving data would be helpful for future experiences. On continued research and improvisation, the model can be updated to even more characteristics of human behavior which may lead to an accident. A statistical analysis could be done through these "Black Boxes". There may be concerns over the privacy of the driver and the data that is collected inside the car. The collected data should only be used for computation of risk. Once a certain amount of calculations is done the visual recordings should no longer be required.

References

1. Auflick, J., Angell, L.S., Kochar, D., Austria, P.A., Biever, W.J., Tijerina, L.: Driver Workload Metrics Task 2 Final Report. National Highway Traffic Safety Administration, Washington, DC (2006)
2. Mohamad, N., Yusuff, H., Yahaya, A.S., Ngah, U.K.: Breast cancer analysis using logistic. IJRRAS 10(1) (2012)
3. Google Road API documentation, Google developers (7 June 2017). https://developers.google.com/maps/documentation/roads/snap
4. Rosebrock, A.: Face Alignment with OpenCV and Python, (retrieved 25 June 2017). http://www.pyimagesearch.com/2017/05/22/face-alignment-with-opencv-and-python/
5. Byun, H., Lee, S.W.: Applications of Support Vector Machines for Pattern Recognition: A Survey. Paper presented at the Pattern Recognition with Support Vector Machines: First International Workshop, SVM 2002, Niagara Falls, Canada (2002)
6. Soukupová, T., Čech, J.: Real-Time Eye Blink Detection Using Facial Landmarks. Center for Machine Perception, Department of Cybernetics Faculty of Electrical Engineering, Czech Technical University in Prague (2016)
7. Zhao, Z., Eick, C., Zeidat, N.: Supervised Clustering—Algorithms and Benefits. Paper presented at the International Conference on Tools with AI (ICTAI) (2004). Asthana, A., Zafeoriou S., Cheng S., Pantic M.: Incremental face alignment in the wild. In: Conference on Computer Vision and Pattern Recognition (2014). (1, 2, 3, 4, 5, 7)
8. Harbluk, J.L., Noy, Y.I., Trbovich, P.L., Eizenman, M.: An on-road assessment of cognitive distraction: impacts on drivers' visual behavior and braking performance. Accid. Anal. Prev. 39(2), 372–379 (2007)
9. Blanco, M., Biever, W.J., Gallagher, J.P., Dingus, T.A.: The impact of secondary task cognitive processing demand on driving performance. Accid. Anal. Prev. 38, 895–906 (2006)
10. Yu, X., Yang, F., Yang, P., Huang, J. Metaxas, D.: Robust Eyelid Tracking for Fatigue Detection. In ICIP (2012)

11. Wu, S., Amari, S.: Improving support vector machine classifiers by modifying kernel functions. Neural Netw. **12**(6), 783–798 (1999)
12. Sigari, M.H., Pourshahabi, M.R., Soryani, M., Fathy, M.: A review on driver face monitoring systems for fatigue and distraction detection. Int. J. Adv. Sci. Technol. **64**, 73–100 (2014). http://dx.doi.org/10.14257/ijast.2014.64.07

Highlight Generation of Cricket Match Using Deep Learning

K. Midhu$^{(\boxtimes)}$ and N. K. Anantha Padmanabhan

Department of Computer Science and Engineering, TKM College of
Engineering, Kollam, India
Midhu997@gmail.com

Abstract. In this paper, we propose various algorithms to generate highlight of a cricket video by detecting important events such as replay, pitch view, boundary view, bowler, batsman, umpire, spectator, player's gathering etc. The proposed method work in two parts. The first part contains five levels. At level one key frame are identified by using hue histogram difference. At level two classify frames to replay frames or real-time frames by detecting the absence of scoreboard. At level three real time frames are classified into field view or non field view based on Dominant Grass Pixel Ration. At level 4a onfield frames are classified into pitch view and boundary view. At level 4b by using edge detection method close-up and crowd frames are detected. Level 5a and 5b again divide each close-up and crowd frames into umpire, bowler, batsman, players gathering and spectator. Concept mining is done on the second part by using the Apriori algorithm and labeled frame events are input. Then combines all the concept detected to form a summarized video. Results at the end of paper show the accuracy of our approach.

Keywords: Histogram · Dominant grass pixel ratio · Concept mining
Apriori algorithm

1 Introduction

The increasing amount of digital video content complicates the users to select the relevant information from the video. Highlight generation process produces the summary of the whole video and it contains all the necessary events in the video. Highlight generation from sports video is an important research area because of its huge viewership and commercial importance [1, 2]. Full length sports video contains so many uninteresting events and it is very boring for people to see the whole events other than exciting events.

Cricket is next to soccer in viewership and fans. Major cricket playing nations are India, Australia, Pakistan, South Africa, Zimbabwe, Bangladesh, Sri Lanka, New Zealand, and England. Cricket is in variable formats such as tests, one day and T20. T20 are approximately 3 h, which is the shortest version of cricket, greater than soccer (approximately 90 min) and hockey (approximately 70 min). Cricket is a long duration

© Springer International Publishing AG 2018
D. J. Hemanth and S. Smys (eds.), *Computational Vision and Bio Inspired Computing*,
Lecture Notes in Computational Vision and Biomechanics 28,
https://doi.org/10.1007/978-3-319-71767-8_79

game, hence its highlight generation process is more difficult than other sports video analysis, such as soccer, basketball, tennis, etc. So many manual highlight generation tools are available, but they cannot take all the interesting events in a cricket video, such as wicket and hit.

The proposed method focuses on highlighting the major three events in the cricket videos like 'wicket', 'six', 'four'. The organization of the paper is as follows: Sect. 2 describes the related work. Section 3 discusses the proposed method along with the details of the techniques used for various event detection from cricket videos and concept mining from those events. Sections 4 and 5 focuses on the experimental results obtained and the conclusion of the work.

2   Literature Survey

Highlight extraction means capturing many interesting events from sports games such as wicket, six and boundary from cricket, goal from football etc. Event based [3, 4] and excitement model based [5, 6] are two techniques to extract highlight from a sports video. Both these methods sometimes contain some events which are not interesting to viewers. Highlight generation by using audio energy is an example for excitement model based system. Here clips containing high audio energy are selected. Sometimes spectator and commentator may cheer even for off-field distractions, and as a results highlight generation based on audio energy may include some false alarm sequences.

Baoxin used replays [7] to detect the highlights. But wide ball, no ball, etc. is very common for cricket and replays are shown also for such unimportant activity. Ekin [8] found the relevant video segments by the detection of "play" and "break". But the main problem is that all the events in a "play" segment cannot be considered as a highlight.

Goyani et al. [9], Bhawarthi et al. [10] and Kolekar [11, 13] detect highlight from the cricket match by considering low level features and then high level concept such as wicket are extracted from that low level features by using the Apriori algorithm. But they all are only concentrated on wicket fall concept mining. Pradeep [12] extract important events from cricket video by the use of audio as a feature. Classify the input sports video into video and audio streams. Peak values from the audio streams are selected and video frame corresponding to that values is selected. The concept is generated by choosing some frames before and after peak frame.

3   Proposed System

In the first level events such as replay, pitch view, boundary view, bowler, batsman, umpire, spectator, player's gathering are identified by analyzing low level features [13]. Concepts such as wicket and hit are extracted in the second level through the proper association of events [13].

For cricket balled out, catch out, run out, stumped out causes wicket fall. Boundary contains four and sixes. Hence this type of important concept contains so many events such as replay, close up of batsman who got out, close-up of bowler who contributed the wicket, crowd gathering, fielders gathering, etc. These types of events are extracted in the first level. In the second level high level concepts such as wicket and hit are extracted by using the Apriori algorithm [14].

Figure 1 describes the complete flow of the proposed work. To reduce the processing time and analysis time first key frames are extracted from the cricket video by considering the hue histogram difference between the adjacent frames. Various events are detected from these key frames by extracting low level features. Following the detection and classification events, we associate these events to form a concept. This is finally fed into the Apriori algorithm [14] to finally achieve the mined concept.

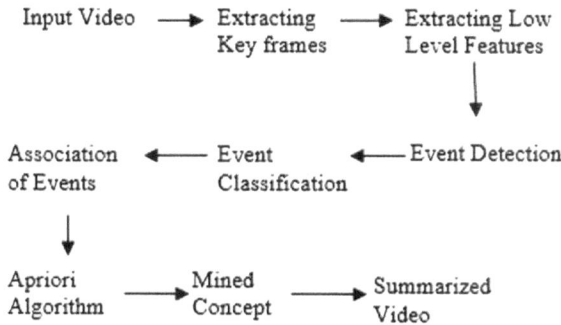

Fig. 1 Block diagram of proposed system

3.1 Event Detection

Figure 2 shows the hierarchical classification method [13]. Number of frames to be processed is reduced in each level. This decreases the processing time and improves computational efficiency.

3.1.1 Level 1: Key Frame Detection

The key frame detection method helps to reduce the analysis time and processing time. It is based on the rule that continuous video frames will have similar features over a length or a little difference between them. Shot boundary detection is the main task in this method. For identifying the shot boundaries hue histogram difference between the adjacent frames are calculated and if it is greater than a threshold we say it is a cut, means that there have been large changes in the content of the video. In the below algorithm N1, N2 are the size of the image. In Fig. 3 a cut is detected between frame 132 and frame 133.

Algorithm

1. Transform the input *RGB* frame into *HSV* image format and plot 256-bins Hue histogram H_n
2. Calculate hue-histogram of the previous frame H_p
3. Calculate Hue-Histogram Difference (*Diff$_H$*) between frame n and the frame n − 1 using the following formula:

$$Diff_H(n) = \frac{1}{N_1 N_2} \sum_{j=1}^{256} H_n(j) - H_p(j) \tag{1}$$

4. If the difference exceeds a threshold produce the current frame as a cut.

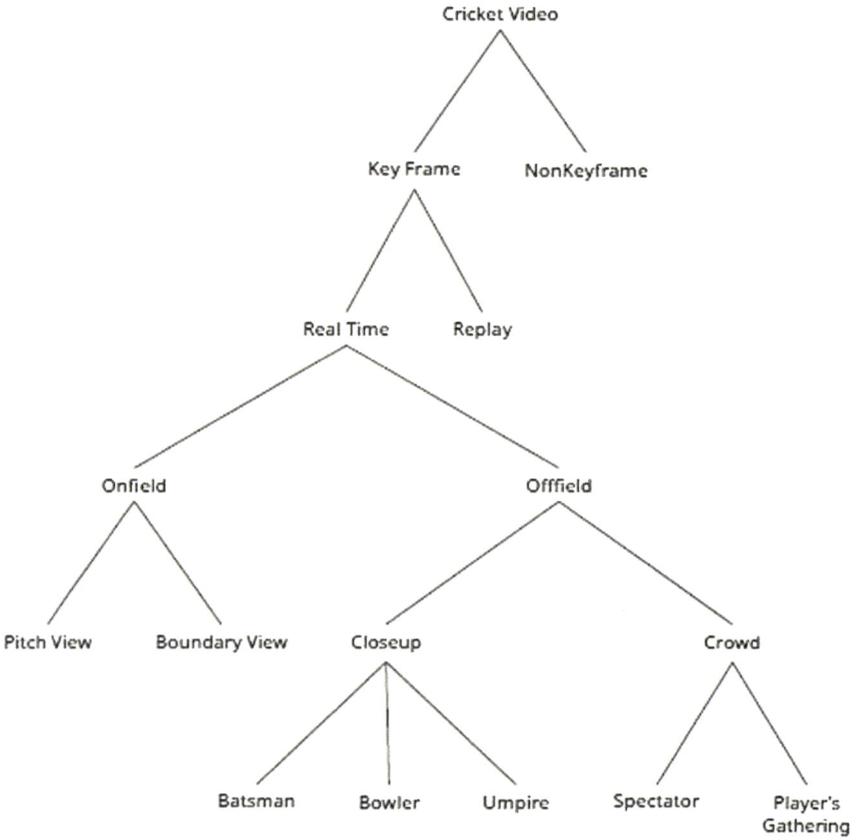

Cricket Video

Key Frame   NonKeyframe

Real Time   Replay

Onfield   Offfield

Pitch View   Boundary View   Closeup   Crowd

Batsman   Bowler   Umpire   Spectator   Player's Gathering

Fig. 2 Hierarchical classification levels

frame131   frame132   frame133   frame134

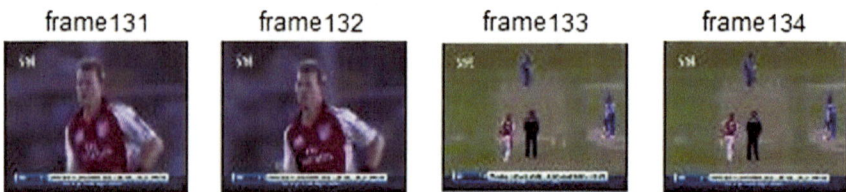

Fig. 3 Cut detection

3.1.2   Level 2: Replay Detection

Cricket videos contain a replay segment and real time segment. Image analysis technique is used to classify the segment into real or replay. A replay segment is always sandwiched between two logo frames. Hence, to find a replay segment we first find out the logo frames by calculating the hue histogram difference of frames with the reference logo template [13]. There is different reference logo for different match. Hence the threshold value used to differentiate each frame to logo frame or a normal frame may differ to match to match and logo to logo. Scoreboard detection method is a good solution to this problem. Scoreboard is not displayed in replay frames. Hence, by checking the absence of the scoreboard in a frame, we conclude that it corresponds to a replay. Hence in this work we identified shots from cricket video and then real time shots are selected for further processing. In the below algorithm N1, N2 are the size of the image.

Algorithm

1. Convert the Reference Scorebar frame into *HSV* image format and plot 256-bins Hue histogram H_n
2. Convert the scoreboard portion of the frame into *HSV* image format and plot 256-bins Hue histogram H_p
3. Compute Hue-Histogram Difference (Diff$_r$) between frame n and the Reference score bar using the following formula:

$$\text{Diff}_r(n) = \frac{1}{N_1 N_2} \sum_{j=1}^{256} H_n(j) - H_p(j) \tag{2}$$

4. If the difference exceeds a threshold Produce the current frame as a replay frame (Fig. 4).

a: Non-Replay Frame   b: Replay Frame

Fig. 4 Non-replay frame and replay frame

3.1.3   Level 3: Onfield View Detection

At level-3, extracting field view and non field view frames from real time frames by considering the grass pixel ratio DGPR [13]. This value helps us to classify frames into onfield view and off field view. To find the green color peak bin number consider 60 field view images in HSV format for training. Then 256-bin histogram of the hue component of these images are plotted. We select the peaks of the hue histogram of these images. By analyzing this 60 images we found that the green color peak occurs between bin $k = 51$ to $k = 61$. The peak of the histogram gives the number of the pixels of the grass in the image. P_g indicate this number. *DGPR* is calculated by the equation P_g/P. Where P indicates the total number of pixels in the frame. For field view images *DGPR* values vary from 0.25 to 0.5. DGPR value is very small for offfield view images.

3.1.4   Level 4a: OnField View Classification

OnField view frames are classified into pitch view, boundary view [15]. We use threshold to classify frames into pitch view and boundary view. But this threshold varies from match to match and it is very tedious task to calculate threshold value for every match. By using the algorithm specified in [15] so many false positive of pitch view may be reported due to the little variations in threshold. Here we are trained naive bayesian classifier on some training images of field view, and then use the trained Naïve Bayesian classifier to classify images as boundary or pitch view.

3.1.5   Level 4b: OffField View Classification

At this level, we are classifying off field view images into close-up and crowd by using the canny edge detection method [13, 14]. Close up and crowd frames are shown after all the exciting events. A canny edge detection algorithm is used to classify image because edginess is more for crowd frames (Fig. 5).

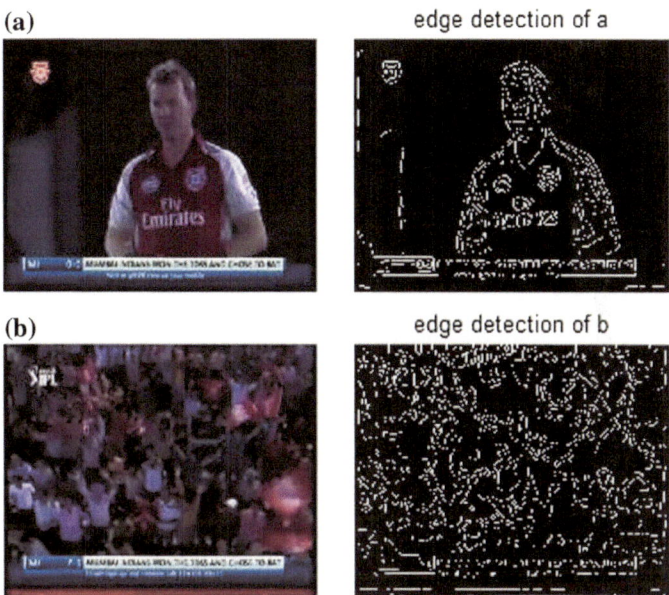

Fig. 5 **a** Close-up image, **b** Crowd image

3.1.6 Level 5a: Closeup Classification

Close up frames contain batsman, bowler and umpire. We can classify close up frames by using skin color and jersey color information [13, 14]. From the cricket videos it can be observed that blocks 6, 7, 10, 11 mostly contain skin blocks. After determining the face block locations, we select corresponding jersey blocks for close up classification. After testing various close up images skin blocks to jersey color table is already determined [14] (Fig. 6).

Fig. 6 **a**, **b** Batsman, **c** Umpire

3.1.7 Crowd Classification

Here we classify crowd frames into spectator and players gathering [13]. Every exciting event contains these two events. The spectator will cheer after all the exciting events. Hence the classifications into these two events are very important (Fig. 7).

Fig. 7 Crowd classification to players gathering and spectator

3.2   Semantic Concept Mining

We found the labels associated with each frame in part1. Now we want to find out the concept associated with each event. Table 1 shows the labels corresponding to each event.

Table 1 Label associated with events

Sl. No	Label	Event
1	V_P	Pitch view
2	V_B	Boundary view
3	F	Fielder
4	B	Batsman
5	U	Umpire
6	S	Spectator
7	G	Player's gathering
8	R	Replay

It is very sure that an exciting clip contains so many events other than normal clip. For example wicket fall is an exciting clip and it contains so many events in a short duration of time.

By proper association of these events, we can extract out the concept like wicket fall and hit. If there is a hit, pitch view, spectators, boundary view, replay etc., events will be detected repeatedly in very short span of time. Using Table 1, we associate a specific label to each classified frame. If a shot does not contain 30 frames the shot is avoided from further processing. After the shots are labeled and then used an Apriori algorithm to get a meaningful concept like wicket fall and hit. The match between Kings XI Punjab versus Mumbai Indians is selected for video V1. The below figure shows the hit concept mining from that video. We have checked 960 frames of V1 which contain fall of wicket. Event detection of frames are shown in Table 2.

Table 2 Event detection

Frame No.	Label
1–65	Pitch view
66–140	Spectator
141–250	Pitch view
251–360	Players gathering
361–440	Spectator
552–960	Replay

3.3 Video Summarization

All the shots are detected and tagged with the name of the event contained in it as mentioned in the above sections. In this module, the method selects all the shots tagged with important events. This includes wicket, six, four and major fouls. Then all these events are stitched together in the same order as they appear in the original video to form the summary of the video.

4 Experimental Result

T20 match videos are used as a dataset. It only contains 20 overs in each innings and includes so many events in each over compared to other matches. First we have divided the whole cricket match into 2 megaslot and each megaslot is again divided into slots. Here in our work each slot contain one over. And then events are extracted from each slot and by using apriori algorithm concepts are mined. Concepts are combined in their temporal order to form highlight of the video. Precision and recall measured by using the below equation. N_c indicates correctly detected frames. N_m indicates missed frames and N_f indicates false positive (Table 3).

$$Recall = \frac{N_c}{N_c + N_m} \qquad Precision = \frac{N_c}{N_c + N_f} \qquad (3)$$

Table 3 Experimental database

Cricket video	Cup	Match (A vs. B)
V1	IPL 2010	Match 41: KXIP versus MI
V2	IPL 2010	Match 21: MI versus CSK
V3	IPL 2010	Match 8: KKR versus CSK

4.1 Level 1: Cut Detection

We have taken segment 1, segment 2 and segment 3 from video V1, V2, V3. Cut detection result is shown in Table 4.

Table 4 Cut detection

Video	Total	Reported	Correct	Precision (%)
Segment 1	58	55	52	89.65
Segment 2	106	103	96	90.56
Segment 3	110	108	100	90.90

4.2 Level 2: Replay Detection

Replay detection testing was done on 10,000 frames. Precision and recall are shown in Table 5.

Table 5 Replay detection

Precision	99%
Recall	100%

4.3 Level 3 Real Time Frames Classification

Classification of onfield and offfield views testing is done on 8000 frames. Precision and recall are shown in Table 6

Table 6 Real time frames classification

	On field (%)	Off field (%)
precision	95.17	96.23
recall	96.05	96.19

4.4 Level 4a: Onfield View Classification

After training Naïve Bayesian Classifier on 4000 images and testing on 4177 images, we observed following precision and recall (Table 7).

Table 7 Field view classification

	Pitch view (%)	Boundary view (%)
Precision	92.21	95
Recall	90.21	90

4.5 Level 4b: Off Field View Classification

Non field view classification testing is done on 4000 frames. Precision and recall are shown in Table 8.

Table 8 Non field view classification

Classification	Precision (%)	Recall (%)
Close up	91.27	93.3
Crowd	93.48	88.12
Umpire	85.36	83.56
Bowler	86.29	87.99
Batsman	84.13	88.59
spectator	93.18	94.67
Player's gathering	82.42	79.17

Figure 8 shows the events such as fielder, pitch view, players gathering and spectator.

These events corresponding to wicket fall.

| fielder | pitch view | players gathering | spectator |

Fig. 8 Events associated with wicket fall

5   Conclusion

In this paper a new method to extract highlight from cricket videos are presented. The system works in 2 steps. In the first step various events are generated from each over of the video. By proper association of the events highlights are generated.

Experimentally all these steps are tested and the result obtained from the steps are verified. In future, umpire gesture detection and audio commentary detection can also be add to increase the accuracy of the system.

References

1. Kokaram, A., Rea, N., Dahyot, R., Tekalp, M., Bouthemy, P., Gros P, Sezan, I.: Browsing sports video: trends in sports-related indexing and retrieval work. IEEE Sig. Process. Mag. **23**(2), 47–58 (2006)
2. Li, Y., Smith, J., Zhang, T., Chang, S.F.: Multimedia database management systems. J. Vis. Commun. Image Represent. **15**(3), 261–264 (2004)
3. Hua, W., Han, M., Gong, Y.: Baseball Scene Classification Using Multimedia Features. In: Proceedings of the IEEE International Conference on Multimedia and Expo, vol. 1, pp. 821–824 (2002)

4. Li, B., Sezan, I.: Semantic Sports Video Analysis: Approaches and New Applications. In: Proceedings 2003 International Conference on Image Processing, 2003. ICIP 2003, vol. 1, pp. I–17. IEEE (2003)

5. Duan, LY., Xu, M, Tian, Q., Xu, CS., Jin, JS.: A unified framework for semantic shot classification in sports video. IEEE Trans. Multimedia **7**(6), 1066–1083 (2005)

6. Hanjalic, A.: Adaptive extraction of highlights from a sport video based on excitement modeling. IEEE Trans Multimedia **7**(6), 1114–1122 (2005)

7. Li, B., Pan, H., Sezan, I.: A General Framework for Sports Video Summarization with its Application to Soccer. In: Proceedings (ICASSP'03) 2003 IEEE International Conference on Acoustics, Speech, and Signal Processing, 2003, vol. 3, pp. III–169. IEEE (2003)

8. Ekin, A., Tekalp, A.M., Mehrotra, R.: Automatic soccer video analysis and summarization. IEEE Trans. Image Process. **12**(7), 796–807 (2003)

9. Goyani, M., Dutta, S., Gohil, G., Naik, S.: Wicket fall concept mining from cricket video using a-priori algorithm. Proc. Int. J. Multimedia Appl. (IJMA) **3**(1) (2011)

10. Bhawarthi, D., Gadage, S.: Enriching feature extraction using a-priory algorithm for cricket video. Int. J. Eng. Res. Appl. (IJERA), ISSN, 2248–9622

11. Kolekar, M.H., Palaniappan, K., Sengupta, S., Seetharaman, G.: Semantic concept mining based on hierarchical event detection for soccer video indexing. J. Multimedia **4**(5) (2009)

12. Pradeep, K.: Significant event detection in sports video using audio cues. Int. J. Innovations Eng. Technol. (IJIET) **3**(1) (2013)

13. Kolekar, M.H., Sengupta, S.: Semantic concept mining in cricket videos for automated highlight generation. Multimedia Tools Appl. **47**(3), 545–579 (2010)

14. Zhu, X., Wu, X., Elmagarmid, A.K., Feng, Z., Wu, L.: Video data mining: semantic indexing and event detection from the association perspective. IEEE Trans. Knowl. Data Eng. **17**(5), 665–677 (2005)

15. Kumar, A., Masulkar, S., Dr. Mukerjee, A.: Automatic Highlight Extraction in Cricket (2014)

Detection of Pulmonary Nodules Using Thresholding and Fractal Analysis

Deepthi Ramesh[(✉)], Deepa Jose[(✉)], R. Keerthana,
and V. Krishnaveni

Department of ECE, KCG College of Technology, Karapakkam, Chennai
600097, Tamil Nadu, India
{deepthi95,deepajosell}@gmail.com

Abstract. Automated detection of pulmonary nodules helps radiologists in early detection of lung cancer from computed tomography (CT) scans. It is very costly computationally because of its complexity of the process. The CT scan has more advantages than other computational algorithms. The preprocessed CT scan is thresholded using Otsu's method and the lung region is segmented using K-means Clustering which is based on geometric features. Texture based feature analysis algorithm is used to identify the major descriptors. The Artificial Neural Networks (ANN) is used for training, testing and validation process takes place to identify nodule and classify in stages i.e. Stage 1 (initial), Stage 2 (middle) and Stage 3 (critical). The results obtained in this method has been checked for accuracy.

Keywords: Median filter · CLAHE · Otsu thresholding · K-means clustering
SFTA · ANN

1 Introduction

Detection of pulmonary nodules is important for diagnosis of lung cancer, but the detection is a painstaking task because its physical appearance varies in a wide range and they have low contrast against neighbouring cells and other lung tissues [1]. One of the important indicators for lung cancer are the presence of pulmonary nodules (focal densities with diameter from 3 mm to 3 cm). Chest computed tomography (CT) scan, especially high resolution CT, has been widely adopted for detection of lung tumors. CT scan is better compared to other imaging modalities like chest radiograph and MRI because it has the unique ability to display tiny nodules. Inspite of this, its difficult to detect the nodule in CT scan using naked eye and to separate nodules from vessels accurately [2]. Small lung nodules noted on CT images make the accurate diagnosis difficult and give wrong outputs in decision-making. Small lung nodules are malignant and not reliably identified by the scan [3].

Pulmonary nodules can be diagnosed as cancer based on the properties of shape. A pulmonary nodule can occur anywhere in the lung region, thus making diagnosis by

© Springer International Publishing AG 2018
D. J. Hemanth and S. Smys (eds.), *Computational Vision and Bio Inspired Computing*,
Lecture Notes in Computational Vision and Biomechanics 28,
https://doi.org/10.1007/978-3-319-71767-8_80

radiologists a complicated task [1]. Symptoms of lung cancer are not identified until the disease is already at an advanced stage [4]. Even if lung cancer causes symptoms, many people may misdiagnose them for other diseases, such as an infection or long-term effects from smoking. This delays the diagnosis. For solving this issue, a Computer Aided Diagnosis (CAD) is implemented for early detection of lung cancer.

2 Literature Survey

This paper proposes an effective method of combining image recognition and thresholding concepts [1]. Rules are learned from prior knowledge of true nodules verified by the radiologists, while the classification parameters for SVM is given as input from features extracted from the labeled dataset. Hierarchical vector quantization (VQ) approach is supposed to overcome the preprocessing and nodules detection issues in an intuitive manner [1]. Juxta-pleural nodules (i. nodules near the lung wall) is included by applying the initial lung mask using morphological closing operation. Then, low-level VQ is used to find and segment the nodules within the extracted region [1].

Results demonstrated that CADe using the paper obtained a high overall sensitivity and specificity [5]. The median portion of candidate nodules is used to calculate the features. The candidate nodules in 3D contains the voxels, are used to calculate the features in 3D. The candidate nodules are detected from the masked lung regions. The candidate nodule's features are extracted and classification is performed. The threshold value is calculated and lung lobes are segmented by methods such as thresholding, lung lobe extraction, hole filling and contour correction [6]. The ROIs are segmented from extracted lung volume. The threshold value of each overlapped circular region is calculated iteratively by using one of the multi-level thresholding methods known as K-means clustering [6]. The lung lobes are separated from the labelled lung CT scan. The actual nodules present in lung lobes are labelled. The candidate nodules are selected from the ROIs. Artificial Neural Network classifier is trained and tested on the dataset [6].

The Automatic detection and distinguishing techniques could be done by analyzing all the images and defining all the nodules with the accuracy required in a less computational time. Distinguishing the cancerous nodules (is a malignant tumor of the lung) from the blood vessels is the main challenge that can be carried out through the proposed approach. The first stage is to segment the lung to define the exact ROI and the final result of the segmentation process is to segregate the lung and vascular tree from the nodule. In order to correct the image and regain all the human tissue including the organs and bones filling of the holes is performed using image fill function. The resulting image is used as a reference pattern over the original image to crop all the unnecessary sections and the final image produced consist of only the human body without any other interfering object such as the foreground and background value. By screening the images produced from the two stages of segmentation, a final result is obtained in the form of isolation and separation of the nodule from all the remaining tissues by the original CT image [7].

3 Methodology

The following steps are done for detection of pulmonary nodules is described in Fig. 1.

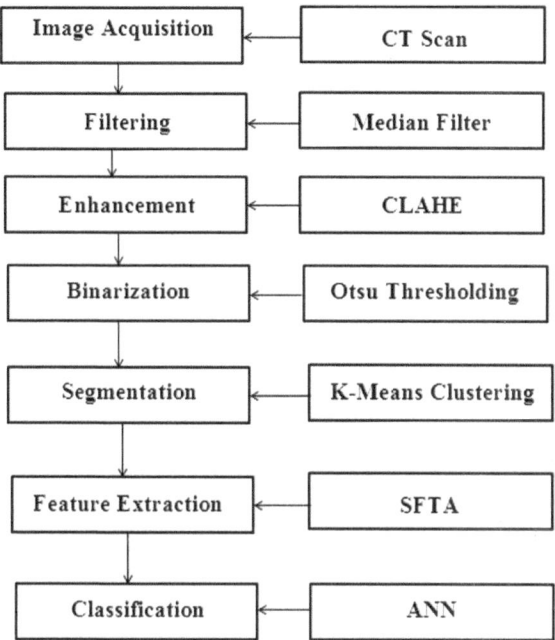

Fig. 1 Process for detection of pulmonary nodules

3.1 Image Acquisition

Multiple imaging modalities like chest radiography, MRI and CT scan can be used to obtain lung region. CT scan is used here due its better clarity and low contrast provided. The CT images are obtained from Parvathy hospital in Chrompet.

3.2 Filtering

Noise removal therefore is necessary to enhance and recover the analysis details that may be hidden in the data [8]. Filtering is applied to clear such images. CT scans are commonly affected by impulse noise or salt and pepper noise which is removed by using median filter [9]. The median filter does a better job of removing noise while preserving edges of the image because the value output of each pixel is determined by the median of the ranking process rather than mean.

3.3   Enhancement

The objective of enhancement techniques is to process an image so that the result is more desirable than the original image for detection of nodules. The main purpose of image enhancement is to identify the detail that is hidden in an image or to increase contrast in a low contrast image. In this method, CLAHE (Contrast Limited Adaptive Histogram Equalization) with command adapthisteq is used. The contrast, especially in homogeneous areas, can be limited to avoid amplifying any noise that might be present in the image [9].

3.4   Binarization

Binarization is used for object recognition and for detection of boundary of image. Otsu's method, named after Nobuyuki Otsu, is used for the reduction of a gray level image to obtaining a binary image and classifying into foreground pixels and background pixels as expressed in Table 2 [10].

3.5   Segmentation

Segmentation is essential and time consuming stage in image analysis which is done in Table 2. K-Means Clustering divides each pixel and groups in terms of their geometry. It will add the pixels into a cluster when the pixel properly satisfies the specified a condition or avoid the pixel [11].

3.6   Feature Extraction

SFTA (Segmentation-based Fractal Texture Analysis) is used here to extract features from the image. The extraction algorithm consists in transforming the input image into a set of binary images from which the fractal dimensions of the ROIs are computed in order to get the segmented texture pattern. The fractal analysis technique is used to characterize the textural features in multiple spatial domain to determine if solitary nodules are malignant by calculating the weight of the ROIs [12].

3.7   Classification

An Artificial Neural Network (ANN) is a mathematical model that tries to simulate the structure and functionalities of brain as shown in Fig. 2. While the features introduced in the classifier have been commonly used on lung nodule detection, advanced features such as texture features for nodule malignancy diagnosis is implemented. Artificial Neural Network classifier is trained and tested on the dataset [13]. The neural network classification in the proposed system classifies the candidate nodules based on geometric and texture features.

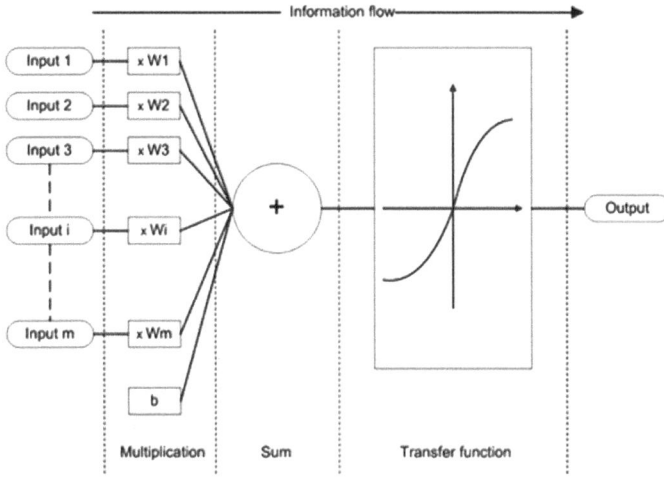

Fig. 2 Working principle of an artificial neuron

4 Experimental Results

The methodology described in the previous section has been executed. The results have been obtained successfully.

4.1 Image Acquisition

The input images are in jpg format which conform to the DIOCOM standards with 512×512 dimensions as given in Table 1.

4.2 Filtering

The values of neighbouring pixels are identified and sorted. Impulse noise or salt and pepper noise is greatly reduced which is shown in Table 1.

Table 1 Preprocessing

Steps	Image 1	Image 2	Image 3
Image acquisition			
Filtering			
Enhancement			

4.3   Enhancement

In adapthisteq command, input image and value given to Neighbouring pixels is given to limit contrast so that the boundary between nodules and other tissues are preserved not distorted as shown in Table 1.

4.4   Binarization

Otsu's thresholding method is used to differentiate between background (black region) and for ground pixels (white region) by using for loop to go through all the possible threshold values and calculating a measure of threshold value, i.e. the pixels that either fall in foreground or background as shown in Table 2. A threshold value is specified where the sum of foreground and background spreads is at its minimum [14].

4.5 Segmentation

K-means clustering concept is used to used to get the lung region border in LAB color format. The lung regions found by repetitive iteration belong to one cluster and the rest of the regions belong to another cluster. Masking of the lung region is done by using region properties command to specify parameters like Orientation, FilledArea, BoundingBox, MajorAxisLength, PixelList, MinorAxisLength, Centroid which represent the geometric features of the nodule. The segmented region and blobs region are combined by assigning black (or zeroes) to the regions that do fit into the constraints imposed. A nodule is usually a localized object that produces a more or less pronounced blob-like pattern, which is (slightly) lighter than the surrounding structures as shown in Table 2.

4.6 ANN

Here the classification process is applied using supervised learning. The type of training images are determined. Training data should consist of images for which the stage of cancer is known and verified by the experts. The training dataset gives instruction to ANN in a form it understands [14]. Learning and testing of performance of the system is done with the test dataset which is the data that has not been introduced to network while learning. The training, testing and validation is a complex and time consuming task [2]. The training of the datasets is done by using Neural Network Training toolbox. The extracted features are compared with other images and classified as Stage 1 (initial), Stage 2 (middle) and Stage 3 (critical). This process is done in Table 3.

Table 2 Segmentation process

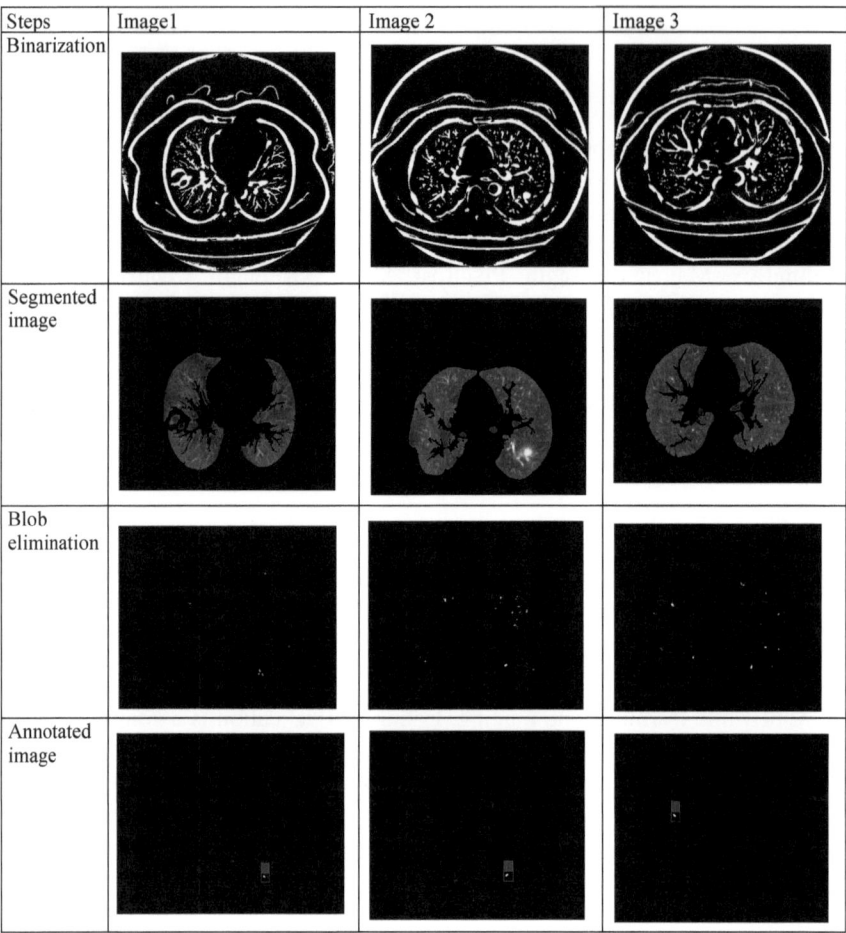

Table 3 Outputs from SFTA and ANN stage

Image	Outputs from SFTA and ANN
Image 1	Command Window ① New to MATLAB? Watch this Video, see Examples, or read Getting Started. Descriptors = 0.2236 255.0000 44.0000 NaN NaN 0 Stage 1 fx
Image 2	Command Window ① New to MATLAB? Watch this Video, see Examples, or read Getting Started. Descriptors = 0.1518 255.0000 53.0000 NaN NaN 0 Stage 2 fx
Image 3	Command Window ① New to MATLAB? Watch this Video, see Examples, or read Getting Started. Descriptors = 0.0626 255.0000 35.0000 NaN NaN 0 fx Stage 3

5 Conclusion

Artificial neural networks has a good future scope in everyday services, products and applications. So, further improvements need to be done to the existing system to optimize the process and achieve higher accuracy. For high-performance and low power environments, double edge triggered flip-flops will be used because of less number of clocked transistors and lowest power [15]. In future, this project will be implemented in Virtex 5 (FPGA) to increase optimization and processing speed of the detection of pulmonary nodules using parallel computing methods [16].

References

1. Poomimadevi, C.S., Helen Sulochana, C.: Automatic Detection of Pulmonary using image processing techniques. In: IEEE WiSPNET 2016 Conference
2. Chen, S., Suzuki, K.: Separation of bones from chest radiographs by means of anatomically specific multiple massive-training ANNs combined with total variation minimization smoothing. IEEE Trans Med Imaging **33**(2) (2014)
3. Han, H., Li, L., Han, F., Song, B., Moore, W., Liang, Z.: Fast and adaptive detection of pulmonary nodules in thoracic ct images using a hierarchical vector quantization scheme. IEEE J. Biomed. Health Inf. **19**(2) (2015)
4. Elsayed, O., Mahar, K., Kholief, M., Khater, H.A.: Automatic Detection of the Pulmonary Nodules from CT Image. In: IEEE SAI Intelligent Systems Conference (2015)
5. Jose, D., Kumar, P.N., Hussain, A.: Fault Tolerant Adaptive neuro-fuzzy based automated cruise controller on FPGA. In: IEEE INDICON Conference, pp. 1–5, IIT Bombay, India (2013)
6. Akram, S., Javed, M.Y., Hussain, A.: Automated thresholding of lung CT scan for artificial neural network based classification of nodules. In: 2015 IEEE ICIS 2015, June 28–July 1 (2015)
7. Wang, J., Cheng, Y.: A New Pulmonary Nodules Computer-aided Detection System in Chest CT Images Based on Adaptive Fuzzy C-Means Technology. In: 2015 7th International Conference on Intelligent Human-Machine Systems and Cybernetics (2015)
8. Li, S., Fang, L., Yin, H.: An efficient dictionary learning algorithm and its application to 3-D medical image denoising. In: IEEE Trans Biomed Eng. **59**(2) (2012)
9. Chhabra, T., Dua, G., Malhotra, T.: Comparative analysis of methods to denoise CT scan images. Int. J. Adv. Res. Electr. Electron. Instrum. Eng. **2**(7), 3363–3369
10. Ilackiya, R., Dr. Perumal, K.: Factal texture based image classification. IJCSMC **4**(9), 192–198 (2015)
11. Oladimeji, M.O., Turkey, M., Ghavami, M. (SMIEE), Dudley, S. (MIEEE): A New Approach for Event Detection using k-means Clustering and Neural Networks. In: 2015 International Joint Conference on Neural Networks (IJCNN)
12. Akila Devi, M.P., Latha, T., Helen Sulochana, C.: Iterative thresholding based image segmentation using 2D improved Otsu algorithm. In: IEEE Proceedings of 2015 Global Conference on Communication Technologies (GCCT 2015)
13. Sun, S., Guo, Y., Guan, Y., Ren, H., Fan, L., Kang, Y.: Juxta-vascular nodule segmentation based on flow entropy and geodesic distance. IEEE J. Biomed. Health Inf. **18**(4) (2014)
14. Wang, J., Cheng, Y.: A new pulmonary nodules computer-aided detection system in chest CT images based on adaptive fuzzy C-means technology. In: 2015 7th International Conference on Intelligent Human-Machine Systems and Cybernetics (IHMSC) (2015)
15. Renganayaki, G., Jeyakumar, V.: Design of an efficient low power shift register using double edge triggered flip flop. Int. J. Innov. Res. Comput. Commun. Eng. **2**(1), 2707–2716 (2014)
16. Jose, D, Kumar, P.N.: Parallel pseudo-exhaustive and low power delay testing of VLSI systems. In: Proceedings of the Third International Conference on Trends in Information, Telecommunication and Computing, vol. 150, pp. 399–405 (2013)

PSO Based Blind Deconvolution Technique of Image Restoration Using Cepstrum Domain of Motion Blur

G. Ramteke Mamta[1] and Maitreyee Dutta[2(✉)]

[1] Chandigarh Engineering College, Chandigarh, India
mamta.cse@cgc.edu.in
[2] Department of Computer Science and Engineering, NITTTR, Chandigarh,
India
d_maitreyee@yahoo.co.in

Abstract. In this paper, blind deconvolution technique is planned based on the Particle Swarm Optimization (PSO) and cepstrum method. Angle and distance is obtained from motion blurred images using cepstrum method. The parameters of cepstrum are optimized through PSO technique. Here, we are optimizing values of theta and length from cepstrum of blurred image, which will help in PSF calculations. To extend the result of our previous work (Almeida and Almeidain in IEEE Trans Image Process 19(1):36–52, 2010 [1]) we have used PSO technique which is giving better result than GA. Also the convergence rate is faster and computational time of an algorithm is also reduced. Hence, the proposed method outperforms the previous method.

Keywords: Image restoration · PSO · Blind · Deconvolution

1 Introduction

Image Restoration is the important part of various image processing applications. Distorted form of image may be obtained in many forms due to error in capturing process such as motion blur, noise and camera mis-focus. Point spread function has a big role in case of image restoration which helps to bring back the image information missing due to the blurring process. The various processing techniques of an image are operationalized either in the spatial domain or the frequency domain. Frequency domain deconvolution is most common technique for image restoration. After computing the Fourier transform of both the image and the PSF, it restores the information loss caused by the blurring factors. As PSF is not known, this deconvolution technique is called blind. Blind deconvolution method is clearly harder than its nonblind correspondent; due to its ill-posed nature. But sometimes restoration becomes immense need to interpret the data or in case of decision making in various applications like remote sensing, microscopy and medical imaging where data has to be interpreted through image. Gradually all applications mostly follow image data than the textual data.

© Springer International Publishing AG 2018
D. J. Hemanth and S. Smys (eds.), *Computational Vision and Bio Inspired Computing*,
Lecture Notes in Computational Vision and Biomechanics 28,
https://doi.org/10.1007/978-3-319-71767-8_81

Recapturing of an image is not the solution of this problem because it will be costly and time consuming process. Blurring in images occurs due to several reasons like camera mis-focus, relative motion between object and camera, atmospheric turbulence and other similar conditions. There are two ways of blind deconvolution methods. Blind deconvolution is impossible for an unique solution to the problem. An image can be represented as a Fourier transformation in terms of spatial frequency. The information about fine details, sharp edges can be represented using high spatial frequency components whereas large area represent low spatial frequency. Since the cepstral features are based on Fourier transform, they are translational invariant and also moreover, produces rotation invariant features resulting from the fft conversion. Once the image is represented in cepstral domain, low frequencies are compared to high-frequencies. Faster calculation can be done in the Fourier space, deconvolution used the Fourier transformation. This filtering is used in the log-spectral domain to separate out the filter effects and it is comprehensive technique for signal and image processing. Computation of the angle or linear distance can be calculated using cepstrum domain. First take fft of the blurred image afterwards, take logarithm of the frequency spectrum and squared it. From output, it has been found that the most powerful frequencies are highlighted. Here, we have suggested an improved scheme for solving the blind image restoration problem. The scheme utilizes cepstrum method within its parameters optimized by particle swarm optimization. The rest of the paper is organised as follows. Section 2 deals with the cepstrum domain. Section 3 gives literature survey. The role of the Particle Swarm Optimization is discussed in Sect. 4. Section 5 presents the results analysis with discussion.

2 Blur Patterns in Cepstrum Domain

The cepstrum techniques are mostly used in digital signal processing used to detect speech signals. But sometimes it is useful in some applications of image processing. Here, it is used to estimate the PSF of the blur image. The cepstrum of an image can be described as a two-dimensional signal which is used to find out the translational variation of two sub-areas. To express the image degradation as a mathematical expression a blurred image g(x, y) can be expressed as follows:

$$g(x, y) = a(x, y) \otimes p(x, y) + n(x, y) \tag{1}$$

where a(x, y) is the original image, p(x, y) is the PSF, n(x, y) is additive noise, and $\otimes$ represents the convolution. We assume the degradation of an image to be an energy-preserving transformation, and indicate it as follows

$$\int_{-\infty}^{\infty} d(x, y) dx dy = 1 \tag{2}$$

The degradation in image processing can also be represented in the form of cepstrum mathematically as follows.

$$c(m, n) = F^{-1}\{\log|g(u, v)|\} \tag{3}$$

3 Literature Survey

Here cepstrum based Blind deconvolution methods [1] are described in detail highlighting different techniques of PSF estimation [2–4] which are relevant with the proposed work like cepstrum based [5] and PSO method. In this technique of image restoration method [6], support vector regression (SVR) is suggested for solving the blind image restoration problem. The method utilizes particle swarm optimization (PSO) with its parameters optimized. This particle swarm optimization (PSO) helps for faster convergence and easier implementation. In this technique of blind deconvolution of motion blur [7], author proposed to estimate blur kernel caused by motion blur. To analyse in details, assumptions had been derived regarding the bluring effects in cepstrum domain. It had been implemented in two phases, initially various estimation of PSF methods from the cepstrum of a blurred image had been done and restored the image with a fast deconvolution algorithm. Afterwards, selection of the method of PSF was done. In the paper [8], there were many comparisions of the different cepstrum-based methods based on their performance. The first and second cepstrum method was the generalized and was based on spectral root respectively. The remaining five methods were based on the complex cepstrum. These all methods have different computational techniques used in the spatial and frequency domain. A novel method [9] was developed to estimate the point spread function (PSF) from a single blurred image caused by motion blur. The approximation about PSF can be done using cepstral analysis in the cepstrum domain. Edge detection method [10], has been used to build the objective function which was utilized for searching the parameters of PSF. Here, PSO was used to investigate the unknown PSF. Edge information and image morphology has been used to devise the objective function. It also used Wiener filter to restore the estimated image. A technique is described for the optimization of multidimensional grayscale soft morphological filters for applications in automatic film archive restoration [11], specific to the problem of film dirt removal. Fergus [12] introduced a BID method that uses natural image statistics to estimate the blur kernel. Author [13] proposed a blind image deconvolution scheme based on soft integration of parametric blur structure. It addressed issue of parametric blur information. The PDR method helps to approximate the parameter blur quite accurately. In PDR method a prior parameter knowledge is available, then it becomes most flexible method. This method aids in estimation of the fuzzy blur structure. The PDR technique consists of optimization of the cost function which consists of all the parameters. Bayesian technique [14] used hierarchical method which simultaneously estimate image and blur. This method is help to remove the blur, which encountered due to long exposure and also low light conditions. Total variation priors are also used on the image to predict the statistical data observed from natural images. Hence it preserved the edges in the image by not overpenalizing discontinuities while applying smoothess. Blur estimation using radon transform [15] provides better results for the most of the natural images which were based on the spectrum of the blurred images and the output for natural images was better. The power-spectrum is approximately isotropic and has a power-law decay with the spatial frequency. A new method for image recovery proposed in [13], where the blur kernel was estimated on the basis of statistical irregularities [16] of the power spectrum. This method incorporates a new model and a spectral whitening formula. It has been used to estimate the power spectrum of the blur and further recovered using a phase retrieval algorithm. Blind deconvolution [17] introduces Bayesian phase unwrapping method using a noncausal Markov random chain model. For the noise removal and smoothness used prior regularizing term. A least mean square optimization is used for phase unwrapping noniteratively using cosine transform.

4 Particle Swarm Optimization

Population based technique is inspired by communal actions of bird flocking which is called as Particle swarm optimization (PSO). PSO has lots of similarities with Genetic Algorithms which is evolutionary computation technique. Initially the random solutions of population is set and searches for optimum solution by updating generations. But PSO has no evolution operators such as crossover and mutation like GA. The movement of each particle is influenced by its local best known position, but is also directed towards the best known positions in the search-space, that is updated by other particles. This is helps to move the swarm toward the best solutions. This method is easy to implement because only few parameters need to be modified in comparison with GA. This technique has been applied in many areas like cloud computing, wireless network, classification of custering, artificial neural network etc. Behaviors of bird flocking is stimulated by PSO. A supposition is made to describe the scenario of PSO, there is randomly search of food in an area by a group of birds. In the area being searched there is only one piece of food. Some birds are unknown of the location of the food. All they are aware of is that how much distant the food is in each iteration. Hence what can be the optimum strategy for locating the food? The most successful strategy is to chase the bird which is closest to the food. The learnt PSO from the scenario is used to resolve the problems of optimization. Here, each single solution corresponds to a "bird" in the search space, called as "particle". The evaluation of the fitness values of all the particles is done by the optimization of the fitness function. The direction of the flying of the particles is decided by the velocities possessed by these fitness function. Initialization of PSO is done with a set of random particles i.e. solution. Later on by updating generations PSO finds the optimum solution. Each particle is modified by following two "best" values in every iteration. The best solution that is achieved so far is first one i.e. the fitness (with stored fitness value known as **pbest**). The most appropriate value achieved so far by any particle in the population is the global best known as the **gbest**. There is another best value, the local best known as lbest which is obtained when the particle participates in the population as its topological neighbor. The following Eqs. (4) and (5) are used to update the position and velocity after obtaining gbest and pbest:

$$v = v + k1 * \text{rand}() * (\text{pbest} - \text{cps}) + k2 * \text{rand}() * (\text{gbest} - \text{cps}) \qquad (4)$$

$$\text{cps} = \text{cps} + v \qquad (5)$$

the particle velocity is **v**, the current particle solution is **cps**, pbest and gbest are present best and global best. Random number is a rand() between (0, 1) k1, k2 are learning factors (usually k1 = k2 = 2).

Each particle represents a potential solution of the optimization in PSO that maintains a population of particles. A randomized velocity is allocated every particle. The PSO intends to locate the particle position that results in the best evaluation of a given fitness function. The following information of each particle in the problem space is: the current position of the particle (xi); the current velocity of the particle (vi); and the personal best position of the particle (yi). Here, yi is the best position that it has achieved so far.

4.1   Proposed PSO-Based Cepstrum Method for Blind Image

In the present work, we improved the cepstrum approach towards blind image deconvolution by optimizing the parameters of cepstrum. PSO based approach has been proposed to optimize the cepstrum parameters. Authors in [18] have also suggested a GA-based approach for optimizing the cepstrum parameters. We have utilized the particle swarm optimization (PSO) for faster convergence and easier implementation.PSO also has fewer adjusting parameters than GA. Here pictorial representation of proposed framework is given which describes sequence of steps to restored image. It signifies the role of PSO in our proposed algorithm which helps to improve the quality of an image. Figure 1 shows the Framework of PSO based cepstrum method of image restoration.

5   Result Analysis

We first apply proposed algorithm on the standard images like Cameraman, Rice and Lena image. Then applied some mathematical computation to calculate theta and length. These angle and distance are input to the PSO to find out the more optimized result of angle and linear distance. The computed value of angle and distance of corrupted image provide the PSF which helps to restored the image of having better PSNR than GA. We have taken readings of cost function of first ten readings of both the methods and compare it to analyse in Table 5. From Fig. 5, it is observed that the PSO converge more faster than GA. Figures 2, 3 and 4 show visual display of blurred image, restored image using GA and restored image of using PSO of Cameraman, Rice and Lena image respectively.

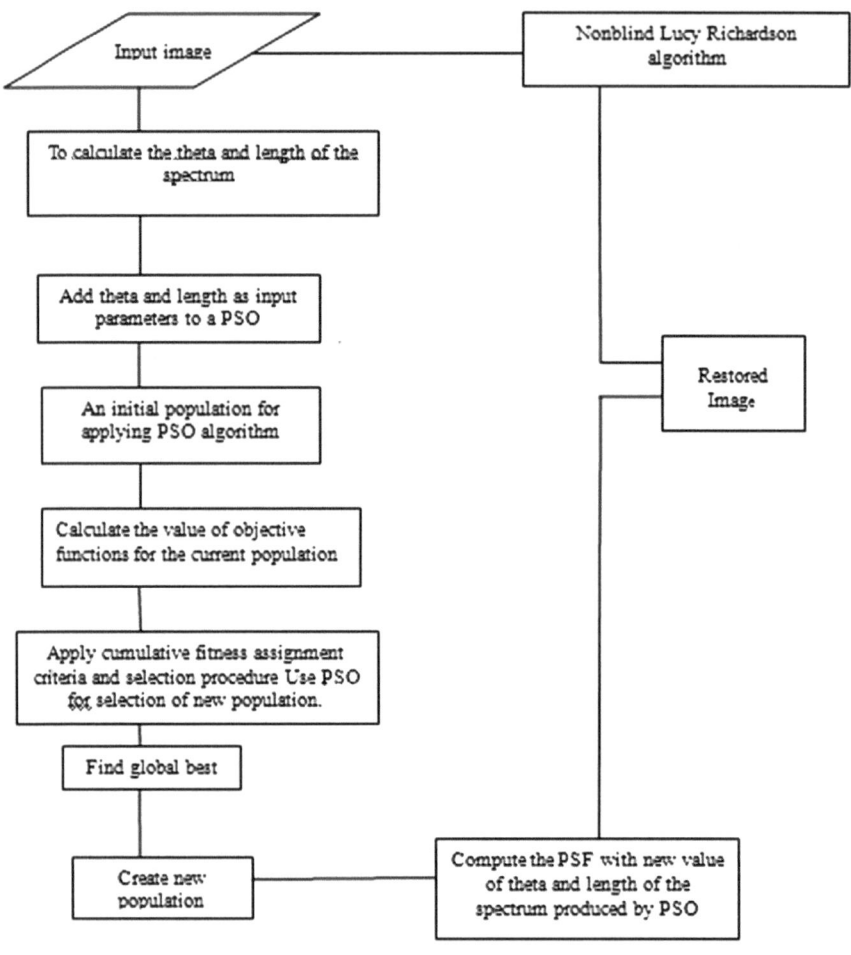

Fig. 1 Framework of PSO based cepstrum method of image restoration

Tables 2, 3 and 4 hows visual display of the motion blurred, cepsrum based restored image and PSO based restored image of cameraman, rice and Lena image respectively and Tables 1, 2 and 3 gives angle and length calculation done using basic cepstrum method, GA bases cepstrum method, PSO based cepstrum method for cameraman, rice and lenna image respectively. Table 4 gives comparative analysis of GA and PSO based cepstrum method in terms of PSNR for cameraman, rice and lena image. From table itself, it proves that is our PSO based method outperforms the GA based cepstrum method.

(a)   (b)   (c)

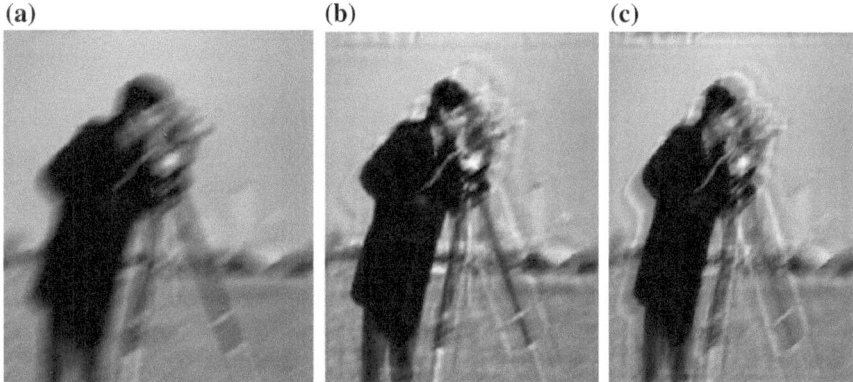

Fig. 2 **a** Degraded motion blurred camera image. **b** Restored using method GA based cepstrum method. **c** Restored using PSO based cepstrum method

(a)   (b)   (c)

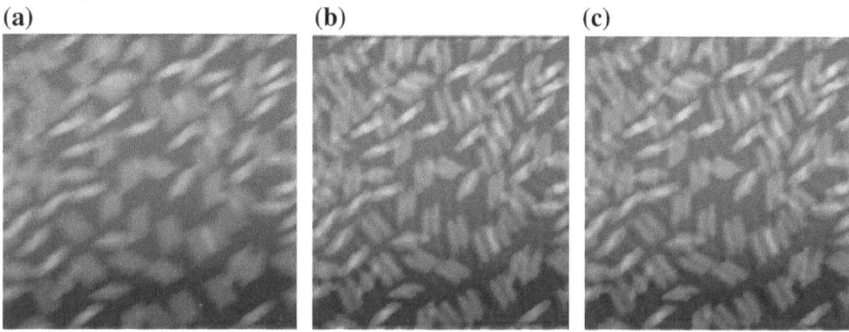

Fig. 3 **a** Degraded Motion blurred rice image. **b** Restored image using GA based cepstrum method. **c** Restored image using PSO based cepstrum method

(a)   (b)   (c)

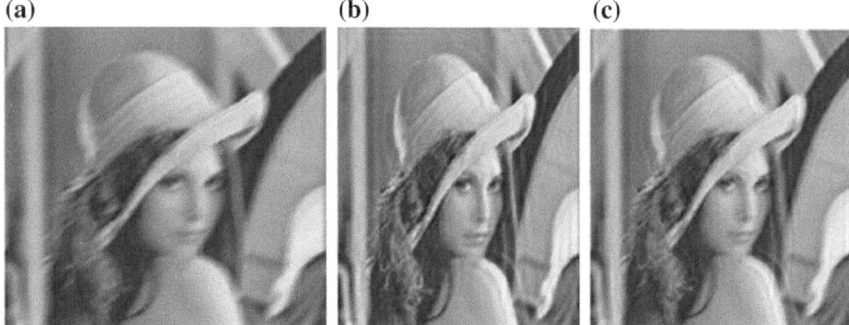

Fig. 4 **a** Motion blurred Lena image. **b** Restored using GA based cepstrum method. **c** Restored using PSO based cepstrum method

Table 1 Comparison of the Theta and Length calculation for cameraman

Parameters of blur	Basic cepstrum method	GA based proposed method	PSO based cepstrum method
Angle	30	32.085	29.02
Distance	20	20	20

Table 2 Assessment of the Angle and Distance calculation for rice image

Parameters of blur	Basic cepstrum method	GA based proposed method	PSO based cepstrum method
Angle	33	34.016	31.998
Distance	11	20	20

Table 3 Assessment of the Angle and Distance calculation for lena image

Parameters of blur	Initial cepstrum method	GA based proposed method	PSO based method
Angle	30.5	32.201	29.583
Distance	20.2	21	19

Table 4 Compartive analysis of GA and PSO based cepstrum method in terms of PSNR

Images	Basic cepstrum based method	GA based cepsrum method	Proposed PSO based cepstrum method
Cameraman	25.093	25.5053	29.0832
Rice	24.4814	24.5756	30.003
Lena	26.8577	29.4015	31.6046

Table 5 gives comparison of Cost function of GA based cepstrum method and PSO based cepstrum method for first ten generations. After plotting iterations versus cost functions of GA and PSO which displayed in Fig. 5. It proves from Fig. 5, PSO converge faster than GA. From Table 6, it is clear that the PSNR is less in lower and higher theta and length. But giving better output if theta is in the middle of the quadrant.

In the next experiment, we apply proposed techniques for different test images Mendrill, bridge, crane, living room, bee by varying number of iterations. The deblurring results on test images with size 256 × 256 pixels are shown in Table 7. Our algorithm was work with various iterations in case of Mendrill image as shown in Fig. 6. We apply it for bee, crane, living room and bridge images. In every case, PSNR is increased by increasing number of iterations as shown in Table 7. Clearly, the PSO based cepstruem technique is much more efficient than GA based cepstrum technique.

Table 5 Comparison of cost function of GA based cepstrum method and PSO based cepstrum method

No. of iteration	Cost function of GA	Cost function of PSO
0	0.39487	0.39894
1	0.39487	0.39894
2	0.39804	0.39894
3	0.39804	0.40073
4	0.39819	0.40338
5	0.39819	0.41024
6	0.39819	0.4179
7	0.39903	0.42066
8	0.40117	0.44009
9	0.40461	0.44331
10	0.40845	0.44844

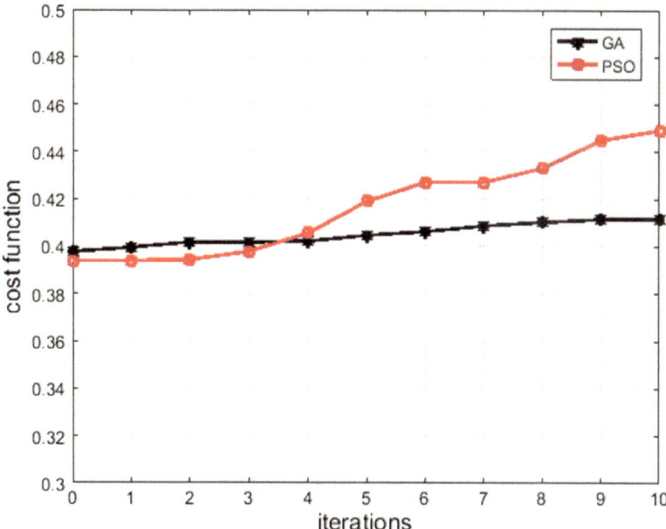

Fig. 5 Comparative analysis of cost function cepstrum method using GA and PSO

Table 6 Comparitive analysis theta and length and their PSNR

Sr. No.	Theta	Length	Optimized theta	Length	PSNR of cepstrum method	PSNRO based cepstrum method
1	5	10	11.256	12	23.4717	25.035
2	20	10	20.918	40	26.09	30.7168
3	30	20	28.738	20	24.9449	28.3947
4	40	30	39.538	27	25.5476	28.5067
5	50	40	49.524	36	25.3399	26.6092

Table 7 Performance measure for different iterations for test images

Image	Trial1	Trial2	Trial3	Trial
Mendrill				
Iteration	10	15	20	25
PSNR	24.51	32.7635	58.16	64.59
Bee				
PSNR	35.037	37.79	47.361	71.85
Living room				
PSNR	29.05	34.14	42.879	64.9464
crane				
PSNR	34.89	43.0459	53.166	90.38
Bridge				
PSNR	31	37.24	43.74	78.034

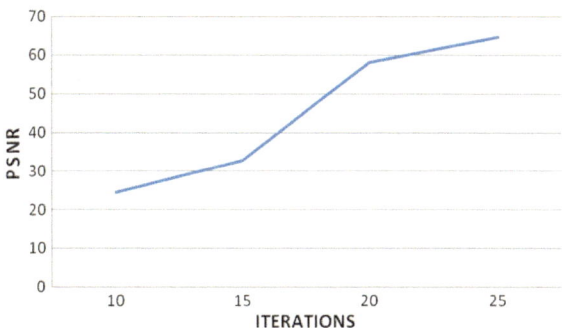

Fig. 6 Performance measure of proposed method in terms of PSNR

Figure 6 indicates that the number of generations in the particle swarm optimization give scope to improvement in the image. So for constant theta and length, increasing the number of iterations PSNR increases. But after certain stage it converges to global best. So from all three experiments it has been clear that the technique will support the motion blur having the range of theta between 0° and 90°. In all the experiments though there are variations in the value of PSNR. But the PSO based cepstrum method gives better performance than basic cepstrum and GA based cepstrum method. In case of test image data, all restored images gives better PSNR using proposed method.

6   Conclusion

In this paper we suggested a new motion blur PSF estimation method based on a new critera-cepstrum. To find theta and angle in normal cepstrum method some mathematical trigonometric functions are used. We have added PSO technique to find optimized value of theta and length of spectrum in a better way. The framework include nonblind deconvolution method. The results obtained show that PSO based cepstrum method in blind deconvolution algorithm is more efficient than GA based method. There is a scope of improvement by postprocessing output image to remove the ringing effect. In addition it should be noted that cepstrum based approach is not limited to optimize length and theta, it can be extended to hybridization.

References

1. Almeida, M., Almeida, L.: Blind and semi-blind deblurring of natural images. IEEE Trans. Image Process. **19**(1), 36–52 (2010)
2. Shi, M., Liu, S.: PSF estimation via gradient cepstrum analysis for image deblurring in hybrid sensor network, Hindawi Publishing Corporation. Int. J. Distrib. Sens. Netw. **11**(10), 11 p, Article ID 758034 (2015)

3. Qin, F.Q., Min, J., Guo, H.R.: A Blind Image Restoration Method Based on PSF Estimation. In: IEEE World Congress on Software Engineering, pp. 173–176 (2009)
4. Hu, W., Xue, J., Zheng, N.: PSF estimation via gradient domain correlation. IEEE Trans. Image Process. **21**(1), 386–392 (2012)
5. Kang, X., Peng, Q., Thomas, G., Yu, C.: Blind Image Restoration Using the Cepstrum Method. Conference, Ottawa (2006)
6. Dash, R., Sa, P.K. (Member, IEEE), Majhi, B. (Member, IEEE): Particle swarm optimization based support vector regression for blind image restoration. J. Comput. Sci. Technol. **27**(5): 989–995 (2012)
7. Asai, H., Oyamada, Y., Pilet, J., Saito, H.: Cepstral Analysis Based Blind Deconvolution for Motion Blur. In: International Conference on Image Processing, 26–29 Sept 2010, Hong Kong
8. Taxt, T.: Comparision of cepstrum based methods for radial blind deconvolution of ultrasound images. IEEE Trans. Ultrason Ferroelectr Freq Control **44**(3) (1997)
9. Asai, H., Oyamada, Y., Pilet, J., Saito, H.: Cepstral Analysis Based Blind Deconvolution of Motion of Blur. In: Proceedings of 2010 IEEE 17th International Conference on Image Processing, 26–29 Sept 2010, Hong Kong
10. Lai, Y.C, Huo, C.L., Yu, Y.H., Sun, T.Y.: PSO-Based Estimation for Gaussian Blur in Blind Image Deconvolution Problem. In: IEEE International Conference on Fuzzy Systems (2011)
11. Hamid, M.S., Harvey, N.R., Marshall, S.: Genetic algorithm optimization of multidimensional grayscale soft morphological filters with applications in film archive restoration. IEEE Tans. Circ. Syst. Video Technol. **13**(5), 406–416 (2003)
12. Fergus, R., Singh, B., Hertzmann, A., Roweis, S.T., Freeman, W.: Removing camera shake from a single photograph. ACM Trans. Graph. SIGGRAPH **25**, 787–794 (2006)
13. Chen, L., Yap, K.H.: A soft double regularization approach to parametric blind image deconvolution. IEEE Trans. Image Process. **14**(5), 624–633 (2005)
14. Babacan, S.D., Wang, J., Molina, R., Katsaggelos, A.K.: Bayesian blind deconvolution from differently exposed image pairs. IEEE Trans. Image Process. **19**(11), 2874–2888 (2010)
15. Oliveira, J.P., Figueiredo, M.A., Bioucas-Dias, J.M.: Blur estimation for blind restoration of natural images: linear motion and out-of-focus. IEEE Trans. Image Process. **23**(1), 466–477 (2014)
16. Goldstein, A., Fattal, R.: Blur-Kernel Estimation from Spectral Irregularities. In Proceedings of the ECCV, pp. 622–635 (2012)
17. Taxt, T., Frolova, G.V.: Noise robust one-dimensional blind deconvolution of medical ultrasound images. IEEE Trans Ultrason Ferroelectr. Freq. Control **46**(2), 291–299 (1999)
18. Mamta, R., Dutta, M.: GA based blind deconvolution technique of image restoration using cepstrum domain of motionblur. Indian J. Sci. Technol. **10**(16), 20 Apr 2017

Traffic Density Analysis Employing Locality Sensitive Hashing on GPS Data and Image Processing Techniques

K. Sowmya and P. N. Kumar[✉]

Department of Computer Science and Engineering, Amrita School of
Engineering, Coimbatore, Amrita Vishwa Vidyapeetham, Amrita University,
Coimbatore, India
cb.en.p2csel5020@cb.students.amrita.edu, pn_kumar@cb.
amrita.edu

Abstract. Recent development of GPS enabled devices helps in tracking the approximate location of any device. Any GPS enabled device with working internet can be tracked at any point of time. The data obtained from GPS serves several purposes, such as tracking lost devices, providing directions to a certain destination, etc. In several public environments, difficulty arises in plugging the rescue operation during any emergency needs. In case of traffic, the raw data about the traffic closure will not help the authority to reach right location. Instead, the information such as, near accurate location and time of traffic collected from GPS helps the authority to reach the destination. In this paper location of vehicle crowd formation is detected by applying similarity detection with locality sensitive hashing to the collected GPS data and two approaches with LSH (on numerical computation and on image processing) are proposed. The LSH technique is used to hash the location data to find the vehicles at similar locations and time. The proposed approaches are lightweight and needs less computational effort, since dual hashing is employed thus making it suitable for real time applications. This paper uses Apache spark for detecting vehicle crowd by applying LSH to the GPS data, since it is very fast and handles enormous data.

Keywords: GPS · Image processing · Locality sensitive hashing · Apache spark

1 Introduction

Tracking devices are becoming preferably and more generally used, for instance, in location based services, auto-system management, navigation-system, congestion-control, vehicle-floating authority and wildlife tracking. The amount of trajectory data is therefore increasing rapidly. In order to support these applications, databases management systems have been specially designed. The increasing availability of enormous quantity of spatio-temporal data emerging from an individual's trajectories has given

© Springer International Publishing AG 2018
D. J. Hemanth and S. Smys (eds.), *Computational Vision and Bio Inspired Computing*,
Lecture Notes in Computational Vision and Biomechanics 28,
https://doi.org/10.1007/978-3-319-71767-8_82

rise to diverse geographic information and systems. It also adds in opportunities and challenges to instantly obtain knowledge from the geographic trajectories. There are several frameworks [1, 2] available for handling these geographic data and extract useful information from them. LSH is a similarity detection technique [3] that is increasingly employed to find data items that are technically (by distance, by property) neighbors. Thus, LSH shall be recognized as well suited technique for finding vehicles whose location data indicates their geographic closeness. LSH drastically reduces the time complexity of computation. LSH uses buckets to hold the similar items, so, at the time of querying, it is enough to search only the relevant bucket to which the query object hashes the item to. Location data are not a single value data. It contains coordinates corresponding to the spatial dimension. Time at which the location was recorded is also integrated to the location data. LSH fits high dimensional data into smaller feature spaces, thus making the processing of GPS data simple and less error prone. When LSH is employed on the image obtained with plotting the GPS data, the algorithm considers the color pixel information to identify the vehicle locations and determine if vehicle crowd is more at any square area in the considered location space.

The sections following the introduction is organized to cover the following, Sect. 2 Locality sensitive hashing, a similarity technique, Sect. 3 Existing methodologies that work on data obtained from GPS enabled devices, Sect. 4 the proposed Work, Sect. 5 Implementation and the results, Sect. 6 Conclusion and Sect. 7 the possible enhancements in future.

1.1 Abbreviation

See Table 1.

Table 1 Abbreviations used

Abbreviations	Full form
GPS	Global positioning system
LSH	Locality sensitive hashing
GPX	GPS exchange format
TXT	Text format
MD	Merge distance
DTW	Dynamic time wrapping

2 Locality Sensitive Hashing

The exponential growth of data over the past twenty years has now created many instances where collecting and retaining all the relevant information and finding similar items is not feasible. Handling unstructured data has become difficult too. Locality sensitive hashing is the better solution for the both of these problems. Locality sensitive hashing is one among the most important nearest neighbor search algorithms. The secret behind this algorithm is that, by using specific hashing functions, it is possible to hash the points pertaining to their similarities. The probability of collision for data points that are similar (close) to each other is much higher comparing to those that are far in distance. It works by reducing the dimensionality of the data to increase the similarity among them and to use them for knowledge discovery. The dimensionality reduction is obtained by applying hashing, so that similar items map to same buckets with increased probability. LSH does not perform the same function as cryptographic hash because it focuses on maximizing the collision probability of similar items. An integral of Locality Sensitive Hashing called Min Hashing reduces feature space size using a group of random hashing functions to hash each individual piece of raw input data retaining only the minimum values resulted from each unique hashing function. A Min-Hash function performs to convert the tokenized data into a set of hash values, now the role of LSH is to break the Min-Hashes into a series of bands comprised of rows. Hashing is performed on each band. If two data have exact Min-Hashes in a band, it is that they shall be hashed to the same bucket. Thus, by reducing the dimensionality it is easy to obtain the similarities between two data points, since the search can be restricted to the bucket that is relevant rather than going through every single data item in the dataset. Employing LSH along with image processing features will address the traffic analysis with minimal computational effort since there are several tools available to visualize GPS data in a computational friendly manner.

2.1 Why Locality Sensitive Hashing?

GPS devices produce location data. Location data need not be accurate. Finding the vehicle density at any location does not require two vehicle that are accurately in the same location. Simply, it is required to find the vehicles that are closer to each other; in other words, to find the set of vehicles with similar location data indicating that they are geographically close by. Locality sensitive hashing also implies the same principle of finding similar items that prove to be alike in whatever criteria they are similar to each other.

3 Literature Survey

Li et al. [1] design a framework called hierarchical-graph-based similarity measurement (HGSM), to mine the similarity between objects with respect to the historical data their location, geographically. This paper explains about a geographic information system to model each object's location consistently and measure the similarity among them effectively. The sequence property of object's movement and the hierarchy property of geographic spaces are taken for consideration. The user similarity is explored by location history extraction and similarity sequence matching. The approach proposed in the paper is compared against three baselines namely similarity by counting, Similarity by 'cosine similarity' and Similarity by Pearson that proved it better in accuracy and efficiency of knowledge discovery. User similarity is important for effectively obtaining the knowledge with more relevance.

A new algorithm for measuring the similarity between trajectories, and in particular between GPS traces was proposed by Ismail et al. [2]. A new distance measure for measuring similarity between two or more trajectories that are suitable for GPS data is designed and used. In particular, robustness are considered for sub-sampling and super-sampling, as GPS devices only provide sampled points along the actual trajectory. The new distance function, called the Merge Distance (MD), based on the length of the shortest trajectory that is a super-sequence of both trajectories. Intuitively, this length should be short when the two trajectories come from the same curve. The Merge Distance can be computed in quadratic time, like DTW. Then an experimental comparison between MD, the Euclidean distance and DTW are performed. The results indicate that MD is robust under sub-sampling and super-sampling, and that it has comparable results with DTW in several other settings. At this point, it is difficult to decide whether MD is a metric or not.

Application of locality-sensitive hashing method to nearest neighbor search problems for the data that is mixture of both numerical and categorical attributes is done by Lee et al. [3]. The method that is proposed in the paper uses dual hashing functions, where one function performs hashing on numerical attributes and the second one on categorical attributes. The design includes, forming an indexing structures corresponding to each of the hashing functions, gathering and combining the candidate sets, and thoroughly examining them to determine the nearest ones. The paper also deals with Locality sensitive hashing for numeric data, categorical data, numeric attributes, and categorical attributes.

Computer forensic workers are extremely overwhelmed by regular increase in the amount of data and the availability of storage devices in mass to store the data. Existing hashing validate media, find out the duplicate values, helps to avoid and escape from trivial conditions. Altering a single bit of data will radically disregard to hash. There is less proof that a cryptographic hash can feature us to the associations of files that are possibly to have minute changes in data. This paper explains a means for fuzzy hashing with a reactionary hashing algorithm to find the data that are closely similar, and could have had the data items that are changed. Fuzzy hashing looks for similarity and not the exactness.

Prabhakar et al. [4] have proposed a system to support the ambulatory services throughout their way to the hospital. In the proposed approach, GPS module replaces the RF transmitter to determine the near accurate location of the ambulance. Several processing is done on the images to avoid disturbances. The concept of processing approaches used in this paper can be taken as base for future implementations.

4   Proposed Works

4.1   Locality Sensitive Hashing on GPS Data in Visual Format

GPS data becomes more useful when they are provided in a visualized format. The coordinates (latitude, longitude) that are obtained from the moving vehicles are taken and passed through a visualization tool and the visual depiction of it is obtained. The depiction of the coordinates are plotted in the (lat-min, long-min), (lat-min, long-max), (lat-max, long-min), (lat-max, long-max) square area. The latitude (min, max) and longitude (min, max) are obtained from the GPS data captured at a point of time in a particular locality. Now, with the available square plot, it is possible to divide them into smaller squares (of area x * y), with each square mapping to a certain coverage by the latitude (x), longitude (y) length. These smaller squares will represent the smallest possible areas on which the traffic density can be obtained with less error in a more optimized and faster way. Here LSH is employed to obtain the smaller squares, where the hashing functions being the length of latitude and longitude of the determined smaller squares. The area to be considered for the smaller squares will remain the same irrespective the locality from where the data is captured. Using any efficient programming languages, code can be written to analyze the density of the color in a square area. The application of LSH is as follows: It is not necessary to apply the color density analysis code on all the smaller squares, here the principle of LSH can be used to reduce the number of squares on which the analysis needs to be made. While dividing the large square plot into smaller squares, initially, divisions must be made only along the longitude (or latitudes), now code runs through these divided rectangular blocks and decides the blocks that have color filled at area greater or equal to the smaller square area (x * y) within the block, only those blocks are selected and further latitudinal (or longitudinal) divisions are made. This reduces the computational effort by at least 0.4 times and time taken to analyze traffic density by 0.5 times.

4.2 Locality Sensitive Hashing on the Raw GPS Data

The flow of the architecture is shown in Fig. 1. GPS is fitted to modern devices. The devices can be attached to any movable or non-movable object like vehicles. The scenario considered focuses on the data obtained from vehicles moving on roads. At any particular point of time like 8.30 a.m., the data from the vehicles at a particular area is obtained. The data shall be obtained in the intervals of microseconds for about 10 s. Since the data collected will be in gpx format, it is suggested to transform the data to a format that makes the processing and viewing of information without much effort. Thereby the *.gpx file is converted to *.txt format and the data is available for knowledge querying. The data contains the latitude, longitude coordinate values, time when the data was captured and altitude details of the vehicle's location. Since, LSH employs hashing to find the similar item, location data is hashed containing two coordinates (latitude, longitude) with certain hashing technique in order to find out the vehicles that are close by, to analyze the traffic density. Therefore, the two-level hashing is performed, one with a function that hashes the latitude values and separating them into buckets containing vehicles falling into respective buckets as their latitude location. Next level of hashing will be performed on the buckets that are obtained after first-level hashing. Finally, the buckets obtained with vehicles falling into a particular area (both latitudinal and longitudinal). The derivation of respective hashing functions is explained in the next section. Following the hashing, it is necessary to find the length of each buckets in order to obtain the regions that is dense packed with vehicles. When this analysis is performed at regular intervals of days, weeks and months, it can be concluded that, when and what time at a city, the traffic is more likely to pack. Therefore, the rescue can arrive in prior. However, in cases of accidents and natural calamities, analysis should be made only at the time of occurrence.

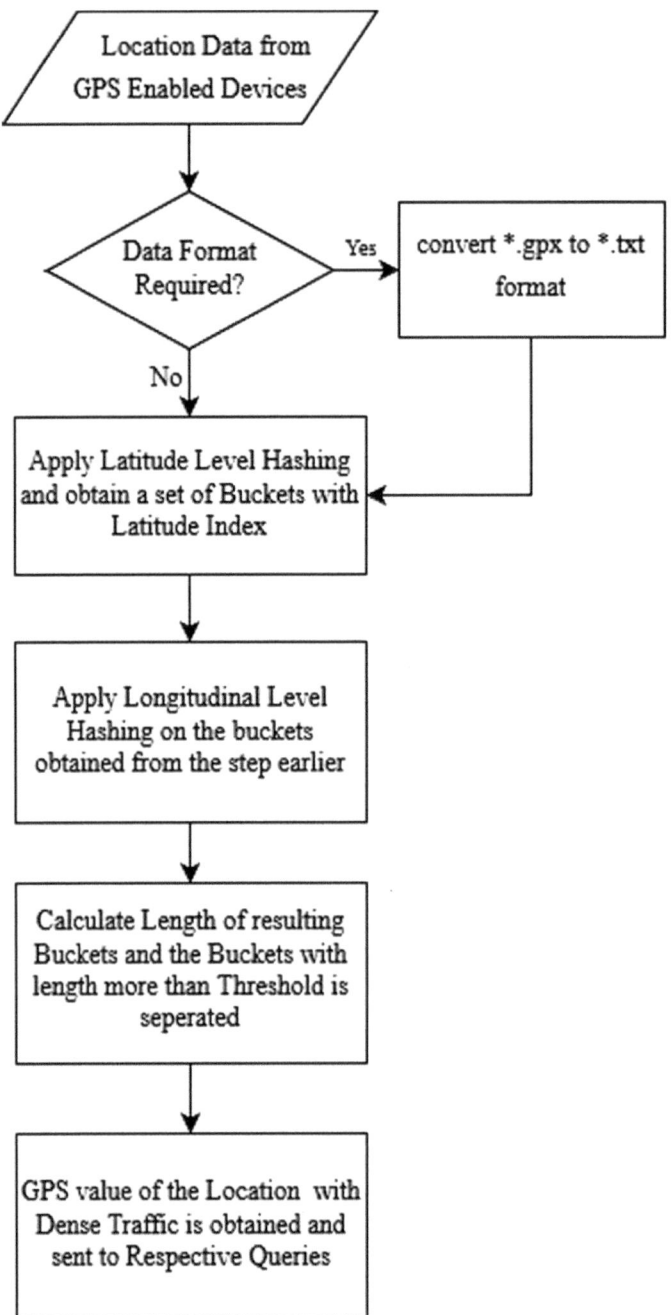

Fig. 1 Proposed work flow

5 Implementation

5.1 Scenario

Traffic jams are either caused by lack of traffic guides or due to accidents. Our scenario revolves around these issues. Since, these days, GPS device is available in all the vehicles, it is possible to get real time data from moving vehicles and analyses as in, where the traffic is possibly to get jammed at what point in time (say 9.00 a.m. on Monday) and at which place (in terms of latitude and longitude).

5.2 Dataset Description

The dataset contains the attributes like Latitude, Longitude, Altitude and Time. Latitude and longitudes are decimal converted degree-minute-second values. One latitude degree = 69 miles, One latitude min = 1.15 miles, One latitude second = 101 ft, One longitude degree = 54.6 miles, One longitude min = 0.91 miles, One longitude second = 80 ft. The data taken for analysis covers the roads between (11.0210° North, 76.9663° East) and (11.0812° North, 76.9416° East), that is between two locations in Coimbatore, Tamil Nadu, India.

5.3 Apache Spark

Apache Spark is a computing framework, which is an open-source and effectively uses cluster computing for all its functionalities. Spark is positioned as a fast and general engine for Big Data. It generalizes the MapReduce model and is poised to replace MapReduce and other runtimes. Choosing Apache Spark as the engine for the taken scenario would be justified because it is faster than all other frameworks especially for employing map reduce. The speed of the engine is high, because it handles all the data in-memory. Since, the considered scenario needs to handle the enormous amount of data and requires solution at real time; the expected processing rate is high.

5.4 Result and Analysis of Approach Proposed in 4.2

The GPS data of moving vehicles are obtained on timely basis and is fed through Apache Spark, where locality sensitive hashing is applied on the data, through the following steps.

Step 1: Conversion of *.gpx to *.txt format (Fig. 2).

Time	Latitude	Longitude	Altitude
2016-12-15 09-45-16	11.0345636	77.61034357	410
2016-12-15 09-45-16	11.030788774	77.610887467	412
2016-12-15 09-45-16	11.036765004	77.610837573	412
2016-12-15 09-45-16	11.0323654	77.61623754	410
2016-12-15 09-45-16	11.033461	77.6102462	411
2016-12-15 09-45-16	11.0345636	77.61034357	412
2016-12-15 09-45-16	11.045788774	77.61016437467	411
2016-12-15 09-45-17	11.0313131	77.61054	412

Fig. 2 Dataset in text format

Step 2: Data are hashed along the latitude and the corresponding signature matrix (set of buckets resulting from hashing) is obtained. Hashing function for this step is the distance that should considered for traffic analysis (say, if we consider 100 vehicles in 1 km * 1 km for more than 30 s as traffic jam, then latitude equivalent of 1 km would be the hashing function under consideration).

In the taken scenario, the number of buckets = 7 (Fig. 3).

Fig. 3 Result of latitudinal hashing (1st bucket)

Let 20 vehicles in x by y area be considered as the area (road) with dense traffic. Let z be the number of road in a certain city. And, p by q be the average area covered by a single road. The total area covered by all the roads in the city z * p * q. So the total number of buckets after first hashing will be equal to (z * p * q)/(x). Assuming that x is the length along the latitude.

The hash function for this first level hashing would be,

$$\frac{(Latitude\ Maximum - Latitude\ Minimum) * x}{p * q}.$$

Step 3: The buckets of signature matrix is taken individually and are further hashed along the longitude to obtain all the vehicles in a particular area.

The function to hash the buckets obtained from Step 2 along longitude will be,

$$\frac{(Longitude\ Maximum - Longitude\ Minimum) * x}{p * q}$$

Finally,

$$Number\ of\ Buckets = \left(\frac{z * p * q}{x}\right) * y$$

In the taken scenario, the number of buckets = 7 * 6 = 42 (Fig. 4).

```
Buckets After longitudinal hashing:

Bucket1 cross Bucket1
=[[[11.0345636,77.61034357],[11.036573648,77.6
1067573],[11.035788774,77.61016437467],[11.037
6701,77.61045867],[11.03888774,77.61013487],[1
1.055788774,77.61016437467],[11.035788774,77.6
1097467],[11.03592774,77.6104437467],

[11.030788774,77.610887467],[11.039999774,77.6
1013267],[11.034288774,77.610217467],[11.03987
264,77.61069897467],[11.034274,77.6104437467],
[11.0323891,77.610666],[11.032034,77.610988],[
11.08564,77.615662],[11.09234444,77.61055532],

[11.033894,77.61016437467],
[11.0347454,77.610321235],[11.03442537,77.6103
2424],[11.03010121,77.610311156],[11.03288424,
77.610621001],[11.035788709,77.61016437467],[1
1.033354335,77.610013125],[11.031233,77.610456
3],

[11.045788774,77.61016437467],[11.035788774,77
.61016437467],[11.035788774,77.61016437467],[1
1.039087,77.61087],[11.034187,77.6108111],[11.
03288774,77.61000342],[11.030788774,77.6108874
67],[11.030788774,77.610887467],

[11.030788774,77.610887467],[11.0313131,77.610
54],[11.03984,77.610613],[11.038181,77.610523]
,[11.031237,77.6104114],[11.03034114,77.610235
],[11.03674587,77.610826523],[11.033461,77.610
2462],[11.033765004,77.610837573],

[11.03365483237,77.6102476613],[11.03200333,77
.61021372]]
```

Fig. 4 Result of longitudinal hashing (1st bucket)

Step 4: Now after second signature hashing, the length of each bucket will give the number of vehicles in a particular area. Setting a threshold [number of vehicles in a given area causing traffic jam (say 20)] can help in concluding if a specific area is crowded or not (Fig. 5).

```
Areas with heavy traffic are:

[11.03112345, 77.6106437467] to [11.03112541,
77.6138342]
[11.0763271, 77.62031848] to [11.077719836,
77.620441529]
[11.08124326, 77,619824319] to [11.085102273,
77.617971638]
```

Fig. 5 Areas with heavy traffic

Step 5: The length of each bucket is calculated and the ones with items more than the threshold is separated and the areas covered by the respective bucket (after dual hashing) is considered to be the areas with higher traffic volumes.

5.5 Verification

The ideal verification for locality sensitive hashing is the Jaccard similarity. Jaccard similarity of any two items is considered to be the true similarity between them. Jaccard similarity is applied between two different datasets collected at same time point but different weeks of a month and the correctness of the algorithm is verified.

$$Jaccard\,Similarity = \frac{A \cap B}{A \cup B},$$

where A and B are two items for which similarity is to be found. Comparing Jaccard similarity against Locality sensitive hashing proves that the results of LSH is 95.66% as the ideal solution. It's less complexity to process the data makes LSH a better solution for real-time scenarios. Apache Spark makes the processing fast and handles large amount of data without much effort, adding to increase the effectiveness of LSH.

6 Conclusion

A Framework for efficiently mining the location dataset to analyze the traffic density is designed using Locality sensitive hashing. The data obtained is hashed into buckets using LSH for quicker analysis and faster results. From the result, it can be concluded that Locality sensitive hashing occupies only 6% of the total space required for storing the dataset, with enormous amount of data arriving regularly during real-time scenarios, LSH can perform knowledge discovery quickly, with less errors and efforts. The vehicle density details at any location on roads are collected with two assumptions, first is by considering that all vehicles are GPS enabled and less technical faults and the second is parked vehicles do not send their data. From the rate at which the analyses was obtained, can used to conclude that Apache Spark is definitely the better tool for real time data analytics. To make the analysis more beneficial during real-time emergencies, the processing time of information can be reduced by feeding the data through streaming framework for spark, thus reducing the time take taken for storage. To make the result more user friendly and to reduce the time required to analyze the location name from the GPS coordinates, an efficient image processing technique with LSH employed, is proposed.

7 Future Work

Following the successful implementation with the raw data and providing the coordinate information about the areas that are populated with vehicles leading to heavy traffic, the next step is to predict, implement, depict the result in visual format using efficient image processing approaches and analyze the result to give useful information in visual format about the vehicle density to the traffic control team. The following are the tasks to successfully implement the proposed approach 4.1: to identify best suited visualization tool, to write optimal code to determine the density of any color, to apply LSH and analyze the result.

References

1. Li, Q., et al.: Mining User Similarity Based on Location History. In: Proceedings of the 16th ACM SIGSPATIAL International Conference on Advances in Geographic Information Systems. ACM, USA (2008)
2. Ismail, A., Vigneron, A.: A new trajectory similarity Measure for GPS data. In: Proceedings of the 6th ACM SIGSPATIAL International Workshop on GeoStreaming. ACM, pp. 19–22 (2015)
3. Lee, K.M., Lee, K.M.: Efficient search for data with numerical and categorical attributes based on dual locality-sensitive hashing. Indian J. Sci. Technol. 9(24) (2016)
4. Prabhakar, Manoj K., Kumar, Manoj S.: GPS tracking system coupled with image processing in traffic signals to enhance life security. Int. J. Comput. Sci. Inf. Technol. 5(4), 131 (2013)

A Novel Transfer Learning Approach upon Hindi, Arabic, and Bangla Numerals Using Convolutional Neural Networks

Abdul Kawsar Tushar, Akm Ashiquzzaman, Afia Afrin,
and Md. Rashedul Islam$^{(\boxtimes)}$

Department of CSE, University of Asia Pacific, Dhaka, Bangladesh
{tushar.kawsar,zamanashiq3,meghlaprottoy,rashed.cse}
@gmail.com

Abstract. Increased accuracy in predictive models for handwritten character recognition will open up new frontiers for optical character recognition. Major drawbacks of predictive machine learning models are headed by the elongated training time taken by some models, and the requirement that training and test data be in the same feature space and consist of the same distribution. In this study, these obstacles are minimized by presenting a model for transferring knowledge from one task to another. This model is presented for the recognition of handwritten numerals in Indic languages. The model utilizes convolutional neural networks with backpropagation for error reduction and dropout for data overfitting. The output performance of the proposed neural network is shown to have closely matched other state-of-the-art methods using only a fraction of time used by the state-of-the-arts.

Keywords: Transfer learning · Indic numerals · Numeral recognition
Convolutional neural networks · Optical character recognition

1 Introduction

Knowing how to cook fish helps while cooking chicken. Solving some mathematical problems enhances the ability to solve other similar problems. Once the basic circuitry of a small cell phone is known, it gets easier to explain the mechanism of similar other electronic devices. Knowledge is a dynamic horizon of the repertoire of human beings. Learning a new technique and applying it to solve different related problems—this is the natural way of learning. For example, when we learn how to differentiate between rotten potato and fresh potato we intuitively learn the basics of identifying a rotten vegetable, be it a tomato or potato!

Such observations lead us to a specialized learning method—where previous experiences and stored knowledge are used to resolve new tasks and problems. This technique is known as *transfer learning*. It is an important subsection of the field of machine learning (ML). Transfer is one of the most important elements of human

© Springer International Publishing AG 2018
D. J. Hemanth and S. Smys (eds.), *Computational Vision and Bio Inspired Computing*,
Lecture Notes in Computational Vision and Biomechanics 28,
https://doi.org/10.1007/978-3-319-71767-8_83

learning mechanism which indicates that one major application of transfer learning is to deploy it in neural networks, as neural network is modeled after the human brain and nervous system. Several experiments and research works agree with this. For example, experimental results obtained from [1] provide evidence that transferring knowledge across related tasks helps the learner experience more and generalize better. An inductive transfer mechanism namely multitask learning has been demonstrated in [2] which improves generalization by learning tasks in parallel while using a shared representation. A similar approach, self-taught learning, has been adopted in [3] which works with unlabeled data sets.

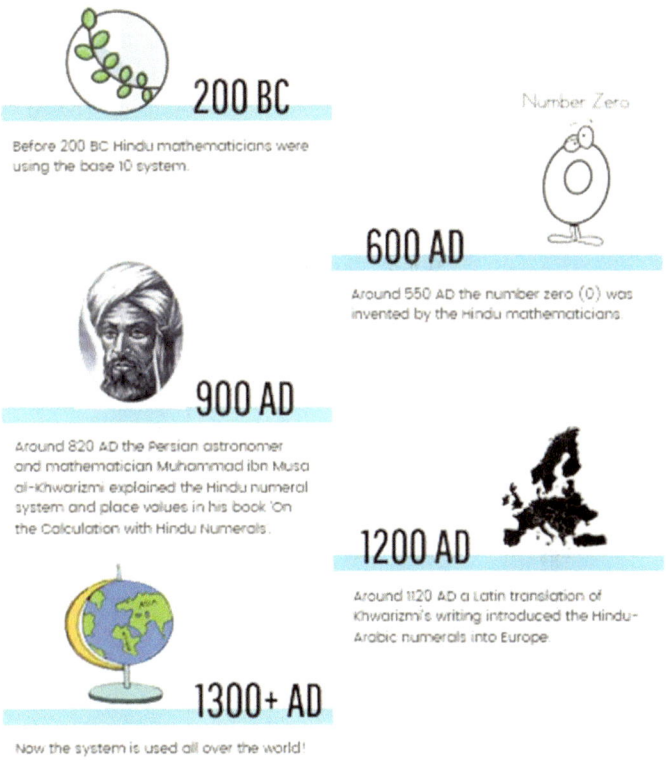

Fig. 1 Brief history of the modern number system

Modern decimal system is a descendant of the Hindu-Arabic numeral system, invented by the Indian mathematicians between the 1st and 4th centuries. Very early use of the place value system is seen in the *Bakhshali manuscript*, the oldest extant manuscript in Indian mathematics [4]. Though the actual date of the composition of the manuscript is a matter of debate, the language used in the manuscript indicates that it could not have been composed any later than 400 AD. The oldest known mention of numeral zero ("0") and the decimal positional system has been found in the

Lokavibhaga, a *Jain* cosmological text translated from a *Prakrit* original internally dated to AD 458 [5, 6]. By the 9th century the system was adopted in Arabic mathematics and later introduced to Europe by the High Middle Ages (around 12th century) [7]. The book *On the Calculation with Hindu Numerals*, written in about 825 AD in Arabic by a Persian astronomer and mathematician, Muhammad Ibn Musa al-Khwarizmi introduced the Indian numeral system to the Arabian Peninsula. Later in the 12th century, the Western world got introduced to this system through the Latin translations of his work [8]. Thus, the Indian numerals form the basis of European number systems which are now broadly used worldwide. However, this long journey of numbers from India to Europe through Arab was not as simple as it sounds. The Eastern and Western parts of the Arabic world adopted the basic Indian numeral system differently and the European digits that we observe nowadays are descendant of the Western Arabic glyphs. On the other hand, the Eastern Arabic numerals (also known as the "Indic numerals") spread among many countries to the East of the Arab world. Figure 1 shows the entire historical journey of the modern number system using a simple timeline. Nonetheless, the history of invention and evaluation of modern number system clearly stipulates that Hindu-Arabic numeral system is the precursor of modern numerals. This observation indicates a high correlation among different numeral systems which is the primary motivation behind our work.

This paper proposes a deep learning model appropriate for training and transferring the knowledge to a later task. Later transfer learning techniques are utilized to compare reduced training time and recognition accuracy where prediction accuracy for image numerals are scored. The model is trained individually with three different existing numeral systems in image forms, namely—Bangla, Urdu, and Hindi; all of these originated from the Indic numerals under Indic languages which are used by a considerable percentage of world's population. All these three numeral systems, having a fairly recent common predecessor, are highly correlated with each other which consequently makes it possible to apply transfer learning, as stated above.

The rest of this paper is structured as follows: Sect. 2 introduces the concept of deep neural networks (DNN) with which the model is formulated. Section 3 outlines the proposed model which follows the explanation for transfer learning. Section 4 provides details about the experimentation and relevant dataset, and discusses about the results. Then Sect. 5 concludes the paper.

2 Deep Neural Network

This section presents an overview of the deep learning system that will be used in the proposed model. Deep neural network is one type of artificial neural network (ANN)— a computing system inspired by the biological neural networks. Whereas a simple predictive algorithm tries to mimic the mapping between input and output variables, ANN has a unique characteristic of creating transient states through the *artificial neurons* which are the basic building blocks of ANN. What constitutes the main difference between human brain and simple machine is the creativity and decision taking capability. ANN is the first step of modern technology to eradicate this "little" difference.

Artificial neurons organized in several layers form an artificial neural network. A DNN will presumably have more layers than a simple ANN. Though the number of layers is not fixed, it is usually no less than four for DNN. Among various types of DNN, the Convolutional neural networks (CNN) are widely used for processing visual and other two-dimensional data [9, 10].

3 Proposed Method

This section outlines a brief technical discussion on transfer learning and then describes the proposed technique that is central to this study.

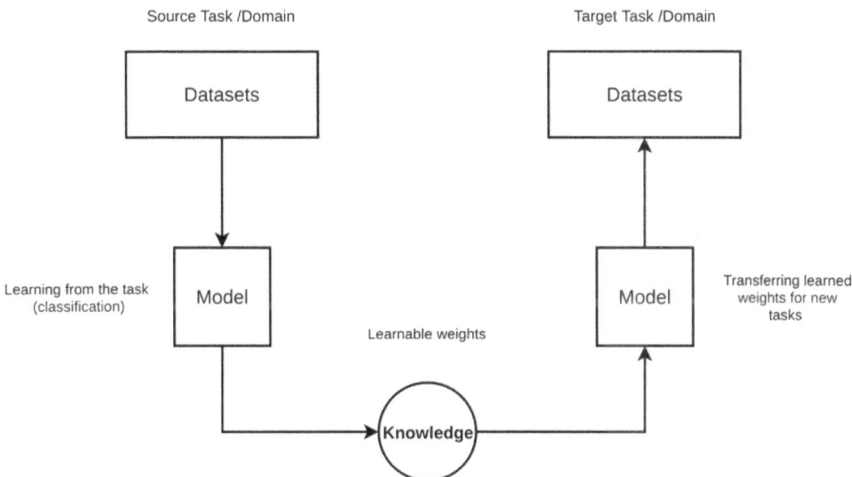

Fig. 2 Transfer learning flow diagram

3.1 Transfer Learning

ML and data mining technologies have introduced a revolutionary change in the field of knowledge engineering including classification, clustering and regression [11, 12]. However, a major drawback of these renowned techniques is most of the ML algorithms perform well only when training and test data are drawn from the same feature space and follow the same distribution. When this *identical distribution* assumption does not hold, most of the ML models need to be rebuilt from scratch with new sets of training data which is undoubtedly a time-consuming approach. Furthermore, large amounts of hand-crafted, structured training data are required for this purpose, which is expensive. Thus, despite being a successful learning technique, prevalent ML algorithms may fail in some real-world situations [13]. In such cases, *knowledge transfer* can improve the learning performance, as transferring knowledge between different but similar problem domains eliminates the extra burden of recollecting training data and rebuilding the model from scratch. One such example is image clustering where heterogeneous data from different feature spaces are used for training purpose. A novel method for image clustering using heterogeneous transfer learning technique has been proposed in [14]. Similar techniques have been devised in [15, 16] for the task of classifying Chinese text documents using another collection of English texts as training data.

Another major application of transfer learning is when there is a chance of training data to be outdated with time. One such example is *Web mining* [17], where the Web data used in training a particular Web-page classification model can be easily outdated afterwards, as the topics on the web change frequently. In general, training is acquired from a source task and infused into the model. The knowledge from the model is then carried over to another model working on a new task in same or different domain. The target task is then re-trained while utilizing the transferred knowledge to its advantage in reduction of time and increase in accuracy. Figure 2 depicts these steps between the source and target tasks.

3.2   Proposed Model Architecture

The whole model can be divided into two main clusters depending on their tasks: [9, 18] use CNN layers as unsupervised feature vector extractors and fully connected layers as classifiers in their studies. The current study follows the same footprint. The feature extractor cluster consists of four separate convolutional layers with variable output as well as kernel size 3×3. Each convolution layer learns increasingly complex patterns or features of the input image data and gradually shifts closer toward recognition of the same. After the last convolution layer, a max pooling layer with 2×2 sized kernel and 2 sized strides in both axes is inserted in order to increase weight of the most important features. The classification cluster has three fully connected neural network layers with same variable output numbers. The fully connected layers are used as classifiers which act upon complex features learned in the convolution layers. All of the layers have exponential linear unit (ELU) as activation functions, except for the last layer; the endmost layer has a Softmax function for output [19]. ELU is used to overcome vanishing gradient problem during training [20]. The convolution and fully connected layers are insured with dropout function to reduce overfitting [21]. Details of the neural network layers are described in Table 1.

The whole model is trained from scratch with numerals from the mentioned three scripts, separately. The first phase of the process sees the models being trained for 300 epochs and the best resulting weights subsequently saved in each case. In the second phase, all of the feature extraction cluster, i.e. the CNN layers are halted for weight update due to overfitting concerns. At the same time, layers in the classification cluster get to learn to detect different digits using feature vectors obtained from recognition process of numerals of another language. All simulations are performed on a computer hosting Intel i3 7th generation processor with 8 GB RAM. Nvidia GTX-1050Ti 4 GB DDR-5 GPU is used for CUDA accelerated computing.

Table 1 Architecture of deep neural network

Layer	Number of neurons	Activation function used	Kernel size
Convolutional	64	ELU	3×3
Convolutional	64	ELU	3×3
Convolutional	64	ELU	3×3
Convolutional	32	ELU	3×3
Max pooling	N/A	N/A	2×2
Fully connected	512	ELU	N/A
Fully connected	256	ELU	N/A
Fully connected	128	ELU	N/A
Output	10	Softmax	N/A

4 Experiments

This section presents a brief discussion on the data sets that are used in the experiments and then examines the results of these experiments.

4.1 Dataset

In this study, all of the images chosen are samples of handwritten numerals collected form a range of writers. Bangla handwritten numerical dataset CMATERDB 3.1.1, Hindi (Devanagari) handwritten numerical dataset CMATERDB 3.2.1, and Urdu (Arabic) handwritten numerical dataset CMATERDB 3.3.1 are chosen for this study [22]. All three of these datasets contain handwritten 32×32 pixels that are binary-scanned and segmented samples of the digits. Table 2 shows brief details of the datasets.

The size and quality of training dataset greatly influence the accuracy and correctness of any ML algorithm. For deep learning, this effect is even more prominent, as convergence rate of any neural network is highly dependent on the model complexity and availability of data. In this study, we have used three datasets containing handwritten digits of different languages sharing a common ancestor. The images are inverted before feeding into CNN. As a result, the numerals are in white foreground on the backdrop of black background. Edges are a very important feature in character recognition, converting the background black along with white foreground makes the edge detection slightly easier [23].

Table 2 Brief details of the dataset

ID	Lingual origin of the numerals	Training samples	Digits/classes
1	Urdu	3000	10
2	Bangla	6000	10
3	Hindi	3000	10

4.2   Results and Discussion

An objective of this study is to verify the hypothesis that numerals of the same origins will be easier of detect as they have shared the morphological common traits. It will be difficult to estimate competitiveness of performance for transfer learning solely by observing similar features for both source and targeted domains. In contrast, key insights will be gained by considering the time of training deep learning models. For this purpose, firstly, Table 3 shows the comparison of the three tasks, i.e. separate simulations on three datasets described in Sect. 3. The table demonstrates the learning times, or the numbers of iterations over the whole dataset the model needs to classify with state-of-the-art correctness. For each of the datasets used, the accuracy is observed to have been achieved at around 200th epochs on average. Note that after the best accuracy is achieved, the model for that epoch is saved and used in subsequent epochs for the same dataset. For another one of the datasets, a fresh model is used where weights and biases are again initialized. Therefore, parameters of these three models are in no way dependent on each other.

Table 3 Standalone training performance on datasets

Dataset used for training	Best accuracy achieved (%)	Number of epochs	Best accuracy in epoch number
Urdu	99.30	300	230
Bangla	99.40	300	216
Hindi	99.26	300	189

After that, weights and biases from the three isolated runs of the model with each of the datasets are saved and re-used for a different task, different from the source task. At this stage, the feature extraction cluster of the models, as mentioned in Sect. 4, are frozen for weight update, otherwise the main objective of transfer learning would not be ensured. 100 epochs are allocated to each of the three runs at this stage. Table 4 shows the task combinations with which transfer learning experiments are executed. We highlight three things in this result: the best accuracy achieved with the transferred model, epoch number where this best accuracy is achieved, and the accuracy of that run as recorded after ten initial epochs are completed. It is observed form Table 4 that, even though the feature extraction layers are trained on a different task in an unsupervised manner, the classification accuracy is on par with state-of-the-art methods mentioned in Table 3. It is also evident that all the transferred models are capable of instantaneous recognition of digits from a related but different numeral script. This power is lent to the transferred model from the intuitive explanation of transfer learning mentioned in Sect. 1. Another remarkable insight gained from this experiment is that, in all the runs, over 92% accuracy is achieved after only 10 epochs. It is a very short time compared to the original model mentioned in Table 3 where it took 300 epochs to train them sufficiently enough. A point of note here is that the accuracy could have improved further, had the models been run for more than the 100 epochs allocated. Furthermore, when a task does not have sufficient labeled training data, the proposed approach will be of significant help if the model is pre-trained on a related but separate dataset.

Table 4 Transfer learning comparisons

Source task	Destination task	Best accuracy achieved (%)	Best accuracy in epoch number	Accuracy after 10 epochs (%)
Urdu	Bangla	96.99	42	93.90
Bangla	Urdu	97.79	48	97.12
Hindi	Bangla	98.66	38	95.45
Urdu	Hindi	95.88	77	92.11
Bangla	Hindi	98.57	52	93.67
Hindi	Urdu	98.57	64	92.44

Results of the transfer learning process can, in part, be explained through the morphological similarities of the digits among the languages under discussion. However, the feature extraction process of the CNN is in itself an unsupervised operation. Moreover, where the model will be applied after transfer is not known from before to the earlier task model. Therefore, it cannot be said that before transferring, the model learns the important features it recognizes in the previous task to better adapt to the task after transfer. Interestingly, this phenomenon partially explains the discrepancy in accuracy in transferring skills from one language to another and vice versa in Table 4. The competitive results obtained from the proposed transfer learning process, taken along with the limited amount of re-training time, can be termed as a very competitive process.

5   Conclusion

In this research, a model for transferring knowledge from one character recognition task to another in proposed. The model is used for recognition of handwritten numerals in Indic scripts used by a considerable percentage of world's population. A novel form of convolutional neural network for transfer learning with curtailed time and competitive accuracy is discussed for this purpose. Independence of model-specific training has helped to reduce the re-training time in the target task significantly. Furthermore, by diminishing the effect of overfitting in the proposed model, very competitive accuracy is achieved via experimentation. As a result, performance for predictive models for numeral recognition can now have prediction scores or performance gains on par with state-of-art methods in shorter time, which can lead to future breakthroughs in numeral recognition tasks in specific and transfer learning in general.

Acknowledgements. The authors would like to thank the department of Computer Science and Engineering, University of Asia Pacific for supporting this research in various ways. Abdul Kawsar Tushar and Akm Ashiquzzaman contributed equally to this work.

References

1. Thrun, S.: Is learning the n-th thing any easier than learning the first? In: Advances in Neural Information Processing Systems, pp. 640–646 (1996)
2. Caruana, R.: Multitask learning. In: Learning to Learn, pp. 95–133. Springer, Berlin (1998)
3. Raina, R., Battle, A., Lee, H., Packer, B., Ng, A.Y.: Self-taught learning: transfer learning from unlabeled data. In: Proceedings of the 24th International Conference on Machine Learning, pp. 759–766 (2007)
4. Hayashi, T.: Bakhshali Manuscript. Encycl. Hist. Sci. Technol. Med. Non-Western Cult. 387–389 (2008)
5. Crump, T.: The Anthropology of Numbers. Cambridge University Press, Cambridge (1992)
6. Ifrah, G., Harding, E.F., Bellos, D., Wood, S., et al.: The Universal History of Computing: From the Abacus to Quantum Computing. Wiley, USA (2000)
7. Smith, D.E., Karpinski, L.C.: The Hindu-Arabic Numerals. Courier Corporation, USA (2013)
8. Struik, D.J.: A Concise History of Mathematics. Courier Corporation, USA (2012)
9. LeCun, Y., Boser, B., Denker, J.S., Henderson, D., Howard, R.E., Hubbard, W., Jackel, L. D.: Backpropagation applied to handwritten zip code recognition. Neural Comput. 1, 541–551 (1989)
10. LeCun, Y., Bengio, Y., Hinton, G.: Deep learning. Nature 521, 439–440 (2015)
11. Wu, X., Kumar, V., Quinlan, J.R., Ghosh, J., Yang, Q., Motoda, H., McLachlan, G.J., Ng, A., Liu, B., Philip, S.Y.: others: top 10 algorithms in data mining. Knowl. Inf. Syst. 14, 1–37 (2008)
12. Pradhan, A.: Support vector machine-A survey. Int. J. Emerg. Technol. Adv. Eng. 2, 82–85 (2012)
13. Hoffmann, A.G., et al.: General limitations on machine learning. In: ECAI, pp. 345–347 (1990)
14. Yang, Q., Chen, Y., Xue, G.-R., Dai, W., Yu, Y.: Heterogeneous transfer learning for image clustering via the social web. In: Proceedings of the Joint Conference of the 47th Annual Meeting of the ACL and the 4th International Joint Conference on Natural Language Processing of the AFNLP, vol. 1, pp. 1–9 (2009)
15. Ling, X., Xue, G.-R., Dai, W., Jiang, Y., Yang, Q., Yu, Y.: Can chinese web pages be classified with english data source? In: Proceedings of the 17th International Conference on World Wide Web, pp. 969–978 (2008)
16. Cireşan, D.C., Meier, U., Schmidhuber, J.: Transfer learning for Latin and Chinese characters with deep neural networks. In: The 2012 International Joint Conference on Neural Networks (IJCNN), pp. 1–6 (2012)
17. Al-Mubaid, H., Umair, S.A.: A new text categorization technique using distributional clustering and learning logic. IEEE Trans. Knowl. Data Eng. 18, 1156–1165 (2006)
18. Maitra, D. Sen, Bhattacharya, U., Parui, S.K.: CNN based common approach to handwritten character recognition of multiple scripts. In: 13th International Conference on Document Analysis and Recognition (ICDAR), 2015, pp. 1021–1025 (2015)
19. Hinton, G.E., Salakhutdinov, R.R.: Replicated softmax: an undirected topic model. In: Advances in Neural Information Processing Systems, pp. 1607–1614 (2009)
20. Clevert, D.-A., Unterthiner, T., Hochreiter, S.: Fast and accurate deep network learning by exponential linear units (elus). arXiv Prepr. arXiv1511.07289. (2015)
21. Srivastava, N., Hinton, G.E., Krizhevsky, A., Sutskever, I., Salakhutdinov, R.: Dropout: a simple way to prevent neural networks from overfitting. J. Mach. Learn. Res. 15, 1929–1958 (2014)

22. Google Code Archieve—Long-term Storage for Google Code Project Hosting. https://code. google.com/archive/p/cmaterdb/downloads
23. Ashiquzzaman, A., Tushar, A.K.: Handwritten Arabic numeral recognition using deep learning neural networks. In: IEEE International Conference on Imaging, Vision & Pattern Recognition (icIVPR), 2017, pp. 1–4 (2017)

Diabetic Retinopathy Detection in Fundus Image Using Cross Sectional Profiles and ANN

M. Smitha[1]([email]), A. K. Nisa[1], and K. Archana[2]

[1] Department of Computer Science and Engineering, TKM College of Engineering, Kollam, India
smithamundath@gmail.com
[2] Department of Computer Science and Engineering, LBS College of Engineering, Kasaragod, India

Abstract. An eye disease which destroys the normal vision ability of Diabetic Patients is known as diabetic retinopathy. Early diagnosis of this disease is necessary because, it is severe in the later stages. The presence of the microaneurysm (MA) is the first clear clinical symptom of this disease. MAs are red dots formed by swelling of the weak part of the capillary wall. The detection of microaneurysms in retinal fundus images is an important task for applications such as diabetic retinopathy screening and early treatment. The proposed method detects MAs by the use of directional cross sectional profiles of some central pixels. The cross sectional profiles of each local maximum pixels of the preprocessed images are drawn and then these profiles are analyzed. For each profile, peak detection step is employed and some attributes which includes the shape, height and size of the peak are computed. The numerical measures of these attributes are included in the feature set. This feature set is used as an input for the classifier. Artificial Neural Network (ANN) is used as the classifier. The results are compared with the well-known classifier Naïve Bayes and obtained good results.

Keywords: Diabetic retinopathy · Color fundus images · Cross sectional profiles microaneurysms · Artificial neural network · Extreme learning machine

1 Introduction

Diabetic Retinopathy (DR) is a retinal disease that can cause the effected patient to loss his/her visual capacity, if left undiagnosed in the initial stage. The presence of micro aneurysms are the initial indications of diabetic retinopathy. They arise due to high sugar levels in the blood. The recent studies indicates a tremendous growth in the number of diabetic patients in the last few years. It is not only a problem in the developed countries, but also in developing countries across the world [1].

Diabetic retinopathy (DR) is one of the complicated stage of diabetes mellitus. The retinopathy refers an undesirable change in the retina. A narrow layer of sensitive tissue that develops at the backside of patient's eye. It happens due to the longstanding of diabetes, and due to, the blood flow pathway may get obstructed. It does not make any

© Springer International Publishing AG 2018
D. J. Hemanth and S. Smys (eds.), *Computational Vision and Bio Inspired Computing*,
Lecture Notes in Computational Vision and Biomechanics 28,
https://doi.org/10.1007/978-3-319-71767-8_84

problem with normal vision in the starting stages. It creates problem only when it reaches a final phase. Therefore, regular check up of eye is very important for the detection of this disease. To build an automatic system which is able to perform the diagnosis task needs the usage of digital retinal images.

The DR can be treated well in the first stages. But in the later stages, it becomes complicated. Vision loss will occur and the cost of treatment will also be high. This paper describes an automatic system that can aid in the diabetic retinopathy detection. As the number of diabetes affected people is increasing day by day, the need for automated detection methods are also necessary. To automatically detect diabetic retinopathy and also for the mass screening of diabetic retinopathy, a computer has to interpret and analyze digital images of the retina. The first computerized approaches to classify retinal MAs was described by Laÿ [2].

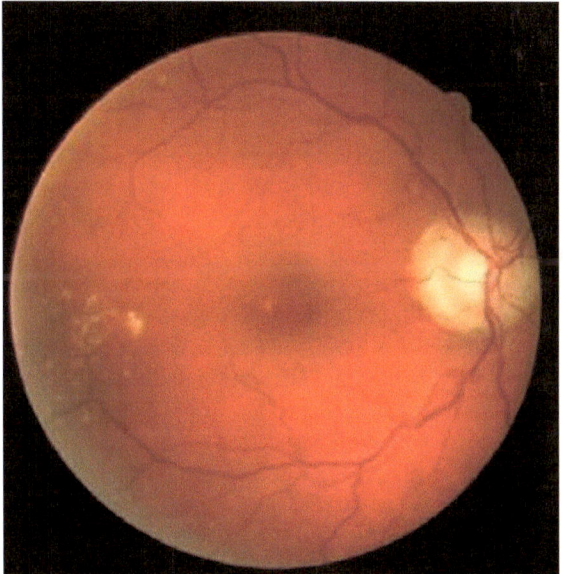

Fig. 1 Presence of MAs in retina

Figure 1 shows retina with the presence of MAs. Automated screening techniques for micro aneurysm detection have importance in reducing the treatment cost and time. The screening of diabetic patients for DR can reduce the risk of vision loss in the patients. Image processing techniques can be used to reduce noises in the input image. Also the Optic Disks and blood vessels can be extracted from the retinal images. Various techniques have been developed so far for the detection of micro aneurysms. Most of the techniques are based on the image processing techniques and uses retinal images from online databases like ROC and E-ophtha ex.

The rest of the paper is organized as follows: related works is discussed in Sect. 2, the proposed methodology and subsequent feature processing, classification steps are discussed in Sect. 3. The results of the proposed method are presented in Sect. 4 and the paper concludes in Sect. 5.

2 Literature Survey

In 2011, Hatanaka et al. [3] used double pass filter for detecting the presence of microaneurysms and exudates in the retina. The first step includes the conversion of input image into RGB color space. Then pixel values are compared with the neighboring pixel values. Green channel is selected because, MAs can be seen as bright objects in green channel. Edges are removed and there is chance that some MAs may be present on the blood vessels also. These are removed by extracting the blood vessels from the image. Once the features are extracted, MAs are classified using 3 layered ANN classifier. Drawbacks contains detection of many false positives on the capillaries.

Giancardo et al. [4] in 2011 proposed a method for detecting diabetic retinopathy. Here, a radon based transform is used for detecting MAs. Gaussian filters are used for normalizing the image in the preprocessing step and the green channel of the input image is used here. Feature vectors are computed with radon transformation and SVM classifier is used for classification. The simple training step the advantage of this method and also segmentation of vessels are not needed.

Antal and Hajdu [5] proposed a method in 2012. The preprocessing techniques used in this method is a combination of previously existing methods and some new methods. Candidate extraction algorithms are used as a second step. An approach known as ensemble creation is also used here. Ensemble creation aids for the selecting the best combination.

Lazar and Hajdu [6] proposed a method which detects MAs by the use of directional cross sectional profiles. Local Maximum Regions (LMR) are computed after the preprocessing. The cross sectional profiles of each local maximum pixels of the pre-processed images are drawn as the third step and then these profiles are analyzed. For each profile, peak detection step is done and a set of attributes such as the size, height, and shape of the peak are calculated. This feature set is given as the input to the naïve bayes classifier. It doesn't worked well with naive bayes classifier.

3 Proposed System

The proposed method starts with preprocessing. The input of the first step is the inverted green channel image of the original color fundus image. Many methods are there to detect the ROI, e.g., the one proposed by Gagnon et al. [7]. In almost all existing MA detection methods, the main input is the green channel image is considered because hemorrhages, MAs can be seen as bright objects in green channel.

Local maximum region extraction step is performed after the preprocessing. In the local maximum region extraction step local intensity maximum regions of the input image are found. In the third step, cross sectional profiles of LMA are drawn and then these profiles are analysed. It will be followed by the peak detection step. After the peak detection step, a set of candidates will be obtained. For each of the candidates, a set of statistical measures are calculated which will form the feature set of the ANN classifier. After the classification step, final candidates are only included by the thresholding step.

For the better understanding of the proposed method, its workflow is shown in Fig. 2.

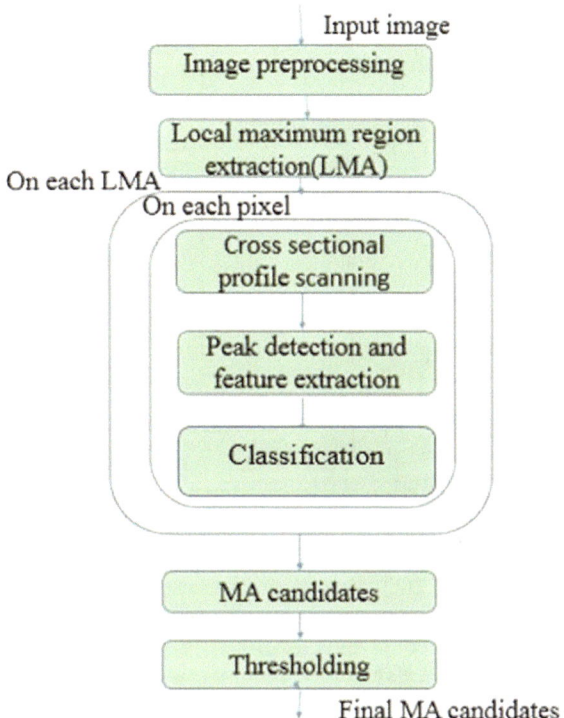

Fig. 2 Workflow of the proposed method

3.1 Preprocessing

Many fundus image dataset constitute images that are available in an irreversible compressed format which will result in the missing of small structures. Our proposed method depends on the local intensity distribution of micro aneurysms. So it is important to reduce the effect of noise. In order to decrease the effect of noise in such images, some image smoothening techniques are used here. In order to suppress the effect of noise, here we convolute the image with a Gaussian mask. For detecting the MAs some image preprocessing techniques are done prior to the original steps.

3.2 Local Maximum Region Extraction

Local maximum region extraction is the second step in the proposed method. In the preprocessed image, MAs are visible as local intensity maximum structures. Gaussian like intensity distribution of MA, conveys the idea that every MA region contains at least one regional maximum.

A local maximum region (LMR), of a gray scale image is an interconnected set of pixels with a fixed intensity value, such that each and every pixels in its neighborhood has a lower intensity value [8].

(a)   (b)

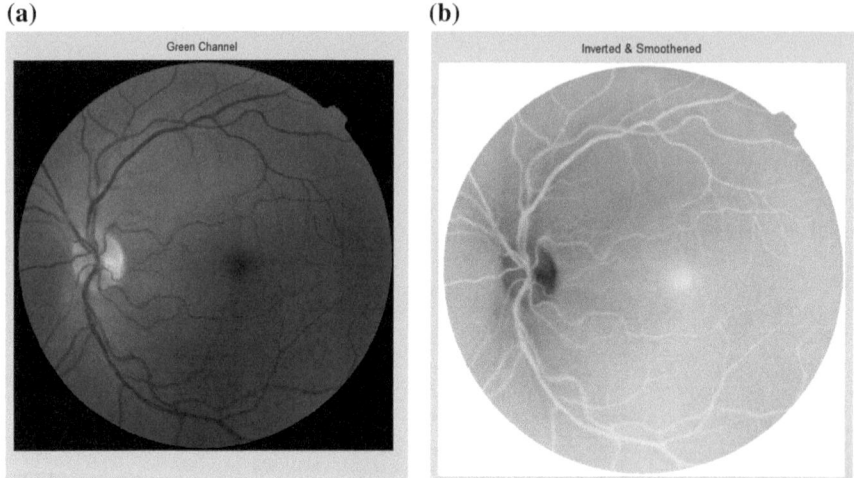

Fig. 3 **a** Green channel of input image, **b** inverted and smoothened image of input image

So we are taking into account only LMRs of the image. A breadth-first search algorithm which resembles similar to the one in [8] will give all the probable LMRS. The pixels are taken one by one, sequentially and then compared with its 8 neighbors. If there is a condition that all neighborhood pixels have a less intensity value than the current processing pixel, then the pixel itself is a LMR. The current pixel cannot be a maximum, If there is a condition that the current processing pixel has a neighbor pixel with greater intensity value.

3.3 Cross Sectional Profile Creation

In order to evaluate the neighbors of one maximum pixel in a MA candidate region, the examination of the intensity values along discrete line segments are taken. These profiles are taken along different orientations and best results are taken. The waveforms generated are recorded and they form the cross sectional intensity profiles. The cross sectional length was chosen as 31 based on the previous methodologies by Lazar and Hajdu [6]. For getting correct results, the difference between consecutive cross section is also important, so larger cross sections have to be chosen. Figure 3 shows the cross sectional profile of a candidate pixel along different orientation. If the pixel is a possible candidate, the cross sectional profile will show a peak at the center of length of the cross section. Similar procedure is described in [9] as a basic line detector.

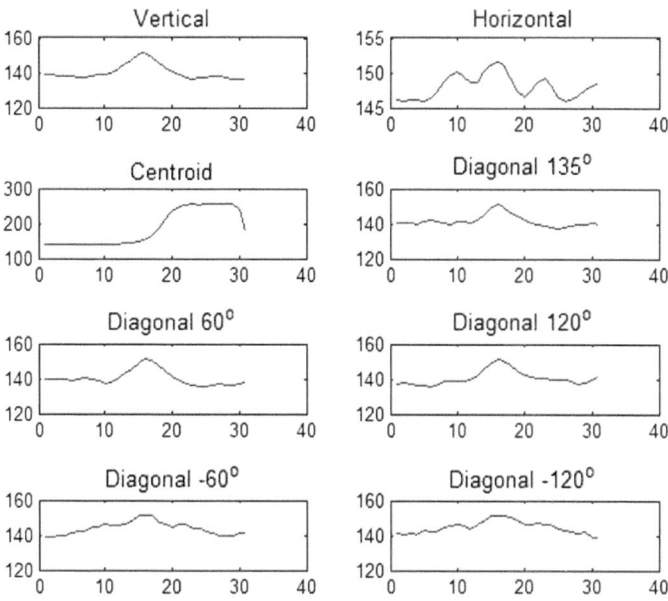

Fig. 4 Cross sectional profile of a candidate pixel

3.4 Peak Detection

The main intension of the peak detection step is to decide whether a peak is present at the center of the profile. For classification several properties of the peak have to be calculated in advance. Here P denotes the cross sectional profile and P[i] is its ith value. The peak detection is an important step in the proposed method as it finds out the presence of a peak on the cross sectional profiles of each profile. Figure 4 shows the ramps and peaks. At last, the feature set consists of a set of numerical measures that show how these values vary when the orientation of the cross-section is changes. After the peak detection step, presence of ramp is also checked. If it is there, then only the properties of the ramp are calculated by the next steps. The peak detection in one dimensional discrete data is a popular problem in many fields of s such as the automatic spectrometric evaluation [10, 11], chromatographic [12, 13].

A ramp is defined as a segment of the profile, where the sign of the difference between the consecutive values is nonzero. Once the ramps are found, it is tested to check if there is a full peak present at the center of the profile, i.e., the ramp left to the central index is increasing, and the one on the right is decreasing. The peak is represented by four values such as inc_s, dec_s, inc_e, dec_e. Among these, the first two values indicates the start index and end index of the increasing ramp and other two indicates the boundaries of the decreasing ramp, respectively.

When a peak is detected at the center of a profile, the following properties are to be calculated.

(1) The peak width is the difference between the peak start and Peak end indices of the peak: $w_{peak} = dec_e - inc_s$
(2) The top width is the size of the gap between the increasing and decreasing ramp: $w_{top} = dec_s - inc_e$
(3) The increasing ramp height: $h_{inc} = P[inc_e] - P[inc_s]$
(4) The decreasing ramp height: $h_{dec} = P[dec_s] - P[dec_e]$
(5) The increasing ramp slope: $s_{inc} = h_{inc}/(inc_e - inc_s)$
(6) The decreasing ramp slope: $s_{dec} = h_{dec}/(dec_e - dec_s)$
(7) The peak height is calculated as the difference between the intensity of the central pixel and a baseline that connects the start and end of the profile

$$h_{peak} = P[center] - (P[dec_e]/w_{peak} \cdot center - inc_s) + P[inc_s].$$

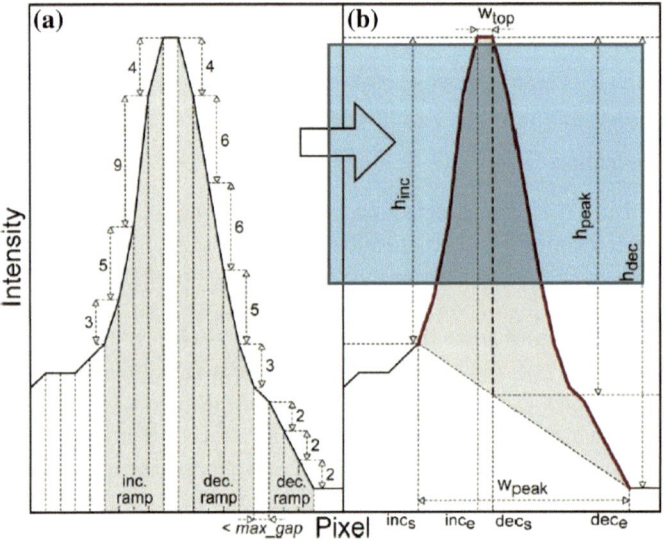

Fig. 5 a Increasing and decreasing ramps on a sample profile, **b** Result of the peak detection and the calculated peak measures

At the *center* of the profile, the peak is detected. The value of *peak width* denotes the pixel extension in the particular direction. Area in the structure is represented *top width* and the heights and slopes of the increasing and decreasing ramps shows the distinction from the neighbor pixels and the intensity transition. The *peak height* value shows the heights of the increasing and decreasing ramps (Fig. 5).

3.5   Feature Set Creation

When the above values have been calculated, next step is to create feature set for the classification. The peak properties are concluded in the following five sets. RHEIGHTS indicates decreasing and increasing height values of the ramp, RSLOPES indicates the ramp slope values., PWIDTHS, PHEIGHTS and TWIDTHS sets contain the peak width, peak height, and top height values. Let cv_T, σ_T, μ_T denotes the coefficient of variation, standard deviation and respective mean of the values in set. The feature set for the classification includes

$$F = \{\mu_{PWIDTH}, cv_{RHEIGHT}, \sigma_{TWIDTH}, \mu_{TWIDTH}, cv_{PHEIGHT}, \sigma_{PWIDTH}, \sigma_{RSLOPES}\}.$$

3.6   Classification

A classification procedure classifies the data into different classes based on some similarity measure. The items which belongs to a single class shares some common Here it is a two class problem, because the presence of only two set classes. The Images with microaneurysm which shows the presence of diabetic retinopathy and images without microaneurysm. For classification, artificial neural network is used. An ANN is defined by three types of parameters:

1. The interconnections between the neurons
2. Weight with respect to each interconnection (Weights are updated in the learning process).
3. The function that converts a neuron's weighted input to its output (known as activation function).

ANN uses data as input and allots them into the class that best fits to the training samples. Here nonlinear regression is uses. Consider Fig. 6 which shows the diagram of ANN i.e. input, hidden layer and output. The type of network which is used here is ELM.

The Extreme learning machine (ELM) is a feed forward neural network based supervised classifier which contains only a single hidden layer. Feed forward network consist of multiple layers. The first layer gets the network input. The last layer creates the networks output. Feed forward networks are used for any input to output mapping. A feed forward network with one hidden layer and enough neurons in the hidden can fit any finite input–output mapping problem [14]. The operation of this network include 2 phases

• Phase for learning
• Phase for classification.

In the learning phase, a particular pattern is fed to the inputs. At the time when it reaches the output layer, the pattern is transformed along the network path. Each unit in the output layer represents different class. The present outputs of the network are compared with the expected output of the pattern that are correctly classified. The output the unit with the correct category will have maximum output value, and the output values of the other output units are very small.

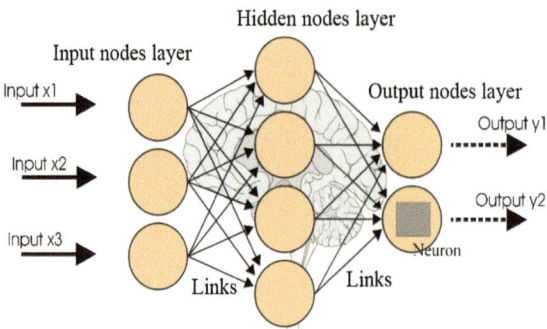

Input nodes layer

Hidden nodes layer

Input x1

Input x2

Input x3

Output nodes layer

Output y1

Output y2

Neuron

Links

Links

Fig. 6 structure of ANN

Based on this comparison, all the weights in each of the connections are modified. It is done in order to guarantee that, when this same pattern is fed as the input, the output should belong to the correct category.

4 Experimental Result

The dataset used is Retinal Online Challenge (ROC) database [15] and E-Ophtha database. The ROC database consists of 50 color retinal images for 50 color retinal images for retinal purpose and 50 images for testing. E-Ophtha is a database of color fundus images useful for research in Diabetic Retinopathy. E-Ophtha is well known data set for retinal images due to its collection of heterogenous retinal images. Performance of the classifier is analyzed by score and accuracy calculation of microaneurysms. The Eq. (1) is used to calculate the scores of MA candidates. Based on the score, the thresholding step is performed. The candidates with less scores are avoided from the MAs.

$$Score = \frac{\min_{PHEIGHT} \mu_{RSLOPES}}{1 + \sigma_{TWIDTHS} + \sigma_{PWIDTHS} + \sigma_{RSLOPES} + \sigma_{RHEIGHTS} + \sigma_{RPHEIGHTS}} \quad (1)$$

To cope up with the requirements of DR screening systems in real life, score values are assigned to the MA candidates that were classified as true MAs. The scores are assigned in such a way that stronger, more visible MAs get higher score. Non maximum suppression and thresholding are optional step and non-maximum suppression refers to the operation of selecting the point with the highest score from all candidates. Therefore, points with non-maximal score in a candidate region are neglected, and the output of the proposed method is a set of coordinates and the corresponding score values.

Table 1 shows the peak feature values of images 7, image 9 and image 16. The values obtained are shown in Table 1.

Table 1 Peak feature values of images in ROC dataset

Images	w_{peak}	w_{top}	h_{inc}	h_{dec}	s_{inc}	s_{dec}	h_{peak}
Image 7	9	2	7	8	1.6	1.76	7.3
Image 9	11	1	14	6	2.3	2.66	15.0
Image 16	4	2	7	8	3.5	4	7.5
Image 27	7	2	5	22	5	3.6	9.8

The performance is evaluated with True positives, True negatives, false positives and false negatives

- True Positives (TP): MA regions which are correctly classified by the classifier.
- False Positives (FP): Non-MA regions which are wrongly classified as lesion regions by the classifier.
- True Negatives (TN): Non-MA regions which are correctly classified by the classifier.
- False Negatives (FN): MA regions which are wrongly classified as non-MA regions by the classifier.

The specificity sensitivity and accuracy can be calculated as

Specificity = TN/(TN + FP)
Sensitivity = TP/(TP + FN)
Accuracy = (TP + TN)/(TP + TN + FP + FN).

These values, the accuracy of the proposed method can be compared with other existing methods. The testing parameters are computed on ROC dataset (Table 2) and also computed on E-ophtha EX dataset (Table 3).

Table 2 Performance analysis in ROC dataset

Classifier	TP	FN	FP	TN	Sensitivity	Specificity	Accuracy
NB	26	4	4	8	86.6	66.6	80.9
ANN	28	2	5	7	93.3	58.3	83.3

Table 3 Performance analysis in E-ophtha EX dataset

Classifier	TP	FN	FP	TN	Sensitivity	Specificity	Accuracy
NB	25	5	3	8	83	75	80.9
ANN	28	3	2	7	90.6	80	88.09

The first clinical data set used for experiments consists of 100 fundus images from ROC dataset. For the training stage 50 fundus images were used (31 with DR effected and 19 without DR). The system performance was tested with 50 images. The total accuracy obtained was 88.09%, the sensitivity was 90.6%, whereas the specificity was 80%.

5 Conclusion

This proposed method uses a cross sectional profile based approach for diabetic retinopathy detection. From the preprocessed image, local maximum regions are selected. Cross sectional profiles of the selected pixels are recorded and analyzed. The peak detection step ends by giving a set preprocessed image. Results shows that performance wise it improves the result compared to the existing method.

The proposed method could be exploited in other medical image processing problems, such as abnormality detection that includes the recognition circular structures in an image.

The obtained results shows that the system meets the requirements needed for the screening of DR and can help the ophthalmologists. It also saves the money for the screening process. The large amount of time needed in laboratories can also be avoided.

References

1. Wild, S., Roglic, G., Green, A., Sicree, R., King, H.: Global prevalence of diabetes estimates: estimates for the year 2000 and Projections for 2030. Diab. Care **27**(5), 1047–1053 (2004)
2. Laÿ, B.: Analyse automatique des images angiofluorographiques au cours de la retinopathie diabetique. Ecole Nationale Superieure des Mines de Paris, Centre de Morphologie Mathematique, Paris, France (1983)
3. Hatanaka, Y., Inoue, T., Okumura, S., Muramatsu, C., Fujita, H.: Automated microaneurysm detection method based on double-ring filter and feature analysis in retinal fundus images. In: 25th International Symposium on Computer-Based Medical Systems (CBMS), 2012, pp. 1–4. IEEE (2012)
4. Giancardo, L., Meriaudeau, F., Karnowski, T.P., Li, Y., Tobin, K.W., Chaum, E.: Microaneurysm detection with radon transform-based classification on retina images. In: Annual International Conference of the IEEE Engineering in Medicine and Biology Society, EMBC, 2011, pp. 5939–5942. IEEE (2011)
5. Antal, B., Hajdu, A.: An ensemble-based system for microaneurysm detection and diabetic retinopathy grading. IEEE Trans. Biomed. Eng. **59**(6), 1720–1726 (2012)
6. Lazar, I., Hajdu, A.: Retinal microaneurysm detection through local rotating cross-section profile analysis. IEEE Trans. Med. Imaging **32**(2), 400–407 (2013)
7. Gagnon, L., Lalonde, M., Beaulieu, M. and Boucher, M.C.: Procedure to detect anatomical structures in optical fundus images. In Proceedings of the SPIE, vol. 4322, pp. 1218–1225 (2001)
8. Cree, M.J., Olson, J.A., McHardy, K.C., Sharp, P.F., Forrester, J.V.: A fully automated comparative microaneurysm digital detection detection system. Eye **11**, 622–628 (1997)
9. Ricci, E., Perfetti, R.: Retinal blood vessel segmentation using line operators and support vector classification. IEEE Trans. Med. Imaging **26**(10), 1357–1365 (2007)
10. Jarman, K.H., Daly, D.S., Anderson, K.K., Wahl, K.L.: A new approach to automated peak detection. Chemometr. Intell. Lab. Syst. **69**(1), 61–76 (2003)
11. Coombes, K.R., Tsavachidis, S., Morris, J.S., Baggerly, K.A., Hung, M.C., Kuerer, H.M.: Improved peak detection and quantification of mass spectrometry data acquired from surface-enhanced laser desorption and ionization by denoising spectra with the undecimated discrete wavelet transform. Proteomics **5**(16), 4107–4117 (2005)

12. Vivó-Truyols, G., Torres-Lapasió, J.R., Van Nederkassel, A.M., Vander Heyden, Y., Massart, D.L.: Automatic program for peak detection and deconvolution of multi-overlapped chromatographic signals: part I: peak detection. J. Chromatogr. A **1096**(1), 133–145 (2005)
13. Peters, S., Vivo-Truyols, G., Marriott, P.J., Schoenmakers, P.J.: Development of an algorithm for peak detection in comprehensive two-dimensional chromatography. J. Chromatogr. A **1156**(1), 14–24 (2007)
14. Huang, G.-B., Zhu, Q.-Y., Siew, C.-K.: Extreme learning machine: theory and applications. Neurocomputing **70**(1), 489–501 (2006)
15. Niemeijer, M., Van Ginneken, B., Cree, M.J., Mizutani, A., Quellec, G., Sánchez, C.I., Zhang, B., Hornero, R., Lamard, M., Muramatsu, C., Wu, X.: Retinopathy online challenge: automatic detection of microaneurysms in digital color fundus photographs. IEEE Trans. Med. Imaging **29**(1), 185–195 (2010)

An Application of Image Processing Technique for Compression of ECG Signals Based on Region of Interest Strategy

T. Shreekanth$^{(\boxtimes)}$ and R. Shashidhar

Department of Electronics and Communication, Sri Jayachamarajendra College
of Engineering, Mysore, India
{shreekanth_t,shashidhar.r}@sjce.ac.in

Abstract. In this paper, a novel Region of Interest (ROI) based 2-Dimensional
(2D) compression of ECG signals using JPEG2000 compression standard is
proposed. Because of its high efficiency JPEG2000 is the global benchmark for
compression of stationary images. This work is to illustrate that the JPEG2000
compression technique is not only restricted to compress images but also it can
be applied to compress ECG signals. First the one dimensional ECG signal is
transformed to 2D representation or image to explore the correlation among the
samples and among the beats. This necessitates few steps that includes QRS
detection and arrangement of QRS complex at relative position, Period sorting,
mean extension, period normalization, amplitude normalization and locating
ROI. Then this resultant 2D ECG data array is compressed using JPEG2000.
The ROI region is extracted by applying the Otsu thresholding and the vertical
projection profile on the resultant 2D ECG data. The core idea in this work is to
compress the ROI region at low compression rate and non-ROI region at high
compression rate. The proposed method is evaluated on the selected data from
MITs Beth Israel hospital and it was conceded that this method surpasses some
of the prevailing methods in the literature by attaining a higher Compression
Ratio (CR) and moderate Percentage root- mean square difference (PRD).

Keywords: ECG · Compression · Connected component labelling · CR
JPEG2000 · PRD · ROI

1 Introduction

The ECG is a standard biomedical signal used for detection and diagnosis of any
abnormalities related to heart. Real time data of the obtained signals needs to be
transmitted over telephone lines which requires high bandwidth. Hence, an efficient
technique to store and retrieve longstanding recording (24 * 7) and a serviceable real
time solution are important considerations in the modern medical applications. The
compression approaches can be widely categorized into three classes: time-domain
approach [1, 2], parametric approach [3], and transform domain approach [4–7]. The

© Springer International Publishing AG 2018
D. J. Hemanth and S. Smys (eds.), *Computational Vision and Bio Inspired Computing*,
Lecture Notes in Computational Vision and Biomechanics 28,
https://doi.org/10.1007/978-3-319-71767-8_85

typical transform domain approach involves the use of diverse wavelet-based compression methods. The marked reason that makes the wavelets appropriate for data compression applications is its time-frequency localization property. In recent times wavelets have been in use for many number of data compression applications producing optimum results. The JPEG 2000 image compression format is the conventional successor of JPEG having added functionalities and complexity than JPEG. Former uses the wavelet as the transform, whereas later uses the DCT. In the past most of the compression methods had presented intra-beat correlation among consecutive samples following which in the current times inter-beat correlation between heart beats is accepted as a solution for obtaining the high compression ratio [8–13].

In this paper a 2-D approach for compression of ECG signals is presented, which take advantage of the redundancies that exist within the beat and between the adjacent. ROI coding is done in order to compress the region with particular clinical significance at lower compression rate and the rest of the region at higher compression rate. ROI coding is the inherent property of JPEG2000 codec [14]. In this method the Region of Interest is the QRS complex i.e. the focus is only on retaining the morphology of QRS complex. The clinical significance of QRS complex is-its prolonged duration indicates bundle branch block, its increased amplitude indicates cardiac hypertrophy. Patient's data pertaining to the above clinical significance can be compressed and reproduces effectively using this approach.

The reminder of the paper is categorized as follows. Section 2 provides in detail the proposed algorithm. The metrics used to evaluate the proposed algorithm is described in Sect. 3. Section 4 provides the results obtained by evaluating the proposed algorithm on the MIT database and comparison of the results of proposed method with that of the existing techniques in the literature and finally Sect. 5 concludes the paper.

2 Proposed Method

Heart beat signals usually exhibit notable similarity among adjacent beats with short-term correlation between subsequent samples. This examination establishes that, adopting temporal beat alignment method leads to ECG compression methods that are highly efficient. Figure 1 depicts the process flow diagram of the proposed method and is executed through the aforementioned steps:

(1) Detection of QRS complex from 1D ECG signal.
(2) Construction of 2D ECG array from 1D signal.
(3) Sorting of beat periods.
(4) Equalization of length through mean extension.
(5) Normalization of Amplitude and Period.
(6) ROI Mapping.
(7) Compression and Decompression using JPEG2000.
(8) Construction of 1D ECG signal from 2D array.

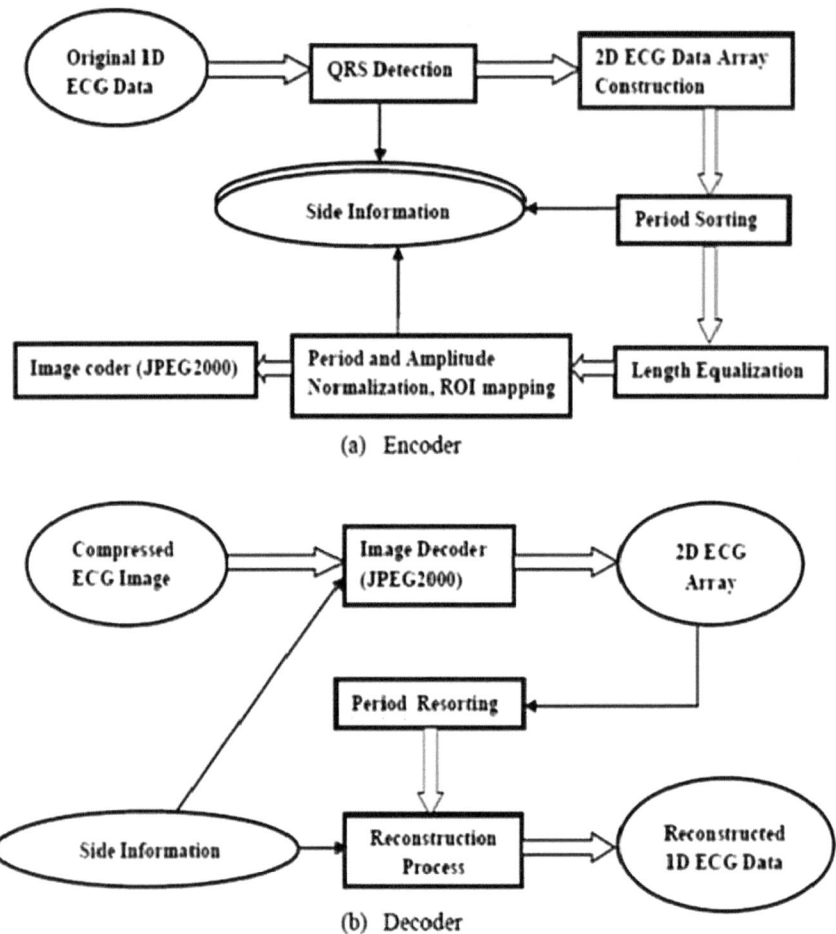

Fig. 1 Process flow of the proposed method

1. Detection of QRS complex from 1D ECG signal

In order to exploit the reciprocity between the successive beats the 1D input ECG data needs to be fragmented and arranged properly. In order to perform this, input 1D signal has to be QRS detected as depicted in Fig. 3a before they can be segmented and aligned based on the result of the QRS detection. Performance of the proposed algorithm will be dependent on the accuracy of the QRS detection method [15].

2. Construction of 2D ECG array from 1D signal

After locating the QRS complex, the ECG is cut periodically and aligned accordingly to form a 2D array. For the sake of removing the redundancies between adjacent heart-beats few pre-processing steps are performed namely: mean removal, amplitude normalization and period normalization. Figure 3b illustrates a sample of cut and aligned beat ECG data array. Gray scale mapping of the same is shown in Fig. 3c. Acceptable alignment of QRS complex is necessary to utilize the beat to beat correlation maximally. To circumvent sharp boundaries and transitions in 2D arrays, existing 1D ECG signal is cropped at 130th sample, that is, prior to each R peak. (A smooth section determined from human physiology [16]) Each heartbeat length is maintained and transmitted to decoder as an additional information. This helps in signal reconstruction. Also, ECG cycle periods are differentially encoded and sent across. Each heart beat pattern is unique and thus the length varies. To facilitate this fact, adequate number of zeros are padded at the end of each heartbeat data sequence.

3. Sorting of beat periods

The matrix obtained from the QRS detection and alignment displays the inter-beat correlation of the original ECG signal. This is followed with certain periodic irregularities which possess a big challenge for 2D compression algorithms. In order to maximize inter-beat correlation and simultaneously obtain high performance, a period sort is performed that sorts the heart beat sequence depending on their periods in an ascending or descending fashion. This is a unique and innovative method of compression as it brings down the period difference between the adjacent heart-beats, resulting in improved CR and PR. This step is redundant for regular ECG as the period difference between the heart beats are negligible for the current purpose, thus period sorting escalates the compression overhead. But, abnormalities in ECG signals are quite significant and period sort effectively addresses this issue by exposing the 2D correlation structure as shown in Fig. 3d.

4. Equalization of length through mean extension

This step is required to equalize the length of each heartbeat fragment to form an acceptable 2D array. There are different ways of extending the signal some of which include zero-extension that pads the short segment with zeros, zero-order extension which extends the segment by concatenating its last element and mean extension which pads short segments with last samples of heart beat sequences. It is as shown in Fig. 3e.

5. Normalization of amplitude and period

It is evident from the obtained data that period of each heart-beat is different which results in variable number of data points in the 2D array. In order to explore the inter-beat dependencies using JPEG2000, normalizing of the data points is done to accommodate in a constant number of columns for each row [9].

Let $x_m = [x_m(1) \, x_m(2) \ldots x_m(N_m)]$ denote the m-th ECG cycle. Then the period-normalized ECG cycle $y_m = [y_m(1) \, y_m(2) \ldots y_m(N)]$ is found using

$$y_m(n) = x_m(t') \tag{1}$$

where $x_m(t')$ is an interpolated representation of the samples of $x_m(n)$, and

$$t' = \frac{(n-1)(N_m-1)}{N-1} + 1 \tag{2}$$

The period of the m-th ECG cycle is represented by N_m and the normalized period by N. To compute $x_m(t')$ Cubic-spline interpolation [10] is employed. N is a constant. The period-normalized data analogous to the data in Fig. 3e is depicted in Fig. 3f.

By dividing each sample value within the beat by the maximum amplitude value of that beat, amplitude normalization ensures that each beat will have the highest amplitude as unity. The normalized period in this experiment, was selected as

$$N = [\alpha \times M] \tag{3}$$

where M is defined as the mean of the detected periods for the given record, and α is used to calculate the number of columns in the period normalized image. It is noticed that the value of α influences the compression performance. For low CR's, $\alpha = 1.0$ seems to be a good choice, whereas for high CR's smaller values of α results in superior performance. The influence of α on the compression performance is as shown in Fig. 2.

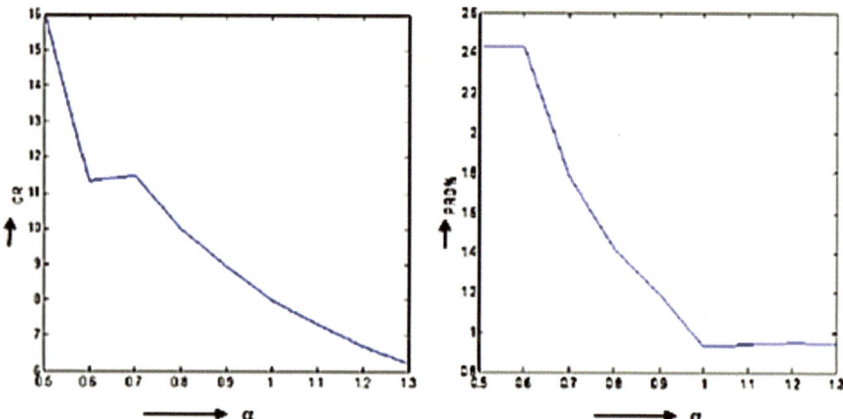

Fig. 2 Effects of α on performance for different CR's and PRD's

6. ROI mapping

Following the equalization of the length of the beats through mean extension and normalizing the period of the 2D ECG array, the ROI mapping is done in order to select only those data that falls in the QRS region. This is done by applying the Otsu thresholding to the 2D ECG data array as shown in Fig. 3g and then finding the vertical projection profile of the thresholded image as depicted in Fig. 3h. Vertical projection profile is nothing but the sum of all the pixels in the column direction [17]. This width of the vertical projection profile locates the Region of Interest which corresponds to the QRS complex as shown in Fig. 3i. This is followed by compressing the QRS region at lower compression rate and the remaining part of the image i.e. Non QRS region, at slightly higher compression rate. Compression rate of both the regions are provided as side information to the image decoder. This results in very high compression ratio, also the morphology of the QRS part is defined very well in the reconstructed signal. ROI mapping of QRS is as shown in Fig. 3j which is indicated by a rectangular box.

7. Compression and decompression using JPEG2000

For this work, Kakadu V64_demo_apps JPEG2000 codec is used with default parameters for compression and code block sizes of 64 × 64, 5 wavelet transform levels, 9–7 filters, etc. The only parameter that changes across records is the image dimension. This dimension is derived from the ECG cycle period data that is received by decoder as side information. The reported Compression Ratios (CR) are from actual compressed files and include all side information required by the decoder.

8. Construction of 1D ECG signal from 2D array

In order to obtain the original signal from the segmented data, first the JPEG2000 decoded data is recorded and restored at its original signal. Now, the restored matrix is period de-normalized to its original data. The number of samples are picked up from the know heart-beat length and multiplied to its corresponding amplitude and mean value. Concatenating all the heart beats data in the form of a series which gives us back the original signal.

3 Performance Measure

The metrics used to evaluate the performance of the proposed technique are namely: Compression Ratio (CR) and Percentage Root-mean-square Difference (PRD). The CR is determined using Eq. 1.

$$CR = \frac{Total\ number\ of\ bits\ original\ signal}{Total\ number\ of\ bits\ in\ Reconstructed\ signal} \tag{1}$$

The PRD is the most commonly used metric for evaluating the quality of any compression technique. The PRD is considered here to examine the difference between the original and reconstructed signal, which is determined using Eq. 2.

$$PRD = \sqrt{\frac{\sum_{i=1}^{n}\left(x_{org}(i) - x_{rec}(i)\right)^2}{\sum_{i=1}^{n} x^2_{org(i)}}} \times 100 \tag{2}$$

where x_{org} represents the original signal, x_{rec} represents the reconstructed signal and n is the number of samples in the data.

1000   T. Shreekanth and R. Shashidhar

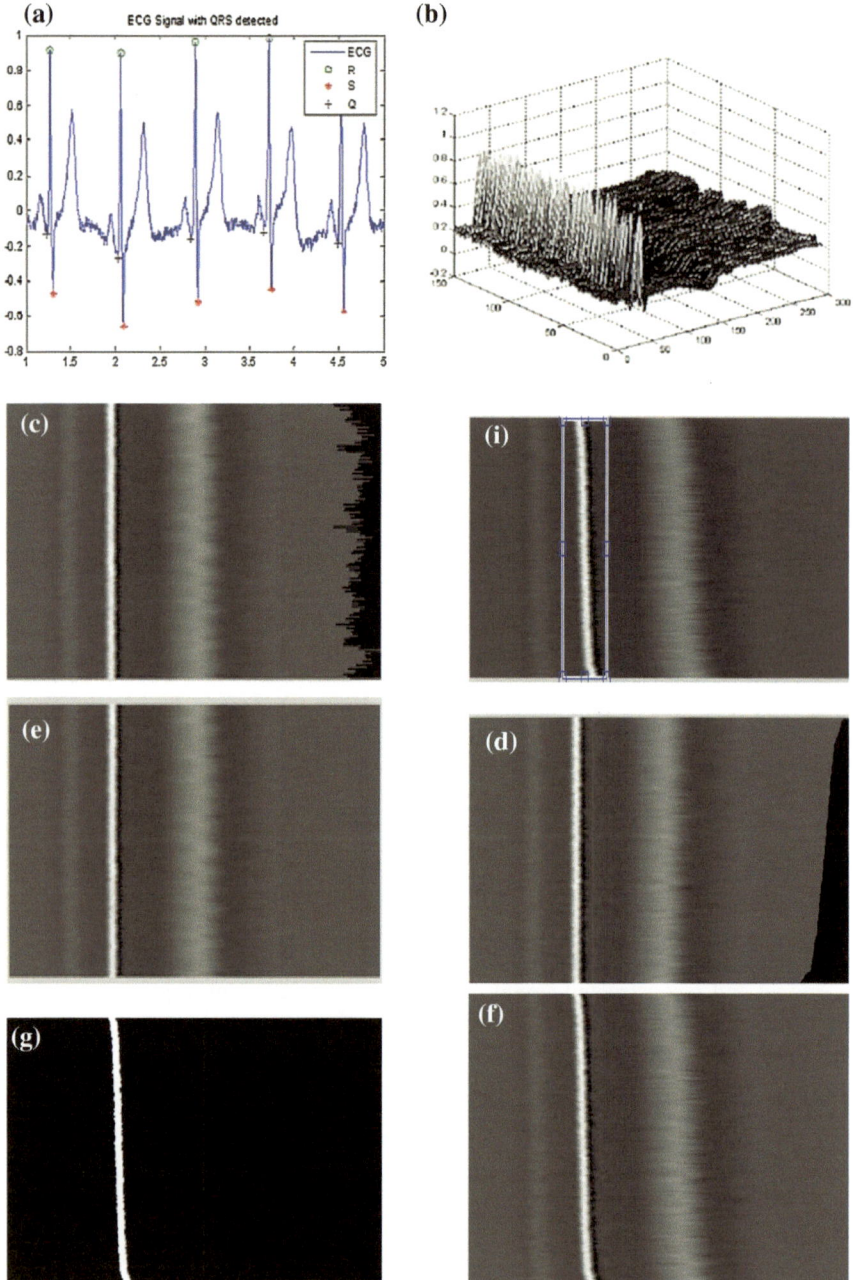

Fig. 3 Construction of a 2D ECG data array. **a** QRS recognized original ECG signal, **b** cut and aligned ECG data, **c** mapping of a 2D ECG gray scale data before period normalization and period sorting, **d** a 2D ECG data array after period sorting, **e** a 2D ECG data array after mean extension, **f** a 2D ECG data array after period normalization, **g** a 2D ECG data array after thresholding using otsu's method, **h** a vertical histogram profile of **g**, **i** detection of ROI cropping region and **j** rectangular box indicates the QRS part as the region of interest

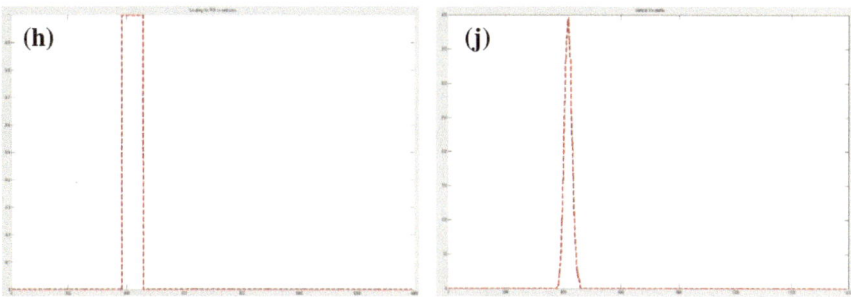

Fig. 3 (*continued*)

4 Experimental Results and Performance Comparision

The above algorithm was implemented to process a 5 min data obtained from the MIT-BIH arrhythmia database and the ECG was sampled at a rate of 360 Hz, 11 bits/s [18]. Varying patterns of ECG signals were considered to evaluate recoverability of the reconstructed waveforms including record numbers 100, 102, 103, 114, 117 and 220. These records contain signals of different rhythms and morphologies. Reported compression ratios are from actual compressed file sizes and PRDs of decompressed files.

In Table 1, shows the PRD comparison of different coding algorithms and in Table 2 shows the performance comparison of the proposed algorithm. Also, the graph depicting the performance comparison of the proposed method with the other algorithms in the literature is shown in Fig. 4. In addition to good results in CR vs. PRD challenge, the significant advantage of the proposed method is that generally vital QRS complexes are recovered with a good fidelity. In addition the distribution of reconstruction error is almost uniform, and thus the morphology of all the components are preserved so, the clinical performance of the method can be considered good because it retains more clinical relevant information with high fidelity.

Clinical performance of this method is considerably good for the reason that it retains high fidelity information that is clinically relevant. Also, morphology of components is preserved as the distribution of reconstruction error is uniform. Figures 5 and 6 show 1000 samples of original signal, reconstructed and error signal pertaining to record number 102 and 117. This reveals the visual quality of reconstructed signals in order that the error signal is uniformly distributed. Main advantage of proposed method is smoothing of background noise along with preservation of characteristic features of signal.

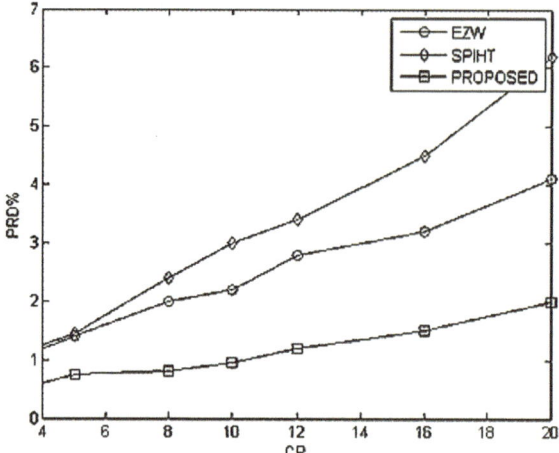

Fig. 4 Comparison of performance of proposed method with other algorithms in the literature

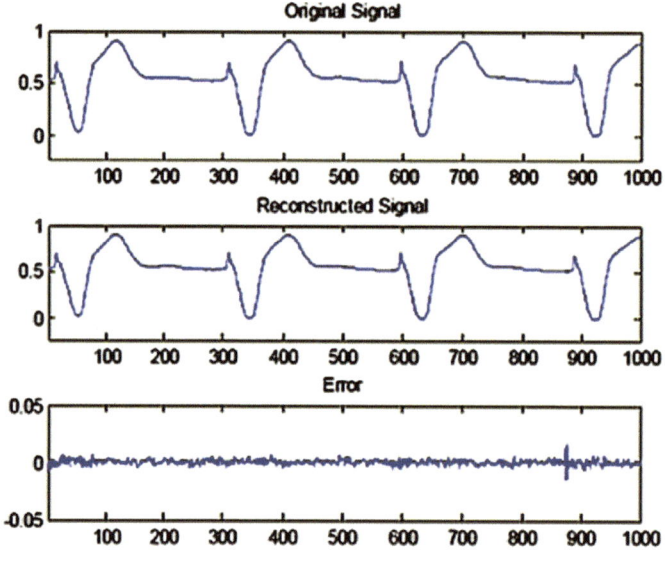

Fig. 5 Record 102, CR = 10, PRD = 0.5%

Table 1 Performance comparison (PRD) of different coding algorithms

Algorithm	Record no.	CR	PRD (%)	Sampling frequency (Hz)	Resolution (bit/sample)
SPHIT	117	8:1	1.18	360	11
Hilton	117	8:1	2.6	360	11
Djohn	117	8:1	3.9	360	11
Lu et al.	117	8:1	1.18	360	11
Proposed	117	10:1	0.6	360	11

Table 2 Performance comparison (PRD) of the proposed algorithm with different records

Record no.	CR	PRD (%)	Sampling frequency(Hz)	Resolution (bit/sample)
117	8:1	0.6	360	11
114	8:1	0.4	360	11
102	8:1	0.5	360	11
100	8:1	1.2	360	11
220	8:1	0.65	360	11

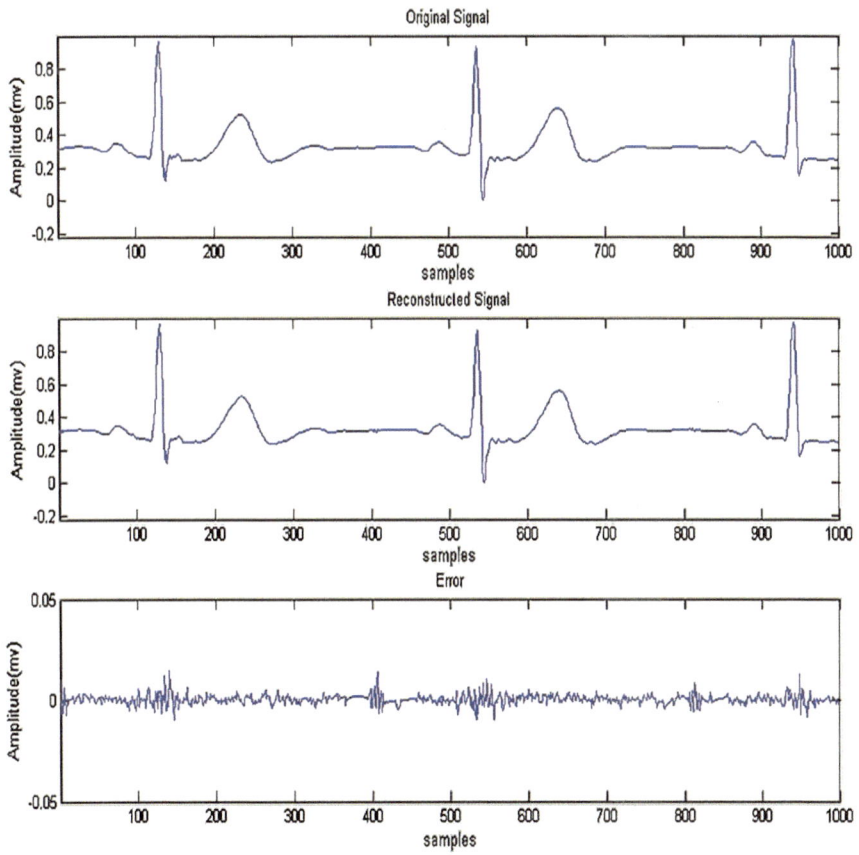

Fig. 6 Record 114, CR = 10, PRD = 0.6%

5   Conclusion

In this work it has been illustrated that apart from images, the JPEG2000 image compression standard can also be utilized to compress the ECG signals. The performance of the proposed algorithm is evaluated by compressing various records from the MIT-BIH arrhythmia database and the results are compared with the existing methods in the literature. The results depicts that the presented algorithm performs reasonably well when compared to other methods found in the literature. Also the 2D-ECG compression with ROI coding is superior to the 1D ECG compression methods, despite of its more computation time to produce 2D-ECG. Future work in this directions will include utilizing the lossless compression capability of JPEG2000 and using the multi-component capabilities of JPEG2000 to compress multichannel ECG data.

References

1. Cox, J., Fozzard, H., Nolle, F.M., Oliver, G.: AZTEC: a preprocessing system for real-time ECG rhythm analysis. IEEE Trans. Biomed. Eng. **15**, 128–129 (1968)
2. Jalaleddine, S., Hutchens, C., Strattan, R., Coberly, W.: ECG data compression techniques— a unified approach. IEEE Trans. Biomed. Eng. **37**, 329–343 (1990)
3. Zigel, Y., Cohen, A., Abu-ful, A., Wagshal, A., Katz, A.: Analysis by synthesis ECG signal compression. Comput. Cardiol. **24**, 279–282 (1997)
4. Cetin, E., Koymen, M., Aydin, M.C.: Multichannel ECG data compression by multirate signal processing and transform domain coding techniques. IEEE Trans. Biomed. Eng. **40**, 495–499 (1993)
5. Djohan, T., Nguyen, Q., Tompkins, W.J.: ECG compression using discrete symmetric wavelet transform. In: Proceedings of 17th International IEEE Medicine and Biology Conference (1995)
6. Bradie, B.: Wavelet packet-based compression of single lead ECG. IEEE Trans. Biomed. Eng. **43**, 493–501 (1996)
7. Hilton, M.L.: Wavelet and wavelet packet compression of electrocardiograms. IEEE Trans. Biomed. Eng. **44**, 394–402 (1997)
8. Lu, Z., Kim, D.Y., Pearlman, W.A.: Wavelet compression of ECG signals by the set partitioning in hierarchical trees algorithm. IEEE Trans. Biomed. Eng. **47**, 849–856 (2000)
9. Wei, J.-J., Chang, C.-J., Chou, N.-K., Jan, G.-J.: "ECG data compression using truncated singular value decomposition. IEEE Trans. Inf. Technol. Biomed. **5**, 290–299 (2001)
10. Bilgin, M., Marcellin, W., Altbach, M.I.: Compression of electrocardiogram signals using JPEG2000. IEEE Trans. Consum. Electron. **49**(4), 833–840 (2003)
11. Lee, H., Buckley, K.M.: ECG data compression using cut and align beats approach and 2-D transform. IEEE Trans. Biomed. Eng. **46**(5), 556–564 (1999)
12. Chou, H.H., Chen, Y.J., Shiau, Y.C., Kuo, T.S.: An effective and efficient compression algorithm for ECG signals with irregular periods. IEEE Trans. Biomed. Eng. **53**(6), 1198–1205 (2006)
13. Tai, S.C., Sun, C.C., Tan, W.C.: 2-D ECG compression method based on wavelet transform and modified SPIHT. IEEE Trans. Biomed. Eng. **52**(6), 999–1008 (2005)
14. Aboy, M., Crespo, C., McNames, J., Bassale, J., Jenkins, L., Goldstein, B.: A biomedical signal processing toolbox. In: Proceedings of Biosignal 2002 (2002)

15. Pan, J., Tompkins, W.J.: A real-time QRS detection algorithm. IEEE Trans. Biomed Eng, BME **32**, 230–236 (1985)
16. Vander, J., Sherman, J.H., Luciano, D.S.: Human Physiology. McGraw-Hill, New York, chap. 14, pp. 393–472 (1994)
17. Gonzalz, R.C., Woods, R.E.: Digital Image Processing, 2nd edn. Prentice Hall, Upper Saddle River, NJ (2002).
18. Goldberger, A.L., Amaral, L.A.N., Glass, L., Hausdorff, J.M., Ivanov, P.Ch., Mark, R.G., Mietus, J.E., Moody, G.B., Peng, C.-K., Stanley, H.E.: PhysioBank, physioToolkit, and physioNet: components of a new research resource for complex physiologic signals. Circulation **101**(23), e215–e220 (2000)

Face Authentication Using Thermal Imaging

S. Athira$^{(\boxtimes)}$ and O. V. Ramana Murthy

Department of Electrical and Electronics Engineering, Amrita School of
Engineering, Coimbatore, Amrita Vishwa Vidyapeetham, Coimbatore, India
athirasasdhar@gmail.com

Abstract. The objective of the paper is to develop an algorithm for face
authentication using thermal imaging. Thermal imaging makes use of thermal
cameras to detect infrared radiation emitted by objects. Thermal images are
especially useful for face detection and authentication because of its low sensitivity to illumination changes and its ability in detecting disguises. The
objective of the project is achieved through three stages. In the first stage, an
existing algorithm is verified by implementing it on the benchmark dataset.
Currently existing face authentication algorithms have only been verified by
implementing on datasets having plain background i.e. images without any
background objects. There is a need to investigate the efficiency of face
authentication algorithms on realistic scenarios. So, in the second stage an effort
had been put forth to create a new dataset of images with realistic background
and analyse the performance of various existing face authentication algorithms
on the new dataset.

Keywords: Face authentication · Thermal imaging · Dataset extension

1 Introduction

Identification systems has a lot of applications spanning from governmental projects
like border control or criminal identification to civil purposes like e-commerce, network access or transport. Out of the existing identification systems, biometrics has
become very popular. It is the most reliable identification system since it depends upon
the physiological characteristics of a person which is unique for every person in this
planet and cannot be stolen or swapped. Different physiological features can be used as
biometrics such as iris, retina, face etc.

Face authentication is one of the most popular biometric identification systems. It is
more direct, user-friendly and convenient compared to other methods. There are many
existing face authentication systems in visible spectrum. But such systems may fail
when there is a variation in illumination. Further, such systems can be easily spoofed
by using disguises or by using photographs. Such systems cannot distinguish a live face
from a non-live face. A secure system needs liveness detection to guard against such
spoofing.

© Springer International Publishing AG 2018
D. J. Hemanth and S. Smys (eds.), *Computational Vision and Bio Inspired Computing*,
Lecture Notes in Computational Vision and Biomechanics 28,
https://doi.org/10.1007/978-3-319-71767-8_86

These challenges can be met with the help of thermal imaging. Thermal imaging makes use of thermal cameras to detect infrared radiation emitted by objects. Since thermal imaging depends on the heat emitted there is no issue of illumination variation. Only a live face will emit thermal radiation. So liveness detection can be easily carried out with thermal imaging. Also the radiation emitted by human face is at a different range compared to other objects thus aiding in detecting disguises.

Previous works done by researchers indicate that many face authentication algorithms such as Principal Component Analysis (PCA) [1], Local Binary Pattern (LBP), Histogram of Oriented Gradients (HOG) etc. which gives very accuracy in face authentication in visible spectrum works very well in the case of thermal images too. However, they were all verified on datasets containing non-realistic backgrounds. There are no other background objects. But that may not always be the case. There may be other objects with similar spectral range in the background. The work presented here analyses the performance of various face authentication algorithms on a newly created database with images having realistic background.

The paper is organized as follows: Sect. 2 provides an overview on related work done previously by other authors. Section 3 explains the method implemented to achieve the objective. Section 4 presents the results obtained. Section 5 gives the conclusion reached and future scope of the work.

2 Literature Survey

Researchers have realized the potential of thermal imaging in the area of human identification. One of the earliest works on the topic was conducted by Chen et al. He had conducted some significant studies on face authentication in visible and infrared spectrum using principal component analysis (PCA) and has presented the results in [2]. The algorithm was implemented on their dataset called C-X1. Database consists of 10,916 images in both visible and IR spectrum from 488 subjects.

Guzman et al. [3] proposed an algorithm that does face authentication based on the vasculature information which is different for every person. The algorithm is implemented by using localization of active contours and morphological operations. The features are then matched using similarity measurements. The proposed algorithm was implemented on C-X1 dataset and on a newly created dataset comprising 13 subjects. When implemented on the Guzman dataset it gave an average recognition rate of 89% and a recognition rate of 68.5% on C-X1 dataset by using different similarity measures. The poor performance of algorithm on C-X1 dataset was claimed to be due to the lack of non uniformity correction in the dataset. Major drawback is that the algorithm was validated only on a small dataset.

Carrapico et al. [4] has taken into consideration several image descriptors and analyzed their performance in face authentication using thermal images. The image descriptors considered include Gabor Bank-Filters, Localized Binary Patterns (LBP), Colour and Edge Directivity Descriptor (CEDD) [5] and Fuzzy Colour and Texture Histogram (FCTH) [6]. The four algorithms were implemented on dataset provided by University of Science and Technology, China along with a k-nn classifier. Out of the four the Localized Binary Patterns (LBP) feature descriptor gave the highest accuracy of 91%.

Xie and Wang [7] has proposed a modified LBP based feature extraction algorithm for face authentication using thermal imaging. The major drawback of LBP is its large dimensionality. The proposed approach tackles this problem by selecting personalized features for each subject. The algorithm was implemented on their own dataset of 400 thermal images of 40 subjects in combination with KNN classifier. Its performance was compared with other LBP based techniques. Their algorithm gave the recognition rate of 98.2%.

Espinosa-Duró et al. [8] created a database of thermal face images. It consist of 41 subjects (32 males and 9 females). Visible and thermal images of all the subjects were captured under three different illumination conditions: IR (infrared illumination), AR (artificial illumination) and NA (Natural illumination). A feature extraction method based on discrete cosine transform (DCT) was implemented.

In all the above mentioned works the feature extraction algorithms were verified only on datasets containing images with plain background. However these algorithms need to be verified on images with realistic background as well.

3   Methodology

Figure 1 depicts the basic flow diagram of the face authentication process.

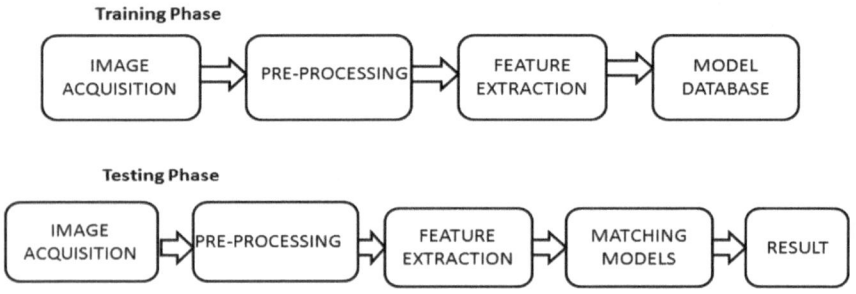

Fig. 1 Block diagram

(a) *Pre-Processing*: A median filter with a kernel size of 5 was used to remove the higher frequencies and the impulse noises while preserving the edges.
(b) *Feature Extraction*: Four different feature extraction algorithms were implemented on the two datasets: Local Binary Pattern (LBP), Histogram of Oriented Gradients (HOG), Pyramid Histogram of Oriented Gradients (PHOG) and Discrete Cosine Transform (DCT).

- **Local Binary Pattern (LBP)** [7]: It is a powerful feature for texture classification. It has two major parameters: P and R. P is the number of neighboring pixels and R is the radius. Here P is taken as 8 and R as 1. The entire image is divided into non-overlapping blocks of uniform size of 16 × 16 pixels. For each pixel in the block, its intensity is compared with 8 neighboring pixels at a radius of 1 pixel. A binary code for centre pixel

$$LBP_{P,R}(g_c) = \sum_{i=0}^{P-1} 2^i \cdot B \tag{1}$$

is generated based on the comparison as per the following equation:

$$B = \begin{cases} 1, & g_i - g_c > 0 \\ 0, & g_i - g_c < 0 \end{cases} \tag{2}$$

where g_c is the gray value of the central pixel, g_i is the gray value of the neighboring pixels. $g_i - g_c$ is the generated binary code for central pixel g_c. A binary value of 1 is assigned if the gray value of central pixel is less than that of the neighboring pixels and a value of 0 is assigned if the gray value of central pixel is greater than that of the neighboring pixels. The number bits depend on P value. Since it is taken as 8 here an 8 bit binary code is generated for each pixel.

For each non-overlapping block histogram is computed based on the following equations:

$$H(r) = \sum_{x_c}^{M-1} \sum_{y_c}^{N-1} f\left(LBP_{P,R}(x,y), r\right) \tag{3}$$

$$f\left(LBP_{P,R}(x,y), r\right) = \begin{cases} 1 & LBP_{P,R}(x,y) = r \\ 0 & otherwise \end{cases} \tag{4}$$

The final feature matrix is formed by concatenating normalized histogram of the binary patterns of all the blocks in the image.

- **Histogram Of Oriented Gradients (HOG)** [9]: It is a scale invariant feature descriptor. Entire image is divided into uniform blocks with 50% overlapping. The gradient of an image is defined by:

$$\Delta f = \begin{pmatrix} g_x \\ g_y \end{pmatrix} = \begin{pmatrix} \frac{\partial f}{\partial x} \\ \frac{\partial f}{\partial y} \end{pmatrix} \tag{5}$$

where $\frac{\partial f}{\partial x}$ is the gradient in x direction and $\frac{\partial f}{\partial y}$ is the gradient in y direction.

The gradient direction is calculated by:

$$\theta = \tan^{-1}\left(\frac{g_y}{g_x}\right) \tag{6}$$

The gradient magnitude and direction is calculated for each pixel in each block. Then a histogram of gradients is computed for each cell (9 bins, 0°–360°. Concatenating the normalized histograms all the blocks forms the final feature vector.

- **Pyramid Histogram of Oriented Gradients** [10]: It is a variant of HOG. HOG features are calculated for three different block sizes. Then final feature vector is formed by concatenating the normalized histograms of all the levels.
- **Discrete Cosine Transform** [8]: It is an invertible linear transform. The image is converted into frequency domain by applying DCT. DCT2 defined by the following equation is used:

$$X[k,l] = \frac{2}{N} \cdot c_k \cdot c_l \cdot \sum_{m=0}^{a-1} \sum_{n=0}^{b-1} x[m,n] \cos\left(\frac{(2n+1)l\pi}{2N}\right)$$

where a and b indicate the number of coefficients in x and y direction, N is the total number of significant coefficients and c_k and c_l are defined as:

$$c_k, c_l = \begin{cases} \frac{1}{\sqrt{2}} \, to \, k = 0, \, l = 0 \\ 1 \, to \, k = 1, 2, \ldots a-1 \quad and \quad l = 1, 2, \ldots b-1 \end{cases}$$

The DCT coefficients indicate the importance of frequencies in it. The first coefficient signifies the lowest frequency component and carries the most significant information. Higher frequencies are mostly noise components. The first N DCT coefficients are used as feature vector.

(c) *Classification*: The features formed were classified using two popular classifiers: K-Nearest Neighbours (KNN) [11] and Support Vector Machine (SVM) [12].

A k-fold cross validation process was carried out to get the classification accuracy. Here k was chosen as 10. The entire dataset is divided into 10 random groups. Each group is taken as the test matrix and the remaining group as training groups in an iterative process. The classification accuracy is calculated as follows:

$$A = \frac{C}{T}$$

where A is the classification accuracy, C is the number of correct classifications and T is the total number of classifications.

(d) *Datasets*: The algorithm was implemented on two datasets: the publicly available Carl database and a newly created database. Carl database consists of thermal, visible and infrared images of 41 subjects (32 males and 9 females) under three different illumination conditions: natural illumination, artificial illumination and infrared illumination. The dataset also contains face localized images of the same. There are many face localization algorithms in existence like level set algorithm [13] etc. The face localized images in Carl database were created using a new algorithm [14] (Figs. 2, 3 and 4).

(a)   (b)

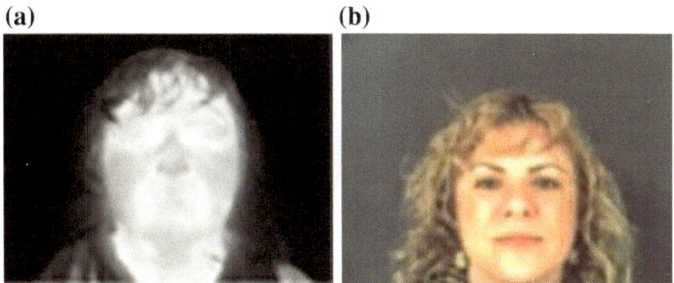

Fig. 2 a Thermal, **b** Visible

(a)   (b)

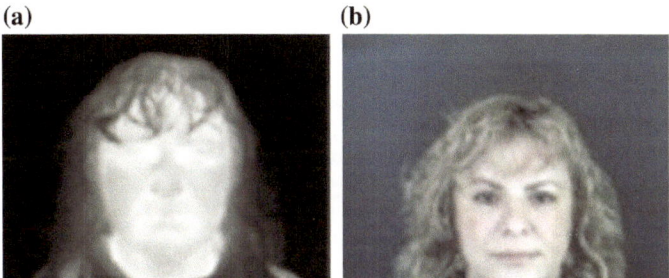

Fig. 3 a Thermal, **b** Visible

(a)   (b)

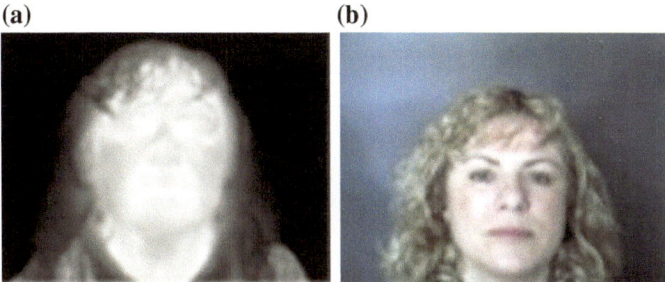

Fig. 4 a Thermal, **b** Visible

The newly created database consists of thermal and near infrared images of 30 subjects (10 males and 20 females) with realistic backgrounds. The backgrounds vary for different subjects. The images were captured at different times of the day from morning 9 a.m to evening 4 p.m to take into account the temperature variations that occur. The thermal images were acquired using FLIR One thermal camera with a temperature range of −20° to 120 °C and thermal sensitivity of 0.1 °C. The images captured are in jpg format with dimensions of 480 × 640 pixels. The near infrared images were captured using a MAPIR NDVI + Red Survey 2 camera with a spectral range of 700 nm–1 mm (Fig. 5).

The captured images were 3 dimensional with the infrared component as the 2nd dimension. So only that was extracted from all the images. The images are in raw jpg format with dimensions of 3840 × 2160 pixels. The images were acquired with an approximate distance of 1 m between the subject and the camera. Unlike the existing thermal face databases, the images were captured with a non-plain realistic background.

(a)   **(b)**

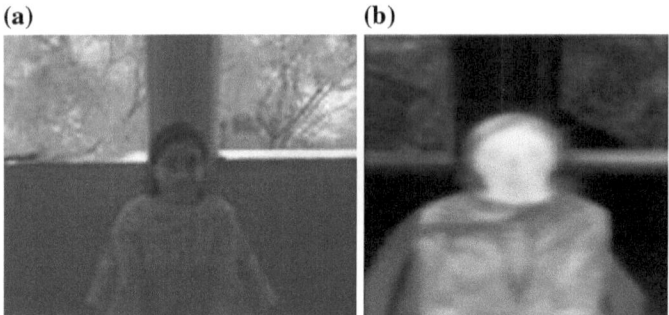

Fig. 5 **a** Near infrared, **b** Thermal

4   Results

The results obtained are presented in Tables 1, 2 and 3.

Table 1 Classification accuracy using Carl and new database

Algorithm	Carl—KNN (%)	Carl—SVM (%)	New—KNN (%)	New—SVM (%)
LBP	41	15	30	10
HOG	38	35	28	25
PHOG	30	25	23	20
DCT	37	56.2	29	45

Table 1 depicts the classification accuracy for KNN and SVM on the different feature extraction algorithms. The tables indicate that the accuracy is highest for LBP features. Also, it is observed that the accuracy value decreases when considering images with non-plain background.

Table 2 Classification accuracy of face localized images

Algorithm	KNN (%)	SVM (%)
LBP	91	88
HOG	90.5	88
PHOG	89	25
DCT	81.7	87.8

Table 2 gives the accuracies when the algorithms were implemented on the face localized images of Carl database. The results indicate the advantage of face localizing the images before feature extraction.

Table 3 Comparison of different illuminations using KNN

Algorithm	NA (%)	AR (%)	IR (%)
LBP	94.5	91	90.7
HOG	91	90.5	91
PHOG	88	89	90
DCT	98	97	98

Table 3 compares the accuracy values for different illuminations. It can be observed that the accuracy does not vary much with change in illumination.

5 Conclusion and Future Work

A new dataset of thermal and near infrared face images with non-plain realistic backgrounds was created. Different feature extraction algorithms (LBP, HOG, PHOG, DCT) were compared by implementing them on a benchmark dataset and on a newly created dataset. The LBP features gave the highest accuracy. Also, the accuracy decreased significantly when implemented on the new dataset because of its non-plain background. Also, accuracy increased dramatically when face localization was done before feature extraction. There was not much change in accuracy when implemented for different illuminations conforming that thermal images remain unaffected by illumination variations.

Face localization algorithm for images with plain background already exists. Such algorithms for non-plain backgrounds should be investigated. The algorithm needs to be verified on a larger dataset.

References

1. Nimmy, K., Sethumadhavan, M.: Biometric authentication via facial recognition (2014)
2. Chen, X., Flynn, P.J., Bowyer, K.W.: Ir and visible light face recognition. Comput. Vis. Image Underst. **99**(3), 332–358 (2005)
3. Guzman, A.M., Goryawala, M., Wang, J., Barreto, A., Andrian, J., Rishe, N., et al.: Thermal imaging as a biometrics approach to facial signature authentication. IEEE J. Biomed. Health Inform. **17**(1), 214–222 (2013)
4. Carrapico, R., Mourão, A., Magalhaes, J., Cavaco, S.: A comparison of thermal image descriptors for face analysis. In: 2015 23rd European Signal Processing Conference (EUSIPCO). IEEE, pp. 829–833 (2015)
5. Chatzichristofis, S., Boutalis, Y.: Ccdd: color and edge directivity descriptor: a compact descriptor for image indexing and retrieval. In: *Computer Vision Systems*, pp. 312–322 (2008)

6. Chatzichristofis, S.A., Boutalis, Y.S.: Fcth: Fuzzy color and texture histogram—a low level feature for accurate image retrieval. In: Ninth International Workshop on Image Analysis for Multimedia Interactive Services, 2008. WIAMIS'08. IEEE, pp. 191–196 (2008)
7. Xie, Z., Wang, Z.: Infrared face recognition based on personalized features selection of lbp. In: 2015 7th International Conference on Intelligent Human-Machine Systems and Cybernetics (IHMSC), vol. 2. IEEE, pp. 228–231 (2015)
8. Espinosa-Duró, V., Faundez-Zanuy, M., Mekyska, J.: A new face database simultaneously acquired in visible, near-infrared and thermal spectrums. Cogn. Comput. **5**(1):119–135 (2013)
9. Dalal, N., Triggs, B.: Histograms of oriented gradients for human detection. In: IEEE Computer Society Conference on Computer Vision and Pattern Recognition, 2005. CVPR 2005, vol. 1. IEEE, pp. 886–893 (2005)
10. Bai, Y., Guo, L., Jin, L., Huang, Q.: A novel feature extraction method using pyramid histogram of orientation gradients for smile recognition. In: 2009 16th IEEE International Conference on Image Processing (ICIP). IEEE, pp. 3305–3308 (2009)
11. Ameur, B., Masmoudi, S., Derbel, A.G., Hamida, A.B.: Fusing gabor and lbp feature sets for knn and src-based face recognition. In: 2016 2nd International Conference on Advanced Technologies for Signal and Image Processing (ATSIP). IEEE, pp. 453–458 (2016)
12. Suykens, J.A.., Vandewalle, J.: Least squares support vector machine classifiers. Neural Process. Lett. **9**(3), 293–300 (1999)
13. Kumaravel, M.S Karthik, P.S., Soman, K.: Human face image segmentation using level set methodology. Int. J. Comput. Appl. **44**(12), 16–22 (2012)
14. Marzec, M., Koprowski, R.: Wróbel, Z., Kleszcz, A., Wilczyński, S.: Automatic method for detection of characteristic areas in thermal face images. Multimedia Tools Appl. **74**(12), 4351–4368 (2015)

Meme Classification Using Textual and Visual Features

E. S. Smitha[1]($\boxtimes$), S. Sendhilkumar[1], and G. S. Mahalaksmi[2]

[1] Department of Information Science and Technology, College Engineering
Guindy, Anna University, 600025 Chennai, India
smithaengoor@gmail.com
[2] Department of Computer Science, College Engineering Guindy, Anna
University, 600025 Chennai, India
gsmaha@annauniv.edu

Abstract. Social networks became a global phenomenon and made an enormous impact in different fields of society in a few years. These days, a huge number of memes have available in social networks. An internet meme is a cultural style that propagates from one to another in social media. It is a unit of information that jumps from place to place with slight modification. These memes play a vital part in expressing emotions of users in social networks and serve as an effective promotional and marketing tool. Visual memes are important because they will show emotion, humor, or portray something that words cannot. This paper recommends a framework that could be utilized to categorize internet memes by certain visual features and textual features.

Keywords: Social networks · Meme · Classify · Internet · Textual
Visual feature · Sentiment analysis

1 Introduction

Internet memes turn out to be a new social phenomenon for previous few years. The main idea behind the internet meme is to convey the social thoughts which can be in the form of actions, voice, manuscripts with an imitated theme. "The meme" is pronounced as meam, which rhymes with beam or team. The term 'meme' had derived by the British ethologist Richard Dawkins. The word 'meme' had come from the Greek word 'mimema' meaning 'something imitated'. As portrayed by Dawkins [1] memes are a form of social engendering or cultural propagation for individuals to pass social recollections and social thoughts to each other. They can be simply say that, an internet meme becomes viral and as part of online and social networks. Memes have the ability to self-replicate, mutate and respond to certain situations, that make memes to resemblance with genes. As like the way that DNA and life will spread from area to area, an image thought will likewise venture out from brain to mind with or without change or modification. A 'meme character' is a figurative entity that normally at the

© Springer International Publishing AG 2018
D. J. Hemanth and S. Smys (eds.), *Computational Vision and Bio Inspired Computing*,
Lecture Notes in Computational Vision and Biomechanics 28,
https://doi.org/10.1007/978-3-319-71767-8_87

core of an internet meme, which shows some sort of articulation that a meme passes on. Image content is normally comes on the upper or base part of the picture. In social media and social networks visuals can improve the quality of communication and business because most of the people are visual learners.

1.1 Role of Memes

The online social networking can enhance people's feeling of connectedness with genuine or potentially online groups and social networking. Memes shared through online social networking are normally humour, fast feed recordings, urban legends, funny pictures or trendy music.

Most of the present day memes are adorned photographs that are proposed to be amusing, frequently as an approach to freely scorn human conduct. Different memes can be recordings and verbal articulations. A few memes have heavier and more philosophical substance. The memes are important for two reasons: meme is an overall social phenomenon, and they will fly out from individual to individual rapidly through online networking.

1.2 Challenges

It is very difficult task to recognise the exact meaning and user intension of a meme contains sarcastic expression. Different people may interpret the same image which has some sarcastic expressions. The same sarcastic image can be interpreted differently by different people. So it will be very difficult for the system to identify the actual intention of the image. It is also difficult for the system to understand the conditional sentences For example: "The mobile phone will be awesome if its camera is good" after considering dependencies the system will interpret it as "phone-awesome" and "camera-good" but here the user's perception is quite different.

Usually the researchers and developers will ignore the pronouns during sentiment analysis, but it is ambiguous to the system to detect what may a noun or pronoun phrase is actually referring to in the sentence. But in some sentences a pronoun may refer something which is so vital in extracting users' intention.

1.3 Motivation

The primary work reported in this work is classification of memes. The application of these classified meme is predicting the trend in the emotion on a particular topic, that will be the future work. The online social networks like twitter, face book, micro blogs such as Weibo produce billions of tex and visual information, which are useful to recognize sentiment indicated by visual and text data. With the growing number of users and activities in online social network services, image sentiment analysis has become an important keyword for psychological study and commercial marketing. This work is to analyze the sentiments from memes by considering both image features and textual features.

The fundamental point of the work is to discover the emotion in the social media meme which may be useful in the fields of marketing, advertising, trend analysis and so on.

This paper suggests a method for analyzing memes for sentiments. Memes highlighting happiness, misery, anger, disgust, fear, neutral kinds of emotions are collected. Text features are extracted using Optical Character Recognition (OCR) and visual features using Matrix Laboratory (MATLAB) functions from the memes. Using these features labels are assigned using Natural Language Tool Kit (NLTK) for memes. The training data thus created is then given to WEKA tool for generation of decision rules. These rules applied to classify any meme from social networks.

Section 2 discussed about a few works identified similar this work. Section 3 is about the structure design utilized as a part of this work. The Sect. 4 is about the outcome and performance analysis. The last section discuss about conclusion and future work.

2   Related Works

Memes have become an important way to convey information in social media. As per Machajdik and Hanbury [2] images will affect people emotionally.

Extracting both text and visual features from memes and using these features, for understanding memes has many challenges. Exhaustive research has happened in the history about the classification about the image data. A classification model for image data by Naïve Bayes classifier is proposed by Park [3]. The resulting classifier provided appropriate classification with limited computational efforts.

The images collected from the Caltech-101 dataset contained only images and those images were classified successfully using random forests and ferns classifiers in [4] by Bosch et al., here the features which describe the shape and appearance of a picture were used for classification. The outputs of both shape and appearance based classification were concatenated and served to describe the image. The authors found similarities between images in the Caltech dataset using a kernel function. The feature descriptors both shape and appearance with the kernel function delivered fixed regions from many training images. A random ferns classifier was used to provide speedy training times.

Deza and Parikh [5] were attempted to gain an understanding on what makes an image viral. They had identified some characteristics of the extensiveness of an image and the features that can be used to predict virality. Reddit suggested a way to determine top rated posters on the basis of reputation. In the context of Reddit, viral images are defined as the images that had many upvotes, few downvotes and had been resubmitted several times by many users. A virality score function also calculated with those factors,4 forms contexts were derived to better understanding of image virality.

As per Machajdik and Hanbury [2] images will affect people emotionally. For object detection Viola and Jones proposed Haar-Cascade classifier. Rosebrock found that the execution of the Haar-course classifier could be hazardous. Rosebrock formulated six step process to achieve peak performance for object identification using Histogram of Oriented Gradients and Linear Support Vector Machines.

Various algorithms for shape detection and texture feature extraction like fractal dimension technique, Edge Detection and Boundary Tracing, Fourier Transform of Boundary, Scale Space fruit were discussed by Drashti Jasani et al. [6] reviewed the advancement of the data and communication technology within the field of agriculture. From these methods, it is inferred that Circular Hough Transform (CHT), Edge Detection and Boundary Tracing will provide better results.

Yuan et al. [7], proposed an image sentiment prediction method, in which the mid-level attributes of an image used to predict its sentiment. Scene descriptor low-level features from the SUN Database are extracted and then four types of mid-level attributes are selected. The features or attributes are used for classification (a) Material: such as metal, vegetation; (b) Function: cooking, playing; (c) Surface property: glossy, rusty; and (d) Spatial Envelope: man-made, natural: For predicting sentiments 102 predefined midlevel attributes are used.

Wang and Wen [8] proposed a robust non paranormal approach to transform all vision and text features into the cumulative density function space. By learning the stochastic dependencies, they show that their model significantly outperforms various competitive baselines in the prediction experiments. In addition, they also proposed a simple pipeline for generating memes from raw images, drawing the wisdom from reverse image search and traditional information retrieval point of view. Wang et al. [9] utilized re-sampling, cost-sensitive learning and one-class classification as supervised learning methods. They are very popular and proved to be effective in sentiment classification. It is difficult to work with supervised methods as they are expensive and time consuming. Researchers are working on this area to find out better techniques. Opinion mining is to identify the polarity as positive or negative. Saleh et al. [10], extracted this using Support Vector Machines (SVM) by using various datasets and weighting schemes. In [11] Nagarjuna Devi et al. have built up a general procedure of 'Aspect or Feature based Sentiment Analysis' using Support ector Machine (SVM) in a novel approach. It is turned out to be a standout amongst the best approaches to analyze and extract the general clients' view about the specific feature as well as whole product.

3   Proposed Architecture

The various components for proposed meme classification model is given in Fig. 1. The steps involved in meme mining are as follows: data collection, visual and text feature extraction, generating training data and classification using Weka tool and NLTK 3.0 using python.

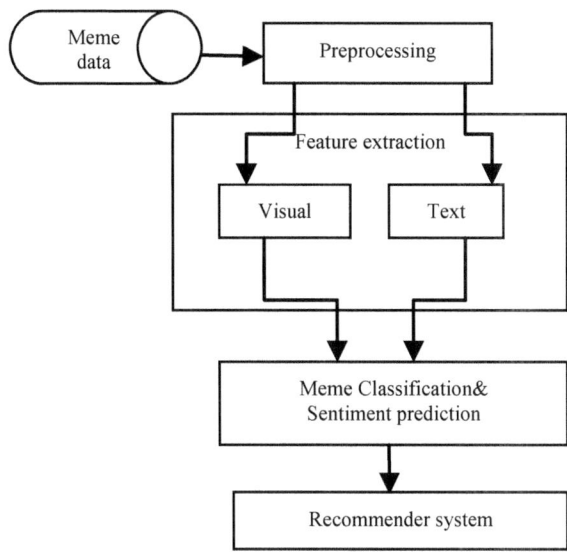

Fig. 1 Meme classification model

3.1 Data Collection

Memes mining is at its initial stage, hence, there is no proper data set available. There are a lot of sites that contain millions of memes but they may not be labelled appropriately for the use of training. Hence for experimental purpose manual collection of memes is done from the following social networking websites: Face book, Instagram, Flickr. The collected memes contain text in the top and bottom and image in the middle. Totally 650 memes with human or animal faces are taken for classification. The emotions or expressions considered are following: 'happy', 'sad', 'anger', 'disgust', 'fear', 'neutral'. The categories and its corresponding number of memes are shown in Table 1. Sample memes are shown in Fig. 2.

Table 1 Categories of memes

Category name	No. of memes collected
Happy	130
Sad	120
Angry	115
Fear	105
Disgust	095
Neutral	085

Fig. 2 Sample memes

3.2   Meme Preprocessing: Resize and Segmentation

The memes collected are of various sizes. They are then resized uniformly into 256 by 256 pixels. So that uniformity in segmentation can be achieved. The memes are then segmented into three region: two text part which is the top and bottom sections of the meme and third central image part. The memes with top and bottom regions are considered for extracting text features and the central (image) region for visual features. Figure 3 shows an example of segmented meme.

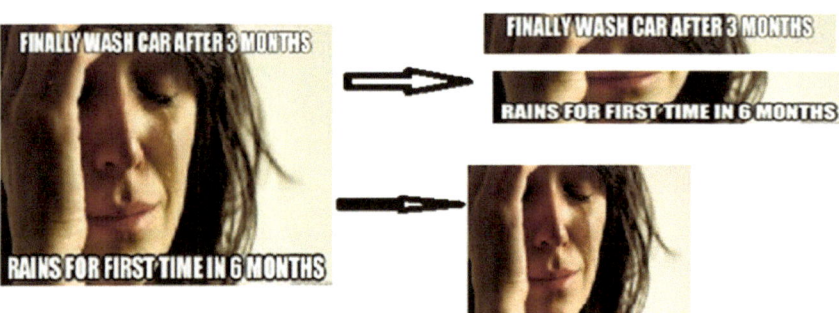

Fig. 3 Segmented meme

3.3   Text Feature Extraction

Text in the meme can be extracted using various techniques. In The segmented text from the meme is fed into an online OCR (http://ocr.space) for character recognition. The output of the OCR will be copied into a word file (Fig. 4). The word file hence obtained is converted into an excel file (Fig. 5) and feed to Natural Language Tool Kit (NLTK 3.0) in Python to perform operations like tokenization, stopwords removal, stemming and part of speech tagging which shown in Fig. 6. It provides a lot of words processing instrument and packages including data mining, machine learning sentiment analysis and other various language processing tasks.

AUDIENCE FACE AFTER COMING OUT OF ANJAAN
The walls we build around us to keep sadness out also keeps out the joy. you choose
The hardest thing is Not talking to someone you used to talk every day & night...(I know, it Kills)
I don't care who you are where you're from, what you did, as long as you love me
HAPPY NEW YEAR HAPPY NEW YEAR EVERYWHERE
We all have that one person in our life, who replies us late but still we wait for their message with a smile
THE SECRET OF HAPPINESS ...IS TO HAVE BAD MEMORY!
FINALLY WASH CAR AFTER 3 MONTHS RAINS FOR FIRST TIME IN 6 MONTHS
It hurts and kills me,when you are in online and you won't reply for my text.
I'M SAD,THE DOG TOLD ME HE WAS CUTER...
After Breakup,Boys are not actually Cruel,They secretly care, for their EX

Fig. 4 Extracted text from OCR

1	AUDIENCE FACE AFTER COMING OUT OF ANJAAN
2	The walls we build around us to keep sadness out also keeps out the joy. you choose
3	The hardest thing is Not talking to someone you used to talk every day & night...(I know, it Kills)
4	I don't care who you are where you're from, what you did, as long as you love me
5	HAPPY NEW YEAR HAPPY NEW YEAR EVERYWHERE
6	We all have that one person in our life, who replies us late but still we wait for their message with a smile
7	THE SECRET OF HAPPINESS ...IS TO HAVE BAD MEMORY!
8	FINALLY WASH CAR AFTER 3 MONTHS RAINS FOR FIRST TIME IN 6 MONTHS
9	It hurts and kills me,when you are in online and you won't reply for my text.
10	I'M SAD,THE DOG TOLD ME HE WAS CUTER...
11	After Breakup,Boys are not actually Cruel,They secretly care, for their EX
12	Just Be Friends"with someone you fell in love with
13	Life become disturbed'when you don't chat with the one,whom you used to chat daily
14	Dear 2017 I haven't met you but I already hate you
15	SADNESS DID THE FLOP,sad Tip for 2017 Don't get emotionally attached to anyone
16	SADNESS IS... MONDAY

Fig. 5 Extracted text in Excel

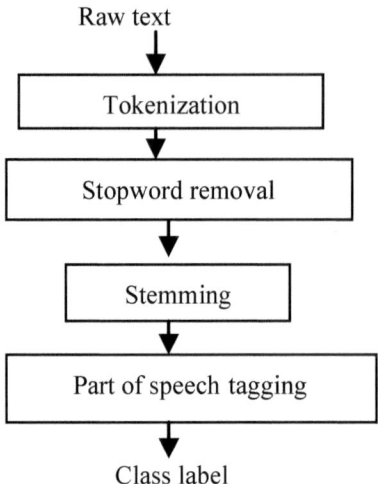

Raw text

Tokenization

Stopword removal

Stemming

Part of speech tagging

Class label

Fig. 6 Text feature extraction

3.3.1 Tokenization

Tokenization is mainly a splitting process that breaks up a chronological sequence of string into tokens such as words, symbols, phrases and other elements. Tokenization divides extracted texts using OCR from the memes into separate words. Figure 7 shows an example of tokenization.

```
>>> import nltk
>>> sentence = """sadness is seeing yourr best friend sadd...."""
>>> tokens = nltk.word_tokenize(sentence)
>>> tokens
['sadness', 'is', "seeing", 'yourr', 'best', 'friend',
'sadd', '.']
```

Fig. 7 Example for tokenization

3.3.2 Stopwords

In computing, stop words are words which are filtered out before or after processing of natural language data (text) [12]. Though stop words generally refer to the most common words in a language, there is no common or general list of stop words used by all natural language processing tools. A few devices particularly abstain from evacuating these stop words to help state seek. Stop words removal is used to remove words from the tokenized meme text such as 'a' and 'the', in order to improve performance. Stop words here used are list of English stopwords. Figure 8 shows an example code for stopwords removal.

```
from nltk.corpus import stopwords
from nltk.tokenize import word_tokenize
example_sent = "This is a sample sentence, showing off the stop words filtration."
stop_words = set(stopwords.words('english'))
word_tokens = word_tokenize(example_sent)
filtered_sentence = [w for w in word_tokens if not w in stop_words]
filtered_sentence = []
for w in word_tokens:
    if w not in stop_words:
    filtered_sentence.append(w)
print(word_tokens)
print(filtered_sentence)
.............................................................................
output here:
['This', 'is', 'a', 'sample', 'sentence', ',', 'showing', 'off', 'the', 'stop', 'words', 'filtration', '.']
['This', 'sample', 'sentence', ',', 'showing', 'stop', 'words', 'filtration', '.']
```

Fig. 8 Example for stopword removal

3.3.3 Stemming

Stemming is the technique of finding stem or root of a word, generally a written word form. The stem not required to indistinguishable to the morphological base of the word. It is typically adequate that related words guide to a similar stem, regardless of the possibility that this stem is not in itself a substantial root. Many web crawlers treat words with an indistinguishable originate from equivalent words as a sort of inquiry extension, a procedure called conflation. For example: happiness and happily are both reduced to their root happy. Figure 9 shows an example of stemming.

```
from nltk.stem import PorterStemmer
from nltk.tokenize import sent_tokenize, word_tokenize

ps = PorterStemmer()

sentence = "gaming, the gamers play games"
words = word_tokenize(sentence)

for word in words:
    print(word + ":" + ps.stem(word))

output:
gaming:game
;:;
the:the
gamers:gamer
play:play
games:game
```

Fig. 9 Example code for stemming

3.3.4 Part of Speech Tagging

In corpus etymology, part-of-speech tagging (POS tagging or POST) is similar to grammatical tagging, or removing the ambiguity of word category. It is the way toward increasing a word in a content (corpus) as relating to a particular grammatical feature, in light of its definition and its unique circumstance. It is the association with adjoining and related words in an expression, sentence, or passage. For instance: paper is labelled as thing and go is labelled as verb.

One of the vital parts of the NLTK module is the Part of Speech labelling. This implies naming words in a phrase or sentence as things, descriptors, verbs and so forth. NLTK has a rundown of POS labels. Figure 10 shows an example of parts of speech.

```
import nltk
from nltk.corpus import state_union
from nltk.tokenize import PunktSentenceTokenizer

document = 'Today the Netherlands celebrates King\'s Day. To honor this tradition,
the Dutch embassy in San Francisco invited me to'
sentences = nltk.sent_tokenize(document)
data = []
for sent in sentences:
    data = data + nltk.pos_tag(nltk.word_tokenize(sent))
for word in data:
    if 'NNP' in word[1]:
        print(word)

output:
('Netherlands','NNP')
('King','NNP')
('Day','NNP')
('San','NNP')
('Francisco','NNP')
```

Fig. 10 Example for parts of speech

3.3.5 Text Feature Classification

The text classification using NLTK classifies the meme text into three classes: Positive, Negative and Neutral. Most probably the sentences related to sad, anger, disgust will be classified as negative and the sentences related to happy and surprise will be in positive classes. The sentences which are not included in that category, like "the game got over" will come under neutral. The sample training and testing set for text classification shown in Figs. 11 and 12 respectively.

H18 fx

	A	B
1	AUDIENCE FACE AFTER COMING OUT OF ANJAAN!	pos
2	"The walls we build around us to keep sadness out also ke	neg
3	The hardest thing is Not talking to someone you used to ta	neg
4	I don't care who you are where you're from, what you did,	neg
5	HAPPY NEW YEAR HAPPY NEW YEAR EVERYWHERE	pos
6	We all have that one person in our life, who replies us late	pos
7	THE SECRET OF HAPPINESS ...IS TO HAVE BAD MEMORY!	pos
8	LOVE FAILURE IS THE MOST PAINFUL FEELING IN THE WORL	neg
9	It hurts and kills me,when you are in online and you won't	neg
10	I'M SAD,THE DOG TOLD ME HE WAS CUTER...	neg
11	After Breakup,Boys are not actually Cruel,They secretly ca	neg
12	"Just Be Friends"with someone you fell in love with	pos
13	'Life become disturbed'when you don't chat with the one,	neg
14	Dear 2017 I haven't met you but I already hate you	neg
15	SADNESS DID THE FLOP,sad Tip for 2017 Don't get emotion	neg
16	SADNESS IS... MONDAY	neg
17	I NEVER FEEL SAD WHEN A COMPANY REJECT ME DURING IN	neg
18	I SAID IM A MECHANICAL ENGINEER NOT A MECHANIC!!! a	neg

Fig. 11 Training set

	A	B
1	DON'T MAKE ME PUT MY ANGRY EYES ON	?
2	Have u ever been so angry thet you kicked your own face	?
3	HAVE YOU EVER BEEN SO ANGRY	?
4	GOOD USE YOUR ANGER	?
5	YOUR FACE WHEN IDIOTS ARE TALKING	?
6	I DID'T CHOOSE ENGINEERING LIFE THE ENGINEERING LIFE CHOSE ME	?

Fig. 12 Test set

3.4 Visual Feature Extraction

Memes exhibit a variety of features: General features like pixel, local and global features, structural and global features, space and frequency domain features. Memes image feature extraction involve reducing the number of resources which are needed for the huge data set description. The main problem in complex data analysis is the involvement of huge number of variables in the data. The requirement of memory and computation power will increase, when the number of variables increases. It might make the characterization calculation over fit the preparation tests and sum up ineffectively to new samples. Feature extraction is a general term for strategies for building combinations of the variables to get around these issues while still describing the data with adequate accuracy. The best results are achieved when we construct a set of application-dependent features, a process called feature engineering. Here we had extracted GLCM features, edge and contour-based features and histogram features. The features are extracted using MATLAB. From the GLCM features, contrast, dissimilarity, cluster prominence, energy, entropy and homogeneity are selected.

Emotion detection using the extracted image features from the segmented meme is done in MATLAB by using an application as shown in Fig. 13. The image features such as energy, entropy, homogeneity, width, height, curvature and so on are extracted and stored in a database. Then the features are assigned with suitable emotion labels.

With a specific end goal to perceive emotions from faces we use AdaBoost algorithm [13] for identifying classes of emotions. Note that we are not trying to propose an algorithm that that beats the best facial emotion detection algorithms.

3.4.1   Emotion Detection and Classification

Image classification is done using extracted features and the application in MATLAB. The pre-processed, segmented image part of the memes will be taken for emotion detection. Emotion detection using the segmented meme is done in MATLAB by using an application (http://matlab-recognition-code.com/facial-expression-recognition-matlab-code) as shown in Fig. 13. The image features such as energy, entropy, homogeneity, width, height, curvature are extracted and stored. Then the features are assigned with suitable emotion labels.

Here we are not attempting to propose an algorithm for facial emotion detection, because the concept is not coming under the boundary of this work.

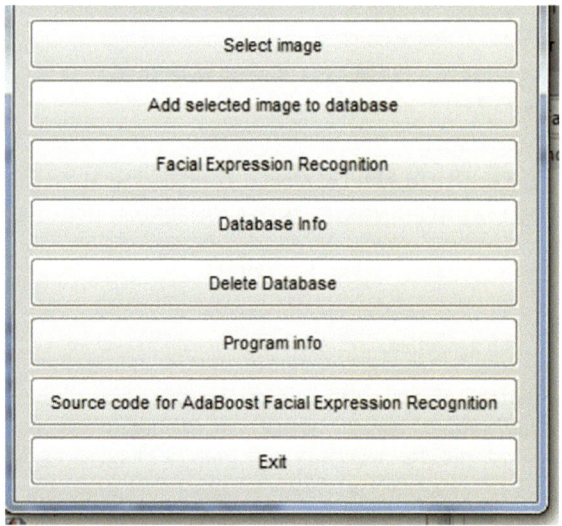

Fig. 13 Adaboost facial expression recognition system

3.4.2   Classification

The Classification is a general process of categorizing ideas and objects that are differentiated, recognized and understood. Here classification is achieved using text and image features. For the text classification, NLTK 3.0 is used. The features of images are extracted using MATLAB and the Adaboost classifier is used for emotion detection in the memes. In text classification using NLTK classifies the meme text into three classes: Negative, Positive and Neutral. Most probably the sentences related to sad, anger, disgust will be classified as negative and the sentences related to happy and surprise will be in positive classes. The sentences like "the game got over" will come under neutral. Memes with faces will classify as angry, disgust, fear, sad, happy and surprise. Weka 3.8 is used for overall classification.

4   Result and Performance Analysis

4.1   Building Model Using Weka Tool

WEKA tool has a set of machine learning algorithms for performing data mining and analysis tasks. The extracted image features, emotion detection application and labels from text classification using NLTK are given as input to WEKA tool. Two famous classification algorithms are used for performance evaluation. The first algorithm, the J48 algorithm allows classification through decision trees. The second one is Naive Bayes algorithm, it is a classification technique based on the assumption, independence among predictors or feature. It is essential to recall that a test instance can count in both the true positive and false positive sets if the classification module outputs more than one label or an unexpected label.

4.2   Decision Tree Algorithm J48

After extracting text and image features from input meme, the given meme is classified using J48 decision tree algorithm in WEKA. J48 algorithm ignores the missing values during the construction of the tree i.e., the value of an item can be predicted based on the attribute values of the other records. The fundamental idea is to partition the data into range based on the attribute values for the items that are found in the training sample. Figure 14 shows the training (a) and test data (b) sets given into the WEKA tool for classification. A tree constructed using J48 algorithm is given Fig. 15. Performance analysis by J48 algorithm and Confusion matrix generated by J48 algorithm are shown in Table 2 and Fig. 16 respectively.

	A	B	C	D	E	F	G	H	I	J
1	imge	cont	Clus.Prom	Diss	Ener	Entro	Homo	F.label	T.label	Img.label
2	10000	0.66	760.5	0.28	0.34	2.05	0.91	angry	neg	angry
3	10001	0.923	940.6	0.31	0.39	1.89	0.91	angry	neg	angry
4	10002	0.986	1480.3	0.28	0.34	1.99	0.91	angry	neutral	angry
5	10003	1.215	387.7	0.4	0.12	2.69	0.87	angry	neutral	disgust
6	10004	0.766	805.4	0.25	0.33	1.95	0.92	angry	neg	angry
7	10005	0.901	1348.5	0.29	0.35	1.96	0.9	angry	neg	angry
8	10006	1.215	387.7	0.4	0.33	1.95	0.87	angry	neg	sad
9	10007	0.923	940.6	0.28	0.34	2.05	0.92	angry	neg	disgust
10	10010	1.896	1301.4	0.416	0.337	2.18	0.868	angry	neg	fear
11	10011	1.03	1601.7	0.326	0.402	1.89	0.891	angry	neg	angry
12	10012	0.812	887.32	0.259	0.341	1.89	0.917	angry	neutral	disgust
13	10013	1.167	1120.2	0.376	0.348	2.102	0.885	angry	neg	angry
14	10014	1.03	712.2	0.314	0.393	1.85	0.905	angry	neg	angry
15	20000	0.706	1154.2	0.23	0.34	1.9	0.93	disgust	pos	disgust

(a) Training data

	A	B	C	D	E	F	G	H	I	J
1	imge	cont	Clus.Prom	Diss	Ener	Entro	Homo	F.label	T.label	Img.label
2	8000	0.66	760.5	0.28	0.34	2.05	0.91	angry	neg	?
3	8001	0.923	940.6	0.31	0.39	1.89	0.91	fear	neg	?
4										
5										

(b) Test data

Fig. 14 Sample data

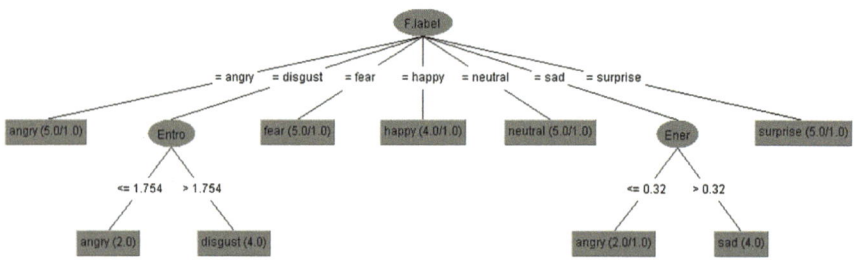

Fig. 15 Tree formed by J48 algorithm

Table 2 Detailed accuracy by class

TP Rate	FP Rate	Precision	Recall	F-measure	MCC	ROC area	PRC area	Class
0.778	0.051	0.778	0.778	0.787	0.726	0.947	0.753	Angry
0.833	0.024	0.833	0.841	0.828	0.810	0.956	0.771	Sad
0.833	0.024	0.836	0.834	0.833	0.814	0.975	0.7778	Disgust
0.833	0.024	0.834	0.832	0.834	0.821	0.984	0.819	Fear
0.875	0.075	0.700	0.875	0.776	0.734	0.969	684	Happy
0.625	0.001	1.00	0.625	0.762	0.762	0.987	0.852	Surprise
1.00	0.023	0.833	1.00	0.909	0.902	0.988	833	Neutral

```
=== Confusion Matrix ===

a b c d e f g   <-- classified as
7 0 1 1 0 0 0 | a = angry
1 5 0 0 0 0 0 | b = sad
0 1 5 0 0 0 0 | c = disgust
1 0 0 5 0 0 0 | d = fear
0 0 0 0 7 0 1 | e = happy
0 0 0 0 3 5 0 | f = surprise
0 0 0 0 0 0 5 | g = neutral
```

Fig. 16 Confusion matrix generated by J48 algorithm

4.3 Naive Bayes Algorithm

Naive Bayes algorithm, is a classification technique based on Bayes theorem, here the features or attributes are assumed to be independent. Naive Bayes model simple and especially useful for very huge amount of data sets. Bayes theorem provides a way of calculating posterior probability $P(c/x)$ from $P(c)$, $P(x)$ and $P(x/c)$ as shown in Eq. (4.1).

$$P(Ci/X) = \frac{P(Xi/Ci) * P(Ci)}{P(Ci)} \tag{4.1}$$

It predicts X belongs to a class label Ci if the probability $P(Ci|X)$ is the highest among all the $P(Ci|X)$ for all the k classes. Here the training set has 3 class labels negative, positive and neutral, so k = 3. First calculate the prior probabilities of each class $P(Ci)$, then conditional probabilities $(P(X/Ci)$ have to be calculated. Then find the posterior probability $P(Ci/X)$ for all 3 classes using Eq. 4.1. Compare the posterior probabilities $P(Ci/X)$ for 3 classes, which value is greater, then the test data has to be assigned with that particular label. Performance analysis by Naive Base algorithm and Confusion matrix generated by Naive Base algorithm are shown in The classification result by Naïve Base algorithm is given in Table 3 and Fig. 17 respectively.

Table 3 Detailed accuracy by class

TP rate	FP rate	Precision	Recall	F-measure	MCC	ROC area	PRC area	Class
0.556	0.011	1.000	0.556	0.711	0.710	0.960	0.922	Angry
0.833	0.048	0.714	0.833	0.736	0.736	0.980	0.877	Sad
1.00	0.051	0.750	0.998	0.845	0.845	0.996	0.976	Disgust
0.834	0.047	0.556	0.833	0.625	0.625	0.972	0.910	Fear
0.250	0.095	0.998	0.250	0.466	0.466	0.964	0.748	Happy
1.000	0.002	0.667	1.000	0.775	0.775	0.976	0.832	Surprise
0.899	0.001	1.000	0.997	1.000	0.998	1.000	1.000	Neutral

```
=== Confusion Matrix ===

a b c d e f g   <-- classified as
5 1 2 1 0 0 0 | a = angry
0 5 0 0 0 1 0 | b = sad
0 0 6 0 0 0 0 | c = disgust
0 1 0 5 0 0 0 | d = fear
0 0 0 3 2 3 0 | e = happy
0 0 0 0 0 8 0 | f = surprise
0 0 0 0 0 0 5 | g = neutral
```

Fig. 17 Confusion matrix generated by Naive Base algorithm

4.4   Performance Analysis

Table 4.

Table 4 Performance evaluation

Summary	J48 (%)	Naïve base (%)
Correctly classified instances	81.25	70.83
Incorrectly classified instances	18.75	29.17
Kappa statistic	00.78	00.67
Mean absolute error	00.07	00.09
Root mean squared error	00.19	00.23
Relative absolute error	32.73	40.44
Root relative squared error	57.26	70.77

Two classification algorithms, J48 and Naive Base are used for performance evaluation. The Table 4 summarizes the result of both algorithms. The J48 algorithm classified the memes more accurately compared to Naïve Base algorithm. The Naïve Base algorithm is a straightforward algorithm which has a well-built assumption of independence of features. If the features are independent accuracy of the result will be more in this algorithm. Here the features are dependent, a tree might be expected to work better. So the decision tree J48 algorithm performs well. Root mean square error is also less in the case of J48.

5   Conclusion and Future Work

The paper focuses on classifying the memes depending on the sentiment of the given meme. Visual sentiment analysis on social network content helps us to understand the user behaviour and provides useful information for related data analysis. A frame work using text and image features had introduced. The model will be useful to analyse social media memes for finding user interest and predictive modelling.

The work can be extended to find the trends and patterns in memes being shared among social media. The domains in which memes are trending during a particular span of time also can be analysed. It can be useful in advertising and promotion of products and services. Trend analysis can also serve to be useful in areas like, finding the popularity of a political party, psychological study and so on.

References

1. Dawkins, R.: The Selfish Gene. Oxford University Press (1989)
2. Machajdik, J., Hanbury, A.: Affective image classification using features inspired by psychology and art theory. In: MM'10 Proceedings of the 18th ACM international conference on Multimedia, pp. 83–92 (2010)
3. Park, D.C.: Image classification using naïve bayes classifier. Int. J. Comp. Sci. Electron. Eng. (IJCSEE) **4**, Issue ISSN 2320–4028, (online), 3 (2016)
4. Bosch, A., Zisserman, A., Munoz, X.: Image classification using random forests and ferns. In: Proceedings of the IEEE 11th International Conference on Computer Vision (ICCV), pp. 20–34 (2007)
5. Deza, A., Parikh, D.: Understanding image virality. In: Proceedings of the IEEE Conference on Computer Vision and Pattern Recognition (CVPR), pp. 25–34 (2015)
6. Drashti Jasani, G., Patel, P., Patel, S., Ahir, B., Patel, K., Dixit, M.: Review of shape and texture feature extraction techniques for fruits. Int. J. **6**, pp. 4851–4854 (2015)
7. Yuan, J., Mcdonough, S., You, Q., Luo, J.: Sentribute: image sentiment analysis from a mid-level perspective. In: Proceedings of the Second International Workshop on Issues of Sentiment Discovery and Opinion Mining (2013)
8. Wang, W.Y., Wen, M.: I Can Has Cheezburger? A nonparanormal approach to combining textual and visual information for predicting and generating popular meme descriptions. In: The 2015 Annual Conference of the North American Chapter of the ACL, pp. 355–365 (2015)
9. Wang, Z., Li, S., Lee, S.Y.M., Zhou, G.: Semi-supervised learning for imbalanced sentiment classification. In: Proceedings of the Twenty-Second International Joint Conference on Artificial Intelligence, pp. 1826–1831 (2011)
10. Rushdi-Saleh, M., Mart´ın-Valdivia, M., R´aez, A., L´opez, L.: Experiments with svm to classify opinions in different domains. Expert Sys. App. **38**, 14799–14804 (2011)
11. Nagarjuna Devi, D.V., Kumar, C., Prasad, S.: A feature based approach for sentiment analysis by using support vector machine. In: IEEE 6th International Conference on Advanced Computing, pp. 44–49 (2016)
12. https://en.wikipedia.org/wiki/Stop_words#cite_note-1
13. Owusu, E., Zhan, Y., Mao, Q.R.: An SVM AdadaBoost facial expression recognition system. Springer Science+Business Media, New York (2013)

Image Based Group Happiness Intensity Analysis

V. Lakshmy$^{(\boxtimes)}$ and O. V. Ramana Murthy

Department of Electrical and Electronics Engineering, Amrita School of
Engineering, Amrita University, Coimbatore, India
`lakshmikoduvayur@gmail.com`

Abstract. The objective of this project is to compute happiness index of an
image containing single or multiple persons. There are situations in social
gatherings such as marriage and festivals, where large number of group photos
is taken. If only few photos, say best ten, have to be selected for representing the
whole occasion, manual selection is based on individual perception. Different
individuals have different perception. In this project the perceptions are ranked
in a 0–5 scale of happiness emotion content; thus trying to automate the process.
Image processing techniques are used for computing happiness index of the
image. Initially, faces in the image are detected, followed by expression classification such as happy, neutral and so on. If the image contains multiple
persons, each person's expression is given an individual happiness index. All
the individual happiness indices are integrated to yield the image's happiness
index. Perceptions of different experts are used to validate the happiness indices.
The different combinations of feature extraction, classification and integration is
experimented so that the validation is as high as possible. This technique is
validated on the benchmark HAPPEI database.

Keywords: Group happiness intensity · Feature extraction · HAPPEI database

1 Introduction

Emotion analysis of a single person has been an active research topic for 2 years. A lot
of methodologies have been developed to recognize human emotions using facial
analysis. But, most of the algorithms are validated using datasets created in lab environments. However, analysis of the overall expression theme conveyed by a group of
people in an image taken in wild environments is still an unexplored area of research.
Automatic emotion analysis of a group image is has a wide variety of applications such
as image retrieval, surveillance, early event prediction, video thumbnail creation, event
summarization, advertisement and content recommendation. Number of photos taken
each day is growing exponentially, a recent research by market research firm info
Trends says that more than a trillion digital images will be taken in this year (2017) [1].
Each and every moment is captured in different angles and everyone wants to share

© Springer International Publishing AG 2018
D. J. Hemanth and S. Smys (eds.), *Computational Vision and Bio Inspired Computing*,
Lecture Notes in Computational Vision and Biomechanics 28,
https://doi.org/10.1007/978-3-319-71767-8_88

best shots of their happy moments. It is the need of today's world to develop a technology which can select the best shot from a set of pictures.

One of the important factor which determines the best shot is the happiness content expressed by the faces in the photo. This project tries to automate the selection of best shots from a set of photos based on the perceived happiness content.

1.1   Challenges

There are several challenges in this regard

1. Face detection in outdoor, unconstrained environments is itself an open challenge owing to variations in orientation or pose, illumination, distance from camera, and so on.
2. Facial expression recognition is the next challenging task. Many factors like individual variations in expressing emotions, occlusion, presence of accessories like spectacles, scarf could contribute to the complexity. Moreover, facial expressions are subtle facial muscle movements, and it is challenge to detect and represent these kinds of slight changes.
3. Human perception of group expression is very subjective. Different people consider different attributes to rank group emotion. Apart from emotion intensity of each person there are other contexts which should be considered such as background, clothing, pose, relative distance of persons and so on. Even the contribution of each face may vary depending on the face size, position and occlusion degree.

While computing happiness index, all these factors have to be considered. This project analyses the performance of different classification and feature extraction algorithms for obtaining an index of an image based on it's happiness content.

The next section is an overview of researches related to this area. Section 3 describes the overall framework of this project. The section is divided into 6 subsection where each section explains different methods used in each step of the project and a subsection describes HAPPEI dataset in which the results were validated. Section 4 compares the result obtained for various combinations of feature extraction, classification and integration methods for the computation of happiness index of an image. Section 5 consists of the conclusions arrived from the various experimentations done in the project. Section 6 gives a light to further improvisations which can be done to improve the result and various applications where this work can be implemented.

2   Related Literature

The research in group emotion analysis is in its starting stage. In 2012 an experiment [2] was conducted to infer the happiness level of the persons who are passing through the cameras installed in MIT campus. They used Shore framework to detect faces and to compute the geometrical features from the faces. They computed individual smile intensity based on geometrical features. They took average of the individual smile intensities to obtain overall frame level happiness.

Dhall et al. [3] formulated a framework for estimating the Group emotion index for a given image, focusing on happiness in a social context. They conducted an opinion survey and used occlusion intensity, happiness intensity as local attributes; relative size and distance between centroids of faces as global attributes for estimating group happiness index. They proposed three integration algorithms and a database (HAPPEI) for inferring the happiness mood intensities of a group of people. The database is created from google images using keyword search. Each image and each face in the image is labelled with happiness intensities in the range of 0–5 by four human labellers. In the first algorithm, they took average of individual happiness indices to obtain group happiness index. In the second algorithm they assigned weight to each individual happiness index based on occlusion intensity, relative size and position of the face. Third algorithm was based on topic models. They proved that proper assignment of weights to individual happiness indices reduces the validation error and topic model performed better than the other two algorithms.

As a continuation of the work in [3], Dhall et al. [4] classified group photos into 3 categories based on group expression positive, neutral and negative. They extended HAPPEI database from Google images using keyword search and human annotation. They modelled framework for inferring affect based on multi-modal fusion using Multiple kernel learning (MKL). They formulated two bag of words representations. First one is based on active facial units (BOWAU) and second one is based on low level features (BOWLL). They performed feature augmentation by computing low-level features such as occlusion intensity, attractiveness on the aligned faces. Pyramid of Histogram of Gradients (PHOG) and Local Phase Quantization (LPQ) descriptors are computed over an aligned face. Then they performed feature fusion to obtain local context values for each face. They used CENTRIST [5] and GIST descriptors to obtain global or scene context. These four modalities were used to determine the affect content of the image. The maximum accuracy (67.64%) was obtained by using BOWLL, BOWAU and CENTRIST features.

Recently Mou et al. [6] classified group images into 5 categories based on valance (positive, neutral, negative) and arousal (high, medium). They used face, body and context features. They used Quantised Local Zernike Moments (QLZM) [7] for extracting global and local features from 49 facial points detected using intraface. They also extracted geometrical features from facial points. PHOG is used to extract upper body features. Bounding boxes are fitted to each individual person; and a group box fitted to all the persons together. The relative location and size of each individual box with respect to the group box are used to compute the context features. Further, the relative location and size of the group box with respect to the full image is used to enhance the context features. Then the images are classified using decision level fusion. They used k-NN classifier and obtained an accuracy of 54% for valence, and 55% arousal. Their experimental results showed that geometrical features computed from facial points yielded highest individual accuracy (53%).

3 Overall Framework and Background

The purpose of the system is to estimate the group happiness index of the given group image. The system consist of four stages. First stage of the system is face detection. After detecting all faces in the image, happiness index of each face is computed. Finally group happiness index is computed from the individual happiness indices using an integration algorithm. The different steps are explained in the following subsections (Fig. 1).

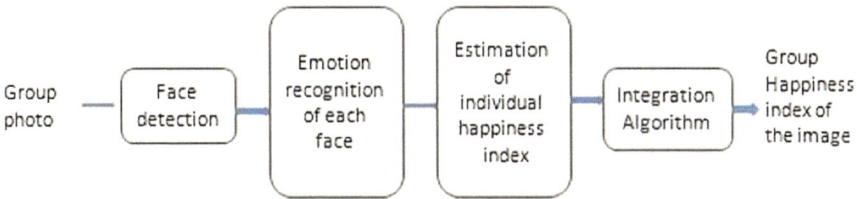

Fig. 1 Overall framework: Happiness indices of individual face are integrated to obtain group happiness index

3.1 Face Detection

Lot of algorithms exists for face detection. The prominent one is Viola Jones algorithm [8]. Matlab implementation of the algorithm is available. This algorithm was used in this to detect faces and to detect mouth from faces. It was observed that it has high false positive rates. The algorithm was tested with various merge threshold values. It was found that false positive rate is decreasing while increasing the merge threshold values. The false positive rates and accuracies for different merge threshold values of Viola Jones algorithm are given in Table 1.

Table 1 Performance of Viola Johns algorithm with different merge threshold values

Merge threshold	False detection (%)	Not detected (%)
8	4.03	7.09
7	5.30	9.49
6	6.75	8.95
4	13.83	3.17
3	29.91	2.56

3.2 Individual Happiness Intensity Computation

The individual happiness intensity is estimated by two ways,

- Feature vector is calculated for the whole face
- Feature vector is calculated only for the mouth.

This stage can be divided into three steps: Pre-processing, Feature extraction and Classification. The various combinations of feature extraction and classification algorithms has implemented.

1. Preprocessing: As a first step, the faces in the image are first detected and cropped to a size of 70 × 70 pixels. The cropped faces are converted to gray scale images before applying feature extraction algorithms.
2. Feature extraction: Feature extraction can be done using different algorithms. Three feature extraction algorithms were experimented for happiness analysis.

- Histogram of Oriented Gradients (HOG) [8]: It is a scale-invariant feature descriptor which is computed on a dense grid of uniformly spaced cells and uses overlapping. HOG feature vector of each cell is the weighted histogram of the gradients (9 bins, 0–3600), where weight of each pixel's orientation is determined by its magnitude. Concatenation of these individual histograms form the final feature vector. The gradient of an image is defined by:

$$\Delta f = \begin{pmatrix} g_x \\ g_y \end{pmatrix} = \begin{pmatrix} \frac{\partial f}{\partial x} \\ \frac{\partial f}{\partial y} \end{pmatrix}$$

where $\frac{\partial f}{\partial x}$ is the gradient in x direction and $\frac{\partial f}{\partial y}$ is the gradient in y direction. The gradient direction is calculated by:

$$\theta = \tan^{-1}\left(\frac{g_y}{g_x}\right)$$

- Pyramid Histogram of gradients (PHOG) [9]: It is another feature extraction algorithm which is similar to HOG. The implementation PHOG consists of the following steps. First, the canny edge detector is applied to the cropped face. Then the face was divided into spatial grids at all pyramid levels. After this a 3 × 3 Sobel mask is applied to the edge contours for calculating the orientation gradients. Then the gradients of each grid are joined together at each pyramid level. The implemented PHOG was of three pyramid levels, 0–360 angle range and bin count of 16.
- Local Binary Pattern (LBP) [10]: It is a texture operator which labels the pixels of an image by threshold the neighbourhood of each pixel and considers the result as a binary number. Then histogram of this LBP image is calculated as feature vector. A label of a pixel gc is generated based on the comparison as per the following equation:

$$LBP_{P,R}(g_c) = \sum_{i=0}^{P-1} 2^i . B$$

where

$$B = \begin{cases} 1, g_i - g_c > 0 \\ 0, g_i - g_c < 0 \end{cases}$$

- PHOG using HOG: It is obtained by concatenating HOG feature calculated in 3 pyramid levels.

3. Classification: The features formed were classified using two popular classifiers: K-Nearest Neighbours (KNN) [11] and Support Vector Machine (SVM) [12].

- K-Nearest Neighbours (KNN): It compares the features of the test image with that of the training images and calculates the distance between each value. The distance can be Euclidean, Manhattan etc. Here Euclidean distance is considered. Using distance calculated it finds k nearest neighbours of the test image and checks their label matrix values and assigns the majority label as the label for the test image.
- Support Vector Machine (SVM): It is a discriminative classifier formally defined by a separating hyperplane. During training phase of the algorithm, it gives an optimal hyperplane for categorizing testing features based on labelled training data. Linear was used for testing and training.

A k-fold cross validation process was carried out to get the classification accuracy. Here k was chosen as 10. The entire dataset is divided into 10 random groups. Each group is taken as the test matrix and the remaining group as training groups in an iterative process. The classification accuracy is calculated as follows:

$$A = \frac{C}{T}$$

where A is the classification accuracy, C is the number of correct classifications and T is the total number of classifications.

3.3 Computation of Group Happiness Index

The group happiness index was computed using individual happiness indices. Two approaches were experimented for this. The first method was to take average of the individual indices. The second one was weighted average. Weight for individual index were based on the size of the face. Two methods were compared based on Root Mean Square error (RMSE). RMSE is calculated using the equation

$$A = \sqrt{(e - o)^2}$$

where e is the happiness intensity given in the dataset and o is the intensity estimated using the method.

3.4 HAPPEI Dataset

HAPPEI dataset is a collection of happy people images. It consists of 2638 images. The images for this database is collected using keyword search in flicker. The keywords include different scenarios such as marriage, graduation day and so on. Each image is labelled with image level happiness intensity in the scale of 0–10. The dataset also contains position, happiness intensity and occlusion intensity of 8500 faces in the images of the dataset. The labelling is done based on human perception. Figure 2 contains the sample images from HAPPEI dataset.

Fig. 2 Few images in HAPPEI dataset

4 Results and Discussion

The performances of different combinations of feature extraction and classification algorithms for estimating individual happiness intensities using features from face and mouth are given in Tables 2 and 3 respectively. From the results, it was observed that features extracted from the whole face has better performance than that from the mouth. It is also observed that PHOG with SVM classifier has more performance than any other combination.

Table 2 Performance of features extracted from face

	KNN classifier (%)	SVM classifier (%)
PHOG	28.65	47.0
HOG	33.1	46.70
LBP	26.65	43
PHOG using HOG	31.95	43.12

The methods for estimating group happiness index were compared using Root Mean Square Error obtained for the images in HAPPEI dataset. Lower the RMSE, better the performance. The Root Mean Square Error (RMSE) obtained using averaging

Table 3 Performance of features extracted from mouth

	KNN classifier (%)	SVM classifier (%)
PHOG	18.65	30.0
HOG	30.1	35.10
LBP	21.65	33
PHOG using HOG	28.86	36.75

for computing group happiness intensity is shown in Table 4. The performance of weighted averaging is shown in Table 5. From the results, it was observed that weighted average has more performance than simple average.

Table 4 Performance (RMSE) of averaging algorithm to compute group happiness index

	KNN classifier	SVM classifier
PHOG	1.13	0.80
HOG	1.23	2.90

Table 5 Performance (RMSE) of weighted averaging to compute group happiness index

	KNN classifier	SVM classifier
PHOG	0.9	0.75
HOG	1.1	2.07

5 Conclusion

Face detection with Viola Johns algorithm is able to detect frontal faces and its false positive rates can be reduced by setting proper merge threshold value. PHOG feature extracted from the face can be used with SVM classifier to estimate individual happiness indices. Group happiness index can be computed from these individual intensities. Weighted average (where weight is based on individual face size) has better performance than simple average for calculating group happiness index.

6 Future Scope

This work was just a beginning in the area where a lot of progress can be achieved. The accuracy can be improved by including feature descriptors such as centrist and gist which can account the background features. The result can further improved by using better face detection methods and classification algorithms. The influence of factors such as number of faces, background, variation of face sizes in an image, orientation of faces, orientation of body and gestures of the persons in the image in estimating happiness index can be evaluated. As mentioned in the introduction, this work has a lot of applications. It can be integrated to various mobile applications which can automatically select best shots and create mini movies or collages to describe an entire event.

References

1. Hernandez, J., Hoque, M.E., Drevo, W., Picard, R.W.: Mood meter: counting smiles in the wild. In: ACM Conference on Ubiquitous Computing, Pittsburgh, pp. 301–310 (2012)
2. Dhall, A., Goecke, R., Gedeon, T.: Automatic group happiness intensity analysis. IEEE Trans. Affect. Comput. **6**(1), 13–26 (2015)
3. Dhall, A., Joshi, J., Sikka, K., Goecke, R., Sebe, N.: The more the merrier: analysing the affect of a group of people in images. In: 11th IEEE International Conference and Workshops on Automatic Face and Gesture Recognition (FG), Slovenia, pp. 1–8 (2015)
4. Mou, W., Celiktutan, O., Gunes, H.: Group-level arousal and valence recognition in static images: face, body and context. In: 11th IEEE International Conference and Workshops on Automatic Face and Gesture Recognition (FG), Slovenia, pp. 1–6 (2015)
5. Wu, J., Rehg, J.M.: CENTRIST: a visual descriptor for scene categorization. IEEE Trans. Pattern Anal. Mach. Intell. **33**(8), 1489–1501 (2011)
6. Viola, P., Jones, M.J.: Robust real-time face detection. Int. J. Compu. Vis. **57**(2), 137 (2004)
7. Sariyanidi, E., Gunes, H., Gökmen, M., Cavallaro, A.: Local zernike moment representation for facial affect recognition. In: BMVC (2013)
8. Bosch, A., Zisserman, A., Munoz, X.: Representing shape with a Spatial Pyramid Kernel. In: 6th ACM International Conference on Image and video Retrieval, Amsterdam, pp. 401–408 (2007)
9. Dalal, N., Triggs, B.: Histograms of oriented gradients for human detection. In: IEEE Computer Society Conference on Computer Vision and Pattern Recognition, San Diego, pp. 886–893 (2005)
10. Ameur, B., Masmoudi, S., Derbel A.G., Hamida, A.B.: Fusing gabor and LBP feature sets for KNN and SRC based face recognition. In: International Conference on Advanced Technologies for Signal and Image Processing, Tunisia, pp. 453–458 (2016)
11. Dhall, M., Ramana, O.V., Goecke, R., Joshi, J., Gedeon, T.: Video and image based emotion recognition challenges in the wild: EmotiW 2015. In: ACM on International Conference on Multimodal Interaction. New York, USA (2015)
12. Suja, P., Tripathi, S., Deepthy, J.: Emotion recognition from facial expressions using frequency domain techniques. In: Advances in signal processing and intelligent recognition systems, pp. 299–310 (2014)

Earthquake Analysis: Visualizing Seismic Data with Python

Saksham Tulsyan[(⊠)], Bharat Bahl, Shivi Kaya, and G. Thippa Reddy

SITE, VIT University, Vellore, India
saksham.tulsyan2015@vit.ac.in

Abstract. Earthquake is one of the most destructive and life wrecking natural calamity that basically happens due to energy released from Earth's crust. In this paper we propose to serve previous years' earthquake data as a parameter and visualize relationships, if any, among various attributes of the data set. In conclusion we aim to discover regions which are most and least prone to earthquake and the probable magnitude of earthquake at various latitudes and longitudes. Studying raw data from excel sheet is old century and very hard so we have used data mining techniques such as clustering and regression to achieve the obtained results.

Keywords: Earthquake · Clustering · Regression · Visualization

1 Introduction

Earthquake by definition is a tremor or the shaking of earth due to sudden release of energy in earth's lithosphere that originates seismic activity. The seismic waves define the nature of the earthquake its type and frequency over a given period of time. Millions of people have lost their lives and property because of the earthquake and no prior warning. Earthquakes are one the most destructive natural disaster which if warned against can save millions of lives. The aim of the paper is to analyse previous years' data and mine some results which might help in drawing conclusions about a pattern in earthquake analysis. Earthquakes can't be averted completely using data mining but a probable warning or analysis might help people prepare for it.

Data mining is a technique in which huge amount of data is studied using various algorithms and some logical patterns which discover some hidden meaning that might be useful in general are mined. Data mining techniques can also be deployed for prediction and analysis of these natural hazards and calamities. For analysis of earthquake data we have used clustering algorithm and linear regression and visualised the results using Orange. In the data available we had records of various earthquakes' attributes like latitude, longitude, magnitude etc. Plots for various columns were made, clusters formed were analysed and the line of best fit was plotted to get the desired points. When the aforementioned approach was applied to latitude and longitude we

D. J. Hemanth and S. Smys (eds.), *Computational Vision and Bio Inspired Computing*,
Lecture Notes in Computational Vision and Biomechanics 28,
https://doi.org/10.1007/978-3-319-71767-8_89

got the regions that were the most and least prone to earthquake. The line of best fit predicted the most probable earthquake prone latitude for a given longitude and vice versa. The results matched the present data. Similarly magnitude was predicted for various latitudes and longitudes.

2 Literature Survey

Otari and Kulkarni in [1] have discussed in their paper various techniques of predicting natural disasters like earthquakes using various logistic and classification models, decision trees and neural networks. They have provided methods to overcome all issues regarding prediction of various seismic activities around the globe. Various other methods are put open for further research by them.

Gupta and Gupta in [2] have discussed their analysis of seismic data using the Hadoop Hive big data analysis tool. Plotting of graphs and tables of magnitude of earthquakes, destruction caused, lives lost has been done by using Tableau and Power Map features and a model was developed.

In [3], authors have used previous earthquake data as a parameter to visualize dependencies among factors that cause earthquakes such as velocity, acceleration and displacement. They have developed a tool in which the data can be entered in raw format and graphs will be generated after application of data mining techniques.

Negarestani et al. [4] aims at calculation of radon concentration in soil using a Layered Neural Network (LNN). This technique can obtain any useful pattern between the radon concentration and the environmental parameters. On comparison with linear computation techniques results state that LNN can give a better understanding of radon levels that may be the cause of various environmental problems. Radon and environmental parameters might have a non-linear relationship which can be better discovered using LNN.

In [5], author tell us that various regression methodologies have been deployed to predict the radon levels in soil gas based on environmental parameters like barometric pressure, soil temperature, air temperature and rainfall. A model has been built from the analysis provided by three stations in the Krkso basin, Slovenia. The studies also stated that trees give more accurate and better results than regression techniques. The model predicts radon concentration with a correlation of 0.8. With seismic activities present the correlation decreases.

In [6], it is explained that seismic data is generally random and very difficult to analyse. From data, it is tough to extract the hidden relationships between earthquake occurrences. In this paper, global earthquake data from 1965 till 2014 is analyzed. Earthquakes are divided according to their magnitude, depth, geographic location and time of occurrence, into two groups: namely according to the Flinn-Engdahl method or using a rectangular grid based on latitude and longitude and then the two portions are compared. The first portion is the earthquake distribution is approximated by a method given by Gutenberg and Richter whereas in the second portion, mutual information is calculated to approximate. Therefore the charts generated due to this are a good alternative to analyze the behavior of earthquakes.

In [7], tools have been proposed which can go along with the various tools that are available for researches to work on large data sets of earthquakes. In this paper, various new analytical techniques have also been discussed but the most important part is where the proposed tool architectures have been stated and described in order to inform the users how the tools actually work. Implementation of the tools has also been described in the paper along with an option to refer to YouTube videos for any help.

In [10], authors have presented a web client-server WEB-IS for remote analysis and visualization of seismic data for both small and large earthquakes. The clustering schemes, feature generation, feature extraction techniques and rendering algorithms form a computational framework for this environment. The have told us about the usefulness of NaradaBrokering (iNtegrated Asynchronous Real-time Adaptive Distributed Architecture) as a flexible middleware for providing a high throughput in remote visualization of geophysical data.

In [11], authors have told about the challenges faced in earthquake prediction. Reliable, practical prediction methods should be developed to save lives. The book contains the rigorous statistical methods and analysis which allow readers to recognize promising approaches.

3   Proposed Methodology

Firstly, the large amount of data used for earthquake analysis was taken and analyzed.

Then data pre-processing was performed on the data set. Data pre-processing is the most important step for analyzing large amounts of data and big data sets. It involves following steps:

- Data Cleaning—The dates for the occurrence of earthquake were given in dd/mm/yyyy format which was converted into yyyy format so that graph plotting could become easier and many values were missing which were replaced by the average of the previous and next available value. This was done using a python script.
- Data Integration—All the cleaned data was integrated together into a one large data set. All the data in different formats are put together and conflicts within the data are removed.
- Data Reduction—The data was available from 1965 BC to 2016 AD so we reduced it to 1965 AD–2016 AD so that the results could be more relevant.

After the data was ready to be used, wee first visualized the data using WEKA tool. Following are the results obtained from the WEKA tool (Fig. 1).

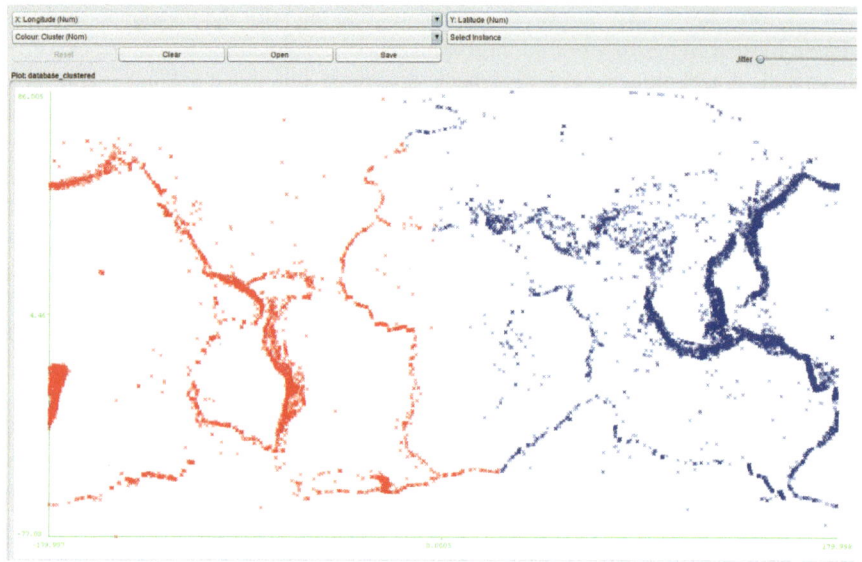

Fig. 1 Longitude versus latitude

After visualizing the data from the WEKA tool, we developed an algorithm for scatter plot, clustering and visualized it using Orange. We plotted all the data points one by one and many clusters were formed signalling earthquake prone areas. To determine the regions that are more prone to earthquakes, we plotted the different data available to us, various regression lines were plotted. For this, we used linear regression techniques. Linear Regression develops relationship between a dependent variable and one or more independent variables using a line of best fit (also known as regression line). Using above stated concepts, we developed an algorithm to visualize the relationship between various attributes of the data set. To implement this, we used csv, numpy and matplotlib libraries of python. Functions used are subplots(), polyfit(), plot () and scatter(). The results are as follows (Fig. 2).

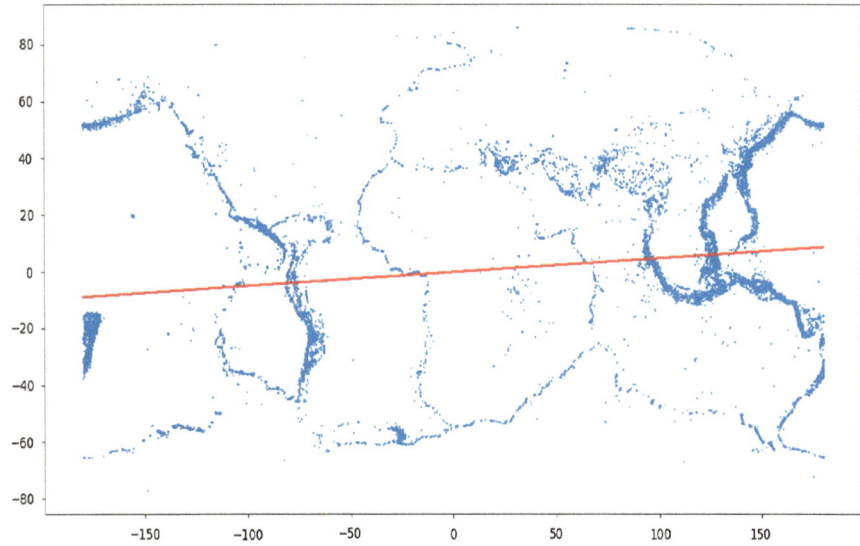

Fig. 2 Longitude versus latitude. Equation of the best fit line: Y = 0.048x − 0.256

When plotted on the world map, following regions could be designated as earth-quake prone areas (Fig. 3).

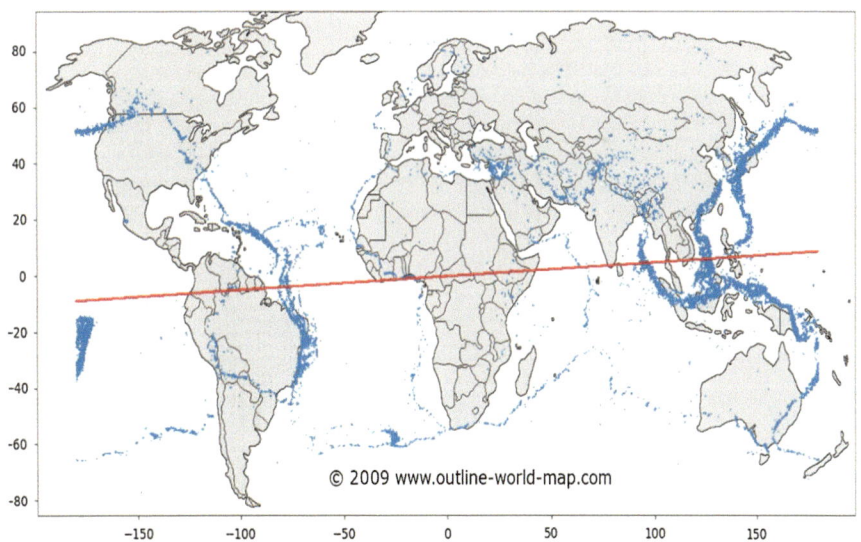

Fig. 3 Longitude versus latitude (World Map)

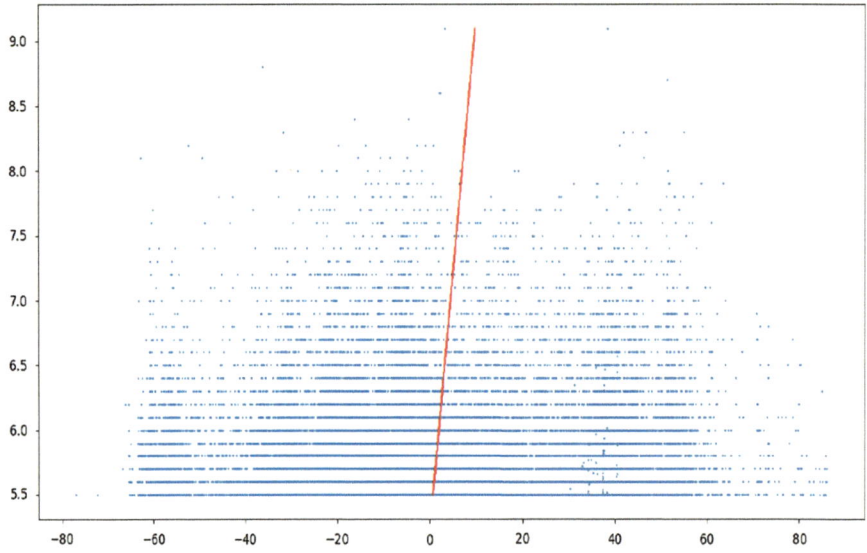

Fig. 4 Magnitude versus latitude. Equation of the best fit line: X = 2.490y − 12.97

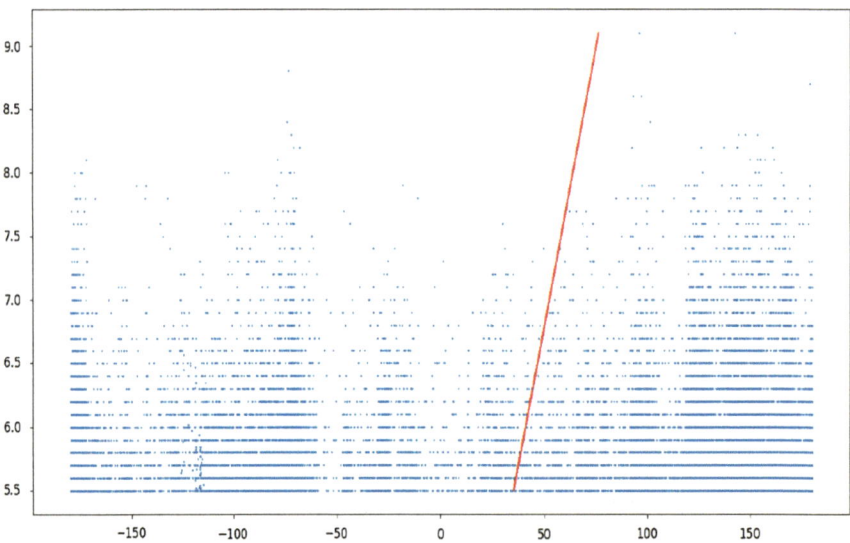

Fig. 5 Magnitude versus longitude. Equation of the best fit line: X = 11.44y − 27.68

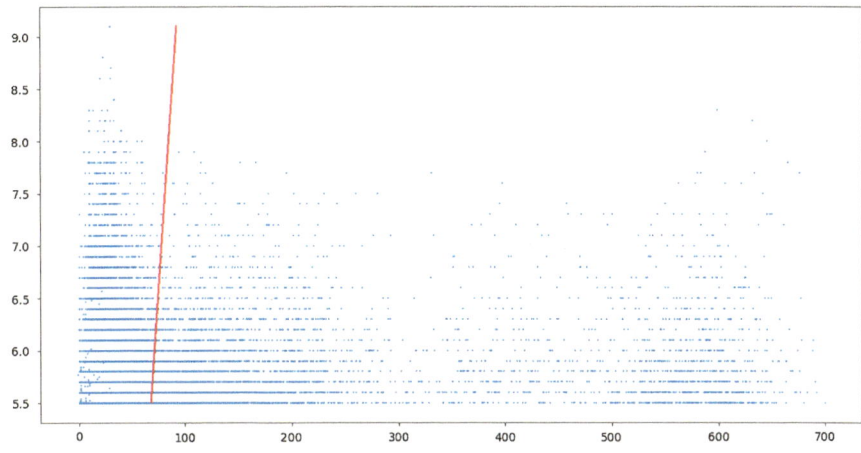

Fig. 6 Magnitude versus depth. Equation of the best fit line: x = 6.8y + 30.763

4   Results

For example, at longitude 113.288°E, on putting this value in the equation: **y = 0.048x − 0.256**, we get latitude of 5.181°N which points to the following location.

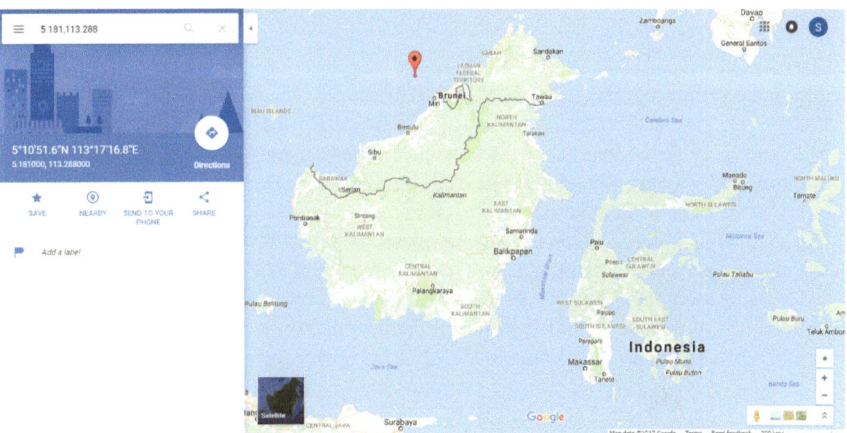

This location is near Indonesia which is an earthquake prone region.

From Fig. 4, at latitude 8°N, on putting this value in the equation: **x = 2.490y − 12.97**, we get magnitude of 6.95.

From Fig. 5, at longitude 50°E, on putting this value in the equation: **x = 11.44y − 27.68**, we get magnitude of 6.79.

From Fig. 6, at a depth of 85 km, on putting this value in the equation: **x = 6.8y + 30.763**, we get magnitude of 7.97.

5 Conclusion

An integral part of any analysis is the building of an efficient classification model from the existing data. In this study, we have visualized seismic activity of the world occurring from 1965 to 2016 using data mining techniques and python. We have visualized data of magnitude, longitude, latitude and depth of all earthquakes on scatter plots and then using regression algorithm to make predictions that can to determine the major risk zones and be prepared for any seismic activity to happen. By doing so, we can heavily minimize the havoc and destruction which can be caused by a future earthquake. The methods and techniques used in this experiment can open new paths for seismic research.

References

1. Otari, G.V., Kulkarni, R.V.: A review of application of data mining in earthquake prediction. Int. J. Comput. Sci. Inform. Technol. 3(2), 3570–3574 (2012)
2. Gupta, G., Gupta, I.S.: Earthquake Data Analysis and Visualization using Big Data Tool
3. Patil, S., Roy M.: Visualization of Earthquake Data using Data Mining
4. Negarestani, A., Setayeshi, S., Ghannadi-Maragheh, M., Akashe, B.: Layered neural networks based analysis of radon concentration and environmental parameters in earthquake prediction. J. Environ. Radioact. 62(3), 225–233 (2002)
5. Zmazek, B., Todorovski, L., Džeroski, S., Vaupotič, J., Kobal, I.: Application of decision trees to the analysis of soil radon data for earthquake prediction. Appl. Radiat. Isot. 58(6), 697–706 (2003)
6. Machado, J.A.T., Lopes, A.M.: Analysis and visualization of seismic data using mutual information. Entropy 15(9), 3892–3909 (2013)
7. Luksys, E., Asimakopoulou, E., Bessis, N.: Development of tools for data analysis of earthquakes. In: 2014 International Conference on Intelligent Networking and Collaborative Systems (INCoS), pp. 406–410. IEEE (2014)
8. https://www.ngdc.noaa.gov/nndc/struts/results
9. https://en.wikipedia.org/wiki/Lists_of_earthquakes
10. Yuen, D.A., Kadlec, B.J., Bollig, E.F., Dzwinel, W., Garbow, Z.A., da Silva, C.R.: Clustering and visualization of earthquake data in a grid environment. Vis. Geosci. 10(1), 1–12 (2005)
11. Lomnitz, C.: Fundamentals of Earthquake Prediction. Wiley, USA (1994)
12. https://www.analyticsvidhya.com/blog/2015/08/comprehensive-guide-regression/

Instill Dental Antennas for Minimally Interfering Bio-medical Devices

T. Gomathi$^{(\boxtimes)}$, S. Maflin Shaby, and B. Priyadharshini

Department of ETCE, Sathyabama University, Chennai, India
{gomes20,maflin.s,priyadharshini8821}@gmail.com

Abstract. Implantable medical devices are mainly in use for the wireless electronic health devices and remote home care applications because of their miniature techniques and low-power consumption integrated circuits. The demand for bio-medical applications that employ wireless telemetry systems has increased significantly. Antennas can be implanted into human bodies to form a bio-communication system between medical devices and exterior instruments for short-range bio-telemetry applications. A compact antenna is being implanted in dental applications. The antenna consists of Archimedean spiral antenna and a Hilbert-based curve gives high gain. In this design two Hilbert based curves and one spiral antenna is embedded. This 3D folded antenna is implanted in teeth. It gives uni-directional radiation pattern and eliminates the back radiation. Due to unidirectional radiation the power loss is reduced. In Archimedean spiral antenna FR4 material is used as a substrate with dielectric constant 2.2. Here probe feed technique is used to excite the current. Simulation is done by using HFSS software. Shortening wall technique is used to enhance the bandwidth. The simulated antenna performs high gain (>6 dB), return loss (<−10 dB) and broad bandwidth. This antenna can be used for bio-medical applications.

Keywords: Hilbert fractal antenna · Implantable antenna · Archimedean spiral antenna · HFSS · Gain

1 Introduction

It emphasizes mainly on the implantable medical devices for dental antenna. It is one of the most important innovations in today's healthcare systems. An implant is a medical device which is manufactured to replace a missing biological structure, support a damaged biological structure, support a damaged biological structure or enhance an existing biological structure. These implantable biomedical devices have founded considerable consciousness for their use in wireless electronic health devices and remote home care applications because of its small size and less usage of power integrated circuits [1]. Antennas are very important component of communication system. Here in dental application antennas play a significant role.

© Springer International Publishing AG 2018
D. J. Hemanth and S. Smys (eds.), *Computational Vision and Bio Inspired Computing*,
Lecture Notes in Computational Vision and Biomechanics 28,
https://doi.org/10.1007/978-3-319-71767-8_90

Here, implantable devices are antennas embedded in system contains biosensors and interface circuits, which enables the exchange of data between implantable devices and external environment. We are using HFSS software. HFSS is a high performance full wave electromagnetic (EM) field simulator. The solution to 3D EM problems can be obtained quickly and accurately using the integration of simulation automation and visualization and solid modeling. Here, implantable devices are antennas embedded in system contains biosensors and interface circuits, which enables the exchange of data between implantable devices and external environment. In this paper, implantable antenna is a combination of square spiral and Hilbert base curve for effective size reduction and gain at a fixed frequency. The implant devices require a radio frequency range to transmit the information collected from the biological devices. There will be a certain range where these implanted antennas will be used and that range of frequency is normalized. The range is termed as MEDRADIO (Medical Device Radio Communicate Service) [2].

The Medical Device Radio Communicate Service (MEDRADIO) is in the 401–406, 413–419, 426–432, 438–444 and 451–457 MHz range. MedRadio spectrum is used for methodical healing purposed in implanted medical devices [3]. In this paper we are designing an antenna for implantable systems that operate at 400.8 MHz. We are operating antenna at 401 MHz frequency because human body allows this frequency. While designing an antenna the first and foremost consideration is patient safety, and then comes biocompatibility, size, and high-quality communication with exterior equipments. This all requirements are fulfilled at 401 MHz frequency at which we are designing and operating the antenna. We have designed an antenna which two Hilbert based curves and one spiral antenna is embedded. This 3-D folded antenna will be implant in teeth [4–6].

2 Antenna Design

Here we are going to design antenna in which Archimedean Spiral and hilbert curves are embedded. The antenna substrate must be selected based on the dentures dimensions and properties of the oral environment. Here we are considering mainly molars because their size is large compared to other teeth and practically support antenna in Medradio band. $8 * 11.5 * 8$ mm^3 is an average molar of a person. We are using a regular cube shape as a easy teeth model for antenna design. We are using ceramic dentures. Alumina and zirconium dioxide are considered as a common materials according to the study. Depending upon the shape of the teeth we are going to design archimedean square spiral antenna. We are using square spiral antenna because when examined we can see that square antenna has an advantage of operating at lower frequencies with the same performances. First radiation band of a spiral antenna occurs when the circumference of the spiral is one wavelength, corresponding to a diameter $D = \lambda/\pi$ for the circular spiral and a width $w = \lambda/4$ for the square spiral. According to this relation square spiral geometries have more advantage in terms of size (Fig. 1).

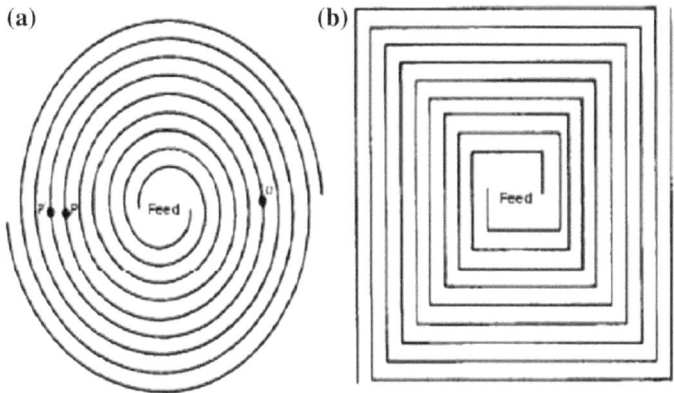

Fig. 1 Archimedean spirals, **a** round, **b** square

The Hilbert Curve will be designed. The Hilbert curve is a space filling curve that visits every point in a square grid. It has advantages in those operations where the coherence between neighbouring pixels is important. The Hilbert curve is also a special version of a quad tree; any image processing functions that benefits from the use of quadtrees may also use a Hilbert curve. Hilbert curve geometry in antenna design has been configured to reduce the size of antenna as well as to get multiple resonances. Total length of the Hilbert curve increases when increasing the iteration stage, while keeping the overall space of the entire geometry fixed [7].

This system model modeled about Hilbert's space-filling curve and some of its properties that are need. For more background, there is monograph Sagan in the year 1994 on the space-filling curves that describes the Hilbert's curve. Then Zumbusch in the year 2003 describes all the multilevel numerical methods, which also includes space-filling curves. Throughout this paper, parameter d is a positive integer, λd is the d-dimensional Lebesgue measure, and II (Figs. 2 and 3). II is the usual Euclidean norm. For integer m $\geq$ 0, define 2 dm intervals

$$Idm(k) = [k/2dm, k + 1/2dm], \quad k = 0, \ldots, 2dm - 1 \tag{1}$$

$$Idm = \{Idm(k) \mid k < 2dm\}. \tag{2}$$

and let Idm = {Idm (k) | k < 2dm}.

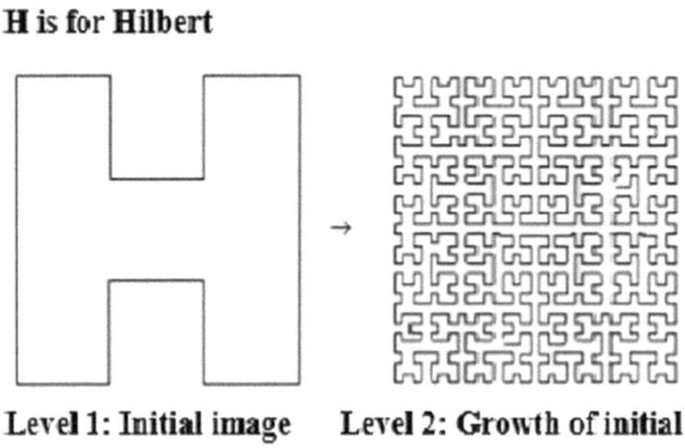

H is for Hilbert

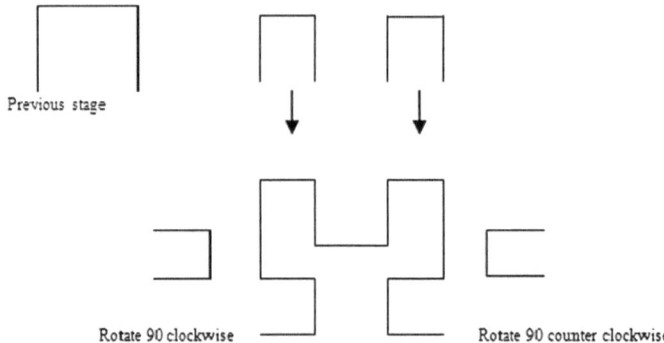

Level 1: Initial image   Level 2: Growth of initial

Fig. 2 The Hilbert curve's stages of growth

Fig. 3 Hilbert curve construction

Next, for k = (k1, ... kd) with kj ∈ {0, 1, ..., 2m − 1} define 2dm sub cubes of [0, 1]d via

$$Edm(k) = \Pi dj = 1[kj/2m, kj + 1/2m \qquad (3)$$

The set of indices κ is Kmd = {0, 1 ... 2m − 1} d and let Emd = {Emd (κ) | κ ∈ Kmd}. Sequence of mapping Hm: Idm dm can be found with the following properties,

- **Bijection**:

$$\text{For } k \neq k', Hm(Idm(k)) \neq Hm(Idm(k')) \qquad (4)$$

- **Adjacency**:

The Hm(Idm(k)) and Hm(Idm(k + 1)) two sub cubes are adjacent. They have one (d − 1)-dimensional face in common.

- **Nesting**:

If we split Imd (k) into the 2d successive sub intervals Im + 1d (kl), kl = 2dk + l, l = 0, …, 2d − 1, then Hm + 1(Im + 1d (kl)) are the sub cubes in which union is Hm (Imd (k)). Hilbert curve is defined by H(x) = limm → ∞ Hm(x). The point x ∈ [0, 1] that is Imd (km) of intervals which compress to x. If x is not having a terminating base 2 representation then the sequence Imd (km) is unique and thus Hm(Imd (km)) is a unique sequence of the sub cubes (Fig. 4).

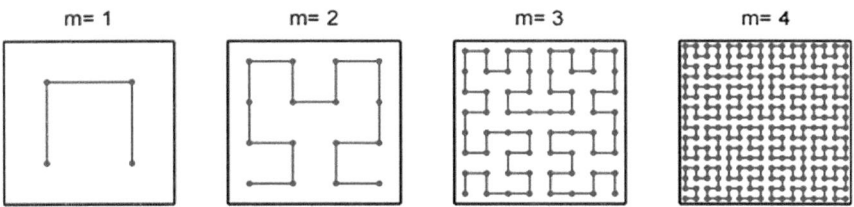

Fig. 4 First 4 stages in the approximation of Hilbert's space-filling curve

Points x = 1/4 = 0.010 = 0.001 with the two binary representations however has uniquely defined H(x). The Hilbert curve passes through the every point in [0, 1] d. It is not the surjective: there are some points where x is not equal to H(x) = H(x′). There are more than one way to define the sequence of |mappings in a Hilbert curve. But any one of those ways produces a mapping H with all these properties:

- P (1): H (Idm(k)) = Hm(Idm(k)).
- P (2): If Ac[0, 1] is measurable, then $\lambda 1(A) = (H (A))$.
- P (3): if x ~ U ([0, 1]) then H (x) ~ U([0,1]d). It admits the change of variables.

The antenna proposed at Medradio band [8] was based on Hilbert-shaped fractal geometry and it consists of grounded shorting pin earth feed port as shown in Fig. 5. A metal shorting pin which has an equivalent shunt inductance, also was grounded to improve impedance of compact radiator. Thus the effective current path of the radiator is extended based on this design, and to allow antenna miniaturization it reduces the antenna operation frequency. To avoid dental problems from tearing caused by chewing or biting actions, the two sides of the molar were observed by the fractal radiator of the reference antenna. The reference antenna is then hidden and protected in the spaces between adjacent teeth with the help of a thin metal wire connecting at both radiators [9].

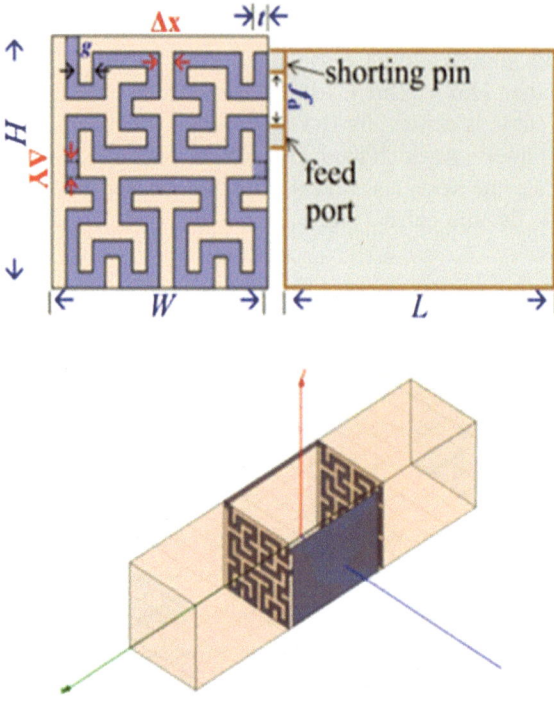

Fig. 5 Antenna designed using Hilbert curve

With the increase in the orders of the Hilbert curves, there is a decrease in operation frequency; however, when the fractal Hilbert curves order is exceeded by four, then the operation frequency insignificantly decreases, and then the implementation limitations of the wire gaps and the width become more precised [10]. After determining the order of the line gaps and the width are also determined in the fixed area [11, 12]. The frequency operation for MedRadio can be further clarified by adjusting ΔX and ΔY. The detailed dimensions of the antenna are as follows: $L = 11.5$, $W = 8$, $H = 8$, $t = 0.45$, $g = 0.45$, and $f_d = 1.4$ mm [10] as shown below in Fig. 6.

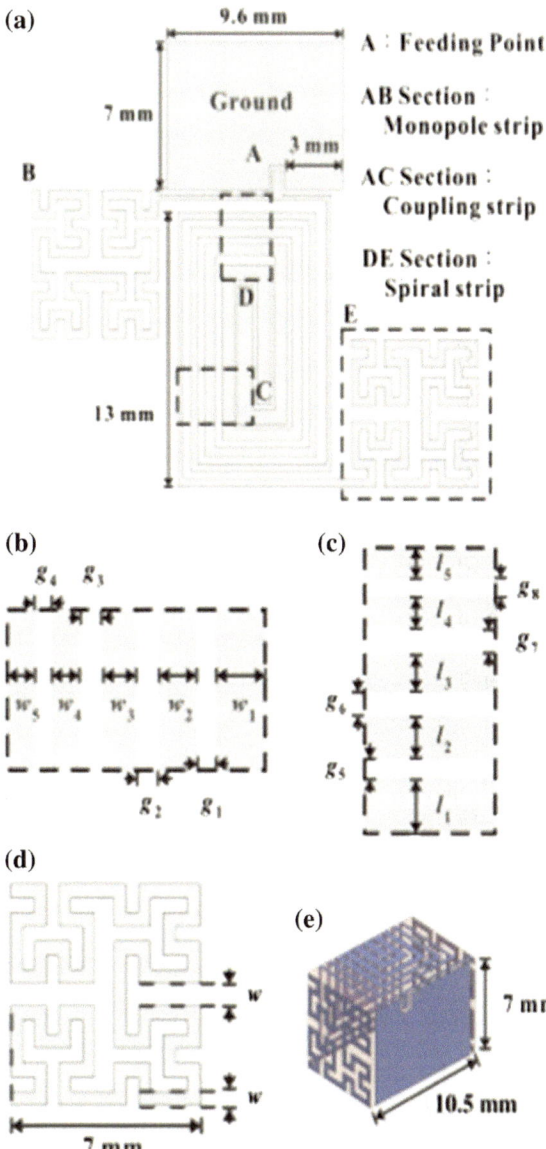

Fig. 6 Structure and dimension of broadband antenna

3 Simulations and Discussions

The figure below shows the simulated 3D model.

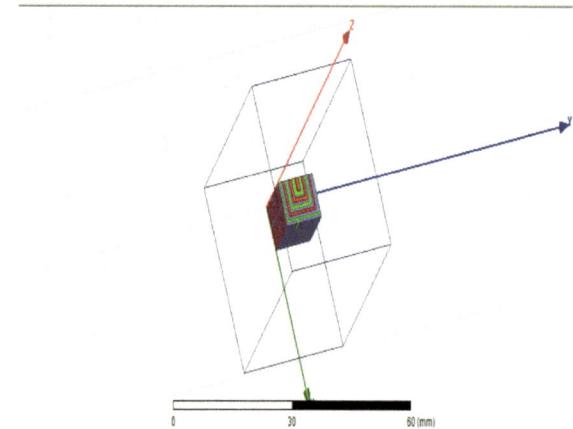

Radiation Pattern

The below result i.e., radiation pattern shows us a visualized picture of where our antenna transmits and receives power, where theta is the horizontal polarization and phi is the vertical (Fig. 7).

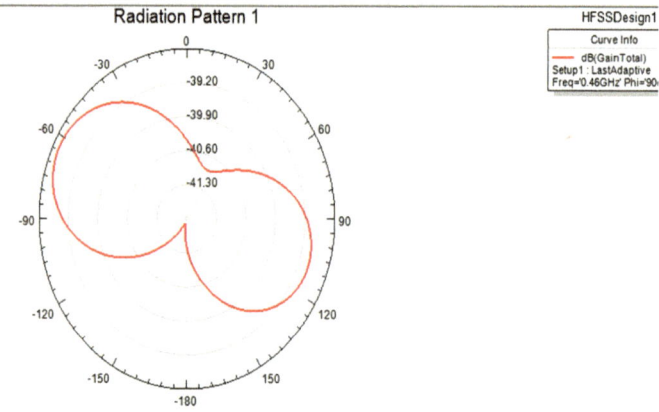

Fig. 7 Radiation pattern

Return loss

Figure 8 shows the return loss, which is of −24 dB at a frequency of 464 MHz and a peak shape is observed in the graph. With the help of the return loss we are able to conclude that whether the antenna which we have designed is propagating in the desired frequency range. The sweep values in the X and Y directions are observed in the graph.

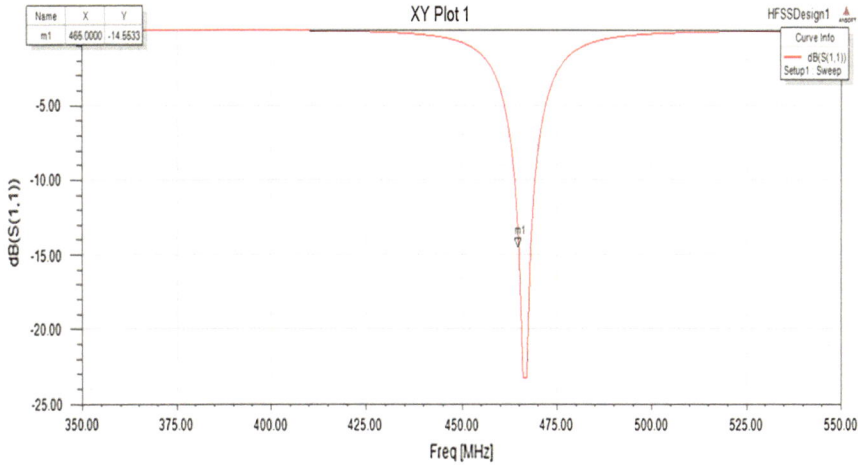

Fig. 8 Return loss

Gain

The proposed antenna gain bandwidth is nearly 12 MHz and the gain −3.7 dBi and a return loss of 474 MHz. The red area shows high directivity. From the above gain graph we can understand at what region our antenna is more efficient (Fig. 9).

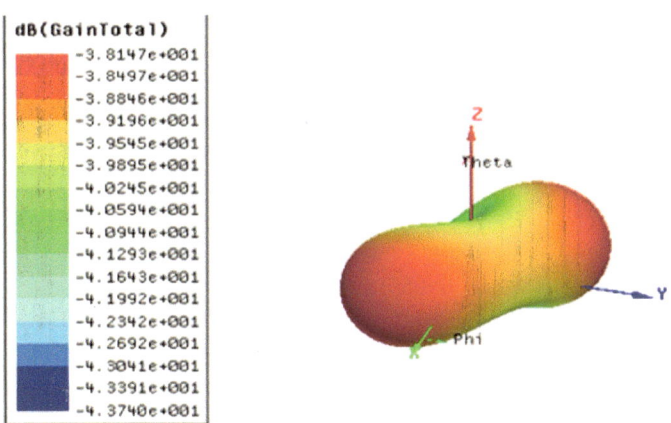

Fig. 9 Gain of the antenna

Fabricated Antenna

The antenna proposed is designed and fabricated on ZrO_2 substrate. Before exhibiting ceramic properties ZrO_2 must be heated to 1200° by using cavity test the dielectric constant and loss tangent of ZrO_2 is remeasured after processing the ceramic (Fig. 10).

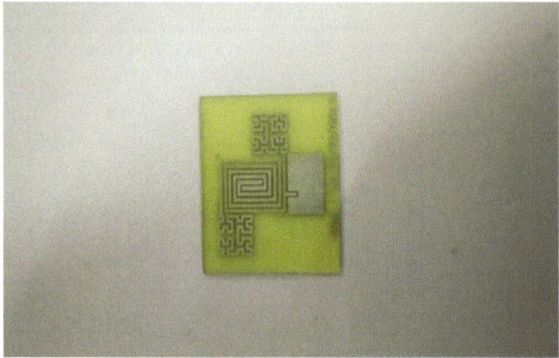

Fig. 10 Fabricated antenna

The FR4 material as a flexible substrate is used in designing the antenna. In the fabrication process, planar fabricated antenna was cut in 5 pieces on a FPCB substrate and the design was attached using the adhesive H20E which is bio-compatible and it is a medical device adhesive and it was used to glue the inner ceramic to the FPCB with the etched antenna radiator. Then, soldering of RF connector was done. The connector is specified as SMA connector. Then the output return loss graph is obtained with the help of Network Analyser (Fig. 11).

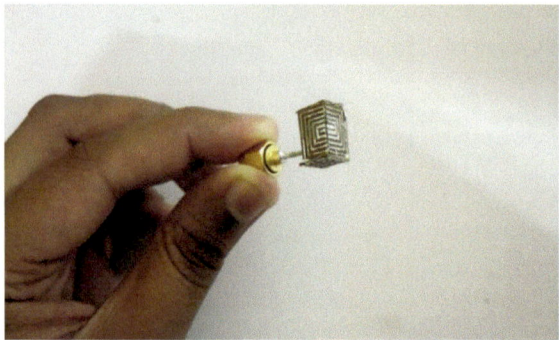

Fig. 11 Etched antenna on the FPCB

Testing Result Analysis

In the proposed model after the fabrication we are getting the return loss of −30 dB at the frequency of 430 MHz at M1 and 442 MHz at M2 at the peak shape observed in the graph in Fig. 12. There is a 5–10% increase in the return loss at the desired frequency. With the help of the return loss we are able to conclude that whether the antenna which we have designed is propagating in the desired frequency range (Fig. 12).

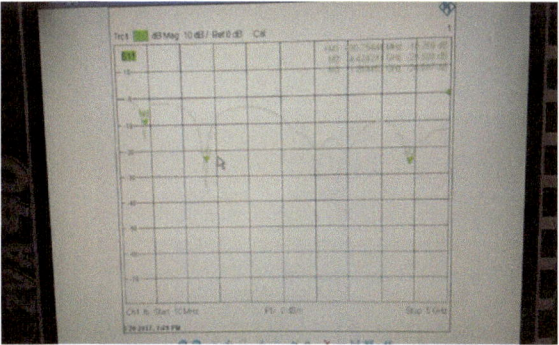

Fig. 12 Tested return loss result

The Reference antenna gives a gain of −6.78 db whereas the proposed antenna gives for about −3.4 db. The bandwidth is given as 60 MHz whereas in the proposed antenna it's 12 MHz. Return Loss is given as −24 db in the existing antenna but in the proposed system it is −30 db. Hence, there is a proper increase in the proposed antenna model.

4 Conclusion

Thus a 3D compact implantable dental antenna that operates in a MedRadio band has been designed. This proposed design is a combination of a Archimedean spirals and a Hilbert based fractal, which is designed to create a minimally interfering implantable devices for remote health care applications. This antenna design characteristics a high antenna gain, compressed area and volume, reasonable 2D origami implementation, and a broad bandwidth to conquer environmental variations. After the combination of Archimedean spiral design along with the fractal Hilbert shapes, two current paths are found in the MedRadio band which generates multiple modes, and achieves a broadband design. Thus the proposed antenna has a total area of less than 245 mm^2. Moreover, the proposed antenna gain bandwidth is nearly 12 MHz. We are getting a gain of −3.7 dBi and a return loss of 474 MHz. Here in this design FR4 material is used because of which an antenna of low cost is designed, and this design is not a complex system. Thus this proposed broadband dental antenna is feasible for dental application and promising in future applications.

References

1. Lee, C.M., Yo, T.C., Tu, C.H., Juang, Y.Z.: Compact broadband stacked implantable antenna for biotelemetry medical devices. Electron. Lett. **43**, 660–662 (2007)
2. Yang, C.L., Tsai, C.L., Chen, S.H.: Implantable high-gain dental antennas for minimally invasive biomedical devices. IEEE Trans. Antennas Propag. **61**(5) (2013)

3. FCC MedRadio Specification. Available: http://wireless.fcc.gov/services/indexhtm?Job= service_home&id=medical_implant

4. Kim, J., Rahmat-Samii, Y.: Implantable antennas inside a human body: simulations, designs and characteristics. IEEE Trans. Microw. Theory. Tech. **52**(8), 1934–1943 (2004)

5. Soontornpipit, P., Furse, C., Chung, Y.C.: Design of implantable microstrip antenna for communication with medical implants. IEEE Trans. Microw. Theory Tech. **52**(8), 1944–1951 (2004)

6. Gosalia, K., Humayun, M.S., Lazzi, G.: Impedance matching and implementation of planar space-filling dipoles as intraocular implanted antennas in a retinal prosthesis. IEEE Trans. Antennas Propag. **53**(8), 2265–2373 (2008)

7. Haga, Y., Esashi, M.: Biomedical microsystems for minimally invasive diagnosis and treatment. Proc. 2004 IEEE Conf. **92**(1), 98–114 (2004). (IEEE Conference)

8. Shaby, S.M., Juliet, A.V.: Performance Analysis and Validation of sensitivity of piezoresistive MEMS pressure sensor. In: Proceedings of International conference IEEE Recent Advances in Intelligent Computational Systems (RAICS2011), organized by IEEE, Thiruvananthapuram, pp. 692–695, ISBN 978-1-4244-9477-4 (2011)

9. Balyan, R.K., Asst. Professor Richa: Improving the gain of high-gain Hilbert space filling dental antenna's for biomedical devices. IJARECE **4**(11), ISSN:2278-909X (2015)

10. Chen, S.-H., Yang, C.-L.: Implantable fractal dental antennas for low invasive biomedical devices. In: IEEE Antennas for Low Invasive Biomedical Devices, IEEE Antennas and Propagation. Society International Symposium (APS), USA (2010)

11. Shaby, S.M., Juliet, A.V.: Analysis and optimization of sensitivity of a MEMS piezoresistive pressure sensor. In: Advanced Materials Research, vol. 548, pp. 652–656, ISSN 1662-8985 (2012)

12. Shaby, S.M., Juliet, A.V.: Analysis of sensitivity and linearity of SiGe MEMS piezoresistive pressure sensor. In: Applied Mechanics and Materials, vol. 241–244, pp. 1024–1027, ISSN 1662-7482 (2013)

A Survey on Multi-feature Hand Biometrics Recognition

E. GokulaKrishnan[(⊠)] and G. Malathi

School of Computing Science and Engineering, VIT University Chennai
Campus, Chennai, India
Gokulakrishnan.e2016@vitstudent.ac.in,
Malathi.g@vit.ac.in

Abstract. Biometrics is the one of the most emerging technology in our day to
day life. Biometrics is going to be a future of security and applications. Why we
need biometrics? The password is not user-friendly, we could not dump all
password in our brain. Sometimes we couldn't remember which application which
password we used. To overcome all those problems, Biometrics provides security
"You are the password for your application". Biometrics provides trustworthiness.
Both behavioral and physiological characteristics of biometric features recognize
the individuality of a person whether the user is genuine or an imposter. Hand
biometrics is one of the traditional biometric systems. Generally, hand biometrics
can be either captured by contact and contactless-based approach. Hand biometrics
comprises of palm textures, knuckle, hand geometry, fingerprints which can be
used for recognition. Hand biometrics consist of more uniqueness and individuality of the person has been identified. In this survey paper, we are going to present
the working of the hand geometry and palm print technically. The primary
objective is to study in-depth of hand biometric system and architecture of hand
biometrics. The secondary objective is to literature survey on hand geometry and
palmprint. This paper also studies the pre-processing and feature extraction of
various systems used in hand and palm print. Finally paper addresses the performance evaluation techniques of hand biometrics.

Keywords: Hand · Palm print · Finger print · SVM · Random transfer
Contour coding · Hand feature · Biometrics · Hand shape

1 Introduction

The term biometrics is defined as to measure a unique behavioral and physiological
characteristics of an individual.2020 vision of future biometrics is genuinely replaced the
smart cards [1]. Smartcards are one of the famous authentications but the main drawback
is the identity of the users. Biometrics is the base plethora of human identification and
verification. Even though there is lot security functionalities in biometrics there is a
security breach and vulnerabilities to evade the human biometrics. To overcome those
problems researches focus on developing the high-end security system to identify the

© Springer International Publishing AG 2018
D. J. Hemanth and S. Smys (eds.), *Computational Vision and Bio Inspired Computing*,
Lecture Notes in Computational Vision and Biomechanics 28,
https://doi.org/10.1007/978-3-319-71767-8_91

suspects. A new innovative biometric that can be integrated into IOT. To ensure the security level "Raspberry Pi" can be used to fetch the data's and transmit data's to the cloud Azure with the secure cryptographic algorithm using RSA and AES-256 [2].

1.1 Classification of Biometric Traits

Biometrics can be classified into two types (Fig. 1)

- **Physiological** consist of Face, Fingerprint, Hand Geometry, Iris, vein, Knuckle.
- **Behavioral** trait has Signature, Keystroke, voice etc.

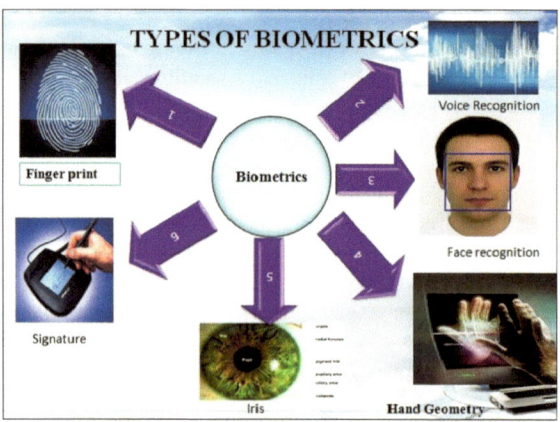

Fig. 1 Types of biometrics

1.2 Characteristics of Biometrics

- **Universality** attribute should be common from all over the world.
- **Invariance of properties** attribute should be common for a longer period.
- **Singularity** attribute must vary from person to person.
- **Acceptance** attribute must be accepted by universally.
- **Performance** attribute performance should be high.

2 Hand Biometrics

Hand Biometrics is one of the traditional methods to identify the persons [3]. In Hand biometrics basically, user can be recognized based on hand shape (Hand Geometry) and a surface of the palm (Palm Print) [4]. Hand biometrics has unique features we can identify and extract the feature. Hand biometrics consist of fingerprint, palmprint, knuckle, vein these are the type of biometrics. In hand palm region consists more than 90 unique hidden features are found in our humans. Hand geometry is used to measure the size of the finger, width etc. In palmprint based the creases, we can able to detect and recognize the person [5].

2.1   Human Anatomy of Hand

Figure 2 describes a human anatomy of hand [6, 7]. Everyone has unique features of hand, the inner surface of hand palm has various unique features hidden inside like

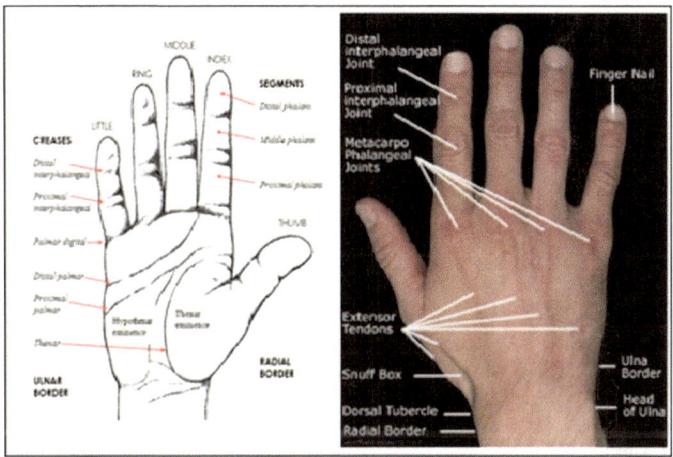

Fig. 2 Anatomy of human hand surface

minutiae, creases, thumb finger length, index finger length, middle finger length, ring finger length, pinkie finger length or ring finger length. The upper portion of the palm region dorsal palmar lines and left side border of the palm lies hypothenar eminence and right side of the border hand has a thenar eminence. Every finger inner surface has a three distinct creases upper creases named as a dorsal phalanx, middle creases named as the middle phalanx, lower finger creases named as a proximal phalanx. In palmar region N-number, unique features reside and it can be used for recognition. Some of the researchers focus only on fingerprint region and fingertip and few of them researcher focuses on palmar region [8]. Metacarpal bones, Radio metacarpal joint, styloid process explains about how they discover the joints and bones and describes differentiate modern people hand and ancient people hand [9]. Cloning the human hand using palm print region the technique they focused are: 3D model of human hand system and hand contours [10]. The hand is working under the brain whenever the brain stimulates the action the hand will do "Think" and "Feel" the hand bones are grouped into carpus which comprises of eight bones used for the root of the hand [11]. In the depth of hand, anatomy consists of fingers, palm, opisthenar, wrist, nails. Every human hand consist of 19 bones, palm consist of five metacarpals, Fingertip has distal phalanx, Melanin couldn't found in palm [12].

Hand Biometrics is classified into two types.

2.2   Contact Based Approach

- **Invariant Pose** the devices with pegs or other guidance parts which instruct a user how to place a hand to capture the certain pattern.
- **Variant Pose** devices without any hand pose or guidance, user have to put the hand on the flat surface.

2.3   Contact Free Approach

A system without any hand shapes guidance by capturing through the standard webcam or CCTV cameras.

2.4   Architecture of Hand Biometrics

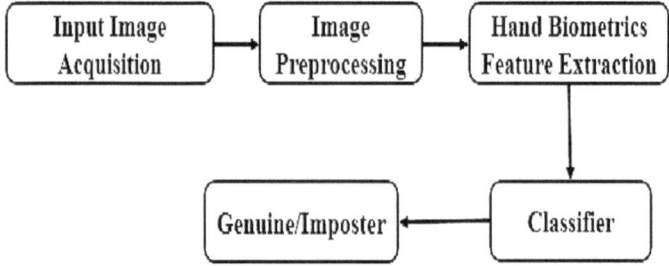

Fig. 3 Work flow of hand biometrics

Figure 3 shows the traditional architecture of hand biometrics. Initially, the input of the image is taken and preprocessing the image samples and extracting the features minutely and the classifier is used for a performance of the system. The result of the classifier score decides whether the user is genuine or imposter [13].

3   Literature Survey

In this survey, we are focusing on hand geometry and palm print. Most of the researchers focus on hand biometrics because it is traditional method for biometrics recognition [3].

3.1   Methods Used in Hand Geometry

In hand geometry calculating the distance, the measure can be done using peg based approach provides a solution to alignment problem while capturing, Alignment and measurement of the hand vary person to person. Peg based hand biometric recognition author used number pegs like 2 pegs, 3 pegs, 4 pegs and 5 pegs and evaluated the study based on the number of pegs. The result shows that the presence of the pegs reduces the variability of the finger spread improving the valley between finger estimation [14]. Peg based system for image capturing and detection they used contour based detection for hand surfaces in different intensity levels. For optimizing the result they used nearest neighbor (NN) classifier to achieve the performance and reducing the error-correction codes [4]. Multimodal biometrics based on hand geometry and palm texture. SVM used as a geometrical verifier for palm print hamming distance is used. Palm print texture is obtained by 2D Gabor phase encoding scheme to increase the performance level they have used the fusion level and score level [15]. Novel method for peg free hand geometry using hand surface with low-resolution images using document scanner. Instead of measuring the length and width author extract the position invariant method for hand shape using the random transform to increase the global performance of the system score level fusion is used [16]. Using 2D and 3D based laser digitizer to acquire the intensity and range of an image in a contact-free manner without using any pegs. The experimental results demonstrate the combination of 3D hand geometry feature with 2D geometry feature can be used to increase the performance [17]. Comparing the verification performance with kids and adult using hand biometrics which comprises of the fingerprint, palmprint, Hand-geometry, and digit-print. This paper mainly focuses, is there any aging difference between child and adults and also recognition of accuracy between child and adult this method also peg free and the user can place a hand on sensors. Feature extraction process consist of five different algorithms used to detect like minutiae, Eigen palms and Eigen's finger, Geometry, Shape and Fingerprint [18]. Multimodal biometrics approach which is a combination of a palm print, fingerprint and finger geometry. Feature extraction for finger tips and finger valley can be obtained by using neighborhood tracking algorithm. LDA algorithm is used to extract the palmprint and fingerprint features. Three decision fusion rule AND rule, OR rule, Majority voting rule can be used to strengthen the performance of recognition [19]. Hand biometrics based on hand shape it can be done using a contactless webcam, used to detect the curvature of the hand shape. Processing the hand shape DCT transform is applied for geometrical hand template. Verifier used in the system is least square SVM for statistical learning and structural risk minimization [20] (Table 1).

Table 1 Comparison of various hand biometrics

Reference	Year	Samples/person	Hand biometric feature	Similarity measure	Algorithm/verifier	Performance analysis
4	2006	109	Hand Geometry and Palm texture	Width, Height of the finger, Palm crease	SVM, 2D Gabor filter	FAR: 0.003 FRR: 0.11
13	2008	150	Hand Geometry and Palm texture	Finger length, hand contour, palm crease	SVM	FRR: 0.66
14	2005	22	Hand geometry	13 features of hand like Finger length, middle finger length	Contour coding, NN classifier	NA
16	2009	136	Hand geometry	Hand surface	Random transform	EER: 5.1%
19	2009	1800	Hand shape	Hand length	LS-SVM	EER: 0.60%

Providing the solution to template security issues on hand biometrics. Fuzzy Analytic hierarchy process is applied with five decision parameter like template generation, template security, fusion overhead, template size overhead, storage complexity. An important observation that template may secure at feature fusion level and indexing technique used to improve the size of the secured templates [22]. Touch based biometric authentication using various touch gestures. The study states that pressure on the sensitivity on the multi touch screen on mobile devices and tablets. Discriminative feature such as distance, angle, touch point used for statistical analysis for verification. The system provides robust protection against unauthorized access [23].

4 Preprocessing

Preprocessing is the process of improving the data that suppress unwanted distortions or enhance some image important feature for further processing. Some preprocessing techniques are cropping the image, thinning the image, Conversion of binary to the gray level. Removing noisy data in that image, Segmentation the image by pixel by pixel. Binarizing the image samples. Contact free hand based geometric and palmprint images can be used for morphology. Preprocessing module consist of stripe regions is used to extract the Hand geometric features. Block segmentation, palmprint feature extraction are in the feature extraction module. In recognition module it has two stage (i) coarse recognition, (ii) fine recognition. Morphing an image can be processed into the different distance the image has to capture [24] (Fig. 4).

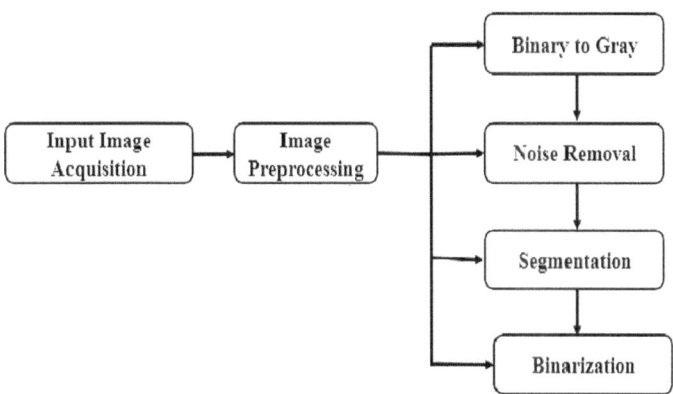

Fig. 4 Hand biometrics preprocessing architecture

5 Feature Extraction

Feature extraction is the process of identifying and reduction of dimensionality as a composite feature vector. Fetching the data's in a particularly marked region. Recognizing hand images using automated hand tracking contact-free method. Conventional Hand geometry is used to detect the hand position in a precise manner and obtaining accuracy. Hand tracking algorithm is used to recognize the person through online. Feature extraction is based on hand length, a width of the fingertip of the finger [25] (Fig. 5).

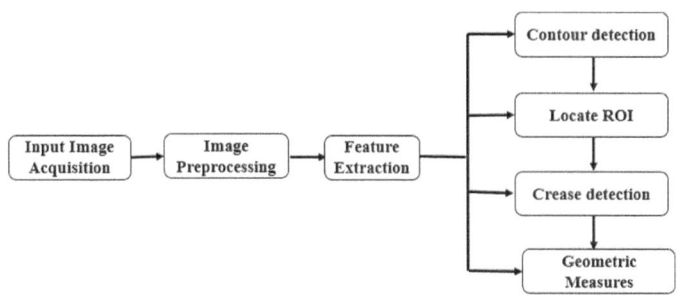

Fig. 5 Hand biometrics feature extraction architecture

6 Performance Analysis

In Hand biometrics performance analysis is based on FAR, FRR and EER.

6.1 False Acceptance Rate (FAR)

Higher the score value provides higher the similarity value. Basically, they use the scores for fixing the threshold. If a score of a trained person (identification) or the person pattern is verified against (verification) is higher than the threshold value [26] and formula for FAR [27] is given in Eq. 1.

$$FAR = \frac{NFA}{NIIA} \tag{1}$$

where,
NFA   Number of false acceptance.
NIIA   Number of imposter identification attempts.

6.2 False Rejection Rate (FRR)

If a score value of the trained person is lower than the threshold value then it is called FRR. The formula for FRR [28] is given in Eq. 2.

$$FAR = \frac{NFR}{NEIA} \tag{2}$$

where,
NFR   Number of false rejection rates.
NEIA   Number of enrollee identification attempt.

6.3 Equal Error Rate (ERR)

If the score distribution overlaps FAR and FRR intersect at one point. FAR and FRR value must be same in certain point it is called ERR. The formula for ERR [29] is given in Eq. 3.

$$ERR = \frac{FAR + FRR}{2} \tag{3}$$

7   Results

Contact based approach (i.e.) using peg based methods provides more efficient than Contactless based approach. Increasing the number of pegs will increase the efficiency recognition. To control the poor recognition if we increase the quality of the image at the capturing stage because template database having the good quality templates. In this survey paper discussed about various hand biometrics such as palmprint, exterior and interior of hand biometrics. Depends upon the requirement various hand biometrics techniques are used.

8   Discussion

Hand biometrics is the traditional method that can be used in the several decades, palm print recognition consist of n number of individuality rich information for high end security we can use this palm print. Contactless biometrics recognition rate can be increased based upon the quality of the camera used at the stage of acquisition. Peg based method provides more acceptance rate because of it is used in under controlled environment.

9   Conclusion

This paper gives in-depth analysis of hand biometrics including hand geometry and palmprint. Liveness detection on hand biometrics plays an important role in identifying the spoofing. It also reviews various bi-modal hand biometric, techniques used and verification methods which were used in the biometric systems. It gives a clear idea about while designing high-end biometric system. This paper also describes in-depth analysis of preprocessing and feature extraction of Hand Biometrics. Finally paper represented about the performance and fusion technique which gives the accuracy of the results.

References

1. 2020 Vision Technology: Biometrics—The Future of Security. Available: http://www.2020cctv.com/news/biometrics-the-future-of-security. Accessed 13 Sept 2016
2. Shah, D., Haradi, V.: IoT based biometrics implementation on raspberry Pi. Procedia Comp. Sci. **79**, 328–336 (2016)
3. El Kalam, A.A., Ibjaoun, S.: Biometric authentication systems based on hand pattern vein, digital certificates and smart cards. In: IEEE on National Security Days, pp. 1–8 (2013)
4. Faundez-Zanuy, M.: Biometric verification of humans by means of hand geometry. In: 39th Annual 2005 International Carnahan Conference on Security Technology CCST, pp. 61–67 (2005)
5. Miguel A.F., Travieso, C.M. Alonso, J.B.: Multimodal biometric system based on hand geometry and palm print texture. In: 40th Annual IEEE International Carnahan Conferences Security Technology, pp. 61–67 (2006)

6. http://image.wikifoundry.com/image/1/Ctd8CFWvudRlq0jFWuTaRA73328/GW480H633 (2016)
7. http://www.sydneyhandsurgeryclinic.com.au/_data/docs/surface-anatomy.gif (2016)
8. "Hand", Wikipedia. Available: https://en.wikipediaorg/wiki/Hand. Accessed 13 Sept 2016 (2016)
9. Ward, C., Tocheri, M., Plavcan, J., Brown, F., Manthi, F.: Early Pleistocene third metacarpal from Kenya and the evolution of modern human-like hand morphology. Proc. Natl. Acad. Sci. **111**(1), 121–124 (2013)
10. Rhee, T.: Human hand modeling from surface anatomy. In: ACM SIGGRAPH Symposium on Interactive 3D Graphics and Games, pp. 1–9 (2006)
11. TAYLOR, C.L.: The Anatomy and Mechanics of the Human Hand, pp. 22–35 (2016)
12. HealthLine: Hand Anatomy, Pictures & Diagram|Body Maps, Healthline.com, 2016. Available: http://www.healthline.com/human-body-maps/hand. Accessed 13 Sept 2016
13. Zheng, G., Wang, C.-J., Boult, T.: Application of projective invariants in hand geometry biometrics. IEEE Trans. Inform. Forensic Secur. **2**(4), 758–768 (2007)
14. Ferrer, M.A., Morales, A., Travieso, C.M.: Influence of the pegs number and distribution on a biometric device based on hand geometry. In: 42nd Annual IEEE International Carnahan Conference on Security Technology ICCST, pp. 221–225 (2008)
15. Ferrer, M.A., Travieso, C.M., Alonso, J.B.: Multimodal biometric system based on hand geometry and palm print texture. In: 40th Annual IEEE International Carnahan Conferences Security Technology, pp. 61–67 (2006)
16. Mostayed, A., Kabir, M.E.: Biometric authentication from low resolution hand images using random transform. In: 12th International Conference on Computers and Information Technology, ICCIT, pp. 587–592 (2009)
17. Kanhangad, V., Kumar, A., Zhang, D.: Combining 2D and 3D hand geometry features for biometric verification. In: IEEE Computer Society Conference on Computer Vision and Pattern Recognition Workshops, pp. 39–44 (2009)
18. Uhl, A., Wild, P.: Comparing verification performance of kids and adults for fingerprint, palmprint, hand-geometry and digitprint biometrics. In: IEEE 3rd International Conference on Biometrics: Theory, Applications, and Systems, BTAS, pp. 1–6 (2009)
19. Yu, P., Xu, D., Zhou, H.: Decision fusion for hand biometric authentication. In: IEEE International Conference on Intelligent Computing and Intelligent Systems, ICIS, pp. 486–490 (2009)
20. González, E., Morales, A., Ferrer, M.A.: Looking for hand shape based biometric devices interoperability. In: IEEE International Carnahan Conference on Security Technology (ICCST), pp. 1–5 (2011)
21. Ahmad, M., Woo, W., Dlay, S.: Non-stationary feature fusion of face and palmprint multimodal biometrics. Neurocomputing **177**, 49–61 (2016)
22. Selwal, A., Gupta, S., Surender, Anubhuti: Template security analysis of multimodal biometric frameworks based on fingerprint and hand geometry. In: Perspectives in Science (2016)
23. Qiao, M., Zhang, S., Sung, A.H.: A novel touchscreen-based authentication scheme using static and dynamic hand biometrics. In: IEEE 39th Annual Computer Software and Applications Conference COMPSAC, vol. 2, pp. 494–503 (2015)
24. Wang, W.-C., Chen, W.-S. Shih, S.-W.: Biometric recognition by fusing palmprint and hand-geometry based on morphology. In: IEEE International Conference on Acoustics, Speech and Signal Processing, pp. 893–896 (2009)
25. Michael, Goh Kah Ong, Connie, Tee: Locating geometrical descriptors for hand biometrics in a contactless environment. IEEE Int. Symp. Inf. Technol. **1**, 1–6 (2010)

26. FRR FAR EER explanations| Griaule Biometrics. Griaulebiometrics.com. Available: http://www.griaulebiometrics.com/en-us/forum/frr-far-eer-explanations. Accessed 13 Sept 2016 (2016)
27. GokulaKrishnan, E., Asha, S.: Subspace based face recognition using clustering. Int. J. Comput. Sci. Netw. IJCSN **3**(5), 321–325 (2014)
28. Biometrics, Wikipedia, 2016. Available: https://en.wikipedia.org/wiki/Biometrics. Accessed 13 Sept 2016
29. Calculate EER from FAR and FRR?, Stats.stackexchange.com. Available: http://stats.stackexchange.com/questions/221562/calculate-eer-from-far-and-frr. Accessed 13 Sept 2016 (2016)

Early Detection and Classification of Diabetic Retinopathy Using Empirical Transform and SVM

Sumandeep Kaur[(⊠)] and Daljit Singh

Electronics and Communication Engineering, GNDEC, Ludhiana, India
sumansaraon44@gmail.com

Abstract. Diabetic Retinopathy is the name given to 'disease of retina'. The objective of this work is for timely diagnosis and classification of diabetic retinopathy using curvelet transforms and SVM. Firstly, retinal images are enhanced using empirical transform. Canny edge detection is applied for extracting eyeball from retinal fundus image. Then morphological operations are applied for locating the imperfections in the images. At the end, images are classified into normal, proliferative or non-proliferative by using SVM. Both accuracy and sensitivity of the images is improved when compared with previous technique in which only k-means and fuzzy classifier is used. The number of exudates detected in present work is more than that of the process without enhancement. The sensitivity, specificity and accuracy of system are calculated as 96.77, 100 and 97.78 respectively.

Keywords: Diabetic retinopathy · Curvelets · Empirical transform · SVM

1 Introduction

Diabetes is a chronic syndrome that arises due to a decrease in production of insulin, or alternatively, due to ineffective use of produced insulin by the body. Diabetes led to the weakness of the blood vessels in the body. The small and delicate blood vessels are more prone to this weakness. The weakness in retinal blood vessels, lead to structural variations within the retina, which is known as diabetic retinopathy. The blood vessels in retina undergo many changes like swelling, leakage or may close off completely [1].

WHO reported that in the year 2012, approximately 347 million people around the world have diabetes. WHO projected that diabetes may become the 7th leading root cause of death by 2030 [2]. WHO also predicted that diabetic population in India will rise to 79.4 million by the year 2030. This digit is largest among any nation in the world.

Diabetic retinopathy usually impacts both the eyes. It remains undetected at early stages, but at later stages, it may lead to total vision loss that can't be reversed. Keeping strict control of blood glucose level and regular eye screening by a doctor are means to avoid diabetic retinopathy and vision loss. The aim of this algorithm is to reduce the

© Springer International Publishing AG 2018
D. J. Hemanth and S. Smys (eds.), *Computational Vision and Bio Inspired Computing*,
Lecture Notes in Computational Vision and Biomechanics 28,
https://doi.org/10.1007/978-3-319-71767-8_92

manual work in diagnosing the diabetic retinopathy. Specialized training of ophthalmologists is required to prevent diabetic retinopathy and blindness in diabetic patients.

2 Methodology

Step 1 Enhance the retinal fundus images from the database by applying curvelet transform on them.

Step 2 Extract eyeball from the background of retinal image by using canny edge detector.

Step 3 Extract the feature from curvelet pre-processed image, by applying morphological operations, for detecting exudates and hemorrhages in fundus images.

Step 4 Select the feature from the extracted features.

Step 5 Classify the retinal fundus images into normal, Proliferative Diabetic Retinopathy (PDR) and Non-Proliferative Diabetic Retinopathy (NPDR) images by using Support Vector Machine (SVM).

Step 6 Compare the results obtained with the previous techniques (Fig. 1).

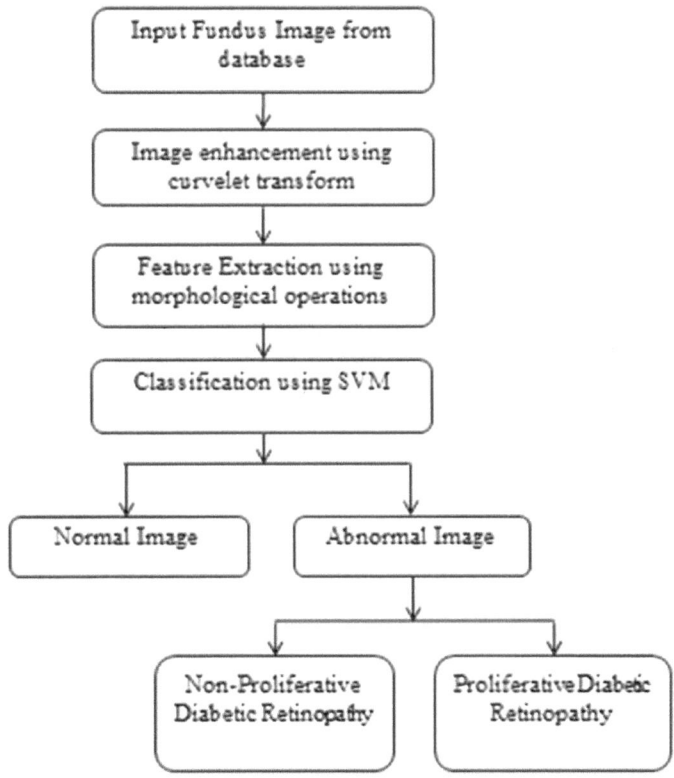

Fig. 1 Methodology of work

3   Present Work

3.1   Image Enhancement Using Curvelet Transform

In the detection process of diabetic retinopathy, digital retinal images of the eyes are capture. Most of these images encompass noise due to some bugs during capturing or transmission process. So, this noise is the main hindrance in the efficient diagnosis of diabetic retinopathy. Therefore, image enhancement is done at the very first step of detection. The finest approximation to the original image is prepared by using the curvelet transforms. The curvelet transforms are capable of eradicating additive white Gaussian noise while conserving edges [3].

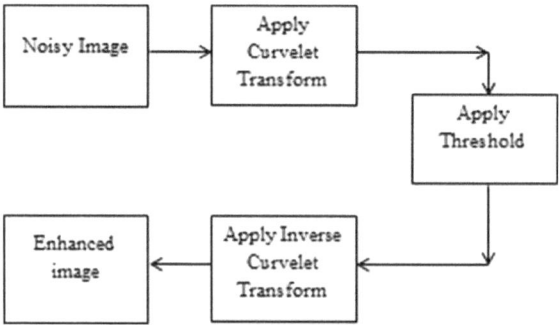

Fig. 2 Image enhancement using curvelet transform

Enhancement of retinal fundus images is done using curvelet transform is shown in Fig. 2. This removes any noisy coefficients such as coefficients of Poisson, Speckle, Gaussian and Random noise. Denoising of the image is done in following steps:

Step 1: Calculation of threshold.
Step 2: Curvelet transform is applied to the image.
Step 3: Measured thresholds are applied to remove noisy coefficients.
Step 4: Apply inverse curvelet transform to reconstruct the image.

3.2   Eyeball Extraction Using Canny Edge Detection

Edge detection of an image is a way of finding edges of the image which is significant in discovering the approximate absolute gradient scale at every point of a gray scale image at an input. A canny edge detector initiate with removal of speckle noise. An operator is used for isolating eyeball from its background by firstly perceiving edges of the eyeball. It is gradient centered edge detection technique. Gradient (G_x, G_y) can be measured by measuring partial derivatives $\frac{\partial f}{\partial x}$ and $\frac{\partial f}{\partial y}$ at each pixel location [4]. Hence, the gradient of an image is specified as:

$$\nabla f = \begin{bmatrix} G_x \\ G_y \end{bmatrix} = \begin{bmatrix} \frac{\partial f}{\partial x} \\ \frac{\partial f}{\partial y} \end{bmatrix} \tag{1}$$

Mostly detection of edges is based on the supposition that the edges exist wherever discontinuity in an intensity function exists or wherever a sharp intensity gradient is present in an image [5]. Then derivative of an intensity value is taken across an image and points were discovered at which derivative is highest. The best pixels for edge are selected from the numerous possible values in local neighborhood. This is called non-maximum suppression. Using this value the edges can be founded.

Canny operation can be measured by shortening the partial derivative if the 3×3 kernel is known. Canny operator is given as:

$$G = |G_x| + |G_y| = \sqrt{G_x^2 + G_y^2} \tag{2}$$

here, l stands for the pixel intensity of the kernel matrix. Hence edges could be found by moving and convolving a kernel throughout the image.

3.3   Feature Extraction Using Morphological Operations

The enhanced retinal fundus images are used for selecting threshold. For this process, a histogram of enhance image is made and the upper and lower threshold for the gray scale image is obtained from this histogram. Using these thresholds, the denoised gray scale image is converted into the binary images by equating the values above threshold as zero i.e. black while equating values below threshold as one i.e. white. On these binary images, morphological operations are applied to extract the imperfections from the images. New binary image is created at the end, which contains non-zero value only at the places in the image where abnormalities are present. Finally, a region based bounding box is used to identify and quantify the exudates.

3.4   Classification Using Support Vector Machine

Support vector machine is a supervised learning technique used for classification. It is used for constructing single hyperplane or set of hyperplanes. These hyperplanes are employed for regression as well as classification purpose. From the set of these hyperplanes, a hyperplane is carefully chosen that has maximum distance from the nearest training data point. The error of classification is inversely proportional to this distance. Therefore, error would be least if the distance is large. MultiSVM is employed here for classification purpose. Since, SVM is a binary classifier as it classifies data into two classes only. So, MultiSVM is used for classifying data into more than two classes. Here retinal fundus images are classified into three classes that is normal images, proliferative and non-proliferative images. Proliferative and non-proliferative are the forms of diabetic retinopathy. MultiSVM efficiently classifies the input image into one of these three classes.

3.5   Comparison with the Previous Technique

At the end, the accuracy and sensitivity of this system i.e. detection and classification using empirical transforms and support vector machine is compared with the previous work in which k-mean clustering and fuzzy classifier were used [5].

4   Result and Discussion

The experimentation for normal and diseased retinal fundus images is done using the standard database available on internet. This database contains the normal as well as abnormal images of an eye. So, detection is done using the presented method and results are compared with the pre-known data.

4.1   Diabetic Retinopathy Detection

The raw image is taken from the database and given to the system. The operations are performed on this retinal image for detecting hemorrhages and exudates. The raw image is shown in Fig. 3.

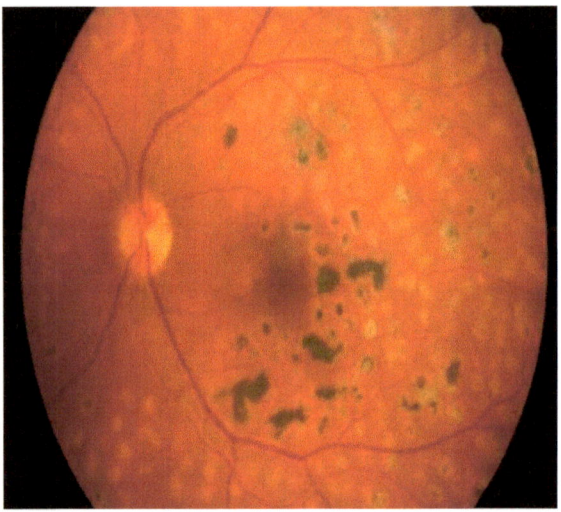

Fig. 3 Retinal fundus image with diabetic retinopathy

The image taken directly from the database may contain some sort of noise. So, enhancement of image is done using the curvelet transform, so that features could be efficiently extracted from the images. First of all, the curvelet transform of noisy image is calculated. Then after the application of hard thresholding, noisy coefficients are removed. At the end, inverse curvelet transform is again carried out for reconstructing the image. The noise in image and curvelet reconstructed image are shown in Fig. 4.

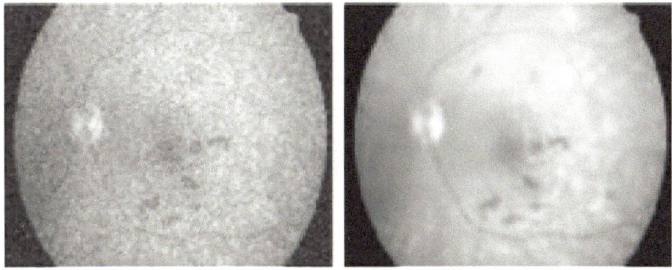

Fig. 4 Noise in image (left) and Reconstructed image (right)

The canny edge detection is done so as to extract eyeball from the retinal fundus image. The output of canny edge detection and extracted eyeball is shown in Fig. 5.

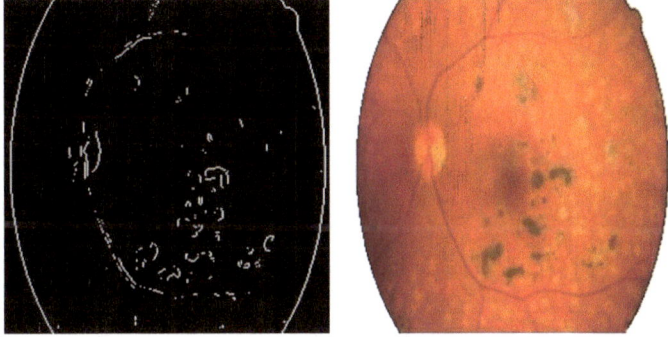

Fig. 5 Canny edge detection (left) and extracted eyeball (right)

Then histogram of both original and reconstructed image is computed. Here difference in the histograms could be clearly seen in Fig. 6. So, enhancement has positive result on the detection process and threshold could be chosen from the reconstructed image for efficient diagnosis of exudates.

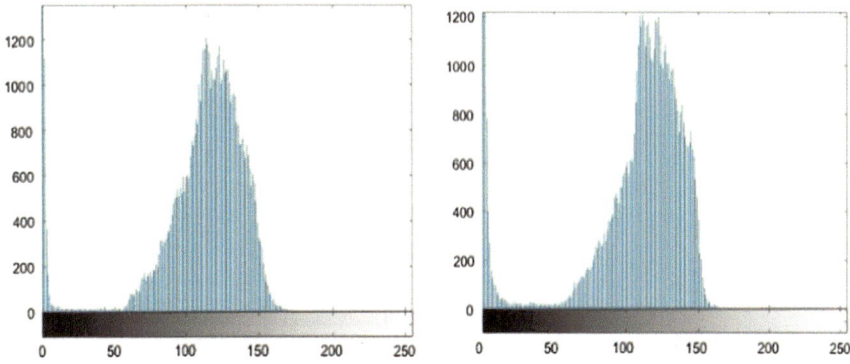

Fig. 6 Histogram of raw (left) and enhanced image (right)

Then the chosen threshold value from the histogram is used for computing binary images by equating the values above threshold as zero i.e. black while equating values below threshold as one i.e. white. On these binary images, morphological operations are applied to extract the imperfections from the images. Numerous morphological methods like erode, dilate, filling etc. are applied to efficiently locate the abnormalities. Figure 7 depicts the changes before and after applying the hole filling morphological operation.

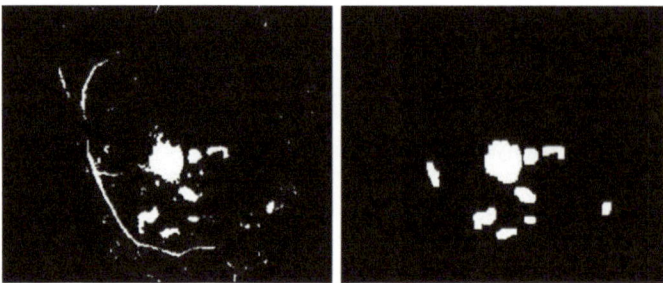

Fig. 7 Exudates before (left) and after (right) morphological filling

New binary image is created at the end, which contains non-zero value only at the places in the image where abnormalities are present. Finally, a region based bounding box is used to identify and quantify the exudates. The image after segmenting exudates is merged with input fundus image for visual comparison. This image is shown in Fig. 8.

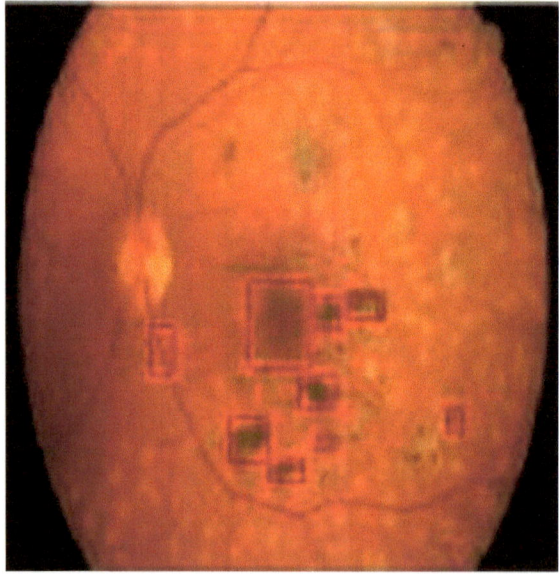

Fig. 8 Merged image and quantified exudates

The sensitivity and accuracy of the system is computed for measuring the performance of the system. The analysis is done using confusion matrix shown in Table 1.

Table 1 Confusion matrix

		Predicted data		Total
		True	False	
Real data	True	TP	FN	T
	False	FP	TN	P
Total		T'	P'	T + P or T' + P'

In this table, TP, FP, TN, FN represents the total of true positive, false positive, true negative and false negative trials respectively. TN and TP indicate that system is receiving right results, whereas FN and FP indicate that system is receiving wrong results. P represents no. of elements in a positive set while N represents no. of elements in a negative set.

Sensitivity: To measure true positive ratio of classified images.

$$Sensitivity = \frac{TP}{TP + FN} \tag{3}$$

Specificity: To measure true negative ratio of images.

$$Specificity = \frac{TN}{TN + FP} \tag{4}$$

Accuracy: To measure percentage of correctly classified images.

$$Accracy = \frac{TP + TN}{TP + FN + TN + FP} \tag{5}$$

4.2   Diabetic Retinopathy Classification

Various features are extracted from the retinal fundus images using empirical transform and morphological operations. Then at the last step, images are classified into three classes namely normal images, non-proliferative and proliferative images. These are given class 0, 1 and 2 respectively. MultiSVM is used for this classification purpose. SVM classifies images into two classes only. So, MultiSVM is used for classifying images into more than two classes. So, Multi-SVM system gives the result after classification.

4.3   Comparison of Present and Previous Work

In the present technique image enhancement is done using empirical transform then abnormalities are located using morphological operations. While in the previous work, k-means and fuzzy logics were used for detecting and classifying diabetic retinopathy. The number of exudates detected by current method is far more than that of previous method in which enhancement was not done.

The number of exudates detected in present work is more than that of the process without enhancement. If the value of PSNR is less, number of exudates detected in noisy image is far less than when detected with the curvelet-denoised images. The results for 10 images are shown in Fig. 9.

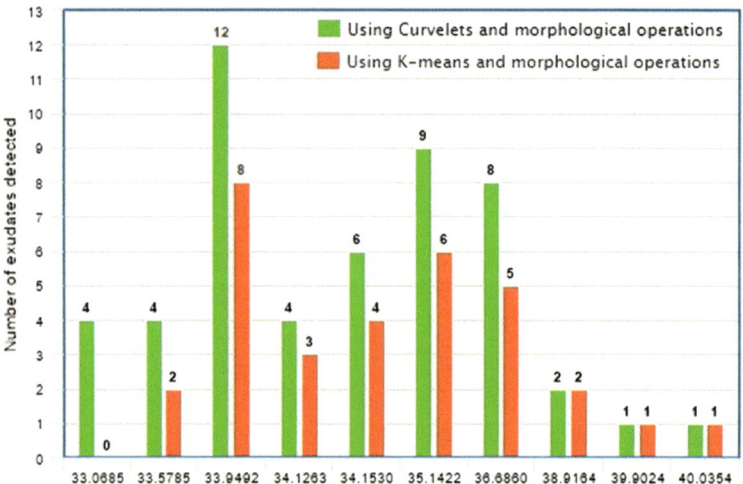

Fig. 9 Variation of number of exudates detected by both methods with PSNR

5   Calculations

5.1   Calculations for Present Work

The curvelet transform and morphological operations are used for detection in present work. After performing the operation total of 44 images were identified accurately while 1 image shows normal class while it was in the diseased class. Therefore 44 out of 45 images were detected accurately.

Therefore, TP = 30

TN = 14

FN = 1

FP = 0

$$\text{Accuracy} = \frac{30+14}{30+1+14+0} = \frac{44}{45} = 97.78\%$$

$$\textit{Sensitivity} = \frac{30}{30+1} = 96.77\%$$

$$\textit{Specificity} = \frac{14}{14+0} = 100\%$$

Therefore, accuracy, specificity and sensitivity of the system is attained as 97.78%, 100% and 96.77% respectively.

5.2   Calculations for Previous Work

The k-means clustering and morphological operations are used for detection in present work. After performing the operation total of 44 images were identified accurately while 3 images shows normal class while it was in the diseased class. Therefore 42 out of 45 images were detected accurately.

Therefore, TP = 28

TN = 14

FN = 2

FP = 1

$$\text{Accuracy} = \frac{28+14}{28+2+14+1} = \frac{42}{45} = 93.33\%$$

$$\textit{Sensitivity} = \frac{28}{28+2} = 93.33\%$$

$$\textit{Specificity} = \frac{14}{14+1} = 93.33\%$$

Therefore, accuracy, specificity and sensitivity of the system is attained as 93.33%, 93.33% and 99.33% respectively. The comparison of both methods are shown in Table 2 and Fig. 10.

Table 2 Comparison of results

	Using curvelet and SVM	Using K-means and FIS
Accuracy	97.78%	93.33%
Specificity	100%	93.33%
Sensitivity	96.77%	93.33%
Approximate execution time	45.854221 s	25.568224 s

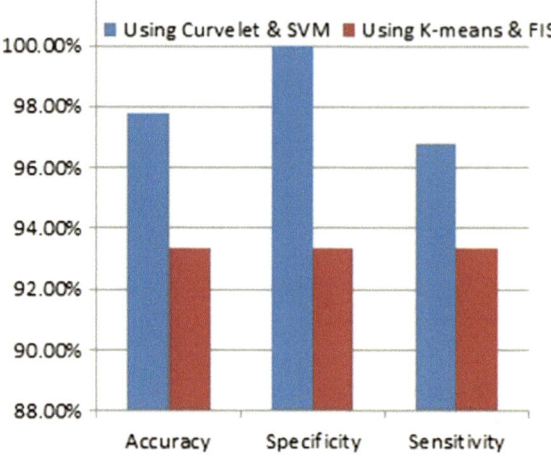

Fig. 10 Comparison of results

6 Conclusions

In this work, a methodology for timely diagnosis and classification of diabetic retinopathy using curvelet transform and support vector machine is developed. Diabetic retinopathy is detected by discovering hemorrhages and exudates in the fundus images. First of all, retinal images are enhanced using empirical transform. Then morphological operations are applied for locating the imperfections in the images. At the end, images are classified into normal, proliferative or non-proliferative by using support vector machine. Hence, enhancement of images using curvelets helps to improve the accuracy and sensitivity of detection and classification. The number of exudates detected in present work is more than that of the process without enhancement. Also, the accuracy of detection system depends upon the PSNR in the image. If the value of PSNR is less, number of exudates detected by noisy image is far less than when detected with the curvelet denoised images. The accuracy, sensitivity and specificity of system is calculated as 97.78, 96.77 and 100% respectively.

References

1. Chauhan, R., Uniyal, A., Dubey, V.P.: Detection of retinal blood vessels and reduction of false microaneurysms for diagnosis of diabetic retinopathy. IEEE Trans. Med. Imaging **30**, 191–196 (2016)
2. World Health Organization.: Global health estimates: deaths by cause, age, sex and country 2000–2012. In: Department of Health Statistics and Information Systems WHO, Geneva (2014)
3. Rakshitha, T.R., Devaraj, D., Prasanna, K.S.C.: Comparative study of imaging transforms on diabetic retinopathy images. In: IEEE International Conference on Recent Trends in Electronics Information Communication Technology, pp. 118–122, India (2016)

4. Manjula, R., Rajesh, V.: Early detection of diabetic retinopathy from retinal fundus images using Eigen value analysis. In: International Conference on Control, Instrumentation, Communication and Computational Technologies, vol. 1, Issue No. 15, pp. 347–351 (2015)
5. Jahiruzzaman, M.D., Hossain, A.: Detection and classification of diabetic retinopathy using K-Means clustering and fuzzy logic. In 18th International Conference on Computing And Information Technology, pp. 534–538 (2015)

A Prototype to Detect Anomalies Using Machine Learning Algorithms and Deep Neural Network

Malathi A.[1,2(⊠)], Amudha J.[1,2], and Puneeth Narayana[3]

[1] Department of Computer Science and Engineering, Amrita School of
Engineering, Bengaluru, India
malathiarumugam4ll@gmail.com, j_amudha@blr.amrita.edu
[2] Amrita Vishwa Vidyapeetham, Amrita University, Coimbatore, India
[3] Cinqueon Technologies, Bengaluru, India
puneeth@cinqueon.com

Abstract. Artificial Intelligence is making a huge impact nowadays in almost all the applications. It is all about instructing a machine to perceive an object like a human. Making the machine to excel at this perception requires training it by feeding large number of examples. In this way machine learning algorithms find many applications for real time problems. In the last five years, Deep Learning techniques have transfigured the field of machine learning, data mining and big data analytics. This research is to investigate the presence of anomalies in given data. This model can be used as a prototype and can be applied in domains like finding abnormalities in medical tests, to segregate fraud applications in banking, insurance records, to monitor IT infrastructure, observing energy consumption and vehicle tracking. The concepts of supervised learning and unsupervised learning are used with the help of machine learning algorithms and deep neural networks. Even though machine learning algorithms are effective for classification problems, the data in question, determines the efficiency of these algorithms. It has been showed in this paper; the traditional machine learning algorithms fail when the data is highly imbalanced and necessitate the use of deep neural networks. One such neural network called deep Autoencoder, is used to detect anomalies present in a large data set which is largely biased. The results derived out of this study, proved to be very accurate.

Keywords: Anomaly · Autoencoder · Dimensionality reduction · One class classification · Random forest · Supervised learning · Unsupervised learning

1 Introduction

Deep learning algorithms make use of huge amount of data to automatically extract intricate description of them. These algorithms are influenced by AI which emulates the human brain's ability to perceive, examine, learn, and make decisions, especially for

D. J. Hemanth and S. Smys (eds.), *Computational Vision and Bio Inspired Computing*,
Lecture Notes in Computational Vision and Biomechanics 28,
https://doi.org/10.1007/978-3-319-71767-8_93

very complicated problems [1]. Artificial Neural Networks (ANN) such as Convolutional Neural Networks, Autoencoders, Restricted Boltzmann's Machines and Recurrent Neural Networks offer accurate results by learning many levels of abstractions of data [2] by making a detailed analysis of features.

With the increase in number of computer networks every day, and huge number of applications running on top of it, providing security to the genuine communication has become a complex problem. The authors [3] have designed a neural network based eye gaze detection. The authors of this paper [4] have scrutinized the structural information of the document detect plagiarism. This paper [5] used random forest algorithm to detect intrusions and achieved good detection rates. Fuzzy Rule based analysis and optimization models is discussed in this paper [6]. Even though the availability of security software makes the system more robust, they do not prevent all types of attacks that come through the networks. The enormous volumes of systems which form a network suffer from cyber security vulnerabilities. Resolving all proved to be very difficult technically, and economically costly. So, the design of an efficient system for finding out anomalies in the network is crucial. Recent analysis shows that several machine learning algorithms have proved to be efficient in detecting anomalies. Data defined by high-dimensions can be converted to low-dimensional data by training a multilayer neural network with a small middle layer [7] to reconstruct high-dimensional input vectors. Autoencoder network is used effectively for initializing the weights that allows networks to learn low-dimensional codes and work better than principal components analysis as a tool to reduce the dimensionality of data. An IOT based distributed structure [8] which contains master and slave servers are implemented. In each slave server, a stacked sparse Autoencoder is used to extract the representation of data and then SoftMax is used for extraction of anomalous points. Ebborth et al. [9] used Autoencoder network for fraud detection and acquired an accuracy rate of 95.4. The authors [10] have used Least Square Support Vector Machine (LS-SVM) and have produced a good accuracy rate. Zhao et al. [11] have applied the ensemble methods which use score and rank aggregation and used sequential methods where one algorithm is followed by another. Their study shows the sequential method performs better. Peter klimek, Yuri Yegorov et al. have developed [12] a parametric statistical model in this paper, finding the essential statistical properties in the election results and produced robust results. Motivated by these results, it is proposed to design an efficient Network Intrusion Detection System in this paper.

The original data set consists of seven weeks of network traffic, of about five million connection records. The test data is of about two weeks data which makes around 2 million connection records. The authors of this paper [13] have performed a detailed statistical analysis on this dataset. The analysis has come out with identification of redundancy in the records. Redundancy in data sets makes the classifier more biased towards the presence of more frequent records. The authors have modified the data set, eliminating redundancy. This dataset is available in the website [14]. This paper [15] discusses about various neural networks and machine learning algorithms for the detection of anomalies in networks. It has been found out that the mixture of machine learning models applied on the Knowledge Discovery Data set (KDD Cup '99) has produced very high accuracy on particular attacks. Deep Belief Networks (DBN) has been trained as a classifier by Gao et al. [16] to detect intrusion on KDD data set. The authors proved DBN can successfully be used as an effective Intrusion Detection System IDS. The results showed that DBN recorded the best accuracy of 93.49%, a True Positive Rate (TPR) value of 92.33 and False Positive Rate (FPR) of 0.76%. Alom et al. [17] exploited the DBNs capabilities to detect intrusion through series of experiments. The authors trained DBN with reduced KDD data set to identify unknown attacks on it. They concluded by proposing DBN as a good Intrusion Detection System (IDS) based on an accuracy of 97.5% achieved in the experiment. The advantage of a Variable Auto Encoder (VAE) over an Autoencoder and a Principal Component Analysis (PCA) is that it provides a probability measure [18] rather than a reconstruction error as an anomaly score, this paper shows that PCA has given poor results.

Many research papers focus on the performance of the algorithms in terms of improving the overall accuracy score. Less research has been done in acquiring perfect segregation of True positive examples. This issue is the key focus in this paper. This paper pays attention on improving the recall value. Series of experiments are done in improving the classification and getting a perfect score. In the first analysis, different machine learning algorithms are compared for their performance. The second analysis highlights how deep Autoencoder network performs better than these algorithms in getting satisfactory results by predicting True Positive rates accurately, apart from getting the highest accuracy score.

2   Methodologies

In this study, the network intrusion problem has been analyzed from Artificial Intelligence perspective and different approaches in giving effective solutions have been proposed to deal with it. All the approaches have given the best results. The first approach uses different machine learning algorithms and found almost all the algorithms equally good in getting excellent accuracy score with respect to class with majority examples. The second approach uses autoencoder deep neural network.

The data used in these experiments is a part of KDD'99 Cup data set. It has 60,839 samples and 43 features including labels. The label for this data set is 'outlier'. Duration, number of failed logins, wrong fragments, number of access files are some of the 43 features provided in this dataset. Even though rule based systems exist in solving anomaly detection problem, they detect known anomalies. The purpose of applying machine learning algorithms is to find out the patterns which are not usual and happen in uncertain situations. So, all the approaches used here find out such patterns which are abnormal and segregate them.

Fig. 1 Distribution of negative and positive examples in the data

In some binary classification problems, the two classes do not have equal presence like the KDD dataset representation shown in the Fig. 1 above. Such a data set is called as unbalanced because one class has more examples than the other. In such problems, the class with few examples is of main concern, because the algorithm will not generalize properly and perform poorly with respect to prediction of minority class [19]. The formal assumption in machine learning problems is that both training and testing samples should be taken from the same distribution. Occurrence of unbalanced data is common in medical diagnostic cases where the negative cases of blood samples are more than positive cases of blood samples for a disease. Machine learning algorithms would still give an accuracy score of 99 percent for this type of unbalanced dataset [20]. Such datasets bias the results towards majority class. The predictions with respect to negative examples will be 99%. But the same cannot be expected from positive examples. So, data sets of this kind need special attention. Unsupervised learning algorithms like clustering algorithms are helpful in learning similar patterns out of these datasets and in acquiring good results. This paper makes a detailed analysis of the data set by applying supervised machine learning algorithms and unsupervised Autoencoder deep neural network. Since the objective is to find anomalies which are very less compared to normal instances, many experiments are made to improve the model performance especially on the positive examples.

2.1 Analysis 1

The data set is divided into a training set and a test set. The training data set consists of 70% of 60,839 instances and the remaining data is the test set. The training data set is again split into multiple folds (K folds where K = 10) for efficient prediction by avoiding over fitting and reducing the variance. The machine learning algorithms, K-Nearest Neighbours (KNN), Classification and Regression Tree (CART), Naïve Bayes (NB) algorithm, Support Vector Machines (SVM), and Random Forest (RF), are applied on the training dataset. The table below shows the accuracy score, Confusion Matrix, Precision, Recall and F1generated by applying each of the algorithms on the data set.

Table 1 Performance comparison of algorithms on the dataset

Algorithm	Accuracy score		Confusion Matrix		Class	Precision	Recall	F1
	Training set	Testing set						
KNN	0.99877	0.99873	18173	1	0.0	1.00	1.00	1.00
			22	56	1.0	0.98	0.72	0.83
NB	0.97846	0.97885	17798	376	0.0	1.00	0.98	0.99
			10	68	1.0	0.15	0.87	0.26
RF	0.99892	0.99890	18170	4	0.0	1.00	1.00	1.00
			16	62	1.0	0.94	0.79	0.86
SVM	0.99746	0.99720	18174	0	0.0	1.00	1.00	1.00
			51	27	1.0	0.83	0.79	0.81
CART	0.99872	0.99841	18161	13	0.0	1.00	1.00	1.00
			16	62	1.0	0.83	0.79	0.81

From the Table 1 all the algorithms produced an accuracy score which is greater than 97%. Recall is a measure of classifier's performance with respect to false negatives while precision is a measure of its performance with respect to false positives. The proportion of examples which are anomalies is given by the bottom row in the confusion matrix. Precision is what proportion of examples that are anomalies and found out by classifier given by the second column of the matrix.

Naïve Bayes gives a good measure of recall whereas all the algorithms are good in classifying the majority examples. So, there is another measure called F1score. This score gives the mean of recall and precision to determine the efficiency of an algorithm. Comparatively RF algorithm produced the best performance compared to all the other algorithms as shown in the Table 1. Random Forest is a bagging model. It selects many subsets from the data called trees. The algorithm randomly chooses some of the features and portions of the dataset for growing each tree. The trees are grown with the replacement of features and records. Each tree is of the same size. The classification of objects is based on the voting process. Decision on classification is based on the support executed by the trees. As seen in the Table 1 above, this analysis shows that if the distribution of the labels is skewed the algorithms try to classify all instances as majority label instances [21].

2.2   Analysis II

In this approach, the deep neural network Autoencoder is utilized for efficient feature representation. It is an unsupervised learning technique. By applying back propagation algorithm, it tries to set the target value to be equal to input value. The autoencoder network has an input layer, an output layer and a hidden layer. Hidden layers can be more than 1. The id field and the labels are removed from the data. The original 41 features which define the data points are represented in a reduced dimensional space of 30 in the hidden layers.

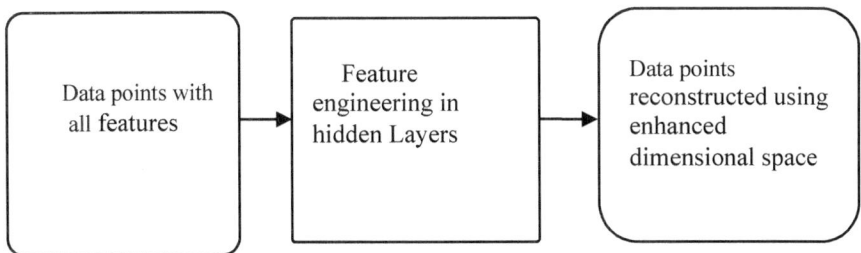

Fig. 2 Block Diagram of a simple autoencoder

A simple Autoencoder is a feed forward, neural network identical to a Multilayer Perceptron (MLP). The block diagram of a simple Autoencoder is shown in the Fig. 2. It has an input layer, one or more than one hidden layer and an output layer. The output layer will have the same number of nodes as the input layer. Autoencoder will not predict target values. But tries to reconstruct its own input. In this way, it is an unsupervised learning model. The structure of a deep Autoencoder is shown in the Fig. 3. For each input, the network does a feed forward flow to determine activations at all hidden layers then at the output layer to derive an output. The deviation of output from the input is measured using a distance measure. This error is propagated backwards. This forms one iteration. Then weights are updated for the second iteration. The Autoencoder used for the analysis has five layers, with three hidden layers apart from the input layer and output layers. It is trained to learn compressed representation of 41 features. In the first experiment 41 features are reduced to 10 hidden nodes in the second layer, then again compressed to 3 nodes in the middle layer. This form of feature reduction forms the encoder network. The 3 nodes in the middle layer is the

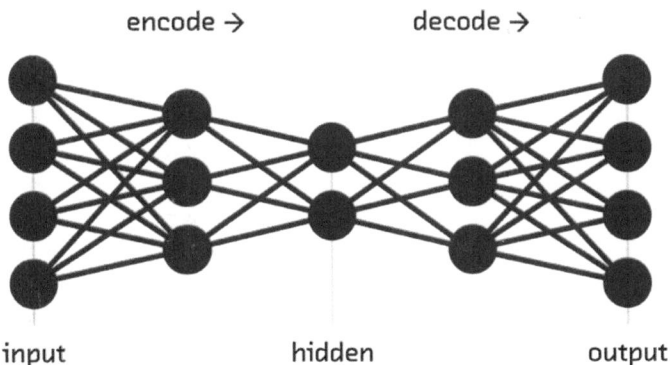

Fig. 3 Structure of a deep autoencoder

encoded version of the data point represented by original 41 features. The decoding part of the deep autoencoder is a feed forward net again with 3 nodes expanded to 10 and then 41 nodes in the output layer. This part learns to reconstruct the data points.

The task of dimensionality reduction would have been very difficult if the features are not correlated. The network tries to find a pattern in the structure of the data and thus represent the features in a low dimensional space. The layers are initialized with weights randomly. But finding the initial weights which leads to a good approximation of the model is another study. Because the errors become very small and insignificant after they pass through the hidden layers. The process of discovering these initial weights is called as Pretraining [20]. The first set of the examples are trained in unsupervised fashion. The autoencoder learns the patterns of the input data without knowing the given class labels. It clusters the patterns of the data samples based on similarity. The mapping of inputs to the outputs follows nonlinearity. The nonlinear activation function used is 'Tanh' and 100 epochs are tried on the training set. The plot of reduced representation of the data is shown in the Fig. 4. From the figure, it can be deduced dimensionality reduction alone is not enough to find the cluster of anomalies. A network model is obtained on the training set and the metric Mean Squared Error (MSE) is used as a loss function. This model is used as a pre-training input for the next 40% of data samples. These data samples are trained using the weights from the previous model and with the labels in a supervised manner. Then this one is used for the testing data. The results are given in the Table 2. The results obtained from this

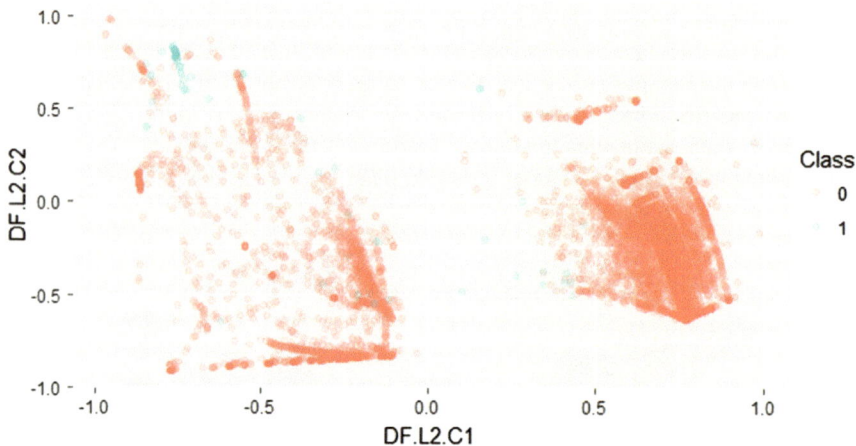

Fig. 4 Two-dimensional feature representation

experiment showed 87% detection of anomalies. The parameters of the input, hidden dropout ratios, L1, L2 regularization are kept at the default values.

The second experiment is done with same data splits but with different hidden layer size. The other three experiments are conducted based on the One Class Classification (OCC) also called as Unary Classification technique. OCC recognizes objects of a class among the entire data objects by learning from a training set which consists of data of that class [22]. In OCC tasks, the negative instances are very less in the entire data distribution. But one side of the boundary can be determined certainly. OCC is harder than regular binary class or multiclass problems. It describes a boundary around the target class, so that it accepts instances from that class reducing the acceptance of instances from the other class. Since only one side of the boundary can be determined, in OCC, it is hard to decide, based on just one class how tightly the boundary should fit in each of the directions around the data. It is difficult to know which attributes contribute the separation of positive and negative classes. Such problems require many training examples than traditional machine learning problems [23]. The other three experiments are conducted based on OCC which has more number of training examples. In the third one, data splits are 70, 10 and 20. The training set has 70 percent of the data but the negative examples are removed from it. This set is trained with Autoencoder with 20 nodes in the first hidden layer, 3 nodes in the second and 20 nodes in the third hidden layer. The model derived from training is applied on the 10 percent of the data which included a response. And this second model is applied on the testing set. This experiment improved the results. The fourth experiment is done in the same way with different data splits. It didn't improve the results much. In fifth one training set consists of 60 percent of data and trained using One Class Classification technique. The model is trained with set of all negative examples and tested on the data

Table 2 The confusion matrix of different autoencoder models

Models	Data splits	Hidden layer size	Actual	Predict	N	Freq.
1	40, 40, 20	10, 3, 10	0	0	12047	0.9947
			0	1	64	0.0052
			1	0	7	0.1296
			1	1	47	0.8703
2	40, 40, 20	15, 3, 15	0	0	12040	0.9890
			0	1	71	0.0109
			1	0	6	0.0555
			1	1	48	0.8944
3	70, 10, 20	20, 3, 20	0	0	12015	0.9920
			0	1	96	0.0079
			1	0	2	0.0370
			1	1	52	0.9629
4	80, 10, 10	20, 4, 20	0	0	6166	0.9953
			0	1	27	0.0043
			1	0	1	0.0384
			1	1	25	0.9616
5	60, 20, 20	20, 4, 20	0	0	11975	0.9978
			0	1	136	0.0021
			1	0	1	0.0185
			1	1	53	0.9814

which consisted of both positive and negative examples. The results are 98 percent accurate with respect to positive examples. It detected all true positive cases except one.

The data is split in various ratios in all these trials to check whether doing it improves the results. In the first and second trial data split ratio is the same. But nodes in the hidden layers are changed. It did not improve the results much. In the third procedure size of the training set is increased. The model derived is applied on the testing set. The nodes in the hidden layer are changed to 20, 4, 20. It improved the classification accuracy of positive examples and it is 96%. In the final model, the training set size is set to 60 percent, validation set is 20%. This division of the data set brought the result which classified almost all the positive examples accurately. The third, fourth and fifth experiments are performed based on the OCC technique. From the results obtained, Analysis II performed better than Analysis I.

Table 3 The metrics and values derived from the fifth model

Metric	Value
maxf1	0.9961
maxf2	0.9984
maxaccuracy	0.9961
maxprecision	1.0000
maxrecall	1.0000
maxspecificity	1.0000
maxmin_per_class_accuracy	0.9961

The percent of data splits, the number of the nodes in the hidden layers and the confusion matrix derived from each model is shown in the Table 2. The results shown in the final model is highly accurate. The confusion matrix correctly classified all the positive examples and misclassified 26 negative examples as positive in the test set. The metrics and values of this model are shown in the Table 3.

3 Conclusion

In this paper two types of analysis are done using artificial neural networks, on a data set which is very much biased to negative examples. In the Analysis I, it has been proved that traditional machine learning algorithms do not give better results in classifying data which has more examples towards one class compared to other class. In the Analysis II, it has been shown deep Autoencoder with unary classification approach offered the best solution in the classification of a data set which is highly biased towards one class. This model can be used as a prototype to detect the presence of anomalies in data obtained from different domains. The quality of these results can be varied by changing the number of hidden neurons, varying activation functions, distance measures and gradient optimization. Deep neural network with many hidden layers is found to be very useful, for this type of complex and real-time problems, and with high dimensional data.

References

1. Najafabadi, M., Villanustre, F., Khoshgoftaar, T.M., Seliya, N., Wald, R., Muharemagic, E.: Deep learning applications and challenges in big data analytics. J. Big Data **2**(1), 1–21 (2015)
2. Nielsen, M.A.: Neural Networks and Deep Learning. Determination Press (2015)
3. Nandhakumar, H., Amudha, J.: A comparative analysis of a neural based remote eye gaze tracker. In: 2014 International Conference on Embedded Systems, pp. 69–75
4. Gupta, D., Vani, K., Leema, L.M.: Plagiarism detection in text documents using sentence bounded stop word N-Grams. J. Eng. Sci. Technol. (JESTEC) **11**(10), 1403–1420 (2016)
5. Aggarwala, P., Sharma, S.K.: Analysis of KDD dataset attributes—class wise for intrusion detection. Procedia Comput. Sci. **57**, 842–851 (2015) (1877-0509© 2015 3rd International Conference on Recent Trends in Computing)
6. Amudha, J., Radha, D.: Analysis of fuzzy rule optimization models. Int. J. Eng. Technol. **7** (5), 1564–1570 (2015)
7. Hinton, G.E., Salakhutdinov, R.R.: Reducing the dimensionality of data with neural networks. Science **313**(5786), 504–507
8. Yuan, Y., Jia, K.: A distributed anomaly detection method or operation energy consumption using smart meter data. In: 2015 International Conference on Intelligent information Hiding and Multimedia Signal Processing
9. Ebborth, M.: Deep learning anomaly detection as support fraud investigation in brazilian exports and anti-money laundering. In: 2016 15th IEEE International Conference on Machine Learning and Applications
10. Ge, S., Jun, L.: Anomaly detection of condition monitoring with predicted uncertainty for aerospace applications. In: 2015 IEEE 12th International Conference on Electronic Measurement & Instruments 248–253
11. Zhao, Z., Mehrotra, K.G., Mohan C.K.: Ensemble algorithms for unsupervised anomaly detection. In: Current Approaches in Applied Artificial Intelligence, pp. 514–525 Springer (2015)
12. Klimek, P.: Statistical detection of systematic election irregularities. In: Proceedings of the National Academy of Sciences of United States of America, vol. 109, Issue No. 41, pp. 16469–16473 (2015)
13. Tavallaee, M., Bagheri, E., Lu, W., Ghorbani, A.A.: A Detailed analysis of the KDD CUP 99 data set. In: Proceedings of the 2009 IEEE Symposium on Computational Intelligence in Security and Defense Applications (CISDA 2009), pp. 53–58 (2009)
14. http://kdd.ics.uci.edu/databases/kddcup99/kddcup99.html
15. Foster, E.: Xavier: arXiv:1701.02145. Shallow and Deep Networks Intrusion Detection System: A Taxonomy and Survey
16. Gao, C.N., Gao, L., Gao, Q., Wang, H.: An intrusion detection model based on deep belief networks. In: 2014 Second International Conference on Advanced Cloud and Big Data, 2014, pp. 247–252. National Center for Biotechnology Information, http://www.ncbi.nlm.nih.gov
17. Alom, M.Z., Bontupalli, V., Taha, T.M.: Intrusion detection using deep belief networks. In: National Aerospace and Electronics Conference (NAECON), pp. 339–344 (2015)
18. An, J., Cho, S.: Variational autoencoder based anomaly detection using reconstruction probability. In: Special Lecture on IE, vol. 2, pp. 1–18. SNU Data Mining Center (2015)
19. Japkowicz, N., Stephen, S.: The class imbalance problem: a systematic study. Intell. Data Anal. **6**(5), 429–449 (2002)

20. Pozzolo, A.Z., Caelen, O., Johnson, R.A., Bontempi, G.: Calibrating probability with under sampling for unbalanced classification. In: Symposium on Computational Intelligence and Data Mining (CIDM), IEEE (2015)
21. Hinton, G.E., Salakhutdinov, R.R.: Reducing the dimensionality of data with neural networks. Science **313**(5786), 504–507 (2006)
22. Khan, S.S., Madden, M.G.: One-class classification: taxonomy of study and review of techniques. Knowl. Eng. Rev. **29**(03), 345–374 (2014)
23. Tax, D., Duin, R.: Uniform object generation for optimizing one-class classifiers. J. Mach. Learn. Res. **2**, 155–173 (2001)

Histogram Modification and Bi-level Moment Preservation Based Reversible Watermarking

R Rajkumar[1(✉)] and A Vasuki[2]

[1] Department of Electronics and Communication Engineering, Dr. NGP Institute of Technology, Coimbatore, Tamilnadu, India
rajkumarramasami@gmail.com
[2] Department of Electronics and Communication Engineering, Kumaraguru College of Technology, Coimbatore, Tamilnadu, India
vasuki.a.ece@kct.ac.in

Abstract. Reversible watermarking is technique to hide the secret information in an image and to reconstruct the original image after the extraction of the watermark. In this paper, we present a reversible watermarking algorithm based on block separation, moment preservation, histogram shifting and embedding secret information. The peak points of the blocks are identified and histogram shifting is performed to improve the embedding capacity of the image. Due to the increase in the embedding capacity of the proposed method, the distortion of the host image increases. The bi-level moment preservation technique is used, in order to improve the image quality. The performance of the proposed algorithm has been evaluated based on the parameters like embedding capacity, Peak Signal to Noise Ratio (PSNR) and Color Peak Signal to Noise Ratio (CPSNR) and compared with the existing histogram modification scheme and it is found to give better results.

Keywords: Histogram shifting · Histogram modification · Bi-level moment preservation · Reversible watermarking · Block separation · Moment preservation

1 Introduction

In recent years, digital images are easily transmitted and received through wired and wireless communication channels. Due to the development of various data processing systems, the properties or the content of the data is easily changed. Watermarking is an information hiding technique that is used to hide proprietary information within digital media like photograph, music, or video. The method of embedding data inside the digital image is known as image watermarking. The data which is embedded is known as the watermark. The classification of watermarking [1] is based on different criteria such as human perception, working domain etc.

© Springer International Publishing AG 2018
D. J. Hemanth and S. Smys (eds.), *Computational Vision and Bio Inspired Computing*,
Lecture Notes in Computational Vision and Biomechanics 28,
https://doi.org/10.1007/978-3-319-71767-8_94

The basic types of watermarking techniques are [2]:

- Spatial Domain and Frequency Domain Watermarking
- Reversible and Irreversible Watermarking
- Perceptible and Imperceptible Watermarking
- Fragile and Robust Watermarking
- Public and Private Watermarking
- Asymmetric and Symmetric Watermarking

Spatial domain watermarking is the method in which the secret data is embedded directly into the host image pixels. But in transform domain [3, 4], the watermark is embedded in frequency domain of the host image. The transformations which are commonly used are Discrete Cosine Transform (DCT) [5], Discrete Sine Transform (DST), Discrete Wavelet Transform (DWT) [6, 7], Singular Value Decomposition (SVD), etc. [8]. In reversible watermarking the original information (host image) can be completely recovered after the extraction of the secret message, but in irreversible technique the original information cannot be completely recovered. Perceptible watermarking is the method in which the embedded information is visible when the original content is viewed [9]. This type of watermarking is mainly used for authentication of the images, whereas in the imperceptible watermarking the information added in the host image cannot be viewed directly. If the secret information is not exactly detected after slight modification of the original signal, it is fragile watermarking. Semi-fragile watermarking [10, 11] resists the transformation but cannot be detected. Digital watermarking is said to be robust since it resists all the transformations and attacks. In public watermarking the secret data can be accessed by all the users, whereas in private watermarking the secret data can be accessed only by the authorized persons. If the watermarking method uses different keys for the embedding and extraction of watermark, it is termed as asymmetric watermarking. In symmetric watermarking, same secret keys are used for both embedding and extraction of watermark.

The main features of watermarking are robustness, imperceptibility, embedding capacity, security. The first three properties play a vital role in the watermarking algorithms and there is trade off between these properties of watermarking [4]. Many image watermarking algorithms have been proposed in the past several years. The algorithms which have been developed introduce some permanent distortions due to watermarking and the original image cannot be retrieved back. In order to overcome it, many new algorithms have been developed and are described in literature review.

A prediction error expansion method for reversible watermarking has been developed to reduce the watermarking distortion [12]. The prediction error is then embedded in both the prediction context and the current pixel. The complex predictors such as median edge detector or the gradient adjust predictor are used to reduce the distortion of watermarking. A tree based parity check method is used to reduce distortion between the cover object and the stego object uses the steganography [13]. It is simple method of hiding the message in the image. The stego object is constructed effectively by using majority vote strategy method. A binary linear stego code is generated which is tree structure model effectively reduces the distortion of the host image.

A recovery packet which contains the information about the watermarked area is used to restore the information [14] which reduces the image distortion. The recovery packet is inserted invisibly to the image region. The original watermarked pattern can be reconstructed using the recovery packet. This packet is encoded before the embedding process. The image recovery process is done by decoding the recovery packet without considering the watermark information embedded in the image. A triple layered embedding [15] of the secret data, which allows the modifications $\{-2, -1, 0, +1, +2\}$ of the cover image pixels. The modifications up to the third LSB of the original image pixel is done, which increases the embedding capacity of the image and distortion is increased due to third LSB modification. In this technique, the performance on distortion can be measured at low embedding capacities and mitigates the capacity control problem and reversible data-embedding technique using prediction-error expansion is used. This technique better exploits the correlation inherent in the neighborhood of a pixel than the difference-expansion scheme. The embedding capacity is maximized by using combined prediction-error expansion and histogram shifting combine to form an effective method for data embedding [16].

The watermarking algorithm for High Dynamic Range (HDR) images has been developed [17]. This HDR images have a wide range of luminance values which are applied to Tone mapping preprocessing technique to reduce the dynamic range of the images. A blind watermarking technique is based on quantization index modulation has been used. The watermarking algorithm which embeds the information in all the three color channels of the host image with tolerable distortion have been developed [18]. This algorithm is based on wavelet transform of the original signal and the color difference equation is used to measure the redundancy of perception of the host signal. In this method, the original signal is not required for the extraction of the watermark and is referred as transparent algorithm. This method is highly robust since it undergo various attacks such as noise addition, cropping, filtering, scaling etc.

An effective recovery of image and tamper detection process is achieved by Cyclic Redundancy Check (CRC) and dual watermarking [19]. The input image blocks are XOR with the watermark are generated using CRC. This dual watermarking scheme ensures the image recovery in an efficient manner with low distortion. A semi-fragile method which embeds the approximation details with the watermark is proposed [20]. The luminance of the image is added in the three color channels of the input image. The luminance addition into the color channels is performed by using mapping function. The embedding process is implemented using the secret key which maps the position of the pixel. The luminance value is taken from the watermark information which is embedded in original signal.

The watermarking scheme for image authentication and tamper proofing has been discussed [21]. The algorithm uses bi level moment preservation technique which consists of the authentication data and the feature information. A two stage dual parity check method and morphological operations are used to improve the tamper proofing process. A data hiding scheme based on histogram modification has been proposed [22] in which the adjacent pixel differences are calculated and these differences are used to embed the data. The histogram shifting technique is used control the overflow and underflow of the pixels. The count of the pixel value which has the maximum pixel difference determines the embedding capacity of the image. In color image the color channels are separated from the host image. The secret data can be embedded in all the three channels and recombined to from the watermarked image.

In this paper, a novel reversible data hiding algorithm for colour images using bi-level moment preservation and histogram shifting technique is proposed. Here, the bi-level moment preservation technique is used for efficient reconstruction of the image. The histogram shifting method after the block separation of the color channels is used for efficient embedding of the watermark data. The bit map generated by using the representative pixels and the secret data are embedded in the original signal and then it is transmitted. At the receiver, with the help of peak point information the watermark data's are extracted. The original image is reconstructed by using the feature information stored in the individual blocks of the images.

The rest of the paper is organised as follows. In Sect. 2, the bi level moment preservation technique is explained in detail. The proposed watermarking technique with the algorithm is explained in Sect. 3. The experimental results are tabulated and discussed in Sect. 4. Finally, Sect. 5 concludes the paper with future work.

2 Bi-level Moment Preservation

The characteristics and shape of the objects can be clearly provided by its moments [8]. The shape of the histogram of an image can be described by its central moments or moments about the mean expressed as:

$$\mu_n = \sum_{i=0}^{L-1} (z_i - m)^n p(z_i) \tag{1}$$

where, n is the order of the moment and m is the mean. The mean value is calculated as:

$$m = \sum_{i=0}^{L-1} z_i p(z_i) \tag{2}$$

Since the histogram is normalized as per the assumption, the sum of all the components is '1'. So the zeroth moment and the first moment is assumed as:

$$\mu_0 = 1$$
$$\mu_1 = 0$$

The second order moment is calculated as:

$$\mu_2 = \sum_{i=0}^{L-1} (z_i - m)^2 p(z_i) \tag{3}$$

In order to calculate the threshold and identify the representative grey values the moment of the input image are calculated before the thresholding process [23]. For the input image I, the bth moment is calculated as:

$$m_b = \frac{\sum_{i=1}^{N} G^b(i)}{N} \qquad (4)$$

where b is the order of the moment G(i) is the gray level value of the ith pixel, N is the total number of pixels in the input image I, by using the histogram of the image I, the moment can also be expressed as:

$$m_b = \frac{\sum_k n_k (g_k)^b}{N} \qquad (5)$$

where n_k is the number of pixels at the gray level g_k.

The pixels of the image are classified into pixels groups and the pixels in each group are assigned with the gray level value and this image thresholding method is known as moment persevering technique. Bi level moment preserving technique is a simple method in which only one threshold is determined to classify the pixels. Here the image pixels are classified into two groups, the first group contains the pixels which are greater than the calculated threshold value and the second group contains the pixels which are equal or less than the threshold value.

In bi-level moment thresholding technique, the threshold is calculated by using the following equivalent expression as follows.

$$p_x = \frac{\sum g_{x \le t} n_k}{N} \qquad (6)$$

In this type of moment preservation [23], the pixel values greater than the threshold is replaced with g_x and the pixels equal or less than the threshold is replaced with g_y. Where the values of g_x and g_y are calculated by using the below expressions.

$$g_x = \left(-b - \sqrt{b^2 - 4a}\right)/2 \qquad (7)$$

$$g_y = \left(-b + \sqrt{b^2 - 4a}\right)/2 \qquad (8)$$

where, a $= (m_1 m_3 - m_2^2)/(m_0 m_2 - m_1^2)$ and b $= (m_1 m_2 - m_0 m_3)/(m_0 m_2 - m_1^2)$
and

$$m_0 = p_x g_x^0 + p_y g_y^0$$
$$m_1 = p_x g_x^1 + p_y g_y^1$$
$$m_2 = p_x g_x^2 + p_y g_y^2$$
$$m_3 = p_x g_x^3 + p_y g_y^3$$

3   Proposed Reversible Watermarking Method

The proposed reversible watermarking technique using bi level moment preservation and histogram shifting is explained in this section. The algorithm is implemented for color images by separating the input RGB image into three different color channels. Each color channel is implemented with histogram shifting, block separation and bi-level moment preservation. Here the histogram shifting method is used to increase the embedding capacity of the image. In order improve the quality of the reconstructed image the bi-level moment preservation technique is used. The frame work of the proposed reversible watermarking algorithm is explained below in detail. Figure 1 shows the block diagram of the watermark embedding method for the proposed technique.

The algorithm for the proposed watermark embedding scheme is as follows.

- Read the input color image.
- Separate the RGB image into three different colour channels.
- Block separation is performed for the R, G, B components.
- Compute the histogram of the individual blocks.
- Perform histogram shifting process.
- The histogram shifting is done on either side of the peak value in the histogram.
- In the existing techniques, only one level of shifting is done for the data embedding process to ensure the originality of the host signal.

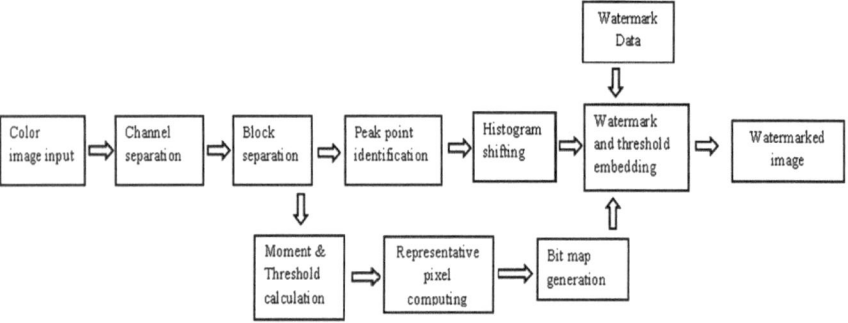

Fig. 1 Block diagram for watermark embedding

- Whereas in the proposed method, multilevel level shifting is done to improve the embedding capacity.
- The increase in the embedding capacity reduces the quality of the original image.
- In order to improve the quality of the original image, the moment preserving technique is adopted.
- For each block, the moments are calculated. By using the moment values the two representative pixels are selected. The moment calculation and the pixel selection is done as explained in bi-level moment preservation (Sect. 2).

- These representative pixels and the watermark bits are embedded to the shifted pixels positions of the original histogram.
- The watermark can be embedded in all the three color channels by using step 4 to step 9.
- Finally, all the color channels are combined to form the RGB image with the watermark data is known as the watermarked image.
- This watermarked image is then transmitted to the receiver along with the side information.

The block diagram for the watermark extraction and the reconstruction of the original image is shown in Fig. 2.

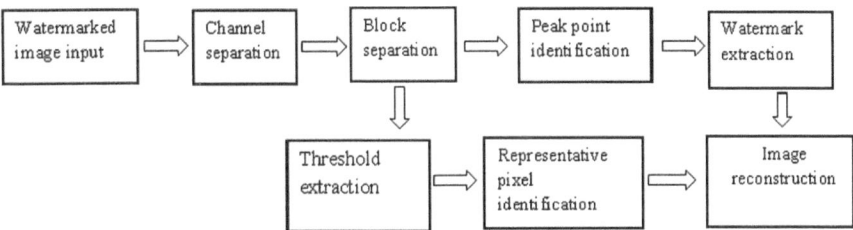

Fig. 2 Block diagram for watermark extraction and image reconstruction

The algorithm for the proposed watermark extraction processes is as follows.

- The received watermarked RGB image is separated into the three color channels.
- The representative pixel values and the watermark is embedded, are identified using the peak point information and the shifted pixels values of the histogram.
- The representative pixel values and the watermark data is extracted separately.
- The watermark data of the entire color channel is finally combined to for the complete watermark information.
- The representative pixels of each block, which are selected from the bi level moment preservation technique is used to the reconstruction of the original image.
- The quality of the reconstructed image is highly improved by the bi-level moment preservation technique.

4   Results and Discussions

The performance of the proposed watermarking scheme on different images has been evaluated. The evaluation parameters are PSNR, CPSNR and embedding capacity, used for comparison of various images.

The embedding capacity of the proposed watermarking scheme is defined as total number of pixels present on the peak point of the histogram in bits.

The logarithmic value of the ratio of the maximum gray level to the mean square error is the peak signal to noise ratio (PSNR) of the image.

$$PSNR = 10 \log_{10} \left(\frac{255^2}{MSE} \right) (\text{dB}) \tag{9}$$

where, Mean Square Error (MSE) is the square value of the difference between the original image and the watermarked image,

$$MSE = \frac{1}{M \times N} \sum_{i=1}^{M} \sum_{j=1}^{N} (x(i,j) - w(i,j))^2 \tag{10}$$

where, x(i, j) is the original image, w(i, j) is the watermarked image.

The Color Peak Signal to Noise Ratio (CPSNR) is given by

$$CPSNR = 10 \log_{10} \left(\frac{255^2}{CMSE} \right) \tag{11}$$

where color mean square error (CMSE) is calculated as

$$CMSE = \frac{1}{3MN} \sum_{i=r,g,b} \sum_{y=1}^{N} \sum_{x=1}^{M} (I_i(x,y,i) - I_r(x,y,i))^2 \tag{12}$$

where M, N—number of rows and columns in the image, I_i is the input image and I_r is the reconstructed image.

The proposed watermarking algorithm is tested on various color images of size 800×600. The original images are shown in Fig. 3. The watermarked images and the reconstructed images are show in Figs. 4 and 5 respectively. The parameters for the proposed algorithm are compared with the existing method based on pixel differences.

The PSNR values of the individual color components of all the images for the proposed method, shows a better improvement due to the moment preservation technique, i.e. the average PSNR of the proposed method for the sample images is 46.01 dB, whereas the existing method attains the PSNR of only 44.93 dB. The average CPSNR of the images for the proposed method is 44.53 dB and fro the existing method it is 43.32 dB. Tables 1 and 2 shows the comparison of PSNR and CPSNR of the proposed watermarking method and the existing method. The proposed method shows the improvement in the embedding capacity of the each color component, due to peak point identification by block separation process, so the overall embedding capacity of the increases. Table 3 shows the comparison of the embedding capacity of the proposed method and the existing method.

(a) Sunset   (b) Blue hill   (c) Water lily   (d) Winter

(e) Nature   (f) Lena   (g) Baboon   (h) Peppers

Fig. 3 Original images

(a) Sunset   (b) Blue hill   (c) Water lily   (d) Winter

(e) Nature   (f) Lena   (g) Baboon   (h) Peppers

Fig. 4 Watermarked images

(a) Sunset   (b) Blue hill   (c) Water lily   (d) Winter

(e) Nature   (f) Lena   (g) Baboon   (h) Peppers

Fig. 5 Reconstructed images after watermark extaction

Table 1 Comparison of PSNR of proposed watermarking method and existing method

Input images	PSNR in dB (Pixel difference method)				PSNR in dB (Proposed method)			
	R	G	B	Avg	R	G	B	Avg
Sunset	44.79	44.91	44.54	44.75	45.23	45.02	45.65	45.30
Bluehill	47.85	47.53	48.51	47.96	48.59	48.32	48.14	48.35
Waterlily	44.61	44.42	44.58	44.54	45.22	44.96	44.73	44.97
Winter	43.41	45.45	44.18	44.35	45.47	45.84	45.45	45.59
Nature	44.58	44.95	43.92	44.48	46.02	45.68	45.76	45.82
Lena	46.98	44.25	45.63	45.62	47.14	46.88	46.03	46.68
Baboon	44.87	42.56	43.45	43.63	45.32	44.98	45.41	45.24
Peppers	44.52	44.01	44.86	44.46	46.11	45.79	46.55	46.15
Overall average	44.93				46.01			

Table 2 Comparison of CPSNR of proposed watermarking method and existing method

Input images	CPSNR in dB (Pixel difference method)	CPSNR in dB (Proposed method)
Sunset	42.53	43.44
Bluehill	47.21	47.98
Waterlily	43.11	44.23
Winter	42.58	43.78
Nature	42.14	43.72
Lena	44.84	45.44
Baboon	42.11	43.88
Peppers	42.02	43.79
Average CPSNR	43.32	44.53

Table 3 Comparison of embedding capacity of proposed watermarking method and existing method

Input images	Embedding capacity in bits (Pixel difference method)				Embedding capacity in bits (Proposed method)			
	R	G	B	Total	R	G	B	Total
Sunset	100,281	113,199	105,975	319,455	115,241	147,832	114,021	377,094
Bluehill	211,054	186,406	224,209	621,669	218,156	220,073	228,932	667,161
Waterlily	95,191	128,804	91,425	315,420	103,259	102,767	125,683	331,709
Winter	95,090	126,754	91,832	313,676	145,287	100,043	98,987	344,317
Nature	89,475	138,721	92,050	320,246	108,248	149,671	99,027	356,946
Lena	245,291	187,352	110,735	543,378	268,775	205,891	145,014	619,680
Baboon	90,980	98,560	99,420	288,960	96,425	105,143	117,761	319,329
Peppers	126,684	98,456	173,204	398,344	139,672	110,078	185,680	435,430

As per the comparison the embedding capacity of the image is increased when compared with the existing scheme. The PSNR and CPSNR values show improvement for the proposed scheme. Generally, the increase in embedding capacity will reduce PSNR value (image quality). But, in this method, the use of bi-level moment preservation maintains the image quality at a better level.

5   Conclusion

In this paper, a reversible watermarking algorithm has been presented by using the histogram shifting and the bi-level moment preservation technique. The block separation and histogram shifting technique improves the embedding capacity of the images greater than the existing scheme. The bi-level moment preserving technique ensures the reconstruction of the original image with better quality when compared to the existing method. The proposed algorithm is tested and compared with existing algorithms on RGB images with the parameters such as embedding capacity, CPSNR and PSNR. The peak point information is shared along with the watermark signal for the efficient retrieval of the watermark. The simulation, experimental verification and measurements are done in MATLAB. In future, the algorithm can be implemented in various color spaces.

References

1. Chaturvedi, N.: Various digital image watermarking techniques and wavelet transforms. Int. J. Emerg. Technol. Adv. Eng. **2**(5), 363–366 (2012)
2. Tao, H., Chongmin, L., Zain, J.M., Abdalla, A.N.: Robust image watermarking theories and techniques: A review. J. Appl. Res. Technol. **12**(1), 122–138 (2014)
3. Mishra, A., Agarwal, C., Sharma, A., Bedi, P.: Optimized gray-scale image watermarking using DWT–SVD and firefly algorithm. Expert Syst. Appl. **41**(17), 7858–7867 (2014)
4. Hu, H.T., Hsu, L.Y.: Exploring DWT–SVD–DCT feature parameters for robust multiple watermarking against JPEG and JPEG2000 compression. Comput. Electr. Eng. **41**(1), 52–63 (2015)
5. Roy, S., Pal, A.K.: A blind DCT based color watermarking algorithm for embedding multiple watermarks. AEU—Int. J. Electron. Commun. **72** (2), 149–161 (2017)
6. Preda, R.O.: Semi-fragile watermarking for image authentication with sensitive tamper localization in the wavelet domain. Meas. **46**(1), 367–373 (2013)
7. Nguyen, T.S., Chang, C.C., Yang, X.Q.: A reversible image authentication scheme based on fragile watermarking in discrete wavelet transform domain. AEU—Int. J. Electron. Commun. **70**(8), 1055–1061 (2016)
8. Jain, A.K.: Fundamentals of Digital Image Processing, pp. 113–180. Prentice Hall International Edition
9. Khan, A., Siddiqa, A., Munib, S., Malik, S.A.: A recent survey of reversible watermarking techniques. Inf. Sci. **279**(20), 251–272 (2014)
10. Su, Q., Niu, Y., Liu, X., Yao, T.: A novel blind digital watermarking algorithm for embedding color image into color image. Optik—Int. J. Light Electron Opt. **124**(18), 3254–3259 (2013)
11. Qi, X., Xin, X.: A singular-value-based semi-fragile watermarking scheme for image content authentication with tamper localization. J. Vis. Commun. Image Represent. **30**, 312–327 (2015)
12. Li, X., Yang, B., Zeng, T.: Efficient reversible watermarking based on adaptive prediction-error expansion and pixel selection. IEEE Trans. Image Process. **20**(12), 3524–3533 (2011)
13. Hou, C.L., Lu, C., Tsai, S.C., Tzeng, W.G.: An optimal data hiding scheme with tree-based parity check. IEEE Trans. Image Process. **20**(3), 880–886 (2011)
14. Yang, Y., Sun, X., Yang, H., Li, C.T., Xiao, R.: A contrast-sensitive reversible visible image watermarking technique. IEEE Trans. Circ. Syst. Video Technol. **19**(5), 656–667 (2009)
15. Zhang, W., Chen, B., Nenghai, Y.: Improving various reversible data hiding schemes via optimal codes for binary covers. IEEE Trans. Image Process. **21**(6), 2991–3003 (2009)
16. Zhang, X.: Efficient data hiding with plus-minus one or two. IEEE Signal Proc. Lett. **17**(7), 635–638 (2010)

17. Guerrini, F., Okuda, M., Adami, N., Leonardi, R.: High dynamic range image watermarking robust against tone-mapping operators. IEEE Trans. Inf. Forensics Secur. **6**(2), 283–295 (2011)
18. Chou, C.H., Liu, K.C.: A perceptually tuned watermarking scheme for color images. IEEE Trans. Image Process. **19**(11), 2966–2982 (2010)
19. Chang, C.C., Tai, W.L., Lin, C.C.: A multipurpose wavelet-based image watermarking. In: Proceedings of First International Conference on Innovative Computing, Information and Control, vol. 3, pp. 70–73. Beijing, China, 30 August to September 1 2006
20. Qian, Z., Feng, G., Ren, Y.: Fragile watermarking for color image recover based on color filter array interpolation. Lect. Notes Comput. Sci., **6184**, 537–543 (2010).
21. Liu, K.C.: Moment preserving based watermarking for color image authentication and recovery. In: Proceedings of International Conference Network and Computer Science, pp. 139–143. Changchun, China, 29 December to 31 December 2012
22. Ramaswamy, R., Arumugam, V.: Lossless data hiding based on histogram modification. Int. Arab J. Inf. Technol. **9**(5), 445–451 (2012)
23. Liu, K.C.: Self-embedding watermarking scheme for colour images by bi-level moment-preserving technique. IET Image Process. **8**(6), 363–372 (2014)

Combining Diffusion Filter Algorithms with Super—Resolution for Abnormality Detection in Medical Images

Shilpa Joshi$^{(\boxtimes)}$ and R. K. Kulkarni

Department of Electronics and Communication Engineering, VESIT, Chembur,
Mumbai, Maharashtra, India
shilpa132205@gmail.com

Abstract. One of the most significant areas of image research is Image Enhancement. The main aspect of image enhancement involves the improvisation of the visual manifestation of an image. Poor contrast and noise affect many kinds of images today, such as satellite images, remote sensing images, medical images, real-life images and electron microscope images. Therefore, noise removal and resolution increment are important as well as necessary to ensure and enhance the quality of images. There are many imaging modalities and each of them performs different functions ranging from the provision of information about human anatomy/structure to the provision of location statistics about specific activities and tasks. Physical constraints of system detectors —which are tuned to signal-to-noise and timing considerations are used to determine the resolution of imaging systems. The hybrid techniques designed n this paper uses algorithms are mostly based on standard diffusion filters and SR algorithms. Results demonstrate the potential in introducing SR techniques into practical medical applications.

Keywords: Image modalities · Medical image analysis · Diffusion filters
Image enhancement · Super-resolution · Malignancy detection

1 Introduction

It is useful and obvious that increasing the image resolution of medical images will result in the betterment of diagnostic ability thereby and accurate treatment [1]. In addition, it is only intuitive that greater resolution will favor the improvement of automatic detection (of diseases) and segmentation results. The digital medical imaging technologies such as Computerized Tomography (CT), Positron Emission Tomography (PET), Magnetic Resonance Imaging (MRI) etc. have revolutionized modern medicine [2]. However, even with advances in acquisition technology and optimized reconstruction algorithms over the two last decades, obtaining an image at a desired resolution is not easy due to imaging environments and the limitations of physical imaging systems. To add on, quality-limiting factors such as Noise and Blur also play a major

© Springer International Publishing AG 2018
D. J. Hemanth and S. Smys (eds.), *Computational Vision and Bio Inspired Computing*,
Lecture Notes in Computational Vision and Biomechanics 28,
https://doi.org/10.1007/978-3-319-71767-8_95

role in image resolution [3]. Super-resolution in important imaging modalities such as MRI, fMRI and PET has been reviewed in this paper. Standard region growing algorithms (K-means Algorithm) [4] are used along with OTSU binary segmentation to identify region of interests (ROIs); then super resolution (SR) algorithm is used to increase the image resolution. Prior to that two stage filtering is applied. Median filtering to remove Gaussian Noise and diffusion filtering is applied to reduce speckle noise [5]. In the next stage objects of an image are identified and feature extraction is done to obtain six features. n this paper, we analyze the performance of the Speckle Reduction Filter (SRDF) and Anisotropic Diffusion along with Super-resolution [6]. The rest of the paper is organized as follows: In Sect. 2, explanation of the methods and materials used in the comparative approach are given, in Sect. 3, we look at the experimental results and discussions. In Sect. 4, we conclude the paper.

1.1   Domain Based Image Enhancement

Image enhancement is a method in which an image is processed in such a way that the resulting image (after processing) is more appropriate for application in comparison to the original image. In this process, the characteristics ad attributes and modified/altered the attributes of the images are modified during enhancement [7]. Most fields that intensively and extensively make use of images and imaging require image enhancement. A few examples are medical images, satellite images, etc. Enhancement methods are of two classes: [8]

- **Spatial Domain Methods** [9]
- **Frequency Domain Methods**

In **spatial domain** methods, the image pixels are directly dealt with and the pixel values are modified to acquire the required enhancement. The high simplicity and low complexity of spatial domain methods support real time implementations. However, these techniques sacrifice robustness [10, 11].

On the other hand, in **frequency domain** methods, the images are transferred into a frequency domain and the Fourier transform of the respective images are performed. The enhancement techniques are applied on these transformed images. Lastly, an inverse Fourier transform is then performed on the images to obtain the final enhanced images. Frequency domain methods are of relatively low complexity but also cannot target all parts on an image [12, 13].

Spatial filtering techniques can be applied to reduce the speckle noise but they work well as long as the underlying signal is smooth [14]. In this paper, we analyze the performance of the Speckle Reduction Filter (SRDF) and Anisotropic Diffusion along with Super-resolution [6]. The rest of the paper is organized as follows: In Sect. 2, explanation of the methods and materials used in the comparative approach are given, in Sect. 3, we look at the experimental results and discussions. In Sect. 4, we conclude the paper.

1.2   Speckle Dominance and Filtering

Speckle noise is a type of granular noise which generally degrades the quality of medical images. It usually results due to the consistent processing of backscattered signals from several distributed targets. Speckle noise also increases the mean grey level of the affected image. This noise creates a lot of difficulty in interpreting the image [15]. Speckle Noise can be analyzed by various mathematical models [16]. In particular study general model described by Jain [17] is adopted which is expressed by Eq. 1

$$S(x,y) = f(x,y)n_{m(x,y)} + n_{a(x,y)} \qquad (1)$$

where $f(x, y)$ is the noise free image to be recovered, $S(x, y)$ is the noisy image, (x, y) and (x, y) are multiplicative and additive noises respectively. For any speckle, the contrast ration ό is defined as

$$\sigma = \frac{standard\ deviation\ of\ I}{mean\ value\ of\ I} \qquad (2)$$

where I is the intensity of the field (Fig. 1).

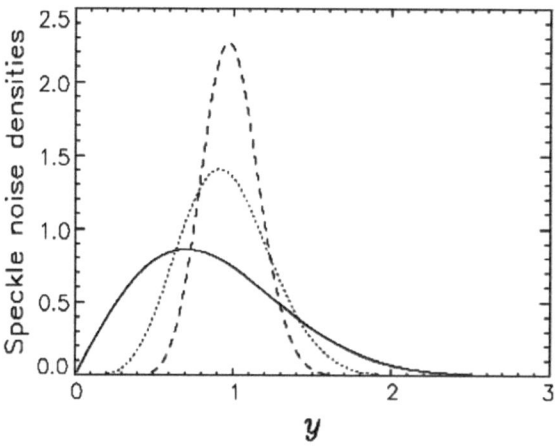

Fig. 1 Speckle noise modeling

Spatial filters [18] provide a convenient way to remove the random noises from the image intensity profile. In this paper, for speckle de noising four non-linear spatial mean filters are studied and out of that the filters from diffusion category are proved to be best. One should choose the technique according to the type and amount of noise present in the image. The diffusion filtering in Spatial domain method is discussed below.

2 Novel Proposed Approach

The proposed filter given in Fig. 2 is based on combination of 3 techniques of image processing which will produce clear, blur-free and super-resolute image. This includes (1) Pre-processing (2) Feature Extraction (3) Abnormality detection. Each stage is described in detail in this section. Figure 2 shows the structure of the proposed method. Each section is described in detail in this segment.

2.1 Preprocessing

To obtain most desirable results and proper probability detection whether the lesions found are malignant or benign, pre-processing step is very important. In the proposed method pre-processing is performed in two stages. One is to improve quality by two stage filtering and by obtaining region of interest. Absolute improvement can be seen in quality of an image after this stage [19].

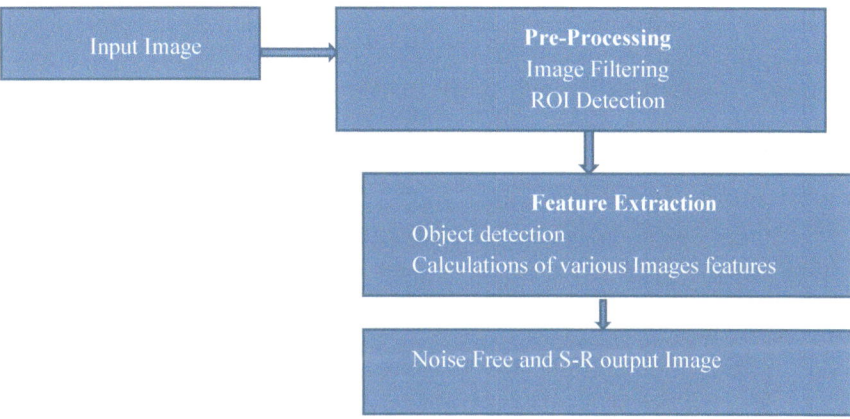

Fig. 2 Proposed algorithm

The first stage of the proposed method is to remove the additional impurities like speckle noise and other types of noise such as random noise and Gaussian Noise [4]. Diffusion Filters [20] proposed by Perona-Malik uses theory of diffusion process as base. It is a technique of reducing image noise without removing significant parts of images. Diffusion process is linear and space-invariant but AIDF (Anisotropic Diffusion Filter) is generalization of diffusion process. In this paper AIDF is used as it gives shape adaptive smoothing and coherence enhancing.

2.2 Region of Interest (*ROIs*) Detection

The additional margins are removed by segmentation. This makes image smaller which results into less computational burden (Fig. 3). To reduce false positive rate it is absolutely necessary to remove the artifacts such as label on the image which contains patient information and as well as some masses or calcifications which are similar to pectoral muscles. To remove this various segmentations methods are used [21]. The proposed technique have used Otsu binary segmentation [4] along with K-means algorithm. This proposes multilevel thresholding. Otsu method works on global thresholding where K-means works on local thresholding. Both the methods together yield a very good result.

2.3   Image Improvements

The quality of medical images need to be preserved for highlighting masses and calcifications. Because it causes edges and some regions of the images should be extracted and studied effectively. As these regions normally contains high frequency so a frequency domain method needs to be applied to achieve them. For this purpose super resolution algorithms are used. The goal of super resolution is to get an image with high resolutions based on low resolution data base. The advantage of S-R algorithm is it gives output image with increased resolution without altering existing hardware. It improves the resolution by considering first high frequency components and then it operates on degradation caused by high frequency components. There are various S-R algorithms exists but here learning based approach is used as it is suitable for single medical image as well as it has capacity to predict high frequency components.

2.4   Final Image Improvement

Most imaging applications fundamentally rely upon high-resolution symbolism. Expanding determination by enhancing indicator cluster determination is not generally a possible way to deal with enhancing determination. Lion's share of imaging applications include making pictures from radiation engendering through three-dimensional protests yet the last pictures are two-dimensional where they speak to some type of projection through a three-dimensional volume. Super-Resolution [22] is a procedure of picture upgrade by which low quality, low resolution (LR) pictures are utilized to create a great, high resolution (HR) picture. Super-Resolution system can be connected to enhance the nature of the picture. The necessities for better resolution in all medicinal imaging modalities as of now speak to a vital and open test. Exact estimation and perception of structure in living tissues is naturally restricted by the imaging framework highlights. Imaging past these cutoff points in therapeutic imaging is alluded to as Super-Resolution. In this paper, another way to deal with single-picture super resolution, is connected subsequent to separating. We propose a novel technique for Super-Resolution in view of scanty portrayal which is inadequate portrayal of a low resolution restorative picture squares (frames) are utilized to produce a high resolution frame. By combining together the low- and high-resolution picture patches, we can uphold the comparability of m between the low-resolution and high-resolution picture fix combine of the fix sets, approaches, which basically test a lot of picture fix sets [23], diminishing the computational cost generously. A few assessments are executed to contrast and past strategy, and the proposed calculation (Diffusion separating and Super Resolution) has its preference on just super-resolution. In this way, the proposed calculation can deal with SR with uproarious contributions to a more brought together system concerning their own word references. The educated lexicon match is a more minimized portrayal.

3 Experimental Results

The performance parameters are most important criteria to justify results through evaluation. Section *Represents* a summary of parameter variation at high noise density before and after super-resolution technique has been applied. Performance of all algorithms is tested on medical images captured with various modalities. The computational result showed that speckle reduction diffusion filter shows best computational results in terms of statistical analysis as well as visual quality is retained at the cost of more runtime at all the noise densities. The parameters considered here are peak signal to noise ratio (PSNR) and mean square error (MSE) by which the objective quality of the reconstructed image is measured. The Universal Quality Index (UQI) and structural similarity index (SSIM) are mainly used for measuring the similarity between two images. The results are *taken* **before and after application of super Resolution technique**. The filters used for implementation are Lee, Kuan Frost from **Linear Category**, Median, Geometric and Holomorphic from **Nonlinear category** and Anisotropic Speckle reduction filters from **Diffusion** Filter. The computational result showed that frost filters from linear category have the best results Due to its nonlinear nature and adaptive property (Table 1).

Table 1 Comparative analysis of measurable parameters before and after S-R

Parameters	Lee	Kuan	Frost	Median	AIDF	SRDF
PSNR (dB)	25.49	24.32	26.59	23.33	**17.04**	**26.32**
PSNR–SR (dB)	42.47	43.93	44.12	42.11	**33.22**	**46.34**
SSIM	.82	.8	8	.71	**.69**	**.7**
SSIM-SR	.34	.36	.33	.38	**.81**	**.87**
Run Time (S)	1.26	1.14	4.48	.98	**34.68**	**13.74**
Run Time-SR (S)	5.44	5.23	18.49	18	**47.84**	**42.57**

It proves to be excellent both speckle reduction and detail preserving properties. The response of the filter varies locally with the coefficient of variation [24]. In case of low coefficient of variation, the filter is more average like, and in cases of high coefficient of variation, the filter attempts to preserve sharp features by not averaging. Figures 3 and 4 respectively represents the graphical and statistical analysis of proposed technique for USG image. The noisy input of USG of liver is processed through three stages proposed method. The visual analysis is given for Low, Medium and High noise density respectively in Figs. 4, 5 and 6. The step wise execution is expressed in terms of 5 sub-figures named as a (Input Noisy Image), b (Two stage filtered Image), c (Region Grown Image), d (Segmented Image), e (Final S-R Image). The amount of contamination which is Noise Density is classified as follows. Low Noise Density (L.N.D)—Noise contamination 10–20%. Medium Noise Density (M.N.D)—Noise contamination 21–40% High Noise Density (H.N.D)—Noise contamination >40%.

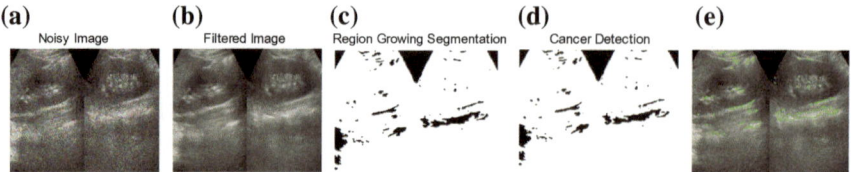

Fig. 3 Low noise density analysis of proposed technique

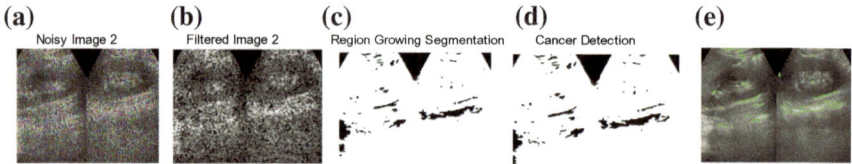

Fig. 4 Medium noise density analysis of proposed technique

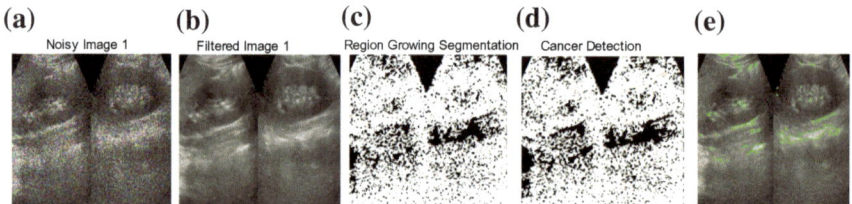

Fig. 5 High noise density analysis of proposed technique

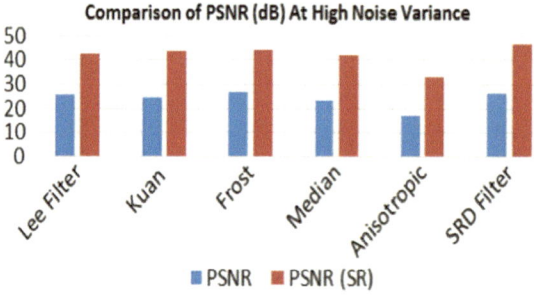

Fig. 6 Graphical representation of PSNR (dB)

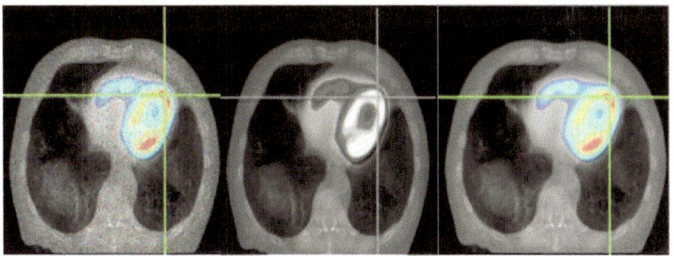

Fig. 7 Output of proposed technique for MRI of brain

Figure 7 indicates the output produced through proposed technique which is clearer and noise free which is the real success of the proposed method.

4 Conclusion

As we can observe, diffusion filters, in conclusion, are more effective in comparison to other standard or non-linear filtering techniques. This is because it has the added advantage of Selective Enhancement. Even though the Diffusion Filter Algorithm shows better performance in comparison to other filtering algorithms, it also has marginally higher run time. However, for various image modalities such as X-Ray, CT, PET, MRI and USG, the technique exhibits better performance in terms of image clarity. USG and MRI proved to be best in terms of both visual as well as statistical Parameters.

References

1. Jeyalakshmi, T.R., Ramar, K.: A modified method for speckle noise removal in ultrasound medical images. Int. J. Comput. Electr. Eng. **2**(1), 1793–8163 (2010)
2. Loizou, C.P., Pattichis, C.S.: Introduction to ultrasound imaging and speckle noise. In: Despeckle Filtering Algorithms and Software for Ultrasound Imaging, pp. 3–23. Morgan & Claypool Publishers (2008)
3. Jain, A.K.: Fundamentals of digital image processing, pp. 86–95. Prentice-Hall, Englewood Cliffs, NJ (1989)
4. Hosotani, F., Inuzuka, Y., Hasegawa, M., Hirobayashi (Member, IEEE), S., Misawa, T.: Image denoising with edge-preserving and segmentation based on mask nha. IEEE Trans. Image Process. **24**(12), 6025–6033 (2015)
5. Gonzalez, R.C., Woods, R.E.: Digital Image Processing, 2nd edn. Prentice-Hall, Englewood Cliffs, NJ (2002)
6. Nicolae, M.C., Moraru, L., Onose, L.: Comparative approach for speckle reduction in medical ultrasound images. Rom. J. biophys. **20**(1), 13–21 (2010)
7. Kaurl, N., Khullar, S.: An approach for speckle noise reduction from US images using enhanced sticks filtering. Int. J. Comput. Sci Commun. Eng. **2**(1), 55–57 (2013)
8. Chopra, N., Anand, M.A.: Despeckling of images using wiener filter in dual wavelet transform domain. Int. J. Comput. Sci. Inf. Technol. **5**(3), 4069–4071 (2014)

9. Singh, I., Neeru, N.: Performance comparison of various image denoising filters under spatial domain. Int. J. Comput. Appl. **96**(19), (2011). (Department of Computer Engineering Punjabi University, Patiala) (0975–8887)
10. Kulkarni (Ph.D. Thesis), R.K.: Thesis novel restoration techniques for images corrupted with high density impulsive noise. NIT Rourkela (2012)
11. Srivastava, C., Mishra, S.K., Asthana, P.: Performance comparison of various filters and wavelet transform for image de-noising. IOSR J. Comput. Eng. **10**(1), 55–63 (2013)
12. Gnanambal Ilango, B., Gowri, S.: Neighbourhood median filters to remove speckle noise from ct–images. Int. J. Appl. Inf. Syst. (IJAIS) **4**(10), 41–46 (2012). ISSN: 2249-0868 Foundation of Computer Science FCS, New York, USA, www.ijais.org
13. Akl, A., Yaacoub, C.: A hybrid wavelet-spatial denoising filter. Digital Signal Processing (DSP), 18th International Conference on1–3 July 2013, pp. 1546–1874 (2013). ISBN: 978-1-4673-5807-1, https://doi.org/10.1109/ICDSP.2013.6622695
14. Krissian (Member, IEEE), K., Westin (Member, IEEE), C.F., Kikinis, R., Vosburgh (Member, IEEE), K. G.: Oriented speckle reducing anisotropic diffusion. IEEE Trans. Image Process. **16**(5) (2007)
15. Weickert, J., Romeny (Member, IEEE), B.T.H., Viergever, M.A.: Efficient and reliable schemes for nonlinear diffusion filtering. IEEE Trans. Image Process. **7**(3) (1998). ISSN Information: Print ISSN: 1057-7149, Electronic ISSN: 1941-0042
16. Hiremath, P.S., Akkasaligar, P.T., Badiger, S.: Noise reduction in medical ultrasound imagesbook edited by Gunti Gunarathne (2013). ISBN 978-953-51-1159-7, INSPEC Accession Number: 5867956, https://doi.org/10.5772/56519
17. Sivakumar, R., Gayathri, M.K., Nedumaran, D.: Speckle filtering of ultrasound b-scan images—a comparative study between spatial and diffusion filters. Int. J. Eng. Technol. **2**(6), 514–523 (2006). ISSN:1793-8236
18. Somasundaram, K., Kalavathi, P.: Performance of spatial mean filters on denoising medical images for edge detection. IJCST **3**(2) (2012), ISSN: 0976-8491 (Online) | ISSN: 2229-4333 (Print). (Department of Computer Science and Applications, Gandhigram Rural Institute Deemed University, Gandhigram, Tamilnadu, India)
19. Samak, A.H.: A new nonlinear anisotropic-wiener method for speckle noise reduction in optical coherence tomography. Int. J. Comput. Appl. **65**(12), (2013). (0975–8887)
20. Yu (Senior Member, IEEE), Y., Acton, S.T.: Speckle reducing anisotropic diffusion. IEEE Trans. Image Process. **11**(11), 1546–1874 (2002)
21. Dalwadi, M.N., Khandhar, D.N., Wandra, K.H.: Automatic boundary detection and generation of region of interest for focal liver lesion ultrasound image using texture analysis. Int. J. Adv. Res. Comput. Eng. Technol. (IJARCET) **2**(7), 2369–2373 (2013). ISSN: 2278–1323
22. Kouame, D.: Super-resolution in medical imaging: An illustrative approach through ultrasound. ISBN 978-1- 4244-3932-4/09/$25.00. IEEE (2009)
23. Yang (Student Member, IEEE), J., Wright (Member, IEEE), J., Huang (Fellow, IEEE), T.S., Ma (Senior Member, IEEE), Y.: Image super-resolution via sparse representation. IEEE Trans. Image Process. **19**(11), 2861–2873 (2010)
24. Sivakumar, R.: Comparative study of speckle noise reduction of ultrasound b-scan images in matrix laboratory environment. Int. J. Comput. Appl. **10**(9), (2010). (0975–8887)

Design of an Algorithm for People Identification Using Facial Descriptors

Pranav V.$^{(\boxtimes)}$, R. Manjusha, and Latha Parameswaran

Department of Computer Science and Engineering, Amrita School
of Engineering, Amrita Vishwa Vidyapeetham, Amrita University, Coimbatore,
India
cb.en.p2cvil5017@cb.students.amrita.edu,
{r_manjusha,p_latha}@cb.amrita.edu

Abstract. The proposed work aims at identifying and greeting people. A new technique is introduced that incorporates the facial features and KNN classifier. Since human face is the one which is said to be the most representative, the features of the face (eyes, nose and mouth) are extracted and are used for training the classifier. The proposed frame work consists of four phases: Facial Features Detection (FFD), Detected Features Positioning (DFP), Descriptive Features Extraction (DFE) and Face Identification (FI). The proposed algorithm can be used in a wide variety of scenarios such as campus, office etc. after being trained with the corresponding dataset. The performance of the system is analyzed for various scenarios. Good average accuracy of 96.05% has been achieved.

Keywords: Facial features detection · Detected features positioning
Descriptive features extraction · Face identification · K-nearest neighbor

1 Introduction

Researches have proved that the greeting/identification by someone, especially in a strange environment can make a person feel very comfortable. Each person has unique features which enable us to distinguish and recognize people. When we see a person whom we already know, we observe his/her prime noticeable features and compare or match with those which are familiar. And finally we identify the person to be the one with whom the features match the most. All these happen in a fraction of a second. This is the motivation behind the implementation of this concept in computers. That's how the people identification system evolved.

Computer Vision based people identification is used in the proposed work. People identification can be done based on features such as height, hairstyle, walking style etc. Face, being the most representative part of a human, contains the maximum number of features. The proposed work focuses on extracting features from face for identifying a person. The facial features are extracted from the image and are used for training and classification. The classifier used for the proposed work is the KNN classifier.

© Springer International Publishing AG 2018
D. J. Hemanth and S. Smys (eds.), *Computational Vision and Bio Inspired Computing*,
Lecture Notes in Computational Vision and Biomechanics 28,
https://doi.org/10.1007/978-3-319-71767-8_96

The important steps of our proposed technique includes detection of facial features, finding the ratios of the distances between them, and finally classifying the face using the classifier which is trained based on the extracted features. This proposed technique is based on the idea that each and every human has a unique set of lengths between the facial features. The ratios of the lengths between the facial features which are independent of the camera distance are considered in the proposed work.

2   Related Work

A lot of research has been conducted in the field of people identification.

In [1] the authors used the stride parameters along with the height as distinguishable features of a person. Here they estimate the height by separating the person and the background using segmentation and after that the evident height is fit to a model which is time dependent.

With the evolution of filters, several facial features extraction was possible because of which people identification became popular. A formal and ancient method of categorizing faces was first suggested in [2], a method in which facial profiles are collected as curves and their norms were calculated. Other profiles were classified according to the deviations from the norm.

The progress in the field of face recognition was rapid due to the vital evolution of algorithms and the obtainability of vast databases of facial images. In the process of people identification, the initial step is to detect the face and extract features like mouth, nose and eyes. There are many popular algorithms such as Viola Jones and other algorithms were developed based on morphological image processing [3] and skin color segmentation [4].

In [3] the author assumes that anterior face image is easily available. At first facial regions were detected based on the eye areas. After identifying the face region other facial feature points such as junctions of eyes, nostril, corners of mouth etc. were detected. The system in [3] comprises of two phases. First, the eye regions were extracted by applying morphological operations method and rule based method separately. The probable eye regions were found by linking both results and using the relative geometrical position between eye pairs. After that, the facial regions were positioned and extracted.

In [4], facial features such as eyes, nose and lips were considered for identification. This algorithm works on the skin color segmentation and this method consists of three basic stages, face detection, region localization and facial features detection. The illumination discrepancy problem of images has been minimized by doing the color space conversion from RGB to YCbCr. The author found the unique range of color values for specific facial features.

The skin colour based segmentation algorithm [4] can be combined with popular algorithms like Viola Jones face detector. Multi-objective difference evolution (MODE) was used in [5] to discover the ideal threshold ranges of color components in a composite color space of RGB and CbCrCg.

After a comparative analysis between several color models, RGB-H-CbCr was the one used in [6] in the first stage. Improved Haar-like features were used in the next stage. Representative features were extracted after the positioning of facial features.

Like almost all other face recognition method, in [6] also Viola-Jones algorithm [7] was used for facial features detection that uses Haar like features which is good enough to detect face components.

In [8], the author proposes a method which involves three steps. First step deals with the conversion of the given color image into different color space models. In the next step eigen values and corresponding eigen vectors are identified from all the color space models. In the third step a nearest neighbor classifier was applied and face images were classified based on the extracted features.

In [9] the authors have experimented an algorithm for face recognition which is invariant to pause and orientation using Gabor wavelets. This algorithm can recognize faces under challenging illumination conditions too. But the limitation is that the algorithm will work only for facial images.

In [10] the authors have proposed an algorithm to understand scenes from a given image. It can perform face detection and estimate the sentiment expressed by each person in the given image. The challenge in this algorithm is that when similarity of the faces in the image is high, it gives lesser accuracy.

Based on the above literature survey, it may be observed that face detection has produced higher accuracy than face recognition. In this work, we propose a feature description based algorithm that has a 96% accuracy in face detection and accuracy.

3 Proposed Work

The proposed framework in detail, shown in Fig. 1 represents the architecture diagram of the proposed system. Figure 2 illustrates the system flow diagram of the proposed work. The proposed work has four phases. Facial Features Detection (FFD), Detected Features Positioning (DFP), Descriptive Features Extraction (DFE) and Face Identification (FI).

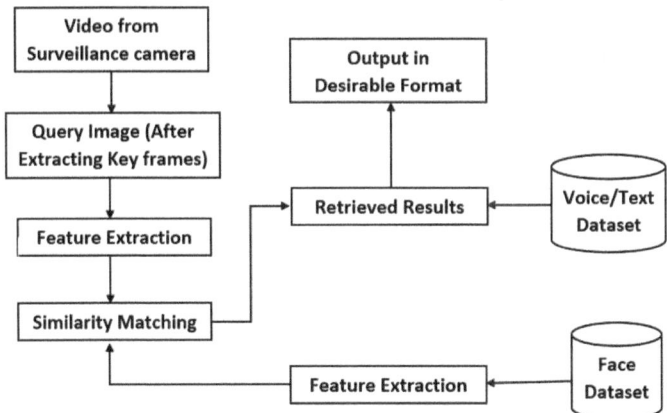

Fig. 1 Architecture diagram of the proposed system

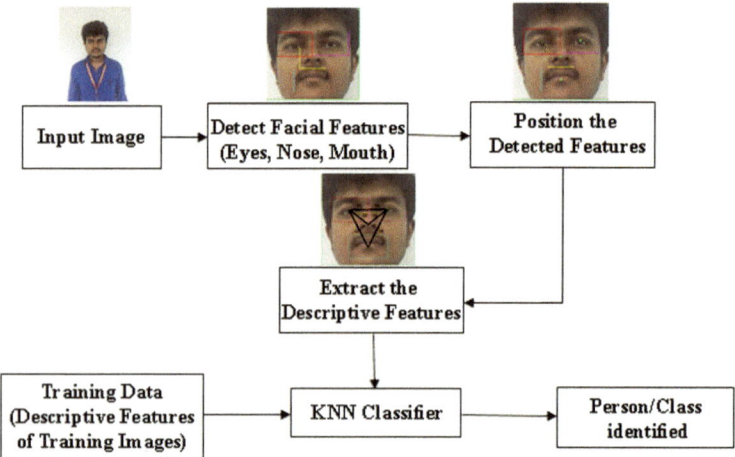

Fig. 2 System flow diagram of the proposed work

3.1   Facial Features Detection (FFD)

Being the most representative part of a human, face helps in identifying people. Eyes, nose and mouth are the important features which makes the face of one person distinguishable from another.

Here we use Viola Jones algorithm [7] for detecting face along with the facial features. The Viola Jones algorithm comprises four steps. Selection of Haar features, Integral image formation, Adaboost training and cascading classifiers.

Haar feature selection is the primary step of viola jones algorithm. Convolution kernels are used for extracting certain features from the image. For example, the convolution kernel shown in Fig. 3 extract the horizontal lines from an image. Similar is the case with the haar features. They act like kernels with a value of +1 in black region and a value of −1 in the white region.

Human faces possess some common properties. For example, the eye area is darker than the upper cheeks. Similarly the nose bridge is brighter than the eye region. These uniformities are matched using haar features. The kernel is traversed throughout the window and the value at each sub region is calculated as

$$\text{Value} = \sum (\text{pixels in dark region}) - \sum (\text{pixels in white region})$$

In the above step it is needed to calculate the sum of values of the black and white regions. For the datasets of high quality it may be difficult to calculate the sum of huge number of pixels. So the integral image representation is used. Integral image is created by filling each pixel position with the sum of all the pixel values which are to the left and top of it (including itself). From the integral image the sum of the pixels in a region is calculated by using the corner pixel values, as shown in Fig. 4.

-1	-1	-1
2	2	2
-1	-1	-1

Fig. 3 A kernel for horizontal line extraction

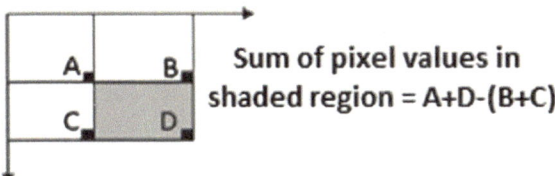

Fig. 4 Integral image calculation

There can be more than 160,000 feature values within a detector which need to be considered. But out of these, only certain features will be useful to identify a face. Adaboost algorithm in which more weights are given to those samples which are misclassified in the previous iteration is used to identify these features.

So far we have seen that the detector is scanned more than once through the same regions in the image, each time with a new size. Most of the considered sub windows would still be non faces. So we need the algorithm to be focused on ignoring non faces rapidly so that it can employ more time on likely face regions. Therefore for efficiency, we need to apply less accurate but faster classifiers first to immediately discard windows that clearly appear to be negative. This is known as cascading.

Finally after going through all these four stages, Viola Jones algorithm gave detected facial features as shown in Fig. 5b.

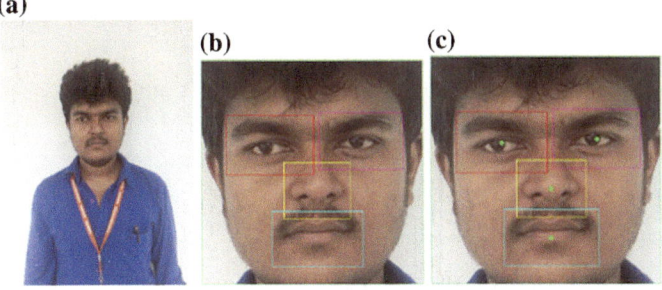

Fig. 5 Results of FFD and DFP

3.2   Detected Features Positioning (DFP)

After applying Viola Jones algorithm on the input image the face and facial feature regions were detected. The next step is to localize these detected facial features to a particular pixel value. The best single point with which we can represent a region is the centroid itself.

Figure 5c shows the features detected in the previous stage (FFD), being positioned at the centroid coordinates (marked with a green color '+').

3.3   Descriptive Features Extraction (DFE)

The objective in this stage is to extract the features which can be used for training and testing phases of the classifier.

First the distance between the positioned facial features are calculated. Euclidean distance metric is used here. The following are the distances that are calculated in this stage. (Here eye center is the center point of the line connecting left eye and right eye.)

$$D1 = \text{Midpoint of left eye} - \text{Midpoint between eyes}$$
$$D2 = \text{Midpoint of right eye} - \text{Midpoint between eyes}$$
$$D3 = \text{Midpoint of left eye} - \text{Midpoint of mouth}$$
$$D4 = \text{Midpoint of right eye} - \text{Midpoint of mouth}$$
$$D5 = \text{Midpoint of left eye} - \text{Midpoint of nose}$$
$$D6 = \text{Midpoint of right eye} - \text{Midpoint of nose}$$
$$D7 = \text{Midpoint of nose} - \text{Midpoint of mouth}$$

All the distances from D1 to D7 are represented in Fig. 6a.

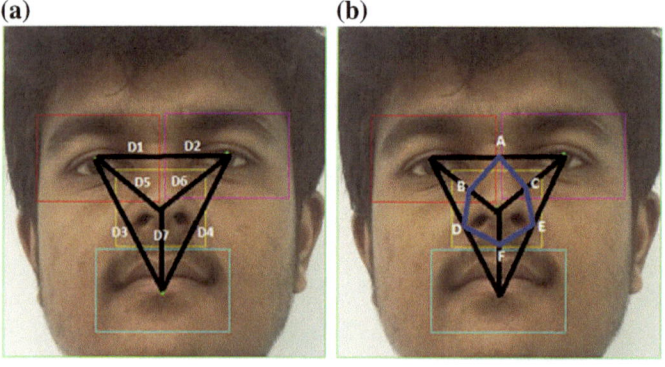

Fig. 6 Distances used for DFE

After calculating these distances, the mid points are identified. Totally seven midpoints are obtained. The next distance measures considered are those which are drawn from each midpoint to the nearest two midpoints. Final distance measures obtained are illustrated in Fig. 6b, in which the midpoints are represented with the alphabets from A to F. M1–M6 represent the distances between the midpoints and are defined as follows.

$$M1 = AB$$
$$M2 = AC$$
$$M3 = DF$$
$$M4 = EF$$
$$M5 = BD$$
$$M6 = CE$$

Now the distances D1–D7 and M1–M6 are found. But these can not be chosen as the descriptive features, since they may vary according to the position and resolution of the camera. But these distances help in extraction of the Descriptive features which remain consistent for all types of photographs. Descriptive Feature extraction is carried out by calculating the ratio between the distances as follows.

$$F1 = \frac{D1 + D2}{D3}$$
$$F2 = \frac{D1 + D2}{D4}$$
$$F3 = \frac{D1 + D2}{D5}$$
$$F4 = \frac{D1 + D2}{D6}$$
$$F5 = \frac{D1 + D2}{D7}$$
$$F6 = \frac{D3}{D4}$$
$$F7 = \frac{D5}{D6}$$

$$F8 = \frac{M1}{M2}$$

$$F9 = \frac{M5}{M6}$$

$$F10 = \frac{D1}{M1 + M3 + M5}$$

$$F11 = \frac{D2}{M2 + M4 + M6}$$

$$F12 = \frac{M1 + M2}{M5 + M6}$$

$$F13 = \frac{M1 + M2}{M3 + M4}$$

$$F14 = \frac{M1 + M3 + M5}{M2 + M4 + M6}$$

3.4   Face Identification (FI)

Face identification, which is the final phase of the proposed framework, uses the descriptive features that are extracted from the previous stage (DFE) for training the classifier. Later, the same descriptive features are extracted from the test image during the process of classification.

K Nearest Neighbors, a popular non-parametric classifier, is the one which is used in this phase. Given a number of N labelled instances $\{p_i, q_i\}$ which are used for training and a test vector p_0, the role of the classifier is to find out the class to which the test vector p_0 belongs to. The K-nearest neighbor (KNN) classification algorithm does this by considering the class label to which the majority of the N nearest neighbors of the test vector p_0 is associated with. The N nearest neighbors of the test vector are those labelled instances which have the minimum distance from it. Euclidean distance is the default distance metric used by the KNN classifier. The descriptive features F1, F2 ... F14 replace the p_i here and the labels which are people id become q_i.

The main contribution in this work is that the proposed algorithm can identify faces from their family photos where the similarity in faces is very high. Many existing algorithms have less accuracy in these conditions.

4   Experimental Results

The proposed algorithm was implemented using MATLAB 2015b. Datasets from our University laboratory and family photographs were used for testing the proposed work.

Table 1 contains the coordinates of the facial features after the application of DFP step on Fig. 5. Similarly the distances and corresponding ratios (Descriptive features) extracted from the same image (Fig. 5) are tabulated in Tables 2 and 3 respectively.

Around 200 images were collected from the University laboratory and the system has been trained with the descriptive features extracted from those. The photographs were taken with different facial expressions and different resolutions. This constituted

Table 1 Coordinate positions of facial features

Facial features	Positioned coordinate	
	x	y
Left eye	248.50	306.00
Right eye	581.00	292.50
Nose	421.00	469.00
Mouth	420.50	643.00

the first dataset. Similarly another dataset was made by collecting the photographs of people who belonged to the same family.

While identifying each person say x, 20 photographs out of 200 were of the same person x and the remaining 180 were not. Table 4 indicates the confusion matrix from which the accuracy o+f the system is calculated. True Positive (T+ve) defines that the person is correctly identified, while True Negative (T−ve) indicates that the person is correctly rejected. Similarly False Positive (F+ve) denotes that the person is wrongly identified and False Negative (F−ve) implies that he is incorrectly rejected.

The accuracy is calculated as

$$Accuracy = \frac{TP + TN}{TP + TN + FP + FN}$$

where TP, TN, FP and FN represent True Positive, True Negative, False Positive and False Negative respectively.

Table 4 leads to the conclusion that the average accuracy of identifying 10 people with various 20 facial expressions among 180 negative (wrong) faces was 96.05%.

Also it could be observed that the number of photographs required for training the classifier is more in the scenario of the same family than that of others. This is because of the fact that people from the same family have more probability of having similar facial features.

5   Conclusions and Future Work

The variations in both horizontal and vertical symmetry of a human face have been extracted as the facial features. The proposed system could be successfully experimented in different scenarios. A good net accuracy of 96.05% obtained from the experimental results confirms the reliability and stability of the proposed system.

The incorporation of various key frame extraction techniques to the proposed work is planned as a future work in order to make the system completely automated.

Table 2 Distance measures

D1	D2	D3	D4	D5	D6	D7	M1	M2	M3	M4	M5	M6
323.35	511.44	378.35	385.50	237.33	238.23	174.00	454.44	512.07	118.66	119.11	104.42	87.00

Table 3 Descriptive features

F1	F2	F3	F4	F5	F6	F7	F8	F9	F10	F11	F12	F13	F14
2.21	2.17	3.52	3.50	4.80	0.98	0.99	0.89	1.20	0.48	0.71	5.05	4.07	0.94

Table 4 Confusion matrix

People id	T+ve	F−ve	T−ve	F+ve	Accuracy
1	19	1	175	5	97.00%
2	17	0	174	9	95.50%
3	16	6	173	5	94.50%
4	15	6	172	7	93.50%
5	18	4	170	8	94.00%
6	18	2	178	2	98.00%
7	19	2	174	5	96.50%
8	20	3	177	0	98.50%
9	17	1	176	6	96.50%
10	18	3	175	4	96.50%

References

1. BenAbdelkader, C., Cutler, R., Davis, L.: Stride and cadence as a biometric in automatic person identification and Verification. Fifth IEEE Int. Conf. Autom. Face Gesture Recognit. **4**, 377–380 (2002)
2. Galton, F.: Personal identification and description. J Anthr. Inst. Great Britain Ireland **18**, 177–191 (1889)
3. Devadethan S., Titus, G., Purushothaman, S.: Face detection and facial feature extraction based on a fusion of knowledge based method and morphological image processing. In: Annual International Conference on Emerging Research Areas: Magnetics, Machines and Drives (AICERA/iCMMD) (2014). https://doi.org/10.1109/AICERA.2014.6908216
4. Hasan, M.M., Hossain, M.F.: Facial features detection in color images based on skin color segmentation. In: 3rd international conference on informatics, electronics & vision (2014). https://doi.org/10.1109/ICIEV.2014.6850763
5. Luh, G.C.: Face detection using combination of skin color pixel detection and viola-jones face detector. In: International Conference on Machine Learning and Cybernetics, vol. I, pp. 364–370 (2014)
6. Semary, N.A., Gad, A.F.: A proposed framework for robust face identification system. In: 9th International Conference on Computer Engineering and Systems, pp. 62–67 (2014)
7. Viola, P., Jones, M.: Rapid object detection using a boosted cascade of simple features. In: Proceedings of the 2001 IEEE Computer Society Conference on Computer Vision and Pattern Recognition, Vol I, pp. 511–518 (2001)
8. Jose, J.P., Poornima, P., Kumar, K.M.: A novel method for color face recognition using KNN classifier. In: International Conference on Computing, Communication and Applications (2012)

9. Karthika, R., Parameswaran, L.: Study of gabor wavelet for face recognition invariant to pose and orientation. In: Proceedings of the International Conference on Soft Computing Systems, Advances in Intelligent Systems and Computing, vol. 397, pp. 501–509 (2016)
10. Athira, S., Manjusha, R., Parameswaran, L.: Scene understanding in images, intelligent systems technologies and applications. Advances in Intelligent Systems and Computing, vol. 530, pp. 261–271 (2016)

Image Tampering Detection Based on Inherent Lighting Fingerprints

Manoj Kumar[1]([⊠]) and Sangeet Srivastava[2]

[1] Department of Computer Science, The Northcap University, Gurugram, India
wss.manojkumar@gmail.com
[2] Department of Applied Science, The Northcap University, Gurugram, India
sangeetsrivastava@ncuindia.edu

Abstract. Digital Imaging experienced unprecedented growth in the past few years with the increase in easily accessible handheld digital devices. This furthers the applications of digital images in many areas. With the amassed recognition and accessibility of low cost editing software, the integrity of images can be easily compromised. Recently, image forensics has gained popularity for such forgery detection. However, these techniques still lag behind to prove the authenticity of the images against counterfeits. Therefore, the issue of credibility of the images as a legal proof of some event or location become imperative. The proposed work showcases a state-of-the art forgery detection methodology based on the lighting fingerprints available within digital images. Any manipulation in the image(s) leaves dissimilar fingerprints, which can be used to prove the integrity of the images after the analysis. This technique performs various operations to obtain the intensity and structural information. Dissimilar features in an image are obtained using Laplacian method followed by surface normal estimation. Applying this information, source of the light direction is estimated in terms of angle ψ. The proposed technique demonstrates an efficient tool of digital image forgery detection by identifying dissimilar fingerprints based on lighting parameters. Evaluation of the proposed technique is successfully done using CASIA1 image dataset.

Keywords: Image tampering detection · Lighting estimation · Edge detection

1 Introduction

In ancient time, a photograph was considered as a piece of *actuality*. With advanced digital image processing and the availability of imaging technology the actuality of a photograph is no longer granted as a sign of truthiness. The basic reason behind the fact is that without any specialist skills anyone can easily manipulate photographs. From magazine to other media industries, journals, courtroom applications, political campaigns and photo pranks in our phones, the manipulated photographs are increasing exponentially [1]. As a result, truthiness and authenticity of images are diminishing.

© Springer International Publishing AG 2018
D. J. Hemanth and S. Smys (eds.), *Computational Vision and Bio Inspired Computing*,
Lecture Notes in Computational Vision and Biomechanics 28,
https://doi.org/10.1007/978-3-319-71767-8_97

Recently, to detect the tampering in images, the concept of image forensics has been developed. Digital image forensics is mainly divided into two parts. Part (a): It determines whether an image gone through under some tampering or post processing operations. In this category of forgery detection, we try to detect the inconsistency in the image processing features and image acquisition processes. Part (b): In this, forgery detection deals with image source identification. The image forensics science does not require any prior information about original image [2]. Forgery in the images is usually accomplished by manipulating statistical, geometrical and physical properties of image data. Pixel based forgeries changed the pixel data during compression or by manipulating wavelet coefficients for image objects [3].

According to a survey by Wall Street Journal, approximately 10% photographs published in the United States are digitally reformed. Even scientific reports are also altered by some ways to change the integrity of original data. As shown in Fig. 1, photographs are easily deleted, reformed using sophisticated editing software's without even leaving the residues of these manipulation operations. The manipulated photographs shown in Fig. 1a, b are recently reported in digital media [4]. Past few years, blind passive digital forgery detection techniques developed a significant interest among researchers compare to active approaches. Active approaches are implemented using watermarking or embedding digital signatures in digital data. Blind forgery detection techniques assess the authenticity or integrity of image without knowing the hidden information present in the image.

The proposed work is inspired by blind approach and therefore we developed a novel technique to detect image tampering based on physics/light elements present in the image. By estimating inherent traces or inconsistencies in the lighting parameters, we can easily state whether a photograph is real or fake.

(a)   **(b)**

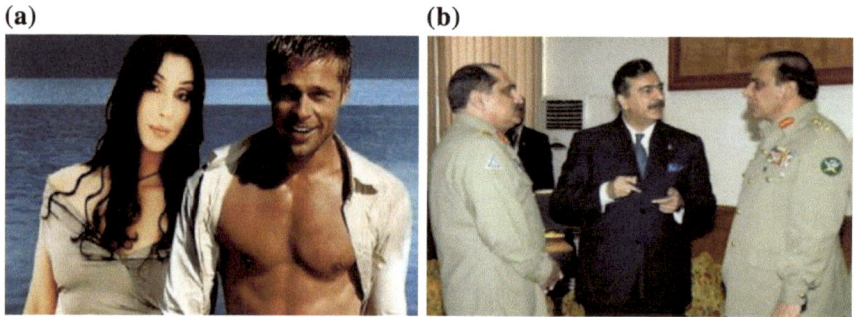

Fig. 1 Some of recent hot photograph forgeries: **a** Photo of Brad Pit is composite with Cher **b** Pakistani President Yousef Gilani in the image [4]

The whole idea of our technique is discussed in the following sections of this paper is as follows: After reviewing related forgery detection techniques in Sect. 2, we discussed our proposed approach in Sect. 3 followed by the result and analysis in Sect. 4. Final conclusion and future work are drawn in Sect. 5.

2   Related Work

Various techniques were developed in the literature to address tampering detection in digital images, but many techniques are remains unidentified to provide an accurate solution till now. Each technique tries to address different aspects to verify the integrity and authenticity of the images. In this Section, we tried to present a glimpse of forgery detection techniques with more focus towards pixel and physics based forgery detection. Blind forgery detection techniques follow many steps such as image pre-processing, feature extraction, classifier or feature preprocessing, classification and post processing. In the preprocessing or feature extraction some operations like cropping, transferring RGB image to Grayscale, DCT, DWT transformation are performed to improve the classification. After completion of pre-processing, some features from the image are extracted to generalize various characteristics of image data sensitive to image tampering. Finally, classifiers are implemented for classifying images into original and forged images. Image forgery detection can be achieved by using following approaches.

2.1   Copy Move and Splicing

Copy move forgery is the simplest technique to perform forgery over images. In this, copy from one part of the image is moved to user desired position and pasted. Texture parts are idyllic for this kind of forgery. Since these areas have similar color and noise properties which are imperceptible for human eyes, therefore, finding inconsistencies in these areas become clumsy. First approach for detecting copy move forgery was implemented by Fridrich et al. [5]. The author proposed copy move detection by using DCT (Discrete Cosine Transformation) of overlapping image regions followed by lexicographical representation to remove computational cost. In addition the technique performance was satisfactory towards tampering detection using block matching algorithm. Popescu and Farid [6], proposed almost similar approach using principal component analysis for image overlapped clocks representation. Using this technique authors succeeded in reduction of computational cost compare to [5]. They obtained forgery detection accuracy of 50% when JPEG quality = 95 with block size 32 × 32 and 100% when JPEG quality = 95 with block size 160 × 160. The accuracy of this technique diminishes with small block size. Myna et al. [7] designed a new approach for detecting copy move forgery. They implemented log polar coordinates and wavelet transforms to detect this kind of forgery. Dimensionality reduction was achieved using wavelet transformation and exhaustive search is applied to find similar blocks in image by mapping with log-polar coordinates. The current work by Zhang et al. [8], image · splicing detection was performed using planer homography to locate fake part of image. Automatic feature extraction method using graph cut with the online feature selection was implemented to segment the fake part. Many researchers implemented the same kind of forgery detection using the photometric consistency of illumination in image object shadows. They formulated color characteristics of shadows by using shadow matte values [9]. More techniques related to such kind of forgery detection can be studied from [10–12].

2.2 Forgery Detection Using Image Compression

JPEG image compression is commonly used by many applications. Even most of the cameras available in the market export JPEG format. Whether an image is compressed or not is an issue for image processing applications which can be used for image tampering detection. Many authors in the literature proposed image forgery detection based on image compression. Popescu [13] developed a technique to detect forgery if image is double compressed. The detection of forgery is examined by using the histogram of DCT coefficients. The manipulation of the image using compression introduces some specific artifacts in these coefficients. By analysing histograms of these coefficients, forgeries can be detected. The major drawback of this technique is that with availability of poor quality of the compressed image. It becomes difficult to detect the traces of forgery. Zhang et al. [14] detected image forgery based on double JPEG2000 compression. They used Fourier transform of DWT coefficient histograms. They analysed specific inconsistencies in the quantization of sub band DWT coefficients for forgery detection. Bianchi and Piva [15] Implemented a reliable method for detecting non-aligned double JPEG compression based on single feature of DCT coefficients. These coefficients were calculated according to the grid of the previous JPEG compression available in the image. Their method estimated quantization step and grid shift of JPEG compression for forgery detection. Recent work related to this domain can be studied from [16–18].

2.3 Forgery Detection Based on Lighting Inconsistency

As we know that photographs are taken under different lighting conditions. Every image contains different illumination map artifacts in it. Combining various objects in a single image becomes difficult to match the lighting conditions for these objects. By analysing the lighting inconsistencies between the image and fake objects we can identify the tampering. Therefore, these lighting inconsistencies are used to prove whether the image is real or doctored one. Johnson and Farid [19] suggested an approach to detect forgery by estimating direction of light source. The authors estimated light source direction for all the objects present in the image. Inconsistencies in the light sources illustrated the presence of forgery. But their technique was not able to detect forgery based on complex lighting environments. Further they [20] designed an approach to find complex lighting environments by modelling lighting coefficients with 9D model. They also proposed forgery detection based on specularity highlights in human eyes. Their technique measures the 3D direction light source from specularity present in the human eye. Inconsistencies in the location of the light source can be used to prove forgery. Peng et al. [21] proposed an optimized light estimation model based on illumination coefficients. They relaxed the assumption of convexity and constant reflection in their approach. The authors considered occlusion geometry and surface texture to analyse the structural profiles followed by lighting estimation. The proposed reflection model shows accuracy for detecting image forgery. Kumar et al. [3] identifies forgery using 3D lighting estimation. With minimum human interaction authors were able to detect image forgery based on light source estimation. The orientation of the image object with respect to light source was calculated to find the forgery. In reflection based forgery detection Riess et al. [22] proposed a method based on intrinsic contour estimation. The technique is able to integrate reflections from multiple materials into a single lighting environment. The method shows an improvement by improving mean error by almost 30% for forgery detection. Forgery detection using lighting, reflections and shadows can be studied from [23–25]. See [4] for a detailed summary of different types of forgery detection methods.

3 Proposed Method

This section explains the details of the proposed approach. The input to our algorithm is a single image (with a fake object in it) in which user selects two patches as an input to the algorithm. The first patch is corresponding to the real part of the image while another one is corresponding to fake part of the image. By analysing both the image patches, our algorithm return angle ψ indicates the incident angle in the image plane. For the calculation of ψ, we need 3D surface and illumination information from the image to detect forgery. The proposed method follows certain assumption such as image surface is lambertian with constant reflectance values. Firstly, preprocessing step is performed to make image more eloquent for obtaining surface and intensity profiles.

3.1 Pre-processing

In order to make the input image more suitable for the process, some pre-processing is required to remove texture, highlights and noise. To achieve this, we implemented various filters to improve the intensity profile and for remove outlier noise from the input image.

The proposed methodology applies adaptive histogram to the Gray pixels of the input selected patches to enhance the visual information. The method performs gray mapping of the pixels using grey operation and transform the histogram towards uniform, smooth and clear gray level to enhance the image for further processing [26]. Mathematically, Gray operation look for a gray value of pixel and map it with a new value and move on. The operation works as a function of $g(v)$ that operates on pixel value $v = f(x, y)$ of image at location (x, y). We applied CLAHE histogram equalization on the selected image patches. The function uses the cumulative distribution function as mapping function and intensity levels are changed for image smoothing. After applying this filter, we require surface texture for getting structural information. To make surface texture more informative Laplacian of Gaussian [27] filter is applied to the image patches. This filter combines Gaussian with Laplacian for edge detection on the image patches. Consider an image with intensity I(x), where each pixel have values $I_1, I_2, I_3 \ldots I_n$. With second order derivative $\left(\delta^2 I / \delta^2 X\right)$ the edges should be better located in the image. Gaussian operator is applied to the selected patch to remove the noise with Gaussian kernel width σ. This is defined as shown in Eq. (1).

$$G_\sigma(x, y) = \frac{1}{\sqrt{2\pi\sigma^2}} \exp\left(-\frac{x^2 + y^2}{2\sigma^2}\right) \tag{1}$$

Calculate the Laplacian of Gaussian $\Delta G_\sigma(x, y)$ and then convolve this with selected patch of image. This is achieved by calculate first and second order derivative of image intensity with Gaussian operators as shown in Eq. (2).

$$\frac{\partial}{\partial_x} G_\sigma(x, y) = \frac{\partial}{\partial_x} e^{-(x^2 + y^2)/2\sigma^2}$$

and   (2)

$$\frac{\partial^2}{\partial^2_x} G_\sigma(x, y) = \frac{x^2 - \sigma^2}{\sigma^4} e^{-(x^2 + y^2)/2\sigma^2}, \frac{\partial^2}{\partial^2_y} G_\sigma(x, y) = \frac{y^2 - \sigma^2}{\sigma^4} e^{-(x^2 + y^2)/2\sigma^2}$$

Equation (2) is applied on selected patch to extract the edge based on a threshold. The modified Equation which is used as LoG is shown in Eq. (3).

$$LoG = \frac{x^2 + y^2 - 2\sigma^2}{\sigma^4} e^{-(x^2 + y^2)/2\sigma^2}$$ (3)

Finally, strong edges by threshold zero-crossings are obtained for further processing.

3.2   Estimating Lighting Source and Angle ψ

To estimate light source direction ψ, geometry of the surface is used. For calculation the known surface geometry is required. In our case we have reliable surface normals on object edges. The selected objects have surface normals perpendicular to the viewing and incident light direction. With an isotropic reflective surface, it is assumed that both the viewing and the incident light source have the same angle of incident i.e. ψ for all objects present in the image. So angle ψ is same as per this assumption, but if we obtain differences in ψ between image objects, then it shows the forgery. It is our preliminary belief for forgery detection and obtained results confirms the validity of the proposed approach. Figure 2 shows the geometry of incident and outgoing light directions in the local coordinate frame. In the proposed work, we are calculating incident light in terms of ψ angle for various patches on the image surface.

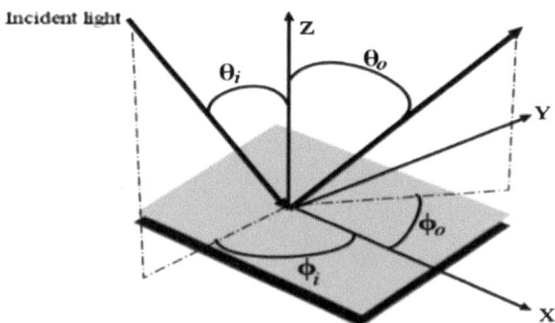

Fig. 2 General geometry of incident light and reflected light directions on a surface

Using the structural information we can obtain surface normal and image radiance on the image patch. Given that illumination in the image is from a distant light source and under this the image intensity I is represent as Eq. (4) [3].

$$I = k(\overrightarrow{n} \bullet \overrightarrow{l}) + a \tag{4}$$

In this $\overrightarrow{n} = (n_x, n_y, n_z)^T$ is a 3D surface normal on the edge, $\overrightarrow{l} = (l_x, l_y, l_z)^T$ states direction of light source. k and a are is surface reflectance parameter and constant indirect illumination. The value for these constant are $0 \leq k, a \leq 1$. The image intensity at the edge can be expressed as Eq. (5).

$$I = k(\overrightarrow{n} \bullet \overrightarrow{l}) + a = k(n_x l_x + n_y l_y + n_z l_z) + a \tag{5}$$

This equation can also be represented as: $I = \begin{pmatrix} n_1^T \\ n_2^T \\ . \\ n_p^T \end{pmatrix} \cdot \begin{pmatrix} s \\ a \end{pmatrix}$

Even implementing the above mentioned conditions there is an error in the light source estimation due to unknowns present in the equation i.e. the exact value of k and a is unknown. To improve the estimation of light source direction ψ, least square is estimated $\hat{L} = (\hat{x}, \hat{y}, \hat{z}, \hat{a})^T$ is calculated as shown in Eq. (6).

$$\hat{L} = (N^T N)^{-1} N^T I \tag{6}$$

Finally, the estimated angle ψ of the light source is calculated using Eq. (7).

$$\psi_i = \cos^{-1}\left(\frac{I - a}{|n|.|s|.k}\right) \tag{7}$$

where $0 \leq \psi \leq \pi$.

4 Results

The proposed technique is tested on images chosen from freely available CASIA1 [28] image database. The database contains real and fake images as its repository. All the images chosen for testing are in JPEG format. From the images, angle ψ is estimated with respect to light source direction. Value of ψ indicated whether the image is fake or real. The estimation is done for both real and fake image patches. The differences between the angles for both the patches indicate the reality of the image. The same ψ values for both the patches indicates that selected patches are consistent with the direction of light source and are original. Inconsistency between these angles indicates that the image is tampered. A sample of images from our test set is shown in Fig. 3.

(a)   (b)   (c)   (d)   (e)

(f)   (g)   (h)   (i)

Fig. 3. A sample of images for evaluating the proposed technique

In image 3a dog and in 3b Girl standing on the beach is inserted in the image scene. In other images the fake parts are men, girl on ship, mug on glass cup, swan on grass, panda, and football respectively. The proposed technique will work better for almost all the selected patches and produce expected results. In the experiment two patches are selected from the real part while one patch is selected from fake part of the image. But in case of loose strong edge detection, this technique produces unexpected results. With more accuracy on the edges, the technique produce better estimation of surface normal and light source direction angle ψ. Figure 4 shows the outcome in terms of edge detection on selected patch and also demonstrates that how surface normals are estimated on strong edges. Yellow part in Fig. 4 shows the basic geometry for normal and ψ calculations as per physics lighting model.

Fig. 4. Edges obtained using LoG on girl hat

In the proposed technique, original ψ must be of difference $\pm 10°$ for different parts of image. More difference i.e. $> 10°$ between ψ values for selected patches shows that these patches are inconsistent and therefore the image is a fake image. Figure 5 shows the results obtained from images in Fig. 3a, b and e. From all the images, two patches are selected from the real parts of the image while third patch is taken from the fake object to estimate the light source direction. The x axis in Fig. 5 is pertaining to both the fake and real image patches while y axis represents the ψ values ranges from $0°$ to $180°$.

Form results, it is clear that real parts are consistence with light source. The difference between ψ values for chosen patches indicates that the fake patches are is inconsistent with real patches values of ψ. This inconsistency shows that the image is a fake image while the rest of the scene is consistent with the image.

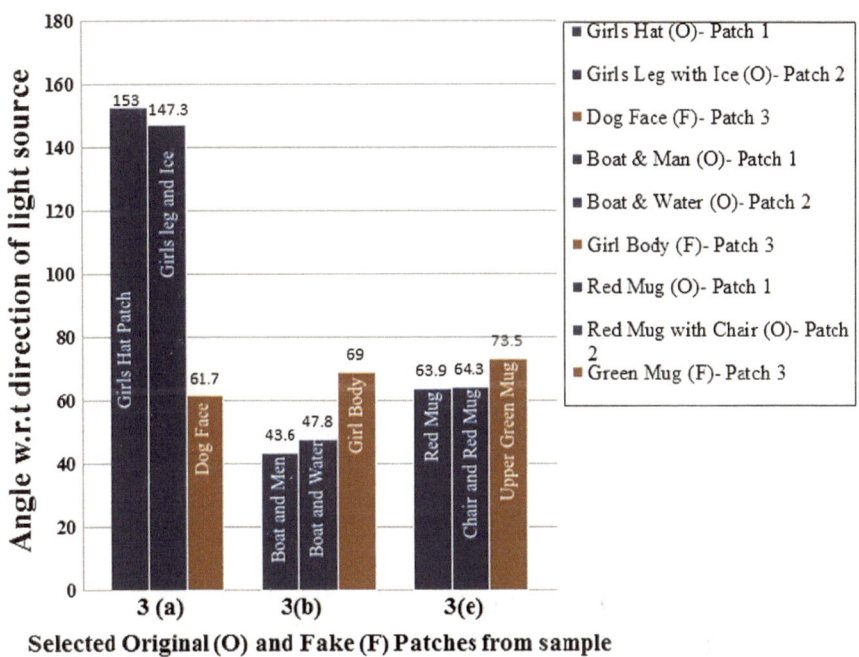

Selected Original (O) and Fake (F) Patches from sample Images

Fig. 5. A sample represenation of estimated ψ values from images 3a, b and e

From Tables 1, 2 and 3, values of ψ clearly indicate the inconsistencies in the direction of light source for selected patches from the image. The experiment is tested against various images chosen from the CASIAv2 dataset.

Table 1 ψ from Fig. 5 (*Image Source* 3a)

Selected object	Obtained ψ angle
Girls hat (O)	153°
Girl and ice (O)	147.3°
Dog face (F)	61.7°

Table 2 ψ from Fig. 5 (*Image Source* 3b)

Selected object	Obtained ψ angle
Boat and water (O)	47.8°
Boat and man (O)	43.6°
Girl body (F)	69°

Table 3 ψ from Fig. 5 (*Image Source* 3e)

Selected object	Obtained ψ angle
Red mug (O)	63.9°
Red mug and chair (O)	64.3°
Upper green mug (F)	73.5°

The obtained results demonstrate that the proposed technique is able to estimate the direction of the light source as per physics property of lighting. Therefore, this method has enough potential to identify tampering. The proposed technique identifies tampering in 8 images out of a test sample of 10 images. For some images, if the patch selection have more pixels from the original image while less from the fake will results in poor estimation of ψ.

The proposed work takes out the human involvement for selecting probes as compare to work done in [29]. In [29] authors selected the 3D shape probe manually to establish the z direction. The accuracy of their results depends on the probe selection, if the direction of the probe is wrong then the estimation of surface normal is wrong which effects the outcomes of the technique. The user assistance is not required in the proposed technique by implementing automatic detection of strong edges for surface normal estimation. The accuracy to estimation the light source is more in this work. As compare to work done in [3] and [30], the proposed technique reduced the estimated error threshold form ±15° to ±10° for light source estimation. We are not considering shadows to identify the light source as done in [30]. Because for estimation of shadows in image, the position of the relative camera and other timestamp metadata is required. Instead of using own created images and known metadata, a standard database is used for testing the proposed technique. The chances of errors for shadow estimation also increases if sharp curvatures on occluding boundaries occur because of z component approaches to zero. These shortcomings are removed in the proposed paper. The obtained results from proposed technique demonstrates the improvement over state of art approaches discussed above.

5   Discussion and Future Work

The proposed work demonstrates a novel technique for detecting forgeries present in the images. This method uses the inherent lighting features for estimation of light source direction by using structural and intensity profiles of the image. To enhance the intensity profile, the preprocessing step the proposed method uses adaptive histogram equalization followed by Laplacian of Gaussian edge detection method for pattern analysis. Further, surface normals are obtained on strong edges so that light direction in terms of angle ψ is obtained. In practice, there may be errors in estimating the angle ψ because the selection of edges is done based on human judgement. To reduce it, we applied least square estimation. Broadly, the obtained results proves that the proposed technique is able to identify forgery based on 3D lighting parameters.

In the contrary, this technique needs improvement for automatic selection of image parts so that human judgement can be completely eliminated for selecting patches on the image surface. This technique can be implemented for complex lighting environments. These shortcomings of the proposed technique can be improved in the future extension of this work.

References

1. Gloe, T., Kirchner, M., Winkler, A., Böhme, R.: Can we trust digital image forensics. In Proceedings of the 15th ACM International Conference on Multimedia, Augsburg, Germany, 25–29 September 2007
2. Ng, T.-T., Chang, S.-F., Lin, C.-Y., Sun, Q.: Passive-blind image forensic. In: Multimedia Security Technologies for Digital Rights Management, pp. 383–412. Academic Press (2006)
3. Kumar, M., Srivastava, S.: Identifying photo forgery using lighting elements. Indian J. Sci. Technol. **9**(48), 1–5 (2016)
4. Birajdar, G.K., Mankar, V.H.: Digital image forgery detection using passive techniques: a survey. Elsevier. Digit. Investig. **10**(2013), 226–245 (2013)
5. Fridrich, J., Soukal, D., Lukas, J.: Detection of copy-move forgery in digital images. In: International Proceedings of Digital Forensic Research Workshop (2003)
6. Popescu, A., Farid, H.: Exposing digital forgeries by detecting duplicated image regions. Department of Computer Science, Dartmouth College (2004)
7. Myna, A., Venkateshmurthy, M., Patil, C.: Detection of region duplication forgery in digital images using wavelets and log-polar mapping. In: Proceedings of the International Conference on Computational Intelligence (2007)
8. Zhang, W., Cao, X., Qu, Y., Hou, Y., Zhao, H., Zhang, C.: Detecting and extracting the photo composites using planar homography and graph cut. IEEE Trans. Inf. Forensics Secur. **5**(10), 544–555 (2010)
9. Liu, Q., Cao, X., Deng, C., Guo, X.: Identifying image composites through shadow matte consistency. IEEE Trans. Inf. Forensics Secur. **6**(3), 1111–1122 (2011)
10. Kakar, P., Sudha, N.: Exposing postprocessed copy-paste forgeries through transform-invariant features. IEEE Trans. Inf. Forensics Secur. **7**(3), 1018–1028 (2012)
11. Yanga, F., Lia, J., Lu, W., Weng, J.: Copy-move forgery detection based on hybrid features. Eng. Appl. Artif. Intell. **59**, 73–83 (2016)
12. Shen, X., Shi, Z., Chen, H.: Splicing image forgery detection using textural features based on the grey level co-occurrence matrices. IET Image Process. **11**(1), 44–53 (2017)
13. Popescu, A.: Statistical tools for digital image forensics Ph.D. thesis, Department of Computer Science, Dartmouth College; Hanover (2004)
14. Zhang, J., Wang, H., Su, Y.: Detection of double-compression in JPEG2000 images for application in image forensics. J. Multimed. **4**(6), 379–388 (2009)

15. Bianchi, T., Piva, A.: Detection of non-aligned double JPEG compression with estimation of primary compression parameters. In: Proceeding of International Conference on Image Processing, (2011)
16. Bhartiya, G., Singh, J.: Forgery detection using feature-clustering in recompressed JPEG images. Multimed. Tools Appl. **75**(20), 1–16 (2016)
17. Tariang, D.B., Naskar, R.,: Re-compressed based JPEG forgery detection and localization through automated quality factor investigation. In: IEEE International Conference on Wireless Communications, Signal Processing and Networking (WiSPNET), Chennai (2016)
18. Kee, E., Johnson, M.K., Farid, H.: Digital image authentication from JPEG headers. IEEE Trans. Inf. Forensics Secur. **6**(3), 1066–1075 (2011)
19. Johnson, M., Farid, H.: Exposing digital forgeries by detecting inconsistencies in lighting. In: Proceeding of ACM Multimedia and Security Workshop (2005)
20. Johnson, M., Farid, H.: Exposing digital forgeries in complex lighting environments. In IEEE Transaction on Information Forensics Security, **3**(2), 450–61 (2007)
21. Peng, B., Wang, W., Dong, J., Tan, T.: Optimized 3D lighting environment estimation for image forgery detection. In: IEEE Transactions on Information Forensics and Security, **12**(2) (2017)
22. Riess, C., Unberath, M., Sven, F.N., Stamminger, P. M., Angelopoulou, E.: Handling multiple materials for exposure of digital forgeries using 2-D lighting environments. Multimed. Tools Appl. **76**(4), 4747–4764 Feb. (2017). doi: 10.1007/s11042-016-3655-0
23. Zhang, W., Cao, X., Zhang, J., Zhu, J., Wang, P.: Detecting photographic composites using shadows. In: IEEE International Conference on Multimedia and Expo (2009)
24. L. Yingda, X. Shen, C. Haipeng: An improved image blind identification based on inconsistency in light source direction. J. Super comput. **58**(1), 50–67 (2011)
25. Kee, J.O.H.F.E.: Exposing photo manipulation from shading and shadows. ACM Trans. Graph. **33**(5), 1–21 (2014)
26. Zhu, Y., Huang, C.: An adaptive histogram equalization algorithm on the image gray level mapping. In: 2012 International Conference on Solid State Devices and Materials Science, Elsevier, Macao, (2012)
27. Marr, D., Hildreth, E.: Theory of edge detection. Proceedings of the Royal Society of London. Series B. Biol. Sci. **207**(1167), 187–217 Feb. 29 (1980)
28. Dong, J., Wang, W.: [Online]. Available: http://forensics.idealtest.org/casiav1/join/ (2009). Accessed 22 December 2016
29. Carvalho, T., Farid, H., Kee, E.: Exposing photo manipulation from user guided 3D lighting analysis. In SPIE Symposium on Electronic Imaging, San Francisco, CA (2015)
30. Roy, A., Mitra, S., Agrawal, R.,: A novel method for detecting light source for digital images forensic. Opto−Electron **19**(2), 211–218 (2011)

Author Index

© Springer International Publishing AG 2018
D. J. Hemanth and S. Smys (eds.), *Computational Vision and Bio Inspired Computing*,
Lecture Notes in Computational Vision and Biomechanics 28,
https://doi.org/10.1007/978-3-319-71767-8